THE PULMONARY ENDOTHELIUM

THE PULMONARY ENDOTHELIUM

Function in health and disease

Editors

Norbert F. Voelkel

Virginia Commonwealth University, Richmond, VA, USA

Sharon Rounds

Alpert Medical School of Brown University, Providence, RI, USA

WILEY-BLACKWELL

A John Wiley & Sons, Ltd., Publication

This edition first published 2009

Wiley-Blackwell is an imprint of John Wiley & Sons, formed by the merger of Wiley's global Scientific, Technical and Medical business with Blackwell Publishing.

Registered office: John Wiley & Sons Ltd, The Atrium, Southern Gate, Chichester, West Sussex, PO19 8SQ, UK

Other Editorial Offices:
9600 Garsington Road, Oxford, OX4 2DQ, UK
111 River Street, Hoboken, NJ 07030-5774, USA

For details of our global editorial offices, for customer services and for information about how to apply for permission to reuse the copyright material in this book please see our website at www.wiley.com/wiley-blackwell

Library of Congress Cataloguing-in-Publication Data

The pulmonary endothelium / [edited by] Norbert F. Voelkel, Sharon Rounds.
p. ; cm.
Includes bibliographical references and index.
ISBN 978-0-470-72361-6
1. Pulmonary endothelium. 2. Pulmonary endothelium–Pathophysiology. I. Voelkel, Norbert F. II. Rounds, Sharon, 1946-
[DNLM: 1. Lung. 2. Endothelium, Vascular. WF 600 P98344 2009]
QP88.45.P847 2009
612.2—dc22

2009011988

ISBN: 978-0-470-72361-6 (HB)

A catalogue record for this book is available from the British Library.

Typeset in 9/11pt Times by Laserwords Private Ltd, Chennai, India
Printed in Singapore by Fabulous Printers Pte Ltd.
First Impression 2009

This book is dedicated to our families and to our mentors. We particularly acknowledge the contributions of Robert Grover, Ivan McMurtry, and the late Jack Reeves to our careers.

Contents

List of Contributors

ABU-BAKR AL-MEHDI

Department of Pharmacology, University of South Alabama College of Medicine, Mobile, AL 36688, USA

ASHOK N. BABU

Cardiovascular Surgery, University of Colorado Health Sciences Center, Aurora, CO 80045, USA

EDWARD R. BLOCK

Department of Medicine, University of Florida-Gainesville School of Medicine, Gainesville, FL 32610, USA

KENNETH E. CHAPMAN

Institute for Environmental Medicine, University of Pennsylvania, Philadelphia, PA 19104, USA

SHAMPA CHATTERJEE

Institute for Environmental Medicine, University of Pennsylvania, Philadelphia, PA 19104, USA

GAURAV CHOUDHARY

Alpert Medical School of Brown University, Vascular Research Laboratory, Providence VA Medical Center, Providence, RI 02908, USA

DONNA L. CIOFFI

Department of Biochemistry and Molecular Biology, Center for Lung Biology, College of Medicine, University of South Alabama, Mobile, AL 36688, USA

ONDINE B. CLEAVER

Assistant Professor Department of Molecular Biology, University of Texas Southwestern Medical Center at Dallas, Dallas, TX, USA

CARLYNE D. COOL

Department of Pathology, National Jewish Health, Denver, CO, USA

NEIL DAVIE

Pulmonary Vascular Business Unit, Pfizer, Tadworth, Surrey, UK

SERPIL C. ERZURUM

Department of Pathobiology and Respiratory Institute, The Cleveland Clinic Foundation, Cleveland, OH 44195, USA

BARRY L. FANBURG

Tufts University School of Medicine, Tufts Medical Center, Pulmonary and Critical Care Division, Boston MA, 02111-1526, USA

ARON B. FISHER

Institute for Environmental Medicine, University of Pennsylvania, Philadelphia, PA 19104, USA

MARIA FRID

Department of Pediatrics, University of Colorado Health Sciences Center, Denver, CO 80262, USA

YUANSHENG GAO

Department of Physiology and Pathophysiology Peking University Health Science Center, Beijing, 100191, China

JOE G.N. GARCIA

Department of Medicine, Pritzker School of Medicine, University of Chicago, Chicago, IL 60637, USA

EVGENIA V. GERASIMOVSKAYA

Department of Pediatrics, University of Colorado Health Sciences Center, Denver, CO 80262, USA

KATIE L. GRINELL

Vascular Research Laboratory, Providence VA Medical Center, Alpert Medical School of Brown University, Providence, RI 02908, USA

ELIZABETH O. HARRINGTON

Vascular Research Laboratory, Providence VA Medical Center, Alpert Medical School of Brown University, Providence, RI 02908, USA

GUOCHANG HU

Department of Pharmacology and Center for Lung and Vascular Biology, University of Illinois College of Medicine, Chicago, IL 60612, USA

JEFF JACOBSON

Department of Medicine, Pritzker School of Medicine, University of Chicago, Chicago, IL 60637, USA

MATTHEW JANKOWICH

Alpert Medical School of Brown University, Vascular Research Laboratory, Providence VA Medical Center, Providence, RI 02908, USA

NIKKI L. JERNIGAN

Vascular Physiology Group, Department of Cell Biology and Physiology, University of New Mexico Health Sciences Center, Albuquerque, NM, USA

USAMAH S. KAYYALI

Tufts University School of Medicine, Tufts Medical Center, Pulmonary and Critical Care Division, Boston MA, 02111-1526, USA

NEBOJSA KNEZEVIC

Center for Lung and Vascular Biology, Department of Pharmacology, University of Illinois at Chicago, Chicago, IL, USA

WOLFGANG M. KUEBLER

University of Toronto, Ontario, Canada

-

Charité - Universitätsmedizin Berlin, Germany

-

The Keenan Research Centre at the Li Ka Shing Knowledge Institute of St. Michael's Hospital, Toronto M5B 1W8, Ontario, Canada

GABRIEL LANG

Department of Medicine, Pritzker School of Medicine, University of Chicago, Chicago, IL 60637, USA

IRENE M. LANG

Department of Internal Medicine II, Division of Cardiology, Medical University of Vienna, 1090 Vienna, Austria

DANIEL TURNER LLOVERAS

Pritzker School of Medicine, Department of Medicine, Section of Pulmonary/Critical Care Medicine, University of Chicago, Chicago, IL 60637, USA

QING LU

Vascular Research Laboratory, Providence VA Medical Center, Providence, RI 02908, USA

SHWU–FAN MA

Department of Medicine, Section of Pulmonary/Critical Care Medicine, University of Chicago, Chicago, IL 60637, USA

KERRI A. MASSEY

Protogenomics Research Institute for Systems Medicine, Sidney Kimmel Cancer Center, San Diego , CA 92121, USA

DOLLY MEHTA

Center for Lung and Vascular Biology, Department of Pharmacology, University of Illinois at Chicago, Chicago, IL, USA

RICHARD D. MINSHALL

Departments of Anesthesiology and Pharmacology and Center for Lung and Vascular Biology, University of Illinois College of Medicine, Chicago, IL 60612, USA

SUMEGHA MITRA

Department of Medicine, Section of Pulmonary/Critical Care Medicine, University of Chicago, Chicago, IL 60637, USA

VLADIMIR R. MUZYKANTOV

Department of Pharmacology and Program in Targeted Therapeutics of the Institute for Translational Medicine Therapeutics, University of Pennsylvania School of Medicine, Institute for Environmental Medicine, University of Pennsylvania Medical Center, Philadelphia, PA 19104-6068, USA

RAMESH NATARAJAN

Pulmonary and Critical Care Medicine Division, Department of Internal Medicine, Virginia Commonwealth University, Richmond VA 23298, USA

MARK R. NICOLLS

Divisions of Pulmonary and Critical Care Medicine, Immunology and Rheumatology, Stanford University School of Medicine, VA Palo Alto Health Care System, Palo Alto CA 94306, USA

KAUSHIK PARTHASARATHI

Departments of Physiology and Biomedical Engineering, The University of Tennessee Health Science Center, Memphis, TN 38163, USA

SADIQA K. QUADRI

Division of Pulmonary, Allergy & Critical Care Medicine, Columbia University College of Physicians & Surgeons, New York, NY, USA

PRADEEP R. RAI

Division of Pulmonary and Critical Care Medicine, University of Colorado Health Sciences Center, Denver, CO, USA

J. USHA RAJ

Department of Pediatrics, University of Illinois at Chicago, Chicago, IL 60612, USA

THOMAS C. RESTA

Vascular Physiology Group, Department of Cell Biology and Physiology, University of New Mexico Health Sciences Center, Albuquerque, NM, USA

ROSANA SOUZA RODRIGUES

Department of Radiology, Federal University of Rio de Janeiro, Rio de Janeiro, Brazil

SHARON ROUNDS

Alpert Medical School of Brown University, Chief, Medical Service, Providence VA Medical Center, Providence, RI 02908, USA

JAN E. SCHNITZER

Protogenomics Research Institute for Systems Medicine, Sidney Kimmel Cancer Center, San Diego , CA 92121, USA

MARGARET A. SCHWARZ

Associate Professor Department of Pediatrics, University of Texas Southwestern Medical Center at Dallas, Dallas, TX, USA

RADU V. STAN

Department of Pathology, Dartmouth Medical School, Lebanon, NH, USA

KURT R. STENMARK

Department of Pediatrics, University of Colorado Health Sciences Center, Denver, CO 80262, USA

TROY STEVENS

Departments of Pharmacology and Medicine, Center for Lung Biology, College of Medicine, University of South Alabama, Mobile, AL 36688, USA

YUNCHAO SU

Department of Pharmacology and Toxicology, Medical College of Georgia, Augusta, GA 30912, USA

RASA TAMOSIUNIENE

Stanford University, Palo Alto Institute of Research Education, VA Palo Alto Health Care System, Palo Alto CA 94306, USA

MOHAMMAD TAUSSEEF

Center for Lung and Vascular Biology, Department of Pharmacology, University of Illinois at Chicago, Chicago, IL, USA

BERNARD THÉBAUD

Department of Pediatrics, Division of Neonatology, University of Alberta, Edmonton, AB T6G 2S2, Canada

RUBIN M. TUDER

Division of Pulmonary and Critical Care Medicine, University of Colorado Denver, School of Medicine, Aurora, CO 80045, USA

NORBERT VOELKEL

The E. Raymond Fenton Professor of Pulmonary Research, Director, Victoria Johnson Center for Pulmonary Obstructive Disease Research, Pulmonary and Critical Care Medicine Division, Virginia Commonwealth University, Richmond, VA 23298, USA

ELIZABETH WAGNER

Department of Medicine, Division of Pulmonary and Critical Care Medicine, Johns Hopkins Asthma and Allergy Center, Baltimore, MD 21224, USA

BENJIMEN R. WALKER

Vascular Physiology Group, Department of Cell Biology and Physiology, University of New Mexico Health Sciences Center, Albuquerque, NM, USA

RACHEL K. WOLFSON

Department of Medicine, Pritzker School of Medicine, University of Chicago, Chicago, IL 60637, USA

SONGWEI WU

Department of Pharmacology, Center for Lung Biology, College of Medicine, University of South Alabama, Mobile, AL 36688, USA

MERVIN C. YODER

Department of Pediatrics, Division of Neonatal-Perinatal Medicine, Indiana University School of Medicine, Indianapolis, IN 46202, USA

GUY A. ZIMMERMAN

Department of Internal Medicine, University of Utah School of Medicine, Salt Lake City, UT, USA

Introduction

Sharon Rounds[1] and Norbert Voelkel[2]

[1]Vascular Research Laboratory, Providence VA Medical Center, Alpert Medical School of Brown University, Providence, RI, USA

[2]Victoria Johnson Center for Pulmonary Obstructive Disease Research, Pulmonary and Critical Care Medicine Division, Virginia Commonwealth University, Richmond, VA, USA

Over the past 40 years there has been an explosion of new knowledge regarding normal and abnormal function of vascular endothelium. In the past, endothelium was regarded as a passive lining of blood vessels with organ-specific variability with regard to its role in filtration of blood or in maintenance of minimal fluid filtration. As the nonrespiratory functions of the lung became recognized, the importance of the endothelium became evident. In his review on this topic in 1969, Fishman stated with prescience "It is clear from the observations and speculations above that the degree to which the pleuripotential [sic] endothelial cells actually fulfill their potential promises to be a rewarding line of investigation" [1]. Indeed, with the advent of recognition of metabolic functions of endothelium, it became clear that the endothelium is critical to maintenance of a thrombosis-free surface, to interactions with circulating blood cells, and to modulation of vasomotor tone. This Introduction and this volume are not intended to enumerate all of the investigators and their contributions to the understanding of lung endothelial pathobiology, but to describe highlights in the field and to describe the current state of understanding.

The lung endothelium is now recognized to have a number of unique functional attributes that are due to its central location in the circulation. The entire cardiac output passes through the lung with every heartbeat. Furthermore, the lung endothelium has a vast surface area, estimated to be $120\,m^2$. Thus, lung endothelium is uniquely positioned to interact with circulating cells and vasoactive mediators. Indeed, it is now clear that the pathogenesis of many lung diseases, such as acute lung injury, is related to this important attribute.

Another unique feature of the lung endothelium is the need for the lung to maintain a relatively dry interstitial and alveolar gas space to facilitate gas exchange. The anatomic features of lung endothelium are critical to fluid and protein filtration, and crucial for normal lung function. The ultrastructural features of the pulmonary capillary endothelium important in maintenance of normal lung vascular permeability [2] and the effects of injury on endothelium have been elegantly described [3]. There has also been an enormous increase in understanding of the cell biology of lung endothelial permeability and the effects of injury on signaling mechanisms, such as increased permeability caused by thrombin [4].

The study of the lung endothelium originally used the study of the metabolism of circulating substances, such as angiotensin I [5], 5-hydroxytryptamine (serotonin) [6], and eicosanoids [7], using passage through isolated perfused lungs [8]. Similarly, isolated perfused lungs were used to assess perturbation of endothelial permeability [9]. The advent of techniques for isolation and culture of endothelial cells (EC)s from umbilical veins [10, 11], the main pulmonary artery [12], and pulmonary microvessels [13–15] has allowed the study of endothelium alone, without confounding factors related to distribution of perfusate. Correlation of results using cultured ECs and intact lungs was an important advance in the field [16]. In addition, the availability of cultured endothelium has allowed elucidation of the interactions of ECs with blood cells and platelets. More recently, with the advent of animal models of disease and genetically manipulated models, emphasis has shifted to the study of endothelium of intact lungs.

Recent research has made clear that the lung ECs are heterogeneous in calcium handling, permeability, and proliferative potential with differences between endothelium of conduit vessels and the microcirculation, as described in Chapters 5 and 9 of this volume. Furthermore, the bronchial and pulmonary circulations differ in their physiology and responses to disease, as discussed in Chapter 14. It is now apparent that the lung endothelium is not a static organ, but is capable of regeneration and repopulation via resident and circulating progenitor cells, as described in Chapter 13.

The pulmonary circulation, unlike the systemic circulation, is a low-pressure, high-volume circulation that responds to hypoxia with vasoconstriction. The lung endothelium is critical to maintenance of normal lung vascular tone and modulation of hypoxic vasoconstriction, reviewed in Chapter 12. In addition, the pulmonary circulation responds to alveolar hypoxia with vascular remodeling and sustained pulmonary hypertension. The lung endothelium again is key in modulation of pulmonary vascular remodeling, as discussed in Chapters 11 and 27.

The most recent group of very exciting advances is the growing recognition that the lung endothelium plays an important role in the pathogenesis of lung diseases and this work is highlighted in this volume in Chapters 23–30. It has become increasingly clear that many lung diseases are directly due to or complicated by pulmonary EC dysfunction.

This volume is a group of essays that describe the state-of-the-art knowledge of lung endothelium. The volume is divided into three sections. The first section describes the Normal Pulmonary Endothelium, including development, structure, cell biology, signaling, functions, heterogeneity, interactions with circulating cells and mesenchymal cells, and the endothelium of the bronchial circulation. The second section of the volume deals with Mechanisms and Consequences of Pulmonary Endothelial Cell Injury, ranging from effects on ECs to organ injury, including protection against lung permeability and drug targeting to pulmonary endothelium. The third section of the volume focuses on the Pulmonary Endothelium in Disease. Although not a diseased state, this includes the transition from the fetal to the newborn lung. Throughout the volume, it will be evident that these sections are somewhat arbitrary since insights into normal function inevitably enhance understanding of pathophysiology and vice versa.

We are grateful to the authors who have contributed outstanding chapters that reflect both their work and overviews of the field. We are also grateful to our colleagues and spouses for their support of this effort. Finally, we thank our publishers, especially Fiona Woods of John Wiley & Sons, Ltd, who has patiently and firmly encouraged the completion of this work.

References

1. Heinemann, H.O. and Fishman, A.P. (1969) Non-respiratory functions of mammalian lung. *Physical Review*, **49**, 1–47.
2. Schneeberger-Keeley, E.E. and Karnovsky, M.J. (1968) The ultrastructural basis of alveolar-capillary membrane permeability to peroxidase used as a tracer. *Journal of Cell Biology*, **37**, 781–93.
3. Bachofen, M. and Weibel, E.R. (1977) Alterations of the gas exchange apparatus in adult respiratory insufficiency associated with septicemia. *American Review of Respiratory Disease*, **116**, 589–615.
4. Mehta, D. and Malik, A.B. (2006) Signaling mechanisms regulating endothelial permeability. *Physical Review*, **86**, 279–367.
5. Fanburg, B.L. and Glazier, J.B. (1973) Conversion of angiotensin 1 to angiotensin 2 in the isolated perfused dog lung. *Journal of Applied Physiology*, **35**, 325–31.
6. Block, E.R. and Fisher, A.B. (1977) Depression of serotonin clearance by rate lungs during oxygen exposure. *Journal of Applied Physiology: Respiratory, Environmental and Exercise Physiology*, **42**, 33–38.
7. Bakhle, Y.S., Jancar, S., and Whittle, B.J.R. (1978) Uptake and inactivation of prostaglandin E_2 methyl analogues in the pulmonary circulation. *British Journal of Pharmacology*, **62**, 275–80.
8. Dawson, C.A., Bongard, R.D., Rickaby, D.A. *et al.* (1989) Effect of transit time on metabolism of a pulmonary endothelial enzyme substrate. *American Journal of Physiology: Heart and Circulatory Physiology*, **257**, H853–65.
9. Schneeberger, E.E. and Neary, B.A. (1982) The bloodless rat: a new model for macromolecular transport across lung endothelium. *American Journal of Physiology: Heart and Circulatory Physiology*, **242**, H890–99.
10. Jaffe, E.A., Nachman, R.L., Becker, C.G., and Minick, R.C. (1973) Culture of human endothelial cells derived from umbilical veins. Identification by morphologic and immunologic criteria. *Journal of Clinical Investigation*, **52**, 2745–56.
11. Gimbrone, M.A. Jr., Cotran, R.S., and Folkman, J. (1974) Human vascular endothelial cells in culture. Growth and DNA synthesis. *Journal of Cell Biology*, **60**, 673–84.

12. Ryan, U.S., Clements, E., Habliston, D., and Ryan, J.W. (1978) Isolation and culture of pulmonary artery endothelial cells. *Tissue and Cell*, **10**, 535–54.
13. Ryan, U.S., White, L.A., Lopez, M., and Ryan, J.W. (1982) Use of microcarriers to isolate and culture pulmonary microvascular endothelium. *Tissue and Cell*, **14**, 597–606.
14. Alvarez, D.F., Huang, L., King, J.A. *et al.* (2008) Lung microvascular endothelium is enriched with progenitor cells with vasculogenic capacity. *American Journal of Physiology: Lung Cellular and Molecular Physiology*, **294**, L419–30.
15. Masri, F.A., Xu, W., Comhair, S.A.A. *et al.* (2007) Hyperproliferative apoptosis-resistant endothelial cells in idiopathic pulmonary hypertension. *American Journal of Physiology: Lung Cellular and Molecular Physiology*, **293**, L548–54.
16. Junod, A.F. and Ody, C. (1977) Amine uptake and metabolism by endothelium of pig pulmonary artery and aorta. *American Journal of Physiology: Cell Physiology*, **232**, C88–94.

SECTION I:

NORMAL PULMONARY ENDOTHELIUM. STRUCTURE, FUNCTION, CELL BIOLOGY

1

Development of the Pulmonary Endothelium in Development of the Pulmonary Circulation: Vasculogenesis and Angiogenesis

Margaret A. Schwarz[1] and Ondine B. Cleaver[2]

[1]Department of Pediatrics, University of Texas Southwestern Medical Center at Dallas, Dallas, TX, USA

[2]Department of Molecular Biology, University of Texas Southwestern Medical Center at Dallas, Dallas, TX, USA

INTRODUCTION

Role of the Pulmonary Vasculature

The cardiovascular system, comprised of the heart and blood vessels, is the first functional organ formed during embryogenesis in higher vertebrates. In the mouse, the heart and first vessels become functional as early as 8 days following fertilization, while in humans the cardiovascular system forms after approximately 3 weeks of development. Cardiovascular function is essential to the survival of higher organisms, because every cell requires nutrition, gas exchange, and elimination of wastes via blood vessels. The primary site of gas exchange is the vascular/alveolar interface, located deep within the lung. Once blood is oxygenated in the lung, pumping of the blood by the heart disperses oxygen-rich blood throughout the body, where exchange of gas within tissues occurs via capillary beds. Then, oxygen-depleted, carbon dioxide-rich blood is returned to the lungs via the vena cava, for the respiratory/circulatory cycle to begin anew. Despite decades of research into the biology of this vascular/pulmonary interface, little is known about how the pulmonary vasculature ensures its proper coordinated growth and intimate development along the tree-like epithelium of the developing lung.

Vascular Development Overview

Morphogenesis of the embryonic vascular system begins with the emergence of angioblasts, or endothelial progenitor cells, which are initially scattered within the mesoderm prior to their incorporation into patent vessels [1]. Angioblasts are fibroblast-like, mesodermal cells capable of migrating, recognizing other angioblasts, adhering, and organizing into vascular structures. Once an angioblast is recruited into forming a vascular "tube," or vessel, it differentiates into a bona fide differentiated endothelial cell (EC). The defining cell type of the established cardiovascular system is thus the EC, which forms the seamless lining of the entire circulatory system. As the vasculature develops, the initial circulatory system is composed of a rather homogeneous system of primitive vessels, or "plexus." However, as the embryo develops, this plexus reshapes and remodels into a hierarchical network of large and small vessels. In large vessels, such as the major arteries and veins, the endothelial inner lining becomes insulated by thick layers of extracellular matrix (ECM) components and smooth muscle. In capillary beds, where vessels taper to very narrow diameters, and gases and nutrients are actively exchanged, the endothelium is relatively more "naked" and in immediate contact with surrounding tissues. Thus, development of the vas-

The Pulmonary Endothelium: Function in health and disease Editors Norbert F. Voelkel, Sharon Rounds

cular system is a step-wise series of dynamic cellular activities, which together shape individual blood vessels, thereby ensuring proper distribution of oxygen-rich blood throughout the body. Interestingly, most key steps in specification and differentiation of vascular cell types are driven by the molecular interaction of vascular endothelial growth factor (VEGF) with its receptor vascular endothelial growth factor receptor VEGFR-2, which is expressed in vascular ECs. In this chapter, we will review the basic steps during systemic and pulmonary vessel development, since they are driven by many analogous mechanisms, and we will present new ideas regarding the molecular basis of their coordinated growth.

ONTOGENY OF VASCULAR CELLS

Endothelial Origin

To fully understand vascular development, it is essential to know where exactly endothelial precursors come from. Although their exact cell of origin has long remained elusive, angioblasts are known to differentiate exclusively from the mesoderm [2, 3]. In addition, it has been demonstrated that angioblasts arise in both extra- and intra-embryonic mesoderm, with their extra-embryonic emergence in the yolk sac preceding their differentiation in embryonic tissues. In mouse, the first extra-embryonic angioblasts can be detected as early as embryonic day (E) 6.5, while those in the embryo proper can be identified later, around E7.0 [4–6]. The first angioblasts identified in the yolk sac can be found within local proliferative foci of extra-embryonic mesoderm. These aggregations of angioblasts progressively take a more definitive shape, either as angioblast "cords" (linear aggregates) or blood islands (see following section) [5, 6]. In all vertebrates examined, these primitive vascular structures precede the formation of a functional and continuous vasculature.

Blood Islands and Hemangioblasts

As mentioned in the previous section, some of the earliest angioblasts identified in vertebrates are those in or near structures called "blood islands" [5, 7]. In mouse, blood islands are scattered in a ring around the distal yolk sac mesoderm [8–10]. In frog and fish, on the other hand, a single blood island is found on the ventral aspect of the gut. Blood islands have been described as "mesodermal cell aggregates," where inner cells consist of blood or hematopoietic stem cells and outer cells comprise a mantle of angioblasts [5]. Thought to represent transitional structures, blood islands have been shown to grow and fuse, creating a continuous network of blood filled vessels [6, 11, 12]. However recent work calls into question this "blood island fusion" mechanism of vascular development, and suggests instead that embryonic vessels are more likely to derive from ECs migrating and enveloping, or "capturing," hematopoietic precursors, as they generate a continuous vasculature [5]. Regardless of the exact dynamics, blood islands have been observed for over a century and are a hallmark of the primitive vertebrate yolk sac vasculature.

The close spatial and temporal association of hematopoietic and EC development in the yolk sac blood islands led to the idea that both lineages originated from common precursor called the "hemangioblast" [1, 13–16]. This possibility is supported by the observation that vessel and blood progenitors express many common markers and mutation of a number of genes affects both lineages [11, 17]. For decades, evidence has accumulated that supports the existence of a hemangioblast [18–20]. However, the isolation of a truly bipotential cell in the embryo, with the capacity to give rise exclusively to both EC and hematopoietic cell types, has yet to be conclusively shown. Recent experiments demonstrate that most intra-embryonic ECs do not emerge from blood islands, and in addition, few blood and ECs actually arise from common progenitors [21–23]. Therefore, the question remains open as to the true nature of the hemangioblast, the breadth of its potential to give rise to different cell types, and its actual frequency within the early vertebrate embryo.

The Endothelial Cell

The fundamental building unit of the blood vessel is the EC. Together, blood vessels of an adult human consist of approximately 1×10^{13} ECs, which stitch together to form the hierarchical network of vessels that carry blood throughout the body [24]. One interesting question that arises is exactly how does one define the EC? Only two shared characteristics have been identified that can be applied to all ECs [25]. The first is anatomical, in that ECs adhere to one another and form the seamless inner lining of all blood vessels. The second is functional, in that ECs create a selectively permeable and active interface, between blood and tissues, which controls the passage of nutrients, gases, and immune cells. Surprisingly, beyond these two traits, no single definition can be applied globally to all ECs. Blood vessels are strikingly different from one tissue to the next. It has been said that there are as many different types of ECs as there are tissues [26]. In the last decade, ECs have been shown to be extremely heterogeneous in their transcriptional profile, structural features, and regionalized functions [27–29]. Therefore, perhaps a more apt definition of ECs is that they can generally be defined as the cells that line the lumen of blood vessels, but display a variable nature that is strikingly heterogeneous, dynamic, and plastic.

ONTOGENY OF THE VASCULATURE

Cellular Mechanisms of Blood Vessel Formation

Blood vessel development occurs via two principal and distinct cellular mechanisms, referred to as vasculogenesis and angiogenesis (Figure 1.1) [15, 30, 31–34]. The initial primitive vascular plexus emerges via vasculogenesis, which describes the *de novo* formation of blood vessels from individual angioblasts. Angiogenesis, in contrast, describes the growth and remodeling of the existing primitive vasculature, and occurs during normal growth of embryonic organs and tissues. Both vasculogenesis and angiogenesis strictly refer to "the genesis of blood vessels"; however, they have been used to describe very different cellular mechanisms of blood vessel formation.

Vasculogenesis

Vasculogenesis refers to the formation of blood vessels via the clustering and organization of individual angioblasts into linear aggregates, or "cords," followed by the formation of a patent lumen (Figure 1.1a) [15, 30, 35, 36]. In addition, the term has also been used to describe the fusion of blood islands into blood-filled tubes within the yolk sac. Vasculogenesis is known to be the primary mechanism by which the first embryonic vessels form [2, 36]. This includes the primordia of most primitive blood vessels, including the dorsal aortae and the endocardium, as well as the relatively homogeneous capillary network found in tissues such as the yolk sac. Vasculogenesis is therefore a term that describes a step-wise developmental process, which includes angioblast migration, proliferation, adhesion, morphogenesis, differentiation, and maturation into ECs. Coalescence of these individual vascular progenitors ultimately leads to the formation of a continuous network of vessels, which circulation depends on. "Vasculogenesis" and "neovascularization" are both terms that refer to this *de novo* formation of blood vessels, and are often used interchangeably. Two types of vasculogenesis have been described, type 1 and type 2, with the distinction being based on the location of angioblast emergence relative to the location of vessel formation. In type 1, angioblasts aggregate into cords, at

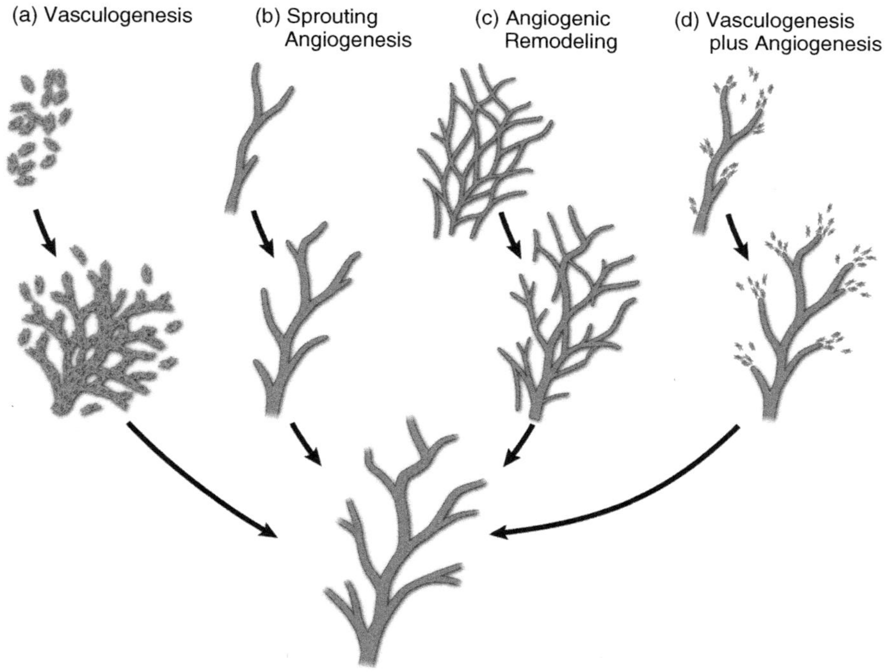

Figure 1.1 Schematic illustrating the different mechanisms of blood vessel formation. (a) Vasculogenesis is the *de novo* formation of vessels via aggregation of angioblasts within the mesoderm. (b) Sprouting angiogenesis is the formation and extension of new sprouts from pre-existing vessels. (c) Angiogenic remodeling is the reorganization and shape change of vessels within an existing vascular plexus. (d) In many tissues, including lung, vasculogenesis and angiogenesis are coordinated to create vascular beds within developing organs and tissues.

the same location where they emerge in the mesoderm. In type 2, angioblasts appear in the mesoderm, but then actively migrate to a different location, where they then coalesce into vessels. During embryonic vascular development, dorsal aortae formation in mouse occurs by vasculogenesis type 1 [37], while the formation of a single dorsal aorta in frog entails vasculogenesis type 2 [38, 39].

Tubulogenesis

Central to the concept of vasculogenesis is the concept of endothelial tubulogenesis. Morphogenesis of a vascular "tube," from a "cord" of angioblasts or within a growing angiogenic sprout, occurs via tubulogenesis. Tubulogenesis has been described as occurring by two distinct mechanisms. In the first mechanism, the vascular lumen forms by the alignment and fusion of "intracellular spaces," such as large vacuoles [40, 41]. Classical observations in the avian embryo suggest this first mechanism, where a lumen can be shown to form from the fusion and expansion of intracellular vacuoles into a long continuous space across many cells, at the center of a cord [40–45]. Alternatively, the lumen can be generated by the enlargement of an "extracellular space" located between adjacent angioblasts [46]. The latter mechanism for vascular "tube" formation primarily involves cellular rearrangements that drive the transformation of a solid cord of cells, into a patent cylinder. Based on zebrafish observations [46], it might be predicted that vacuole fusion-based tubulogenesis is likely to be predominantly used in angiogenic sprouting as discussed below, whereas rearrangement-based tubulogenesis is likely to occur primarily during vasculogenesis.

Angiogenesis

Following the formation of the initial primitive vascular plexus via vasculogenesis, the simple circulatory system is then elaborated and extended via angiogenesis. Two fundamentally distinct angiogenic mechanisms have been identified: "sprouting angiogenesis" and "angiogenic remodeling." Sprouting angiogenesis is defined as the sprouting and extension of new vessels from pre-existing vessels. Quiescent cells within the walls of vessels proliferate, branch, and extend new sprouts into avascular tissues. Angiogenic remodeling encompasses the multiple gross changes that pre-existing vessels can undergo in their basic size or pattern, including the splitting or fusion of the vessel and the enlargement or shrinking of vessel diameter [47–49]. Often these changes in vessel size or shape occur in response to hemodynamic forces. Here, we describe the general features distinguishing each type of angiogenesis.

Sprouting Angiogenesis

Sprouting angiogenesis involves sprouting of new capillaries from the walls of pre-existing blood vessels (Figure 1.1b). Quiescent cells at a specific point along the vessel wall initiate a cascade of targeted cellular activities, all aimed at building an entirely new vessel branch from a pre-existing parent vessel. To create a new sprout, proteolytic degradation of the ECM surrounding the parent vessel is coordinated with proliferation of the sprouting ECs. Together these cellular activities generate a new growing vascular branch, which will eventually fuse with the wall of an adjacent vessel.

Cells at the distal tip of extending angiogenic sprouts, termed "tip" cells, have attracted recent attention. New capillary sprouts grow into the interstitium by the ameboid migration of distal tip ECs. These invade surrounding avascular tissue, migrate as the sprout extends, fuse with the endothelium of an adjacent vessel, and open up a new connecting lumen [14]. Interestingly, the growth of new sprouts is not believed to occur by proliferation of the tip cells. As the angiogenic sprouts extend, it is within the growing stalk that new cells are added by mitotic proliferation of pre-existing ECs [50]. Classical observations of neural angiogenesis demonstrated that ECs located at the tip of sprouts exhibited a number of distinctive "filiform" processes, hypothesized to function in seeking out and fusing with other growing vessels [51]. More recent studies on endothelial tip cell filopodia in growing retinal vessels have shown that filopodia are the primary target of VEGF signaling and function to drive vessel growth and extension [52, 53].

Remodeling Angiogenesis

Another angiogenic process that generates basic morphogenetic changes in the vascular network architecture is "remodeling angiogenesis," or "angiogenic remodeling." In this angiogenic process, pre-existing vessels change in shape, size, and fundamental organization (Figure 1.1c). Generally, these changes involve a wide range of cellular modifications that dynamically alter blood vessel size or architecture. During remodeling, vessels of an initial embryonic plexus either enlarge or regress during development, accommodating the coordinated growth and differentiation of other tissues. Once the vascular system is mature, the vascular network becomes relatively stable and undergoes angiogenic remodeling only in select tissues, such as in female reproductive organs, wound healing, or during pathological processes (e.g., tumor growth).

A dramatic example of angiogenic remodeling involves the primary capillary plexus of the early murine yolk sac. Initially, this plexus presents as a relatively homogeneous network of vessels, resembling a fisherman's net, with most vessels being of equal size, length, and similar appearance. However, this primitive plexus is rapidly remodeled and modified into the familiar hierarchical, tree-like array of larger and smaller blood vessels. These transformations occur via "angiogenic remodeling" [31, 54]. Angiogenic remodeling remains poorly understood, despite the fact many mouse mutants display clear failure of vascular remodeling.

A wide variety of cellular mechanisms underlie angiogenic remodeling, causing either an increase or decrease in vessel density. Here, we describe intussusception, regression, and pruning. Intussusception is the process of splitting and reorganizing pre-existing vessels, resulting in the expansion of a capillary network [55, 56]. During intussusception, proliferation of ECs within a vessel results in the formation of a large lumen that is subsequently split by intervening endothelial walls (thus resulting in the splitting of one vessel into two). Another mechanism of vascular remodeling, which in contrast decreases capillary density, involves endothelial regression [57]. Key steps in vessel regression include changes in EC shape, lumen narrowing, increased vacuolation, cessation of blood flow, detachment from the basement membrane, and cell death. Regression of vessels often occurs as a result of either a reduction of blood flow, cessation of VEGF-mediated maintenance, or other genetically determined processes, such as changes in expression of angiogenic cues in surrounding tissues. Yet another type of vascular remodeling, which also decreases vessel density and does not involve cell death, has been termed "pruning," as it resembles the process of thinning out excess branches on a tree [31]. Pruning was first observed in the embryonic retinal vasculature and involves the regression of redundant, parallel channels [58]. In these vessels, blood flow ceases, their lumens collapse and ECs retract out of the regressing vessel. In all cases of angiogenic remodeling described above, the principal goal is to fine tune the vasculature so that it perfuses tissues at the required density, satisfying local oxygen demands, by trimming excessive, unneeded vessels or reorganizing vessels to meet physiological demands.

Vasculogenesis and Angiogenesis within Organs

Vascularization of most developing embryonic organs has long been thought to occur primarily via angiogenic invasion of vessels. This was a sensible supposition, given that growing organs appeared to be vascularized by ingrowth of vessels that originated and sprouted from the pre-existing primary vascular plexus. However, improved technology for visualization of the vasculature and its precursors, using newly identified molecular markers and new vascular reporters, has revealed that most organs develop at least part of their vasculature via *in situ* aggregation of local mesenchymal angioblasts or vasculogenesis [34]. This holds true for the growing vasculature of the lung, liver, stomach, spleen, pancreas, intestine, and kidney [32, 59–63]. During embryonic development of these organs, it is known that angiogenic sprouting from existing vessels also contributes to maintenance and extension of the primitive organ vasculature [34]. New observations have demonstrated that peripheral vasculogenic vessels often fuse with invading angiogenic vessels [64]. Thus, it seems likely that building a continuous vasculature within most organs is a coordinated joining of both vasculogenic beds with angiogenic ingrowth of sprouting vessels.

ARTERIAL VERSUS VENOUS DIFFERENTIATION

Once blood flow begins within the circulatory system, the immature vascular plexus becomes segregated into recognizable arteries and veins (Figure 1.2). Vessels can be categorized as either veins or arteries by a number of parameters, including the direction of blood flow within their lumens, anatomical and functional differences, as well as by the expression of several markers. For instance, the expression of ephrin B2 (Efnb2) ligand is enriched in arteries, while expression of the B4 ephrin receptor (EphB4) is enriched in veins. In addition, a variety of other markers are specific for arteries, including Dll4 [65, 66], Jag1 [67], Notch1 [68], Hey1 and Hey2 [69], activin receptor-like kinase 1 [70], and EPAS1/hypoxia-inducible factor (HIF) [71].

The mechanisms underlying the specification of arterial and venous cell fate are largely unknown. Previously, circulatory dynamics were thought to be the driving cause of arteries and veins developing into structurally and functionally different vessels. However, growing evidence points to a genetic program underlying this fundamental distinction. Indeed, labeling experiments in zebrafish suggest that arterial and venous EC fate may be determined before the formation of blood vessels [72]. Similarly, work in chicks has demonstrated that segregation of arterial and venous markers has already occurred in subpopulations of blood islands long before vessel formation [73]. Therefore, growing evidence points to hard-wired genetic cues specifying arteriovenous cell fate extremely early during vascular development.

Interestingly though, it also seems likely that different vascular beds experience artery/vein specification at different times. For instance, arteriovenous markers in certain organs, such as myocardium [74] and pancreas (Cleaver, unpublished), appear to acquire their identities much later during development. In addition, it is

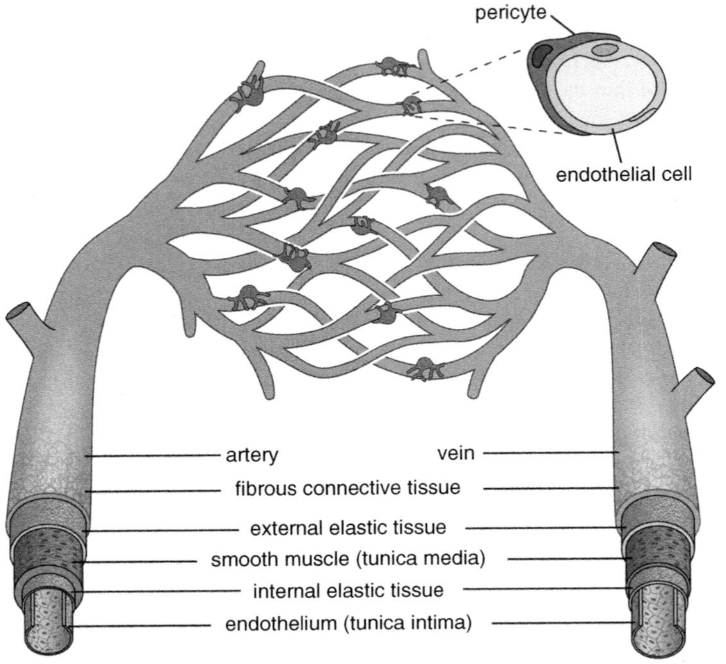

Figure 1.2 Fundamental architecture of blood vessels. Capillary beds perfuse tissues. Capillaries are small caliber vessels, the lumen often forming from single ECs. Capillaries are largely devoid of supportive cells, except for sparse coverage by pericytes. Capillaries are connected in a hierarchical fashion to larger arterioles and venules, which in turn connect to arteries and veins. Arteries and veins are insulated by thick layers of elastic, smooth muscle and fibrous tissues. A color version of this figure appears in the plate section of this volume.

known that arteriovenous cell fate is highly plastic and reversible. In grafting experiments in chicks, vascular ECs were shown to be plastic with respect to their arteriovenous fate [75]. In these experiments, fragments of arteries were heterotopically transplanted to different embryonic sites. Strikingly, cells from the grafted arteries would quickly colonize either host arteries or veins. When they colonized veins, arterial ECs turned off arterial markers and upregulated venous markers. Thus, EC fate remains plastic with respect to arteriovenous differentiation, at least for a period of time during early development.

KEY MOLECULES IN VASCULAR DEVELOPMENT

VEGF [76, 77], and its receptors VEGFR-1 (also called Flt-1) and VEGFR-2 (also called KDR or Flk-1) [78] have long been known to be critical regulators of endothelial differentiation, as well as blood vessel formation and morphogenesis [79]. VEGF-A is essential for proper vessel formation and selective expression of VEGF-A isoforms (murine 120, 164, 188; human 121, 145, 165, 189, 206) drives different aspects of vessel formation in many different organs, including the lung [80]. Here, we introduce the principal vascular developmental factors and outline their roles in vessel formation.

VEGF-A and its Isoforms

The VEGF family of growth factors consists of VEGF-A, B, C, D, and E, and placental growth factor (PlGF). All family members regulate at least some aspect of EC proliferation, migration, and/or survival [79, 81]. Gene targeting demonstrates that VEGF-A plays an essential role in early vessel development. VEGF-A expression is dynamic throughout embryonic development and is often expressed in tissues immediately adjacent to developing blood vessels [38, 77, 82, 83]. VEGF-mediated signaling drives both vessel formation by vasculogenesis, as well as angiogenic invasion of developing tissues. Mice lacking a single VEGF allele die early during embryogenesis (around E10.5). These VEGF-null embryos show a range of vascular defects, including severe abnormalities in EC differentiation, sprouting angiogenesis, vessel lumen formation, and in the overall patterning of the vasculature [84, 85]. The profound vascular phenotype that results from the loss of a single allele of VEGF demonstrates that tight regulation of VEGF levels is critical for proper vascular development. However, given that angioblasts are present in the VEGF knockout embryos, it can be inferred that VEGF signaling is not required for initial specification of angioblasts [86], but is critical for their proper differentiation and morphogenesis.

VEGF-A presents a number of alternate isoforms, which are generated by alternative splicing of the VEGF-A mRNA. Resulting isoforms differ in their biological activities, as a direct result of differences in their receptor binding affinities and in their ability to diffuse within the extracellular environment. The larger forms of VEGF (VEGF164, 188, and 205 in mouse) possess a motif that tethers them to various ECM components and thus decreases their diffusibility. The smallest isoform of VEGF lacks this domain and can freely diffuse. This form has been shown to drive chemotaxis of migrating angioblasts [39]. Gene targeting of these different isoforms results in a range of vascular defects [87]. Therefore, it seems likely the coordination of different isoforms is critical for the generation of a continuous and functional embryonic vasculature.

VEGFRs

The principal receptor for VEGF is the receptor tyrosine kinase VEGFR-2. VEGFR-2 has been shown to be critical for both vasculogenesis and angiogenesis, and is one of the most reliable markers of angioblasts and differentiated ECs. Expression of VEGFR-2 has been shown to be high during embryonic blood vessel formation and in tumor vessels [38, 77, 78, 88]. Mice lacking VEGFR-2 function die early during development, between E8.5 and E10.5, from almost total failure of vascular development [17]. Mutant animals lack almost all angioblast differentiation and either cord or vessel formation. In addition, these mice lack all hematopoietic cells. Thus, VEGFR-2 is a key regulator of both angioblast specification and differentiation. In this chapter, we will review its role during pulmonary vascular development in detail (see "Vascular Growth Factors in Lung Morphogenesis").

VEGFR-1 displays structural and expression similarities to VEGFR-2, but appears to play a distinct role during vessel formation. VEGFR-1 is a high-affinity receptor for VEGF and PlGF, much like VEGFR-2 [89]. In contrast to VEGFR-2-null mutants however, loss of VEGFR-1 function does not affect early angioblast development, but it does affect their ability to assemble and organize into vessels [90]. In addition, VEGFR-1-deficient embryos actually show an increase, rather than a decrease, in the number of EC precursors throughout the embryo [91]. While VEGFR-1, like VEGFR-2, possesses an intracellular tyrosine kinase domain, mutation of this domain does not impede normal vessel formation. This suggests that the intracellular portion of the receptor may not transduce active intracellular signaling. Instead, it has been proposed that VEGFR-1 normally functions to sequester excess VEGF ligand, which may regulate the number of differentiated angioblasts and subsequent EC proliferation.

FORMATION OF PULMONARY VASCULATURE

Once the embryo has established a rudimentary circulatory system capable of providing oxygen and nutrients to growing tissues, organ development begins, driven by genetic cues. Coordinately, organ vascular beds also begin to emerge and grow. Although a significant amount is known regarding the forces that drive embryonic vessel formation and lung branching morphogenesis, the angiogenic and vasculogenic mechanisms that establish the pulmonary circulation remain poorly understood. This is in part a result of the complexity of distal pulmonary development, where intimate association of alveolar and vascular tissues must be coordinated to create a functional interface that allows proper oxygen exchange in the mature lung. Given this interdependent relationship between alveolar and vascular development, it has proven difficult to distinguish the mechanisms underlying vascular emergence from those driving distal epithelial morphogenesis. In the second half of this chapter, we review the stages of pulmonary branching morphogenesis and place these in context with what is known regarding pulmonary vascular development. In addition, we also introduce new ideas regarding the molecular basis of their close association and coordinated growth.

ORIGIN OF THE LUNG

Lung morphogenesis initiates on the ventral aspect of the foregut. The first signs of lung formation are a thickening of the foregut epithelium and the subsequent evagination of the laryngotracheal groove. The groove then separates from the esophagus posteriorly, giving rise to the laryngotracheal tube. This parallel tube then grows distally into the underlying splanchnopleuric mesoderm. Morphogenetic changes of the endodermal epithelium result in the formation of two small lung buds, composed of inner epithelial pouches surrounded by a thick layer of mesoderm. This mesodermal layer consists of undifferentiated mesenchyme, vascular, and neuronal cells, surrounded by a thin layer of mesothelium. Following initial embryonic lung budding, early lung morphogenesis then involves a stereotypic pattern of reproducible budding and branching events, that generate a complex, tree-like system of epithelial branches, which maintain medial–lateral and left–right axes and form the mature lung organ [92–96].

STAGES OF LUNG DEVELOPMENT

Lung development, including pulmonary neovascularization, can be divided into five classic chronological stages based on the growth and differentiation of specific pulmonary epithelial structures (Figure 1.3) [97–99].

(i) **Embryonic stage**, when the evaginating foregut endodermal epithelium invades the adjacent primitive mesoderm (murine: 9.5–11.5 days; human: 3.5–7 weeks).

(ii) **Pseudoglandular stage**, during which epithelial-lined airways (pre-acinar bronchi) undergo repeated dichotomous branching (murine: 11.5–16 days; human: 7–17 weeks).

(iii) **Canalicular (or vascular) stage**, is marked by proliferation of the vasculature, emergence of capillaries, epithelial thinning, and differentiation of the alveolar type 1 and 2 cells (murine: 16.5–17.4 days; human: 17–27 weeks).

(iv) **Saccular stage**, when vascularization and the number of terminal sacs increases, concurrent with formation of crests and cup-shaped alveoli (murine: 17.4–5 + days; human: 28–36 weeks).

(v) **Alveolar stage**, during which the alveolar ducts and alveoli develop, mature, and proliferate two to threefold before reaching their adult number (murine: 5+ days; human: 36 weeks gestation onwards).

Progression of lung development through these five distinct stages is consistent across mammalian species.

ORIGIN OF LUNG VASCULATURE

Similar to vessel formation within the developing embryo [100], lung neovascularization is governed by complex interactions between ECs, endodermal and mesodermal cells, mural cells, the ECM, and the cellular microenvironment, as well as by epigenetics [28, 101]. Consistent with vessel formation in other tissues, angiogenesis and vasculogenesis are considered to work in concert to form the pulmonary vascular system [64, 99, 102–104]. Identifying the mechanisms underlying formation of the pulmonary circulation poses many challenges. Initial

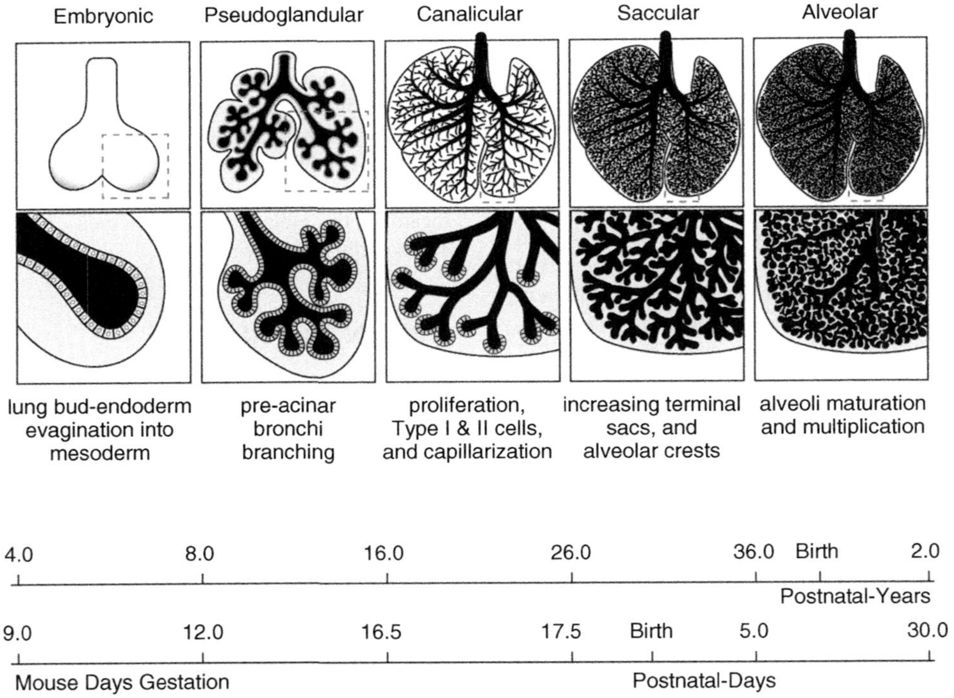

Figure 1.3 Diagram illustrating the stages of lung development that are consistent across mammalian species.

observations using staining for von Willebrand factor suggested that vessel formation in the emerging lung was predominately limited to the canalicular stage [105]. However, more recent observations using *in situ* hybridization and transgenic mouse studies that examined VEGFR-2 expression, generally considered to be a marker of primitive angioblasts and developing vessels, indicate that vessel formation occurs throughout all stages of lung development [106]. Thus, the evolution of available tools and reagents has resulted in an improved anatomical understanding of lung vessel location.

ANGIOGENESIS AND VASCULOGENESIS IN THE DEVELOPING LUNG

Serial histological reconstruction of human embryonic fetal lungs has provided significant insight into the developing lung vasculature. These histological studies indicate that during the embryonic stage of lung development, cells expressing the CD34 antigen (hematopoietic progenitor cell marker) coalesce and form the pulmonary arteries via vasculogenesis within the mesoderm [98, 107, 108]. As lung morphogenesis proceeds to the pseudoglandular stage, pulmonary arteries are believed to continue to be formed via vasculogenesis, while later, during the canalicular and alveolar stages, extension of these vessels occurs via angiogenic mechanisms [98, 107, 108]. Thus, based on these histological studies, it would appear that the development of pulmonary circulation employs sequentially the distinct mechanisms of vasculogenesis and angiogenesis.

In contrast to these histological findings, electron microscopy and methacrylate vessel-casting studies suggests that two independent vascular networks, one angiogenic and one vasculogenic, actually form in parallel and only later connect with each other to generate a continuous circulatory network within the lung [61]. Indeed, these studies suggest that these two networks, which arise simultaneously but independently from each other, have only rare anatomical communication between them during early lung development. Electronic microscopy studies identified vasculogenic pools of clustered angioblasts throughout the embryonic stage, as separate and peripherally located within the lung mesenchyme. To characterize angiogenic vessel formation, vessel casting was performed. The earliest point at which vessel casting could be accomplished, E12 – at the beginning of the pseudoglandular stage – indicated that arterial and venous vessels sprout at this stage from central pulmonary trunk vessels. Communication between the two networks was found to then gradually increase, until a complete vascular circuit is established by E17 just before term in the mouse embryo (term = E18.5) [61]. One complication is that the vessel casting technique is limited, as the location of growing vessels in relationship to the mesenchyme and bronchi is not effectively revealed. Emerging angiogenic vessels are fragile making identification difficult, and casting at earlier stages prior to E12 of fetal development is limited by embryo size. However despite these limitations, casting studies were the first to identify the simultaneous development of the two parallel pulmonary vascular networks.

Analysis of the expression of EC-specific reporter genes has further expanded our understanding of vasculogenesis and angiogenesis during lung vascular development. Utilizing transgenic reporter mouse lines, both vasculogenic and angiogenic derived emergence of vessels has been observed. Distribution of Tie2 receptor expression in Tie2–lacZ transgenic mice suggests that vessels do not originate *de novo* in the lung bud mesenchyme, but are instead attracted to the lung bud and grow into the lung mesenchyme by angiogenic sprouting [109]. Indeed, vessels expressing Tie2 are observed extending from the medial gut tube toward the distal tip of the lung buds. Vessel emergence via vasculogenesis within the lung mesenchyme is supported by observations of VEGFR-2 reporter expression. VEGFR-2–lacZ transgenic mice, in contrast to the Tie2–lacZ pattern, reveal the presence of an intact vascular plexus within the mesenchyme in E10.5 mouse lungs [106]. Therefore vascular identification studies carried out with different markers reveal endothelial heterogeneity, indicating that different types of ECs are found in the proximal versus the distal lung bud mesenchyme. Alternatively, as VEGFR-2 is a more primitive EC marker Tie2/platelet-endothelial cell adhesion molecule (PECAM)-1 (CD31) [110], it is possible that observed differences may be based on the distinct stages of EC commitment in different regions of the bud. Nonetheless, these studies indicate that vessels are present within the distal mesoderm early, but do little to delineate the exact origin of the different vessel populations. Although initial studies suggested sequential vasculogenesis and angiogenesis, recent evidence continues to accumulate supporting the notion that separate parallel angiogenic and vasculogenic processes work coordinately to form the pulmonary vasculature throughout lung development.

In addition to the alveolar endothelial interface that supports oxygen exchange, central vessels are also found in close proximity to the central bronchi of the lung. Interestingly, bronchial circulation and the interface between the central bronchi and vasculature are poorly understood. To date, observations suggest that although arteries are adjacent to the bronchi extending into the peripheral airways in the mature lung, during early pulmonary development there is little contact between the vasculature and the central or peripheral airways [98, 107, 108]. However, there is histological evidence demonstrating that by the canalicular stage bronchi and vessels are in close proximity and that an intact vascular network is found by casting at the saccular stage [61]. The contrast between these studies highlight a persistent void in our knowledge of the mechanisms that mediate formation of the bronchi/bronchial circulation interface.

Pulmonary Arterial and Venous Differentiation

The pulmonary circulation is composed of arterial and venous vessels that coordinate vascular flow to and from the distal oxygen exchanging alveolar cells. As mentioned in "Arterial versus Venous Differentiation," recent studies have identified the endothelial marker EphB4 tyrosine kinase receptor and its membrane-bound ligand EfnB2 as specific venous and arterial vessels markers, respectively [111]. Interestingly, in contrast to other regions throughout the body, the pulmonary arteries carry un-oxygenated blood to the distal capillaries where the EC/alveolar interface facilitates oxygen exchange. Pulmonary veins then return oxygen-rich blood to the left side of the heart. Histological analysis of human fetal lungs (84–98 days gestation) suggests that while a subset of the vascular population expresses EfnB2, all pulmonary EC populations, venous and arterial, express EphB4 [98, 107, 108]. Furthermore, at this stage (E13.5) ECs lack fate specificity as they express both surface markers. It is only at E15.5 that EC arteriovenous cell fate specificity begins to emerge [112]. What is unclear is the stimulus that dictates pulmonary EC specification to either an arterial or venous fate. As oxygen levels *in utero* are relatively low in the developing fetus and the fetal lung is protected from high arterial flow pressures, it is not readily evident that a mechanical or oxidative stress mechanism is involved. An alternative possibility is signaling from smooth muscle cells (SMCs) that are known to line arterial but not the venous system [98, 107, 108]. The paucity of studies that examine arterial and venous EC fate specification highlight our lack of understanding of the mechanisms that regulate the emerging pulmonary vasculature and remain a challenge to pulmonary vascular biologist.

Extension of Primary Pulmonary Vascular Plexus to the Epithelial/Mesenchymal Interface

In light of previous studies on lung vascularization and our recent identification of blood flow in the early lung bud (before E10.5) [112], we set forth a novel proposal for the etiology of lung vascular network formation. We propose that a functional, blood-filled primitive vascular network is present in the mesoderm prior to the evagination of the endodermal lung epithelium (Figure 1.4a). Initially, the relatively homogenous web-like plexus lies within the gut tube mesodermal layer, and runs along the entire length of the foregut and beyond. As the endoderm buds into the mesoderm, the vascular plexus and mesodermal layers are pushed out with it, forming a vascular network that surrounds the budding epithelium like a fish net (Figure 1.4b). This can be seen in a number of studies

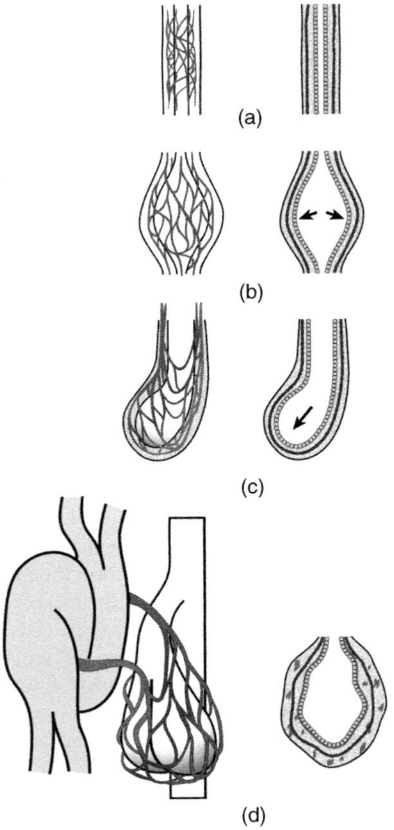

Figure 1.4 Proposal for the sequential progression of lung vascular development. A primitive blood filled vascular network, present within the mesoderm (a), is pushed outward by the invading endodermal bud epithelium (b). Progression of the endodermal epithelial invasion and distal lung bud expansion results in vascular plexus forming a purse-like pouch that narrows at the proximal neck (c). The growing vasculature of the lung bud always maintains a vascular connection with the central circulation system, and the proximal vessels remodel into fewer and larger vessels (d). As the bud grows and the lung vasculature extends and remodels, vasculogenic pools are also present in the distal mesoderm (d). Vascular remodeling of this plexus and the establishment of communication with the vasculogenic clusters completes a multilayered pulmonary vascular network. A color version of this figure appears in the plate section of this volume.

that describe early lung vasculature [109]. However, importantly, the vasculature at these early stages remains sandwiched within the middle of the mesodermal layer and is not in immediate contact with the underlying endodermal epithelium.

As budding continues, we propose that the lung bud extends distally with minimal proximal lung growth. This causes the distal vascular plexus to extend, while the proximal vascular plexus remains in relative close proximity to its origin within the foregut. As the bud tips grow out, proximal vessels remodel into fewer and larger vessels and both the arterial (anterior) and returning venous (posterior) systems take shape. Since there is minimal proximal growth relative to distal proliferation, the vascular plexus comes to form a purse-like pouch, with constriction of the proximal plexus around the thinning neck of the lung bud (Figure 1.4c). Simultaneously, in the distal mesenchyme of the lung bud, vasculogenic pools of angioblasts are also emerging (Figure 1.4d).

Around E12 in the mouse, vessels extend centripetally from their position within the mesenchyme toward the epithelial/mesenchymal interface by angiogenic sprouting. In addition, this same plexus also extends in the opposite direction, centrifugally outwards, and establishes communication with the vasculogenic clusters. Overall, remodeling of this plexus completes a multilayered pulmonary vascular network, within the lung bud, by embryonic day 17. This proposed mechanism is consistent with observed vessel formation in other organs where the vasculature is initially confined to a single layered plexus within the mesoderm, while adjacent endoderm and ectoderm layers are initially avascular. Similarly, lymphatic vessels in skin develop from a simple flat array of vessels, to a multilayered array [113]. In both cases, an initial plexus must grow out of a two-dimensional net-like network, and create a more three-dimensional array. Still to be determined is whether type 1 and/or 2 vasculogenic mechanisms are used in lung vascularization, and the timing and mechanisms underlying pulmonary vascular tubulogenesis and angiogenic remodeling during lung development.

Further complicating our understanding of pulmonary neovascularization has been the difficulty in pinpointing the stage at which the lung vasculature comes in contact with the epithelium. Early studies indicate that cells expressing VEGFR-2 mRNA are present in conjunction with pulmonary epithelium during much of lung development [59, 106]. Although adult murine and human lungs have vessels adjacent to the epithelium of bronchi, branching epithelium, and distal alveoli cells, this does not appear to be the case in embryonic lungs. Serial reconstruction of human embryonic fetal lungs [107] and identification of perfused vessels in the mesenchyme [112] indicates that primitive vessels are present in the mesenchyme but are not immediately adjacent to the evaginating epithelium. Lack of consensus surrounding the stage during lung development where vessel emergence is observed and how vessels develop their interface with the alveolar cell elucidates the difficulty presented in dissecting out the pulmonary circulatory system.

VASCULAR GROWTH FACTORS IN LUNG MORPHOGENESIS

VEGF-A and its Isoforms

Similar to their roles in embryonic vasculogenesis, VEGF-A and its receptors, VEGFR-1, and VEGFR-2, are also essential for pulmonary vessel formation. Indirectly regulated by both fibroblast growth factor (FGF)-9 and "sonic hedgehog" signaling in the mesenchyme, VEGF-A expression mediates distal capillary density and plexus formation [114]. This is supported by the correlation of VEGF-A isoform-specific expression patterns with regional pulmonary vessel formation at different developmental timepoints. VEGF-A isoform distribution and timing suggests that different VEGF-A isoforms facilitate specific aspects of vessel formation. During the early pseudoglandular stages, when vessel formation is confined to the middle mesenchymal cell layer, initial expression of the 120 and 164 VEGF-A isoforms is distributed throughout the mesenchyme [115–118]. At this stage, primitive vessel building and recruitment occurs, and the vascular plexus surrounds the emerging lung bud. This is consistent with the fact that VEGF-A 120 is highly diffusible, allowing it to chemotactically recruit vessels from the plexus or from surrounding vasculogenic pools, while doing little to increase the vascular density within the region [80]. In contrast to VEGF-A 120's highly diffusible properties, VEGF-A 164 exhibits only moderate diffusion capacity, and is therefore capable of both vessel recruitment and increases in vascular density. The presence of both VEGF-A isoforms 120 and 164 would suggest that during early stages of lung development, vessel formation in the mesenchyme occurs by both vessel recruitment (angiogenesis) and *de novo* differentiation (vasculogenesis). Taken together these findings suggest that in the mesenchyme VEGF120 expression is stimulating angiogenesis while VEGF164 facilitates simultaneous angiogenesis and vasculogenesis.

During the later part of the pseudoglandular stage, the VEGF-A 188 isoform that is notable for developing vascular density is found to be tightly associated with ECM and is found at the epithelial tips of the lung buds. Its expression initiates midway during lung development and gives rise to high local concentrations at the distal tips of the lung buds, which increase distal capillary network density [80, 116–118]. However, it is unclear whether the increase in vascular network density results from vasculogenic or angiogenic mechanisms. It is worth noting that at E14, the overall expression of VEGF-A is markedly increased in epithelial cells at the tips of the expanding airways, which coincides strikingly with a dramatic increase in vessel density and vascular ingression into the epithelial/mesenchymal interface [112, 115–118]. The corresponding timing of increased VEGF expression in focal epithelial tip cells and the proximity of the epithelium to the extending vasculature are consistent with the facilitation of distal vessel formation. We propose that the burst in epithelial expression of VEGF-A in the lung is likely to attract the filopodia of the angiogenic tip cells toward the epithelial/mesenchymal interface (Figure 1.5). Differential VEGF-A isoform distribution and focal epithelial expression suggests that VEGF-A regulation is critical to vascular growth in pulmonary development.

The potent effect of VEGF on both the formation of pulmonary vessels and on developing airway epithelium can be demonstrated experimentally. Overexpression of the VEGF-A 164 isoform under the control of the human SP-C promoter results in increased vascularization, as expected, but also in marked lung abnormalities characterized by large dilated tubules, disrupted branching morphogenesis, and inhibition of type 1 epithelial cell differentiation [119]. Selective expression of the VEGF-A 120 isoform (which lacks heparin-binding capacity and

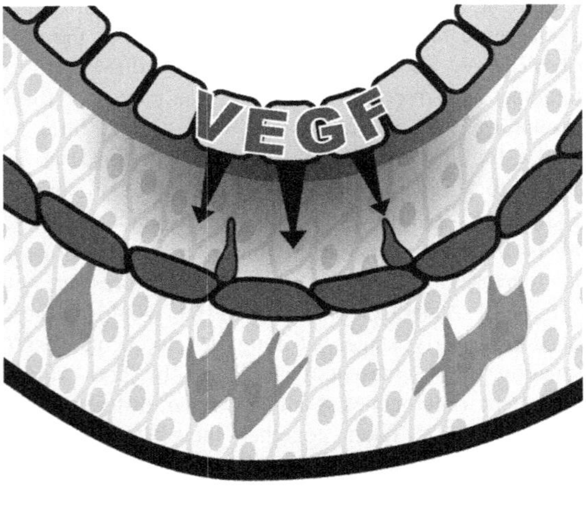

Figure 1.5 Vascular remodeling and establishment of intervascular connections is in part due to the interactions between epithelial VEGF gradients, the vasculogenic pools, and angiogenic extensions from the growing lung plexus. These forces work in concert to develop a functional gas-exchanging vascular/alveolar cell interface. A color version of this figure appears in the plate section of this volume.

therefore lacks ECM interaction domains in mice) resulted in impaired vascular development. Expression of only the VEGF-A 120 isoform resulted in the lack of directed extension of endothelial filopodia and a decrease in vascular branching [120]. Importantly, in addition to defects in lung microvasculature, these mutant mice also displayed a marked delay of airspace maturation [121]. The lack of branching and diminished EC filopodia was attributed to the disruption of the proper VEGF-A concentration gradient. Despite whether expression of one or all VEGF isoforms was altered during development, distal alveolar formation was altered. Together, these experiments suggest that all of the VEGF-A isoforms are necessary for normal alveolar/vascular air–blood barrier formation and confirms that the different VEGF-A isoforms have specific roles in lung morphogenesis [121].

VEGFRs

The influence of VEGF-A on neovascularization is not only regulated by local control of expression levels, but also by the selective expression of its receptors. VEGF-A binds multiple receptors including VEGFR-1, -2, and -3 (also known as Flt-4), and neuropilins 1 and 2. While it has been shown that each member of this family of closely related tyrosine kinases performs very different functions during blood vessel development, little is known about their different roles during development of the pulmonary vasculature. What has been demonstrated is that VEGFR-1 and VEGFR-2 both regulate EC proliferation and differentiation and are therefore essential for development of the pulmonary vasculature [122]. As during initial embryonic vessel formation, VEGFR-2 is more likely to mediate EC proliferation and differentiation, while VEGFR-1 plays a greater role in vessel branching and remodeling [122]. VEGFR-2 mRNA-expressing cells during lung development have been correlated with regions in which endothelial precursors are emerging within the mesenchyme via vasculogenesis [106].

Although all precursor EC express VEGFR-2, recent studies indicate that VEGFR-2 is also expressed on precursors to SMCs or pericytes. Presentation of either a VEGF or platelet-derived growth factor ligand to the precursor cell dictates the cell fate to either an EC or pericyte/SMC, respectively [123]. This observation thus limits the usefulness of engineered VEGFR-2 mRNA and VEGFR-2 reporter mice as a sole means to identify emerging vessels. However, by taking advantage of colocalization using antibodies against phosphorylated VEGR2 and its "endothelial-differentiating" ligand VEGF, one can deduce that a cell positive for both would represent a cell committed to an endothelial fate. Studies that examine colocalization of phosphorylated VEGFR-2 in association with VEGF confirmed that the vasculature is confined to the mesenchymal cells prior to E14.5–15.5 [112]. While most studies have associated VEGFR-2 expression with vascular and perivascular cells, a study by Ahlbrecht *et al.* determined that epithelial cells in later stages of development also initiate VEGFR-2 expression and simultaneously secrete VEGF [124]. In contrast, neuronal cells lack VEGFR and only express VEGF and neuropilin receptors [125]. Clearly these studies point to the growing need to examine the role of the different tyrosine kinases in response to the VEGF ligand in different cell types.

Environmental Influences on VEGF Expression

Although cell-autonomous factors, like receptor availability and composition of intracellular signaling mediators, are strong determinants of VEGF signaling, tissue interactions, ECM, and environmental factors also play an important role in VEGF regulation. Explant experiments demonstrate that epithelial/mesenchymal interactions are required for induction or maintenance of vascular precursors [59]. Specifically, fetal lung mesenchyme isolated and grown in culture in the absence of lung epithelium maintains few VEGFR-2 cells. In contrast, lung mesenchyme recombined with lung epithelium develops abundant VEGFR-2-positive cells. The necessity of both lung rudimentary tissues suggests that during early pulmonary development epithelial/mesenchymal signaling is essential for the proper emergence of vascular precursors and subsequent development of lung vasculature [59].

Oxygen tension, a mediator of the transcription factor HIF, has been shown to regulate VEGF expression levels. Signaling through the HIF–VEGF–VEGFR system in fact actively participates in lung alveolarization and maturation [126]. Genetic ablation of HIF-2α resulted in the development of fatal respiratory distress syndrome in neonatal mice [127]. Associated with the reduction in HIF-2α were lowered alveolar VEGF levels. This resulted in alveolar capillaries that failed to remodel properly and a concomitant insufficient surfactant production by alveolar type 2 cells. However, this could be rescued by either intra-uterine or postnatal intratracheal instillation of VEGF [127]. Further demonstrating the profound impact of HIF on VEGF protein expression, hyperoxia exposure (>95% O_2 days 4–14) resulted in depressed HIF-2α and VEGF mRNA levels [128, 129] resulting in not only a reduction in vessel density, but also arrested lung alveolarization [130]. Mediation of environmental oxygen tension is observed in the premature newborn where fetal lungs are exposed to relatively high oxygen levels compared to what they would experience *in utero*. This premature oxygenation results in the onset of pathologic lung hypoplasia or bronchopulmonary dysplasia (BPD). Studies examining lung development in premature infants using a baboon model of BPD determined that there was a marked and selective downregulation of HIFs [131]. Inhibition of HIF degradation augmented distal alveolar angiogenesis and ameliorated the pathological alveolar dysplasia and physiological consequences of BPD [132, 133]. These studies suggest that environmental influences on VEGF expression play a significant role in the evolution of neonatal lung disease.

ENDOTHELIAL-SPECIFIC FACTORS

In addition to the VEGF and VEGFR family, ECs themselves generate factors that contribute to the regulation of their behaviors during vessel formation. For example, angiopoietin-1 protein is likely to be required for pulmonary vessel integrity and quiescence. High angiopoietin-1 levels in nitrofen-induced hypoplastic lungs were associated with a significant reduction in peripheral capillaries [134]. Further supporting a role for endothelial-selective growth factors in pulmonary vascular development, transgenic mice with an endothelial nitric oxide synthase mutation exhibit capillary hypoperfusion, misaligned pulmonary veins and also display a paucity of distal arteriolar branches [135]. These endothelial-specific factors, while not characterized as endothelial growth factors, directly impact vessel formation during development and warrant further studies to better understand their contribution to lung pulmonary vascular development.

NON-ENDOTHELIAL-SPECIFIC GROWTH FACTORS

In contrast to factors that have endothelial-specific receptors, growth factors secreted from other cell types also contribute to vessel formation. For example, secreted factors such as FGFs influence vessel formation by altering vascular integrity [136] and distal alveolar formation [137]. However, the effects of FGFs are not limited to vessel formation, as lung branching and distal alveolar cell differentiation are directly impacted by FGF levels. Although these studies are beyond the scope of this chapter, review articles by Cardoso and Maeda nicely elaborate in greater detail on the interactions between transcriptional factors and lung morphogenesis [93, 138]. Further examination of nonendothelial-specific growth factors and their contribution to overall lung growth, including vessel formation, is important in broadening our understanding of pulmonary vascular development.

ECM

The ECM has also been shown to be critical in modulating embryonic organ and tissue development, including blood vessel formation. Interactions between adhesion molecules mediate cell–cell cohesion and facilitate establishment of epithelial cell polarity [139]. Despite our growing understanding of secreted growth factors in lung vascularization, the role of the ECM in this process is poorly understood.

One abundant pulmonary ECM component is laminin. In the developing lung, laminin is the predominant ECM molecule found at the epithelial/mesenchymal interface [140–144]. Owing to its proximity to the developing vasculature, laminin is ideally positioned to influence lung vessel formation. Recent experiments have shown that laminin regulates vessel lumen diameter, but overall has little impact on vessel emergence [145]. In these studies, deletion of laminin from embryoid bodies due to a laminin $\gamma 1$ deletion results in minimal impact on vessel emergence and organization, but does increase the frequency of vessels with wide lumens [145].

In contrast to the relatively minor role of laminin on vessel construction, recent studies suggest that the ECM protein tenascin-C is required for pulmonary vessel network formation. Tenascin-C is known to be downstream of the paired-related homeobox gene (*Prx1*) and *Prx1*-null mice die soon after birth from respiratory failure. Histological analysis of *Prx1*-null mice reveals hypoplastic lungs with a marked reduction in both vessel number and tenascin-C expression as compared to control littermates. Ihida-Stansbury *et al.* suggest that not only is *Prx1* required for tenascin-C expression, but that tenascin-C is required for *Prx1*-dependent differentiation of fetal pulmonary EC precursors and vascular network formation [146, 147]. Together, these studies demonstrate that ECM molecules are important mediators of vessel formation in the developing lung.

ANTIANGIOGENIC FACTORS

In contrast to the positive role of many growth factors on vascular development, negative/inhibitory vascular factors provide a counterbalance to vessel formation during lung development. The antiangiogenic protein endothelial-monocyte activating polypeptide (EMAP) II, which is activated by its cleavage from p43 [148–150], plays a significant role during lung vascular development.

EMAP II temporal/spatial expression during lung development is consistent with a role in maintaining specific avascular regions during lung development. During the early, pseudoglandular stage (E14.5–15.5), prior to vascularization of the epithelial/mesenchymal interface, EMAP II was found to be highly expressed. Strikingly, its expression is downregulated coincident with the canalicular stage (E16.5), as this region becomes vascularized. EMAP II expression is limited to the perivascular expression into adulthood [151]. Exogenous delivery of the endogenous antiangiogenic protein EMAP II in a fetal lung allograft model [150] markedly decreased lung vasculature, it induced lung dysplasia and it inhibited distal epithelial cell differentiation. Conversely, and as predicted, delivery of an EMAP II-blocking antibody significantly enhanced vasculature and accelerated differentiation of the distal lung [150].

Early postnatal lung development is profoundly influenced by experimental vascular inhibition, demonstrating the requirement for tight regulation of pulmonary angiogenic factors. Thalidomide, fumagillin, the VEGFR-2 inhibitor SU5416, or PECAM-1-blocking antibodies delivered in the early postnatal period result not only in vascular interruption, but in coincident gross abnormalities in lung development [152, 153]. For example, delivery of the VEGFR-2 inhibitor in the early postnatal period initiates an attenuation of lung development noted by a concomitant decrease vessel formation and alveolarization [152, 153]. Similar results are noted when PECAM-1 is inhibited resulting in the disruption of alveolar septation and reduced endothelium [152, 153]. These studies provide support a role for tight regulation of vascular regulators during lung morphogenesis.

Cross-Talk between Pulmonary Vasculature and Epithelium

As pulmonary morphogenesis progresses, the distal alveoli and ECs have a greater influence on each other's development. This is evident as disruption of either the emerging distal air sacs composed of alveolar clefts or vasculature results in an alteration in the normal morphogenesis of the other. In contrast to embryonic and early pseudoglandular stages, where the vasculature and branching airways are separated by several cell layers, the later pseudoglandular and canalicular stages are characterized by thinning of the mesenchyme and increasing proximity of the lung epithelial and ECs. The close proximity of the two cell types is critical for the facilitation of oxygen exchange across the epithelial/endothelial interface during later development.

The mutual dependence of vasculature and the organs they perfuse is exemplified when vessels are experimentally disrupted. For instance, in lung, vessel inhibition is associated with alterations in epithelial cell morphogenesis. Inhibition of VEGF using a soluble receptor in lung renal capsule grafts [154] inhibited vascular development and epithelial development supporting a role for VEGF in the coordination of epithelial and vascular development [155]. Whereas blockade of vessel growth using the antiangiogenic protein EMAP II in a lung allograft model [150] inhibits epithelial morphogenesis [148]. Furthermore, studies indicate that endogenous VEGF induces fetal epithelial proliferation *in vitro* fetal human lung explants [156], while conversely VEGF blockade interrupts alveolar structural integrity [157]. In addition, transgenic studies where pulmonary blood vessel formation is altered by overexpression of VEGF164 isoform using the SP-C promoter results in concomitant disruption of branching morphogenesis and inhibition of alveolar type 1 cell differentiation [119]. It is important to note that VEGFRs are not found on the epithelium, suggesting that the vasculature is the target, and that the epithelium responds secondarily. On the other hand, inhibition of lung structural maturation by inhibition of transforming growth factor-β1, thyroid transcriptional factor-1, or Wnt7b resulted in vascular malformations in conjunction with severe alterations in distal lung alveolar morphogenesis [96, 158–160]. Taken together, these studies indicate that there is a direct and mutually dependent relationship between vessel formation and epithelial morphogenesis.

It has become increasingly apparent that an intimate and reciprocal relationship between epithelial and ECs is fostered throughout distal lung development, likely via cell–cell signaling mediated by VEGF-A. This theory is supported by several key observations: (i) VEGF-A distribution in development, (ii) EC facilitation of distal epithelial cell differentiation, and (iii) the strikingly evident reciprocal influence that alveolar and vascular development have on each other. First, during lung development the epithelial cells generate VEGF that is deposited in the subepithelial matrix within the lung branches. This results in a clear proximal-to-distal VEGF gradient, with VEGF epithelial expression being highest at the tips of the branching distal airways at E13.5 and lowest at the proximal epithelium [117]. Corresponding to the epithelial VEGF gradient, phosphorylated VEGFR-2 signal can be found on the tips of the pulmonary ECs that are extending toward the epithelial/mesenchymal interface during the pseudoglandular stage [112]. Taken together, this suggests that the epithelial basilar VEGF gradient serves as a guidance and endothelial differentiation signal [123].

The basilar epithelial location of VEGF also suggests a morphologic role where a cross-talk interaction between the VEGF expressing basilar epithelial surface and the ECs initiate distal epithelial differentiation. Previous studies have shown that ECs contribute important paracrine signals that influence the development of surrounding organs. For example, during pancreatic development, key events of endocrine differentiation occur only in close association with ECs [161, 162]. In liver, hepatocyte migration and differentiation require similar signals from blood vessel ECs [60]. Similarly, in lung development, VEGF also patterns and coordinates epithelial/vascular morphogenesis [155, 163]. These studies indicate that without VEGF-A tightly coordinating distal epithelial differentiation and vascular development, progression of epithelial proliferation and sacculation are altered. Interestingly, distal lung differentiation does progress, but the epithelial cell numbers and structure are limited. This suggests that VEGF-A has a broad influence on distal lung formation. Importantly, these studies reinforce the fundamental concept that vascular and epithelial cell cross-talk are essential in the formation of the alveolar/vascular interface that is essential for oxygen exchange.

CONCLUSIONS AND PERSPECTIVES

Lung vascular development is clearly a complex process. Guided by both pro- and antiangiogenic factors, the ECM, epithelial/mesenchymal interactions, and angiogenic and vasculogenic mechanisms work together to establish a functional site of gas exchange at the alveolar/endothelial interface. Mediated by a wide array of vascular growth factors, receptors, and arterial/venous guidance cues, vessel formation is derived by vasculogenic and angiogenic forces. Furthermore, during development vascular growth factors mediate not only endothelial morphogenesis, but also influence directly and indirectly affect a broader cellular community. This results in the close association and coordination of vascular formation and epithelial differentiation, where alteration in either system inevitably and dramatically influences the formation of the other. The intimate relationship between these two interconnected processes makes it exceedingly difficult to identify the individual contributions to either component. Thus, designing methods to distinguish the contribution and regulation of vascularization from epithelial morphogenesis, development of an in-depth understanding of the angiogenic and vasculogenic progression during the early stages of lung formation, and identification of the arterial and venous contributions all remain exciting challenges for future studies.

ACKNOWLEDGMENTS

We are grateful to Dr. Philip Shaul for critical reading of the manuscript and helpful advice. We are also indebted to Jose Cabrera for artistic rendition of complex vascular concepts. This work was supported by Juvenile Diabetes Research Foundation award 99-2007-472, National Institutes of Health R01 grant DK079862-01, American Heart Association award 0755054Y, and the Basil O'Connor March of Dimes award to O.C., and National Institutes of Health R01 grants HL-60061 and HL-75764 to M.S.

References

1. His, W. (1900) Lecithoblast und Angiobalst der Wirbeltiere. *Abhandlungen der Mathematisch-Physischen Klassen der Sachsischen Gesellschaft*, **26**, 171–328.
2. Coffin, J.D. and Poole, T.J. (1988) Embryonic vascular development: immunohistochemical identification of the origin and subsequent morphogenesis of the major vessel primordia in quail embryos. *Development*, **102**, 735–48.
3. Noden, D.M. (1989) Embryonic origins and assembly of blood vessels. *The American Review of Respiratory Disease*, **140**, 1097–103.
4. Drake, C.J. and Fleming, P.A. (2000) Vasculogenesis in the day 6.5 to 9.5 mouse embryo. *Blood*, **95**, 1671–79.
5. Ferkowicz, M.J. and Yoder, M.C. (2005) Blood island formation: longstanding observations and modern interpretations. *Experimental Hematology*, **33**, 1041–47.
6. Haar, J.L. and Ackerman, G.A. (1971) A phase and electron microscopic study of vasculogenesis and erythropoiesis in the yolk sac of the mouse. *The Anatomical Record*, **170**, 199–223.
7. Kinder, S.J., Loebel, D.A., and Tam, P.P. (2001) Allocation and early differentiation of cardiovascular progenitors in the mouse embryo. *Trends in Cardiovascular Medicine*, **11**, 177–84.
8. Wilt, F.H. (1974) The beginnings of erythropoiesis in the yolk sac of the chick embryo. *Annals of the New York Academy of Sciences*, **241**, 99–112.
9. Pardanaud, L., Altmann, C., Kitos, P. *et al.* (1987) Vasculogenesis in the early quail blastodisc as studied with a monoclonal antibody recognizing endothelial cells. *Development*, **100**, 339–49.
10. Peault, B., Coltey, M., and Le Douarin, N.M. (1988) Ontogenic emergence of a quail leukocyte/endothelium cell surface antigen. *Cell Differentiation*, **23**, 165–74.
11. Eichmann, A., Yuan, L., Moyon, D. *et al.* (2005) Vascular development: from precursor cells to branched arterial and venous networks. *The International Journal of Developmental Biology*, **49**, 259–67.
12. Risau, W., Sariola, H., Zerwes, H.G. *et al.* (1988) Vasculogenesis and angiogenesis in embryonic-stem-cell-derived embryoid bodies. *Development*, **102**, 471–78.
13. Sabin, F.R. (1920) Studies on the origin of the blood vessels and of red blood corpuscles as seen in the living blastoderm of chick during the second day of incubation. *Carnegie Contribution to Embryology*, **9**, 215–62.
14. Wagner, R.C. (1980) Endothelial cell embryology and growth. *Advances in Microcirculation*, **9**, 45–75.
15. Risau, W. and Flamme, I. (1995) Vasculogenesis. *Annual Review of Cell and Developmental Biology*, **11**, 73–91.
16. Murray, P.D.F. (1932) The development *in vitro* of the blood of the early chick embryo. *Proceedings of the Royal Society of London*, **111**, 497–521.
17. Shalaby, F., Rossant, J., Yamaguchi, T.P. *et al.* (1995) Failure of blood-island formation and vasculogenesis in Flk-1-deficient mice. *Nature*, **376**, 62–66.
18. Vogeli, K.M., Jin, S.W., Martin, G.R., and Stainier, D.Y. (2006) A common progenitor for haematopoietic and endothelial lineages in the zebrafish gastrula. *Nature*, **443**, 337–39.
19. Park, C., Ma, Y.D., and Choi, K. (2005) Evidence for the hemangioblast. *Experimental Hematology*, **33**, 965–70.
20. Ribatti, D. (2008) Hemangioblast does exist. *Leukemia Research*, **32**, 850–54.
21. Huber, T.L., Kouskoff, V., Fehling, H.J. *et al.* (2004) Haemangioblast commitment is initiated in the primitive streak of the mouse embryo. *Nature*, **432**, 625–30.
22. Ueno, H. and Weissman, I.L. (2006) Clonal analysis of mouse development reveals a polyclonal origin for yolk sac blood islands. *Developmental Cell*, **11**, 519–33.
23. Kinder, S.J., Tsang, T.E., Quinlan, G.A. *et al.* (1999) The orderly allocation of mesodermal cells to the extraembryonic structures and the anteroposterior axis during gastrulation of the mouse embryo. *Development*, **126**, 4691–701.
24. Sumpio, B.E., Riley, J.T., and Dardik, A. (2002) Cells in focus: endothelial cell. *The International Journal of Biochemistry and Cell Biology*, **34**, 1508–12.
25. Aird, W.C. (2003) Endothelial cell heterogeneity. *Critical Care Medicine*, **31**, S221–30.
26. Suter, E.R. and Majno, G. (1965) Passage of lipid across vascular endothelium in newborn rats. An electron microscopic study. *The Journal of Cell Biology*, **27**, 163–77.
27. Garlanda, C. and Dejana, E. (1997) Heterogeneity of endothelial cells. Specific markers. *Arteriosclerosis, Thrombosis, and Vascular Biology*, **17**, 1193–202.
28. Aird, W.C. (2007) Phenotypic heterogeneity of the endothelium: II. Representative vascular beds. *Circulation Research*, **100**, 174–90.

29. Cleaver, O. and Melton, D.A. (2003) Endothelial signaling during development. *Nature Medicine*, **9**, 661–68.
30. Risau, W., Sariola, H., Zerwes, H.G. *et al.* (1988) Vasculogenesis and angiogenesis in embryonic-stem-cell-derived embryoid bodies. *Development*, **102**, 471–78.
31. Risau, W. (1997) Mechanisms of angiogenesis. *Nature*, **386**, 671–74.
32. Pardanaud, L., Yassine, F., and Dieterlen-Lievre, F. (1989) Relationship between vasculogenesis, angiogenesis and haemopoiesis during avian ontogeny. *Development*, **105**, 473–85.
33. Wilting, J. and Christ, B. (1996) Embryonic angiogenesis: a review. *Naturwissenschaften*, **83**, 153–64.
34. Drake, C.J. (2003) Embryonic and adult vasculogenesis. *Birth Defects Research, Part C: Embryo Today*, **69**, 73–82.
35. Sabin, F.R. (1917) Origin and development of the primitive vessels of the chick and the pig. *Carnegie Contribution to Embryology*, **6**, 61–124.
36. Poole, T.J. and Coffin, J.D. (1989) Vasculogenesis and angiogenesis: two distinct morphogenetic mechanisms establish embryonic vascular pattern. *The Journal of Experimental Zoology*, **251**, 224–31.
37. Walls, J.R., Coultas, L., Rossant, J., and Henkelman, R.M. (2008) Three-dimensional analysis of vascular development in the mouse embryo. *PLoS One*, **3**, e2853.
38. Cleaver, O., Tonissen, K.F., Saha, M.S., and Krieg, P.A. (1997) Neovascularization of the *Xenopus* embryo. *Developmental Dynamics*, **210**, 66–77.
39. Cleaver, O. and Krieg, P.A. (1998) VEGF mediates angioblast migration during development of the dorsal aorta in *Xenopus*. *Development*, **125**, 3905–14.
40. Houser, J.W., Ackerman, G.A., and Knouff, R.A. (1961) Vasculogenesis and erythropoiesis in the living yolk sac of the chick embryo. A phase microscopic study. *The Anatomical Record*, **140**, 29–43.
41. Kamei, M., Saunders, W.B., Bayless, K.J. *et al.* (2006) Endothelial tubes assemble from intracellular vacuoles *in vivo*. *Nature*, **442**, 453–56.
42. Folkman, J. and Haudenschild, C. (1980) Angiogenesis *in vitro*. *Nature*, **288**, 551–56.
43. Meyer, G.T., Matthias, L.J., Noack, L. *et al.* (1997) Lumen formation during angiogenesis *in vitro* involves phagocytic activity, formation and secretion of vacuoles, cell death, and capillary tube remodelling by different populations of endothelial cells. *The Anatomical Record*, **249**, 327–40.
44. Lubarsky, B. and Krasnow, M.A. (2003) Tube morphogenesis: making and shaping biological tubes. *Cell*, **112**, 19–28.
45. Davis, G.E., Bayless, K.J., and Mavila, A. (2002) Molecular basis of endothelial cell morphogenesis in three-dimensional extracellular matrices. *The Anatomical Record*, **268**, 252–75.
46. Jin, S.W., Beis, D., Mitchell, T. *et al.* (2005) Cellular and molecular analyses of vascular tube and lumen formation in zebrafish. *Development*, **132**, 5199–209.
47. Folkman, J. and Klagsbrun, M. (1987) Angiogenic factors. *Science*, **235**, 442–47.
48. Klagsbrun, M. and D'Amore, P.A. (1991) Regulators of angiogenesis. *Annual Review of Physiology*, **53**, 217–39.
49. Patan, S., Munn, L.L., and Jain, R.K. (1996) Intussusceptive microvascular growth in a human colon adenocarcinoma xenograft: a novel mechanism of tumor angiogenesis. *Microvascular Research*, **51**, 260–72.
50. Gerhardt, H., Golding, M., Fruttiger, M. *et al.* (2003) VEGF guides angiogenic sprouting utilizing endothelial tip cell filopodia. *The Journal of Cell Biology*, **161**, 1163–77.
51. Klosovskii, B.N. and Zhukova, T.P. (1963) Effect of colchicine on remote phases of growing capillaries in the brain. *Arkhiv Patologii*, **35** (3), 38–44.
52. Hellstrom, M., Phng, L.K., Hofmann, J.J. *et al.* (2007) Dll4 signalling through Notch1 regulates formation of tip cells during angiogenesis. *Nature*, **445**, 776–80.
53. Gerhardt, H. and Betsholtz, C. (2005) How do endothelial cells orientate? in *Mechanisms of Angiogenesis* (eds M. Clauss and G. Breier), Experientia Supplementum, Birkhäuser, Basel, pp. 3–15.
54. Beck, L. Jr. and D'Amore, P.A. (1997) Vascular development: cellular and molecular regulation. *The FASEB Journal*, **11**, 365–73.
55. Patan, S. (2004) Vasculogenesis and angiogenesis. *Cancer Treatment and Research*, **117**, 3–32.
56. Burri, P.H., Hlushchuk, R., and Djonov, V. (2004) Intussusceptive angiogenesis: its emergence, its characteristics, and its significance. *Developmental Dynamics*, **231**, 474–88.
57. Im, E. and Kazlauskas, A. (2006) New insights regarding vessel regression. *Cell Cycle*, **5**, 2057–59.
58. Ashton, N. (1966) Oxygen and the growth and development of retinal vessels. *In vivo* and *in vitro* studies. The XX Francis I. Proctor lecture. *American Journal of Ophthalmology*, **62**, 412–35.
59. Gebb, S.A. and Shannon, J.M. (2000) Tissue interactions mediate early events in pulmonary

vasculogenesis. *Developmental Dynamics*, **217**, 159–69.

60. Matsumoto, K., Yoshitomi, H., Rossant, J., and Zaret, K.S. (2001) Liver organogenesis promoted by endothelial cells prior to vascular function. *Science*, **294**, 559–63.
61. deMello, D.E., Sawyer, D., Galvin, N., and Reid, L.M. (1997) Early fetal development of lung vasculature. *American Journal of Respiratory Cell and Molecular Biology*, **16**, 568–81.
62. Tufro, A., Norwood, V.F., Carey, R.M., and Gomez, R.A. (1999) Vascular endothelial growth factor induces nephrogenesis and vasculogenesis. *Journal of the American Society of Nephrology*, **10**, 2125–34.
63. Robert, B., St. John, P.L., and Abrahamson, D.R. (1998) Direct visualization of renal vascular morphogenesis in Flk1 heterozygous mutant mice. *American Journal of Physiology: Renal Physiology*, **275**, F164–72.
64. Galambos, C. and deMello, D.E. (2007) Molecular mechanisms of pulmonary vascular development. *Pediatric and Developmental Pathology*, **10**, 1–17.
65. Krebs, L.T., Xue, Y., Norton, C.R. *et al.* (2000) Notch signaling is essential for vascular morphogenesis in mice. *Genes and Development*, **14**, 1343–52.
66. Shutter, J.R., Scully, S., Fan, W. *et al.* (2000) Dll4, a novel Notch ligand expressed in arterial endothelium. *Genes and Development*, **14**, 1313–18.
67. Xue, Y., Gao, X., Lindsell, C.E. *et al.* (1999) Embryonic lethality and vascular defects in mice lacking the Notch ligand Jagged1. *Human Molecular Genetics*, **8**, 723–30.
68. Lawson, N.D., Scheer, N., Pham, V.N. *et al.* (2001) Notch signaling is required for arterial–venous differentiation during embryonic vascular development. *Development*, **128**, 3675–83.
69. Nakagawa, O., Nakagawa, M., Richardson, J.A. *et al.* (1999) HRT1, HRT2, and HRT3: a new subclass of bHLH transcription factors marking specific cardiac, somitic, and pharyngeal arch segments. *Developmental Biology*, **216**, 72–84.
70. Seki, T., Yun, J., and Oh, S.P. (2003) Arterial endothelium-specific activin receptor-like kinase 1 expression suggests its role in arterialization and vascular remodeling. *Circulation Research*, **93**, 682–89.
71. Tian, H., Mcknight, S.L., and Russell, D.W. (1997) Endothelial PAS domain protein 1 (EPAS1), a transcription factor selectively expressed in endothelial cells. *Genes and Development*, **11**, 72–82.
72. Zhong, T.P., Childs, S., Leu, J.P., and Fishman, M.C. (2001) Gridlock signalling pathway fashions the first embryonic artery. *Nature*, **414**, 216–20.
73. Herzog, Y., Guttmann-Raviv, N., and Neufeld, G. (2005) Segregation of arterial and venous markers in subpopulations of blood islands before vessel formation. *Developmental Dynamics*, **232**, 1047–55.
74. Lavine, K.J., Long, F., Choi, K. *et al.* (2008) Hedgehog signaling to distinct cell types differentially regulates coronary artery and vein development. *Development*, **135**, 3161–71.
75. Moyon, D., Pardanaud, L., Yuan, L. *et al.* (2001) Plasticity of endothelial cells during arterial–venous differentiation in the avian embryo. *Development*, **128**, 3359–70.
76. Drake, C.J. and Little, C.D. (1995) Exogenous vascular endothelial growth factor induces malformed and hyperfused vessels during embryonic neovascularization. *Proceedings of the National Academy of Sciences of the United States of America*, **92**, 7657–61.
77. Flamme, I., Breier, G., and Risau, W. (1995) Vascular endothelial growth factor (VEGF) and VEGF receptor 2 (flk-1) are expressed during vasculogenesis and vascular differentiation in the quail embryo. *Developmental Biology*, **169**, 699–712.
78. Yamaguchi, T.P., Dumont, D.J., Conlon, R.A. *et al.* (1993) *flk-1*, an *flt*-related receptor tyrosine kinase is an early marker for endothelial cell precursors. *Development*, **118**, 489–98.
79. Ferrara, N., Gerber, H.P., and Lecouter, J. (2003) The biology of VEGF and its receptors. *Nature Medicine*, **9**, 669–76.
80. Grunstein, J., Masbad, J.J., Hickey, R. *et al.* (2000) Isoforms of vascular endothelial growth factor act in a coordinate fashion To recruit and expand tumor vasculature. *Molecular and Cellular Biology*, **20**, 7282–91.
81. Roy, H., Bhardwaj, S., and Yla-Herttuala, S. (2006) Biology of vascular endothelial growth factors. *FEBS Letters*, **580**, 2879–87.
82. Breier, G., Albrecht, U., Sterrer, S., and Risau, W. (1992) Expression of vascular endothelial growth factor during embryonic angiogenesis and endothelial cell differentiation. *Development*, **114**, 521–32.
83. Dumont, D.J., Fong, G.H., Puri, M.C. *et al.* (1995) Vascularization of the mouse embryo: a study of *flk-1*, *tek*, *tie*, and vascular endothelial growth factor expression during development. *Developmental Dynamics*, **203**, 80–92.
84. Carmeliet, P., Ferreira, V., Breier, G. *et al.* (1996) Abnormal blood vessel development and lethality

in embryos lacking a single VEGF allele. *Nature*, **380**, 435–39.
85. Ferrara, N., Carver-Moore, K., Chen, H. *et al.* (1996) Heterozygous embryonic lethality induced by targeted inactivation of the VEGF gene. *Nature*, **380**, 439–42.
86. Duan, L.J., Nagy, A., and Fong, G.H. (2003) Gastrulation and angiogenesis, not endothelial specification, is sensitive to partial deficiency in vascular endothelial growth factor-a in mice. *Biology of Reproduction*, **69**, 1852–58.
87. Stalmans, I., Ng, Y.S., Rohan, R. *et al.* (2002) Arteriolar and venular patterning in retinas of mice selectively expressing VEGF isoforms. *The Journal of Clinical Investigation*, **109**, 327–36.
88. Millauer, B., Wizigmann-Voos, S., Schnurch, H. *et al.* (1993) High affinity VEGF binding and developmental expression suggest Flk-1 as a major regulator of vasculogenesis and angiogenesis. *Cell*, **72**, 835–46.
89. de Vries, C., Escobedo, J.A., Ueno, H. *et al.* (1992) The fms-like tyrosine kinase, a receptor for vascular endothelial growth factor. *Science*, **255**, 989–91.
90. Fong, G.H., Rossant, J., Gertsenstein, M., and Breitman, M.L. (1995) Role of the Flt-1 receptor tyrosine kinase in regulating the assembly of vascular endothelium. *Nature*, **376**, 66–70.
91. Fong, G.H., Zhang, L., Bryce, D.M., and Peng, J. (1999) Increased hemangioblast commitment, not vascular disorganization, is the primary defect in *flt-1* knock-out mice. *Development*, **126**, 3015–25.
92. Millien, G., Beane, J., Lenburg, M. *et al.* (2008) Characterization of the mid-foregut transcriptome identifies genes regulated during lung bud induction. *Gene Expression Patterns*, **8**, 124–39.
93. Cardoso, W.V. and Lu, J. (2006) Regulation of early lung morphogenesis: questions, facts and controversies. *Development*, **133**, 1611–24.
94. Copland, I. and Post, M. (2004) Lung development and fetal lung growth. *Paediatric Respiratory Reviews*, **5** (Suppl A), S259–64.
95. Maeda, Y., Dave, V., and Whitsett, J.A. (2007) Transcriptional control of lung morphogenesis. *Physiological Reviews*, **87**, 219–44.
96. Defelice, M., Silberschmidt, D., Dilauro, R. *et al.* (2003) TTF-1 phosphorylation is required for peripheral lung morphogenesis, perinatal survival, and tissue-specific gene expression. *The Journal of Biological Chemistry*, **278**, 35574–83.
97. Tenhave-Opbroek, A. (1981) The development of the lung in mammals: an analysis of concepts and findings. *The American Journal of Anatomy*, **152**, 201–19.
98. Hislop, A. (2005) Developmental biology of the pulmonary circulation. *Paediatric Respiratory Reviews*, **6**, 35–43.
99. Kimura, J. and Deutsch, G.H. (2007) Key mechanisms of early lung development. *Pediatric and Developmental Pathology*, **10**, 335–47.
100. Jain, R.K. (2003) Molecular regulation of vessel maturation. *Nature Medicine*, **9**, 685–93.
101. Aird, W.C. (2007) Phenotypic heterogeneity of the endothelium: I. Structure, function, and mechanisms. *Circulation Research*, **100**, 158–73.
102. deMello, D.E. and Reid, L.M. (2000) Embryonic and early fetal development of human lung vasculature and its functional implications. *Pediatric and Developmental Pathology*, **3**, 439–49.
103. Kumar, V.H. and Ryan, R.M. (2004) Growth factors in the fetal and neonatal lung. *Frontiers in Bioscience*, **9**, 464–80.
104. Pauling, M.H. and Vu, T.H. (2004) Mechanisms and regulation of lung vascular development. *Current Topics in Developmental Biology*, **64**, 73–99.
105. Sundell, C. and Roman, J. (1994) Control of lung vasculogenesis by extracellular matrix composition. *Molecular Biology of the Cell*, **5**, 179a.
106. Schachtner, S.K., Wang, Y., and Scott Baldwin, H. (2000) Qualitative and quantitative analysis of embryonic pulmonary vessel formation. *American Journal of Respiratory Cell and Molecular Biology*, **22**, 157–65.
107. Hall, S.M., Hislop, A.A., and Haworth, S.G. (2002) Origin, differentiation, and maturation of human pulmonary veins. *American Journal of Respiratory Cell and Molecular Biology*, **26**, 333–40.
108. Hislop, A.A. (2002) Airway and blood vessel interaction during lung development. *Journal of Anatomy*, **201**, 325–34.
109. Parera, M.C., Van Dooren, M., Van Kempen, M. *et al.* (2005) Distal angiogenesis: a new concept for lung vascular morphogenesis. *American Journal of Physiology: Lung Cellular and Molecular Physiology*, **288**, L141–49.
110. Nishikawa, S.I., Nishikawa, S., Hirashima, M. *et al.* (1998) Progressive lineage analysis by cell sorting and culture identifies $FLK1^+VE^-$ cadherin$^+$ cells at a diverging point of endothelial and hemopoietic lineages. *Development*, **125**, 1747–57.
111. Adams, R.H., Wilkinson, G.A., Weiss, C. *et al.* (1999) Roles of ephrinB ligands and EphB receptors in cardiovascular development: demarcation

of arterial/venous domains, vascular morphogenesis, and sprouting angiogenesis. *Genes and Development*, **13**, 295–306.
112. Schwarz, M.A., Caldwell, L., Cafasso, D., and Zheng, H. (2009) Emerging pulmonary vasculature lacks fate specification. *American Journal of Physiology: Lung Cellular and Molecular Physiology*, **296**, L71–81.
113. Dellinger, M., Hunter, R., Bernas, M. *et al.* (2008) Defective remodeling and maturation of the lymphatic vasculature in angiopoietin-2 deficient mice. *Developmental Biology*, **319**, 309–20.
114. White, A.C., Lavine, K.J., and Ornitz, D.M. (2007) FGF9 and SHH regulate mesenchymal *Vegfa* expression and development of the pulmonary capillary network. *Development*, **134**, 3743–52.
115. Greenberg, J.M., Thompson, F.Y., Brooks, S.K. *et al.* (2002) Mesenchymal expression of vascular endothelial growth factors D and A defines vascular patterning in developing lung. *Developmental Dynamics*, **224**, 144–53.
116. Ng, Y.S., Rohan, R., Sunday, M.E. *et al.* (2001) Differential expression of VEGF isoforms in mouse during development and in the adult. *Developmental Dynamics*, **220**, 112–21.
117. Healy, A.M., Morgenthau, L., Zhu, X. *et al.* (2000) VEGF is deposited in the subepithelial matrix at the leading edge of branching airways and stimulates neovascularization in the murine embryonic lung. *Developmental Dynamics*, **219**, 341–52.
118. Voelkel, N.F., Vandivier, R.W., and Tuder, R.M. (2006) Vascular endothelial growth factor in the lung. *American Journal of Physiology: Lung Cellular and Molecular Physiology*, **290**, L209–21.
119. Zeng, X., Wert, S.E., Federici, R. *et al.* (1998) VEGF enhances pulmonary vasculogenesis and disrupts lung morphogenesis *in vivo*. *Developmental Dynamics*, **211**, 215–27.
120. Ruhrberg, C., Gerhardt, H., Golding, M. *et al.* (2002) Spatially restricted patterning cues provided by heparin-binding VEGF-A control blood vessel branching morphogenesis. *Genes and Development*, **16**, 2684–98.
121. Galambos, C., Ng, Y.S., Ali, A. *et al.* (2002) Defective pulmonary development in the absence of heparin-binding vascular endothelial growth factor isoforms. *American Journal of Respiratory Cell and Molecular Biology*, **27**, 194–203.
122. Yamamoto, Y., Shiraishi, I., Dai, P. *et al.* (2007) Regulation of embryonic lung vascular development by vascular endothelial growth factor receptors, Flk-1 and Flt-1. *The Anatomical Record (Hoboken)*, **290**, 958–73.
123. Yamashita, J., Itoh, H., Hirashima, M. *et al.* (2000) Flk1-positive cells derived from embryonic stem cells serve as vascular progenitors. *Nature*, **408**, 92–96.
124. Ahlbrecht, K., Schmitz, J., Seay, U. *et al.* (2008) Spatiotemporal expression of flk-1 in pulmonary epithelial cells during lung development. *American Journal of Respiratory Cell and Molecular Biology*, **39**, 163–70.
125. Kutcher, M.E., Klagsbrun, M., and Mamluk, R. (2004) VEGF is required for the maintenance of dorsal root ganglia blood vessels but not neurons during development. *The FASEB Journal*, **18**, 1952–54.
126. Wagner, K.F., Hellberg, A.K., Balenger, S. *et al.* (2004) Hypoxia-induced mitogenic factor has antiapoptotic action and is upregulated in the developing lung: coexpression with hypoxia-inducible factor-2alpha. *American Journal of Respiratory Cell and Molecular Biology*, **31**, 276–82.
127. Compernolle, V., Brusselmans, K., Acker, T. *et al.* (2002) Loss of HIF-2alpha and inhibition of VEGF impair fetal lung maturation, whereas treatment with VEGF prevents fatal respiratory distress in premature mice. *Nature Medicine*, **8**, 702–10.
128. Hosford, G.E. and Olson, D.M. (2003) Effects of hyperoxia on VEGF, its receptors, and HIF-2alpha in the newborn rat lung. *American Journal of Physiology: Lung Cellular and Molecular Physiology*, **285**, L161–68.
129. Grover, T.R., Asikainen, T.M., Kinsella, J.P. *et al.* (2007) Hypoxia-inducible factors HIF-1alpha and HIF-2alpha are decreased in an experimental model of severe respiratory distress syndrome in preterm lambs. *American Journal of Physiology: Lung Cellular and Molecular Physiology*, **292**, L1345–51.
130. Kunig, A.M., Balasubramaniam, V., Markham, N.E. *et al.* (2005) Recombinant human VEGF treatment enhances alveolarization after hyperoxic lung injury in neonatal rats. *American Journal of Physiology: Lung Cellular and Molecular Physiology*, **289**, L529–35.
131. Asikainen, T.M., Ahmad, A., Schneider, B.K., and White, C.W. (2005) Effect of preterm birth on hypoxia-inducible factors and vascular endothelial growth factor in primate lungs. *Pediatric Pulmonology*, **40**, 538–46.
132. Asikainen, T.M., Chang, L.Y., Coalson, J.J. *et al.* (2006) Improved lung growth and function through hypoxia-inducible factor in primate chronic lung disease of prematurity. *The FASEB Journal*, **20**, 1698–700.

133. Asikainen, T.M., Waleh, N.S., Schneider, B.K. *et al.* (2006) Enhancement of angiogenic effectors through hypoxia-inducible factor in preterm primate lung *in vivo*. *American Journal of Physiology: Lung Cellular and Molecular Physiology*, **291**, L588–95.
134. Chinoy, M.R., Graybill, M.M., Miller, S.A. *et al.* (2002) Angiopoietin-1 and VEGF in vascular development and angiogenesis in hypoplastic lungs. *American Journal of Physiology: Lung Cellular and Molecular Physiology*, **283**, L60–66.
135. Han, R.N., Babaei, S., Robb, M. *et al.* (2004) Defective lung vascular development and fatal respiratory distress in endothelial NO synthase-deficient mice: a model of alveolar capillary dysplasia? *Circulation Research*, **94**, 1115–23.
136. Murakami, M., Nguyen, L.T., Zhang, Z.W. *et al.* (2008) The FGF system has a key role in regulating vascular integrity. *The Journal of Clinical Investigation*, **118**, 3355–66.
137. Cardoso, W.V., Itoh, A., Nogawa, H. *et al.* (1997) FGF-1 and FGF-7 induce distinct patterns of growth and differentiation in embryonic lung epithelium. *Developmental Dynamics*, **208**, 398–405.
138. Maeda, S., Suzuki, S., Suzuki, T. *et al.* (2002) Analysis of intrapulmonary vessels and epithelial–endothelial interactions in the human developing lung. *Laboratory Investigation*, **82**, 293–301.
139. Byzova, T.V., Goldman, C.K., Pampori, N. *et al.* (2000) A mechanism for modulation of cellular responses to VEGF: activation of the integrins. *Molecular Cell*, **6**, 851–60.
140. Schuger, L., O'shea, S., Rheinheimer, J., and Varani, J. (1990) Laminin in lung development: effects of anti-laminin antibody in murine lung morphogenesis. *Developmental Biology*, **137**, 26–32.
141. Mollard, R. and Dziadek, M. (1998) A correlation between epithelial proliferation rates, basement membrane component localization patterns, and morphogenetic potential in the embryonic mouse lung. *American Journal of Respiratory Cell and Molecular Biology*, **19**, 71–82.
142. Schuger, L. (1997) Laminins in lung development. *Experimental Lung Research*, **23**, 119–29.
143. Schuger, L., Skubitz, A.P., O'Shea, K.S. *et al.* (1991) Identification of laminin domains involved in branching morphogenesis: effects of anti-laminin monoclonal antibodies on mouse embryonic lung development. *Developmental Biology*, **146**, 531–41.
144. Yang, Y., Palmer, K.C., Relan, N. *et al.* (1998) Role of laminin polymerization at the epithelial mesenchymal interface in bronchial myogenesis. *Development*, **125**, 2621–29.
145. Jakobsson, L., Domogatskaya, A., Tryggvason, K. *et al.* (2008) Laminin deposition is dispensable for vasculogenesis but regulates blood vessel diameter independent of flow. *The FASEB Journal*, **22**, 1530–39.
146. Ihida-Stansbury, K., Mckean, D.M., Gebb, S.A. *et al.* (2004) Paired-related homeobox gene *Prx1* is required for pulmonary vascular development. *Circulation Research*, **94**, 1507–14.
147. Jones, P.L. (2003) Homeobox genes in pulmonary vascular development and disease. *Trends in Cardiovascular Medicine*, **13**, 336–45.
148. Schwarz, M.A., Zhang, F., Lane, J.E. *et al.* (2000) Angiogenesis and morphogenesis of murine fetal distal lung in an allograft model. *American Journal of Physiology: Lung Cellular and Molecular Physiology*, **278**, L1000–7.
149. Schwarz, M.A., Wan, Z., Liu, J., and Lee, M.K. (2004) Epithelial–mesenchymal interactions are linked to neovascularization. *American Journal of Respiratory Cell and Molecular Biology*, **30**, 784–92.
150. Schwarz, M.A., Zhang, F., Gebb, S. *et al.* (2000) EMAP II inhibits lung neovascularization and airway epithelial morphogenesis. *Mechanisms of Development*, **95**, 123–32.
151. Schwarz, M., Lee, M., Zhang, F. *et al.* (1999) EMAP II: a modulator of neovascularization in the developing lung. *American Journal of Physiology: Lung Cellular and Molecular Physiology*, **276**, L365–75.
152. Delisser, H.M., Helmke, B.P., Cao, G. *et al.* (2006) Loss of PECAM-1 function impairs alveolarization. *The Journal of Biological Chemistry*, **281**, 8724–31.
153. Jakkula, M., Le Cras, T.D., Gebb, S. *et al.* (2000) Inhibition of angiogenesis decreases alveolarization in the developing rat lung. *American Journal of Physiology: Lung Cellular and Molecular Physiology*, **279**, L600–7.
154. Vu, T.H., Alemayehu, Y., and Werb, Z. (2003) New insights into saccular development and vascular formation in lung allografts under the renal capsule. *Mechanisms of Development*, **120**, 305–13.
155. Zhao, L., Wang, K., Ferrara, N., and Vu, T.H. (2005) Vascular endothelial growth factor co-ordinates proper development of lung epithelium and vasculature. *Mechanisms of Development*, **122**, 877–86.

156. Brown, K.R., England, K.M., Goss, K.L. *et al.* (2001) VEGF induces airway epithelial cell proliferation in human fetal lung *in vitro*. *American Journal of Physiology: Lung Cellular and Molecular Physiology*, **281**, L1001–10.
157. Kasahara, Y., Tuder, R.M., Taraseviciene-Stewart, L. *et al.* (2000) Inhibition of VEGF receptors causes lung cell apoptosis and emphysema. *The Journal of Clinical Investigation*, **106**, 1311–19.
158. Shu, W., Jiang, Y.Q., Lu, M.M., and Morrisey, E.E. (2002) Wnt7b regulates mesenchymal proliferation and vascular development in the lung. *Development*, **129**, 4831–42.
159. Dickson, M.C., Martin, J.S., Cousins, F.M. *et al.* (1995) Defective haematopoiesis and vasculogenesis in transforming growth factor-beta 1 knock out mice. *Development*, **121**, 1845–54.
160. Zhou, L., Dey, C.R., Wert, S.E., and Whitsett, J.A. (1996) Arrested lung morphogenesis in transgenic mice bearing an SP-C–TGF-beta 1 chimeric gene. *Developmental Biology*, **175**, 227–38.
161. Jacquemin, P., Yoshitomi, H., Kashima, Y. *et al.* (2006) An endothelial–mesenchymal relay pathway regulates early phases of pancreas development. *Developmental Biology*, **290**, 189–99.
162. Lammert, E., Cleaver, O., and Melton, D. (2001) Induction of pancreatic differentiation by signals from blood vessels. *Science*, **294**, 564–67.
163. Chen, R.R., Silva, E.A., Yuen, W.W., and Mooney, D.J. (2007) Spatio-temporal VEGF and PDGF delivery patterns blood vessel formation and maturation. *Pharmaceutical Research*, **24**, 258–64.

2

Anatomy of the Pulmonary Endothelium

Radu V. Stan

Departments of Pathology and Microbiology & Immunology, Dartmouth Medical School, Lebanon, NH, USA

INTRODUCTION

Vascular endothelium is a highly differentiated cellular monolayer with the organization of a simple squamous epithelium. It lines the entire cardiovascular system, and thus constitutes a quasi-ubiquitous presence in all organs and tissues throughout the body. In the lung, as in other organs, endothelium is a critical participant in several processes such as vascular permeability, coagulation and anticoagulation cascades, regulation of vascular tone, interactions with the immune system, and formation of new vessels by vasculogenesis and angiogenesis. The lung has an additional function of matching perfusion with ventilation for optimal gas exchange.

A dual circulation supplies the lung: (i) the pulmonary circulation that is involved in gas exchange, and (ii) the bronchial circulation that supplies the airways down to the terminal bronchioles (depending on the species), and participates in the thermoregulation and humidification of the air (see Chapter 14). Both types of circulation have important roles in mediating host defense mechanisms. The lung has also a well-developed lymphatic system [1, 2] with the critical function of drainage of the fluid from interstitial space in order to maintain an efficient diffusion barrier. The lymphatic network starts at the pleura, continuing in collecting vessels in the interlobular and interlobar septa, and finally collecting into the hilar lymphatics.

This chapter is intended to provide a brief overview of the anatomy of the pulmonary circulation, pointing out the advances contributed by electron microscopy (EM) studies.

HISTORIC PERSPECTIVE

Many aspects of the lung architecture and circulation were established by light microscopy at the beginning of the last century. The few controversies due to the limited resolution of the light microscope, especially on the structure of alveolar capillary barrier, were solved by studies employing EM in 1950s starting with Low [3, 4], who was the first to examine lung sections by EM. The normal vasculature of the lung was examined in the 1950s by Scklipkoeter [5], Clemens [6], Bargmann and Knoop [7–10], Karer [8–10] Policard and Collet [11–13], De Groodt [14], and Takahashi *et al.* [15]. These studies showed that the air–blood tissue barrier was consistently composed of a capillary endothelium separated from the epithelium by a very narrow interstitial space. The following decades witnessed an explosion of studies of different aspects of normal lungs at higher resolution due to advances in specimen preparation and methodology of study (i.e., stereology) and in different species [16–23] completing the picture of the lung structure from an evolutionary standpoint (see [22–24] for reviews). Numerous other EM studies subsequently established our current picture of fine-structural organization of the interalveolar septum in mammals and other species. Of these, some studies deserve special mention, including the studies of Bensch and Dominguez [25–27], Weibel [20, 28, 29], Ryan [30–32], Lauweryns [2, 33], and Palade and Simonescu [34–38]. By far, most reports deal with the lung in different pathological conditions. This work is highlighted in the chapters throughout this volume.

PULMONARY CIRCULATION

The pulmonary circulation has a large surface area (120 m^2) with the main function of gas exchange as well as important roles in host defense, monitoring, and maintenance of blood homeostasis. This vascular bed is unique as it receives all cardiac output, maintains a low blood pressure, and is exposed to high mechanical stress and

The Pulmonary Endothelium: Function in health and disease Editors Norbert F. Voelkel, Sharon Rounds

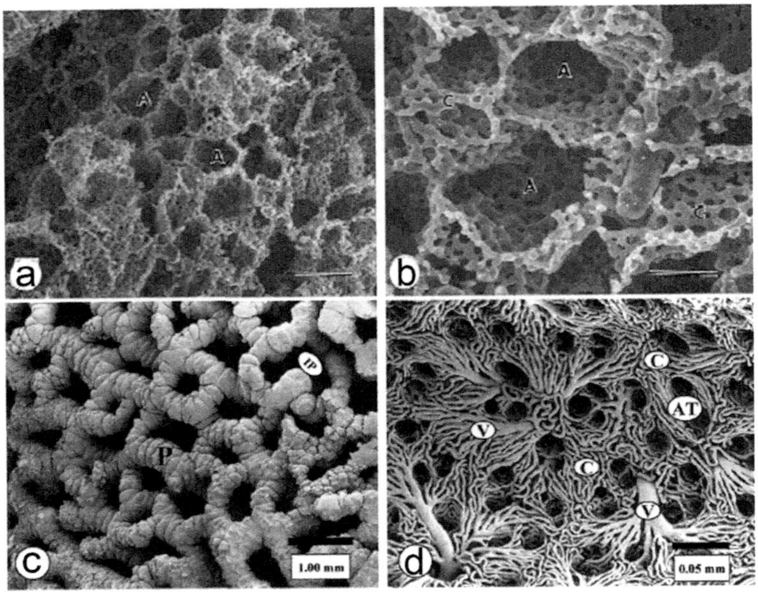

Figure 2.1 Scanning EM of corrosion casts of mammalian lung (a and b) and avian lung (c and d). (a) Vascular corrosion cast of a normal inflated hamster lung demonstrating the disposition of capillaries in the alveolar septa. A, alveolae; C, capillaries. Bar = 100 µm. (b) A higher power micrograph from (a). Bar = 50 µm. (c) Resin corrosion cast of the avian lung parenchyma demonstrating the parabronchi that anastomose and form a complicated network. P, parabronchi; IP, interparabronchial part containing capillaries. Bar = 1 mm. (d) Vascular corrosion cast of the capillaries of an avian lung demonstrating the arrangement of the capillaries around the atria of parabronchi (AT). V, venules; C, blood capillaries; AT, air capillaries. Bar = 50 µm. (a) & (b) reprinted from Hossler, F.E. and Douglas, J.E. (2001) *Microscopy and Microanalysis*, **7** (3), 253–64, copyright © 2001, Cambridge University Press. (c) & (d) reprinted from [75], copyright © 2005, the Japanese Society of Veterinary Science.

to the highest oxygen tension of all vascular beds. The pulmonary circulation starts at the right ventricle with the pulmonary artery that hierarchically branches out following the airways. Thus, blood with high content in CO_2 and low O_2 (pO_2 ~40 mmHg) is taken from the right ventricle by the pulmonary artery (the precapillary segment of pulmonary circulation) to a web-like capillary network (Figures 2.1a,b and 2.2a) forming a net through which the blood is thought to seep as a sheet or a film [39–42] the holes of the net being constituted by alveolar spaces. Finally, from the capillaries the oxygenated blood (pO_2 ~100 mmHg) is collected by the venous tree (or the postcapillary segment of pulmonary circulation) and drained into the left atrium of the heart.

Arteries (Precapillary Segment)

The walls of arteries and veins closely resemble each other, having similar thickness due to thinner smooth muscle layers in the arteries [43]. This is different from the high-pressure systemic circulation where arterial walls are much thicker. The arterial endothelial cells (ECs) rest on a thick basement membrane, and form tight junctions with up to six adjacent cells and are aligned in the direction of flow [44, 45]. The EC nucleus is situated centrally, surrounded by the "organelle zone" [34] then by the attenuated "peripheral" zone, thicker than in capillaries.

Capillary Network

The lung capillary endothelium is of the continuous type, forming a complete barrier between the blood and the tissues [46, 47]. It is composed of highly attenuated ECs resting on a thin basement membrane (Figure 2.2b) [28, 35]. Their cytoplasm surrounds the nucleus as a thin layer and the perinuclear zone of the capillary ECs is characteristically poor in organelles (Figure 2.2e). In all zones, the plasma membrane features large numbers of membrane invaginations, such as caveolae and other non-coated vesicles [48–51], and is poor in clathrin-coated pits and vesicles (Figure 2.2d) [48, 52]. The caveolae are provided with stomatal diaphragms containing the diaphragm protein PV-1 (Figure 2.3) [51, 53]. This is

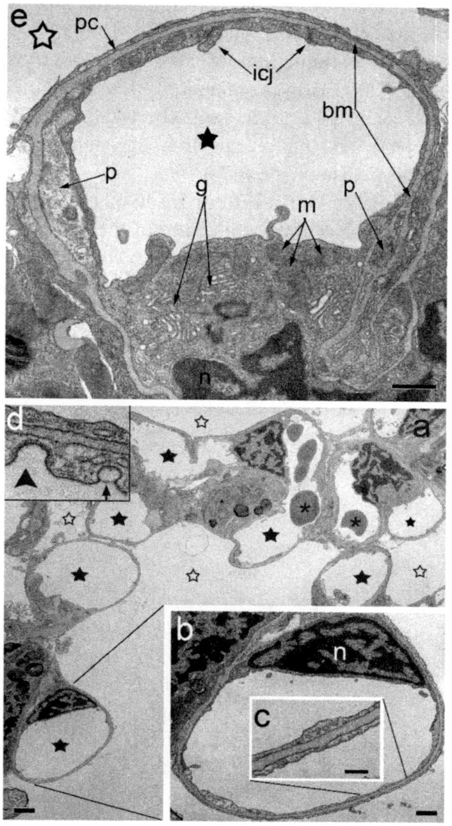

Figure 2.2 Transmission electron micrographs demonstrating different aspects of the pulmonary capillary network. (a) Low-power field showing the capillary loops (filled stars) and alveolar airspaces (open stars). Asterisks show red blood cells. Bar = 1 μm. (b) A transverse section through a capillary from (a) indicated by lines. It demonstrates the nucleus (n), the relative paucity in organelles of the perinuclear region of EC as well as the extensive attenuated parts of alveolo-capillary unit involved in the gas exchange. Bar = 500 nm. (c) A higher power micrograph of the gas exchange unit demonstrating its components: the attenuated EC and type I pneumocyte separated by a thin basement membrane. Bar = 100 nm. (d) Detail of an EC demonstrating a clathrin-coated pit (arrowhead) as well as a caveola with a stomatal diaphragm (arrow). (e) Transverse section through a pulmonary capillary loop at the level of the perinuclear zone. It demonstrates several of the endothelial organelles (m, mitochondria; g, golgi) as well as the intercellular junctions (icjs). The very thin basement membrane (bm) separates the endothelium from pneumocytes type I (pc) or pericytes (p). Capillary loops (filled stars) and alveolar airspaces (open stars).

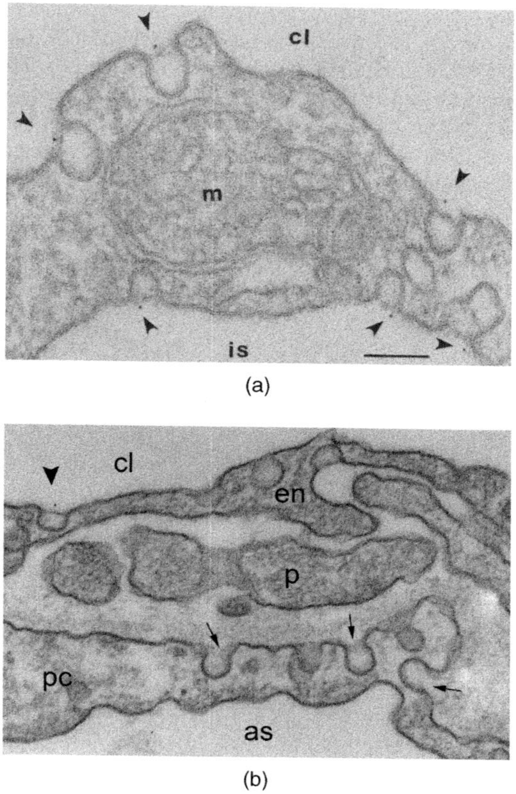

Figure 2.3 Pulmonary capillary ECs have caveolae with stomatal diaphragms. (a) High magnification of a capillary EC showing caveolae whose stomatal diaphragms are labeled with anti-PV1 antibodies (arrowheads). (b) Micrograph demonstrating the lack (arrows) of stomatal diaphragms on pulmonary epithelial cells (pc); m, mitochondrion; p, pericyte; en, EC; cl, capillary lumen; is, interstitial space; as, air space.

a feature that differentiates the capillary ECs from those of the arterial tree and large veins. Venules and small collecting veins do have caveolae with stomatal diaphragms. The periphery of the lung capillary ECs is much thinner than that of other continuous endothelia (i.e., heart or muscle). The areas facing the type I pneumocytes in the alveolus form an extremely attenuated (down to 20 nm thick) "avesicular" zone [35] consisting of the luminal and abluminal plasma membranes separated by a minute amount of cytoplasm and devoid of membrane invaginations and organelles (Figure 2.2b,c,e). These areas are thought to be directly involved in gas exchange [22, 24, 29] and the proper function of the lung depends on their maintenance. The thickness and extent of such regions seem to depend on lung size: they are extremely rare in human or dog lungs, but become more frequent in lungs of rats and mice, and are a predominant feature in the smallest mammal – the Etruscan shrew [20]. The intercellular junctions of capillary ECs (Figure 2.2e and Figure 2.4) seem to be tighter and to confer a better barrier function than in the pre- or postcapillary segment of the pulmonary circulation. In addition to VE-cadherin (cadherin-5), the junctions contain other molecules such as E-cadherin and N-cadherin, which is also a difference from the pre- and postcapillary segments [45, 54, 55]. Finally, the pulmonary capillary ECs are in contact with sparse pericytes [56] as well as fibroblasts that might provide a link with type 2 pneumocytes [57].

The normal endothelium of alveolar capillaries has no fenestrae [59]. However, under certain pathological conditions, such as fibrotic lung disease [60–62], leptospirosis [63], and neoplasms [64, 65], fenestrae may develop (for a review, see [47]).

The mammalian lung differs from other vertebrates and birds in terms of the architecture of the gas exchange units. In frogs and fish lungs, the alveolar epithelium is made of a single cell type. The "interalveolar" septa

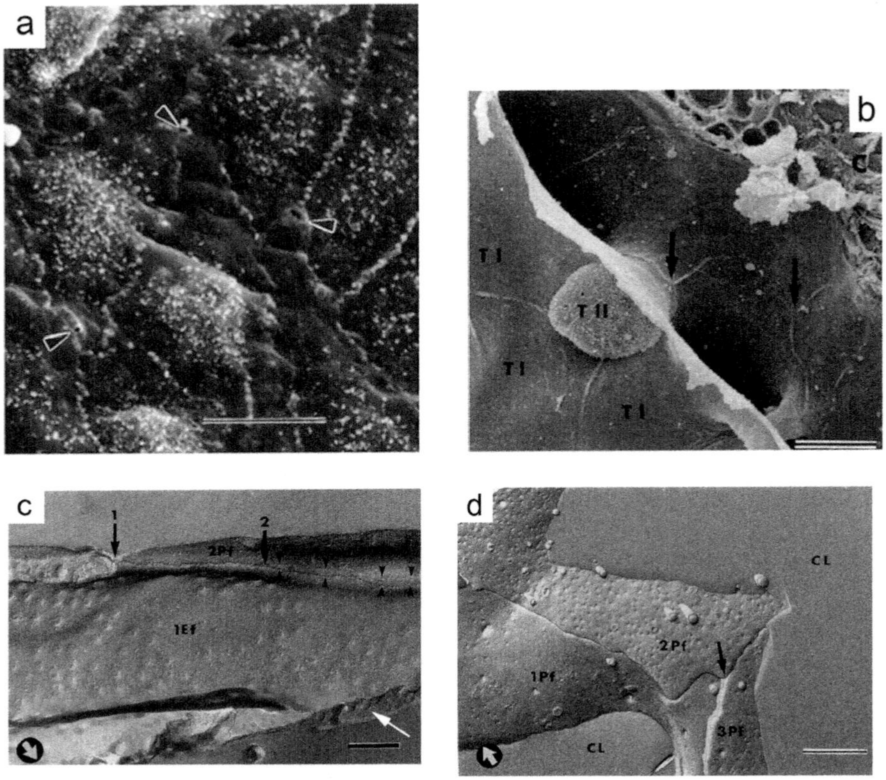

Figure 2.4 Inter-EC junctions in large vessels (a) and capillaries (b–d) as demonstrated by scanning EM (a and b) as well as freeze-fracture (c and d). (a) Scanning electron micrographs showing the endothelial monolayer in pulmonary large vessels in the rat. Sparse discontinuities in the junctions are demonstrated by arrowheads. Bar = 10 μm. (b) Scanning electron micrographs showing the endothelial monolayer in pulmonary capillaries in the mouse. The junctions in the capillary as well as in the air space appear as fine discolored lines. Arrows point to the junctional points between three cells, thought to contain discontinuities which may be pores. C, capillary EC body; T I, type I pneumocyte; T II, type II pneumocyte. (c) Transmission EM of a platinum carbon freeze-fracture replica of mouse lung. It demonstrates the junction between two adjacent ECs (1 and 2). Arrows and arrowheads point to the junctional complexes. The dimples show the introits of vesicular carrier attached to the plasma membrane. White arrow points to caveolae obviated at a site where the fracture plane cut across the EC. Pf, P face of the replica containing the inner leaflet of the plasma membrane; Ef, E face of the replica containing the external leaflet of the plasma membrane bilayer. Arrow in a dark background shows the direction of metal shadowing. (d) An example of the intersection point of three ECs by freeze-fracture. Reproduced from (a) [54] and (b–d) [58], with permission of the American Physiological Society.

contain a continuous broad connective tissue sheet, or septum, that contains a separate capillary on either side [17, 18, 20, 21].

The general anatomy of the avian lung is fundamentally different from that of the mammalian lung [16, 66]. The gas exchange units contained in parabronchi are so designed that are continuously perfused with a unidirectional stream of air (Figure 2.1c,d) [19, 67]. In the actual gas exchange apparatus the airways consist of air capillaries of about 10 µm diameter that are densely interwoven with blood capillaries. The air–blood barrier again is composed of epithelium, interstitium, and endothelium, but all three layers are extremely thin, making a total barrier only 0.1 µm thick over the major part. Avian lungs are considered to be superior to the mammalian lungs in terms of efficiency [23, 24].

Pulmonary Venous System (Postcapillary Segment)

As noted, the pulmonary veins collect the blood from capillaries. Their branching orders are similar to those of arteries and they can be recognized by their location [45]. Pulmonary veins do not contain valves, which also discriminates them from the bronchial veins [68]. Venules and small veins seem to feature venous sphincters to aid in the progression of the blood [69]. Ultrastructurally, pulmonary vein ECs resemble those in the arteries.

BRONCHIAL CIRCULATION

The bronchial circulation receives around 3% of systemic blood flow and originates from the aorta or intercostal arteries. These vessels might have been discovered by Leonardo da Vinci [70] (see Chapter 14). They are classified as either intrapulmonary or extrapulmonary. The intrapulmonary bronchial arteries perfuse the vasa vasorum of large pulmonary arteries and veins, the airways to the terminal bronchioles, and visceral pleura. The intrapulmonary capillaries drain into the pulmonary vein, whereas their extrapulmonary counterparts drain in bronchiolar veins. These veins contain valves and resemble the systemic veins in their architecture. There are species-specific differences in the territory supplied by the bronchial circulation [71].

The ECs in the capillary segment of the bronchial vasculature are more permeable to solutes and have a far greater capacity for angiogenesis compared with ECs from the pulmonary vasculature [72]. In some conditions the capillary ECs can be fenestrated [73].

CONCLUSIONS AND PERSPECTIVES

It is evident that the ultrastructure of pulmonary ECs has features that facilitate gas exchange and minimize fluid flux from the blood into the lung interstitium (see Chapter 8). Future research is needed to characterize the endothelial structures participating in nonrespiratory pulmonary functions and to ascertain the structural bases of heterogeneity among lung vessel endothelium (see Chapter 9).

References

1. Kato, F. (1966) The fine structure of the lymphatics and the passage of China ink particles through their walls. 1. The fine structure of the lymphatics of the cattle lung and the passage of China ink through their walls. *Nagoya Medical Journal*, **12** (4), 221–36.
2. Lauweryns, J.M. (1971) The blood and lymphatic microcirculation of the lung. *Pathology Annual*, **6**, 365–415.
3. Low, F.N. (1952) Electron microscopy of the rat lung. *The Anatomical Record*, **113** (4), 437–49.
4. Low, F.N. (1954) The electron microscopy of sectioned lung tissue after varied duration of fixation in buffered osmium tetroxide. *The Anatomical Record*, **120** (4), 827–51.
5. Schlipkoeter, H.W. (1954) Elektronenoptische Untersuchungen ultradiinner Lungenschnitte. *Deutsche Medizinische Wochenschrift*, **79**, 1658–9.
6. Clemens, H.J. (1954) [Electron microscopical studies on the structure of the alveolar wall of the rat lung]. *Verhandlungen der Anatomischen Gesellschaft*, 204–5.
7. Bargmann, W. and Knoop, A. (1956) Comparative electron microscopic research of pulmonary capillaries. *Zeitschrift fur Zellforschung und mikroskopische Anatomie*, **44** (3), 263–81.
8. Karrer, H.E. (1956) An electron microscopic study of the fine structure of pulmonary capillaries and alveoli of the mouse; preliminary report. *Bulletin of the Johns Hopkins Hospital*, **98** (2), 65–83.
9. Karrer, H.E. (1956) The ultrastructure of mouse lung fine structure of the capillary endothelium. *Experimental Cell Research*, **11** (3), 542–47.
10. Karrer, H.E. (1956) The ultrastructure of mouse lung; general architecture of capillary and alveolar walls. *The Journal of Biophysical and Biochemical Cytology*, **2** (3), 241–52.
11. Policard, A. and Collet, A. (1956) Contribution of electron microscopy to the knowledge of the histophysiology of the alveolar wall. *Journal of Physiology (Paris)*, **48** (3), 687–90.
12. Policard, A., Collet, A. *et al.* (1956) Inframicroscopic arrangements of the endothelium of the pulmonary capillaries in mammals. *Comptes Rendus*

Hebdomadairies des Seances de l'Academie des Sciences, **243** (1), 8–11.
13. Policard, A., Collet, A. *et al.* (1959) Research with the electron microscope on the parietal alveolar cells of the lung of mammals. *Zeitschrift fur Zellforschung und Mikroskopische Anatomie*, **50**, 561–87.
14. De Groodt, M., Lagasse, A. *et al.* (1958) Structure of the pulmonary alveoli, as seen with the electron microscope. *Archives d'Anatomie Pathologique (Paris)*, **47** (1), 67–84.
15. Takahashi, K., Kawano, A. *et al.* (1961) Electron microscopic studies on the lung tissue. *The Tokushima Journal of Experimental Medicine*, **8**, 149–55.
16. Irving, L. (1964) Comparative anatomy and physiology of gas transport mechanisms, in *Handbook of Physiology. 3: Respiration*, vol. **I**, American Physiological Society, Bethesda, MD, pp. 177–212.
17. Klika, E. and Petrik, P. (1965) A study of the structure of the lung alveolus and bronchiolar epithelium. (A histological and histochemical study using the method of membranous preparations). *Acta Histochemica*, **20** (5), 331–42.
18. Stehbens, W.E. (1965) Ultrastructure of vascular endothelium in the frog. *Quarterly Journal of Experimental Physiology and Cognate Medical Sciences*, **50** (4), 375–84.
19. Bretz, W.L. and Schmidt-Nielsen, K. (1971) Bird respiration: flow patterns in the duck lung. *The Journal of Experimental Biology*, **54** (1), 103–18.
20. Weibel, E.R. (1972) Morphometric estimation of pulmonary diffusion capacity. V. Comparative morphometry of alveolar lungs. *Respiration Physiology*, **14** (1), 26–43.
21. Hughes, G.M. and Weibel, E.R. (1978) Visualization of layers lining the lung of the South American lungfish (*Lepidosiren paradoxa*) and a comparison with the frog and rat. *Tissue and Cell*, **10** (2), 343–53.
22. West, J.B. and Mathieu-Costello, O. (1999) Structure, strength, failure, and remodeling of the pulmonary blood–gas barrier. *Annual Review of Physiology*, **61**, 543–72.
23. West, J.B., Watson, R.R. *et al.* (2007) The human lung: did evolution get it wrong? *The European Respiratory Journal*, **29** (1), 11–17.
24. Weibel, E.R., Sapoval, B. *et al.* (2005) Design of peripheral airways for efficient gas exchange. *Respiratory Physiology and Neurobiology*, **148** (1-2), 3–21.
25. Bensch, K.G. (1967). Electron microscopy in the study of the lung, in *The Lung* (eds A.A. Liebow and D.E. Smith), International Academy of Pathology Monographs, Williams & Wilkins, Baltimore, MD, pp. 323.
26. Bensch, K.G., Dominguez, E. *et al.* (1967) Absorption of intact protein molecules across the pulmonary air–tissue barrier. *Science*, **157** (793), 1204–6.
27. Dominguez, E.A., Liebow, A.A. *et al.* (1967) Studies on the pulmonary air-tissue barrier. I. Absorption of albumin by the alveolar wall. *Laboratory Investigation*, **16** (6), 905–11.
28. Weibel, E.R. (1969) The ultrastructure of the alveolar–capillary membrane barrier, in *The Pulmonary Circulation and Interstitial Space* (eds A.P. Fishman and H.H. Hecht), University of Chicago Press, Chicago, IL, p. 9.
29. Weibel, E.R. (1973) Morphological basis of alveolar–capillary gas exchange. *Physiological Reviews*, **53** (2), 419–95.
30. Ryan, S.F. (1969) The structure of the interalveolar septum of the mammalian lung. *The Anatomical Record*, **165** (4), 467–83.
31. Ryan, S.F. (1973) The structure of the primary lung lobule. *Annals of Clinical and Laboratory Science*, **3** (3), 147–55.
32. Smith, U. and Ryan, J.W. (1973) Electron microscopy of endothelial and epithelial components of the lungs: correlations of structure and function. *Federation Proceedings*, **32** (9), 1957–66.
33. Lauweryns, J.M. and Boussauw, L. (1973) The ultrastructure of lymphatic valves in the adult rabbit lung. *Zeitschrift fur Zellforschung und Mikroskopische Anatomie*, **143** (2), 149–68.
34. Simionescu, M., Simionescu, N. *et al.* (1974) Morphometric data on the endothelium of blood capillaries. *The Journal of Cell Biology*, **60** (1), 128–52.
35. Simionescu, M. (1980) Ultrastructural organization of the alveolar–capillary unit. *Ciba Foundation Symposium*, **78**, 11–36.
36. Simionescu, D. and Simionescu, M. (1983) Differentiated distribution of the cell surface charge on the alveolar–capillary unit. Characteristic paucity of anionic sites on the air-blood barrier. *Microvascular Research*, **25** (1), 85–100.
37. Simionescu, M. (1983) Alveolar capillary endothelium, functional morphology. *Current Problems in Clinical Biochemistry*, **13**, 74–81.
38. Simionescu, N., Lupu, F. *et al.* (1983) Rings of membrane sterols surround the openings of vesicles and fenestrae, in capillary endothelium. *The Journal of Cell Biology*, **97** (5 Pt 1), 1592–600.
39. Fung, Y.C. and Sobin, S.S. (1969) Theory of sheet flow in lung alveoli. *Journal of Applied Physiology*, **26** (4), 472–88.

40. Sobin, S.S., Tremer, H.M. *et al.* (1970) Morphometric basis of the sheet-flow concept of the pulmonary alveolar microcirculation in the cat. *Circulation Research*, **26** (3), 397–414.
41. Fung, Y.C. and Sobin, S.S. (1972) Pulmonary alveolar blood flow. *Circulation Research*, **30** (4), 470–90.
42. Sobin, S.S. and Fung, Y.C. (1992) Response to challenge to the Sobin–Fung approach to the study of pulmonary microcirculation. *Chest*, **101** (4), 1135–43.
43. Sasaki, S., Kobayashi, N. *et al.* (1995) Structural organization of pulmonary arteries in the rat lung. *Anatomy and Embryology (Berlin)*, **191** (6), 477–89.
44. Virchow, R. (1847) Uber die Standpunkte in den Wissenschaftlichen Medizin. *Virchows Archiv*, **1**, 1–19.
45. Stevens, T. (2007) Microheterogeneity of lung endothelium, in *Endothelial Biomedicine* (ed. W.C. Aird), Cambridge University Press, Cambridge, pp. 1161–77.
46. Stan, R.V. (2005) *Channels Across Endothelial Cells*, Landes Bioscience, Austin, TX.
47. Stan, R.V. (2007) Endothelial stomatal and fenestral diaphragms in normal vessels and angiogenesis. *Journal of Cellular and Molecular Medicine*, **11** (4), 621–43.
48. Palade, G.E. (1991) Role of plasmalemmal vesicles, in *The Lung: Scientific Foundations* (eds R.G. Crystal and J.B. West), Raven Press, New York, p. 359–67.
49. Moldovan, N.I., Heltianu, C. *et al.* (1995) Ultrastructural evidence of differential solubility in Triton X-100 of endothelial vesicles and plasma membrane. *Experimental Cell Research*, **219** (1), 309–13.
50. Stan, R.V., Roberts, W.G. *et al.* (1997) Immunoisolation and partial characterization of endothelial plasmalemmal vesicles (caveolae). *Molecular Biology of the Cell*, **8** (4), 595–605.
51. Stan, R.V. (2005) Structure of caveolae. *Biochimica et Biophysica Acta*, **1746** (3), 334–48.
52. King, J., Hamil, T. *et al.* (2004) Structural and functional characteristics of lung macro- and microvascular endothelial cell phenotypes. *Microvascular Research*, **67** (2), 139–51.
53. Stan, R.V., Ghitescu, L. *et al.* (1999) Isolation, cloning, and localization of rat PV-1, a novel endothelial caveolar protein. *The Journal of Cell Biology*, **145** (6), 1189–98.
54. Chetham, P.M., Babal, P. *et al.* (1999) Segmental regulation of pulmonary vascular permeability by store-operated Ca^{2+} entry. *American Journal of Physiology: Lung Cellular and Molecular Physiology*, **276** (1 Pt 1), L41–50.
55. Ofori-Acquah, S.F., King, J. *et al.* (2008) Heterogeneity of barrier function in the lung reflects diversity in endothelial cell junctions. *Microvascular Research*, **75** (3), 391–402.
56. Weibel, E.R. (1974) On pericytes, particularly their existence on lung capillaries. *Microvascular Research*, **8** (2), 218–35.
57. Sirianni, F.E., Chu, F.S. *et al.* (2003) Human alveolar wall fibroblasts directly link epithelial type 2 cells to capillary endothelium. *American Journal of Respiratory and Critical Care Medicine*, **168** (12), 1532–37.
58. Walker, D.C., MacKenzie A. Hosford, S. (1994) The structure of the tricellular region of endothelial tight junctions of pulmonary capillaries analyzed by freeze-fracture. *Microvasc Res*, Nov; **48** (3), 259–81.
59. Palade, G.E. and Bruns, R.R. (1968) Structural modulations of plasmalemmal vesicles. *The Journal of Cell Biology*, **37** (3), 633–49.
60. Suzuki, Y. (1969) Fenestration of alveolar capillary endothelium in experimental pulmonary fibrosis. *Laboratory Investigation*, **21** (4), 304–8.
61. Kawanami, O., Matsuda, K. *et al.* (1992) Endothelial fenestration of the alveolar capillaries in interstitial fibrotic lung diseases. *Acta Pathologica Japonica*, **42** (3), 177–84.
62. Kawanami, O., Jiang, H.X. *et al.* (1995) Alveolar fibrosis and capillary alteration in experimental pulmonary silicosis in rats. *American Journal of Respiratory and Critical Care Medicine*, **151** (6), 1946–55.
63. De Brito, T., Bohm, G.M. *et al.* (1979) Vascular damage in acute experimental leptospirosis of the guinea-pig. *The Journal of Pathology*, **128** (4), 177–82.
64. Jin, E., Fujiwara, M. *et al.* (2001) Aerogenous spread of primary lung adenocarcinoma induces ultrastructural remodeling of the alveolar capillary endothelium. *Human Pathology*, **32** (10), 1050–58.
65. Jin, E., Ghazizadeh, M. *et al.* (2001) Angiogenesis and phenotypic alteration of alveolar capillary endothelium in areas of neoplastic cell spread in primary lung adenocarcinoma. *Pathology International*, **51** (9), 691–700.
66. Duncker, H.R. (1971) The lung air sac system of birds. A contribution to the functional anatomy of the respiratory apparatus. *Ergebnisse der Anatomie und Entwicklungsgeschichte*, **45** (6), 7–171.
67. Scheid, P. and Piiper, J. (1972) Cross-current gas exchange in avian lungs: effects of reversed

parabronchial air flow in ducks. *Respiration Physiology*, **16** (3), 304–12.

68. Yen, R.T., Zhuang, F.Y. *et al.* (1983) Morphometry of cat pulmonary venous tree. *Journal of Applied Physiology*, **55** (1 Pt 1), 236–42.
69. Aharinejad, S., Bock, P. *et al.* (1992) Scanning and transmission electron microscopy of venous sphincters in the rat lung. *The Anatomical Record*, **233** (4), 555–68.
70. Mitzner, W. and Wagner, E. (1992) On the purported discovery of the bronchial circulation by Leonardo da Vinci. *Journal of Applied Physiology*, **73** (3), 1196–201.
71. Mitzner, W. and Wagner, E.M. (2004) Vascular remodeling in the circulations of the lung. *Journal of Applied Physiology*, **97** (5), 1999–2004.
72. Wagner, E. and Moldobaeva, A. (2007) Bronchial endothelium, in *Endothelial Biomedicine* (ed. W.C. Aird), Cambridge University Press, Cambridge, p. 1171–77.
73. Hirabayashi, M. and Yamamoto, T. (1984) An electron-microscopic study of the endothelium in mammalian bronchial microvasculature. *Cell and Tissue Research*, **236** (1), 19–25.
74. Hossler, F.E. and Douglas, J.E. (2001) Vascular corrosion casting: review of advantages and limitations in the application of some simple quantitative methods. *Microscopy and Microanalysis*, **7** (3), 253–64.
75. Nasu, T. (2005) Scanning electron microscopic study on the microarchitecture of the vascular system in the pigeon lung. *The Journal of Veterinary Medical Science*, **67** (10), 1071–74.

3

Cadherins and Connexins in Pulmonary Endothelial Function

Kaushik Parthasarathi[1] and Sadiqa K. Quadri[2]

[1]Departments of Physiology and Biomedical Engineering, University of Tennessee Health Science Center, Memphis, TN, USA

[2]Division of Pulmonary, Allergy and Critical Care Medicine, Columbia University College of Physicians & Surgeons, New York, NY, USA

INTRODUCTION

The interendothelial junction contains several proteins including cadherins and connexins that are constituent proteins of adherens junctions (AJs) and gap junctions (GJs), respectively. In addition, resident at the junction are the junctional adhesion molecules (JAMs), and claudins and occludins that are constituent proteins of tight junctions (TJs). Recent reviews detail the molecular structures of these proteins [1–4]. These junctional proteins provide structural support to the microvasculature (cadherins), regulate junctional permeability (claudins and occludins), mediate intercellular communication (connexins), and facilitate leukocyte migration (JAMs).

Here, we review the primary functions of AJ and GJ proteins as relevant to the pulmonary circulation. Recent studies not only redefine our existing understanding of cadherin and connexin function, but also reveal their novel roles in the lung microvasculature [5–8]. While it was thought that VE-cadherin mediated barrier functions of both macro- and microvessels in lung, recent reports reveal that E-cadherin regulates primarily the microvascular barrier. Similarly, only recently have connexins been implicated in interendothelial signaling in pulmonary circulation [5]. However, this contrasts with reports that systemic capillaries and venules do not support connexin-dependent communication – findings that may have contributed to the reduced focus until now on pulmonary endothelial connexins [9]. These exciting new findings provide new paradigms for the role of cadherins and connexins in pulmonary vasculature and are elaborated below.

CADHERINS

Cadherin Subtypes

AJs are located at cell–cell contact sites and link the actin cytoskeleton of adjacent cells [10, 11]. Cadherins, the major constituent of AJs [12], are a family of single-chain transmembrane proteins [13–16] that support homophilic cell–cell binding between similar molecules on opposing cells. Classical cadherins, number more than 15, and can be subgrouped into type I and type II cadherins, based on variations in amino acid sequence [12, 17]. They share a similar structure that has five extracellular homologous domains and one transmembrane region. The type I subgroup includes B-, E-, EP-, M-, N-, P-, and R-cadherin, and cadherin-4; while the type II subgroup includes cadherin-5 through -12. Cadherin-13 lacks the sequence corresponding to the cytoplasmic domain of typical cadherins. (The letters in the prefix indicate the tissue in which the corresponding cadherin was first detected, e.g., B, brain; E, endothelium; M, muscle; N, neuron; P, placenta;, R, retina; VE, vascular endothelium, etc.; EP-cadherin is a novel *Xenopus* Ca^{2+}-dependent adhesion molecule that shares comparable homology with mouse E- and P-cadherins.)

In the vascular endothelium, the three major cadherins include VE-, E-, and N-cadherin [18, 19]. VE-cadherin (cadherin-5) is located at intercellular junctions of all endothelial types, and its expression has been confirmed both *in vitro and in vivo* [18, 20]. In the intact pulmonary vasculature, large vessels primarily express VE-cadherin [21–23], while microvessels express E-cadherin [22, 24,

The Pulmonary Endothelium: Function in health and disease Editors Norbert F. Voelkel, Sharon Rounds

25]. As VE-cadherin belongs to the type II subgroup, only 23% of its sequence is identical with the classical cadherins, E-, N-, and P-cadherin from the type I subgroup [26]. N-cadherin, the other major endothelial cadherin, is not clustered at cell–cell junctions, but diffusely distributed on the cell membrane [27].

Classical cadherins consist of five extracellular domains (ΔEXDs) of around 110 amino acids each with internal sequence homology and conserved Ca^{2+}-binding motifs, a transmembrane region, and a highly conserved cytoplasmic region that interacts with actin filaments via catenins [28, 29]. The N-terminal ΔEXDs on cadherins mediate homotypic cell–cell contact on opposing membranes. Based on crystal lattice contacts in the structure of an N-cadherin molecule, a two-step association mechanism was defined [30]. First, a cis-interaction pair ("strand dimer") is formed between two parallel molecules by mutual exchange of a β-strand due to binding of a tryptophan (Trp2) from one molecule into a hydrophobic pocket of the partner molecule. Secondly, a cis-dimerized pair undergoes a trans-interaction ("adhesion dimer") with a complementary antiparallel unit. Alternating cis- and trans-interactions then form an "endless" zipper-like superstructure. In addition, the crystal structures also reveal that different sites are involved in forming the interface between two adjacent cadherin molecules [30–32].

Cadherins depend on Ca^{2+} for their function and removal of Ca^{2+} reduces adhesive activity. X-ray crystallography studies show that Ca^{2+} is essential for the stabilization of elongated rod-like structures of E-cadherin [33] and for protection against proteases [34]. The high concentration of Ca^{2+} necessary for saturating all Ca^{2+}-binding sites on E-cadherin and for effective cis-dimerization ($[Ca^{2+}] \sim 0.5 \pm 1$ mM) and trans-interactions of E-cadherin molecules ($[Ca^{2+}] >$ 1 mM) points to a possible physiological regulation of cadherin-mediated adhesive interactions [35].

Cadherins form a complex with cytosolic catenins, suggesting that the cadherin–catenin complex may play a role in mediating endothelial permeability and cell adhesion to the matrix (Figure 3.1). β-Catenin, a structural component of AJs in endothelial cells (ECs), consists of a N-terminal region with 140 amino acids, followed by a 524-residue domain that contains 12 repeats of 42 amino acids known as armadillo (*arm*) repeats and a 119-residue C-terminal tail [36]. The *arm* repeats are required for association with cadherins [37]. The β-catenin-binding site on E-cadherin is critical for chaperoning E-cadherin out of the endoplasmic reticulum, and therefore plays a major role in processing and targeting E-cadherin [38].

Cadherin Phosphorylation

AJs are dynamic structures that vary cell–cell binding strength according to cellular requirements. Inflammatory agents modulate the integrity of endothelial junctions through phosphorylation of tyrosine residues on AJ proteins. In human umbilical vein ECs (HUVECs), histamine increases the phosphorylation state of AJs in long-confluent cultures and induces VE-cadherin dissociation from the actin cytoskeleton. The cAMP agonist, dibutyryl cAMP, inhibits these responses [39]. However, activation of cAMP-specific Epac1 may reverse histamine-induced compromise of barrier integrity of HUVEC monolayers and concomitantly tighten the barrier [40, 41]. In pulmonary artery EC monolayers, thrombin induces disassembly of the cell–cell junction and augments permeability by increasing phosphorylation of VE-cadherin and p120, and correspondingly dephosphorylating β-catenin [42]. In confluent monolayers of both pulmonary artery and human lung microvascular ECs, tumor necrosis factor (TNF)-α increases permeability by phosphorylating tyrosine residues on VE-cadherin, β-catenin, and γ-catenin [43].

In primary endothelial cultures, activation of the Ca^{2+}-dependent, redox-sensitive, proline-rich tyrosine kinase-2 (Pyk2) phosphorylates tyrosine on cadherins. Pyk2 activation and its subsequent translocation to cell–cell junctions initiates catenin tyrosine phosphorylation and results in a loss of VE-cadherin homotypic adhesion. Endothelial expression of the Pyk2 (calcium-dependent tyrosine kinase)-related non-kinase CRNK – a N-terminal deletion mutant that is dominant negative – abolishes the Pyk2-induced increase in β-catenin tyrosine phosphorylation and blocks the loss of cell–cell contact [44].

There are six tyrosine residues in the cytoplasmic domain of VE-cadherin. In HUVECs, tyrosine phosphorylation of VE-cadherin on Tyr658 and Tyr731, which correspond to the p120-catenin- and β-catenin-binding sites, respectively, requires activation of both Src and Pyk2 [45]. Mutation of either Tyr851 or Tyr883, or both (Tyr to Phe), decreases binding of the adaptor protein Shc to cadherin, as determined by Sepharose bead-binding and gel-overlay assays [46]. These mutations also decrease Src phosphorylation and the capacity of cadherin to act as a Src substrate. Mutation of Tyr851 and/or Tyr883 does not alter the capacity of the cytoplasmic domain of cadherin to bind β-catenin *in vitro*. However, Shc binding to cadherin negatively influences β-catenin binding to the same molecule [46]. Since the capacity of Shc to interact with cadherin and tyrosine phosphorylation of Src and Pyk2 is dependent on the tyrosine phosphorylation of cadherin, it is possible that agonist induced permeability changes involve cadherin phosphorylation through calcium-sensitive activation of the Pyk2 pathway. The cytoskeletal signaling most probably include interactions between β- and α-catenin, increased phosphorylation of catenins, and Src and Pyk2 activation-dependent increased opening of cell–cell junction and permeability.

Role of Cytoskeleton in AJ Stability

Cadherin-mediated cell–cell interactions are regulated by protein interactions at the cytoplasmic face of the membrane (Figure 3.1). The interaction of cadherin with cytoplasmic proteins and the actin cytoskeleton is thought to mediate many aspects of cell–cell adhesion [47], including clustering of cadherin, strengthening of adhesive contacts, and downstream effects on membrane and cell organization. Cadherin–cytoskeleton interaction is only beginning to be understood, primarily from studies in epithelial cells. The general assumption is that cadherins are linked to the actin cytoskeleton through the β-catenin–α-catenin complex and that this complex participates in transmembrane signaling [10]. Moreover, cadherins may be involved in regulating actin filament assembly – indicating the bidirectional nature of this interaction [48]. An intact circumferential cortical actin network is required for retaining the cadherin–catenin complex at the cell surface [49]. Agents that disrupt actin microfilaments perturb cell–cell adhesion [50]. It has also been suggested that α-catenin interacts with the E-cadherin–β-catenin complex only in the monomeric form. In the dimer form, α-catenin may directly bind and regulate actin filaments [51]. In intact microvessels, disruption of the actin cytoskeleton reduces adhesion of VE-cadherin-coated microbeads to the EC surface, suggesting that this disruption leads to untethering of VE-cadherin and disassembly of endothelial AJs [52]. Thus, disruption of the link between cadherins and the actin cytoskeleton, and actin depolymerization may both separately lead to microvascular barrier compromise.

E-cadherin Dynamics

E-cadherin is a major AJ component in both epithelial cells and ECs. Thus, an understanding of E-cadherin comes from studies in native epithelial cells and ECs, and cells transfected with exogenous E-cadherin. These studies suggest that E-cadherin is delivered to the cell surface and recycled from there through active internalization via various endocytic carriers and pathways. Small GTPases mediate the internalization of E-cadherin

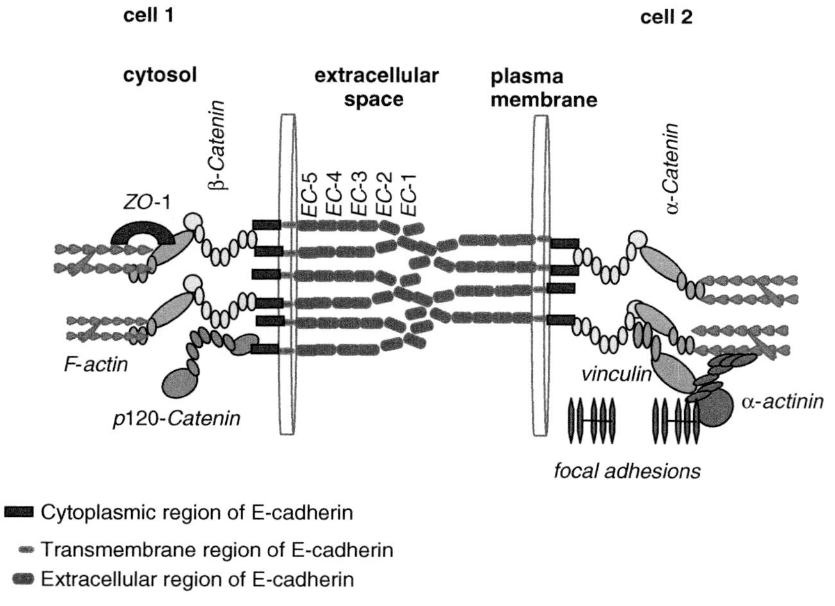

Figure 3.1 Cytosolic domains of E-cadherin bind direct or indirectly to multiple proteins and participate in intracellular signaling pathways.

at cell–cell contact sites through a clathrin-independent mechanism [53]. F-actin depolymerization is a necessary step for E-cadherin endocytosis [54]. Tracking E-cadherin movements through transfected E-cadherin–Green Fluorescent Protein (GFP) reveals that newly synthesized E-cadherin–GFP appears at the perinuclear Golgi region 3 h post-transfection and is subsequently transported in Pleiomorphic tubulovesicular carriers via the Rab11-positive endosome toward fusion sites on the cell surface [55]. The carriers range from spheres (250 nm diameter) to tubules (1–20 μm length). Golgin-97 is a selective and essential component of these carriers. Expression of Golgin-97 facilitates efficient trafficking of E-cadherin–GFP out of the trans-Golgi network (TGN) to the cell surface [56].

After internalization only a portion of the endocytosed E-cadherin is degraded, while the remainder is recycled back to the cell surface [55]. Preconfluent cultures exhibit increased E-cadherin recycling and a greater proportion of intracellular E-cadherin than fully confluent cultures. Continuous recycling of E-cadherin may be essential to form cell–cell contacts [55]. β-Catenin binds to E-cadherin early in the biosynthetic pathway, while p120 binds to the catenin–cadherin complex, at or near the cell membrane [38, 57]. There is reduced association of the internalized pool of E-cadherin with β-catenin indicating that this association may be dependent on whether E-cadherin is undergoing recycling or stabilized at the cell surface. The extent of E-cadherin expression on the cell surface determines both its adhesiveness [10, 58] and the recruitment of the intracellular cadherin pool to the cell surface upon cell–cell contact [47]. E-cadherin molecules on the cell surface exist as oligomers of different sizes, thereby suggesting that the oligomerization occurred prior to E-cadherin assembly at the cell adhesion site [59]. However, mechanisms that underlie the regulation of the barrier by cell surface cadherins are still not clear.

Clustering of the intracellular domain of the E-cadherin–β-catenin complex does not affect binding of β-catenin to α-catenin and with α-catenin that is bound to F-actin [60]. Vinculin and α-actinin bind to β-catenin or α-catenin. Vinculin also binds to the E-cadherin–β-catenin complex or actin, but does not bind simultaneously. The intracellular domain of cadherins has high-affinity binding with β-catenin [61] and β-catenin has a lower-affinity interaction with α-catenin [62]. It is widely accepted that α-catenin is bound to the cadherin–β-catenin complex bridges. Since α-catenin and β-catenin also bind to the actin-binding proteins including vinculin and α-actinin [60], these are also known as focal adhesion-interacting protein, suggesting that cadherin–catenin complex could also link to the focal adhesion complex. Figure 3.1 shows the binding schema for cadherin-associated proteins.

From our recent studies using rat lung microvascular ECs, it is evident that cadherin function is a dynamic process and that the distribution of AJs is an active process that requires the activity of focal adhesion kinase (FAK) [25]. A 15-min exposure of confluent monolayers to a hyperosmolar solution strengthened the barrier as determined by increases in transendothelial resistance. Concomitantly, focal adhesion formation, FAK activity, and E-cadherin accumulation at the cell periphery also increased. These hyperosmolarity-induced increases were blunted in monolayers expressing the kinase-deficient mutant of FAK. These studies point to an E-cadherin-dependent mechanism in that E-cadherin acts as a switch to either increase or decrease barrier strength through FAK signaling, which in turn regulates cadherin accumulation or clustering [6]. Moreover, H_2O_2 exposure induces an immediate loss of surface E-cadherin that then progressively increases with time (Figure 3.2). This response may be due to focal adhesions driving E-cadherin toward the surface. Thus, inhibition of FAK activation may block the signal for E-cadherin translocation to the surface, thereby compromising the integrity of the microvascular barrier.

Cadherin and GTPases

The general assumption is that Rho GTPase activity is involved in the formation and development of cadherin-dependent cell–cell contacts [63, 64]. GTPases of the Rho family are known to mediate cadherin–actin signaling and actin reorganization [11, 65]. Rac regulates endothelial barrier properties both in intact microvessels and in culture by regulation of actin filament polymerization and acting directly on the tether between VE-cadherin and the cytoskeleton [52]. Mutations in the small GTPase Rac1 disrupt the circumferential, cortical actin filament network and the targeting of cadherin–catenin complex components to the cell surface [49]. Expressing the constitutively active form of Rac, RacV12, in ECs increases tyrosine phosphorylation of α-catenin and a loss of VE-cadherin-mediated cell–cell adhesion [66]. Intact mouse lungs transfected with a VE-cadherin mutant lacking the Δ*EXD* exhibit a fivefold increase in vascular permeability. However, coexpression of a dominant-negative Rho GTPase, Cdc42, blocks this response, suggesting a role of Rho GTPases in the maintenance of the lung microvascular barrier [67].

Cadherin Function

Protein and fluid flux across the endothelial barrier occur through a paracellular pathway or by a transcellular route

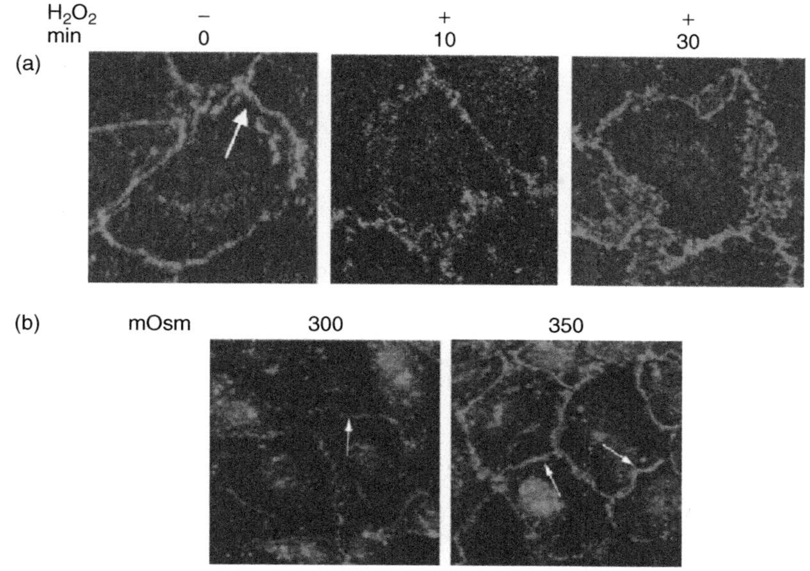

Figure 3.2 Dynamic regulation of E-cadherin at cell junctions. Confocal images show E-cadherin–GFP fluorescence in rat lung microvascular ECs. (a) H_2O_2 (100 μM) induces a transient reduction in the E-cadherin expression. (b) Hyperosmolarity (350 mOsm) increases junctional E-cadherin expression. Reproduced from [6, 25], by permission of the American Physiological Society and with permission © 2003 The American Society for Biochemistry and Molecular Biology.

involving vesicular transport. AJs play a critical role in regulating the paracellular pathway and, thus, in maintaining the integrity of the endothelial barrier. Cell culture studies using pulmonary artery and lung microvascular ECs separately have demonstrated that downregulation of cadherin expression at the endothelial junction leads to increases in endothelial permeability [39, 42]. These studies also point to a concomitant increase in VE- and E-cadherin phosphorylation [24], indicating the critical role of both proteins in the lung vasculature.

Comparison of cadherin contents using monlayers suggests that VE- and E-cadherin expression predominate in pulmonary artery and lung microvascular ECs, respectively [22, 25]. In both lung microvascular endothelial cultures and in intact vessels, E-cadherin is primary to maintaining the endothelial barrier and in mediating junction resealing [6, 22, 23]. It is of interest to note that lung endothelial barrier may be tighter in the microvessels than in macrovessels [22]. Thus, the cadherin subtype may play a critical role in determining the strength of the pulmonary vascular barrier.

VE-cadherin plays an important role in the establishment and maintenance of endothelial monolayer integrity [68, 69]. In addition, VE-cadherin has been implicated in the regulation of leukocyte migration [70, 71]. Leukocyte migration is impaired across endothelial monolayers overexpressing nonphosphorylatable mutants of VE-cadherin [45]. Moreover, intact animal studies reveal a role for VE-cadherin in angiogenesis in the lung [8]. Thus, VE-cadherin may play a significant role not only in the maintenance of the lung vascular barrier, but also in other endothelium-dependent responses.

The range of cellular processes mediated by cadherins in pulmonary microvessels is only beginning to be established. Although VE- and E-cadherin are currently the most studied, it appears that other cadherins may also play hereto unknown roles. For example, while the heterogenous expression pattern of N-cadherin in pulmonary vessels is described in a recent report [72], its function in lung vasculature remains unknown. Future studies may yet reveal more exciting roles of endothelial cadherins in the pulmonary vasculature.

CONNEXINS

Most mammalian cell types communicate with each other, thus coordinating their actions. Intercellular GJs facilitate this communication, through formation of channels that allow transfer of small molecules (<1 kDa) and ions (Figure 3.3). In the vasculature, GJs play a role in coordinating vasodilation, cancer cell metastasis, leukocyte

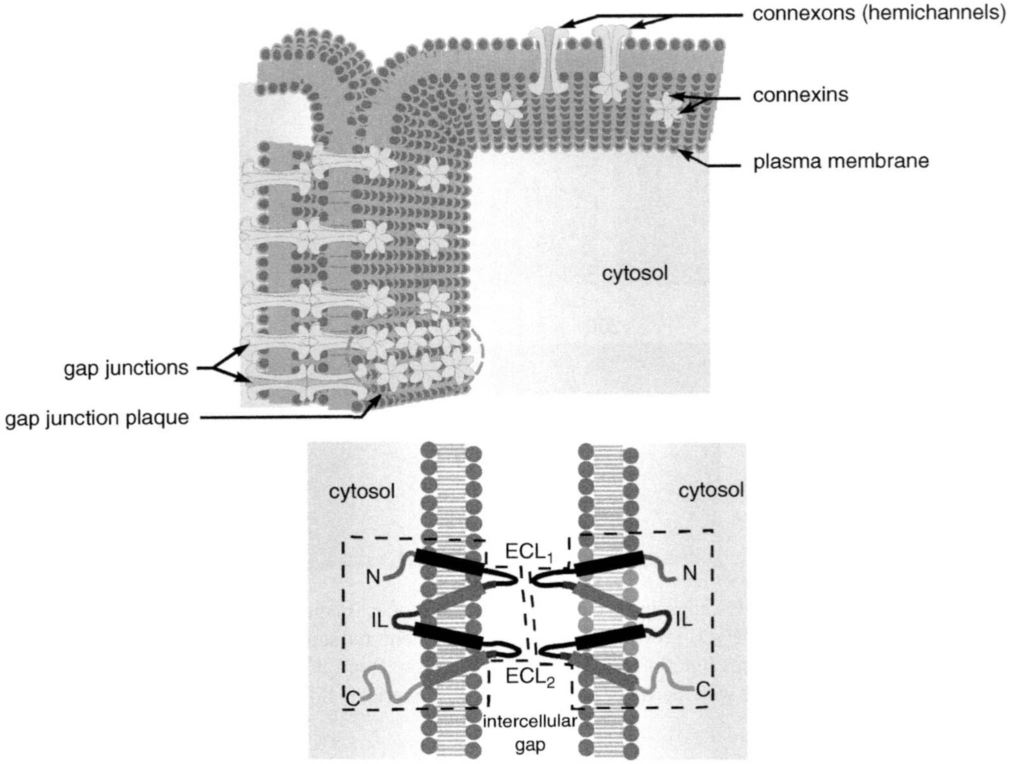

Figure 3.3 (a) GJ organization and structure. (b) Juxtaposed connexin molecules at a cell–cell junction. ECL, extracellular loop; IL, intracellular loop.

migration, and inflammation [5, 73–75]. We discuss recent reports that address these functions.

Connexin Subtypes

Functional characteristics of GJs derive from the composition of their protein subunits, the connexins. The number of connexin genes in the mouse and human genome, respectively are 19 and 20 [76–80]. Within the cell, connexins assemble into hexameric units called "hemichannels" or "connexins" (Figure 3.3). The hemichannel can be composed of the same or different connexin subtype, making them either homomeric or heteromeric. The assembled hemichannels are transported to the plasma membrane, where they dock with opposing hemichannels on the plasma membrane of adjacent cells to form functional GJs. As adjacent cells can contribute either identical or different hemichannels, the resulting GJs can be either homotypic or heterotypic channels, respectively. The implications for this wide array of channel subtypes are not yet clear.

In channels composed of a single connexin subtype, the connexin in itself determines the channel characteristics. For example, a decrease in intracellular pH to 7.1 is most likely to close a Cx38-containing channel, but not a Cx50-containing channel [81]. Thus, the connexin make-up of a GJ channel may render a particular functional characteristic to the cell. However, the functional characteristic of the heteromeric or heterotypic channels may depend on the connexin make-up of that channel.

Vascular endothelial cells consistently express Cx37, Cx40, and Cx43 [82–86]. However, the distribution of these connexins is not uniform. In the systemic circulation, endothelia from straight aortic segments express Cx37 and Cx40, while those from branch points express Cx43, suggesting that shear stress may influence connexin expression. In skeletal muscle, all arteries express Cx37 and Cx40, while only the large arteries express Cx43. Moreover, no connexin expression is evident in either systemic capillaries or venules [9], suggesting that in the systemic circulation, connexin expression is vessel specific. In addition, the magnitude of connexin expression is also heterogeneous, with Cx37 expressed the most and Cx40 the least frequently [83, 87, 88].

In contrast to the systemic vasculature, arteries in the pulmonary vasculature express Cx37, Cx40, and Cx43 [87], with Cx43 expression the maximum [89]. Moreover, we have shown that *in situ* pulmonary capillaries and venules of both rat and mouse express Cx43 [5]. In addition, Cx43 expression may be greater in the pulmonary artery compared to microvessels [89]. Thus, it is clear that the pulmonary vasculature differs significantly from the systemic vasculature in both the connexin expression in itself and the magnitude of the expression in different vessels subtypes. How these differences bear upon the possible dissimilarities in endothelial function in the two vasculatures needs to be explored.

Endothelial connexin expression is, however, not static and changes in response to inflammatory or injurious stimuli. Mechanical injury increases endothelial Cx43 expression in pulmonary microvessels, but not in the pulmonary artery [89]. Increased shear stress augments Cx43 expression in ECs [90, 91]. However, both the magnitude and duration of this response is heterogeneous [91]. While ECs derived from high shear vessels exhibit a sustained increase in Cx43 expression, those from low shear vessels exhibit only a transient response. Thus, it appears that injury and stress both alter endothelial connexin expression, with the caveat that these changes are heterogeneous. In addition, the responses may depend on the *in situ* location of the cell, and thus differ between micro- and macrovessels.

Connexin Trafficking and GJ Regulation

Regulated trafficking of connexin hemichannels to the plasma membrane and a rapid turnover of GJs from the plasma membrane both facilitate changes in intercellular GJ channels [92]. Trafficking of hemichannels begins with their assembly at intracellular sites that may be connexin subtype-specific. However, the exact sites of assembly are only beginning to be understood. For example, Cx32 is assembled in the TGN, while Cx43 may be assembled in either the endoplasmic reticulum or the TGN [92, 93]. However, hemichannel assembly is required prior to their transport to the plasma membrane. The Golgi play a major role in transporting the assembled hemichannels to the plasma membrane [94]. This process is also dependent on actin filaments and microtubules [94–96].

The hemichannels dock randomly into the plasma membrane and then move laterally to cluster at GJ plaques [97]. Newer hemichannels are added to the periphery of the existing plaques, while the older ones are removed from the center [98]. It is also suggested that connexins are directly delivered to the GJ plaques [95]. The plasma membrane target for connexins in this transport type appears to be cadherins [95], indicating an interaction between connexins and cadherins. The turnover of GJs is rapid with a half life of about 5 h [99]. Cx43-containing GJs are removed more rapidly from plasma membrane sites at an exponential rate with a half-life of about 1 h [100]. Both the lysosomal and endosomal compartments of the cell are involved in degrading the GJs [100].

Endogenous regulators of connexin trafficking and GJ formation include cAMP, protein kinase A (PKA), and Ca^{2+}/calmodulin protein kinase (CaMK). cAMP regulates Cx43 GJ formation either through increased trafficking of the protein to the intercellular junction or by facilitating assembly of the GJ itself [101, 102]. Thus, cAMP augments intercellular dye transfer [101], relaxation of systemic arteries [103], and interendothelial Ca^{2+} communication [104]. PKA augments both Cx43 expression and mRNA levels, and increases functional GJ formation [105, 106]. Different from cAMP and PKA, CaMK has been reported to increase gap junctional coupling, without increasing Cx43 expression [107]. Thus, cAMP, known to augment the lung microvascular barrier [39], may also induce a concomitant increase in interendothelial GJ communication in microvessels. Therefore, an intact microvascular barrier may be a key requirement for synchronizing or coordinating endothelial function.

Exogenous regulators include pharmacological agents and synthetic peptides. Pharmacological agents, including 18α-glycyrrhetinic acid, oleamide, carbenoxolone, heptanol, and octonol, decrease GJ communication [108–111]. However, these agents may have other nonspecific effects on the endothelium, including mitochondrial depolarization [109, 112, 113]. In addition, these agents may provide complete inhibition only at high toxic concentrations. Recently other agents including mefloquine [114] and 2-aminoethoxydiphenyl borate (2-APB) [115] have been shown to regulate GJs. The advantage of these agents is that their connexin target is more specific. For example, mefloquine efficiently blocks GJs with Cx36 and Cx50, but not those with Cx26, Cx32, or Cx43 [114]. In contrast, 2-APB blocks Cx26, Cx30, Cx36, Cx40, Cx45, and Cx50, but not Cx32, Cx43, and Cx46 [115].

Other regulators of GJ communication include synthetic peptides that mimic portions of the first or second extracellular loop of the connexin subunits. These connexin mimetic peptides block intercellular communication-mediated rhythmic contractile activity and vasorelaxation in arteries [116, 117], and propagation of intercellular calcium waves in alveolar epithelium [118]. In intact pulmonary capillaries, the peptides gap26 and gap27 block the interendothelial Ca^{2+} communication and attendant expression of leukocyte adhesion molecules in adjacent venules [5]. While the mechanism of inhibition is not clear, it is speculated that the peptides inhibit assembly of newly formed GJs through impairing hemichannel docking to plaques or by inducing conformational changes in GJs, thereby closing intercellular channels [109, 116]. The advantage of the connexin mimetic peptides over pharmacological agents is the easy reversibility of their inhibitory effect with a Ringer's wash [5]. Their main drawback may be the long exposure period (>45 min) required for complete inhibition of GJ communication. However, short exposures (~30 min) inhibit ATP secretion and dye uptake from extracellular region, while having no effect on the intercellular communication [119].

Connexins in Tumor Cell Metastasis

The role of GJs in tumor cell metastasis is not entirely clear (see Chapter 30). Several studies indicate that GJs may facilitate tumor cell metastasis. Adhesion of tumor cells to the pulmonary endothelium may initiate an assembly of GJs at points of contact between the cells, which may serve to establish a metabolic coupling between the two cells [120].

In human melanoma lesions, Cx26 expression is low in noninvasive melanoma cells compared to invasive cells [121]. Metastasizing mouse melanoma cells *in vitro* exhibit greater Cx26 GJ communication with ECs [122]. In addition, intravenous infusions of agents that inhibit Cx26 GJ communication block spontaneous lung metastasis of melanoma cells in mice [123, 124]. However, inhibiting Cx43 GJ communication had no effect on the extent of lung metastasis [123]. Metastatic lung cancer cells and ECs form Cx43-containing GJs *in vitro*, with the extent of gap junctional coupling defined by the amount of Cx43 protein in both cells [120]. Expressing Cx43 in tumor cells increases their transendothelial migration more than twofold in microvascular endothelial monolayers [125]. In patients with lung squamous cell carcinoma, significantly higher numbers of Cx26-positive cells are present in both the primary tumor and metastatic foci in lymph nodes. In addition, these patients have a significantly lower survival rate [122]. More interestingly, Cx37-derived peptides from lung carcinomas in mice effectively reduced metastatic loads in mice carrying pre-established micrometastases and decreased spontaneous metastasis in mice [126, 127]. This effect was found to occur through peptide activation of antitumor cytotoxic T lymphocytes [127].

In contrast to these studies, other reports point to the role of connexins as tumor suppressors. Cx32-deficient mice exhibit increased proliferation of lung tumors, suggesting a role for Cx32 in suppressing lung tumors [128]. In addition, stable transfection of tumorigenic lung cancer cells with Cx43 renders the cell line nontumorigenic [129].

We think that these conflicting findings suggest that in tumor cells, a lack of connexins and, thus, intercellular communication may influence the spread of the primary tumor itself. In contrast, the presence of connexins and the establishment of GJ channels with ECs may be a primary requirement for transendothelial migration of the tumor cells and establishment of a secondary site for tumor development in the lung. Establishment of secondary sites in the lung may be enhanced by the fact that intercellular communication between tumor and ECs induces angiogenesis-like mechanisms in tumor cells [130]. Pulmonary arterioles and capillaries may be favored sites for establishment of secondary tumor sites, as they both express connexins and are preferred sites for tumor cell attachment [131].

One possible mechanism that increases adhesion and metastatic potential of cancer cells may be changes in their cadherin levels. Metastatic bone cancer cells stably transfected with Cx43 DNA, showed reduced adhesion to HUVEC monolayers [132]. These cells also expressed reduced levels of OB-cadherin, indicating that increased connexin expression may competitively inhibit cadherin expression in metastatic cancer cells. In addition, human mesothelioma tumor cells, expressed both Cx43 and N-cadherin [133]. Well differentiated lung cancer cells expressed both E-cadherin and Cx43, while poorly differentiated cells did not, indicating that interactions between gap and AJ proteins is essential in cancer progression [134]. The interaction of connexins with cadherins and the up- or downregulation of connexins may well be interconnected. The observation that a loss of E-cadherin increases tumor cell invasion in the lung [135] supports the above possibility. Thus, regulation of both gap and AJs may be critical in tumor cell metastasis in the lung and tumor progression, and thus need tested.

Connexins in Endothelial–Leukocyte Communication

The expression of connexins in freshly isolated leukocytes from untreated animals and patients is unclear. Some reports indicate that leukocytes (monocytes, lymphocytes, and granulocytes) do not express connexins prior to activation [136, 137]. In contrast, other reports reveal that unstimulated neutrophils express Cx37, Cx40, and Cx43 [138]. Cx37 expression in neutrophils predominates in pseudopodia, while Cx40 and Cx43 localize to the membrane [138]. This distinct localization suggests that Cx37 may play a role in migration, while Cx40 and Cx43 might regulate the initial neutrophil–endothelial communication. In addition, both T and B lymphocytes from peripheral blood express Cx43 and support inter-lymphocyte GJ communication [139]. However, lymphocytes do not express Cx26, Cx32, Cx37, and Cx45 [139]. Moreover, lymphocytes form functional GJ channels with ECs in culture [140].

The expression levels of connexins increase in response to leukocyte activation. Leukocyte connexin expression increases in response to agonists such as lipopolysaccharide (LPS) [136] and phorbol myristate acetate [138]. While TNF-α alone decreases gap junctional coupling between neutrophils and ECs [138], TNF-α in combination with interferon-γ increased monocyte–endothelial communication [137]. Both immunofluorescence and ultrastructural studies support this hypothesis by indicating the possible presence of "GJ-like channels" between endothelium and leukocytes [136]. These GJ channels exhibit bidirectional communication [138]. An increase in leukocyte Cx43 expression was also observed in leukocytes adherent to endothelium in inflamed tissue [136].

The data thus far indicate that the role of GJs in endothelial–leukocyte interaction may be varied. Thus, inhibition of GJs facilitates polymorphonuclear neutrophil transendothelial migration [138, 141], but inhibits monocyte migration [137]. Lymphocytes retain their ability to migrate across the endothelium even under GJ-inhibited conditions [140]. In acute lung inflammation, inflammatory cells in the bronchoalveolar lavage fluid from Cx40 mice remain unchanged compared to wild-type mice [142]. Thus, the type and function of GJs involved in endothelial–leukocyte interaction may be specific to the particular pathological process involved (see also Chapter 10).

Connexins in Inflammation

While the role of connexins in the development of atherosclerosis has received the bulk of the attention, several recent studies demonstrate connexin-dependent mechanisms in other inflammatory responses. A decrease in Cx37 may contribute to the establishment of the pathophysiological features of allergic airway disease, including increased expression of intercellular and vascular cell adhesion molecules [143]. Inflammatory conditions such as sepsis and agents such as TNF-α lead to a reduction in expression of Cx40 and Cx43, respectively [144, 145], in the myocardium. These cause loss of intercalated disk structural integrity, resulting in myocardial depression [144] and development of atrial arrhythmias [145]. LPS decreases Cx43 expression in nasal epithelial cells [87]. In aortic segments, LPS treatment decreases endothelial dye coupling, concomitant to a reduction in endothelial Cx40 and Cx37 expression levels [146]. Inflammatory responses associated with multiple sclerosis decrease Cx43 expression in glial cells [147]. Heterozygous Cx43-null mice brains subjected to an ischemic insult exhibit increased apoptosis and inflammation [148]. Acinar cell injury in acute pancreatitis is exacerbated in mice with deleted Cx32 gene [149]. In intact alveoli, increases in epithelial cytosolic Ca^{2+} spreads spatially to adjacent alveoli in a Cx43-dependent manner [150]. The spatial spread serves to coordinate alveolar surfactant secretion [150]. The above studies show that connexins and, thus, gap junctional communication may play a critical role in inflammation and disease.

Determination of responses in the lung microvascular network in septic mice indicates that endothelial expression of Cx40 is decreased compared to that in control mice [142]. GJ communication mediates interendothelial Ca^{2+} waves in lung venular capillaries [111]. In confluent endothelial monolayers, mechanical perturbation of a single cell induces an interendothelial Ca^{2+} wave [151] that is mediated by GJs. Focal release of endothelial Ca^{2+} by Ca^{2+} uncaging in intact lung capillaries spreads spatially to adjacent capillaries and venules. Cx43 gap peptides inhibit the spatial spread of Ca^{2+} in capillaries [5] (Figure 3.4) (see also Chapters 5 and 9).

Data obtained using endothelial-specific Cx43 mice further established the role of endothelial Cx43 in mediating spatially extensive Ca^{2+} responses in lung capillaries [5]. Embryonic lethality of Cx43 knockout (KO) mice [152] precludes their use and necessitates the use of mice with targeted endothelial Cx43 deletion. It is interesting to note that Cx40 and Cx37 KO animals survive into adulthood. This may be due to a two- to threefold compensatory upregulation of endothelial Cx37 and Cx43 mRNA levels in Cx40 KO animals [153]. Uncaging-induced Ca^{2+} communication is inhibited in endothelial-specific Cx43 KO mice [5]. The downstream signaling effect of the Cx43-dependent Ca^{2+} communication is the increase in expression of the proinflammatory leukocyte adhesion molecule, P-selectin in venules adjacent to Ca^{2+} uncaging site [5]. Though the bulk of the data on the role of connexins in inflammation pertains to the systemic vasculature as evident from the discussion in this section, it is now emerging that Cx43 may play a major role in mediating proinflammatory signaling in the pulmonary microvasculature. Additional studies are required to establish further this emerging concept.

Interactions among Junctional Proteins

Connexins may also be involved in the regulation of pulmonary microvascular permeability (see Chapter 8). Inhibiting GJ communication with gap peptides blocks thrombin-induced increases in lung microvascular permeability determined by quantifying microvascular filtration coefficient [5] (Figure 3.4). Heretofore, cadherins and TJ

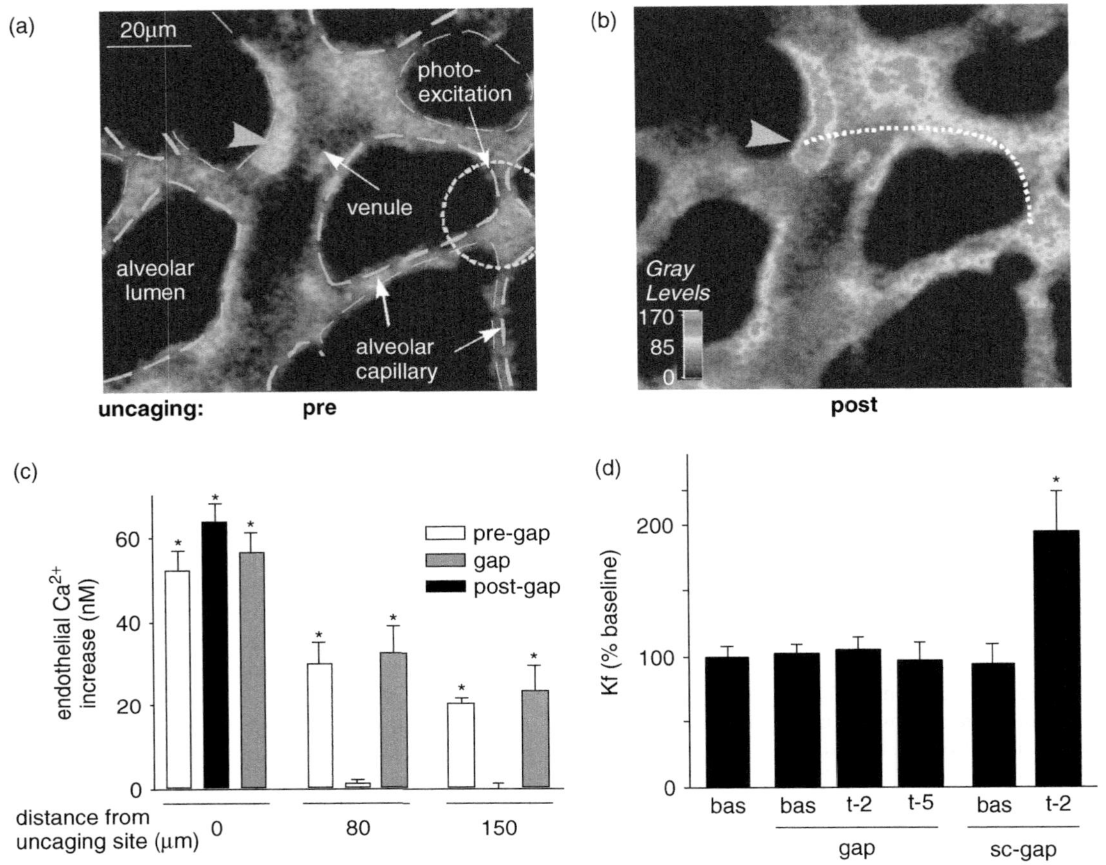

Figure 3.4 GJ-dependent responses in lung microvessels. Images show fluorescence of the Ca^{2+} indicator Fluo4 at baseline (a) and in response to Ca^{2+} uncaging (b). Note the increase in Fluo4 fluorescence in a venule located 150 μm (arrowhead) from the uncaging site (circle). (c) Gap peptides specific to Cx43 (gap) reversibly blocked the spatial spread of Ca^{2+}. (d) Gap peptides (gap) inhibited thrombin-induced increases in microvascular permeability (K_f), while scrambled gap peptides (sc-gap) failed to block the thrombin response. bas, baseline; t-2, thrombin 2 U/ml; t-5, thrombin 5 U/ml. A color version of this figure appears in the plate section of this volume. Reproduced from [5] by permission of the American Society for Clinical Investigation.

proteins have been implicated as primary regulators of microvascular permeability. Emerging evidence indicates that connexins may play a role in microvascular permeability through their association with both cadherins and TJ proteins. It is even possible that connexins themselves regulate vascular permeability. Many recent reports point to close morphological association and shared functions between connexins and cadherins [154]. Cx43 localize to intercellular junctions that predominantly express similar cadherin subtypes [95]. Post-transcriptional downregulation of E-cadherin using E-cadherin antisense oligonucleotides concomitantly decreased connexin localization at cell–cell borders and increased their levels in the cytosol [155]. In addition, N- or E-cadherin knockdown decreased intercellular communication, indicating that coassembly of cadherins and connexins regulates GJ formation [154, 155].

In addition, connexins also associate with TJ proteins. Immunofluorescence studies reveal that Cx43 colocalizes with both zona occludens ZO-1 and -2 zona occludens proteins on plasma membrane GJ plaques in lung epithelial cells [156, 157]. The C-terminal tail of Cx43 interacts with ZO-1, and microtubules consisting of α- and β-tubulin dimers [158]. ZO-1 association with Cx43 controls the size and distribution of the GJ plaque [159]. In human airway epithelium, connexin expression blocks ouabain-induced barrier disruption, and loss of the TJ proteins, occludin, ZO-1, and claudin [160]. These reports suggest that connexins may work in tandem with TJ proteins in barrier regulation. Our data on the inhibition of thrombin-induced increases in lung microvessel permeability in intact lungs of endothelial-specific Cx43 KO mice lends support to this possibility [5]. Thus, connexins in association with other junctional proteins may play a critical role in mediating microvascular barrier functions.

CONCLUSIONS AND PERSPECTIVES

In this chapter, we have brought to the fore recent developments that establish a major functional role for cadherins and connexins in the lung vasculature. Cadherins are primary junctional proteins involved in maintenance of the vascular barrier. Recent studies have established that Rho GTPases, actin, catenins, and other AJ proteins participate in tandem with cadherins in barrier regulation. As evinced in recent publications, cadherins also play a major role in leukocyte migration and angiogenesis in the lung. Connexins, once thought of as mere conduits, are now shown to mediate inflammatory signaling in the lung. In addition, it is becoming more clear that connexins may be major players in cancer metastasis in the lung. Interestingly, it is now emerging that junctional proteins interact amongst themselves and work in synchrony more often than not in the maintenance of the pulmonary vascular barrier.

The studies discussed clearly establish that both junctional proteins, either individually or acting in tandem, play important roles in microvascular function. However, the role of these proteins as relevant to pathological contexts and the attendant signaling mechanisms needs to be further defined. While the role of connexins and cadherins as modulators of local inflammatory responses is established, the role of these proteins in modulating the severity and spatial extent of lung injury remains to be determined. In addition, several signaling mediators such as reactive oxygen species and nitric oxide are implicated in endothelial inflammatory responses, possibly through alterations in the activity of signaling intermediates, Rho and Rac. It is unclear how these signaling intermediates modify responses mediated by the junctional proteins. Moreover, the functional interactions among the junctional proteins needs defined. Future studies may resolve these issues and establish further the role of cadherins and connexins in pulmonary endothelial function.

ACKNOWLEDGMENTS

We thank Dr. Jahar Bhattacharya for his suggestions and critical reading of this chapter. Funding support: HL75503 (K.P.) and HL36024 (S.K.Q.; Principal Investigator: Jahar Bhattacharya).

References

1. Pokutta, S. and Weis, W.I. (2007) Structure and mechanism of cadherins and catenins in cell–cell contacts. *Annual Review of Cell and Developmental Biology*, **23**, 237–61.
2. Mese, G., Richard, G., and White, T.W. (2007) Gap junctions: basic structure and function. *The Journal of Investigative Dermatology*, **127**, 2516–24.
3. Schneeberger, E.E. and Lynch, R.D. (2004) The tight junction: a multifunctional complex. *American Journal of Physiology: Cell Physiology*, **286**, C1213–28.
4. Weber, C., Fraemohs, L., and Dejana, E. (2007) The role of junctional adhesion molecules in vascular inflammation. *Nature Reviews: Immunology*, **7**, 467–77.
5. Parthasarathi, K., Ichimura, H., Monma, E. *et al.* (2006) Connexin 43 mediates spread of Ca^{2+}-dependent proinflammatory responses in lung capillaries. *The Journal of Clinical Investigation*, **116**, 2193–200.
6. Quadri, S.K. and Bhattacharya, J. (2007) Resealing of endothelial junctions by focal adhesion kinase. *American Journal of Physiology: Lung Cellular and Molecular Physiology*, **292**, L334–42.
7. Rao, R.M., Yang, L., Garcia-Cardena, G., and Luscinskas, F.W. (2007) Endothelial-dependent mechanisms of leukocyte recruitment to the vascular wall. *Circulation Research*, **101**, 234–47.
8. Lampugnani, M.G. and Dejana, E. (2007) The control of endothelial cell functions by adherens junctions, in *Vascular Development*, Novartis Foundation Symposium 283, John Wiley & Sons, Ltd, Chichester, pp. 4–13; discussion 13–17, 238–241.
9. Looft-Wilson, R.C., Payne, G.W., and Segal, S.S. (2004) Connexin expression and conducted vasodilation along arteriolar endothelium in mouse skeletal muscle. *Journal of Applied Physiology*, **97**, 1152–58.
10. Yap, A.S., Brieher, W.M., and Gumbiner, B.M. (1997) Molecular and functional analysis of cadherin-based adherens junctions. *Annual Review of Cell and Developmental Biology*, **13**, 119–46.
11. Yap, A.S. and Kovacs, E.M. (2003) Direct cadherin-activated cell signaling: a view from the plasma membrane. *The Journal of Cell Biology*, **160**, 11–16.
12. Takeichi, M. (1995) Morphogenetic roles of classic cadherins. *Current Opinion in Cell Biology*, **7**, 619–27.

13. Goodwin, M. and Yap, A.S. (2004) Classical cadherin adhesion molecules: coordinating cell adhesion, signaling and the cytoskeleton. *Journal of Molecular Histology*, **35**, 839–44.
14. Kemler, R. (1992) Classical cadherins. *Seminars in Cell Biology*, **3**, 149–55.
15. Takeichi, M. (1991) Cadherin cell adhesion receptors as a morphogenetic regulator. *Science*, **251**, 1451–55.
16. Nagafuchi, A., Tsukita, S., and Takeichi, M. (1993) Transmembrane control of cadherin-mediated cell–cell adhesion. *Seminars in Cell Biology*, **4**, 175–81.
17. Tanihara, H., Sano, K., Heimark, R.L. *et al.* (1994) Cloning of five human cadherins clarifies characteristic features of cadherin extracellular domain and provides further evidence for two structurally different types of cadherin. *Cell Adhesion and Communication*, **2**, 15–26.
18. Liaw, C.W., Cannon, C., Power, M.D. *et al.* (1990) Identification and cloning of two species of cadherins in bovine endothelial cells. *The EMBO Journal*, **9**, 2701–8.
19. Corada, M., Mariotti, M., Thurston, G. *et al.* (1999) Vascular endothelial-cadherin is an important determinant of microvascular integrity *in vivo*. *Proceedings of the National Academy of Sciences of the United States of America*, **96**, 9815–20.
20. Dejana, E., Bazzoni, G., and Lampugnani, M.G. (1999) Vascular endothelial (VE)-cadherin: only an intercellular glue? *Experimental Cell Research*, **252**, 13–19.
21. Gao, X., Kouklis, P., Xu, N. *et al.* (2000) Reversibility of increased microvessel permeability in response to VE-cadherin disassembly. *American Journal of Physiology: Lung Cellular and Molecular Physiology*, **279**, L1218–25.
22. Parker, J.C., Stevens, T., Randall, J. *et al.* (2006) Hydraulic conductance of pulmonary microvascular and macrovascular endothelial cell monolayers. *American Journal of Physiology: Lung Cellular and Molecular Physiology*, **291**, L30–37.
23. Safdar, Z., Wang, P., Ichimura, H. *et al.* (2003) Hyperosmolarity enhances the lung capillary barrier. *The Journal of Clinical Investigation*, **112**, 1541–49.
24. Godzich, M., Hodnett, M., Frank, J.A. *et al.* (2006) Activation of the stress protein response prevents the development of pulmonary edema by inhibiting VEGF cell signaling in a model of lung ischemia–reperfusion injury in rats. *The FASEB Journal*, **20**, 1519–21.
25. Quadri, S.K., Bhattacharjee, M., Parthasarathi, K. *et al.* (2003) Endothelial barrier strengthening by activation of focal adhesion kinase. *The Journal of Biological Chemistry*, **278**, 13342–49.
26. Breier, G., Breviario, F., Caveda, L. *et al.* (1996) Molecular cloning and expression of murine vascular endothelial-cadherin in early stage development of cardiovascular system. *Blood*, **87**, 630–41.
27. Salomon, D., Ayalon, O., Patel-King, R. *et al.* (1992) Extrajunctional distribution of N-cadherin in cultured human endothelial cells. *Journal of Cell Science*, **102** (Pt 1), 7–17.
28. Kemler, R. and Ozawa, M. (1989) Uvomorulin–catenin complex: cytoplasmic anchorage of a Ca^{2+}-dependent cell adhesion molecule. *BioEssays*, **11**, 88–91.
29. Aberle, H., Schwartz, H., and Kemler, R. (1996) Cadherin–catenin complex: protein interactions and their implications for cadherin function. *Journal of Cellular Biochemistry*, **61**, 514–23.
30. Shapiro, L., Fannon, A.M., Kwong, P.D. *et al.* (1995) Structural basis of cell–cell adhesion by cadherins. *Nature*, **374**, 327–37.
31. Nagar, B., Overduin, M., Ikura, M., and Rini, J.M. (1996) Structural basis of calcium-induced E-cadherin rigidification and dimerization. *Nature*, **380**, 360–64.
32. Koch, A.W., Bozic, D., Pertz, O., and Engel, J. (1999) Homophilic adhesion by cadherins. *Current Opinion in Structural Biology*, **9**, 275–81.
33. Pokutta, S., Herrenknecht, K., Kemler, R., and Engel, J. (1994) Conformational changes of the recombinant extracellular domain of E-cadherin upon calcium binding. *European Journal of Biochemistry*, **223**, 1019–26.
34. Hyafil, F., Babinet, C., and Jacob, F. (1981) Cell–cell interactions in early embryogenesis: a molecular approach to the role of calcium. *Cell*, **26**, 447–54.
35. Pertz, O., Bozic, D., Koch, A.W. *et al.* (1999) A new crystal structure, Ca^{2+} dependence and mutational analysis reveal molecular details of E-cadherin homoassociation. *The EMBO Journal*, **18**, 1738–47.
36. Peifer, M., Berg, S., and Reynolds, A.B. (1994) A repeating amino acid motif shared by proteins with diverse cellular roles. *Cell*, **76**, 789–91.
37. Hulsken, J., Birchmeier, W., and Behrens, J. (1994) E-cadherin and APC compete for the interaction with beta-catenin and the cytoskeleton. *The Journal of Cell Biology*, **127**, 2061–69.
38. Chen, Y.T., Stewart, D.B., and Nelson, W.J. (1999) Coupling assembly of the E-cadherin/beta-catenin complex to efficient endoplasmic reticulum exit and basal-lateral membrane targeting of E-cadherin in polarized

MDCK cells. *The Journal of Cell Biology*, **144**, 687–99.

39. Andriopoulou, P., Navarro, P., Zanetti, A. *et al.* (1999) Histamine induces tyrosine phosphorylation of endothelial cell-to-cell adherens junctions. *Arteriosclerosis, Thrombosis, and Vascular Biology*, **19**, 2286–97.
40. Kooistra, M.R., Corada, M., Dejana, E., and Bos, J.L. (2005) Epac1 regulates integrity of endothelial cell junctions through VE-cadherin. *FEBS Letters*, **579**, 4966–72.
41. Fukuhara, S., Sakurai, A., Sano, H. *et al.* (2005) Cyclic AMP potentiates vascular endothelial cadherin-mediated cell–cell contact to enhance endothelial barrier function through an Epac–Rap1 signaling pathway. *Molecular and Cellular Biology*, **25**, 136–46.
42. Konstantoulaki, M., Kouklis, P., and Malik, A.B. (2003) Protein kinase C modifications of VE-cadherin, p120, and beta-catenin contribute to endothelial barrier dysregulation induced by thrombin. *American Journal of Physiology: Lung Cellular and Molecular Physiology*, **285**, L434–42.
43. Angelini, D.J., Hyun, S.W., Grigoryev, D.N. *et al.* (2006) TNF-alpha increases tyrosine phosphorylation of vascular endothelial cadherin and opens the paracellular pathway through fyn activation in human lung endothelia. *American Journal of Physiology: Lung Cellular and Molecular Physiology*, **291**, L1232–45.
44. van Buul, J.D., Anthony, E.C., Fernandez-Borja, M. *et al.* (2005) Proline-rich tyrosine kinase 2 (Pyk2) mediates vascular endothelial-cadherin-based cell–cell adhesion by regulating beta-catenin tyrosine phosphorylation. *The Journal of Biological Chemistry*, **280**, 21129–36.
45. Allingham, M.J., van Buul, J.D., and Burridge, K. (2007) ICAM-1-mediated, Src- and Pyk2-dependent vascular endothelial cadherin tyrosine phosphorylation is required for leukocyte transendothelial migration. *Journal of Immunology*, **179**, 4053–64.
46. Xu, Y. and Carpenter, G. (1999) Identification of cadherin tyrosine residues that are phosphorylated and mediate Shc association. *Journal of Cellular Biochemistry*, **75**, 264–71.
47. Adams, C.L., Nelson, W.J., and Smith, S.J. (1996) Quantitative analysis of cadherin–catenin-actin reorganization during development of cell–cell adhesion. *The Journal of Cell Biology*, **135**, 1899–911.
48. Kovacs, E.M., Goodwin, M., Ali, R.G. *et al.* (2002) Cadherin-directed actin assembly: E-cadherin physically associates with the Arp2/3 complex to direct actin assembly in nascent adhesive contacts. *Current Biology*, **12**, 379–82.
49. Quinlan, M.P. and Hyatt, J.L. (1999) Establishment of the circumferential actin filament network is a prerequisite for localization of the cadherin–catenin complex in epithelial cells. *Cell Growth and Differentiation*, **10**, 839–54.
50. Hirano, S., Nose, A., Hatta, K. *et al.* (1987) Calcium-dependent cell–cell adhesion molecules (cadherins): subclass specificities and possible involvement of actin bundles. *The Journal of Cell Biology*, **105**, 2501–10.
51. Drees, F., Pokutta, S., Yamada, S. *et al.* (2005) Alpha-catenin is a molecular switch that binds E-cadherin–beta-catenin and regulates actin-filament assembly. *Cell*, **123**, 903–15.
52. Waschke, J., Baumgartner, W., Adamson, R.H. *et al.* (2004) Requirement of Rac activity for maintenance of capillary endothelial barrier properties. *American Journal of Physiology: Heart and Circulatory Physiology*, **286**, H394–401.
53. Akhtar, N. and Hotchin, N.A. (2001) RAC1 regulates adherens junctions through endocytosis of E-cadherin. *Molecular Biology of the Cell*, **12**, 847–62.
54. Le, T.L., Joseph, S.R., Yap, A.S., and Stow, J.L. (2002) Protein kinase C regulates endocytosis and recycling of E-cadherin. *American Journal of Physiology: Cell Physiology*, **283**, C489–99.
55. Le, T.L., Yap, A.S., and Stow, J.L. (1999) Recycling of E-cadherin: a potential mechanism for regulating cadherin dynamics. *The Journal of Cell Biology*, **146**, 219–32.
56. Lock, J.G., Hammond, L.A., Houghton, F. *et al.* (2005) E-cadherin transport from the trans-Golgi network in tubulovesicular carriers is selectively regulated by golgin-97. *Traffic*, **6**, 1142–56.
57. Miranda, K.C., Joseph, S.R., Yap, A.S. *et al.* (2003) Contextual binding of $p120^{ctn}$ to E-cadherin at the basolateral plasma membrane in polarized epithelia. *The Journal of Biological Chemistry*, **278**, 43480–88.
58. Angres, B., Barth, A., and Nelson, W.J. (1996) Mechanism for transition from initial to stable cell–cell adhesion: kinetic analysis of E-cadherin-mediated adhesion using a quantitative adhesion assay. *The Journal of Cell Biology*, **134**, 549–57.
59. Iino, R., Koyama, I., and Kusumi, A. (2001) Single molecule imaging of green fluorescent proteins in living cells: E-cadherin forms oligomers on the free cell surface. *Biophysical Journal*, **80**, 2667–77.

60. Yamada, S., Pokutta, S., Drees, F. *et al.* (2005) Deconstructing the cadherin–catenin–actin complex. *Cell*, **123**, 889–901.
61. Huber, A.H., Stewart, D.B., Laurents, D.V. *et al.* (2001) The cadherin cytoplasmic domain is unstructured in the absence of beta-catenin. A possible mechanism for regulating cadherin turnover. *The Journal of Biological Chemistry*, **276**, 12301–9.
62. Pokutta, S. and Weis, W.I. (2000) Structure of the dimerization and beta-catenin-binding region of alpha-catenin. *Molecular Cell*, **5**, 533–43.
63. Noren, N.K., Niessen, C.M., Gumbiner, B.M., and Burridge, K. (2001) Cadherin engagement regulates Rho family GTPases. *The Journal of Biological Chemistry*, **276**, 33305–8.
64. Ehrlich, J.S., Hansen, M.D., and Nelson, W.J. (2002) Spatio-temporal regulation of Rac1 localization and lamellipodia dynamics during epithelial cell–cell adhesion. *Developmental Cell*, **3**, 259–70.
65. Braga, V.M. (2002) Cell–cell adhesion and signalling. *Current Opinion in Cell Biology*, **14**, 546–56.
66. van Wetering, S., van Buul, J.D., Quik, S. *et al.* (2002) Reactive oxygen species mediate Rac-induced loss of cell–cell adhesion in primary human endothelial cells. *Journal of Cell Science*, **115**, 1837–46.
67. Broman, M.T., Kouklis, P., Gao, X. *et al.* (2006) Cdc42 regulates adherens junction stability and endothelial permeability by inducing alpha-catenin interaction with the vascular endothelial cadherin complex. *Circulation Research*, **98**, 73–80.
68. Ali, J., Liao, F., Martens, E., and Muller, W.A. (1997) Vascular endothelial cadherin (VE-cadherin): cloning and role in endothelial cell–cell adhesion. *Microcirculation*, **4**, 267–77.
69. Minagar, A., Long, A., Ma, T. *et al.* (2003) Interferon (IFN)-beta 1a and IFN-beta 1b block IFN-gamma-induced disintegration of endothelial junction integrity and barrier. *Endothelium*, **10**, 299–307.
70. Hordijk, P. (2003) Endothelial signaling in leukocyte transmigration. *Cell Biochemistry and Biophysics*, **38**, 305–22.
71. Shaw, S.K., Bamba, P.S., Perkins, B.N., and Luscinskas, F.W. (2001) Real-time imaging of vascular endothelial-cadherin during leukocyte transmigration across endothelium. *Journal of Immunology*, **167**, 2323–30.
72. Ofori-Acquah, S.F., King, J., Voelkel, N. *et al.* (2008) Heterogeneity of barrier function in the lung reflects diversity in endothelial cell junctions. *Microvascular Research*, **75**, 391–402.
73. Chadjichristos, C.E. and Kwak, B.R. (2007) Connexins: new genes in atherosclerosis. *Annals of Medicine*, **39**, 402–11.
74. Saito-Katsuragi, M., Asada, H., Niizeki, H. *et al.* (2007) Role for connexin 26 in metastasis of human malignant melanoma: communication between melanoma and endothelial cells via connexin 26. *Cancer*, **110**, 1162–72.
75. Elias, L.A., Wang, D.D., and Kriegstein, A.R. (2007) Gap junction adhesion is necessary for radial migration in the neocortex. *Nature*, **448**, 901–7.
76. Willecke, K., Jungbluth, S., Dahl, E. *et al.* (1990) Six genes of the human connexin gene family coding for gap junctional proteins are assigned to four different human chromosomes. *European Journal of Cell Biology*, **53**, 275–80.
77. Willecke, K., Hennemann, H., Dahl, E. *et al.* (1991) The diversity of connexin genes encoding gap junctional proteins. *European Journal of Cell Biology*, **56**, 1–7.
78. Hennemann, H., Suchyna, T., Lichtenberg-Frate, H. *et al.* (1992) Molecular cloning and functional expression of mouse connexin40, a second gap junction gene preferentially expressed in lung. *The Journal of Cell Biology*, **117**, 1299–310.
79. Nielsen, P.A., Beahm, D.L., Giepmans, B.N. *et al.* (2002) Molecular cloning, functional expression, and tissue distribution of a novel human gap junction-forming protein, connexin-31.9. Interaction with zona occludens protein-1. *The Journal of Biological Chemistry*, **277**, 38272–83.
80. Saez, J.C., Berthoud, V.M., Branes, M.C. *et al.* (2003) Plasma membrane channels formed by connexins: their regulation and functions. *Physiological Reviews*, **83**, 1359–400.
81. Peracchia, C. (2004) Chemical gating of gap junction channels; roles of calcium, pH and calmodulin. *Biochimica et Biophysica Acta*, **1662**, 61–80.
82. Beny, J.L. and Connat, J.L. (1992) An electron-microscopic study of smooth muscle cell dye coupling in the pig coronary arteries. Role of gap junctions. *Circulation Research*, **70**, 49–55.
83. Reed, K.E., Westphale, E.M., Larson, D.M. *et al.* (1993) Molecular cloning and functional expression of human connexin37, an endothelial cell gap junction protein. *The Journal of Clinical Investigation*, **91**, 997–1004.

84. Larson, D.M., Haudenschild, C.C., and Beyer, E.C. (1990) Gap junction messenger RNA expression by vascular wall cells. *Circulation Research*, **66**, 1074–80.
85. Little, T.L., Beyer, E.C., and Duling, B.R. (1995) Connexin 43 and connexin 40 gap junctional proteins are present in arteriolar smooth muscle and endothelium *in vivo*. *American Journal of Physiology: Heart and Circulatory Physiology*, **268**, H729–39.
86. Van Rijen, H., van Kempen, M.J., Analbers, L.J. *et al.* (1997) Gap junctions in human umbilical cord endothelial cells contain multiple connexins. *American Journal of Physiology: Cell Physiology*, **272**, C117–30.
87. Yeh, H.I., Rothery, S., Dupont, E. *et al.* (1998) Individual gap junction plaques contain multiple connexins in arterial endothelium. *Circulation Research*, **83**, 1248–63.
88. Ko, Y.S., Yeh, H.I., Rothery, S. *et al.* (1999) Connexin make-up of endothelial gap junctions in the rat pulmonary artery as revealed by immunoconfocal microscopy and triple-label immunogold electron microscopy. *The Journal of Histochemistry and Cytochemistry*, **47**, 683–92.
89. Pepper, M.S., Montesano, R., el Aoumari, A. *et al.* (1992) Coupling and connexin 43 expression in microvascular and large vessel endothelial cells. *American Journal of Physiology: Cell Physiology*, **262**, C1246–57.
90. Gabriels, J.E. and Paul, D.L. (1998) Connexin43 is highly localized to sites of disturbed flow in rat aortic endothelium but connexin37 and connexin40 are more uniformly distributed. *Circulation Research*, **83**, 636–43.
91. DePaola, N., Davies, P.F., Pritchard, W.F. Jr. *et al.* (1999) Spatial and temporal regulation of gap junction connexin43 in vascular endothelial cells exposed to controlled disturbed flows *in vitro*. *Proceedings of the National Academy of Sciences of the United States of America*, **96**, 3154–59.
92. Martin, P.E. and Evans, W.H. (2004) Incorporation of connexins into plasma membranes and gap junctions. *Cardiovascular Research*, **62**, 378–87.
93. Das Sarma, J., Meyer, R.A., Wang, F. *et al.* (2001) Multimeric connexin interactions prior to the trans-Golgi network. *Journal of Cell Science*, **114**, 4013–24.
94. Martin, P.E., Blundell, G., Ahmad, S. *et al.* (2001) Multiple pathways in the trafficking and assembly of connexin 26, 32 and 43 into gap junction intercellular communication channels. *Journal of Cell Science*, **114**, 3845–55.
95. Shaw, R.M., Fay, A.J., Puthenveedu, M.A. *et al.* (2007) Microtubule plus-end-tracking proteins target gap junctions directly from the cell interior to adherens junctions. *Cell*, **128**, 547–60.
96. Thomas, T., Jordan, K., and Laird, D.W. (2001) Role of cytoskeletal elements in the recruitment of Cx43–GFP and Cx26–YFP into gap junctions. *Cell Communication and Adhesion*, **8**, 231–36.
97. Thomas, T., Jordan, K., Simek, J. *et al.* (2005) Mechanisms of Cx43 and Cx26 transport to the plasma membrane and gap junction regeneration. *Journal of Cell Science*, **118**, 4451–62.
98. Gaietta, G., Deerinck, T.J., Adams, S.R. *et al.* (2002) Multicolor and electron microscopic imaging of connexin trafficking. *Science*, **296**, 503–7.
99. Fallon, R.F. and Goodenough, D.A. (1981) Five-hour half-life of mouse liver gap-junction protein. *The Journal of Cell Biology*, **90**, 521–26.
100. Laing, J.G., Tadros, P.N., Westphale, E.M., and Beyer, E.C. (1997) Degradation of connexin43 gap junctions involves both the proteasome and the lysosome. *Experimental Cell Research*, **236**, 482–92.
101. Burghardt, R.C., Barhoumi, R., Sewall, T.C., and Bowen, J.A. (1995) Cyclic AMP induces rapid increases in gap junction permeability and changes in the cellular distribution of connexin43. *The Journal of Membrane Biology*, **148**, 243–53.
102. Popp, R., Brandes, R.P., Ott, G. *et al.* (2002) Dynamic modulation of interendothelial gap junctional communication by 11,12-epoxyeicosatrienoic acid. *Circulation Research*, **90**, 800–6.
103. Griffith, T.M., Chaytor, A.T., Taylor, H.J. *et al.* (2002) cAMP facilitates EDHF-type relaxations in conduit arteries by enhancing electrotonic conduction via gap junctions. *Proceedings of the National Academy of Sciences of the United States of America*, **99**, 6392–97.
104. D'Hondt, C., Srinivas, S.P., Vereecke, J., and Himpens, B. (2007) Adenosine opposes thrombin-induced inhibition of intercellular calcium wave in corneal endothelial cells. *Investigative Ophthalmology and Visual Science*, **48**, 1518–27.
105. Zhang, F., Cheng, J., Lam, G. *et al.* (2005) Adenovirus vector E4 gene regulates connexin 40 and 43 expression in endothelial cells via PKA and PI3K signal pathways. *Circulation Research*, **96**, 950–57.
106. Somekawa, S., Fukuhara, S., Nakaoka, Y. *et al.* (2005) Enhanced functional gap junction neoformation by protein kinase A-dependent and Epac-dependent signals downstream of cAMP

in cardiac myocytes. *Circulation Research*, **97**, 655–62.

107. De Pina-Benabou, M.H., Srinivas, M., Spray, D.C., and Scemes, E. (2001) Calmodulin kinase pathway mediates the K^+-induced increase in gap junctional communication between mouse spinal cord astrocytes. *The Journal of Neuroscience*, **21**, 6635–43.
108. Taylor, H.J., Chaytor, A.T., Evans, W.H., and Griffith, T.M. (1998) Inhibition of the gap junctional component of endothelium-dependent relaxations in rabbit iliac artery by 18-alpha glycyrrhetinic acid. *British Journal of Pharmacology*, **125**, 1–3.
109. Chaytor, A.T., Evans, W.H., and Griffith, T.M. (1998) Central role of heterocellular gap junctional communication in endothelium-dependent relaxations of rabbit arteries. *The Journal of Physiology*, **508** (Pt 2), 561–73.
110. Isakson, B.E. and Duling, B.R. (2005) Heterocellular contact at the myoendothelial junction influences gap junction organization. *Circulation Research*, **97**, 44–51.
111. Ying, X., Minamiya, Y., Fu, C., and Bhattacharya, J. (1996) Ca^{2+} waves in lung capillary endothelium. *Circulation Research*, **79**, 898–908.
112. Goldberg, G.S., Moreno, A.P., Bechberger, J.F. *et al.* (1996) Evidence that disruption of connexon particle arrangements in gap junction plaques is associated with inhibition of gap junctional communication by a glycyrrhetinic acid derivative. *Experimental Cell Research*, **222**, 48–53.
113. Zundorf, G., Kahlert, S., and Reiser, G. (2007) Gap-junction blocker carbenoxolone differentially enhances NMDA-induced cell death in hippocampal neurons and astrocytes in co-culture. *Journal of Neurochemistry*, **102**, 508–21.
114. Cruikshank, S.J., Hopperstad, M., Younger, M. *et al.* (2004) Potent block of Cx36 and Cx50 gap junction channels by mefloquine. *Proceedings of the National Academy of Sciences of the United States of America*, **101**, 12364–69.
115. Bai, D., del Corsso, C., Srinivas, M., and Spray, D.C. (2006) Block of specific gap junction channel subtypes by 2-aminoethoxydiphenyl borate (2-APB). *The Journal of Pharmacology and Experimental Therapeutics*, **319**, 1452–58.
116. Chaytor, A.T., Evans, W.H., and Griffith, T.M. (1997) Peptides homologous to extracellular loop motifs of connexin 43 reversibly abolish rhythmic contractile activity in rabbit arteries. *The Journal of Physiology*, **503**, 99–110.
117. Dora, K.A., Martin, P.E., Chaytor, A.T. *et al.* (1999) Role of heterocellular Gap junctional communication in endothelium-dependent smooth muscle hyperpolarization: inhibition by a connexin-mimetic peptide. *Biochemical and Biophysical Research Communications*, **254**, 27–31.
118. Boitano, S. and Evans, W.H. (1998) Inhibition of intercellular Ca^{2+} signalling with sequence-specific antibodies and peptides made from gap junctions. *Molecular and Cellular Biology*, **9**, 324a.
119. Leybaert, L., Braet, K., Vandamme, W. *et al.* (2003) Connexin channels, connexin mimetic peptides and ATP release. *Cell Communication and Adhesion*, **10**, 251–57.
120. el-Sabban, M.E. and Pauli, B.U. (1994) Adhesion-mediated gap junctional communication between lung-metastatatic cancer cells and endothelium. *Invasion and Metastasis*, **14**, 164–76.
121. Ito, A., Katoh, F., Kataoka, T.R. *et al.* (2000) A role for heterologous gap junctions between melanoma and endothelial cells in metastasis. *The Journal of Clinical Investigation*, **105**, 1189–97.
122. Ito, A., Koma, Y., Uchino, K. *et al.* (2006) Increased expression of connexin 26 in the invasive component of lung squamous cell carcinoma: significant correlation with poor prognosis. *Cancer Letters*, **234**, 239–48.
123. Ito, A., Morita, N., Miura, D. *et al.* (2004) A derivative of oleamide potently inhibits the spontaneous metastasis of mouse melanoma BL6 cells. *Carcinogenesis*, **25**, 2015–22.
124. Miura, D., Kida, Y., and Nojima, H. (2007) Camellia oil and its distillate fractions effectively inhibit the spontaneous metastasis of mouse melanoma BL6 cells. *FEBS Letters*, **581**, 2541–48.
125. Pollmann, M.A., Shao, Q., Laird, D.W., and Sandig, M. (2005) Connexin 43 mediated gap junctional communication enhances breast tumor cell diapedesis in culture. *Breast Cancer Research*, **7**, R522–34.
126. Seo, N., Tokura, Y., Nishijima, T. *et al.* (2000) Percutaneous peptide immunization via corneum barrier-disrupted murine skin for experimental tumor immunoprophylaxis. *Proceedings of the National Academy of Sciences of the United States of America*, **97**, 371–76.
127. Mandelboim, O., Vadai, E., Fridkin, M. *et al.* (1995) Regression of established murine carcinoma metastases following vaccination with tumour-associated antigen peptides. *Nature Medicine*, **1**, 1179–83.

128. King, T.J. and Lampe, P.D. (2004) The gap junction protein connexin32 is a mouse lung tumor suppressor. *Cancer Research*, **64**, 7191–96.
129. Peebles, K.A., Duncan, M.W., Ruch, R.J., and Malkinson, A.M. (2003) Proteomic analysis of a neoplastic mouse lung epithelial cell line whose tumorigenicity has been abrogated by transfection with the gap junction structural gene for connexin 43, *Gja1*. *Carcinogenesis*, **24**, 651–57.
130. Bazarbachi, A., Abou Merhi, R., Gessain, A. *et al.* (2004) Human T-cell lymphotropic virus type I-infected cells extravasate through the endothelial barrier by a local angiogenesis-like mechanism. *Cancer Research*, **64**, 2039–46.
131. Al-Mehdi, A.B., Tozawa, K., Fisher, A.B. *et al.* (2000) Intravascular origin of metastasis from the proliferation of endothelium-attached tumor cells: a new model for metastasis. *Nature Medicine*, **6**, 100–2.
132. Li, Z., Zhou, Z., and Donahue, H.J. (2008) Alterations in Cx43 and OB-cadherin affect breast cancer cell metastatic potential. *Clinical and Experimental Metastasis*, **25**, 265–72.
133. Pelin, K., Hirvonen, A., and Linnainmaa, K. (1994) Expression of cell adhesion molecules and connexins in gap junctional intercellular communication deficient human mesothelioma tumour cell lines and communication competent primary mesothelial cells. *Carcinogenesis*, **15**, 2673–75.
134. Jinn, Y., Ichioka, M., and Marumo, F. (1998) Expression of connexin32 and connexin43 gap junction proteins and E-cadherin in human lung cancer. *Cancer Letters*, **127**, 161–69.
135. Nawrocki-Raby, B., Gilles, C., Polette, M. *et al.* (2003) Upregulation of MMPs by soluble E-cadherin in human lung tumor cells. *International Journal of Cancer*, **105**, 790–95.
136. Jara, P.I., Boric, M.P., and Saez, J.C. (1995) Leukocytes express connexin 43 after activation with lipopolysaccharide and appear to form gap junctions with endothelial cells after ischemia–reperfusion. *Proceedings of the National Academy of Sciences of the United States of America*, **92**, 7011–15.
137. Eugenin, E.A., Branes, M.C., Berman, J.W., and Saez, J.C. (2003) TNF-alpha plus IFN-gamma induce connexin43 expression and formation of gap junctions between human monocytes/macrophages that enhance physiological responses. *Journal of Immunology*, **170**, 1320–28.
138. Zahler, S., Hoffmann, A., Gloe, T., and Pohl, U. (2003) Gap-junctional coupling between neutrophils and endothelial cells: a novel modulator of transendothelial migration. *Journal of Leukocyte Biology*, **73**, 118–26.
139. Oviedo-Orta, E., Hoy, T., and Evans, W.H. (2000) Intercellular communication in the immune system: differential expression of connexin40 and 43, and perturbation of gap junction channel functions in peripheral blood and tonsil human lymphocyte subpopulations. *Immunology*, **99**, 578–90.
140. Oviedo-Orta, E., Errington, R.J., and Evans, W.H. (2002) Gap junction intercellular communication during lymphocyte transendothelial migration. *Cell Biology International*, **26**, 253–63.
141. Doring, B., Pfitzer, G., Adam, B. *et al.* (2007) Ablation of connexin43 in smooth muscle cells of the mouse intestine: functional insights into physiology and morphology. *Cell and Tissue Research*, **327**, 333–42.
142. Rignault, S., Haefliger, J.A., Waeber, B. *et al.* (2007) Acute inflammation decreases the expression of connexin 40 in mouse lung. *Shock*, **28**, 78–85.
143. Park, S.J., Lee, K.S., Kim, S.R. *et al.* (2007) Change of connexin 37 in allergen-induced airway inflammation. *Experimental and Molecular Medicine*, **39**, 629–40.
144. Celes, M.R., Torres-Duenas, D., Alves-Filho, J.C. *et al.* (2007) Reduction of gap and adherens junction proteins and intercalated disc structural remodeling in the hearts of mice submitted to severe cecal ligation and puncture sepsis. *Critical Care Medicine*, **35**, 2176–85.
145. Sawaya, S.E., Rajawat, Y.S., Rami, T.G. *et al.* (2007) Downregulation of connexin40 and increased prevalence of atrial arrhythmias in transgenic mice with cardiac-restricted overexpression of tumor necrosis factor. *American Journal of Physiology: Heart and Circulatory Physiology*, **292**, H1561–67.
146. Simon, A.M., McWhorter, A.R., Chen, H. *et al.* (2004) Decreased intercellular communication and connexin expression in mouse aortic endothelium during lipopolysaccharide-induced inflammation. *Journal of Vascular Research*, **41**, 323–33.
147. Brand-Schieber, E., Werner, P., Iacobas, D.A. *et al.* (2005) Connexin43, the major gap junction protein of astrocytes, is down-regulated in inflamed white matter in an animal model of multiple sclerosis. *Journal of Neuroscience Research*, **80**, 798–808.
148. Nakase, T., Sohl, G., Theis, M. *et al.* (2004) Increased apoptosis and inflammation after focal brain ischemia in mice lacking connexin43 in astrocytes. *American Journal of Pathology*, **164**, 2067–75.

149. Frossard, J.L., Rubbia-Brandt, L., Wallig, M.A. *et al.* (2003) Severe acute pancreatitis and reduced acinar cell apoptosis in the exocrine pancreas of mice deficient for the Cx32 gene. *Gastroenterology*, **124**, 481–93.
150. Ichimura, H., Parthasarathi, K., Lindert, J., and Bhattacharya, J. (2006) Lung surfactant secretion by interalveolar Ca^{2+} signaling. *American Journal of Physiology: Lung Cellular and Molecular Physiology*, **291**, L596–601.
151. Thuringer, D. (2004) The vascular endothelial growth factor-induced disruption of gap junctions is relayed by an autocrine communication via ATP release in coronary capillary endothelium. *Annals of the New York Academy of Sciences*, **1030**, 14–27.
152. Liao, Y., Day, K.H., Damon, D.N., and Duling, B.R. (2001) Endothelial cell-specific knockout of connexin 43 causes hypotension and bradycardia in mice. *Proceedings of the National Academy of Sciences of the United States of America*, **98**, 9989–94.
153. Kruger, O., Beny, J.L., Chabaud, F. *et al.* (2002) Altered dye diffusion and upregulation of connexin37 in mouse aortic endothelium deficient in connexin40. *Journal of Vascular Research*, **39**, 160–72.
154. Wei, C.J., Francis, R., Xu, X., and Lo, C.W. (2005) Connexin43 associated with an N-cadherin-containing multiprotein complex is required for gap junction formation in NIH3T3 cells. *The Journal of Biological Chemistry*, **280**, 19925–36.
155. Nishimura, M., Saito, T., Yamasaki, H., and Kudo, R. (2003) Suppression of gap junctional intercellular communication via 5' CpG island methylation in promoter region of E-cadherin gene in endometrial cancer cells. *Carcinogenesis*, **24**, 1615–23.
156. Singh, D., Solan, J.L., Taffet, S.M. *et al.* (2005) Connexin 43 interacts with zona occludens-1 and -2 proteins in a cell cycle stage-specific manner. *The Journal of Biological Chemistry*, **280**, 30416–21.
157. Giepmans, B.N. and Moolenaar, W.H. (1998) The gap junction protein connexin43 interacts with the second PDZ domain of the zona occludens-1 protein. *Current Biology*, **8**, 931–34.
158. Giepmans, B.N., Verlaan, I., and Moolenaar, W.H. (2001) Connexin-43 interactions with ZO-1 and alpha- and beta-tubulin. *Cell Communication and Adhesion*, **8**, 219–23.
159. Hunter, A.W., Barker, R.J., Zhu, C., and Gourdie, R.G. (2005) Zonula occludens-1 alters connexin43 gap junction size and organization by influencing channel accretion. *Molecular Biology of the Cell*, **16**, 5686–98.
160. Go, M., Kojima, T., Takano, K. *et al.* (2006) Connexin 26 expression prevents down-regulation of barrier and fence functions of tight junctions by Na^+/K^+-ATPase inhibitor ouabain in human airway epithelial cell line Calu-3. *Experimental Cell Research*, **312**, 3847–56.

4

Pulmonary Endothelial Cell Interactions with the Extracellular Matrix

Katie L. Grinnell and Elizabeth O. Harrington

Vascular Research Laboratory, Providence VA Medical Center, Alpert Medical School of Brown University, Providence, RI, USA

INTRODUCTION

The endothelium serves an essential role throughout the circulation as the first interface between blood and interstitium. Lining the inner surface of all vasculature, the endothelium coordinates numerous functions, including platelet adhesion, immune function, and the volume and electrolyte content of the intravascular and extravascular spaces [1–3]. This multi-fold ability is mediated, in part, by the interactions between the endothelial cells (ECs) themselves and the extracellular matrix (ECM) upon which the cells are anchored. Through physical interactions, the ECM provides a protein scaffold upon which ECs migrate, proliferate, apoptose, and regulate blood vessel stabilization – events critical for vascularization [4–7].

COMPONENTS OF CELL–ECM INTERACTIONS

Basement Membrane

The basement membrane is a complex arrangement of fibrillar and nonfibrillar protein molecules, referred to as the ECM, that surrounds and supports the cells of all mammalian tissues, including the endothelium [4]. The basement membrane appears in a cross-section electron micrograph as an amorphous band roughly 40–60 nm thick [8]. A combination of structural glycoproteins, proteoglycans, water, and nonmatrix proteins, including growth factors and cytokines, give the ECM its strength and resilience [4, 6]. The most abundant ECM components are the various members of the collagen family, consisting of at least 16 isoforms [4, 6]. Collagen types I and IV are the primary isoforms found in the ECM underlying endothelium, with types III and V found in smaller amounts [4, 6, 9]. Its triple-helical structure and staggered lateral assembly of adjacent collagen molecules confer tensile strength to the ECM [10].

Fibronectin, laminin, and elastin are the other major protein components of the ECM [10]. Fibronectin is a large glycoprotein comprised of a series of modular domains that are able to undergo conformational changes in response to tension. The multiple domains of fibronectin also contain numerous binding sites for interactions with other ECM molecules [11, 12]. The laminins are heterotrimeric and form independent networks that play an integral role in the tensile strength of the ECM. These glycoproteins are also involved in cell adhesion and are closely associated with collagen IV [13, 14]. In the ECM of the endothelium, the predominant laminin isoforms expressed are laminin-8/laminin 411 and laminin-10/laminin 511 [15]. Finally, elastin, with its amorphous elastic core surrounded by microfibrils, is widely expressed in the ECM of tissues that undergo a high degree of deformation or contraction, such as the skin; thus, elastin is less abundant in the ECM of the endothelium [6, 16, 17].

While collagen, fibronectin, laminin, and elastin serve roles important for the structure and adhesion of the cells, additional ECM proteins are responsible for controlling other cellular functions, including migration, replication, differentiation, and apoptosis. The nidogens, or entactins as they are also known, are a family of highly conserved glycoproteins ubiquitously expressed throughout the ECM [18, 19]. Until recently, little was known regarding their function. It is now known that the nidogens serve to stabilize the ECM during periods of rapid

The Pulmonary Endothelium: Function in health and disease Editors Norbert F. Voelkel, Sharon Rounds

growth or turnover, such as during development and angiogenesis [20, 21]. Tenascin-C is also a glycoprotein that has been implicated in modulating angiogenesis and vessel sprouting [22, 23]. Interestingly, tenascin-C knockout mice display no significant vascular anomalies [24]. An additional ECM associated glycoprotein is referred to as secreted protein, acidic and rich in cysteine (SPARC). SPARC upregulation has been associated with tissue remodeling and angiogenesis, serving to disrupt cell adhesion, inhibit proliferation, and regulate the synthesis of several ECM proteins, including laminin and fibronectin [25, 26]. The proteoglycans, chondroitin sulfate and heparin sulfate, function largely to sequester cytokines and growth factors in the vicinity of responsive cells. In addition, their net negative charge attracts water molecules, keeping the ECM and resident cells hydrated [27]. Once metabolized, the glycosaminoglycan side-chains of these proteoglycans are released and contribute to the viscosity and resistance of the ECM [28]. Two additional proteoglycan molecules, perlecan and syndecan, play an important role in determining the pore size and charge density of the matrix and, hence, contribute to the function of the ECM as a selective filter controlling which substances reach the cellular surface of the endothelium [29]. Additionally, agrin is a heparin sulfate proteoglycan expressed primarily in central nervous system and muscle cells, which plays an important role in the aggregation of the nicotinic acetylcholine receptors.

Formation of the ECM begins with secretion of laminin polymers by the endothelium [30, 31]. Once the native laminin network has been laid down, collagen is produced and interacts with the laminin to form a scaffold upon which the other ECM components are assembled [6]. Laminin and collagen are unique in that they contain the necessary information within their protein sequence so as to mediate their self-assembly into sheet-like structures [32]. Nidogen/entactins and perlecans are then secreted, bridging the laminin and collagen networks. The other ECM components interact with this combined network to facilitate the functional needs of the endothelium over time [6].

Since distinct ECM molecules regulate a variety of cellular functions, it is not surprising that ECM remodeling has been shown to occur in the lung in experimental settings of either acute or chronic hyperoxia, bleomycin-induced injury, or whole-body irradiation [33, 34]. Further, *in vitro* studies have shown altered production of selected ECM proteins by pulmonary artery endothelial cell (PAECs) in response to exposure to lipopolysaccharide (LPS), with an enhancement of SPARC deposition and a concomitant diminution of procollagen III and V, as well as fibronectin protein levels [35]. Likewise, increased fibronectin production was noted in ECs of hyperoxic lungs [36, 37]. Altered adhesion to fibronectin and disruption of the fibronectin matrix are events that are important in tumor necrosis factor (TNF)-α-induced endothelial monolayer permeability [38, 39]. Similar to fibronectin, tenascin-C synthesis is upregulated in the lungs of human subjects with acute respiratory distress syndrome (ARDS) or bronchopulmonary dysplasia [40]. Data have suggested that the levels of tenascin-C protein influences fetal lung branching and vascularization during lung development [41]. Finally, PAECs are protected against the induction of apoptosis upon exposure to bleomycin or LPS if grown on selected ECM proteins; such as collagen IV, laminin, fibronectin, or gelatin [42, 43]. Thus, the cell–ECM interactions of the lung endothelium serve a yin–yang relationship, whereby the biological state of the lung endothelium can promote remodeling of the surrounding ECM and the ECM composition can affect the functional state of the lung endothelium.

ECM Remodeling

The composition of the ECM is dependent upon the balance between matrix protein synthesis and degradation. The most commonly found matrix degrading enzymes are the matrix metalloproteinases (MMPs) [44–46]. These matrix-degrading enzymes can be produced by stromal cells or ECs and, in some instances, by tumor cells. In the pulmonary endothelium, the primary MMP responsible for remodeling during normal angiogenesis or in response to pulmonary edema are the zinc-dependent MMP, MMP-2 (gelatinase A), and MMP-9 (gelatinase B) [47–49]. MMPs are regulated at the transcriptional and post-translational levels, as well as by direct binding to competitive, reversible inhibitors, termed tissue inhibitors of metalloproteinases (TIMPs). Degradation of the ECM constituents by MMPs releases ECs from their cell surface anchors, integrins, facilitating a more migratory and proliferative phenotype [44–46]. The actions of MMPs also result in the liberation of ECM-sequestered growth factors and release of proteolytic ECM cleavage byproducts which can affect local cell function [44–46]. For example, MMP proteolysis of collagen results in the formation of various peptides, referred to as endostatin, arrestin, canstatin, and tumstatin, which induce EC apoptosis [6, 50, 51]. Also, MMP-mediated cleavage of perlecan or fibronectin produces peptides, endorepellin or anastellin, that mediate antiangiogenic effects through the disruption of focal adhesion complex formation and cell migration or induction of cell cycle arrest, respectively [52, 53].

Pulmonary Disease and Dysregulated ECM

As the ECM serves as the three-dimensional surface on which cells adhere and contribute to tissue structure and function, it is not surprising that the composition of the ECM can affect normal lung function through regulation of the tensile and compressive strength and elasticity [54, 55], modulation of interstitial fluid dynamics and gas exchange [56], and regulation of availability of signaling molecules and/or cellular surface receptors [57]. ECM remodeling commonly occurs within the lung during the progression of chronic obstructive pulmonary disease (COPD), asthma, fibrosis, and cancer [44, 45]. Some common changes noted in the settings of COPD and fibrosis include an increased deposition of ECM molecules, as well as an altered ECM protein composition of the lung tissue. In contrast, MMP-mediated degradation of the ECM is thought to contribute to asthma and cancer progression by providing a setting within the lung interstitial space for infiltration and migration of inflammatory cells, cancer cells, or ECs involved in tumor-associated angiogenesis. Ongoing studies are investigating whether modulation of ECM turnover by MMPs and/or TIMPs may prove to be efficacious therapies for attenuating the pathogenesis of a variety of chronic lung diseases.

Types of Cell–ECM Junctions

Several types of cellular junctions control the adhesive interactions of the EC monolayer and the ECM; these include focal adhesions, focal contacts, fibrillar adhesions, podosomes, dystroglycan (DG) contacts, and hemidesmosomes (Table 4.1). Each of the many ECM proteins has a particular EC surface marker to which it binds, conveying distinct functional properties to the cells. These EC–ECM structures are reviewed, with a brief description of their respective role(s) in the pulmonary endothelium (Tables 4.1 and 4.2).

Focal Adhesions/Focal Contacts/Fibrillar Adhesions

The strongest and most well-studied EC–ECM interaction is focal adhesion (Figure 4.1). The four main constituents of this adhesion complex are the transmembrane

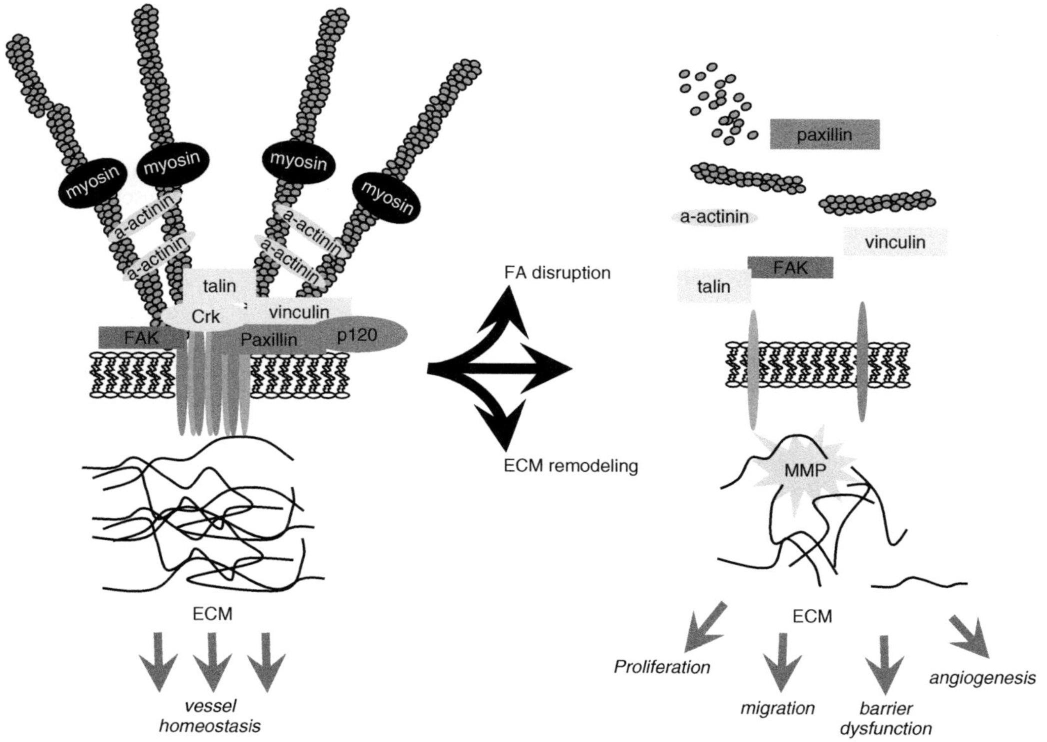

Figure 4.1 Schematic representation of the effects of cell–ECM interactions on pulmonary EC function. Normally, the cell–ECM interactions within the lung endothelium are stable, providing a protein scaffold on which the vessels are homeostatic. In settings of injury, disease, or environmental stresses, the ECM can become remodeled, which, in turn, may promote fewer cell–ECM interactions signaling the pulmonary endothelium to become more migratory, undergo apoptosis, induce angiogenesis, or modulate the vessel barrier function.

Table 4.1 Characterization of cell–ECM structures

Cell–ECM interaction	Characteristics	Cell types
Focal adhesions	strongest cell–ECM adhesions located at cell periphery large, rod-shaped protein complexes provide reciprocal communication between ECM and actin cytoskeleton direct cell shape, fate, and motion acutely responsive to shear stress, cyclic stretch, angiogenic signals, and proinflammatory stimuli	endothelium cardiomyocytes fibroblasts platelets vascular smooth muscle cells
Focal complexes/contacts	similar to focal adhesions but smaller in size (<1 μm in diameter) located along cell periphery more dynamic than focal adhesion, displaying increased motility abundant in lamellipodia maintenance is not tension dependent	endothelium cardiomyocytes fibroblasts platelets vascular smooth muscle cells
Fibrillar adhesions	elongated or punctuate structures located centrally within the cell relatively immobile can generate considerable contractile force function largely in migration and lamellipodial protrusion associated with tissue repair involved in matrix reorganization translocate centripetally in response to actomyosin pulling function in fibrillogenesis of ECM maintenance does not depend on actomyosin contractility	cardiac valvular interstitial cells fibroblasts endothelium smooth muscle cells
Podosomes	similar in structure to fibroblast invadopodia finger-like invaginations of the ventral cell membrane directed toward center of cell, toward the substratum consist of F-actin core and actin-associated proteins, surround by plaque proteins diameter of 0.5–1 μm and a depth of 0.2–0.4 μm extremely dynamic play major role in diapedesis, migration, bone resorption, ECM degradation and remodeling, angiogenesis, and vasculogenesis responsive to inflammation and shear stress do not require protein synthesis	monocyte-derived hematopoietic cells, including macrophages, leukocytes, and osteoclasts endothelium smooth muscle cells transformed fibroblasts and carcinoma cells

Table 4.1 (*continued*)

Cell–ECM interaction	Characteristics	Cell types
DG contacts	characterized in the inherited disease Duchenne muscular dystrophy absence/mutation of dystrophin leads to muscle wasting, respiratory difficulties, poor coordination essential for maintenance of vessel barrier function transducers of shear stress play a role in myelinogenesis of peripheral nerves help maintain polarity of epithelial cells	skeletal muscle smooth muscle neurons endothelium epithelial cells
Hemidesmosomes	best characterized in epithelial tissues firmly attach keratinocytes to ECM a new type (type II) has been characterized in ECs May function to connect the vimentin (IF cytoskeleton to the plasma membrane at sites of ECM adhesion	type I: epithelial cells type II: endothelium

integrins, ECM protein ligands, the cytoskeletal microfilament actin, and intracellular anchor proteins, including talin, vinculin, α-actinin, and filamin [58]. Focal adhesion formation is initiated when specific ligands in the ECM bind to their specific integrin receptors. This "adhesive interface" then undergoes a maturation phase during which additional integrins and cytoskeletal components are recruited [59]. According to a recent review by Romer *et al.*, four major factors influence the assembly, rate, size, and specific constituency of focal adhesion: (i) the biophysical and biochemical properties of the ECM, (ii) the degree of integrin activation and avidity for the available ligands, (iii) the contraction state of the cytoskeleton, and (iv) the specific cellular and tissue environment [60].

The most widely expressed endothelial adhesion molecules are the members of the integrin family. These obligate heterodimers consist of one α-chain and one β-chain. Each subunit is a transmembrane glycoprotein composed of a large ectodomain and a smaller cytoplasmic domain. Of the 19 α-subunits and eight β-subunits, the following combinations are found in ECs: $\alpha_1\beta_1$ and $\alpha_2\beta_1$ (that bind to collagen), $\alpha_3\beta_1$, $\alpha_6\beta_1$, and $\alpha_6\beta_4$ (that bind to laminin), $\alpha_4\beta_1$ and $\alpha_5\beta_1$ (that serve as fibronectin receptors), and $\alpha_v\beta_3$ and $\alpha_v\beta_5$ (that selectively bind vitronectin) [30, 61–63]. The extracellular domains of integrins interact with the amino acid motif, Arg–Gly–Asp (RGD), within ECM proteins. The short cytoplasmic domains are in turn linked to actin-binding proteins [64].

Upon integrin binding to ECM protein RGD domains, the cytoplasmic tail becomes associated with the actin-binding proteins, vinculin, α-actinin, paxillin, talin, zyxin, tensin, and filamin [65]. Each of these primary anchor proteins then forms their own extensive signaling complexes with specific target molecules. Vinculin binds to actin-related protein (Arp)-2/3 and phosphatidylinositol 4,5-kinase (PIP5K) [66, 67]. α-Actinin binds to neighboring zyxin, which in turn associates with the vasodilator-stimulated phosphoprotein (VASP) and profilin [68, 69]. Paxillin forms interactions with p21-activated kinase (PAK) and PAK-interacting exchange factor (PIX), *Abl* protooncogene (Abl), and p120 GTPase-activating protein ($p120^{RasGAP}$) [70, 71]. Through its c-Src homology-2 domain, tensin initiates interactions with multiple phosphotyrosine signaling molecules, including Src and p130 Crk-associated substrate ($p130^{Cas}$) ([71, 72], while filamin stimulates the activity of the small Rho GTPases, RhoA, Rac, and Cdc42, as well as Ral1, RhoA-associated kinase (ROCK), and calveolin-1 [73–75].

Perhaps the most important of the focal adhesion signaling components, focal adhesion kinase (FAK), binds to multiple primary anchor proteins (e.g., paxillin, talin, vinculin), uniting all of their individual signaling complexes. FAK is a nonreceptor protein tyrosine kinase with an internal catalytic domain interposed between its N- and C-terminal domains [76]. The N-terminal domain of FAK binds to a protein called Trio, which itself is comprised of three domains [77]. Together, FAK and Trio regulate actin dynamics and RhoA activity [78, 79]. The FAK N-terminus is also involved in mediating the phosphorylation and activation of Wiskott–Aldrich syndrome

Table 4.2 Signaling molecules associated with cell–ECM structures

Cell–ECM interaction	Intracellular components	Intercellular components	ECM components
Focal adhesions	actin α-actinin ezrin filamin fimbrin paladin parvin profilin radixin talin tensin VASP vinculin Abl Csk FAK Pyk2 c-Src Fyn PAK PAX PKC SHP2 PTP1B $p130^{Cas}$ Crk DOCK180 paxillin zyxin Arp-2/3 Cas GIT1/GIT2 PI3K PIP5K $p120^{GAP}$ Cdc42 Rac Rho ROCK Shc Trio WASP	integrins (heterodimers) $\alpha_1\beta_1$ $\alpha_2\beta_1$ $\alpha_3\beta_1$ $\alpha_6\beta_1$ $\alpha_6\beta_4$ $\alpha_4\beta_1$ $\alpha_5\beta_1$ $\alpha_5\beta_3$ $\alpha_v\beta_3$ $\alpha_v\beta_5$	collagen fibronectin heparan sulfate laminin proteoglycan vitronectin
Focal complexes/contacts	actin α-actinin Talin FAK VASP exhibit enriched levels of vinculin and paxillin demonstrate decreased expression of zyxin and tensin	same integrin isoforms as focal adhesion highly enriched in $\alpha_v\beta_3$	collagen fibronectin heparan sulfate laminin proteoglycan enriched expression of vitronectin

Table 4.2 (*continued*)

Cell–ECM interaction	Intracellular components	Intercellular components	ECM components
Fibrillar adhesions	high levels of tensin actin myosin parvin/actopaxin contain little or no phosphotyrosine	enriched in $\alpha_5\beta_1$ integrin	Fibronectin
Podosomes	Actin α-actinin WASP Arp-2/3 cortactin dynamin gelsolin paxillin talin vinulin PI3K Pyk2 FAK Cdc42 RhoA PKC c-Src	enriched in integrins $\alpha_2\beta 1$ $\alpha_3\beta_1$ $\alpha_5\beta_1$ $\alpha_v\beta_3$ $\alpha_M\beta_2$ (CD18)	fibronectin collagen type I osteopontin vitronectin
DG contacts	dystrophin actin utrophin rapsyn Grb-2	$\alpha_1\beta_1$ $\alpha_3\beta_1$ $\alpha_6\beta_1$, $\alpha_6\beta_4$ DG α-DG β-DG	laminin agrin perlecan
Hemidesmosomes	Type I plectin BP230 Type II plectin	Type I integrin $\alpha_6\beta_4$ CD151 keratin Type II integrin $\alpha_6\beta_4$ CD151 vimentin	Type I type XVII collagen BP180 laminin-5 Type II laminin-5

protein (WASP), a downstream effector of the small Rho GTPase, Cdc42, which in turn induces actin polymerization through the binding to Arp-2/3 [80–82]. The FAK C-terminal domain controls cellular localization of the protein through its focal adhesion targeting (FAT) motif [76]. Autophosphorylation at Tyr397 within its central catalytic domain provides a signaling platform for members of the Src family of tyrosine kinases, including c-Src and Fyn [83–85]. This catalytic activation of c-Src is traditionally considered to be the primary means by which focal adhesions are involved in dynamic regulation of the endothelial barrier [76]; changes in c-Src kinase activation have been shown to contribute to increased endothelial permeability in response to numerous stimuli, including production of oxygen radicals, thrombin, and vascular endothelial growth factor (VEGF) [86–90].

The focal adhesion complexes serve as the primary mechanosensors for many cellular signaling cascades and the specific combination of focal adhesion constituents differs according to the state of the endothelium. Under conditions of shear stress, cyclic stretch, angiogenesis, or inflammation, the predominant integrin combinations may change, along with the anchor-associated signaling molecules [60]. In the pulmonary endothelium, the following focal adhesion components have been observed to undergo dynamic regulation: $\alpha_1\beta_1$, $\alpha_2\beta_1$, $\alpha_5\beta_1$, $\alpha_v\beta_3$, $\alpha_v\beta_5$, $\alpha_6\beta_4$, FAK, Fyn, proline-rich tyrosine kinase-2 (Pyk2), GIT1, and GIT2, $p130^{Cas}$, paxillin, phosphoinositol-3 kinase (PI3K), Shc, c-Src, VASP, and vinculin [91–95]. The constant bidirectional signaling between the focal adhesion complex and the ECM allows the endothelium to respond to the constantly changing vascular environment.

Smaller, more nascent adhesive structures that are often considered the predecessors to focal adhesion complex are known as focal complexes/contacts. These structures, which usually measure less than 1 μm in diameter, are more dynamic than focal adhesion, displaying motility even in stationary cells [96]. Focal contacts do not contain the same diversity of signaling molecules contained within focal adhesions, with enriched levels of vinculin and paxillin, and comparatively low expression of zyxin and tensin [97]. Maturation of focal contacts into focal adhesions is thought to be dependent upon EC interactions with the actin cytoskeleton and levels of Rho-modulated actomyosin tension [98].

Fibrillar adhesions are another variant of cell–ECM interaction arising from focal contacts. Unlike focal contacts that are located at the cell periphery, fibrillar adhesions are elongated or punctate structures located centrally within the cell. They contain high levels of $\alpha_5\beta_1$ integrin, tensin, and a member of a relatively novel family of adhesion proteins, parvin/actopaxin [99]. Fibrillar adhesions are primarily attached to fibronectin fibrils through their interaction with $\alpha_5\beta_1$ integrin molecules, whereas focal contacts are bound largely to vitronectin, through $\alpha_v\beta_3$ integrins [100]. Due to the rigid nature of vitronectin, focal contacts can remain relatively immobile despite considerable contractile forces. Fibrillar adhesions, in contrast, being bound to the considerably more pliable fibronectin, translocate centripetally in response to slight variations in actomyosin pulling [100]. The resultant movement of the fibronectin receptors stretches the underlying fibronectin matrix, promoting fibrillogenesis [101].

Podosomes

One of the more recently discovered cell–ECM adhesions is the podosome. Although these modules are sites of integrin-associated actin polymerization like focal adhesions, focal contacts, and fibrillar adhesions, they are more similar in composition and structure to invadopodia, traditionally observed in fibroblasts [102, 103]. While the podosomes were originally thought to serve as "cellular feet," podosomes are actually finger-like invaginations of the ventral membrane, directed towards the cell center, perpendicular to the substratum [104, 105]. Accordingly, they contain several unique markers, including the actin-binding proteins, gelsolin, dynamin, and cortactin [105, 106]. Perhaps the most intriguing characteristic of these extremely dynamic structures is that their formation does not require protein synthesis [105]. Podosomes are found in monocyte-derived hematopoietic cells, including macrophages, leukocytes, and osteoclasts, where they play a major role in diapedesis, migration, and bone resorption [107–109]. It is now known that podosomes are present within a wide variety of other cell types, including the endothelium. It is currently believed that they may also play a role in ECM degradation and cell remodeling. Given the importance of these two processes in angiogenesis and vasculogenesis, as well as in response to inflammation and shear stress, recent studies have begun to address the particular role of podosomes in ECs and have revealed a link between podosome formation and the small Rho GTPases, Cdc42 and RhoA, protein kinase C (PKC), and the tyrosine kinase, c-Src [110].

DG Contacts

Another of the more newly characterized cell–ECM adhesion complexes is the DG contact. Initially discovered while studying the inherited disease Duchenne muscular dystrophy, the DG contact is centered around dystrophin, a cortical cytoskeletal protein [111]. In muscle, dystrophin binds actin as part of a multiprotein complex named the dystrophin-associated glycoprotein complex (DGC). DG is one of the central components of the DGC, serving as the link between dystrophin and the ECM. The mature DG protein is composed of an α- and a β-subunit [112]. α-DG is a heavily glycosylated peripheral membrane protein that binds to the ECM molecules laminin, agrin, and perlecan, as well as to β-DG. β-DG, which spans the plasma membrane, interacts intracellularly with the C-terminus of dystrophin, as well as utrophin, rapsyn, and the cytoskeletal adaptor protein, growth factor receptor-bound protein 2 (Grb-2) [113].

It has now been established that DG and its associated contacts are expressed not only in muscle, but also in nervous tissue and endothelium, particularly in the cerebral microvasculature [111]. The DG complex shares its primary ECM ligand, laminin, with a number of integrin receptors, including $\alpha_1\beta_1$, $\alpha_3\beta_1$, $\alpha_6\beta_1$, and $\alpha_6\beta_4$. DG, however, is expressed in a different pattern than these integrin receptors, demonstrating presence on blood vessels of all diameters [111]. Adhesion of ECs to the ECM is essential for maintenance of an impermeable barrier and within the central nervous system, alterations in DG contacts have been noted following cerebral ischemia [114]. More recently, studies in mice have revealed that lack of functional dystrophin, and hence DG complexes, displayed a defect in transduction of shear stress into vessel dilation through the nitric oxide–cGMP pathway [115]. Additionally, adhesion complexes centered around the DGC protein utrophin, referred to as utrophin-associated protein complexes (UAPCs), have been shown to play a role in the regulation of vascular tone in human umbilical vein ECs [116].

Hemidesmosomes

The final cell–ECM adhesion complex that has been characterized to date is the hemidesmosome. Best characterized in epithelial tissues, hemidesmosomes function to firmly attach keratinocytes in the basal layer of the skin to the underlying ECM. In the skin, these adhesion complexes contain the integrin, $\alpha_6\beta_4$, the type XVII collagen, BP180, the tetraspanin CD151, and the two plakin family members, plectin and BP230. The $\alpha_6\beta_4$ integrin serves to connect cells to laminin-5, while the cytoplasmic proteins plectin and BP230 connect to keratin intermediate filaments (IFs) [117, 118]. Until recently, it was thought that these ECM adhesions occurred only in epithelial cells. However, integrin $\alpha_6\beta_4$ and plectin have been found assembled into structures resembling hemidesmosomes in ECs [118]. Notably, these structures lack BP180 and BP230, and as such have been classified as type II hemidesmosomes [118], in order to distinguish them from type I hemidesmosomes found in stratified epithelial cells. The function of type II hemidesmosomes in endothelia remains elusive. Studies suggest that in these complexes, $\alpha_6\beta_4$ may function to connect the vimentin IF cytoskeleton to the plasma membrane at sites of ECM adhesion [119].

FUNCTIONAL EFFECT OF EC–ECM INTERACTIONS IN THE PULMONARY VASCULATURE

The concerted functions of these numerous cell–ECM adhesion complexes serve to direct EC fate, shape, migration, differentiation, and apoptosis. In the pulmonary endothelium, they play an essential role in maintaining barrier function and re-endothelialization following injury or insult to vessel walls. In cases of malignant transformation, there is a downregulation of ECM matrix proteins, particularly fibronectin, and alterations in the expression levels of integrins and proteoglycans. This leads to a decrease in matrix adhesion and a more migratory phenotype. In the pulmonary endothelium, this causes an increase in vessel permeability and subsequent formation of pulmonary edema.

Regulation of Pulmonary EC Cycle Progression, Proliferation, and Apoptosis

In quiescent endothelium, the primary signals generated by the ECM inhibit proliferation and promote cell adhesion. As with many cell types, integrin-mediated adhesion of ECs regulates the cell cycle progression through the G_1 phase through altered expression and/or activity of cyclins A and D1, as well as several cyclin-dependent kinases. In addition, studies have suggested that tension-dependent changes in the shape of the cell or actin cytoskeleton, as well as cell–ECM interactions, regulate EC cycle progression into S phase [120, 121]. Furthermore, interactions of distinct integrin pairs with the ECM may direct the cell to proliferate or undergo growth arrest. For example, EC interactions with fibronectin through the $\alpha_5\beta_1$ integrin promoted growth factor-mediated cell proliferation, whereas $\alpha_2\beta_1$ integrin binding to laminin caused EC growth arrest in G_1 phase [122]. Additionally, studies demonstrated that $\alpha_v\beta_3$ integrin ligation, but not the ligation of $\alpha_5\beta_1$ or $\alpha_2\beta_1$, promoted a greater proliferative response of ECs to VEGF [123, 124] – effects mediated through multimeric protein complex formation between VEGF receptor-2, $\alpha_v\beta_3$, and c-Src [123]. Thus, many intracellular and extracellular factors regulate cell cycle progression and proliferation through cell adhesion to the ECM.

Apoptosis (or anoikis) occurs in numerous cells, including the endothelium, upon loss of cell–ECM interactions (see Chapter 16). Indeed, ECs grown in a single-cell suspension have been shown to undergo apoptosis [125]. Also, ECs deprived of serum, activated by Fas ligand, or exposed to TNF-α were protected from apoptosis when grown on distinct ECM proteins [126–129]. Furthermore, soluble ligand antagonists that disrupted selected integrin–ECM interactions or select ECM proteolytic fragments can induce apoptosis in proliferating endothelium [6, 50, 51]. Endothelial apoptotic bioactive functions have been attributed to endostatin and canstatin, tumstatin, and arrestin, proteolytic fragments of collagen XVIII and collagen IV, respectively. One mode of action of these soluble matrix proteolytic products is to block integrin receptors, $\alpha_v\beta_3$, $\alpha_5\beta_1$, or $\alpha_1\beta_1$, resulting in antiangiogenic signaling, in part, by promoting EC apoptosis [6]. Further data have demonstrated that laminar shear stress upregulates the expression of EC integrins and integrin-associated proteins [130, 131], suggesting that antiapoptotic signaling pathways induced by these biomechanical forces are mediated, in part, at the level of cell–ECM interactions. Finally, exposure of ECs to angiogenesis-inducing agents promotes upregulation of $\alpha_v\beta_3$ integrins [132], further demonstrating a role for cell–ECM interactions to regulate survival/antiapoptotic pathways.

As mentioned above, cell–ECM signaling through various mediators is important in the progression of the cell cycle and proliferation or induction of apoptosis; including crosstalk with growth factor stimulated receptor tyrosine kinases, activation of small GTPases, clustering at lipid rafts, and activation of integrin-associated proteins. Indeed, binding of human umbilical vein ECs to fibronectin can promote proliferation and cell cycle progression by G_1 transition into the S phase in both Rac-1- and RhoA-dependent manners through modulation of cyclins D1 and E, respectively [122, 133]. Also, disruption of the integrin complex associated protein, FAK, promotes EC apoptosis in response to serum deprivation [134]. Survival signals conveyed by FAK, in ECs, are mediated through various signaling molecules, including $p130^{Cas}$, Rac-1, PKC, and mitogen-activated protein kinases [134, 135]. We have also found that increased degradation of focal adhesion-associated proteins, FAK, $p130^{Cas}$, and paxillin, was associated with the induction of apoptosis in pulmonary-derived ECs upon exposure to adenosine/homocysteine (Figure 4.2); apoptotic effects were attenuated upon protein tyrosine phosphatase or caspase inhibition or overexpression of FAK [136, 137]. We have also shown a member of the ubiquitous intracellular signaling family, PKC, to regulate EC proliferation, possibly through the modulation of EC–ECM interactions. Indeed, overexpression of the PKCδ isoform promoted EC adhesion to the ECM protein, vitronectin [138], increased the number of focal adhesions within the ECs (Figure 4.3) [139], and caused delayed S-phase transition, and hence diminished proliferation, through the upregulation of $p27^{Kip1}$ [140]. Thus, cell–ECM interactions serve as potent, complex signaling cues dictating life or death response of many cell types, including the endothelium.

The pathological progression of emphysema is thought to occur in response to chronic airway inflammation, proteolytic degradation of the lung parenchymal ECM, and apoptotic-mediated turnover of various lung cells, including the endothelium (see Chapter 26). Wiebe *et al.* demonstrated diminished vascular bed volume, relative to the total lung surface, in patient lungs with emphysema and COPD [141]. While several groups have noted an increased number of apoptotic ECs in emphysematous lungs [142–144], the mechanism by which apoptosis was induced in the progression of the disease is not clear. It is well recognized that airway inflammation and enhanced oxidative stress in the settings of emphysema lead to an imbalance in the level of proteolytic MMPs and their inhibitors, leading to the overall degradation of the lung ECM and airway space enlargement.

It has similarly been shown that altered ECM protein deposition [145] and increased pulmonary EC apoptosis [146] occur in early stages of systemic scleroderma (see Chapter 27). In addition, autoimmune antibodies directed against the vascular endothelium [147], and against MMP-1 and MMP-3 [148, 149], have been identified in the sera of patients with systemic scleroderma, suggesting additional mechanisms for the vascular dysfunction and concomitant development of pulmonary fibrosis in this disease. Thus, data suggest that pathological progression of several diseases of the lung occur, in part, from an imbalance between angiogenic and apoptotic homeostasis in the lung vasculature, possibly through altered cell–ECM interactions [150].

Angiogenesis in the Pulmonary Vasculature

In quiescent, uninjured blood vessels, the cell–ECM interactions signal to inhibit cellular proliferation and to facilitate cell adhesion. However, in instances of angiogenesis, vascular remodeling, or repair, the various ECM components are rearranged, proteolyzed, and/or newly synthesized resulting in the exposure of different functional protein domains to the EC surface; such changes induce the normally quiescent endothelium to become more proliferative, migratory, and/or adhesive (see Chapter 13). Thus, the integrity of the surrounding ECM plays a critical role in regulating blood vessel survival, and in regulating proliferation, migration, and tubulogenesis of the endothelium.

Unlike most other organs, angiogenesis occurs infrequently within the healthy adult lung; however, it is noted in lungs during the pathological progression of pulmonary arterial hypertension (PAH) [150], asthma [151–153], cancer, transplantation [154], and in response to pneumonectomy [155]. Neovessel formation is noted in each lung disease listed above, with the exception of PAH. Instead, in patients with PAH, plexiform lesions are noted within the pulmonary arteries that are characterized as clusters of ECs, without tubule sprouting and differentiation [156] (see Chapter 21). Angiogenesis and vascular remodeling of the bronchial circulation (see Chapter 14) has also been shown to occur in settings of pulmonary embolism or pulmonary artery obstruction [157–160], high altitude [161], and hypoxia [162, 163], hence enhancing the blood supply to the lung. Finally, during acute lung injury, angiogenesis can occur during the later, reparative phase of the injury and may be mediated, in part, via a VEGF-dependent pathway [164, 165] (see Chapter 23).

Maintenance of Pulmonary Endothelial Barrier Function

Vascular permeability occurs through transcellular and paracellular pathways and is a primary role of the en-

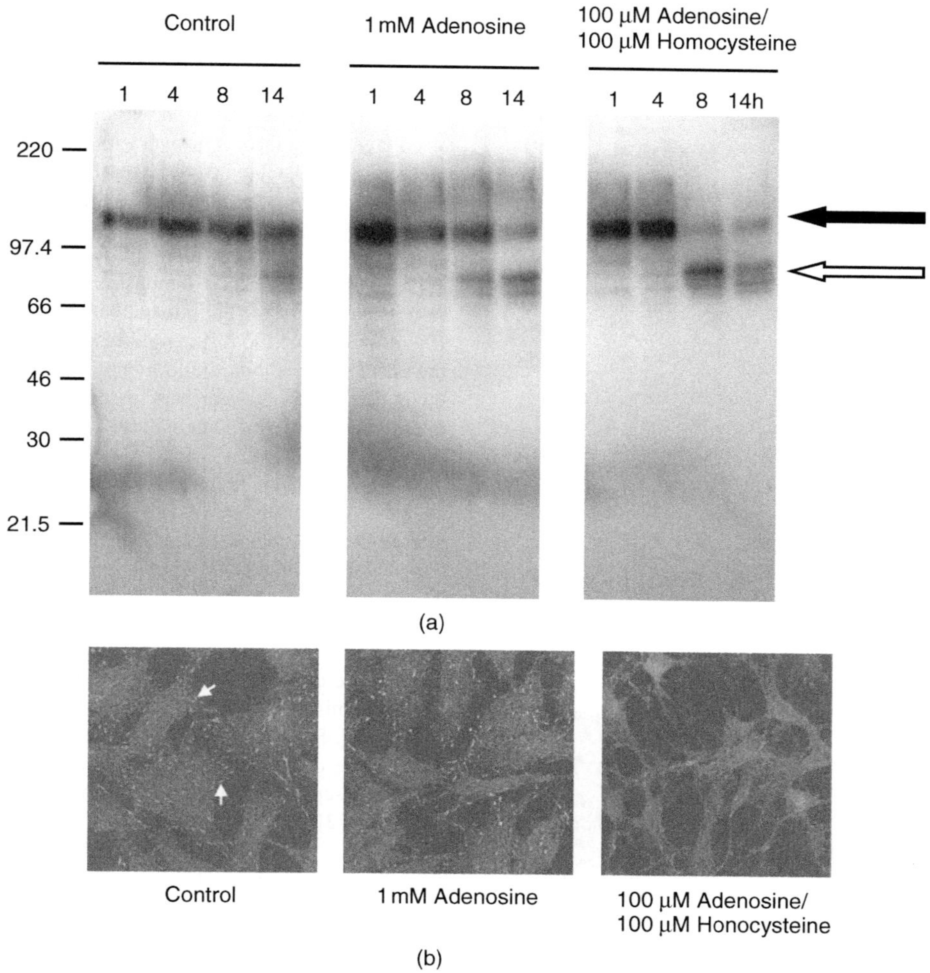

Figure 4.2 FAK proteolysis and focal adhesion disruption precedes apoptosis in pulmonary artery ECs. PAECs were incubated with HEPES buffer alone (Control) or with 1 mM adenosine or 100 μM adenosine and 100 μM homocysteine for the time indicated (a) or for 4 h (b). In (a), cells were harvested, equivalent amounts of protein were resolved by sodium dodecylsulfate–polyacrylamide gel electrophoresis, transferred to nitrocellulose, and immunoblotted for FAK. Open arrows indicate proteolytic fragments; solid arrows indicate intact protein. Proteolysis of a central focal adhesion-associated protein, FAK, occurs in PAECs as early as 8 h following exposure to adenosine or adenosine/homocysteine. In (b), the PAECs were immunofluorescently stained for FAK and visualized with laser scanning confocal microscopy. Arrows indicate focal adhesion. Fewer FAK-containing focal adhesion were noted in adenosine- or adenosine/homocysteine-treated PAECs. Thus, focal adhesion disruption through proteolysis of key protein components may be early steps in the onset of PAEC apoptosis upon exposure to adenosine or adenosine and homocysteine. Images were modified from the originally published images from [136].

dothelium (see Chapter 8). Modulation of blood vessel barrier function via the paracellular pathway is tightly regulated at the level of both inter-EC junctions and EC–ECM interactions; protein complexes transmit signals in response to environmental, biochemical, and mechanical cues. Microarray analyses have shown differential gene expression of both proteins which compose the ECM and of ECM-modifying proteins in ECs isolated from the macrovasculature, relative to the microvasculature [166], suggesting that the ECM is specifically tailored within the vascular tree for the function the ECs are serving, such as vascular permeability. For

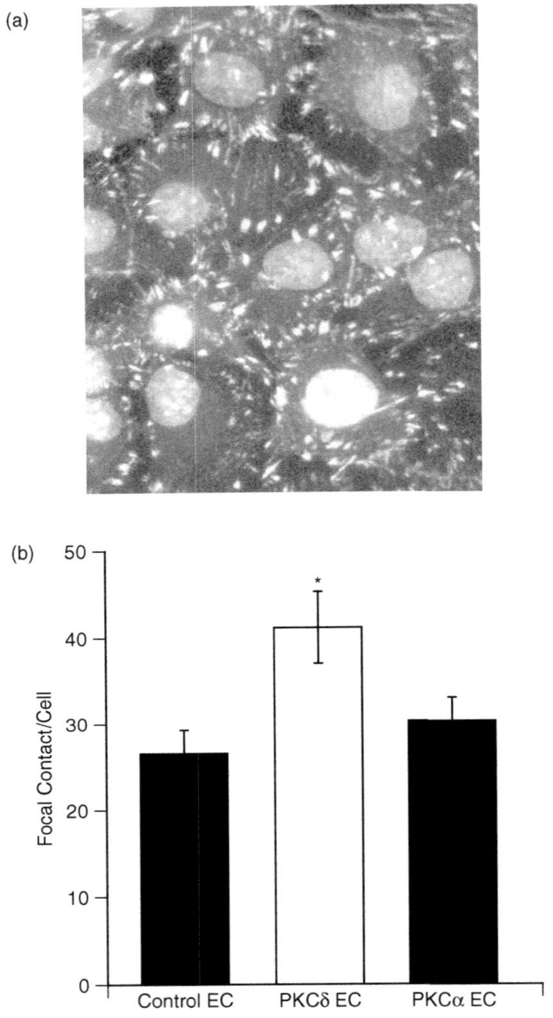

Figure 4.3 (a) PKCδ overexpression promotes focal adhesion formation in microvascular ECs. Microvascular ECs derived from rat epididymis stably overexpressing an eukaryotic vector encoding PKCα cDNA (PKCα EC), PKCδ cDNA (PKCδ EC), or empty (Control EC) were fixed and immunofluorescently stained for a focal adhesion-associated protein, vinculin. The nuclei of the cells were counterstained with 4',6-diamidino-2-phenylindole. Images were obtained via immunofluorescence microscopy and a representative image of PKCδ cDNA is shown in (b). The focal adhesions were quantitated and are expressed as the mean ± standard error of the number of focal adhesions per EC. $n = 3$; $*p < 0.01$. PKCδ may play an important role in focal adhesion formation and/or stabilization in ECs. Images were modified from the originally published images from [139].

example, microvessel-derived ECs displayed increased levels of α1 and α2 type IV collagen, laminin, LIMK ("lin-1, isl-1, and mec-3 kinase"), myosin light chain kinase, Vav, and myosin – proteins shown to be important in barrier function regulation. Whereas α1 and α2 type V collagen and fibronectin were found more highly expressed in ECs isolated from large, thicker walled vessels [166]. Recent *in vitro* and *ex vivo* studies have revealed that the ECs isolated from the extra-alveolar vessels or lung capillary vessels differentially respond to edematogenic agents [167, 168], demonstrating a heterogeneous responsiveness of the endothelial barrier (see Chapter 9). It is possible that the differential response of ECs within the pulmonary vasculature to monolayer permeability inducing agents is due, in part, to differential cell–ECM interactions; however, data is lacking at this time to support this hypothesis.

Not surprisingly, pulmonary edema has also been associated with an increased activity of several MMPs and altered ECM synthesis. Indeed, MMP-2 and MMP-9 were elevated in edema fluids of patients with ARDS [169]. These authors similarly noted increased level of procollagen III in the edema fluid of patients with ARDS – a marker of collagen synthesis [169]. MMP-2 and MMP-9 levels were also increased in the bronchoalveolar lavage fluid of animals using experimental models of sepsis [170, 171]. Similarly, MMP inhibition was shown to attenuate the degree of lung injury and edema in animal models of sepsis, ischemia–reperfusion injury, and ARDS [47–49, 172].

Much work has been done to elucidate signaling pathways important in regulating the endothelial barrier function in the lung at the level of cell–ECM, with ample attention paid to focal adhesions and focal contacts. Indeed, disruption of cell–ECM interactions via disruption of integrin ligation to the ECM has been shown to increase monolayer permeability [39, 173–176]. In response to edematogenic agents, pulmonary ECs display enhanced stress fiber formation and a reorganization of focal adhesions to clusters at the ends of stress fibers [177, 178]; EC–ECM restructuring is thought to serve as an anchoring platform for the generation of centripetal forces during cellular retraction and barrier dysfunction [179, 180]. Multiple signaling molecules have been shown to play key roles in regulating the cell–ECM interactions during the disruption and the restoration phases of agonist-induced increases in monolayer permeability, and are summarized in Table 4.2. We have shown that overexpression of the PKCδ isoform increased the number of focal adhesions in the microvascular endothelium – an effect that correlated with enhanced barrier function [139]. Additionally, we have demonstrated that chemical or molecular inhibition of PKCδ caused endothelial barrier dysfunction in lung microvascular endothelial monolayers and in isolated, perfused lungs Figure 4.4 [181, 182]. These changes correlated with decreased stress fiber and focal adhesion formation, and diminished RhoA and FAK activities [182]. Thus, the cell–ECM interactions are critical in the maintenance of a competent barrier, as well as the response to edematous agents within the pulmonary vasculature.

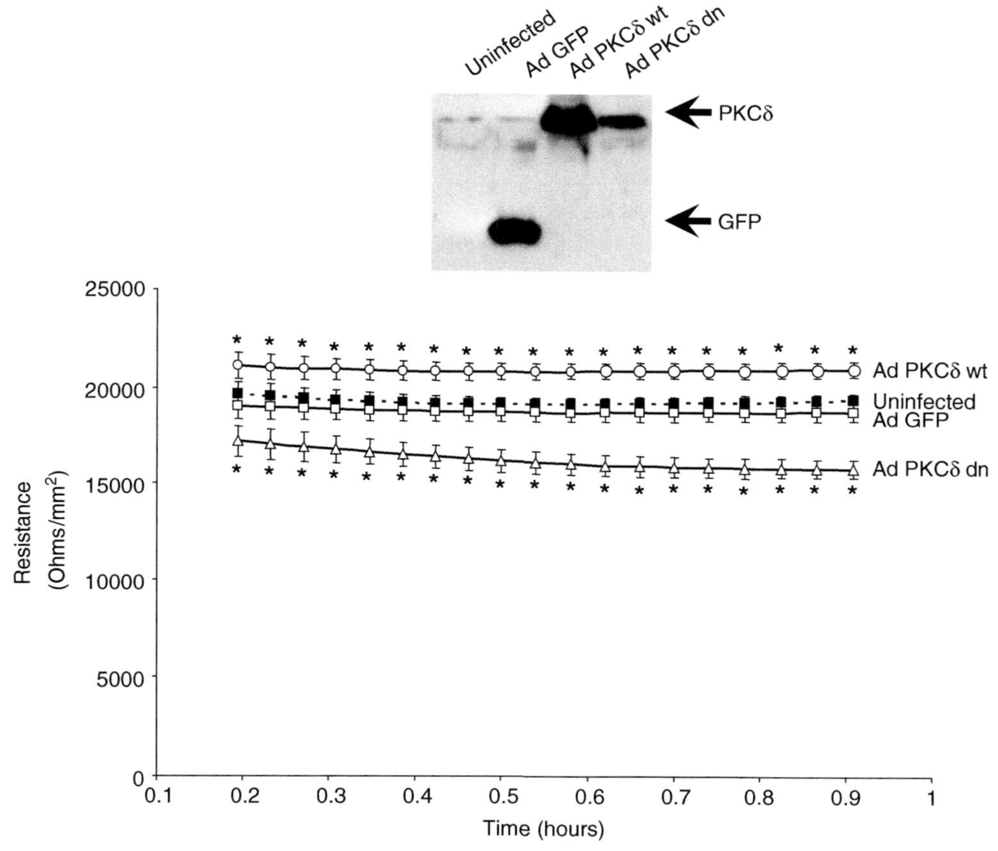

(a)

Figure 4.4 Modulation of PKCδ alters barrier function of vasculature. (a) Monolayer of microvascular ECs derived from rat epididymis were either uninfected or infected with adenoviral particles encoding Green Fluorescent Protein (Ad GFP) or cDNA encoding wild-type PKCδ (Ad PKCδ wt) or dominant-negative PKCδ (Ad PKCδ dn). Protein overexpression was confirmed by immunoblot analysis (inset) and the effect of the overexpressed protein on monolayer permeability was determined by measuring the electrical resistance across the monolayers 24 h postinfection. The mean ± standard error are presented. $n = 16$; $*p < 0.05$ versus infected with adenoviral particles encoding Green Fluorescent Protein or uninfected. Images were modified from the originally published images from [182]. (b) Filtration coefficients (K_f) were obtained from isolated, perfused rat lungs at baseline (solid bars) and following a 45-min exposure to vehicle (dimethylsulfoxide) or the PKCδ chemical inhibitor, 50 μM rottlerin (open bars). $n = 3 - 4$, $*p < 0.05$. (c) Anesthetized rats were injected with a bolus of vehicle (dimethylsulfoxide), 5 μM rottlerin, 250 nM Ro-31-7549, or 10 nM Gö6976 in 1 ml 0.9% NaCl. The rats were then injected with 20 mg/kg Evan's blue dye after 5 min and sacrificed after an additional 45 min. The lungs were harvested and the amount of dye in the lungs was determined spectrophotometrically. Vehicle, $n = 16$; rottlerin, $n = 9$; Ro-31-7549, $n = 11$; Gö6976, $n = 8$. $*p < 0.05$ versus vehicle. Images were modified from the originally published images from [181]. Thus, disruption of PKCδ activity causes pulmonary barrier dysfunction. The data suggests that PKCδ activity is important in maintaining endothelial barrier function in the lung vasculature, possibly for maintenance of EC–ECM interactions.

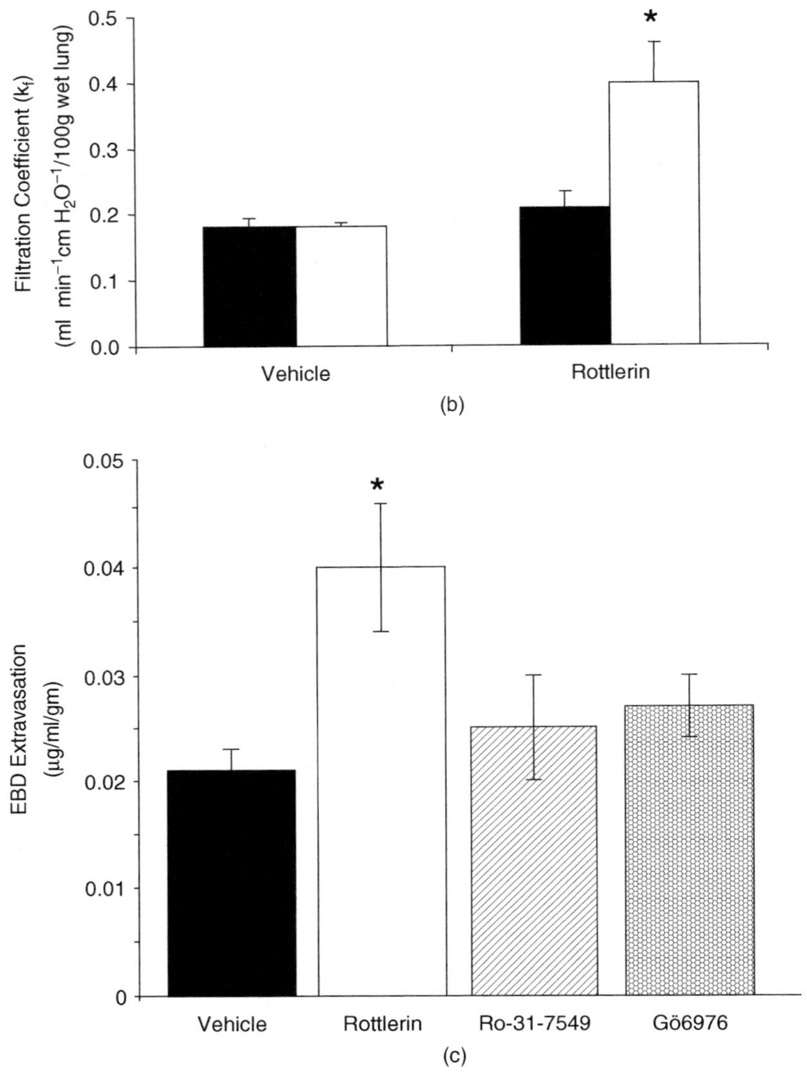

Figure 4.4 *(continued)*

CONCLUSIONS AND PERSPECTIVES

The EC–ECM interactions play many roles in pulmonary vasculature from regulating cell proliferation, migration, and adhesion to angiogenesis and edema formation (Figure 4.1). Not only does the composition of the ECM differ across the vascular tree within the healthy lung, it is modified in settings of pulmonary disease and acute lung injury, and in turn affecting normal EC function and responsiveness to environmental cues. While much work has been accomplished to elucidate molecular mechanisms critical in cell–ECM interactions, much is still unknown. In addition, whether the observations made in *in vitro* culture settings, in which two-dimensional EC–ECM interactions are created, translates into what is occurring in a three-dimensional EC–ECM arrangement of the lung tissue is not yet clear. Future studies should delineate the role of the ECM composition and/or remodeling in the function of the pulmonary endothelium and how disruption of this balance may lead to pathogenesis.

References

1. Eggermont, J., Trouet, D., Carton, I., and Nilius, B. (2001) Cellular function and control of

volume-regulated anion channels. *Cell Biochemistry and Biophysics*, **35**, 263–74.
2. Knolle, P.A. (2006) Cognate interaction between endothelial cells and t cells. *Results and Problems in Cell Differentiation*, **43**, 151–73.
3. Rao, R.M., Yang, L., Garcia-Cardena, G., and Luscinskas, F.W. (2007) Endothelial-dependent mechanisms of leukocyte recruitment to the vascular wall. *Circulation Research*, **101**, 234–47.
4. Berrier, A.L. and Yamada, K.M. (2007) Cell–matrix adhesion. *Journal of Cellular Physiology*, **213**, 565–73.
5. Stupack, D.G. and Cheresh, D.A. (2003) Apoptotic cues from the extracellular matrix: regulators of angiogenesis. *Oncogene*, **22**, 9022–29.
6. Kalluri, R. (2003) Basement membranes: structure, assembly and role in tumour angiogenesis. *Nature Reviews: Cancer*, **3**, 422–33.
7. Rongish, B.J., Hinchman, G., Doty, M.K. *et al.* (1996) Relationship of the extracellular matrix to coronary neovascularization during development. *Journal of Molecular and Cellular Cardiology*, **28**, 2203–15.
8. Majno, G. and Palade, G.E. (1961) Studies on inflammation. 1. The effect of histamine and serotonin on vascular permeability: an electron microscopic study. *Journal of Biophysical and Biochemical Cytology*, **11**, 571–605.
9. Paulsson, M. (1992) Basement membrane proteins: structure, assembly, and cellular interactions. *Critical Reviews in Biochemistry and Molecular Biology*, **27**, 93–127.
10. Vakonakis, I. and Campbell, I.D. (2007) Extracellular matrix: from atomic resolution to ultrastructure. *Current Opinion in Cell Biology*, **19**, 578–83.
11. Lucena, S., Arocha Pinango, C.L., and Guerrero, B. (2007) Fibronectin. Structure and functions associated to hemostasis [Review]. *Investigación Clínica*, **48**, 249–62.
12. Dallas, S.L., Chen, Q., and Sivakumar, P. (2006) Dynamics of assembly and reorganization of extracellular matrix proteins. *Current Topics in Developmental Biology*, **75**, 1–24.
13. Aumailley, M., El Khal, A., Knoss, N., and Tunggal, L. (2003) Laminin 5 processing and its integration into the ECM. *Matrix Biology*, **22**, 49–54.
14. Paez, M.C., Gonzalez, M.J., Serrano, N.C. *et al.* (2007) Physiological and pathological implications of laminins: from the gene to the protein. *Autoimmunity*, **40**, 83–94.
15. Hallmann, R., Horn, N., Selg, M. *et al.* (2005) Expression and function of laminins in the embryonic and mature vasculature. *Physiological Review*, **85**, 979–1000.
16. Gupta, V. and Grande-Allen, K.J. (2006) Effects of static and cyclic loading in regulating extracellular matrix synthesis by cardiovascular cells. *Cardiovascular Research*, **72**, 375–83.
17. Ju, H. and Dixon, I.M. (1996) Extracellular matrix and cardiovascular diseases. *Canadian Journal of Cardiology*, **12**, 1259–67.
18. Hohenester, E. and Engel, J. (2002) Domain structure and organization in extracellular matrix proteins. *Matrix Biology*, **21**, 115–28.
19. Dziadek, M. (1995) Role of laminin–nidogen complexes in basement membrane formation during embryonic development. *Experientia*, **51**, 901–13.
20. Kohling, R., Nischt, R., Vasudevan, A. *et al.* (2006) Nidogen and nidogen-associated basement membrane proteins and neuronal plasticity. *Neurodegenerative Diseases*, **3**, 56–61.
21. Sephel, G.C., Kennedy, R., and Kudravi, S. (1996) Expression of capillary basement membrane components during sequential phases of wound angiogenesis. *Matrix Biology*, **15**, 263–79.
22. Schenk, S., Chiquet-Ehrismann, R., and Battegay, E.J. (1999) The fibrinogen globe of tenascin-C promotes basic fibroblast growth factor-induced endothelial cell elongation. *Molecular Biology of the Cell*, **10**, 2933–43.
23. Paik, D.C., Fu, C., Bhattacharya, J., and Tilson, M.D. (2004) Ongoing angiogenesis in blood vessels of the abdominal aortic aneurysm. *Experimental and Molecular Medicine*, **36**, 524–33.
24. Jones, P.L. and Jones, F.S. (2000) Tenascin-C in development and disease: gene regulation and cell function. *Matrix Biology*, **19**, 581–96.
25. Tremble, P.M., Lane, T.F., Sage, E.H., and Werb, Z. (1993) SPARC, a secreted protein associated with morphogenesis and tissue remodeling, induces expression of metalloproteinases in fibroblasts through a novel extracellular matrix-dependent pathway. *Journal of Cell Biology*, **121**, 1433–44.
26. Brekken, R.A. and Sage, E.H. (2001) SPARC, a matricellular protein: at the crossroads of cell–matrix communication. *Matrix Biology*, **19**, 819–27.
27. Yung, S. and Chan, T.M. (2007) Glycosaminoglycans and proteoglycans: overlooked entities? *Peritoneal Dialysis International: Journal of the International Society for Peritoneal Dialysis*, **27** (Suppl 2), S104–9.
28. Pelosi, P., Rocco, P.R., Negrini, D., and Passi, A. (2007) The extracellular matrix of the lung and its role in edema formation. *Anais da Academia Brasileira de Ciências*, **79**, 285–97.

29. Hassell, J., Yamada, Y., and Arikawa-Hirasawa, E. (2002) Role of perlecan in skeletal development and diseases. *Glycoconjugate Journal*, **19**, 263–67.
30. Albelda, S.M., Daise, M., Levine, E.M., and Buck, C.A. (1989) Identification and characterization of cell-substratum adhesion receptors on cultured human endothelial cells. *The Journal of Clinical Investigation*, **83**, 1992–2002.
31. Languino, L.R., Gehlsen, K.R., Wayner, E. *et al.* (1989) Endothelial cells use $\alpha_2\beta_1$ integrin as a laminin receptor. *Journal of Cell Biology*, **109**, 2455–62.
32. Timpl, R. and Brown, J.C. (1996) Supramolecular assembly of basement membranes. *BioEssays*, **18**, 123–32.
33. Adamson, I.Y., King, G.M., and Bowden, D.H. (1988) Collagen breakdown during acute lung injury. *Thorax*, **43**, 562–68.
34. Durr, R.A., Dubaybo, B.A., and Thet, L.A. (1987) Repair of chronic hyperoxic lung injury: changes in lung ultrastructure and matrix. *Experimental and Molecular Pathology*, **47**, 219–40.
35. Sawhney, R.S. and Bone, R.C. (1993) Endotoxin alters the expression of extracellular matrix proteins by cultured endothelial cells. *Cellular and Molecular Biology Research*, **39**, 589–99.
36. Maniscalco, W.M., Watkins, R.H., Chess, P.R. *et al.* (1998) Cell-specific expression of fibronectin and eiiia and eiiib splice variants after oxygen injury. *The American Journal of Physiology: Lung Cellular and Molecular Physiology*, **274**, L599–609.
37. Sinkin, R.A., Roberts, M., LoMonaco, M.B. *et al.* (1998) Fibronectin expression in bronchopulmonary dysplasia. *Pediatric and Developmental Pathology*, **1**, 494–502.
38. Rotundo, R.F., Curtis, T.M., Shah, M.D. *et al.* (2002) TNF-α disruption of lung endothelial integrity: reduced integrin mediated adhesion to fibronectin. *The American Journal of Physiology: Lung Cellular and Molecular Physiology*, **282**, L316–29.
39. Curtis, T.M., McKeown-Longo, P.J., Vincent, P.A. *et al.* (1995) Fibronectin attenuates increased endothelial monolayer permeability after RGD peptide, anti-$\alpha_5\beta_1$, or TNF-α exposure. *American Journal of Physiology: Lung Cellular and Molecular Physiology*, **269**, L248–60.
40. Kaarteenaho-Wiik, R., Kinnula, V.L., Herva, R. *et al.* (2002) Tenascin-C is highly expressed in respiratory distress syndrome and bronchopulmonary dysplasia. *Journal of Histochemistry and Cytochemistry*, **50**, 423–31.
41. Ihida-Stansbury, K., McKean, D.M., Gebb, S.A. *et al.* (2004) Paired-related homeobox gene *prx1* is required for pulmonary vascular development. *Circulation Research*, **94**, 1507–14.
42. Hoyt, D.G., Rizzo, M., Gerritsen, M.E. *et al.* (1997) Integrin activation protects pulmonary endothelial cells from the genotoxic effects of bleomycin. *The American Journal of Physiology: Lung Cellular and Molecular Physiology*, **273**, L612–17.
43. Hoyt, D.G., Mannix, R.J., Gerritsen, M.E. *et al.* (1996) Integrins inhibit LPS-induced DNA strand breakage in cultured lung endothelial cells. *The American Journal of Physiology: Lung Cellular and Molecular Physiology*, **270**, L689–94.
44. Demedts, I.K., Brusselle, G.G., Bracke, K.R. *et al.* (2005) Matrix metalloproteinases in asthma and COPD. *Current Opinion in Pharmacology*, **5**, 257–63.
45. Gueders, M.M., Foidart, J.M., Noel, A., and Cataldo, D.D. (2006) Matrix metalloproteinases (MMPs) and tissue inhibitors of MMPs in the respiratory tract: potential implications in asthma and other lung diseases. *European Journal of Pharmacology*, **533**, 133–44.
46. Mott, J.D. and Werb, Z. (2004) Regulation of matrix biology by matrix metalloproteinases. *Current Opinion in Cell Biology*, **16**, 558–64.
47. Steinberg, J., Halter, J., Schiller, H.J. *et al.* (2003) Metalloproteinase inhibition reduces lung injury and improves survival after cecal ligation and puncture in rats. *Journal of Surgical Research*, **111**, 185–95.
48. Soccal, P.M., Gasche, Y., Miniati, D.N. *et al.* (2004) Matrix metalloproteinase inhibition decreases ischemia–reperfusion injury after lung transplantation. *American Journal of Transplantation*, **4**, 41–50.
49. Carney, D.E., McCann, U.G., Schiller, H.J. *et al.* (2001) Metalloproteinase inhibition prevents acute respiratory distress syndrome. *Journal of Surgical Research*, **99**, 245–52.
50. Kim, S., Bakre, M., Yin, H., and Varner, J. (2002) Inhibition of endothelial cell survival and angiogenesis by protein kinase A. *The Journal of Clinical Investigation*, **110**, 933–41.
51. Stupack, D.G., Puente, X.S., Boutsaboualoy, S. *et al.* (2001) Apoptosis of adherent cells by recruitment of caspase-8 to unligated integrins. *Journal of Cell Biology*, **155**, 459–70.
52. Ambesi, A., Klein, R.M., Pumiglia, K.M., and McKeown-Longo, P.J. (2005) Anastellin, a fragment of the first type III repeat of fibronectin, inhibits extracellular signal-regulated kinase and

causes G_1 arrest in human microvessel endothelial cells. *Cancer Research*, **65**, 148–56.

53. Bix, G., Fu, J., Gonzalez, E.M. *et al.* (2004) Endorepellin causes endothelial cell disassembly of actin cytoskeleton and focal adhesions through $\alpha_2\beta_1$ integrin. *Journal of Cell Biology*, **166**, 97–109.
54. Rocco, P.R.M., Negri, E.M., Kurtz, P.M. *et al.* (2001) Lung tissue mechanics and extracellular matrix remodeling in acute lung injury. *American Journal of Respiratory and Critical Care Medicine*, **164**, 1067–71.
55. Rocco, P.R.M., Souza, A.B., Faffe, D.S. *et al.* (2003) Effect of corticosteroid on lung parenchyma remodeling at an early phase of acute lung injury. *American Journal of Respiratory and Critical Care Medicine*, **168**, 677–84.
56. Miserocchi, G., Passi, A., Negrini, D. *et al.* (2001) Pulmonary interstitial pressure and tissue matrix structure in acute hypoxia. *The American Journal of Physiology: Lung Cellular and Molecular Physiology*, **280**, L881–87.
57. Johnson, Z., Proudfoot, A.E., and Handel, T.M. (2001) Interaction of chemokines and glycosaminoglycans: a new twist in the regulation of chemokine function with opportunities for therapeutic intervention. *Cytokine and Growth Factor Reviews*, **16**, 625–36.
58. Wu, M.H. (2005) Endothelial focal adhesions and barrier function. *The Journal of Physiology*, **569**, 359–66.
59. Zaidel-Bar, R., Cohen, M., Addadi, L., and Geiger, B. (2004) Hierarchical assembly of cell–matrix adhesion complexes. *Biochemical Society Transactions*, **32**, 416–20.
60. Romer, L.H., Birukov, K.G., and Garcia, J.G. (2006) Focal adhesions: paradigm for a signaling nexus. *Circulation Research*, **98**, 606–16.
61. Cheng, Y.F. and Kramer, R.H. (1989) Human microvascular endothelial cells express integrin-related complexes that mediate adhesion to the extracellular matrix. *Journal of Cellular Physiology*, **139**, 275–86.
62. Dejana, E., Colella, S., Conforti, G. *et al.* (1988) Fibronectin and vitronectin regulate the organization of their respective Arg–Gly–Asp adhesion receptors in cultured human endothelial cells. *Journal of Cell Biology*, **107**, 1215–23.
63. Mechtersheimer, G., Barth, T., Hartschuh, W. *et al.* (1994) In situ expression of β_1, β_3 and β_4 integrin subunits in non-neoplastic endothelium and vascular tumours. *Virchows Archiv*, **425**, 375–84.
64. Tarone, G., Stefanuto, G., Mascarello, P. *et al.* (1990) Expression of receptors for extracellular matrix proteins in human endothelial cells. *Journal of Lipid Mediators*, **2**, S45–S53.
65. Geiger, B., Bershadsky, A., Pankov, R., and Yamada, K.M. (2001) Transmembrane crosstalk between the extracellular matrix–cytoskeleton crosstalk. *Nature Reviews: Molecular Cell Biology*, **2**, 793–805.
66. Ezratty, E.J., Partridge, M.A., and Gundersen, G.G. (2005) Microtubule-induced focal adhesion disassembly is mediated by dynamin and focal adhesion kinase. *Nature Cell Biology*, **7**, 581–90.
67. Fukami, K., Endo, T., Imamura, M., and Takenawa, T. (1994) A-actinin and vinculin are PIP2-binding proteins involved in signaling by tyrosine kinase. *The Journal of Biological Chemistry*, **269**, 1518–22.
68. Crawford, A.W., Michelsen, J.W., and Beckerle, M.C. (1992) An interaction between zyxin and α-actinin. *Journal of Cell Biology*, **116**, 1381–93.
69. Reinhard, M., Jouvenal, K., Tripier, D., and Walter, U. (1995) Identification, purification, and characterization of a zyxin-related protein that binds the focal adhesion and microfilament protein VASP (vasodilator-stimulated phosphoprotein). *Proceedings of the National Academy of Sciences*, **92**, 7956–60.
70. Turner, C.E., Brown, M.C., Perrotta, J.A. *et al.* (1999) Paxillin LD4 motif binds PAK and PIX through a novel 95-kD ankyrin repeat, ARF-GAP protein: a role in cytoskeletal remodeling. *Journal of Cell Biology*, **145**, 851–63.
71. Salgia, R., Uemura, N., Okuda, K. *et al.* (1995) CRKl links $p210^{bcr/abl}$ with paxillin in chronic myelogenous leukemia cells. *The Journal of Biological Chemistry*, **270**, 29145–50.
72. Lo, S.H. (2004) Tensin. *International Journal of Biochemistry and Cell Biology*, **36**, 31–4.
73. Ohta, Y., Suzuki, N., Nakamura, S. *et al.* (1999) The small GTPase rala targets filamin to induce filopodia. *Proceedings of the National Academy of Sciences*, **96**, 2122–28.
74. Ueda, K., Ohta, Y., and Hosoya, H. (2003) The carboxy-terminal pleckstrin homology domain of rock interacts with filamin-A. *Biochemical and Biophysical Research Communications*, **301**, 886–90.
75. Stahlhut, M. and van Deurs, B. (2000) Identification of filamin as a novel ligand for caveolin-1: evidence for the organization of caveolin-1-associated membrane domains by the actin cytoskeleton. *Molecular Biology of the Cell*, **11**, 325–37.
76. Schlaepfer, D.D. and Mitra, S.K. (2004) Multiple connections link FAK to cell motility and invasion.

Current Opinion in Genetics and Development, **14**, 92–101.

77. Medley, Q.G., Buchbinder, E.G., Tachibana, K. *et al.* (2003) Signaling between focal adhesion kinase and Trio. *The Journal of Biological Chemistry*, **278**, 13265–70.
78. Hildebrand, J.D., Taylor, J.M., and Parsons, J.T. (1996) An SH3 domain-containing GTPase-activating protein for Rho and Cdc42 associates with focal adhesion kinase. *Molecular Cell Biology*, **16**, 3169–78.
79. Zhai, J., Lin, H., Nie, Z. *et al.* (2003) Direct interaction of focal adhesion kinase with $p190^{RhoGEF}$. *The Journal of Biological Chemistry*, **278**, 24865–73.
80. Wu, X., Suetsugu, S., Cooper, L.A. *et al.* (2004) Focal adhesion kinase regulation of N-WASP subcellular localization and function. *The Journal of Biological Chemistry*, **279**, 9565–76.
81. Erickson, J.W. and Cerione, R.A. (2001) Multiple roles for Cdc42 in cell regulation. *Current Opinion in Cell Biology*, **13**, 153–57.
82. Millard, T.H., Sharp, S.J., and Machesky, L.M. (2004) Signalling to actin assembly via the WASP (Wiskott–Aldrich syndrome protein)-family proteins and the Arp2/3 complex. *The Biochemical Journal*, **380**, 1–17.
83. Polte, T.R. and Hanks, S.K. (1995) Interaction between focal adhesion kinase and Crk-associated tyrosine kinase substrate $p130^{cas}$. *Proceedings of the National Academy of Sciences*, **92**, 10678–82.
84. Schaller, M.D., Hildebrand, J.D., Shannon, J.D. *et al.* (1994) Autophosphorylation of the focal adhesion kinase, $pp125^{fak}$, directs SH2-dependent binding of $pp60^{src}$. *Molecular Cell Biology*, **14**, 1680–88.
85. Xing, Z., Chen, H.C., Nowlen, J.K. *et al.* (1994) Direct interaction of v-Src with the focal adhesion kinase mediated by the Src SH2 domain. *Molecular Biology of the Cell*, **5**, 413–21.
86. Garcia, J.G., Verin, A.D., Schaphorst, K. *et al.* (1999) Regulation of endothelial cell myosin light chain kinase by Rho, cortactin, and $p60^{src}$. *The American Journal of Physiology: Lung Cellular and Molecular Physiology*, **276**, L989–98.
87. Shi, S., Garcia, J.G., Roy, S. *et al.* (2000) Involvement of c-Src in diperoxovanadate-induced endothelial cell barrier dysfunction. *The American Journal of Physiology: Lung Cellular and Molecular Physiology*, **279**, L441–51.
88. Tiruppathi, C., Naqvi, T., Sandoval, R. *et al.* (2001) Synergistic effects of tumor necrosis factor-α and thrombin in increasing endothelial permeability. *The American Journal of Physiology: Lung Cellular and Molecular Physiology*, **281**, L958–68.
89. Becker, P.M., Verin, A.D., Booth, M.A. *et al.* (2001) Differential regulation of diverse physiological responses to VEGF in pulmonary endothelial cells. *The American Journal of Physiology: Lung Cellular and Molecular Physiology*, **281**, L1500–11.
90. Weis, S., Cui, J., Barnes, L., and Cheresh, D.A. (2004) Endothelial barrier disruption by VEGF-mediated Src activity potentiates tumor cell extravasation and metastasis. *Journal of Cell Biology*, **167**, 223–29.
91. Hoyt, D.G., Rizzo, M., Gerritsen, M.E. *et al.* (1997) Integrin activation protects pulmonary endothelial cells from the genotoxic effects of bleomycin. *The American Journal of Physiology: Lung Cellular and Molecular Physiology*, **273**, L612–17.
92. Su, G., Hodnett, M., Wu, N. *et al.* (2007) Integrin $\alpha_v\beta_5$ regulates lung vascular permeability and pulmonary endothelial barrier function. *American Journal of Respiratory Cell and Molecular Biology*, **36**, 377–86.
93. Janardhan, K.S., Appleyard, G.D., and Singh, B. (2004) Expression of integrin subunits α_v and β_3 in acute lung inflammation. *Histochemistry and Cell Biology*, **121**, 383–90.
94. Shikata, Y., Birukov, K.G., and Garcia, J.G. (2003) S1P induces FA remodeling in human pulmonary endothelial cells: role of Rac, GIT1, FAK, and paxillin. *Journal of Applied Physiology*, **94**, 1193–203.
95. Lee, Y.H., Kayyali, U.S., Sousa, A.M. *et al.* (2007) Transforming growth factor-β1 effects on endothelial monolayer permeability involve focal adhesion kinase/Src. *American Journal of Respiratory Cell and Molecular Biology*, **37**, 485–93.
96. Nobes, C.D. and Hall, A. (1995) Rho, Rac, and Cdc42 GTPases regulate the assembly of multimolecular focal complexes associated with actin stress fibers, lamellipodia, and filopodia. *Cell*, **81**, 53–62.
97. Zaidel-Bar, R., Ballestrem, C., Kam, Z., and Geiger, B. (2003) Early molecular events in the assembly of matrix adhesions at the leading edge of migrating cells. *Journal of Cell Science*, **116**, 4605–13.
98. Ren, X.D., Kiosses, W.B., Sieg, D.J. *et al.* (2000) Focal adhesion kinase suppresses Rho activity to promote focal adhesion turnover. *Journal of Cell Science*, **113**, 3673–78.

99. Zamir, E., Katz, B.Z., Aota, S. *et al.* (1999) Molecular diversity of cell–matrix adhesions. *Journal of Cell Science*, **112**, 1655–69.
100. Zamir, E., Katz, M., Posen, Y. *et al.* (2000) Dynamics and segregation of cell–matrix adhesions in cultured fibroblasts. *Nature Cell Biology*, **2**, 191–96.
101. Pankov, R., Cukierman, E., Katz, B.Z. *et al.* (2000) Integrin dynamics and matrix assembly: tensin-dependent translocation of $\alpha_5\beta_1$ integrins promotes early fibronectin fibrillogenesis. *Journal of Cell Biology*, **148**, 1075–90.
102. Tarone, G., Cirillo, D., Giancotti, F.G. *et al.* (1985) Rous sarcoma virus-transformed fibroblasts adhere primarily at discrete protrusions of the ventral membrane called podosomes. *Experimental Cell Research*, **159**, 141–57.
103. David-Pfeuty, T. and Singer, S.J. (1980) Altered distributions of the cytoskeletal proteins vinculin and α-actinin in cultured fibroblasts transformed by rous sarcoma virus. *Proceedings of the National Academy of Sciences*, **77**, 6687–91.
104. Gimona, M. and Buccione, R. (2006) Adhesions that mediate invasion. *International Journal of Biochemistry and Cell Biology*, **38**, 1875–92.
105. Spinardi, L. and Marchisio, P.C. (2006) Podosomes as smart regulators of cellular adhesion. *European Journal of Cell Biology*, **85**, 191–94.
106. Buccione, R., Orth, J.D., and McNiven, M.A. (2004) Foot and mouth: podosomes, invadopodia and circular dorsal ruffles. *Nature Reviews: Molecular Cell Biology*, **5**, 647–57.
107. Hidalgo, A. and Frenette, P.S. (2007) Leukocyte podosomes sense their way through the endothelium. *Immunity*, **26**, 753–55.
108. Gimona, M., Buccione, R., Courtneidge, S.A., and Linder, S. (2008) Assembly and biological role of podosomes and invadopodia. *Current Opinion in Cell Biology*, **20**, 235–41.
109. Carman, C.V., Sage, P.T., Sciuto, T.E. *et al.* (2007) Transcellular diapedesis is initiated by invasive podosomes. *Immunity*, **26**, 784–97.
110. Linder, S. and Kopp, P. (2005) Podosomes at a glance. *Journal of Cell Science*, **118**, 2079–82.
111. Adams, J.C. (2002) Molecular organization of cell–matrix contacts: essential multiprotein assemblies in cell and tissue function. *Expert Reviews in Molecular Medicine*, **4**, 1–24.
112. Endo, T. and Manya, H. (2006) Defect in glycosylation that causes muscular dystrophy. *Methods in Enzymology*, **417**, 137–52.
113. Durbeej, M., Henry, M.D., and Campbell, K.P. (1998) Dystroglycan in development and disease. *Current Opinion in Cell Biology*, **10**, 594–601.
114. Del Zoppo, G.J., Milner, R., Mabuchi, T. *et al.* (2006) Vascular matrix adhesion and the blood–brain barrier. *Biochemical Society Transactions*, **34**, 1261–66.
115. Loufrani, L., Dubroca, C., You, D. *et al.* (2004) Absence of dystrophin in mice reduces NO-dependent vascular function and vascular density: total recovery after a treatment with the aminoglycoside gentamicin. *Arteriosclerosis and Thrombosis: A Journal of Vascular Biology*, **24**, 671–76.
116. Ramirez-Sanchez, I., Ceballos-Reyes, G., Rosas-Vargas, H. *et al.* (2007) Expression and function of utrophin associated protein complex in stretched endothelial cells: dissociation and activation of eNOS. *Frontiers in Bioscience: A Journal and Virtual Library*, **12**, 1956–62.
117. Litjens, S.H., de Pereda, J.M., and Sonnenberg, A. (2006) Current insights into the formation and breakdown of hemidesmosomes. *Trends in Cell Biology*, **16**, 376–83.
118. Homan, S.M., Mercurio, A.M., and LaFlamme, S.E. (1998) Endothelial cells assemble two distinct $\alpha_6\beta_4$-containing vimentin-associated structures: roles for ligand binding and the β_4 cytoplasmic tail. *Journal of Cell Science*, **111** (Pt 18), 2717–28.
119. Gonzales, M., Weksler, B., Tsuruta, D. *et al.* (2001) Structure and function of a vimentin-associated matrix adhesion in endothelial cells. *Molecular Biology of the Cell*, **12**, 85–100.
120. Huang, S., Chen, C.S., and Ingber, D.E. (1998) Control of cyclin D1, $p27^{Kip1}$, and cell cycle progression in human capillary endothelial cells by cell shape and cytoskeletal tension. *Molecular Biology of the Cell*, **9**, 3179–93.
121. Mammoto, A., Huang, S., Moore, K. *et al.* (2004) Role of RhoA, mDia, and ROCK in cell shape-dependent control of the Skp2–$p27^{kip1}$ pathway and the G_1/S transition. *The Journal of Biological Chemistry*, **279**, 26323–30.
122. Mettouchi, A., Klein, S., Guo, W. *et al.* (2001) Integrin-specific activation of Rac controls progression through the G_1 phase of the cell cycle. *Molecular Cell*, **8**, 115–27.
123. Mahabeleshwar, G.H., Feng, W., Reddy, K. *et al.* (2007) Mechanisms of integrin vascular endothelial growth factor receptor cross-activation in angiogenesis. *Circulation Research*, **101**, 570–80.
124. Soldi, R., Mitola, S., Strasly, M. *et al.* (1999) Role of alphavbeta3 integrin in the activation of vascular endothelial growth factor receptor-2. *The EMBO Journal*, **18**, 882–92.

125. Meredith, J.E. Jr., Fazeli, B., and Schwartz, M.A. (1993) The extracellular matrix as a cell survival factor. *Molecular Biology of the Cell*, **4**, 953–61.
126. Hoyt, D.G., Mannix, R.J., Rusnak, J.M. *et al.* (1995) Collagen is a survival factor against LPS-induced apoptosis in cultured sheep pulmonary artery endothelial cells. *The American Journal of Physiology: Lung Cellular and Molecular Physiology*, **269**, L171–77.
127. Aoudjit, F. and Vuori, K. (2001) Matrix attachment regulates Fas-induced apoptosis in endothelial cells: a role for c-FLIP and implications for anoikis. *Journal of Cell Biology*, **152**, 633–43.
128. Isik, F.F., Gibran, N.S., Jang, Y.C. *et al.* (1998) Vitronectin decreases microvascular endothelial cell apoptosis. *Journal of Cellular Physiology*, **175**, 149–55.
129. Bieler, G., Hasmim, M., Monnier, Y. *et al.* (2007) Distinctive role of integrin-mediated adhesion in TNF-induced PKB/AKT and NF-κB activation and endothelial cell survival. *Oncogene*, **26**, 5722–32.
130. Freyberg, M.A., Kaiser, D., Graf, R. *et al.* (2001) Proatherogenic flow conditions initiate endothelial apoptosis via thrombospondin-1 and the integrin-associated protein. *Biochemical and Biophysical Research Communications*, **286**, 141–49.
131. Urbich, C., Walter, D.H., Zeiher, A.M., and Dimmeler, S. (2000) Laminar shear stress upregulates integrin expression: role in endothelial cell adhesion and apoptosis. *Circulation Research*, **87**, 683–89.
132. Brooks, P.C., Clark, R.A., and Cheresh, D.A. (1994) Requirement of vascular integrin $\alpha_v\beta_3$ for angiogenesis. *Science*, **264**, 569–71.
133. Bourdoulous, S., Orend, G., MacKenna, D.A. *et al.* (1998) Fibronectin matrix regulates activation of Rho and Cdc42 GTPases and cell cycle progression. *Journal of Cell Biology*, **143**, 267–76.
134. Ilic, D., Almeida, E.A.C., Schlaepfer, D.D. *et al.* (1998) Extracellular matrix survival signals transduced by focal adhesion kinase suppress p53-mediated apoptosis. *Journal of Cell Biology*, **143**, 547–60.
135. Almeida, E.A.C., Ili, D., Han, Q. *et al.* (2000) Matrix survival signaling: from fibronectin via focal adhesion kinase to c-Jun NH_2-terminal kinase. *Journal of Cell Biology*, **149**, 741–54.
136. Harrington, E.O., Smeglin, A., Newton, J. *et al.* (2001) Protein tyrosine phosphatase-dependent proteolysis of focal adhesion complexes in endothelial cell apoptosis. *The American Journal of Physiology: Lung Cellular and Molecular Physiology*, **280**, L342–53.
137. Bellas, R.E., Harrington, E.O., Sheahan, K.L. *et al.* (2002) FAK blunts adenosine-homocysteine-induced endothelial cell apoptosis: requirement for PI 3-kinase. *The American Journal of Physiology: Lung Cellular and Molecular Physiology*, **282**, L1135–42.
138. Harrington, E.O., Loffler, J., Nelson, P.R. *et al.* (1997) Enhancement of migration by protein kinase Cα and inhibition of proliferation and cell cycle progression by protein kinase Cδ in capillary endothelial cells. *The Journal of Biological Chemistry*, **272**, 7390–97.
139. Harrington, E.O., Brunelle, J.L., Shannon, C.J. *et al.* (2003) Role of protein kinase C isoforms in rat epididymal microvascular endothelial barrier function. *American Journal of Respiratory Cell and Molecular Biology*, **28**, 626–36.
140. Ashton, A.W., Watanabe, G., Albanese, C. *et al.* (1999) Protein kinase Cδ inhibition of S-phase transition in capillary endothelial cells involves the cyclin-dependent kinase inhibitor $p27^{kip1}$. *The Journal of Biological Chemistry*, **274**, 20805–11.
141. Wiebe, B.M. and Laursen, H. (1998) Lung morphometry by unbiased methods in emphysema: bronchial and blood vessel volume, alveolar surface area and capillary length. *Acta Pathologica Microbiologica, et Immunologica Scandinavica*, **106**, 651–56.
142. Segura-Valdez, L., Pardo, A., Gaxiola, M. *et al.* (2000) Upregulation of gelatinases A and B, collagenases 1 and 2, and increased parenchymal cell death in COPD. *Chest*, **117**, 684–94.
143. Kasahara, Y., Tuder, R.M., Cool, C.D. *et al.* (2001) Endothelial cell death and decreased expression of vascular endothelial growth factor and vascular endothelial growth factor receptor 2 in emphysema. *American Journal of Respiratory and Critical Care Medicine*, **163**, 737–44.
144. Imai, K., Mercer, B.A., Schulman, L.L. *et al.* (2005) Correlation of lung surface area to apoptosis and proliferation in human emphysema. *European Respiratory Journal*, **25**, 250–58.
145. LeRoy, E.C. (1974) Increased collagen synthesis by scleroderma skin fibroblasts *in vitro*: a possible defect in the regulation or activation of the scleroderma fibroblast. *The Journal of Clinical Investigation*, **54**, 880–89.
146. Schachna, L. and Wigley, F.M. (2002) Targeting mediators of vascular injury in scleroderma. *Current Opinion in Rheumatology*, **14**, 686–93.
147. Carvalho, D., Savage, C.O., Black, C.M., and Pearson, J.D. (1996) IgG antiendothelial cell

autoantibodies from scleroderma patients induce leukocyte adhesion to human vascular endothelial cells *in vitro*. Induction of adhesion molecule expression and involvement of endothelium-derived cytokines. *The Journal of Clinical Investigation*, **97**, 111–19.
148. Nishijima, C., Hayakawa, I., Matsushita, T. *et al.* (2004) Autoantibody against matrix metalloproteinase-3 in patients with systemic sclerosis. *Clinical and Experimental Immunology*, **138**, 357–63.
149. Sato, S., Hayakawa, I., Hasegawa, M. *et al.* (2003) Function blocking autoantibodies against matrix metalloproteinase-1 in patients with systemic sclerosis. *Journal of Investigative Dermatology*, **120**, 542–47.
150. Voelkel, N.F., Douglas, I.S., and Nicolls, M. (2007) Angiogenesis in chronic lung disease. *Chest*, **131**, 874–79.
151. Li, X. and Wilson, J.W. (1997) Increased vascularity of the bronchial mucosa in mild asthma. *American Journal of Respiratory and Critical Care Medicine*, **156**, 229–33.
152. Kuwano, K., Bosken, C.H., Paré, P.D. *et al.* (1993) Small airways dimensions in asthma and in chronic obstructive pulmonary disease. *American Review of Respiratory Disease*, **148**, 1220–25.
153. Salvato, G. (2001) Quantitative and morphological analysis of the vascular bed in bronchial biopsy specimens from asthmatic and non-asthmatic subjects. *Thorax*, **56**, 902–6.
154. Zheng, L., Orsida, B.E., Ward, C. *et al.* (1999) Airway vascular changes in lung allograft recipients. *Journal of Heart and Lung Transplantation*, **18**, 231–38.
155. Le Cras, T.D., Fernandez, L.G., Pastura, P.A., and Laubach, V.E. (2005) Vascular growth and remodeling in compensatory lung growth following right lobectomy. *Journal of Applied Physiology*, **98**, 1140–48.
156. Tuder, R.M., Groves, B., Badesch, D.B., and Voelkel, N.F. (1994) Exuberant endothelial cell growth and elements of inflammation are present in plexiform lesions of pulmonary hypertension. *American Journal of Pathology*, **144**, 275–85.
157. Wagner, E.M., Petrache, I., Schofield, B., and Mitzner, W. (2006) Pulmonary ischemia induces lung remodeling and angiogenesis. *Journal of Applied Physiology*, **100**, 587–93.
158. Karsner, H.T. and Ghoreyeb, A.A. (1913) Studies in infarction: III. The circulation in experimental pulmonary embolism. *Journal of Experimental Medicine*, **18**, 507–11.
159. Fadel, E., Riou, J.Y., Mazmanian, M. *et al.* (1999) Pulmonary thromboendartectomy for chronic thromboembolic obstruction of the pulmonary artery in piglets. *Journal of Thoracic and Cardiovascular Surgery*, **117**, 787–93.
160. Charan, N.B. and Carvalho, P. (1997) Angiogenesis in bronchial circulatory system after unilateral pulmonary artery obstruction. *Journal of Applied Physiology*, **82**, 284–91.
161. Stenmark, K.R., Fasules, J., Hyde, D.M. *et al.* (1987) Severe pulmonary hypertension and arterial adventitial changes in newborn calves at 4,300 m. *Journal of Applied Physiology*, **62**, 821–30.
162. Davie, N.J., Crossno, J.T. Jr., Frid, M.G. *et al.* (2004) Hypoxia-induced pulmonary artery adventitial remodeling and neovascularization: contribution of progenitor cells. *The American Journal of Physiology: Lung Cellular and Molecular Physiology*, **286**, L668–78.
163. Heistad, D.D., Armstrong, M.L., and Amundsen, S. (1986) Blood flow through vasa vasorum in arteries and veins: effects of luminal pO_2. *The American Journal of Physiology: Heart and Circulatory Physiology*, **250**, H434–42.
164. Mura, M., dos Santos, C.C., Stewart, D., and Liu, M. (2004) Vascular endothelial growth factor and related molecules in acute lung injury. *Journal of Applied Physiology*, **97**, 1605–17.
165. Thickett, D.R., Armstrong, L., and Millar, A.B. (2002) A role for vascular endothelial growth factor in acute and resolving lung injury. *American Journal of Respiratory and Critical Care Medicine*, **166**, 1332–37.
166. Chi, J.T., Chang, H.Y., Haraldsen, G. *et al.* (2003) Endothelial cell diversity revealed by global expression profiling. *Proceedings of the National Academy of Sciences*, **100**, 10623–28.
167. Lowe, K., Alvarez, D., King, J., and Stevens, T. (2007) Phenotypic heterogeneity in lung capillary and extra-alveolar endothelial cells. Increased extra-alveolar endothelial permeability is sufficient to decrease compliance. *Journal of Surgical Research*, **143**, 70–77.
168. Parker, J.C., Stevens, T., Randall, J. *et al.* (2006) Hydraulic conductance of pulmonary microvascular and macrovascular endothelial cell monolayers. *The American Journal of Physiology: Lung Cellular and Molecular Physiology*, **291**, L30–37.
169. Pugin, J., Verghese, G., Widmer, M.C., and Matthay, M.A. (1999) The alveolar space is the site of intense inflammatory and profibrotic reactions in the early phase of acute respiratory distress syndrome. *Critical Care Medicine*, **27**, 304–12.

170. Lois, M., Brown, L.A.S., Moss, I.M. *et al.* (1999) Ethanol ingestion increases activation of matrix metalloproteinases in rat lungs during acute endotoxemia. *American Journal of Respiratory and Critical Care Medicine*, **160**, 1354–60.
171. D'Ortho, M.P., Jarreau, P.H., Delacourt, C. *et al.* (1994) Matrix metalloproteinase and elastase activities in LPS-induced acute lung injury in guinea pigs. *The American Journal of Physiology: Lung Cellular and Molecular Physiology*, **266**, L209–16.
172. Steinberg, J., Halter, J., Schiller, H. *et al.* (2005) Chemically modified tetracycline prevents the development of septic shock and acute respiratory distress syndrome in a clinically applicable porcine model. *Shock*, **24**, 348–56.
173. Lampugnani, M.G., Resnati, M., Dejana, E., and Marchisio, P.C. (1991) The role of integrins in the maintenance of endothelial monolayer integrity. *Journal of Cell Biology*, **112**, 479–90.
174. Wu, M.H., Ustinova, E., and Granger, H.J. (2001) Integrin binding to fibronectin and vitronectin maintains the barrier function of isolated porcine coronary venules. *The Journal of Physiology*, **532**, 785–91.
175. Qiao, R.L., Yan, W., Lum, H., and Malik, A.B. (1995) Arg–Gly–Asp peptide increases endothelial hydraulic conductivity: comparison with thrombin response. *The American Journal of Physiology: Cell Physiology*, **269**, C110–17.
176. Tsukada, H., Ying, X., Fu, C., and (1995) Ligation of endothelial $\alpha_v\beta_3$ integrin increases capillary hydraulic conductivity of rat lung. *Circulation Research*, **77**, 651–59.
177. Schaphorst, K.L., Pavalko, F.M., Patterson, C.E., and Garcia, J.G.N. (1997) Thrombin-mediated focal adhesion plaque reorganization in endothelium: role of protein phosphorylation. *American Journal of Respiratory Cell and Molecular Biology*, **17**, 443–55.
178. van Nieuw Amerongen, G.P., Natarajan, K., Yin, G. *et al.* (2004) GIT1 mediates thrombin signaling in endothelial cells: role in turnover of RhoA-type focal adhesions. *Circulation Research*, **94**, 1041–49.
179. Ingber, D.E. (2002) Mechanical signaling and the cellular response to extracellular matrix in angiogenesis and cardiovascular physiology. *Circulation Research*, **91**, 877–87.
180. Dudek, S.M. and Garcia, J.G.N. (2001) Cytoskeletal regulation of pulmonary vascular permeability. *Journal of Applied Physiology*, **91**, 1487–500.
181. Klinger, J.R., Murray, J.D., Casserly, B. *et al.* (2007) Rottlerin causes pulmonary edema *in vivo*: a possible role for PKCδ. *Journal of Applied Physiology*, **103**, 2084–94.
182. Harrington, E.O., Shannon, C.J., Morin, N. *et al.* (2005) PKCδ regulates endothelial basal barrier function through modulation of RhoA GTPase activity. *Experimental Cell Research*, **308**, 407–21.

5

Pulmonary Endothelial Cell Calcium Signaling and Regulation of Lung Vascular Barrier Function

Nebojsa Knezevic, Mohammad Tauseef and Dolly Mehta

Department of Pharmacology, and Center for Lung and Vascular Biology, University of Illinois College of Medicine, Chicago, IL, USA

INTRODUCTION

Calcium ion (Ca^{2+}) is a universal second messenger that triggers endothelial cell (EC) contraction [1], expression of genes encoding inflammatory proteins [2], and activation of adhesion molecules [3], subsequently disrupting endothelial barrier function. Resting ECs that form a stable barrier maintain intracellular Ca^{2+} in nanomolar quantities (40–100 nM) compared to 10 000-fold higher levels in the bloodstream. Extracellular calcium has an important influence on endothelial barrier stability owing to the fact that VE-cadherin complexes located in inter-EC junctions bind homophilically in a calcium-dependent manner [1]. Extracellular Ca^{2+} also influences the physiological function of cell surface proteins such as gap junction hemichannels, which permit intercellular communication between ECs [4]. By contrast, a rise in free intracellular cytosolic Ca^{2+}, $[Ca^{2+}]_i$, is established as a key inducer of signaling events that disrupt endothelial barrier function [1, 2]. However, an increase in the $[Ca^{2+}]_i$ level secondary to activation of ECs with sphingosine 1-phosphate (S1P) elicits pathways that strengthen barrier function [5]. Thus, molecular mechanisms regulating the increase in $[Ca^{2+}]_i$ remain an area of active research for development of therapeutic targets that can prevent disease caused by loss of lung vascular barrier function such as acute lung injury. In this chapter, we will describe endothelial mechanisms of intracellular Ca^{2+} regulation and their impact on lung vascular permeability.

CALCIUM RESPONSE

The earliest signal that appears in ECs following their stimulation with inflammatory mediators or with certain barrier enhancing agonists is an increase in $[Ca^{2+}]_i$ concentration [1, 2, 5]. There is an initial transient peak of cytosolic Ca^{2+} as a result of Ca^{2+} release from the endoplasmic reticulum (ER) store, which is followed by a more sustained phase of Ca^{2+} entry via plasmalemmal ion channels (Figure 5.1). Real-time imaging of capillaries *in situ* demonstrated that this increase in $[Ca^{2+}]_i$ occurs as a result of cytosolic Ca^{2+} oscillations in specialized ECs known as "pacemaker cells" [6], which is propagated to adjacent or neighboring ECs, thereby reinforcing Ca^{2+}-dependent changes in endothelial barrier function.

Calcium Release

Role of Phospholipase C

Phosphoinositide-specific phospholipase C (PLC) serves as a common effector crucial for inducing Ca^{2+} release downstream of EC surface G-protein-coupled receptors (GPCRs), growth factor receptors, and cytokine receptors [1, 2, 5]. In addition, reactive oxygen species (ROS), mechanical stress, and lipid mediators such as arachidonic acid (AA) can elicit an increase in intracellular Ca^{2+} by mechanisms that require PLC activity [2, 7]. PLC generates inositol 1,4,5-triphosphate (IP_3) and diacylglycerol (DAG) from

The Pulmonary Endothelium: Function in health and disease Editors Norbert F. Voelkel, Sharon Rounds

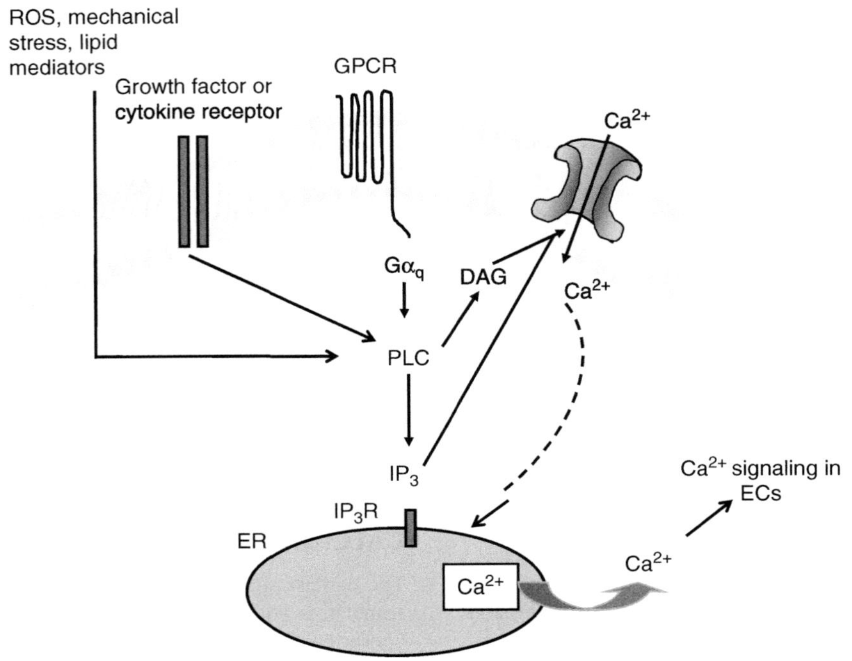

Figure 5.1 Ca^{2+} release mechanisms in ECs. PLC is a common effector for inducing Ca^{2+} release. PLC can be activated downstream of GPCRs, growth factor, or cytokine receptors. In addition, ROS, mechanical stress, and lipid mediators can activate PLC. PLC generates IP_3 and DAG from PIP_2. IP_3 binds to its receptor, IP_3R, on the ER and leads to release of Ca^{2+} into the cytosol. DAG and IP_3 activate plasmalemmal nonselective Ca^{2+} channels that by mediating Ca^{2+} influx increases intracellular Ca^{2+} leading to EC activation.

phosphatidylinositol (4,5)-bisphosphate (PIP_2). IP_3 diffuses to the ER and binds with its tetrameric receptor inositol 1,4,5-triphosphate receptor (IP_3R) leading to release of sequestered Ca^{2+} into the cytosol (Figure 5.1). DAG directly activates specific Ca^{2+} channels, thereby inducing Ca^{2+} influx (see "Ca^{2+} Entry").

Around 10 mammalian PLC isoforms have been described, which include four β and δ isoforms and two γ and ε isoforms [8]. The molecular mass of various PLC isoforms ranges from 85 to 210 kDa. Common domains that are shared by each PLC isoform include a catalytic α/β barrel, a hydrophobic rim, X/Y spanning sequence, pleckstrin homology, EF hand, and C2 [8]. Additional domains that are unique to each PLC isoform are also present. For example, only PLCβ which induces IP_3 and DAG formation downstream of GPCRs linked with $G_{\alpha q}$ as well as $G_{\beta\gamma}$ has a GTPase-activating domain specific for the heterotrimeric G-protein $G_{\alpha q}$ [8]. PLCγ isozymes contain Src homology-2 and -3 (Src homology-2SH2 and Src homology-3SH3) domains that allow PLCγ to bind tyrosine and phosphatidylinositol-3-kinases, and to generate IP_3 and DAG downstream of growth factor and cytokine receptors [8]. Interestingly, PLCγ through its SH3 domain also interacts with transient receptor potential (TRP) channel 3 (TRPC3, see "Role of TRP Channels" for nomenclature) – a receptor-operated Ca^{2+} channel (ROC) [9]. The PLCε isoform contains a Ras-binding domain and two RA domains, and thereby acts as a guanine nucleotide exchange factor (GEF) for Ras GTPase [10]. $G_{\alpha 12}$ and $G_{\alpha 13}$, known to regulate RhoA activity [11] and cadherin localization [12], enhanced the lipase activity of PLCε [13]. These findings indicate that in addition to their catalytic activity multifunctional PLCs may potentiate Ca^{2+} signaling in the endothelium by forming a signalplex with Ca^{2+} channels and Ras GTPases. However, the isoform-specific contribution of PLCs in mediating the increase in intracellular Ca^{2+} levels following activation of ECs with various agonists remains to be delineated.

Role of Intracellular Ca^{2+} Stores

ER and mitochondria are two well-known intracellular calcium stores in ECs. Several Ca^{2+}-binding proteins including calreticulin are known to localize in ER, which can accumulate Ca^{2+} in concentrations approaching

3 mM [14]. The ER is generally organized as a meshwork, but can rearrange into mobile vesicles that move along microtubules by means of kinesin motors [15], thus indicating that ER trafficking to plasmalemma may link Ca^{2+} mobilization with Ca^{2+} entry. In contrast to ER, mitochondria can accumulate only a limited amount of Ca^{2+} (~25% of the total EC Ca^{2+} reserve) due to the toxic effects of Ca^{2+} on mitochondrial membrane potential and energy production [14]. Interestingly, the close proximity between ER and mitochondria, which in fibroblasts is only around 50 nm [16], suggests that coupled mobilization of Ca^{2+} from ER and mitochondrial stores may amplify total cytosolic Ca^{2+} levels during EC stimulation. In fact, mechanical stimulation of lung capillaries was shown to increase the amplitude of mitochondrial Ca^{2+} oscillations following elicitation of IP_3R-mediated Ca^{2+} transients [17].

ER IP_3R, an approximately 270-kDa complex, is responsible for release of stored intracellular Ca^{2+}, which occurs upon interaction of IP_3 with IP_3R [14]. Type I, II, and III isoforms of IP_3R are expressed in ECs [14]. These isoforms are 65% homologous with each other, but significantly differ in their sensitivity to IP_3 as well as intracellular Ca^{2+}. Whereas type II is most sensitive to IP_3, type III is the least sensitive [18]. The type I receptor is regulated by intracellular Ca^{2+} concentration, whereas the type III isoform is Ca^{2+}-independent [18]. Thus, the open probability of the IP_3R channel depends on both $[IP_3]$ and local $[Ca^{2+}]_i$ [14]. Whereas cytosolic Ca^{2+} levels between 100 and 300 nM sensitize IP_3R to IP_3, the receptor becomes desensitized as cytosolic Ca^{2+} levels approach micromolar concentrations. In contrast, when the ER luminal Ca^{2+} concentration exceeds the buffering capacity of the ER, the IP_3R complex becomes more sensitive to agonist stimulation [14].

Structural analysis of the IP_3R indicates that it is organized as a tetramer, contains multiple cavities, dynamically changes shape, and can even function after fragmentation by proteases [15]. The IP_3-binding pocket in IP_3R consists of an N-terminal β-trefoil domain and a C-terminal α-helical domain [15]. IP_3 generation may position IP_3Rs at the ER surface, thus enabling a global change in cytosolic Ca^{2+} [15]. Specific interactions between IP_3R and accessory modulatory proteins such as cytochrome *c* [19], homer [19], and RhoA [20] have been described, and it is possible that these regulatory proteins may cluster with IP_3R at a specific ER site or facilitate IP_3R interaction with TRP channels (see "Role of TRP Channels") for regulating Ca^{2+} increase within the cell. IP_3R also contains consensus phosphorylation sites for protein kinase A, Ca^{2+}/calmodulin protein kinase (CaMK) II, protein kinase C (PKC), and tyrosine kinases, indicating that phosphorylation by these kinases may further modulate IP_3R-mediated Ca^{2+} release [2].

Ryanodine receptors, which respond to a plant alkaloid, ryanodine, are shown to be variably expressed in ECs [21] and also induce Ca^{2+} mobilization [22]. However, the relative contribution of ryanodine receptors in the mechanism of endothelial Ca^{2+} signaling remains to be established.

The refilling of ER Ca^{2+} stores is an active process that repletes stored Ca^{2+} to millimolar levels required for normal ER stress response and for subsequent EC activation [2, 23]. The Ca^{2+}-ATPase (sarco/endoplasmic reticulum Ca^{2+}-ATPaseSERCA) of ER actively sequesters Ca^{2+} against a concentration gradient. However, Malli *et al.* showed that ER Ca^{2+} refilling was apparent even in the presence of IP_3-induced Ca^{2+} release, indicating SERCA and IP_3R activation to be closely linked [24]. Interestingly, in this study, mitochondria also contributed to ER Ca^{2+} refilling as inhibition of mitochondrial Ca^{2+} flux abrogated ER refilling.

Ca^{2+} Entry

Role of TRP Channels

A 10 000-fold Ca^{2+} gradient exists across the cell membrane in naïve ECs. Thus, Ca^{2+} channels opening in response to stimulation of ECs by any of several agonists experience a large driving force favoring Ca^{2+} entry into cytosol. The Ca^{2+} entry activates several EC functions and also refills the ER for further EC activation [25–28]. Although the molecular identity of EC Ca^{2+} channels that mediate Ca^{2+} entry remains to be established, proteins of the *Drosophila* TRP superfamily have been shown to exist in ECs and have been thought to regulate Ca^{2+} entry in these cells. Based on structural homology, the TRP superfamily is categorized into seven groups: TRPC (canonical), TRPV (vanilloid), TRPM (melastatin), TRRP (polycystin), TRPML (mucolipin), TRPA (ankyrin), and TRPN ("no mechanoreceptor potential C") [29, 30]. Around 27 genes specifying various TRP channels are described in human and murine tissues. Each TRP channel contains six putative transmembrane domains with cytosolic N- and C-termini that vary in length depending on the family of TRP channel [29] (Figure 5.2). The pore-forming region is believed to be located between fifth and sixth transmembrane domain. TRPC, TRPV, TRPN, and TRPA channels also contain ankyrin repeats at their N-terminus [29, 30]. TRPM does not contain ankyrin repeats in its N-terminus, but has an enzymatically active domain at C-terminus [30] (Figure 5.2). This property makes these channels unique among TRP family members, but physiological significance of kinase activity in TRPM channels remains to be established. A conserved TRP domain which follows the sixth transmembrane domain is present in the C-terminus of TRPC, TRPM, and TRPN channels

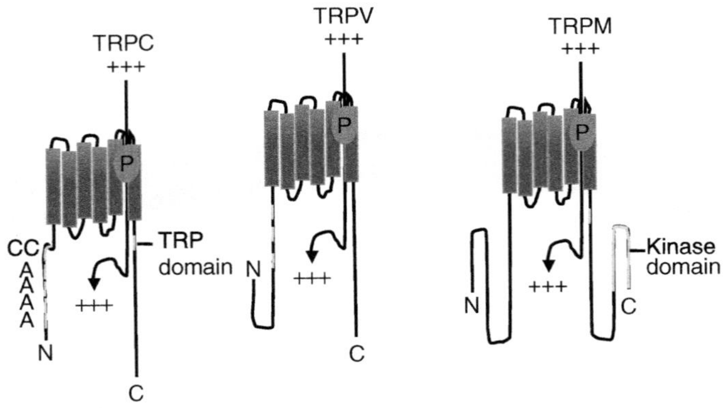

Figure 5.2 Domain representation of three groups of the TRP superfamily. TRP consists of six transmembrane domains. The pore loop (P), which allow the passage of cations (+++), is predicted to be located between fifth and sixth transmembrane domain. The TRPC and TRPV N-terminus contains ankyrin repeats (A,---), a coiled-coil domain (CC), and a TRP domain. TRPM lacks ankyrin repeats, but has a C-terminus protein kinase domain. Adapted from [30].

(Figure 5.2). The TRPC family members have been extensively studied in ECs [25–28]. Progress has been made in investigating the role of TRPM and TRPV in mediating Ca^{2+} entry in ECs. The role of TRPC, TRPV, and TRPM in mediating Ca^{2+} entry in ECs is described in the following subsections (Table 5.1).

TRPC Seven TRPC (1–7) family members are reported thus far. However, human tissues express six of them because TRPC2 is a pseudogene [30]. Considerable heterogeneity has been noted in the expression profile of the various TRPCs among ECs from different species and also within EC types from the same species [25, 26, 28]. For example, the mRNA profile of TRPCs showed TRPC1 and TRPC6 to be expressed at higher levels in human ECs compared to TRPC4, TRPC5, or TRPC7 [2]. In contrast, the mRNA transcript profile of mouse lung microvascular ECs showed TRPC4 to be expressed to a greater extent than TRPC1. Whether these differences in expression levels are reflected at the level of TRPC proteins remains to be established. TRPC requires PLC-induced IP_3 and DAG generation for their activation [25–28] (Figure 5.3). IP_3 activates Ca^{2+} entry by activating store-operated Ca^{2+} channels (SOCs) secondary to depletion of ER stores. Whereas DAG directly activates Ca^{2+} entry by activating ROCs. The subunit stoichiometry of SOCs and ROCs, and whether such a distinction can be apparent following activation of TRPC by physiological agonist remains controversial, but in general the mode of activation of TRPC has been used to classify TRPC channels into two subfamilies. TRPC1, TRPC4, and TRPC5 are presumed to form the SOCs as they are generally activated by depletion of ER stores but not by the released Ca^{2+} *per se* [25–28, 31]. TRPC3, TRPC6, and TRPC7 form ROCs as these channels are activated by DAG independently of store depletion. Interestingly, multiple heterologous combinations of TRPCs (TRPC1 with TRPC4 or TRPC5, and TRPC6 with TRPC3 and TRPC7) may combine to form tetrameric channels with unique properties [32]. A recent study demonstrated that when TRPC6 is expressed at a lower level it heteromerize with TRPC4 and functions as a SOC [33]. In pulmonary ECs TRPC4 was coimmunoprecipitated with TRPC1 [34], at least providing a clue that these channels can form heteromers in an endogenous system [34]. Evidence also indicates that glycosylation, phosphorylation as well as nitrosylation of SOCs such as TRPC1/TRPC5 can promote Ca^{2+} entry through them [35–37]. The interaction of TRPC with cytoskeletal proteins [20, 38, 39], protein components of the endocytic machinery [40–43], adherens junction (AJ) proteins [44], the Na^+/H^+ exchanger regulatory factor NHERF [45], and Na^+/Ca^{2+} exchanger (NCX) [46] has been shown to modulate TRPC activity. Together, these findings suggest that TRPC regulation of endothelial barrier function is likely to be modulated by heteromerization, post-translational modification, and protein–protein interaction.

Single-channel conductance and channel selectivity for Ca^{2+} relative to other permeating cations such as Na^+ can vary substantially among different TRPCs [30]. Although PLC activation should lead to both ROC and SOC activation, it has been difficult to parse the Ca^{2+} entry response into separate SOC and ROC components. Thus, nonphysiological agonists such as thapsigargin (TG), which can directly activate SOCs by passive depletion of ER stores, have been used to assess the function of TRPC1, TRPC4, and TRPC5 [25–28,

Table 5.1 Expression of TRP channels in ECs.

Gene name	Modification of activity	Expression in ECs [28]	Constitute
TRPC subfamily			
TRPC1	store depletion, IP_3, TG, mechanical stress	BAEC[a], HPAEC[ab], BPAEC, MPAEC, RPAEC, HCAEC[b], HCerAEC[b], HDMEC, HMAEC, HUVEC[a], RSSEC	SOC
TRPC3	DAG, store depletion	BAEC, PAEC, HPAEC, MPAEC, RPAEC, HCAEC[b], HCerAEC[b], HDMEC, HMAEC, HUVEC	SOC
TRPC4	store depletion, IP_3, TG	BAEC, MAEC[a], HPAEC[a], BPAEC, MPAEC, HCAEC[b], HCerAEC[b], HDMEC, HUVEC	SOC
TRPC5	store depletion, IP_3, TG, S1P	BAEC, RPAEC, HCAEC[b], HCerAEC[b], HUVEC	SOC
TRPC6	DAG, PIP_3	BAEC, HPAEC, MPAEC, HCAEC[b], HCerAEC[b], HDMEC, HUVEC	ROC
TRPC7	DAG, store depletion	HPAEC, HCAEC, HCerAEC, HDMEC, HUVEC	ROC
TRPV subfamily			
TRPV4	4α-PDD, EETs, mechanical stress, alteration of extracellular temperature	MAEC, HCerAEC, HPAEC	vanilloid
TRPM subfamily			
TRPM2	ADP-ribose, H_2O_2, NAD, AA	HPAEC[a]	melastatin
TRPM5	PIP_2, voltage modulation, heat	HPAEC	melastatin

[a]Demonstrated by immunoblotting;
[b]also demonstrated by immunostaining of cultured cells.
EC cell prefixes: BA, bovine aortic; BPA, bovine pulmonary artery; HCA, human coronary artery; HCerA, human cerebral artery; HDM, human dermal microvascular; HMA, human mesenteric artery; HPA, human pulmonary artery; HUV, human umbilical vein; MA, mice aortic; MPA, mouse pulmonary artery; PA, porcine aortic; RPA, rat pulmonary artery; RSS, rat splenic sinus.

34]. Likewise, 1-oleoyl-2-acetyl-*sn*-glycerol (OAG), a cell-permeable analog of DAG, is often used to assess TRPC3 and TRPC6 activity [25–27]. A great deal of work has been carried out to investigate the role of SOCs in mediating Ca^{2+} entry in ECs. TRPC1, TRPC4, and TRPC5 function is important for mediating Ca^{2+} entry following stimulation of ECs with thrombin, vascular endothelial growth factor (VEGF), or nitric oxide (NO) [2, 20, 47, 48]. Antisense depletion of TRPC1 [49] or inhibition of TRPC1 with a TRPC1-blocking antibody [37] reduced Ca^{2+} entry by 50%. Overexpression of TRPC1 in ECs increased Ca^{2+} entry [50]. Microvessel ECs isolated from TRPC4$^{-/-}$ mice also showed inhibition of Ca^{2+} entry in response to thrombin [51]. Consistent with the expression of ROC channels in ECs, the DAG analog OAG induced Ca^{2+} entry in human microvascular and pulmonary arterial ECs independently of IP_3-dependent store depletion or PKC activity [48, 52]. VEGF also directly activated cation current in cells expressing recombinant TRPC3 and TRPC6 [47]. However, small interfering RNA-induced "knockdown" of TRPC6 prevented OAG-induced Ca^{2+} entry and significantly reduced thrombin-induced Ca^{2+} entry, demonstrating that TRPC6 is the primary mediator of receptor-operated Ca^{2+} entry in ECs [52] (Figure 5.4). Evidence from studies of vascular smooth muscle cells (SMCs) suggests that TRPC6 may also control intracellular Na^+ [53], but whether this occurs in ECs remains unclear.

TRPV The vanilloid subfamily of TRP channels, TRPV, consists of six members (TRPV1–TRPV6) and mediates Ca^{2+} entry in response to osmolar, thermal, and mechanical stress, chemical stimuli (vannilloids,

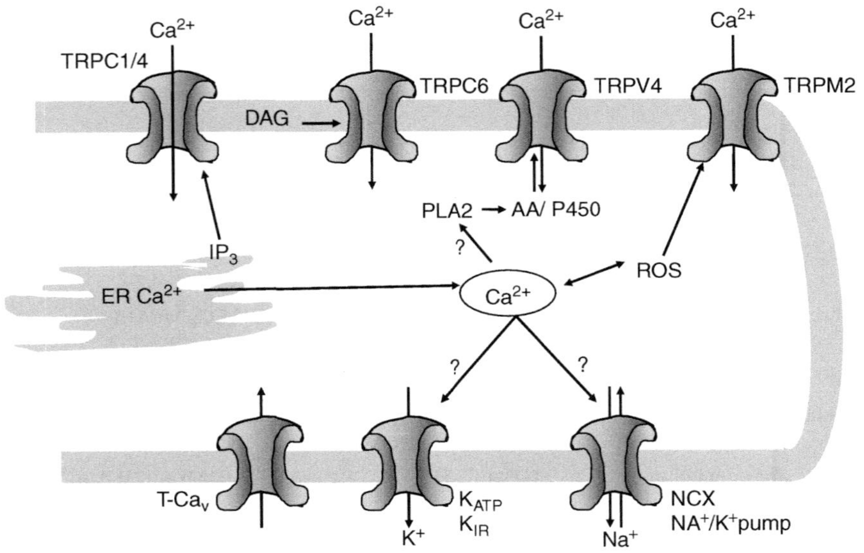

Figure 5.3 Model showing Ca^{2+} entry pathways in an EC. Upon generation, IP_3 induces Ca^{2+} release from ER followed by activation of Ca^{2+} entry through the SOC (TRPC1/4). DAG directly activates the ROC (TRPC6) and increases intracellular Ca^{2+} concentration. A rise in intracellular Ca^{2+} leads to PLA_2 activation, inducing the generation of AA and cytochrome P450, which in turn activates TRPV4 and induces Ca^{2+} entry in ECs. Alternatively, TRPV4 can be activated by osmotic stress or Ca^{2+}-independent PLA_2. Oxidant generation in the cells activates TRPM2, which mediates Ca^{2+} entry. Increased intracellular Ca^{2+} concentration by inducing membrane depolarization activate other membrane channels such as T-Ca_v, K^+ channels (K_{ATP} and K_{ir}), NCX, and the Na^+/K^+ pump to potentiate Ca^{2+} entry.

anandamide, camphor, piperine, allicin, etc.), and even store depletion [30]. Of various TRPVs, TRPV4 function has been investigated in lungs and in ECs [54] (Figure 5.3). The synthetic phorbol ester 4α-phorbol 12,13-didecanoate (4α-PDD) and AA-derived epoxyeicosatrienoic acids (EETs) are endogenous lipid mediators that activate TRPV4 [54, 55]. Cytochrome P450 epoxygenase activity is required for generating EETs from AA. Application of 4α-PDD, mechanical stress, or alteration of extracellular temperature induced large calcium signals in ECs in a TRPV4-sensitive manner [56]. It appears, however, that cell swelling activated TRPV4 by means of the phospholipase A_2 (PLA_2)-dependent formation of AA, whereas 4α-PDD and heat activated TRPV4 by inducing post-translational modification of an aromatic residue at the N-terminus of the third transmembrane domain [57]. Interestingly, ECs isolated from mouse aorta lacking $TRPV4^{-/-}$ showed no Ca^{2+} responses to 4α-PDD, whereas SOC activation remained unaltered, suggesting the involvement of this channel in specifically mediating endothelial Ca^{2+} signaling by AA metabolites or osmotic stress [57]. However, a physical interaction between TRPV4 and IP_3 receptor sensitizes TRPV4 to EET [58], indicating that a functional cross-talk may exist between the SOC and TRPV4, which may amplify TRPV4 activity. Evidence also indicates that Ca^{2+} entry through endothelial TRPV4 channels triggers NO- and endothelium-derived hyperpolarizing factor-dependent vasodilatation [59, 60]. Moreover, TRPV4 appears to be mechanistically important in endothelial mechanosensing of shear stress [59]. Studies also show that activators of TRPV4, such as 4α-PDD and hypotonicity, inhibited aquaporin5 expression in lung epithelial cells [61], raising the possibility that TRPV4 may also be an important regulator of aquaporin1, the predominant isoform in lung ECs [1, 61]. It is thus possible that TRPV4 may be important in maintaining endothelial water permeability and hence water homeostasis.

TRPM The TRPM family is comprised of eight members [30]. These channels possess variable permeability to cations like Ca^{2+} and Mg^{2+}. For example, TRPM4 and TRPM5 are impermeable to Ca^{2+}, while TRPM6 and TRPM7 show high permeability to Ca^{2+} and Mg^{2+} [27, 62].

TRPM2 and TRPM5 are known to occur in lung ECs [30, 63]. Whereas TRPM2 is activated by intracellular ADP-ribose, hydrogen peroxide, AA, and NAD,

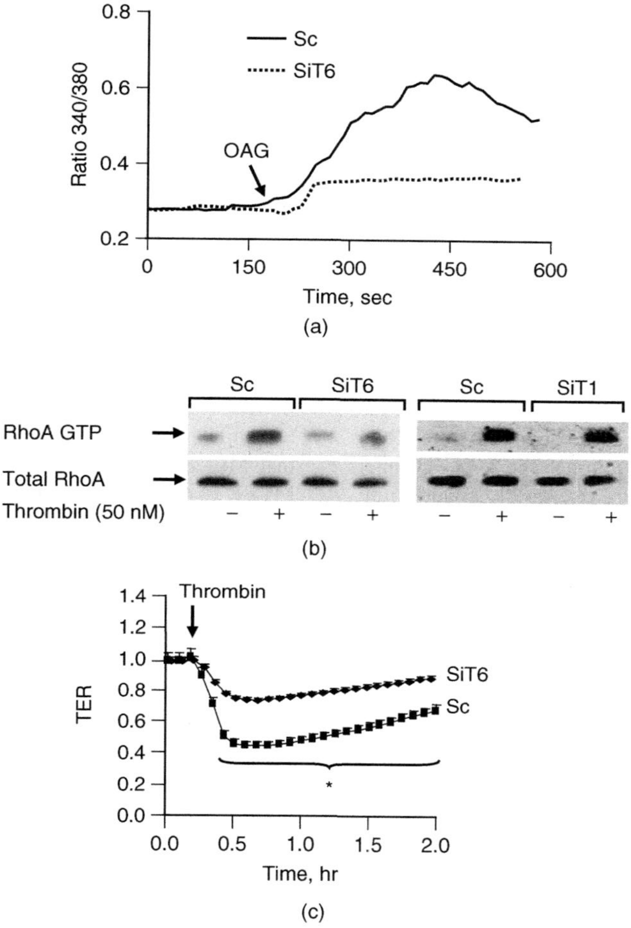

Figure 5.4 TRPC6-dependent ROC entry, RhoA activation, and barrier dysfunction in response to OAG and thrombin, respectively. (a) Intracellular Ca^{2+} increase in response to OAG is predominantly regulated by TRPC6. ECs transfected with TRPC6 siRNA (siT6) or control siRNA (siSc) were stimulated with OAG in the presence of extracellular Ca^{2+}. Inhibition of TRPC6 expression prevented OAG-induced Ca^{2+} entry as compared to cells transfected with control siRNA. (b) RhoA activity in response to a 4-min OAG stimulation in ECs transfected with TRPC6 siRNA (siT6) or control siRNA (siSc). RhoA activation is evident by the increased amount of GTP-bound RhoA compared with total amount of RhoA in whole-cell lysates. (c) Changes in TER in response to thrombin in cells transfected with TRPC6 siRNA (siT6) or control siRNA (Sc). Reproduced from [52] with permission © 2007 The American Society for Biochemistry and Molecular Biology.

TRPM5 responded to voltage modulation, PIP_2, and heat [64]. Hecquet *et al.* recently showed that H_2O_2 in a concentration-dependent manner increased Ca^{2+} entry and cationic currents in ECs [63]. Inhibiting TRPM2 function by either suppressing endogenous expression of TRPM2, or by pretreatment of ECs with TRPM2 blocking antibody, expression of dominant-negative splice variant of TRPM2 or inhibition of ADP-ribose formation inhibited the cationic current and Ca^{2+} entry elicited by H_2O_2, indicating that TRPM2 mediates H_2O_2-induced Ca^{2+} entry [63]. Thus, TRPM2 may be linked with NADPH oxidase to serve as a cellular redox sensor in ECs.

Role of T-Type Calcium Channels

A few studies have reported that the T-type calcium channel, a low-voltage activated Ca^{2+} channel (T-type calcium channel T-Ca_v), is present in pulmonary microvascular ECs [54, 65]. Similar to other Ca_V channels, the α-, β-, γ-, and α2δ-subunits constitute T-Ca_V. However, the α1G-subunit forms the functional T-Ca_V [66]. The α-subunit consists of four homologous membrane-spanning domains, which in turn contain six transmembrane segments (S1–S6). The S4 segment acts as a voltage sensor [67]. Analysis of T-Ca_V-induced Ca^{2+} currents using the whole-cell variation of the patch clamp technique indicated that T-Ca_V in pulmonary microvascular ECs has a low threshold for voltage activation, which is not affected by high concentrations of tetrodotoxin, a potent inhibitor of voltage-gated fast sodium channels. T-Ca_V activity is induced by voltages less negative than −60 mV with maximal current activation occurring at −10 mV [65] (Figure 5.3). The T-Ca_V is rapidly inactivated during depolarization [65]. Zhou *et al.* showed that Ca^{2+} entry through the T-Ca_V channel augmented the inflammatory cascade in endothelium by potentiating the release of von Willebrand factor and membrane surface expression of P-selectin [65]. Although the resting membrane potential of freshly dissociated bovine and cultured human capillary cells is shown to range between −50 and −58 mV [68], it remains unclear whether depolarization to −10 mV occurs in pulmonary microvascular endothelium *in situ*.

Regulation of Ca^{2+} Entry

TRP channel binding and activation mechanisms remain to be established for many permeability-increasing agonists. Interestingly, the mechanism that facilitates SOC activation is progressively becoming clear. It has been shown that, in ECs, the ER and plasma membrane are separated by a distance of around 200–400 nm [34]. These findings prompted investigation of mechanisms that integrate store depletion with TRPC1/4 activation. Although the identification of several mechanisms revealed a complex regulation of SOC activity, considerable attention has been focused on the possible roles of cytoskeletal remodeling and caveolin (Cav) in the mechanism inducing coupling between ER store depletion and SOC activation [1]. For example, alteration of actin remodeling from monomeric G-actin to polymeric F-actin and vice versa has been shown to inhibit SOC activity in ECs [20, 39, 69]. Actin was shown to mediate the interaction between IP_3R and TRPC1 [20].

Wu *et al.* recently demonstrated that perturbing microtubule organization with nocadozole (which depolymerizes microtubules) or taxol (which aligns microtubules) inhibited SOC activity in ECs [38]. These authors further showed that inhibition of kinesin, a retrograde microtubule motor, prevented TG-induced SOC activity in pulmonary arterial ECs. Since kinesin also interacts with actin [1], it is possible that actin and microtubules via kinesin facilitate the coupling between ER and TRPC1/4 to regulate SOC activity.

Small GTPase RhoA and the long isoform of myosin light chain (MLC) kinase (myosin light chain kinase MLCK)-L, a multifunctional 210-kDa Ca^{2+}/calmodulin-dependent enzyme, control actin and microtubule remodeling, and thereby may contribute to the promotion of Ca^{2+} entry. However, few studies have described the importance of these cytoskeletal regulators in altering SOC activity. Coimmunoprecipitation assays showed that RhoA interacted with the IP_3 receptor and TRPC1 [20]. Inhibition of RhoA interfered with IP_3R–TRPC1 interaction and impaired SOC activity induced by thrombin and TG. Pharmacologic inhibitors of MLCK such as wortmannin or ML-9 were also shown to inhibit SOC activity [70]. Interestingly, we found that MLCK-L also interacted with TRPC1 [71]. Using MLCK-L mutants either lacking the N- or C-terminus, we observed that TRPC1 interacted with the N-terminus of MLCK-L. These findings indicate that kinase activity of MLCK-L, which resides in the C-terminus of MLCK, is not required for TRPC1 activation. We further found that the actin binding domain in the N-terminus of MLCK-L promoted SOC activation, since N-MLCK-L mutants lacking an actin-binding site inhibited calcium entry. Since RhoA-mediated SOC activity also required actin remodeling, it is possible that RhoA may utilize MLCK-L as an effector arm to enable SOC activity. Evidence indicates that protein-4.1 by forming a complex with spectrin, an integral component of the membrane cytoskeleton that cross-links actin filaments, and TRPC4 also contribute to regulation of SOC activity [31, 72, 73]. Disruption of the spectrin–protein-4.1 interaction using antibodies against the protein-4.1 binding domain on spectrin, reduced SOC activity by around 50% [74], confirming the critical role of the protein-4.1, actin, and spectrin signaling complex in regulating SOC activity.

Cav-1 coats the flask-shaped plasma membrane invaginations known as caveolae, which account for about 95% of vesicles in ECs [1, 75]. Cav-1 contains a scaffolding domain (Cav-1 scaffold domain CSD), located between residues 82 and 101, through which it binds many signaling molecules such as heterotrimeric G-proteins, Src, PKC, IP_3R, TRPC, and endothelial NO synthase (endothelial nitric oxide synthaseeNOS) [76]. Using a Ca^{2+} sensor, Isshiki *et al.* first showed that shear stress induced SOC activity which preferentially localized to caveolae [40]. Kwaitek *et al.* identified a CSD sequence motif at the C-terminus of TRPC1 [41]. A cell-permeable peptide raised against the CSD sequence markedly reduced ER stored Ca^{2+} release as well as SOC activity in response to thrombin challenge of ECs [41]. CSD interacted with TRPC1 [41]. Arteries as well as microvascular ECs isolated from Cav-1-null mice showed diminished acetylcholine-induced SOC activity [43]. Although TRPC1 and TRPC4 expression were not altered, deletion of Cav-1 grossly impaired TRPC4 localization at the intercellular junctions [43]. Cav-1 deletion did not alter the cytosolic localization of TRPC1 [43]. Interestingly, rescuing Cav-1 expression in ECs restored SOC activity and TRPC4 localization at the interendothelial junctions [43]. Collectively these findings support the view that RhoA, MLCK-L, spectrin, and Cav-1 interact with TRPC1/TRPC4 to facilitate SOC activity. However, whether each of these players acts together or in parallel to regulate endothelial SOC remains to be investigated.

Stromal interacting molecule (STIM) is a recently described single-pass transmembrane "Ca^{2+} sensor" that detects changes of Ca^{2+} content in the ER [77]. STIM1 has been shown to form homomultimers and heteromultimers with itself and TRPC1/4 as well as TRPC6 [33]. These interactions are shown to be mediated by a coiled-coil ezrin, radixin, and moesin (ERM) domain but may involve other domains of STIM1 including a sterile α-motif, a serine- and proline-rich region, and a lysine-rich region [33]. Although the possible contribution of STIM1 in ECs is not yet verified, the possibility exists that RhoA, MLCK-L, spectrin, and Cav may merge at STIM1 to regulate coupling between ER and TRPC1/TRPC4, and thereby SOC activation.

Recent evidence also indicates that actin interacts with TRP channels such as TRPV4 and TRPC6. For example, polymerized actin was shown to directly associate with TRPV4 in living cells and perturbing actin filaments altered TRPV4 activity [78], indicating that actin may act as a mechanosensor affecting the activity of TRPV4 in regulating cell volume. TRPC6 also interacts with actin but the functional significance of this interaction in regulating TRPC6 activity remains unclear.

CALCIUM SIGNALING AND ENDOTHELIAL BARRIER FUNCTION

General Role

A widely accepted view is that the rise in intracellular Ca^{2+} induced by inflammatory mediators such as thrombin, VEGF, and oxidants is an early pivotal signal that induces actin–myosin-based cell contraction leading to increased endothelial permeability [1] (Figures 5.5 and 5.6). However, as mentioned and discussed in the following section, an increase in cytosolic Ca^{2+} induced

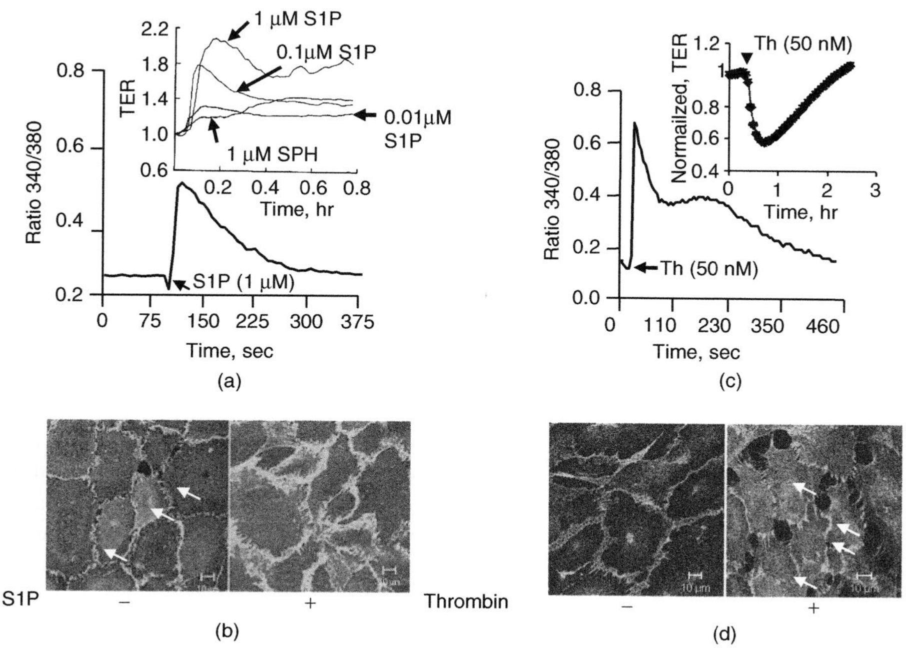

Figure 5.5 Contrasting effects of S1P and thrombin on intracellular Ca^{2+} concentration, AJ organization, and endothelial barrier function. Both S1P (a) and thrombin (c) increase intracellular Ca^{2+}. Whereas S1P-mediated Ca^{2+} increase is accompanied by annealing of AJ (b) and enhancement of TER (a, inset), thrombin-mediated Ca^{2+} increase disrupts AJs (d) and decreases TER (c, inset). Data in (a) and (b) reproduced from [5] with permission © 2005 The American Society for Biochemistry and Molecular Biology. (c) and (d) use unpublished data.

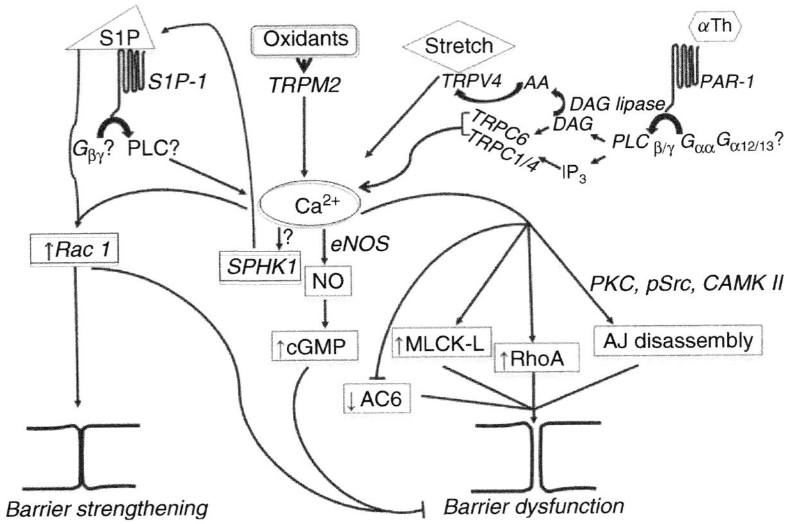

Figure 5.6 Intracellular signaling mechanisms regulating endothelial permeability. Inflammatory mediators such as thrombin by generating IP_3 and DAG lead to activation of TRPC1/4 and TRPC6. DAG catalysis via lipase may lead to AA generation that may stimulate TRPV4. Alternatively, stretch (or osmotic swelling, not shown) can directly activate TRPV4. Oxidants generated post ECs activation stimulates TRPM2. Upon activation, TRPC1/4, TRPC6, TRPV4, and TRPM2 mediate Ca^{2+} entry that stimulates RhoA and MLCK-L activities but inhibits adenylate cyclase 6 (AC6) activity. Together, these events lead to increased actin–myosin-induced EC contraction resulting in disruption of barrier function. A rise in intracellular Ca^{2+} also induces AJ disassembly via PKC, pSrc and CaMK II-mediated phosphorylation of AJ components. Ca^{2+} increase also trigger the activation of SPHK1 and eNOS, causing generation of NO and S1P. NO via cGMP negatively suppress endothelial barrier dysfunction. S1P by binding to its receptor S1P1 activates Gβ → PLC pathway that by activating Rac1 promotes barrier repair.

by the barrier protective agent, S1P has been shown to strengthen barrier function [5] (Figure 5.5). Intriguingly, S1P-mediated strengthening of barrier function required Ca^{2+} release but not Ca^{2+} entry, indicating that the two components of the Ca^{2+} response differentially regulate endothelial permeability [5]. Thus, further investigation into the mechanisms by which increases in cytosolic Ca^{2+} induced by permeability-increasing or permeability-decreasing mediators modify endothelial contraction or relaxation is likely to be useful in delineating the role of calcium responses in regulating endothelial permeability.

Studies have demonstrated that TRPC1/4-, TRPC6-, TRPV4-, and TRPM2-induced rise in cytosolic Ca^{2+} increases endothelial permeability [37, 50, 52, 63, 79] (Figure 5.6). For example, antisense depletion of TRPC1 in ECs [49] or inhibition of TRPC1 with TRPC1-blocking antibody [37] decreased endothelial permeability in response to thrombin [37, 49]. Conversely, overexpression of TRPC1 in ECs potentiated the increase in endothelial permeability due to tumor necrosis factor-α [50, 79]. Lung microvessel ECs lacking TRPC4 also showed attenuated permeability response following thrombin stimulation [51]. Singh *et al.* have demonstrated that the activity of TRPC6, a highly expressed ROC in ECs and lung, is also required for mediating the loss of endothelial barrier function [52]. They showed that suppressing endogenous expression of TRPC6 in ECs markedly attenuated the increase in permeability by thrombin [52] (Figure 5.4). Evidence that TRPM2 regulates oxidant-mediated increase in endothelial permeability was recently demonstrated [63]. Hecquet *et al.* showed that H_2O_2 in a concentration-dependent manner increased endothelial permeability [63]. However, inhibiting TRPM2 function by either suppressing endogenous expression of TRPM2, or using TRPM2 blocking antibody attenuated H_2O_2-induced increases in endothelial permeability, indicating that TRPM2-mediated Ca^{2+} entry contributes to the mechanism of increased endothelial permeability.

The importance of Ca^{2+} signaling in regulating endothelial barrier function *in vivo* has been further demonstrated using real-time optical imaging of lung venular capillaries and mouse models [6]. Kubeler *et al.* elegantly showed *in situ* that increases in the amplitude of Ca^{2+} oscillations and Ca^{2+} influx through gadolinium-sensitive channels were responsible for the augmentation of pulmonary microvascular permeability induced by elevation of lung capillary pressure (P_{pc}) [6] (see also Chapter 20). Following these studies, Ichimura *et al.* reported that increases in P_{pc}-coupled Ca^{2+} release from the ER to increases in amplitude of mitochondrial calcium oscillations [17]. Recent studies showed that TRPV4 activity was in part responsible for mediating the P_{pc}-induced increase in lung microvascular permeability [59], since edema formation in response to elevation of P_{pc} was significantly reduced in mice lacking TRPV4 [59].

Direct activation of SOC by TG also increased the microvessel filtration coefficient ($K_{f,c}$) – a measure of liquid permeability across the pulmonary microvascular barrier [44]. Pocock *et al.* showed that activation of TRPC6 either by OAG or flufenamic acid increased the hydraulic conductivity of individually perfused frog mesenteric microvessels [47]. Our preliminary data also point to the possibility that ROC activation increases endothelial permeability in isolated perfused mouse lungs [80]. We further found that TRPC6 is obligatory for mediating ROC-induced increase in lung microvessel permeability since OAG failed to induce the permeability response in mice lacking TRPC6 [80]. The perfusion of activators of TRPV channels such as 4α-PDD and 5,6- or 14,15-EET in mouse lung also increased $K_{f,c}$ [81]. The permeability response to 4α-PDD was absent in $TRPV4^{-/-}$ mice [81], indicating that TRPV4 in the intact microcirculation predominantly mediates increased endothelial permeability induced by AA products [81].

Collectively the above findings demonstrate that an increase in cytosolic Ca^{2+} following direct activation of SOCs, ROCs, or TRPV4 plays a key role in increasing endothelial permeability in the lung microcirculation. However, the above studies do not provide a complete understanding of the individual contribution by each of these channels to the Ca^{2+} entry regulating microvascular permeability in response to physiological and clinically relevant agonists, which by generating several second messengers, may induce cross-talk between these channels. For example, thrombin, which is a known permeability increasing mediator and is known to be released during vascular injury [82], triggers signaling events that lead to generation of IP_3, DAG, and oxidants [1] as well as PLA_2 activation [83], and by inference should activate TRPC, TRPV, and TRPM in endothelium. Likewise, the increase in endothelial permeability induced by elevation of the lung microvascular pressure may involve stretch-induced activation of signaling events that should lead to the opening of other membrane ionic channels in addition to TRPV4. Clearly, strategies that mimic the increase in lung microvascular permeability in clinical settings, like pulmonary venous hypertension, would be useful in delineating the role of Ca^{2+} entry through these channels in altering lung microvascular permeability. Some progress has been made in this direction. Tiruppathi *et al.* assessed the role of TRPC4 in regulating the thrombin-induced increase in pulmonary microvessel permeability by selectively activating protease-activating receptor (PAR)-1, which predominantly regulates lung microvascular permeability [51]. They demonstrated that the increase in lung microvessel permeability induced by a PAR-1-specific peptide was reduced by 50% in TRPC4-null mice [51], indicating that thrombin-activated Ca^{2+} store depletion and the subsequent Ca^{2+} entry via TRPC4 account for a component of increased endothelial permeability. Intriguingly, our preliminary data indicate that lungs isolated from TRPC6-null mice are completely protected against PAR-1 as well as endotoxin (lipopolysaccharide)-induced lung edema [80]. These findings raise the possibility that TRPC6 serves as the critical influx pathway for Ca^{2+} required for lipopolysaccharide-related increase in endothelial permeability.

Downstream targets of TRP channel activity, which mediate the increase in endothelial permeability, remain an area of active investigation. It is known that EC contraction, disruption of intercellular adhesion, and remodeling of endothelial attachments with the underlying matrix precede gap formation between cells [1]. Actin–myosin-induced stress fiber formation leading to EC contraction may be the predominant pathway regulating the Ca^{2+}-dependent increase in endothelial permeability [84]. Both MLCK-L and RhoA activities are required to induce endothelial contraction [84]. Ca^{2+} binding to calmodulin induces a conformational change in MLCK-L leading to its activation [84] (Figure 5.6). Upon activation, MLCK phosphorylates MLCs that, subsequently, increase interaction with filamentous actin resulting in cytoskeletal rearrangement and stress fiber formation [84]. RhoA, through its downstream effector, Rho kinase, stimulates phosphorylation of the regulatory subunit of MLC phosphatase, PP1, which attenuates the phosphatase activity resulting in an overall increase in MLC phosphorylation [85]. Thus RhoA and MLCK-L may serve as effectors of TRP channels in mediating Ca^{2+}-dependent endothelial contraction. Singh *et al.* showed that Ca^{2+} entry mediated RhoA activation in response to thrombin [52]. Using small interfering RNA (siRNA) that inhibited the endogenous expression of TRPC6 or TRPC1, they further demonstrated that the activity of TRPC6, but not that of TRPC1, was required for thrombin induction of RhoA activity [52] (Figure 5.4). Interestingly, both RhoA and MLCK-L (discussed in "Regulation of Ca^{2+} Entry") regulated TRPC1 activity [20, 71]. These findings raise the possibility that RhoA and

MLCK-L may functionally couple TRPC6 with TRPC1/4 to regulate the endothelial barrier. TRPV4-mediated Ca^{2+} entry was shown to increase lung microvascular permeability in a MLCK-dependent manner. Studies also showed that P_{pc}-induced increases in cytosolic Ca^{2+} enhanced surface expression of P-selectin and release of ROS [17, 86], indicating that increased cytosolic Ca^{2+} is sufficient for inducing proinflammatory activation of endothelium dependent on RhoA and MLCK-L, resulting in loss of endothelial barrier function.

Findings that TRPC4 is colocalized with intercellular junctions [31, 43] raise another possibility – that an increase in cytosolic Ca^{2+} entry may trigger junctional disassembly by directly influencing association between VE-cadherin, β-catenin, or p120-catenin [2]. In support of this notion is the finding that thrombin modulated the phosphorylation state of cadherin, β-catenin, and p120 by either activating kinases such as PKC and Src, or by impairing the association of the protein tyrosine phosphatase SHP2 with AJs [1] (Figure 5.6). Evidence also indicates that increased cytosolic Ca^{2+} may induce loss of barrier function by altering levels of intracellular cAMP – a well-known barrier protective substance [87]. cAMP synthesis in ECs is regulated by adenylate cyclase (AC) and phosphodiesterases [88]; therefore, it is possible that agents may increase endothelial permeability by inhibiting AC activity. The AC_6 isoform, which is expressed in ECs [89], was colocalized with TRPC4 near the plasma membrane [90] and propitiously at sites of cell–cell contact [90]. AC_6 exhibits both high and low affinity Ca^{2+} binding [91]. Permeability-increasing mediators were shown to transiently inhibit AC_6 activity. Overexpression of another AC isoform, AC_8, that reverses the Ca^{2+}-induced inhibition of cAMP production in ECs, was able to prevent the inter-endothelial junction gap formation caused by thrombin [90] (Figure 5.6). Therefore, the overall effect of actin–myosin contraction, phosphorylation of AJs, activation of proinflammatory signals in ECs, and Ca^{2+}-induced inhibition of AC_6 by TRP channels may be to induce cytoskeletal rearrangement leading to intercellular gap formation and increased permeability.

In contrast to thrombin (and many other permeability increasing mediators), S1P-induced increases in Ca^{2+} were shown to be required for endothelial barrier strengthening [5] (Figure 5.5). S1P-induced Ca^{2+} increase activated Rac1 GTPase, AJ assembly, and enhancement of endothelial barrier function by the pathway, $G_i \rightarrow$ PLC $\rightarrow$ IP_3R. Although Mehta *et al.* did not identify the S1P receptor mediating Ca^{2+}-induced barrier strengthening [5], it is possible that this barrier strengthening depends on G_i-receptor coupling to S1P1. Interestingly, inhibition of Ca^{2+} influx by lanthanum, or of TRPC1 by a TRPC1-blocking antibody, failed to prevent S1P-induced Rac1 translocation to junctions and AJ assembly, suggesting that Ca^{2+} release from ER rather than Ca^{2+} influx regulated the observed barrier enhancement. Rac1 activation requires upstream effectors that convert Rac-GDP (inactive state) to Rac-GTP (active state) [92]. Rac1 is held inactive by guanine nucleotide dissociation inhibitor (GDI)-1. Dissociation of GDI-1 from the Rac-GDP–GDI complex is required for GTP exchange by GEFs such as Tiam-1 and Vav2 [93, 94]. Evidence suggests that Ca^{2+} signaling can activate Rac1 by stimulating dissociation of GDI-1 from the Rac–GDI complex [95]. In addition, Tiam-1 activation may occur by a Ca^{2+}-dependent pathway [96], which may lead to Rac1 activation. Sphingosine kinase (SPHK), phosphatase, and lyase regulate S1P levels in plasma and cells [97]. SPHK, by phosphorylating sphingosine, leads to formation of S1P in the cell and maintains the vascular S1P gradient [97]. Our findings suggest that SPHK1 predominantly regulates S1P levels in ECs [98]. Whether a rise in cytosolic Ca^{2+} induces SPHK1 activation which, by generating S1P strengthens barrier function via autocrine/paracrine mechanisms is yet to be elucidated. Evidence also indicates that pressure-induced endothelial Ca^{2+} influx activates cGMP, which by generating NO suppressed endothelial $[Ca^{2+}]_i$ responses, thereby protecting the vascular barrier [6]. Thus, S1P and NO generated as a result of increased intracellular Ca^{2+} may serve to redirect the net effect of Ca^{2+} from barrier disruption to barrier restoration (Figure 5.6).

Role in Segmental Variation of Endothelial Permeability

Phenotypic differences in barrier function of conduit and microvascular ECs are well established (see Chapter 9). Studies in monolayers of cultured ECs from pulmonary microvessels showed that this segment was the most restrictive to albumin [1]. Permeability of $[^{125}I]$albumin was found to be three- to fourfold less in confluent monolayers of pulmonary microvessel ECs than those from mainstem arteries or veins [1]. Transendothelial electrical resistance (TER) was also 10-fold greater in pulmonary microvessel endothelia than in large-vessel endothelia [99]. Although the basis for segmental variations in liquid and albumin permeability is not completely clear, evidence does suggest that Ca^{2+} signaling may in part contribute to this response [100, 101]. Studies showed that SOC activity differs between ECs isolated from rat pulmonary microvessel ECs (rat pulmonary microvessel endothelial cell RPMVECs) and rat pulmonary macrovascular ECs (rat pulmonary macrovascular endothelial cell RPAECs) in several ways: (i) the presence of extracellular Ca^{2+}, TG or ATP produced a smaller increase in cytosolic Ca^{2+} in RPMVECs, which had a shorter duration, and (ii) repletion of extracellular Ca^{2+} evoked Ca^{2+}

entry in both RPMVECs and RPAECs, but SOC activity induced a comparatively greater rise in cytosolic Ca^{2+} in RPMVECs than RPAECs. Surprisingly, TG-induced SOC activity did not increase endothelial permeability in PMVECs to dextran [100]. Likewise, TG increased lung microvascular permeability in rat lungs by inducing intercellular gap formation in the extralaveolar sites [100]. In contrast, Tiruppathi *et al*. demonstrated that TG can induce Ca^{2+} influx in vascular ECs isolated from mouse lungs (mouse lung vascular endothelial cells MLVECs). These authors further showed that MLVECs lacking TRPC4 exhibited significantly less Ca^{2+} influx in response to TG. Thrombin-induced increase in endothelial permeability response was also compromised in TRPC4-deficient cells [51]. Moreover, TRPC4-null mice were partially protected from thrombin-induced increase in lung microvascular permeability. However, in these studies the sites where TRPC4 deletion altered lung vascular leak were not determined. Alvarez *et al.* [81] recently demonstrated that TRPV4-induced increase in lung microvascular permeability occurred at the level of alveolar-capillary septa not venules. Since TRPC6, TRPV4, and TRPM2 are also important in mediating Ca^{2+} entry, future studies are needed to investigate the contribution of these Ca^{2+} channels and their effectors in modulating segmental variation in endothelial permeability.

ENDOTHELIAL HANDLING OF OTHER IONS

Potassium (K^+) Ion

The selective membrane permeability of ECs to K^+ ions is crucial for maintaining the resting membrane potential and for normal function of the Na^+/K^+ pump [68, 102]. Several types of K^+ channels have been shown to be expressed in ECs: inward rectifier (K_{ir}), ATP-sensitive K channels (K_{ATP}), flow-activated potassium currents (K_s), Ca^{2+} activated K^+ channels (K_{Ca}), and transient (A-type) K channels (voltage-sensitive K^+ channels, K_v) [68]. Ca^{2+}-activated K^+ channels are further subgrouped into large conductance (BK_{Ca}), intermediate conductance (IK_{Ca}), and small conductance (SK_{Ca}) Ca^{2+}-activated K^+ channels. Both physiologic stimuli (such as shear stress) and pathophysiologic stimuli (such as hyperosmolarity, vasoactive substances, hypoxia) can activate these channels [68, 103]. Although the role of various K^+ channels in regulating endothelial function in lung microcirculation has not been parsed out, studies point to a crucial role of K_{ATP} channels as a sensor of acute ischemic response in the pulmonary circulation [104]. It was shown that during flow cessation, which mimicked acute ischemia, ECs rapidly depolarized leading to generation of ROS [104]. However, pulmonary microvascular ECs isolated from K_{ATP}-null mice failed to depolarize and generate ROS [104], demonstrating that K_{ATP} channels act as an ischemic sensor in the lung microcirculation [105, 106]. Since K_{ir} predominantly maintains the membrane potential in ECs, it can be speculated that K_{ir} activation may increase the electrochemical driving force for Ca^{2+} via TRP channels [107], subsequently potentiating Ca^{2+}-dependent loss of endothelial barrier function (Figure 5.3). In addition, activation of K_{Ca}-dependent Ca^{2+} entry, as a result of TRP channels or ER-induced increase in intracellular Ca^{2+}, may contribute to the increase in cytosolic calcium [107, 108] (Figure 5.3). Since in guinea-pig endocardial endothelium K_{ir} was shown to be predominantly localized at the luminal surface of endothelium [109], it is also possible that K channels may contribute to the spatial regulation of lung vascular endothelium permeability. ECs and SMCs are coupled electrically via myo-endothelial gap junctions [110]. Thus, it is possible that opening of K channels in endothelium through generation of endothelium-derived relaxing factor (e.g., NO) may modulate the increase in endothelial permeability by activating cross-talk between SMCs and ECs [111].

Sodium (Na^+) Ion

The intracellular Na^+ concentration regulates several cellular functions such as intracellular ion activity via Na^+K^+-ATPase [112], pH via the Na^+/H^+ exchanger, intracellular Ca^{2+} via NCX, and cell volume [53, 113]. However, only a few studies have been performed so far to investigate the role of intracellular Na^+ in regulating endothelial barrier function. The basal intracellular Na^+ in ECs was shown to vary between 9 and 20 mM, which increased following inhibition of Na^+K^+-ATPase with ouabain [112]. Stimulation of bovine pulmonary arterial ECs with H_2O_2 and xanthine/xanthine oxidase increased the Na^+/K^+ pump activity [112]. Since changes in cell shape lead to barrier disruption [1], the possibility exists that increased Na^+/K^+ pump activity, by altering cell volume and Na^+ homeostasis, may be important in maintaining endothelial barrier integrity. TRPC3/6 interacted with NCX and coimmunoprecipitated with Na^+/K^+-ATPase [114, 115]. In SMCs, TRPC6 activation was linked with activation of voltage-dependent Na^+ channels [53, 116], indicating that a functional coupling between NCX and the TRPC6 could drive NCX to prolong Ca^{2+} influx in ECs (Figure 5.3).

CONCLUSIONS AND PERSPECTIVES

Based on the notion that Ca^{2+} is a key second messenger and thus the cellular carrier of information transfer,

Ca^{2+} signaling is the most intensely studied area in basic and clinical research on the endothelium. As discussed in this chapter, a great deal of recent work has been carried out using mouse models in which genes such as TRPC4, TRPV4, or TRPC6 are deleted. These studies have for the first time provided clues concerning the *in vivo* role of the TRP channel-mediated Ca^{2+} entry in mediating increased endothelial permeability. However several questions remain unclear. For example, the isoform specific roles of PLC, endogenous agonists which activate TRP channels, and their effectors that regulate endothelial permeability are still incompletely understood. Another dilemma in the field is to investigate whether differences in spatial increase in Ca^{2+} regulate the enhancement or disruption of barrier function. The role of segmental variations in cytosolic Ca^{2+} accompanying changes in endothelial permeability requires further investigation in order to fully characterize the contribution of specific (or perhaps localized) Ca^{2+} domains in modulating EC permeability. Furthermore, there is still much to learn about the contribution of K^+ and Na^+ in polarizing Ca^{2+} signaling. We hope that the development of practical methods that can combine real-time imaging of cytosolic Ca^{2+} increase in response to relevant physiological agonists such as S1P, thrombin, mechanical stretch, and oxidants *in situ* along with assessment of endothelial barrier function will potentially resolve the Ca^{2+}-dependent regulation of microvessel endothelial permeability in the normal state and during inflammation.

References

1. Mehta, D. and Malik, A.B. (2006) *Physiological Reviews*, **86**, 279–367.
2. Tiruppathi, C., Minshall, R.D., Paria, B.C. *et al.* (2002) *Vascular Pharmacology*, **39**, 173–85.
3. True, A.L., Rahman, A., and Malik, A.B. (2000) *American Journal of Physiology: Lung Cellular and Molecular Physiology*, **279**, L302–11.
4. Thimm, J., Mechler, A., Lin, H. *et al.* (2005) *The Journal of Biological Chemistry*, **280**, 10646–54.
5. Mehta, D., Konstantoulaki, M., Ahmmed, G.U., and Malik, A.B. (2005) *The Journal of Biological Chemistry*, **280**, 17320–28.
6. Kuebler, W.M., Ying, X., and Bhattacharya, J. (2002) *American Journal of Physiology: Lung Cellular and Molecular Physiology*, **282**, L917–23.
7. Usatyuk, P.V., Fomin, V.P., Shi, S. *et al.* (2003) *American Journal of Physiology: Lung Cellular and Molecular Physiology*, **285**, L1006–17.
8. Rebecchi, M.J. and Pentyala, S.N. (2000) *Physiological Reviews*, **80**, 1291–335.
9. van Rossum, D.B., Patterson, R.L., Sharma, S. *et al.* (2005) *Nature*, **434**, 99–104.
10. Wing, M.R., Bourdon, D.M., and Harden, T.K. (2003) *Molecular Interventions*, **3**, 273–80.
11. Kozasa, T., Jiang, X., Hart, M.J. *et al.* (1998) *Science*, **280**, 2109–11.
12. Meigs, T.E., Fedor-Chaiken, M., Kaplan, D.D. *et al.* (2002) *The Journal of Biological Chemistry*, **277**, 24594–600.
13. Hains, M.D., Wing, M.R., Maddileti, S. *et al.* (2006) *Molecular Pharmacology*, **69**, 2068–75.
14. Tran, Q.K., Ohashi, K., and Watanabe, H. (2000) *Cardiovascular Research*, **48**, 13–22.
15. Mikoshiba, K. (2007) *Biochemical Society Symposium*, **74**, 9–22.
16. Wang, H.J., Guay, G., Pogan, L. *et al.* (2000) *The Journal of Cell Biology*, **150**, 1489–98.
17. Ichimura, H., Parthasarathi, K., Quadri, S. *et al.* (2003) *The Journal of Clinical Investigations*, **111**, 691–99.
18. Patel, S., Joseph, S.K., and Thomas, A.P. (1999) *Cell Calcium*, **25**, 247–64.
19. Choe, C.U. and Ehrlich, B.E. (2006) *Science's STKE*, **2006**, re15.
20. Mehta, D., Ahmmed, G.U., Paria, B.C. *et al.* (2003) *The Journal of Biological Chemistry*, **278**, 33492–500.
21. Lesh, R.E., Marks, A.R., Somlyo, A.V. *et al.* (1993) *Circulation Research*, **72**, 481–88.
22. Kohler, R., Brakemeier, S., Kuhn, M. *et al.* (2001) *Cardiovascular Research*, **51**, 160–68.
23. Nilius, B. and Droogmans, G. (2001) *Physiological Reviews*, **81**, 1415–59.
24. Malli, R., Frieden, M., Trenker, M., and Graier, W.F. (2005) *The Journal of Biological Chemistry*, **280**, 12114–22.
25. Tiruppathi, C., Ahmmed, G.U., Vogel, S.M., and Malik, A.B. (2006) *Microcirculation*, **13**, 693–708.
26. Ahmmed, G.U. and Malik, A.B. (2005) *Pflugers Archiv*, **451**, 131–42.
27. Nilius, B., Owsianik, G., Voets, T., and Peters, J.A. (2007) *Physiological Reviews*, **87**, 165–217.
28. Yao, X. and Garland, C.J. (2005) *Circulation Research*, **97**, 853–63.
29. Nilius, B. and Mahieu, F. (2006) *Molecular Cell*, **22**, 297–307.
30. Venkatachalam, K. and Montell, C. (2007) *Annual Review of Biochemistry*, **76**, 387–417.
31. Cioffi, D.L., Wu, S., Alexeyev, M. *et al.* (2005) *Circulation Research*, **97**, 1164–72.
32. Hofmann, T., Schaefer, M., Schultz, G., and Gudermann, T. (2002) *Proceedings of the National Academy of Sciences of the United States of America*, **99**, 7461–66.

33. Yuan, J.P., Zeng, W., Huang, G.N. *et al.* (2007) *Nature Cell Biology*, **9**, 636–45.
34. Cioffi, D.L. and Stevens, T. (2006) *Microcirculation*, **13**, 709–23.
35. Dietrich, A., Mederos y Schnitzler, M., Emmel, J. *et al.* (2003) *The Journal of Biological Chemistry*, **278**, 47842–52.
36. Yoshida, T., Inoue, R., Morii, T. *et al.* (2006) *Nature Chemical Biology*, **2**, 596–607.
37. Ahmmed, G.U., Mehta, D., Vogel, S. *et al.* (2004) *The Journal of Biological Chemistry*, **279**, 20941–49.
38. Wu, S., Chen, H., Alexeyev, M.F. *et al.* (2007) *The Journal of Biological Chemistry*, **282**, 34801–8.
39. Holda, J.R. and Blatter, L.A. (1997) *FEBS Letters*, **403**, 191–96.
40. Isshiki, M., Ying, Y.S., Fujita, T., and Anderson, R.G. (2002) *The Journal of Biological Chemistry*, **277**, 43389–98.
41. Kwiatek, A.M., Minshall, R.D., Cool, D.R. *et al.* (2006) *Molecular Pharmacology*, **70**, 1174–83.
42. Ambudkar, I.S., Brazer, S.C., Liu, X. *et al.* (2004) In *Mammalian TRP Channels as Molecular Targets*, Novartis Foundation Symposium 258, John Wiley & Sons, Ltd, Chichester, pp. 63–70; discussion 70–74, 98–102, 263–66.
43. Murata, T., Lin, M.I., Stan, R.V. *et al.* (2007) *The Journal of Biological Chemistry*, **282**, 16631–43.
44. Wu, S., Cioffi, E.A., Alvarez, D. *et al.* (2005) *Circulation Research*, **96**, 856–63.
45. Tang, Y., Tang, J., Chen, Z. *et al.* (2000) *The Journal of Biological Chemistry*, **275**, 37559–64.
46. Rosker, C., Graziani, A., Lukas, M. *et al.* (2004) *The Journal of Biological Chemistry*, **279**, 13696–704.
47. Pocock, T.M., Foster, R.R., and Bates, D.O. (2004) *American Journal of Physiology: Heart and Circulatory Physiology*, **286**, H1015–26.
48. Cheng, H.W., James, A.F., Foster, R.R. *et al.* (2006) *Arteriosclerosis, Thrombosis, and Vascular Biology*, **26**, 1768–76.
49. Brough, G.H., Wu, S., Cioffi, D. *et al.* (2001) *The FASEB Journal*, **15**, 1727–38.
50. Paria, B.C., Vogel, S.M., Ahmmed, G.U. *et al.* (2004) *American Journal of Physiology: Lung Cellular and Molecular Physiology*, **287**, L1303–13.
51. Tiruppathi, C., Freichel, M., Vogel, S.M. *et al.* (2002) *Circulation Research*, **91**, 70–76.
52. Singh, I., Knezevic, N., Ahmmed, G.U. *et al.* (2007) *The Journal of Biological Chemistry*, **282**, 7833–43.
53. Lemos, V.S., Poburko, D., Liao, C.H. *et al.* (2007) *Biochemical and Biophysical Research Communications*, **352**, 130–34.
54. Townsley, M.I., King, J.A., and Alvarez, D.F. (2006) *Microcirculation*, **13**, 725–39.
55. Watanabe, H., Davis, J.B., Smart, D. *et al.* (2002) *The Journal of Biological Chemistry*, **277**, 13569–77.
56. Wu, L., Gao, X., Brown, R.C. *et al.* (2007) *American Journal of Physiology: Renal Physiology*, **293**, F1699–713.
57. Vriens, J., Owsianik, G., Fisslthaler, B. *et al.* (2005) *Circulation Research*, **97**, 908–15.
58. Fernandes, J., Lorenzo, I.M., Andrade, Y.N. *et al.* (2008) *The Journal of Cell Biology*, **181**, 143–55.
59. Yin, J., Hoffmann, J., Kaestle, S.M. *et al.* (2008) *Circulation Research*, **102**, 966–74.
60. Kohler, R., Heyken, W.T., Heinau, P. *et al.* (2006) *Arteriosclerosis, Thrombosis, and Vascular Biology*, **26**, 1495–502.
61. Sidhaye, V.K., Schweitzer, K.S., Caterina, M.J. *et al.* (2008) *Proceedings of the National Academy of Sciences of the United States of America*, **105**, 3345–50.
62. Ramsey, I.S., Delling, M., and Clapham, D.E. (2006) *Annual Review of Physiology*, **68**, 619–47.
63. Hecquet, C.M., Ahmmed, G.U., Vogel, S.M., and Malik, A.B. (2008) *Circulation Research*, **102**, 347–55.
64. Talavera, K., Yasumatsu, K., Voets, T. *et al.* (2005) *Nature*, **438**, 1022–25.
65. Zhou, C. and Wu, S. (2006) *Microcirculation*, **13**, 645–56.
66. Serrano, J.R., Perez-Reyes, E., and Jones, S.W. (1999) *The Journal of General Physiology*, **114**, 185–201.
67. Talavera, K. and Nilius, B. (2006) *Cell Calcium*, **40**, 97–114.
68. Adams, D.J. and Hill, M.A. (2004) *Journal of Cardiovascular Electrophysiology*, **15**, 598–610.
69. Ribeiro, C.M., Reece, J., and Putney, J.W. Jr. (1997) *The Journal of Biological Chemistry*, **272**, 26555–61.
70. Watanabe, H., Tran, Q.K., Takeuchi, K. *et al.* (2001) *The FASEB Journal*, **15**, 282–84.
71. Kini, V., Knezevic, N., Vogel, S. *et al.* (2008) *American Journal of Respiratory and Critical Care Medicine*, **177**, A846.
72. Bennett, V. and Gilligan, D.M. (1993) *Annual Review of Cell Biology*, **9**, 27–66.
73. Hartwig, J.H. (1995) *Protein Profile*, **2**, 703–800.
74. Wu, S., Sangerman, J., Li, M. *et al.* (2001) *The Journal of Cell Biology*, **154**, 1225–33.
75. Razani, B., Woodman, S.E., and Lisanti, M.P. (2002) *Pharmacological Reviews*, **54**, 431–67.
76. Schlegel, A. and Lisanti, M.P. (2001) *Journal of Cellular Physiology*, **186**, 329–37.

77. Hewavitharana, T., Deng, X., Soboloff, J., and Gill, D.L. (2007) *Cell Calcium*, **42**, 173–82.
78. Suzuki, M., Hirao, A., and Mizuno, A. (2003) *The Journal of Biological Chemistry*, **278**, 51448–53.
79. Paria, B.C., Malik, A.B., Kwiatek, A.M. *et al.* (2003) *The Journal of Biological Chemistry*, **278**, 37195–203.
80. Tauseef, M., Kini, V., Knezevic, N. *et al.* (2008) *The FASEB Journal*, **22**, 1213.1.
81. Alvarez, D.F., King, J.A., Weber, D. *et al.* (2006) *Circulation Research*, **99**, 988–95.
82. Coughlin, S.R. (2000) *Nature*, **407**, 258–64.
83. Cheng, J., Baldassare, J.J., and Raben, D.M. (1999) *The Biochemical Journal*, **337** (Pt 1), 97–104.
84. Dudek, S.M. and Garcia, J.G. (2001) *Journal of Applied Physiology*, **91**, 1487–500.
85. Verin, A.D., Birukova, A., Wang, P. *et al.* (2001) *American Journal of Physiology: Lung Cellular and Molecular Physiology*, **281**, L565–74.
86. Parthasarathi, K., Ichimura, H., Quadri, S. *et al.* (2002) *Journal of Immunology*, **169**, 7078–86.
87. Moore, T.M., Chetham, P.M., Kelly, J.J., and Stevens, T. (1998) *American Journal of Physiology: Lung Cellular and Molecular Physiology*, **275**, L203–22.
88. Stevens, T., Creighton, J., and Thompson, W.J. (1999) *American Journal of Physiology: Lung Cellular and Molecular Physiology*, **277**, L119–26.
89. Stevens, T., Nakahashi, Y., Cornfield, D.N. *et al.* (1995) *Proceedings of the National Academy of Sciences of the United States of America*, **92**, 2696–700.
90. Cioffi, D.L., Moore, T.M., Schaack, J. *et al.* (2002) *The Journal of Cell Biology*, **157**, 1267–78.
91. Mons, N., Guillou, J.L., and Jaffard, R. (1999) *Cellular and Molecular Life Sciences*, **55**, 525–33.
92. Takai, Y., Sasaki, T., and Matozaki, T. (2001) *Physiological Reviews*, **81**, 153–208.
93. Crespo, P., Schuebel, K.E., Ostrom, A.A. *et al.* (1997) *Nature*, **385**, 169–72.
94. Mertens, A.E., Roovers, R.C., and Collard, J.G. (2003) *FEBS Letters*, **546**, 11–16.
95. Price, L.S., Langeslag, M., ten Klooster, J.P. *et al.* (2003) *The Journal of Biological Chemistry*, **278**, 39413–21.
96. Buchanan, F.G., Elliot, C.M., Gibbs, M., and Exton, J.H. (2000) *The Journal of Biological Chemistry*, **275**, 9742–48.
97. Spiegel, S. and Milstien, S. (2003) *Biochemical Society Transactions*, **31**, 1216–19.
98. Tauseef, M., Kini, V., Knezevic, N. *et al.* (2008) *Circulation Research*, **103**, 1164–72.
99. Blum, M.S., Toninelli, E., Anderson, J.M. *et al.* (1997) *American Journal of Physiology: Heart and Circulatory Physiology*, **273**, H286–94.
100. Chetham, P.M., Babal, P., Bridges, J.P. *et al.* (1999) *American Journal of Physiology: Lung Cellular and Molecular Physiology*, **276**, L41–50.
101. Gebb, S. and Stevens, T. (2004) *Microvascular Research*, **68**, 1–12.
102. Faraci, F.M. and Heistad, D.D. (1998) *Physiological Reviews*, **78**, 53–97.
103. Sobey, C.G. (2001) *Arteriosclerosis, Thrombosis, and Vascular Biology*, **21**, 28–38.
104. Milovanova, T., Chatterjee, S., Manevich, Y. *et al.* (2006) *American Journal of Physiology: Cell Physiology*, **290**, C66–76.
105. Al-Mehdi, A.B., Zhao, G., and Fisher, A.B. (1998) *American Journal of Respiratory Cell and Molecular Biology*, **18**, 653–61.
106. Song, C., Al-Mehdi, A.B., and Fisher, A.B. (2001) *American Journal of Physiology: Lung Cellular and Molecular Physiology*, **281**, L993–1000.
107. Nilius, B., Droogmans, G., and Wondergem, R. (2003) *Endothelium*, **10**, 5–15.
108. Brenner, R., Perez, G.J., Bonev, A.D. *et al.* (2000) *Nature*, **407**, 870–76.
109. Manabe, K., Ito, H., Matsuda, H. *et al.* (1995) *The Journal of Physiology*, **484**, 41–52.
110. Dora, K.A., Sandow, S.L., Gallagher, N.T. *et al.* (2003) *Journal of Vascular Research*, **40**, 480–90.
111. Woodman, O.L., Wongsawatkul, O., and Sobey, C.G. (2000) *Clinical and Experimental Pharmacology and Physiology*, **27**, 34–40.
112. Meharg, J.V., McGowan-Jordan, J., Charles, A. *et al.* (1993) *American Journal of Physiology: Lung Cellular and Molecular Physiology*, **265**, L613–21.
113. Schelling, J.R. and Abu Jawdeh, B.G. (2008) *American Journal of Physiology: Renal Physiology*, **295**, F625–32.
114. Goel, M., Sinkins, W., Keightley, A. *et al.* (2005) *Pflugers Archiv*, **451**, 87–98.
115. Goel, M., Zuo, C.D., Sinkins, W.G., and Schilling, W.P. (2007) *American Journal of Physiology: Heart and Circulatory Physiology*, **292**, H874–83.
116. Soboloff, J., Spassova, M., Xu, W. *et al.* (2005) *The Journal of Biological Chemistry*, **280**, 39786–94.

6

Pulmonary Endothelium and Nitric Oxide

Yunchao Su[1] and Edward R. Block[2,3]

[1] Department of Pharmacology and Toxicology, Medical College of Georgia, Augusta, GA, USA
[2] Department of Medicine, University of Florida College of Medicine, Gainesville, FL, USA
[3] Research Service, Malcom Randall Veterans Affairs Medical Center, Gainesville, FL, USA

INTRODUCTION

Nitric oxide (NO) is a biologically active gas that functions as a potent signaling molecule in a number of physiological and pathophysiological processes such as neuronal communication, host defense, the regulation of vascular tone, platelet aggregation, and angiogenesis [1]. Pulmonary endothelium generates NO from L-arginine via the catalytic action of NO synthase (nitric oxide synthase NOS) [2]. There are three isoforms of NOS: neuronal (neuronal nitric oxide synthase nNOS, or NOS-1), inducible (inducible nitric oxide synthase iNOS, or NOS-2), and endothelial (endothelial nitric oxide synthase eNOS, or NOS-3). Although nNOS and iNOS are expressed in pulmonary endothelium [3–5], the contribution of NO produced from these two isoforms of NOS is minimal [6]. The principal NOS isoform in pulmonary endothelium is eNOS [6], which is constitutively expressed, and is regulated by calcium and calmodulin. The synthesis of NO depends on the availability of its substrate, L-arginine, which is delivered to eNOS via an L-arginine transporter located on the plasma membrane [7, 8] or synthesized in endothelium [9]. The cofactors required include NADPH, tetrahydrobiopterin (BH_4), flavin adenine dinucleotide (FAD), flavin mononucleotide (FMN), and oxygen (Figure 6.1). The coproduct L-citrulline is simultaneously produced when NO is synthesized. The half-life of NO is only 2–3 s. From its site of production in the endothelial cell (EC) to its targets in the vascular smooth muscle cell (SMC) and in blood cells, NO's bioavailability is affected by biological scavengers such as hemoglobin (Hb) and reactive oxygen species (ROS).

FUNCTION OF LUNG ENDOTHELIAL NO SIGNALING

NO in the Regulation of Pulmonary Vascular Tone

NO produced from eNOS in pulmonary endothelium is a major endogenous vasodilator that contributes to the low vascular resistance in the pulmonary circulation [10, 11]. Inhibition of NO synthesis by infusion of NOS inhibitors, N^G-monomethyl-L-arginine (L-NMMA) or nitro-L-arginine methyl ester (L-NAME), to human volunteers [10], rabbits [12], and isolated perfused lungs [13] induces significant increases in pulmonary artery pressure and pulmonary vascular resistance, and in hypoxic pulmonary vasoconstriction. eNOS knockout mice lack endothelium-dependent vasodilation, and have elevated pulmonary artery pressure and enhanced hypoxic pulmonary vasoconstriction [6, 14]. Adenovirus-mediated overexpression of eNOS in pulmonary endothelium counteracts the increases in pulmonary artery pressure induced by vasoconstrictive agents and hypoxia [15, 16], and restores normal pulmonary artery pressure in eNOS knockout mice [17]. Inhaled NO selectively dilates pulmonary vessels and reverses hypoxic pulmonary vasoconstriction [18, 19]. These experimental data indicate that NO is continuously released from eNOS in pulmonary endothelium and plays an essential role in modulating pulmonary vascular tone.

NO in Fetal Lung Vascular Development and Lung Angiogenesis

NO is an important mediator in fetal lung vascular development and lung angiogenesis [20] (see Chapter 1). eNOS knockout mice display major abnormalities in pul-

The Pulmonary Endothelium: Function in health and disease Editors Norbert F. Voelkel, Sharon Rounds

monary vascular development and lung alveolarization [21, 22]. The compensatory lung growth after partial pneumonectomy is severely impaired in eNOS knockout mice [23]. Inhaled NO restores lung growth in bronchopulmonary dysplasia caused by eNOS knockout [24], hyperoxia [25], and SU5416 – a vascular endothelial growth factor (VEGF) receptor inhibitor [26]. The importance of NO in fetal lung vascular development and morphogenesis is due to its role as a mediator of angiogenesis involving VEGF [20]. Lung angiogenesis is essential for fetal lung development [27]. VEGF upregulates eNOS gene expression [28], and increases eNOS activity and NO production in pulmonary artery ECs [29]. NO mediates VEGF-induced migration, proliferation, tube formation, and survival of pulmonary ECs.

NO in Leukocyte–and Platelet–Endothelium Interactions in Pulmonary Vessels

eNOS-derived NO is an endogenous homeostatic regulator of leukocyte–and platelet–endothelium interactions in the pulmonary microcirculation (see Chapter 10). Upregulation of eNOS inhibits leukocyte migration following lung ischemia–reperfusion in mice [30]. Inhibition of NO synthesis using L-NMMA and L-NAME induces increases in leukocyte rolling and migration, and its adhesion to endothelium, which can be inhibited by L-arginine, but not D-arginine [31, 32]. Reduced expression of intercellular adhesion molecule-1, vascular cell adhesion molecule-1, and P-selectin is responsible for the inhibitory effect of eNOS-derived NO on leukocyte–endothelium interaction [33]. NO also inhibits platelet aggregation and adhesion to endothelium under both static and flow conditions [34, 35]. The anti-inflammatory and antiplatelet aggregation properties of NO are associated with the beneficial effects of inhaled NO on arterial oxygenation and pulmonary circulation in acute respiratory distress syndrome and acute lung injury [36].

NO in Ventilation/Perfusion Matching and Blood Gas Transport

NO released from eNOS in pulmonary endothelium also diffuses into red blood cells where it *S*-nitrosylates cysteine thiols to form *S*-nitrosothiols (SNOs). SNO in erythrocytes undergoes a transnitrosylation reaction in which a conserved cysteine residue at position 93 of the β-subunit of Hb is *S*-nitrosylated resulting in formation of SNO-Hb [37]. Sufficient amounts of SNO and SNO-Hb are required for optimal hypoxic pulmonary vasoconstriction which maintains ventilation/perfusion (V/Q) matching [38]. SNO-deficient erythrocytes produce impaired vasodilator responses and enhance hypoxic pulmonary vasoconstriction resulting in defects in blood oxygenation [38]. Repletion of SNO improves V/Q matching and oxygenation [38, 39].

The effect of NO on blood gas transport is mediated by the interplay between SNO and SNO-Hb in the processes of oxygen uptake and delivery. In the pulmonary microcirculation, the transnitrosylation reaction for the formation of SNO-Hb is favored in the oxygenated (R, relaxed) state of Hb [40, 41]. Formation of SNO-Hb increases the affinity of Hb to oxygen thus enhancing oxygen uptake in the lung [37, 40]. In the periphery, deoxygenation of the Hb with an R to T (tense) conformational change results in release of SNO to acceptor thiols [42]. This further decreases the affinity of Hb to oxygen, thereby enhancing oxygen delivery. Moreover, SNO released from Hb in the presence of low oxygen saturation causes hypoxic vasodilation that directs microvascular blood flow from well-perfused tissue to hypoxic tissue [43]. The depleted SNO-Hb is replenished in the lung by transnitrosylation due to Hb reoxygenation and by NO released from eNOS in the pulmonary endothelium.

L-ARGININE IN LUNG ENDOTHELIAL NO SIGNALING

Endothelial L-Arginine Availability is Required for NO Production

The rate of NO production in the lung endothelium is critically dependent on the availability of L-arginine [44]. In physiological conditions, the concentration of circulating arginine is approximately 100 μM and the tissue concentration ranges from 100 to 1000 μM. Several studies have shown that the K_m for eNOS is less than 10 μM. eNOS should be saturated in ECs and therefore increasing extracellular arginine should not increase NO production any further. However, a number of *in vitro* and *in vivo* studies indicate that NO production by vascular ECs under physiological conditions can be increased by extracellular arginine despite a saturating intracellular arginine concentration [44, 45]. This observation has been termed the "arginine paradox." Attempting to understand this paradox, the existence of two arginine pools has been demonstrated in ECs [46, 47]. Pool-I can be depleted by extracellular lysine through an exchange mechanism mediated by membrane transporters such as the cationic amino acid transporter (CAT)-1. Pool-II is not freely exchangeable with extracellular lysine, but accessible to eNOS, thereby rendering eNOS independent of extracellular arginine. Pool-II consists of recycled arginine from citrulline (pool-IIA) and protein breakdown (pool-IIB) [47]. Other evidence for the existence of two arginine pools comes from the observation that a caveolar complex between CAT-1 and eNOS exists in pulmonary ECs [8]. This caveolar complex provides a mechanism for the directed delivery of extracellular L-arginine to eNOS in pool-I. The existence of a CAT-1-eNOS complex suggests that the K_m of the arginine transporter may be more important than the K_m of NOS. The K_m of NO production by ECs is approximately 73–150 μM [48], which is in the range of physiological arginine concentrations and the K_m values of the CAT-1 transporters [49].

L-Arginine availability in pulmonary ECs represents a balance between supply and utilization pathways. The sources for intracellular L-arginine include membrane L-arginine transport and endogenous generation (synthesis and protein turnover). The L-arginine utilization pathways include NO synthesis (eNOS), ornithine synthesis (arginase), creatine synthesis (arginine/glycine amidinotransferase), and agmatine synthesis (arginine decarboxylase) [50] (Figure 6.1). Any alterations in the activity of these pathways can affect intracellular L-arginine content and endothelial NO synthesis.

Endothelial L-Arginine Transporters

L-Arginine is transported across the plasma membrane by four different transport systems (y^+, $b^{0,+}$, y^+L, and $B^{0,+}$) [49, 51]. The transport activity of system y^+ is characterized by a high affinity for cationic amino acids, sodium independence, and stimulation of transport by substrate on the opposite (trans) site of the membrane. The other three systems accept a wider range of substrates, including cationic and neutral amino acids. Systems $b^{0,+}$ is Na^+-independent and system $B^{0,+}$ is Na^+-dependent. The y^+ transport system is the main transporter that is responsible for 60–95% of carrier-mediated L-arginine delivery into lung vascular ECs [52]. System y^+ transporter activity is attributed to a family of CATs. The K_m of system y^+ (100–250 μM) lies within the physiological concentration range for circulating L-arginine. Four related CAT proteins have been identified and referred to as CAT-1, CAT-2 (A and B), CAT-3, and CAT-4 [53, 54]. These CAT proteins constitute a subfamily of the solute carrier family 7 (SLC7). The gene names of SLC7A1, A2, A3, and A4 have been assigned to CAT-1, CAT-2, CAT-3, and CAT-4 respectively [53, 54].

CAT-1 is the most extensively studied system y^+ protein. It was originally cloned and identified as the

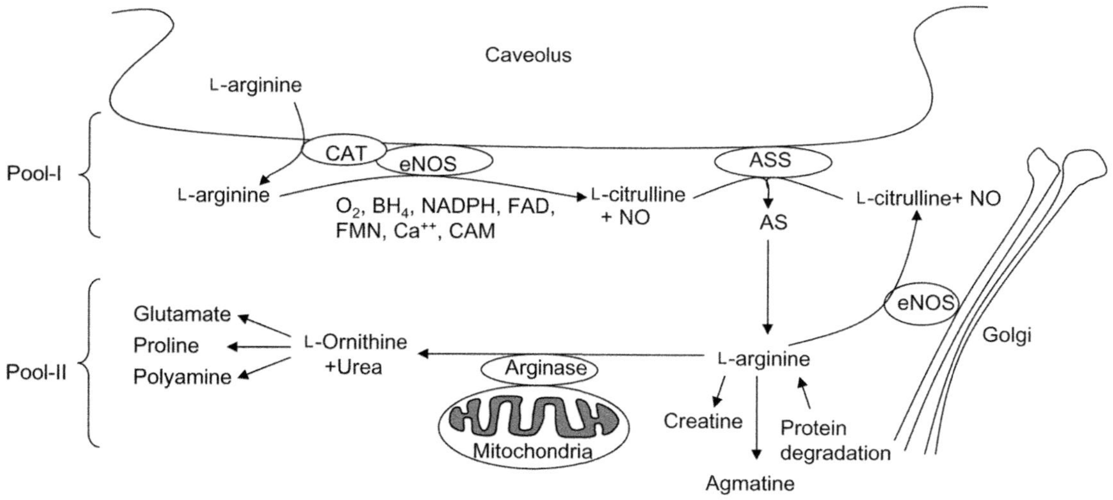

Figure 6.1 Schematic model of the L-arginine/NO pathway in pulmonary ECs. eNOS catalyzes the reaction to produce NO and L-citrulline from L-arginine, which is delivered to eNOS via an L-arginine transporter located on the plasma membrane. The cofactors required include oxygen, BH_4, NADPH, FAD, FMN, calcium, and calmodulin (CAM). eNOS localized in caveolae is associated with CAT-1 which directly delivers extracellular L-arginine to pool-I. L-Arginine pool-II consists of recycled arginine from citrulline and protein breakdown. The L-citrulline recycling pathway is catalyzed by ASS. The L-arginine utilization pathways include NO synthesis (eNOS), ornithine synthesis (arginase), creatine synthesis (arginine/glycine amidinotransferase), and agmatine synthesis (arginine decarboxylase).

receptor for Moloney murine leukemia virus (MMLV). Amino acid similarities observed between the MMLV receptor and L-histidine and L-arginine permeases from *Saccharomyces cerevisiae* led to the discovery of its physiological function as a Na^+-independent CAT. Amino acid starvation is a strong stimulus for the transcription of SLC7A1 gene (CAT-1) and for increased mRNA stability and translation of CAT-1 mRNA. Transcription of CAT-1 mRNA occurs from a TATA-less promoter that extends into the first exon of the CAT-1 mRNA. An amino acid-responsive element located in the first exon is required for the stimulation of transcription. The starvation-induced increase in the stability of CAT-1 mRNA is dependent on the presence of an AU-rich element within a 217-bp fragment in the distal part of the CAT-1 3'-untranslated region (UTR) that binds to the embryonic lethal abnormal vision protein HuR [55]. Translation of CAT-1 mRNA caused by amino acid starvation is initiated from an internal ribosomal entry sequence (IRES) within the 5'-UTR of the CAT-1 mRNA. The IRES requires the phosphorylation of translational initiation factor eIF-2α by either GCN2, double-stranded RNA-dependent protein kinase (PKR), or PKR-like endoplasmic reticulum kinase [53]. Post-translationally, CAT-1 activity is regulated by phosphorylation, trans-stimulation, and membrane potential. Activation of protein kinase C (PKC) induces phosphorylation of the CAT-1 transporter, which leads to inhibition of its transport activity in lung ECs. In contrast, depletion of PKC after long-term treatment with phorbol myristate acetate or thymeleatoxin promotes dephosphorylation of the CAT-1 transporter and activation of its transport activity [56]. In cells with initially low concentrations of intracellular cationic amino acids, a rise in the level of these amino acids (e.g., by protein breakdown) can cause an increase in CAT-1-mediated L-arginine uptake. Hyperpolarization of the plasma membrane increases influx and decreases efflux rates of CAT-1 [57].

CAT-1 activity is required for NO production in lung ECs. CAT-1 and eNOS proteins can be coimmunoprecipitated in the lysates of pulmonary ECs. Immunohistochemical studies demonstrated that CAT-1, eNOS, and caveolin are colocalized in caveolae, suggesting that CAT directly delivers extracellular L-arginine to eNOS [8]. Pertussis toxin-induced activation of CAT-1 in pulmonary ECs increased NO production without affecting eNOS activity [58], indicating that the caveolar CAT-1–eNOS complex is interacting with L-arginine pool-I in ECs.

Endothelial L-Arginine Generation

Pulmonary ECs have an L-arginine biosynthetic pathway that converts L-citrulline to L-arginine [9]. This L-citrulline recycling pathway consists of a two-step enzymatic process [48]. In the first and rate-limiting step, L-citrulline is converted to argininosuccinate by argininosuccinate synthase (ASS) in the presence of L-aspartate and ATP. In the second step, argininosuccinate is converted to arginine by the action of argininosuccinate lyase (ASL). A gradient fractionation study has shown that ASS and ASL colocalize with eNOS in the caveolar fraction [59], suggesting that L-arginine pool-IIA is closely associated with caveolae [47].

L-Glutamine and hypoxia are physiological regulators of the L-citrulline recycling pathway in lung ECs [9]. Glutamine induces inhibition of endothelial arginine synthesis and this appears to occur via competitive inhibition of L-citrulline uptake and a decrease in ASS activity [60, 61]. Hypoxia reduces ASS gene expression [62] and increases intracellular L-glutamine content [9]. These two factors contribute to hypoxia-induced inhibition of arginine synthesis in pulmonary endothelium [9].

The L-citrulline recycling pathway is closely coupled to endothelial NO production. Knockdown of argininosuccinate expression by argininosuccinate small interfering RNA reduced NO production in ECs [63]. Inhibition of ASS activity by L-glutamine and hypoxia also reduces NO production [64]. Likewise, activation of eNOS by bradykinin increases endothelial L-arginine synthesis from L-citrulline [65].

Another important source for intracellular L-arginine generation is protein turnover that contributes L-arginine to pool-IIB in ECs [47]. Even under conditions where the exchangeable arginine is depleted by extracellular L-lysine and the recycling of citrulline to arginine is inhibited by L-glutamine, ECs still retain a NO-producing capacity of 30–50% of control capacity [47]. However, NO production is completely abolished when cells are incubated in medium containing L-lysine, L-glutamine, and the proteasome inhibitor MG132, and this inhibition is completely reversed by addition of extracellular arginine, suggesting that L-arginine in pool-IIB generated from protein turnover contributes to NO synthesis [47].

Arginase in Lung Endothelial NO Signaling

L-Arginine in pulmonary ECs is catabolized by eNOS, arginase, arginine/glycine amidinotransferase, and arginine decarboxylase for the synthesis of NO, ornithine, creatine, and agmatine, respectively. Besides eNOS, arginase is the most active enzyme in lung endothelial NO signaling. The products of arginase are L-ornithine and urea. L-Ornithine is subsequently metabolized into polyamines, proline, and glutamate. There are two isoforms of arginase in lung ECs: cytosolic arginase I and mitochondrial arginase II. Arginase II is the major isoenzyme in ECs [50, 66].

Inflammatory cytokines, ROS, and cyclic strain have been found to upregulate arginase gene expression in lung ECs [66, 67]. Potential redox-sensitive response elements have been identified in the arginase I promoter region [68]. Cyclic strain is a potent inducer of arginase I mRNA expression [66].

Arginase is one of the major regulators of intracellular L-arginine bioavailability for endothelial NO synthesis. The depletion of freely exchangeable L-arginine pools with extracellular L-lysine does not prevent the competitive arginase inhibitor N^{ω}-hydroxy-nor-L-arginine from increasing NO release, suggesting that L-arginine in pool-II is accessible to arginase [69]. Based on the biochemical properties of arginase, it is not surprising that arginase inhibits NO synthesis by competing with eNOS for L-arginine. The V_{max} of arginase for L-arginine at physiological pH (approximately 1400 µmol/min/mg) is more than 1000 times that of eNOS (approximately 1 µmol/min/mg), although the K_m for arginine is in the 2–20 mM range for arginases and is in the 2–20 µM range for eNOS [48]. Indeed, inhibition of arginase has been shown to stimulate NO synthesis in lung ECs [70]. Moreover, overexpression of arginase I or arginase II suppresses NO generation associated with a significant decrease in intracellular L-arginine content in ECs [71]. Similarly, constitutive expression of arginase in microvascular ECs counteracts NO-mediated vasodilation [72]. In addition, arginase activity is higher in serum from pulmonary arterial hypertension (PAH) patients than in controls, and pulmonary artery ECs derived from the lungs of patients with PAH have higher arginase II expression and produce lower NO than control cells *in vitro*, suggesting that arginase subserves a tonic vasoconstrictor function [73].

eNOS IN LUNG ENDOTHELIAL NO SIGNALING

Transcriptional Regulation of eNOS

The eNOS promoter has been cloned from ECs of several species, and there is a high degree of promoter sequence homology [74]. The 5' regulatory region of the eNOS gene contains a "TATA-less" promoter and a variety of cis elements for the putative binding of transcription factors. Specific sites that may influence eNOS transcription found in the eNOS gene include a CCAT box, Sp1, and GATA sites, a sterol regulatory element, activator protein (AP)-1 and -2 elements, a nuclear factor (NF)-1 element, acute-phase reactant regulatory elements, partial estrogen-responsive elements, a cAMP response element, and a putative shear stress response element (SSRE) (Figure 6.2). The human eNOS proximal core promoter has two regulatory regions involved in basal eNOS transcription: positive regulatory domain (PRD I; −104 to −95 relative to transcription initiation), which binds Sp1 and two variants of Sp3, and PRD II (−144 to −115), which binds transcription factors Ets-1, Elf-1, YY1, Sp1, and MYC-associated zinc finger protein.

eNOS promoter activity is controlled by DNA methylation and histone modifications. Methylation of eNOS promoter-reporter regions PRD I and PRD II is associated with a marked impairment of promoter activity. The eNOS promoter is more heavily methylated in non-ECs

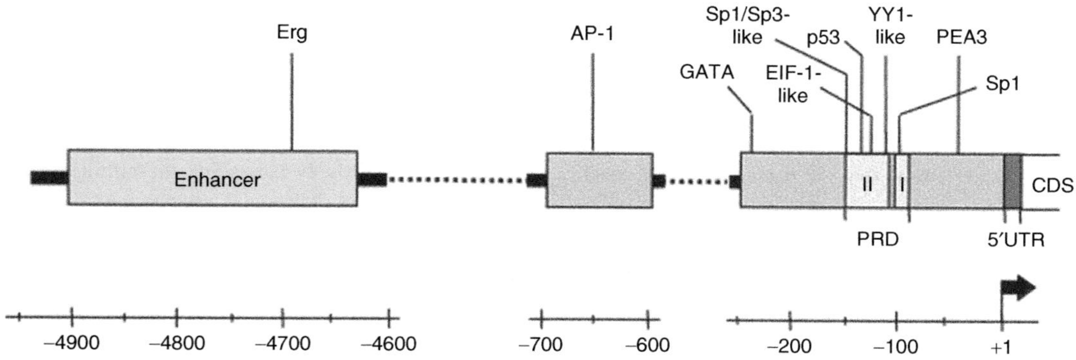

Figure 6.2 Structure of the human eNOS promoter with relevant transcription factor binding sites. The figure shows transcription factor binding sites: PEA3 binding site (position −24 to −40 bp), Sp1 binding site (position −95 to −104 bp), YY1-like binding site (position −117 to −121 bp), Elf-1-like binding site (position −126 to −129 bp), p53-like binding site (position −120 to −143 bp), Sp1/Sp3-like binding site (position −141 to −146 bp), GATA binding site (position −225 to −230 bp), and AP-1 binding site (position −656 to −662 bp). Also shown is the Erg binding site (position −4687 to −4697 bp) in the 269-bp enhancer element (position −4638 to −4907 bp) of the human eNOS promoter. CDS, coding sequence; bp, base pair. Reproduced from [75], with permission of Elsevier Ltd.

than in ECs, explaining why a variety of non-ECs do not express appreciable steady-state levels of eNOS mRNA [76]. Histone acetylation and Lys4 methylation of histone H3 in the eNOS proximal promoter is necessary for eNOS mRNA expression [77]. The eNOS core promoter and proximal downstream coding regions are highly enriched in acetylated histones H3 and H4 and methylated Lys4 of histone H3 which are selectively associated with functionally competent RNA polymerase II complexes.

Several physiological and pathophysiogical stimuli, such as transforming growth factor (TGF)-β, estrogen, and shear stress, influence eNOS gene transcription [75, 78]. TGF-β1 increases eNOS transcription via recruitment of multiple transcription factors (Smad2 and NF-1) to distinct cis-acting sequences [74]. Estrogen-induced increase in eNOS gene transcription is caused by enhanced activity of the estrogen response element and of the binding activity of transcription factor Sp1 [75]. Shear stress-induced upregulation of eNOS gene transcription is due to activation of NF-κB leading to translocation of p50/p65 heterodimers to the nucleus and binding to SSRE [79]. The integrity of F-actin is required for shear stress-induced activation of the eNOS promoter because of the role of F-actin in mechanotransduction [80]. Interestingly, a 27-nucleotide DNA sequence (5'-GAAGTCTAGACCTGCTGCAGGGGTGAG-3') in eNOS intron 4 was shown to bind β-actin in human aortic ECs. It was subsequently shown that silencing β-actin expression decreased eNOS gene transcription and that β-actin overexpression increased eNOS gene transcription in human aortic ECs [81]. However, in pulmonary ECs, silencing of β-actin expression did not affect eNOS gene transcription [82], suggesting fundamental differences in the contribution of the actin cytoskeleton to eNOS gene transcription between pulmonary and systemic ECs.

Post-transcriptional Regulation of eNOS

eNOS mRNA levels represent the balance between gene transcription and mRNA degradation. eNOS mRNA degradation is a major mechanism for post-transcriptional control of eNOS mRNA levels. eNOS is a very stable mRNA species with measured half-lives that, after transcriptional arrest, average 24–48 h [74]. The kinetics of mRNA degradation are dependent in part on nucleotide sequence motifs, which are usually located in the 3'-UTR of mRNA. Possible interactions of specific proteins with these sequences may render the mRNA more or less susceptible to endonucleolytic cleavage. Two motifs often implicated in mRNA destabilization are present at the 3'-end of the eNOS mRNA. Tumor necrosis factor-α destabilizes eNOS mRNA, which is suggested to be mediated by the increased binding of regulatory cytosolic proteins to the 3'-UTR of the eNOS mRNA. Other stimuli that have been reported to decrease eNOS mRNA stability include lipopolysaccharides, hypoxia, and oxidized low-density lipoprotein [75]. VEGF- and H_2O_2-induced eNOS upregulation are dependent on an enhanced stability of eNOS mRNA [83]. eNOS mRNA stability is also regulated by the polymerization state of actin [84, 85]. Disruption of actin filaments by the Rho inhibitor, *Clostridium botulinum* C3 transferase, cytochalasin D, swinholide, or statins or a decrease in actin stress fiber formation by overexpressing a dominant-negative Rho mutant results in increases in eNOS mRNA stability, eNOS protein content, and eNOS activity [84–86]. However, increased binding of G-actin to a 43-nucleotide cis-element in the proximal portion of eNOS 3'-UTR due to a higher G- to F-actin ratio was associated with destabilization of eNOS mRNA [87].

Post-Translational Regulation of eNOS

At the post-translational level, eNOS is regulated by protein–protein interactions, by fatty acylation with myristate and palmitate, by phosphorylation, and by *S*-nitrosylation [88–90].

Protein–Protein Interactions

It has been shown that calmodulin serves as an allosteric activator for eNOS and that caveolin directly interacts with and inhibits eNOS [91]. Bradykinin B2 receptor has been shown to reside in endothelial caveolae and to interact with eNOS in a ligand- and calcium-dependent manner via its C-terminal intracellular domain 4. The binding of Ca^{2+}/calmodulin to eNOS disrupts the inhibitory eNOS–caveolin or eNOS–bradykinin B2 complex, leading to enzyme activation [92]. Heat shock protein 90 (HSP90) serves as an allosteric activator of eNOS [93, 94]. HSP90 binding stimulates eNOS activity by cooperatively enhancing the affinity of eNOS for calmodulin, by balancing output of NO versus superoxide, by facilitating heme binding, and by increasing the rate of Akt (protein kinase B)-dependent eNOS phosphorylation [95]. Endoglin is enriched in caveolae and stabilizes eNOS by promoting its association with HSP90. NOSIP competes with caveolin to bind a site on the oxygenase domain of eNOS and uncouples eNOS from its caveolar attachments, thereby reducing eNOS activity. Increased interaction of NOSTRIN with eNOS promotes the translocation of eNOS from the plasma membrane to intracellular vesicles with a concomitant reduction in eNOS enzyme activity. Dynamin-2 binds the eNOS reductase domain and increases eNOS activity by potentiating electron transfer [96]. CAT-1 forms a caveolar CAT-1–eNOS complex, serving as an optimal substrate delivery channel for eNOS [8]. eNOS is also associated with microtubules [97]. Modifications of tubulin polymerization by either taxol or nocodazole do not influence eNOS–tubulin association but do affect eNOS–HSP90 interactions. Pharmacological stabilization of microtubules increases the association of eNOS with HSP90, eNOS activity, and NO production. Disruption of microtubules decreases the association of eNOS with HSP90, eNOS activity, and NO production [97]. eNOS localized to the plasma membrane is colocalized with cortical F-actin. eNOS that is located in the perinuclear area (probably Golgi) is colocalized with G-actin [86]. The actin-interacting site on eNOS is located in the oxygenase domain [82]. eNOS also associates indirectly with β-actin through other eNOS interacting proteins, such as caveolin, calmodulin, HSP90, dynamin-2, CAT-1, NOSTRIN, and NOSIP [80, 98] (Figure 6.3). The association of β-actin with eNOS increases eNOS activity [86].

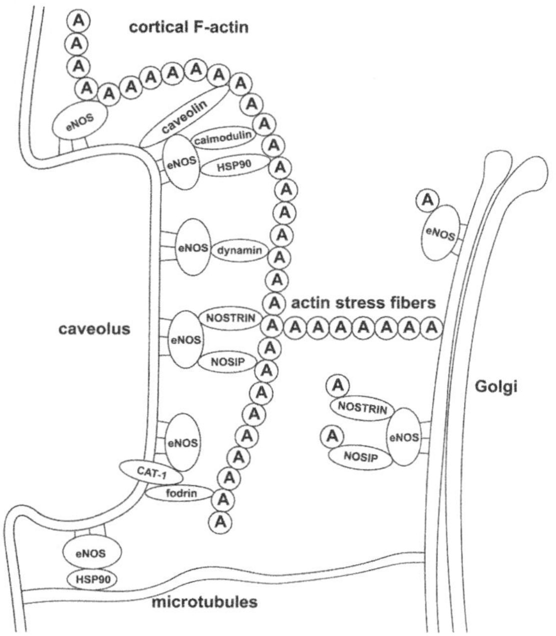

Figure 6.3 Schematic model of eNOS association with the cytoskeleton and other cellular proteins in ECs. eNOS interacts with microtubules, cortical F-actin, and with G-actin in the Golgi directly or indirectly through other eNOS-interacting proteins such as caveolin, calmodulin, HSP90, dynamin-2, CAT-1, NOSTRIN, and NOSIP "A" represents G-actin. The chain "A" of represents F-actin. Reproduced from [98], with permission of Humana Press.

Fatty Acylation with Myristate and Palmitate

eNOS is localized to specific cellular domains in ECs, including Golgi and plasmalemmal caveolae. Localization of eNOS to these membrane domains is dependent on irreversible myristoylation of its N-terminal glycine. Without this modification, eNOS is almost completely cytosolic and lacks palmitoyl moieties. eNOS is palmitoylated on two cysteine residues near the N-terminus (Cys15 and Cys26). This modification is reversible, requires eNOS myristoylation, stabilizes the association of eNOS with the membrane, and is required for proper intracellular localization of eNOS. The NO-generating activity of the myristoyl/palmitoyl-deficient eNOS *in vitro* is not impaired. However, the cellular NO generated by the myristoyl/palmitoylation-deficient enzyme within the EC is significantly reduced, suggesting that fatty acylation-dependent intracellular localization of eNOS is critical for optimal NO production [89].

eNOS Phosphorylation

eNOS undergoes phosphorylation at residues Ser1177 (primary sequence numbering corresponds to human eNOS), Ser635, Ser617, Thr495, and Ser116 [89, 90]. Phosphorylation at Ser1177, Ser635, and Ser617 is stimulatory, whereas phosphorylation at Thr495 and Ser116 is inhibitory. The activation of eNOS catalytic function by Ser1177 phosphorylation is due to inhibition of calmodulin dissociation from eNOS and also enhancement of eNOS electron transfer. Ser1177 phosphorylation is catalyzed by several kinases, including kinase Akt as well as cyclic protein kinase A (PKA), AMP-activated protein kinase (AMPK), protein kinase G, and Ca^{2+}/calmodulin-dependent protein kinase II. Phosphorylation at Ser635 is responsive to PKA and increases eNOS activity. Phosphorylation at Ser617 is caused by PKA or Akt, which sensitize eNOS to calmodulin binding. Akt-mediated eNOS phosphorylation is responsible for eNOS activation induced by shear stress, VEGF, and insulin. Phosphorylation at Thr495 is downstream of PKC and AMPK, and attenuates the binding of calmodulin by eNOS. Phosphorylation of eNOS at Ser116 inhibits enzyme activity, and dephosphorylation of eNOS at this site is promoted by VEGF.

S-Nitrosylation

eNOS is active only as a homodimer because electrons need to transfer from the reductase domain of one monomer to the oxygenase domain of another monomer. Nitrosylation of cysteine residues Cys94 and Cys99 in eNOS inhibits eNOS enzymatic activity due to monomerization of eNOS protein via the destruction of a zinc tetrathiolate cluster at the dimeric interface [99]. Stimulation of ECs with eNOS agonists such as VEGF promotes rapid and reversible denitrosylation of eNOS, temporally associated with enzyme activation [100]. Receptor-regulated eNOS nitrosylation may represent an important determinant of NO signaling in the vascular wall.

BIOLOGICAL FATE OF NO FROM LUNG ENDOTHELIUM

The biological half-life of NO is only 2–3 s. As NO diffuses from its site of production in the EC to its targets in the vascular SMC and in blood cells, NO is rapidly oxidized to nitrite and nitrate [101]. NO can also react with O_2^- and form $ONOO^-$, reducing the biological availability of NO. O_2^- could be generated from eNOS itself at lower levels of BH_4. eNOS catalysis involves an electron transfer from the reductase domain of one monomer to the oxygenase domain of the other monomer where ferric heme-superoxy species are formed. Upon receiving the second electron from BH_4, the species react with L-arginine to form NO and L-citrulline. In the absence of BH_4, the ferric heme-superoxy species decay to generate O_2^- [102] (see Chapter 17).

Importantly, NO interacts with heme-containing proteins such as soluble guanylate cyclase (sGC), Hb, myoglobin (Mb), and enzymes containing iron–sulfur centers. Direct binding of NO to the heme prosthetic group of sGC forms the nitrosyl heme adduct, which induces a conformational change in sGC and a subsequent increase in its enzymatic activity. Activation of sGC leads to cGMP production which mediates a number of physiological and pathophysiological effects of NO. Hb and Mb not only scavenge NO, but also play an important role in the maintenance of NO homeostasis [103]. The reactions of Hb and Mb with NO form SNO-Hb and SNO-Mb which can protect the cells against any possible oxidative damage of NO. SNO-Hb and SNO-Mb can induce SMC relaxation through activation of sGC and function as a store of vasoactive NO [104].

NO SIGNALING IN PATHOPHYSIOLOGY OF PULMONARY DISEASES

NO in Hypoxic Pulmonary Arterial Hypertension

NO produced from eNOS in pulmonary endothelium is essential for maintaining the low pulmonary vascular resistance and the inactive proliferation status of pulmonary vascular SMCs [10, 11, 105]. The increase in pulmonary arterial pressure and remodeling of the pulmonary vasculature associated with hypoxic PAH are attributable to reduction of NO release from hypoxic pulmonary endothelium (see Chapter 18). The studies from cultured pulmonary ECs, animal models, and human subjects have confirmed that hypoxia reduces NO production from lung endothelium [106]. Several mechanisms are involved. (i) Substrate (L-arginine) delivery to eNOS is compromised under hypoxic conditions. Hypoxia decreases CAT-1-mediated L-arginine uptake by inducing membrane depolarization and by disrupting CAT-1-cytoskeleton association [7, 107]. (ii) Oxygen is a cofactor for NO synthesis by eNOS, and the K_m for oxygen is in the physiological range (5–20 μM), suggesting that a decrease in oxygen attenuates NO production by ECs [108]. (iii) eNOS activity is reduced under hypoxic conditions. There are conflicting reports on the effects of hypoxia on eNOS gene expression. Liao *et al.* [109] and Fike *et al.* [110] demonstrated that hypoxia decreased eNOS gene expression in cell culture and animal models, respectively. Xu *et al.* [111] and Arnet *et al.* [112] reported that exposure to hypoxia for up to 6 h increased eNOS expression. Our data indicate that exposure of pulmonary artery ECs to hypoxia for 24 h did not alter eNOS protein contents [113]. However, in nearly all models of hypoxia, both *in vivo* and *in vitro*, where eNOS activity has been directly measured, a decrease in activity has been observed even when eNOS expression is increased [106]. Hypoxia-induced decrease in eNOS activity is caused by post-translational mechanisms. The interactions of eNOS with its activating proteins, HSP90 and β-actin, are disrupted [80, 86, 113, 114], and with its inhibiting protein, caveolin, is tightened [114]. eNOS Ser1177 phosphorylation is also decreased during hypoxia [114]. In addition, hypoxia increases O_2^- formation and the O_2^- may react with NO to produce peroxynitrite and reduce bioavailable NO [115].

NO in PAH of Sickle Cell Disease

Pulmonary arterial hypertension is a common complication of sickle cell disease. The pathophysiology of PAH in sickle cell disease is associated with pulmonary vasoconstriction, vascular smooth muscle hyperplasia, and *in situ* pulmonary arterial thrombosis [116]. These processes are caused by impaired NO synthesis and decreased NO bioavailability. Arginase activity and arginine/ornithine ratio in the plasma of patients with sickle cell disease are elevated and tend to increase with the level of pulmonary hypertension [117]. Increased arginase activity due to inflammation, chronic end-organ damage, and hemolysis decrease plasma L-arginine leading to reduction of substrate availability for eNOS and shift arginine metabolism toward L-ornithine production [73]. The bioavailability of L-arginine is further decreased by increased L-ornithine levels because L-ornithine and L-arginine compete for CAT-1 transporter uptake by ECs. Impaired eNOS dimerization in the lungs of patients with sickle cell disease decreases eNOS activity even in the case of increased eNOS protein expression observed in mouse models of sickle cell disease [118]. Moreover, xanthine oxidase levels in the lungs of sickle cell disease patients are higher [119]. O_2^- generated from xanthine oxidase not only decreases the association of eNOS with its activating protein HSP90, but also consumes NO [119]. Furthermore, plasma hemoglobin released from erythrocytes due to intravascular hemolysis reacts with NO and severely reduces NO bioavailability in patients with sickle cell disease [60, 118] (Figure 6.4).

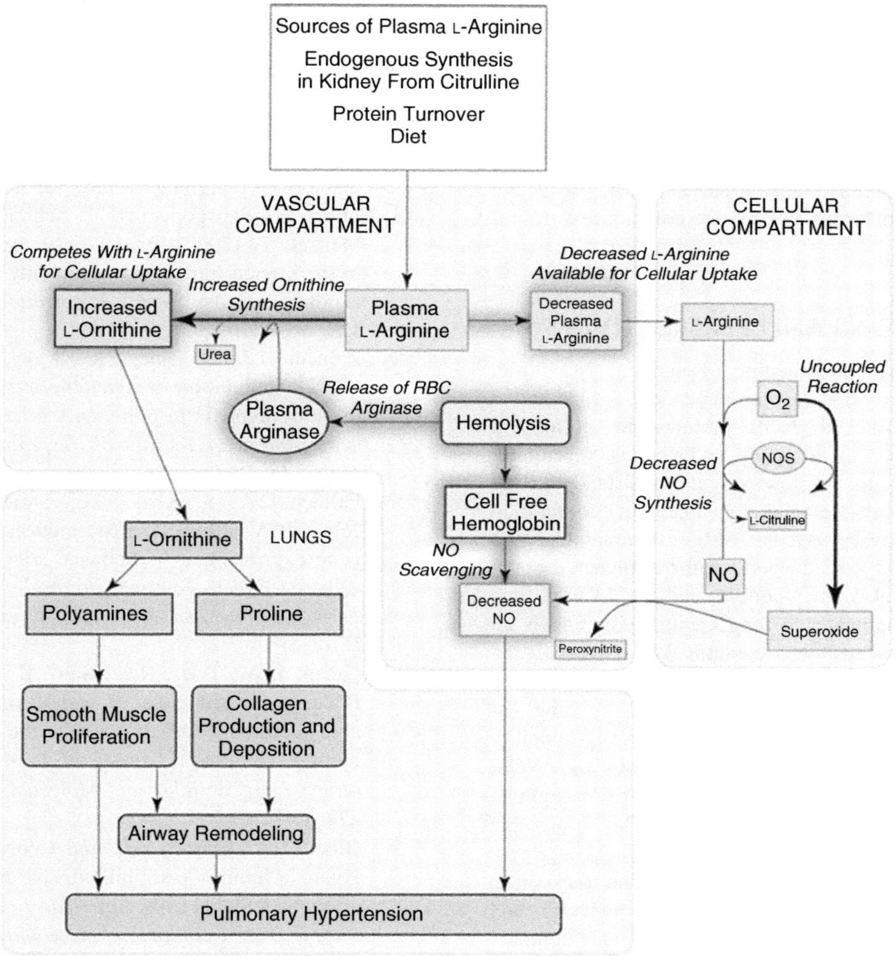

Figure 6.4 Altered L-arginine metabolism in sickle cell disease. Arginine is synthesized endogenously from citrulline, primarily in the kidney via the intestinal–renal axis. Arginase and eNOS compete for L-arginine, their common substrate. In sickle cell disease, bioavailability of L-arginine and NO are decreased by several mechanisms linked to hemolysis. The release of erythrocyte arginase during hemolysis increases plasma arginase levels and shifts arginine metabolism toward L-ornithine production, decreasing the amount available for NO synthesis. The bioavailability of arginine is further decreased by increased L-ornithine levels because L-ornithine and L-arginine compete for the same transporter system for cellular uptake. NO bioavailability in sickle cell disease is low due to low substrate availability, NO scavenging by cell free hemoglobin released during hemolysis, and through reactions with free radicals such as superoxide. Endothelial dysfunction resulting from NO depletion and increased levels of the downstream products of L-ornithine metabolism (polyamines and proline) likely contributes to the pathogenesis of lung injury and pulmonary hypertension. Reproduced from [117], with permission of the American Medical Association.

NO in Chronic Obstructive Pulmonary Diseases

Decreased endothelial NO production has been associated with the pathogenesis of chronic obstructive pulmonary disease (COPD). eNOS polymorphism is involved in the susceptibility to the development of certain phenotypes of COPD such as emphysema associated with α_1-antitrypsin deficiency [120]. eNOS expression has also been reported to be reduced in the pulmonary arteries of smokers [121], in the lungs of smoking animal models [122], and in smoke-exposed pulmonary artery ECs *in vitro* [123]. Oxidative stress in COPD causes oxidation of BH_4 [124], a cofactor for eNOS, further exaggerating the reduction of NO release [125]. eNOS deficiency leads to impairment of compensatory post-pneumonectomy lung growth [23] and developmental alveolarization [21, 126]. Ineffective repair of alveoli results in airspace enlargement in emphysema.

CONCLUSIONS AND PERSPECTIVES

The pulmonary endothelium generates NO from L-arginine via a calcium-dependent constitutive eNOS – the principal NOS isoform in pulmonary endothelium. NO is a major endogenous vasodilator that contributes to the low vascular resistance in the pulmonary circulation and the inactive proliferation status of pulmonary vascular SMCs. Maintaining V/Q match and blood gas transport require sufficient amount of NO. NO is also an important mediator in fetal lung vascular development and lung angiogenesis and is an endogenous homeostatic regulator of leukocyte–and platelet–endothelium interactions in the pulmonary microcirculation. The production of NO by lung endothelium is critically dependent on the availability of L-arginine and the enzymatic activity of eNOS. L-Arginine availability is orchestrated by L-arginine transporter and enzymes for intracellular L-arginine generation and metabolism such as ASS and arginase. The activity of CAT-1, the major L-arginine transporter, is regulated by phosphorylation, trans-stimulation, and membrane potential. CAT-1 forms a caveolar complex with eNOS, facilitating directly delivery of extracellular L-arginine to eNOS in lung ECs. eNOS gene transcription is modulated by TGF-β, estrogen, and shear stress. eNOS activity is also controlled by post-translational mechanisms including protein–protein interactions, fatty acylation, phosphorylation, *S*-nitrosylation. NO's bioavailability is affected by biological scavengers such as Hb and ROS. Abnormality of NO signaling is associated with hypoxic PAH, sickle cell disease-associated PAH, and COPD. The mechanism responsible for aberrant NO signaling in these lung diseases involves reduction of NO release from the pulmonary endothelium and decrease in NO bioavailability. Modification of the NO signaling pathway provides therapeutic options for a variety of pulmonary diseases characterized by impaired endothelial NO production.

References

1. Moncada, S., Palmer, R.M., and Higgs, E.A. (1991) Nitric oxide: physiology, pathophysiology, and pharmacology. *Pharmacological Reviews*, **43**, 109–42.
2. Palmer, R.M., Ashton, D.S., and Moncada, S. (1988) Vascular endothelial cells synthesize nitric oxide from L-arginine. *Nature*, **333**, 664–66.
3. Luhrs, H., Papadopoulos, T., Schmidt, H.H., and Menzel, T. (2002) Type I nitric oxide synthase in the human lung is predominantly expressed in capillary endothelial cells. *Respiration Physiology*, **129**, 367–74.
4. Zulueta, J.J., Sawhney, R., Kayyali, U. *et al.* (2002) Modulation of inducible nitric oxide synthase by hypoxia in pulmonary artery endothelial cells. *American Journal of Respiratory Cell and Molecular Biology*, **26**, 22–30.
5. Palmer, L.A., Semenza, G.L., Stoler, M.H., and Johns, R.A. (1998) Hypoxia induces type II NOS gene expression in pulmonary artery endothelial cells via HIF-1. *American Journal of Physiology: Lung Cellular and Molecular Physiology*, **274**, L212–19.
6. Fagan, K.A., Tyler, R.C., Sato, K. *et al.* (1999) Relative contributions of endothelial, inducible, and neuronal NOS to tone in the murine pulmonary circulation. *American Journal of Physiology: Lung Cellular and Molecular Physiology*, **277**, L472–78.
7. Block, E.R., Herrera, H., and Couch, M. (1995) Hypoxia inhibits L-arginine uptake by pulmonary artery endothelial cells. *American Journal of Physiology: Lung Cellular and Molecular Physiology*, **269**, L574–80.
8. McDonald, K.K., Zharikov, S., Block, E.R., and Kilberg, M.S. (1997) A caveolar complex between the cationic amino acid transporter 1 and endothelial nitric-oxide synthase may explain the "arginine paradox". *The Journal of Biological Chemistry*, **272**, 31213–16.
9. Su, Y. and Block, E.R. (1995) Hypoxia inhibits L-arginine synthesis from L-citrulline in porcine pulmonary artery endothelial cells. *American Journal of Physiology: Lung Cellular and Molecular Physiology*, **269**, L581–87.

10. Stamler, J.S., Loh, E., Roddy, M.A. *et al.* (1994) Nitric oxide regulates basal systemic and pulmonary vascular resistance in healthy humans. *Circulation*, **89**, 2035–40.
11. Fagan, K.A., McMurtry, I., and Rodman, D.M. (2000) Nitric oxide synthase in pulmonary hypertension: lessons from knockout mice. *Physiological Research*, **49**, 539–48.
12. Vaughan, D.J., Brogan, T.V., Kerr, M.E. *et al.* (2003) Contributions of nitric oxide synthase isozymes to exhaled nitric oxide and hypoxic pulmonary vasoconstriction in rabbit lungs. *American Journal of Physiology: Lung Cellular and Molecular Physiology*, **284**, L834–43.
13. Cremona, G., Wood, A.M., Hall, L.W. *et al.* (1994) Effect of inhibitors of nitric oxide release and action on vascular tone in isolated lungs of pig, sheep, dog and man. *The Journal of Physiology*, **481**, 185–95.
14. Steudel, W., Ichinose, F., Huang, P.L. *et al.* (1997) Pulmonary vasoconstriction and hypertension in mice with targeted disruption of the endothelial nitric oxide synthase (NOS 3) gene. *Circulation Research*, **81**, 34–41.
15. Jiang, L., Quarck, R., Janssens, S. *et al.* (2006) Effect of adenovirus-mediated gene transfer of nitric oxide synthase on vascular reactivity of rat isolated pulmonary arteries. *Pflugers Archiv*, **452**, 213–21.
16. Champion, H.C., Bivalacqua, T.J., D'Souza, F.M. *et al.* (1999) Gene transfer of endothelial nitric oxide synthase to the lung of the mouse *in vivo*. Effect on agonist-induced and flow-mediated vascular responses. *Circulation Research*, **84**, 1422–32.
17. Champion, H.C., Bivalacqua, T.J., Greenberg, S.S. *et al.* (2002) Adenoviral gene transfer of endothelial nitric-oxide synthase (eNOS) partially restores normal pulmonary arterial pressure in eNOS-deficient mice. *Proceedings of the National Academy of Sciences of the United States of America*, **99**, 13248–53.
18. Frostell, C., Fratacci, M.D., Wain, J.C. *et al.* (1991) Inhaled nitric oxide. A selective pulmonary vasodilator reversing hypoxic pulmonary vasoconstriction. *Circulation*, **83**, 2038–47.
19. Ichinose, F., Roberts, J.D. Jr., and Zapol, W.M. (2004) Inhaled nitric oxide: a selective pulmonary vasodilator: current uses and therapeutic potential. *Circulation*, **109**, 3106–11.
20. Han, R.N. and Stewart, D.J. (2006) Defective lung vascular development in endothelial nitric oxide synthase-deficient mice. *Trends in Cardiovascular Medicine*, **16**, 29–34.
21. Han, R.N., Babaei, S., Robb, M. *et al.* (2004) Defective lung vascular development and fatal respiratory distress in endothelial NO synthase-deficient mice: a model of alveolar capillary dysplasia? *Circulation Research*, **94**, 1115–23.
22. Balasubramaniam, V., Tang, J.R., Maxey, A. *et al.* (2003) Mild hypoxia impairs alveolarization in the endothelial nitric oxide synthase-deficient mouse. *American Journal of Physiology: Lung Cellular and Molecular Physiology*, **284**, L964–71.
23. Leuwerke, S.M., Kaza, A.K., Tribble, C.G. *et al.* (2002) Inhibition of compensatory lung growth in endothelial nitric oxide synthase-deficient mice. *American Journal of Physiology: Lung Cellular and Molecular Physiology*, **282**, L1272–78.
24. Balasubramaniam, V., Maxey, A.M., Morgan, D.B. *et al.* (2006) Inhaled NO restores lung structure in eNOS-deficient mice recovering from neonatal hypoxia. *American Journal of Physiology: Lung Cellular and Molecular Physiology*, **291**, L119–27.
25. Lin, Y.J., Markham, N.E., Balasubramaniam, V. *et al.* (2005) Inhaled nitric oxide enhances distal lung growth after exposure to hyperoxia in neonatal rats. *Pediatric Research*, **58**, 22–29.
26. Tang, J.R., Markham, N.E., Lin, Y.J. *et al.* (2004) Inhaled nitric oxide attenuates pulmonary hypertension and improves lung growth in infant rats after neonatal treatment with a VEGF receptor inhibitor. *American Journal of Physiology: Lung Cellular and Molecular Physiology*, **287**, L344–51.
27. Stenmark, K.R. and Abman, S.H. (2005) Lung vascular development: implications for the pathogenesis of bronchopulmonary dysplasia. *Annual Review of Physiology*, **67**, 623–61.
28. Hood, J.D., Meininger, C.J., Ziche, M., and Granger, H.J. (1998) VEGF upregulates ecNOS message, protein, and NO production in human endothelial cells. *American Journal of Physiology: Heart and Circulatory Physiology*, **274**, H1054–58.
29. Chen, Y., Medhora, M., Falck, J.R. *et al.* (2006) Mechanisms of activation of eNOS by 20-HETE and VEGF in bovine pulmonary artery endothelial cells. *American Journal of Physiology: Lung Cellular and Molecular Physiology*, **291**, L378–85.
30. Kaminski, A., Pohl, C.B., Sponholz, C. *et al.* (2004) Up-regulation of endothelial nitric oxide synthase inhibits pulmonary leukocyte migration following lung ischemia–reperfusion in mice. *American Journal of Pathology*, **164**, 2241–49.

31. Hickey, M.J. and Kubes, P. (1997) Role of nitric oxide in regulation of leucocyte–endothelial cell interactions. *Experimental Physiology*, **82**, 339–48.
32. Kubes, P., Suzuki, M., and Granger, D.N. (1991) Nitric oxide: an endogenous modulator of leukocyte adhesion. *Proceedings of the National Academy of Sciences of the United States of America*, **88**, 4651–55.
33. Lefer, D.J., Jones, S.P., Girod, W.G. *et al.* (1999) Leukocyte–endothelial cell interactions in nitric oxide synthase-deficient mice. *American Journal of Physiology: Heart and Circulatory Physiology*, **276**, H1943–50.
34. de Graaf, J.C., Banga, J.D., Moncada, S. *et al.* (1992) Nitric oxide functions as an inhibitor of platelet adhesion under flow conditions. *Circulation*, **85**, 2284–90.
35. Radomski, M.W., Palmer, R.M., and Moncada, S. (1987) The role of nitric oxide and cGMP in platelet adhesion to vascular endothelium. *Biochemical and Biophysical Research Communications*, **148**, 1482–89.
36. Payen, D.M. (2000) Inhaled nitric oxide and acute lung injury. *Clinics in Chest Medicine*, **21**, 519–29, ix.
37. Gaston, B., Singel, D., Doctor, A., and Stamler, J.S. (2006) S-nitrosothiol signaling in respiratory biology. *American Journal of Respiratory and Critical Care Medicine*, **173**, 1186–93.
38. McMahon, T.J., Ahearn, G.S., Moya, M.P. *et al.* (2005) A nitric oxide processing defect of red blood cells created by hypoxia: deficiency of *S*-nitrosohemoglobin in pulmonary hypertension. *Proceedings of the National Academy of Sciences of the United States of America*, **102**, 14801–6.
39. Moya, M.P., Gow, A.J., McMahon, T.J. *et al.* (2001) *S*-Nitrosothiol repletion by an inhaled gas regulates pulmonary function. *Proceedings of the National Academy of Sciences of the United States of America*, **98**, 5792–97.
40. McMahon, T.J., Exton, S.A., Bonaventura, J. *et al.* (2000) Functional coupling of oxygen binding and vasoactivity in *S*-nitrosohemoglobin. *The Journal of Biological Chemistry*, **275**, 16738–45.
41. McMahon, T.J., Moon, R.E., Luschinger, B.P. *et al.* (2002) Nitric oxide in the human respiratory cycle. *Nature Medicine*, **8**, 711–17.
42. Gow, A.J. and Stamler, J.S. (1998) Reactions between nitric oxide and haemoglobin under physiological conditions. *Nature*, **391**, 169–73.
43. Coggins, M.P. and Bloch, K.D. (2007) Nitric oxide in the pulmonary vasculature. *Arteriosclerosis, Thrombosis, and Vascular Biology*, **27**, 1877–85.
44. Aisaka, K., Gross, S.S., Griffith, O.W., and Levi, R. (1989) L-Arginine availability determines the duration of acetylcholine-induced systemic vasodilation *in vivo*. *Biochemical and Biophysical Research Communications*, **163**, 710–17.
45. Eddahibi, S., Adnot, S., Carville, C. *et al.* (1992) L-Arginine restores endothelium-dependent relaxation in pulmonary circulation of chronically hypoxic rats. *American Journal of Physiology: Lung Cellular and Molecular Physiology*, **263**, L194–200.
46. Closs, E.I., Scheld, J.S., Sharafi, M., and Forstermann, U. (2000) Substrate supply for nitric-oxide synthase in macrophages and endothelial cells: role of cationic amino acid transporters. *Molecular Pharmacology*, **57**, 68–74.
47. Simon, A., Plies, L., Habermeier, A. *et al.* (2003) Role of neutral amino acid transport and protein breakdown for substrate supply of nitric oxide synthase in human endothelial cells. *Circulation Research*, **93**, 813–20.
48. Wu, G. and Morris, S.M. Jr. (1998) Arginine metabolism: nitric oxide and beyond. *The Biochemical Journal*, **336**, 1–17.
49. Hallemeesch, M.M., Lamers, W.H., and Deutz, N.E. (2002) Reduced arginine availability and nitric oxide production. *Clinical Nutrition*, **21**, 273–79.
50. Morris, S.M. Jr. (2005) Arginine metabolism in vascular biology and disease. *Vascular Medicine*, **10** (Suppl 1), S83–S87.
51. Deves, R. and Boyd, C.A. (1998) Transporters for cationic amino acids in animal cells: discovery, structure, and function. *Physiological Reviews*, **78**, 487–545.
52. Zharikov, S.I. and Block, E.R. (1998) Characterization of L-arginine uptake by plasma membrane vesicles isolated from cultured pulmonary artery endothelial cells. *Biochimica et Biophysica Acta*, **1369**, 173–83.
53. Closs, E.I., Boissel, J.P., Habermeier, A., and Rotmann, A. (2006) Structure and function of cationic amino acid transporters (CATs). *The Journal of Membrane Biology*, **213**, 67–77.
54. Hammermann, R., Brunn, G., and Racke, K. (2001) Analysis of the genomic organization of the human cationic amino acid transporters CAT-1, CAT-2 and CAT-4. *Amino Acids*, **21**, 211–19.
55. Hatzoglou, M., Fernandez, J., Yaman, I., and Closs, E. (2004) Regulation of cationic amino acid transport: the story of the CAT-1 transporter. *Annual Review of Nutrition*, **24**, 377–99.
56. Krotova, K.Y., Zharikov, S.I., and Block, E.R. (2003) Classical isoforms of PKC as regulators

of CAT-1 transporter activity in pulmonary artery endothelial cells. *American Journal of Physiology: Lung Cellular and Molecular Physiology*, **284**, L1037–44.
57. Zharikov, S.I., Herrera, H., and Block, E.R. (1997) Role of membrane potential in hypoxic inhibition of L-arginine uptake by lung endothelial cells. *American Journal of Physiology: Lung Cellular and Molecular Physiology*, **272**, L78–84.
58. Zharikov, S.I., Krotova, K.Y., Belayev, L., and Block, E.R. (2004) Pertussis toxin activates L-arginine uptake in pulmonary endothelial cells through downregulation of PKC-alpha activity. *American Journal of Physiology: Lung Cellular and Molecular Physiology*, **286**, L974–83.
59. Flam, B.R., Hartmann, P.J., Harrell-Booth, M. *et al.* (2001) Caveolar localization of arginine regeneration enzymes, argininosuccinate synthase, and lyase, with endothelial nitric oxide synthase. *Nitric Oxide*, **5**, 187–97.
60. Reiter, C.D., Wang, X., Tanus-Santos, J.E. *et al.* (2002) Cell-free hemoglobin limits nitric oxide bioavailability in sickle-cell disease. *Nature Medicine*, **8**, 1383–89.
61. Sessa, W.C., Hecker, M., Mitchell, J.A., and Vane, J.R. (1990) The metabolism of L-arginine and its significance for the biosynthesis of endothelium-derived relaxing factor: L-glutamine inhibits the generation of L-arginine by cultured endothelial cells. *Proceedings of the National Academy of Sciences of the United States of America*, **87**, 8607–11.
62. Su, Y. and Block, E.R. (1997) Hypoxia inhibits the induction of argininosuccinate synthetase by endotoxin in lung endothelial cells. *American Journal of Physiology: Lung Cellular and Molecular Physiology*, **272**, L934–38.
63. Flam, B.R., Eichler, D.C., and Solomonson, L.P. (2007) Endothelial nitric oxide production is tightly coupled to the citrulline–NO cycle. *Nitric Oxide*, **17**, 115–21.
64. Arnal, J.F., Munzel, T., Venema, R.C. *et al.* (1995) Interactions between L-arginine and L-glutamine change endothelial NO production. An effect independent of NO synthase substrate availability. *The Journal of Clinical Investigation*, **95**, 2565–72.
65. Solomonson, L.P., Flam, B.R., Pendleton, L.C. *et al.* (2003) The caveolar nitric oxide synthase/arginine regeneration system for NO production in endothelial cells. *The Journal of Experimental Biology*, **206**, 2083–87.
66. Durante, W., Johnson, F.K., and Johnson, R.A. (2007) Arginase: a critical regulator of nitric oxide synthesis and vascular function. *Clinical and Experimental Pharmacology and Physiology*, **34**, 906–11.
67. Nelin, L.D., Nash, H.E., and Chicoine, L.G. (2001) Cytokine treatment increases arginine metabolism and uptake in bovine pulmonary arterial endothelial cells. *American Journal of Physiology: Lung Cellular and Molecular Physiology*, **281**, L1232–39.
68. Kawamoto, S., Amaya, Y., Murakami, K. *et al.* (1987) Complete nucleotide sequence of cDNA and deduced amino acid sequence of rat liver arginase. *The Journal of Biological Chemistry*, **262**, 6280–83.
69. Topal, G., Brunet, A., Walch, L. *et al.* (2006) Mitochondrial arginase II modulates nitric-oxide synthesis through nonfreely exchangeable L-arginine pools in human endothelial cells. *The Journal of Pharmacology and Experimental Therapeutics*, **318**, 1368–74.
70. Chicoine, L.G., Paffett, M.L., Young, T.L., and Nelin, L.D. (2004) Arginase inhibition increases nitric oxide production in bovine pulmonary arterial endothelial cells. *American Journal of Physiology: Lung Cellular and Molecular Physiology*, **287**, L60–68.
71. Li, H., Meininger, C.J., Hawker, J.R. Jr. *et al.* (2001) Regulatory role of arginase I and II in nitric oxide, polyamine, and proline syntheses in endothelial cells. *American Journal of Physiology: Endocrinology and Metabolism*, **280**, E75–82.
72. Zhang, C., Hein, T.W., Wang, W. *et al.* (2001) Constitutive expression of arginase in microvascular endothelial cells counteracts nitric oxide-mediated vasodilatory function. *The FASEB Journal*, **15**, 1264–66.
73. Xu, W., Kaneko, F.T., Zheng, S. *et al.* (2004) Increased arginase II and decreased NO synthesis in endothelial cells of patients with pulmonary arterial hypertension. *The FASEB Journal*, **18**, 1746–48.
74. Searles, C.D. (2006) Transcriptional and posttranscriptional regulation of endothelial nitric oxide synthase expression. *American Journal of Physiology: Cell Physiology*, **291**, C803–16.
75. Li, H., Wallerath, T., and Forstermann, U. (2002) Physiological mechanisms regulating the expression of endothelial-type NO synthase. *Nitric Oxide*, **7**, 132–47.
76. Chan, Y., Fish, J.E., D'Abreo, C. *et al.* (2004) The cell-specific expression of endothelial nitric-oxide synthase: a role for DNA methylation. *The Journal of Biological Chemistry*, **279**, 35087–100.
77. Fish, J.E., Matouk, C.C., Rachlis, A. *et al.* (2005) The expression of endothelial nitric-oxide synthase

is controlled by a cell-specific histone code. *The Journal of Biological Chemistry*, **280**, 24824–38.

78. MacRitchie, A.N., Jun, S.S., Chen, Z. *et al.* (1997) Estrogen upregulates endothelial nitric oxide synthase gene expression in fetal pulmonary artery endothelium. *Circulation Research*, **81**, 355–62.
79. Davis, M.E., Grumbach, I.M., Fukai, T. *et al.* (2004) Shear stress regulates endothelial nitric-oxide synthase promoter activity through nuclear factor kappaB binding. *The Journal of Biological Chemistry*, **279**, 163–68.
80. Su, Y., Kondrikov, D., and Block, E.R. (2007) β-Actin: a regulator of NOS-3. *Science STKE*, **2007**, e52-1–e52-3.
81. Ou, H., Shen, Y.H., Utama, B. *et al.* (2005) Effect of nuclear actin on endothelial nitric oxide synthase expression. *Arteriosclerosis, Thrombosis, and Vascular Biology*, **25**, 2509–14.
82. Kondrikov, D., Han, H.R., Block, E.R., and Su, Y. (2006) Growth and density-dependent regulation of NO synthase by the actin cytoskeleton in pulmonary artery endothelial cells. *American Journal of Physiology: Lung Cellular and Molecular Physiology*, **290**, L41–50.
83. Bouloumie, A., Schini-Kerth, V.B., and Busse, R. (1999) Vascular endothelial growth factor up-regulates nitric oxide synthase expression in endothelial cells. *Cardiovascular Research*, **41**, 773–80.
84. Laufs, U., Endres, M., Stagliano, N. *et al.* (2000) Neuroprotection mediated by changes in the endothelial actin cytoskeleton. *The Journal of Clinical Investigation*, **106**, 15–24.
85. Takemoto, M., Sun, J., Hiroki, J. *et al.* (2002) Rho-kinase mediates hypoxia-induced downregulation of endothelial nitric oxide synthase. *Circulation*, **106**, 57–62.
86. Su, Y., Edwards-Bennett, S., Bubb, M.R., and Block, E.R. (2003) Regulation of endothelial nitric oxide synthase by the actin cytoskeleton. *American Journal of Physiology: Cell Physiology*, **284**, C1542–49.
87. Searles, C.D., Ide, L., Davis, M.E. *et al.* (2004) Actin cytoskeleton organization and posttranscriptional regulation of endothelial nitric oxide synthase during cell growth. *Circulation Research*, **95**, 488–95.
88. Kone, B.C. (2000) Protein–protein interactions controlling nitric oxide synthases. *Acta Physiologica Scandinavica*, **168**, 27–31.
89. Govers, R. and Rabelink, T.J. (2001) Cellular regulation of endothelial nitric oxide synthase. *American Journal of Physiology: Renal Physiology*, **280**, F193–206.
90. Dudzinski, D.M. and Michel, T. (2007) Life history of eNOS: partners and pathways. *Cardiovascular Research*, **75**, 247–60.
91. Feron, O., Saldana, F., Michel, J.B., and Michel, T. (1998) The endothelial nitric-oxide synthase–caveolin regulatory cycle. *The Journal of Biological Chemistry*, **273**, 3125–28.
92. Ju, H., Venema, V.J., Marrero, M.B., and Venema, R.C. (1998) Inhibitory interactions of the bradykinin B2 receptor with endothelial nitric-oxide synthase. *The Journal of Biological Chemistry*, **273**, 24025–29.
93. Harris, M.B., Ju, H., Venema, V.J. *et al.* (2000) Role of heat shock protein 90 in bradykinin-stimulated endothelial nitric oxide release. *General Pharmacology*, **35**, 165–70.
94. Garcia-Cardena, G., Fan, R., Shah, V. *et al.* (1998) Dynamic activation of endothelial nitric oxide synthase by HSP90. *Nature*, **392**, 821–24.
95. Fontana, J., Fulton, D., Chen, Y. *et al.* (2002) Domain mapping studies reveal that the M domain of hsp90 serves as a molecular scaffold to regulate Akt-dependent phosphorylation of endothelial nitric oxide synthase and NO release. *Circulation Research*, **90**, 866–73.
96. Cao, S., Yao, J., and Shah, V. (2003) The proline-rich domain of dynamin-2 is responsible for dynamin-dependent *in vitro* potentiation of endothelial nitric-oxide synthase activity via selective effects on reductase domain function. *The Journal of Biological Chemistry*, **278**, 5894–901.
97. Su, Y., Zharikov, S.I., and Block, E.R. (2002) Microtubule-active agents modify nitric oxide production in pulmonary artery endothelial cells. *American Journal of Physiology: Lung Cellular and Molecular Physiology*, **282**, L1183–89.
98. Su, Y., Kondrikov, D., and Block, E.R. (2005) Cytoskeletal regulation of nitric oxide synthase. *Cell Biochemistry and Biophysics*, **43**, 439–50.
99. Ravi, K., Brennan, L.A., Levic, S. *et al.* (2004) *S*-Nitrosylation of endothelial nitric oxide synthase is associated with monomerization and decreased enzyme activity. *Proceedings of the National Academy of Sciences of the United States of America*, **101**, 2619–24.
100. Erwin, P.A., Lin, A.J., Golan, D.E., and Michel, T. (2005) Receptor-regulated dynamic *S*-nitrosylation of endothelial nitric-oxide synthase in vascular endothelial cells. *The Journal of Biological Chemistry*, **280**, 19888–94.
101. Moncada, S. and Higgs, A. (1993) The L-arginine–nitric oxide pathway. *The New England Journal of Medicine*, **329**, 2002–12.

102. Xia, Y. (2007) Superoxide generation from nitric oxide synthases. *Antioxidants and Redox Signalling*, **9**, 1773–78.
103. Andriambeloson, E. and Witting, P.K. (2002) Chemical regulation of nitric oxide: a role for intracellular myoglobin? *Redox Report*, **7**, 131–36.
104. Rayner, B.S., Wu, B.J., Raftery, M. *et al.* (2005) Human *S*-nitroso oxymyoglobin is a store of vasoactive nitric oxide. *The Journal of Biological Chemistry*, **280**, 9985–93.
105. Ozaki, M., Kawashima, S., Yamashita, T. *et al.* (2001) Reduced hypoxic pulmonary vascular remodeling by nitric oxide from the endothelium. *Hypertension*, **37**, 322–27.
106. Le Cras, T.D. and McMurtry, I.F. (2001) Nitric oxide production in the hypoxic lung. *American Journal of Physiology: Lung Cellular and Molecular Physiology*, **280**, L575–82.
107. Zharikov, S.I. and Block, E.R. (2000) Association of L-arginine transporters with fodrin: implications for hypoxic inhibition of arginine uptake. *American Journal of Physiology: Lung Cellular and Molecular Physiology*, **278**, L111–17.
108. Rengasamy, A. and Johns, R.A. (1996) Determination of K_m for oxygen of nitric oxide synthase isoforms. *The Journal of Pharmacology and Experimental Therapeutics*, **276**, 30–33.
109. Liao, J.K., Zulueta, J.J., Yu, F.S. *et al.* (1995) Regulation of bovine endothelial constitutive nitric oxide synthase by oxygen. *The Journal of Clinical Investigation*, **96**, 2661–66.
110. Fike, C.D., Kaplowitz, M.R., Thomas, C.J., and Nelin, L.D. (1998) Chronic hypoxia decreases nitric oxide production and endothelial nitric oxide synthase in newborn pig lungs. *American Journal of Physiology: Lung Cellular and Molecular Physiology*, **274**, L517–26.
111. Xu, X.P., Pollock, J.S., Tanner, M.A., and Myers, P.R. (1995) Hypoxia activates nitric oxide synthase and stimulates nitric oxide production in porcine coronary resistance arteriolar endothelial cells. *Cardiovascular Research*, **30**, 841–47.
112. Arnet, U.A., McMillan, A., Dinerman, J.L. *et al.* (1996) Regulation of endothelial nitric-oxide synthase during hypoxia. *The Journal of Biological Chemistry*, **271**, 15069–73.
113. Su, Y. and Block, E.R. (2000) Role of calpain in hypoxic inhibition of nitric oxide synthase activity in pulmonary endothelial cells. *American Journal of Physiology: Lung Cellular and Molecular Physiology*, **278**, L1204–12.
114. Murata, T., Sato, K., Hori, M. *et al.* (2002) Decreased endothelial nitric-oxide synthase (eNOS) activity resulting from abnormal interaction between eNOS and its regulatory proteins in hypoxia-induced pulmonary hypertension. *The Journal of Biological Chemistry*, **277**, 44085–92.
115. Liu, J.Q., Zelko, I.N., Erbynn, E.M. *et al.* (2006) Hypoxic pulmonary hypertension: role of superoxide and NADPH oxidase (gp91^{phox}). *American Journal of Physiology: Lung Cellular and Molecular Physiology*, **290**, L2–10.
116. Kato, G.J., Onyekwere, O.C., and Gladwin, M.T. (2007) Pulmonary hypertension in sickle cell disease: relevance to children. *Pediatric Hematology and Oncology*, **24**, 159–70.
117. Morris, C.R., Kato, G.J., Poljakovic, M. *et al.* (2005) Dysregulated arginine metabolism, hemolysis-associated pulmonary hypertension, and mortality in sickle cell disease. *The Journal of the American Medical Association*, **294**, 81–90.
118. Hsu, L.L., Champion, H.C., Campbell-Lee, S.A. *et al.* (2007) Hemolysis in sickle cell mice causes pulmonary hypertension due to global impairment in nitric oxide bioavailability. *Blood*, **109**, 3088–98.
119. Pritchard, K.A. Jr., Ou, J., Ou, Z. *et al.* (2004) Hypoxia-induced acute lung injury in murine models of sickle cell disease. *American Journal of Physiology: Lung Cellular and Molecular Physiology*, **286**, L705–14.
120. Novoradovsky, A., Brantly, M.L., Waclawiw, M.A. *et al.* (1999) Endothelial nitric oxide synthase as a potential susceptibility gene in the pathogenesis of emphysema in alpha$_1$-antitrypsin deficiency. *American Journal of Respiratory Cell and Molecular Biology*, **20**, 441–47.
121. Barbera, J.A., Peinado, V.I., Santos, S. *et al.* (2001) Reduced expression of endothelial nitric oxide synthase in pulmonary arteries of smokers. *American Journal of Respiratory and Critical Care Medicine*, **164**, 709–13.
122. Lee, J.H., Lee, D.S., Kim, E.K. *et al.* (2005) Simvastatin inhibits cigarette smoking-induced emphysema and pulmonary hypertension in rat lungs. *American Journal of Respiratory and Critical Care Medicine*, **172**, 987–93.
123. Su, Y., Han, W., Giraldo, C. *et al.* (1998) Effect of cigarette smoke extract on nitric oxide synthase in pulmonary artery endothelial cells. *American Journal of Respiratory Cell and Molecular Biology*, **19**, 819–25.

124. Lowe, E.R., Everett, A.C., Lee, A.J. *et al.* (2005) Time-dependent inhibition and tetrahydrobiopterin depletion of endothelial nitric-oxide synthase caused by cigarettes. *Drug Metabolism and Disposition*, **33**, 131–38.

125. Clini, E., Cremona, G., Campana, M. *et al.* (2000) Production of endogenous nitric oxide in chronic obstructive pulmonary disease and patients with cor pulmonale. Correlates with echo-Doppler assessment. *American Journal of Respiratory and Critical Care Medicine*, **162**, 446–50.

126. Balasubramaniam, V., Tang, J.R., Maxey, A. *et al.* (2003) Mild hypoxia impairs alveolarization in the endothelial nitric oxide synthase (eNOS) deficient mouse. *American Journal of Physiology: Lung Cellular and Molecular Physiology*, **284**, L964–71.

7

Pulmonary Endothelial Cell Surface Metabolic Functions

Usamah S. Kayyali and Barry L. Fanburg
Tufts University School of Medicine, Tufts Medical Center, Boston, MA, USA

INTRODUCTION

By virtue of its interface with the pulmonary blood, the endothelium of the pulmonary circulation is positioned to provide unique regulatory control over both soluble and cellular components of systemic venous blood before its return to the systemic circulation [1]. Any constituent of the systemic circulation may be conditioned by actions at the surface of the pulmonary endothelium, which constitutes the largest component of the total body endothelium. It has long been known that composition of the systemic arterial blood differs from that of the venous return. The pulmonary endothelial layer expresses a variety of plasma membrane proteins that possess enzymatic, transport, and binding properties (Figure 7.1). Some of these proteins on the cell surface may be in equilibrium with their counterparts in the plasma. Many of the proteins partner with glycoaminoglycans and lipid components of the endothelial cell (EC) membrane to stabilize their attachment and produce their actions. Although they have been identified to be present on the pulmonary EC surface, the physiologic function and pathologic roles of many of these proteins are presently unknown. In addition to altering substances in the venous return to the right side of the heart, these membrane proteins may be important in transmitting communications from the EC to other vascular cells including smooth muscle cells (SMCs), fibroblasts, and pericytes. Ligands for these cell surface proteins such as serotonin [5-hydroxytryptamine (5-HT)] have been shown to alter gene expression of other pulmonary cells such as SMCs [2, 3]. Permeability of the endothelial surface also may be regulated by cell surface proteins. Many of these actions of the endothelium will be discussed in greater detail in other chapters of this book.

Although there are many hydrolytic proteins on the EC surface, one of the most well-characterized pulmonary endothelial plasma membrane proteins is angiotensin I-converting enzyme (ACE). ACE hydrolyzes the terminal dipeptide from angiotensin I to form the vasoactive octapeptide, angiotensin II, and also degrades bradykinin. Similarly, lipoprotein lipase enzymatically interacts with circulating lipids and may also affect vascular physiology. Proteins with active transport properties such as the serotonin transporter (SERT) are present on the EC surface and clear circulating molecules, such as 5-HT, from the venous return [4]. There are other proteins with transport function, such as the system L neutral amino acid transporters [5, 6], that are in need of study.

Binding properties of endothelial surface membrane proteins have only begun to be characterized, but these proteins may participate in important pathological processes as diverse as thrombus formation, leukocyte accumulation resulting in membrane injury and alterations in permeability (see Chapter 10), and attachment of circulating tumor cells resulting in metastatic disease in the lung (see Chapter 30). Information currently known about these components of the endothelial surface will be discussed in this chapter, but it is recognized that many more remain to be discovered and may play important physiologic and pathologic roles in the pulmonary circulation.

PROTEINS WITH HYDROLYTIC PROPERTIES ON THE EC SURFACE

One of the best-characterized enzymes on the surface of the pulmonary endothelium is ACE – a large glycoprotein composed of a single polypeptide chain that was first discovered over 50 years ago. About three decades

The Pulmonary Endothelium: Function in health and disease Editors Norbert F. Voelkel, Sharon Rounds

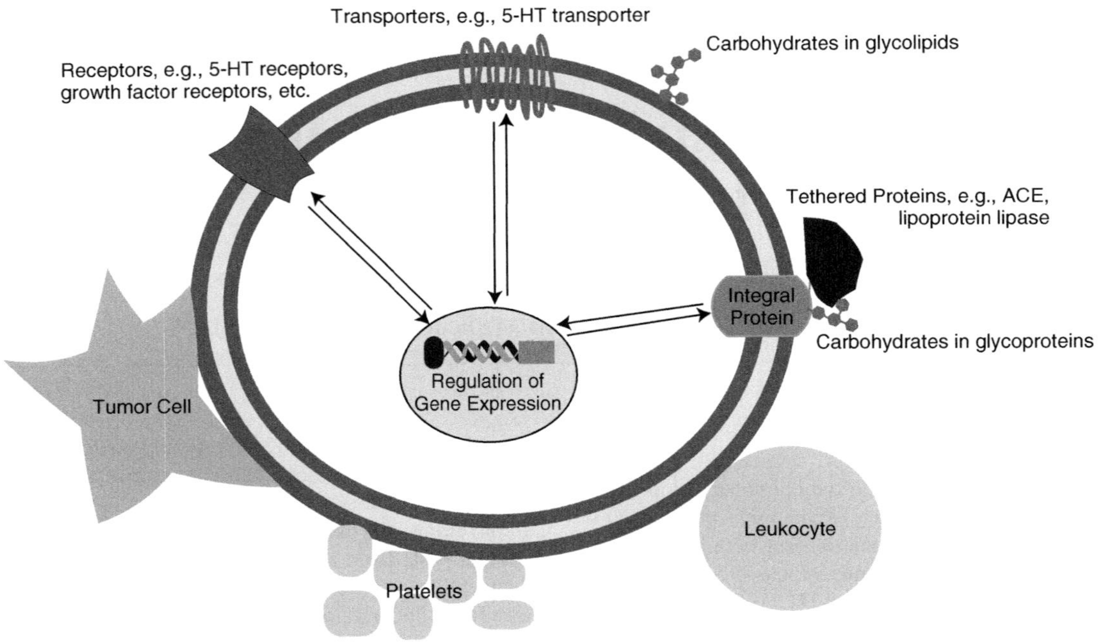

Figure 7.1 In addition to small molecules, protein growth factors, ECM proteins, and formed elements of blood attach to glycoproteins and lipoproteins on the EC surface.

ago this enzyme was localized on the surface of ECs of the pulmonary circulation [7, 8]. Since ACE is a metalloproteinase that hydrolyzes the circulating decapeptide angiotensin I to the vasoactive form, angiotensin II, and degrades the vasodilator bradykinin, it likely has significance in the regulation of the renin–angiotensin system and systemic blood pressure. However, ACE is also present in a soluble form in the serum [9] and on the surface of ECs from the systemic circulation. The relative contributions of the surface-bound and soluble ACE in participation in the renin–angiotensin–aldosterone system and regulation of systemic blood pressure remain unknown. Furthermore, the kinetics of interchange between the enzyme anchored to membrane through a hydrophobic domain near its C-terminal region [10, 11] and soluble ACE are incompletely understood [12]. Endothelial cellular ACE is clearly under regulatory control by factors such as hepatocyte growth factor [13], hormones [14, 15], a factor produced by SMCs [16], cAMP, and hypoxia [17]. In addition, a familial elevation of serum ACE has been identified [18]. Another important point to consider about ACE function is that the enzyme itself has been proposed to mediate signaling via c-Jun N-terminal kinase, which might explain the observation that ACE inhibitors exert actions beyond those mediated by inhibiting its enzymatic activity on peptide substrates [19].

ACE may have functions and utilities other than regulation of systemic circulatory pressure through the renin–angiotensin–aldosterone system [20]. Although angiotensin II causes proliferation and hypertrophy of vascular SMCs [21–23], there is no clear indication that angiotensin II influences pressures in the pulmonary circulation. Of interest, serum ACE has been noted to be elevated in sarcoidosis, but here too the relationship of this elevation in serum to that present on the endothelial surface is unknown. It is possible that the elevated serum ACE in sarcoidosis is derived from macrophages in granulomas [24]. Pulmonary endothelial ACE has been targeted by Danilov *et al.* with agents ligated to its antibody as a method to deliver imaging and therapeutic substances to the pulmonary vasculature [25, 26] (see Chapter 22). Actions of ACE also have been used to assess pulmonary vascular surface area and blood flow [27, 28]. Although consideration has been given to the possibility that release of ACE in the pulmonary circulation might serve as a marker of pulmonary vascular injury [29, 30], a practical use of this approach has never been brought to fruition. Inhibitors of ACE have been used in attempts to prevent vascular injury [31]. Polymorphisms of ACE have been associated with systemic hypertension and coronary heart disease [32–34], but there is no known relationship to pulmonary vascular disease. In conclusion, although many studies have been carried out on ACE in the pulmonary circulation, its role and potential targeting warrant further investigation.

Lipoprotein lipase is another example of an EC surface protein that has catalytic properties. This enzyme participates in the hydrolysis of plasma lipoproteins to form triglycerides with subsequent uptake of their fatty acids into tissues [35]. This process has been studied extensively in relation to atherosclerosis [36]. Studies have been carried out to access the binding of lipoprotein lipase to the endothelial surface and have suggested attachment to heparin-like glycoaminoglycans [37]. Work by Saxena *et al.* indicates that this attachment occurs through a 220-kDa heparin sulfate proteoglycan [38]. These findings are important as the binding of lipoprotein lipase to the EC surface may be prototypic of that of other plasma components such as platelet and clotting factors, antithrombin III, and lipoproteins. Similarly to ACE, lipoprotein lipase exists in the circulation both soluble in plasma and attached to the endothelium. Its release from the endothelial surface may be regulated by a variety of factors, including tumor necrosis factor [39]. Whether lipoprotein lipase in the pulmonary circulation mediates unique functions that are distinct from the systemic circulation is an area worthy of investigation.

ACTIVE TRANSPORT AT THE EC SURFACE

Several cell surface proteins that participate in active transport have been specifically identified, but one that has been most fully characterized is SERT. This transporter is a 12-transmembrane serpentine loop molecule [40], similar to many G-protein receptors, that is capable of coupling with the active transport of sodium and removing the circulating tryptophan derivative, 5-HT, from the blood. Systemically administered 5-HT is removed from the blood by pulmonary ECs during the first passage [41]. One possible function of 5-HT uptake by pulmonary endothelium is to protect the left side of the heart from high concentrations of 5-HT that may produce cardiac heart valve fibrosis. Valvular abnormalities on the left side of the heart have been described in the carcinoid syndrome where the pulmonary 5-HT transport system is probably overwhelmed by high concentrations of circulating 5-HT. Polymorphisms of the 5-HT transporter have been found in pulmonary hypertension [42] and 5-HT itself has been noted to possibly participate in the pathogenesis of pulmonary hypertension [43]. Similarly, 5-HT may be important in the pathogenesis of pulmonary hypertension associated with anorexigens [44]. The mechanism by which 5-HT produces pulmonary hypertension and the method by which the 5-HT transporter may participate in the process is still under considerable study. A schematic of the actions of 5-HT on both the endothelium and adjacent SMCs and fibroblasts is shown in Figure 7.2. 5-HT has been shown to trigger pulmonary artery smooth muscle hypertrophy and proliferation [45] as well as fibroblast proliferation [46]. To further complicate the process, a variety of 15 or more 5-HT receptors are present on the endothelial surface in addition to the 5-HT transporter.

As opposed to SMCs where the functional role may be directly related to muscle contractility and growth, the role of the 5-HT transporter on the EC surface other than for purposes of clearing 5-HT from blood is currently unknown. However, SERT may play a role in some of the reported effects of 5-HT on ECs such as increased endothelial permeability *in vivo* [47–49] and in cultured monolayers [50, 51]. Other reports indicate that 5-HT decreases endothelial permeability and inflammatory infiltration [52]. The fact that 5-HT induced changes in endothelial permeability and actin stress fiber formation [50, 53] can be blocked by 5-HT receptor antagonists suggests an important function for 5-HT receptors in EC physiology that warrants further studies. The role of the endothelial 5-HT transporter, if any, in regulating the permeability barrier is also worthy of investigation. Depending on what part of the pulmonary vasculature is affected, 5-HT modulation of endothelial permeability might have implications for lung injury and pulmonary edema, as well as for vascular remodeling in conditions such as pulmonary hypertension.

BINDING PROPERTIES OF THE EC SURFACE

There is long-standing recognition that circulating leukocytes interact with the pulmonary endothelial surface and that this interaction may result in the release of mediators that alter lung function or capillary permeability [54, 55]. Various models have been proposed to study the process of rolling and adhesion of leukocytes in the pulmonary microvasculature [56] and certain members of the selectin family have been implicated in the process (see Chapter 10). Small amounts of circulating lipopolysaccharide in the presence of leukocyte chemotactic factor produce leukopenia and lung injury in experimental animals suggesting that sequestration of leukocytes within the pulmonary circulation is injurious to the lung [57]. Certain adhesion-promoting glycoproteins such as CD11/CD18 on the surface of leukocytes have been suggested to be important in the sequestration process [58]. Similarly, it has been suggested that shock induced by various methods enhances leukocyte–endothelial interactions in the pulmonary microcirculation, resulting in organ injury and the respiratory distress syndrome [59]. These and other similar observations emphasize the importance of developing more knowledge about adherence of proteins to the surface of the pulmonary vascular endothelium.

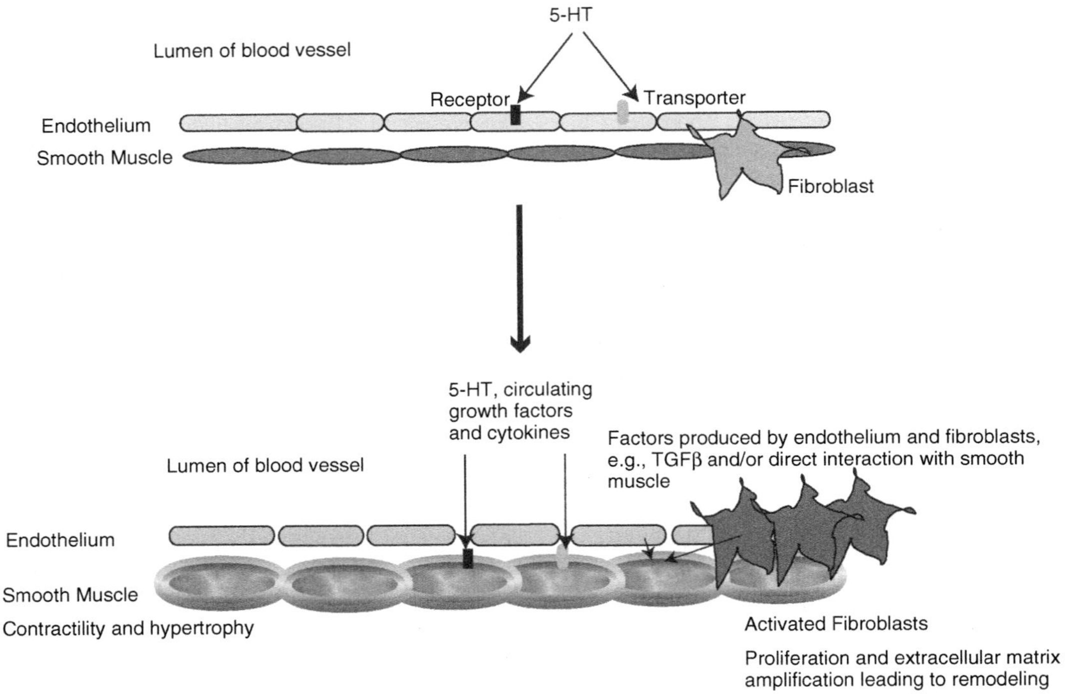

Figure 7.2 In addition to the interaction of endothelium at the vascular fluid interface, ECs may regulate vascular permeability or produce substances that influence vascular SMCs or fibroblasts.

Similar to leukocytes, platelets at times aggregate in the pulmonary microcirculation and this aggregation may lead to the formation of pulmonary thrombi that interfere with circulation (see Chapters 10 and 25). A well-known initiator of such events is thrombin and the specifics of this process have been studied extensively [60, 61]. Similarly, ischemia–reperfusion has been demonstrated to result in platelet adhesion in subpleural arterioles in rabbits [62]. Like the adherence of leukocytes, that of platelets is not a passive process but rather requires specific binding proteins. For example, platelet-endothelial cell adhesion molecule-1 (CD31) has been demonstrated to be required for interaction of platelets with ECs after irradiation [63].

Selective adherence to the pulmonary EC surface may also be important in hematogenous metastases to the lung (see Chapter 30). Although the potential for such an occurrence has been recognized [64, 65], few studies have yet addressed this experimentally.

Like leukocytes, ECs bind a variety of lectins. Characterization of different lectin-binding proteins has shown that there is considerable heterogeneity of ECs among vascular beds (see Chapter 9). For example, macrovascular ECs express a different lectin-binding profile from that of microvascular ECs. These differences may be due to different progenitors for these cell populations, but also may reflect variations in microenvironments in which the ECs develop. In the pulmonary circulation, microvascular ECs preferentially bind the lectin of *Griffonia simplicifolia* that has affinity toward galactose-enriched moieties [66]. On the other hand, ECs from the pulmonary macrovasculature preferentially bind *Helix pomatia* that has affinity towards α- and β-*N*-acetyl-galactosamine [66]. These differences offer possibilities for identification and isolation of different pulmonary vascular EC populations. Moreover, they suggest the possibility that EC populations from different locations interact variably with inflammatory and other types of cells as well as with similar ECs. It is possible that signaling in different EC populations varies even when the expression of receptors and transporters is similar. Thus, different pulmonary vascular ECs might express similar levels of ACE and 5-HT receptors and transporters, but respond differently to stimuli because heterogeneity of surface proteins affects receptor coupling. As examples, ACE dimerization is known to be important for ACE inhibitor action and ACE shedding is believed to involve a moiety that exhibits lectin-like behavior with affinity for galactose [67, 68]. Also, the 5-HT transporter is a glycoprotein that has been reported to bind to mitogen lectins such as conconavalin A [69], which also modulate its function in lymphocytes [70]. It would be of interest to investigate the correlation of the lectin-binding profile of different vascular beds with surface protein function.

There has been ample evidence that ECs from variable sources respond differently to stimuli. For example, cyclic stretch increases β-catenin in rat pulmonary macrovascular, but decreases it in microvascular ECs [71]. Since mechanical stretch activates signaling via focal adhesions and adherens junctions (AJs), it would be of interest to test the role played by lectin-binding proteins in modulating the function of cell surface proteins such as VE-cadherin, integrins, and focal adhesion kinase (FAK). All of these cell surface proteins coordinate the communication of information from the exterior of the EC to the interior via complex signaling that involves the cytoskeleton and results in modification of the cell structure and biophysical properties, as well as its expression profile.

EC SURFACE PROTEINS THAT PARTICIPATE IN BARRIER REGULATION AND INTERCELLULAR COMMUNICATION

In addition to receptors that endocytose specific molecules for transcellular transport, several endothelial surface proteins are believed to play a major role in regulation of endothelial barrier properties [72]. While these proteins are discussed in detail elsewhere in the book (see Chapters 3 and 8), molecules that mediate cell–cell adhesion, such as cadherens and occludins, are believed to mediate bidirectional signaling and communication between the outside of the cell and the inside via their interaction with the cytoskeleton. Similar communication is also mediated by integrin complex proteins that regulate interaction with the extracellular matrix (ECM; see Chapter 4). Together these proteins regulate the number and size of intercellular gaps and hence endothelial barrier permeability. Differences in these proteins across pulmonary EC types are an area of active investigation. The cell's surface proteins that are involved in endothelial barrier function are part of different types of cellular junctions. Integrins bind to the ECM as part of focal adhesions, which link to actin filaments via a complex of proteins including FAK, paxillin, and vinculin, and coordinate communication in both directions. Another type of junction that coordinates communication between ECs is the AJ. AJs are formed by homophilic interaction between VE-cadherin molecules in different ECs. These proteins also coordinate communication and intercellular gap formation by communicating with the interior of the cell through the actin cytoskeleton via complexes that involve proteins such as β-catenin. Tight junctions represent yet another type of intercellular junction that is composed of homophilic binding of intermembrane occludins. In ECs, claudin is the major occludin, which also coordinates communication to the interior of the cell through the actin cytoskeleton via other proteins, such as zona occludens. Finally, there are gap junctions (GJs) that constitute channels that form between cells through channel forming connexins (see Chapter 3). GJs can allow signals and second messengers to travel between cells such as in the case of observed Ca^{2+} fluxes in pulmonary ECs.

CONCLUSIONS AND PERSPECTIVES

In summary, the pulmonary EC surface is an interactive interface that mediates a variety of physiological and pathological responses. Membrane-associated components on the luminal side of the endothelium regulate interactions with blood cells and circulating factors. Receptors and transporters for ligands such as 5-HT might have wider roles that are unique to the pulmonary circulation, and the complexity of the signaling in relation to other ligands and pathways is only beginning to be appreciated. Our understanding of isolated effects of the interaction of ligands and formed blood components with EC surface needs to be considered in the wider context of pursuant interaction of circulating components or other factors produced by the endothelium with other cells in the pulmonary vasculature such as SMCs and fibroblasts.

References

1. Gillis, C.N. and Pitt, B.R. (1982) The fate of circulating amines within the pulmonary circulation. *Annual Review of Physiology*, **44**, 269–81.
2. Lee, S.L., Wang, W.W., Lanzillo, J.J., and Fanburg, B.L. (1994) Regulation of serotonin-induced DNA synthesis of bovine pulmonary artery smooth muscle cells. *American Journal of Physiology: Lung Cellular and Molecular Physiology*, **266**, L53–60.
3. Simon, A.R., Severgnini, M., Takahashi, S. *et al.* (2005) 5-HT induction of c-*fos* gene expression requires reactive oxygen species and Rac1 and Ras GTPases. *Cell Biochemistry and Biophysics*, **42**, 263–76.
4. Fanburg, B.L. and Lee, S.L. (1997) A new role for an old molecule: serotonin as a mitogen. *American Journal of Physiology: Lung Cellular and Molecular Physiology*, **272**, L795–806.
5. Takabe, W., Kanai, Y., Chairoungdua, A. *et al.* (2004) Lysophosphatidylcholine enhances cytokine

production of endothelial cells via induction of L-type amino acid transporter 1 and cell surface antigen 4F2. *Arteriosclerosis, Thrombosis, and Vascular Biology*, **24**, 1640–45.

6. Verrey, F. (2003) System L: heteromeric exchangers of large, neutral amino acids involved in directional transport. *Pflugers Archiv*, **445**, 529–33.
7. Ryan, U.S., Ryan, J.W., Whitaker, C., and Chiu, A. (1976) Localization of angiotensin converting enzyme (kininase II). II. Immunocytochemistry and immunofluorescence. *Tissue and Cell*, **8**, 125–45.
8. Johnson, A.R. and Erdos, E.G. (1977) Metabolism of vasoactive peptides by human endothelial cells in culture. Angiotensin I converting enzyme (kininase II) and angiotensinase. *The Journal of Clinical Investigation*, **59**, 684–95.
9. Lanzillo, J.J., Stevens, J., Tumas, J., and Fanburg, B.L. (1983) Spontaneous change of human plasma angiotensin I converting enzyme isoelectric point. *Archives of Biochemistry and Biophysics*, **227**, 434–39.
10. Beldent, V., Michaud, A., Bonnefoy, C. *et al.* (1995) Cell surface localization of proteolysis of human endothelial angiotensin I-converting enzyme. Effect of the amino-terminal domain in the solubilization process. *The Journal of Biological Chemistry*, **270**, 28962–69.
11. Ramchandran, R., Sen, G.C., Misono, K., and Sen, I. (1994) Regulated cleavage-secretion of the membrane-bound angiotensin-converting enzyme. *The Journal of Biological Chemistry*, **269**, 2125–30.
12. Silberstein, L.E., Berkman, E.M., and Fanburg, B.L. (1983) Angiotensin-1-converting enzyme (ACE) activity in human serum following plasma exchange. *Proceedings of the Society for Experimental Biology and Medicine*, **173**, 222–26.
13. Day, R.M., Thiel, G., Lum, J. *et al.* (2004) Hepatocyte growth factor regulates angiotensin converting enzyme expression. *The Journal of Biological Chemistry*, **279**, 8792–801.
14. Dasarathy, Y., Lanzillo, J.J., and Fanburg, B.L. (1992) Stimulation of bovine pulmonary artery endothelial cell ACE by dexamethasone: involvement of steroid receptors. *American Journal of Physiology: Lung Cellular and Molecular Physiology*, **263**, L645–49.
15. Krulewitz, A.H., Baur, W.E., and Fanburg, B.L. (1984) Hormonal influence on endothelial cell angiotensin-converting enzyme activity. *American Journal of Physiology: Cell Physiology*, **247**, C163–68.
16. Yu, F.S., Lee, S.L., and Fanburg, B.L. (1989) Smooth muscle cell conditioned medium elevates angiotensin-converting enzyme of bovine pulmonary artery endothelial cells. *American Journal of Respiratory Cell and Molecular Biology*, **1**, 401–5.
17. Krulewitz, A.H. and Fanburg, B.L. (1984) The effect of oxygen tension on the *in vitro* production and release of angiotensin-converting enzyme by bovine pulmonary artery endothelial cells. *The American Review of Respiratory Disease*, **130**, 866–69.
18. Okabe, T., Fujisawa, M., Yotsumoto, H. *et al.* (1985) Familial elevation of serum angiotensin converting enzyme. *The Quarterly Journal of Medicine*, **55**, 55–61.
19. Kohlstedt, K., Brandes, R.P., Muller-Esterl, W. *et al.* (2004) Angiotensin-converting enzyme is involved in outside-in signaling in endothelial cells. *Circulation Research*, **94**, 60–67.
20. Ehlers, M.R. and Riordan, J.F. (1989) Angiotensin-converting enzyme: new concepts concerning its biological role. *Biochemistry*, **28**, 5311–18.
21. Berk, B.C., Vekshtein, V., Gordon, H.M., and Tsuda, T. (1989) Angiotensin II-stimulated protein synthesis in cultured vascular smooth muscle cells. *Hypertension*, **13**, 305–14.
22. Geisterfer, A.A., Peach, M.J., and Owens, G.K. (1988) Angiotensin II induces hypertrophy, not hyperplasia, of cultured rat aortic smooth muscle cells. *Circulation Research*, **62**, 749–56.
23. Morishita, R., Gibbons, G.H., Ellison, K.E. *et al.* (1994) Evidence for direct local effect of angiotensin in vascular hypertrophy. *In vivo* gene transfer of angiotensin converting enzyme. *The Journal of Clinical Investigation*, **94**, 978–84.
24. Gilbert, S., Steinbrech, D.S., Landas, S.K., and Hunninghake, G.W. (1993) Amounts of angiotensin-converting enzyme mRNA reflect the burden of granulomas in granulomatous lung disease. *The American Review of Respiratory Disease*, **148**, 483–86.
25. Danilov, S.M., Faerman, A.I., Printseva, O. *et al.* (1987) Immunohistochemical study of angiotensin-converting enzyme in human tissues using monoclonal antibodies. *Histochemistry*, **87**, 487–90.
26. Danilov, S., Jaspard, E., Churakova, T. *et al.* (1994) Structure–function analysis of angiotensin I-converting enzyme using monoclonal antibodies. Selective inhibition of the amino-terminal active site. *The Journal of Biological Chemistry*, **269**, 26806–14.
27. Fanburg, B.L. (1988) Relationship of the pulmonary vascular endothelium to altered pulmonary vascular resistance. State of the art. *Chest*, **93**, 101S–105S.

28. Pitt, B.R., Lister, G., Davies, P., and Reid, L. (1987) Correlation of pulmonary ACE activity and capillary surface area during postnatal development. *Journal of Applied Physiology*, **62**, 2031–41.
29. Nukiwa, T., Matsuoka, R., Takagi, H. *et al.* (1982) Responses of serum and lung angiotensin-converting enzyme activities in the early phase of pulmonary damage induced by oleic acid in dogs. *The American Review of Respiratory Disease*, **126**, 1080–86.
30. Rakugi, H., Kim, D.K., Krieger, J.E. *et al.* (1994) Induction of angiotensin converting enzyme in the neointima after vascular injury. Possible role in restenosis. *The Journal of Clinical Investigation*, **93**, 339–46.
31. Powell, J.S., Clozel, J.P., Muller, R.K. *et al.* (1989) Inhibitors of angiotensin-converting enzyme prevent myointimal proliferation after vascular injury. *Science*, **245**, 186–88.
32. Higashimori, K., Zhao, Y., Higaki, J. *et al.* (1993) Association analysis of a polymorphism of the angiotensin converting enzyme gene with essential hypertension in the Japanese population. *Biochemical and Biophysical Research Communication*, **191**, 399–404.
33. Cambien, F., Poirier, O., Lecerf, L. *et al.* (1992) Deletion polymorphism in the gene for angiotensin-converting enzyme is a potent risk factor for myocardial infarction. *Nature*, **359**, 641–44.
34. Oike, Y., Hata, A., Ogata, Y. *et al.* (1995) Angiotensin converting enzyme as a genetic risk factor for coronary artery spasm. Implication in the pathogenesis of myocardial infarction. *The Journal of Clinical Investigation*, **96**, 2975–79.
35. Saxena, U., Witte, L.D., and Goldberg, I.J. (1989) Release of endothelial cell lipoprotein lipase by plasma lipoproteins and free fatty acids. *The Journal of Biological Chemistry*, **264**, 4349–55.
36. Tabas, I., Williams, K.J., and Boren, J. (2007) Subendothelial lipoprotein retention as the initiating process in atherosclerosis: update and therapeutic implications. *Circulation*, **116**, 1832–44.
37. Shimada, K., Gill, P.J., Silbert, J.E. *et al.* (1981) Involvement of cell surface heparin sulfate in the binding of lipoprotein lipase to cultured bovine endothelial cells. *The Journal of Clinical Investigation*, **68**, 995–1002.
38. Saxena, U., Klein, M.G., and Goldberg, I.J. (1991) Identification and characterization of the endothelial cell surface lipoprotein lipase receptor. *The Journal of Biological Chemistry*, **266**, 17516–21.
39. Saxena, U., Witte, L.D., and Goldberg, I.J. (1990) Tumor necrosis factor induced release of endothelial cell lipoprotein lipase. *Arteriosclerosis*, **10**, 470–76.
40. Blakely, R.D., Berson, H.E., Fremeau, R.T. Jr. *et al.* (1991) Cloning and expression of a functional serotonin transporter from rat brain. *Nature*, **354**, 66–70.
41. Strum, J.M. and Junod, A.F. (1972) Radioautographic demonstration of 5-hydroxytryptamine-^{3}H uptake by pulmonary endothelial cells. *The Journal of Cell Biology*, **54**, 456–67.
42. Eddahibi, S., Humbert, M., Fadel, E. *et al.* (2001) Serotonin transporter overexpression is responsible for pulmonary artery smooth muscle hyperplasia in primary pulmonary hypertension. *The Journal of Clinical Investigation*, **108**, 1141–50.
43. Sato, K., Webb, S., Tucker, A. *et al.* (1992) Factors influencing the idiopathic development of pulmonary hypertension in the fawn hooded rat. *The American Review of Respiratory Disease*, **145**, 793–97.
44. Rothman, R.B., Ayestas, M.A., Dersch, C.M., and Baumann, M.H. (1999) Aminorex, fenfluramine, and chlorphentermine are serotonin transporter substrates. Implications for primary pulmonary hypertension. *Circulation*, **100**, 869–75.
45. Lee, S.L., Wang, W.W., Lanzillo, J.J., and Fanburg, B.L. (1994) Serotonin produces both hyperplasia and hypertrophy of bovine pulmonary artery smooth muscle cells in culture. *American Journal of Physiology: Lung Cellular and Molecular Physiology*, **266**, L46–52.
46. Welsh, D.J., Harnett, M., MacLean, M., and Peacock, A.J. (2004) Proliferation and signaling in fibroblasts: role of 5-hydroxytryptamine2A receptor and transporter. *American Journal of Respiratory and Critical Care Medicine*, **170**, 252–59.
47. Bergman, R.K. and Munoz, J.J. (1975) Effect of *Bordetella pertussis* extract and vasoactive amines on vascular permeability. *The Journal of Allergy and Clinical Immunology*, **55**, 378–85.
48. Thurston, G., Suri, C., Smith, K. *et al.* (1999) Leakage-resistant blood vessels in mice transgenically overexpressing angiopoietin-1. *Science*, **286**, 2511–14.
49. Delaunois, A., Gustin, P., and Ansay, M. (1993) Role of neuropeptides in acetylcholine-induced edema in isolated and perfused rabbit lungs. *The Journal of Pharmacology and Experimental Therapeutics*, **266**, 483–91.
50. Lee, H.-Z. (1997) Inhibitory effect of 2-phenyl-4-quinolone on serotonin-mediated changes in the

morphology and permeability of endothelial monolayers. *European Journal of Pharmacology*, **335**, 245–54.

51. Nomura, A., Zhang, M., Sakamoto, T. *et al.* (2003) Anti-inflammatory activity of creatine supplementation in endothelial cells *in vitro*. *British Journal of Pharmacology*, **139**, 715–20.
52. Mineau-Hanschke, R., Wiles, M.E., Morel, N. *et al.* (1990) Modulation of cultured pulmonary microvessel and arterial endothelial cell barrier structure and function by serotonin. *Microvascular Research*, **39**, 140–55.
53. Shepro, D. and Hechtman, H.B. (1985) Endothelial serotonin uptake and mediation of prostanoid secretion and stress fiber formation. *Federation Proceedings*, **44**, 2616–19.
54. Brigham, K.L. and Meyrick, B. (1984) Interactions of granulocytes with the lungs. *Circulation Research*, **54**, 623–35.
55. Hogg, J.C., Coxson, H.O., Brumwell, M.L. *et al.* (1994) Erythrocyte and polymorphonuclear cell transit time and concentration in human pulmonary capillaries. *Journal of Applied Physiology*, **77**, 1795–800.
56. Sikora, L., Rao, S.P., and Sriramarao, P. (2003) Selectin-dependent rolling and adhesion of leukocytes in nicotine-exposed microvessels of lung allografts. *American Journal of Physiology: Lung Cellular and Molecular Physiology*, **285**, L654–63.
57. Worthen, G.S., Haslett, C., Rees, A.J. *et al.* (1987) Neutrophil-mediated pulmonary vascular injury. Synergistic effect of trace amounts of lipopolysaccharide and neutrophil stimuli on vascular permeability and neutrophil sequestration in the lung. *The American Review of Respiratory Disease*, **136**, 19–28.
58. Yoder, M.C., Checkley, L.L., Giger, U. *et al.* (1990) Pulmonary microcirculatory kinetics of neutrophils deficient in leukocyte adhesion-promoting glycoproteins. *Journal of Applied Physiology*, **69**, 207–13.
59. Nakagawa, N.K., Nogueira, R.A., Correia, C.J. *et al.* (2006) Leukocyte–endothelium interactions after hemorrhagic shock/reperfusion and cecal ligation/puncture: an intravital microscopic study in rat mesentery. *Shock*, **26**, 180–86.
60. Vanhauwe, J.F., Thomas, T.O., Minshall, R.D. *et al.* (2002) Thrombin receptors activate G_o proteins in endothelial cells to regulate intracellular calcium and cell shape changes. *The Journal of Biological Chemistry*, **277**, 34143–49.
61. Vogel, S.M., Gao, X., Mehta, D. *et al.* (2000) Abrogation of thrombin-induced increase in pulmonary microvascular permeability in PAR-1 knockout mice. *Physiological Genomics*, **4**, 137–45.
62. Roberts, A.M., Ovechkin, A.V., Mowbray, J.G. *et al.* (2004) Effects of pulmonary ischemia–reperfusion on platelet adhesion in subpleural arterioles in rabbits. *Microvascular Research*, **67**, 29–37.
63. Gaugler, M.H., Vereycken-Holler, V., Squiban, C., and Aigueperse, J. (2004) PECAM-1 (CD31) is required for interactions of platelets with endothelial cells after irradiation. *Journal of Thrombosis and Haemostasis*, **2**, 2020–26.
64. Lafrenie, R., Shaughnessy, S.G., and Orr, F.W. (1992) Cancer cell interactions with injured or activated endothelium. *Cancer Metastasis Reviews*, **11**, 377–88.
65. Glinsky, V.V., Glinsky, G.V., Glinskii, O.V. *et al.* (2003) Intravascular metastatic cancer cell homotypic aggregation at the sites of primary attachment to the endothelium. *Cancer Research*, **63**, 3805–11.
66. King, J., Hamil, T., Creighton, J. *et al.* (2004) Structural and functional characteristics of lung macro- and microvascular endothelial cell phenotypes. *Microvascular Research*, **67**, 139–51.
67. Kohlstedt, K., Gershome, C., Friedrich, M. *et al.* (2006) Angiotensin-converting enzyme (ACE) dimerization is the initial step in the ACE inhibitor-induced ACE signaling cascade in endothelial cells. *Molecular Pharmacology*, **69**, 1725–32.
68. Kost, O.A., Balyasnikova, I.V., Chemodanova, E.E. *et al.* (2003) Epitope-dependent blocking of the angiotensin-converting enzyme dimerization by monoclonal antibodies to the N-terminal domain of ACE: possible link of ACE dimerization and shedding from the cell surface. *Biochemistry*, **42**, 6965–76.
69. Helmeste, D.M., Chudzik, J., and Tang, S.W. (1994) Interaction of lectins with human platelet serotonin transporter. *European Journal of Pharmacology*, **266**, 327–31.
70. Cedeno, N., Urbina, M., Obregon, F., and Lima, L. (2005) Characterization of serotonin transporter in blood lymphocytes of rats. Modulation by *in vivo* administration of mitogens. *Journal of Neuroimmunology*, **159**, 31–40.
71. Adkison, J.B., Miller, G.T., Weber, D.S. *et al.* (2006) Differential responses of pulmonary endothelial phenotypes to cyclical stretch. *Microvascular Research*, **71**, 175–84.
72. Mehta, D. and Malik, A.B. (2006) Signaling mechanisms regulating endothelial permeability. *Physiological Reviews*, **86**, 279–367.

8

Cell Biology of Lung Endothelial Permeability

Guochang Hu and Richard D. Minshall

Departments of Anesthesiology and Pharmacology, and Center for Lung and Vascular Biology, University of Illinois College of Medicine, Chicago, IL, USA

INTRODUCTION

The pulmonary vascular endothelium is uniquely located to filter the entire blood before it enters the systemic circulation, and consequently the integrity of this endothelial barrier is essential for the maintenance of adequate homeostasis in both the pulmonary and systemic circulations. Under physiological conditions, the pulmonary endothelium forms a semipermeable dynamic barrier that controls the exchange of fluid and macromolecules between the blood and the interstitium. This "basal" endothelial permeability plays a crucial role in regulation of normal tissue homeostasis and the maintenance of pulmonary function. During inflammatory responses, the integrity of the vascular endothelial barrier is compromised mainly in postcapillary venules leading to increased permeability [1, 2]. Increased pulmonary endothelial permeability is a hallmark of acute lung injury (ALI) and acute respiratory distress syndrome (ARDS) (see Chapters 21 and 24). Vascular leak of plasma components into pulmonary interstitial tissues and alveoli results in protein-rich pulmonary edema, atelectasis, hypoxemia, and respiratory failure. The degree of increased pulmonary vascular permeability correlates clinically with the severity of lung injury and the amount of neutrophil traffic in the injured lung. A persistent increase in endothelial permeability contributes to the high morbidity and mortality of patients with ALI [3].

There are two cellular pathways that have been identified that control endothelial barrier function (Figure 8.1). Normally, the integrity of the endothelial barrier depends on endothelial cell–cell contact (junctions) and cell–matrix contacts (focal adhesions). This keeps the "paracellular leak" of macromolecules such as albumin low. However, there is basal endothelial albumin permeability which is maintained by movement of these macromolecules through the "transcellular" pathway via caveolae (Figure 8.2). Accumulating evidence has shown convincingly that active transport of albumin across the endothelial barrier *in situ* can be fully accounted for by the formation, fission, and transport of caveolae [4–7]. Paracellular permeability is regulated by protein complexes that make up adherens junctions (AJs), tight junctions (TJs) (Figure 8.3), and focal adhesions, in part through direct interaction of these complexes with the actin cytoskeleton [8]. The increase in vascular permeability is characterized by disruption of endothelial cell–cell and cell–matrix contacts, and opening of paracellular junctions between adjacent cells as well as enhancement of basal permeability (i.e., caveolae-mediated transendothelial transport of albumin). This chapter will focus on the signaling mechanisms and proteins involved in regulation of lung endothelial permeability.

CHARACTERISTICS OF LUNG ENDOTHELIAL PERMEABILITY

With regard to normal barrier function, endothelial cell (EC)s display organ-specific heterogeneity (see Chapter 9). Physical forces and local physiological needs of the specific organ lead to heterogeneity and segmental variation of endothelial junctions. For example, the microvascular endothelial barriers in the lung and brain are much "tighter" than those in the kidney, liver, heart, and

The Pulmonary Endothelium: Function in health and disease Editors Norbert F. Voelkel, Sharon Rounds

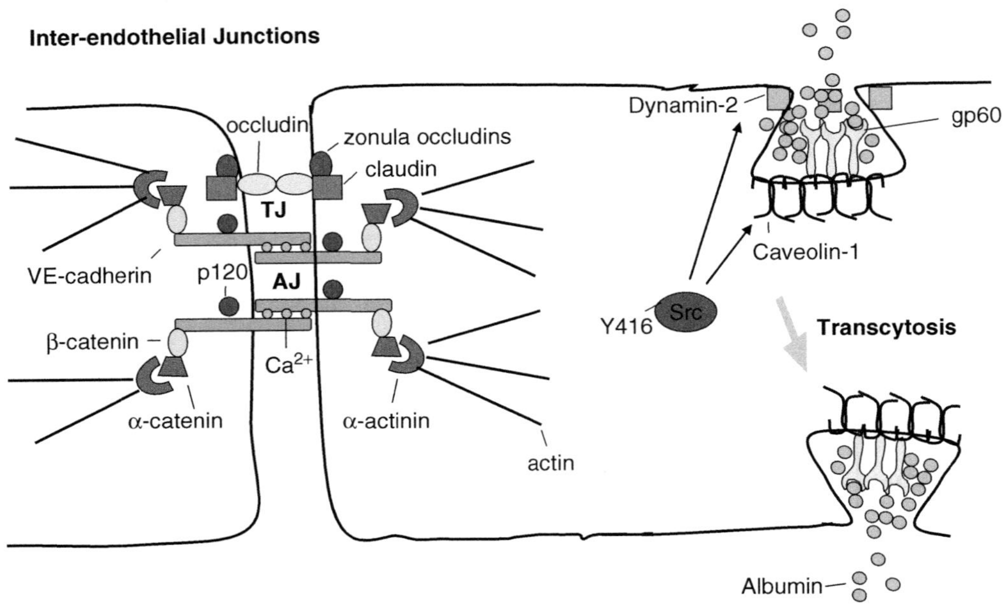

Figure 8.1 Endothelial permeability pathways. The restrictive endothelial barrier is formed by occluding TJs comprised of occludins, claudins, and ZO proteins, and adhesive AJs made up of membrane-spanning VE-cadherin and associated catenins (α, β, and p120-catenin) that are linked to the actin cytoskeleton via α-catenin. Destabilization of interendothelial junctional complexes leads to increased paracellular permeability. Transcellular permeability via caveolae mediates constitutive transport of macromolecules, primarily albumin and associated cargo and fluid-phase molecules, from the luminal to abluminal compartments. Albumin binding to EC surface gp60 induces gp60 clustering and physical interaction with caveolin-1, followed by $G_{i\beta\gamma}$-linked Src tyrosine kinase signaling. Activated Src phosphorylates dynamin-2 and caveolin-1, which initiate plasmalemmal vesicle fission (endocytosis) and trafficking (transcytosis) to the opposing membrane where vesicular contents are deposited (exocytosis).

intestines [9]. Both sites require a restrictive endothelial barrier to ensure vital organ function. The tight barrier of the pulmonary endothelium effectively restricts the leakage of fluid and solutes required for maintenance of normal gas exchange.

Endothelial permeability varies not only in different organs, but also in vessels of different caliber within a single organ [10]. For example, pulmonary microvascular endothelium provides a more restrictive barrier than pulmonary macrovascular endothelium at baseline [11]. In the lung, the special structure of the alveolar–capillary barrier exists for gas exchange between the alveolar air and the blood flowing through the pulmonary capillaries. The pulmonary endothelial barrier is continuously exposed to biophysical forces in the form of shear stress and mechanical strain imposed by both blood flow and respiratory cycles. Owing to the enormous EC surface area of the pulmonary vasculature, the pulmonary endothelium is particularly sensitive to dynamic barrier regulation [12]. Endothelial permeability is also not uniform throughout the pulmonary vasculature, with greater macromolecule diffusion in postcapillary venules compared with pulmonary arterioles in whole lung models [13–15].

Structural Features of Lung Endothelial Barrier

All endothelial barriers share several common features. Anatomically, the structural components responsible for endothelial permeability include the following: (i) TJs composed of occludins, claudins, zonula occludens (ZO)-1, -2, and -3, and cingulin; (ii) AJs composed of cadherins, catenins, actinin, and actin (Figure 8.1); (iii) junctional adhesion molecules (JAMs) (see Chapter 3); and (iv) focal adhesions (see Chapter 4). The presence and organization of interendothelial junctions varies in different organs (see Chapter 9). The pulmonary microvessel endothelium is of the continuous, nonfenestrated type, and is the primary site of fluid and solute exchange. The brain is lined by continuous endothelium connected by TJs that help to maintain the blood–brain barrier. The liver, spleen, and bone marrow sinusoids are lined by a discontinuous endothelium that allows

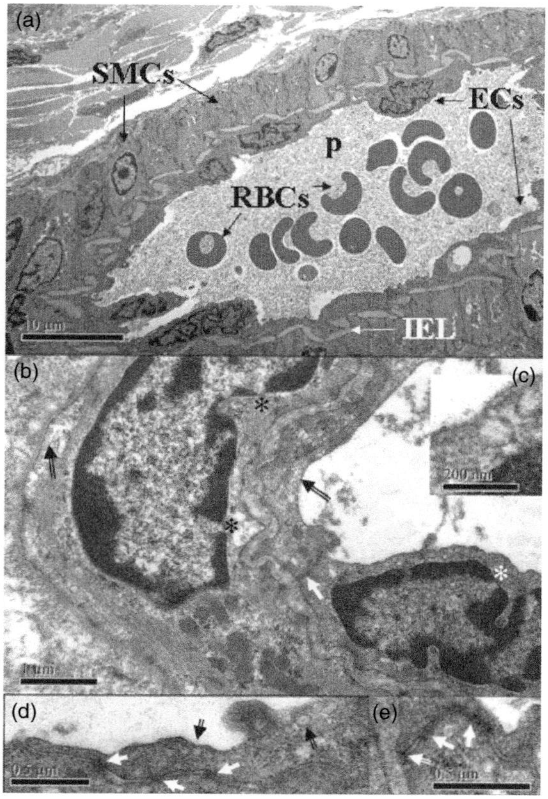

Figure 8.2 Transmission electron micrographs (TEMs) of small arterial blood vessel in the mouse mesentery. (a) Low magnification TEM shows vessel lumen with red blood cells (RBCs) and plasma (p) surrounded by EC monolayer (ECs), internal elastic lamina (IEL), and vascular smooth muscle cell layer (SMCs). Scale bar = 10 μm. (b) Higher magnification image demonstrates nuclear pores (*), junctions between ECs (solid white arrow), and caveolae in both ECs and smooth muscle cells (open black arrows). Scale bar = 1 μm. (c) EC caveolae; scale bar = 200 nm. (d and e) Overlapping ECs highlight abundance of AJs (solid white arrows) and caveolae (open black arrows) with much rarer TJ complex (open white arrow in e). Scale bars = 0.5 μm.

cellular trafficking via intercellular gaps, while the intestinal villi, endocrine glands, and kidneys are lined by a fenestrated endothelium that facilitates selective permeability required for efficient absorption, secretion, and filtering [16]. ECs also possess an abundance of caveolae (Figures 8.2 and 8.3) for transendothelial transport of albumin between the blood and underlying interstitial tissues.

Expression of Junction-Related Proteins

In most of the systemic circulation and in the conduit pulmonary circulation, VE-cadherin is the dominant cadherin isoform [17]. E-cadherin is expressed in bovine brain microvessel endothelium and identified in rat pulmonary capillary endothelium [18–20], and may therefore be the dominant cadherin in rat pulmonary microvascular ECs rather than VE-cadherin [20]. Occludin is expressed strongly in brain endothelium and in testis [21], whereas it is not detected in lung microvessels [22]. In the brain and lung, claudin-5 is expressed in all vessels including arteries, capillaries, and veins, and its expression pattern overlaps with that of VE-cadherin [22]. However, in some vascular beds, such as in the kidney, claudin-5 is expressed in arteries, but not capillaries or veins [23].

Expression of Inflammation-Related Proteins

Vascular endothelial growth factor (VEGF) expression has been identified in perivascular cells of many organs. The lung has the highest level of VEGF gene expression among normal tissues, reflecting the critical role of VEGF in the structural integrity of the lung [24, 25]. Although mesenchymal and alveolar type II epithelial cells are a major source of VEGFs [26], VEGF can slowly diffuse across the alveolar epithelium to adjacent vascular endothelium and act in a paracrine fashion [27]. Expression of adhesion molecules such as intercellular adhesion molecule (ICAM)-1 [28] and P-selectin [29] is also higher in the lungs as compared to other organs (see Chapter 10). Proinflammatory cytokines strongly increase the level of expression of ICAM-1 and the lung and small intestine exhibit the largest responses to lipopolysaccharide challenge [29]. Overexpression of ICAM-1 in cultured human dermal microvascular ECs can cause vascular leakiness as well as EC shape change, cytoskeletal reorganization, and junctional protein alterations in the absence of cross-linking or leukocyte binding [30]. Endothelial P-selectin blockade also reduces acid aspiration-induced lung permeability [31].

Caveolae

Caveolae – small, flask-shaped endocytic structures – are ubiquitous features of ECs, and comprise around 95% of the cell surface vesicles and around 15% of the EC volume [32] (Figure 8.3). The existence of a transcytotic pathway suggests that vascular permeability may also be regulated by caveolae. The selective permeability of the endothelial transcytosis pathway to specific molecules is controlled not only by the size of caveolar vesicles, but also by the presence of specific receptors

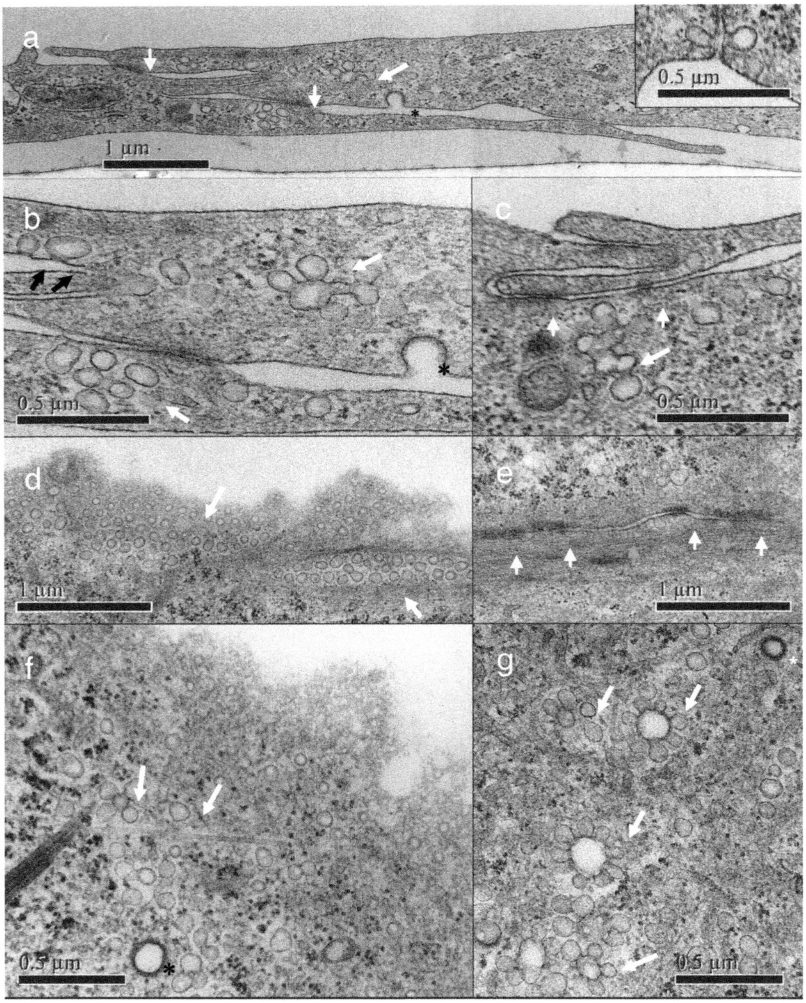

Figure 8.3 TEMs showing cell–cell junctions and abundance of caveolae in human lung microvascular ECs. Cross-sections (a–c) and horizontal sections (d–g) (~80 nm thick sections from paraformaldehyde-fixed and Epon-embedded cultured ECs at passage 5) with noted AJs (solid white arrow), TJs (gold arrow), GJs (green arrow), caveolae and caveosomes (open white arrows), and clathrin-coated pits and vesicles (*). (a) ECs grown to confluence show interdigitating protrusion from each cell joined by AJs and TJs. Individual caveolae and their clusters are also prominent, along side the less frequently observed clathrin-coated pit. Inset shows apposing caveolae open to the junctional cleft. (b) Higher magnification of TEM shown in (a). Note caveolae docked at points of cell–cell contact (black open arrows) and clusters of caveolae (open white arrow). (c) Cell–cell AJs and GJs are observed in this cross-section within the interwoven cell membranes, along with typical caveolae and caveosome profiles. (d) Horizontal section shows numerous caveolae lined-up in apparent rows, and (e) relative abundance of AJs (white arrows) and GJs (green arrows) compared to TJs (gold arrow). (f and g) Horizontal sections dramatically demonstrate abundance of caveolae and especially caveosomes (rosette-like structures marked by open arrows) relative to clathrin-coated vesicles (*) in pulmonary ECs. A color version of this figure appears in the plate section of this volume.

within caveolae. Lung microvascular ECs have an abundance of caveolae compared to brain capillaries [33, 34]. Whether subclasses of caveolae can be distinguished by the specific expression of receptors that exist in the lung endothelium is unclear; however, a recent study demonstrated that aminopeptidase P is concentrated in the caveolae of lung endothelium [35]. This finding suggests that tissue-specific caveolae may contribute to the differential regulation of endothelial permeability via transendothelial transport.

Properties of Lung Endothelial Permeability

Heterogeneity in endothelial permeability is evident along all segments of the lung vascular tree including the arterial, capillary, and postcapillary venule segments (see Chapter 9). Basal permeability properties of lung macrovascular and microvascular monolayers of the endothelium differ in a manner opposite to those of systemic vasculatures. Within the microvascular endothelium, the intercellular sealing via TJs is stronger in arterioles than in capillaries and it is loose in venules; about 30% of venular junctions are open and measure 3–6 nm. The intercellular communication via gap junctions (GJs) is more developed in arterioles (Figure 8.3c,e) than in venules and bona fide GJs seem to be absent between capillary ECs [36]. In the intact circulation, constitutive fluid filtration is 28- and 56-fold greater in pulmonary arterial and venous segments, respectively, compared to pulmonary microvascular segments [37]. Furthermore, cultured microvascular ECs exhibit 10-fold higher barrier properties than macrovascular ECs as measured by electrical resistance across monolayers [15]. This phenotypic heterogeneity equates to distinct regulation of endothelial barriers in the pulmonary artery and microvascular endothelium. Rat pulmonary microvascular endothelium exhibits tight intercellular connections, whereas the macrovascular endothelium has visible gaps between cells [22]. In monolayers of pulmonary microvascular ECs, the basal indices of permeability are lower, including hydraulic conductivity, solute permeability coefficients, and transendothelial electrical resistance. Also, Ca^{2+} transients induced by store depletion are attenuated and the increment in solute permeability induced by store depletion is minimal compared to that observed in pulmonary artery endothelium [9, 23, 38].

Heterogeneity in endothelial permeability correlates with site-specific protein expression patterns [9, 39]. The pulmonary arterial endothelium expresses a greater amount of the endothelial form of nitric oxide (NO) synthase (endothelial nitric oxide synthase eNOS) and produces more NO than capillary ECs [40]. The unique molecular anatomy of junctional proteins also contributes to differences of endothelial permeability between pulmonary artery and microvascular ECs [37, 38]. Capillary ECs express more VE-cadherin, N-cadherin, and adhesion molecules than do pulmonary artery ECs [39, 40].

ECs in arterial and capillary segments exhibit different responses to inflammatory mediators. Oxidants preferentially target postcapillary segments, whereas mechanical perturbation principally increases capillary endothelial permeability [9]. The activation of store-operated Ca^{2+} entry increases EC permeability in arterial and venule segments while typically sparing capillary EC segments [41]. In the lung, not all ECs within a vascular segment respond to inflammatory mediators with a change in cell shape. The endothelium in postcapillary venules responds to histamine by forming intercellular gaps that increase permeability, whereas immediately adjacent ECs exhibited little to no discernible response. Thapsigargin, a calcium entry activator, increases pulmonary vascular permeability in isolated perfused rat lungs, but inter-EC gap formation was only observed in intermediate to large arteries and veins, not in capillary ECs. These findings suggest that activation of store-operated Ca^{2+} entry may only increase macro- and not microvascular permeability [41] (see Chapters 5 and 9.).

BASAL LUNG ENDOTHELIAL PERMEABILITY

Determinants of Basal Lung Endothelial Permeability

In the normal lung, basal lung permeability depends mainly upon transport of the most abundant plasma protein, albumin. Because the severely restricted interendothelial junctions keeps paracellular permeability to macromolecules very low, albumin transport via the transcellular pathway plays a key role in maintenance of basal permeability. There are two pathways for transcellular transport of albumin: transcytosis (the shuttling of vesicles between the luminal and abluminal surfaces of ECs) (Figure 8.1) [7] and "pores" through which the macromolecules are carried by convection [42]. In the pore theory, molecular diameter rather than molecular weight determines the ability of molecules to pass through microvascular walls. An ultrastructural study demonstrated that plasmalemmal vesicles represent the large pore system of continuous microvascular endothelium [32] (Figure 8.3d,f,g). Accumulating evidence has also demonstrated that active transport of albumin across the endothelial barrier via caveolae is likely responsible for maintaining basal lung endothelial permeability [1]. Clathrin-coated pits may also contribute to albumin transport, however, less than 5% of vesicles in ECs are clathrin-coated while more than 95% are of caveolar origin [43, 44] (Figure 8.3a,b).

Physiological Significance of Basal Lung Endothelial Permeability

Regulation of Oncotic Pressure

The fluid components of the blood remain within the vessels of the microcirculation due to the oncotic pressure difference, which is the result of a greater protein concentration in the plasma compared to the interstitial fluid. Albumin, the primary plasma protein, is responsible for 75–80% of the plasma oncotic pressure and plays a pivotal role in modulating the distribution of fluids between compartments [45]. Under physiological circumstances, there is a net movement of albumin from the intravascular to the interstitial space and back, via the lymphatic vessels. Albumin transport from the blood to the interstitium is determined by the capillary and interstitial free albumin concentrations, the capillary permeability to albumin, solvent, and solute movements, and to a much lesser degree, the electrical charges across the capillary wall. Albumin is also the predominant protein in the interstitium, contributing to the interstitial oncotic pressure. The oncotic pressure gradient across the capillary membrane is the primary determinant of fluid leakage into the interstitium. Water and solutes, such as small ions (e.g., Na^+, K^+ Cl^-, HCO_3^-), glucose, urea, and lactic acid, can exchange freely through the pulmonary endothelium. Their transport is characterized in terms of coupling between protein and fluid fluxes through the vessel wall. In the absence of fluid filtration, the concentration of albumin in the fluid immediately outside the vessel will rise as the albumin equilibrates across the vessel wall. If fluid is absorbed into the vessel, the concentration of albumin immediately outside it may rise even more rapidly. Thus, the oncotic pressure gradient across the vessel wall diminishes with time if there is fluid reabsorption or no filtration. If fluid is filtered from plasma to tissues, however, the oncotic pressure gradient across the vessel wall increases [46].

Transport of Metabolites

Albumin binds weakly and reversibly to both cations and anions by its net negative charge. It can thus function as a carrier and circulating depot for a large number of molecules including cholesterol, NO, fatty acids, ions (especially Ca^{2+} and Cu^{2+}), thyroxine, bilirubin, and amino acids. It also binds covalently and irreversibly with Ag^{2+}, Hg^{2+}, D-glucose, and D-galactose. The glycosylation of albumin has effects upon its charge and can have significant effects on its subsequent permeability characteristics. There are four discrete binding sites on the albumin molecule with varying specificity for different substances. Ligands can compete for a single binding site or may compete by altering the affinity of remote sites by conformational changes to the tertiary structure of albumin. Albumin binding is also important for drug delivery as well as stabilization *in vivo*. Drugs binding at the same site will compete for occupancy and are likely to displace one another, whilst drugs binding at separate sites will not compete or alter the "free" concentration of other drugs simultaneously in the circulation.

INCREASED LUNG ENDOTHELIAL PERMEABILITY

Inflammatory mediators such as thrombin and tumor necrosis factor (TNF)-α, generated in response to sepsis and intravascular coagulation, disrupt the endothelial barrier by forming intercellular gaps formed in part from actin–myosin-regulated contraction of ECs [12]. These gaps permit the passage of albumin and other plasma proteins in an unrestricted manner, resulting in an increase in the protein concentration in the interstitial space. In addition, enhanced caveolae-mediated transcytosis may also contribute to increased lung endothelial permeability in response to inflammatory mediators.

Paracellular Permeability

Structural Basis of Paracellular Permeability

The microvascular barrier consists of the endothelial monolayer, intercellular contacts between adjacent ECs, and focal adhesions anchoring the EC basal membrane to surrounding matrices in the vascular wall (Figure 8.2a,b). The integrity of these structural elements is necessary to maintain normal permeability. Paracellular permeability of the pulmonary vascular barrier is regulated by the contractile forces generated by the endothelial cytoskeleton, and the adhesive forces produced at cell–cell junctions and cell–matrix contacts [12]. The disintegration of endothelial cell–cell contacts (junctions) and cell–matrix contacts (focal adhesions) leads to increased endothelial permeability through open paracellular pathways [1, 12].

Interendothelial Cell Contacts ECs form junctional complexes consisting of TJs and AJs, which are the sites of diffusional transport of solutes (Figure 8.1). At TJs, endothelial adhesion is mediated by claudins, occludins, JAM family members, and EC-selective adhesion molecule. The core of the AJ complex is comprised of homotypic and heterotypic interactions among transmembrane glycoproteins of the cadherin superfamily, such as VE-cadherin, and the catenin family members including p120-catenin, β-catenin, and α-catenin. Although cell–cell adhesion in pulmonary ECs may be due to both TJs and AJs, VE-cadherin plays a prominent role in cell tethering [47]. The extracellular domains of VE-cadherin mediate homophilic binding and adhesion between adjacent cells. The VE-cadherin extracellular domain consists of five cadherin-like repeats that form a rigid, rod-like structure that is stabilized by the binding of Ca^{2+} to the intervening sequences located at the base of each domain [48]. Also, the actin cytoskeleton within ECs associates with the cytoplasmic tail of VE-cadherin molecules thereby linking contractile elements to cell–cell junctions. Actin and myosin are the major contractile components in the cytoskeleton. Phosphorylation of myosin light chain (MLC) by Ca^{2+}/calmodulin-dependent MLC kinase (myosin light chain kinase MLCK) is required for actin–myosin interactions and engagement of the endothelial contractile apparatus [49].

EC–Matrix Contact The EC–matrix interaction is dynamically controlled through assembly and disassembly of focal adhesions [50] (see Chapter 4). Focal adhesions are contact points enabling the actin cytoskeletal network to connect to the extracellular matrix (ECM) through a complex of proteins that include vinculin, talin, paxillin, α-actinin, and focal adhesion kinase (FAK). Interaction of the focal adhesion complex with proteins of the ECM is mediated mainly by transmembrane α- and β-integrins that not only function as adhesion receptors, but also transmit chemical signals and mechanical forces between the matrix and cytoskeleton [51, 52]. The adhesive interactions between integrins and their extracellular ligands at focal adhesion complexes regulate EC shape and thereby serve to maintain endothelial barrier properties [8]. FAK is a protein tyrosine kinase that is recruited at an early stage to focal adhesions, and mediates many of the downstream signaling reactions leading to integrin engagement and focal adhesion assembly that ultimately affects barrier function [53, 54]. The activity of FAK is stimulated by Src family tyrosine kinases which are recruited to Src homology-2 (SH2) binding sites generated on FAK by autophosphorylation. FAK subsequently interacts with a number of downstream signaling proteins, including the adaptor protein Grb2 and the p85 subunit of phosphatidylinositol 3-kinase. Under basal conditions, the constitutive activity of FAK and associated adhesive interactions of integrins with their matrix ligands plays an essential role in the maintenance of microvascular barrier function by providing a tethering force for anchoring ECs to the extracellular matrix.

Mechanisms of Increased Paracellular Permeability

Disruption of Cell–Cell Junctions Increased paracellular permeability may result from increased contractile forces and/or decreased adhesive forces between neighboring ECs. The alteration in tethering and contractile forces mediate EC retraction and change in cell shape which leads to the opening of paracellular pathways of macromolecular transport [1]. There are several signal transduction pathways that disrupt interendothelial junctions but in general, they all ultimately result in the rapid and sustained phosphorylation of MLC and simultaneous inhibition of MLC-associated phosphatase, which thereby increases and prolongs the contractile response. EC retraction is thus due to disruption of endothelial AJs secondary to actinomyosin-mediated contractile forces [12].

Dissociation of VE-cadherin and catenins can lead to a decrease in adhesive forces that results in intercellular gap formation and increased paracellular permeability. Several lines of evidence now point to the essential role of VE-cadherin containing AJs in regulating paracellular permeability [16, 55]. Interendothelial cell–cell adhesive forces may also be affected by the phosphorylation status of β-catenin, whereas tyrosine phosphorylation of p120-catenin may promote dissociation from cytoskeletal anchors [47].

Ca^{2+} has an important regulatory role in the maintenance of endothelial barrier function (see Chapter 5). The increase of cytosolic Ca^{2+} induces the uncoupling of endothelial junctions by promoting phosphorylation of MLC and dissociation of cis and trans homotypic and homophilic VE-cadherin-mediated binding in AJs [56]. Many proinflammatory mediators such as oxidants, thrombin, bradykinin, and histamine induce an increase in paracellular permeability by activating Ca^{2+}-sensitive signaling pathways in ECs. The mechanism by which cytosolic Ca^{2+} increases is thought to involve inositol 1,4,5-triphosphate-induced stored Ca^{2+} release from the endoplasmic reticulum (transient rise) and subsequent capacitive Ca^{2+} entry from the extracellular medium in an effort to refill the stores [57].

Disassembly of Focal Adhesion Complexes The strength and stability of cell attachments to ECM components is regulated by interactions between the clustering of integrins and proteins that link integrins to the actin cytoskeleton at sites of cell–matrix adhesion (see Chapter 4). The interruption of integrin–matrix binding leads to microvascular leakage [58]. Integrin adhesion receptors constitute a cell-signaling system whereby interactions in the small cytoplasmic domains of the heterodimeric α- and β-subunits provoke major functional alterations in the large extracellular domains. Inhibition of $\alpha_5\beta_1$ and $\alpha_v\beta_3$ integrin binding with an Arg–Gly–Asp (RGD) peptide has been shown to cause EC detachment from subendothelial matrices and an increase in endothelial permeability [59, 60].

The regulatory role of FAK activity in endothelial paracellular permeability is still controversial. Activation of FAK enhances endothelial barrier function in bovine and rat pulmonary artery EC monolayers [61, 62]. Also, expression of a kinase-deficient FAK mutant was shown to blunt the barrier-strengthening effect of hyperosmolarity in rat lung microvascular ECs [63]. In contrast, other studies demonstrated that activated FAK in response to inflammatory mediators contributes to increased endothelial permeability [64, 65]. In addition, FAK activation has been demonstrated to promote barrier recovery after transient hyperpermeability responses in human pulmonary artery ECs [66], perhaps indicating context-dependent or time-dependent roles of FAK in barrier regulation. FAK activity is mainly regulated by Src phosphorylation. Association of Src with FAK may facilitate Src-mediated phosphorylation of tyrosine residues in FAK, some of which serve as binding sites for additional SH2 domain-containing proteins [67, 68]. Src-dependent tyrosine phosphorylation is also a critical requirement for integrin-dependent focal adhesion attachment to F-actin stress fibers [69]. Thus, interruption of interactions between Src and FAK affects vascular permeability by interfering with integrin adhesion and signaling [68].

Transcellular Permeability

Caveolae-Mediated Transcytosis

Caveolae are cholesterol-rich membrane microdomains characterized as flask-shaped invaginations of the plasma membrane and as free cytoplasmic vesicles that are notably abundant in pulmonary microvascular ECs (Figures 8.2b–e and 8.3). Caveolae, which are approximately 50–80 nm in diameter (Figures 8.2c and 8.3a–d), are distinct in morphology from the more electron-dense and much larger (about 200–300 nm in diameter) clathrin-coated vesicles (Figure 8.3a,b) [70]. Functionally, caveolae are motile organelles that have been implicated in major processes such as endothelial transcytosis, signal transduction, and as docking sites for glycolipids, cholesterol, and glycosylphosphatidylinositol-linked proteins. Current evidence supports the concept that basal albumin permeability can be fully accounted for by the formation, fission, and transport of caveolae [4, 6, 43].

Caveolae-mediated transendothelial transport is rapid (~30 s), the cargo is predominantly in the fluid phase of the vesicle but may also be receptor-bound, and requires Src family kinase signaling to activate fission and shuttling between apical and basal surfaces [43, 71, 72]. Caveolin-1, an integral membrane protein (20–22 kDa), is a specific marker and the primary structural component of endothelial caveolae. The evidence that indicates that caveolin-1 regulates endothelial transcellular transport of albumin is as follows. First, the recent generation of caveolin-1-null mice has revealed the absence of caveolae and defective uptake and transport of albumin, which could be reversed by reintroduction of caveolin-1 cDNA [73–75]. Furthermore, we [71, 76, 77] and others [78, 79] have demonstrated that phosphorylation of caveolin-1 on Tyr14 by Src family kinases initiates plasmalemmal vesicle fission and transendothelial vesicular transport, and that this facilitates the uptake and transport of albumin through ECs. Thus, caveolae are the predominant vesicular carriers in pulmonary ECs that transport albumin and other blood constituents to the tissue.

Mechanisms of Increased Transcellular Permeability

The mechanism(s) by which ECs internalize and transport albumin from the luminal to abluminal side of cells are not completely understood. Studies demonstrated that phosphorylation of caveolin-1 Tyr14 and dynamin-2 at Tyr231 and Tyr597 by Src kinase are key switches initiating caveolar fission from the plasma membrane [71, 76–80]. Phosphorylation of caveolin-1 promotes profound changes in caveolin-1 localization, and induces aggregation and fusion of caveolae and/or caveolae-derived vesicles [81, 82]. Phosphorylation of dynamin-2 may enable its localization to caveolae, specifically at the neck region, thereby forming a collar that "pinches" caveolae from the membrane [76, 78]. It is known that albumin binding to the 60-kDa glycoprotein (gp60) on the EC surface induces clustering of gp60 and its physical interaction with caveolin-1 [71]. Src tyrosine kinase is bound to the caveolin-1 scaffolding domain [83], palmitoylated C-terminal cysteine residue, and N-terminal phosphorylated tyrosine residue, and is activated upon albumin binding to gp60 [71, 76, 77]. Activated Src, in turn, phosphorylates caveolin-1, gp60, and dynamin-2 to initiate plasmalemmal vesicle (containing albumin) fission and transendothelial vesicular transport [76, 77].

The heterotrimeric GTP-binding protein, G_i, which also binds to caveolin-1, has been shown to play a fundamental role in the mechanism of caveolae-mediated endocytosis by activation of Src kinases [43, 71]. Caveolae-mediated endocytosis is pertussis toxin- and $G_{\alpha i}$-minigene peptide-sensitive [71], and $G_{\beta\gamma}$ signaling of Src activation induces caveolae-mediated transcytosis of albumin [77]. Although the role of actin in caveolae-mediated endocytosis remains unclear, both Src and dynamin are known to participate in actin cytoskeletal remodeling by regulating cortactin [84–86]. It is therefore possible that Src controls the function of actin or associated binding proteins and thereby regulates caveolar movement along the actin filaments or microtubule "tracks" [86]. This would be an additional control exerted by Src beyond $G_{\beta\gamma}$-dependent Src activation and the subsequent phosphorylation of caveolin-1 and dynamin-2 [76, 77].

Increased Lung Endothelial Permeability during Inflammation

Oxidants

Tissue injury and inflammation induce the formation of toxic reactive oxygen species (ROS), including superoxide anion (O_2^-), hydroxyl radical (OH^-), hypochlorous acid (HOCl), hydrogen peroxide (H_2O_2), and peroxynitrite ($OONO^-$) upon association with NO (see Chapters 6 and 17). Oxidants have been shown to increase pulmonary vascular endothelial permeability by inducing intercellular gap formation, cell shape change, and actin filament reorganization [87–89]. The mechanism of oxidant-induced endothelial hyperpermeability involves the interaction of oxidants with endothelial barrier-related proteins. Oxidants have also been reported to increase Src-dependent phosphorylation of MLC, which would promote stress fiber formation and cell contraction [90, 91]. Exposure of ECs to H_2O_2 increases the activation of Src and other members of the Src kinase family, which results in an increase in endothelial permeability [91]. H_2O_2 also causes an increase in endothelial permeability by inducing loss of VE-cadherin junctional staining along with concomitant gap formation and dissociation of β-catenin from the EC cytoskeleton [88]. In addition, oxidants may functionally upregulate adhesivity of adhesion molecules (ICAM-1, P-selectin) by altering the conformation and/or clustering of existing adhesion molecules through interaction with cortical actin filaments [1]. These adhesion molecules promote ROS generation and neutrophil adhesion to endothelium, which in turn can further increase endothelial permeability.

TNF-α

TNF-α is a cytokine produced by a variety of immune and nonimmune cells in response to inflammatory stimuli. TNF-α can induce an increase in endothelial permeability via intercellular gap formation [92] as it can activate Src kinases resulting in the tyrosine phosphorylation and redistribution of VE-cadherin and thereby gap formation [93]. Confocal imaging studies indicate that the Src inhibitor PP2 can prevent TNF-α-induced phosphorylation of VE-cadherin and intercellular gap formation, suggesting that a Src family tyrosine kinase activated by TNF-α acts upstream of VE-cadherin to affect changes in endothelial permeability [94]. The mechanism of Src activation may relate to TNF-α-mediated oxidant generation in ECs [95, 96]. TNF-α also induces the activation of protein kinase C (PKC) α and/or β isotypes in pulmonary artery ECs which promote actin stress fiber formation, AJ disassembly, and endothelial barrier disruption by activating RhoA [97]. In addition, activation of p38 mitogen-activate protein kinase and zinc-dependent matrix metalloproteinase gelatinase also contribute to TNF-α-induced increase in endothelial permeability by AJ disassembly and focal adhesion disruption [98]. Finally, TNF-α can increase endothelial permeability by enhancing thrombin-induced Ca^{2+} influx [57].

VEGF

VEGF, also known as vascular permeability factor, is the founding member of a family of closely related cytokines that exert critical functions in angiogenesis and vasculogenesis. Increased endothelial permeability by VEGF has been reported in pulmonary arterial endothelial monolayers [99, 100]. Overexpression of VEGF in murine lungs also resulted in high-permeability edema *in vivo*. Inhibition of the biological effect of VEGF markedly reduced pulmonary vascular protein extravasation and edema formation [27]. VEGF-induced increased vascular permeability also requires Src family kinase activity [68]. Activated Src following VEGF stimulation induces the phosphorylation of VE-cadherin and β-catenin [101]. VEGF also promotes VE-cadherin endocytosis by regulating β-arrestin2 and Vav2 through Src [101, 102]. These processes induce the disassembly of endothelial cell–cell junctions, resulting in the enhanced permeability of the blood vessel wall. In addition, an increase in intracellular Ca^{2+} through increasing Ca^{2+} release from intracellular stores and Ca^{2+} influx also contributes to VEGF-induced increase in endothelial permeability [57].

While both *in vitro* and *in vivo* studies have shown that VEGF can increase endothelial permeability, controversy, and uncertainty exists about the role of VEGF in the regulation of pulmonary vascular permeability. Recent animal studies and clinical data support a protective role for VEGF in ALI and ARDS patients [27, 103]. Whether VEGF is actively involved in promoting repair of the alveolar–capillary membrane remains unclear.

Thrombin

Thrombin is a multifunctional serine protease that is involved not only in mediating the cleavage of fibrinogen to fibrin in the coagulation cascade but also in activating ECs as a potential proinflammatory mediator. Emerging evidence indicates that thrombin increases pulmonary endothelial permeability by causing EC retraction and shape change [1]. This thrombin-induced increase in endothelial permeability occurs within minutes and recovers within 2 h. Thrombin signaling in the regulation of endothelial permeability is mediated by protease-activated receptor-1. Thrombin increases the intracellular Ca^{2+} concentration in ECs by mobilizing Ca^{2+} from intracellular stores and via its influx into the cell. Ca^{2+} signaling is critical in the mechanism of thrombin-induced MLC phosphorylation and subsequent actin–myosin cross-bridging (which induces actin stress fiber formation) [57]. Thrombin also increases Src-dependent tyrosine phosphorylation of VE-cadherin-associated β-, γ-, and p120-catenin by modulating the level of SHP-2 associated with VE-cadherin complexes [68]. This event promotes cell–cell junction disassembly and intercellular gap formation detected in EC monolayers after thrombin treatment, resulting in an increase in monolayer permeability [104]. PKCα, downstream of Ca^{2+}, is also involved in the thrombin-induced permeability increase via a cadherin-dependent mechanism [56].

Neutrophils

It is well known that activated neutrophils increase the permeability of the endothelium to albumin, thus promoting fluid loss into the interstitial space (see Chapter 10). This increase in albumin transport is thought to be mediated by disruption of the paracellular pathway due to the opening of interendothelial junctions. Although the mechanisms have not been completely described, a fairly detailed picture has emerged [1]. The interaction between activated polymorphonuclear neutrophils (PMNs) and ECs is crucial in regulating neutrophil-induced barrier dysfunction. Activation of neutrophils by a number of proinflammatory mediators involves an increase in the surface expression of CD11/CD18 complexes on PMNs. CD11/CD18-mediated ligation of ICAM-1 on ECs can directly initiate signaling events in ECs and trigger reorganization of endothelial actin filaments that leads to the opening of the endothelial junctions [8, 105]. Activated neutrophils produce oxygen free radicals that not only cause membrane lipid peroxidation and cell disintegration, but also activate signaling pathways leading to an elevation of intracellular free Ca^{2+}, the activation of MLCK, and reorganization of junctional proteins [12]. Oxygen free radicals also promote the release of proinflammatory mediators from ECs and other sources that increase the expression of ICAM-1 that, in turn, further promote neutrophil adhesion [8, 106]. In addition, proteinases released by activated neutrophils can cause structural rearrangements in adjacent ECs via proteolytic activity-dependent and -independent pathways, which then leads to the formation of interendothelial gaps and increased plasma fluid filtration [105].

Recently, it was demonstrated that activation of neutrophils stimulates caveolae-mediated transport of albumin via the transendothelial permeability pathway [107]. These findings show that ICAM-1-dependent Src activation and Src phosphorylation of caveolin-1 following the binding of neutrophils to ECs enhances caveolae-mediated transcytosis of albumin. The increase in albumin permeability of pulmonary microvessels by means of caveolae was also shown to be an important mechanism of pulmonary edema formation. These results suggest that caveolae-mediated transport of albumin and the opening of interendothelial junctions following neutrophil activation can both contribute to the formation of pulmonary edema in patients with ARDS.

CONCLUSIONS AND PERSPECTIVES

Lung vascular permeability is regulated by caveolae-mediated protein transport within individual ECs (transcellular pathway) and solute transport between adjacent ECs (paracellular pathway). Under normal physiological conditions, transport of proteins via caveolae and convective fluid flux through cell–cell junctions accounts for the basal endothelial permeability properties of the lung microvasculature important for maintaining vascular homeostasis. Transcellular protein permeability via caveolae, predominantly controlled by Src-dependent phosphorylation of caveolin-1 and dynamin-2, and paracellular solute and fluid permeability via interendothelial junctions are both significantly enhanced by pathological stimuli. Paracellular permeability of the pulmonary vascular barrier is regulated by increased contractile forces generated by the endothelial cytoskeleton and decreased adhesive forces at cell–cell junctions and cell–matrix contacts. Connections between the actin cytoskeleton and cell–cell junctions and focal adhesions are therefore essential for both maintenance as well as dysfunction of the endothelial barrier. Various inflammatory mediators increase pulmonary vascular permeability via formation of junctional gaps between contiguous ECs through a number of different signaling pathways [1, 68] generally culminating in contraction of the actin cytoskeleton upon sustained phosphorylation of myosin. In addition, recent evidence also points to the contribution of stimulated transcellular albumin transport through enhanced caveolae trafficking in response to pathologic insults in the development of increased endothelial permeability [107–109].

Although several molecular determinants that alter paracellular junctional integrity and transcellular protein transport under both physiological and pathological conditions have been identified, many important questions remain unanswered. Interactions between junctional and cytoskeletal proteins have been mapped; however, additional studies are necessary not only to identify the precise role of these interactions in regulation of junctional assembly, but also to determine which specific signals and protein modifications control endothelial permeability in response to different pathological insults.

In addition, a regulatory interplay between increased endothelial permeability and recovery of endothelial barrier exists at several levels. However, there is relatively little information regarding the signaling mechanisms that mediate reannealing of interendothelial gaps after an inflammatory insult. Although considerable advances have been made in our understanding of the signaling mechanisms of caveolae-mediated albumin transport, relatively little is known about the precise role of this pathway in edema formation and tissue injury. There is also little information as to whether and how transcellular and paracellular pathways interact with each other, and whether these pathways are compensatory or cooperative in their actions. Finally, numerous studies utilizing cultured endothelial monolayers and animal models have provided unique insights into the regulatory mechanisms of pulmonary endothelial permeability, often revealing novel targets for directed treatments. It is fundamental that we accurately translate and apply these mechanisms into viable treatment strategies in the clinic. In summary, further elucidation of the cellular regulatory mechanisms of endothelial junctions and caveolae-mediated transcytosis will help identify novel therapeutic targets for treatment of lung inflammation and injury such as seen in ALI and ARDS.

ACKNOWLEDGMENTS

Transmission electron microscopy (sectioning and digital image acquisition) was provided by Oleg Chaga, PhD (Department of Pharmacology) and Linda Juarez, PhD (Research Resources Center, Electron Microscopy Services), University of Illinois at Chicago.

References

1. Mehta, D. and Malik, A.B. (2006) Signaling mechanisms regulating endothelial permeability. *Physiological Reviews*, **86**, 279–367.
2. Michel, C.C. and Curry, F.E. (1999) Microvascular permeability. *Physiological Reviews*, **79**, 703–61.
3. Sinclair, D.G., Braude, S., Haslam, P.L., and Evans, T.W. (1994) Pulmonary endothelial permeability in patients with severe lung injury. Clinical correlates and natural history. *Chest*, **106**, 535–39.
4. Stan, R.V. (2002) Structure and function of endothelial caveolae. *Microscopy Research and Technique*, **57**, 350–64.
5. Pelkmans, L. and Helenius, A. (2002) Endocytosis via caveolae. *Traffic*, **3**, 311–20.
6. Rippe, B., Rosengren, B.I., Carlsson, O., and Venturoli, D. (2002) Transendothelial transport: the vesicle controversy. *Journal of Vascular Research*, **39**, 375–90.
7. Minshall, R.D. and Malik, A.B. (2007) Transport across the endothelium: regulation of endothelial permeability. *Handbook of Experimental Pharmacology*, **176**, 107–44.
8. Lum, H. and Malik, A.B. (1994) Regulation of vascular endothelial barrier function. *American Journal of Physiology: Lung Cellular and Molecular Physiology*, **267**, L223–41.
9. Gebb, S. and Stevens, T. (2004) On lung endothelial cell heterogeneity. *Microvascular Research*, **68**, 1–12.
10. Cines, D.B., Pollak, E.S., Buck, C.A. *et al.* (1998) Endothelial cells in physiology and in the pathophysiology of vascular disorders. *Blood*, **91**, 3527–61.
11. Del Vecchio, P.J., Siflinger-Birnboim, A., Belloni, P.N. *et al.* (1992) Culture and characterization of pulmonary microvascular endothelial cells. *In Vitro Cellular and Developmental Biology*, **28A**, 711–15.
12. Dudek, S.M. and Garcia, J.G. (2001) Cytoskeletal regulation of pulmonary vascular permeability. *Journal of Applied Physiology*, **91**, 1487–500.
13. Lamm, W.J., Luchtel, D., and Albert, R.K. (1988) Sites of leakage in three models of acute lung injury. *Journal of Applied Physiology*, **64**, 1079–83.
14. Lin, W., Jacobs, E., Schapira, R.M. *et al.* (1998) Stop-flow studies of the distribution of filtration in rat lungs. *Journal of Applied Physiology*, **84**, 47–52.
15. Qiao, R.L. and Bhattacharya, J. (1991) Segmental barrier properties of the pulmonary microvascular bed. *Journal of Applied Physiology*, **71**, 2152–59.
16. Dejana, E. (1996) Endothelial adherens junctions: implications in the control of vascular permeability and angiogenesis. *The Journal of Clinical Investigation*, **98**, 1949–53.
17. Dejana, E. (2004) Endothelial cell–cell junctions: happy together. *Nature Reviews: Molecular Cell Biology*, **5**, 261–70.
18. Pal, D., Audus, K.L., and Siahaan, T.J. (1997) Modulation of cellular adhesion in bovine brain microvessel endothelial cells by a decapeptide. *Brain Research*, **747**, 103–13.
19. Safdar, Z., Wang, P., Ichimura, H. *et al.* (2003) Hyperosmolarity enhances the lung capillary barrier. *The Journal of Clinical Investigation*, **112**, 1541–49.
20. Parker, J.C., Stevens, T., Randall, J. *et al.* (2006) Hydraulic conductance of pulmonary microvascular and macrovascular endothelial cell monolayers.

American Journal of Physiology: Lung Cellular and Molecular Physiology, **291**, L30–37.

21. Hirase, T., Staddon, J.M., Saitou, M. *et al.* (1997) Occludin as a possible determinant of tight junction permeability in endothelial cells. *Journal of Cell Science*, **110**, 1603–13.
22. Morita, K., Sasaki, H., Furuse, M., and Tsukita, S. (1999) Endothelial claudin: claudin-5/TMVCF constitutes tight junction strands in endothelial cells. *The Journal of Cell Biology*, **147**, 185–94.
23. Shasby, D.M. (2007) Cell–cell adhesion in lung endothelium. *American Journal of Physiology: Lung Cellular and Molecular Physiology*, **292**, L593–607.
24. Kaner, R.J., Ladetto, J.V., Singh, R. *et al.* (2000) Lung overexpression of the vascular endothelial growth factor gene induces pulmonary edema. *American Journal of Respiratory Cell and Molecular Biology*, **22**, 657–64.
25. Lahm, T., Crisostomo, P.R., Markel, T.A. *et al.* (2007) The critical role of vascular endothelial growth factor in pulmonary vascular remodeling after lung injury. *Shock*, **28**, 4–14.
26. Kaner, R.J. and Crystal, R.G. (2001) Compartmentalization of vascular endothelial growth factor to the epithelial surface of the human lung. *Molecular Medicine*, **7**, 240–46.
27. Mura, M., dos Santos, C.C., Stewart, D., and Liu, M. (2004) Vascular endothelial growth factor and related molecules in acute lung injury. *Journal of Applied Physiology*, **97**, 1605–17.
28. Panés, J., Perry, M.A., Anderson, D.C. *et al.* (1995) Regional differences in constitutive and induced ICAM-1 expression *in vivo*. *American Journal of Physiology: Heart and Circulatory Physiology*, **269**, H1955–64.
29. Eppihimer, M.J., Wolitzky, B., Anderson, D.C. *et al.* (1996) Heterogeneity of expression of E- and P-selectins *in vivo*. *Circulation Research*, **79**, 560–69.
30. Clark, P.R., Manes, T.D., Pober, J.S., and Kluger, M.S. (2007) Increased ICAM-1 expression causes endothelial cell leakiness, cytoskeletal reorganization and junctional alterations. *The Journal of Investigative Dermatology*, **127**, 762–74.
31. Kyriakides, C., Austen, W. Jr., Wang, Y. *et al.* (2000) Endothelial selectin blockade attenuates lung permeability of experimental acid aspiration. *Surgery*, **128**, 327–31.
32. Predescu, D. and Palade, G.E. (1993) Plasmalemmal vesicles represent the large pore system of continuous microvascular endothelium. *American Journal of Physiology: Heart and Circulatory Physiology*, **265**, H725–33.
33. Simionescu, M., Simionescu, N., and Palade, G.E. (1974) Morphometric data on the endothelium of blood capillaries. *The Journal of Cell Biology*, **60**, 128–52.
34. Stewart, P.A. (2000) Endothelial vesicles in the blood–brain barrier: are they related to permeability? *Cellular and Molecular Neurobiology*, **20**, 149–63.
35. Oh, P., Borgström, P., Witkiewicz, H. *et al.* (2007) Live dynamic imaging of caveolae pumping targeted antibody rapidly and specifically across endothelium in the lung. *Nature Biotechnology*, **25**, 327–37.
36. Simionescu, M., Simionescu, N., and Palade, G.E. (1975) Segmental differentiations of cell junctions in the vascular endothelium. The microvasculature. *The Journal of Cell Biology*, **67**, 863–85.
37. Parker, J.C. and Yoshikawa, S. (2002) Vascular segmental permeabilities at high peak inflation pressure in isolated rat lungs. *American Journal of Physiology: Lung Cellular and Molecular Physiology*, **283**, L1203–9.
38. Kelly, J.J., Moore, T.M., Babal, P. *et al.* (1998) Pulmonary microvascular and macrovascular endothelial cells: differential regulation of Ca^{2+} and permeability. *American Journal of Physiology: Lung Cellular and Molecular Physiology*, **274**, L810–19.
39. Garlanda, C. and Dejana, E. (1997) Heterogeneity of endothelial cells: specific markers. *Arteriosclerosis, Thrombosis, and Vascular Biology*, **17**, 1193–202.
40. Stevens, T. (2005) Molecular and cellular determinants of lung endothelial cell heterogeneity. *Chest*, **128**, 558S–64S.
41. Chetham, P.M., Babal, P., Bridges, J.P. *et al.* (1999) Segmental regulation of pulmonary vascular permeability by store-operated Ca^{2+} entry. *American Journal of Physiology: Lung Cellular and Molecular Physiology*, **276**, L41–50.
42. Moore, T.M., Chetham, P.M., Kelly, J.J., and Stevens, T. (1998) Signal transduction and regulation of lung endothelial cell permeability. Interaction between calcium and cAMP. *American Journal of Physiology: Lung Cellular and Molecular Physiology*, **275**, L203–22.
43. Minshall, R.D., Tiruppathi, C., Vogel, S.M., and Malik, A.B. (2002) Vesicle formation and trafficking in endothelial cells and regulation of endothelial barrier function. *Histochemistry and Cell Biology*, **117**, 105–12.
44. Oh, P., McIntosh, D.P., and Schnitzer, J.E. (1998) Dynamin at the neck of caveolae mediates their budding to form transport vesicles by GTP-driven

fission from the plasma membrane of endothelium. *The Journal of Cell Biology*, **141**, 101–14.
45. Michel, C.C. (1996) Transport of macromolecules through microvascular walls. *Cardiovascular Research*, **32**, 644–53.
46. Michel, C.C. (1997) Starling: the formulation of his hypothesis of microvascular fluid exchange and its significance after 100 years. *Experimental Physiology*, **82**, 1–30.
47. Aberle, H., Schwartz, H., and Kemler, R. (1996) Cadherin–catenin complex: protein interactions and their implications for cadherin function. *Journal of Cellular Biochemistry*, **61**, 514–23.
48. Leckband, D. and Sivasankar, S. (2000) Mechanism of homophilic cadherin adhesion. *Current Opinion in Cell Biology*, **12**, 587–92.
49. Dejana, E., Bazzoni, G., and Lampugnani, M.G. (1999) Vascular endothelial (VE)-cadherin: only an intercellular glue? *Experimental Cell Research*, **252**, 13–19.
50. Riveline, D., Zamir, E., Balaban, N.Q. *et al.* (2001) Focal contacts as mechanosensors: externally applied local mechanical force induces growth of focal contacts by an mDia1-dependent and ROCK-independent mechanism. *The Journal of Cell Biology*, **153**, 1175–86.
51. Luscinskas, F.W. and Lawler, J. (1994) Integrins as dynamic regulators of vascular function. *The FASEB Journal*, **8**, 929–38.
52. Sheetz, M.P. (2001) Cell control by membrane–cytoskeleton adhesion. *Nature Reviews: Molecular Cell Biology*, **2**, 392–96.
53. Schlaepfer, D.D., Hauck, C.R., and Sieg, D.J. (1999) Signaling through focal adhesion kinase. *Progress in Biophysics and Molecular Biology*, **71**, 435–78.
54. Schaller, M.D. (2001) Biochemical signals and biological responses elicited by the focal adhesion kinase. *Biochimica et Biophysica Acta*, **1540**, 1–21.
55. Corada, M., Mariotti, M., Thurston, G. *et al.* (1999) Vascular endothelial-cadherin is an important determinant of microvascular integrity *in vivo*. *Proceedings of the National Academy of Sciences of the United States of America*, **96**, 9815–20.
56. Sandoval, R., Malik, A.B., Minshall, R.D. *et al.* (2001) Ca^{2+} signalling and PKCα activate increased endothelial permeability by disassembly of VE-cadherin junctions. *The Journal of Physiology*, **533**, 433–45.
57. Tiruppathi, C., Ahmmed, G.U., Vogel, S.M., and Malik, A.B. (2006) Ca^{2+} signaling, TRP channels, and endothelial permeability. *Microcirculation*, **13**, 693–708.
58. Wu, M.H., Ustinova, E., and Granger, H.J. (2001) Integrin binding to fibronectin and vitronectin maintains the barrier function of isolated porcine coronary venules. *The Journal of Physiology*, **532**, 785–91.
59. Dejana, E., Lampugnani, M.G., Giorgi, M. *et al.* (1990) Fibrinogen induces endothelial cell adhesion and spreading via the release of endogenous matrix proteins and the recruitment of more than one integrin receptor. *Blood*, **75**, 1509–17.
60. Yuan, S.Y. (2002) Protein kinase signaling in the modulation of microvascular permeability. *Vascular Pharmacology*, **39**, 213–23.
61. Garcia, J.G.N., Schaphorst, K.L., and Verin, A.D. *et al.* (2000) Diperoxovanadate alters endothelial cell focal contacts and barrier function: role of tyrosine phosphorylation. *Journal of Applied Physiology*, **89**, 2333–43.
62. Wu, M.H. (2005) Endothelial focal adhesions and barrier function. *The Journal of Physiology*, **569**, 359–66.
63. Quadri, S.K., Bhattacharjee, M., Parthasarathi, K. *et al.* (2003) Endothelial barrier strengthening by activation of focal adhesion kinase. *The Journal of Biological Chemistry*, **278**, 13342–49.
64. Shikata, Y., Birukov, K.G., Birukova, A.A. *et al.* (2003) Involvement of site-specific FAK phosphorylation in sphingosine-1 phosphate- and thrombin-induced focal adhesion remodeling: role of Src and GIT. *The FASEB Journal*, **17**, 2240–49.
65. van Nieuw Amerongen, G.P., Natarajan, K., Yin, G. *et al.* (2004) GIT1 mediates thrombin signaling in endothelial cells: role in turnover of RhoA-type focal adhesions. *Circulation Research*, **94**, 1041–49.
66. Mehta, D., Tiruppathi, C., Sandoval, R. *et al.* (2002) Modulatory role of focal adhesion kinase in regulating human pulmonary arterial endothelial barrier function. *The Journal of Physiology*, **539**, 779–89.
67. Abu-Ghazaleh, R., Kabir, J., Jia, H. *et al.* (2001) Src mediates stimulation by vascular endothelial growth factor of the phosphorylation of focal adhesion kinase at tyrosine 861, and migration and anti-apoptosis in endothelial cells. *The Biochemical Journal*, **360**, 255–64.
68. Hu, G., Place, A.T., and Minshall, R.D. (2008) Regulation of endothelial permeability by Src kinase signaling: vascular leakage versus transcellular transport of drugs and macromolecules. *Chemico-Biological Interactions*, **171**, 177–89.
69. Miyamoto, S., Teramoto, M., Coso, O.A. *et al.* (1995) Integrin function: molecular hierarchies of

cytoskeletal and signaling molecules. *The Journal of Cell Biology*, **131**, 791–805.

70. Razani, B. and Lisanti, M.P. (2001) Caveolin-deficient mice: insights into caveolar function human disease. *The Journal of Clinical Investigation*, **108**, 1553–61.
71. Minshall, R.D., Tiruppathi, C., Vogel, S.M. *et al.* (2000) Endothelial cell–surface gp60 activates vesicle formation and trafficking via G_i-coupled Src kinase signaling pathway. *The Journal of Cell Biology*, **150**, 1057–70.
72. Tuma, P.L. and Hubbard, A.L. (2003) Transcytosis: crossing cellular barriers. *Physiological Reviews*, **83**, 871–932.
73. Drab, M., Verkade, P., Elger, M. *et al.* (2001) Loss of caveolae, vascular dysfunction, and pulmonary defects in caveolin-1 gene-disrupted mice. *Science*, **293**, 2449–52.
74. Razani, B., Engelman, J.A., Wang, X.B. *et al.* (2001) Caveolin-1 null mice are viable but show evidence of hyperproliferative and vascular abnormalities. *The Journal of Biological Chemistry*, **276**, 38121–38.
75. Schubert, W., Frank, P.G., Razani, B. *et al.* (2001) Caveolae-deficient endothelial cells show defects in the uptake and transport of albumin *in vivo*. *The Journal of Biological Chemistry*, **276**, 48619–22.
76. Shajahan, A.N., Timblin, B.K., Sandoval, R. *et al.* (2004) Role of Src-induced dynamin-2 phosphorylation in caveolae-mediated endocytosis in endothelial cells. *The Journal of Biological Chemistry*, **279**, 20392–400.
77. Shajahan, A.N., Tiruppathi, C., Smrcka, A.V. *et al.* (2004) $G_{\beta\gamma}$ activation of Src induces caveolae-mediated endocytosis in endothelial cells. *The Journal of Biological Chemistry*, **279**, 48055–62.
78. Conner, S.D. and Schmid, S.L. (2003) Regulated portals of entry into the cell. *Nature*, **422**, 37–44.
79. Parton, R.G., Joggerst, B., and Simons, K. (1994) Regulated internalization of caveolae. *The Journal of Cell Biology*, **127**, 1199–215.
80. Sverdlov, M., Shajahan, A.N., and Minshall, R.D. (2007) Tyrosine phosphorylation-dependence of caveolae-mediated endocytosis. *Journal of Cellular and Molecular Medicine*, **11**, 1239–50.
81. Geiger, B., Bershadsky, A., Pankov, R., and Yamada, K.M. (2001) Transmembrane crosstalk between the extracellular matrix–cytoskeleton crosstalk. *Nature Reviews: Molecular Cell Biology*, **2**, 793–805.
82. Birukov, K.G., Csortos, C., Marzilli, L. *et al.* (2001) Differential regulation of alternatively spliced endothelial cell myosin light chain kinase isoforms by $p60^{Src}$. *The Journal of Biological Chemistry*, **276**, 8567–73.
83. Li, S., Seitz, R., and Lisanti, M.P. (1996) Phosphorylation of caveolin by Src tyrosine kinases. The α-isoform of caveolin is selectively phosphorylated by v-Src *in vivo*. *The Journal of Biological Chemistry*, **271**, 3863–68.
84. McNiven, M.A., Kim, L., Krueger, E.W. *et al.* (2000) Regulated interactions between dynamin and the actin-binding protein cortactin modulate cell shape. *The Journal of Cell Biology*, **151**, 187–98.
85. Cao, H., Orth, J.D., Chen, J. *et al.* (2003) Cortactin is a component of clathrin-coated pits and participates in receptor-mediated endocytosis. *Molecular and Cellular Biology*, **23**, 2162–70.
86. Krueger, E.W., Orth, J.D., Cao, H., and McNiven, M.A. (2003) A dynamin–cortactin–Arp2/3 complex mediates actin reorganization in growth factor-stimulated cells. *Molecular Biology of the Cell*, **14**, 1085–96.
87. Barnard, M.L. and Matalon, S. (1992) Mechanisms of extracellular reactive oxygen species injury to the pulmonary microvasculature. *Journal of Applied Physiology*, **72**, 1724–29.
88. Johnson, A., Phillips, P., Hocking, D. *et al.* (1989) Protein kinase C inhibitor prevents pulmonary edema in response to H_2O_2. *American Journal of Physiology: Heart and Circulatory Physiology*, **256**, H1012–22.
89. Seeger, W., Hansen, T., Rossig, R. *et al.* (1995) Hydrogen peroxide-induced increase in lung endothelial and epithelial permeability – effect of adenylate cyclase stimulation and phosphodiesterase inhibition. *Microvascular Research*, **50**, 1–17.
90. Hardwick, J.S. and Sefton, B.M. (1995) Activation of the Lck tyrosine protein kinase by H_2O_2 requires the phosphorylation of Tyr-394. *Proceedings of the National Academy of Sciences of the United States of America*, **92**, 4527–31.
91. Kevil, C.G., Okayama, N., and Alexander, J.S. (2001) H_2O_2-mediated permeability II: importance of tyrosine phosphatase and kinase activity. *American Journal of Physiology: Cell Physiology*, **281**, C1940–47.
92. Mark, K.S. and Miller, D.W. (1999) Increased permeability of primary cultured brain microvessel endothelial cell monolayers following TNF-α exposure. *Life Sciences*, **64**, 1941–53.
93. Aikawa, R., Komuro, I., Yamazaki, T. *et al.* (1997) Oxidative stress activated extracellular signal-regulated kinases through Src and Ras in

cultured cardiac myocytes of neonatal rats. *The Journal of Clinical Investigation*, **100**, 1813–21.

94. Nwariaku, F.E., Liu, Z., Zhu, X. *et al.* (2002) Tyrosine phosphorylation of vascular endothelial cadherin and the regulation of microvascular permeability. *Surgery*, **132**, 180–85.

95. Nishida, M., Maruyama, Y., Tanaka, R. *et al.* (2000) $G_{\alpha i}$ and $G_{\alpha o}$ are target proteins of reactive oxygen species. *Nature*, **408**, 492–95.

96. Rahman, A., Bando, M., Kefer, J. *et al.* (1999) Protein kinase C-activated oxidant generation in endothelial cells signals intracellular adhesion molecule-1 gene transcription. *Molecular Pharmacology*, **55**, 575–83.

97. Ferro, T.J., Parker, D.M., Commins, L.M. *et al.* (1993) Tumor necrosis factor-α activates pulmonary artery endothelial protein kinase C. *American Journal of Physiology: Lung Cellular and Molecular Physiology*, **264**, L7–14.

98. Partridge, C.A., Jeffrey, J.J., and Malik, A.B. (1993) A 96-kDa gelatinase induced by TNF-α contributes to increased microvascular endothelial permeability. *American Journal of Physiology: Lung Cellular and Molecular Physiology*, **265**, L438–47.

99. Hippenstiel, S., Krull, M., Ikemann, A. *et al.* (1998) VEGF induces hyperpermeability by a direct action on endothelial cells. *American Journal of Physiology: Lung Cellular and Molecular Physiology*, **274**, L678–84.

100. Becker, P.M., Verin, A.D., Booth, M.A. *et al.* (2001) Differential regulation of diverse physiological responses to VEGF in pulmonary endothelial cells. *American Journal of Physiology: Lung Cellular and Molecular Physiology*, **281**, L1500–11.

101. Chiariello, M., Marinissen, M.J., and Gutkind, J.S. (2001) Regulation of c-myc expression by PDGF through Rho GTPases. *Nature Cell Biology*, **3**, 580–86.

102. Luttrell, L.M., Ferguson, S.S., Daaka, Y. *et al.* (1999) β-arrestin-dependent formation of β_2 adrenergic receptor–Src protein kinase complexes. *Science*, **283**, 655–61.

103. Thickett, D.R., Armstrong, L., and Millar, A.B. (2002) A role for vascular endothelial growth factor in acute and resolving lung injury. *American Journal of Respiratory and Critical Care Medicine*, **166**, 1332–37.

104. Tinsley, J.H., Ustinova, E.E., Xu, W., and Yuan, S.Y. (2002) Src-dependent, neutrophil-mediated vascular hyperpermeability and β-catenin modification. *American Journal of Physiology: Cell Physiology*, **283**, C1745–51.

105. Lindbom, L. (2003) Regulation of vascular permeability by neutrophils in acute inflammation. *Chemical Immunology and Allergy*, **83**, 146–66.

106. Lo, S.K., Everitt, J., Gu, J., and Malik, A.B. (1992) Tumor necrosis factor mediates experimental pulmonary edema by ICAM-1 and CD18-dependent mechanisms. *The Journal of Clinical Investigation*, **89**, 981–88.

107. Hu, G., Vogel, S.M., Schwartz, D.E. *et al.* (2008) Intercellular adhesion molecule-1-dependent neutrophil adhesion to endothelial cells induces caveolae-mediated pulmonary vascular hyperpermeability. *Circulation Research*, **20** (102), e120–31.

108. Chen, J., Braet, F., Brodsky, S. *et al.* (2002) VEGF-induced mobilization of caveolae and increase in permeability of endothelial cells. *American Journal of Physiology: Cell Physiology*, **282**, C1053–63.

109. Childs, E.W., Udobi, K.F., Hunter, F.A. *et al.* (2005) Evidence of transcellular albumin transport after hemorrhagic shock. *Shock*, **23**, 565–70.

9

Lung Endothelial Phenotypes: Insights Derived from the Systematic Study of Calcium Channels

Donna L. Cioffi[1,4], Songwei Wu[2,4] and Troy Stevens[2,3,4]

[1]Department of Biochemistry and Molecular Biology, University of South Alabama College of Medicine, Mobile AL, USA
[2]Department of Pharmacology, University of South Alabama College of Medicine, Mobile AL, USA
[3]Department of Medicine, University of South Alabama College of Medicine, Mobile AL, USA
[4]Center for Lung Biology, University of South Alabama College of Medicine, Mobile AL, USA

INTRODUCTION

Endothelium is a thin cell layer that forms a contiguous lining interconnecting all vascular structures [1–3]. Endothelial cells (ECs) were once considered a metabolically inactive cell lining and were described in the classic literature as a simple nucleated membrane [1–3]. More recent work, however, has demonstrated that this early perception was incorrect. Indeed, these cells are highly dynamic, metabolically active regulators of vascular and tissue homeostasis. ECs are sufficient to generate new vascular structures, with lumens capable of blood perfusion. They produce a number of powerful vasoregulatory substances, such as the vasodilators nitric oxide (NO) and prostacyclin, and the vasoconstrictor endothelin-1, and they possess an endocrine function by controlling the synthesis of blood proteins such as angiotensin II and bradykinin. Moreover, circulating hormones must first cross the EC barrier to access target tissues. Hormone delivery may be due to the passive diffusion of macromolecules between adjacent cells or due to the directed delivery of hormones by transcytosis through endocytotic pathways. While maintaining a functional permeability barrier to water, solutes, and macromolecules, ECs are key determinants of coagulation, hemostasis, and white blood cell recruitment to sites of infection.

While ECs line all vascular (and lymphatic) structures, not all ECs are the same [1–3]. Morphological studies have resolved key structural differences among these cells in different organs. Whereas continuous ECs line blood vessels in the brain, heart, and lung, discontinuous ECs line blood vessels in sinusoids, such as those found in liver. Fenestrated ECs line blood vessels whose role is to filter blood, such as that seen in the glomerulus. Structural differences in EC phenotype accompany the unique functional roles of these cells within highly specialized vascular structures. What has been less apparent is the structural and functional EC heterogeneity that exists within a population of "like" cells in a particular organ, such as endothelia of the continuous type.

Continuous ECs line blood vessels within the pulmonary circulation. In recent years the remarkable heterogeneity in structure and function that exists among these cells, ranging from pulmonary arteries to capillaries and, ultimately, veins, has been examined in detail [4, 5]. With the development of novel experimental tools that enable systematic study of the molecular, cellular, and physiological EC attributes, a greater appreciation of

The Pulmonary Endothelium: Function in health and disease Editors Norbert F. Voelkel, Sharon Rounds

the extent of heterogeneity, even within adjacent cells lining a common vascular structure, has emerged.

HETEROGENEITY IN PULMONARY ARTERY, CAPILLARY, AND VEIN ENDOTHELIUM

The pulmonary circulation arises from the right ventricle, is highly specialized among all vascular beds, and is the only one that receives 100% of the cardiac output. Despite accommodating a blood flow of approximately 5 L/min at rest (in humans) and in excess of 35 L/min during exercise (in humans), the pulmonary systolic and diastolic blood pressure remains well below that of the systemic circulation. Indeed, the mean pulmonary pressure is 18 mmHg, whereas the mean systemic blood pressure is 100 mmHg, in resting humans. While the vasculature maintains low pressures in the face of high flows, it is simultaneously impacted by mechanical perturbations of breathing. Arteries and bronchioles are aligned from the proximal to the distal lung segments, and as bronchioles branch ultimately into alveolar sacs, the alveoli become surrounded by capillary loops. Thus, even tidal volume breathing creates a mechanical stress in the bronchi, bronchioles, and alveoli that is sensed in arteries and capillaries, respectively. In addition, the pulmonary circulation is unique in that its arteries carry mixed venous blood, with a $PO_2 \sim 40$ mmHg and $PCO_2 \sim 46$ mmHg. Gas exchange occurs across the large alveolar–capillary surface area, and the venous blood then carries oxygenated blood, with a $PO_2 \sim 100$ mmHg and $PCO_2 \sim 40$ mmHg. Thus, the pulmonary circulation is adapted to accommodate a unique complex of biophysical and environmental challenges.

Endothelium within the pulmonary circulation is similarly specialized, not only in comparison to other circulations, but also along the pulmonary vascular tree [4–10]. Pulmonary artery endothelium sees the highest bulk blood flow in the body. High blood flow imposes a mechanical stress on endothelium that causes these cells to align in the direction of blood flow. Each of these cells interacts with up to six neighboring cells, and possess multiple contacts with both adjacent cells and the underlying matrix. Pulmonary artery ECs (PAECs) contain a full complement of intracellular organelles, including endoplasmic reticulum (ER), Golgi apparatus, mitochondria, and the highly specialized endothelial-specific Weibel–Palade body (Figure 9.1). Weibel–Palade body exocytosis delivers von Willebrand factor to the blood, and inserts P-selectin in the plasma membrane, important for hemostasis and leukocyte trafficking, respectively [11]. Thus, pulmonary artery endothelium is specialized to accommodate the unique demands of the conduit vasculature.

Pulmonary microvascular endothelium is similarly adapted to meet the demands of its unique niche [4–10]. Unlike conduit vessels, capillary loops are comprised of single ECs that contact only one or two adjacent ECs and form lumens that are 5–10 μm in diameter [12]. The capillary surface area is enormous. Blood flow through pulmonary capillaries can be intermittent, as not all capillaries are perfused with red blood cells under resting conditions (e.g., when venous pressure is not elevated). Red blood cells align in single file as they transit through capillary loops, optimizing the surface area for efficient gas exchange. Endothelial nuclei protrude into the capillary lumen, but because the luminal diameter is small, nuclei from different cells are offset, so as not to obstruct blood flow. Capillary ECs closely abut type I pneumocytes, and these two cell types form the alveolar–capillary membrane [13]. The alveolar–capillary membrane forms thick and thin sections. The thick section includes cellular regions rich in intracellular organelles, including the nucleus, ER, Golgi apparatus, and mitochondria (Figure 9.1). The thin section is comprised merely of cytoplasmic extensions largely devoid of organelles; it is this thin section, which is only around 100 nm in thickness, that is principally responsible for gas exchange. While capillary ECs are highly metabolically active and possess a full complement of intracellular organelles, they do not typically possess Weibel–Palade bodies. Despite the absence of Weibel–Palade bodies, capillary ECs express von Willebrand factor and P-selectin, suggesting they organize these key hemato-regulatory proteins in unique ways [14]. Capillary ECs possess highly restrictive cell–cell junctions, and are greatly enriched with caveolae and endocytotic pathways. These cells produce vasoregulatory molecules, such as NO and prostacyclin, but to a lesser degree than do their macrovascular counterparts. Indeed, capillary ECs are highly specialized in their structure and function.

Less is known about the specialized attributes of pulmonary vein ECs (PVECs), when compared with pulmonary artery and capillary endothelium [4–10]. Similar to other conduit segments, vein endothelium contacts multiple neighboring cells. The conduit veins return nearly 100% of the cardiac output to the left ventricle, and consequently this vascular segment transports large blood volumes. This biophysical stress only modestly orients vein ECs in the direction of flow, to a lesser degree than is seen in pulmonary artery endothelium. Vein ECs possess vasoregulatory molecules, similar to the arterial segment, and possess a full complement of organelles, including Weibel–Palade bodies. Although vein endothelium forms

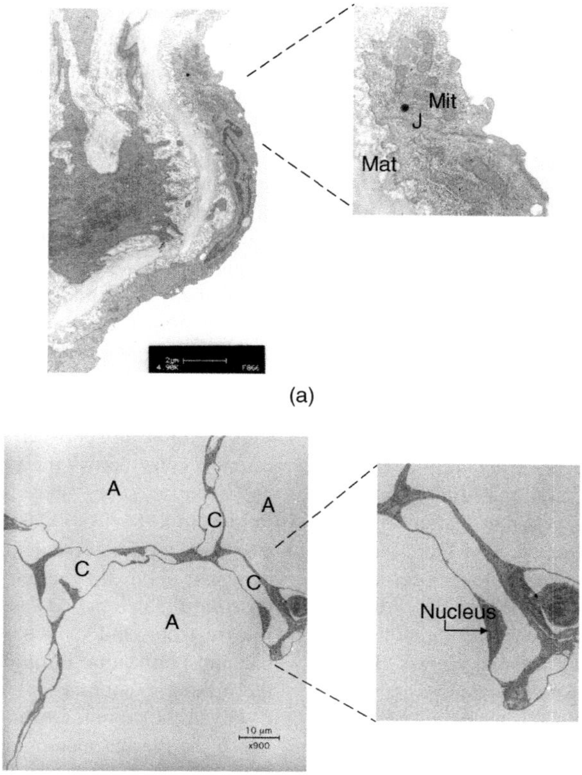

Figure 9.1 Pulmonary artery and capillary ECs display a wide range of unique anatomical features. (a) Transmission electron micrograph illustrating that PAECs reside on a complex matrix composition (Mat), form complex EC–EC junctions (J), and possess numerous organelles, such as mitochondria (Mit). (b) Transmission electron micrograph showing that pulmonary capillary ECs form, with type I epithelial cells, the alveolar–capillary membrane separating the alveoli (A) from the capillary lumen (C). Endothelial nuclei protrude into the lumen in the thick portion of the capillary, whereas the thin cellular extensions form the surface area for gas exchange and are largely devoid of organelles. Electron micrographs courtesy of Dr. Judy King (Center for Lung Biology, University of South Alabama).

a semirestrictive barrier, the permeability across this segment is higher than it is across arterial and microvascular segments. It is clear that PVECs are specialized, when compared with arterial and capillary cells, yet the unique structure and function of endothelium within the vascular segment remains incompletely understood.

While EC heterogeneity among the arterial, capillary, and vein segments can be appreciated by a systematic study of cell (and vascular) morphology, it is further discriminated based upon the interaction of ECs with lectins. Indeed, whereas pulmonary ECs from all vascular segments interact with *Ulex europaeus* lectin, only PAECs and PVECs interact with *Helix pomatia*, and only pulmonary microvascular ECs (PMVECs) interact with *Griffonia simplicifolia* [4–6, 15]. The border between *H. pomatia*-positive cells and *G. simplicifolia*-positive cells is demarcated, and resides at around 25 μm vessel diameters.

Lectin binding has therefore been used to enrich for cultures of purified PAECs and PMVECs *in vitro*. Just as is observed *in vivo*, PAECs interact with *H. pomatia in vitro*, and this behavior is retained irrespective of cell passage. Similarly, PMVECs interact with *G. simplicifolia in vitro* and this behavior is retained irrespective of cell passage. Using lectin criteria as a guide to specify cell phenotypes, systematic study of PAECs and PMVECs has revealed fundamentally unique functions of these cell types. Indeed, PAECs and PMVECs display a remarkable range of phenotypic heterogeneity, reflecting their *in vivo* behaviors, that is stable in culture.

EXQUISITE FIDELITY IN CALCIUM SIGNALING REVEALS LUNG EC HETEROGENEITY

Integrated *in vivo* and *in vitro* studies have demonstrated that PMVECs possess a more restrictive permeability barrier than do PAECs and PVECs [16–19]. *In vivo*, the capillary EC barrier is approximately 26- and 58-fold more restrictive than that of the PAEC and PVEC barriers, respectively [16]. This finding has led Parker *et al.* to conclude that the inherently tight capillary EC barrier is the single most important safety factor preventing alveolar edema, replacing all of the previously recognized safety factors from the Starling equation, including hydrostatic and oncotic pressure gradients [17].

Evidence for such a profoundly different barrier function among pulmonary EC segments brings into question whether each vascular segment is similarly targeted by, and responsive to, inflammatory mediators. Many of the neurohumoral inflammatory mediators that induce gaps between adjacent ECs initiate this physiological response by increasing cytosolic calcium [20]. When these classical observations were first made, the intracellular signal necessary to trigger gap formation was not known. However, in subsequent years, classic inflammatory agonists such as histamine [21], serotonin [22], platelet-activating factor [23], and many others were all shown to increase cytosolic calcium, and it is this rise in cytosolic calcium that initiates the cytoskeletal reorganization necessary to induce inter-EC gaps and increase macromolecular permeability [24, 25].

Whereas extracellular calcium concentrations approach 2 mM [26], ECs maintain a low basal cytosolic calcium concentration near 100 nM [27]. Intracellular calcium is stored within organelles, most notably the ER. Estimates vary as to the exact calcium concentration within the ER, although it appears to reside between 100 μM and 2 mM [28–31]. Thus, a large calcium concentration gradient exists between both the plasma membrane (approximate 20 000 : 1) and the ER membrane (at least 1000 : 1), and the cytosolic compartment. Inflammatory agonists transiently release calcium from the ER, but interestingly, this calcium source is not sufficient to increase EC permeability [19]. In addition, inflammatory mediators promote calcium entry across the plasma membrane, and this calcium source is sufficient to increase EC permeability. Evidence that calcium entry across the plasma membrane increases permeability has therefore led to extensive studies seeking to identify the plasma membrane calcium channels that regulate barrier function (see Chapter 5).

It is remarkable to consider that as recently as 1990, the molecular identity of not even a single EC calcium channel was known. Studies at that time largely focused on determining how first messenger signaling molecules coordinate complex intracellular responses, such as inter-EC gap formation. Results from this work revealed that inflammatory mediators, which act on distinct transmembrane receptors, activate common intracellular G-proteins [32]. Thus, histamine and platelet activating factor (and others) bind to different receptors, but each activate G_q-proteins that cause phospholipase C (PLC)-dependent generation of inositol 1,4,5-triphosphate (IP_3). IP_3 diffuses into the cytosol and binds to its receptor on the ER. IP_3 binding to its receptor triggers a transient calcium release, and importantly, the transient depletion of ER calcium opens calcium channels on the plasma membrane and promotes calcium influx (Figure 9.2). This relationship between calcium store depletion and calcium entry across the plasma membrane was termed "capacitative", or "store-operated", calcium entry by Putney in 1986 [33]. Calcium resequesteration into the ER then terminates calcium influx. Calcium reuptake depends upon the activity of the sarco/endoplasmic reticulum calcium ATPase (SERCA). Thus, calcium release from, and reuptake into, the ER finely controls calcium permeation through a subset of ion channels on the plasma membrane.

Given the complex nature of signaling events that occur at the plasma membrane, studies were undertaken to more directly address the impact of calcium permeation through store-operated Ca^{2+} channels (SOCs) on EC barrier function. In 1990, Thastrup *et al.* [34] discovered that the plant alkaloid thapsigargin irreversibly inhibits SERCA function. SERCA inhibition decreases ER calcium and activates SOCs. Thapsigargin could thus be used to specifically address how calcium permeation through SOCs impacts EC permeability, without the confounding influences of simultaneous G-protein activation. Thapsigargin induced a dose-dependent increase in EC permeability, both in the isolated perfused lung and in cultured PAECs [35, 36]. However, lung histology revealed that increased endothelial permeability resulted in the appearance of large perivascular cuffs, without alveolar edema (Figure 9.3). These findings could have been explained by the idea that fluid escapes from the circulation in capillaries and is drawn into perivascular cuffs by negative interstitial pressure in order to be cleared by lymphatics. However, scanning and transmission electron micrographs did not detect a disruption in capillary–EC junctional integrity, while prominent gaps were observed in PAECs and PVECs. Moreover, fluid could be seen coursing through adjacent smooth muscle cells (SMCs) underlying conduit vessels, resulting in an increase in the distance between adjacent SMCs. These findings were therefore most consistent with the idea that the site of fluid and macromolecular permeation

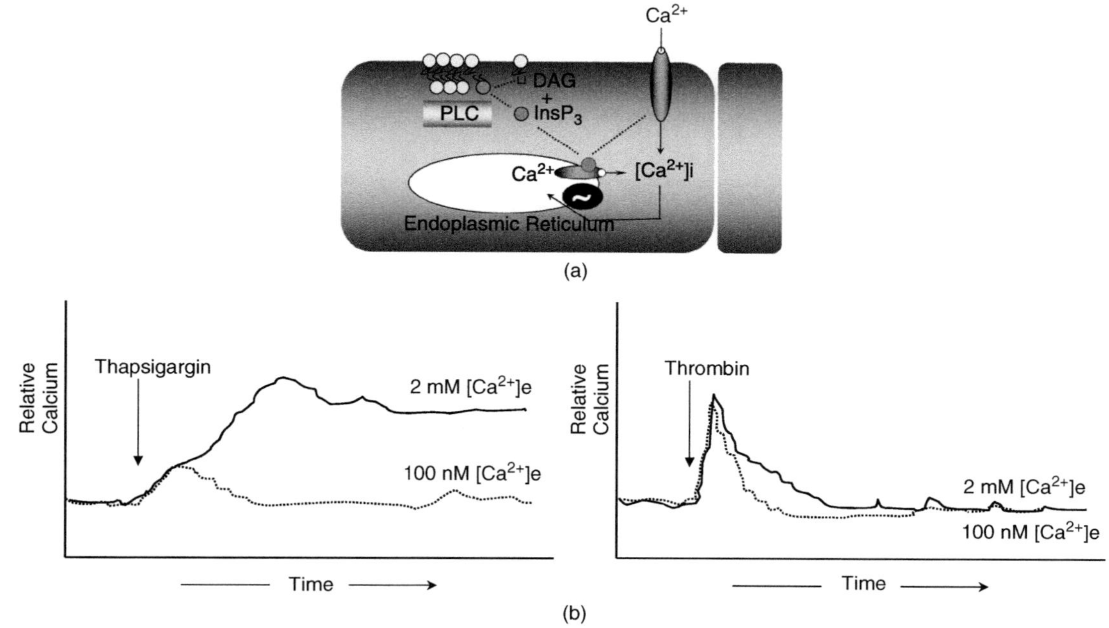

Figure 9.2 Depletion of ER calcium activates SOCs on the plasma membrane. (a) Multiple inflammatory mediators bind membrane receptors that activate heterotrimeric G_q proteins, which in turn activate PLC. PLC dissociates phosphatidylinositol 4,5-bisphosphate ( ) into diacylglycerol (DAG) and IP_3. IP_3 diffuses into the cytosol and binds IP_3 receptors on the ER. Calcium permeates through IP_3-bound receptors, from the ER into the cytosol, and transiently increases cytosolic calcium ($[Ca^{2+}]_i$). Depletion of stored calcium also activates SOCs on the plasma membrane, which allows calcium to permeate across the plasma membrane causing a sustained $[Ca^{2+}]_i$ rise. The sustained $[Ca^{2+}]_i$ rise is terminated by SERCA ( ), which utilizes ATP to pump calcium against its electrochemical gradient, from the cytosol into the ER. Like generation of IP_3, SERCA inhibition depletes stored calcium and activates SOCs. (b) Schematic of the rise in $[Ca^{2+}]_i$ following thapsigargin (left panel), which inhibits SERCA, and thrombin (right panel), which increases intracellular IP_3. Solid lines represent the $[Ca^{2+}]_i$ response when the agonists are added with 2 mM extracellular calcium ($[Ca^{2+}]_e$) and dotted lines represent the $[Ca^{2+}]_i$ response when agonists are added with 100 nM $[Ca^{2+}]_e$.

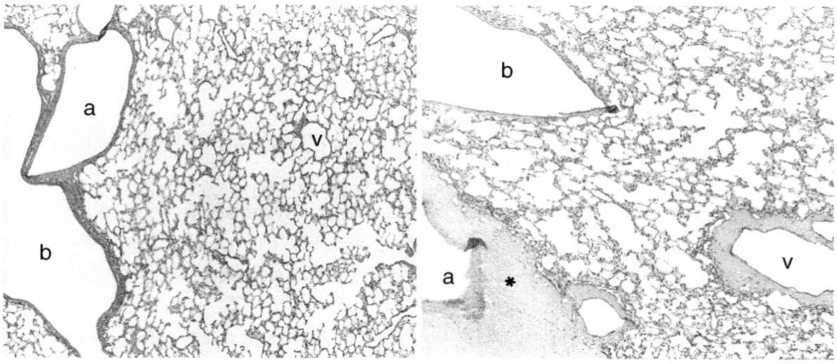

Figure 9.3 Thapsigargin induces perivascular cuffs surrounding arteries and veins. Histology of a control lung shows a large artery (a) and bronchiole (b), with a vein (v) in the parenchyma (left panel). Following thapsigargin infusion, large perivascular fluid cuffs accumulate surrounding arteries and veins (right panel). For experimental details, see [36].

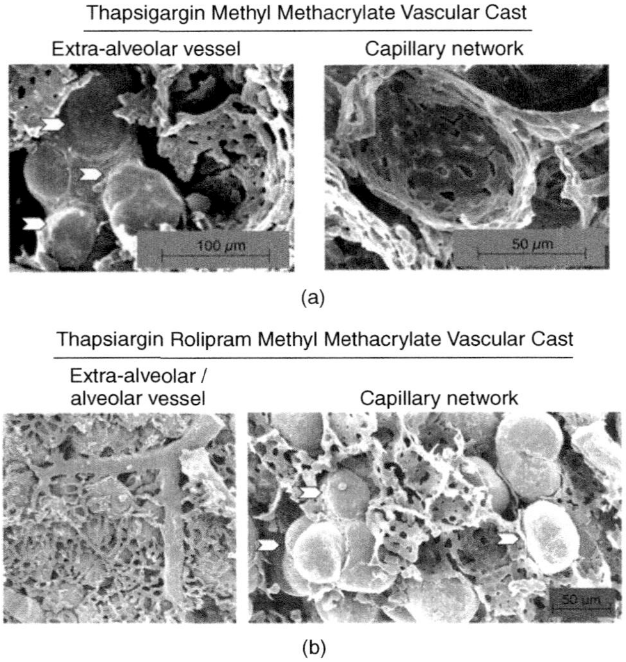

Figure 9.4 Perfusion casts reveal that thapsigargin induces extra-alveolar leak sites, whereas rolipram/thapsigargin induce alveolar leak sites. Thapsigargin perfusion into the lung circulation results in inter-EC gaps in extra-alveolar vessels, resulting in bulging of casting material out of the circulation (a). Whereas large bulges of casting material can be seen in large and intermediate vessel sizes (arrowheads), capillaries architecture remains intact. Rolipram treatment before thapsigargin perfusion abolishes extra-alveolar leak sites, but reveals new leak sites in the microcirculation (arrowheads; b). For experimental details, see [37].

was across extra-alveolar, conduit vessel endothelium. Studies utilizing cultured PMVECs further supported this idea. Whereas thapsigargin increased cytosolic calcium in these microvascular ECs, the calcium rise was not sufficient to induce EC gaps and increase macromolecular permeability [19]. Thus, it appeared that thapsigargin, and activation of SOCs, only increased permeability across conduit vessel ECs, revealing an unanticipated level of EC heterogeneity.

In subsequent studies, we would come to learn that capillary ECs possess the SOC that is capable of inducing gap formation, but due to a unique ER compartmentalization in these cells, thapsigargin does not directly activate this channel [15, 37, 38]. Clues supporting this conclusion first came from studies undertaken in the isolated perfused lung, in which the type 4 phosphodiesterase inhibitor rolipram was used to increase cAMP and reduce the permeability response. Rolipram decreased whole lung permeability in response to thapsigargin by 50%. However, histology and electron microscopy experiments revealed that rolipram had abolished the permeability response across extra-alveolar segments, and revealed new leak sites in the microcirculation (Figure 9.4). Perfusion casts of the lung circulation further supported this contention, as thapsigargin induced large leak sites in extra-alveolar segments that were abolished by rolipram, yet in rolipram-pretreated lungs, the capillary leak sites were not observed. Thus, activation of SOC entry was sufficient to increase permeability across both extra-alveolar and alveolar ECs, but mechanisms controlling this response were unique among the vascular segments.

Measurements of whole-cell calcium demonstrate that thapsigargin induces a slowly developing and sustained increase in cytosolic calcium that is due to the activation of multiple different ion channels [39–42]. Given the disparate effects of rolipram on extra-alveolar and alveolar EC permeability, electrophysiology studies were performed to more discretely identify a calcium entry pathway responsible for the thapsigargin-induced increase in permeability. Using the whole cell patch clamp approach, thapsigargin activated a nonselective cationic conductance (e.g., one that is permeable to multiple divalent cations) in PAECs and PMVECs. Since this current was similarly activated in both cell types, it was not consistent with *in vivo* data illustrating that thapsigargin increases extra-alveolar EC permeability [40]. A second current was also identified which displayed calcium selectivity and was not equally permeable to other divalent cations. Thapsigargin only activated this current in PAECs [37, 38, 41, 42]. Importantly, this current was inhibited by rolipram in PAECs, whereas rolipram enabled thapsigargin to activate the current in PMVECs. Thus, the thapsigargin activated calcium selective current, called I_{SOC}, was regulated by rolipram in a manner entirely consistent with the way that rolipram influences EC permeability *in vivo*.

Search for the molecular identity of the I_{SOC} channel advanced rapidly. Discovery of the transient receptor potential (TRP) protein in *Drosophila melanogaster* retina [43–46], and its mammalian orthologs (TRPC1–7) [32, 47–52], provided molecular candidates for SOC entry channels. TRPC1, TRPC4, TRPC5, and potentially TRPC3 proteins are considered subunits of endogenous SOC entry channels [53], whereas other members in this family may form the molecular basis of receptor-operated channels [54]. Separate groups provided evidence that TRPC1 and TRPC4 contribute to the I_{SOC}. TRPC1 antisense inhibition reduced the global cytosolic calcium response to thapsigargin [41]. However, the I_{SOC} was only reduced by approximately 50%, and the current's reversal potential was not left shifted, suggesting other subunits contributed to the I_{SOC} channel pore. Additionally, in ECs isolated from TRPC4 knockout mice, the thapsigargin-activated I_{SOC} was abolished, and the current's reversal potential was left shifted, indicating that TRPC4 plays a key role in forming the channel's pore [55]. Thus, both TRPC1 and TRPC4 appear to comprise the I_{SOC} channel's pore, in some presently undetermined stoichiometry. Pulmonary artery and microvascular ECs both express TRPC1 and TRPC4, and I_{SOC} permeation characteristics through the TRPC1/TRPC4 channel are similar in these cells. Moreover, TRPC1 and TRPC4 coimmunoprecipitate in a larger channel complex, consistent with the idea that these proteins form the molecular basis of the I_{SOC} channel in both PAECs and PMVECs [39, 56–58].

Further support for the importance of TRPC1 and TRPC4 in regulating EC permeability came from studies in an animal model of chronic heart failure. Placement of an infrarenal aortocaval fistula resulted in a high flow vascular adaptation, that over several weeks caused heart failure reminiscent of the human condition [59, 60]. Whereas increased pulmonary vascular pressure results in a hydrostatic edema in these animals, the endothelium adapts to this high-pressure environment by strengthening its barrier function. Indeed, baseline permeability is reduced in heart failure animals. Both TRPC1 and TRPC4 proteins are downregulated in extra-alveolar endothelium, and the permeability response to thapsigargin is abolished in these animals (Figure 9.5). Remarkably, whereas thapsigargin-induced perivascular cuff formation is abolished, the permeability response to other inflammatory agonists, such as 14,15-epoxyeicosatrienoic acid (EET), is retained (Figure 9.5). Close inspection of the vascular response to 14,15-EET revealed an increase in capillary, and not extra-alveolar, permeability. Thus, 14,15-EET targeting to capillary endothelium revealed yet another level of unexpected endothelial heterogeneity, illustrating the exquisite selectivity with which inflammatory agonists act.

Exactly how 14,15-EET targets the capillary endothelium, and not the extra-alveolar endothelium, was initially unclear. However, evidence that 14,15-EET activates an ion channel in the vanilloid family of TRP proteins, the TRPV4 channel [61], provided a potential molecular explanation for such selective targeting to a discrete vascular segment [59]. Indeed, TRPV4 expression is highly enriched in the lung's microcirculation. Unlike the TRPC1/TRPC4 channel, TRPV4 is not down-regulated in animals with heart failure. Moreover, 4α-phorbol 12,13-didecanoate (4α-PDD), which directly activates TRPV4 channels, similarly increases lung capillary permeability and the permeability response to 4α-PDD is abolished in TRPV4 knockout mice (Figure 9.6) [62]. Thus, TRPV4 expression is prominent in lung capillary segments and the TRPV4 channel is activated by inflammatory stimuli that do not cross-activate the TRPC1/TRPC4 channel.

These findings provide insight into the exquisite organization of ion channels along the arterial-capillary-venous axis, and the regulation of TRPC1/TRPC4 and TRPV4 by discrete inflammatory stimuli. This work also illustrates that thapsigargin and 4α-PDD can be used to increase permeability in discrete vascular segments, resulting in perivascular cuffs or alveolar flooding, respectively. Alveolar flooding inactivates surfactant, decreases compliance [63], and causes hypoxemia, but less is known about the physiological consequences of perivascular cuffing. To address this issue, thapsigargin and 4α-PDD were applied to the intact pulmonary circulation at concentrations eliciting identical rises in the filtration coefficient (K_f) [64]. Whereas thapsigargin caused perivascular cuff formation, 4α-PDD produced alveolar edema (Figure 9.7). Surprisingly, thapsigargin decreased dynamic compliance by nearly 20%. Recently, the idea that perivascular cuffs decrease lung compliance (both static and dynamic) has been substantiated in intact, sedated animals (Stevens, unpublished). Decreased compliance is due to a reduction in the efficiency of mechanical coupling between the bronchovascular bundle that is engorged with fluid and the lung parenchyma.

Such decrements in mechanical coupling efficiency occur without concomitant alveolar flooding, and hence, without surfactant inactivation. Thus, perivascular cuffs and alveolar flooding each decrease compliance, although they do so through different mechanisms.

It may appear from this collective work as though calcium permeation through membrane ion channels increases EC permeability by necessity. However, this is not the case. In 2003, Wu *et al.* [65] discovered that lung capillary ECs express a voltage-gated T-type calcium channel. This was a surprising result since ECs are nonexcitable cells. It was unclear how cells lacking action potentials could activate a channel controlled by membrane potential. Reverse transcription polymerase chain reaction (RT-PCR) cloning revealed the channel was the α_{1G} subtype ($Ca_v3.1$) and electrophysiology studies revealed that the channel activated at approximately -60 mV, with peak currents seen at -10 mV (Figure 9.8).

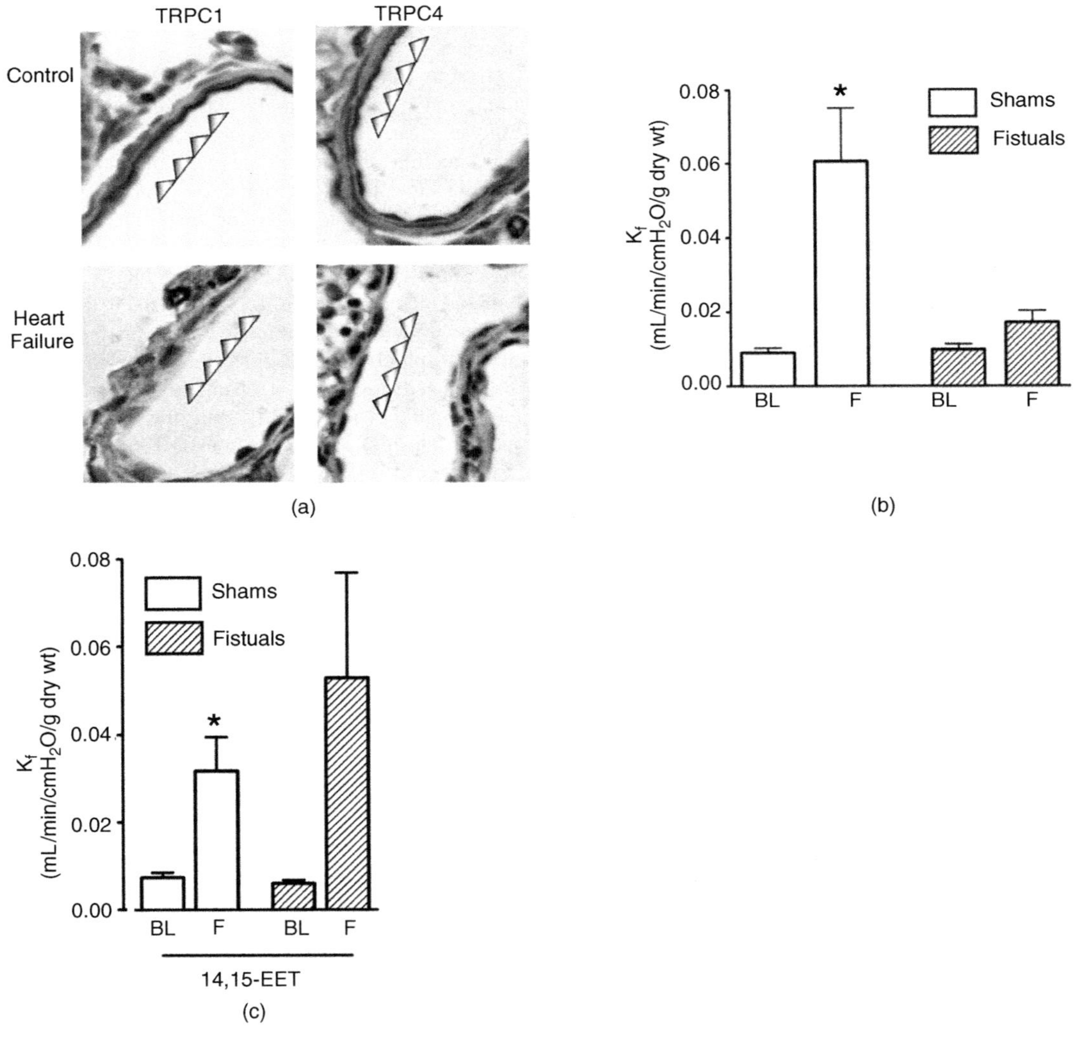

Figure 9.5 TRPC1 and TRPC4 proteins contribute to a thapsigargin-activated calcium entry pathway important for disrupting the EC barrier. (a) Extra-alveolar endothelium express TRPC1 and TRPC4 proteins *in vivo*; these channel proteins are downregulated in animals with heart failure. (b) Thapsigargin increases whole lung permeability (K_f: filtration coefficient) in control animals, but not in animals with heart failure. (c) 14,15-EETs increase permeability in control animals and in animals with heart failure. BL, baseline measurements; F, final measurements. Sham denotes a surgical manipulation and fistula denotes placement of an aortocaval shunt. For experimental details, see [60].

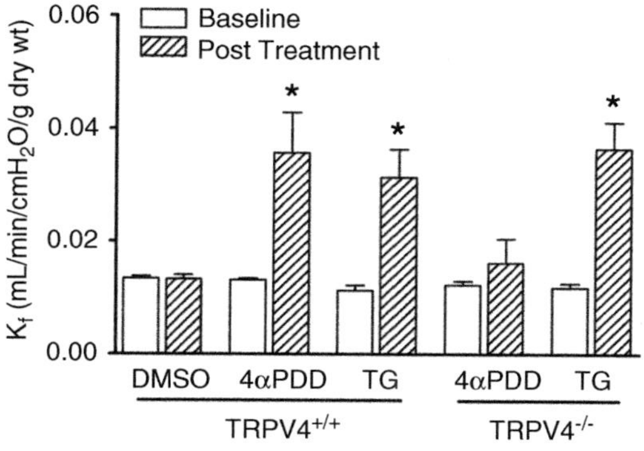

Figure 9.6 4α-PDD increases permeability in control animals, but not in TRPV4$^{-/-}$ animals. Infusion of 4α-PDD increases K_f (filtration coefficient) in wild-type mice, but not in TRPV4$^{-/-}$ mice. In contrast, thapsigargin (TG) increases permeability in both wild-type and TRPV4$^{-/-}$ mice. DMSO, dimethyl sulfoxide (vehicle control). For experimental details, see [62].

In electrophysiology studies, window currents represent the physiologically relevant membrane potentials at which calcium permeates voltage-gated channels. The EC T-type calcium channel's window current was identified, and interestingly, the window current ranged from −60 to −30 mV, which is a more depolarized range than in most other cell types expressing this channel. Thrombin and other inflammatory agonists cause membrane depolarization in ECs into this range of voltages, providing a mechanism by which inflammatory agonists activate the T-type calcium channel. Thus, the α_{1G} T-type calcium channel represents a novel type of channel for endothelium, as it is activated by membrane depolarization.

It is interesting that α_{1G} ($Ca_v3.1$) channel expression is restricted to lung capillary segments and microvessels approximately 10–20 μm in diameter. Studies on the physiological significance of this channel have now demonstrated that its activation results in von Willebrand factor secretion [66] and P-selectin membrane translocation (Wu, unpublished), processes essential for controlling hemostasis and neutrophil trafficking, respectively. These processes are highly specialized in lung capillary endothelium, as von Willebrand factor and P-selectin are contained within Weibel–Palade bodies in extra-alveolar vessels, but not in capillary endothelium [67] (Wu, unpublished). The intracellular locale of von Willebrand factor and P-selectin in capillary endothelium, and mechanisms controlling their secretion and membrane translocation, respectively, remain poorly understood in capillary ECs. Functionally, α_{1G} channel activation promotes

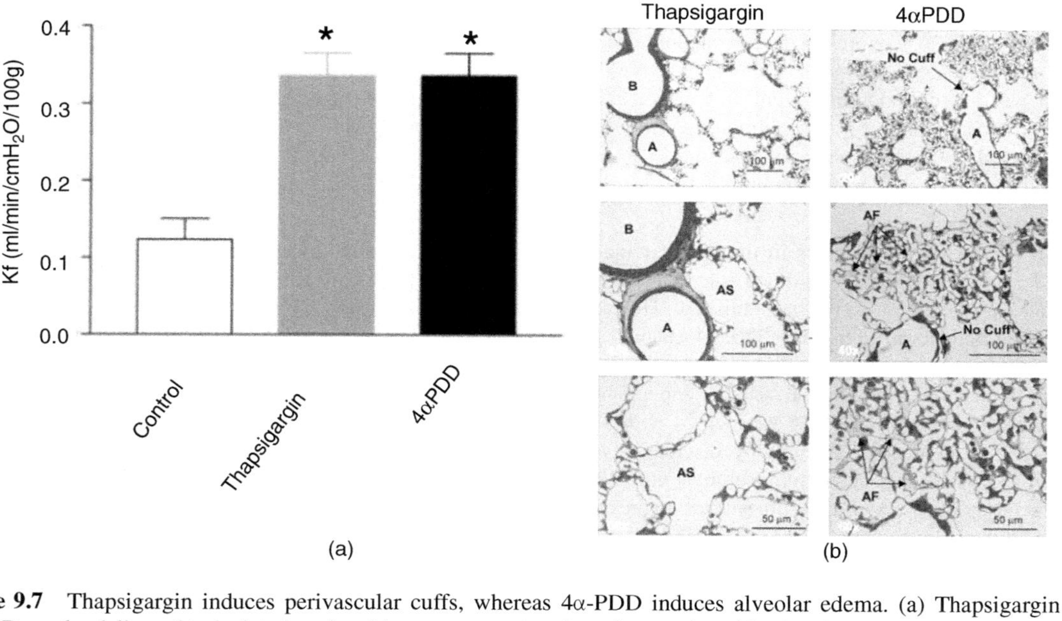

Figure 9.7 Thapsigargin induces perivascular cuffs, whereas 4α-PDD induces alveolar edema. (a) Thapsigargin and 4α-PDD can be delivered to isolated perfused lungs at concentrations that produce identical increases in permeability. (b) Whereas thapsigargin induces extra-alveolar leak sites that cause perivascular cuff formation, 4α-PDD induces capillary leak sites that result in alveolar flooding without concurrent perivascular cuff formation. A, artery; B, bronchiole; AS, alveolar space; AF, alveolar fluid. For experimental details, see [64].

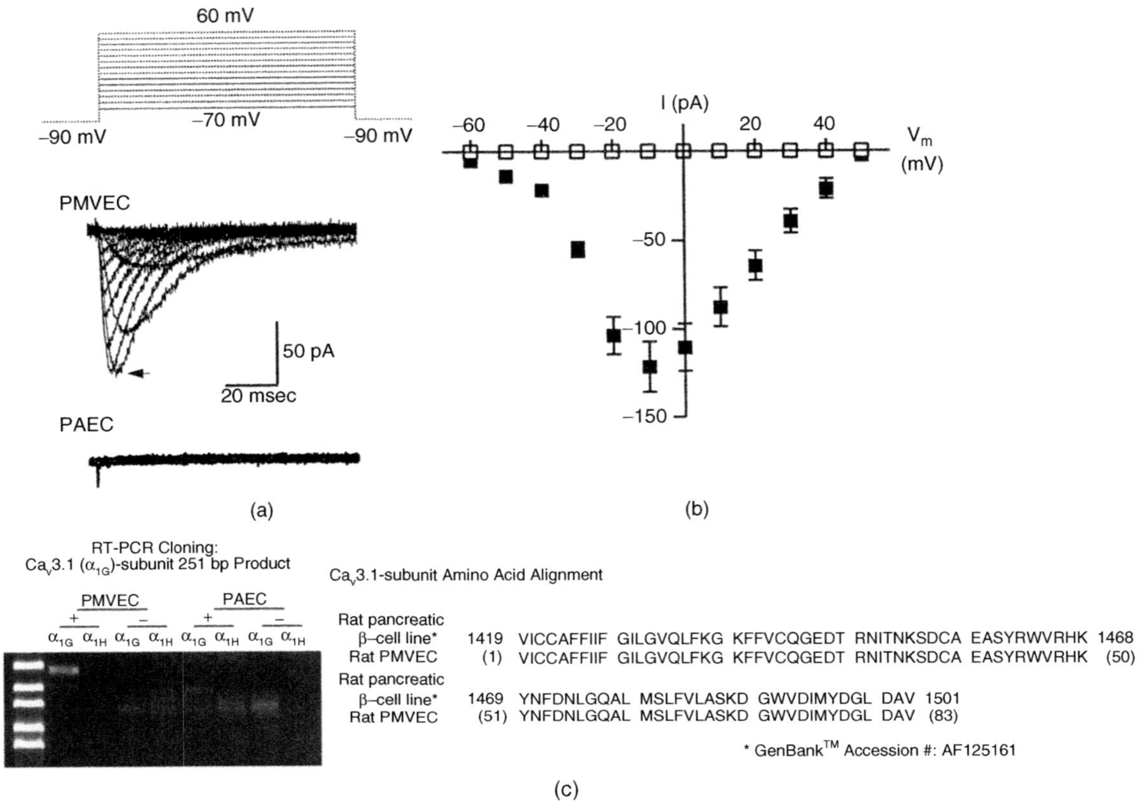

Figure 9.8 PMVECs express a T-type calcium channel that is not present in PAECs. (a) Whole-cell electrophysiology reveals the presence of a voltage-activated, T-type calcium channel. Single ECs were held at – 90 mV and a ramp protocol performed. Calcium current was detected at – 60 mV, peaking at – 10 mV. This prototypical T current was resolved in PMVECs, but not in pulmonary artery ECs. (b) The current (I)–voltage (V) plot reveals a T-type calcium current in PMVECs (filled squares), but not in PAECs (open squares). (c) RT-PCR cloning illustrates the T channel expressed in microvascular endothelium is the α_{1G}, or $Ca_v3.1$, subtype. For experimental details, see [65].

retention of sickled erythrocytes in the lung circulation [65] and appears to play an important role in neutrophil recruitment to the airways (Wu, unpublished). However, α_{1G} channel activation, unlike the TRPV4 channel that is similarly expressed in the capillary, does not increase capillary permeability (Wu and Townsley, unpublished). Thus, two calcium channels that are located within the same vascular segment display highly discrete physiological roles, where α_{1G} activation promotes von Willebrand factor release and P-selectin surface expression, and TRPV4 activation increases EC permeability.

CONCLUSIONS AND PERSPECTIVES

Perhaps it should not be surprising that endothelium displays such finely tuned heterogeneity, even within a single circulation, such as the pulmonary circulation. After all, conduit vessels and capillary segments each fulfill highly specialized physiological functions. Nonetheless, the precision with which endothelium controls expression of ion channels, such as the TRPC1/TRPC4, TRPV4, and α_{1G} channels, and the fidelity whereby these channels are activated, have revealed a much greater complexity than previously recognized.

The increasing appreciation for such functional EC heterogeneity presents exciting new challenges that stand to prominently shape the future of pulmonary medicine. In considering the basis of pulmonary vascular disease, one must now recognize the afflicted vascular sites with a greater degree of accuracy. As one example, in the case of idiopathic pulmonary hypertension with prominent obliterative vascular lesions, what is the specific vascular site from which the lesion emanates? Which EC phenotypes are involved in lesion progression, and in what way? What unique cellular attributes render a specific EC phenotype susceptible or resistant to injury? What unique phenotypic attributes may be pharmacologically targeted? While we know little about the answers to these questions in complex pulmonary vascular diseases, such as idiopathic pulmonary arterial hypertension, Kaposi's sarcoma, and hereditary hemorrhagic telangiectasia, we now appreciate that the ECs so well adapted to perform a site-specific function may also make them susceptible to disease.

Whereas idiopathic pulmonary arterial hypertension is a predominately precapillary disorder, acute lung injury afflicts all vascular segments. Therapeutic strategies targeting endothelium have universally failed in acute lung injury, with the possible exception of activated protein C [68], which shows promise in sepsis and the sepsis syndrome. Now appreciating the heterogeneity among lung vascular segments, the diversity of ion channels that are expressed among EC populations, and how these channels impact on the EC response to inflammation, we recognize that a pan-endothelial approach to therapy is doomed to fail. Yet, this also provides exciting possibilities for exploiting novel molecular targets for therapeutic benefits that are restricted to a specific vascular segment. Inhibiting TRPC1/TRPC4 channel(s) may limit both perivascular cuff formation and alveolar flooding, whereas inhibiting TRPV4 channels may limit alveolar flooding. Strategies to inhibit α_{1G} function will not prevent edema formation, but could limit coagulation and potentially decrease neutrophil recruitment to the airways in pneumonia. Although the therapeutic potential of TRPC1/TRPC4, TRPV4, and α_{1G} channels remains to be fully tested, the prospect of such targeted therapy is exciting.

In addition, the manner in which TRPC1/TRPC4, TRPV4, and α_{1G} channels impact cell function stand to provide keen insight into basic EC biology. For example, TRPC1/TRPC4 activation results in inter-EC gap formation, in a manner originally described by Majno and Palade [69, 70] regarding the endothelial response to histamine and serotonin in postcapillary venules of the systemic circulation. However, to date, TRPV4 activation has not been shown to cause gaps in capillary ECs. Rather, TRPV4 activation disrupts cell–matrix tethering and causes EC sluffing off the basement membrane [59, 62], which may represent a mechanism of apoptosis. How calcium permeation through these two ion channels causes such disparate responses remains unclear. Subcellular compartmentalization of proteins, protein complexes, and second messenger signaling modalities is becoming increasingly well studied, and may contribute to the distinct responses to the TRPC1/TRPC4- and TRPV4-dependent calcium signals.

The α_{1G} channel possesses its own level of subcellular compartmentalization, enabling calcium that permeates its channel to trigger von Willebrand secretion and P-selectin membrane translocation without increasing permeability. The complexity of this signaling modality is further complicated by evidence that capillary ECs do not possess Weibel–Palade bodies. Indeed, it is of interest that capillaries do not possess this organelle, since expression of von Willebrand factor is typically sufficient to promote organelle biogenesis. Lung capillary ECs clearly possess von Willebrand factor, and heterologous expression of the full-length von Willebrand factor induces Weibel–Palade body biogenesis in the capillary cells [66]. It is therefore not presently clear why the endogenously expressed von Willebrand factor fails to induce organelle biogenesis. Moreover, von Willebrand factor and P-selectin do not colocalize in capillary endothelium, as they are not partitioned into a common organelle. The intracellular structure that houses P-selectin remains undetermined, as do the cellular events responsible for responding to the α_{1G}-dependent calcium signal and inserting P-selectin in the plasma membrane. In the aggregate, the intracellular events that are responsible for transducing channel-specific calcium signals into a meaningful physiological response largely remain enigmatic.

EC biologists have made great strides forward in recent years, better defining the structure and function of this intriguing cell type. From this work has emerged a new understanding of the segment-restricted phenotypes of pulmonary artery, capillary, and venous endothelium, with novel molecular fingerprints responsible for highly discrete vascular control. Not only has this insight provided new ways of considering normal vascular homeostasis, the cause of vascular disease, and mechanisms underlying vascular repair, but it has also provided new molecular targets with therapeutic potential. In the foreseeable future, this work will be translated into meaningful practice in pulmonary medicine, while setting the groundwork for an ever more detailed understanding of the basic cell biology that is responsible for maintaining vascular homeostasis.

References

1. Aird, W.C. (ed.) (2007) *Endothelial Biomedicine*, Cambridge University Press, New York.
2. Aird, W.C. (2007) Phenotypic heterogeneity of the endothelium: II. Representative vascular beds. *Circulation Research*, **100**, 174–90.
3. Aird, W.C. (2007) Phenotypic heterogeneity of the endothelium: I. Structure, function, and mechanisms. *Circulation Research*, **100**, 158–73.

4. Gebb, S. and Stevens, T. (2004) On lung endothelial cell heterogeneity. *Microvascular Research*, **68**, 1–12.
5. Stevens, T. (2005) Molecular and cellular determinants of lung endothelial cell heterogeneity. *Chest*, **128**, 558S–564S.
6. Stevens, T., Phan, S., Frid, M.G. *et al.* (2008) Lung vascular cell heterogeneity: endothelium, smooth muscle, and fibroblasts. *Proceedings of the American Thoracic Society*, **5**, 783–91.
7. Thorin, E., Stevens, T., and Patterson, C.E. (2005) Heterogeneity of lung endothelial cells, in *Perspectives on Lung Endothelial Barrier Function* (ed. C.E. Patterson), Elsevier Science, Amsterdam, pp. 277–310.
8. Stevens, T. (2007) Microheterogeneity of lung endothelium, in *Endothelial Biomedicine* (ed. W.C. Aird), Cambridge University Press, New York, pp. 1161–70.
9. Stevens, T., Kasper, M., Cool, C., and Voelkel, N.F. (2005) Pulmonary circulation and pulmonary hypertension, in *Endothelial Cells in Health and Disease* (ed. W.C. Aird), Informa Healthcare, New York, pp. 417–38.
10. Al-Mehdi, A., Schaphorst, K.L., and Stevens, T. (2005) Lung endothelial heterogeneity, in *Encyclopedia of the Microvasculature* (ed. J.G. Garcia), Elsevier Science, Amsterdam, pp. 465–70.
11. Lowenstein, C.J., Morrell, C.N., and Yamakuchi, M. (2005) Regulation of Weibel–Palade body exocytosis. *Trends in Cardiovascular Medicine*, **15**, 302–8.
12. Doerschuk, C.M., Beyers, N., Coxson, H.O. *et al.* (1993) Comparison of neutrophil and capillary diameters and their relation to neutrophil sequestration in the lung. *Journal of Applied Physiology*, **74**, 3040–45.
13. Weibel, E.R. (1973) Morphological basis of alveolar–capillary gas exchange. *Physiological Reviews*, **53**, 419–95.
14. Craig, L.E., Spelman, J.P., Strandberg, J.D., and Zink, M.C. (1998) Endothelial cells from diverse tissues exhibit differences in growth and morphology. *Microvascular Research*, **55**, 65–76.
15. King, J., Hamil, T., Creighton, J. *et al.* (2004) Structural and functional characteristics of lung macro- and microvascular endothelial cell phenotypes. *Microvascular Research*, **67**, 139–51.
16. Parker, J.C. and Yoshikawa, S. (2002) Vascular segmental permeabilities at high peak inflation pressure in isolated rat lungs. *American Journal of Physiology: Lung Cellular and Molecular Physiology*, **283**, L1203–9.
17. Parker, J.C. (2007) Hydraulic conductance of lung endothelial phenotypes and starling safety factors against edema. *American Journal of Physiology: Lung Cellular and Molecular Physiology*, **292**, L378–80.
18. Parker, J.C., Stevens, T., Randall, J. *et al.* (2006) Hydraulic conductance of pulmonary microvascular and macrovascular endothelial cell monolayers. *American Journal of Physiology: Lung Cellular and Molecular Physiology*, **291**, L30–37.
19. Kelly, J.J., Moore, T.M., Babal, P. *et al.* (1998) Pulmonary microvascular and macrovascular endothelial cells: differential regulation of Ca^{2+} and permeability. *American Journal of Physiology: Lung Cellular and Molecular Physiology* **274**, L810–19.
20. Moore, T.M., Chetham, P.M., Kelly, J.J., and Stevens, T. (1998) Signal transduction and regulation of lung endothelial cell permeability. Interaction between calcium and cAMP. *American Journal of Physiology: Lung Cellular and Molecular Physiology*, **275**, L203–22.
21. Rotrosen, D. and Gallin, J.I. (1986) Histamine type I receptor occupancy increases endothelial cytosolic calcium, reduces F-actin, and promotes albumin diffusion across cultured endothelial monolayers. *The Journal of Cell Biology*, **103**, 2379–87.
22. Ullmer, C., Boddeke, H.G., Schmuck, K., and Lubbert, H. (1996) 5-HT_{2B} receptor-mediated calcium release from ryanodine-sensitive intracellular stores in human pulmonary artery endothelial cells. *British Journal of Pharmacology*, **117**, 1081–88.
23. Korth, R.M., Hirafuji, M., Benveniste, J., and Russo-Marie, F. (1995) Human umbilical vein endothelial cells: specific binding of platelet-activating factor and cytosolic calcium flux. *Biochemical Pharmacology*, **49**, 1793–99.
24. Lum, H., Del Vecchio, P.J., Schneider, A.S. *et al.* (1989) Calcium dependence of the thrombin-induced increase in endothelial albumin permeability. *Journal of Applied Physiology*, **66**, 1471–76.
25. Goligorsky, M.S., Menton, D.N., Laszlo, A., and Lum, H. (1989) Nature of thrombin-induced sustained increase in cytosolic calcium concentration in cultured endothelial cells. *The Journal of Biological Chemistry*, **264**, 16771–75.
26. Clapham, D.E. (1995) Calcium signaling. *Cell*, **80**, 259–68.
27. Buchan, K.W. and Martin, W. (1991) Bradykinin induces elevations of cytosolic calcium through mobilisation of intracellular and extracellular pools in bovine aortic endothelial cells. *British Journal of Pharmacology*, **102**, 35–40.

28. Yu, R. and Hinkle, P.M. (2000) Rapid turnover of calcium in the endoplasmic reticulum during signaling. Studies with cameleon calcium indicators. *The Journal of Biological Chemistry*, **275**, 23648–53.
29. Meldolesi, J. and Pozzan, T. (1998) The endoplasmic reticulum Ca^{2+} store: a view from the lumen. *Trends in Biochemical Sciences*, **23**, 10–14.
30. Montero, M., Brini, M., Marsault, R. *et al.* (1995) Monitoring dynamic changes in free Ca^{2+} concentration in the endoplasmic reticulum of intact cells. *The EMBO Journal*, **14**, 5467–75.
31. Varadi, A. and Rutter, G.A. (2002) Dynamic imaging of endoplasmic reticulum Ca^{2+} concentration in insulin-secreting MIN6 Cells using recombinant targeted cameleons: roles of sarco(endo)plasmic reticulum Ca^{2+}-ATPase (SERCA)-2 and ryanodine receptors. *Diabetes*, **51** (Suppl 1), S190–201.
32. Parekh, A.B. and Putney, J.W. Jr. (2005) Store-operated calcium channels. *Physiological Reviews*, **85**, 757–810.
33. Putney, J.W. Jr. (1986) A model for receptor-regulated calcium entry. *Cell Calcium*, **7**, 1–12.
34. Thastrup, O., Cullen, P.J., Drobak, B.K. *et al.* (1990) Thapsigargin, a tumor promoter, discharges intracellular Ca^{2+} stores by specific inhibition of the endoplasmic reticulum Ca^{2+}-ATPase. *Proceedings of the National Academy of Sciences of the United States of America*, **87**, 2466–70.
35. Chetham, P.M., Guldemeester, H.A., Mons, N. *et al.* (1997) Ca^{2+}-inhibitable adenylyl cyclase and pulmonary microvascular permeability. *American Journal of Physiology: Lung Cellular and Molecular Physiology*, **273**, L22–30.
36. Chetham, P.M., Babal, P., Bridges, J.P. *et al.* (1999) Segmental regulation of pulmonary vascular permeability by store-operated Ca^{2+} entry. *American Journal of Physiology: Lung Cellular and Molecular Physiology*, **276**, L41–50.
37. Wu, S., Cioffi, E.A., Alvarez, D. *et al.* (2005) Essential role of a Ca^{2+}-selective, store-operated current (I_{SOC}) in endothelial cell permeability: determinants of the vascular leak site. *Circulation Research*, **96**, 856–63.
38. Wu, S., Chen, H., Alexeyev, M.F. *et al.* (2007) Microtubule motors regulate I_{SOC} activation necessary to increase endothelial cell permeability. *The Journal of Biological Chemistry*, **282**, 34801–8.
39. Cioffi, D.L., Wu, S., and Stevens, T. (2003) On the endothelial cell I_{SOC}. *Cell Calcium*, **33**, 323–36.
40. Wu, S., Moore, T.M., Brough, G.H. *et al.* (2000) Cyclic nucleotide-gated channels mediate membrane depolarization following activation of store-operated calcium entry in endothelial cells. *The Journal of Biological Chemistry*, **275**, 18887–96.
41. Brough, G.H., Wu, S., Cioffi, D. *et al.* (2001) Contribution of endogenously expressed Trp1 to a Ca^{2+}-selective, store-operated Ca^{2+} entry pathway. *The FASEB Journal*, **15**, 1727–38.
42. Wu, S., Sangerman, J., Li, M. *et al.* (2001) Essential control of an endothelial cell I_{SOC} by the spectrin membrane skeleton. *The Journal of Cell Biology*, **154**, 1225–33.
43. Montell, C. (2005) *Drosophila* TRP channels. *Pflugers Archiv*, **451**, 19–28.
44. Montell, C. (2005) The TRP superfamily of cation channels. *Science's STKE*, **2005**, re3.
45. Montell, C. (2001) Physiology, phylogeny, and functions of the TRP superfamily of cation channels. *Science's STKE*, **2001**, RE1.
46. Montell, C. (2004) Molecular genetics of *Drosophila* TRP channels, in *Mammalian TRP Channels as Molecular Targets*, Novartis Foundation Symposium 258, John Wiley & Sons, Ltd, Chichester, pp. 3–12; discussion 12–7, 98–102, 263–6.
47. Putney, J.W. Jr. (2007) New molecular players in capacitative Ca^{2+} entry. *Journal of Cell Science*, **120**, 1959–65.
48. Putney, J.W. (2005) Physiological mechanisms of TRPC activation. *Pflugers Archiv*, **451**, 29–34.
49. Vazquez, G., Wedel, B.J., Aziz, O. *et al.* (2004) The mammalian TRPC cation channels. *Biochimica et Biophysica Acta*, **1742**, 21–36.
50. Clapham, D.E. (2007) SnapShot: mammalian TRP channels. *Cell*, **129**, 220.
51. Ramsey, I.S., Delling, M., and Clapham, D.E. (2006) An introduction to TRP channels. *Annual Review of Physiology*, **68**, 619–47.
52. Clapham, D.E. (2003) TRP channels as cellular sensors. *Nature*, **426**, 517–24.
53. Nilius, B. and Voets, T. (2005) TRP channels: a TR(I)P through a world of multifunctional cation channels. *Pflugers Archiv*, **451**, 1–10.
54. Trebak, M., Vazquez, G., Bird, G.S., and Putney, J.W. Jr. (2003) The TRPC3/6/7 subfamily of cation channels. *Cell Calcium*, **33**, 451–61.
55. Freichel, M., Suh, S.H., Pfeifer, A. *et al.* (2001) Lack of an endothelial store-operated Ca^{2+} current impairs agonist-dependent vasorelaxation in $TRP4^{-/-}$ mice. *Nature Cell Biology*, **3**, 121–27.
56. Cioffi, D.L., Wu, S., Alexeyev, M. *et al.* (2005) Activation of the endothelial store-operated I_{SOC} Ca^{2+} channel requires interaction of protein 4.1 with TRPC4. *Circulation Research*, **97**, 1164–72.

57. Cioffi, D.L. and Stevens, T. (2006) Regulation of endothelial cell barrier function by store-operated calcium entry. *Microcirculation*, **13**, 709–23.
58. Cioffi, D.L., Lowe, K., Alvarez, D.F. *et al.* (2009) TRPing on the lung endothelium. Calcium channels that regulate barrier function. *Antioxidants and Redox Signaling*, **11**, 765–76.
59. Townsley, M.I., King, J.A., and Alvarez, D.F. (2006) Ca^{2+} channels and pulmonary endothelial permeability: insights from study of intact lung and chronic pulmonary hypertension. *Microcirculation*, **13**, 725–39.
60. Alvarez, D.F., King, J.A., and Townsley, M.I. (2005) Resistance to store depletion-induced endothelial injury in rat lung after chronic heart failure. *American Journal of Respiratory and Critical Care Medicine*, **172**, 1153–60.
61. Vriens, J., Owsianik, G., Fisslthaler, B. *et al.* (2005) Modulation of the Ca^{2+} permeable cation channel TRPV4 by cytochrome P450 epoxygenases in vascular endothelium. *Circulation Research*, **97**, 908–15.
62. Alvarez, D.F., King, J.A., Weber, D. *et al.* (2006) Transient receptor potential vanilloid 4-mediated disruption of the alveolar septal barrier: a novel mechanism of acute lung injury. *Circulation Research*, **99**, 988–95.
63. Grossman, R.F., Jones, J.G., and Murray, J.F. (1980) Effects of oleic acid-induced pulmonary edema on lung mechanics. *Journal of Applied Physiology*, **48**, 1045–51.
64. Lowe, K., Alvarez, D., King, J., and Stevens, T. (2007) Phenotypic heterogeneity in lung capillary and extra-alveolar endothelial cells. Increased extra-alveolar endothelial permeability is sufficient to decrease compliance. *The Journal of Surgical Research*, **143**, 70–77.
65. Wu, S., Haynes, J. Jr., Taylor, J.T. *et al.* (2003) Cav3.1 (α_{1G}) T-type Ca^{2+} channels mediate vaso-occlusion of sickled erythrocytes in lung microcirculation. *Circulation Research*, **93**, 346–53.
66. Zhou, C., Chen, H., Lu, F. *et al.* (2007) Cav3.1 (alpha1G) controls von Willebrand factor secretion in rat pulmonary microvascular endothelial cells. *American Journal of Physiology: Lung Cellular and Molecular Physiology*, **292**, L833–44.
67. Fuchs, A. and Weibel, E.R. (1966) Morphometric study of the distribution of a specific cytoplasmatic organoid in the rat's endothelial cells. *Zeitschrift fur Zellforschung und Mikroskopische Anatomie*, **73**, 1–9.
68. Finigan, J.H., Dudek, S.M., Singleton, P.A. *et al.* (2005) Activated protein C mediates novel lung endothelial barrier enhancement: role of sphingosine 1-phosphate receptor transactivation. *The Journal of Biological Chemistry*, **280**, 17286–93.
69. Majno, G. and Palade, G.E. (1961) Studies on inflammation. 1. The effect of histamine and serotonin on vascular permeability: an electron microscopic study. *The Journal of Biophysical and Biochemical Cytology*, **11**, 571–605.
70. Majno, G., Palade, G.E., and Schoefl, G.I. (1961) Studies on inflammation. II. The site of action of histamine and serotonin along the vascular tree: a topographic study. *The Journal of Biophysical and Biochemical Cytology*, **11**, 607–26.

10

Pulmonary Endothelial Interactions with Leukocytes and Platelets

Rosana Souza Rodrigues[1] and Guy A. Zimmerman[2]

[1] Department of Radiology, Federal University of Rio de Janeiro, Rio de Janeiro, Brazil
[2] Department of Internal Medicine, Program in Human Molecular Biology and Genetics, University of Utah School of Medicine, Salt Lake City, UT, USA

INTRODUCTION

When we turn to lung we find polynuclear cells present in disproportionate numbers ... Failing any evidence that the narrowness of the capillaries mechanically arrests them, one is tempted to assume that they voluntarily tarry.

F. W. Andrewes, 1910 [1]

Leukocytes and the lungs have a complicated and interesting relationship. There are transient and resident populations of leukocytes in intravascular and extravascular compartments of the lungs, and these immune effector cells participate in lung homeostasis, defense, repair, and injury. This chapter is focused on the intravascular compartment, and on interactions of pulmonary endothelial cells (ECs) with white blood cells (WBC) and platelets. Unique interactions between the lung vascular network and leukocytes have been recognized for over a century. A key point is that leukocytes and pulmonary ECs interact in a dynamic fashion continuously in normal physiologic conditions – in addition to states of defensive inflammation (as in responses to infection and wound surveillance), and in pathologic inflammation and injury. In this chapter we focus on the basal physiologic state and leukocyte–endothelial interactions in regulated inflammation, although we refer frequently to lung injury. We also briefly discuss interactions of pulmonary endothelium with platelets (see also Chapter 25). Interactions of endothelium with red blood cells (RBCs) are touched on briefly in this chapter. There is a very large literature on leukocyte–endothelial interactions and related topics, and our bibliography is not comprehensive but instead includes a mix of liberally cited reviews [2–21] and primary articles that report key findings an/or investigative approaches. We begin with studies of margination and sequestration of leukocytes and platelets in human subjects and patients, and then move to experimental models and mechanisms. Mechanisms of leukocyte–endothelial interactions in systemic vascular beds heavily influence concepts of these cell–cell interactions in the lung and are discussed in detail. We attempt to integrate classic observations [1, 22], with interval discoveries and more recent studies in a variety of systems.

OBSERVATIONS IN PATIENTS AND THE "HUMAN MODEL"

Bierman *et al.*, in studies at the Laboratory of Experimental Oncology of the National Cancer Institute and the University of California at San Francisco, examined transfusion of leukocytes in patients with neoplastic diseases and found evidence that large numbers of transfused WBCs are removed by the lungs. Based on preliminary observations they proposed that a mechanism for accumulation of leukocytes in the lungs exists, that it is under physiologic control, that leukocytes can be rapidly delivered from the lungs to systemic vessels, and that this is paralleled by a mechanism for rapid delivery of platelets from the lungs into the peripheral circulation [23–25]. They also concluded that these mechanisms are impaired in leukemic patients. Building on these initial findings, they conducted additional clinical investigations [26, 27] using intravascular catheters to selectively sample the right and left circulations (Figure 10.1a) that were later summarized in a seminal review [28]. One of the most

The Pulmonary Endothelium: Function in health and disease Editors Norbert F. Voelkel, Sharon Rounds

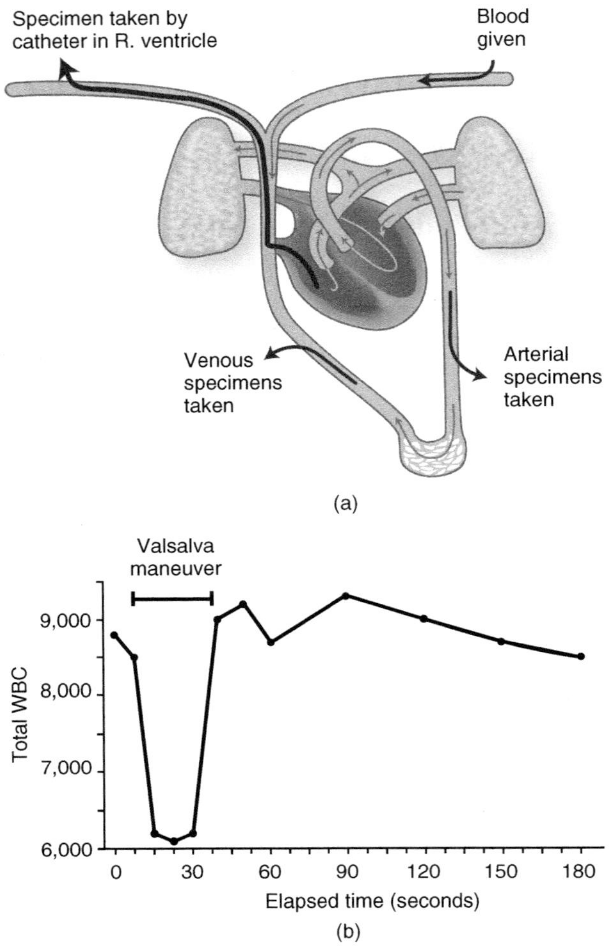

Figure 10.1 Studies in human subjects identified a marginated pool of leukocytes in the lungs. (a) Bierman *et al.* sampled pulmonary and systemic blood using intravascular catheters, and determined the concentrations of leukocytes before and after administration of transfused leukocytes, forced ventilatory maneuvers (see Figure 10.1b), and infusion of epinephrine or histamine. (b) A Valsalva maneuver resulted in rapid and reversible decreases in the total WBCs in arterial blood, whereas a Muller maneuver (not shown) had the opposite effect. The greatest impact of forced ventilatory maneuvers was on PMN number in the arterial blood. This research was originally published in *Blood* and the figure is redrawn from [28].

interesting observations in this series of reports was that leukocytes accumulate in the lungs with a decrease in arterial numbers during inspiration and are released into the arterial blood on expiration, with marked changes when subjects performed Valsalva or Muller maneuvers [27], (Figure 10.1b). Parallel measurements indicated that the changes in leukocyte numbers were not due to hemoconcentration or hemodilution, leading them to suggest that there is a significant tidal ebb and flow of leukocytes into and from the pulmonary circulation during breathing that can be accentuated by forced ventilatory actions [27]. Comparison of leukocyte numbers and time-dependent variations in arterial and venous blood indicated that the lungs can more rapidly alter circulating leukocyte numbers than all other organs combined. Bierman *et al.* further reported that the greatest changes with respiratory maneuvers are in numbers of circulating polymorphonuclear leukocytes (PMNs; also interchangeably termed neutrophils and granulocytes in original reports and in this chapter) and suggested that adhesiveness of these cells might play a role in addition to, or exclusive of, mechanical retention of these cells in the lungs [27]. In parallel, they also reported that infusion of epinephrine causes an increase in leukocyte and platelet numbers in the arterial circulation followed by increases in the venous blood [26], and that, conversely, intravenous histamine causes rapid accumulation of leukocytes in the pulmonary vascular bed accompanied by transient peripheral leukopenia [24, 28]. The greatest effect of histamine was on the granulocyte series, but other leukocyte subtypes were also involved. Although many of their studies were conducted in patients with cancer, their conclusions regarding the marginated pool of leukocytes in the lungs and leukocyte sequestration in response to inflammatory agonists have largely been validated in healthy subjects and experimental animals. (Both "marginated" and "marginal" have been used to identify the localized population of leukocytes present in lung vessels and other microvascular beds. "Marginated" is more common and is used in most places in this chapter.)

In their studies of transfusion, Bierman *et al.* reported that the pulmonary circulation of nonleukemic subjects is capable of removing large numbers of leukemic myeloid or nonmyeloid leukocytes delivered by infusion or cross-transfusion [28]. Interestingly, they also described two episodes of pulmonary edema that occurred within minutes of transfusion of leukocytes [28]. These may be some of the earliest cases of what is now called transfusion-associated acute lung injury [29] and illustrate the concept that accumulation of leukocytes in the pulmonary vascular bed may sometimes have pathologic consequences.

In a classic series of experiments in the "human model," Athens *et al.* at the University of Utah devised methods to radiolabel leukocytes and reported that infused labeled granulocytes are distributed to a total blood pool that is made up of a circulating, freely accessible pool and a "marginal" pool that are in dynamic equilibrium in normal male subjects [30, 31] (Figure 10.2). A similar analysis in women and children has not been reported. Athens *et al.* found that physical exercise or infusion of epinephrine resulted in a shift of granulocytes from the marginal pool to the circulating pool without altering the total blood granulocyte number. The latter observation was consistent with Bierman's report of changes in distribution of unlabeled PMNs in response to epinephrine [26]. Athens *et al.* also reported that injection of *Salmonella* lipopolysaccharide (LPS; endotoxin) 4 h after infusion of labeled granulocytes resulted in a shift of cells from the circulating pool to the marginal pool 1.5 h later, with transient leukopenia in the venous blood in some, but not all, subjects (Figure 10.2). This

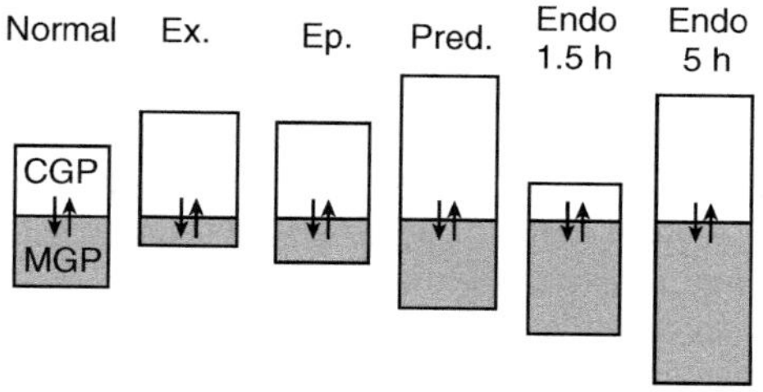

Figure 10.2 Observations using radiolabeled leukocytes characterized the marginal and circulating pools of PMNs in human volunteers. Athens *et al.* labeled granulocytes (PMNs) and returned them to the venous blood of male volunteers. They found the labeled cells partitioned into a circulating granulocyte pool (CGP) and a "marginal" granulocyte pool (MGP) that were in dynamic equilibrium (arrows). Exercise (Ex), epinephrine (Ep.), prednisone (Pred.), and endotoxin (Endo) dramatically altered the numbers of PMNs in the MGP and CGP. In some cases (prednisone; endotoxin) the total blood granulocyte pool (MGP + CGP) was increased. Redrawn from [31].

was followed by expansion of both the marginal and the circulating pools of granulocytes 5 h after challenge with LPS [31]. These studies generated insights regarding variations in granulocyte number in response to physiologic and pathologic stimuli (Figure 10.2) that remain clinically useful today. Athens *et al.* did not specifically explore the contributions of the lung to the marginal pool, but nevertheless commented that the work of Bierman and others suggested that the "... vascular system of the lungs is of particular significance in the margination of granulocytes" [30]. Athens later suggested that the lungs are a particular destination for leukocytes newly released from the bone marrow in patients rebounding from cancer chemotherapy and observed that this is often temporally accompanied by lung injury syndromes (J. Athens, personal communication).

Hogg *et al.* subsequently reported that increased pulmonary blood flow contributes to release of PMNs into the circulation of human subjects [32], and that leukocytes are retained in the lung in part because of the effects of differential alveolar and capillary pressures in lung zones 3, 2, and 1 [33]. The latter observations were made in spontaneously breathing volunteers who also performed forced expiratory maneuvers and are consistent with the earlier studies by Bierman (Figure 10.1). Hogg's group contributed additional important studies that defined characteristics of the lung marginated pool of leukocytes (see "Cellular and Molecular Features that Influence the Marginated Pool of Leukocytes in Lung Capillaries") in human subjects [34–38]. Parallel studies in dogs and rabbits provided comparisons to human data and, suggested relationships between neutrophil kinetics and lung injury [37, 39] (reviewed in [40]).

Clinical studies between 1980 and 1985 identified a previously unrecognized heritable disorder of leukocyte–endothelial interactions characterized by recurrent infections of the soft tissues including the lungs, impaired wound healing, and defective leukocyte accumulation at extravascular sites [41]. This syndrome, ultimately termed leukocyte adhesion deficiency (LAD) type I, was subsequently shown to be due to absent or severely reduced numbers of β_2 integrins (also called CD11/CD18 integrins and "leukocyte integrins") on the surface of PMNs, monocytes, and other leukocytes – a deficiency that causes impaired adhesion of these cells to ECs and to other surfaces; later clinical studies identified two additional molecular mechanisms of defective leukocyte adhesion to endothelium (reviewed in [12]). The β_2 integrins are $\alpha\beta$ heterodimers with highly restricted expression on leukocytes that mediate adhesion to ligands on the surfaces of resting or inflamed EC. The β_2 integrins also bind to structures on other cells such as platelets or neighboring WBCs, and to matrix molecules [8]. This mechanism of adhesion requires activation of the leukocytes and a change in the affinity and/or avidity of the β_2 integrins that makes them competent to bind to the counter-receptors on ECs and other targets [8, 12]. Studies to define the molecular basis of the LAD syndromes and characterize regulation of adhesive mechanisms mediated by β_2 integrins contributed directly to development of the concept of a multistep cascade of leukocyte–endothelial interactions in the systemic circulation in infection and inflammation (reviewed in [12, 15]). This paradigm is important in understanding current views of leukocyte–endothelial interactions in the normal lung (See Sticky controversies: Physiologic margination of leukocytes in pulmonary arterioles and venules, and adhesive margination in alveolar capillaries). The β_2 integrins also contribute to accumulation of leukocytes in lung vessels and alveoli in a variety of physiologic and pathologic inflammatory conditions [9, 16].

EVOLUTION OF SURROGATE EXPERIMENTAL SYSTEMS

Andrewes reviewed the roles of leukocytes in host defense and the distribution of leukocytes in different vascular beds of rabbits and other experimental animals based on his own work and that of others [1, 22]. He concluded that the lung contains a disproportionate number of PMNs in its microvascular network and suggested that this has some particular physiologic significance, possibly related to oxygenation of these cells. This opinion unwittingly anticipated molecular relationships between oxygen sensing, innate immunity, and PMN function that are currently being examined [42]. Later, Vejlens reviewed the distribution of leukocytes in the vascular system and noted that as early as 1867, Conheim reported leukocytes in a marginal location along the endothelial surfaces of venules of experimental animals [43]. This influenced the views of Athens *et al.* in interpretation of their studies of the marginal granulocyte pool [30]. Bierman discussed the hematologic activities of the lung in man in the context of earlier observations in experimental animals [28]. Other investigators built on Bierman's studies in human subjects with experiments in dogs that yielded similar findings [44]. Experimental animals of a variety of species continue to be important in studies of leukocyte interactions in the lung, and have been utilized extensively in investigations of the lung marginated pool and the process of inflammation-induced leukocyte sequestration in the lung [4, 9, 14, 40].

Renewed interest in intravital microscopic approaches provided significant technical advances in the analysis of animal models. Intravital techniques have been extensively utilized in studies of systemic leukocyte–endothelial interactions (Figure 10.3). Building on studies by Malphigi in the seventeenth century and Terry in the 1930s (reviewed in [14]), Wagner *et al.* examined leukocyte transit in lung vessels of dogs using intravital microscopy [45–47]. Kuebler *et al.*, working

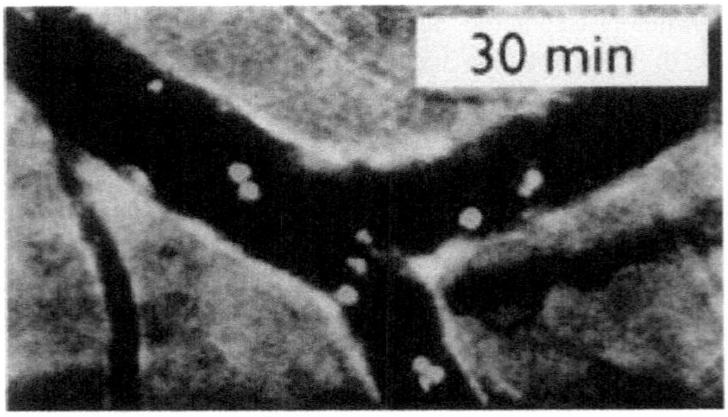

Figure 10.3 Intravital microscopy allows direct examination of cell–cell interactions in systemic vessels. Intravital microscopic evaluation of leukocyte rolling and adhesion in vessels was examined in dorsal skinfold chambers in hamsters. Animals were exposed to cigarette smoke, leukocytes were stained with Acridine orange by intravascular injection, and rolling and adhesion were visualized by fluorescence microscopy. Cigarette smoke induced slow rolling and adhesion of leukocytes, and also caused the formation of intravascular platelet–leukocyte aggregates (not shown). Reproduced from [127] with permission of The American Society for Clinical Investigation.

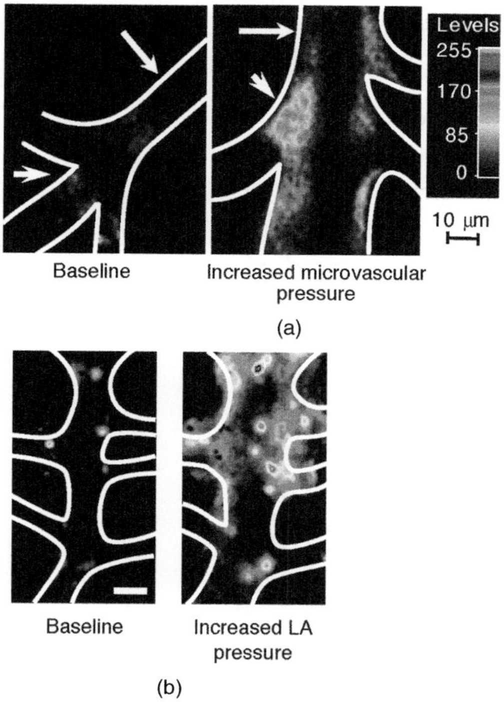

Figure 10.4 Intravital microscopy of the lung provides key insights regarding margination and sequestration of leukocytes. Many investigators have applied intravital microscopic techniques to studies of margination and sequestration of leukocytes in the lungs (see text). This figure illustrates examples in which intravital imagining was used to detect surface display of P-selectin (a) and accumulation of fluorescently labeled leukocytes (b) in microvessels of isolated, blood-perfused rat lungs under conditions of increased microvascular pressure. See [50, 51] for experimental details. The panels in this figure were provided by Jahar Bahattacharya and are reproduced with permission of the author and publishers [50, 51].

in Munich, later developed intravital fluorescence microscopic technologies for studies of the lungs of rabbits, and reported measurements of microvascular hemodynamics and leukocyte kinetics [48, 49] (reviewed in [14]). More recently, Bhattacharya's group in New York has applied elegant intravital approaches that allow imaging of endothelial signal transduction events and display of selectins as well as the local behavior of leukocytes in pulmonary microvessels of rats and mice [50–52] (Figure 10.4).

In the interval between the human physiologic studies in the 1950s and early 1960s [28, 31], and adaptation of intravital microscopy to lung analysis in the 1980s (see previous paragraph), modern vascular biology was revolutionized by development of techniques for reproducible culture of human endothelium with phenotypic characteristics similar to those found *in situ* [53, 54], (reviewed in [55]). Subsequently, cultured human endothelium has been used to characterize many known functions and responses of ECs at the molecular level, to identify novel endothelial activities, and to characterize roles of ECs under physiologic conditions and in disease [7, 17, 18]. Among these biologic features are changes in phenotype and function in response to cues from the environment – including signals delivered by inflammatory, thrombotic, and microbial agonists – a group of processes that have been termed EC "activation" [17, 56]. While there is evidence for phenotypic heterogeneity in endothelium in different vascular beds, the ability to sense and respond to experimental signals may be a defining feature of EC across the spectrum of organs and tissues (reviewed in [17]).

Cultured human endothelial models provided the first clear evidence that activated EC present molecules that mediate adhesion and signaling of leukocytes [57–60] (Figure 10.5). This was a new biologic role for EC, dramatically altering understanding of the functional capabilities of these cells in inflammation and hemostasis (reviewed in [3, 7]). A variety of observations demonstrated that EC are not exclusively inert cellular "landing sites" for adherent PMNs and other leukocytes stimulated by inflammatory mediators or chemotactic factors in the flowing blood [3], although this is one mechanism by which cell–cell interactions between leukocytes and endothelium occurs [61–63]. Cultured EC models were subsequently modified to allow study of leukocyte–endothelial interactions under conditions of flow and shear, providing additional refinement and important insights relevant to both the systemic and pulmonary circulations. These early studies also contributed the initial observations leading to evolution of the "multistep paradigm" concept of leukocyte interaction with the systemic endothelium (see following section). Leukocytes from subjects with LAD were studied in cultured EC models [64], providing new basic insights into molecular mechanisms of endothelial–leukocyte interactions that have clinical relevance [12].

In the 1970s and 1980s, molecular approaches to alteration of the murine genome by homologous recombination [65] broadly revolutionized experimental biology and medicine. Mice with targeted deletions of leukocyte integrins, selectins, and ligands for these adhesion molecules were developed and used in studies of endothelial–leukocyte interactions in the systemic circulation [66] (reviewed in [67, 68]). Although there are significant differences in innate immune and inflammatory responses of mice and humans [69, 70], murine studies often provided *in vivo* evidence for mechanisms

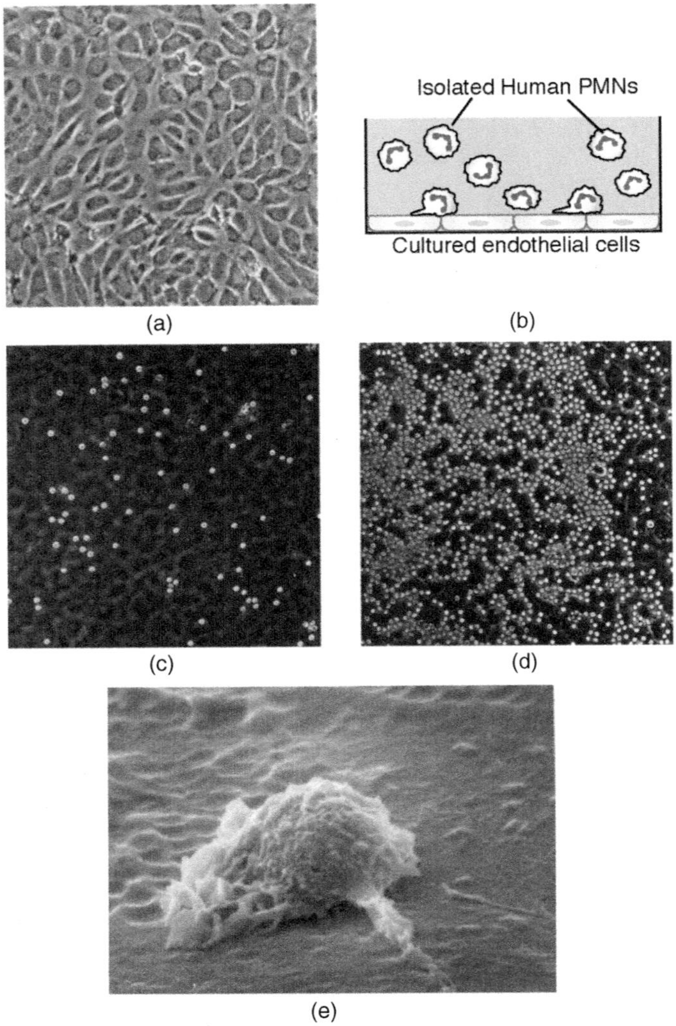

Figure 10.5 Human EC models: key experimental systems in mechanistic studies of endothelial–leukocyte interactions. (a) Primary cultures of HUVECs are highly informative models of EC phenotype, activation, and function *in vivo*. Cultured ECs from other vascular beds have also been used in mechanistic studies. (b) Models employing HUVECs and isolated human PMNs were used to examine molecular mechanisms of tethering, signaling, and adhesion in endothelial–leukocyte interactions. See [141] for additional details. (c and d) Treatment of EC monolayers with thrombin (d) or histamine (not shown) induced rapid adhesion of PMNs when compared to EC monolayers treated with control buffer alone (c) [58]. The mechanism for PMN adhesion, which peaked within minutes, was subsequently shown to depend on translocation of P-selectin to the EC surface resulting in tethering of the PMNs, rapid synthesis and signaling by PAF presented in juxtacrine fashion by the stimulated EC, and binding of β_2 integrins on the activated PMNs to ligands on the EC plasma membrane [3, 64]. These studies contributed to the concept of a multistep paradigm of molecular events in endothelial–PMN interactions (see Figure 10.6). (e and a) PMN adherent to the surface of a thrombin-stimulated EC monolayer (d) was imaged by scanning election microscopy. The leukocyte has an anterior spreading lamellopodium and a trailing tail-like uropod demonstrating polarization and indicating directional migration. Points of intimate adhesion to the EC plasma membrane are also obvious. Reproduced from [58] with permission of The American Society for Clinical Investigation.

of EC–leukocyte interaction that were initially identified in *in vitro* models employing human EC and leukocytes. Genetically manipulated mice were subsequently utilized to examine determinants of leukocyte adhesion and emigration in the lung in models of inflammation, infection, and injury (reviewed in [13, 19, 70]). Some studies of lung inflammation in mice examined gene knockouts in parallel with alternative approaches such as treatment of wild-type animals with blocking antibodies or antisense constructs, at times yielding differing results depending on the experimental perturbation [71].

INTERACTION OF PMNs WITH SYSTEMIC ENDOTHELIUM AND THE MULTISTEP PARADIGM

The cellular and molecular bases for endothelial–leukocyte interactions were first studied in the context of systemic endothelium, and were then later extended to questions relevant to the pulmonary circulation. A multistep process for spatially regulated targeting of leukocytes to specific regions of the systemic vasculature was identified based on a variety of observations that, in one instance or another, utilized all of the experimental approaches outlined in previous sections. The steps include activation of ECs in a localized fashion, initial tethering events mediated by members of the selectin family of adhesion molecules, localized signaling and activation of the leukocytes, tight adhesion and arrest of the WBCs on the inflamed endothelial surface, and subsequent emigration of the leukocytes from the blood to the extravascular space. As the linked molecular events occur in apparent sequence this has been called a multistep cascade or the "multistep paradigm" [2, 6]. The multistep paradigm has been extensively reviewed (e.g., [3, 5, 7, 10, 11, 15, 16, 21]), and this concept is now included in the discussions of vascular function and inflammation in general textbooks of clinical medicine, pulmonary science, and cell and molecular biology. A grasp of the multistep cascade is necessary in order to understand concepts and controversies regarding endothelial–leukocyte interactions in the lung.

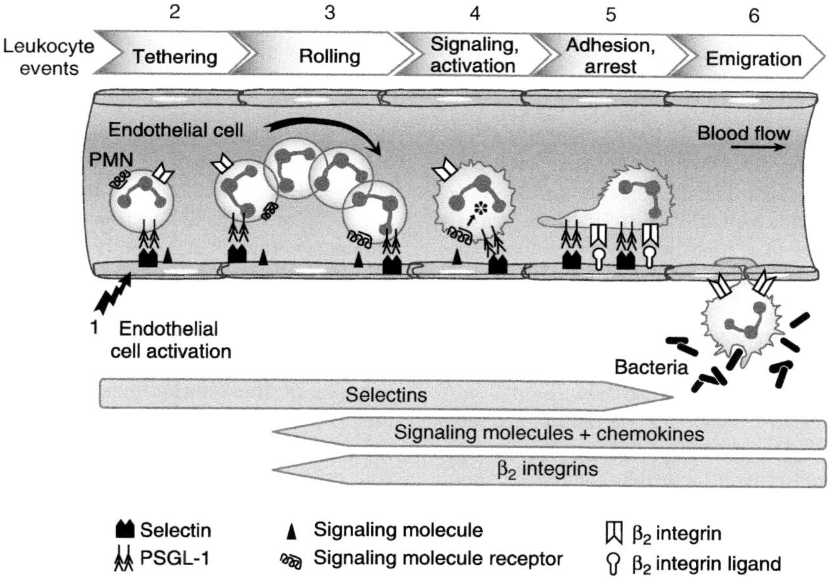

Figure 10.6 Endothelial interactions with PMNs in the systemic circulation involve a multistep cascade of molecular events. Studies with human cell models (Figure 10.5), *in vivo* animal models and intravital microscopic techniques (Figure 10.3), cells from humans with leukocyte adhesion deficiency syndromes, and genetically altered mice indicate that a sequence of molecular events mediates PMN tethering, adhesion, and emigration in postcapillary venules and at other systemic vascular sites in response to infection, pathologic inflammation, and tissue injury. See text for details of each step, and variations in systemic vascular beds. The molecular sequence may not be strictly linear and has overlapping features depending on the inflammatory context, as indicated by the bars below the figure. The systemic multistep paradigm has had a major influence on concepts of margination and sequestration of PMNs in the lungs (Figures 10.7 and 10.8). Modified by permission of Lippincott, Williams and Wilkins from Bunting, M., Harris E.S. *et al*, (2002) *Current Opinion in Hematology*, **10**(2) 150–158.

Central aspects of the multistep cascade will be outlined using PMN–endothelial interactions (Figure 10.6) [12, 15], but the paradigm is also relevant to interactions of endothelium with other leukocyte subtypes [6, 72]. Multiple steps in PMN interactions with inflamed endothelium were identified using human EC cell models and isolated adhesion molecules [64, 73], and further dissected using genetically modified mice and other surrogate systems (reviewed in [11, 15, 67]).

Under basal conditions, systemic ECs and PMNs are not adhesive for one another or are minimally adhesive [14] – a feature that allows free circulation of the leukocytes in the blood. In an initial step leading to targeted adhesion of PMNs at inflamed sites, stimulation of ECs with inflammatory mediators or hemostatic signaling molecules triggers endothelial activation (Step 1 in Figure 10.6). As a consequence, the ECs become adhesive for unactivated PMNs in an agonist- and time-dependent fashion. Thrombin, histamine, and certain other rapidly acting agonists trigger endothelial-dependent adhesion of neutrophils within minutes in cultured human EC models, whereas LPS, interleukin (IL)-1, and tumor necrosis factor (TNF)-α induce adhesion that requires approximately 4 h to become maximal, but that is then sustained for 18–24 h *in vitro* (reviewed in [3]). Time-dependent differential mechanisms that allow rapid and transient or, in contrast, delayed and sustained targeting of PMNs depending on the requirements of the host provide significant biologic advantages in physiologic inflammation and host defense [3], and have been observed *in vivo* [67]. Biochemical and molecular events that contribute to EC activation, adhesion factor expression, and signaling molecule synthesis and surface display continue to be defined (reviewed in [7, 17, 18]) and are discussed elsewhere in this volume.

The next step in the multistep cascade (Step 2 in Figure 10.6) is mediated by selectins – a family of three glycoproteins with high structural homology in extracellular lectin, epidermal growth factor-like, and consensus repeat domains [10, 20]. Display of a specific selectin on the plasma membranes of activated ECs mediates tethering and initial capture of the PMN from the flowing blood. Activation of ECs by thrombin, histamine, leukotriene C_4, oxidants, and other agonists results in translocation of P-selectin to the endothelial surface. P-selectin (which was called GMP-140 and PADGEM in early reports) is constitutively present in subcellular storage granules termed Weibel–Palade bodies and is transferred to the plasma membrane by regulated exocytosis that is dependent on intracellular calcium (Ca^{2+}) transients and signaling cascades [10]. Translocation and surface expression of P-selectin was first shown in platelets and cultured human umbilical vein ECs (HUVECs) [10, 74, 75], but has also been reported in postalveolar venular capillaries of the lung (Figure 10.4) and in other pulmonary vessels in animals and man [14, 76] (see "Sticky Controversies"). P-selectin binds to a glycoprotein ligand on PMNs termed P-selectin glycoprotein ligand-1 (PSGL-1) – a molecular interaction that loosely tethers the two cells together in a Ca^{2+}-dependent fashion. PSGL-1 is basally present on PMNs and this initial adhesion event does not require activation of the leukocyte. In *in vitro* human cell models in which the endothelial monolayer is stimulated with thrombin or histamine, P-selectin is translocated to the EC plasma membrane within seconds, its surface display peaks within a few minutes, and it is then reinternalized, providing a mechanism for initial tethering of PMNs with rapid kinetics [3, 58, 64, 75] (Figure 10.5). In contrast to this rapidly induced mechanism, LPS, TNF-α, IL-1, and additional agonists stimulate display of a different selectin, E-selectin, which is not stored in ECs but, instead, is synthesized under transcriptional control [3, 59]. E-selectin is also recognized by PSGL-1 on PMNs, again mediating tethering of the leukocytes to stimulated endothelium [15]. Evidence from knockout mice deficient in single or multiple selectins or in fucosyltransferases that catalyze glycation of selectin ligands indicates that selectin-mediated interactions are essential for tethering and rolling of PMNs (Step 3 in Figure 10.6) and cannot be replaced by alternative adhesion molecules, at least in the murine system [10, 13, 15]. Genetic defects in fucose metabolism that impair selectin ligand function cause the extremely rare human syndrome, LAD type II, and knockout of PSGL-1 in mice impairs selectin-dependent events [12, 15].

Surface topography, dimerization, and intracellular interactions of the cytoplasmic tails of selectins and PSGL-1 influence in complex fashions initial tethering, strength of the tethers under flow, and longevity of the selectin–PSGL-1 bond [10, 15]. Although selectins bind to many different glyconjugates with variable affinity – particularly fucosylated and sialylated *O*-glycans – experiments with blocking antibodies and targeted deletion of PSGL-1 in mice indicate that it is the principal binding partner for P-selectin on PMNs and that it is a ligand for E-selectin; there also appears to be one or more as yet unidentified ligands for E-selectin in the murine system [10, 15, 77]. Fucoidin, a reagent that has been used to examine selectin-mediated events *in vitro* and *in vivo* [78], interrupts binding of P- and L-selectins to PSGL-1 and potentially other glycan ligands and sulfated structures.

Tethering of PMNs leads to rolling of the leukocytes under the influence of flow and shear forces in postcapillary venules and other systemic vessels (Step 3 in

Figure 10.6). This component of the multistep cascade was shown to involve selectins in an *in vitro* system using purified P-selectin immobilized in a model membrane and isolated PMNs under flow conditions [73], providing a molecular mechanism for observations of rolling made much earlier by Dutrocet and Conheim [14] and in later studies [79]. P- and E-selectin on EC and PSGL-1 on neutrophils have been shown to mediate both tethering and rolling of PMNs *in vitro* and *in vivo* [10, 13, 15, 21, 77]. A third member of the selectin family, L-selectin, also mediates rolling of PMNs on unidentified endothelial counterligands [67] and participates in "leukocyte-assisted" capture of flowing PMNs by engaging PSGL-1 on target WBCs [10, 13, 80]. Integrins on leukocytes, including heterodimers of the β_2 family and integrin $\alpha_4\beta_1$ (VLA4), can also contribute to rolling of PMNs and other leukocyte subtypes on inflamed endothelium [15, 21]. Adherent platelets and leukocytes themselves can substitute for ECs as surfaces for tethering and rolling of PMNs *in vitro* [10, 15, 80].

In the next step (Step 4 in Figure 10.6), outside-in signals, delivered via receptors on the surfaces of rolling PMNs, trigger activation of the leukocytes. One of the functional consequences is conversion of surface β_2 integrins on the PMN to a conformation that allows them to recognize ligands on the endothelial plasma membrane. Binding of β_2 integrins, principally $\alpha_m\beta_2$ and $\alpha_L\beta_2$, on activated PMNs then mediates tight adhesion of the leukocytes and their arrest in the next phase of the multistep cascade (Step 5, in Figure 10.6; also see the following paragraphs). This was shown in a system in which a soluble chemoattractant was added to isolated human PMNs flowing over purified P-selectin immobilized together with purified intercellular adhesion molecule (ICAM)-1 – a ligand for activated β_2 integrins [73]. Tight adhesion mediated by β_2 integrins has also been demonstrated in a variety of other *in vitro* and *in vivo* models [12, 13, 15]. Ligands on EC for β_2 integrins on activated PMNs include ICAM-1 and ICAM-2, and others are likely to be identified [15].

In separate studies, it was shown that stimulated ECs can rapidly synthesize and display on their surfaces a phospholipid signaling molecule for PMNs, platelet-activating factor (PAF), triggering juxtacrine activation of the leukocytes and cooperative adhesion mediated by P-selectin/PSGL-1 and β_2 integrins [3, 57, 64]. This provided the first evidence that inflamed EC can contribute endogenous signaling molecules that trigger PMN activation. Subsequently, it was also demonstrated that IL-8 and several other polypeptide chemokines synthesized by inflamed ECs, or released by activated platelets or extravascular cells such as macrophages, can activate tethered and rolling PMNs [3, 11, 15, 21]. In addition to acting in a paracrine fashion in solution, locally released chemokines can bind to glycosaminoglycans or to the Duffy receptor on the apical EC surface, allowing them to be presented to the leukocytes in a spatially restricted fashion at the tips of plasma membrane microvilli [3, 21, 81]. This is similar to the spatially restricted signaling by PAF that is synthesized by stimulated ECs [3]. Chemokines including IL-8 and ENA-78 are expressed by pulmonary ECs in inflammatory lung syndromes [82], indicating that signaling of this nature may occur in the lungs of human subjects. Specific "arrest chemokines" that mediate Step 4 and trigger Step 5 in the multistep cascade (Figure 10.6) have been identified and/or are proposed for PMNs, monocytes, and lymphocyte subsets [11]. "Postarrest" steps in the multistep cascade appear to be important for β_2 integrin clustering, resistance to detachment, and adhesion strengthening [83].

PAF, "arrest" chemokines, and other PMN agonists that induce β_2 integrin-mediated adhesiveness – such as the activated fifth component of complement (C5a) and leukotriene B_4 – bind to seven-transmembrane G-protein-coupled receptors on PMNs (Step 4 in Figure 10.6). Outside-in signals delivered by these receptors trigger cellular activation via intracellular signal transduction pathways that may be organized in preformed "signalosomes" [11, 15, 21, 84]. In early steps in the multistep cascade, tethering and rolling not only accomplish initial capture of PMNs from the flowing blood, but also establish close apposition of the leukocytes with the EC surface, facilitating signaling by endothelial-bound agonists [10, 11]. Engagement of PSGL-1 by P-selectin enhances outside-in signals delivered via the PAF receptor [64] and receptors for IL-8 [85] on human PMNs – a mechanism that provides signal integration and enhancement; there may be differences in human and murine PMNs in these responses [11, 15]. There is evidence that signaling by apically displayed endothelial factors, including PAF and IL-8, can take place extremely rapidly, within seconds, under *in vivo* and *in vitro* conditions [84, 86].

Tightly adherent, polarized PMNs migrate over the endothelial surface and then transmigrate to extravascular sites (Steps 5 and 6 in Figure 10.6) – a process that can occur within minutes after arrest [21]. Polarization of the leukocytes, which is required for directed migration, is an early signal-dependent response of PMNs adherent to inflamed EC (Figure 10.5e). Emigration of PMNs from the apical surfaces of ECs requires ultimate release from selectin and integrin bonds [87], processes that are largely uncharacterized [15]. PMNs then migrate via paracellular pathways between adjacent ECs although, remarkably, a transcellular route has also been reported [16, 17]. The molecular basis for paracellular emigration of PMNs and other leukocytes between ECs (Step 6 in Figure 10.6), which are linked by adherens and tight junctions [88], is the least well-characterized component of the multistep cascade. Nevertheless, there is evidence from human and murine systems that this route of transmigration involves additional sequential events and key endothelial junctional molecules that include (PECAM-1; CD31), junctional adhesion molecule (JAM)-A, other members of the JAM family, and CD99 [21, 88–90].

The events in the linear cascade outlined in this section (Figure 10.6) have additional important molecular, biochemical, and biophysical features that will not be detailed here, but that are discussed in reviews and other articles that we include in the bibliography and the reports that they cite. The linear multistep paradigm, while useful and consistent with a large body of experimental information, may also be an oversimplification in many *in vivo* conditions, where networks of events with overlapping contributions to tethering, activation, and arrest of leukocyte subclasses operate in concert with additional molecular checks and balances [11, 15]. This concept is illustrated by the overlapping steps in Figure 10.6.

The multistep cascade was originally conceived as a mechanism that mediates spatial targeting of PMNs and other leukocytes in systemic vessels and inflamed tissue, but it has additional important facets that were initially unrecognized. "Information transfer" occurs between the interacting cells, in addition to spatial localization of the leukocytes [3, 11, 15, 87]. There is evidence, largely from *in vitro* models, that PMNs and monocytes receive complex outside-in signals via PSGL-1, L-selectin, and β_2 integrins – in addition to signaling molecules such as PAF and chemokines – while adherent to activated EC. These signals induce functional responses in addition to tight adhesion and polarization (Figures 10.5 and 10.6), including changes in patterns of expressed genes that may influence subsequent inflammatory events (reviewed in [11, 15]). Recent studies suggest that interactions of PMNs with activated EC also modulate apoptosis and lifespan of the leukocytes [21]. In turn, there is evidence that activated leukocytes deliver outside-in signals to EC, leading to alterations in intracellular Ca^{2+} levels, cytoskeletal rearrangement, altered activity of small GTPases and intracellular kinases, and modulation of junctional interactions of the endothelial monolayer. These events may be critical in leukocyte emigration (reviewed in [21]). Activated PMNs can also alter the display of signal-transducing receptors on the EC plasma membrane and the pattern of inflammatory genes expressed by EC [91–93]. Signaling of EC by PMNs may occur if the leukocytes are activated in the blood by soluble inflammatory agonists and become sequestered in microvascular beds (see "Leukocyte Sequestration") or in a reciprocal fashion when PMNs are first activated by juxtacrine signaling molecules at the EC surface (Figure 10.6). Inhibitory factors such as prostacyclin and nitric oxide, which are generated by EC [7], can modulate adhesiveness, activation, and biophysical properties of PMNs [94], and thereby influence margination and sequestration in the lungs (see the following three sections).

CELLULAR AND MOLECULAR FEATURES THAT INFLUENCE THE MARGINATED POOL OF LEUKOCYTES IN LUNG CAPILLARIES: SIZE MATTERS

There are similarities, and key variations, in leukocyte–endothelial interactions in the pulmonary and systemic circulations. There are also differences in the sinusoidal circulation of the liver and in brain vessels compared to other regions of the systemic vasculature [95, 96], and likely in other microvascular beds such as kidney glomeruli [1]. These variations indicate organ-specific specializations. Early and more recent reviews highlight characteristics of the alveolar capillaries – which receive the entire blood flow, in contrast to the microvascular beds of other organs – in differentially influencing physiologic localization of PMNs and other leukocytes in the lungs [4, 9, 13, 14, 16, 40]. The size and dynamic nature of the marginated pool of PMNs in the lungs [14] indicate a unique relationship between leukocytes and the pulmonary vasculature. Such unique interactions were initially suggested by early investigation of the circulating and marginated pools in humans [28, 30] (Figures 10.1 and 10.2). Subsequent studies of human subjects and human lungs *ex vivo* further demonstrated and characterized the marginated pool of leukocytes in the pulmonary vessels [32, 33, 35–38, 40]. Observations in experimental animals have refined our understanding of the cellular basis for the lung marginated pool.

The pulmonary "marginated" pool has been defined physiologically as the increased concentration of leukocytes in the blood in normal, uninflamed lungs compared to leukocyte numbers in the blood in peripheral vessels [9]. It has also been defined as the subset of leukocytes that are in physical contact with ECs as detected by intravital microscopy [4]. Most studies of the pulmonary marginated pool of leukocytes have focused on PMNs, extending classic observations indicating that PMNs are disproportionately present in the blood compartment of the lungs, and are dominantly influenced by variables that alter the distribution of leukocytes between the lungs and peripheral circulation [1, 28].

Studies by many laboratories indicate that the pulmonary capillaries of human and experimental animals harbor neutrophils in numbers that are 30- to 100-fold greater than their concentration in systemic and venous conduit vessels, and that monocytes and lymphocytes are also concentrated in the pulmonary capillary blood [9, 14, 16, 39, 97]. Environmental variables may alter the pulmonary marginated pool (e.g., cigarette smoke induces retention of PMNs in lung vessels in humans [35]). In addition, variables including alterations in pulmonary blood flow, elevation of airway pressure, and positive pressure ventilation increase retention of leukocytes in the lungs of experimental animals and humans [14, 33]. These observations are of potential clinical significance in the operating room and intensive care unit.

A key finding in studies of the mechanisms that determine the lung marginated pool is that leukocytes are retained for varying periods when passing through alveolar capillaries, resulting in transit times that are prolonged compared to those of RBCs [9, 14, 46]. Capillary retention of leukocytes is under the influence of intravascular pressure and flow and alveolar pressure [32, 47, 98–100]. Analysis of transit times of fluorescent PMNs in subpleural capillaries by video microscopy indicated that PMNs require a mean of approximately 6 s and a median of 26 s to move through capillaries in the canine lung [46]. This contrasts with RBC or plasma transit times, which are of the order of 1–4 s [9, 101]. Delayed pulmonary capillary transit of PMNs compared to RBCs, resulting in concentration of PMNs in the capillary blood, has also been documented in the human lung; longer transit times of leukocytes were proposed to establish "an opportunity for PMN–endothelial interactions" [35, 36, 38]. PMN transit in lung capillaries is not simply slowed, but occurs in "stops" that can last from less than 1 s to more than 1 min [46, 47, 100]. Greater than 90% of PMNs are retained in capillaries at least once during transit through the lungs of experimental animals, resulting in occlusion of up to 15% of alveolar capillaries [14, 100]. The geometry and network organization of the pulmonary microvascular bed prevents this leukocyte retention from significantly altering pulmonary microvascular resistance and flow, allowing dynamic accumulation of PMNs in this compartment under physiologic circumstances.

The evolutionary pressures that resulted in a concentrated pool of leukocytes in lung capillaries have not been defined, but Doerschuk has proposed that prolonged transit of PMNs in the lungs is biologically advantageous to the host and facilitates detection of infectious agents by intravascular PMNs and responses in homeostatic inflammation [9]. Others also identify potential biologic advantages of the lung marginated pool, but emphasize that concentration of leukocytes in pulmonary capillaries is a liability in syndromes of inflammatory lung injury [4, 14, 35].

Disproportionate retention of a large number of marginated leukocytes in pulmonary capillaries is most commonly explained by anatomic features of the lung microvessels and size and biophysical features of the WBCs [9, 13, 16]. The majority of the information on this point again comes from studies of PMNs [9, 37]. The adult human lung contains approximately 300 million alveoli, each of which made up of approximately 1000 capillary segments with the entire network connecting to 300 million arterioles and venules [16]. Thus, the pulmonary capillary bed is a huge reservoir, with a surface area of around 80 m^2. The alveolar capillary beds of other mammals are also extensive and intricate. The diameter of spherical neutrophils is 6–8 μm, whereas the diameter of many lung capillary segments is 2–15 μm [9], depending on the species [37]. Estimates from a variety of studies indicate that approximately 50% of the capillary segments will not allow PMNs to transit without deformation of the shape of the leukocytes because of these differences in diameter [9, 16]. By this analysis, PMN movement and retention in lung capillaries can be reduced to issues of physical constraint as illustrated in Figure 10.7, although this has not been the uniform interpretation [1] (see Sticky controversies). A key feature is that PMNs cannot roll in capillary segments that have diameters equal to, or smaller than, the leukocyte diameter [9, 14, 16], as they can in larger microvessels in the systemic circulation (Figure 10.6), and in lung arterioles and venules (Figure 10.8). It seems likely, however, that leukocyte rolling may occur in some alveolar capillaries with larger diameters (~10–15 μm).

Additional important features contribute to the lung marginated pool and establish a dynamic equilibrium between PMNs and the lung microvasculature. The alveolar capillary bed is a complex of short interconnecting segments of approximately 8 μm in length. A leukocyte must traverse 40–100 capillary segments in transit from precapillary arterioles to postcapillary venules [9]. If only a small number of capillaries, estimated to be approximately 1–2%, mechanically retard leukocyte movement it is estimated that 55–60% of transiting PMNs will be slowed and retained in the lungs based on direct observations and computational models [9, 14]. Studies of PMN diameter and shape in human, canine, and rabbit lungs demonstrated that the leukocytes deform from spherical to ellipsoid shape in all three species, and indicated that deformation is a time dependent event during transit [37, 102, 103]. Neutrophils initially deform almost as quickly as RBCs, but after a threshold (approximately a diameter of 5 μm) change shape much more slowly, resulting in a net ability to deform approximately 1000 times slower than that of normal RBCs [9, 102, 104]. Taken together, these features lead to the conclusion that the increased numbers of PMNs in the alveolar capillary bed under physiologic conditions is due to the diameters of capillary segments and biomechanical properties of the leukocytes that slow their deformation relative to

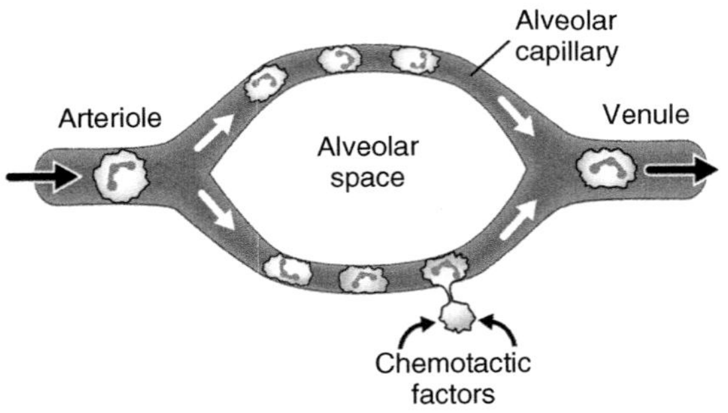

Figure 10.7 Alveolar capillaries are the dominant sites of margination of PMNs under physiologic conditions and are a major site of sequestration of PMNs in response to proinflammatory stimuli. Under physiologic conditions, the marginated pool of PMNs is primarily located in the alveolar capillary bed. Dynamic retention of leukocytes in these vessels is thought to result from differences in diameter of PMNs and alveolar capillaries that slow transit of the leukocytes through the capillaries compared to the transit times of RBCs and platelets. Biomechanical deformation of the PMNs is required for passage through capillaries with diameters smaller than those of the leukocytes. In some studies PMNs remain partially deformed as they exit the alveolar capillary bed into pulmonary venules. Initial biomechanical "stiffening" of PMNs followed by adhesive interactions further slows transit of these leukocytes through alveolar capillaries in response to inflammatory stimuli. This process, known as sequestration, further increases the concentration of PMNs in the alveolar capillary reservoir above that in the basal state. Sequestration is an early event in physiologic and pathologic lung inflammation.

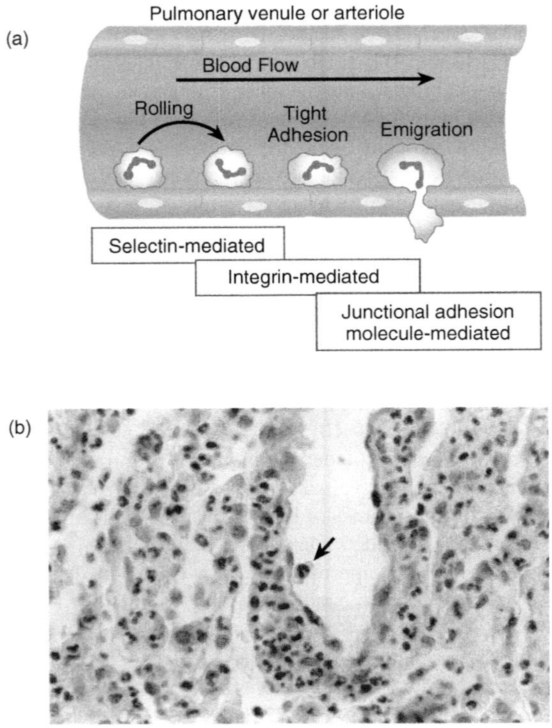

Figure 10.8 Tethering, rolling, and tight adhesion are mechanisms of leukocyte margination and sequestration in pulmonary arterioles and venules. (a) Experiments utilizing several animal species and intravital microscopy indicate that leukocytes tether, roll, and arrest on pulmonary arteriolar and venular endothelium. Leukocytes rolling in arterioles and venules to be a part of the marginated pool under physiologic circumstances, although they are a much smaller fraction of lung leukocytes than those in alveolar capillaries under basal conditions. The mechanisms for margination in arteriolar and venular beds in the lung have not been completely established, but appear to involve adhesive events that parallel those in the systemic circulation (Figure 10.6). Increased tethering, rolling, and tight adhesion contribute to sequestration of leukocytes in the lungs in response to inflammatory stimuli. See text for details. (b) A monocyte in close apposition to ECs in a venule in the lung of a patient dying with bacterial pneumonia is indicated (arrow). The leukocyte may have been rolling along the vascular surface (see a). The vessel is surrounded by many extravascular PMNs and monocytes. Lung venules and arterioles may be important sites of leukocyte sequestration and trafficking in clinical lung inflammation, and in experimental models with sufficiently long time courses. Panel b reproduced from [56] by permission of The American College of Chest Physicians.

that of RBCs, causing much slower transit through lung capillaries [9, 16]. The large number of short interacting alveolar capillary segments allows the more rapidly moving RBCs to "stream" around segments transiently occluded by slowly moving, deformed leukocytes [35]. PMNs "... squeezed into an elongated sausage form in the act of passage..." were observed in rabbit lungs, liver, and kidneys in early studies [1].

Gravitational and other differences in pulmonary blood flow [40, 98], and alveolar pressure [33], influence capillary diameter and perfusion and by these mechanisms contribute to regional differences in the numbers of leukocytes that are marginated in capillaries by mechanical constraints [14, 33]. Changes in the pattern of ventilation can rapidly alter the size of the marginated pool of PMNs in the alveolar capillaries, as noted previously (Figure 10.1). Infusion of agents such as epinephrine increases blood flow in regions of the lung with relatively low perfusion, resulting in decreases in the number of marginated PMNs [32] and providing a mechanism for earlier observations of leukocytosis following catecholamine administration and exercise [26, 31].

Based on these observations, establishment of the alveolar capillary pool of marginated PMNs under normal physiologic circumstances does not require activation of either the capillary EC or the leukocyte – or signaling and information transfer between the two interacting cells – in contrast to margination in inflamed systemic microvessels (Figure 10.6). Close apposition of leukocytes and ECs in pulmonary capillaries and long transit times of leukocytes (Figure 10.7) does, however, make it possible that intercellular signaling may occur [38] if one cell or the other receives an activating stimulus (see "Interaction of PMNs with Systemic Endothelium and the Multistep Paradigm" and [105, 106]).

STICKY CONTROVERSIES: PHYSIOLOGIC MARGINATION OF LEUKOCYTES IN PULMONARY ARTERIOLES AND VENULES, AND ADHESIVE MARGINATION IN ALVEOLAR CAPILLARIES

Owing to the fact that the surface area of the alveolar capillary microvascular bed exceeds the combined vascular surface areas of pulmonary arterioles and venules by 20- to 30-fold, the capillaries are the dominant sites of the marginated pool of leukocytes in the lungs under physiologic conditions (reviewed in [9, 14, 16]). Nevertheless, intravital microscopic visualization of leukocytes labeled *in vivo* indicated that the intravascular concentration of fluorescent cells was significantly elevated in all compartments of the rabbit pulmonary microcirculation compared to leukocyte numbers in aortic blood [49, 107]. This method of labeling does not distinguish leukocyte subclasses from one another, so the identities of the WBCs in individual compartments are unknown. Labeled leukocytes were observed to roll and stick on arteriolar and venular surfaces, and to stop in transit through alveolar capillaries. The numbers of leukocytes were greatest in capillaries followed by venules and arterioles with relative concentrations of pulmonary leukocytes in these compartments compared to peripheral blood of 36-, 24-, and 8-fold, respectively [107]. Rolling velocities of leukocytes in arterioles and venules were correlated with shear forces, and indicated that microvascular blood flow is a major force that releases leukocytes from the arteriolar and venular, as well as capillary, compartments of the total lung marginated pool [14, 107]. These investigators concluded that rolling of leukocytes in pulmonary macrovessels (Figure 10.8) could not be explained by surgical manipulation – which induces margination in systemic vessels [108] and is likely to also induce margination in the lung under some circumstances – or by hemodynamic variables alone, and proposed that adhesive factors of the classes involved in rolling and adhesion of leukocytes in the systemic circulation (Figure 10.6) are involved.

Studies by others also indicate that leukocytes marginate in macrovessels in the lung. Doerschuk reported increased numbers of leukocytes in arterioles and venules of the rabbit lung compared to leukocyte concentrations in peripheral blood [39]. Albertine *et al.* observed that the distribution of marginated PMNs in the lungs of sheep is greatest in capillaries, arteries, and veins in that order, and suggested that adhesion proteins and their ligands contribute to pulmonary margination (reviewed in [4]). Differences in relative proportions of PMNs in lung arterial and venular pools in experiments using rodents [107] and sheep [4] suggest species differences and/or differential inflammatory stimulation by operative manipulation or other experimental procedures. PMN margination in vessels larger than capillaries was also observed in canine [101] and rabbit [109] lungs in earlier studies, but in each case this was thought to be of minimal physiologic significance. Observations of vessels in the exposed canine lung by high magnification *in vivo* video microscopy identified leukocytes that rolled slowly or "skidded" along the surfaces of pulmonary arterioles and venules in addition to elongated, deformed leukocytes that stopped in alveolar capillary segments [103]. As leukocyte rolling and adhesion are generally considered to be absent in systemic arterioles, and absent or present at only low levels in systemic venules under basal conditions in the absence in infection or inflammation, Kuebler and Goetz proposed that leukocyte margination in arterioles and venules of the uninflamed lung is a unique feature of the pulmonary circulation [14].

While adhesive interactions similar to those that occur in systemic vessels (Figure 10.6) have been implicated in physiologic margination of leukocytes in the uninflamed lung [1, 4, 27, 103, 107], this has been a controversial question [9, 14, 17, 20]. One issue is expression of selectins in lung vessels. Early studies indicated that neither P-selectin nor E-selectin are expressed in pulmonary capillary ECs (reviewed in [9]), although sensitivity of the detection methods has been questioned [16]. An antibody against P-selectin stained endothelium of canine lung arterioles and venules, but not lung capillaries, under basal conditions [103]. More recently, pulmonary microvascular P-selectin was implicated in rodent lung injury models [110] and basal and stimulated surface display of P-selectin was detected in postalveolar venular capillaries – a site of significant neutrophil accumulation – in excised rat lungs [50, 111] (Figure 10.4). In lung samples from humans without evidence for inflammation or injury, P- and E- selectin were detected in arterioles and venules, but not alveolar capillaries [76]. Thus, selectins are present in lung macrovessels and some microvessels depending on the species and conditions. A nonspecific selectin inhibitor, fucoidin, which interrupts binding of P- and L-selectins to glycoconjugate ligands [78], interrupted leukocyte rolling in arterioles and venules in the lungs of instrumented rabbits under basal conditions [112]. Together with evidence that P-selectin is present in pulmonary vessels, these observations were taken to suggest that basal leukocyte rolling in the lungs is mediated by P-selectin [14]. In contrast to these results, venular rolling was blocked by fucoidin but not by an anti-P-selectin antibody in exposed canine lungs – a finding interpreted as demonstrating L-selectin-mediated rolling adhesion [103]. It is not clear whether differences in these studies were due to species variations or to differences in experimental procedures. Efficacy of the anti-P-selectin antibody as an inhibitor of tethering and rolling was not documented under conditions of the *in vivo* experiments in the canine study [103].

Selectin-mediated adhesion is also implicated in margination of leukocytes in alveolar capillaries. In the same *in vivo* experiments in which fucoidin inhibited leukocyte rolling in pulmonary arterioles and venules, retention of leukocytes in capillaries and their transit times were also significantly reduced by treatment with this agent [112]. This led to the conclusion that L-selectin contributes to physiologic accumulation of leukocytes in the alveolar capillary bed [14]. Changes in capillary retention of leukocytes induced by fucoidin treatment were not commented on in the earlier study in dogs [103], however. To add to the conundrum, genetically modified mice deficient in P- and E-selectin had no apparent alteration in their marginated pool size [9]. The latter observations would not exclude a role for L-selectin and its ligands in capillary margination, however.

In addition to selectins, leukocyte integrins were suggested to participate in basal margination in lung venular, arteriolar, and capillary compartments [14]. Rat lung endothelium is reported to constitutively express high levels of ICAM-1 – a major counterligand for β_2 integrins [113]. Leukocytes have increased rolling velocities in systemic vessels of knockout mice deficient in β_2 integrins [11, 13], and leukocyte integrins participate in slow rolling in addition to tight adhesion, arrest, and transmigration [114]. Low velocity rolling facilitates leukocyte recruitment by increasing transit times along the endothelial surface [10, 11, 72]. Slow leukocyte rolling was observed in lung arterioles and venules under basal conditions (noted earlier in this section), suggesting the possibility that β_2 integrins might participate in this leukocyte–endothelial interaction (Figure 10.8a). It was also suggested that β_2 integrins may mediate tight adhesion in lung venules, arterioles, and capillaries under physiologic conditions [14]. A critical consideration, however, is that PMNs and monocytes must become activated in order for β_2 integrins to change conformation and recognize ICAM-1 and other ligands on EC as outlined in previous sections [8, 12]. Recent observations indicate that β_2 integrins on human PMNs can do this rapidly and, potentially, reversibly in a localized fashion when appropriately stimulated [86]. Nevertheless, no mechanism for the requisite outside-in signaling events that trigger β_2 integrin adhesiveness (Step 4 in Figure 10.6) has yet been identified in vessels of the uninflamed lung. A related observation is that PMNs from a dog with genetic deficiency of β_2 integrins had pulmonary vascular transit times similar to those of wild-type canine neutrophils under basal conditions [115]. In limited human studies, infusion of epinephrine increased circulating leukocyte numbers in two children with complete deficiency of β_2 integrins, suggesting that subjects with the LAD I syndrome have a basal marginated pool (but not defining its compartments) [9]. The impact of targeted deletion of β_2 integrins on the pulmonary marginated pool of PMNs in mice is not yet clear.

LEUKOCYTE SEQUESTRATION: INFLAMMATORY SIGNALING AND NEWLY DISCOVERED MECHANISMS

"Margination" and "sequestration" have frequently been used interchangeably in describing leukocyte interactions in the pulmonary vasculature. This has the potential to cause confusion with respect to the lung pool of intravascular leukocytes in the basal physiologic state, and increases in this marginated population in host defense responses and in pathologic conditions. For this reason sequestration will be defined here as an active process resulting from a functional change in the WBCs, ECs, or both cell types that further increases the number of leukocytes in lung vessels over the number marginated in the lung in the basal state. Definitions similar to this have been used by others [4, 9, 16, 20, 116]. Sequestration can occur at any of the vascular sites at which leukocytes marginate – arterioles, venules, and capillaries (Figures 10.7, 10.8) – but the alveolar capillary is traditionally identified as the dominant location [9]. It is possible, however, that macrovascular sites represent major locales of sequestration and routes of leukocyte transmigration in clinical lung inflammation and injury (Figure 10.8). Clinical inflammatory lung syndromes have much longer time courses than usual experimental models and mechanisms of sequestration may be quite different in injured or diseased human lungs.

One mechanism of sequestration that has been studied extensively is activation of PMNs in the blood, or in suspension *ex vivo* followed by infusion into experimental animals. Agonists for PMN activation in these experiments include chemokines, inflammatory proteins of other classes such as C5a, proinflammatory lipids, *N*-formyl-methionyl-leucyl-phenylalanine (fMLP – a mimetic of bacterial peptides), and LPS [16]. Sequestration resulting from activation of leukocytes in the circulating blood may be one mechanism by which PMNs further accumulate in the lungs in inflammatory conditions [9]. Intravascular activation of PMNs by infusion of a selective agonist, fMLP, also leads to sequestration of platelets in pulmonary microvessels [117]. This demonstrated complex cell–cell interactions between leukocytes and platelets in the lung that have pathophysiologic significance in disease models [118].

An assumption of many experimental studies is that PMNs activated by soluble agonists interact with resting, *unactivated* pulmonary endothelium, in contrast to the EC-dependent mechanism outlined previously (Figure 10.6). Activation of PMNs followed by their binding to resting ECs is clearly a mechanism for PMN adhesion [12, 61–63]. LPS, cytokines such as TNF-α, and certain other agonists can activate both cell types; however, this is dependent on species, time of analysis, and other aspects of the model (see "Interaction of PMNs with Systemic Endothelium and the Multistep Paradigm").

A variety of observations indicate that activation of PMNs by soluble agonists further retards transit through normal lung vessels, leading to sequestration of increased numbers of the leukocytes above those basally present in the lung (reviewed in [16]). Intravascular infusion of IL-8, PAF, leukotriene B_4, C5a, TNF-α, or LPS results in rapid sequestration (within 1 min or less), increasing the number of PMNs retained in lung capillaries in this fashion (reviewed in [9, 13]). Biophysical responses are again implicated in sequestration of activated leukocytes in lung capillaries. Activation of PMNs by soluble chemotactic factors or other stimuli was reported to induce transient cell "stiffening" and impaired deformation, increasing their retention in micropore devices and rabbit lungs [119, 120]. The rapid phase of sequestration in experimental models is in general not dependent on neutrophil L-selectin or members of the β_2 integrin family [9] even though agonist-induced stiffening and decreased deformability are temporally correlated with activation of β_2 integrins, which also occurs within 1 min or less [8, 16]. Thus, rapid stimulus-induced leukocyte retention, and consequent delay in capillary transit time, are thought to be largely dependent on cellular stiffening and further impairment in dynamic change in shape of the WBCs [9]. These biophysical changes are in part due to regulated alterations in the subcortical actin cytoskeleton [9, 120]. Viscous, granule-laden cytoplasm and a multilobed nucleus are also thought to contribute to impaired deformation of activated PMNs [120, 121]. Deformability as a determinant of PMN sequestration has been studied in normal subjects and patients with stable and decompensated chronic obstructive pulmonary disease (COPD), together with variations in accumulation of radiolabeled PMNs in the lungs [121].

Although adhesion molecules are not required for initial sequestration of activated PMNs and further delay of their transit through alveolar capillary segments, both L-selectin and β_2 integrins appear to contribute to a second phase of retention of the leukocytes that lasts for 4–7 min or more [9, 14, 122]. These studies imply binding of leukocyte integrins and/or L-selectin to ligands on the alveolar capillary EC. PMNs released from the bone marrow in response to inflammatory mediators have enhanced sequestration in the lungs under experimental conditions, a feature that may in part depend on L-selectin [123]. Mechanisms of sequestration that are both dependent on, and independent of, β_2 integrins have been reported in animal models [9]. Engagement of selectin or integrin ligands on other sequestered leukocytes or on platelets in complex homotypic and/or heterotypic cell–cell interactions [10, 15] may potentially occur in the pulmonary microvessels [9], in addition to adhesion of sequestered leukocytes to microvascular ECs. A mechanism in which microvascular transit of PMNs is first retarded by agonist-induced changes in deformability and then by β_2 integrin has also been reported in the liver [16, 124].

PMN sequestration triggered by initial stimulus-induced stiffening and subsequent enhanced adhesiveness is thought to be the first step in a sequence of events that leads to emigration of the leukocytes out of pulmonary capillaries and to subsequent transmigration across the alveolar epithelium in both defensive and pathologic inflammatory conditions [9][1]. The events beyond initial sequestration of leukocytes – which requires intimate interaction with endothelium in alveolar capillaries, as outlined here – are complex and incompletely defined, and have recently been reviewed

---

[1]Transmigration across the alveolar epithelium has recently been reviewed. See Zemans R.L., Colgan S.P., Downey G.P. (2009) Transepithelial migration of neutrophils. Mechanisms and implications for acute lung injury. American Journal of Respiratory Cell and Molecular Biology **40** (5) 519–535.

[16, 19]. While enhanced alveolar accumulation of PMNs initiated by intravascular sequestration may be requisite and protective in defense of the host against microbial invasion via the distal airspaces [9, 125], it contributes to pathologic inflammation and pulmonary damage in experimental models [19]. Sequestration of PMNs is triggered not only by mediators in the blood but also by agonists delivered via the airways [126]. Of interest, pathologic stimuli delivered via the alveoli, such as cigarette smoke, can induce systemic activation of leukocytes and their accumulation in peripheral vessels [127] (Figure 10.3) in addition to sequestration of leukocytes in the lungs [35].

Sequestration of PMNs in lung vessels can also occur by endothelial-dependent mechanisms, resulting in retention of initially unactivated PMNs. This may, for example, be the mechanism for shifts of PMNs from the circulating to the marginated pool in response to histamine observed in early studies [24, 28]. Histamine can activate ECs and trigger rapid endothelial-dependent PMN adhesion (Figure 10.4) mediated by surface display of P-selectin, synthesis PAF, and triggering of β_2 integrin-dependent adhesiveness [3, 64, 75]. PAF can stimulate both PMN stiffening and β_2 integrin activation, and therefore could mediate sequestration of PMNs in pulmonary capillaries and/or macrovessels. Histamine infusion was reported to trigger adhesion of leukocytes in pulmonary macrovessels *in vivo* in a rodent model [16].

LPS, which induces rapid and prolonged increases in the number of PMNs in the pulmonary vascular beds of experimental animals and humans [31] (Figure 10.2), may mediate sequestration by activating both neutrophils and lung EC. LPS triggers actin reorganization and increased stiffness of human PMNs within minutes *in vitro* [120]. LPS delivered via the airways also causes glucose uptake by lung PMNs, a process that can be detected by [^{18}F]2-fluoro-deoxy-D-glucose positron emission tomography imaging in humans [126]. In addition, however, LPS causes delayed and sustained activation of human ECs and consequent EC-dependent PMN adhesion in cultured human cell models [3, 128]. In mice, pulmonary endothelial activation mediated by Toll-like receptor 4 appears to be a central event in LPS-induced sequestration of PMNs in the lungs, questioning the interpretation that this primarily occurs via a direct effect on the leukocytes [129].

In addition to histamine and LPS, TNF-α is a third example of an inflammatory mediator that can trigger pulmonary sequestration of PMNs and other leukocytes by activating lung ECs. TNF-α induces EC-dependent adhesion of PMNs mediated by E-selectin in human models, in addition to directly activating leukocytes [3, 60, 128]. TNF-α triggers P-selectin expression by microvascular ECs in isolated, perfused rat lungs [111]. In a murine model, TNF-α induced leukocyte rolling and adhesion in venules and arterioles of revascularized lung allografts studied by intravital microscopy in dorsal skin fold chambers; P- L-, and E- selectin each contributed to rolling, based on inhibition by anti-selectin antibodies [130]. Thus this cytokine, like histamine and LPS, may alter the number of leukocytes sequestered in lung vessels by acting directly on EC. It should be noted that there are differences in expression of P-selectin in response to cytokines and LPS in mice and humans [69].

Recent experimental observations suggest that leukocyte sequestration in lung vessels may also occur via previously unrecognized mechanisms that are not simply accounted for by "traditional" inflammatory stimuli. In isolated, blood-perfused rat lungs, increased left atrial pressure resulted in intracellular Ca^{2+} transients and exocytosis of P-selectin in digitally imaged venular capillaries [50]. Low basal display of P-selectin was greatest at branch points, but increased throughout the vessel segment in response to pressure (Figure 10.4). This suggests that pressure "stress" is proinflammatory in the lung microcirculation under some conditions [50] (see also Chapter 20). In this model, alveolar epithelial cells can induce surface display of P-selectin in lung capillaries via unexpected biochemical signaling pathways [111, 116]. In the same experimental preparation, pressure elevation in lung postcapillary venules induced leukocyte sequestration that was blocked by an anti-P-selection antibody and by fucoidin [51].

In more recent observations using isolated lungs from rats and genetically modified mice, oxygen radicals generated by RBCs under hypoxic conditions induced increased intracellular Ca^{2+} and surface expression of P-selectin in venular capillaries [52], identifying a complex interaction between RBCs, ECs, and leukocytes (Figure 10.9). This was associated with increased rolling and sticking of leukocytes in these vascular segments.

These experiments suggest previously unrecognized mechanisms by which lung microvascular ECs may become proadhesive and contribute to leukocyte sequestration in the absence of "traditional" humoral inflammatory agonists or infectious stimuli [50, 52]. It is possible that these mechanisms are relevant to clinical conditions in which patients with normal lungs are subjected to increased intravascular pressures, global or regional hypoxemia, or transfusion of large volumes of aged RBCs [131]. Finally, the observations also suggest ways in which experimental lung preparations or instrumented experimental animals may be perturbed in unintended fashions (inadvertent hypoxia, alterations in intravascular pressure, etc.) that yield surface expression of selectins and local leukocyte accumulation not present in the truly basal state. Experimental manipulations such as

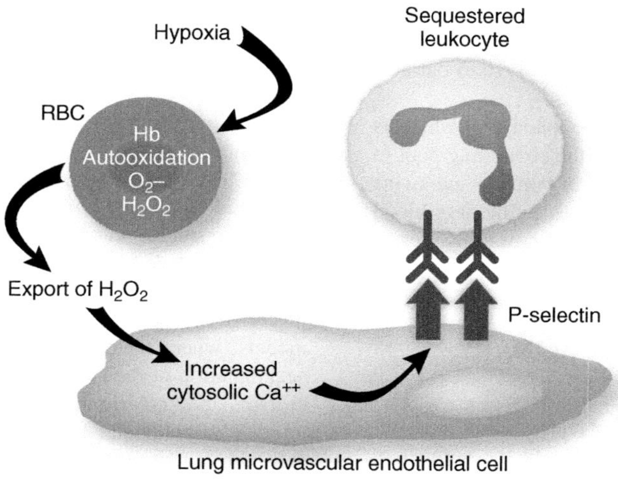

Figure 10.9 Previously unrecognized cell–cell interactions and proinflammatory mechanisms may contribute to microvascular sequestration of leukocytes in the lung. In hypoxic rat lungs, H_2O_2 generated by auto-oxidation of membrane-bound hemoglobin (Hb) in RBCs is exported to microvascular ECs. This induces increased cytosolic calcium, translocation of P-selectin to the EC plasma membrane, and leukocyte adhesion in venules and venular capillaries. This may be one of several previously unrecognized mechanisms for leukocyte sequestration in the lungs. See text and [52, 131] for details. This figure was originally published in *Blood* and is reproduced from [131].

simple placement of intravascular catheters have previously been reported to induce leukocyte sequestration in the lungs [4].

OTHER LUNG VESSELS AND OTHER LEUKOCYTES

This chapter focuses on ECs in vessels of the pulmonary circulation. The lung is also invested with the bronchial circulation – a systemic vascular system. Bronchial arteries perfuse microvascular beds in the airways from the level of the carina to the terminal bronchioles as well as capillary beds in the pleura, walls of the pulmonary artery and vein, esophagus, and hilar lymph nodes [18]. There are anastamoses between the bronchial and pulmonary circulations that vary among experimental animals [132], as well as other species differences. For example, mice do not have a functional bronchial circulation beyond the main stem bronchi [18], implying that leukocyte delivery to distal airways and, potentially, proximal alveolar units is different from that in man. ECs in bronchial vessels are reported to be more responsive to inflammatory stimuli [18]. As the bronchial vessels are part of the systemic circulation, endothelial–leukocyte interactions might be predicted to follow paradigms common to other systemic vessels (Figure 10.6), although this has not been extensively studied [40]. Bronchial ECs may also participate in unique leukocyte homing pathways [20, 133] (see Chapter 14).

We have also focused heavily on margination and sequestration of PMNs because of the bulk of information on this leukocyte subtype and its particular roles in lung defense and injury. Nevertheless, monocytes and lymphocytes also have substantial marginated populations in the lungs. Although initially controversial, monocytes are distributed between marginated and circulating pools [97, 134]. In rabbits, the lung marginated pool of monocytes was seven times larger than the circulating pool, whereas that of PMNs was three times larger than the circulating pool [97]. Basal monocyte margination in pulmonary vessels likely contributes to physiological trafficking to extravascular sites, which maintains alveolar macrophage and dendritic cell number, and can be enhanced in response to inflammatory signaling [135]. Marginated monocytes may also "patrol" lung compartments perfused by pulmonary and bronchial vessels [136].

Alterations in stiffness and increased adhesiveness mediated by β_2 integrins resulted in monocyte sequestration in the lungs of rabbits, and parallel changes were detected in human monocytes [137]. Monocytes that sequestered in the pulmonary circulation of mice appeared to signal EC, amplifying local inflammation [138]. The specific vascular locations and mechanisms of monocyte margination and sequestration in lung models have not been extensively explored; they are likely to have features similar to those that govern PMN interactions with systemic and pulmonary endothelium (Figures 10.6–10.8) but also cell-specific variations.

The lung marginated pool of lymphocytes, which is approximately 10 times larger that the circulating pool in rabbits and has also been reported in humans and swine [97], is poorly characterized. In dogs, pulmonary blood flow influences accumulation and release of lymphocytes in lung vessels as it does with PMNs [99].

PLATELETS AND THE PULMONARY ENDOTHELIUM

In human volunteers, infusion of epinephrine was followed within 30–60 s by an increase in the number of platelets in arterial blood that persisted for approximately 5 min and then declined in eight of 10 subjects tested [26]. It had been suggested earlier that epinephrine induces thrombocytosis because of splenic contraction, but Bierman *et al.* concluded that the increase in platelet number was due to release from the lung and that the pulmonary circulation is a significant reservoir for platelets as well as leukocytes [23, 26, 28]. Platelet interactions with the pulmonary endothelium have, however, not been explored as extensively as have those of leukocytes. In instrumented human volunteers, a forced expiratory maneuver resulted in a decrease in left ventricular PMNs and lymphocyte numbers relative to counts in pulmonary arterial blood, but platelet and RBCs numbers were unchanged in the left ventricular samples [33]. This difference may be due to the smaller diameters of platelets, which would be expected to preserve their transit times in the face of decreases in capillary diameter that further retarded transit of leukocytes. The difference in behavior of leukocytes and platelets could also be due to variations in deformability of platelets, or to both size and biomechanical factors. In studies of radiolabeled PMNs, monocytes, lymphocytes, and platelets in spontaneously breathing instrumented rabbits, each blood cell type was distributed to a lung marginated pool [97]. The estimated size of the platelet marginated pool was 13 times smaller than the circulating pool. The marginated pool of platelets in the lungs was also much smaller than the marginated pools of the leukocyte subtypes, a variation that was attributed to the smaller size of the platelet. While platelets appear to have the smallest lung marginated pool, studies in several animal models indicate that platelets contribute to normal pulmonary vascular integrity and alveolar–capillary barrier function [139, 140]. Thus, platelets may have intimate interactions with the lung endothelium under basal conditions (see Chapter 25). In addition, there is evidence that megakarocytes are present in the pulmonary microvascular bed and that the normal lung is a site of thrombopoiesis [23, 140].

In studies in rabbits, sequestration of PMNs and platelets was examined in response to infusion of complement fragments. Platelets accumulated in the lungs following sequestration of PMNs, a pattern that suggested that the platelets were adhering to neutrophils that were first sequestered in lung microvessels [123]. In earlier studies in rabbits, sequestration of PMNs in the lungs induced by fMLP or PAF was also accompanied by platelet sequestration [117]. These findings are potentially important in the interpretation of lung injury models, where interactions of PMNs and platelets and accumulation of the two cell types in the pulmonary microcirculation contribute to alveolar–capillary membrane injury [118]. Platelet interactions in lung inflammation and injury were recently reviewed [140].

CONCLUSIONS AND PERSPECTIVES

As outlined in the Introduction and developed in subsequent sections of this chapter, leukocytes and the vascular beds of the lung have unique relationships under physiologic conditions. These interactions are the basis for additional intricate interplay in pathologic lung inflammation, injury, and repair. Characterization of interactions between the pulmonary endothelium and leukocytes began with seminal observations in the "human model" and in experimental animals, and have now been extended using the tools of modern cell and molecular biology and investigative pathology. While mechanical and biophysical forces were thought by many observers to predominately regulate leukocyte interactions with lung endothelium under basal conditions, pathways of cellular adhesion and intercellular signaling have been proposed and, in some cases, experimentally dissected. Mechanisms of endothelial–leukocyte interaction vary in the alveolar and macrovascular divisions of the pulmonary circulation and there are similarities and differences in these cell–cell interactions in the pulmonary and systemic vascular beds, including the bronchial circulation. As with interactions of endothelium and leukocytes, early studies in humans suggested that there are unique intercellular relationships as platelets transit lung vessels. This issue has been studied much less intensively than have contacts and dynamic encounters between leukocytes and the lung endothelium, however.

Many additional questions remain. How basal interactions of leukocytes, platelets, and ECs change during exercise, sleep, extremes of body temperature, aging, and in response to environmental variations (altitude, hyperbaric conditions, air pollution, etc.) are largely unexplored. There are similar outstanding questions when one considers iatrogenic manipulations such as variable modes of mechanical ventilation or perfusion, transfusion, hematopoietic reconstitution after chemotherapy or bone marrow transplantation, lung transplantation, and so on. In addition, most of what we believe we know now is based on studies of interactions of PMNs with the lung endothelium in short-term human studies and abbreviated surrogate models; there is much less information available regarding other leukocyte types, and specific subtypes, and platelets. Furthermore, time as a variable has not been completely explored. These represent gaps in our knowledge. How findings in instrumented animal models apply to uninstrumented humans continues to evoke controversy. New approaches to molecular imaging and noninvasive *in vivo* cell biology may be on the horizon, and may provide avenues to address these and other questions that will then generate new discoveries in the field.

ACKNOWLEDGMENTS

The authors thank many postdoctoral fellows, students, technical associates, and colleagues who contributed to work cited here. We particularly thank Tom McIntyre, PhD, and Jahar Bhattacharya, PhD, for sending original images to be included in the figures. We greatly appreciate the efforts of Sharren Brewer and Linn Steele in the preparation of the manuscript, and creative contributions by Diana Lim in drafting the figures, including new drawings of illustrations from the original literature. Work cited in this chapter was supported by a grant from the National Institutes of Health (R37R01 HL44525), a previous National Institutes of Health/ National Heart, Lung, and Blood Institute Specialized Centers of Research in Acute Respiratory Distress Syndrome grant (P50 HL50153), and earlier grants from the American Heart Association to G.A.Z. R.S.R. was the recipient of an award from Coordenação de Aperfeiçoamento de Pessoal de Nivel Superior of the Ministerio da Educação, Brazil.

References

1. Andrewes, F.W. and Oxon, M.A. (1910) The behaviour of the leucocytes in infection and immunity. *The Lancet*, **ii**, 8–16.
2. Butcher, E.C. (1991) Leukocyte–endothelial cell recognition: three (or more) steps to specificity and diversity. *Cell*, **67** (6), 1033–36.
3. Zimmerman, G.A., Prescott, S.M. *et al.* (1992) Endothelial cell interactions with granulocytes: tethering and signaling molecules. *Immunology Today*, **13** (3), 93–100.
4. Gee, M.H. and Albertine, K.H. (1993) Neutrophil–endothelial cell interactions in the lung. *Annual Review of Physiology*, **55**, 227–48.
5. Hynes, R.O. and Lander, A.D. (1992) Contact and adhesive specificities in the associations, migrations, and targeting of cells and axons. *Cell*, **68** (2), 303–22.
6. Springer, T.A. (1994) Traffic signals for lymphocyte recirculation and leukocyte emigration: the multistep paradigm. *Cell*, **76**, 301–14.
7. Cines, D.B., Pollak, E.S. *et al.* (1998) Endothelial cells in physiology and in the pathophysiology of vascular disorders. *Blood*, **91** (10), 3527–61.
8. Harris, E.S., McIntyre, T.M. *et al.* (2000) The leukocyte integrins. *The Journal of Biological Chemistry*, **275** (31), 23409–12.
9. Doerschuk, C.M. (2001) Mechanisms of leukocyte sequestration in inflamed lungs. *Microcirculation*, **8** (2), 71–88.
10. McEver, R.P. (2001) Adhesive interactions of leukocytes, platelets, and the vessel wall during hemostasis and inflammation. *Thrombosis and Haemostasis*, **86** (3), 746–56.
11. Ley, K. (2002) Integration of inflammatory signals by rolling neutrophils. *Immunological Reviews*, **186**, 8–18.
12. Bunting, M., Harris, E.S. *et al.* (2002) Leukocyte adhesion deficiency syndromes: adhesion and tethering defects involving beta 2 integrins and selectin ligands. *Current Opinion in Hematology*, **9** (1), 30–35.
13. Forlow, S.B., Rose, C.E. *et al.* (2002) Leukocyte adhesion and emigration in the lung, in *Interactions of Blood and Pulmonary Circulation* (eds E. Weir, H.L. Reeve, and J.T. Reeves), Futura, New York, pp. 255–75.
14. Kuebler, W.M. and Goetz, A.E. (2002) The marginated pool. *European Surgical Research*, **34** (1-2), 92–100.
15. McIntyre, T.M., Prescott, S.M. *et al.* (2003) Cell–cell interactions: leukocyte–endothelial interactions. *Current Opinion in Hematology*, **10** (2), 150–58.
16. Burns, A.R., Smith, C.W. *et al.* (2003) Unique structural features that influence neutrophil emigration into the lung. *Physiological Reviews*, **83** (2), 309–36.
17. Aird, W.C. (2007) Phenotypic heterogeneity of the endothelium: I. Structure, function, and mechanisms. *Circulation Research*, **100** (2), 158–73.
18. Aird, W.C. (2007) Phenotypic heterogeneity of the endothelium: II. Representative vascular beds. *Circulation Research*, **100** (2), 174–90.

19. Reutershan, J. and Ley, K. (2004) Bench-to-bedside review: acute respiratory distress syndrome – how neutrophils migrate into the lung. *Critical Care*, **8** (6), 453–61.
20. Bock, D., Aydt, E.M. *et al.* (2006) The role of selectins during lung inflammation and their potential impact for innovative therapeutic strategies. *Current Respiratory Medicine Review*, **2**, 339–54.
21. Vielhaur, V., Cullere, X. *et al.* (2007). Leukocyte–endothelial cell interactions, in *Endothelial Biomedicine* (ed. W.C. Aird), Cambridge University Press, New York, pp. 576–86.
22. Andrewes, F.W. and Oxon, M.A. (1910) The behaviour of the leucocytes in infection and immunity. *The Lancet*, **i**, 1737–43.
23. Bierman, H.R., Kelly, K.H. *et al.* (1951) The pulmonary circulation as a source of leucocytes and platelets in man. *Science*, **114** (2959), 276–77.
24. Bierman, H.R., Kelly, K.H. *et al.* (1951c) The effect of intravenous histamine administration on the level of the white blood count in the peripheral blood. *Blood*, **6** (10), 926–35.
25. Bierman, H.R., Byron, R.L. *et al.* (1951) The lung removal mechanism for leukocytes. *Blood*, **6**, 770–71.
26. Bierman, H.R., Kelly, K.H. *et al.* (1952) The release of leukocytes and platelets from the pulmonary circulation by epinephrine. *Blood*, **7** (7), 683–92.
27. Bierman, H.R., Kelly, K.H. *et al.* (1952) The influence of respiratory movements upon the circulating leukocytes. *Blood*, **7** (5), 533–44.
28. Bierman, H.R. (1955) The hematologic role of the lung in man. *American Journal of Surgery*, **89** (1), 130–40.
29. Looney, M.R., Su, X. *et al.* (2006) Neutrophils and their Fc gamma receptors are essential in a mouse model of transfusion-related acute lung injury. *The Journal of Clinical Investigation*, **116** (6), 1615–23.
30. Athens, J.W., Raab, S.O. *et al.* (1961) Leukokinetic studies. III. The distribution of granulocytes in the blood of normal subjects. *The Journal of Clinical Investigation*, **40**, 159–64.
31. Athens, J.W., Haab, O.P. *et al.* (1961) Leukokinetic studies. IV. The total blood, circulating and marginal granulocyte pools and the granulocyte turnover rate in normal subjects. *The Journal of Clinical Investigation*, **40**, 989–95.
32. Muir, A.L., Cruz, M. *et al.* (1984) Leukocyte kinetics in the human lung: role of exercise and catecholamines. *Journal of Applied Physiology*, **57** (3), 711–19.
33. Markos, J., Hooper, R.O. *et al.* (1990) Effect of raised alveolar pressure on leukocyte retention in the human lung. *Journal of Applied Physiology*, **69** (1), 214–21.
34. MacNee, W., Martin, B.A. *et al.* (1989) Regional pulmonary transit times in humans. *Journal of Applied Physiology*, **66** (2), 844–50.
35. MacNee, W., Wiggs, B. *et al.* (1989) The effect of cigarette smoking on neutrophil kinetics in human lungs. *The New England Journal of Medicine*, **321** (14), 924–28.
36. Hogg, J.C., Doerschuk, C.M. *et al.* (1992) Neutrophil retention during a single transit through the pulmonary circulation. *Journal of Applied Physiology*, **73** (4), 1683–85.
37. Doerschuk, C.M., Beyers, N. *et al.* (1993) Comparison of neutrophil and capillary diameters and their relation to neutrophil sequestration in the lung. *Journal of Applied Physiology*, **74** (6), 3040–45.
38. Hogg, J.C., Coxson, H.O. *et al.* (1994) Erythrocyte and polymorphonuclear cell transit time and concentration in human pulmonary capillaries. *Journal of Applied Physiology*, **77** (4), 1795–800.
39. Doerschuk, C.M., Allard, M.F. *et al.* (1987) Marginated pool of neutrophils in rabbit lungs. *Journal of Applied Physiology*, **63** (5), 1806–15.
40. Hogg, J.C. and Doerschuk, C.M. (1995) Leukocyte traffic in the lung. *Annual Review of Physiology*, **57**, 97–114.
41. Anderson, D.C., Schmalsteig, F.C. *et al.* (1985) The severe and moderate phenotypes of heritable Mac-1, LFA-1 deficiency: their quantitative definition and relation to leukocyte dysfunction and clinical features. *The Journal of Infectious Diseases*, **152** (4), 668–89.
42. Walmsley, S.R., McGovern, N.N. *et al.* (2008) The HIF/VHL pathway: from oxygen sensing to innate immunity. *American Journal of Respiratory Cell and Molecular Biology*, **38** (3), 251–55.
43. Vejlens, G. (1938) The distribution of leukocytes in the vascular system. *Acta Pathologica et Microbiologica Scandinavica*, Suppl 33, 1–239.
44. Ambrus, C.M., Ambrus, J.L. *et al.* (1954) Role of the lungs in regulation of the white blood cell level. *The American Journal of Physiology*, **178** (1), 33–44.
45. Wagner, W.W. Jr., Latham, L.P. *et al.* (1982) Direct measurement of pulmonary capillary transit times. *Science*, **218** (4570), 379–81.
46. Lien, D.C., Wagner, W.W. Jr. *et al.* (1987) Physiological neutrophil sequestration in the lung: visual evidence for localization in capillaries. *Journal of Applied Physiology*, **62** (3), 1236–43.

47. Lien, D.C., Worthen, G.S. *et al.* (1990) Neutrophil kinetics in the pulmonary microcirculation. Effects of pressure and flow in the dependent lung. *The American Review of Respiratory Disease*, **141** (4 Pt 1), 953–59.
48. Kuhnle, G.E., Leipfinger, F.H. *et al.* (1993) Measurement of microhemodynamics in the ventilated rabbit lung by intravital fluorescence microscopy. *Journal of Applied Physiology*, **74** (3), 1462–71.
49. Kuebler, W.M., Kuhnle, G.E. *et al.* (1994) Leukocyte kinetics in pulmonary microcirculation: intravital fluorescence microscopic study. *Journal of Applied Physiology*, **76** (1), 65–71.
50. Kuebler, W.M., Ying, X. *et al.* (1999) Pressure is proinflammatory in lung venular capillaries. *The Journal of Clinical Investigation*, **104** (4), 495–502.
51. Ichimura, H., Parthasarathi, K. *et al.* (2005) Pressure-induced leukocyte margination in lung postcapillary venules. *The American Journal of Physiology: Lung Cellular and Molecular Physiology*, **289** (3), L407–12.
52. Kiefmann, R., Rifkind, J.M. *et al.* (2008) Red blood cells induce hypoxic lung inflammation. *Blood*, **111** (10), 5205–14.
53. Jaffe, E.A., Nachman, R.L. *et al.* (1973) Culture of human endothelial cells derived from umbilical veins. Identification by morphologic and immunologic criteria. *Journal of Clinical Investigation*, 52, 2745–56.
54. Gimbrone, M.A. Jr., Cotran, R.S. *et al.* (1974) Human vascular endothelial cells in culture. Growth and DNA synthesis. *The Journal of Cell Biology*, **60** (3), 673–84.
55. Nachman, R.L. and Jaffe, E.A. (2004) Endothelial cell culture: beginnings of modern vascular biology. *The Journal of Clinical Investigation*, **114** (8), 1037–40.
56. Zimmerman, G.A., Albertine, K.H. *et al.* (1999) Endothelial activation in ARDS. *Chest*, **116** (Suppl 1), 18S–24S.
57. Prescott, S.M., Zimmerman, G.A. *et al.* (1984) Human endothelial cells in culture produce platelet-activating factor (1-alkyl-2-acetyl-*sn*-glycero-3-phosphocholine) when stimulated with thrombin. *Proceedings of the National Academy of Sciences of the United States of America*, **81** (11), 3534–38.
58. Zimmerman, G.A., McIntyre, T.M. *et al.* (1985) Thrombin stimulates the adherence of neutrophils to human endothelial cells *in vitro*. *The Journal of Clinical Investigation*, **76** (6), 2235–46.
59. Bevilacqua, M.P., Pober, J.S. *et al.* (1985) Interleukin 1 acts on cultured human vascular endothelium to increase the adhesion of polymorphonuclear leukocytes, monocytes, and related leukocyte cell lines. *The Journal of Clinical Investigation*, **76** (5), 2003–11.
60. Gamble, J.R., Harlan, J.M. *et al.* (1985) Stimulation of the adherence of neutrophils to umbilical vein endothelium by human recombinant tumor necrosis factor. *Proceedings of the National Academy of Sciences of the United States of America*, **82** (24), 8667–71.
61. Hoover, R.L., Briggs, R.T. *et al.* (1978) The adhesive interaction between polymorphonuclear leukocytes and endothelial cells *in vitro*. *Cell*, **14** (2), 423–28.
62. Gimbrone, M.A. Jr., Brock, A.F. *et al.* (1984) Leukotriene B4 stimulates polymorphonuclear leukocyte adhesion to cultured vascular endothelial cells. *The Journal of Clinical Investigation*, **74** (4), 1552–55.
63. Tonnesen, M.G., Smedly, L.A. *et al.* (1984) Neutrophil-endothelial cell interactions. Modulation of neutrophil adhesiveness induced by complement fragments C5a and C5a des arg and formylmethionyl-leucyl-phenylalanine *in vitro*. *The Journal of Clinical Investigation*, **74** (5), 1581–92.
64. Lorant, D.E., Patel, K.D. *et al.* (1991) Coexpression of GMP-140 and PAF by endothelium stimulated by histamine or thrombin: a juxtacrine system for adhesion and activation of neutrophils. *The Journal of Cell Biology*, **115** (1), 223–34.
65. Capecchi, M.R. (1989) Altering the genome by homologous recombination. *Science*, **244** (4910), 1288–92.
66. Mayadas, T.N., Johnson, R.C. *et al.* (1993) Leukocyte rolling and extravasation are severely compromised in P selectin-deficient mice. *Cell*, 74, 541–54.
67. Ley, K. (1995) Gene-targeted mice in leukocyte adhesion research. *Microcirculation*, **2** (2), 141–50.
68. Etzioni, A. Doerschuk, C.M., *et al.* (1999) Of man and mouse; leukocyte and endothelial adhesion molecule deficiencies. *Blood*, **94** (15), 3281–3288.
69. Mestas, J. and Hughes, C.C. (2004) Of mice and not men: differences between mouse and human immunology. *Journal of Immunology*, **172** (5), 2731–38.
70. Mizgerd, J.P. and Skerrett, S.J. (2008) Animal models of human pneumonia. *The American Journal of Physiology: Lung Cellular and Molecular Physiology*, **294** (3), L387–98.

71. Kumasaka, T., Quinlan, W.M. *et al.* (1996) Role of the intercellular adhesion molecule-1 (ICAM-1) in endotoxin-induced pneumonia evaluated using ICAM-1 antisense oligonucleotides, anti-ICAM-1 monoclonal antibodies, and ICAM-1 mutant mice. *The Journal of Clinical Investigation*, **97** (10), 2362–69.
72. Steeber, D.A., Venturi, G.M. *et al.* (2005) A new twist to the leukocyte adhesion cascade: intimate cooperation is key. *Trends in Immunology*, **26** (1), 9–12.
73. Lawrence, M.B., Springer, T.A. (1991) Leukocytes roll on a selectin at physiologic flow rates: distinction from and prerequisite for adhesion through integrins. *Cell*, **65** (5), 859–73.
74. Hattori, R., Hamilton, K.K. *et al.* (1989) Stimulated secretion of endothelial von Willebrand factor is accompanied by rapid redistribution to the cell surface of the intracellular granule membrane protein GMP-140. *The Journal of Biological Chemistry*, **264** (14), 7768–71.
75. Geng, J.G., Bevilacqua, M.P. *et al.* (1990) Rapid neutrophil adhesion to activated endothelium mediated by GMP-140. *Nature*, **343** (6260), 757–60.
76. Feuerhake, F., Fuchsl, G. *et al.* (1998) Expression of inducible cell adhesion molecules in the normal human lung: immunohistochemical study of their distribution in pulmonary blood vessels. *Histochemistry and Cell Biology*, **110** (4), 387–94.
77. Xia, L., Sperandio, M. *et al.* (2002) P-selectin glycoprotein ligand-1-deficient mice have impaired leukocyte tethering to E-selectin under flow. *The Journal of Clinical Investigation*, **109** (7), 939–50.
78. Ley, K., Linnemann, G. *et al.* (1993) Fucoidin, but not yeast polyphosphomannan PPME, inhibits leukocyte rolling in venules of the rat mesentery. *Blood*, **81** (1), 177–85.
79. Atherton, A. and Born, G.V. (1972) Quantitative investigations of the adhesiveness of circulating polymorphonuclear leucocytes to blood vessel walls. *The Journal of Physiology*, **222** (2), 447–74.
80. Walcheck, B., Moore, K.L. *et al.* (1996) Neutrophil-neutrophil interactions under hydrodynamic shear stress involve L-selectin and PSGL-1. A mechanism that amplifies initial leukocyte accumulation of P-selectin *in vitro*. *The Journal of Clinical Investigation*, **98** (5), 1081–87.
81. Middleton, J., Neil, S. *et al.* (1997) Transcytosis and surface presentation of IL-8 by venular endothelial cells. *Cell*, **91** (3), 385–95.
82. Imaizumi, T., Albertine, K.H. *et al.* (1997) Human endothelial cells synthesize ENA-78: relationship to IL-8 and to signaling of PMN adhesion. *American Journal of Respiratory Cell and Molecular Biology*, **17** (2), 181–92.
83. Ley, K. and Zarbock, A. (2006) Hold on to your endothelium: postarrest steps of the leukocyte adhesion cascade. *Immunity*, **25** (2), 185–87.
84. Laudanna, C. and Alon, R. (2006) Right on the spot. Chemokine triggering of integrin-mediated arrest of rolling leukocytes. *Thrombosis and Haemostasis*, **95** (1), 5–11.
85. DiVietro, J.A., Smith, M.J. *et al.* (2001) Immobilized IL-8 triggers progressive activation of neutrophils rolling *in vitro* on P-selectin and intercellular adhesion molecule-1. *Journal of Immunology*, **167** (7), 4017–25.
86. Alon, R., Aker, M. *et al.* (2003) A novel genetic leukocyte adhesion deficiency in subsecond triggering of integrin avidity by endothelial chemokines results in impaired leukocyte arrest on vascular endothelium under shear flow. *Blood*, **101** (11), 4437–45.
87. Lorant, D.E., McEver, R.P. *et al.* (1995) Activation of polymorphonuclear leukocytes reduces their adhesion to P-selectin and causes redistribution of ligands for P-selectin on their surfaces. *The Journal of Clinical Investigation*, **96** (1), 171–82.
88. Dejana, E. (2004) Endothelial cell–cell junctions: happy together. *Nature Reviews: Molecular Cell Biology*, **5** (4), 261–70.
89. Bradfield, P.F., Nourshargh, S. *et al.* (2007) JAM family and related proteins in leukocyte migration (Vestweber series). *Arteriosclerosis, Thrombosis, and Vascular Biology*, **27** (10), 2104–12.
90. Lou, O., Alcaide, P. *et al.* (2007) CD99 is a key mediator of the transendothelial migration of neutrophils. *Journal of Immunology*, **178** (2), 1136–43.
91. Modur, V., Li, Y. *et al.* (1997) Retrograde inflammatory signaling from neutrophils to endothelial cells by soluble interleukin-6 receptor alpha. *Journal of Clinical Investigation*, 100, 2752–6.
92. Romano, M., Sironi, M. *et al.* (1997) Role of IL-6 and its soluble receptor in induction of chemokines and leukocyte recruitment. *Immunity*, 6, 315–25.
93. Lindemann, S.W., Yost, C.C. *et al.* (2004) Neutrophils alter the inflammatory milieu by signal-dependent translation of constitutive messenger RNAs. *Proceedings of the National Academy of Sciences of the United States of America*, **101** (18), 7076–81.
94. Sato, Y., Walley, K.R. *et al.* (1999) Nitric oxide reduces the sequestration of polymorphonuclear leukocytes in lung by changing deformability and

CD18 expression. *American Journal of Respiratory and Critical Care Medicine*, **159** (5 Pt 1), 1469–76.

95. Kerfoot, S.M. and Kubes, P. (2005) Local coordination verses systemic disregulation: complexities in leukocyte recruitment revealed by local and systemic activation of TLR4 *in vivo*. *Journal of Leukocyte Biology*, **77** (6), 862–67.
96. McDonald, B., McAvoy, E.F. *et al.* (2008) Interaction of CD44 and hyaluronan is the dominant mechanism for neutrophil sequestration in inflamed liver sinusoids. *The Journal of Experimental Medicine*, **205** (4), 915–27.
97. Doerschuk, C.M., Downey, G.P. *et al.* (1990) Leukocyte and platelet margination within microvasculature of rabbit lungs. *Journal of Applied Physiology*, **68** (5), 1956–61.
98. Martin, B.A., Wright, J.L. *et al.* (1982) Effect of pulmonary blood flow on the exchange between the circulating and marginating pool of polymorphonuclear leukocytes in dog lungs. *The Journal of Clinical Investigation*, **69** (6), 1277–85.
99. Thommasen, H.V., Martin, B.A. *et al.* (1984) Effect of pulmonary blood flow on leukocyte uptake and release by dog lung. *Journal of Applied Physiology*, **56** (4), 966–74.
100. Kuebler, W.M., Kuhnle, G.E. *et al.* (1999) Leukocyte margination in alveolar capillaries: interrelationship with functional capillary geometry and microhemodynamics. *Journal of Vascular Research*, **36** (4), 282–88.
101. Hogg, J.C., McLean, T. *et al.* (1988) Erythrocyte transit and neutrophil concentration in the dog lung. *Journal of Applied Physiology*, **65** (3), 1217–25.
102. Wiggs, B.R., English, D. *et al.* (1994) Contributions of capillary pathway size and neutrophil deformability to neutrophil transit through rabbit lungs. *Journal of Applied Physiology*, **77** (1), 463–70.
103. Gebb, S.A., Graham, J.A. *et al.* (1995) Sites of leukocyte sequestration in the pulmonary microcirculation. *Journal of Applied Physiology*, **79** (2), 493–97.
104. Downey, G.P., Doherty, D.E. *et al.* (1990) Retention of leukocytes in capillaries: role of cell size and deformability. *Journal of Applied Physiology*, **69** (5), 1767–78.
105. Wang, Q. and Doerschuk, C.M. (2000) Neutrophil-induced changes in the biomechanical properties of endothelial cells: roles of ICAM-1 and reactive oxygen species. *Journal of Immunology*, **164** (12), 6487–94.
106. Wang, Q., Chiang, E.T. *et al.* (2001) Changes in the biomechanical properties of neutrophils and endothelial cells during adhesion. *Blood*, **97** (3), 660–68.
107. Kuhnle, G.E., Kuebler, W.M. *et al.* (1995) Effect of blood flow on the leukocyte–endothelium interaction in pulmonary microvessels. *American Journal of Respiratory and Critical Care Medicine*, **152** (4 Pt 1), 1221–28.
108. Ley, K., Bullard, D.C. *et al.* (1995) Sequential contribution of L- and P-selectin to leukocyte rolling *in vivo*. *The Journal of Experimental Medicine*, **181** (2), 669–75.
109. Downey, G.P., Worthen, G.S. *et al.* (1993) Neutrophil sequestration and migration in localized pulmonary inflammation. Capillary localization and migration across the interalveolar septum. *The American Review of Respiratory Disease*, **147** (1), 168–76.
110. Bless, N.M., Tojo, S.J. *et al.* (1998) Differing patterns of P-selectin expression in lung injury. *American Journal of Pathology*, **153** (4), 1113–22.
111. Parthasarathi, K., Ichimura, H. *et al.* (2002) Mitochondrial reactive oxygen species regulate spatial profile of proinflammatory responses in lung venular capillaries. *Journal of Immunology*, **169** (12), 7078–86.
112. Kuebler, W.M., Kuhnle, G.E. *et al.* (1997) Contribution of selectins to leucocyte sequestration in pulmonary microvessels by intravital microscopy in rabbits. *The Journal of Physiology*, **501** (Pt 2), 375–86.
113. Panes, J., Perry, M.A. *et al.* (1995) Regional differences in constitutive and induced ICAM-1 expression *in vivo*. *The American Journal of Physiology: Heart and Circulatory Physiology*, **269** (6 Pt 2), H1955–64.
114. Jung, U., Norman, K.E. *et al.* (1998) Transit time of leukocytes rolling through venules controls cytokine-induced inflammatory cell recruitment *in vivo*. *The Journal of Clinical Investigation*, **102** (8), 1526–33.
115. Yoder, M.C., Checkley, L.L. *et al.* (1990) Pulmonary microcirculatory kinetics of neutrophils deficient in leukocyte adhesion-promoting glycoproteins. *Journal of Applied Physiology*, **69** (1), 207–13.
116. Kuebler, W.M., Borges, J. *et al.* (2000) Role of L-selectin in leukocyte sequestration in lung capillaries in a rabbit model of endotoxemia. *American Journal of Respiratory and Critical Care Medicine*, 161, 36–43.
117. Issekutz, A.C. and Ripley, M. (1986) The effect of intravascular neutrophil chemotactic factors on

blood neutrophil and platelet kinetics. *American Journal of Hematology*, **21** (2), 157–71.

118. Zarbock, A., Singbartl, K. *et al.* (2006) Complete reversal of acid-induced acute lung injury by blocking of platelet-neutrophil interactions. *Journal of Clinical Investigation*, 116, 3211–9.

119. Worthen, G.S., Schwab, B. *et al.*, Mechanics of stimulated neutrophils: cell stiffening induces retention in capillaries. *Science*, 245, 183–6.

120. Erzurum, S.C., Downey, G.P. *et al.* (1992) Mechanisms of lipopolysaccharide-induced neutrophil retention. Relative contributions of adhesive and cellular mechanical properties. *Journal of Immunology*, **149** (1), 154–62.

121. Selby, C., Drost, E. *et al.* (1991) *In vivo* neutrophil sequestration within lungs of humans is determined by *in vitro* "filterability". *Journal of Applied Physiology*, **71** (5), 1996–2003.

122. Olson, T.S., Singbartl, K. *et al.* (2002) L-selectin is required for fMLP- but not C5a-induced margination of neutrophils in the pulmonary circulation. *American Journal of Physiology: Regulatory, Integrative and Comparative Physiology*, 282, R1245–52.

123. Kubo, H., Graham, L. *et al.* (1998) Complement fragment-induced release of neutrophils from bone marrow and sequestration within pulmonary capillaries in rabbits. *Blood*, **92** (1), 283–90.

124. Liu, L. and Kubes, P. (2003) Molecular mechanisms of leukocyte recruitment: organ-specific mechanisms of action. *Thrombosis and Haemostasis*, 89, 213–20.

125. Mizgerd, J.P. (2008) Acute lower respiratory tract infection. *The New England Journal of Medicine*, **358** (7), 716–27.

126. Chen, D.L., Rosenbluth, D.B. *et al.* (2006) FDG-PET imaging of pulmonary inflammation in healthy volunteers after airway instillation of endotoxin. *Journal of Applied Physiology*, **100** (5), 1602–9.

127. Lehr, H.A., Weyrich, A.S. *et al.* (1997) Vitamin C blocks inflammatory platelet-activating factor mimetics created by cigarette smoking. *The Journal of Clinical Investigation*, **99** (10), 2358–64.

128. Dobrina, A., Schwartz, B.R. *et al.* (1989) CD11/CD18-independent neutrophil adherence to inducible endothelial–leucocyte adhesion molecules (E-LAM) *in vitro*. *Immunology*, **67** (4), 502–8.

129. Andonegui, G., Bonder, C.S. *et al.* (2003) Endothelium-derived toll-like receptor-4 is the key molecule in LPS-induced neutrophil sequestration into lungs. *Journal of Clinical Investigation*, 111, 1011–20.

130. Sikora, L., Johansson, A.C. *et al.* (2003) A murine model to study leukocyte rolling and intravascular trafficking in lung microvessels. *American Journal of Pathology*, **162** (6), 2019–28.

131. Zimmerman, G.A. (2008) Hypoxic erythrocytes spark lung leukocyte adhesion. *Blood*, **111** (10), 4831–32.

132. Deffebach, M.E., Charan, N.B. *et al.* (1987) The bronchial circulation. Small, but a vital attribute of the lung. *American Review of Respiratory Disease*, **135** (2), 463–81.

133. Ainslie, M.P., McNulty, C.A. *et al.* (2002) Characterisation of adhesion receptors mediating lymphocyte adhesion to bronchial endothelium provides evidence for a distinct lung homing pathway. *Thorax*, **57** (12), 1054–59.

134. van Furth, R. and Sluiter, W. (1986) Distribution of blood monocytes between a marginating and a circulating pool. *The Journal of Experimental Medicine*, **163** (2), 474–79.

135. Srivastava, M., Jung, S. *et al.*(2005). The inflammatory versus constitutive trafficking of mononuclear phagocytes into the alveolar space of mice is associated with drastic changes in their gene expression profiles. *The Journal of Immunology*, 175, 1884–93.

136. Auffray, C., Fogg, D. *et al.* (2007) Monitoring of blood vessels and tissues by a population of monocytes with patrolling behavior. *Science*, **317** (5838), 666–70.

137. Doherty, D.E., Downey, G.P. *et al.* (1994) Lipopolysaccharide-induced monocyte retention in the lung. Role of monocyte stiffness, actin assembly, and CD18-dependent adherence. *Journal of Immunology*, **153** (1), 241–55.

138. O'Dea, K.P., Young, A.J. *et al.*, (2005) Lung-marginated monocytes modulate pulmonary microvascular injury during early endotoxemia. *American Journal of Respiratory and Critical Care Medicine*, 172, 1119–27.

139. Heffner, J.E. and Repine, J.E. (1997) Platelets, in *The Lung: Scientific Foundations*, 2nd edn (eds R.G. Crystal and J.B. West, E.R. Weibel and P.J. Barnes), Lippincott-Raven, Philapdelphia, PA, p. 947–60.

140. Bozza, F.A., Shah, A.M. *et al.* (2009). Amicus or adversary: platelets in lung biology: acute lung injury, and inflammatory pulmonary syndromes. *American Journal of Respiratory Cell and Molecular Biology*, **40** (2), 123–34.

141. Zimmerman, G.A., McIntyre, T.M. *et al.* (1996) Adhesion and signaling in vascular cell–cell interactions. *Journal of Clinical Investigation*, 98, 1699–1702.

11

Mesenchymal–Endothelial Interactions in the Control of Angiogenic, Inflammatory, and Fibrotic Responses in the Pulmonary Circulation

Kurt R. Stenmark[1], Evgenia V. Gerasimovskaya[1], Neil Davie[2] and Maria Frid[1]

[1]Developmental Lung Biology and Cardiovascular Pulmonary Research Laboratories, University of Colorado Health Sciences Center, Denver, CO, USA
[2]Pulmonary Vascular Business Unit, Pfizer, Tadworth, Surrey, UK

INTRODUCTION

All forms of chronic pulmonary hypertension are characterized by vascular remodeling (see Chapters 27–29). Structural changes in the intimal, medial, and adventitial compartments have been described, and collectively all contribute to the alterations in vascular resistance and compliance that are observed in chronic pulmonary hypertension [1–3]. It is also increasingly appreciated that many forms of chronic pulmonary hypertension are associated with significant inflammatory responses in the pulmonary vessel wall, which are suggested to contribute significantly to the vascular remodeling process [1, 4–7]. Chronic inflammation usually results in or is associated with fibrosis [8, 9]. Fibrotic changes have been well described in the remodeled pulmonary circulation of animals with experimental pulmonary hypertension, and increases in accumulation of extracellular matrix (ECM) proteins and in the expression of a number of metalloproteinases and proteases have been reported in pulmonary vessels from patients with various forms of pulmonary hypertension [1, 10–12]. Since chronic inflammation and fibrosis are known to be associated with angiogenic responses, the possibility exists that the vascular remodeling associated with at least some forms of pulmonary hypertension would involve angiogenic responses in or around the vessel wall (see Chapter 13). In the systemic circulation, the adventitial vasa vasorum microcirculation undergoes marked neovascularization in a number of vasculopathies, including atherosclerosis, type 2 diabetes, metabolic syndrome, restenosis, and vasculitis [13–17]. This neovascularization process is thought to play a direct role in the remodeling. In the setting of pulmonary hypertension, expansion of the vasa vasorum network in the adventitia and media has also been described [18–20]. Expansion of the pulmonary artery vasa vasorum is commonly observed in the setting of pulmonary artery obstruction [20]. For instance, Kimura *et al.* reported in patients with chronic thromboembolic obstruction of the pulmonary arteries that the volume of pulmonary adventitia vasa vasorum increases and the core of the nonresolving clots is recannulized by neovascular endothelialized structures that originate from the vasa vasorum [21]. Increased density of capillaries in the adventitial and peri-adventitial regions of pulmonary arteries in patients with severe idiopathic pulmonary fibrosis and pulmonary hypertension has also been described [10] (Figure 11.1). In an animal model of severe hypoxia-induced pulmonary hypertension (i.e., the hypoxic neonatal calf) marked expansion of the vasa vasorum network in the adventitia and within the outer aspects of the media of vessels all along the longitudinal axis of the pulmonary circulation have been described [18] (Figures 11.2 and 11.3).

The mechanisms controlling expansion of the vasa vasorum network in the pulmonary and systemic circulations are not well understood. However, it is increasingly appreciated that activation of what are considered the

The Pulmonary Endothelium: Function in health and disease Editors Norbert F. Voelkel, Sharon Rounds

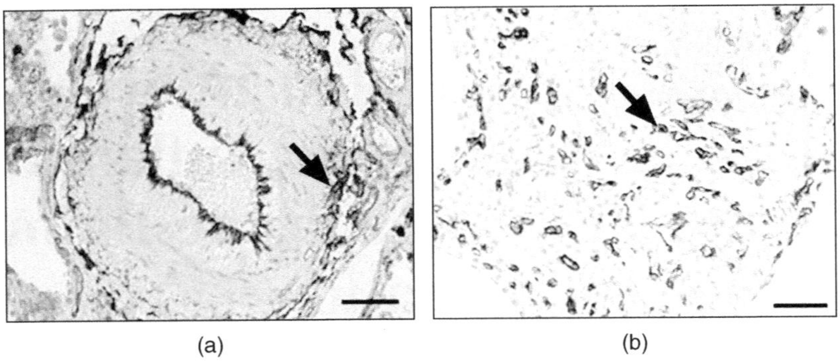

Figure 11.1 Angiogenic responses in the perivascular region of a patient with pulmonary fibrosis and associated pulmonary hypertension. (a) Medium-sized pulmonary artery stained with CD31, demonstrating capillary network expansion in the perivascular area (medial/adventitial region, arrow). (b) CD31 immunohistochemical evidence of capillary proliferation in the fibrotic foci (arrow). Bar = 25 μm. A color version of this figure appears in the plate section of this volume.

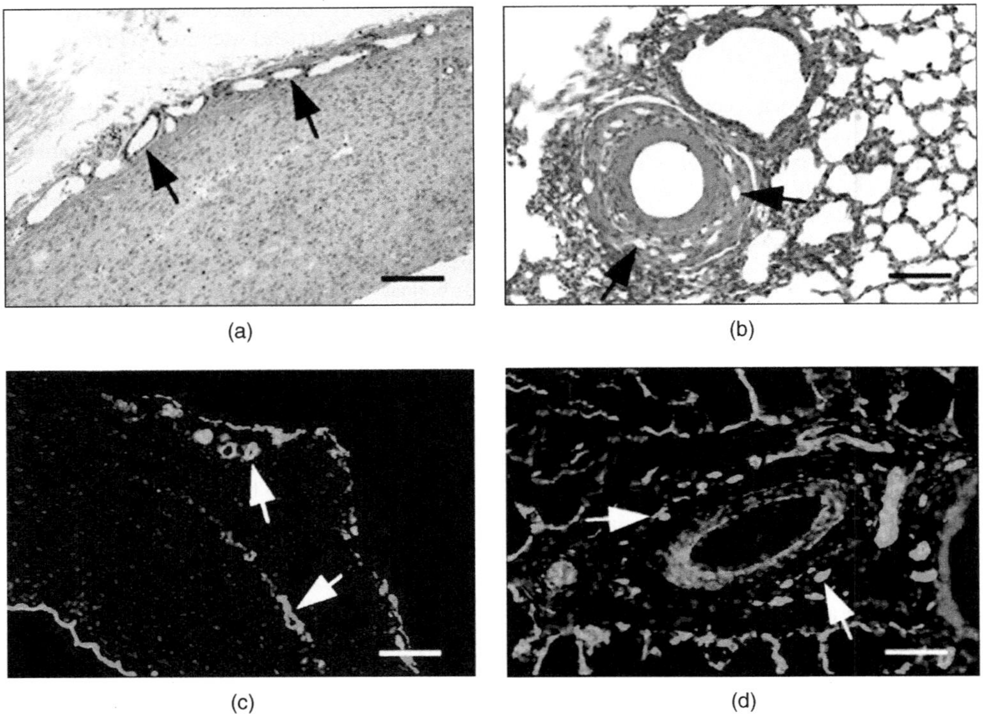

Figure 11.2 Angiogenic responses in the pulmonary arteries of calves with severe hypoxia-induced pulmonary hypertension. Histopathology of large (a and c) and small (b and d) pulmonary arteries. Both histological hematoxylin & eosin (a and b) and immunofluorescent PECAM-1/CD31 (c and d) stainings demonstrate marked expansion of vasa vasorum capillary network in adventitial perivascular regions. Bar = (a and c) 500 and (b and d) 100 μm. A color version of this figure appears in the plate section of this volume.

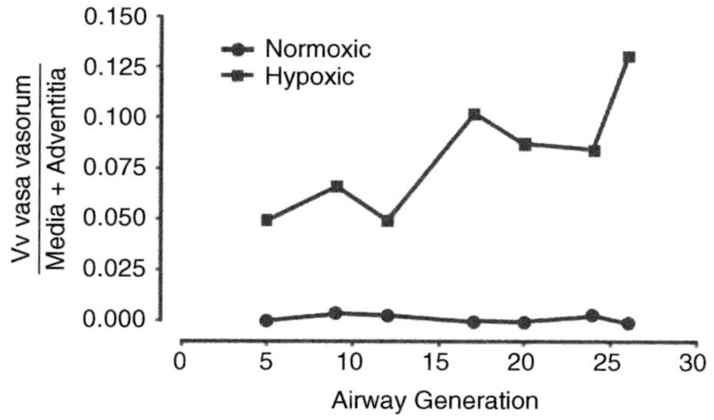

Figure 11.3 Expansion of the vasa vasorum in the severe hypoxic pulmonary hypertension. Quantitative morphometric analyses demonstrated that the volume density (Vv) of vasa vasorum is significantly greater in pulmonary arteries from neonatal calves with severe experimental hypoxia-induced pulmonary hypertension compared with normoxic controls.

stromal constituents of the blood vessel wall (i.e., cells within the adventitial compartment) must play a critical role, since stromal cells have been clearly implicated in the angiogenesis that accompanies tumor progression in cancers of epithelial origin and in chronic inflammatory diseases such as rheumatoid arthritis (RA) [22–26]. The stromal microenvironment consists of several cell types including fibroblasts, macrophages, vascular components [i.e., resident endothelial cells (ECs) of the vasa vasorum (vasa vasorum endothelial cells, VVECs)], inflammatory cells of the innate and acquired immune systems, as well as the ECM that these cells collectively elaborate and all the molecules that are concentrated and immobilized on it. The purpose of this brief chapter is to present evidence that activation of cells within the stromal (adventitial) environment occurs and regulates the inflammatory, fibrotic, and angiogenic responses observed in the pulmonary vessel wall in at least some forms of pulmonary hypertension.

ADVENTITIAL STROMAL CELLS: DEFINITIONS AND ROLES IN INFLAMMATION AND ANGIOGENESIS

A paradigm shift has occurred over the past 10 years in the fields of inflammation and cancer cell research. It has become clear that hematopoietic inflammatory cells can no longer be thought of and analyzed in isolation but need to be considered in the context of organ-specific stromal microenvironments. These environments are composed of tissue-specific cells including, but not limited to, fibroblasts, ECs, and resident macrophages along with their highly specialized ECM components.

Collectively, stromal cells (in the pulmonary circulation we refer to cells principally located in the adventitia) are responsible for defining the specialized architecture of organs and tissues via secretion of ECM components and characteristic cytokine/chemokine combinations [22–26]. Stromal cells have been defined in terms of their embryological origins and lineage relationships, and are generally considered to be mesenchymal in origin. However, the embryological mesenchyme, from which fibroblasts are derived, is not itself a germ layer (usually defined as ectoderm, endoderm, and mesoderm), but is variably considered to be either wholly composed of embryonic mesoderm, or a combination of the mesoderm and either ectoderm or endoderm layers. Complicating the picture, especially in the setting of chronic inflammation, is the fact that blood cell lineages are also derived from the mesoderm, which means that hematopoietic stem cells are technically "mesenchymal" stem cells. Importantly, cell populations have now been identified that appear to blur the distinction between hematopoietic and nonhematopoietic stromal cell populations. An example is the fibrocyte, which has been implicated in the pathogenesis of many fibrotic conditions including hypoxia-induced pulmonary hypertension [27, 28] (see "Monocytes as Stromal Cell Intermediates"). Other unexpected shifts in cell lineage have been reported, including a transition or transdifferentiation of epithelial and ECs into mesenchymal cells or even myofibroblasts [29]. Thus, embyoligic origin, as the standard approach to cellular classification, is becoming increasingly unreliable.

There is accumulating evidence that tissue stromal cells, especially fibroblasts or myofibroblasts, determine the type and duration of leukocytic infiltrates during an inflammatory response [30, 31]. Furthermore, during the resolution of chronic inflammatory responses, stromal cells contribute to the withdrawal of survival signals for the inflammatory cells and ultimately to the normalization of chemokine gradients that allow infiltrating cells to either undergo apoptosis or leave through draining vessels and lymphatics [30, 32].

Thus, as we begin to consider new drug therapies for pulmonary arterial hypertension (PAH), their effects on stromal cells should be considered, especially given the fact that therapies directed at stromal cells are showing increasing promise in cancer and rheumatoid arthritis (RA) [24, 26, 30, 31]. It will be important to take issues of cell origin and tissue-specific functions of cell subpopulations into account.

Roles of Specific Stromal Cells in Inflammation, Fibrosis, and Angiogenesis

Fibroblasts

Fibroblasts are the most abundant cells in the stromal microenvironment, and are responsible for the synthesis and maintenance of the ECM found in specific stromal microenvironments. Unfortunately, a specific definition of the fibroblast phenotype using biochemical markers has not been universally accepted [33]. This is a significant problem, which has retarded progress in the quest to define the role of the fibroblast in chronic inflammation and fibrosis. At present, descriptions or definitions of fibroblasts generally rely upon a combination of morphology (spindle shape) and expression (or lack thereof) of phenotypic markers [e.g., vimentin-positive, α-smooth muscle-actin (α-SM-actin)-negative], and functional features such as production of type I collagen. The lack of specific fibroblast markers is quite troubling because recent work has demonstrated that fibroblasts comprise a very diverse class of distinct cell types [34, 35]. Transcriptional profiling studies have convincingly demonstrated that fibroblasts from different sites can be consistently separated by differences in their transcriptomes [35]. The highly diverse nature of transcriptomes in distinct fibroblast populations suggests very specialized functions in specific tissues. Thus, when investigating angiogenic and inflammatory responses in the pulmonary circulation, it is essential to utilize specifically pulmonary artery adventitial fibroblasts (AdvFBs) in studies evaluating fibroblast interactions with other cells (such as ECs or leukocytes).

In addition to production of ECM proteins, fibroblasts produce and respond to growth factors in their microenvironment, allowing reciprocal interactions with themselves as well as other stromal cell types (ECs, resident macrophages) and with adjacent smooth muscle cells (SMCs) or epithelial cells. As a consequence, fibroblasts play a critical role during tissue development and in maintenance of homeostasis [34, 36]. They have also recently been described as having "sentinel" functions during pathophysiologic processes [24, 30, 34]. Thus, fibroblasts myotibroblasts contribute to the pathology of inflammatory/fibrotic diseases both directly, by production of ECM components and expression of α-SM-actin, as well as indirectly, by influencing the behavior of neighboring cell types through production and release of soluble factors/mediators. We believe both functions contribute to the vascular remodeling observed in pulmonary hypertension.

A good example of how fibroblasts act to cause a persistence of inflammation and thus contribute to fibrotic and angiogenic responses comes from studies in RA models [30–32]. RA synovial fibroblasts have been shown to regulate tissue injury and remodeling *in vivo*, and to display an "imprinted phenotype," which is significantly different from normal synovial fibroblasts and which is stable under *ex vivo* culture conditions. This phenotype results in functionally important outcomes, such as cartilage invasion and bone erosion, and determines disease outcome for the majority of RA patients [37, 38]. Cytokines including monocyte chemotactic protein-1 (MCP-1), stromal cell-derived factor (SDF)-1 (CXCL12), interleukin (IL)-8, RANTES, Fractaline, vascular endothelial growth factor (VEGF), and transforming growth factor (TGF)-β produced by RA fibroblasts play a key role in the persistence of inflammation through recruitment and retention of effector cells of the immune system [39]. Type I interferons have also been shown to be produced by RA synovial fibroblasts (as well as by stromal macrophages) and can block apoptotic signals, which normally result in T lymphocyte death at the conclusion of an inflammatory response, thus contributing to perpetuation of inflammation [40]. Interestingly, the unique imprinted phenotype of RA synovial fibroblasts bears a remarkable phenotypic similarity to that of the bone marrow stromal cells, which are involved in the accumulation and support of hematopoietic cells. Recent studies have suggested that the distinct phenotype of at least some RA synovial fibroblasts may be due to the accumulation of bone marrow-derived stromal (mesenchymal) progenitor cells [31, 41].

Another example of how stromal fibroblasts play critical roles in regulating tissue responses comes from the cancer literature. Stromal fibroblasts surrounding a carcinoma have been referred to as "reactive fibroblasts" or also as "carcinoma-associated fibroblasts" (CAFs) [22, 23]. The CAFs differ from normal fibroblasts by abnormally high expression of α-SM-actin, and increased expression of proteolytic enzymes and ECM proteins, such as tenacin and fibrillar collagens. The importance of CAFs in carcinogenesis has been established in elegant recombination experiments. When immortalized nontumorigenic human prostate epithelial cells were mixed with fibroblasts from human prostate carcinomas and grafted to immune-deficient animals, large carcinomas developed. In contrast, mixing the epithelial cells with fibroblasts from a normal prostate gland did not result in carcinomas [23, 42]. Some of the molecules potentially responsible for the tumorigenic effect of fibroblasts have been identified. Among these is S100A4 (also called mts1 or fibroblast-specific protein-1), which is expressed in both CAFs and in carcinoma cells during tumor progression [43]. S100A4 is a calcium-binding protein with both intracellular and extracellular protein-binding partners. Intracellularly, it interacts with and possibly inactivates p53. S100A4 also interacts with nonmuscle myosin heavy chain, actin filaments and nonmuscle tropomysin, thereby

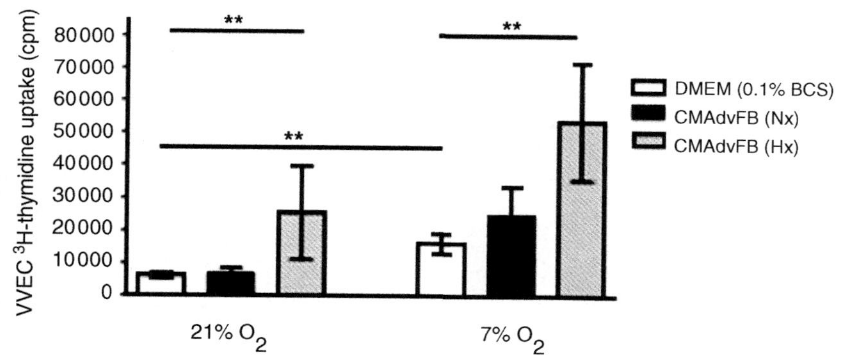

Figure 11.4 DNA synthesis of VVECs was significantly stimulated by conditioned media (CM) from hypoxia-activated (Hx) AdvFBs but not from normoxic (Nx) AdvFBs. The graphs are presented as mean values (mean ± standard deviation) of five separate cell isolates (obtained from five different animals) measured in triplicate. DMEM, Dulbecco's modified Eagle's medium; BCS, bovine calf serum. $^{**}p < 0.01$.

potentially influencing the cytoskeleton and regulating cell motility [44]. The extracellular binding partners are unknown although it may interact with Annexin-II. S100A4 binds to this coreceptor for the serine proteinase plasminogen, which results in increased activation of plasminogen [45]. S100A4 is proangiogenic and this action is possibly mediated by activation of plasminogen or through transcriptional upregulation of the matrix metalloproteinase (MMP)-13 [46]. Both of these proteases are thought to play a role in endothelial invasion. It also appears that S100A4 play a role in epithelial–mesenchymal transitions (EMTs). Thus, production of S100A4 as well as molecules such as SDF-1 (CXCL12), VEGF, MMPs, and TGF-β, can influence the local environment, and result in inflammation, angiogenesis, tumor progression, and metastasis.

Concurrent with the observations of distinct properties of fibroblasts in RA and cancer, we sought to examine whether fibroblasts derived from the adventitia of chronically hypoxic hypertensive animals would exhibit distinct functional properties compatible with inflammation, fibrosis, and angiogenesis [1, 34]. Since it has been suggested that the expanding vasa vasorum plays an important role in the inflammatory and remodeling process, we examined the possibility that pulmonary artery Adventitial Fibroblasts (AdvFBs) may regulate the adaptive expansion of the vasa vasorum by exhibiting proangiogenic capabilities and interacting with resident or recruited VVECs [18, 19]. Further, if this were the case, we wanted to determine potential mechanisms involved.

First, VVECs and AdvFBs were isolated from the same adventitial compartment of intralobar pulmonary arteries from both normoxic and hypoxic neonatal calves. Concurrently, ECs from the pulmonary artery and aorta of the same animals were isolated to ultimately examine the possibility that VVECs exhibit distinct functional properties. Using both coculture and conditioned media approaches, we found that AdvFBs were capable of stimulating VVECs proliferation [19]. Furthermore, conditioned media from AdvFBs was capable of augmenting both the self-assembly and the integrity of cord-like networks formed by VVECs in Matrigel (Figure 11.4). Importantly, exposure to hypoxia (7% O_2) augmented all these responses (Figure 11.4). These observations are in accordance with a number of previous studies, in which hypoxic conditions have been shown to induce angiogenic phenotypes in a number of stromal cell types [47–49]. Moreover, these observations are consistent with the well-established paradigm that hypoxia is a common feature of many pathological conditions associated with neovascular growth. Importantly, previous studies in our laboratory have established that AdvFBs exhibit the earliest and most dramatic activation responses among all cells in the vessel wall [34].

To further examine the fibroblast–endothelial interactions, we performed three-dimensional coculture experiments in Matrigel and found that coculturing AdvFBs and VVECs facilitated assembly of heterotypic cord-like networks (Figure 11.5). These interactions were augmented by hypoxia (Figure 11.5). We were interested in the factors, secreted by AdvFBs under hypoxic conditions that might stimulate or regulate these responses. We have previously shown increased expression of a number of proangiogenic molecules in the pulmonary artery adventitia in the neonatal bovine model of chronic hypoxic PAH, including VEGF, fibronectin, thrombin, and S100A4 (Figure 11.6). Interestingly, all these molecules have been found to be upregulated in the pulmonary arteries of human patients with various forms of PAH [6,

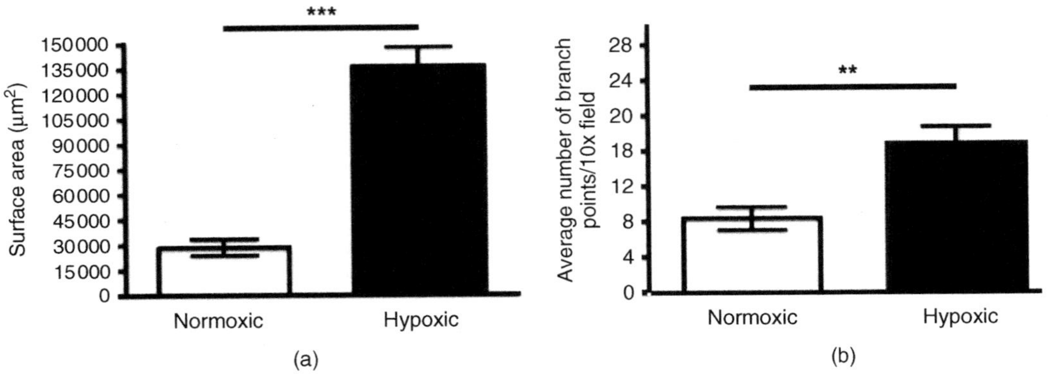

Figure 11.5 Quantitative image analysis illustrating increases in both surface area (a) and number of branch points (b) of cord-like structures in VVEC/AdvFBs Matrigel cocultures in response to hypoxia (7% O_2).

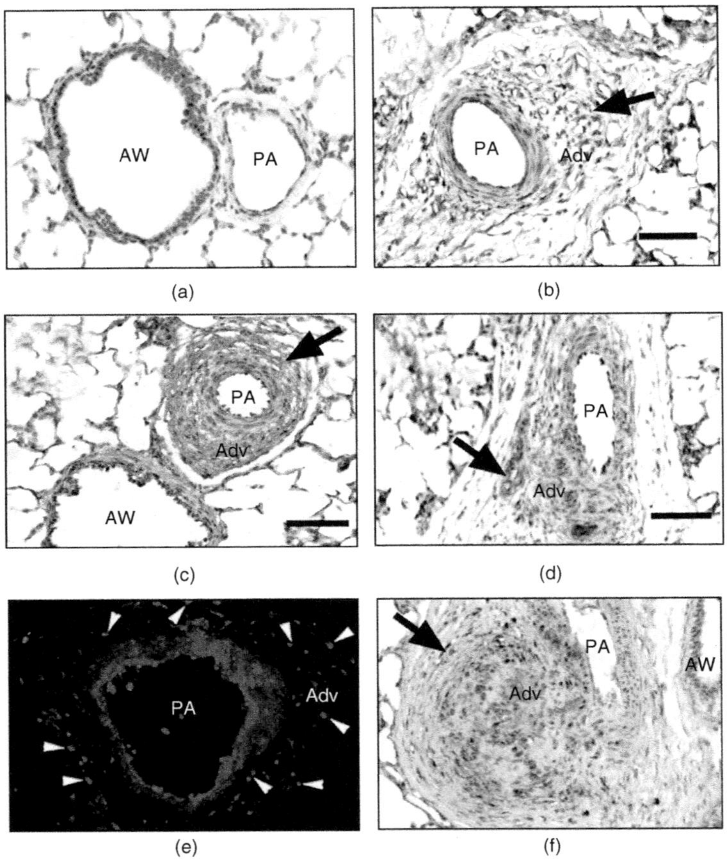

Figure 11.6 Proteins described as having proangiogenic potential [VEGF (b), fibronectin (c), thrombin (d), TGF-β1 (e), and S100A4 (f)], are all expressed in the remodeled adventitia of neonatal calves with severe hypoxia-induced pulmonary hypertension. (a) Control vessel. Adv, adventitia; AW, airway; PA, pulmonary artery. A color version of this figure appears in the plate section of this volume.

7]. In addition, studies in the systemic circulation have suggested that endothelin (ET-1), a factor well known to be upregulated in pulmonary hypertension, plays a critical role in coronary vasa vasorum neovascularization in the setting of experimental hypercholesterolemia through local upregulation of VEGF. We were thus interested in determining if VEGF could play a role in AdvFBs-directed angiogenic responses [16].

We found that ET-1 was released from hypoxia-activated AdvFBs, thus raising the possibility that ET-1 regulates proliferation of VVECs and facilitates their assembly into cord-like networks in coculture with AdvFBs. Using complimentary inhibitory strategies (pharmacologic inhibitors and antisense oligonucleotides), we found that blockade of endogenously produced ET-1 attenuated the assembly of cord-like structures in cocultures of VVECs and AdvFBs (Figure 11.7). These data complement and extend the data of others who have found that ET-1 stimulates an angiogenic phenotype in cultured ECs and regulates the assembly of capillary networks in the systemic circulation [50, 51].

Further, the data are consistent with the emerging concept that the endothelium of *de novo* forming microvessels receives and integrates proangiogenic signals from a number of non-ECs, including fibroblasts [14, 52–55].

Fibroblasts, cultured on ECM proteins, have been shown to secrete cytokines and proangiogenic growth factors that regulate the formation of capillary-like networks by human umbilical vein ECs and systemically derived microvascular ECs [14, 53, 56, 57]. Other studies have shown that stromal cells, including fibroblast-like cells, not only provide initial stimuli for the angiogenic cascade but also provide a stabilizing force to newly formed vessels [14, 52–57]. Tissue fibroblasts have also been described to exhibit proangiogenic capabilities at sites of wound healing and inflammation. These cells respond to chemotactic cytokines released in the tissue environment and are frequently the first cell type to migrate to the wound site where they orchestrate reparative neovascularization [52]. Thus, our studies support the concept that activated AdvFBs regulate angiogenic responses of the resident ECs in the adventitia and stimulate a process of neovascular growth, be it normal or disordered.

ECs

As mentioned above in "Adventitial Stromal Cells: Definitions and Roles in Inflammation and Angiogenesis," ECs are an important component of the stromal environment. VVECs, observed in the adventitia of large pulmonary arteries, are thought to be derived from the bronchial circulation. As the concept of endothelial heterogeneity is well established in the literature [58], it would not be unlikely that VVECs would exhibit different phenotypical characteristics including angiogenic capabilities than pulmonary artery ECs (PAECs) or aortic ECs (AOECs). Supporting the possibility that pulmonary adventitial VVECs exhibit unique angiogenic responses to environmental stimuli, are observations demonstrating that selective expansion of the bronchial, rather than the pulmonary, circulation in the lung takes place under a variety of pathophysiologic conditions [20]. This response seems more consistent and vigorous than angiogenesis of the pulmonary circulation, which, however, has been described to take place in lung cancers and following pneumonectomy [20]. Expansion of the bronchial circulation is well known to take place in the setting of chronic inflammation [59, 60] (see Chapter 14). As noted in the Introduction, ischemia or complete occlusion of the pulmonary artery also leads to tremendous systemic vascular responses in the lung with expansion of the bronchial circulation [20]. Bronchial arteriograms in patients with chronic thromboembolic disease demonstrate the unique capacity of systemic vessels to proliferate and to invade the ischemic lung parenchyma and

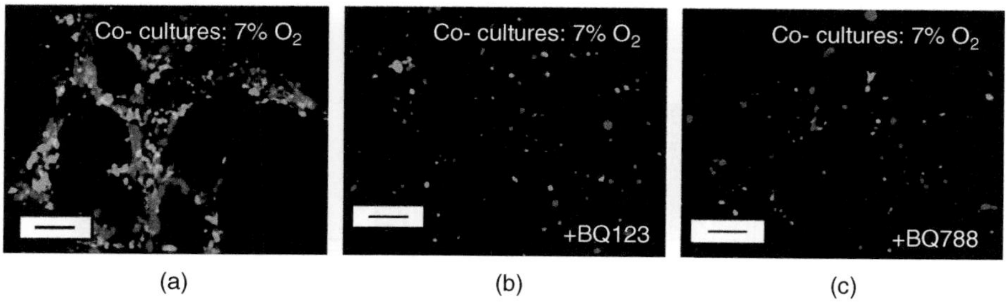

Figure 11.7 Fluorescence microscopy showing that cord-like networks, formed in hypoxic VVEC-AdvFBs Matrigel co-cultures (a), were markedly attenuated when cells were incubated with either the ET_A receptor antagonist BQ123 (b) or the ET_B receptor antagonist BQ788 (c).

even the pulmonary arteries themselves [61]. Neovascularization of the systemic circulation in the lung after pulmonary artery obstruction has been confirmed and studied in humans, sheep, dogs, pigs, guinea pigs, and rats [20]. Recent studies in Wagner's laboratory have suggested that ischemia leads to upregulation of several CXC chemokines which appear to play a role in the generation of new systemic vessels in the lung [62]. These observations are supported by data from Strieter *et al.* in several lung pathologies, showing that the Glu–Leu–Arg $(ELR)^+$ CXC chemokines and their G-protein-coupled CXCR receptors are important proangiogenic factors in the lung [63, 64]. However, to our knowledge few studies have directly examined and contrasted angiogenic capabilities of bronchial circulation-derived ECs and PAECs.

We therefore performed studies examining potential differences in angiogenic capabilities between VVECs (isolated from the pulmonary artery adventitia), luminal PAECs, and AOECs. We found that VVECs exhibited unique proangiogenic capabilities compared to the PAECs and AOECs [65]. Hypoxia exerted far greater proliferative and migratory effects on VVECs than on PAECs or AOECs [65] (see Chapter 18). Further, since it has been previously shown that hypoxia stimulates ATP release from ECs, fibroblasts, and circulating cells, we sought to determine if extracellular ATP would exert angiogenic effects on VVECs (or PAECs and AOECs) and/or if it would augment the effects of other angiogenic factors [VEGF, fibroblast growth factor (FGF)-2] known to be present in the hypoxic microenvironment [65]. We found that extracellular ATP markedly modulated VVECs capabilities by increasing their DNA synthesis, migration, and rearrangement into tube-like networks on Matrigel, but did not affect these characteristics in PAECs or AOECs obtained from the same animals. Extracellular ATP also potentiated the effects of both VEGF and FGF-2 to stimulate DNA synthesis in VVECs, but not in PAECs or AOECs (Figure 11.8).

We also investigated the possibility that differences in postreceptor signaling events in VVECs, AOECs, and PAECs are involved in conferring these distinct proliferative and migratory phenotypes in different cell types. We found significant differences in the magnitude, duration and activation of extracellular signal-regulated mitogen-activated protein kinase (ERK) 1/2, Akt (protein kinase B), mammalian target of rapamycin (mTOR), and p70S6 kinase between VVECs, on the one hand, and AOECs and PAECs, on the other (Figure 11.9). Greater and more prolonged activation of ERK1/2 p70S6 kinase was found in VVECs, and blockade of these pathways inhibited VVECs proliferation and migration, thus suggesting the potential contribution of these pathways to proliferative and migratory advantages observed in these cells as compared to PAECs and AOECs.

Our observations also suggest that VVECs may be specifically susceptible to the effects of mitogenic growth factors and nucleotides, released from resident stromal cells, such as the AdvFBs, or from circulating blood

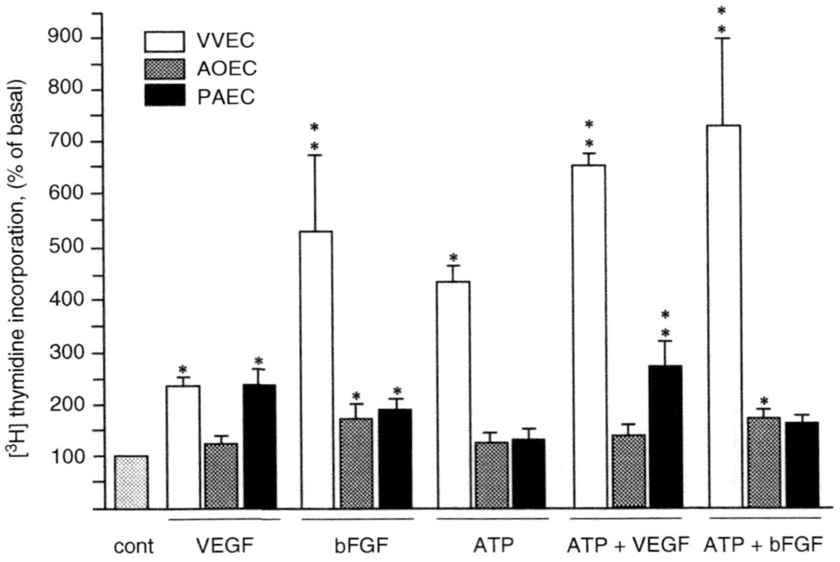

Figure 11.8 ATP exerts proangiogenic effect on cultured VVECs, but not on luminal PAECs (PAEC) and AOECs, and also augments the effects of other angiogenic factors [VEGF and basic FGF (bFGF)] known to be present in the hypoxic microenvironment. (cont = control).

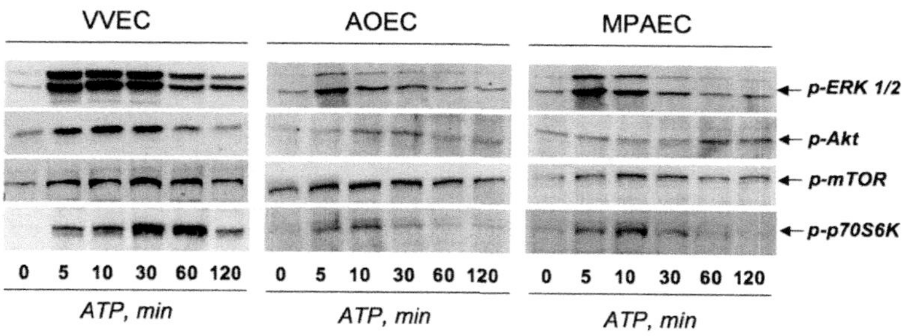

Figure 11.9 Extracellular ATP robustly activates ERK1/2, phosphatidylinositol 3-kinase/Akt, and mTOR/p70S6K signaling pathways in VVECs, but only modestly in AOECs and PAECs (PAEC). Growth arrested cells were stimulated with ATP (100 μM) for the indicated times. Equivalent amounts of total cell protein (25–40 μg) were subjected to Western blot analysis with antibodies against phospho-ERK1/2 (Thr202/Tyr204), phospho-Akt (Ser473), phospho-mTOR (Thr2448), and phospho-p70S6K (Thr421/Ser424). Data shown on each panel illustrate representative experiments for each cell type. Similar results were reproduced in at least three independent experiments.

cells, such as erythrocytes or monocytes that accumulate in the adventitial microenvironment (see Figure 11.10). This may result in an additive stimulatory effects on proliferation and migration of VVECs. Collectively, our observations suggest that the presence of specific cells and cell–cell interactions in the adventitial stromal microenvironment of the pulmonary vessel wall could uniquely contribute to the neovascularization responses observed in certain forms of chronic pulmonary hypertension.

Alternative Sources of Mesenchymal (Stromal) Cells in Inflammatory, Fibrotic, and Angiogenic Responses

Monocytes as Stromal Cell Intermediates

Upon stimulation, monocytes have been shown to differentiate into a wide variety of cells, including macrophages, dendritic cells, and osteoclasts [66]. However, emerging data suggests that cells of a monocytic lineage may also differentiate down a number of stromal cell lineages to produce both fibroblast-like and ECs [67]. The presence of such plasticity in the monocyte lineage helps to explain what appears to be hematopoietic origins of some of the stromal cells that accumulate in inflammatory fibrotic and wound healing responses. A population of circulating peripheral blood leukocytic cells, termed fibrocytes, was described by Bucala *et al.* more than a decade ago which were capable of giving rise to connective tissue fibroblast-like cells in a model of wound injury [68]. These cells were identified by their rapid and specific recruitment from the blood to subcutaneously implanted chambers in mice where they proliferated [68]. Fibrocytes purified from peripheral blood and grown in culture express vimentin and collagens I and III [27, 68]. However these cells are molecularly distinct from resident tissue fibroblasts and exhibit, at least transiently, a profile of leukocytic cell surface markers, including CD34, CD45, CD80, CD86, and major histocompatibility complex class II. These cells do not express typical surface antigens of dendritic cells (CD11a, CD25, CD10, CD38), macrophages (CD14, CD16), B cells (CD19), or ECs (von Willebrand factor). These cells rapidly enter sites of tissue injury where they elaborate a collagen-based matrix and differentiate along fibroblast or potentially even myofibroblast lineages under the influence of cytokines, including TGF-β1, ET-1, and cysteine-containing leukotrienes [27, 68, 69]. In lung fibrosis models, the SDF-1 (CXCL12)–CXCR4 chemokine–receptor pair has been identified as essential in the recruitment of fibrocytes, and targeting of CXCR4 with monoclonal antibodies was shown to ameliorate bleomycin-induced fibrotic lung disease [70]. Fibrocytes have also been demonstrated to support angiogenesis of ECs both *in vivo* and *in vitro* by producing a number of proangiogenic factors, including FGF-2, VEGF, granulocyte macrophage colony-stimulating factor, IL-1, IL-8, and macrophage colony-stimulating factor [27, 68].

We have documented the presence of large numbers of fibrocytes in the pulmonary vessel wall of both rats and calves with severe hypoxia-induced pulmonary hypertension [71]. The fibrocytes accumulated predominately in the adventitia where the active process of vasa vasorum angiogenesis was taking place. Most importantly, *in vivo* depletion or reduction in the number of circulating monocytes and fibrocytes resulted in attenuation of vascular remodeling, as evidenced by reduced adventitial thickening and inhibition of α-SM-actin, collagen, fibronectin, and tenascin expression [71]. Our observations, when combined with others in the setting of RA and tumors, suggest that the stromal environment in pathologic conditions is composed of both activated resident cells and

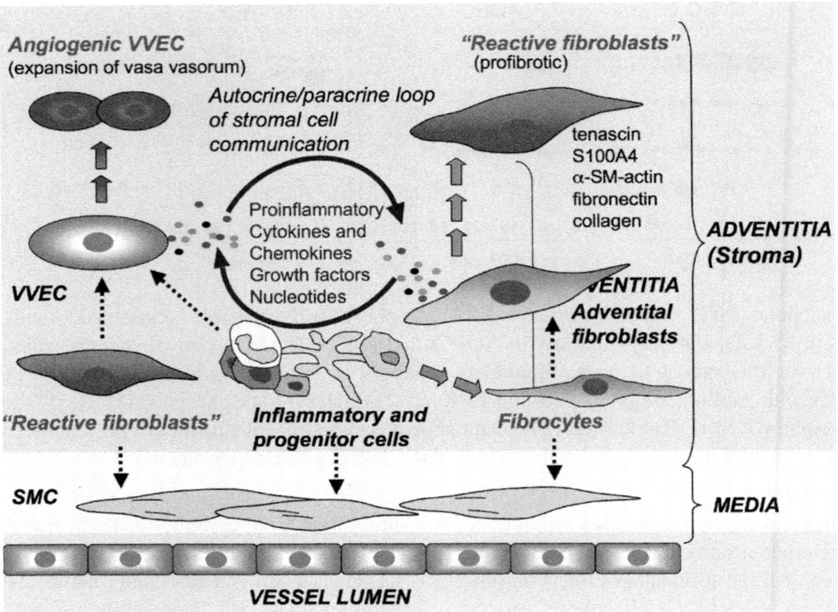

Figure 11.10 Schematic illustrating stromal cell communication in the pulmonary artery vascular wall. Under conditions of chronic hypoxia, adventitial fibroblasts, VVECs, and resident macrophages/dendritic cells release various autocrine/paracrine factors, including proinflammatory cytokines, chemokines, growth factors, and cytokines, which both recruit inflammatory and progenitor cells and "activate" resident fibroblasts, ECs, and SMCs, thus creating micro-environmental circuits that support continued inflammation and remodeling through angiogenic and profibrotic pathways.

circulating, bone marrow-derived, hematopoietic stromal precursors, at least one population of which is fibrocytes. These recruited stromal cells contribute to vascular remodeling both directly, by depositing ECM proteins such as collagen, fibronectin, and tenascin, and indirectly, by producing paracrine factors capable of stimulating growth of VVECs, resident AdvFBs, and adjacent medial SMCs.

Endothelial–Mesenchymal Transition as a Source of Stromal Mesenchymal Cells

Another potential source of fibroblast-like cells in the remodeled vessel wall that could participate in inflammatory, fibrotic, and angiogenic processes, is the resident stromal EC. It has been increasingly appreciated in a number of organ systems that ECs, including microvascular endothelium, are capable of undergoing a process of transition or transdifferentiation into mesenchymal cells of a fibroblast or even a myofibroblast phenotype [29, 72, 73].

Increasing *in vivo* experimental evidence suggests that endothelial–mesenchymal transition (EnMT) plays an important role during cardiovascular development and in various vascular pathologies (see Chapter 1). In the developing vertebrate embryo, the heart is the first organ to be formed, and its endocardial cells, initially expressing an endothelial phenotype [VE-cadherin, Tie-1/2, VEGF receptor 1/2, platelet-endothelial cell adhesion molecule (PECAM)-1/CD31] in the atrio-ventricular canal, give rise to the mesenchymal heart cushion cells through a process of EnMT, and form the mesenchymal portion of cardiac septa and valves [74–76]. Several groups have reported that, in addition to its role in heart development, EnMT is critical in early vascular development. Using ultrastructural, immunohistochemical and cell culture approaches, Arciniegas *et al.* have shown that EnMT is an important event in aortic and pulmonary artery development [77, 78]. De Ruiter *et al.* reported that ECs, "tagged" by immuno-gold labeling at an early stage of development, later (at the onset of SMC differentiation) appeared in the subendothelial space of the developing aorta and by then expressed α-SM-actin [79]. Development of the normal arterial intima, a process which begins relatively early in life, also appears to involve EnMT [77, 80]. Morphologic studies in human embryos suggest that endothelial-like cells give rise to SMCs during the maturation of pulmonary arteries and veins [81, 82]. Findings in experimental wound repair have suggested that EnMT may also take place in the adult, showing that capillary ECs undergo conversion into interstitial connective tissue mesenchymal-like cells in granulation tissue [83]. Microvascular ECs have been reported to transition into mesenchymal cells in response to chronic inflammatory stimuli [84, 85]. A possible role for EnMT in the neointimal thickening in transplant atherosclerosis and restenosis has also been suggested [86–88]. Furthermore, it has recently been demonstrated that transdifferentiation of PAECs into smooth muscle-like cells occurs in hypoxia-induced pulmonary vascular remodeling, and is regulated by myocardin [89].

A number of *in vitro* studies have demonstrated that ECs from a variety of vascular beds retain the capability of transitioning into mesenchymal or even smooth muscle-like cells under a variety of culture conditions [77, 90–92]. ECs derived from the adult bovine aorta convert to spindle-shaped α-SM-actin-expressing cells when treated with TGF-β1 [93]. Certain murine endothelial-like cell lines irreversibly transform into mesenchymal cells upon overgrowth in culture [94]. One concern raised regarding the early reports describing *in vitro* EnMT was the possibility that the primary EC cultures were contaminated with mesenchymal cells. The studies by Frid *et al.* addressed this problem in more detail, by using both fluorescent activated cell sorting (FACS) and magnetic bead techniques (Dynabeads) to "purify" ECs from primary cultures based on their elevated uptake of acetylated low-density lipoprotein, labeled with 1,1′-dioctadecyl-3,3,3′,3′-tetramethyl-indocarbocyanine perchlorate (DiI-AcLDL) or on PECAM-1/CD31 expression (both absent in mesenchymal cells) [91]. These "purified" ECs, however, were still capable of giving rise to colonies of mesenchymal and even smooth muscle-like cells, under the condition that the aforementioned selection of ECs was performed soon (within day 1) after initiating primary cultures. On the contrary, when EC selection was performed 5 days or more after establishing primary cultures, no mesenchymal cells appeared. The possibility was thus raised that the endothelial–mesenchymal phenotypic switch was initiated early in primary culture (due to injury- or stress-related activation of ECs) and would have been concluded by or before day 5. The resulting endothelial-derived mesenchymal cells (appeared via EnMT) would be therefore eliminated by Dynabeads or FACS. Interestingly, investigators employing endothelial cultures derived from large arteries commonly use these techniques to eliminate mesenchymal contaminants, and these methods are usually applied after certain expansion of cells in culture. In contrast, in microvascular ECs, described to be very susceptible to transitioning into mesenchymal cells, the need for selective sorting has been noted as an essential requirement [85, 95–97]. Thus, these studies supported the idea that mesenchymal cells often observed in primary endothelial cultures can arise through the process of EnMT.

An important question is whether the EnMT process is reversible. This is especially important as one begins to consider therapies aimed at reducing inflammatory and fibrotic disease. During embryogenesis, a process that is essentially the reverse of EMT, termed mesenchymal–epithelial transition (MET), occurs during organ morphogenesis [98–102]. Control or upregulation of cadherin expression through developmentally and spatially regulated expression of molecules, such as bone morphogenetic protein (BMP)-7 and hepatocyte growth factor, appears critical for this process [100, 101]. These observations demonstrated that the EMT process in mature organs might be reversible. Indeed, exciting new studies in the kidney suggest that a regenerative program can be initiated in the injured kidney that bears resemblance to renal development [100, 101, 103]. It has been demonstrated that remission, as well as substantial regression, can be achieved by exogenous recombinant human BMP-7 in the fibrotic kidney [100, 101, 104]. BMP-7 reverses EMT, inhibits the release of proinflammatory/fibrogenic cytokines from tubular epithelial cells, and prevents apoptosis, thus resulting in repair and restoration of normal renal architecture. BMP-7 binds to Alk3 and Alk6, which function via phosphorylation of Smad1, 5, and 8, to increase E-cadherin expression, counteracting the TGF-β-induced, Alk5- and Smad2,3-mediated decrease in E-cadherin expression. Thus, strategies aimed at reversing EMT may have a significant positive effect on fibrotic organ injuries.

Less is known regarding the reversibility of EnMT though some data is emerging. Zeisberg *et al.* recently showed that cardiac fibrosis is associated with the appearance of fibroblasts originating from ECs through an EnMT [105]. As in studies with ECs from other organs, TGF-β1 induced ECs to undergo this transition, whereas treatment with BMP-7 preserved the endothelial phenotype. Using a novel Tie-1, CRE:R26RSTOP LacZ mouse, in which cells of endothelial origin are permanently marked by LacZ expression, as well as S100A4–Green Fluorescent Protein (GFP) transgenic mice, in which GFP is expressed under the control of the promoter S100A4, the authors convincingly showed *in vivo* that cardiac ECs undergo EnMT in response to pressure overload and chronic allograft rejection and contribute to the total pool of cardiac fibroblasts. Importantly, systemic administration of recombinant BMP-7 significantly inhibited EnMT and the progression of cardiac fibrosis in both mouse models. These findings provide strong support for the idea that ECs may contribute to fibrotic diseases and that there may be new approaches to ameliorate fibrotic responses in the setting of several forms of pulmonary hypertension.

CONCLUSIONS AND PERSPECTIVES

Stromal cells (including but not limited to, fibroblasts, tissue-resident macrophages, ECs) play a significant role in orchestrating both the recruitment and maintenance of inflammatory cell infiltrates and, as such, of regulating fibrotic and angiogenic responses in a variety of pathologic conditions. Excellent experimental evidence supporting this concept is found in chronic inflammatory diseases (e.g., RA) and in cancer [22, 24–26, 30, 31]. Based on presently available data, the stromal cells of the pulmonary artery adventitia also appear to play an important contributory role in the vascular remodeling observed in several forms of chronic pulmonary hypertension. As in other chronic inflammatory and fibrotic diseases, increased neovascularization within or around the vessel wall might promote the egress of leukocytes to sites of injury or stress which might, in turn, lead to the maintenance and even progression of the disease. As such, it seems that stromal cells have great potential as therapeutic targets in the setting of pulmonary hypertension. Examples of success with therapies targeting stromal cells or the stromal environment can be found in the cancer and chronic inflammatory disease literature. In fact, the concept of tumor stromal "normalization" has now become an accepted aspect of oncologic therapies [22, 25]. Clinical studies of angiogenesis inhibitors and antibodies against ECM components, such as tenascin, have been favorable, and inhibitors of MMPs, overexpression of tissue inhibitor of metalloproteinases (TIMPs), and blockade of integrin signaling have all shown promises in preclinical trials [22, 25].

It should be expected that successful treatments will need to consider cell–cell interactions within the stromal microenvironment. Studies examining the interactions between vascular ECs and their associated pericytes or mesenchymal support cells emphasize the importance of targeting the stroma as a whole. In response to many stresses, ECs have been shown to release platelet-derived growth factor (PDGF), which induces VEGF production from pericytes, leading to bidirectional "conversations" between the two cell types [106]. Interrupting this communication by using PDGF inhibitors proved to be more effective therapy than using VEGF inhibitors alone. Interestingly, although VEGF inhibitors lose their inhibitory effects in late stage tumors, targeting the pericytes helped even late-stage tumors to regress [24, 106]. It has also been demonstrated that at least some of the perivascular progenitor cells in tumors are recruited from the bone marrow, findings similar to those in neovascularized pulmonary artery adventitia [18, 35, 71, 97, 107]. These findings emphasize the importance of knowing that some stromal cells in pathologic process may be derived from nonresident (including bone marrow) or nontraditional (EnMT and EMT) sources. It is becoming increasingly clear that the functional phenotype of mesenchymal cells derived from the circulation or through the process of EnMT is different than that of resident stromal cells. These "nonresident" (circulating or EnMT-derived) cells exhibit different transcriptomes than resident cells [35, 48, 106, 107]. They secrete angiogenic factors and growth factors that influence behavior of resident cells. Importantly, the growth and differentiation of these cells are controlled by signaling pathways distinct from those used by normal resident stromal cells suggesting they will have different responses to therapeutic interventions.

The importance of specific stromal cells in perpetuating inflammatory and fibrotic conditions also raises the possibility that targeted depletion of these cells may open an alternative and exciting new avenue for research and ultimately therapeutic approaches. Abrogating the recruitment of nonresident fibroblasts to the vessel wall may be important in regulating the degree of vascular remodeling that occurs in severe pulmonary hypertension. However, it is important to recognize that recruitment of nonresident leukocytic cells may not always be pathologic but could be important for resolution of injury processes. For instance, Duffield *et al.* have performed studies which conclusively demonstrate that the timing of monocyte/macrophage depletion during a model of liver fibrosis determined whether fibrosis persisted or resolved [108]. Thus, timing of any therapies, directed at deletion of specific cell phenotypes, must be carefully considered. Preventing or reversing EnMT also appears to be of potential benefit and deserves study in animal models. Lastly, because the stromal environment is complex in chronic inflammatory and fibrotic disease states, it appears that multipotent agents (i.e., agents targeting several signaling pathways), or multidrug therapies, rather than a single agent targeting a single factor or a signaling pathway, will be the most effective approach to attenuate angiogenesis, inflammation, and fibrosis. A better understanding of stromal cell interactions will, hopefully soon, lead to an improved approach to the treatment of chronic pulmonary hypertension.

References

1. Stenmark, K.R., Fagan, K.A., and Frid, M.G. (2006) Hypoxia-induced pulmonary vascular remodeling: cellular and molecular mechanisms. *Circulation Research*, **99**, 675–91.
2. Voelkel, N.F. and Cool, C. (2004) Pathology of pulmonary hypertension. *Cardiology Clinics*, **22**, 343–51, v.
3. Humbert, M., Morrell, N.W., Archer, S.L. *et al.* (2004) Cellular and molecular pathobiology of pulmonary arterial hypertension. *Journal of the American College of Cardiology*, **43**, 13S–24S.

4. Denton, C.P., Humbert, M., Rubin, L., and Black, C.M. (2006) Bosentan treatment for pulmonary arterial hypertension related to connective tissue disease: a subgroup analysis of the pivotal clinical trials and their open-label extensions. *Annals of Rheumatic Diseases*, **65**, 1336–40.
5. Ito, T., Okada, T., Miyashita, H. *et al.* (2007) Interleukin-10 expression mediated by an adeno-associated virus vector prevents monocrotaline-induced pulmonary arterial hypertension in rats. *Circulation Research*, **101**, 734–41.
6. Tuder, R.M., Marecki, J.C., Richter, A. *et al.* (2007) Pathology of pulmonary hypertension. *Clinics in Chest Medicine*, **28**, 23–42, vii.
7. Voelkel, N.F., Cool, C., Lee, S.D. *et al.* (1998) Primary pulmonary hypertension between inflammation and cancer. *Chest*, **114**, 225S–230S.
8. Wallace, W.A., Fitch, P.M., Simpson, A.J., and Howie, S.E. (2007) Inflammation-associated remodelling and fibrosis in the lung – a process and an end point. *International Journal of Experimental Pathology*, **88**, 103–10.
9. Wynn, T. (2008) Cellular and molecular mechanisms of fibrosis. *The Journal of Pathology*, **214**, 199–210.
10. Nathan, S., Noble, P., and Tuder, R. (2007) Idiopathic pulmonary fibrosis and pulmonary hypertension: connecting the dots. *American Journal of Respiratory and Critical Care Medicine*, **175**, 875–80.
11. Patel, N.M., Lederer, D.J., Borczuk, A.C., and Kawut, S.M. (2007) Pulmonary hypertension in idiopathic pulmonary fibrosis. *Chest*, **132**, 998–1006.
12. Ryu, J., Krowka, M., Pellikka, P. *et al.* (2007) Pulmonary hypertension in patients with interstitial lung diseases. *Mayo Clinic Proceedings*, **82**, 342–50.
13. Barger, A.C., Beeuwkes, R. III, Lainey, L.L., and Silverman, K.J. (1984) Hypothesis: vasa vasorum and neovascularization of human coronary arteries. A possible role in the pathophysiology of atherosclerosis. *The New England Journal of Medicine*, **310**, 175–77.
14. Hartlapp, I., Abe, R., Saeed, R.W. *et al.* (2001) Fibrocytes induce an angiogenic phenotype in cultured endothelial cells and promote angiogenesis *in vivo*. *The FASEB Journal*, **15**, 2215–24.
15. Hayden, M.R. and Tyagi, S.C. (2004) Vasa vasorum in plaque angiogenesis, metabolic syndrome, type 2 diabetes mellitus, and atheroscleropathy: a malignant transformation. *Cardiovascular Diabetology*, **3**, 1.
16. Herrmann, J., Best, P.J., Ritman, E.L. *et al.* (2002) Chronic endothelin receptor antagonism prevents coronary vasa vasorum neovascularization in experimental hypercholesterolemia. *Journal of the American College of Cardiology*, **39**, 1555–61.
17. Moulton, K.S., Vakili, K., Zurakowski, D. *et al.* (2003) Inhibition of plaque neovascularization reduces macrophage accumulation and progression of advanced atherosclerosis. *Proceedings of the National Academy of Sciences of the United States of America*, **100**, 4736–41.
18. Davie, N.J., Crossno, J.T. Jr., Frid, M.G. *et al.* (2004) Hypoxia-induced pulmonary artery adventitial remodeling and neovascularization: contribution of progenitor cells. *American Journal of Physiology: Lung Cellular and Molecular Physiology*, **286**, L668–78.
19. Davie, N.J., Gerasimovskaya, E.V., Hofmeister, S.E. *et al.* (2006) Pulmonary artery adventitial fibroblasts cooperate with vasa vasorum endothelial cells to regulate vasa vasorum neovascularization: a process mediated by hypoxia and endothelin-1. *American Journal of Pathology*, **168**, 1793–807.
20. Mitzner, W. and Wagner, E.M. (2004) Vascular remodeling in the circulations of the lung. *Journal of Applied Physiology*, **97**, 1999–2004.
21. Kimura, H., Okada, O., Tanabe, N. *et al.* (2001) Plasma monocyte chemoattractant protein-1 and pulmonary vascular resistance in chronic thromboembolic pulmonary hypertension. *American Journal of Respiratory and Critical Care Medicine*, **164**, 319–24.
22. Bhowmick, N.A., Neilson, E.G., and Moses, H.L. (2004) Stromal fibroblasts in cancer initiation and progression. *Nature*, **432**, 332–37.
23. Egeblad, M., Littlepage, L.E., and Werb, Z. (2005) The fibroblastic coconspirator in cancer progression. *Cold Spring Harbor Symposia on Quantitative Biology*, **70**, 383–88.
24. Kalluri, R. and Zeisberg, M. (2006) Fibroblasts in cancer. *Nature Reviews: Cancer*, **6**, 392–401.
25. Mueller, M.M. and Fusenig, N.E. (2004) Friends or foes – bipolar effects of the tumour stroma in cancer. *Nature Reviews: Cancer*, **4**, 839–49.
26. Radisky, E. and Radisky, D. (2007) Stromal induction of breast cancer: inflammation and invasion. *Reviews in Endocrine and Metabolic Disorders*, **8**, 279–87.
27. Bellini, A. and Mattoli, S. (2007) The role of the fibrocyte, a bone marrow-derived mesenchymal progenitor, in reactive and reparative fibroses. *Laboratory Investigation*, **87**, 858–70.
28. Lama, V.N. and Phan, S.H. (2006) The extrapulmonary origin of fibroblasts: stem/progenitor cells

and beyond. *Proceedings of the American Thoracic Society*, **3**, 373–76.

29. Arciniegas, E., Frid, M., Douglas, I., and Stenmark, K. (2007) Perspectives on endothelial-to-mesenchymal transition: potential contribution to vascular remodeling in chronic pulmonary hypertension. *American Journal of Physiology: Lung Cellular and Molecular Physiology*, **293**, L1–8.
30. Filer, A., Pitzalis, C., and Buckley, C.D. (2006) Targeting the stromal microenvironment in chronic inflammation. *Current Opinion in Pharmacology*, **6**, 393–400.
31. Filer, A., Raza, K., Salmon, M., and Buckley, C.D. (2008) The role of chemokines in leucocyte–stromal interactions in rheumatoid arthritis. *Frontiers in Bioscience*, **13**, 2674–85.
32. Parsonage, G., Filer, A.D., Haworth, O. *et al.* (2005) A stromal address code defined by fibroblasts. *Trends in Immunology*, **26**, 150–56.
33. Stevens, T., Phan, S., Frid, M. *et al.* (2008) Lung vascular cell heterogeneity: endothelium, smooth muscle and fibroblasts. *Proceedings of the American Thoracic Society*, **5**, 783–91.
34. Stenmark, K.R., Davie, N., frid, M. *et al.* (2006) Role of the adventitia in pulmonary vascular remodeling. *Physiology (Bethesda)*, **21**, 134–45.
35. Chang, H.Y., Chi, J.T., Dudoit, S. *et al.* (2002) Diversity, topographic differentiation, and positional memory in human fibroblasts. *Proceedings of the National Academy of Sciences of the United States of America*, **99**, 12877–82.
36. Kimura, J. and Deutsch, G.H. (2007) Key mechanisms of early lung development. *Pediatric and Developmental Pathology*, **10**, 335–47.
37. Firestein, G.S. (1996) Invasive fibroblast-like synoviocytes in rheumatoid arthritis. Passive responders or transformed aggressors? *Arthritis and Rheumatism*, **39**, 1781–90.
38. Muller-Ladner, U., Kriegsmann, J., Franklin, B.N. *et al.* (1996) Synovial fibroblasts of patients with rheumatoid arthritis attach to and invade normal human cartilage when engrafted into SCID mice. *American Journal of Pathology*, **149**, 1607–15.
39. Ritchlin, C. (2000) Fibroblast biology. Effector signals released by the synovial fibroblast in arthritis. *Arthritis Research*, **2**, 356–60.
40. Pilling, D., Akbar, A.N., Girdlestone, J. *et al.* (1999) Interferon-beta mediates stromal cell rescue of T cells from apoptosis. *European Journal of Immunology*, **29**, 1041–50.
41. Edwards, J.C. (2000) Fibroblast biology. Development and differentiation of synovial fibroblasts in arthritis. *Arthritis Research*, **2**, 344–47.
42. Olumi, A.F., Grossfeld, G.D., Hayward, S.W. *et al.* (1999) Carcinoma-associated fibroblasts direct tumor progression of initiated human prostatic epithelium. *Cancer Research*, **59**, 5002–11.
43. Ambartsumian, N.S., Grigorian, M.S., Larsen, I.F. *et al.* (1996) Metastasis of mammary carcinomas in GRS/A hybrid mice transgenic for the *mts1* gene. *Oncogene*, **13**, 1621–30.
44. Helfman, D.M., Kim, E.J., Lukanidin, E., and Grigorian, M. (2005) The metastasis associated protein S100A4: role in tumour progression and metastasis. *British Journal of Cancer*, **92**, 1955–58.
45. Semov, A., Moreno, M.J., Onichtchenko, A. *et al.* (2005) Metastasis-associated protein S100A4 induces angiogenesis through interaction with Annexin II and accelerated plasmin formation. *The Journal of Biological Chemistry*, **280**, 20833–41.
46. Schmidt-Hansen, B., Ornas, D., Grigorian, M. *et al.* (2004) Extracellular S100A4 (mts1) stimulates invasive growth of mouse endothelial cells and modulates MMP-13 matrix metalloproteinase activity. *Oncogene*, **23**, 5487–95.
47. Hlatky, L., Tsionou, C., Hahnfeldt, P., and Coleman, C.N. (1994) Mammary fibroblasts may influence breast tumor angiogenesis via hypoxia-induced vascular endothelial growth factor up-regulation and protein expression. *Cancer Research*, **54**, 6083–86.
48. Popovici, R.M., Irwin, J.C., Giaccia, A.J., and Giudice, L.C. (1999) Hypoxia and cAMP stimulate vascular endothelial growth factor (VEGF) in human endometrial stromal cells: potential relevance to menstruation and endometrial regeneration. *The Journal of Clinical Endocrinology and Metabolism*, **84**, 2245–48.
49. Sharkey, A.M., Day, K., McPherson, A. *et al.* (2000) Vascular endothelial growth factor expression in human endometrium is regulated by hypoxia. *The Journal of Clinical Endocrinology and Metabolism*, **85**, 402–9.
50. Smith, P.J., Teichert-Kuliszewska, K., Monge, J.C., and Stewart, D.J. (1998) Regulation of endothelin-B receptor mRNA expression in human endothelial cells by cytokines and growth factors. *Journal of Cardiovascular Pharmacology*, **31** (Suppl 1), S158–60.
51. Wilasrusmee, C., Ondocin, P., Bruch, D. *et al.* (2003) Amelioration of cyclosporin A effect on microvasculature by endothelin inhibitor. *Surgery*, **134**, 384–89.
52. Martin, T.A., Harding, K., and Jiang, W.G. (2001) Matrix-bound fibroblasts regulate angiogenesis by

modulation of VE-cadherin. *European Journal of Clinical Investigation*, **31**, 931–38.

53. Nakatsu, M.N., Sainson, R.C., Aoto, J.N. *et al.* (2003) Angiogenic sprouting and capillary lumen formation modeled by human umbilical vein endothelial cells (HUVEC) in fibrin gels: the role of fibroblasts and angiopoietin-1. *Microvascular Research*, **66**, 102–12.
54. Velazquez, O.C., Snyder, R., Liu, Z.J. *et al.* (2002) Fibroblast-dependent differentiation of human microvascular endothelial cells into capillary-like 3-dimensional networks. *The FASEB Journal*, **16**, 1316–18.
55. Kroon, M.E., Koolwijk, P., van der Vecht, B., and van Hinsbergh, V.W. (2001) Hypoxia in combination with FGF-2 induces tube formation by human microvascular endothelial cells in a fibrin matrix: involvement of at least two signal transduction pathways. *Journal of Cell Science*, **114**, 825–33.
56. Martin, T.A., Harding, K.G., and Jiang, W.G. (1999) Regulation of angiogenesis and endothelial cell motility by matrix-bound fibroblasts. *Angiogenesis*, **3**, 69–76.
57. Tille, J.C. and Pepper, M.S. (2002) Mesenchymal cells potentiate vascular endothelial growth factor-induced angiogenesis *in vitro*. *Experimental Cell Research*, **280**, 179–91.
58. Aird, W.C. (2007) Phenotypic heterogeneity of the endothelium: II. Representative vascular beds. *Circulation Research*, **100**, 174–90.
59. Chu, H.W., Kraft, M., Rex, M.D., and Martin, R.J. (2001) Evaluation of blood vessels and edema in the airways of asthma patients: regulation with clarithromycin treatment. *Chest*, **120**, 416–22.
60. Li, X. and Wilson, J.W. (1997) Increased vascularity of the bronchial mucosa in mild asthma. *American Journal of Respiratory and Critical Care Medicine*, **156**, 229–33.
61. Endrys, J., Hayat, N., and Cherian, G. (1997) Comparison of bronchopulmonary collaterals and collateral blood flow in patients with chronic thromboembolic and primary pulmonary hypertension. *Heart*, **78**, 171–76.
62. Srisuma, S., Biswal, S.S., Mitzner, W.A. *et al.* (2003) Identification of genes promoting angiogenesis in mouse lung by transcriptional profiling. *American Journal of Respiratory Cell and Molecular Biology*, **29**, 172–79.
63. Keane, M.P., Belperio, J.A., Burdick, M.D. *et al.* (2001) ENA-78 is an important angiogenic factor in idiopathic pulmonary fibrosis. *American Journal of Respiratory and Critical Care Medicine*, **164**, 2239–42.
64. Strieter, R.M., Gomperts, B.N., and Keane, M.P. (2007) The role of CXC chemokines in pulmonary fibrosis. *The Journal of Clinical Investigation*, **117**, 549–56.
65. Gerasimovskaya, E.V., Woodward, H.N., Tucker, D.A. *et al.* (2007) Extracellular ATP a pro-angiogenic factor for pulmonary artery vasa vasorum endothelial cells. *Angiogenesis*, **11**, 169–82.
66. Komano, Y., Nanki, T., Hayashida, K. *et al.* (2006) Identification of a human peripheral blood monocyte subset that differentiates into osteoclasts. *Arthritis Research and Therapy*, **8**, R152.
67. Kuwana, M., Okazaki, Y., Kodama, H. *et al.* (2003) Human circulating $CD14^+$ monocytes as a source of progenitors that exhibit mesenchymal cell differentiation. *Journal of Leukocyte Biology*, **74**, 833–45.
68. Quan, T.E., Cowper, S.E., and Bucala, R. (2006) The role of circulating fibrocytes in fibrosis. *Current Rheumatology Reports*, **8**, 145–50.
69. Vannella, K.M., McMillan, T.R., Charbeneau, R.P. *et al.* (2007) Cysteinyl leukotrienes are autocrine and paracrine regulators of fibrocyte function. *Journal of Immunology*, **179**, 7883–90.
70. Phillips, R.J., Burdick, M.D., Hong, K. *et al.* (2004) Circulating fibrocytes traffic to the lungs in response to CXCL12 and mediate fibrosis. *The Journal of Clinical Investigation*, **114**, 438–46.
71. Frid, M.G., Brunetti, J.A., Burke, D.L. *et al.* (2006) Hypoxia-induced pulmonary vascular remodeling requires recruitment of circulating mesenchymal precursors of a monocyte/macrophage lineage. *American Journal of Pathology*, **168**, 659–69.
72. Karasek, M.A. (2007) Does transformation of microvascular endothelial cells into myofibroblasts play a key role in the etiology and pathology of fibrotic disease? *Medical Hypotheses*, **68**, 650–55.
73. Chaudhuri, V., Zhou, L., and Karasek, M. (2007) Inflammatory cytokines induce the transformation of human dermal microvascular endothelial cells into myofibroblasts: a potential role in skin fibrogenesis. *Journal of Cutaneous Pathology*, **34**, 146–53.
74. Armstrong, E.J. and Bischoff, J. (2004) Heart valve development: endothelial cell signaling and differentiation. *Circulation Research*, **95**, 459–70.
75. Eisenberg, L.M. and Markwald, R.R. (1995) Molecular regulation of atrioventricular valvuloseptal morphogenesis. *Circulation Research*, **77**, 1–6.
76. Liebner, S., Cattelino, A., Gallini, R. *et al.* (2004) Beta-catenin is required for

endothelial–mesenchymal transformation during heart cushion development in the mouse. *Journal of Cell Biology*, **166**, 359–67.

77. Arciniegas, E., Neves, C.Y., Carrillo, L.M. *et al.* (2005) Endothelial–mesenchymal transition occurs during embryonic pulmonary artery development. *Endothelium*, **12**, 193–200.
78. Arciniegas, E., Servin, M., Arguello, C., and Mota, M. (1989) Development of the aorta in the chick embryo: structural and ultrastructural study. *Atherosclerosis*, **76**, 219–35.
79. DeRuiter, M.C., Poelmann, R.E., Vanmunsteren, J.C. *et al.* (1997) Embryonic endothelial cells transdifferentiate into mesenchymal cells expressing smooth muscle actins *in vivo* and *in vitro*. *Circulation Research*, **80**, 444–51.
80. Arciniegas, E., Ponce, L., Hartt, Y. *et al.* (2000) Intimal thickening involves transdifferentiation of embryonic endothelial cells. *The Anatomical Record*, **258**, 47–57.
81. Hall, S.M., Hislop, A.A., and Haworth, S.G. (2002) Origin, differentiation, and maturation of human pulmonary veins. *American Journal of Respiratory Cell and Molecular Biology*, **26**, 333–40.
82. Hall, S.M., Hislop, A.A., Pierce, C.M., and Haworth, S.G. (2000) Prenatal origins of human intrapulmonary arteries: formation and smooth muscle maturation. *American Journal of Respiratory Cell and Molecular Biology*, **23**, 194–203.
83. Sarkisov, D.S., Kolokol'chikova, E.G., Kaem, R.I., and Pal'tsyn, A.A. (1988) Vascular changes in maturing granulation tissue. *Biulleten' Eksperimental'noi Biologii i Meditsiny*, **105**, 501–3.
84. Lipton, B.H., Bensch, K.G., and Karasek, M.A. (1992) Histamine-modulated transdifferentiation of dermal microvascular endothelial cells. *Experimental Cell Research*, **199**, 279–91.
85. Romero, L.I., Zhang, D.N., Herron, G.S., and Karasek, M.A. (1997) Interleukin-1 induces major phenotypic changes in human skin microvascular endothelial cells. *Journal of Cellular Physiology*, **173**, 84–92.
86. Beranek, J.T. (1995) Vascular endothelium-derived cells containing smooth muscle actin are present in restenosis. *Laboratory Investigation*, **72**, 771.
87. Beranek, J.T. and Cavarocchi, N.C. (1990) Undifferentiated vascular endothelial cells in coronary allograft atherosclerosis. *International Journal of Cardiology*, **28**, 127–28.
88. Mironov, Y., Hill, C., Starcher, B. *et al.* (1995) Endothelial–mesenchymal transformation in atherogenesis: a recapitulation of embryonic heart cushion tissue morphogenesis. *Annals of Biomedical Engineering*, **23**, S29A.
89. Zhu, P., Huang, L., Ge, X. *et al.* (2006) Transdifferentiation of pulmonary arteriolar endothelial cells into smooth muscle-like cells regulated by myocardin involved in hypoxia-induced pulmonary vascular remodelling. *International Journal of Experimental Pathology*, **87**, 463–74.
90. Arciniegas, E., Parada, D., Graterol, A., and Rec, A. (2003) Mechanically altered embryonic chicken endothelial cells change their phenotype to an epithelioid phenotype. *The Anatomical Record, Part A: Discoveries in Molecular, Cellular, and Evolutionary Biology*, **270**, 67–81.
91. Frid, M.G., Kale, V.A., and Stenmark, K.R. (2002) Mature vascular endothelium can give rise to smooth muscle cells via endothelial–mesenchymal transdifferentiation: *in vitro* analysis. *Circulation Research*, **90**, 1189–96.
92. Ishisaki, A., Hayashi, H., Li, A.J., and Imamura, T. (2003) Human umbilical vein endothelium-derived cells retain potential to differentiate into smooth muscle-like cells. *The Journal of Biological Chemistry*, **278**, 1303–9.
93. Arciniegas, E., Sutton, A.B., Allen, T.D., and Schor, A.M. (1992) Transforming growth factor beta 1 promotes the differentiation of endothelial cells into smooth muscle-like cells *in vitro*. *Journal of Cell Science*, **103**, 521–29.
94. Kohler, A., Jostarndt-Fogen, K., Rottner, K. *et al.* (1999) Intima-like smooth muscle cells: developmental link between endothelium and media? *Anatomy and Embryology (Berlin)*, **200**, 313–23.
95. Hewett, P.W. and Murray, J.C. (1993) Immunomagnetic purification of human microvessel endothelial cells using Dynabeads coated with monoclonal antibodies to PECAM-1. *European Journal of Cell Biology*, **62**, 451–54.
96. Springhorn, J.P., Madri, J.A., and Squinto, S.P. (1995) Human capillary endothelial cells from abdominal wall adipose tissue: isolation using an anti-PECAM antibody. *In Vitro Cellular and Developmental Biology, Animal*, **31**, 473–81.
97. Voyta, J.C., Via, D.P., Butterfield, C.E., and Zetter, B.R. (1984) Identification and isolation of endothelial cells based on their increased uptake of acetylated-low density lipoprotein. *Journal of Cell Biology*, **99**, 2034–40.
98. Dahl, U., Sjodin, A., Larue, L. *et al.* (2002) Genetic dissection of cadherin function during nephrogenesis. *Molecular and Cellular Biology*, **22**, 1474–87.

99. Hay, E.D. (2005) The mesenchymal cell, its role in the embryo, and the remarkable signaling mechanisms that create it. *Developmental Dynamics*, **233**, 706–20.

100. Okada, H. and Kalluri, R. (2005) Cellular and molecular pathways that lead to progression and regression of renal fibrogenesis. *Current Molecular Medicine*, **5**, 467–74.

101. Okada, H. and Kalluri, R. (2005) Recapitulation of kidney development paradigms by BMP-7 reverses chronic renal injury. *Clinical and Experimental Nephrology*, **9**, 100–1.

102. Peinado, H., Portillo, F., and Cano, A. (2004) Transcriptional regulation of cadherins during development and carcinogenesis. *The International Journal of Developmental Biology*, **48**, 365–75.

103. Lee, J.M., Dedhar, S., Kalluri, R., and Thompson, E.W. (2006) The epithelial–mesenchymal transition: new insights in signaling, development, and disease. *The Journal of Cell Biology*, **172**, 973–81.

104. Zeisberg, M., Hanai, J., Sugimoto, H. *et al.* (2003) BMP-7 counteracts TGF-beta1-induced epithelial-to-mesenchymal transition and reverses chronic renal injury. *Nature Medicine*, **9**, 964–68.

105. Zeisberg, E.M., Tarnavski, O., Zeisberg, M. *et al.* (2007) Endothelial-to-mesenchymal transition contributes to cardiac fibrosis. *Nature Medicine*, **13**, 952–61.

106. Bergers, G., Song, S., Meyer-Morse, N. *et al.* (2003) Benefits of targeting both pericytes and endothelial cells in the tumor vasculature with kinase inhibitors. *The Journal of Clinical Investigation*, **111**, 1287–95.

107. Song, S., Ewald, A.J., Stallcup, W. *et al.* (2005) PDGFRbeta$^+$ perivascular progenitor cells in tumours regulate pericyte differentiation and vascular survival. *Nature Cell Biology*, **7**, 870–79.

108. Duffield, J.S., Forbes, S.J., Constandinou, C.M. *et al.* (2005) Selective depletion of macrophages reveals distinct, opposing roles during liver injury and repair. *The Journal of Clinical Investigation*, **115**, 56–65.

12

Pulmonary Endothelium and Vasomotor Control

Nikki L. Jernigan, Benjimen R. Walker and Thomas C. Resta

Department of Cell Biology and Physiology, Vascular Physiology Group, University of New Mexico Health Sciences Center, Albuquerque, NM, USA

INTRODUCTION

The endothelium from each vascular bed has unique structural and functional properties, giving rise to endothelial cell (EC) heterogeneity. An intriguing aspect of pulmonary vascular responses is the marked differences with respect to the systemic circulation. In this chapter, these differences will be highlighted as we discuss mechanisms by which the pulmonary endothelium regulates vascular tone. Topics to be addressed include the endothelium as the primary site for the generation, metabolism, and degradation of a variety of vasoactive factors. We will also examine the importance of ion channels and membrane potential in regulation of pulmonary vascular tone through their influence on the cytosolic free Ca^{2+} concentration ($[Ca^{2+}]_i$), which subsequently drives the production and release of many vasoactive substances. Finally, we will discuss the physiological and pathophysiological significance of the endothelium in controlling pulmonary vascular tone.

The pulmonary circulation is a low-pressure, low-resistance vascular bed with little or no resting tone under normoxic conditions. One of the most fundamental physiological stimuli for pulmonary vasoconstriction is alveolar hypoxia (see Chapter 18). Although the systemic and pulmonary circulations are both reactive to hypoxia, they respond in opposing manners. Hypoxia in the systemic circulation results in vasodilation, which increases oxygen delivery to hypoxic tissues. This vasodilation is largely mediated by endothelium-derived nitric oxide (NO), prostacyclin, and cytochrome P450 products depending on the vascular bed and level of hypoxia [1, 2], although direct vasorelaxant effects of hypoxia on systemic vascular smooth muscle also likely contribute to the response. In contrast, hypoxia elicits pulmonary vasoconstriction – an important physiological response that diverts blood flow to better ventilated areas of the lung. Hypoxia acts predominantly on pulmonary vascular smooth muscle cells (SMCs) to induce contraction. However, the resulting vasoconstrictor response can be modified by release of endothelium-derived vasoactive factors, such as NO, prostacyclin, and endothelin (ET)-1 [3, 4].

Endothelial dysfunction and/or an imbalance between production of vasoconstrictors and vasodilators are common in pulmonary hypertension and may contribute to elevated vascular tone. In addition, although reactive oxygen species (ROS) are essential physiological signaling molecules that mediate basic cellular function [5, 6], endothelial dysfunction occurring in pulmonary hypertension may be further complicated by increased production of ROS [7, 8]. We will address these mechanisms of altered endothelial function and their associated consequences on vascular tone in greater detail below.

GENERATION, METABOLISM, AND DEGRADATION OF VASOACTIVE SUBSTANCES

Although gas exchange is widely considered to be the primary function of the pulmonary circulation, the vast surface area provided by the pulmonary endothelium also facilitates an important metabolic role of this tissue in the uptake, conversion, and removal of vasoactive substances from mixed venous blood (see Chapter 7). The pulmonary vasculature represents a crucial site of metabolic regulation since it filters the entire cardiac output before delivery to the systemic circulation. The pulmonary endothelium is involved in the production, storage, and

The Pulmonary Endothelium: Function in health and disease Editors Norbert F. Voelkel, Sharon Rounds

release of substances acting locally within the lung as well as systemically. In addition, several circulating factors are inactivated, altered, or removed from the blood by actions of pulmonary ECs. For example, prostaglandins (prostaglandin PGE_1, PGE_2, and $PGF_{2\alpha}$), leukotrienes (LTs), and serotonin are almost completely cleared from the blood following a single pass through the lung [9]. The pulmonary circulation is also an important site for both the production and clearance of ET-1. Approximately 50% of ET-1 is cleared from the circulation via ET_B receptor-mediated endocytosis as it passes through the lung [10]. Pulmonary vascular ECs also take up and degrade vasoactive amines such as norepinephrine and serotonin. In addition, bradykinin is largely inactivated in the pulmonary endothelium following degradation by angiotensin-converting enzyme (ACE). This inactivation occurs mainly in the pulmonary endothelium due to the abundant expression of ACE present in *all* alveolar capillary ECs compared to the systemic capillary ECs, where only about 10% of cells express ACE [11]. This predominant expression of ACE in the pulmonary endothelium is additionally responsible for approximately 70% of the conversion of angiotensin I to angiotensin II [9]. Thus, the pulmonary endothelium provides an extensive surface area for the metabolism and alteration of many vasoactive substances as they pass through the lung.

As in the systemic circulation, pulmonary ECs respond to a variety of endocrine, paracrine, and autocrine factors, which bind to receptors on their luminal surface and cause the release of several vasoconstrictors/dilators. Many factors, such as bradykinin, ET, histamine, serotonin, and ATP have dual effects on pulmonary vascular tone. Receptor activation by these substances can either result in vasoconstriction or vasodilation depending on the preexisting vascular tone, site(s) of receptors (endothelial or smooth muscle), and concentration of agonist. Moreover, certain factors such as arginine vasopressin produce endothelium-derived NO-mediated pulmonary vasodilation [12]. With the exception of the cerebral and coronary vasculature [13], this dilation is in contrast to the potent arginine vasopressin-mediated vasoconstriction that occurs in most systemic vascular beds. In addition to responding to a variety of humoral factors, the pulmonary endothelium can also release vasoactive mediators in response to mechanical stimuli such as fluid shear stress and pulsatile stretch of the vascular wall. These endothelium-derived factors typically diffuse to the underlying vascular SMCs and regulate both $[Ca^{2+}]_i$ and myofilament Ca^{2+} sensitivity to effect changes in vascular tone. Although endothelium-dependent vasodilatory influences predominate in the normal pulmonary circulation, endothelial dysfunction characteristic of lung disease creates an imbalance between the synthesis of relaxant and contractile factors (e.g., decreased NO and prostacyclin production, and increased synthesis of ET-1 and ROS), leading to vasoconstriction (Figure 12.1) and resultant pulmonary hypertension. The signaling pathways and regulation of the release of some of these vasoactive factors are discussed in greater detail below.

NO

Furchgott and Zawadzki [14] first demonstrated that the presence of ECs is essential for acetylcholine-induced relaxation of isolated rabbit aorta and initially named this endothelial mediator endothelium-derived relaxing factor (EDRF), which was subsequently identified as NO [15, 16] (see Chapter 6). NO has a relatively short half-life due to rapid conversion to peroxynitrite, nitrite, and then nitrate, but mediates several important physiological functions such as preventing platelet aggregation, cell adhesion, and cell growth. In addition, NO diffuses across the EC membrane and enters vascular smooth muscle where it elicits relaxation by activating soluble guanylyl cyclase (sGC), increasing cGMP levels, and stimulating protein kinase G [17]. Many humoral factors such as bradykinin, ATP, histamine, serotonin, ET-1, arginine vasopressin, and insulin have been reported to induce pulmonary vascular dilation through endothelium-derived NO release (Figure 12.2).

For many years, NO biosynthesis was thought to be relatively simple; that is, an elevation in EC $[Ca^{2+}]_i$ increases Ca^{2+}/calmodulin association which binds to and activates the enzyme endothelial NO synthase (endothelial nitric oxide synthase eNOS). eNOS utilizes L-arginine as a substrate and yields NO and L-citrulline as products [18]. However, in recent years a far more complex model of NO synthesis and eNOS regulation has emerged. Not only does eNOS require several cofactors for NO biosynthesis, including, but not limited to, NADPH, flavin mononucleotide, flavin adenine dinucleotide, tetrahydrobiopterin, O_2, and Ca^{2+}/calmodulin, but eNOS is also subject to multiple levels of post-translational regulation that include subcellular localization, numerous protein–protein interactions, and tightly regulated multisite phosphorylation involving several kinases and phosphatases [19]. Although the activity of eNOS is greatly enhanced with increased $[Ca^{2+}]_i$, it has recently been noted that eNOS can be activated by certain stimuli without an increase in $[Ca^{2+}]_i$ [20]. These new advances reveal alterative possibilities for the regulation of pulmonary vascular tone.

Many serine/threonine protein kinases and phosphatases participate in the regulation of eNOS activity by phosphorylation and dephosphorylation, respectively. To date, there are at least five known eNOS serine/threonine phosphorylation sites. Phosphorylation at these sites results in opposing effects on eNOS activity,

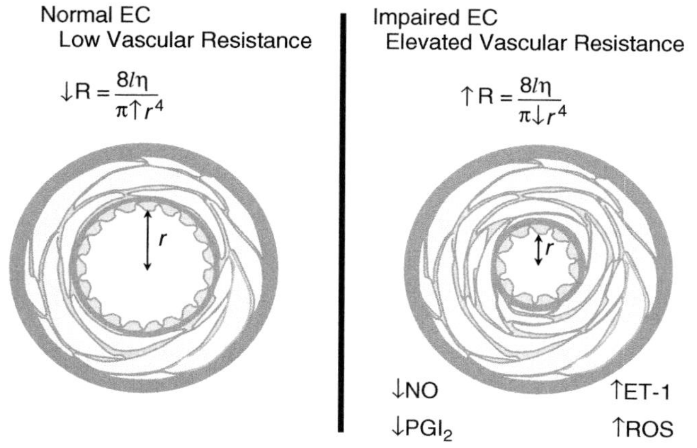

Figure 12.1 Regulation of pulmonary vascular tone by the endothelium. Under normal conditions pulmonary vascular resistance is low. Resistance to blood flow can be increased by vascular smooth muscle contraction and a consequent reduction in vessel radius. The relationship between vessel internal radius (r), blood viscosity (η), vessel length (l), and resistance (R) is approximated by Poiseuille's equation, which states that $R = (8l\eta)/(\pi r^4)$. Pulmonary hypertension is characterized by EC dysfunction, which results in elevated vascular resistance through decreased production of vasodilator products [NO and prostacyclin (PGI_2)] and increased generation of vasoconstrictors (ET-1 and ROS).

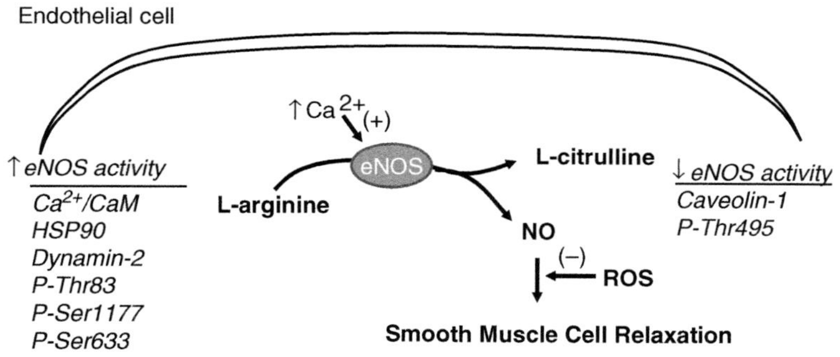

Figure 12.2 NO pathway. L-Arginine is converted to L-citrulline and NO by eNOS. eNOS activity is increased by Ca^{2+}/calmodulin, HSP90, dynamin-2, and phosphorylation of Thr83, Ser1177, or Ser633. In contrast, eNOS activity is diminished by association with caveolin-1 and phosphorylation of Thr495. Once formed, NO can be scavenged by ROS, thereby limiting NO-dependent vasodilation.

which is summarized in Figure 12.2. eNOS is activated by phosphorylation at Thr83, Ser1177, and Ser633 [19]. In contrast, phosphorylation at Thr495 inhibits eNOS activity. Interestingly, agonists that elicit NO synthesis, such as bradykinin, often have a dual effect resulting in simultaneous phosphorylation at Ser1177 and dephosphorylation at Thr495 [19]. The location of eNOS within the ECs is also an important point of regulation. Caveolae are cholesterol-rich, microdomain signaling platforms localized primarily to the cytoplasmic membrane of ECs. Caveolin-1, an integral constituent of caveolae, can associate with eNOS in caveolae and inhibit NO production. This eNOS/caveolin-1 association can be disrupted by binding of protein complexes to eNOS, such as Ca^{2+}/calmodulin, heat shock protein 90 (HSP90), and dynamin-2, leading to increased eNOS activity and NO production [21]. This highly regulated control over NO production plays an important role in modulating pulmonary vascular tone.

Normal and Pathophysiological Role of NO

Although, the normal pulmonary circulation has little inherent tone under normoxic conditions, the importance of NO in regulation of vascular tone is revealed under conditions of increased pulmonary vascular resistance, including that resulting from hypoxic pulmonary vasoconstriction (HPV). As oxygen is required for NO synthesis, hypoxia has been hypothesized to cause constriction by inhibiting basal NO synthesis. However, whether acute hypoxia reduces NO production is controversial [22, 23]. Whereas Hampl *et al.* [24] found that hypoxia increases $[Ca^{2+}]_i$ and augments NO synthesis in bovine pulmonary artery ECs (pulmonary artery endothelial cell PAECs), several other studies have reported opposite effects of acute hypoxia on NO synthesis in cultured pulmonary ECs [22]. The mechanisms whereby hypoxia decreases eNOS activity are not well defined, but may result from reduced O_2 substrate availability for eNOS [22] or alternatively to a reduction in cellular levels of HSP90 [25]. More recently, hypoxia has been reported to decrease NO synthesis in fetal PAECs by lowering $[Ca^{2+}]_i$ [26] or promoting the interaction of eNOS with caveolin-1 [27] – responses that may contribute to the maintenance of high pulmonary vascular resistance in the fetal pulmonary circulation. A similar discrepancy exists regarding effects of acute hypoxia on pulmonary NO production in whole lungs, intact animals and humans. For example, whereas ventilation of anesthetized piglets or isolated piglet lungs with hypoxic gas consistently decreases levels of both exhaled NO and plasma/perfusate NO reaction products (NO_x^-) [28, 29], studies in intact rabbits or perfused rabbit lungs have revealed either no change or a slight reduction in exhaled NO or perfusate NO_x^- production in response to acute hypoxia [22]. In contrast to these observations, preliminary data from our laboratory have demonstrated significant increases in exhaled NO in adult rats in response to ventilation with hypoxic gas (Figure 12.3). However, the relative contribution of NO from vascular versus airway sources is not clear from these studies. Furthermore, it remains to be defined whether this effect of hypoxia to increase exhaled NO is due to direct effects of hypoxia on eNOS activity, or rather to increases in shear stress resulting from pulmonary vasoconstriction or vascular recruitment. Nevertheless, apart from effects of hypoxia on NO production, it is clear that NO synthesis inhibition markedly augments HPV [30–32], thus demonstrating a critical role for NO to limit the degree of vasoconstriction.

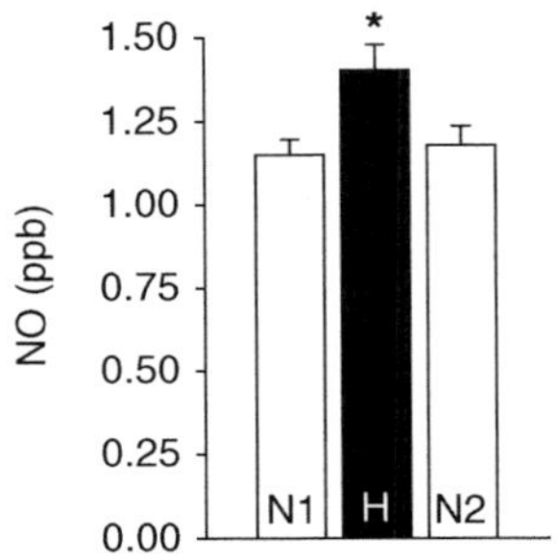

Figure 12.3 Acute hypoxia increases exhaled NO levels in anesthetized rats. Following anesthesia [ketamine (91 mg/kg; intramuscularly) and acepromazine (0.9 mg/kg; intramuscularly)], rats were intubated and ventilated with either a normoxic (N; 21% O_2, balance N_2) or hypoxic (H; 12% O_2, balance N_2) gas mixture at a frequency of 55 breaths/min and a tidal volume of 2.5 ml. A catheter was placed in the left femoral artery and advanced into the abdominal aorta for continuous measurements of mean arterial blood pressure (MABP) and heart rate (HR). Rats ($n = 6$) were initially ventilated with a normoxic gas mixture (N1) for 15 min (arterial $PO_2 = 86.5 \pm 3.1$ mmHg; $PCO_2 = 26.7 \pm 0.8$ mmHg; pH 7.48 $\pm$ 0.01). Exhaled air was collected in a Mylar balloon during the final 5 min of the 15-min interval and immediately analyzed for content of NO (N1; white bar) using an NO chemiluminescence analyzer (Sievers 270B). Rats were then switched to hypoxic ventilation (arterial $PO_2 = 34.8 \pm 1.4$* mmHg; $PCO_2 = 29.5 \pm 1.2$ mmHg; pH 7.43 $\pm$ 0.02) for 15 min with exhaled NO measured from the final 5 min (H; black bar). Animals were then returned to normoxic ventilation (arterial $PO_2 = 74.4 \pm 2.1$ mmHg; $PCO_2 = 29.6 \pm 1.3$ mmHg; pH 7.44 $\pm$ 0.02) for an additional 15 min before exhaled NO was measured in the last 5 min (N2; white bar). Neither MABP (in mmHg; N1 = 100 $\pm$ 6; H = 90 $\pm$ 6; N2 = 106 $\pm$ 5) nor HR (in beats/min; N1 = 372 $\pm$ 18; H = 379 $\pm$ 17; N2 = 409 $\pm$ 22) was significantly altered with hypoxic ventilation. Values are expressed as means $\pm$ standard error. $*p < 0.05$ compared to normoxic ventilation.

An important regulator of NO signaling is the generation of ROS (see Chapter 17). The vascular endothelium appears to have multiple potential sources of ROS production, including xanthine oxidase, NAD(P)H oxidases, cyclooxygenases (COXs), lipoxygenases, cytochrome P450 enzymes, uncoupled NOS, and the mitochondrial electron transport chain. Superoxide anion reacts with NO to produce the cytotoxic oxidant, peroxynitrite ($ONOO^-$), thereby decreasing NO bioavailability [33, 34]. Although ROS may be produced in response to several stimuli, hypoxia and/or reoxygenation following hypoxia appear to play an important role in ROS production [35–39]. Indeed, studies by our laboratory demonstrate that chronic hypoxia increases ROS levels in small pulmonary arteries [40] and that endothelium-derived ROS attenuate NO-dependent pulmonary vasodilation following chronic hypoxia-induced pulmonary hypertension [8]. Consequently, increases in ROS generation following long-term hypoxic exposure may contribute to the development of pulmonary hypertension by limiting the protective influences of endogenous NO.

The role of NO to modulate pulmonary vascular tone is evident not only with physiological stimuli, but is also apparent with lung diseases characterized by pulmonary hypertension. Although acute administration of eNOS inhibitors has little effect in normal lungs, it causes marked vasoconstriction in lungs from animals with chronic hypoxia-induced pulmonary hypertension [23]. In addition, Fagan *et al.* [41], found that targeted disruption of the eNOS gene is not sufficient to produce pulmonary hypertension in otherwise normal animals. However, this deletion results in greater pulmonary vascular remodeling and hypertension in response to chronic hypoxic exposure [41], suggesting that endogenous NO provides an important mechanism to reduce the degree of pulmonary hypertension.

Pulmonary arterial eNOS expression is upregulated following exposure to chronic hypoxia [42–46]. However, it is not clear if this increased expression corresponds to increased production of NO. Early studies demonstrated diminished vasodilatory responses to endothelium-derived NO-dependent agonists [47, 48]. Consistent with an upregulation of eNOS, however, several studies report that NO production is elevated [42, 46, 49, 50] and endothelium-derived NO-dependent vasodilation is enhanced following chronic hypoxia [44–46, 49, 51, 52]. However, it is important to note these measurements were taken in normoxic conditions. Sato *et al.* [53] provided evidence that basal NO production in lungs from chronically hypoxic rats is markedly diminished during hypoxic compared to normoxic ventilation, resulting in levels of NO synthesis similar to normoxic normotensive lungs. Although it appears that PO_2 can influence NO production, abnormal interactions between eNOS and the regulatory proteins caveolin-1 and HSP90 may also decrease NO production in pulmonary ECs following chronic hypoxia [54]. Despite the controversy over the level of NO production seen with pulmonary hypertension, NO remains a key mediator in reducing the severity of pulmonary hypertension [41].

Arachidonic Acid Metabolites

In resting cells, arachidonic acid (AA) is stored in the cell membrane as an esterified glycerol. AA can be released from these stores by receptor-dependent stimulation of the phospholipases PLA_2, PLC, or PLD. PLA_2 acts directly to catalyze the hydrolysis of phospholipids to release AA. In contrast, PLC and PLD generate the lipid products diacylglycerol (DAG) and phosphatidic acid, respectively, which can subsequently be converted to AA by lipases. Once formed, the free AA can then be metabolized by three major pathways: (i) COX, (ii) lipoxygenase, or (iii) cytochrome P450 pathways, which are summarized in Figure 12.4.

COX Pathway

The COX pathway results in the production of PGD_2, PGE_1, PGE_2, and $PGF_{2\alpha}$, thromboxane, and prostacyclin. Generally PGD_2, PGE_2, $PGF_{2\alpha}$, and thromboxane A_2 are pulmonary vasoconstrictors, whereas PGE_1 and prostacyclin are vasodilators. Although basal release of

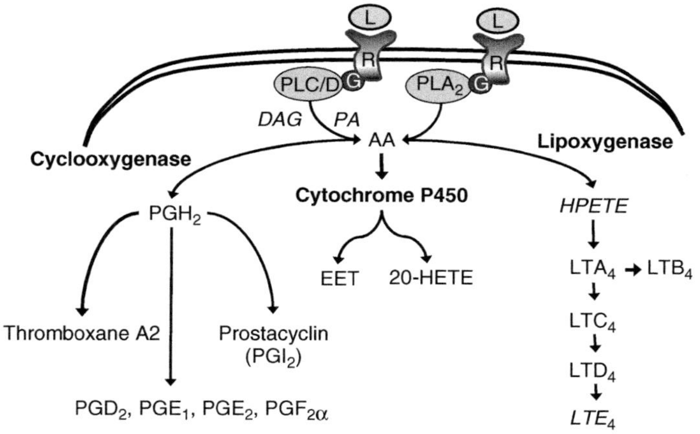

Figure 12.4 AA metabolism. AA can be released from phospholipids by stimulation of PLA_2 or generated from DAG and phosphatidic acid (PA) by PLC and PLD, respectively. Once formed, AA can be metabolized by three major pathways: (i) the COX pathway producing thromboxane, PGD_2, PGE_2, and $PGF_{2\alpha}$, and prostacyclin (PGI_2); (ii) the cytochrome P450 pathway producing EETs and 20-HETE; or (iii) the lipoxygenase pathway producing LTs (LTA_4, LTB_4, LTC_4, LTD_4, and LTE_4). L, ligand; R, receptor; G, G-protein; HPETE, hydroperoxyeicosatetraenoate.

these COX products appears to be quite low under normal conditions [55], their production increases in response to activation of PLs by a variety of stimuli via both Ca^{2+}-dependent and -independent pathways. Cogolludo *et al.* [56] suggest that thromboxane binds to $G_{q/11}$-protein-coupled receptors (TP receptors) and inhibits vascular smooth muscle K_v channels via activation of protein kinase Cζ. In addition, increased production of thromboxane has been associated with pulmonary hypertension [57]. In contrast, prostacyclin binds to G_s protein-coupled receptors on vascular SMCs leading to stimulation of adenylyl cyclase, which catalyzes the formation of cAMP and activates protein kinase A [55]. Prostacyclin is not only a potent vasodilator, but also an effective antiplatelet aggregatory agent [58].

Normal and Pathophysiological Role of Prostacyclin Prostacyclin levels are normally low, but substantially elevated during HPV [59]. Furthermore, inhibition of prostacyclin synthesis increases the pressor response to hypoxia [59, 60], suggesting prostacyclin acts as a functional antagonist of HPV. Although exogenous prostacyclin has little effect on normal pulmonary vascular reactivity [55], the synthesis of prostacyclin is considered a fundamental regulator of pulmonary vascular tone since decreased pulmonary prostacyclin synthesis has been implicated in the pathogenesis of severe pulmonary hypertension [61]. Indeed, one of the most effective treatments for severe pulmonary hypertension is inhalation or intravenous infusion of synthetic prostacyclin analogs [62].

Lipoxygenase Pathway

Products of the lipoxygenase pathway, LTs, and lipoxins, play an important role in mediating increases in vascular permeability that accompany asthma and pulmonary inflammation. However, they have also been shown to induce vasoconstriction mediated by specific LT receptors and through the release of COX products [63], leading to increased pulmonary arterial pressure [64]. 5-Lipoxygenase converts AA to LTA_4. In the presence of downstream enzymes, LTA_4 can be converted into LTB_4, a potent chemotactic agent, or into LTC_4, LTD_4, and LTE_4 (Figure 12.4), a group of smooth muscle contractile agonists [64]. In addition, lipid peroxide radicals generated by 5-lipoxygenase react with and inactivate NO, thereby impairing cGMP-mediated vasorelaxation and inhibiting platelet aggregation [65]. Although 5-lipoxygenase products have been suggested to play a role in HPV [64], later studies by the same group of investigators showed neither an effect of hypoxia on LT release from perfused rat lungs nor inhibition of HPV by lipoxygenase inhibitors [66]. Consequently, the definitive roles of LTs in regulation of pulmonary vascular tone remain unclear.

Cytochrome P450 Pathway

P450 metabolism of AA forms a series of fatty acid epoxides, epoxyeicosatrienoic acids (5,6- 8,9- 11,12- and 14,15-EETs) and hydroxyeicosatetraenoic acids (20-HETEs). The effects of these products on pulmonary vascular reactivity are discussed below.

EETs In the systemic circulation, EETs appear to activate large-conductance, Ca^{2+}-activated potassium channels (BK_{Ca}) in vascular SMCs, thereby promoting hyperpolarization of the resting membrane potential and causing vasorelaxation [67]. Indeed, EETs are considered primary candidates for the endothelium-derived hyperpolarizing factor (EDHF) (see below). Whereas endogenous EETs appear to mediate pulmonary vasodilation in isolated perfused lungs from control rats [68] and following monocrotaline-induced pulmonary hypertension [69], the contribution of EETs to regulation of pulmonary vascular reactivity under normal conditions and in the setting of pulmonary hypertension remains controversial. In addition, Zhu *et al.* [70] demonstrate that all EET isomers induce endothelium- and COX-dependent vasoconstriction in rabbit pulmonary arteries, suggesting EETs release a COX-dependent vasoconstrictor from the pulmonary endothelium, rather than acting as a vasodilator as in the systemic circulation.

20-HETE Elevated levels of 20-HETE contribute to the myogenic response of renal, cerebral, mesenteric, and skeletal muscle arterioles, and play a significant role in autoregulation of renal and cerebral blood flow [67]. 20-HETE produced in *systemic* vascular smooth muscle potently inhibits BK_{Ca} channels resulting in activation of L-type Ca^{2+} channels and vasoconstriction. In addition, there appears to be an important inhibitory effect of NO on the formation of 20-HETE in both renal and cerebral arteries [71–73]. In contrast to vasoconstrictor effects in the systemic circulation, 20-HETE induces pulmonary vasodilation [67]. This response is thought to occur through increased EC $[Ca^{2+}]_i$ and resultant activation of eNOS [74, 75]. Interestingly, whereas the enzyme responsible for 20-HETE synthesis (CYP4A) appears to be localized to vascular smooth muscle in many systemic beds, it is expressed in both vascular smooth muscle and endothelium in pulmonary arteries [76], suggesting a potential role for endothelial 20-HETE in pulmonary vasoregulation [74]. Although responses to EETs and 20-HETE in the pulmonary circulation are in direct contrast to the effects in the systemic circulation, the actions of these cytochrome P450 products may have important implications for the regulation of pulmonary vascular tone that remain unclear.

EDHF

Although both NO and prostacyclin may elicit vascular smooth muscle relaxation in part through membrane hyperpolarization, EDHF-type vasodilatory responses collectively represent mechanisms of endothelium-dependent smooth muscle hyperpolarization that are independent of these endothelial products. This hyperpolarization and subsequent smooth muscle relaxation are postulated to be mediated by increased K^+ conductance through activation of a number of K^+ channels, including voltage-gated (K_v), ATP-sensitive (K_{ATP}), and Ca^{2+}-actived K^+ (K_{Ca}) channels. Several diffusible mediators have been shown to act as EDHFs in different tissues and species and include K^+ ions, EETs, ROS (H_2O_2), AMP, and C-type natriuretic peptide [77]. In addition, heterocellular electrical coupling between endothelium and smooth muscle has been implicated in mediating EDHF activity in many vascular beds; in which case EDHF-type responses may simply be due to a conducted hyperpolarization from the endothelium to the smooth muscle, and not to the activity of a factor, *per se* (Figure 12.5). Such EDHF responses are mediated by an increase in endothelial $[Ca^{2+}]_i$ that results in activation of small-conductance endothelial K_{Ca} channel (SK_{Ca}) and intermediate-conductance endothelial K_{Ca} channel (IK_{Ca}), respectively, hyperpolarization of ECs, and transmittance of this hyperpolarization directly to the vascular smooth muscle through myoendothelial gap junctions [77, 79] (Figure 12.5). Although myoendothelial gap junctions have been characterized in many systemic vascular beds, their presence in the pulmonary circulation has not been directly assessed. Gap junction blockers attenuate bradykinin-induced vasodilatory responses in pulmonary arteries [79], suggesting the existence of myoendothelial coupling in the pulmonary circulation; however, only gap junctions between ECs have been observed microscopically [81, 82]. Alternatively, efflux of K^+ from endothelial SK_{Ca} and IK_{Ca} may result in locally elevated concentrations of K^+ (K^+ cloud) that activate inward rectifier K^+ (K_{ir}) channels on vascular SMCs to induce hyperpolarization (Figure 12.5).

Normal and Pathophysiological Role of EDHF

Although NO- and prostacyclin-independent vasodilation occurs in the pulmonary circulation [44, 68, 80, 83], controversy remains regarding the role of EDHF in this bed.

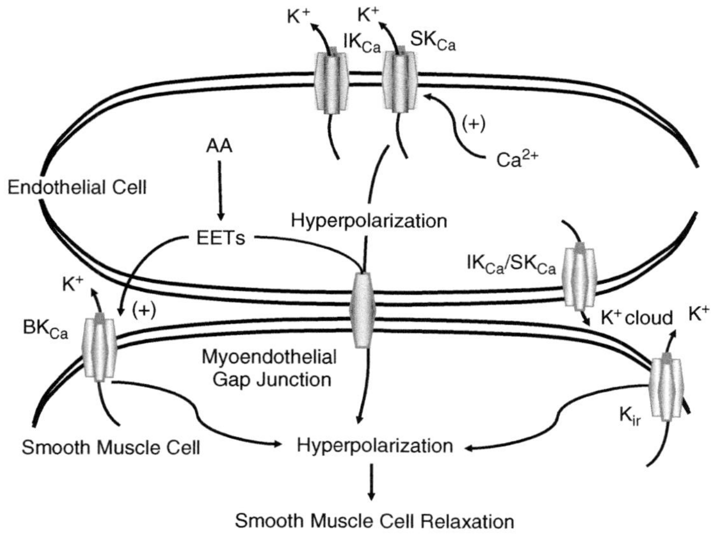

Figure 12.5 Potential EDHF signaling pathways in pulmonary ECs. EETs generated from AA can diffuse to the SMC, potentially through myoendothelial gap junctions, and stimulate large conductance Ca^{2+}-activated K^+ channels (BK_{Ca}) on the vascular smooth muscle. Alternatively increases in intracellular Ca^{2+} stimulate EC intermediate-conductance (IK_{Ca}) and small-conductance (SK_{Ca}) Ca^{2+}-activated K^+ channels, which results in EC hyperpolarization. This hyperpolarization can be transmitted to the SMCs through myoendothelial gap junctions. In addition, endothelial IK_{Ca} and SK_{Ca} in close proximity to inward rectifying K^+ channels (K_{ir}) may activate smooth muscle K_{ir} channels by generating a K^+ cloud.

Morio *et al.* [68] demonstrated that inhibition of EDHF responses (with K_{Ca} channel blockers) was without effect on basal tone and HPV in lungs isolated from either normotensive rats or from rats with chronic hypoxia-induced pulmonary hypertension. Interestingly, this same study identified a role for cytochrome P450 products in mediating thapsigargin-induced EDHF-mediated vasodilation in normotensive, but not hypertensive lungs. These findings are in contrast to a more recent study demonstrating that inhibition of K_{Ca} channels with a combination of charybdotoxin and apamin increased baseline perfusion pressure in monocrotaline-induced hypertensive, but not normotensive lungs, following both NOS and COX inhibition [69]. Interestingly, this response was associated with upregulated mRNA expression of SK_{Ca} and IK_{Ca} channels in lungs from rats with monocrotaline-induced hypertension. These findings support a role for endothelial SK_{Ca} and IK_{Ca} channels in mediating the EDHF response following monocrotaline-induced, but not chronic hypoxia-dependent pulmonary hypertension. Further investigation is needed to delineate the physiological role of EETs and conducted hyperpolarization in regulation of vascular tone in the normal pulmonary circulation, and the mechanism by which EDHF-type vasodilatory responses are differentially regulated in various forms of pulmonary hypertension.

ET-1

Reminiscent of the discovery of EDRF, Hickey *et al.* [83] found that ECs release an endothelium-derived constricting factor (EDCF) that was later identified as ET [84]. There are three isoforms of ET (ET-1, ET-2, and ET-3); however, ET-1 is the primary ET produced by ECs and is a potent endogenous vasoconstrictor peptide. ET-1 is synthesized from a relatively inactive precursor, big ET-1, which is then cleaved by ET-converting enzyme into functional ET-1. Acting through a paracrine or autocrine manner, ET-1 has complex vasoactive properties mediated by at least two distinct G-protein-coupled receptors, ET_A and ET_B. ET_A receptors are located on vascular SMCs and mediate vasoconstriction. ET_B receptors are found in both ECs (ET_{B1}) and SMCs (ET_{B2}) where they mediate endothelium-dependent vasorelaxation and vascular smooth muscle contraction, respectively. The vasodilator effect of ET-1 results from the release of NO and prostacyclin. In addition, the ET_B receptor in pulmonary vascular endothelium is an important regulator of ET-1 clearance [10, 86].

Normal and Pathophysiological Role of ET-1

It is well documented that ET-1 plays an important role in regulating pulmonary vascular tone, cell proliferation, mucus secretion, and EC permeability [55]. Inhibition of ET-1 does not significantly affect pulmonary arterial pressure under normal conditions [87, 88], but can decrease systemic pressure [88]. Indeed, ET-1 appears to be a more potent vasoconstrictor in the systemic circulation than in the pulmonary vascular bed [88]. ET-1 also plays a central role in HPV [4], Although the mechanism by which ET-1 contributes to HPV is not well defined, it is postulated that ET-1 primes the SMCs by either increasing myofilament Ca^{2+} sensitivity or by causing depolarization of SMCs, which increases the open probability of Ca^{2+} channels [89]. In addition, ET-1 may contribute to HPV by inhibition of K_{ATP} channels [90].

Several lines of evidence suggest that ET-1 is an important mediator of chronic hypoxia-induced pulmonary hypertension. Administration of ET-1 receptor antagonists prevents and reverses pulmonary hypertension in chronically hypoxic rats [92–94]. In addition, chronic hypoxia is associated with increases in ET-1, ET_A, and ET_B expression in the lung [95, 96] and augments ET-1-dependent pulmonary vasoconstriction [97–100] through a ROS-dependent mechanism [97]. The effects of both selective ET_A and dual ET_A/ET_B receptor inhibition on the development and treatment of pulmonary hypertension have been evaluated experimentally and clinically. Whereas selective ET_A receptor inhibition has been unambiguously shown to reduce pulmonary hypertension [92–94, 101], the effects of dual ET_A/ET_B receptor inhibition are more complex. Both ET_A/ET_B receptors mediate vasoconstriction in the pulmonary circulation with strong evidence that the two receptors form functional heterodimers [101]. Therefore, whereas combined ET_A/ET_B receptor blockade provides greater inhibition of ET-1-induced vasoconstriction than ET_A receptor inhibition alone [8], inhibition of endothelial ET_B receptors is likely to be disadvantageous considering their role in mediating endothelium-dependent vasodilation and ET-1 clearance. Indeed, Ivy *et al.* [102] reported marked elevations in plasma ET-1 levels and exacerbation of hypoxic pulmonary hypertension in ET_B receptor-deficient rats, supporting a protective role for endothelial ET_B receptors in limiting the severity of pulmonary hypertension. Further studies are needed to clearly define the relative importance of endothelial versus smooth muscle ET_B receptors in the development of pulmonary hypertension.

ENDOTHELIAL ION CHANNELS, MEMBRANE POTENTIAL, AND REGULATION OF CALCIUM ENTRY

Endothelial ion channels and membrane potential are important regulators of pulmonary vascular tone through influences on $[Ca^{2+}]_i$, which subsequently drive the production and release of many vasoactive substances (see

Chapter 5). Ca^{2+} entry is regulated by membrane permeability to Ca^{2+} and the driving force for Ca^{2+} across the membrane (i.e., the difference between membrane potential and the equilibrium potential for Ca^{2+}). Thus, Ca^{2+} entry is influenced by ion channels that modulate both membrane potential and permeability. ECs are conventionally considered "electrically nonexcitable," lacking the ability to rapidly and drastically change membrane potential; and until recently, were not thought to express voltage-gated ion channels. However, voltage-gated Na^+, K^+, and Ca^{2+} channels have all been detected in ECs and contribute to resting membrane potential [103]. Figure 12.6 provides an overview of Ca^{2+} entry pathways and some of the most important ion channels responsible for modulation of membrane potential.

Calcium Entry Pathways

ECs express several Ca^{2+}-permeable, nonselective cation channels. In addition, voltage-gated T-type Ca^{2+} channels have been recently identified in pulmonary microvascular endothelium and their activity may lead to increased $[Ca^{2+}]_i$. A reduction of the Na^+ gradient may provide alternative means to increase $[Ca^{2+}]_i$ by activating the reverse mode of the Na^+/Ca^{2+} exchanger . These Ca^{2+} entry mechanisms are controlled by various signaling cascades, and are important for the synthesis and release of vasoactive substances from the pulmonary endothelium.

Ca^{2+}-Permeable Nonselective Cation Channels

Ligand binding of vasoactive substances to G protein-coupled receptors activates the phosphoinositide (PI) pathway which results in Ca^{2+} entry by two primary mechanisms [104] (Figure 12.6). The downstream second messenger of the PI pathway, DAG, can directly activate Ca^{2+} permeable nonselective cation channels in the plasma membrane; these channels are referred to as receptor-operated channels (ROCs). Inositol 1,4,5-triphosphate (IP_3), also generated from activation of PI pathway, binds to receptors on the endoplasmic reticulum and triggers the release of Ca^{2+}, thus leaving the endoplasmic reticulum in a state of Ca^{2+} depletion. This depletion activates Ca^{2+} entry through store-operated channels (SOCs). In ECs, this Ca^{2+} entry does not appear to be affected by mitochondrial Ca^{2+} buffering, but is offset by the plasma membrane Ca^{2+} pump [105]. Evidence supports a role for canonical transient receptor potential (TRP) C family members to mediate both receptor- and store-operated Ca^{2+} entry. in ECs. Several TRP isoforms have been shown to be involved in EC Ca^{2+} homeostasis [106]. The strongest evidence for TRPC in regulating pulmonary EC Ca^{2+} comes from studies of TRPC4 knockout animals. Store-operated calcium entry is drastically reduced in lung microvascular ECs from TRPC4 knockout mice [107]. Further investigation is required to determine the precise makeup of ROCs and SOCs in pulmonary endothelium, and their role in regulation of endothelial $[Ca^{2+}]_i$ and vasoreactivity.

T-Type Voltage-Gated Ca^{2+} Channels

The expression of voltage-gated T-type Ca^{2+} channels ($Ca_v3.1$) has recently been identified in pulmonary microvascular ECs [108]. T-type Ca^{2+} channels activate rapidly upon slight depolarization of the plasma membrane and inactivate over a similar voltage range, resulting in a transient inward current during sustained membrane depolarization. Since these channels have only recently been described in ECs, our current knowledge and

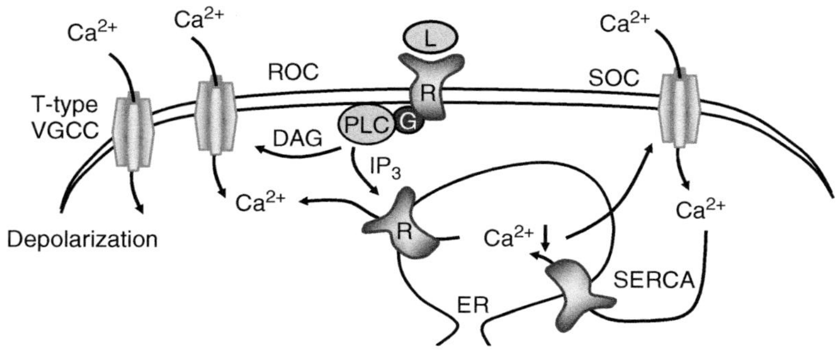

Figure 12.6 Ca^{2+} entry pathways in pulmonary ECs. Ligand (L)–receptor (R) binding activates G-protein-coupled receptors and the PI pathway. DAG activates ROCs and IP_3 causes release of Ca^{2+} from the endoplasmic reticulum (ER). Depletion of Ca^{2+} from the ER activates Ca^{2+} entry through SOCs. In addition, depolarization of the EC may activate Ca^{2+} entry through T-type voltage-gated Ca^{2+} channels (VGCCs). SERCA, sarco/endoplasmic reticulum Ca^{2+}-ATPase.

understanding of a role for T-type Ca^{2+} channels in pulmonary ECs is limited. It has been proposed that a variety of stimuli such as stimulation of G-protein coupled receptors, vascular distension, and shear stress induce membrane depolarization sufficient to activate T-type Ca^{2+} channels [108]. This transient activation may contribute to increases in $[Ca^{2+}]_i$ resulting in the generation and release of vasoactive substances.

Cyclic Nucleotide-Gated Channels

An increase in $[Ca^{2+}]_i$ upon agonist stimulation of ROC and SOC activates eNOS and NO synthesis, which stimulates sGC-mediated increases in cGMP levels as well as cyclic nucleotide-gated channels. The activation of these channels leads to membrane depolarization which decreases the driving force for Ca^{2+} leading to inhibition of Ca^{2+} entry [103]. This is thought to be an important negative feedback mechanism for EC Ca^{2+} entry.

Ion Channels Regulating Membrane Potential

The membrane potential of pulmonary ECs is mainly controlled by K^+, Cl^-, and possibly nonselective cation channels, and modulates the driving force for transmembrane Ca^{2+} flux. With the exception of T-type channel expression in pulmonary microvascular endothelium [109–112], voltage-gated Ca^{2+} channels are generally not observed in ECs. Ca^{2+} entry occurs primarily through nonselective cation channels and is determined by the electrochemical driving force for Ca^{2+}, which is strongly influenced by membrane potential. Hyperpolarizing stimuli enhance Ca^{2+} entry, whereas depolarizing stimuli are associated with reduced $[Ca^{2+}]_i$ levels [103]. In addition, ECs are electrically coupled to one another and possibly to adjacent SMCs by gap junctions [112]. Thus, changes in EC membrane potential can be transmitted along the endothelium and smooth muscle to influence vascular tone. Moreover, it is becoming increasingly evident that membrane potential modulates several cellular processes, including membrane depolarization-induced superoxide generation in human umbilical vein ECs [113]. The role of ion channels in regulating membrane potential is discussed in greater detail below.

K^+ Channels

Potassium channels can regulate Ca^{2+} influx through their influence on EC membrane potential. ECs express a variety of K^+ channels, including K_{ir}, K_{ATP}, K_{Ca}, and K_v channels. Activation of K^+ channels results in EC hyperpolarization, thereby increasing Ca^{2+} influx [114]. The roles of the various K^+ channels are discussed below.

Inward Rectifier K^+ Channels (K_{ir}) K_{ir} channels are characterized by their inward-rectifying current which allows K^+ to flow inward at membrane potentials negative to the K^+ equilibrium potential (E_K). At membrane potentials slightly more depolarized than E_K, these channels conduct a small outward K^+ current [114]. Consequently, K_{ir} channels are thought to be one of the most important channels for the control of the resting potential in ECs [103] and may therefore be involved in regulating the driving force for Ca^{2+} entry in the pulmonary endothelium. K_{ir} channels also function as a K^+ sensor. Activation of K_{ir} channels by modest increases in extracellular $[K^+]$ leads to membrane hyperpolarization [104, 115]. Although the significance of this response in ECs to regulation of vascular tone is not clear, endothelial K_{ir} channels contribute to K^+-induced vasodilation in rat mesenteric artery [115], possibly through the spread of hyperpolarization to the vascular smooth muscle via myoendothelial gap junctions. Given the sensitivity of K_{ir} channels to increases in extracellular $[K^+]$, it is additionally possible that these channels amplify hyperpolarizing responses to activation of SK_{Ca} or IK_{Ca} channels, and may therefore contribute to EDHF responses [77]. Interestingly, vasoactive substances such as angiotensin II, arginine vasopressin, ET-1, and histamine inhibit K_{ir} current in some ECs by a G-protein-coupled mechanism [117–119]. Blocking these channels may depolarize the cell membrane and thereby reduce the driving force for Ca^{2+} influx. In addition, K_{ir} channels may be involved in shear-stress-induced hyperpolarization and vasorelaxation [119]. Although K_{ir} channels are expressed in pulmonary endothelium [121, 122], their role in regulating pulmonary vascular tone is not well characterized.

ATP-Sensitive K^+ Channels (K_{ATP}) K_{ATP} channels are composed of a novel association of K_{ir} (K_{ir} 6.1 and K_{ir} 6.2) and the sulfonylurea receptor and inhibited by increased ATP concentrations. K_{ATP} channels in the pulmonary endothelium are responsible for maintaining membrane potential with normal shear and are inactivated by loss of shear (e.g., ischemia), leading to depolarization [122]. Therefore, these channels may play an important role in sensing alterations in shear stress in pulmonary ECs.

Ca^{2+}-Activated K^+ Channels Three different types of Ca^{2+}-activated K^+ channels have been described including BK_{Ca}, IK_{Ca}, and SK_{Ca} channels. Although BK_{Ca} channels have been found in ECs [103], their functional role is uncertain. The primary K_{Ca} channels localized to ECs are IK_{Ca} and SK_{Ca} [77]. Furthermore, unlike BK_{Ca}, these channels are not voltage-sensitive [103]. To date, strong evidence supports a role for IK_{Ca} and SK_{Ca} channels in mediating EDHF responses [79, 124, 125], as discussed above.

Cl^- Channels

Pulmonary ECs express both volume-regulated anion channels (VRACs) and Ca^{2+}-activated Cl^- channels, but not voltage-gated (ClC-type) or cAMP-gated (cystic fibrosis transmembrane receptor-type) Cl^- channels [103]. These Cl^- channels play an important modulatory role on membrane potential. For example, blockade of Cl^- channels results in hyperpolarization thereby increasing the driving force for Ca^{2+} entry, whereas activation of these channels has the opposite effect [103].

VRACs Classically, VRAC are studied using "cell-swelling" protocols; swelling of the cell activates these channels, whereas shrinkage reduces basal current. Physiologically, VRAC are thought to be activated primarily by mechanical forces, such as shear stress, that induce changes in cell shape [103]. Inhibition of myosin light chain phosphatases via activation of RhoA and Rho/Rho-associated kinase activates VRACs in cultured bovine pulmonary ECs [125], which might be involved in the shear stress-induced mechanosensing pathway of VRACs [103]. In addition, VRACs contribute to intracellular pH homeostasis in pulmonary ECs [126]. Activation of VRACs decreases intracellular pH, which leads to PG synthesis in cerebral ECs [127]. These studies suggest that, in addition to effects on membrane potential, VRAC may also directly contribute to the release of vasoactive compounds.

Ca^{2+}-Activated Cl^- Channels Nilius *et al.* [128] have characterized Ca^{2+}-activated Cl^- channels in bovine pulmonary ECs. These channels are activated by agonists that increase $[Ca^{2+}]_i$, which results in depolarization and negative feedback regulation of Ca^{2+} influx.

Nonselective Cation Channels

Nonselective cation channels are thought to play a minimal role in regulating resting membrane potential, but may become more important following stimulation of ECs. These channels are often more permeable to Na^+ than Ca^{2+} and provide negative feedback regulation of Ca^{2+} entry by inducing depolarization of ECs [103]. In addition the TRP family of cation channels described above, amiloride-sensitive Na^+ channels, or epithelial Na^+ channels (ENaCs), have been described in the microvascular endothelium [129]. Although the functional role of ENaCs in the endothelium is not presently clear, recent studies conducted in cultured bovine corneal ECs undergoing wound-healing show a rise in amiloride-sensitive Na^+ influx and associated depolarization [130]. In addition, cAMP increases both cell membrane-associated ENaC expression and open probability leading to depolarization of alveolar epithelial cells [131]. As in epithelial tissue, ENaCs in vascular ECs are regulated under direct, receptor-mediated signaling of aldosterone [133–135], suggesting a functional regulatory pathway for ENaCs in the endothelium.

Hypoxic Regulation of EC Ca^{2+} Homeostasis

Relatively little is known regarding effects of either acute or chronic hypoxia on endothelial Ca^{2+} homeostasis, despite the potential implications for Ca^{2+} signaling mechanisms in regulation of pulmonary vascular tone in response to these physiological stimuli. In response to acute hypoxia, Hampl *et al.* [24] observed an immediate and transient increase in $[Ca^{2+}]_i$ and stimulated NO synthesis in cultured bovine PAECs that was likely due to Ca^{2+} release from intracellular stores. Later studies by Hogg and Kozlowski [135] identified a nonselective cation current that was increased by acute hypoxia in acutely isolated rat PAECs. However, others have shown an opposite effect of hypoxia to reduce $[Ca^{2+}]_i$ through membrane depolarization and decreased Ca^{2+} influx in this same cell type [136]. Whereas the above studies focus on effects of acute hypoxia on endothelial $[Ca^{2+}]_i$, a recent investigation by Paffett *et al.* [137] demonstrated that chronic hypoxic exposure depolarizes ECs and reduces both basal $[Ca^{2+}]_i$ and store-operated Ca^{2+} entry in intact pulmonary arteries from rats. These findings are consistent with an observation by Murata *et al.* [54] that NO production is impaired in pulmonary arteries from chronically hypoxic rats. This impairment is explained not by a change in eNOS expression, but rather by diminished Ca^{2+} mobilization and abnormal eNOS–caveolin-1 interaction. In contrast to these observations, Fantozzi *et al.* [138], found that 72-h exposure to hypoxia resulted in an increase in TRPC4 expression, elevated basal $[Ca^{2+}]_i$, and enhanced store-operated Ca^{2+} entry in cultured human PAECs. Such differential responses to chronic hypoxia on endothelial $[Ca^{2+}]_i$ between cultured cells and intact arteries from chronically hypoxic animals may reflect changes in endothelial phenotype that occur *in vivo* in response to arterial remodeling or to altered mechanical forces resulting from elevated vascular resistance and hypertension. Whether such alterations in endothelial Ca^{2+} homeostasis in response to acute or chronic hypoxia influence the synthesis of endothelial vasoactive factors and contribute to regulation of pulmonary vascular tone remain important topics of future investigation.

CONCLUSIONS AND PERSPECTIVES

The endothelium is an essential regulator of pulmonary vascular tone. It is responsible for the generation, metabolism, and removal of many vasoactive compounds important in the pulmonary circulation as well as throughout the systemic circulation. Although endothelial control of pulmonary vasomotor tone is largely determined by NO bioavailability, accumulating evidence supports significant roles for other endothelium-derived factors in both normal and disease states, including AA metabolites, ET-1, and various forms of ROS. The generation of many such vasoactive compounds is controlled by $[Ca^{2+}]_i$ levels, determined by the balance of EC Ca^{2+} entry, efflux, and sequestration mechanisms. Ca^{2+} entry, in turn, is a function of both membrane permeability to Ca^{2+} and the electrical driving force for Ca^{2+} across the membrane, and is therefore influenced by ion channels that modulate both membrane potential and Ca^{2+} conductance.

While it is clear that endothelial dysfunction and associated alterations in the production of vasoactive and mitogenic factors play a central role in the etiology of chronic lung diseases characterized by pulmonary hypertension, the underlying mechanisms of this dysregulation are not well understood. Challenges of future studies include identifying processes by which endothelial Ca^{2+} homeostasis is altered under conditions of hypoxic or oxidant stress to promote vasoconstriction and associated pulmonary hypertension, including potential contributions of ROS, impairment of membrane ion channel trafficking, changes in gene expression, and alterations in cell signaling resulting from disruption of lipid raft microdomains.

References

1. Frisbee, J.C., Maier, K.G., Falck, J.R. *et al.* (2002) Integration of hypoxic dilation signaling pathways for skeletal muscle resistance arteries. *American Journal of Physiology: Regulatory, Integrative and Comparative Physiology*, **283**, R309–19.
2. Fredricks, K.T., Liu, Y., Rusch, N.J., and Lombard, J.H. (1994) Role of endothelium and arterial K^+ channels in mediating hypoxic dilation of middle cerebral arteries. *American Journal of Physiology: Heart and Circulatory Physiology*, **267**, H580–86.
3. Aaronson, P.I., Robertson, T.P., Knock, G.A. *et al.* (2006) Hypoxic pulmonary vasoconstriction: mechanisms and controversies. *The Journal of Physiology*, **570**, 53–58.
4. Shimoda, L.A., Sham, J.S., Liu, Q., and Sylvester, J.T. (2002) Acute and chronic hypoxic pulmonary vasoconstriction: a central role for endothelin-1? *Respiratory Physiology and Neurobiology*, **132**, 93–106.
5. Kamata, H. and Hirata, H. (1999) Redox regulation of cellular signalling. *Cellular Signalling*, **11**, 1–14.
6. Natarajan, V. (1995) Oxidants and signal transduction in vascular endothelium. *The Journal of Laboratory and Clinical Medicine*, **125**, 26–37.
7. Lum, H. and Roebuck, K.A. (2001) Oxidant stress and endothelial cell dysfunction. *American Journal of Physiology: Cell Physiology* **280**, C719–41.
8. Jernigan, N.L., Walker, B.R., and Resta, T.C. (2004) Endothelium-derived reactive oxygen species and endothelin-1 attenuate NO-dependent pulmonary vasodilation following chronic hypoxia. *American Journal of Physiology: Lung Cellular and Molecular Physiology*, **287**, L801–8.
9. Levitzky, M.G. (2003) *Pulmonary Physiology*, McGraw-Hill, New York.
10. Dupuis, J., Goresky, C.A., and Fournier, A. (1996) Pulmonary clearance of circulating endothelin-1 in dogs *in vivo*: exclusive role of ETB receptors. *Journal of Applied Physiology*, **81**, 1510–15.
11. Balyasnikova, I.V., Metzger, R., Visintine, D.J. *et al.* (2005) Selective rat lung endothelial targeting with a new set of monoclonal antibodies to angiotensin I-converting enzyme. *Pulmonary Pharmacology and Therapeutics*, **18**, 251–67.
12. Walker, B.R., Haynes, J. Jr., Wang, H.L., and Voelkel, N.F. (1989) Vasopressin-induced pulmonary vasodilation in rats. *American Journal of Physiology: Heart and Circulatory Physiology*, **257**, H415–22.
13. Katusic, Z.S., Shepherd, J.T., and Vanhoutte, P.M. (1984) Vasopressin causes endothelium-dependent relaxations of the canine basilar artery. *Circulation Research*, **55**, 575–79.
14. Furchgott, R.F. and Zawadzki, J.V. (1980) The obligatory role of endothelial cells in the relaxation of arterial smooth muscle by acetylcholine. *Nature*, **288**, 373–76.
15. Palmer, R.M., Ferrige, A.G., and Moncada, S. (1987) Nitric oxide release accounts for the biological activity of endothelium-derived relaxing factor. *Nature*, **327**, 524–26.
16. Ignarro, L.J., Buga, G.M., Wood, K.S. *et al.* (1987) Endothelium-derived relaxing factor produced and released from artery and vein is nitric oxide. *Proceedings of the National Academy of Sciences of the United States of America*, **84**, 9265–69.
17. Carvajal, J., Germain, A., Huidobro-Toro, J., and Weiner, C. (2000) Molecular mechanism

of cGMP-mediated smooth muscle relaxation. *Journal of Cellular Physiology*, **184**, 409–20.
18. Moncada, S. and Higgs, A. (1993) The L-arginine–nitric oxide pathway. *The New England Journal of Medicine*, **329**, 2002–12.
19. Mount, P.F., Kemp, B.E., and Power, D.A. (2007) Regulation of endothelial and myocardial NO synthesis by multi-site eNOS phosphorylation. *Journal of Molecular and Cellular Cardiology*, **42**, 271–79.
20. Maniatis, N.A., Brovkovych, V., Allen, S.E. *et al.* (2006) Novel mechanism of endothelial nitric oxide synthase activation mediated by caveolae internalization in endothelial cells. *Circulation Research*, **99**, 870–77.
21. Minshall, R.D., Sessa, W.C., Stan, R.V. *et al.* (2003) Caveolin regulation of endothelial function. *American Journal of Physiology: Lung Cellular and Molecular Physiology*, **285**, L1179–83.
22. Le Cras, T.D. and McMurtry, I.F. (2001) Nitric oxide production in the hypoxic lung. *American Journal of Physiology: Lung Cellular and Molecular Physiology*, **280**, L575–82.
23. Hampl, V. and Herget, J. (2000) Role of nitric oxide in the pathogenesis of chronic pulmonary hypertension. *Physiological Reviews*, **80**, 1337–72.
24. Hampl, V., Cornfield, D.N., Cowan, N.J., and Archer, S.L. (1995) Hypoxia potentiates nitric oxide synthesis and transiently increases cytosolic calcium levels in pulmonary artery endothelial cells. *The European Respiratory Journal*, **8**, 515–22.
25. Su, Y. and Block, E.R. (2000) Role of calpain in hypoxic inhibition of nitric oxide synthase activity in pulmonary endothelial cells. *American Journal of Physiology: Lung Cellular and Molecular Physiology*, **278**, L1204–12.
26. Tirosh, R., Resnik, E.R., Herron, J. *et al.* (2006) Acute normoxia increases fetal pulmonary artery endothelial cell cytosolic Ca^{2+} via Ca^{2+}-induced Ca^{2+} release. *Pediatric Research*, **60**, 258–63.
27. John, T.A., Ibe, B.O., and Raj, J.U. (2006) Oxygen alters caveolin-1 and nitric oxide synthase-3 functions in ovine fetal and neonatal lung microvascular endothelial cells. *American Journal of Physiology: Lung Cellular and Molecular Physiology*, **291**, L1079–93.
28. Nelin, L.D., Thomas, C.J., and Dawson, C.A. (1996) Effect of hypoxia on nitric oxide production in neonatal pig lung. *American Journal of Physiology: Heart and Circulatory Physiology*, **271**, H8–H14.
29. Pearl, J.M., Nelson, D.P., Wellmann, S.A. *et al.* (2000) Acute hypoxia and reoxygenation impairs exhaled nitric oxide release and pulmonary mechanics. *The Journal of Thoracic and Cardiovascular Surgery*, **119**, 931–38.
30. Archer, S.L., Tolins, J.P., Raij, L., and Weir, E.K. (1989) Hypoxic pulmonary vasoconstriction is enhanced by inhibition of the synthesis of an endothelium derived relaxing factor. *Biochemical and Biophysical Research Communications*, **164**, 1198–205.
31. Liu, S.F., Crawley, D.E., Barnes, P.J., and Evans, T.W. (1991) Endothelium-derived relaxing factor inhibits hypoxic pulmonary vasoconstriction in rats. *The American Review of Respiratory Disease*, **143**, 32–37.
32. Sprague, R.S., Thiemermann, C., and Vane, J.R. (1992) Endogenous endothelium-derived relaxing factor opposes hypoxic pulmonary vasoconstriction and supports blood flow to hypoxic alveoli in anesthetized rabbits. *Proceedings of the National Academy of Sciences of the United States of America*, **89**, 8711–15.
33. Beckman, J.S., Beckman, T.W., Chen, J. *et al.* (1990) Apparent hydroxyl radical production by peroxynitrite: implications for endothelial injury from nitric oxide and superoxide. *Proceedings of the National Academy of Sciences of the United States of America*, **87**, 1620–24.
34. Rubanyi, G.M. and Vanhoutte, P.M. (1986) Superoxide anions and hyperoxia inactivate endothelium-derived relaxing factor. *American Journal of Physiology: Heart and Circulatory Physiology*, **250**, H822–27.
35. Killilea, D.W., Hester, R., Balczon, R. *et al.* (2000) Free radical production in hypoxic pulmonary artery smooth muscle cells. *American Journal of Physiology: Lung Cellular and Molecular Physiology*, **279**, L408–12.
36. Liu, J.Q., Sham, J.S., Shimoda, L.A. *et al.* (2003) Hypoxic constriction and reactive oxygen species in porcine distal pulmonary arteries. *American Journal of Physiology: Lung Cellular and Molecular Physiology*, **285**, L322–33.
37. Marshall, C., Mamary, A.J., Verhoeven, A.J., and Marshall, B.E. (1996) Pulmonary artery NADPH-oxidase is activated in hypoxic pulmonary vasoconstriction. *American Journal of Respiratory Cell and Molecular Biology*, **15**, 633–44.
38. Eddahibi, S., Raffestin, B., Tayarani, Y. *et al.* (1996) Hypoxia-reoxygenation impairs NO-mediated vasodilation in rat lungs. *American Journal of Physiology: Lung Cellular and Molecular Physiology*, **271**, L441–49.

39. Wanstall, J.C., Kaye, J.A., and Gambino, A. (1997) The *in vitro* pulmonary vascular effects of FK409 (nitric oxide donor): a study in normotensive and pulmonary hypertensive rats. *British Journal of Pharmacology*, **121**, 280–86.
40. Jernigan, N.L., Resta, T.C., and Walker, B.R. (2004) Contribution of oxygen radicals to altered NO-dependent pulmonary vasodilation in acute and chronic hypoxia. *American Journal of Physiology: Lung Cellular and Molecular Physiology*, **286**, L947–55.
41. Fagan, K.A., McMurtry, I.F., and Rodman, D.M. (2000) Nitric oxide synthase in pulmonary hypertension: lessons from knockout mice. *Physiological Research*, **49**, 539–48.
42. Fagan, K.A., Morrissey, B., Fouty, B.W. *et al.* (2001) Upregulation of nitric oxide synthase in mice with severe hypoxia-induced pulmonary hypertension. *Respiratory Research*, **2**, 306–13.
43. Le Cras, T., Xue, C., Rengasamy, A., and Johns, R. (1996) Chronic hypoxia upregulates endothelial and inducible NO synthase gene and protein expression in rat lung. *American Journal of Physiology: Lung Cellular and Molecular Physiology*, **270**, L164–70.
44. Resta, T.C., Gonzales, R.J., Dail, W. *et al.* (1997) Selective upregulation of arterial endothelial nitric oxide synthase in pulmonary hypertension. *American Journal of Physiology: Heart and Circulatory Physiology*, **272**, H806–13.
45. Resta, T.C., Chicoine, L.G., Omdahl, J.L., and Walker, B.R. (1999) Maintained upregulation of pulmonary eNOS gene and protein expression during recovery from chronic hypoxia. *American Journal of Physiology: Heart and Circulatory Physiology*, **276**, H699–708.
46. Tyler, R.C., Muramatsu, M., Abman, S.H. *et al.* (1999) Variable expression of endothelial NO synthase in three forms of rat pulmonary hypertension. *American Journal of Physiology: Lung Cellular and Molecular Physiology*, **276**, L297–L303.
47. Adnot, S., Raffestin, B., Eddahibi, S. *et al.* (1991) Loss of endothelium-dependent relaxant activity in the pulmonary circulation of rats exposed to chronic hypoxia. *The Journal of Clinical Investigation*, **87**, 155–62.
48. Eddahibi, S., Adnot, S., Carville, C. *et al.* (1992) L-Arginine restores endothelium-dependent relaxation in pulmonary circulation of chronically hypoxic rats. *American Journal of Physiology: Lung Cellular and Molecular Physiology*, **263**, L194–200.
49. Isaacson, T.C., Hampl, V., Weir, E.K. *et al.* (1994) Increased endothelium-derived NO in hypertensive pulmonary circulation of chronically hypoxic rats. *Journal of Applied Physiology*, **76**, 933–40.
50. Muramatsu, M., Tyler, R., Rodman, D., and McMurtry, I. (1996) Thapsigargin stimulates increased NO activity in hypoxic hypertensive rat lungs and pulmonary arteries. *Journal of Applied Physiology*, **80**, 1336–44.
51. Resta, T.C. and Walker, B.R. (1996) Chronic hypoxia selectively augments endothelium-dependent pulmonary arterial vasodilation. *American Journal of Physiology: Heart and Circulatory Physiology*, **270**, H888–96.
52. Roos, C., Frank, D., Xue, C. *et al.* (1996) Chronic inhaled nitric oxide: effects on pulmonary vascular endothelial function and pathology in rats. *Journal of Applied Physiology*, **80**, 252–60.
53. Sato, K., Rodman, D.M., and McMurtry, I.F. (1999) Hypoxia inhibits increased ETB receptor-mediated NO synthesis in hypertensive rat lungs. *American Journal of Physiology: Lung Cellular and Molecular Physiology*, **276**, L571–81.
54. Murata, T., Sato, K., Hori, M. *et al.* (2002) Decreased endothelial nitric-oxide synthase (eNOS) activity resulting from abnormal interaction between eNOS and its regulatory proteins in hypoxia-induced pulmonary hypertension. *The Journal of Biological Chemistry*, **277**, 44085–92.
55. Chen, Y.F. and Oparil, S. (2000) Endothelial dysfunction in the pulmonary vascular bed. *The American Journal of the Medical Sciences*, **320**, 223–32.
56. Cogolludo, A., Moreno, L., Bosca, L. *et al.* (2003) Thromboxane A_2-induced inhibition of voltage-gated K^+ channels and pulmonary vasoconstriction: role of protein kinase Czeta. *Circulation Research*, **93**, 656–63.
57. Zamora, C.A., Baron, D.A., and Heffner, J.E. (1993) Thromboxane contributes to pulmonary hypertension in ischemia–reperfusion lung injury. *Journal of Applied Physiology*, **74**, 224–29.
58. Korbut, R. and Gryglewski, R.J. (1985) Anti-aggregatory prostanoids. Part 1. *Acta Physiologica Polonica*, **36**, 293–302.
59. Voelkel, N.F., Gerber, J.G., McMurtry, I.F. *et al.* (1981) Release of vasodilator prostaglandin, PGI_2, from isolated rat lung during vasoconstriction. *Circulation Research*, **48**, 207–13.
60. Weir, E.K., McMurtry, I.F., Tucker, A. *et al.* (1976) Prostaglandin synthetase inhibitors do not decrease hypoxic pulmonary vasoconstriction. *Journal of Applied Physiology*, **41**, 714–18.

61. Tuder, R.M., Cool, C.D., Geraci, M.W. *et al.* (1999) Prostacyclin synthase expression is decreased in lungs from patients with severe pulmonary hypertension. *American Journal of Respiratory and Critical Care Medicine*, **159**, 1925–32.
62. Puri, A., McGoon, M.D., and Kushwaha, S.S. (2007) Pulmonary arterial hypertension: current therapeutic strategies. *Nature Clinical Practice Cardiovascular Medicine*, **4**, 319–29.
63. Voelkel, N.F., Stenmark, K.R., Reeves, J.T. *et al.* (1984) Actions of lipoxygenase metabolites in isolated rat lungs. *Journal of Applied Physiology*, **57**, 860–67.
64. Voelkel, N.F., Morganroth, M., and Feddersen, O.C. (1985) Potential role of arachidonic acid metabolites in hypoxic pulmonary vasoconstriction. *Chest*, **88** (4 Suppl), 245S–248S.
65. Zhang, Y.Y., Walker, J.L., Huang, A. *et al.* (2002) Expression of 5-lipoxygenase in pulmonary artery endothelial cells. *The Biochemical Journal*, **361**, 267–76.
66. McDonnell, T.J., Westcott, J.Y., Czartolomna, J., and Voelkel, N.F. (1990) Role of peptidoleukotrienes in hypoxic pulmonary vasoconstriction in rats. *American Journal of Physiology: Heart and Circulatory Physiology*, **259**, H751–58.
67. Roman, R.J. (2002) P-450 metabolites of arachidonic acid in the control of cardiovascular function. *Physiological Reviews*, **82**, 131–85.
68. Morio, Y., Carter, E.P., Oka, M., and McMurtry, I.F. (2003) EDHF-mediated vasodilation involves different mechanisms in normotensive and hypertensive rat lungs. *American Journal of Physiology: Heart and Circulatory Physiology*, **284**, H1762–70.
69. Morio, Y., Homma, N., Takahashi, H. *et al.* (2007) Activity of endothelium-derived hyperpolarizing factor is augmented in monocrotaline-induced pulmonary hypertension of rat lungs. *Journal of Vascular Research*, **44**, 325–35.
70. Zhu, D., Bousamra, M. II, Zeldin, D.C. *et al.* (2000) Epoxyeicosatrienoic acids constrict isolated pressurized rabbit pulmonary arteries. *American Journal of Physiology: Lung Cellular and Molecular Physiology*, **278**, L335–43.
71. Alonso-Garcia, M., Sun, C.W., Falck, J.R. *et al.* (1998) Contribution of 20-HETE to the vasodilator actions of nitric oxide in renal arteries. *American Journal of Physiology: Renal Physiology*, **275**, F370–78.
72. Sun, C.W., Alonso-Galicia, M., Taheri, M.R. *et al.* (1988) Nitric oxide–20-hydroxyeicosatetraenoic acid interaction in the regulation of K^+ channel activity and vascular tone in renal arterioles. *Circulation Research*, **83**, 1069–79.
73. Sun, C.W., Falck, J.R., Okamoto, H. *et al.* (2000) Contribution of cGMP versus 20-HETE on the vasodilator response to nitric oxide in rat cerebral arteries. *American Journal of Physiology: Heart and Circulatory Physiology*, **279**, H339–50.
74. Jacobs, E.R., Zhu, D., Gruenloh, S. *et al.* (2006) VEGF-induced relaxation of pulmonary arteries is mediated by endothelial cytochrome P-450 hydroxylase. *American Journal of Physiology: Lung Cellular and Molecular Physiology*, **291**, L369–77.
75. Chen, Y., Medhora, M., Falck, J.R. *et al.* (2006) Mechanisms of activation of eNOS by 20-HETE and VEGF in bovine pulmonary artery endothelial cells. *American Journal of Physiology: Lung Cellular and Molecular Physiology*, **291**, L378–85.
76. Zhu, D., Zhang, C., Medhora, M., and Jacobs, E.R. (2002) CYP4A mRNA, protein, and product in rat lungs: novel localization in vascular endothelium. *Journal of Applied Physiology*, **93**, 330–37.
77. Köhler, R. and Hoyer, J. (2007) The endothelium-derived hyperpolarizing factor: insights from genetic animal models. *Kidney International*, **72**, 145–50.
78. Coleman, H.A., Tare, M., and Parkington, H.C. (2004) Endothelial potassium channels, endothelium-dependent hyperpolarization and the regulation of vascular tone in health and disease. *Clinical and Experimental Pharmacology and Physiology*, **31**, 641–49.
79. Tracey, A., MacDonald, A., and Shaw, A.M. (2002) Involvement of gap junctions in bradykinin-induced relaxation of bovine pulmonary supernumerary arteries before and after inhibition of nitric oxide/guanylate cyclase. *Clinical Science (London)*, **103**, 553–57.
80. Ko, Y.S., Yeh, H.I., Rothery, S. *et al.* (1999) Connexin make-up of endothelial gap junctions in the rat pulmonary artery as revealed by immunoconfocal microscopy and triple-label immunogold electron microscopy. *The Journal of Histochemistry and Cytochemistry*, **47**, 683–92.
81. Parthasarathi, K., Ichimura, H., Monma, E. *et al.* (2006) Connexin 43 mediates spread of Ca^{2+}-dependent proinflammatory responses in lung capillaries. *The Journal of Clinical Investigation*, **116**, 2193–200.
82. Hasunuma, K., Yamaguchi, T., Rodman, D.M. *et al.* (1991) Effects of inhibitors of EDRF and EDHF on vasoreactivity of perfused rat lungs. *American Journal of Physiology: Lung Cellular and Molecular Physiology*, **260**, L97–104.

83. Hickey, K.A., Rubanyi, C.M., Paul, R.J., and Highsmith, R.F. (1985) Characterization of a coronary vasoconstrictor produced by cultured endothelial cells. *American Journal of Physiology: Cell Physiology*, **248**, C550–56.
84. Yanagisawa, M., Kurihara, H., Kimura, S. *et al.* (1988) A novel potent vasoconstrictor peptide produced by vascular endothelial cells. *Nature*, **332**, 411–15.
85. Fukuroda, T., Fujikawa, T., Ozaki, S. *et al.* (1994) Clearance of circulating endothelin-1 by ETB receptors in rats. *Biochemical and Biophysical Research Communications*, **199**, 1461–65.
86. Chen, S.J., Chen, Y.F., Meng, Q.C. *et al.* (1995) Endothelin-receptor antagonist bosentan prevents and reverses hypoxic pulmonary hypertension in rats. *Journal of Applied Physiology*, **79**, 2122–31.
87. Chen, S.J., Chen, Y.F., Opgenorth, T.J. *et al.* (1997) The orally active nonpeptide endothelin A-receptor antagonist A-127722 prevents and reverses hypoxia-induced pulmonary hypertension and pulmonary vascular remodeling in Sprague-Dawley rats. *Journal of Cardiovascular Pharmacology*, **29**, 713–25.
88. Raffestin, B., Adnot, S., Eddahibi, S. *et al.* (1991) Pulmonary vascular response to endothelin in rats. *Journal of Applied Physiology*, **70**, 567–74.
89. Liu, J.Q., Sham, J.S., Shimoda, L.A., and Sylvester, J.T. (2001) Hypoxic constriction of porcine distal pulmonary arteries: endothelium and endothelin dependence. *American Journal of Physiology: Lung Cellular and Molecular Physiology*, **280**, L856–65.
90. Sato, K., Morio, Y., Morris, K.G. *et al.* (2000) Mechanism of hypoxic pulmonary vasoconstriction involves ET_A receptor-mediated inhibition of K_{ATP} channel. *American Journal of Physiology: Lung Cellular and Molecular Physiology*, **278**, L434–42.
91. Bonvallet, S.T., Zamora, M.R., Hasunuma, K. *et al.* (1994) BQ123, an ET_A-receptor antagonist, attenuates hypoxic pulmonary hypertension in rats. *American Journal of Physiology: Heart and Circulatory Physiology*, **266**, H1327–31.
92. DiCarlo, V.S., Chen, S.J., Meng, Q.C. *et al.* (1995) ET_A-receptor antagonist prevents and reverses chronic hypoxia-induced pulmonary hypertension in rat. *American Journal of Physiology: Lung Cellular and Molecular Physiology*, **269**, L690–97.
93. Eddahibi, S., Raffestin, B., Clozel, M. *et al.* (1995) Protection from pulmonary hypertension with an orally active endothelin receptor antagonist in hypoxic rats. *American Journal of Physiology: Heart and Circulatory Physiology*, **268**, H828–35.
94. Elton, T.S., Oparil, S., Taylor, G.R. *et al.* (1992) Normobaric hypoxia stimulates endothelin-1 gene expression in the rat. *American Journal of Physiology: Regulatory, Integrative and Comparative Physiology*, **263**, R1260–64.
95. Li, H., Elton, T.S., Chen, Y.F., and Oparil, S. (1994) Increased endothelin receptor gene expression in hypoxic rat lung. *American Journal of Physiology: Lung Cellular and Molecular Physiology*, **266**, L553–60.
96. Eddahibi, S., Raffestin, B., Braquet, P. *et al.* (1991) Pulmonary vascular reactivity to endothelin-1 in normal and chronically pulmonary hypertensive rats. *Journal of Cardiovascular Pharmacology*, **17** (Suppl 7), S358–61.
97. Jernigan, N.L., Walker, B.R., and Resta, T.C. (2008) Reactive oxygen species mediate RhoA/Rho kinase-induced Ca^{2+} sensitization in pulmonary vascular smooth muscle following chronic hypoxia. *American Journal of Physiology: Lung Cellular and Molecular Physiology*, **295**, L515–29.
98. MacLean, M.R., McCulloch, K.M., and Baird, M. (1995) Effects of pulmonary hypertension on vasoconstrictor responses to endothelin-1 and sarafotoxin S6C and on inherent tone in rat pulmonary arteries. *Journal of Cardiovascular Pharmacology*, **26**, 822–30.
99. Wanstall, J. and O'Donnell, S. (1990) Endothelin and 5-hydroxytryptamine on rat pulmonary artery in pulmonary hypertension. *European Journal of Pharmacology*, **176**, 159–68.
100. Michel, R.P., Langleben, D., Dupuis, J., and Can, J. (2003) The endothelin system in pulmonary hypertension. *Canadian Journal of Physiology and Pharmacologyhysiol Pharmacol*, **81**, 542–54.
101. Sauvageau, S., Thorin, E., Caron, A., and Dupuis, J. (2007) Endothelin-1-induced pulmonary vasoreactivity is regulated by ET_A and ET_B receptor interactions. *Journal of Vascular Research*, **44**, 375–81.
102. Ivy, D.D., Yanagisawa, M., Gariepy, C.E. *et al.* (2002) Exaggerated hypoxic pulmonary hypertension in endothelin B receptor-deficient rats. *American Journal of Physiology: Lung Cellular and Molecular Physiology*, **282**, L703–12.
103. Nilius, B. and Droogmans, G. (2001) Ion channels and their functional role in vascular endothelium. *Physiological Reviews*, **81**, 1415–59.
104. McFadzean, I. and Gibson, A. (2002) The developing relationship between receptor-operated and store-operated calcium channels in smooth muscle. *British Journal of Pharmacology*, **135**, 1–13.

105. Klishin, A., Sedova, M., and Blatter, L.A. (1998) Time-dependent modulation of capacitative Ca^{2+} entry signals by plasma membrane Ca^{2+} pump in endothelium. *American Journal of Physiology: Cell Physiology*, **274**, C1117–28.
106. Yao, X. and Garland, C.J. (2005) Recent developments in vascular endothelial cell transient receptor potential channels. *Circulation Research*, **97**, 853–63.
107. Tiruppathi, C., Freichel, M., Vogel, S.M. *et al.* (2002) Impairment of store-operated Ca^{2+} entry in TRPC4$^{-/-}$ mice interferes with increase in lung microvascular permeability. *Circulation Research*, **91**, 70–76.
108. Zhou, C. and Wu, S. (2006) T-type calcium channels in pulmonary vascular endothelium. *Microcirculation*, **13**, 645–56.
109. De Proost, I., Brouns, I., Pintelon, I. *et al.* (2007) Pulmonary expression of voltage-gated calcium channels: special reference to sensory airway receptors. *Histochemistry and Cell Biology*, **128**, 301–16.
110. Wu, S., Haynes, J. Jr., Taylor, J.T. *et al.* (2003) Ca_v31 (alpha1G) T-type Ca^{2+} channels mediate vaso-occlusion of sickled erythrocytes in lung microcirculation. *Circulation Research*, **93**, 346–53.
111. Wei, Z., Manevich, Y., Al Mehdi, A.B. *et al.* (2004) Ca^{2+} flux through voltage-gated channels with flow cessation in pulmonary microvascular endothelial cells. *Microcirculation*, **11**, 517–26.
112. de Wit, C. and Wölfle, S.E. (2007) EDHF and gap junctions: important regulators of vascular tone within the microcirculation. *Current Pharmaceutical Biotechnology*, **8**, 11–25.
113. Sohn, Y., Keller, M., Gloe, T. *et al.* (2000) The small G-protein Rac mediates depolarization-induced superoxide formation in human endothelial cells. *The Journal of Biological Chemistry*, **275**, 18745–50.
114. Jackson W.F. (2005) Potassium channels in the peripheral microcirculation. *Microcirculation*, **12**, 113–27.
115. Crane, G.J., Walker, S.D., Dora, K.A., and Garland C.J. (2003) Evidence for a differential cellular distribution of inward rectifier K channels in the rat isolated mesenteric artery. *Journal of Vascular Research*, **40**, 159–68.
116. Nilius, B., Schwarz, G., and Droogmans, G. (1993) Modulation by histamine of an inwardly rectifying potassium channel in human endothelial cells. *The Journal of Physiology (London)*, **472**, 359–71.
117. Hoyer, J., Popp, R., Meyer, J. *et al.* (1991) Angiotensin II, vasopressin and GTPγS inhibit inward-rectifying K^+ channels in porcine cerebral capillary endothelial cells. *The Journal of Membrane Biology*, **123**, 55–62.
118. Zhang, H., Inazu, M., Weir, B., and Daniel, E. (1994) Endothelin-1 inhibits inward rectifier potassium channels and activates nonspecific cation channels in cultured endothelial cells. *Pharmacology*, **49**, 11–22.
119. Olesen, S.P., Clapham, D.E., and Davies, P.F. (1988) Haemodynamic shear stress activates a K^+ current in vascular endothelial cells. *Nature*, **331**, 168–70.
120. Michelakis, E.D., Weir, E.K., Wu, X. *et al.* (2001) Potassium channels regulate tone in rat pulmonary veins. *American Journal of Physiology: Lung Cellular and Molecular Physiology*, **280**, L1138–47.
121. Leung, Y.M., Kwan, C.Y., Daniel, E.E., and Can, J. (2000) Block of inwardly rectifying K^+ currents by extracellular Mg^{2+} and Ba^{2+} in bovine pulmonary artery endothelial cells. *Canadian Journal of Physiology and Pharmacologyhysiol Pharmacol*, **78**, 751–56.
122. Chatterjee, S., Levitan, I., Wei, Z., and Fisher, A.B. (2006) KATP channels are an important component of the shear-sensing mechanism in the pulmonary microvasculature. *Microcirculation*, **13**, 633–44.
123. Edwards, G., Dora, K.A., Gardener, M.J. *et al.* (1998) K^+ is an endothelium-derived hyperpolarizing factor in rat arteries. *Natura*, **396**, 269–72.
124. Doughty, J.M., Plane, F., and Langton, P.D. (1999) Charybdotoxin and apamin block EDHF in rat mesenteric artery if selectively applied to the endothelium. *American Journal of Physiology: Heart and Circulatory Physiology*, **276**, H1107–12.
125. Nilius, B., Voets, T., Prenen, J. *et al.* (1999) Role of Rho and Rho kinase in the activation of volume-regulated anion channels in bovine endothelial cells. *The Journal of Physiology*, **516**, 67–74.
126. Nilius, B., Prenen, J., and Droogmans, G. (1998) Modulation of volume-regulated anion channels by extra- and intracellular pH. *Pflugers Archiv*, **436**, 742–48.
127. Parfenova, H., Haffner, J., and Leffler, C.W. (1999) Phosphorylation-dependent stimulation of prostanoid synthesis by nigericin in cerebral endothelial cells. *American Journal of Physiology: Cell Physiology*, **277**, C728–38.
128. Nilius, B., Prenen, J., Szucs, G. *et al.* (1997) Calcium-activated chloride channels in bovine pulmonary artery endothelial cells. *The Journal of Physiology (London)*, **498**, 381–96.

129. Benos, D.J. and Stanton, B.A. (1999) Functional domains within the degenerin/epithelial sodium channel (Deg/ENaC) superfamily of ion channels. *The Journal of Physiology (London)*, **520**, 631–44.

130. Chifflet, S., Hernández, J.A., and Grasso, S. (2005) A possible role for membrane depolarization in epithelial wound healing. *American Journal of Physiology: Cell Physiology*, **288**, C1420–30.

131. Lazrak, A. and Matalon, S. (2003) cAMP-induced changes of apical membrane potentials of confluent H441 monolayers. *American Journal of Physiology: Lung Cellular and Molecular Physiology*, **285**, L443–50.

132. Chen, W., Valamanesh, F., Mirshahi, T. *et al.* (2004) Aldosterone signaling modifies capillary formation by human bone marrow endothelial cells. *Vascular Pharmacology*, **40**, 269–77.

133. Golestaneh, C., Klein, F., Valamanesh, G. *et al.* (2001) Mineralocorticoid receptor-mediated signaling regulates the ion gated sodium channel in vascular endothelial cells and requires an intact cytoskeleton. *Biochemical and Biophysical Research Communications*, **280**, 1300–6.

134. Kusche-Vihrog, K., Sobczak, K., Bangel, N. *et al.* (2008) Aldosterone and amiloride alter ENaC abundance in vascular endothelium. *Pflugers Archiv*, **455**, 849–57.

135. Hogg, D.S. and Kozlowski, R.Z. (2002) Hypoxia activates a background conductance in freshly isolated pulmonary arterial endothelial cells of the rat. *FEBS Letters*, **522**, 125–29.

136. Stevens, T., Cornfield, D.N., McMurtry, I.F., and Rodman, D.M. (1994) Acute reductions in pO_2 depolarize pulmonary artery endothelial cells and decrease $[Ca^{2+}]i$. *American Journal of Physiology: Heart and Circulatory Physiology*, **266**, H1416–21.

137. Paffett, M.L., Naik, J.S., Resta, T.C., and Walker, B.R. (2007) Reduced store-operated Ca^{2+} entry in pulmonary endothelial cells from chronically hypoxic rats. *American Journal of Physiology: Lung Cellular and Molecular Physiology*, **293**, L1135–42.

138. Fantozzi, I., Zhang, S., Platoshyn, O. *et al.* (2003) Hypoxia increases AP-1 binding activity by enhancing capacitative Ca^{2+} entry in human pulmonary artery endothelial cells. *American Journal of Physiology: Lung Cellular and Molecular Physiology*, **285**, L1233–45.

13

Pulmonary Endothelial Progenitor Cells

Bernard Thébaud[1] and Mervin C. Yoder[2]

[1]Department of Pediatrics, Division of Neonatology, University of Alberta, Edmonton, Canada
[2]Department of Pediatrics, Division of Neonatal-Perinatal Medicine, Indiana University School of Medicine, Indianapolis, IN, USA

INTRODUCTION

The endothelial lining of blood vessels forms a near-continuous barrier between the circulating blood cells and tissues, and maintenance of this barrier is critical for tissue homeostasis. However, the endothelial cells (ECs) constitute so much more than just a physical constraint to the distribution of soluble molecules, suspended cells, and microvesicular particles. As noted in other chapters in this volume and in recent review articles [1, 2], ECs display heterogenous functions (control fluid permeability, leukocyte traffic, hemostasis, vasomotor tone, thermoregulation, sieve function, immune tolerance, and vessel number) that largely depend on their temporal and spatial distribution throughout tissues and organs (see Chapter 9). Furthermore, the vascular ECs are developmentally and homeostatically regulated, and the vessels lined by these ECs, frequently remodeled depending upon the changing demands of those tissues and organs throughout ontogeny. This chapter will focus on recent investigations that have begun to explore pulmonary vascular endothelial heterogeneity with respect to EC proliferative potential, which may have important implications for understanding normal pulmonary growth and development, as well as responses of the lung vasculature to injury or disease.

CIRCULATING ENDOTHELIAL PROGENITOR CELLS

The concept of a distinct progenitor cell for the endothelial lineage is not new. Developmental biologists have long known that ECs are products of angioblast precursor cells derived from the mesodermal germ layer and that these cells constitute the first blood vessels formed in the developing embryo in a process termed vasculogenesis. Upon establishment of a nascent capillary plexus, multiple cell types participate in remodeling the vascular network (particularly following the onset of the systemic circulation) through angiogenesis and arteriogenesis. Formation of new vessels from these preexisting vessels occurs via endothelial sprouting and is termed angiogenesis, whereas the construction of the intricate cellular and matrix composite of the arterial wall is termed arteriogenesis.

ECs are continuously shed into the blood circulation due to a variety of probable causes, including aging, disease, and trauma [3]. The circulating ECs are thought to represent senescent end-stage cells lacking in proliferative potential [4]. Thus, isolation of a circulating cell that displayed potential to give rise to cells appearing endothelial-like *in vitro* and with the potential to incorporate at sites of neoangiogenesis *in vivo* spawned a whole new field of investigation into defining the basic and translational properties of these presumed bone marrow-derived circulating endothelial progenitor cells (EPCs) [5]. The large number of preclinical studies in animal models suggested a high probability for successful clinical translation of EPCs as a cell therapy to treat ischemic tissue disorders via new vessel formation. However, a limitation in realizing the potential of EPCs in human clinical trials to date has been the lack of a clear definition of how to unambiguously identify the circulating EPC and, thus, deciding which population of cells to infuse into the patients.

While no single unique cell surface molecule has been demonstrated to define an EPC in human subjects, CD34 expression is reported to serve as a fundamental marker [6]. CD34 is a sialomucin that is expressed by a host

The Pulmonary Endothelium: Function in health and disease Editors Norbert F. Voelkel, Sharon Rounds

of cells throughout the body, including hematopoietic stem and progenitor cells, and approximately 1% of human bone marrow mononuclear cells express this cell surface molecule [7]. Other antigens such as CD133 and vascular endothelial growth factor (VEGF) receptor (vascular endothelial growth factor receptor VEGFR)-2 (also known as vascular endothelial growth factor receptor-2 KDR) are also thought to define the putative EPCs. In fact, some investigators propose that the $CD133^+ CD34^+ KDR^+$ cell serves as a precursor to the more frequent $CD133^- CD34^+ KDR^+$ circulating EPC [8]. Some evidence has also been presented that human EPCs may be derived from the $CD34^+$ hematopoietic stem cell, although analysis of additional informative human patients would be important to support or refute the published work as only a limited number of *in vivo* studies have been published [9]. Numerous papers have correlated the circulating concentration of $CD133^+ CD34^+ KDR^+$, $CD34^+ KDR^+$, or $CD34^+$ cells with the risk for development of adverse cardiovascular outcomes, and in general an inverse correlation with the circulating concentration of each of these EPC subsets and the highest risk category exists [10, 11].

EPCs have also been defined using *in vitro* colony-forming assays. In one assay, peripheral blood mononuclear cells are plated on fibronectin coated tissue culture wells and 2 days later, the nonadherent colony-forming unit (CFU)-Hill cells are removed and replated into fresh wells coated with fibronectin. Within 5–9 days, one can identify colonies emerging from the plated cells. The number of CFU-Hill-derived colonies that are present in the peripheral blood of patients correlates with the severity of cardiovascular dysfunction and with the risk of developing worsening cardiovascular disease [10].

A second assay identifies another type of circulating endothelial colony-forming cell (ECFC). In this assay, peripheral blood mononuclear cells are plated on collagen type 1-coated tissue culture wells in an endothelial growth media that contains several growth factors including, VEGF, epidermal growth factor, insulin-like growth factor-1, and fibroblast growth factor-2 [12]. In this case, cobblestone-like adherent colonies emerge within 6 days if umbilical cord blood mononuclear cells are plated or 14–21 days if adult peripheral blood mononuclear cells are plated [12]. The ECFCs display varying levels of proliferative potential. Those ECFCs with the highest proliferative potential possess high telomerase activity similar to many embryonic and adult stem cells, despite concomitantly expressing a variety of cell surface proteins similar to mature differentiated vascular ECs. It is important to note that ECFCs never express CD45 (common leukocyte antigen), CD14 [lipopolysaccharide (LPS) receptor], or CD115 (macrophage colony-stimulating factor receptor). In contrast, CFU-Hill cells express these molecules, suggesting that CFU-Hill cells are progeny of the myeloid lineage that expresses endothelial antigens during the process of alternative macrophage differentiation [13].

By definition, EPCs are described as cells that possess the ability to form vessels *de novo in vivo*. A variety of animal models of vascular injury have been developed, and the ability of EPCs to home, engraft, and function as vascular ECs has been reported [14–16]. In these injury models, EPCs clearly play a role in restoring blood flow to the injured and ischemic area. However, the models used do not demonstrate high levels of *de novo* vessel formation from the infused EPCs. To develop a more direct assay, Schechner *et al.* [17] examined the role of human umbilical vein ECs (human umbilical vein endothelial cell HUVECs) to form human blood vessels in collagen/fibronectin gels upon implantation in immunodeficient mice. This approach proved successful only if the HUVECs overexpressed antiapoptotic molecules. We speculated that human EPCs implanted *in vivo* in the same gel constructs may possess *de novo* vessel forming ability. When both CFU-Hill cells and ECFCs were plated in collagen/fibronectin gels and implanted subcutaneously in nonobese diabetic/severe combined immunodeficiency mice, ECFCs gave rise to numerous human blood vessels that had gained access to murine vessels and were perfused with murine red blood cells [18]. While CFU-Hill cells survived in the gels upon implantation, they did not contribute to human–murine chimeric vessel formation. Thus, only ECFCs display direct vessel-forming ability and display the primary defining property of an EPCs.

EPCs are rare circulating cells. Thus, EPCs must display robust proliferative potential if they are to play a major role in neoangiogenesis. When directly comparing the replating potential of ECFCs and CFU-Hill, only ECFCs possess the ability to give rise to secondary and tertiary colonies [18]. To best determine how to define the proliferative potential that resides in the ECs emerging from the colony-forming cells, we developed a method for clonal analysis of endothelial proliferation [12]. In fact, plating of individual ECs derived from the ECFCs demonstrates that the proliferative potential of single ECs is very heterogenous. Some single ECs never divided *in vitro*, while other cells gave rise to a secondary colony of 2–50, 51–500, 501–2000, or more than 2000 endothelial progeny while cultured for only 14 days *in vitro*. These data suggest that circulating ECFCs possess varying levels of proliferative potential and the progeny of ECFCs at a clonal level display a hierarchy of EPC activity (Figure 13.1). In contrast, CFU-Hill cells do not replate at a single-cell level [18]. Thus, the majority of evidence suggests that CFU-Hill cells are not EPCs, but are myeloid lineage cells taht participate in

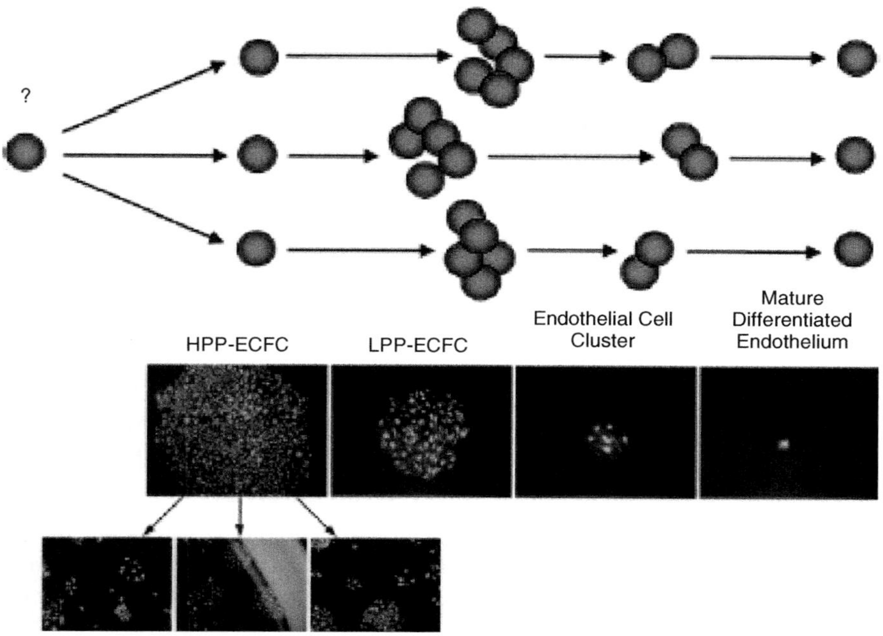

Figure 13.1 Model of an EPC hierarchy based on the proliferative and clonogenic potential of discrete populations of progenitor cells. The high proliferative potential HPP-ECFCs are large colonies that form secondary and tertiary colonies on replating. HPP-ECFCs give rise to all subsequent stages of endothelial progenitors in addition to replating into secondary HPP-ECFCs. The low proliferative potential LPP-ECFCs form colonies that contain more than 50 cells, but do not form secondary colonies or LPP-ECFCs on replating. EC clusters can arise from a single cell, but contain fewer than 50 cells that are typically larger than the smaller cells found in HPP-ECFC and LPP-ECFC colonies. Mature, terminally differentiated ECs do not divide. Whether HPP-ECFCs serve as stem cells for the endothelial lineage remains undetermined [12]. A color version of this figure appears in the plate section of this volume.

neoangiogenesis, although they do not directly form the endothelial monolayer of new vessels. In contrast, ECFCs display all the features of an EPC: clonal proliferative capacity, hierarchy of proliferative potential, and *de novo* vessel-forming ability *in vivo*. Validation of the role of ECFC in human clinical disorders and as a therapy for these disorders will determine the real potential of these EPCs as a means to promote neovascularization.

RESIDENT EPCs

As previously noted, the vascular endothelium displays a host of functions that require a near-continuous monolayer with highly regulated tissue-specific permeability characteristics. Thus, any damage or senescence of vascular ECs must be quickly addressed to maintain the endothelial continuity and avoid thrombus formation with compromised flow. While it would seem obvious that the ECs of most vessels must undergo significant turnover to protect the monolayer, ECs are largely described as quiescent cells and the average lifespan of an EC projected to be as long as 1 year [1]. Surprisingly, little is known about the specific proliferative potential of individual ECs. Numerous animal studies have demonstrated that the rate of aortic EC turnover is less than 1% daily, but that the rate of cell turnover is increased in the growing fetus *in utero*, at sites of vascular injury, near areas of disturbed blood flow, or in hypertensive animals [19–22]. Schwartz and Benditt [23] reported that the distribution of proliferating endothelium is not homogenous throughout the aorta, but in fact is clustered with some areas of the aortic endothelium showing high turnover and other areas devoid of turnover. Using a tritiated thymidine labeling approach, these authors [23] reported that clusters of replicating cells could be detected throughout the length of the rodent aorta under homeostatic conditions. Of interest, HUVECs and human aortic ECs (human aortic endothelial cell HAECs) are frequently used as control differentiated (nonproliferative) ECs in various assays of EPC functions *in vitro* [24–26]. However, it is well known that HUVECs and HAECs, as well as umbilical cord blood ECFCs, can be passaged for at least 40 population doublings *in vitro* [12, 26]. A previously unresolved question was whether all vascular ECs possess the same level of proliferative potential that is somehow released upon *in vitro* culture or if there were different levels of proliferative potential within individual ECs comprising the endothelial monolayer of each blood vessel. Recently, we have provided evidence that a complete hierarchy of ECFCs exists in the individual ECs isolated from umbilical vessels and the human aorta [27]. Thus, a diversity of resident EPCs resides within the human aortic and umbilical arterial and venous endothelium. This concept of resident EPCs provides a new paradigm for understanding yet one additional heterogenous feature of vascular endothelium, and is a stimulus to examine the content and function of resident EPCs in different organ vascular beds, including the lung.

PULMONARY EC DEVELOPMENTAL HETEROGENEITY

The process of pulmonary vascular development is complex and sustained from the embryonic stage of development to early adulthood (when alveolar branching and increases in surface area are completed) (see Chapter 1). Vessels are repeatedly formed and then regress as the vascular bed is remodeled to coincide with specific stages of lung epithelial differentiation that culminates in the finely tuned interface of the capillary endothelial monolayer in close apposition to the alveolar epithelium so as to precisely match ventilation and perfusion [28]. In the end, there are numerous EC subtypes that coat the pulmonary artery, arterioles, alveolar capillaries, pulmonary veins, bronchial arteries, veins, and capillaries. As discussed in other chapters in this volume, EC heterogeneity is prominent in the lung, reflecting differences in cell origin, and the spatial and temporal specificities that evolve as the lung develops. For example, while both the arterial and venous vessels branch throughout the lung, arterial and capillary ECs tend to display more complex interactions and even some codependence with adjacent airway epithelial cells that are not displayed by venous ECs.

The organization of the primary lung vascular plexus is a composite of endothelium from numerous sites. The upper regions of the right and left lung lobes are vascularized from ECs sprouting from the pulmonary trunk (derived from the truncus arteriosus) and the pulmonary arch arteries (derived from the sixth branchial arch) [28]. The lower lobes of the lung are vascularized from ECs sprouting from the intersegmental arteries (penetrating through the diaphragm from the dorsal aorta). Several models have been proposed to attempt to explain how the entire vasculature is derived from these formative sites. Both angiogenesis and vasculogenesis (ECs emerging from mesodermal precursors within the distal lung mesenchyme) are implicated in completing the vast network of pulmonary arterial, capillary, and venous vessels. While numerous genetic and epigenetic factors have been identified that either promote or inhibit growth of the pulmonary vasculature as a whole [28], little is known about how the proliferative potential is regulated in individual ECs of the lung microvasculature or macrovasculature. This is an important question since several recent reports have documented a proproliferative phenotype of rodent pulmonary microvascular ECs (pulmonary microvascular endothelial cell PMVECs) [29, 30].

COMPARISON OF PULMONARY MICROVASCULAR AND MACROVASCULAR ENDOTHELIAL PROLIFERATION

ECs isolated from rat pulmonary artery and pulmonary vessels less than 25 μm, display numerous differences in lectin-binding patterns *in vitro* and *in vivo* [30], physiological responses to a variety of molecules including thapsigargin *in vivo* [31], and permeability characteristics at baseline and with a physiologic stress (see Chapter 9) [32]. More recently, large differences in proliferative potential have been documented between the macro- and microvascular ECs. Recently, Solodushko and Fouty [29] reported that rat PMVECs proliferate at approximately two times the rate of rat pulmonary artery ECs (pulmonary artery endothelial cell PAECs). Even when plated in low serum concentrations (0.1% serum for 72 h), PMVECs displayed an enhanced growth rate compared to PAECs (Figure 13.2). Analysis of the cell cycle status of both cell populations during low serum culture revealed more than one-third of the PMVECs

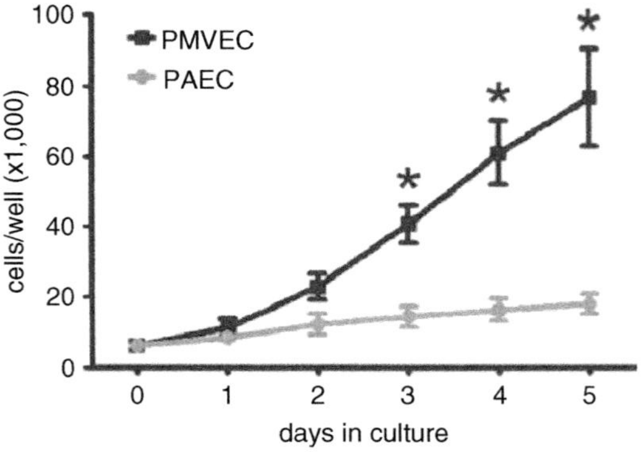

Figure 13.2 PMVECs proliferate faster than PAECs in 0.1% serum. Cells were plated at 20 000 cells/well and cell growth determined by daily cell counts [29].

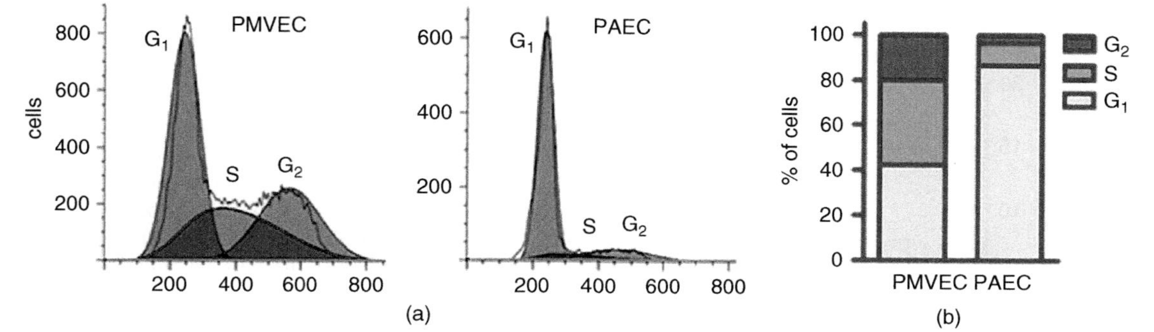

Figure 13.3 Serum withdrawal does not induce G_0/G_1 cell cycle arrest in PMVECs. Cells were plated in 0.1% serum for 3 days, harvested with trypsin, stained with propidium iodide, and analyzed by flow cytometry. (a) Representative cell cycle profiles for PMVECs versus PAECs. (b) There was a statistically significant increase in PMVECs in S and G_2/M and statistically fewer PMVECs in G_0/G_1 compared with PAECs [29].

to still be in S phase, while nearly all PAECs were in the G_0/G_1 phase of the cycle (Figure 13.3a,b). Since it is well known that the tumor suppressor retinoblastoma (Rb) protein must be inactivated (hyperphosphorylated) for cells to pass through the G_1/S transition of the cell cycle, it was perhaps somewhat surprising that the PMVECs displayed hyperphosphorylated Rb even in the presence of low serum culture and was accompanied by enhanced cyclin D–cyclin-dependent kinase (Cdk) 4 and cyclin E–Cdk2 activity (these kinases normally hyperphosphorylate Rb in the presence of mitogens). Cyclin D1 expression was also significantly higher in PMVECs than PAECs in low serum culture. Typically, cyclin D1 transcription and protein half-life are lowered by serum withdrawal leading to G_1 arrest, but this did not occur in the PMVECs. Nonetheless, by inhibiting the activity of Cdk2 *in vitro*, Rb remained active (hypophosphorylated) and the PMVEC became arrested in G_0/G_1. While a low serum concentration did not appear to growth arrest the PMVECs, permitting the cells to grow to complete confluence *in vitro* did. Of interest, this confluence-induced growth arrest was associated with a significant increase in the Cdk inhibitor $p27^{Kip}$ [1] as might be expected, but cyclin D1 expression remained elevated. In sum, these authors suggest that the proproliferative phenotype of the PMVECs likely relates to the significantly higher levels of cyclin D1 expression observed, although the upstream signaling pathways regulating this expression remain unclear.

ISOLATION OF PULMONARY RESIDENT MICROVASCULAR EPCs

Given the published evidence shown in the preceeding section, that rat PMVECs demonstrate higher proliferative potential *in vitro* than PAECs, we collaborated with the laboratory of Dr. Troy Stevens and questioned whether this difference was displayed in each individual PMVEC or was a heterogenous property displayed by some PMVECs and not others [33]. To address differences in proliferative potential at a clonal level, we first isolated fresh PMVECs and PAECs as previously described [29, 30], and demonstrated the purity of the cell populations obtained using cell surface antigen expression and transmembrane electrical resistance measurements known to be specific for PMVECs or PAECs [29, 30]. As predicted, rat PMVECs exhibited a 3.5-fold increase in cell growth for up to 1 week of *in vitro* culture compared to PAECs. During the logarithmic growth phase, PMVECs displayed a population doubling time of 39 h compared to 58 h for PAECs [33]. In subsequent studies, we plated individual PMVEC and PAEC deposited into rat tail collagen type 1-coated tissue culture wells using a flow cytometer, and cultured the cells in standard growth conditions for these cells [33]. Culture wells were examined after 2 weeks of culture and the percent of wells (originally containing a single cell after 24 h of deposition) that contained two or more cells were counted, as well as, the total number of cells in each well. More than 60% of the single PAECs plated in this assay remained as single nondividing cells, whereas less than 30% of the PMVECs remained as single nondividing cells (Figure 13.4). More than 50% of the single plated PMVECs gave rise to colonies of more than 2000 cells within the 14 day culture period in contrast to single plated PAECs, where 20% of the clones gave rise to colonies of more than 2000 cells (Figure 13.4 as above). In fact, some PMVEC clones gave rise to more than 100 000 progeny, whereas the largest colonies derived from PAEC clones rarely exceeded 10 000 cells. Perhaps the most important element in the analysis was that the progeny derived from these rapidly dividing

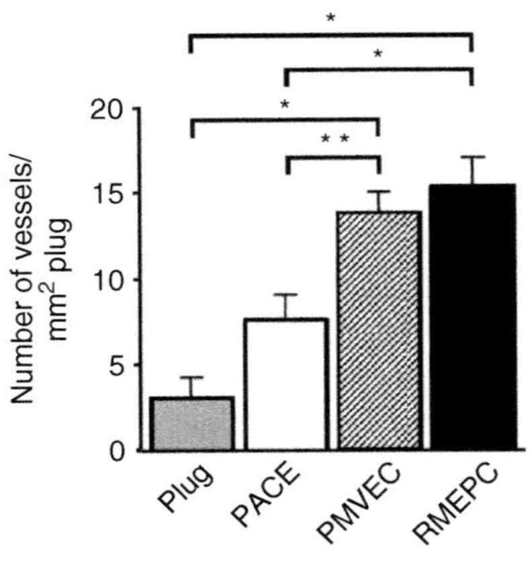

Figure 13.4 RMEPCs exhibit rapid neovasculogenesis in *in vivo* Matrigel plugs. Quantitative analysis of the *de novo* vessels (10 days postinjection) illustrates the high vasculogenic capacity of RMEPCs as compared to the other cell phenotypes or plugs injected with no cells (plug). $*p < 0.001$, $**p < 0.01$ compared to plug and PAECs, respectively, one-way analysis of variance [33].

PMVEC clones when replated at a single-cell level also displayed evidence that some replated clones could give rise to colonies of more than 2000 cells, whereas other individually replated PMVEC clones produced secondary colonies of 2–2000 cells and some replated clones did not divide. Thus, the primary individually plated PMVECs contained some ECs that displayed the potential to give rise to the complete hierarchy of ECFCs, suggesting the presence of resident EPCs with the microvascular endothelium of the rodent lung and we have termed these cells, resident microvascular EPCs (resident microvascular endothelial progenitor cell RMEPCs) [33].

Although highly enriched in proliferative potential, the RMEPCs expressed the same cell surface phenotype as the original pool of PMVECs, and could be easily discriminated from the PAECs using a panel of monoclonal antibodies and/or lectins and fluorescence-activated cell sorting. Furthermore, the RMEPCs displayed greater barrier functional properties than the PAECs, which is a characteristic of freshly isolated PMVECs. The rapidly proliferating RMEPCs displayed long telomeres – a typical feature observed in embryonic and adult stem cells, and a feature previously described in human umbilical cord ECFCs with the highest proliferative potential [12]. Both PMVECs and RMEPCs spontaneously formed elaborate capillary tube-like networks *in vitro* within 8–24 h of plating on Matrigel, whereas PAECs required a longer culture period for tube formation and the tubes regressed more quickly compared to PMVEC or RMEPC cultures [33]. Thus, the highly proliferative PMVECS and the enriched RMEPCs displayed many similar phenotypic and functional properties that differed from the properties of PAECs.

As noted above, perhaps the most stringent test to assay for the presence of an EPC is the ability of a putative EPC population to produce blood vessels *de novo* when implanted *in vivo*. We suspended PAECs, PMVECs, and RMEPCs (at similar cell densities) in Matrigel, and implanted the gels within the peritoneal cavity of anesthetized CD40 rats. Subsequently, animals were sacrificed at 4 or 10 days for analysis of the Matrigel plugs for evidence of neovascularization. Blood filled vessels within 4 days of implantation were noted in the plugs from all three cell populations, but not in the cell-deficient plugs, with vessel sizes ranging from less than 10 μm to arterioles and venules of nearly 50 μm in diameter. At the 10-day postimplantation timepoint, PMVEC- and RMEPC-containing plugs formed twofold more blood vessels than plugs containing PAECs (Figure 13.5). Since a few vessels also formed in the cell-free control plugs, additional studies were carried out where Green Fluorescence Protein (GFP)-expressing RMEPCs were suspended in Matrigel and implanted *in vivo*. The vast majority of blood vessels formed in these plugs *in vivo* were expressing GFP in ECs lining the vessels, proving that RMEPCs generate endothelial lined vessels *de novo* and thus function as EPCs [33]. These

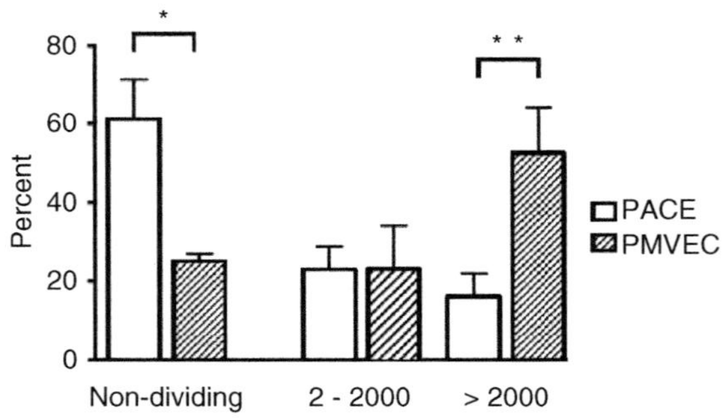

Figure 13.5 PMVECs are enriched with endothelial progenitors. PAECs (opened bars) or PMVECs (striped bars) were seeded at a single cell per well and the percentage of cells undergoing cell division was measured after 14 days in culture. PMVECs exhibited a proproliferative behavior where less that 25% of the cells remain as single nondividing cells–a difference that was around threefold lower compared to PAECs. Further, PMVECs displayed a high proliferative potential, where more than 50% of the plated single cells formed colonies larger than 2000 ECs/colony. $*p < 0.01$, $**p = 0.03$, compared to PAECs, respectively, two-way analysis of variance [33].

results suggest that the rat pulmonary microvasculature possesses ECs with resident EPC activity and that this activity is significantly greater than that observed in the pulmonary artery ECs.

Given the high proliferative potential of the RMEPCs, the capacity for self-renewal, and the high vasculogenic activity *in vivo*, we also conducted additional studies to be assured that the cells had not been transformed *in vitro*. Serum deprivation diminished RMEPC, PMVEC, and PAEC growth *in vitro*, and all three populations of cells were growth arrested upon reaching cell confluence *in vitro*. Plating of PMVECs and RMEPCs in agar resulted in some small colonies that were not observed when PAECs were plated; however, only breast cancer cells gave rise to numerous large colonies in these cultures. Thus, RMEPCs constitute a fast-growing population of ECs that form vessels *de novo* upon implantation but do not display features of transformed cells.

PRELIMINARY OBSERVATIONS ON A MOLECULE THAT REGULATES PULMONARY RMEPC ACTIVITY *IN VITRO*

With evidence that the pulmonary microvasculature contains resident EPC activity, we have recently pursued additional studies in collaboration with the laboratory of Dr. Troy Stevens, to begin to identify molecular pathways that may regulate the EPC activity *in vitro*. In preliminary studies, mRNA profiling of rat PMVECs and PAECs revealed a number of molecules differentially expressed. In our recent studies, we noted that nucleosome assembly protein (NAP)-1 was overexpressed 3.5-fold in PMVECs compared to PAECs at the mRNA level. This protein was of interest since NAP-1 had previously been reported to play an essential role in hematopoietic stem and progenitor cell differentiation [34].

NAP-1 is a highly conserved protein in eukaryotes, and plays an important proproliferative role in yeast, *Xenopus*, and *Arabidopsis thaliana*. NAP-1 was originally described for its ability to remodel chromatin, but more recently has been reported to chaperone the nuclear import of histones H2A and H2B, interact with p300 and regulate gene transcription, facilitate the exchange of histone variants in nucleosomes, and promote nucleosome sliding along DNA. While all of these functions may help explain the important role of NAP-1 in cell proliferation, the specific effects of NAP-1 in PMVECs has only recently been studied [35].

We have determined that NAP-1 is expressed at significantly higher levels in both the cytosol and nuclear fractions in PMVECs compared to PAECs using Western blot analysis [35]. To more directly examine the role

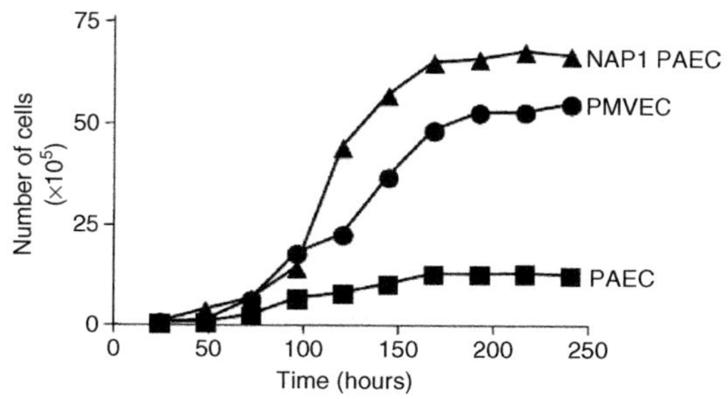

Figure 13.6 NAP-1 overexpression increases PAEC proliferation to the level displayed by PMVECs (a significant increase over that of normal PAECs) [35].

of NAP-1 in determining the proliferative phenotype of endothelium, we overexpressed NAP-1 in PAECs using retroviral infection of a full-length *NAP-1* plasmid. By Western analysis, PAECs infected with the *NAP-1* construct expressed the protein at levels resembling the endogenous NAP-1 expression observed in PMVECs. To examine cell growth, normal PAECs and PMVECs, and the NAP-1-overexpressing PAECs were cultured in low serum (0.1%) concentrations for 2 days and then switched to normal (10%) serum conditions for a 10-day culture period. Whereas PMVECs grew faster than PAECs (as previously observed), the NAP-1-overexpressing PAECs proliferated at the same rate as the PMVECs (Figure 13.6). Despite a change in the proliferative capacity of the NAP-1-overexpressing PAECs, no change in the cell surface phenotype or the barrier properties were observed in these cells compared to freshly isolated PAECs. When PMVECs were infected with a retrovirus carrying a short hairpin RNA that reduced NAP-1 protein expression, the proliferative rate of the infected PMVECs was reduced to the level normally observed for PAECs. Thus, highly proliferative cells express high levels of NAP-1, while cells with limited proliferative potential express low levels of NAP-1, and regulating the level of NAP-1 expression appears to regulate the proliferative potential of pulmonary macro- and microvascular ECs. Further studies will be required to specifically identify the molecular pathways downstream of NAP-1 that mediate changes in the proliferative potential of pulmonary ECs *in vivo*.

CLINICAL DISORDERS ASSOCIATED WITH DIMINISHED LUNG GROWTH OR REMODELING

The recent evidence that the lung contains resident EPCs may have clinical and therapeutic relevance for life-threatening and debilitating lung diseases characterized by arrested alveolar growth or loss of alveoli. Indeed, contemporary observations suggest that blood vessels in the lung actively promote alveolar growth during development and contribute to the maintenance of alveolar structures throughout postnatal life and several lung diseases are associated with a rarefaction of the pulmonary capillary bed (reviewed in [36–38]).

The first observation suggesting a link between blood vessels and impaired lung structure was made in 1959, when Liebow noticed that the alveolar septa in centrilobular emphysema were remarkably thin and almost avascular [39]. Human adult onset emphysema is a disease characterized by airspace enlargement distal to terminal bronchioles and is a major component of chronic obstructive pulmonary disease. Liebow postulated that a reduction in the blood supply of the small precapillary blood vessels might induce the disappearance of alveolar septa [40]. Despite this early observation, pulmonary vessels were long thought to be passive bystanders in lung development, with their development simply following the branching pattern of the airways. A role for diminished vascular development in the pathophysiology of another lung disorder, bronchopulmonary dysplasia (BPD), the chronic lung disease that follows ventilator and oxygen therapy for acute respiratory failure after premature birth, was recently proposed. Described in 1967, BPD was characterized by intense lung fibrosis emerging in the lungs of infants born around 34 weeks gestation that required pulmonary support for acute respiratory failure [40]. Major advances in perinatal medicine have since allowed for the survival of prematurely born infants that are considerably more immature than the 34- to 40-week gestation infants previously at risk for BPD. Thus, BPD now predominantly occurs in infants born at less than 28 weeks gestation–during the late canalicular stage of lung development, just when the blood vessels become juxtaposed to the developing airways [41]. As a result, the disease is now characterized by an arrest in alveolar development *and* capillary rarefaction [42–44], the long-term consequences of which are yet unknown. Consequently, a better understanding of how alveoli and the underlying capillary network develop and how these mechanisms are disrupted in the injured lung is critical to develop efficient therapies to prevent lung injury or regenerate established lung injury in prematurely born infants. The following sections describe evidence accumulated over the past 10 years suggesting the crucial role of lung angiogenesis and angiogenic growth factors during normal lung development, and in response to lung injury and repair, and form a strong rationale to further explore the biology of lung RMEPCs in health and disease.

VASCULAR EC GROWTH PROMOTES LUNG DEVELOPMENT

Histological observations demonstrate that endothelial tubes line up around the terminal buds of the airways [45], suggesting an inductive influence on the part of the epithelium (for endothelial growth) and that a coordinated and timely release of vascular-specific growth factors from respiratory epithelial cells may promote alveolar capillary development (see Chapter 1). For example, VEGF is a highly specific mitogen and survival factor for vascular ECs. The absolute requirement of VEGF for vascular development is well established, and demonstrated by the embryonic lethality and deficient organization of ECs in inactivation studies of VEGF alleles and knockouts of VEGFR-1 and -2 in the murine system (reviewed in [46]). In the lung, VEGF mRNA and protein are localized to distal airway epithelial cells and the basement membrane subjacent to the airway epithelial cells [47]. VEGF is present in alveolar type II cells in the developing mouse lung, and its expression peaks during the canalicular stage, when most of the lung vessel growth occurs, then decreases until day 10 postnatally (P10) when it plateaus at levels maintained throughout adulthood [48]. VEGFR-1 and -2 mRNA expression also increase during normal mouse lung development [49, 50], and these receptors are localized on pulmonary ECs [48], closely apposed to the developing respiratory epithelium. This spatial relationship suggests that VEGF plays a role in the development of the alveolar capillary bed.

Consistent with these observations, pharmacological and genetic VEGF inhibition results in decreased lung capillary growth and arrested alveoli formation [51–54] – features encountered in clinical BPD. In addition, prolonged treatment of adult rats with the VEGFR-1 and -2 blocker SU5416 leads to enlargement of the air spaces [55], indicative of emphysema, suggesting that VEGF is required not only for the formation, but also the maintenance of the pulmonary vasculature and alveolar structures throughout adulthood (see Chapter 25).

PATHOPHYSIOLOGY OF BPD AND ROLE OF VEGF IN LUNG INJURY AND REPAIR

VEGF signaling is disrupted in BPD. The proposed link between alveolarization and angiogenesis is suggested by the secondary abnormalities that occur in one process when the other is primarily affected. Arrested alveolar and lung vascular development are consistent findings in BPD. The first evidence that abnormal vascular development may contribute to postnatal lung disease came from autopsy studies showing reduced pulmonary microvascularization in infants dying from BPD [56]. A more recent postmortem study of newborns dying after short and prolonged durations of mechanical ventilation confirmed the reduction in vascular branching arteries, but interestingly lung platelet/EC adhesion molecule-1 protein content was decreased in infants dying after brief ventilation, but was increased after prolonged ventilation [57]. These findings suggest a transient decrease in endothelial proliferation, followed by a brisk proliferative response, despite an overall reduction in vessel number in infants with BPD. This observation suggests that dysmorphic lung vascular growth in BPD may not necessarily result simply from a reduction in the number of ECs.

Hypoxia is a major stimulator of VEGF expression (see Chapter 18). Premature exposure of the developing lung to a hyperoxic environment may downregulate VEGF expression. Even ambient O_2 levels (21%), that is, premature birth *per se*, may interfere with normal lung development [58]. Accordingly, animal models of impaired alveolar development mimicking BPD also display abnormal lung vascular development [59–63], and this is associated with decreased expression of VEGF and its receptors in lung tissue of hyperoxic neonatal rabbits [64] and rats [53], in ventilated neonatal mice [65], and in chronically ventilated preterm baboons [66], as well as in sheep with antenatal endotoxin exposure [67]. In premature infants with severe respiratory failure, tracheal aspirate VEGF levels are lower in nonsurvivors and in those that subsequently develop BPD as compared to survivors without chronic lung disease, and histopathological findings of arrested alveolarization in preterm infants dying with BPD is associated with decreased VEGF expression [68, 69].

In summary, inhibition of VEGF-driven angiogenesis in animal models causes capillary rarefaction and arrests alveolar development, reminiscent of BPD in human subjects. In infants affected with BPD with lung morphology characterized by arrested alveolar and vascular development, VEGF signaling is known to be decreased. These data form the rationale for testing the therapeutic potential of angiogenic growth factor replacement in lung diseases characterized by arrested alveolar growth.

THERAPEUTIC POTENTIAL OF ANGIOGENIC GROWTH FACTORS IN LUNG DISEASE

Recombinant human VEGF treatment of newborn rats during or after exposure to hyperoxia enhances vessel growth and improves alveolarization [70, 71]. Likewise, intratracheal adenovirus-mediated VEGF gene therapy improves survival, promotes lung capillary formation, preserves alveolar development, and regenerates new alveoli in this same model of irreversible lung injury [37]. However, in both of these animal studies, VEGF therapy induced immature, leaky capillaries, and lung edema – side-effects that might prove to be contributory to further lung injury in human infants. Combined intrapulmonary VEGF and angiopoietin-1 (which promotes vascular maturation) gene transfer preserves lung alveolarization and enhances angiogenesis with more mature capillaries that are less permeable, reducing the vascular leakage seen in VEGF-induced capillaries [37].

These observations highlight the tightly orchestrated process of angiogenesis and point toward the need to closely recapitulate this process to warrant efficient and safe angiogenesis during lung development. Hypoxia-inducible factor (HIF) is a master transcription factor modulating O_2-sensitive gene expression (including VEGF and angiopoietin-1) and vessel growth [72] (see Chapter 18). HIF is activated in hypoxic cells and inhibited by increased O_2 tissue levels. However, because HIF deficiency is lethal during embryonic and/or immediately in the postnatal subject, the role of HIF during lung alveolarization remains unknown [73]. Nonetheless, HIF activation via inhibition of prolyl hydroxylase domain-containing proteins prevents lung injury in the premature baboon model of BPD and further supports a potential role for angiogenic growth factor administration in promoting alveolar development [74].

These observations provide proof of concept for the crucial role of the lung vasculature in what is traditionally thought of as an airspace disease, and opens new therapeutic avenues to protect or regenerate new alveoli through the modulation of angiogenic growth factors and promotion of neoangiogenesis. However, much more needs to be learned about the tightly orchestrated process of angiogenesis if this mechanism is to be exploited therapeutically. The recent excitement in stem cell biology as an approach to tissue regeneration has sparked the interest in the reparative potential of EPCs. If angiogenic growth factors contribute to alveolar homeostasis, then EPCs are appealing candidate cells that may be recruited to participate in new vessel formation.

POTENTIAL ROLE OF RMEPCs IN BPD

The recent discovery of RMEPCs in the lung, as described in the preceeding section promises exciting new insights into the pathophysiology of numerous lung diseases and may present new therapeutic options. Indeed, recent experimental and clinical observations suggest that EPCs contribute to lung repair. LPS-induced murine lung injury is associated with a rapid release of EPCs into the circulation and the collaboration with other bone marrow-derived progenitor cells to lung repair [75]. In elastase-induced emphysematous lung injury, cells derived from the bone marrow develop characteristics of ECs and contribute to repair the alveolar capillary wall [76, 77]. Patients with acute lung injury have twofold higher numbers of circulating EPCs than healthy control subjects [78], suggesting some biological role for the mobilization of these cells during lung disease. More interestingly, and similar to the prognostic role of EPCs in ischemic vascular diseases, improved patient survival in acute lung injury correlates with increased circulating EPCs [79] and severity of illness [78]. Likewise, the number of circulating EPCs is significantly increased in patients with pneumonia and patients with low EPC counts tend to have persistent fibrotic changes in their lungs even after recovery from pneumonia. EPCs are also decreased in patients with restrictive and chronic lung disease [80]. In both diseases, there was a correlation between EPC circulating counts and disease severity. Finally, more recent findings in the developing lung suggest that arrested alveolar growth in experimental hyperoxic-induced BPD in neonatal mice is associated with decreased circulating, lung, and bone marrow EPCs [81]. Interestingly, hyperoxic adult mice did not display alveolar damage and had increased circulating EPCs, implying that decreased EPCs may contribute to the arrested lung growth seen in the neonatal animals [81].

These observations suggest that EPCs contribute to the repair of injured endothelium and help restore lung integrity and are consistent with the beneficial effect of angiogenic growth factors in experimental BPD described above. They also underscore the therapeutic potential of promoting lung angiogenesis to repair the lung. However, as mentioned before, a limitation of these studies has been the lack of a clear definition of how to unambiguously identify the true circulating and resident EPC. The recent discovery of RMEPCs [33] may be a crucial step forward to understanding the contribution of these cells to normal lung development, and determining their role during lung injury and repair with the hope to develop potent cell-based therapies for lung and other organ diseases.

CONCLUSIONS AND PERSPECTIVES

Defining an EPC has been difficult, in large part because of a paucity of distinct assays required to delineate specific cellular functions. Most recently, evidence for a circulating cell that possesses clonal proliferative potential for generating ECs and *in vitro* and *in vivo* vessel-forming ability has led to the recognition that only the rare circulating ECFC is an EPC. The majority of cells previously labeled as EPCs represent a variety of hematopoietic progenitor and lineage committed cells that participate in neoangiogenesis, but do not become functional ECs in those new vessels. Of interest, ECFCs are also found in the resident endothelial intima of vessels throughout the body, including the lung. The pulmonary microvasculature endothelium possesses a higher frequency of cells displaying high proliferative potential than the pulmonary arterial endothelium. Some preliminary studies suggest that certain proteins, such as NAP-1, may play a critical role in modulating the proliferative potential of the resident pulmonary ECFCs. Future studies will assess whether the differences in microvascular and macrovascular endothelial proliferative potential are restricted to the lung or are present in other tissues and organs. It will also be important to determine if the circulating ECFCs arise from resident vascular ECFCs and if the circulating ECFCs can migrate to sites where new vessels are required to serve as a useful cell therapy in vascular disorders in human subjects.

References

1. Aird, W. (2007) Phenotypic heterogeneity of the endothelium I. Structure, function, and mechanisms. *Circulation Research*, **100**, 158–73.
2. Aird, W.C. (2005) Spatial and temporal dynamics of the endothelium. *Journal of Thrombosis and Haemostasis*, **3**, 1392–406.
3. Spaet, T. and Gaynor, E. (1970) Vascular endothelial damage and thrombosis. *Advances in Cardiology*, **4**, 47–66.
4. Dignat-Gorge, F. and Sampol, J. (2000) Circulating endothelial cells in vascular disorders: new insights into an old concept. *European Journal of Haematology*, **65**, 215–20.
5. Asahara, T., Murohara, T., Sullivan, A. *et al.* (1997) Isolation of putative progenitor endothelial cells for angiogenesis. *Science*, **275**, 964–67.
6. Peichev, M., Maiyer, A., Pereira, D. *et al.* (2000) Expression of VEGFR-2 and AC133 be circulating human $CD34^+$ cells identifies a population of functional endothelial precursors. *Blood*, **95**, 952–58.
7. Shizuru, J.A., Negrin, R.S., and Weissman, I.L. (2005) Hematopoietic stem and progenitor cells: clinical and preclinical regeneration of the hematolymphoid system. *Annual Review of Medicine*, **56**, 509–38.
8. Khakoo, A.Y. and Finkel, T. (2005) Endothelial progenitor cells. *Annual Review of Medicine*, **56**, 79–101.
9. Cogle, C.R., Wainman, D.A., Jorgensen, M.L. *et al.* (2004) Adult human hematopoietic cells provide

functional hemangioblast activity. *Blood*, **103**, 133–35.

10. Hill, J.M., Zalos, G., Halcox, J.P. *et al.* (2003) Circulating endothelial progenitor cells, vascular function, and cardiovascular risk. *The New England Journal of Medicine*, **348**, 593–600.
11. Werner, N., Kosiol, S., Schiegl, T. *et al.* (2005) Circulating endothelial progenitor cells and cardiovascular outcomes. *The New England Journal of Medicine*, **353**, 999–1007.
12. Ingram, D.A., Mead, L.E., Tanaka, H. *et al.* (2004) Identification of a novel hierarchy of endothelial progenitor cells using human peripheral and umbilical cord blood. *Blood*, **104**, 2752–60.
13. Ingram, D.A., Caplice, N.M., and Yoder, M.C. (2005) Unresolved questions, changing definitions, and novel paradigms for defining endothelial progenitor cells. *Blood*, **106**, 1525–31.
14. Uemura, A., Kusuhara, S., Katsuta, H., and Nishikawa, S. (2006) Angiogenesis in the mouse retina: a model system for experimental manipulation. *Experimental Cell Research*, **312**, 676–83.
15. Naumov, G.N., Akslen, L.A., and Folkman, J. (2006) Role of angiogenesis in human tumor dormancy: animal models of the angiogenic switch. *Cell Cycle*, **5**, 1779–87.
16. Bonheur, J.A., Albadawi, H., Patton, G.M., and Watkins, M.T. (2004) A noninvasive murine model of hind limb ischemia–reperfusion injury. *The Journal of Surgical Research*, **116**, 55–63.
17. Schechner, J., Nath, A., Zheng, L. *et al.* (2000) *In vivo* formation of complex microvessels lined by human endothelial cells in an immunodeficient mouse. *Proceedings of the National Academy of Sciences of the United States of America*, **97**, 9191–96.
18. Yoder, M., Mead, L., Prater, D. *et al.* (2007) Redefining endothelial progenitor cells via clonal analysis and hematopoietic stem/progenitor cell principles. *Blood*, **109**, 1801–9.
19. Caplan, B.A. and Schwartz, C.J. (1973) Increased endothelial cell turnover in areas of *in vivo* Evans blue uptake in the pig aorta. *Atherosclerosis*, **17**, 401–17.
20. Schwartz, S.M. and Benditt, E.P. (1973) Cell replication in the aortic endothelium: a new method for study of the problem. *Laboratory Investigation*, **28**, 699–707.
21. Schwartz, S.M. and Benditt, E.P. (1977) Aortic endothelial cell replication. I. Effects of age and hypertension in the rat. *Circulation Research*, **41**, 248–55.
22. Schwartz, S.M., Stemerman, M.B. and Benditt, E.P. (1975) The aortic intima. II. Repair of the aortic lining after mechanical denudation. *American Journal of Pathology*, **81**, 15–42.
23. Schwartz, S.M. and Benditt, E.P. (1976) Clustering of replicating cells in aortic endothelium. *Proceedings of the National Academy of Sciences of the United States of America*, **73**, 651–53.
24. Bompais, H., Chagraoui, J., Canron, X. *et al.* (2004) Human endothelial cells derived from circulating progenitors display specific functional properties compared with mature vessel wall endothelial cells. *Blood*, **103**, 2577–84.
25. Gulati, R., Jevremovic, D., Peterson, T. *et al.* (2003) Diverse origin and function of cells with endothelial phenotype obtained from adult human blood. *Circulation Research*, **93**, 1023–5.
26. Auerbach, R., Lewis, R., Shinners, B. *et al.* (2003) Angiogenesis assays: a critical overview. *Clinical Chemistry*, **49**, 32–40.
27. Ingram, D.A., Mead, L.E., Moore, D.B. *et al.* (2005) Vessel wall-derived endothelial cells rapidly proliferate because they contain a complete hierarchy of endothelial progenitor cells. *Blood*, **105**, 2783–86.
28. Jones, P.L. and Aird W.C. (eds) (2007) *Pulmonary Vascular Development*, 1st edn, vol. **1**, Cambridge University Press, New York.
29. Solodushko, V. and Fouty, B. (2007) Proproliferative phenotype of pulmonary microvascular endothelial cells. *American Journal of Physiology: Lung Cellular and Molecular Physiology*, **292**, L671–77.
30. King, J., Hamil, T., Creighton, J. *et al.* (2004) Structural and functional characteristics of lung macroan microvascular endothelial cell phenotypes. *Microvascular Research*, **67**, 139–51.
31. Chetham, P., Babal, P., Bridges, J. *et al.* (1999) Segmental regulation of pulmonary vascular permeability by store-operated Ca^{2+} entry. *American Journal of Physiology: Lung Cellular and Molecular Physiology*, **276**, L41–50.
32. Kelly, J., Moore, T., Babal, P. *et al.* (1998) Pulmonary microvascular and macrovascular endothelial cells: differential regulation of Ca^{2+} and permeability. *American Journal of Physiology: Lung Cellular and Molecular Physiology*, **274**, L810–19.
33. Alvarez, D.F., Huang, L., King, J.A. *et al.* (2007) Lung microvascular endothelium is enriched with progenitor cells that exhibit vasculogenic capacity. *American Journal of Physiology: Lung Cellular and Molecular Physiology*, **294**, L419–30.
34. Stopka, T. and Skoultchi, A.I. (2003) The ISWI ATPase Snf2h is required for early mouse develop-

ment. *Proceedings of the National Academy of Sciences of the United States of America*, **100**, 14097–102.

35. Clark, J., Alvarez, D.F., Alexeyev, M.F. *et al.* (2007) Regulatory role for nucleosome assembly protein-1 in the proliferative and vasculogenic phenotype of pulmonary endothelium. *American Journal of Physiology: Lung Cellular and Molecular Physiology,*, **294**, L431–9.
36. Stenmark, K.R. and Abman, S.H. (2005) Lung vascular development: implications for the pathogenesis of bronchopulmonary dysplasia. *Annual Review of Physiology*, **67**, 623–61.
37. Thebaud, B. and Abman, S.H. (2007) Bronchopulmonary dysplasia: where have all the vessels gone? Roles of angiogenic growth factors in chronic lung disease. *American Journal of Respiratory and Critical Care Medicine*, **175**, 978–85.
38. Voelkel, N.F., Vandivier, R.W., and Tuder, R.M. (2006) Vascular endothelial growth factor in the lung. *American Journal of Physiology: Lung Cellular and Molecular Physiology*, **290**, L209–21.
39. Liebow, A.A. (1959) Pulmonary emphysema with special reference to vascular changes. *The American Review of Respiratory Disease*, **80**, 67–93.
40. Northway, W.H. Jr., Rosan, R.C., and Porter, D.Y. (1967) Pulmonary disease following respirator therapy of hyaline-membrane disease. Bronchopulmonary dysplasia. *The New England Journal of Medicine*, **276**, 357–68.
41. Kinsella, J.P., Greenough, A., and Abman, S.H. (2006) Bronchopulmonary dysplasia. *Lancet*, **367**, 1421–31.
42. Abman, S.H. (2001) Bronchopulmonary dysplasia: "a vascular hypothesis". *American Journal of Respiratory and Critical Care Medicine*, **164**, 1755–56.
43. Jobe, A.H. and Bancalari, E. (2001) Bronchopulmonary dysplasia. *American Journal of Respiratory and Critical Care Medicine*, **163**, 1723–29.
44. Jobe, A.J. (1999) The new BPD: an arrest of lung development. *Pediatric Research*, **46**, 641–43.
45. Hall, S.M., Hislop, A.A., and Haworth, S.G. (2002) Origin, differentiation, and maturation of human pulmonary veins. *American Journal of Respiratory Cell and Molecular Biology*, **26**, 333–40.
46. Yancopoulos, G.D., Davis, S., Gale, N.W. *et al.* (2000) Vascular-specific growth factors and blood vessel formation. Nature, **407**, 242–48.
47. Healy, A.M., Morgenthau, L., Zhu, X. *et al.* (2000) VEGF is deposited in the subepithelial matrix at the leading edge of branching airways and stimulates neovascularization in the murine embryonic lung. *Developmental Dynamics*, **219**, 341–52.
48. Ng, Y.S., Rohan, R., Sunday, M.E. *et al.* (2001) Differential expression of VEGF isoforms in mouse during development and in the adult. *Developmental Dynamics*, **220**, 112–21.
49. Gebb, S.A. and Shannon, J.M. (2000) Tissue interactions mediate early events in pulmonary vasculogenesis. *Developmental Dynamics*, **217**, 159–69.
50. Kalinichenko, V.V., Lim, L., Stolz, D.B. *et al.* (2001) Defects in pulmonary vasculature and perinatal lung hemorrhage in mice heterozygous null for the Forkhead Box f1 transcription factor. *Developmental Biology*, **235**, 489–506.
51. Galambos, C., Ng, Y.S., Ali, A. *et al.* (2002) Defective pulmonary development in the absence of heparin-binding vascular endothelial growth factor isoforms. *American Journal of Respiratory Cell and Molecular Biology*, **27**, 194–203.
52. Jakkula, M., Le Cras, T.D., Gebb, S. *et al.* (2000) Inhibition of angiogenesis decreases alveolarization in the developing rat lung. *American Journal of Physiology: Lung Cellular and Molecular Physiology*, **279**, L600–7.
53. Thebaud, B., Ladha, F., Michelakis, E.D. *et al.* (2005) Vascular endothelial growth factor gene therapy increases survival, promotes lung angiogenesis, and prevents alveolar damage in hyperoxia-induced lung injury: evidence that angiogenesis participates in alveolarization. *Circulation*, **112**, 2477–86.
54. Zhao, L., Wang, K., Ferrara, N., and Vu, T.H. (2005) Vascular endothelial growth factor coordinates proper development of lung epithelium and vasculature. *Mechanisms of Development*, **122**, 877–86.
55. Kasahara, Y., Tuder, R.M., Taraseviciene-Stewart, L. *et al.* (2000) Inhibition of VEGF receptors causes lung cell apoptosis and emphysema. *The Journal of Clinical Investigation*, **106**, 1311–19.
56. Tomashefski, J.F. Jr., Oppermann, H.C., Vawter, G.F., and Reid, L.M. (1984) Bronchopulmonary dysplasia: a morphometric study with emphasis on the pulmonary vasculature. *Pediatric Pathology*, **2**, 469–87.
57. De Paepe, M.E., Mao, Q., Powell, J. *et al.* (2006) Growth of pulmonary microvasculature in ventilated preterm infants. *American Journal of Respiratory and Critical Care Medicine*, **173**, 204–11.
58. Hjalmarson, O. and Sandberg, K. (2002) Abnormal lung function in healthy preterm infants. *American*

Journal of Respiratory and Critical Care Medicine, **165**, 83–87.

59. Han, R.N., Babaei, S., Robb, M. *et al.* (2004) Defective lung vascular development and fatal respiratory distress in endothelial NO synthase-deficient mice: a model of alveolar capillary dysplasia? *Circulation Research*, **94**, 1115–23.
60. Randell, S.H., Mercer, R.R., and Young, S.L. (1990) Neonatal hyperoxia alters the pulmonary alveolar and capillary structure of 40-day-old rats. *American Journal of Pathology*, **136**, 1259–66.
61. Roberts, R.J., Weesner, K.M., and Bucher, J.R. (1983) Oxygen-induced alterations in lung vascular development in the newborn rat. *Pediatric Research*, **17**, 368–75.
62. Shaffer, S.G., O'Neill, D., Bradt, S.K., and Thibeault, D.W. (1987) Chronic vascular pulmonary dysplasia associated with neonatal hyperoxia exposure in the rat. *Pediatric Research*, **21**, 14–20.
63. Wilson, W.L., Mullen, M., Olley, P.M., and Rabinovitch, M. (1985) Hyperoxia-induced pulmonary vascular and lung abnormalities in young rats and potential for recovery. *Pediatric Research*, **19**, 1059–67.
64. Maniscalco, W.M., Watkins, R.H., D'Angio, C.T., and Ryan, R.M. (1997) Hyperoxic injury decreases alveolar epithelial cell expression of vascular endothelial growth factor (VEGF) in neonatal rabbit lung. *American Journal of Respiratory Cell and Molecular Biology*, **16**, 557–67.
65. Bland, R.D., Mokres, L.M., Ertsey, R. *et al.* (2007) Mechanical ventilation with 40% oxygen reduces pulmonary expression of genes that regulate lung development and impairs alveolar septation in newborn mice. *American Journal of Physiology: Lung Cellular and Molecular Physiology*, **293**, L1099–110.
66. Maniscalco, W.M., Watkins, R.H., Pryhuber, G.S. *et al.* (2002) Angiogenic factors and alveolar vasculature: development and alterations by injury in very premature baboons. *American Journal of Physiology: Lung Cellular and Molecular Physiology*, **282**, L811–23.
67. Kallapur, S.G., Bachurski, C.J., Le Cras, T.D. *et al.* (2004) Vascular changes after intra-amniotic endotoxin in preterm lamb lungs. *American Journal of Physiology: Lung Cellular and Molecular Physiology*, **287**, L1178–85.
68. Lassus, P., Ristimaki, A., Ylikorkala, O. *et al.* (1999) Vascular endothelial growth factor in human preterm lung. *American Journal of Respiratory and Critical Care Medicine*, **159**, 1429–33.
69. Lassus, P., Turanlahti, M., Heikkila, P. *et al.* (2001) Pulmonary vascular endothelial growth factor and Flt-1 in fetuses, in acute and chronic lung disease, and in persistent pulmonary hypertension of the newborn. *American Journal of Respiratory and Critical Care Medicine*, **164**, 1981–87.
70. Kunig, A., Balasubramaniam, V., Markham, N.E. *et al.* (2006) Recombinant human VEGF treatment transiently increases lung edema but enhances lung structure after neonatal hyperoxia. *American Journal of Physiology: Lung Cellular and Molecular Physiology*, **291**, L1068–78.
71. Kunig, A.M., Balasubramaniam, V., Markham, N.E. *et al.* (2005) Recombinant human VEGF treatment enhances alveolarization after hyperoxic lung injury in neonatal rats. *American Journal of Physiology: Lung Cellular and Molecular Physiology*, **289**, L529–35.
72. Shweiki, D., Itin, A., Soffer, D., and Keshet, E. (1992) Vascular endothelial growth factor induced by hypoxia may mediate hypoxia-initiated angiogenesis. *Nature*, **359**, 843–45.
73. Compernolle, V., Brusselmans, K., Acker, T. *et al.* (2002) Loss of HIF-2alpha and inhibition of VEGF impair fetal lung maturation, whereas treatment with VEGF prevents fatal respiratory distress in premature mice. *Nature Medicine*, **8**, 702–10.
74. Asikainen, T.M., Chang, L.Y., Coalson, J.J. *et al.* (2006) Improved lung growth and function through hypoxia-inducible factor in primate chronic lung disease of prematurity. *The FASEB Journal*, **20**, 1698–700.
75. Yamada, M., Kubo, H., Kobayashi, S. *et al.* (2004) Bone marrow-derived progenitor cells are important for lung repair after lipopolysaccharide-induced lung injury. *Journal of Immunology*, **172**, 1266–72.
76. Ishizawa, K., Kubo, H., Yamada, M. *et al.* (2004) Bone marrow-derived cells contribute to lung regeneration after elastase-induced pulmonary emphysema. *FEBS Letters*, **556**, 249–52.
77. Ishizawa, K., Kubo, H., Yamada, M. *et al.* (2004) Hepatocyte growth factor induces angiogenesis in injured lungs through mobilizing endothelial progenitor cells. *Biochemical and Biophysical Research Communications*, **324**, 276–80.
78. Burnham, E.L., Taylor, W.R., Quyyumi, A.A. *et al.* (2005) Increased circulating endothelial progenitor

cells are associated with survival in acute lung injury. *American Journal of Respiratory and Critical Care Medicine*, **172**, 854–60.

79. Yamada, M., Kubo, H., Ishizawa, K. *et al.* (2005) Increased circulating endothelial progenitor cells in patients with bacterial pneumonia: evidence that bone marrow derived cells contribute to lung repair. *Thorax*, **60**, 410–13.

80. Fadini, G.P., Schiavon, M., Cantini, M. *et al.* (2006) Circulating progenitor cells are reduced in patients with severe lung disease. *Stem Cells*, **24**, 1806–13.

81. Balasubramaniam, V., Mervis, C.F., Maxey, A.M. *et al.* (2007) Hyperoxia reduces bone marrow, circulating, and lung endothelial progenitor cells in the developing lung: implications for the pathogenesis of bronchopulmonary dysplasia. *American Journal of Physiology: Lung Cellular and Molecular Physiology*, **292**, L1073–84.

14

Bronchial Vasculature: The Other Pulmonary Circulation

Elizabeth Wagner

Department of Medicine, Division of Pulmonary and Critical Care Medicine, Johns Hopkins Asthma and Allergy Center, Baltimore, MD, USA

INTRODUCTION

Frequently forgotten, the bronchial circulation provides lung structures with oxygenated blood at systemic arterial perfusion pressure and comprises 1–3% of the cardiac output. Compared to pulmonary arterial endothelium, bronchial arterial endothelial cells (ECs) are exposed to a much higher perfusion pressure, oxygen tension, and lower blood flow which likely impact functional characteristics. However, both vasculatures share a common drainage pathway, the mechanical stresses imposed by the ventilating lung, and proximity to the external environment. Despite these similarities and differences, relatively little is known specifically about the biology and functional importance of the bronchial vasculature compared to the pulmonary circulation. Largely due to the difficulty in accessing this circulatory bed, attributes of the endothelium lining these systemic vessels, which are dwarfed by the pulmonary vasculature, are still being defined. However, the gross pathologic outcome of bronchial endothelial activation has long been recognized. Specifically, the unique proliferative capacity of bronchial endothelium compared to pulmonary endothelium, in a variety of lung diseases, is well established. Probably the first illustrated example of bronchial EC proliferation during an inflammatory condition was shown by Leonardo da Vinci, in his anatomical drawings of about 1513 (reviewed by Cudkowicz [1]). Close scrutiny of a tuberculous cavity in a terminal lung unit shows bronchial vessels supplying the walls of the cavity. Whether da Vinci used artistic license in his depiction of the vasculature has been argued over the years [2, 3]. However, an additional major discovery in 1847 by Virchow confirmed the proliferative phenotype of bronchial endothelium during chronic pulmonary artery obstruction [4]. These studies demonstrated that under conditions of pulmonary vascular ischemia, tissue responses led to the proliferation and recruitment of systemic bronchial endothelium to ischemic lung regions. Despite the lack of extensive investigation into the proliferative capacity of bronchial ECs *per se*, human lung pathologic reports have shown the propensity of bronchial endothelium to proliferate during conditions of chronic inflammation. Although the specific signals driving this proliferation are still not clearly defined, the lack of responsiveness of pulmonary arterial endothelium under these pathologic conditions underscores the endothelial heterogeneity within the lung.

Another prominent bronchial endothelial attribute, associated with the pathology of asthma, is the alteration of barrier function contributing to airway wall edema. Sir William Osler stated in his original textbook (1892) "the hyperaemia and swelling of the mucosa... explain well the hindrance to inspiration and expiration" during an asthma attack [5]. Although mechanisms regulating airway EC barrier function were not known at the time, the clinical consequences were recognized. Thus, despite a lack of specific understanding of the contribution of ECs within the bronchial vasculature, human lung pathologic conditions have long been recognized as a result of changes in bronchial EC properties.

STRUCTURE

Although few studies of the embryonic development of the bronchial vasculature exist, the bronchial artery arises as an outgrowth from the aorta between week 9 and 12 of gestation in humans [6, 7]. In some species, the

The Pulmonary Endothelium: Function in health and disease Editors Norbert F. Voelkel, Sharon Rounds

bronchial artery originates from the intercostal arteries. The bronchial artery courses to the dorsal aspect of the carina where it bifurcates and sends branches down the mainstem bronchi. This vascular bed perfuses the airways from the level of the carina to the terminal bronchioles [8, 9]. The bronchial arteries send arterioles throughout the airway adventitia that connect with capillaries that are prominent in both the adventitia as well as the mucosa of the airway wall. Thus, the bronchial vasculature forms parallel vascular plexuses, situated on either side of the airway smooth muscle. In the intraparenchymal airways, postcapillary venules collect bronchial venous drainage, which flows into pulmonary venules and/or alveolar capillaries that drain into pulmonary veins and the left atrium. Additionally the bronchial artery sends arteriole branches to large pulmonary vessels as vasa vasorum, to nerves, lymph nodes, and the visceral pleura. Although these represent unique thoracic structures, the bronchial endothelium likely regulates nutrient flow and recruitment of circulating cells to each.

At the microscopic level, relatively little is known specifically about the morphology or unique site-specific markers of the endothelium that lines the vessels of the bronchial vasculature. Capillaries are predominantly continuous and in healthy mammals, lacking in fenestrations. Electron micrographs of the luminal surface of rat tracheal venules (Figure 14.1) show uniform EC borders that are narrow and closely apposed [10, 11]. Whether these tracheal venular EC characteristics are also representative of lower airways is not clear. Only recently have investigators isolated bronchial ECs and studied them *in vitro* [12, 13]. In these studies, functional heterogeneity has, as in the pulmonary vasculature, been observed between macrovascular and microvascular bronchial endothelium. Further heterogeneity may also exist between tracheal and bronchial endothelium, thus contributing regional complexity.

PHYSIOLOGICAL FUNCTION

Given that the bronchial circulation perfuses a variety of structures within the lung, presumably each endothelial component contributes specifically to unique homeostatic functions. However, support for these functional attributes are largely from whole-animal studies and only a few delve into EC-specific responses at the molecular level. As in other vascular beds, the ECs lining the bronchial vasculature have been shown to influence vascular smooth muscle and thereby regulate blood flow, exhibit angiogenesis, interact with circulating cells, provide a barrier function, and influence substrate metabolism. The evidence for each of these functions follows.

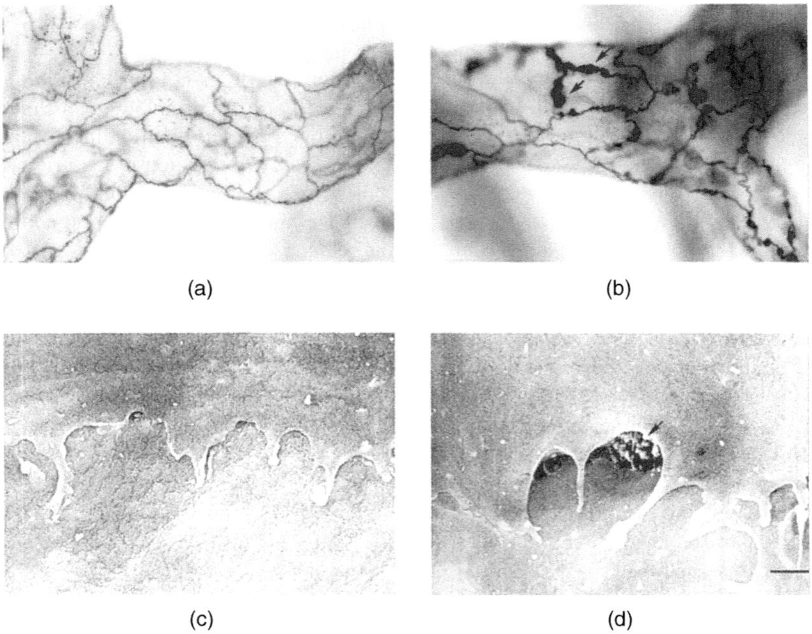

Figure 14.1 The scanning electron micrograph of the normal rat tracheal endothelium (c) shows overlapping, tight borders. Silver nitrate staining shows uniform, regular borders (a). In pathogen-infected rats, endothelial gaps are manifest as focal deposits of silver nitrate (arrows; b) and widely separated EC gaps (d). Bar = 15 μ (a and b) and 1 μm (c and d). Relabeled from [11] by permission of The American Physiological Society.

Bronchial Endothelial-Dependent Vasodilation

Several physiologic studies have demonstrated the importance of endothelial-derived substances in the modulation of bronchial vascular resistance. Table 14.1 lists a variety of bronchial vasodilator and vasoconstrictor substances. Systemic infusion of inhibitors of endothelial nitric oxide (NO) synthase has demonstrated a decrease in bronchial blood flow. Bronchial vasodilators such as bradykinin, acetylcholine, inhaled β-agonists, and ionic and nonionic contrast media all function partially through the release of endothelial-derived NO [14–17]. Cyclooxygenase inhibitors likewise have been shown to elicit endothelium-dependent vasoconstriction [17]. Exogenous delivery of the potent EC-derived constrictor endothelin caused a significant increase in bronchial vascular resistance [18]. These *in vivo* responses support predictions based on EC responses in other systemic organs. However, the endothelial-derived modulation of blood flow through the bronchial vasculature provides little information concerning the functional impact of changes in blood flow on the structures perfused by the bronchial vasculature. Several studies have demonstrated that the level of blood flow through the bronchial circulation can affect both the magnitude and time course of agonist-induced airway smooth muscle constriction by contributing to the passive wash-out of agonist [19–21]. Thus, changes in airway wall perfusion due in part to airway EC activation may ultimately modulate airway smooth muscle reactivity. Interestingly, many of the substances that cause airway smooth muscle constriction also cause an increase in bronchial blood flow. Thus, an increase in flow may contribute to the passive wash-out of airway smooth muscle agonists thereby limiting bronchoconstriction.

Bronchial EC Proliferation and Migration

Vascular proliferation in the adult lung is a bronchial EC phenomenon. The prominent pathologic feature of angiogenesis takes place in the lung during conditions of chronic inflammation such as cystic fibrosis [22], asthma [23], pulmonary fibrosis [24], lung cancer [25], chronic obstructive pulmonary disease (COPD) [26], pulmonary atresia [27], and chronic pulmonary thromboembolic disease [28]. Within the lung, however, the regional extent of neovascularization appears to be somewhat differentiated based on whether the underlying pathology is predominantly an airway disease (asthma, COPD, cystic fibrosis) versus a lung parenchymal disorder (chronic thromboembolism, pulmonary atresia). It is interesting that the pulmonary circulation appears not to participate in angiogenesis except under conditions of pneumonectomy [29] and chronic hypoxia [30]. Numerous studies have shown that systemic vascular beds undergoing angiogenesis are functionally abnormal with a proinflammatory phenotype, demonstrating increased protein transudation, fluid flux, leukocyte recruitment and vasodilation [31, 32], all of which are suggestive of abnormal EC function. Thus, bronchial angiogenesis represents a very active EC responsiveness that is well documented in human lung pathology. However, relatively little is known regarding the mechanisms responsible for systemic neovascularization of the lung or the reasons for the relative resistance of the pulmonary endothelium within the same lung environment. Recent correlative studies demonstrate an increase in lavaged growth factors such as vascular endothelial growth factor (VEGF), monocyte chemotactic protein-1, and angiogenin from asthmatic subjects that cause enhanced *in vitro* tube formation [33]. Additionally, lavaged VEGF was correlated with asthma severity [34]. Animal models of lung angiogenesis allow for more direct assessment of growth factors important in the initiation of neovascularization in the adult mammal (see Chapter 11). Perhaps the best-studied animal models of angiogenesis in the lung are the neovascularization that occurs in the trachea after chronic bacterial infection [35, 36] and in the lung parenchyma after chronic left pulmonary artery ligation [37–39].

Table 14.1 Bronchial blood flow responses.

Vasoconstrictors	Vasodilators
α-Adrenergic agonists	β-adrenergic agonists
Endothelin	acetylcholine
Glucocorticoids	adenosine
Increased airway pressure	antigen
Increased left atrial pressure	bradykinin
Tumor necrosis factor-α	cold, dry air
Vasopressin	histamine
	hypercarbia
	hypoxia
	NO
	prostanoids

Inflammation-Induced Neovascularization

Rodent models have been used to show changes in airway vascularity during inflammation, and are characterized by increased vessel numbers, vessel size, and permeability [36]. Airway remodeling following *Mycoplasma pulmonis* infection showed time- and strain-dependent changes in tracheal vascularity in mice (see Figure 14.2 [40]). New and remodeled vessels form days after infection along with capillaries, which display the phenotype of venules with enhanced plasma leakage and leukocyte influx. In this model, airway lymphatics also showed a proliferative phenotype. Sustained alterations in the tracheal vasculature could be reversed with corticosteroid

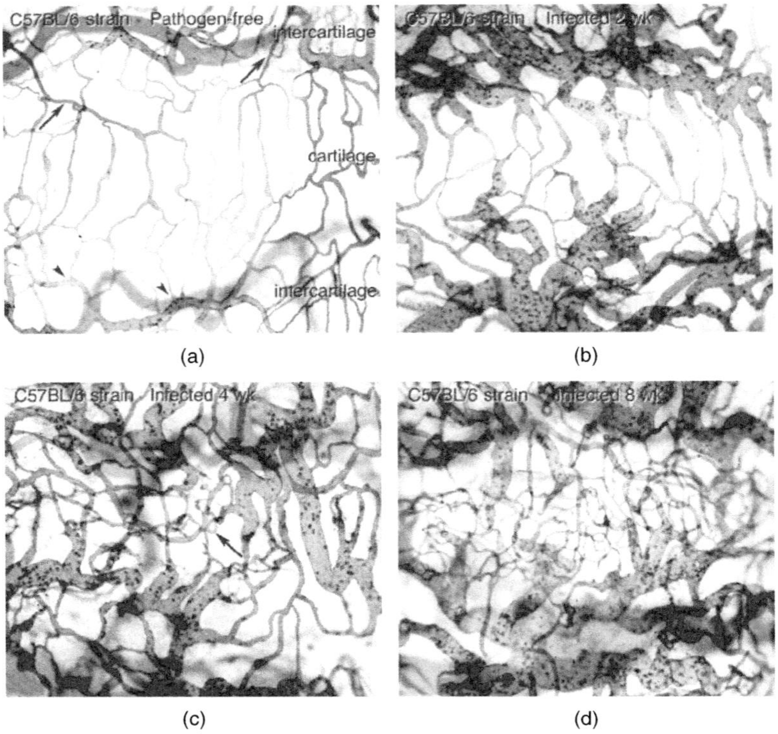

Figure 14.2 Whole-mount sections (perfusion stained with biotinylated *Lycopersicon esculentum* lectin) of mouse tracheal vasculature before and after neovascularization due to *M. pulmonis* infection. (a) The normal vasculature and (b–d) the increase in size and number of vessels, 2, 4, and 8 weeks after infection. Reproduced from Thurston *et al.*, (1998) [40] by permission of the American Society for Investigative Pathology.

treatment [41]. Although VEGF-C and VEGF-D appear to be necessary for lymphangiogenesis, the other VEGF isoforms were not essential for neovascularization [42]. Increased expression of the angiopoietins signaling through the Tie-2 receptor appear to replicate this tracheal model of angiogenesis, thus suggesting an important role for this family of growth factors [43, 44]. Whether these models of tracheal neovascularization are representative of changes that take place in diseased human airways is not clear. However, they provide important model systems to delineate basic mechanisms of systemic EC proliferation.

Left Pulmonary Artery Obstruction

Neovascularization of the systemic circulation into the lung after pulmonary artery obstruction has been confirmed and studied in humans [45, 46]. Bronchial arteriograms in patients with chronic thromboembolic disease demonstrate the unique capacity of systemic vessels to proliferate and to invade the ischemic lung parenchyma [46, 28]. Both a dilated bronchial artery as well as a fine meshwork of vessels distal to the pulmonary occlusion can be seen. As an experimental model of chronic pulmonary thromboembolism, the left pulmonary artery has been obstructed in a variety of research animals, and subsequent systemic neovascularization of the ischemic lung has been studied in sheep [47], dog [48], pig [37], guinea pig [39], rat [49], and mouse [38]. Systemic blood flow to the lung has been shown to increase to as much as 30% of the original pulmonary blood flow after pulmonary artery occlusion [50]. Figure 14.3 shows a methacrylate cast of the angiogenic bronchial vasculature in the rat, 28 days after left pulmonary artery ligation and demonstrates remarkable proliferation of this vasculature. In general, this model demonstrates rapid invasion of systemic vessels (bronchial, intercostal arteries) into ischemic lung parenchyma while the pulmonary vasculature remains relatively inert. In this regard, the bronchial circulation has been compared to "Mother or the Red Cross; normally accepted and unsung, but capable of giving vital help when needed" [8]. To determine growth factors essential to this process, Srisuma *et al.* have identified

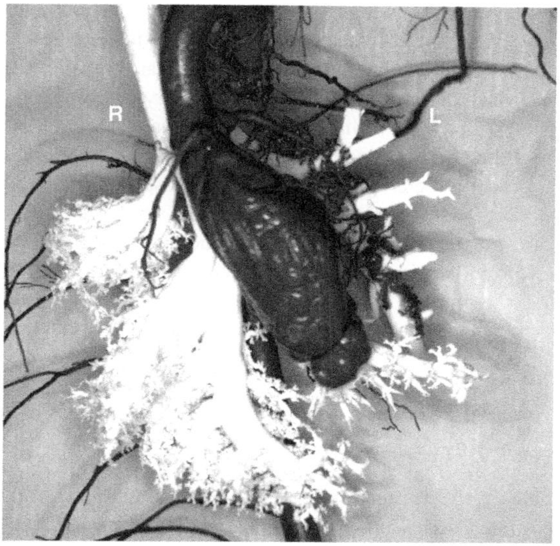

Figure 14.3 Difference in bronchial vascularity of the left (L) airway tree compared to the right (R) lung, 28 days after chronic left pulmonary artery ligation in the rat. Large, tortuous upstream bronchial vessels of the left lung are in stark contrast to the right bronchus, which is essentially devoid of large vessels. Note the aortic arch and coronary arteries filled with methacrylate casting material, the chamber of the left ventricle, and the extensive proliferation of the bronchial vasculature associated with the left bronchus and bronchioles. From unpublished observations of Wagner and Sukkar.

the proangiogenic CXC chemokines to be significantly upregulated in the ischemic left lung [51]. Additionally, neutralizing antibodies to $CXCR_2$, the G-protein-coupled receptor through which these chemokines signal, significantly limited neovascularization in both mice [52] and rats [53] after pulmonary ischemia. These observations add further support to a growing body of evidence reported by Strieter *et al.* in several lung pathologies, that the CXC chemokines and $CXCR_2$ are important proangiogenic factors within the lung [54]. Based on this information, the difference in proangiogenic potential of systemic ECs relative to pulmonary endothelium, was proposed to be related to chemokine responsiveness [55]. The effects of macrophage inflammatory protein (MIP)-2, one of the CXC chemokines, on primary culture mouse systemic arterial and pulmonary artery EC migration were determined. Although no basal differences were apparent, only systemic arterial endothelium demonstrated an increase in migration when exposed to MIP-2. Expression of $CXCR_2$ was not different among cell types. However, cathepsin S, a proteolytic enzyme important for EC motility [56], was shown to be elevated and demonstrate increased activity in systemic arterial ECs relative to pulmonary ECs after chemokine treatment. These results demonstrate one mechanism that may contribute to the difference in proangiogenic potential of these different lung endothelial subtypes. Furthermore, assessing the *in vivo*/*in vitro* responses provide effective model systems to test other growth factors important for selective systemic arterial EC proliferation.

Bronchial Endothelial Barrier Function

Numerous studies in a variety of models, as well as autopsy specimens of human lungs, demonstrate the propensity for the systemic airway circulation to contribute to fluid accumulation within and around airways. Airway wall edema is a prominent feature of the asthmatic airway. We have shown that direct infusion of bradykinin through the bronchial artery resulted in a significant and sustained increase in airway wall area determined by quantitative morphometry, airway wall thickness determined *in vivo* by high-resolution computed tomography, and an increase in lung lymph flow (see Figure 14.4 [57–60]). Both the Hales and Traber laboratories have shown that prostanoids and other inflammatory substances released in the airway after smoke inhalation cause increases in lung lymph flow and protein transudation [61–63]. These responses were eliminated when bronchial arterial flow was impeded, thus implicating the bronchial vasculature in contributing to fluid leak during smoke inhalation. A variety of inflammatory cytokines cause postcapillary venular endothelial gap formation, and allow transudation of plasma and protein into the airway interstitium [64, 10]. Figure 14.1(d) shows an electron micrograph of tracheal EC gap formation in a chronically infected rat. Long finger-like processes bridge gaps. In another study, angiopoietin-1 was shown to reduce endothelial gaps and prevent plasma leakage [65]. To characterize *in vitro* mechanisms responsible for gap formation, ECs isolated from bronchial conduit vessels from sheep were compared to bronchial subepithelial microvessels [12]. Assessment of the permeability coefficient to fluorescein isothiocyanate-conjugated dextran (molecular weight 9500) showed bronchial microvascular endothelium to be more permeable at baseline, and more responsive to both thrombin and bradykinin than bronchial artery endothelium. This result was obtained despite similar levels of bradykinin receptors and intracellular calcium responses to agonists. Thus, unlike the pulmonary vasculature where the microvascular endothelial barrier has been shown to be more restrictive than conduit vessel endothelium, the reverse is true for systemic airway endothelium. Although the mechanisms

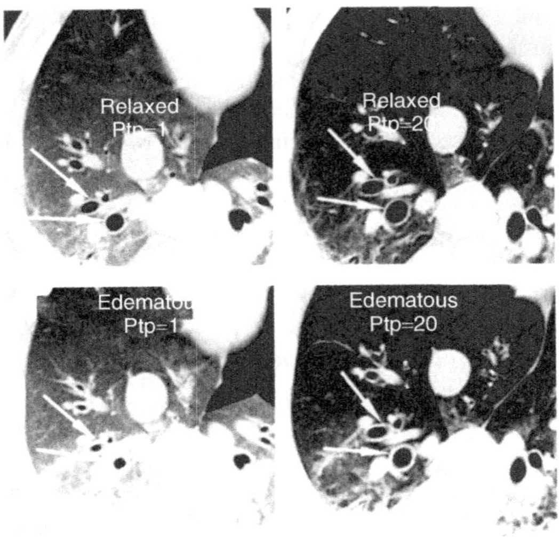

Figure 14.4 High-resolution computed tomography scans from one sheep showing the change in airway wall area and edema after infusion of bradykinin directly into the bronchial artery. Arrows indicate the same two airways in each scan before and after edematous conditions were induced, at low lung volume (Ptp = 1 cmH_2O) and high lung volume (Ptp = 20 cmH_2O). Ptp, transpulmonary pressure. From Brown *et al.*, (1997) [60] used with permission of The American Physiological Society.

responsible for these differences are not currently understood, the results underscore the importance of studying the appropriate EC subtype.

A novel aspect of EC barrier function within the lung relates to the protective effects of bronchial perfusion on pulmonary endothelial barrier function. Although the presence of bronchial artery-derived, pulmonary vasa vasorum has long been recognized, the functional significance of these vessels has been unexplored. The existence of vasa vasorum to pulmonary arterioles of 200 μm diameter has been confirmed [66]. Furthermore, in an ischemia–reperfusion lung injury model, perfusion of the bronchial artery during pulmonary artery ischemia and reperfusion, or pulmonary artery reperfusion alone, significantly attenuated the increase in pulmonary arterial endothelial permeability that occurred in the absence of bronchial artery perfusion [67]. Since bronchial artery perfusion during the reperfusion phase alone was effective at reducing injury, this result excludes the possibility that mere nutrient flow to pulmonary endothelium accounted for the preserved barrier function. These results raised the hypothesis that the bronchial endothelium generates a protective, antioxidant substance to the pulmonary endothelium during pulmonary artery reperfusion. NO produced by bronchial endothelium is a likely, shear stress-induced, antioxidant candidate. Subsequent preliminary studies demonstrated that perfusion of the bronchial circulation during pulmonary artery reperfusion with a NO synthase inhibitor attenuated the protective effect in this model. These results further highlight the potential heterogeneity of EC responses to ischemia with regard to barrier function as well as basic enzymatic function during ventilated ischemia. Furthermore, they provide evidence that the systemic circulation of the lung can impact functional attributes of the pulmonary vasculature, both of which are endothelium-dependent.

Bronchial EC Interaction with Circulating Cells

Leukocyte Recruitment

Leukocyte recruitment in systemic organs involves an orchestrated series of molecular events to occur between rolling leukocytes and postcapillary venular endothelium (see Chapter 10). Since the tracheal and bronchial circulations are systemic circulatory beds, the same EC adhesion molecules likely are responsible for the well-documented leukocyte recruitment of inflammatory airways diseases. Although this is generally assumed, only recently have studies using intravital microscopy to document specific molecules and mechanisms of recruitment been performed in airways *in vivo* [68–70]. In addition to general inflammatory stimuli such as lipopolysaccharide, formyl-Met–Leu–Phe, bradykinin, and thrombin, recent work suggests that excessive ventilatory stress imposed on airway venular endothelium may exert additional EC stimulation that is not predicted by static cell culture conditions. Application of positive end-expiratory pressure (PEEP) was shown to result in stimulus-dependent increases in the number of firmly adherent neutrophils, which could be blocked by P-selectin inhibition (see Figure 14.5 [70]). When cultured systemic venular ECs were studied after excessive cyclic stretch (20% elongation) in an attempt to model *in vivo* conditions, increased P-selectin expression was confirmed. Furthermore, this response was shown to be calcium-dependent and requiring T-type calcium channels [71]. This work is consistent with results obtained in pulmonary microvascular endothelium demonstrating the importance of T-type calcium channels for subsequent activation of endothelial Weibel–Palade bodies [72], which store proteins such as P-selectin, von Willebrand factor, and interleukin-8 [73]. These observations demonstrate overall that excessive lung mechanical stress, in addition to influencing the pulmonary vascular endothelium, also stimulates the systemic airway endothelium to become proinflammatory. However, this limited information regarding proinflammatory activation of bronchial endothelium highlights

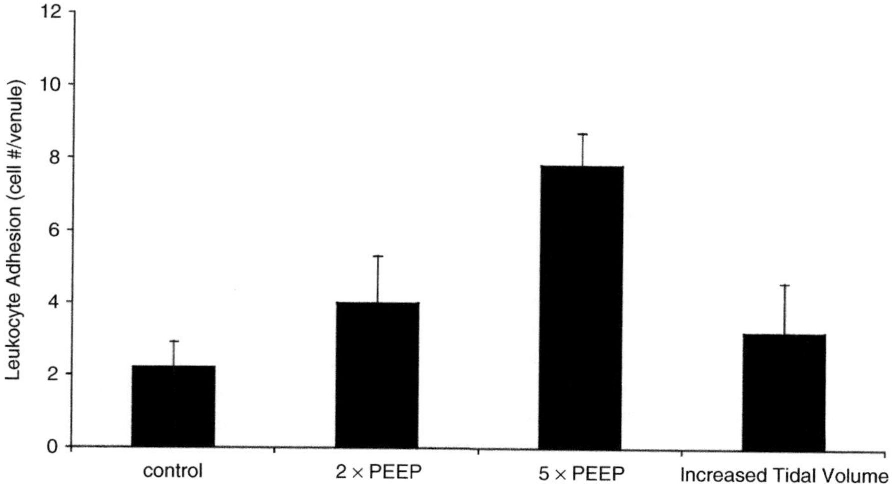

Figure 14.5 The effects of airway distension induced by repetitive exposures to PEEP on leukocyte adhesion in mouse tracheal postcapillary venules. A significant increase in leukocyte adhesion was observed after five 1-min periods of 8 cmH_2O of PEEP were applied. From Wagner and Jenkins (2007) [70] used with permission of The American Physiological Society.

the need for further experimentation to define the relevant stimuli and the essential EC proteins that broker the communication of circulating cells with this specific endothelium.

Progenitor Cells

Recent work by Davie *et al.* demonstrates an association of circulating bone marrow-derived progenitor cells (c-kit$^+$ cells) with expanded vasa vasorum of the pulmonary artery adventitia after chronic hypoxia [74]. Since the bronchial artery normally sends small arterioles to the walls of large pulmonary arteries and veins as vasa vasorum, this increase in vascular density confirms the growth of the systemic vasculature of the lung during these experimental conditions. Furthermore, it suggests that bronchial endothelium interacts with circulating progenitor cells to result in vascular growth and expansion (see Chapter 13). Additional support for the concept that bronchial endothelium may interact with circulating progenitor cells was provided in asthmatic subjects – a group that has in general shown an increase in systemic airway vascularity [23]. Asosingh *et al.* showed that bone marrow-derived endothelial progenitor cells from asthmatic subjects were highly proliferative in an *in vitro* angiogenesis assay [75]. These authors concluded that endothelial progenitor cells interacted directly with bronchial endothelium to promote neovascularization. These observations provide important information suggesting that blood-borne cells home to sites of injury or repair and are associated with bronchial endothelium. However, the overall role of circulating progenitor cells for repair needs further confirmation and definition.

Bronchial Endothelial Metabolism

Few studies have focused directly on the metabolic functions of the bronchial endothelium (see Chapter 7). Work by Grantham and colleagues in an in situ perfused bronchial artery sheep preparation showed that an angiotensin-converting enzyme (ACE) inhibitor significantly depressed metabolism of a synthetic peptide substrate for ACE [76]. These authors concluded that the bronchial circulation is pharmacokinetically and metabolically active with respect to bradykinin and that the enzymes responsible for this metabolic activity line the vascular lumen. The importance of ACE activity was confirmed *in vivo* by work showing the inhibition of bradykinin vasodilation after treatment with an ACE inhibitor [13]. Results of this study also showed that bronchial microvascular ECs *in vitro* expressed levels of ACE activity equivalent to pulmonary ECs. These experiments suggest an important regulatory role for bronchial endothelial ACE in the metabolism of kinin peptides known to contribute to airway pathology [77].

CONCLUSIONS AND PERSPECTIVE

The endothelium of the bronchial circulation shares characteristic features of other systemic vascular beds. Namely, leukocyte recruitment and vascular leak occur at postcapillary sites, and there is a vigorous angiogenesis response to tissue ischemia. This is in direct contrast to the pulmonary vasculature. However, both bronchial endothelium and pulmonary endothelium are exposed to mechanical stress imposed by lung ventilation. Thus, the lung likely imposes unique conditions on the systemic vasculature within the airways and parenchyma. However, the difficulty in isolating bronchial ECs has limited our ability to characterize this vasculature, and relatively little is known specifically about the bronchial endothelium in health and disease. Whether bronchial ECs respond similarly to human umbilical vein ECs, which have been used extensively to establish leukocyte–endothelial interactions, has not been adequately studied. Likewise, the mechanism responsible for the greater proliferative capacity of bronchial ECs relative to pulmonary artery ECs remains poorly understood. The limited available data suggest that bronchial ECs and pulmonary ECs respond differently to *in vitro* stimuli with distinct signaling pathways. Information obtained regarding lung-specific functions using surrogate, readily obtainable ECs may not apply to bronchial endothelium. Lung pathologies associated with the systemic circulation of the lung require focused investigation into basic cellular mechanisms using relevant EC subtypes.

References

1. Cudkowicz, L. (1992) The bronchial circulation in the human, in *The Bronchial Circulation* (ed. J. Butler), Marcel Dekker, New York, pp. 3–41.
2. Mitzner, W. and Wagner, E.M. (1992) On the purported discovery of the bronchial circulation by Leonardo da Vinci. *Journal of Applied Physiology*, **73**, 1196–201.
3. Charan, N.B., Carvalho, P., and Agostoni, P.G. (1994) On the purported discovery of the bronchial circulation by Leonardo da Vinci: a rebuttal. *Journal of Applied Physiology*, **76**, 1836–38.
4. Virchow, V. (1847) Uber die Standpunkte in den Wissenschaftlichen Medizin. *Virchows Archiv*, **1**, 1–19.
5. Osler, W. (1892) *The Principles and Practices of Medicine*, Appleton, New York.
6. Boyden, E. (1970) The time lag in the development of bronchial arteries. *The Anatomical Record*, **166**, 611–14.
7. Hislop, A. (2002) Airway and blood vessel interaction during lung development. *Journal of Anatomy*, **201**, 325–34.
8. Deffebach, M.E., Charan, N.B., Lakshminarayan, S., and Butler, J. (1987) The bronchial circulation – small, but a vital attribute of the lung. *American Review of Respiratory Disease*, **135**, 463–81.
9. Charan, N.B., Turk, G.M., and Dhand, R. (1984) Gross and subgross anatomy of the bronchial circulation in sheep. *Journal of Applied Physiology*, **57**, 658–64.
10. Baluk, P., Hirata, A., Thurston, G. *et al.* (1997) Endothelial gaps: time course of formation and closure in inflamed venules of rats. *American Journal of Physiology: Lung Cellular and Molecular Physiology*, **272**, L155–70.
11. Kwan, M.L., Gomez, A.D., Baluk, P. *et al.* (2001) Airway vasculature after mycoplasma infection: chronic leakiness and selective hypersensitivity to substance P. *American Journal of Physiology: Lung Cellular and Molecular Physiology*, **280**, L286–97.
12. Moldobaeva, A. and Wagner, E. (2002) Heterogeneity of bronchial endothelial cell permeability. *American Journal of Physiology: Lung Cellular and Molecular Physiology*, **283**, L520–7.
13. Moldobaeva, A. and Wagner, E.M. (2003) Angiotensin-converting enzyme activity in ovine bronchial vasculature. *Journal of Applied Physiology*, **95**, 2278–84.
14. Baile, E.M., Wang, L., Verburgt, L., and Pare, P.D. (1997) Bronchial vasodilatory response to ionic and nonionic contrast media. *Journal of Applied Physiology*, **82**, 841–45.
15. Sasaki, F., Pare, P., Ernest, D. *et al.* (1995) Endogenous nitric oxide influences acetylcholine-induced bronchovascular dilation in sheep. *Journal of Applied Physiology*, **78**, 539–45.
16. Carvalho, P., Johnson, S.R., and Charan, N.B. (1998) Non-cAMP-mediated bronchial arterial vasodilation in response to inhaled beta-agonists. *Journal of Applied Physiology*, **84**, 215–21.
17. Pearse, D.B., Dahms, T.E., and Wagner, E.M. (1998) Microsphere-induced bronchial artery vasodilation: role of adenosine, prostacyclin, and nitric oxide. *American Journal of Physiology: Heart and Circulatory Physiology*, **274**, H760–68.
18. Wagner, E.M. (2000) TNF-alpha induced bronchial vasoconstriction. *American Journal of Physiology: Heart and Circulatory Physiology*, **279**, H946–51.
19. Kelly, L., Kolbe, J., Mitzner, W. *et al.* (1986) Bronchial blood flow affects recovery from constriction in dog lung periphery. *Journal of Applied Physiology*, **60**, 1954–59.
20. Wagner, E.M. and Mitzner, W.A. (1990) Bronchial circulatory reversal of methacholine-induced airway constriction. *Journal of Applied Physiology*, **69**, 1220–24.

21. Csete, M.E., Chediak, A.D., Abraham, W.M., and Wanner, A. (1991) Airway blood flow modifies allergic airway smooth muscle contraction. *American Review of Respiratory Disease*, **144**, 59–63.
22. Brinson, G.M., Noone, P.G., Mauro, M.A. *et al.* (1998) Bronchial artery embolization for the treatment of hemoptysis in patients with cystic fibrosis. *American Journal of Respiratory and Critical Care Medicine*, **157**, 1951–58.
23. Li, X. and Wilson, J.W. (1997) Increased vascularity of the bronchial mucosa in mild asthma. *American Journal of Respiratory and Critical Care Medicine*, **156**, 229–33.
24. Turner-Warwick, M. (1963) Precapillary systemic-pulmonary anastomoses. *Thorax*, **18**, 225–37.
25. Muller, K. and Meyer-Schwickerath, M. (1978) Bronchial arteries in various stages of bronchogenic carcinoma. *Pathology, Research and Practice*, **163**, 34–46.
26. Bergeron, C. and Boulet, L.P. (2006) Structural changes in airway diseases: characteristics, mechanisms, consequences, and pharmacologic modulation. *Chest*, **129**, 1068–87.
27. Kaufman, S.L., Kan, J.S., Mitchell, S.E. *et al.* (1986) Embolization of systemic to pulmonary artery collaterals in the management of hemoptysis in pulmonary atresia. *The American Journal of Cardiology*, **58**, 1130–32.
28. Remy-Jardin, M., Duhamel, A., Deken, V. *et al.* (2005) Systemic collateral supply in patients with chronic thromboembolic and primary pulmonary hypertension: assessment with multi-detector row helical CT angiography. *Radiology*, **235**, 274–81.
29. Hsia, C.C., Herazo, L.F., Fryder-Doffey, F., and Weibel, E.R. (1994) Compensatory lung growth occurs in adult dogs after right pneumonectomy. *The Journal of Clinical Investigation*, **94**, 405–12.
30. Hyvelin, J.M., Howell, K., Nichol, A. *et al.* (2005) Inhibition of Rho-kinase attenuates hypoxia-induced angiogenesis in the pulmonary circulation. *Circulation Research*, **97**, 185–91.
31. McDonald, D.M. and Baluk, P. (2002) Significance of blood vessel leakiness in cancer. *Cancer Research*, **62**, 5381–85.
32. Middleton, J., Americh, L., Gayon, R. *et al.* (2004) Endothelial cell phenotypes in the rheumatoid synovium: activated, angiogenic, apoptotic and leaky. *Arthritis Research and Therapy*, **6**, 60–72.
33. Simcock, D.E., Kanabar, V., Clarke, G.W. *et al.* (2007) Proangiogenic activity in bronchoalveolar lavage fluid from patients with asthma. *American Journal of Respiratory and Critical Care Medicine*, **176**, 146–53.
34. Feltis, B.N., Wignarajah, D., Zheng, L. *et al.* (2006) Increased vascular endothelial growth factor and receptors: relationship to angiogenesis in asthma. *American Journal of Respiratory and Critical Care Medicine*, **173**, 1201–7.
35. Ezaki, T., Baluk, P., Thurston, G. *et al.* (2001) Time course of endothelial cell proliferation and microvascular remodeling in chronic inflammation. *American Journal of Pathology*, **158**, 2043–55.
36. McDonald, D.M. and Baluk, P. (2005) Imaging of angiogenesis in inflamed airways and tumors: newly formed blood vessels are not alike and may be wildly abnormal: Parker B. Francis lecture. *Chest*, **128**, 602S–608S.
37. Fadel, E., Riou, J.Y., Mazmanian, M. *et al.* (1999) Pulmonary thromboendarterectomy for chronic thromboembolic obstruction of the pulmonary artery in piglets. *The Journal of Thoracic and Cardiovascular Surgery*, **117**, 787–93.
38. Mitzner, W., Lee, W., Georgakopoulos, D., and Wagner, E. (2000) Angiogenesis in the mouse lung. *American Journal of Pathology*, **157**, 93–101.
39. Shi, W., Hu, F., Kassouf, W., and Michel, R.P. (2000) Altered reactivity of pulmonary vessels in postobstructive pulmonary vasculopathy. *Journal of Applied Physiology*, **88**, 17–25.
40. Thurston, G., Murphy, T.J., Baluk, P. *et al.* (1998) Angiogenesis in mice with chronic airway inflammation: strain-dependent differences. *American Journal of Pathology*, **153**, 1099–112.
41. Thurston, G., Maas, K., Labarbara, A. *et al.* (2000) Microvascular remodelling in chronic airway inflammation in mice. *Clinical and Experimental Pharmacology and Physiology*, **27**, 836–41.
42. Baluk, P., Tammela, T., Ator, E. *et al.* (2005) Pathogenesis of persistent lymphatic vessel hyperplasia in chronic airway inflammation. *The Journal of Clinical Investigation*, **115**, 247–57.
43. Kim, K.E., Cho, C.H., Kim, H.Z. *et al.* (2007) *In vivo* actions of angiopoietins on quiescent and remodeling blood and lymphatic vessels in mouse airways and skin. *Arteriosclerosis, Thrombosis, and Vascular Biology*, **27**, 564–70.
44. McDonald, D.M. (2001) Angiogenesis and remodeling of airway vasculature in chronic inflammation. *American Journal of Respiratory and Critical Care Medicine*, **164**, S39–S45.
45. Endrys, J., Hayat, N., and Cherian, G. (1997) Comparison of bronchopulmonary collaterals and collateral blood flow in patients with chronic thromboembolic and primary pulmonary hypertension. *Heart*, **78**, 171–76.

46. Remy-Jardin, M., Bouaziz, N., Dumont, P. *et al.* (2004) Bronchial and nonbronchial systemic arteries at multi-detector row CT angiography: comparison with conventional angiography. *Radiology*, **233**, 741–49.
47. Charan, N.B. and Carvalho, P. (1997) Angiogenesis in bronchial circulatory system after unilateral pulmonary artery obstruction. *Journal of Applied Physiology*, **82**, 284–91.
48. Michel, R.P., Hakim, T.S., and Petsikas, D. (1990) Segmental vascular resistance in postobstructive pulmonary vasculopathy. *Journal of Applied Physiology*, **69** (3), 1022–32.
49. Weibel, E.R. (1960) Early stages in the development of collateral circulation to the lung in the rat. *Circulation Research*, **8**, 353–76.
50. Michel, R.P. and Hakim, T.S. (1991) Increased resistance in postobstructive pulmonary vasculopathy: structure–function relationships. *Journal of Applied Physiology*, **71**, 601–10.
51. Srisuma, S., Biswal, S.S., Mitzner, W.A. *et al.* (2003) Identification of genes promoting angiogenesis in mouse lung by transcriptional profiling. *American Journal of Respiratory Cell and Molecular Biology*, **29**, 172–79.
52. Sánchez, J., Moldobaeva, A., McClintock, J. *et al.* (2007) The role of CXCR2 in systemic neovascularization of the mouse lung. *Journal of Applied Physiology*, **103**, 594–99.
53. Sukkar, A., Jenkins, J., Sánchez, J., and Wagner, E. (2008) Inhibition of CXCR2 attenuates bronchial angiogenesis in the ischemic rat lung. *Journal of Applied Physiology*, **104**, 1470–75.
54. Strieter, R.M., Polverini, P.J., Kunkel, S.L. *et al.* (1995) The functional role of the ELR motif in CXC chemokine-mediated angiogenesis. *The Journal of Biological Chemistry*, **270**, 27348–57.
55. Moldobaeva, A. and Wagner, E.M. (2005) Difference in proangiogenic potential of systemic and pulmonary endothelium: role of CXCR2. *American Journal of Physiology: Lung Cellular and Molecular Physiology*, **288**, L1117–23.
56. Shi, G.P., Sukhova, G.K., Kuzuya, M. *et al.* (2003) Deficiency of the cysteine protease cathepsin S impairs microvessel growth. *Circulation Research*, **92**, 493–500.
57. Wagner, E.M. (1997) Effects of edema on small airway narrowing. *Journal of Applied Physiology*, **83**, 784–91.
58. Wagner, E.M., Blosser, S., and Mitzner, W. (1998) Bronchial vascular contribution to lung lymph flow. *Journal of Applied Physiology*, **85**, 2190–95.
59. Brown, R., Mitzner, W., and Wagner, E. (1997) Interaction between airway edema and lung inflation on responsiveness of individual airways *in vivo*. *Journal of Applied Physiology*, **83**, 366–70.
60. Brown, R.H., Mitzner, W., Bulut, Y., and Wagner, E.M. (1997) Effect of lung inflation *in vivo* on airways with smooth muscle tone or edema. *Journal of Applied Physiology*, **82**, 491–99.
61. Sakurai, H., Johnigan, R., Kikuchi, Y. *et al.* (1998) Effect of reduced bronchial circulation on lung fluid flux after smoke inhalation in sheep. *Journal of Applied Physiology*, **84**, 980–86.
62. Efimova, O., Volokhov, A.B., Iliaifar, S., and Hales, C.A. (2000) Ligation of the bronchial artery in sheep attenuates early pulmonary changes following exposure to smoke. *Journal of Applied Physiology*, **88**, 888–93.
63. Hales, C.A., Barkin, P., Jung, W. *et al.* (1989) Bronchial artery ligation modifies pulmonary edema after exposure to smoke with acrolein. *Journal of Applied Physiology*, **67**, 1001–6.
64. Persson, C.G. (1993) Plasma exudation from tracheobronchial microvessels in health and disease, in *The Bronchial Circulation*. (ed. J.B. Butler), Marcel Dekker, New York, pp. 443–73.
65. Baffert, F., Le, T., Thurston, G., and McDonald, D.M. (2006) Angiopoietin-1 decreases plasma leakage by reducing number and size of endothelial gaps in venules. *American Journal of Physiology: Heart and Circulatory Physiology*, **290**, H107–18.
66. Schraufnagel, D.E., Pearse, D.B., Mitzner, W.A., and Wagner, E.M. (1995) Three-dimensional structure of the bronchial microcirculation in sheep. *The Anatomical Record*, **243**, 357–66.
67. Pearse, D.B. and Wagner, E.M. (1994) Role of the bronchial circulation in ischemia–reperfusion lung injury. *Journal of Applied Physiology*, **76**, 259–65.
68. Lim, L.H., Bochner, B.S., and Wagner, E.M. (2002) Leukocyte recruitment in the airways: an intravital microscopic study of rat tracheal microcirculation. *American Journal of Physiology: Lung Cellular and Molecular Physiology*, **282**, L959–67.
69. Lim, L.H. and Wagner, E.M. (2003) Airway distension promotes leukocyte recruitment in rat tracheal circulation. *American Journal of Respiratory and Critical Care Medicine*, **168**, 1068–74.
70. Wagner, E.M. and Jenkins, J. (2007) Effects of airway distension on leukocyte recruitment in the mouse tracheal microvasculature. *Journal of Applied Physiology*, **102**, 1528–34.
71. Moldobaeva, A., Jenkins, J., and Wagner, E. (2008) Effects of distension on airway inflammation and venular P-selectin expression. *American Journal of*

Physiology: Lung Cellular and Molecular Physiology, **295**, L941–8.

72. Zhou, C., Chen, H., Lu, F. *et al.* (2007) Cav3.1 (alpha1G) controls von Willebrand factor secretion in rat pulmonary microvascular endothelial cells. *American Journal of Physiology: Lung Cellular and Molecular Physiology*, **292**, L833–44.
73. Rondaij, M.G., Bierings, R., Kragt, A. *et al.* (2006) Dynamics and plasticity of Weibel-Palade bodies in endothelial cells. *Arteriosclerosis, Thrombosis, and Vascular Biology*, **26**, 1002–7.
74. Davie, N.J., Crossno, J.T. Jr., Frid, M.G. *et al.* (2004) Hypoxia-induced pulmonary artery adventitial remodeling and neovascularization: contribution of progenitor cells. *American Journal of Physiology: Lung Cellular and Molecular Physiology*, **286**, L668–78.
75. Asosingh, K., Swaidani, S., Aronica, M., and Erzurum, S.C. (2007) T_h1- and T_h2-dependent endothelial progenitor cell recruitment and angiogenic switch in asthma. *Journal of Immunology*, **178**, 6482–94.
76. Grantham, C.J., Jackowski, J.T., Wanner, A., and Ryan, U.S. (1989) Metabolic and pharmacokinetic activity of the isolated sheep bronchial circulation. *Journal of Applied Physiology*, **67**, 1041–47.
77. Abraham, W.M., Scuri, M., and Farmer, S.G. (2006) Peptide and non-peptide bradykinin receptor antagonists: role in allergic airway disease. *European Journal of Pharmacology*, **533**, 215–21.

15

Mapping Protein Expression on Pulmonary Vascular Endothelium

Kerri A. Massey and Jan E. Schnitzer

Protogenomics Research Institute for Systems Medicine, Sidney Kimmel Cancer Center, San Diego, CA, USA

INTRODUCTION

All blood vessels are lined by a single layer of highly attenuated endothelial cells (ECs) called the endothelium. These cells form a barrier between the circulating blood and the underlying cells inside the tissue. The endothelium plays a significant role in controlling the passage of blood-borne molecules and cells into the tissue, and is important in many functions such as vasoregulation, coagulation, and inflammation as well as tissue nutrition, growth, survival, repair, and overall organ homeostasis and function. Disruption of the vascular endothelium and its normal barrier function can lead rapidly to tissue edema, hypoxia, pathology, and even organ death [1, 2].

Vascular ECs are highly adapted to meet the needs of local tissue, and exhibit molecular and functional variation according to their location in the body [3–5]. Electron microscopy of tissues reveals three categories of endothelium lining the microvascular beds of different organs. Sinusoidal endothelium is minimally restrictive. Large intercellular gaps exist between ECs and the basement membrane is lacking, allowing for rapid and relatively nonselective flow from blood into the tissue. Liver, spleen, and bone marrow all exhibit sinusoidal endothelium. Fenestrated endothelium is defined by the presence of fenestrae (60- to 80-nm circular transcellular openings). This endothelium is generally found in organs that need to rapidly exchange small molecules, such as the kidney, endocrine glands, and intestine. Continuous endothelium (Figure 15.1) forms the most restrictive barrier via a monolayer of attenuated cells linked by intercellular junctions with various degrees of tightness [6].

The lung is an example of continuous endothelium. The lungs are large, well-vascularized organs that showcase a variety of different structural and functional requirements. Pressure and function differ sharply between pulmonary and bronchial vessels. Additionally, the environment changes dramatically at birth when the lungs begin respiring air and receiving 100% of the cardiac output [7]. Lung endothelium is structurally similar to other continuous endothelium, such as the heart and skeletal muscles, but is distinct from the very tight, nonpermissive barriers found in the continuous endothelium of the brain and testes, in part because of the large number of caveolae found in lung.

SEGMENTAL DIFFERENCES OF EC

Although segmental differences have been extensively studied in the lung, we have yet to comprehensively define the structural, molecular, and functional differences between pulmonary and systemic vasculature. Ultrastructural differences can be seen between different types of vasculature. Larger arteries are more likely to have organelles called Weibel–Palade bodies that express both von Willebrand factor and P-selectin. When released, these factors recruit platelets and leukocytes to the site of injuries. Functional differences between vascular types can also be used to predict specific markers. Lung capillaries are adapted for gas exchange and have relatively tight junctions between ECs. Thus, capillaries can be identified by the presence of VE-cadherin as well as E- and N-cadherin to form complex adherens junctions between cells. Differences between tissues can also be predicted from known functions. The lungs can respond to vasoactive compounds such as bradykinin and angiotensin I. Angiotensin-converting enzyme (ACE) can bind and inactivate both of these compounds and is found at high levels in lung capillaries [7–9].

The Pulmonary Endothelium: Function in health and disease Editors Norbert F. Voelkel, Sharon Rounds

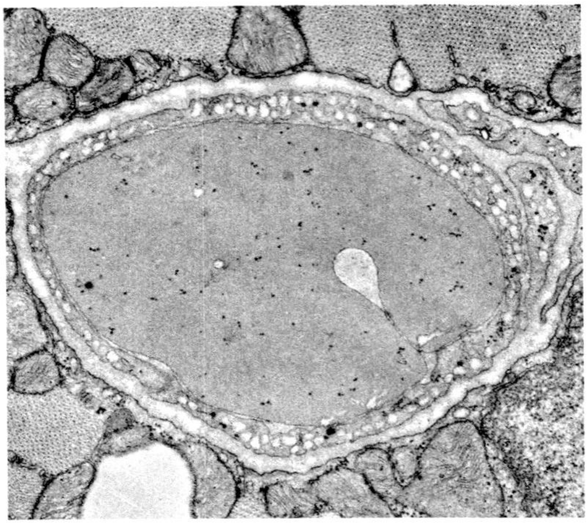

Figure 15.1 A capillary from heart is shown. A red blood cell can be seen to fill the entire capillary. A monolayer of ECs forms a barrier at the perimeter of the vessel. Photo courtesy of G.E. Palade.

Specific and definitive markers differentiating segments of vasculature have been difficult and controversial to define. Noted differences are seen between types of tissue. Though the lung expresses high levels of ACE in capillaries, this enzyme is absent from capillaries in many other organs and is instead a marker of larger arteries [7]. The age of tissue can also have a profound impact on the type of makers seen. Ephrin B2 is a specific marker for arteries while its receptor tyrosine kinase, eprhin B4, clearly marks veins in the developing embryo. These distinctions are lost as the animal ages. In the adult, ephrin B2 is found in ECs as well as the surrounding smooth muscle and pericytes. Ephrin B2 is still found expressed at high levels in capillaries within areas of active angiogenesis, such as within tumor tissue or the female reproductive system [10]. Even species differences can complicate the search for specific markers. Extensive research has shown that CD34 ubiquitously stains vasculature in mice and humans [11–13]. In our hands, expression in rat tissue is limited largely to the lung and, even there, does not consistently stain all vasculature. Experimentally identified markers provide an essential means to identify different vasculature; however, these approaches are far from comprehensive. Clearly, a more global approach is needed.

DIFFICULTIES IN STUDYING ECs

Although there is little question that the microenvironment of the tissue surrounding the blood vessels can significantly influence the phenotype of ECs, there currently is very little molecular information about vascular endothelium and the degree to which EC expression is modulated within different organs, across ages, or across species *in vivo*. Clinically significant tissue-modulated and possibly even organ-specific molecules may be useful as targets for site-directed delivery of drugs, genes, or imaging agents (see Chapter 22). Proteins at the luminal EC surface are directly exposed to the blood and thus inherently accessible to agents circulating in the blood [14, 15]. Comprehensively defining proteins at the interface between the blood and tissue can help define the functions of those cells, environmentally induced changes, and differences between tissues.

Although it is clearly important to define the proteins at the EC surface, the endothelium is a single layer of cells and forms only a tiny part of any tissue homogenate. Even proteins that are highly enriched in ECs can be missed when the total organ homogenate is analyzed [16]. Initial studies focused on electron microscopy of tissue slices to define structural differences between types of endothelium (see Chapter 2) and to identify subdomains of ECs, such as caveolae. This work showed that ECs exhibit enormous diversity, but electron microscopy was not able to explore the function of these cells [9].

ECs IN CULTURE

In the late 1970s, two groups independently isolated ECs [17, 18] and successfully grew these cells in culture. This allowed tremendous insights into the molecular components and functions of these cells, particularly the changes the cells undergo when exposed to changes in vasoactive compounds, cytokines, shear stress, or inflammatory compounds. This enabled the discovery of novel adhesion molecules and receptors [2, 19–21].

At that time, one of the popular ways to discover novel proteins that were specific to ECs was to generate monoclonal antibodies. ECs can be injected directly into mice as an immunogen [22–24]. B cells from the spleen are then purified and used to form hybridomas to produce monoclonal antibodies in culture. Monoclonal antibodies are used to immunoprecipitate the protein of interest while direct amino acid sequencing can then be used to identify the proteins. This is a slow, laborious process.

In the late 1980s and early 1990s, numerous chemical techniques were developed to label surface proteins of ECs. Cultured cells are particularly amenable to these approaches because reagents can simply be added to the media of intact cells and allowed to interact with the surface plasma membrane. Initial studies used radio-iodinated compounds to identify surface components through radioactivity [25]. This allowed the identification of albumin-binding proteins [26, 27]. Different lectins can bind to a unique set of surface glycoproteins, which has been used to determine differences between

ECs from different segments of vasculature [28], from different organs [29], and to compare proteins found *in vitro* with proteins found *in situ* [30]. Radiolabeled proteins are difficult to separate from unlabeled proteins, making identification of unknown proteins challenging. To overcome this obstacle, biotin was used to label surface proteins, which could then be isolated through interaction with avidin (see Chemical Labeling of Surface Proteins *in Vivo*).

As this work advanced, it became apparent that ECs in culture were not the same as those *in vivo*. Once in culture, ECs from unique vascular beds de-differentiate into a more common phenotype, resulting in loss of native function and protein expression as the cells adapt their metabolism to the unnatural cell culture conditions [31–33]. They lose many of their distinctive characteristics found *in vivo* including expression of tissue-specific proteins as well as the usual abundance of caveolae which decreases 30- to 100-fold in cultured ECs [32].

Very recent mass spectrometry (MS) analysis shows large differences in proteins expression *in vitro* versus *in vivo*. Approximately 40% of the proteins expressed *in vivo* are not found *in vitro* [34]. A number of biological mechanisms may account for differences in EC protein expression *in vivo* versus *in vitro*. For example, hemodynamic forces can greatly influence the EC phenotype, including induction and redistribution of protein expression with translocation from intracellular compartments to the cell surface [35]. In addition, cultured cells lack the natural cues from circulating blood, the basement membrane, perivascular cells, and the tissue parenchyma to maintain the protein expression, structure, and phenotype normally found in the native tissue.

DEFINING EC EXPRESSION *IN VIVO*

Several approaches have been taken to identify the molecular components of ECs *in vivo* [36], often building on techniques used for cultured cells. Traditionally, these approaches have focused on validating the expression of proteins of interest. Specific ligands, antibodies, and peptides have been used to stain tissue sections, for Western-type analysis, and, more recently, to follow *in vivo* trafficking after intravenous injection [37]. This approach requires specific probes that may not be available for many proteins, especially those that are unknown or unexpected. This type of experimentally driven discovery of protein expression is unlikely to yield a comprehensive catalogue of proteins present at the EC surface.

PHAGE LIBRARIES

One way to discover novel probes that specifically bind EC proteins is to use phage libraries. With this method, large numbers of peptides or antibody fragments are expressed on the surface of bacterial phage [15, 38]. Phage can then be incubated with cultured cells for direct panning or selection. Phage that bind to available proteins can be purified and used as probes themselves. Phage can also be used to isolate the binding partner for identification via peptide affinity chromatography or immunoprecipitation [39, 40]. Phage can also be intravenously injected and allowed to circulate and presumably bind the EC surface *in vivo*. Then, they can be isolated from each organ or tissue of interest. Several iterations create the opportunity for selection of specific peptides or antibodies with defined tissue tropism. As the peptides are random, this is a nonbiased approach. However, phage are rapidly scavenged from the blood by the liver and spleen, preventing sufficient equilibrium to bind to EC surface proteins *in vivo*. Additionally, short peptides can lack specificity and may bind a large range of proteins in a multitude of organs. Many of the groups that have identified targeting peptides through *in vivo* panning are now doing *ex vivo* screens, for obvious reasons.

Problems with *in vivo* targeting can be partially overcome by using phage that recognize specific proteins on EC extracts to create antibody-like fusion proteins. These can successfully immunotarget *in vivo* [41]. Phage display libraries have revealed promising targets but it is unlikely that this approach can be used as a high-throughput means to identify large numbers of proteins. More robust methods are needed to comprehensively identify and analyze the proteins found on the luminal surface of ECs.

LARGE-SCALE APPROACHES

Genomics and proteomics approaches theoretically provide a means to comprehensively define expression patterns and identify differences among samples in a relatively rapid manner. Genomic analysis can be used to compare global changes in gene expression between tissues or states. Large changes are needed to detect differences between tissues. Additionally, changes in gene expression do not always correlate with changes in protein expression. Proteins may move between different locations within a cell, altering both their accessibility and function. Post-translational changes can also alter protein location and function. To truly define the proteins present at the EC luminal surface and to use this identification as a means to assess function, protein expression itself must be characterized.

Two-dimensional gels are one simple and rapid way to visualize differences between tissues. In these gels, mixtures of proteins are first separated by one property. The gels are then rotated 90°, and separated by a different property. Two-dimensional gels provide better separation between proteins and molecular "fingerprints" are formed. Spots that are unique to one tissue can be excised and identified by MS. For these approaches, successful identification of proteins requires that the proteins migrate onto the gel. Many proteins simply do not separate well on such gels and can be underrepresented or lost altogether.

MS-based techniques can identify proteins based on the presence of digested peptides, but highly complex samples are difficult to separate. Similarly, protein arrays use antibodies or peptides to identify the proteins present in the sample and are limited by affinity of the probes and the complexity of the sample.

PURIFYING ECs

Each of the above methods is limited by the complexity of the starting sample. Identification of proteins using global genomic and proteomic analyses has been limited by the overwhelming molecular complexity of the whole tissues used as the starting material [42]. Sample complexity can be reduced by focusing on subsets of cells or even subdomains of cells. This requires an efficient means to isolate cells. As discussed above, ECs show dramatic changes when isolated and even more so when grown in culture. Up to 40% of the proteins detected *in vivo* are lost *in vitro* [34].

LASER CAPTURE MICRODISSECTION

Purification of ECs has been problematic due to the rapidity in which the cells change once removed from the tissue microenvironment. Laser capture microdissection can be used to surgically isolate ECs from surrounding vasculature in thin tissue sections or from cultured cells [43–46]. This can then be paired with genomic or proteomic analysis. Although laser capture microdissection is labor-intensive and only yields small amounts of material, it may provide a means to separate endothelium based on segmental differences such as artery or vein [47]. With smaller blood vessels such as capillaries, surrounding material will likely be isolated as well. Technical limitations include, but are not limited to, defining microvessel location in whole tissue and surgically isolating each individual microvascular EC with a laser. These challenges must be overcome for the technology to yield the desired information.

CHEMICAL LABELING OF SURFACE PROTEINS *IN VIVO*

As a natural derivative of cell culture studies, several groups have attempted to chemically label proteins at the luminal EC surface. As these proteins are exposed directly to the culture media or blood, they can be radiolabeled or biotinylated by reagents included in the media, perfused through the vasculature or even injected intravenously [48]. Radioiodination of proteins *in situ* results in the fraction of proteins that are exposed to the luminal surface being labeled with ^{125}I [29, 30, 49, 50]. *In situ* labeling requires high amounts of ^{125}I, often exceeding 10 mCi, making this process difficult [30]. Radiolabeling was most often used to verify the presence of known proteins as there is no simple way to separate and identify radiolabeled proteins, significantly limited the utility of this approach.

In vivo biotinylation also chemically labels proteins at the luminal surface of vascular ECs. The strong interaction between biotin and avidin can be used to purify the biotinylated proteins. This method has helped to identify components of EC junctions [51], to determine differences between luminal and abluminal surfaces of cells [52, 53], to identify signaling components present at the cell surface [54], and to determine the location of proteins at different membrane subdomains [55, 56]. The diffusion of labeling reagents through the ECs (especially through junctions) and into surrounding tissue is only somewhat restricted. When biotinylated proteins were compared between *in vivo* perfusions and total tissue homogenates, a unique pattern of proteins was seen in each method, showing that this method identifies a subset of proteins in the tissue [57]. The strong interaction between streptavidin and biotin can be used to purify biotinylated proteins, which can be identified with MS [58–60] or antibodies [51, 53–55, 61]. When these studies were first performed, this was a significant advancement in the field. However, it is difficult to control the degree of biotinylation. Some surface proteins may be missed because biotin might not have equal access to all parts of the cell surface and all proteins might not be biotinylated to a similar degree. Additionally, biotinylation is not limited to the luminal surface of ECs. Small biotin compounds can readily permeate throughout tissue and identify not only luminal EC proteins, but also proteins within ECs (cytoskeletal and others) and surrounding tissue, especially the perivascular space. As a means to label targets that are immediately accessible to small molecules, biotin has clear advantages, but antibodies and other larger molecules simply do not have this type of access. Although *in vivo* biotinylation is clearly an advance over earlier efforts, the degree of specificity remains a serious question and a less permeable molecule is clearly needed not only to identify targets that are accessible to larger molecules but also to define the EC proteome *in vivo* without considerable contamination from other sites being labeled *in situ*.

COLLOIDAL SILICA NANOPARTICLES

Another method to identify proteins exposed to the circulating blood is the colloidal silica coating procedure. A solution of colloidal silica nanoparticles is flushed through the vasculature to selectively coat the luminal surface of all perfused blood vessels. These nanoparticles do not penetrate into the tissue and only adhere to the EC surface. When the tissue is homogenated and centrifuged through a high-density media gradient, the silica-coated luminal EC plasma membranes are easy to separate from other components of the tissue, and even other components of the ECs. This has been confirmed by electron microscopy and western analysis. Known EC surface markers are highly enriched (>15-fold) while markers of blood, other tissue cells and subcellular organelles are markedly depleted (15-fold). Cytoskeletal and other plasma membrane proteins are abundant and quite enriched in these isolated membranes.

The endothelial plasma membrane can be further subfractionated to study functional microdomains such as caveolae. The silica coat stabilizes the surface membrane but silica nanoparticles are large enough that they rarely enter into the caveolae. Caveolae can be separated mechanically from the luminal membrane by shear stress and then isolated by buoyant density centrifugation [62]. GTP can also cause caveolae to bud through activation of dynamin [63, 64] allowing a more physiological isolation technique. Electron micrographs of the isolated membranes showed a homogenous population of appropriately sized, 60- to 80-nm vesicles that frequently retain the distinctive omega-shape of caveolae. Protein biochemistry shows ample enrichment for caveolae markers such as caveolin whereas markers for other subcellular organelles are markedly depleted. Additionally, using magnetic immuno-beads labeled with caveolin antibodies to further isolate caveolae showed that the population was highly pure (>95%). Western blots and gel electrophoresis readily reveal that caveolae concentrate a subset of proteins found in the plasma membrane including caveolin-1, dynamin, *N*-ethylmaleimide-sensitive factor (NSF), soluble NSF attachment proteins (SNAPs), and SNAP receptors (SNAREs) [16, 63, 65, 66].

These methods are especially valuable to separate caveolae from lipid rafts. Much like caveolae, lipid rafts are cholesterol-rich microdomains of the EC plasma membrane. Although they are functionally distinct, they are difficult to isolate away from caveolae because both domains have a similar buoyant density. As lipid rafts are flat domains at the surface, they are coated by silica and retained with the rest of the plasma membrane when caveolae are separated [66]. After removing caveolae, the remaining membrane can be separated from the silica particles, allowing lipid rafts to be isolated using standard techniques [13].

TOWARDS COMPREHENSIVE PROTEIN IDENTIFICATION

Though far simpler than total tissue homogenate, the membrane isolate is still a complex mixture of proteins. Recent advances in proteomics techniques have made it possible to analyze complex samples. Multidimensional protein identification technology (MudPIT) links two-dimensional high-pressure liquid chromatography to MS. For MudPIT, a sample of proteins is first solubilized and digested. The peptides are then separated by hydrophobicity on a reverse phase column. In a second separation step, peptides are passed through an ion-exchange column to separate peptides based on charge. Proteins are then identified through MS. Two-dimensional chromatography offers better separation of proteins in such complex samples. Even with multiple separation steps, each MS measurement only identifies a fraction of the proteins present in a sample. Repeating these analyses measurements multiple times seems essential to identifying the majority of proteins present. Comprehensive measurement is clearly necessary to define the EC proteins within a given tissue and to identify differences between tissues.

Four hundred and fifty proteins were identified when MudPIT analysis was applied to samples of the luminal surface of ECs isolated with colloidal silica. As expected, many plasma membrane associated proteins were identified. Thirty five percent of these proteins were known to peripherally associate with the inner leaflet of the plasma membrane, 31% were integral membrane proteins or proteins with lipid anchors, 25% were cytoskeletal or junctional proteins, while 8% were externally bound, often proteins that are secreted or found in blood. Consistent with the known functions of the plasma membrane, structural and signaling proteins made up more than half of the proteins found associated with the luminal plasma membrane. Trafficking and adhesion proteins, extracellular enzymes and transporter enzymes were all found in abundance as well [34]. It is unlikely that this is the full complement of membrane proteins. Integral membrane proteins are difficult to resolubilize for MudPIT analysis and are likely underrepresented in this sample. Better techniques must be developed to fully sample all proteins present at the plasma membrane. Further analysis must also be performed to fully identify the proteins present in subdomains of the endothelial plasma membrane, such as caveolae and lipid rafts.

CAVEOLAE AS TRANSPORT VESICLES

Although comprehensive proteomics on purified endothelial caveolae have not yet been published, more traditional approaches have been used to determine the presence of proteins of interest using electron microscopy with immuno-gold labeling as well as Western blots of isolated cultures. Caveolae have been thought of as static structures by many investigators [67]. The discovery of proteins classically associated with transport vesicles, including vesicle SNAREs, NSF, and SNAP, suggested that caveolae might function as transport vesicles [65]. Additionally, several GTPases known to play roles in vesicle budding have been found within caveolae [65]. The machinery necessary for vesicle fusion is associated with the caveolae membrane under basal conditions. No further activation is needed to translocate machinery into the caveolae or to recruit intracellular components to the plasma membrane. This suggests that caveolae are ready to bud under basal conditions and may be primed for rapid transport.

Isolated EC membranes were used to definitively show that caveolae can bud and form free vesicles. When high concentrations of GTP are added to isolated EC membranes, caveolae budding is induced. Caveolae are lost from the membrane and pure populations of budded caveolae can be isolated. This reconstituted cell-free assay was used to identify dynamin as the GTPase mediating this fission. Dynamin forms a ring around the neck of caveolae, likely acting as a pinchase to form free vesicles [63].

CAVEOLAE IN MECHANOTRANSDUCTION

One of the major physiological functions of the endothelium is the conversion of hemodynamic forces into a series of adaptive biological responses that minimize mechanical stress as well as injury to the blood vessel and the tissue itself. Clearly the endothelial surface must function in mechanotransduction. We proposed over a decade ago that caveolae themselves may be acute mechanosensing organelles [68–71], either as pressure-transducing structures and/or through concentrating components of the mechanotransduction system. Caveolae are indeed responsive to changes in pressure or shear stress. Increasing hemodynamic stressors including flow and pressure in the pulmonary circulation very rapidly induces the tyrosine phosphorylation of lung EC surface proteins located primarily in caveolae [71]. Such stressors also rapidly activate endothelial nitric oxide synthase (eNOS) that resides concentrated in caveolae [35, 70]. Other signaling molecules that can potentially be activated mechanically also reside concentrated in caveolae including heterotrimeric G-proteins and tyrosine kinases. eNOS appears to bind caveolin directly and is released from caveolin during activation by increased flow in the lung [70]. At higher vascular pressures, caveolae can become distorted at the EC surface and then even disappear or "pop" at defined high threshold pressures [72]. In cultured ECs, increasing flow rates recruit caveolin to the luminal surface and lead to an increase in the number of caveolae at the luminal cell surface [35, 73, 74].

Caveolin is a major structural component of caveolae that oligomerizes to form a shell around caveolae. This may allow caveolae to act like loaded tension bearing springs or coils that sense forces transmitted at or into the cell surface membrane [70]. Caveolin itself can be affected by hemodynamic changes. In addition to a possible tension bearing function, caveolin may serve as a mechanosensitive scaffold, concentrating and inhibiting key signaling molecules in caveolae. Molecular mapping studies show that caveolae are enriched in various signaling molecules including specific G-proteins, select nonreceptor tyrosine kinases, Ras, Raf, and eNOS [35, 69–71]. Many of these are enriched in caveolae under basal conditions and may be functionally inhibited by interaction with caveolin [75–77]. Overexpression of caveolin decreases activation of the p42/44 mitogen-activated protein kinase (MAPK) signaling pathway known to participate in mechanotransduction [78]. Caveolin inhibits eNOS activity, likely through a direct interaction. Increased flow *in situ* rapidly dissociates eNOS from caveolin and concomitantly increases eNOS association with key positive modulators such as calmodulin [70, 79].

Caveolae appear to be necessary for at least some aspects of mechanotransduction. Cholesterol binding agents such as filipin cause disassembly of caveolae and disperse the molecules normally concentrated in this microdomain generally over the cell surface [80]. Filipin causes the cell surface density of caveolae to decrease by 90% and inhibits the flow-induced response, both tyrosine phosphorylation of plasmalemmal proteins as well as downstream activation of the Ras/Raf/MAPK pathway [71, 74]. Mice lacking caveolin-1 lack caveolae. These mice have impaired mechanotransduction, showing a decreased ability to regulate blood vessel diameter in response to changes in flow rates as well as a decreased activation of eNOS [81].

IDENTIFICATION OF LUNG-SPECIFIC PROTEINS

Identification of the components on the luminal face of endothelium and within subdomains such as caveolae will continue to enhance our understanding of the function of these structures (Figure 15.2). Additionally, proteins that are accessible to the blood stream and expressed in the

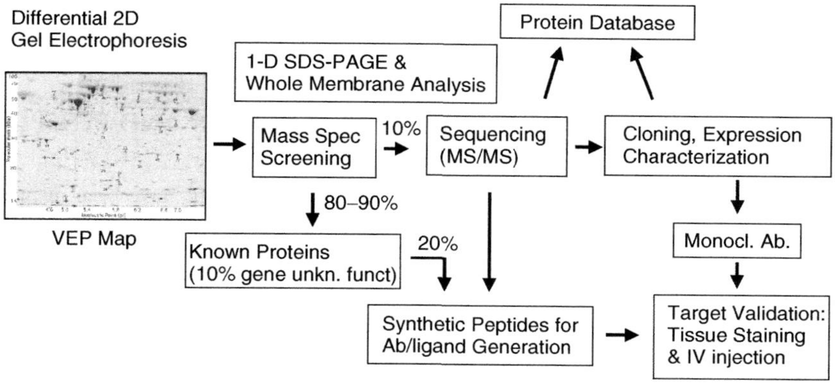

Figure 15.2 Schematic of process used for protein identification. Ab, antibody; IV, intravenous; SDS-PAGE, sodium dodecylsulfate–polyacrylamide gel electrophoresis; VEP, vascular endothelial protein.

vasculature of a single organ can be used as a "molecular address." Identification of tissue-specific markers has been slow due to the difficulty of studying ECs *in vivo*. Although *in vitro* experiments offer valuable insight into EC function, the rapid de-differentiation of ECs in culture has long confounded attempts to study these cells. Surprisingly, 42% of the EC proteins identified in lung tissue using the silica-coating technique were not found in cultured ECs, even when less stringent identification requirements were used [34]. This difference led to the discovery of differential expression of endothelial surface proteins in organs, including lung-specific proteins that were validated and used to target imaging agents *in vivo* [34, 37, 42].

Though frustrating, these rapid changes suggests that ECs are exquisitely dependent on the tissue microenvironment to maintain structure and protein expression. Alterations in this environment, such as those found between different vascular beds, may lead to characteristic changes in protein expression. Proteins that are expressed both *in vivo* and in culture may not depend on the specific microenvironment, and may, therefore, be similar across vascular beds. Similarly, protein expression that is lost when cells are cultured possibly indicates a dependence on the tissue microenvironment and may be a specific marker for a subset of endothelium.

When EC membranes from major organs have been analyzed on two-dimensional gels, each organ showed a specific and characteristic array of proteins, strongly suggesting that tissue-specific markers could be identified. Eliminating proteins also found in culture focuses attention on the proteins that might be tissue specific. The sequences of these proteins were analyzed and those likely to have extracellular domains were compared between tissues. Of the 450 proteins identified at the luminal surface of lung endothelium, only aminopeptidase P (APP) and OX-45 were both specific to lung, and likely to extend into the luminal space [42].

APP is indeed accessible to proteins circulating in the blood flow. Electron microscopy studies showed that intravessel antibodies against APP not only bind at the luminal EC surface, but actually bind directly in caveolae of lung tissue. They concentrated in the caveolae and then were internalized and transcytosed across the EC layer [15]. Transcytosis across the endothelium is very rapid. Dynamic intravital microscopy of engrafted lung tissue revealed that intravenously injected antibodies bound to lung vascular endothelium and rapidly accumulated in the surrounding lung tissue within minutes. Antibodies were pumped across the EC barrier to accumulate at higher concentration in the tissue than the blood. This movement against a concentration gradient is the definition of active transport. The blood vessels were not simply leaky, as fluorescent IgG was maintained within the vasculature.

APP is very enriched in caveolae. It is likely that targeting caveolae is necessary to see this sort of rapid transport. Antibodies against lung endothelium-specific proteins found outside of caveolae bound to the vasculature, but were not transcytosed. If caveolae were depleted by knocking down expression levels of the structural protein caveolin, APP antibodies could still bind to the endothelium, but were not transported out of the vessels and into tissue. Whole-body imaging as well as biodistribution analysis showed that APP antibodies specifically target the lungs in the context of the whole animal. Antibody labeled with ^{125}I rapidly accumulated in the lungs and was maintained in the tissue [37]. Lung uptake of antibodies targeting caveolae is much faster and greater than antibodies targeting other surface proteins not found in caveolae. Also, the APP antibodies remained in lung tissue many days after intravenous injection.

THERAPEUTIC IMPLICATIONS

Rapid, tissue-specific vascular targeting and transcytosis have numerous therapeutic implications. Therapeutic and imaging agents can be delivered to specific sites, decreasing harmful side-effects through interaction with other tissue. Rapid transport into tissue further decreases the risk of side-effects. Drugs are effectively concentrated such that lower doses might be needed. Targeted antiangiogenic compounds can be used to specifically destroy vasculature, effectively destroying all the cells fed by that vasculature [82]. Active pumping of antibodies across the EC layer and into tissue may prevent degradation [37], and increase the therapeutic potential.

Therapeutic efficacy depends on having specific endothelial markers. EC are extremely sensitive to the tissue environment and can rapidly change when removed from their normal environment [34, 42], even undergoing drastic changes when transplanted to different tissue environments [5]. These are drastic changes, but inflammation and other diseases may also alter tissue enough for changes to be reflected in the surrounding vasculature. Cancer is known to alter the surrounding vasculature and the endothelium and also induces a unique "fingerprint" of proteins at the luminal EC surface [34, 42].

When injected into mice bearing lung tumors, APP antibodies did not traffic to the tumors, but instead concentrated in surrounding, tumor-free lung tissue. When the luminal endothelial surface of vasculature from lung tumors was isolated and run on a two-dimensional blot, a distinct pattern of proteins was seen, suggesting that the solid tumors might form a distinct type of tissue. When this extract was further analyzed with MS and Western analysis, APP expression was no longer seen. As expected, several known tumor angiogenesis markers were upregulated.

One surprising result was the induction of a novel protein at the EC surface. Annexin-A1 (AnnA1) is normally found intracellularly. In tumor tissue, AnnA1 could be detected at the luminal surface of ECs. This protein was specific for tumor endothelium and did not appear in the endothelium of normal organs. Immunohistochemistry showed that AnnA1 was also found in diverse human solid tumors (prostate, liver, breast, lung), but not matched normal tissue. AnnA1 was indeed accessible to the blood. Intravenously injected AnnA1 antibodies specifically targeted tumor vasculature. Though tumor-bearing mice treated with radiolabeled, isotype-matched control IgG died, injections of radiolabeled AnnA1 antibodies drastically increased animal survival and led to the virtual elimination of the tumors even many times after just one injection [42]. Many other diseases, inflammation, and hypertension may also lead to changes in protein expression, though these changes may be graded and have yet to be studied.

CONCLUSIONS AND PERSPECTIVES

Nearly 500 proteins associated with the lung endothelial plasma membrane have been identified. Many more proteins likely exist. Additionally, many proteins undergo extensive posttranslational changes, which alter their functional properties and cellular location and may not be detected by MS-based identification techniques. Many of the proteins already identified may exist as multiple isoforms that serve distinct roles. A complete analysis of the EC proteins will likely depend on further advances in MS-based techniques.

Comprehensive analysis of the vascular proteome is essential to identifying tissue-specific protein expression, and identifying differences between normal and diseased tissue. It is essential to first catalog the proteins present in the EC proteome and then to functionally define these proteins. Once this exists for normal tissue, this can be applied to disease states and to understanding the chronology of disease progression. Understanding the functional changes in the EC plasma membrane between tissues and between healthy tissue and disease states provides insight into the needs of each tissue and the changes that develop with disease. Identifying unique markers may allow the targeting of gene therapy, drugs, and imaging agents.

References

1. Fajardo, L.F. (1989) The complexity of endothelial cells. A review. *American Journal of Clinical Pathology*, **92** (2), 241–50.
2. Jaffe, E.A. (1987) Cell biology of endothelial cells. *Human Pathology*, **18** (3), 234–39.
3. Aird, W.C. *et al.* (1997) Vascular bed-specific expression of an endothelial cell gene is programmed by the tissue microenvironment. *The Journal of Cell Biology*, **138** (5), 1117–24.
4. Gumkowski, F. *et al.* (1987) Heterogeneity of mouse vascular endothelium. In vitro studies of lymphatic, large blood vessel and microvascular endothelial cells. *Blood Vessels*, **24** (1–2), 11–23.
5. Janzer, R.C. and Raff, M.C. (1987) Astrocytes induce blood–brain barrier properties in endothelial cells. *Nature*, **325** (6101), 253–57.
6. Jennings, M.A. and Florey, L. (1967) An investigation of some properties of endothelium related to capillary permeability. *Proceedings of the Royal Society of London. Series B: Biological Sciences*, **167** (6), 39–63.

7. Aird, W.C. (2007) Phenotypic heterogeneity of the endothelium: II. Representative vascular beds. *Circulation Research*, **100** (2), 174–90.
8. Stevens, T. (2007) Microheterogeneity of lung endothelium, in *Endothelial Biomedicine* (ed. W.C. Aird), Cambridge University Press, New York, pp. 1161–70.
9. Aird, W.C. (2007) Phenotypic heterogeneity of the endothelium: I. Structure, function, and mechanisms. *Circulation Research*, **100** (2), 158–73.
10. Gale, N.W. *et al.* (2001) Ephrin-B2 selectively marks arterial vessels and neovascularization sites in the adult, with expression in both endothelial and smooth-muscle cells. *Developmental Biology*, **230** (2), 151–60.
11. Baumhueter, S. *et al.* (1994) Global vascular expression of murine CD34, a sialomucin-like endothelial ligand for L-selectin. *Blood*, **84** (8), 2554–65.
12. Fina, L. *et al.* (1990) Expression of the CD34 gene in vascular endothelial cells. *Blood*, **75** (12), 2417–26.
13. Pusztaszeri, M.P., Seelentag, W., and Bosman, F.T. (2006) Immunohistochemical expression of endothelial markers CD31, CD34, von Willebrand factor, and Fli-1 in normal human tissues. *The Journal of Histochemistry and Cytochemistry*, **54** (4), 385–95.
14. Carver, L.A. and Schnitzer, J.E. (2003) Caveolae: mining little caves for new cancer targets. *Nature Reviews: Cancer*, **3** (8), 571–81.
15. McIntosh, D.P. *et al.* (2002) Targeting endothelium and its dynamic caveolae for tissue-specific transcytosis *in vivo*: a pathway to overcome cell barriers to drug and gene delivery. *Proceedings of the National Academy of Sciences of the United States of America*, **99** (4), 1996–2001.
16. Schnitzer, J.E. (2001) Caveolae: from basic trafficking mechanisms to targeting transcytosis for tissue-specific drug and gene delivery *in vivo*. *Advanced Drug Delivery Reviews*, **49** (3), 265–80.
17. Gimbrone, M.A. Jr., Cotran, R.S., and Folkman, J. (1973) Endothelial regeneration: studies with human endothelial cells in culture. *Series Haematologica*, **6** (4), 453–55.
18. Jaffe, E.A. *et al.* (1973) Culture of human endothelial cells derived from umbilical veins. Identification by morphologic and immunologic criteria. *The Journal of Clinical Investigation*, **52** (11), 2745–56.
19. Nagel; T. *et al.* (1994) Shear stress selectively upregulates intercellular adhesion molecule-1 expression in cultured human vascular endothelial cells. *The Journal of Clinical Investigation*, **94** (2), 885–91.
20. Resnick, N. and Gimbrone, M.A. Jr. (1995) Hemodynamic forces are complex regulators of endothelial gene expression. *The FASEB Journal*, **9** (10), 874–82.
21. Topper, J.N. and Gimbrone, M.A. Jr. (1999) Blood flow and vascular gene expression: fluid shear stress as a modulator of endothelial phenotype. *Molecular Medicine Today*, **5** (1), 40–46.
22. Kaplan, K.L. *et al.* (1983) Monoclonal antibodies to E92, an endothelial cell surface antigen. *Arteriosclerosis*, **3** (5), 403–12.
23. Kohler, G. and Milstein, C. (1975) Continuous cultures of fused cells secreting antibody of predefined specificity. *Nature*, **256** (5517), 495–97.
24. Parks, W.M. *et al.* (1985) Identification and characterization of an endothelial, cell-specific antigen with a monoclonal antibody. *Blood*, **66** (4), 816–23.
25. Ghinea, N., Hasu, M., and Popov, D. (1989) Selective radioiodination of the apical and luminal cell surfaces: *in vitro* and *in situ* experiments on vascular endothelial cells with Iodogen-coated Sephadex. *Analytical Biochemistry*, **179** (2), 274–79.
26. Ghinea, N. *et al.* (1989) Endothelial albumin binding proteins are membrane-associated components exposed on the cell surface. *The Journal of Biological Chemistry*, **264** (9), 4755–58.
27. Schnitzer, J.E., Carley, W.W., and Palade, G.E. (1988) Albumin interacts specifically with a 60-kDa microvascular endothelial glycoprotein. *Proceedings of the National Academy of Sciences of the United States of America*, **85** (18), 6773–77.
28. Schnitzer, J.E. *et al.* (1994) Segmental differentiation of permeability, protein glycosylation, and morphology of cultured bovine lung vascular endothelium. *Biochemical and Biophysical Research Communications*, **199** (1), 11–19.
29. Belloni, P.N. and Nicolson, G.L. (1988) Differential expression of cell surface glycoproteins on various organ-derived microvascular endothelia and endothelial cell cultures. *Journal of Cellular Physiology*, **136** (3), 398–410.
30. Schnitzer, J.E., Shen, C.P., and Palade, G.E. (1990) Lectin analysis of common glycoproteins detected on the surface of continuous microvascular endothelium *in situ* and in culture: identification of sialoglycoproteins. *European Journal of Cell Biology*, **52** (2), 241–51.
31. Madri, J.A. and Williams, S.K. (1983) Capillary endothelial cell cultures: phenotypic modulation by matrix components. *The Journal of Cell Biology*, **97** (1), 153–65.

32. Schnitzer, J.E. (1997) The endothelial cell surface and caveolae in health and disease, in *Vascular Endothelium: Physiology, Pathology and Therapeutic Opportunities* (ed. G.V.R. Born and C. V. Schwartz), Schattauer, Stuttgart, pp. 77–95.
33. Thum, T., Haverich, A., and Borlak, J. (2000) Cellular dedifferentiation of endothelium is linked to activation and silencing of certain nuclear transcription factors: implications for endothelial dysfunction and vascular biology. *The FASEB Journal*, **14** (5), 740–51.
34. Durr, E. *et al.* (2004) Direct proteomic mapping of the lung microvascular endothelial cell surface *in vivo* and in cell culture. *Nature Biotechnology*, **22** (8), 985–92.
35. Rizzo, V. *et al.* (2003) Recruitment of endothelial caveolae into mechanotransduction pathways by flow conditioning *in vitro*. *American Journal of Physiology: Heart and Circulatory Physiology*, **285** (4), H1720–29.
36. Carver, L.A. and Schmitzer, J.E. (2007) Proteomic mapping of endothelium and vascular targeting *in vivo*, in *Endothelial Biomedicine* (ed. W.C. Aird), Cambridge University Press, New York, pp. 881–97.
37. Oh, P. *et al.* (2007) Live dynamic imaging of caveolae pumping targeted antibody rapidly and specifically across endothelium in the lung. *Nature Biotechnology*, **25** (3), 327–37.
38. Hamburger, A.W. *et al.* (1985) Isolation and characterization of monoclonal antibodies reactive with endothelial cells. *Tissue and Cell*, **17** (4), 451–59.
39. Pasqualini, R. and Ruoslahti, E. (1996) Organ targeting *in vivo* using phage display peptide libraries. *Nature*, **380** (6572), 364–66.
40. Rajotte, D. *et al.* (1998) Molecular heterogeneity of the vascular endothelium revealed by *in vivo* phage display. *The Journal of Clinical Investigation*, **102** (2), 430–37.
41. Valadon, P. *et al.* (2006) Screening phage display libraries for organ-specific vascular immunotargeting *in vivo*. *Proceedings of the National Academy of Sciences of the United States of America*, **103** (2), 407–12.
42. Oh, P. *et al.* (2004) Subtractive proteomic mapping of the endothelial surface in lung and solid tumours for tissue-specific therapy. *Nature*, **429** (6992), 629–35.
43. Buckanovich, R.J. *et al.* (2007) Tumor vascular proteins as biomarkers in ovarian cancer. *Journal of Clinical Oncology*, **25** (7), 852–61.
44. Espina, V. *et al.* (2006) Laser-capture microdissection. *Nature Protocols*, **1** (2), 586–603.
45. Lu, Q. *et al.* (2008) Analysis of mouse brain microvascular endothelium using immuno-laser capture microdissection coupled to a hybrid linear ion trap with Fourier transform-mass spectrometry proteomics platform. *Electrophoresis*, **29** (12), 2689–95.
46. Macdonald, J.A., Murugesan, N., and Pachter, J.S. (2008) Validation of immuno-laser capture microdissection coupled with quantitative RT-PCR to probe blood–brain barrier gene expression *in situ*. *Journal of Neuroscience Methods*, **174**, 219–26.
47. Asgeirsdottir, S.A. *et al.* (2002) Analysis of *in vivo* endothelial cell activation applying RT-PCR following endothelial cell isolation by laser dissection microscopy. *Annals of the New York Academy of Sciences*, **973**, 586–89.
48. Cole, S.R., Ashman, L.K., and Ey, P.L. (1987) Biotinylation: an alternative to radioiodination for the identification of cell surface antigens in immunoprecipitates. *Molecular Immunology*, **24** (7), 699–705.
49. Merker, M.P., Carley, W.W., and Gillis, C.N. (1990) Molecular mapping of pulmonary endothelial membrane glycoproteins of the intact rabbit lung. *The FASEB Journal*, **4** (12), 3040–48.
50. Schnitzer, J.E. (1992) gp60 is an albumin-binding glycoprotein expressed by continuous endothelium involved in albumin transcytosis. *American Journal of Physiology*, **262** (1Pt 2), H246–54.
51. Alexander, J.S., Blaschuk, O.W., and Haselton, F.R. (1993) An N-cadherin-like protein contributes to solute barrier maintenance in cultured endothelium. *Journal of Cellular Physiology*, **156** (3), 610–18.
52. Roberts, L.M. *et al.* (2008) Subcellular localization of transporters along the rat blood–brain barrier and blood–cerebral-spinal fluid barrier by *in vivo* biotinylation. *Neuroscience*, **155** (2), 423–38.
53. Sargiacomo, M. *et al.* (1989) Integral and peripheral protein composition of the apical and basolateral membrane domains in MDCK cells. *The Journal of Membrane Biology*, **107** (3), 277–86.
54. Walker, J. *et al.* (1993) The adenosine 5′,5‴,P1,P4-tetraphosphate receptor is at the cell surface of heart cells. *Biochemistry*, **32** (50), 14009–14.
55. Fridlich, R., David, A., and Aviram, I. (2006) Membrane proteinase 3 and its interactions within microdomains of neutrophil membranes. *Journal of Cellular Biochemistry*, **99** (1), 117–25.
56. Singh, N. *et al.* (1996) A novel class of cell surface glycolipids of mammalian cells. Free glyco-

syl phosphatidylinositols. *The Journal of Biological Chemistry*, **271** (22), 12879–84.

57. De La Fuente, E.K. *et al.* (1997) Biotinylation of membrane proteins accessible via the pulmonary circulation in normal and hyperoxic rats. *American Journal of Physiology: Lung Cellular and Molecular Physiology*, **272** (3 Pt 1), L461–70.
58. Rybak, J.N. *et al.* (2005) In vivo protein biotinylation for identification of organ-specific antigens accessible from the vasculature. *Nature Methods*, **2** (4), 291–98.
59. Scheurer, S.B. *et al.* (2005) A comparison of different biotinylation reagents, tryptic digestion procedures, and mass spectrometric techniques for 2-D peptide mapping of membrane proteins. *Proteomics*, **5** (12), 3035–39.
60. Scheurer, S.B. *et al.* (2005) Identification and relative quantification of membrane proteins by surface biotinylation and two-dimensional peptide mapping. *Proteomics*, **5** (11), 2718–28.
61. Fujimoto, T. *et al.* (1992) Localization of inositol 1,4,5-trisphosphate receptor-like protein in plasmalemmal caveolae. *The Journal of Cell Biology*, **119** (6), 1507–13.
62. Schnitzer, J.E. *et al.* (1995) Caveolae from luminal plasmalemma of rat lung endothelium: microdomains enriched in caveolin, Ca^{2+}-ATPase, and inositol trisphosphate receptor. *Proceedings of the National Academy of Sciences of the United States of America*, **92** (5), 1759–63.
63. Oh, P., McIntosh, D.P., and Schnitzer, J.E. (1998) Dynamin at the neck of caveolae mediates their budding to form transport vesicles by GTP-driven fission from the plasma membrane of endothelium. *The Journal of Cell Biology*, **141** (1), 101–14.
64. Schnitzer, J.E., Oh, P., and McIntosh, D.P. (1996) Role of GTP hydrolysis in fission of caveolae directly from plasma membranes. *Science*, **274** (5285), 239–42.
65. Schnitzer, J.E., Liu, J., and Oh, P. (1995) Endothelial caveolae have the molecular transport machinery for vesicle budding, docking, and fusion including VAMP, NSF, SNAP, annexins, and GTPases. *The Journal of Biological Chemistry*, **270** (24), 14399–404.
66. Schnitzer, J.E. *et al.* (1995) Separation of caveolae from associated microdomains of GPI-anchored proteins. *Science*, **269** (5229), 1435–39.
67. Severs, N.J. (1988) Caveolae: static inpocketings of the plasma membrane, dynamic vesicles or plain artifact? *Journal of Cell Science*, **90** (Pt 3), 341–48.
68. Liu, J. *et al.* (1997) Organized endothelial cell surface signal transduction in caveolae distinct from glycosylphosphatidylinositol-anchored protein microdomains. *The Journal of Biological Chemistry*, **272** (11), 7211–22.
69. Oh, P. and Schnitzer, J.E. (2001) Segregation of heterotrimeric G proteins in cell surface microdomains. G_q binds caveolin to concentrate in caveolae, whereas G_i and G_s target lipid rafts by default *Molecular Biology of the Cell*, **12** (3), 685–98.
70. Rizzo, V. *et al.* (1998) In situ flow activates endothelial nitric oxide synthase in luminal caveolae of endothelium with rapid caveolin dissociation and calmodulin association. *The Journal of Biological Chemistry*, **273** (52), 34724–29.
71. Rizzo, V. *et al.* (1998) Rapid mechanotransduction *in situ* at the luminal cell surface of vascular endothelium and its caveolae. *The Journal of Biological Chemistry*, **273** (41), 26323–29.
72. Lee, J. and Schmid-Schonbein, G.W. (1995) Biomechanics of skeletal muscle capillaries: hemodynamic resistance, endothelial distensibility, and pseudopod formation. *Annals of Biomedical Engineering*, **23** (3), 226–46.
73. Boyd, N.L. *et al.* (2003) Chronic shear induces caveolae formation and alters ERK and Akt responses in endothelial cells. *American Journal of Physiology: Heart and Circulatory Physiology*, **285** (3), H1113–22.
74. Park, H. *et al.* (2000) Caveolin-1 regulates shear stress-dependent activation of extracellular signal-regulated kinase. *American Journal of Physiology: Heart and Circulatory Physiology*, **278** (4), H1285–93.
75. Li, S., Couet, J., and Lisanti, M.P. (1996) Src tyrosine kinases, Galpha subunits, and H-Ras share a common membrane-anchored scaffolding protein, caveolin. Caveolin binding negatively regulates the auto-activation of Src tyrosine kinases. *The Journal of Biological Chemistry*, **271** (46), 29182–90.
76. Li, S. *et al.* (1995) Evidence for a regulated interaction between heterotrimeric G proteins and caveolin. *The Journal of Biological Chemistry*, **270** (26), 15693–701.
77. Song, K.S. *et al.* (1996) Co-purification and direct interaction of Ras with caveolin, an integral membrane protein of caveolae microdomains. Detergent-free purification of caveolae microdomains. *The Journal of Biological Chemistry*, **271** (16), 9690–97.

78. Engelman, J.A. *et al.* (1998) Caveolin-mediated regulation of signaling along the p42/44 MAP kinase cascade *in vivo*. A role for the caveolin-scaffolding domain. *FEBS Letters*, **428** (3), 205–11.
79. Feron, O. and Balligand, J.L. (2006) Caveolins and the regulation of endothelial nitric oxide synthase in the heart. *Cardiovascular Research*, **69** (4), 788–97.
80. Schnitzer, J.E. *et al.* (1994) Filipin-sensitive caveolae-mediated transport in endothelium: reduced transcytosis, scavenger endocytosis, and capillary permeability of select macromolecules. *The Journal of Cell Biology*, **127** (5), 1217–32.
81. Yu, J. *et al.* (2006) Direct evidence for the role of caveolin-1 and caveolae in mechanotransduction and remodeling of blood vessels. *The Journal of Clinical Investigation*, **116** (5), 1284–91.
82. Red-Horse, K. and Ferrara, N. (2007) Vascular targeting via caveolae. *Nature Biotechnology*, **25** (4), 431–32.

SECTION II:

MECHANISMS AND CONSEQUENCES OF PULMONARY ENDOTHELIAL CELL INJURY

16

Pulmonary Endothelial Cell Death: Implications for Lung Disease Pathogenesis

Qing Lu and Sharon Rounds

Vascular Research Laboratory, Alpert Medical School of Brown University, Providence VA Medical Center, Providence, RI, USA

OVERVIEW OF CELL DEATH

Apoptosis

Apoptosis is a Greek word meaning "dropping off" or "falling away", like leaves falling from trees in the fall. In 1972, Kerr *et al.* used the term "apoptosis" to describe an energy-dependent, genetically determined, active form of programmed cell death – a process by which cells commit suicide in order to eliminate unwanted cells [1]. Apoptosis is characterized by well-ordered morphologic and molecular alterations, including cell surface exposure of phosphatidylserine, cytoskeletal rearrangement, cell shrinkage, plasma membrane blebbing, nuclear membrane collapse, chromatin condensation, internucleosomal DNA fragmentation into 180- to 200-bp fragments (DNA "laddering"), and formation of apoptotic bodies [1]. Cell surface-exposed phosphatidylserine acts as a chemoattractant for phagocytes to engulf apoptotic cells – a process termed "efferocytosis" [2]. Apoptosis is generally considered not to incite inflammatory responses, due to limited release of intracellular contents [3, 4]. Efferocytosis of apoptotic cells also promotes survival of neighboring cells [5]. If the apoptotic cell clearance system is impaired, the apoptotic cells are subjected to secondary necrosis and cytolysis, resulting in inflammation and autoimmunity [6]. Anoikis is apoptosis caused by loss of adhesion to underlying substratum by anchorage-dependent cells, such as pulmonary endothelial cells (ECs). Apoptosis is important in development, maintenance of tissue homeostasis, and tissue remodeling. Apoptosis also plays a fundamental role in genesis of various diseases (see "Lung Diseases Associated with Abnormal Pulmonary EC Apoptosis").

Necrosis

Necrosis is a consequence of acute metabolic perturbation with declined ATP generation and/or disregulated ATP consumption, as in ischemia–reperfusion injury. Necrosis, a caspase-independent cell death, is characterized by cell and organelle swelling and rupture, mitochondrial swelling and degeneration, impaired ATP generation, lysosomal disruption and leakage, disruption of plasma membrane integrity, random fragmentation and degradation of DNA without laddering, and leakage of cellular contents into the surrounding environment [7]. Due to release of potentially injurious, proinflammatory, and proimmunogenic contents into tissues, necrosis often induces inflammation and autoimmune responses. Necrosis has been viewed as a passive and essentially accidental form of cell death, but now is also considered to be an active, regulated, and controllable process [7]. Necrosis is often seen concomitant with apoptosis. Significantly increased necrosis often leads to organ dysfunction.

Morphologic features of apoptosis and necrosis in pulmonary microvascular ECs are illustrated in Figure 16.1.

Autophagy

Autophagy, meaning "to eat oneself," is a mechanism for maintaining cellular homeostasis. Autophagy acts to degrade unwanted cellular proteins (aggregates)

The Pulmonary Endothelium: Function in health and disease Editors Norbert F. Voelkel, Sharon Rounds

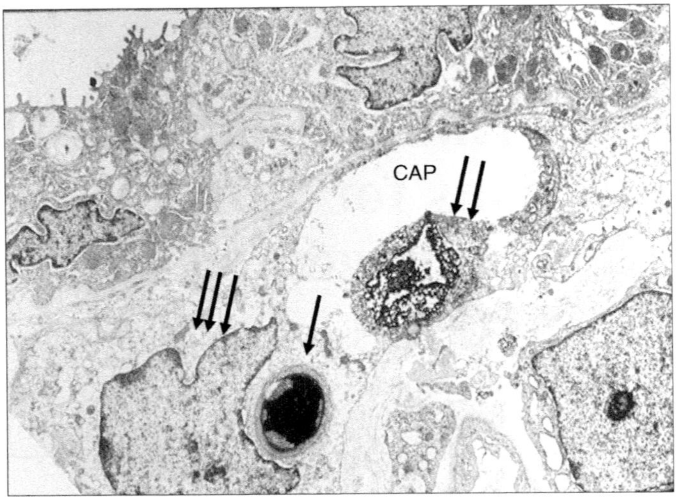

Figure 16.1 Transmission electron micrograph of a lung in the fibroproliferative phase of acute lung injury demonstrating a capillary lumen lined by an apoptotic EC (single arrow) with a pycnotic nucleus, adjacent to an EC undergoing necrosis (double arrow) with hypodense cytoplasm and nuclear chromatin. A third EC (triple arrow) is engulfing the apoptotic cell, which is presumably being phagocytosed. Photograph courtesy of the late Charles Kuhn, MD.

and damaged organelles (in particular, mitochondria) via self-digestion to promote cell survival. Under normal conditions, mammalian target of rapamycin (mTOR) inhibits autophagy by suppressing expression of autophagy-related genes (ATGs). Upon external or internal stimulation, such as nutrient depletion or ischemia, mTOR is inhibited, leading to formation of double-membrane vesicles in the cytosol, sequestering those unwanted proteins and organelles in autophagosomes. Most of the ATG genes, in particular ATG5, ATG12, and light chain 3 (LC3; ATG8), are critical for the formation of autophagosomes. Newly formed autophagosomes subsequently fuse with endosomes or lysosomes to form amphisomes or autolysosomes, respectively. Following fusion of these two vesicular bodies, the autophagosome membrane is broken down by the endosomal or lysosomal proteases, leading to release of the contents of the autophagosomes into the endosome or lysosome for degradation by their proteases. Autophagy is an evolutionarily conserved event existing in all eukaryotic cells. Autophagy is active at a basal level in most cells, acting as a housekeeping process that allows recycling of aged proteins and organelles (mitochondria), thus contributing to the routine turnover of cytoplasmic components. Autophagy is enhanced in response to numerous stresses, such as nutrient starvation, growth factor deprivation, hypoxia, DNA damage, mitochondrial dysfunction, and infection. Autophagy is also associated with various disorders, such as ischemia–reperfusion injury, cardiac diseases, neurodegenerative diseases, and cancer [8]. Autophagy has been considered to be a survival response to stresses, as well as a form of cell death [9]. However, the fundamental question remaining to be resolved is whether autophagy is a mechanism of cell protection or cell death. Autophagy can be chemically inhibited by 3-methyladenine, a class III phosphoinositide 3-kinase inhibitor, and molecularly blocked by knocking down proteins (such as ATG5 and ATG7) important in autophagosome formation. Inhibitors may provide useful tools to investigate the biological roles of autophagy in a particular setting.

Apoptosis, necrosis, and autophagy share a common mechanism of altered mitochondrial permeability and subsequent mitochondrial dysfunction [10]. In this chapter, we focus on apoptosis. However, as biomarkers for different types of cell death become available, their roles in different lung diseases will become an important subject for research.

CELLULAR AND MOLECULAR EVENTS OF APOPTOSIS

Signaling Pathways of Apoptosis

Extrinsic Pathway of Apoptosis

Apoptosis is an energy-dependent, active form of cell death. During development the timing of apoptosis is genetically determined. Apoptosis is also triggered by external and internal stimuli and is mediated through

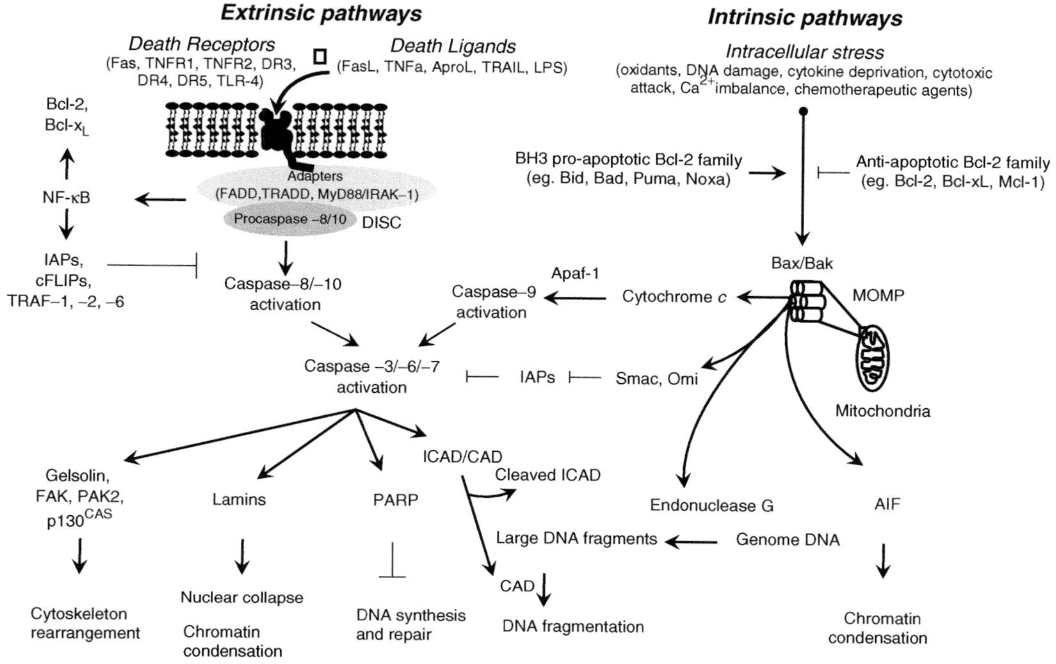

Figure 16.2 Signaling pathways to apoptosis and necrosis. See text for details. CAD, caspase-activated DNase; DISC, death-inducing signaling complex; FAK, focal adhesion kinase; ICAD, inhibitor of caspase-activated DNase; PAK, p21-activated kinase; PARP: poly(ADP-ribose) polymerase.

two fundamental signaling pathways, the extrinsic and intrinsic pathways, as described in Figure 16.2. The extrinsic pathway is triggered by the ligation of cell surface death receptors [Fas, tumor necrosis factor (TNF) receptor (tumor necrosis factor receptorTNFR)-1 and -2, DR3, DR4, DR5, and Toll-like receptor (TLR)-4] with their respective death ligands [Fas ligand (FasL), TNF-α, AproL, TNF-related apoptosis-inducing ligand (tumor necrosis factor-related apoptosis-inducing ligand TRAIL), and lipopolysaccharide (LPS)], resulting in recruitment and activation of initiator caspases, caspase-8 and -10, which subsequently cleave and activate effector caspases, caspase-3, -6, and -7, and apoptosis ensues. The death ligands are often released by inflammatory cells. This pathway to apoptosis may be important in the development of acute lung injury (ALI) (see "Pulmonary Endothelial Apoptosis and ALI/Acute Respiratory Distress Syndrome").

Intrinsic Pathway of Apoptosis

The intrinsic pathway is triggered by apoptosis-inducing signals, such as oxidants, ultraviolet and γ radiation, growth factor and cytokine deprivation, cytotoxic attack, Ca^{2+} imbalance, and chemotherapeutic agents (see Figure 16.2). These intracellular stresses cause mitochondrial outer membrane permeabilization (MOMP), resulting in release of several apoptosis-inducing proteins, including cytochrome *c*, Smac, Omi, apoptosis-inducing factor (AIF), and endonuclease G from mitochondria to the cytosol [11]. The released cytochrome *c* rapidly binds to apoptotic peptidase-activating factor (Apaf)-1, leading to activation of caspase-9, with subsequent activation of caspase-3, -6, and -7, culminating in apoptosis [12, 13]. The released Smac and Omi activate effector caspases by removal of inhibitor of apoptosis proteins (IAPs) [14, 15]. The released AIF and endonuclease G translocate to the nucleus, and initiate chromatin condensation [16] and DNA fragmentation [17], respectively. AIF- and endonuclease G-mediated apoptosis is independent of caspase activity. In addition, MOMP can lead to necrosis as a consequence of reactive oxygen species (ROS) production and abrogation of mitochondrial functions that are indispensable for cell survival [10]. Of note, extrinsic pathway-activated caspase-8 truncates and activates tBid, leading to activation of the mitochondrial-mediated intrinsic pathway [18, 19].

In lung ECs, oxidant-induced injury and LPS-induced injury is mediated via the intrinsic pathway [20, 21]. Gillespie *et al.* have demonstrated that oxidant-induced mitochondrial DNA damage also results in altered mitochondrial membrane potential and apoptosis [20].

Endoplasmic Reticulum Stress-Induced Apoptosis Pathways

Apoptosis can occur as a result of endoplasmic reticulum (ER) stress, as depicted in Figure 16.3. The ER is the site of post-translational modification and folding of proteins. Various insults can disrupt ER protein folding, leading to accumulation of unfolded or misfolded proteins in the ER, thus eliciting ER stress [22, 23]. These insults include chemical inhibitors that deplete ER Ca^{2+} (e.g., A23187, thapsigargin, ionomycin, and EGTA), reductive stress (e.g., homocysteine, β-captoethanol, and dithiothreitol), glycosylation antagonists (e.g., tunicamycin, glucosamine, and 2-deoxyglucose), glucose starvation, hypoxia/ischemia, mutant ER proteins, viral infection, and ATP depletion. Cells may respond to ER stress by a process termed the unfolded protein response (UPR). ER stress is sensed by three sensors, pancreatic endoplasmic reticulum-like kinase (PERK), activating transcription factor (ATF) 6, and inositol-requiring enzyme (IRE) 1. These three sensors are activated in order, with PERK being first, followed by ATF6 and, finally, IRE1. Activated PERK not only inhibits general protein translation by phosphorylation of eukaryotic initiation factor (eIF)-2α, but also promotes ATF4 translation independent of eIF-2α. As a transcription factor, ATF4 induces transcription of genes required to restore ER homeostasis. In addition, ATF4 also induces expression of proapoptotic transcription factor, C/EBP homologous protein (CHOP) (also known as GADD153). Activated ATF6 regulates expression of ER chaperones (such as glucose-regulated protein GRP78, GRP94, and protein disulfide isomerase), X box-binding protein (XBP) 1, and CHOP. Active IRE1 serves as an endoribonuclease to splice XBP1 mRNA to produce an active form, sXBP1. sXBP1 translocates to the nucleus and controls transcription of ER chaperones, genes involved in protein degradation, and P58IPK, which inhibits PERK activation as a negative feedback loop. In addition, active IRE activates c-Jun N-terminal kinase (JNK), leading to phosphorylation of Bim (a proapoptotic protein) and Bcl-2 (an antiapoptotic protein), resulting in subsequent activation of Bim and inactivation of Bcl-2. Through the UPR, cells restore ER function by blocking ER client protein loading for folding, enhancing ER protein folding

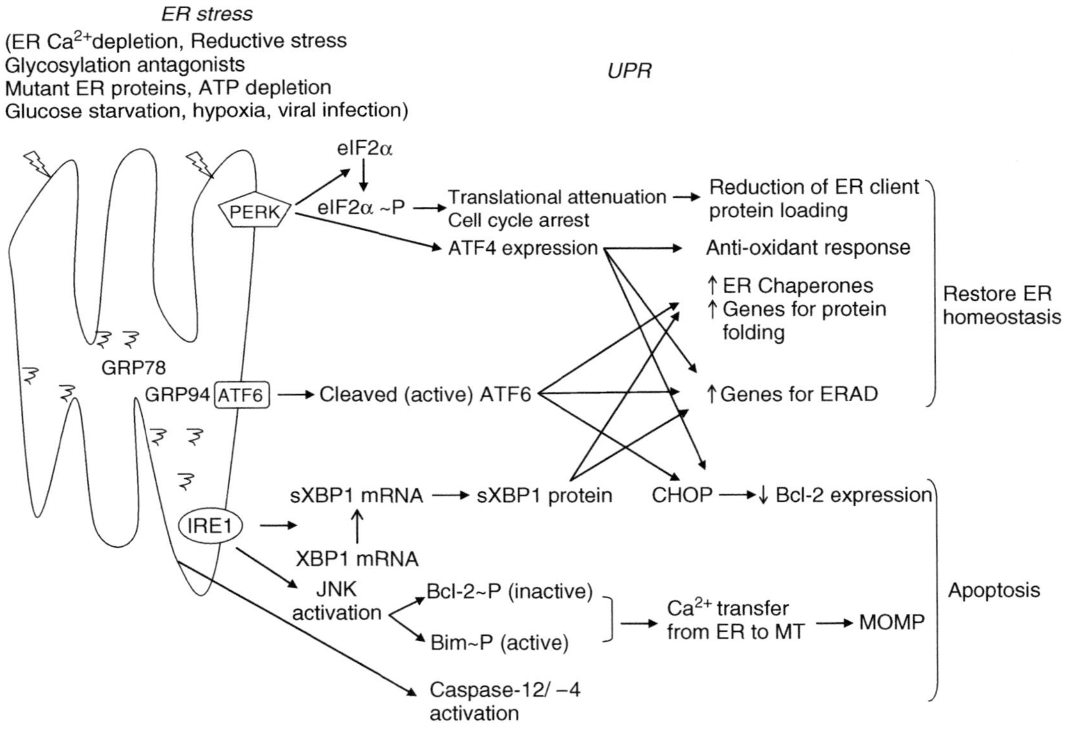

Figure 16.3 Signaling pathways of ER stress, UPR, and apoptosis. See text for details. ERAD, ER-associated protein degradation; MT, mitochondria.

capacity, and promoting ER-associated protein degradation (ERAD) [23, 24]. However, the UPR also activates multiple apoptotic pathways. As stated above, CHOP, which suppresses Bcl-2 expression, is transcriptionally upregulated by ATF4 and ATF6. Additionally, Bcl-2 is inactivated and Bim is activated through IRE1 pathway. These changes allow activation of Bax/Bak in both ER and mitochondrial membrane. Activation of Bax/Bak in the ER membrane leads to transmission of a death signal (such as Ca^{2+}) from the ER to the mitochondria through Ca^{2+}-dependent activation of mitochondrial pathway of apoptosis [24]. Activation of Bax/Bak in the mitochondrial membrane directly activates the mitochondrial apoptotic pathway [24]. In addition, caspase-12 (rodent) and caspase-4 (human), ER membrane-localized caspases, have been proposed to mediate ER stress-induced apoptosis [25, 26]. Caspase-12 and -4 are cleaved and activated through the Ca^{2+}-dependent protease m-calpain exclusively by ER stress triggers [27]. However, recent studies have suggested that ER stress-induced apoptosis is independent of caspase-12 and -4 [28, 29].

Since cells can simultaneously activate both adaptive (survival) and apoptotic pathways in response to ER stress, what decides cell fate? Lin *et al.* have recently reported that while activation of PERK/eIF-2α signaling and induction of CHOP was maintained, IRE1 and ATF6 activities were attenuated by persistent ER stress in human cells [30]. They further showed that maintenance of IRE1 activity enhanced cell survival, suggesting that instability of IRE1 contributes to cell death upon severe ER stress [30]. Rutkowski *et al.* have also demonstrated that adaptation to ER stress is mediated by differential stabilities of prosurvival (adaptive) and proapoptotic mRNAs and proteins [31]. Cells survive mild ER stress because of the short life of proapoptotic proteins compared to those that facilitate protein folding and adaptation [31]. During robust and persistent ER stress, the proapoptotic proteins, particular CHOP, are expressed at a level excessive for degradation and apoptosis occurs [31].

In lung ECs, adenosine-induced apoptosis appears to be via the ER stress pathway (see "ER Stress and Pulmonary EC Apoptosis"). Since ECs may be exposed to oxidant stress (see Chapter 17) under conditions of hyperoxia and upon exposure to environmental conditions such as cigarette smoke, it is tempting to speculate that the ER stress pathway may be important in the genesis of lung diseases.

Control of Apoptosis

As seen in Figure 16.1, the mitochondrial-mediated intrinsic pathway of apoptosis is tightly regulated by Bcl-2 family members. MOMP is controlled by the balance of the activities of antiapoptotic Bcl-2 family members (such as Bcl-2, Bcl-x_L, Mcl-1, Bfl-1, Bcl-w, and Boo) and proapoptotic Bcl-2 family members (such as Bax, Bak, Bok, Bid, Bik, Bad, Bim, Puma, and Noxa) [11, 32]. Normally, proapoptotic Bcl-2 proteins are sequestered in the cytoplasm in an inactive state by antiapoptotic Bcl-2 proteins through nonspecific heterodimerization. During cellular stress, increased availability of proapoptotic Bcl-2 family proteins causes Bax translocation to the mitochondrial outer membrane, and activation of Bax and Bak, leading to Bax/Bak pore formation to allow release of apoptosis-inducing proteins to the cytosol, culminating in apoptosis [11]. The external pathway of apoptosis is regulated by Fas-associated death domain (FADD)-like inhibitory proteins (Fas-associated death domain-like inhibitory proteinFLIPs). As dominant-negative homologs of procaspase-8, FLIPs prevent apoptosis by competing with procaspase-8 for binding to adapter protein FADD [33]. Downregulation of cellular (c) FLIP is implicated in anoikis in ECs [34]. In addition, IAPs directly bind to caspases thus preventing their activation [35]. The nuclear factor nuclear factor-κBNF-κB is an important antiapoptotic transcription factor. NF-κB is activated by TNF-α, AproL, and LPS. Activated NF-κB promotes expression of various antiapoptotic proteins, such as Bcl-2, Bcl-x_L, c-FLIP, c-IAP-1, c-IAP-2, and TNFR-associated factor (tumor necrosis factor receptor-associated factorTRAF)-1 and -2 [36, 37].

Thus, it is possible that lung endothelial apoptosis can be increased both by increased stimuli for apoptosis and by abnormal control of apoptosis.

METHODS TO DETECT APOPTOTIC, NECROTIC, AND AUTOPHAGIC CELLS

Assessment of Apoptosis

Based on the characteristics of different forms of dying cells, apoptotic, necrotic, and autophagic cells can be distinguished by various methods (see Table 16.1). In general, live cells exclude Trypan blue and propidium iodide (PI), thus cell viability can be assessed by Trypan blue or PI exclusion. However, Trypan blue does not distinguish apoptotic versus necrotic cells. Early-stage apoptotic cells that maintain an intact plasma membrane, are PI-negative, while necrotic cells are PI-positive due to disruption of the cellular membrane. Thus, PI staining is useful for distinguishing early apoptotic cells versus necrotic cells in tissue culture.

Evaluation of apoptotic cells in organs, especially in lung tissue, is a challenge, compared to assessment of apoptosis in cultured cells, for the following reasons. (i) Cell-type-specific markers are often lost during progression of apoptosis. Thus, it is difficult to identify the cell types of the apoptotic cells. (ii) The rate of clearance of

Table 16.1 Methods for assessment of apoptosis.

Characteristics	Methodology	Cell culture	Lung sections	References
Changes in cell morphology	electronic microscopy	−	++	[1, 76]
Externalization of phosphatidylserine	Annexin-V assay with PI/7-AAD (IF or flow cytometry)	++	−	[114]
	intravenously administered biotinylated Annexin-V	−	+	[2]
Bcl-2 family members	IB or IHC	++	++	[11, 32, 76]
Apoptotic protein (cytochrome *c*, AIF) translocation	IB, IF	++	−	[11, 32]
Changes in mitochondrial membrane potential	IF or flow cytometry	++	−	[115]
Caspase-3 activation	caspase-3 cleavage by IB, IF, or IHC	++	++	[55, 57]
	DEVDase activity (enzyme activity assay)	++	−	[57]
Targets of caspase-3 [lamin, poly(ADP-ribose) polymerase, cytokeratin-18]	IB, IF, or IHC	++	++	[76]
DNA condensation	Hoechst 33342 or 4′,6-diamidino-2-phenylindole staining	++	−	[116]
DNA fragmentation	DNA laddering	++	−	[117]
	TUNEL staining	++	++	[54, 116]
	in situ ligation of a labeled DNA fragment	+	++	[55]
	single-stranded DNA assay by IHC	+	++	[54]

IB, immunoblotting; IF, immunofluorescence microscopy; IHC, immunohistochemistry; ++, useful; +, potentially useful; −, not useful.

apoptotic cells from tissue is unknown. The net observation at a given time point is a combination of apoptosis induction and apoptotic cell removal. Therefore, an increased value of detected apoptosis may mean either a true increase in apoptosis or a decrease in clearance of apoptotic cells. On the other hand, this may underestimate the true induction of apoptosis, due to a rapid clearance of apoptotic cells. (iii) Apoptosis may increase with age, thus age-matched controls are critical for interpreting experimental results. This makes human studies a challenge due to the low availability of normal human lung tissue. (iv) It is a challenge to distinguish different types of cell death *in vivo*. Despite these challenges, there are many methods available for assessment of apoptosis in cultured cells and lung tissue, as summarized in Table 16.1. Some methods, such as Annexin-V and terminal deoxynucleotidyl transferase-mediated dUTP nick end-labeling (TUNEL) staining can recognize both apoptotic and necrotic cells. In addition, live cells with active transcription and DNA repair are TUNEL-positive. Thus, the combination of Annexin-V with PI or 7-amino actinomycin D (7-AAD) staining for membrane permeability is recommended for assessing apoptotic versus necrotic cells. Apoptotic cells are Annexin-V-positive and PI- or 7-AAD-negative, whereas necrotic cells are double positive. Cell-specific markers are also lost during processing of some apoptosis assays. TUNEL staining is probably the most useful assay currently available to document apoptosis of specific lung cells. Therefore, a complementary approach with at least two or more assays to detect different aspects of apoptosis will provide better precise evidence of lung cell apoptosis.

Assessment of Necrosis

Electron microscopic evidence of necrosis includes disrupted membranes, dilated organelles, and nuclear degradation without condensation. DNA assays reveal DNA smears, but not laddering. Necrotic cells can also be identified using the necrosis marker, high-mobility group box 1 (HMGB1). HMGB1 is tightly bound to chromatin during apoptosis, but released to the extracellular environment during necrosis. Thus HMGB1 can be detected in the culture medium of necrotic cells. In addition, the membrane impermeable dye, SYTOX, stains necrotic cells, but not early-stage apoptotic cells [46].

Assessment of Autophagy

Electron microscopic evidence of autophagy includes lack of condensed nuclei and lack of organelle swelling with appearance of multiple-membrane vesicles. Immunofluorescence microscopy and immunoblot analysis can be used to detect the autophagy marker, microtubule-associated LC3. Normally, LC3 is found as its cytosolic form (LC3-I, 18 kDa, diffuse pattern). Upon induction of autophagy, LC3-I converts to an autophagosome membrane-bound form (LC3-II, 16 kDa, punctuate pattern). Finally, immunoblotting can be used to detect ATG5 and ATG7.

MEDIATORS OF PULMONARY EC APOPTOSIS

EC form a monolayer lining the vasculature. Due to the location of the endothelium at the interface between the blood and surrounding tissue, EC are exposed to multiple stresses. One pathological consequence of these stresses in blood vessels is the induction of EC apoptosis. A variety of biomechanical and biochemical stimuli are involved in endothelial pro- and antiapoptotic processes. Factors that maintain EC survival include physiological levels of shear stress and cyclic strain, integrin interactions with extracellular matrix, activated protein C (APC), and sphingosine 1-phosphate (S1P), whereas EC apoptosis is induced by LPS, TNF-α, ceramide, angiotensin II, and oxidative stress. The pro- and antiapoptotic effects of these biological mediators and underlying signaling pathways have been reviewed in detail elsewhere [47]. This chapter will focus on the review of the current understanding of apoptosis mediated through ER stress pathways, vascular endothelial growth factor (VEGF), transforming growth factor (TGF)-β1, LPS, and ceramide.

ER Stress and Pulmonary EC Apoptosis

Increased levels of ATP and adenosine are released into the blood stream or tissue upon platelet degranulation or cytolytic release from necrotic cells [48]. Our studies have demonstrated that either ATP or adenosine cause pulmonary vascular EC apoptosis, an effect exacerbated by homocysteine [49, 50]. We have also demonstrated that adenosine plus homocysteine cause pulmonary EC apoptosis by inhibiting isoprenylcysteine-*O*-carboxyl methyltransferase (ICMT) [51] – a membrane protein localized to the ER [52, 53] catalyzing post-translational carboxyl methylation of proteins encoding a C-terminal CAAX motif (C, cysteine; A, an aliphatic amino acid; X, any amino acid). Our recent studies have revealed that inhibition of ICMT caused pulmonary EC apoptosis likely through changes in post-translational modification, subcellular relocalization, and a decrease in the level of GRP94 [54] (see Figure 16.4). GRP94 is an ER molecular chaperone important in protein folding and export of the folded proteins [55]. Decrease in GRP94 protein levels has been correlated with apoptosis upon ER stress [56, 57]. Suppression of GRP94 expression accelerates ER stress-induced apoptosis [58, 59], whereas overexpression of GRP94 protects against ER stress-induced neuronal cell death *in vitro* and neuronal cell death in ischemia–reperfusion injury *in vivo* [58, 60]. These findings suggest that GRP94 protects against ER stress-induced apoptosis. Our studies have suggested that not only the GRP94 protein level, but also its post-translational modification and subcellular localization are important in preventing pulmonary EC apoptosis upon ER stress induced by ICMT inhibition [54]. We speculate that the decrease in GRP94 protein levels results from its abnormal post-translational modification and subsequent subcellular relocalization [54]. We also speculate that abnormal post-translational modification and subcellular relocalization and aggregation, and reduction in protein content of GRP94 may reduce ER protein folding capacity, leading to accumulation of unfolded or misfolded proteins in the ER, resulting in ER stress. This in turn activates the UPR, as indicated

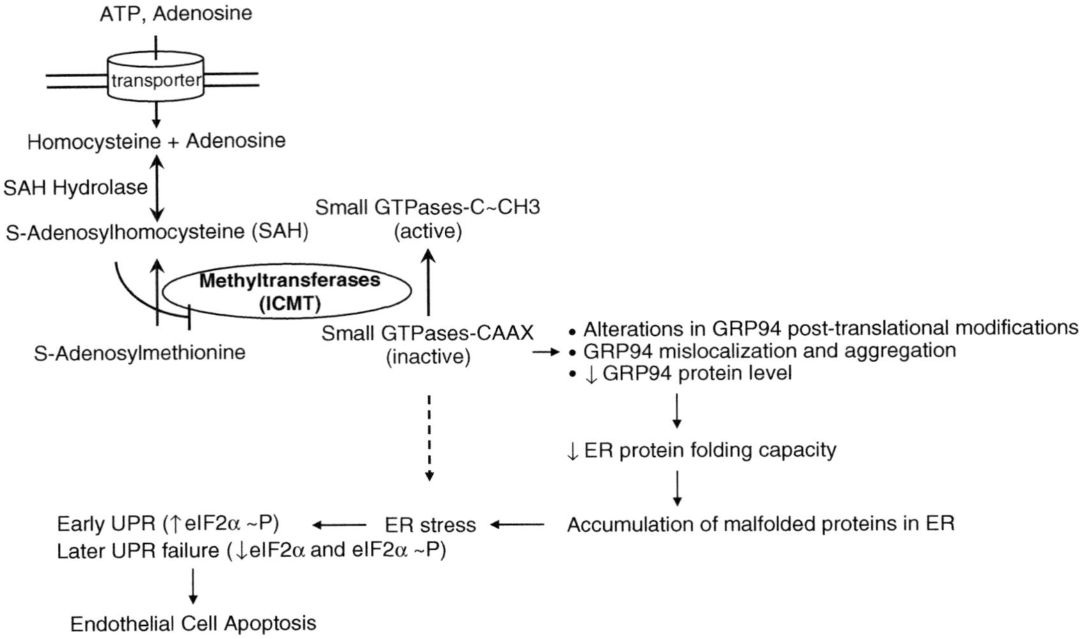

Figure 16.4 Mechanism of pulmonary EC apoptosis caused by extracellular ATP or adenosine via inhibition of intracellular ICMT. See text for details.

by early activation of eIF-2α upon ICMT inhibition (unpublished data). However, prolonged ICMT inhibition (24 h) caused this adaptation to fail, as indicated by decrease in both activated (phosphorylated) and total eIF-2α levels, and apoptosis occurred (unpublished data). Our proposed model is shown in Figure 16.4. Our data suggest that GRP94 may play a central role in mediating ATP- and adenosine-induced ER stress and apoptosis.

VEGF/TGF-β1 Signaling and Pulmonary EC Apoptosis

EC express and secrete abundant VEGF. In addition to promoting EC growth, VEGF is recognized as an EC survival factor important for maintenance of alveolar structures [61]. Kasahara *et al.* have reported a significant decrease in expression of VEGF and VEGF receptor type 2 (vascular endothelial growth factor receptorVEGFR-2) in the lungs of patients with emphysema [45] (see Chapter 26). This decreased expression was associated with apoptosis of both EC and epithelial cells of the alveolar septa [45]. These findings suggest that lung cell apoptosis and the development of emphysema may occur as a result of reduced VEGF signaling. This concept is supported by additional observations that impairment of VEGF signaling by either chronic inhibition of VEGFR-2 or by the genetic deletion of lung VEGF gene caused alveolar septal cell apoptosis and emphysema; effects that were prevented by caspase inhibitors [41, 62]. Our studies have suggested that pulmonary EC apoptosis induced by inhibition of VEGF signaling occurs through Bcl-2 downregulation-mediated activation of the intrinsic pathway [42].

TGF-β1 plays an important role in lung tissue repair and fibrosis. The effect of TGF-β1 on EC apoptosis is controversial. Several studies suggest that TGF-β1 causes EC apoptosis [63–65]. Others have demonstrated that TGF-β1 promotes EC survival [66–68] and protects against EC apoptosis induced by hypoxia [69]. These apparently contradictory findings may be due to studies of different experimental conditions and EC from different organs. We have demonstrated opposite effects of TGF-β1 on apoptosis of EC derived from different pulmonary vascular beds [42, 70]. TGF-β1 caused apoptosis of pulmonary microvascular ECs [70], but protected against pulmonary artery EC apoptosis induced by serum deprivation and by VEGFR-2 blockade [42]. In addition, we have noted that the protective effects of TGF-β1 against pulmonary artery EC apoptosis are dependent upon the type of injury and subsequent activation of apoptotic pathways [42]. TGF-β1 was incapable of protecting pulmonary artery ECs from apoptosis induced by ultraviolet radiation exposure, and TNF-α and ICMT inhibition [42], in which apoptosis occurs by the extrinsic pathway and ER stress pathway, respectively. We further demonstrate that TGF-β1 increased Bcl-2 protein expression and prevented caspase-9 activation induced by VEGFR-2 blockade, but did not alter c-FLIP protein levels or promote the activation of caspase-8 or caspase-12 [42]. These findings suggest that TGF-β1 is effective in inhibiting mitochondrial dysfunction-mediated apoptosis, but ineffective in preventing apoptosis mediated through the extrinsic pathway or ER stress pathway in pulmonary macrovessels. As pulmonary conduit vessels and microvessels have heterogeneous apoptotic responses to TGF-β1, it is important to evaluate apoptosis of EC from different pulmonary vascular beds in various disease states (see Chapter 9).

VEGF has been shown to induce expression and secretion of TGF-β1 in intestinal epithelial cells and glomerular EC [71, 72]. On the other hand, TGF-β1 enhanced expression of VEGF and VEGFR-1 [73, 74]. VEGF is an EC survival factor. However, recent studies have reported that upon TGF-β1 stimulation, VEGF was induced and acted as an apoptosis inducer by mediating the effect of TGF-β1 on apoptosis induction in human umbilical vein ECs (HUVECs) [75]. These findings suggest that VEGF and TGF-β1 may interact to form a positive feedback loop to regulate EC apoptosis. Our data have shown that VEGFR-2 blockage caused pulmonary artery EC apoptosis – an effect associated with downregulation of Bcl-2 [42]. TGF-β1 completely prevented VEGFR-2 blockage-induced pulmonary artery EC apoptosis, likely through upregulation of Bcl-2 [42]. Our data suggest that VEGF and TGF-β1 may control some common gene targets, such as Bcl-2, to maintain cell survival at least in pulmonary artery EC. If one pathway is impaired, another pathway will have the redundant effect to maintain cell survival. Understanding the interaction between VEGF and TGF-β1 at the molecular level will be important for not only understanding their roles in pulmonary EC apoptosis, but also the pathogenesis of emphysema (see Chapter 26).

Ceramide and Pulmonary EC Apoptosis

As a second messenger lipid, ceramide is induced by cigarette smoke extract in cultured bovine pulmonary EC [76]. Ceramide is also upregulated in emphysematous lungs of both patients and animal models based on VEGFR-2 blockade. The increase in ceramide is associated with enhanced alveolar cell apoptosis [76]. Inhibition of *de novo* ceramide synthesis significantly attenuated lung cell apoptosis and emphysema induced by VEGFR-2 blockade [76]. Furthermore, intratracheal instillation of C_{12} ceramide increases lung long-chain ceramide levels, triggers apoptosis of ECs and type II alveolar epithelial cells, and causes emphysema [76]. In addition, ceramide-induced airspace enlargement was prevented by pretreatment with a general caspase inhibitor [76]. These results suggest that ceramide may be a critical mediator of alveolar cell apoptosis and development of emphysema [76].

LPS and Pulmonary EC Apoptosis

LPS is a component of the outer membrane of Gram-negative bacteria and is released into the circulation as microorganisms replicate or die. It has been well documented that LPS is a potent inducer of bacterial sepsis by promoting EC activation, dysfunction, and apoptosis. In the blood, LPS binds to soluble CD14 through LPS-binding protein. The LPS/soluble CD14 complex then binds to the transmembrane protein TLR-4, resulting in the recruitment of adaptor myeloid differentiation factor 88 (MyD88) and subsequent activation of NF-κB. Activation of NF-κB counteracts apoptosis by transcriptional upregulation of several antiapoptotic genes, such as IAP and c-FLIP, as seen in a TNF-α model. However, activation of NF-κB is not protective against LPS-induced apoptosis. Emerging evidence demonstrates that FADD, an adaptor for Fas/FasL system, negatively regulates LPS-induced NF-κB activation by interaction with MyD88. Martin *et al.* have proposed that when Fas is clustered and FADD is bound to the intracellular tail of Fas, MyD88 activation and subsequent enhancement of LPS/TLR4/NF-κB signaling are allowed. In contrast, when Fas is blocked, allowing FADD to bind to and inactivate MyD88, LPS/TLR4/NF-κB signaling declines [77]. Whether occupation of FADD by LPS/MyD88 affects the apoptotic pathway mediated through FADD is not known. Recently, Wang *et al.* have demonstrated that Bid-mediated activation of the intrinsic pathway, but not generation of ROS, is responsible for LPS-induced lung EC apoptosis and lung injury [21].

Mitochondrial DNA Damage and Pulmonary EC Apoptosis

Due to open structure and relatively limited repair capacity, the mitochondrial genome is more vulnerable to oxidative stress than nuclear DNA. Oxidative damage to the mitochondrial DNA leads to altered or impaired expression of mitochondrial genes encoding proteins important in the electron transport chain, which may increase ROS production, thus enhancing oxidative stress, resulting in cell death through apoptosis, necrosis, or autophagy. Recently, Ruchko *et al.* have demonstrated that oxidative stress-induced mitochondrial DNA damage can directly trigger apoptosis via mitochondrial dysfunction, as indicated by loss of mitochondrial membrane potential in lung vascular ECs [20]. This study highlights an important role of mitochondrial DNA damage in pulmonary EC apoptosis.

LUNG DISEASES ASSOCIATED WITH ABNORMAL PULMONARY EC APOPTOSIS

Apoptosis can either ameliorate or exacerbate lung injury, depending upon the cell type. Pulmonary EC apoptosis plays an important role in physiological processes, such as vasculogenesis and angiogenesis during lung development. Pulmonary EC apoptosis is also associated with the initiation and progression of various lung diseases. In the following we focus on the effects of pulmonary EC apoptosis on emphysema, ALI, pulmonary fibrosis, lung ischemia–reperfusion injury, and pulmonary hypertension.

Pulmonary Endothelial Apoptosis and Emphysema

Emphysema is a common and debilitating lung disease characterized by alveolar airspace enlargement and loss of alveolar capillary septa, resulting in impaired gas exchange. Cigarette smoking, air pollution, and genetic deficiency of α_1-antitrypsin (AAT) are known primary risk factors for human emphysema. Currently, there is no specific treatment available to reverse the pathogenesis of emphysema. Protease/antiprotease imbalance has been accepted as a major mechanism for emphysematous lung destruction [78], based on the demonstrations that patients with genetic deficiency of AAT develop emphysema [79] and that intratracheal instillation of proteases caused emphysema in rats [80]. However, most emphysema patients do not have AAT deficiency. Inflammatory cell infiltration is often seen in human emphysema. It has been proposed that neutrophil elastase and macrophage matrix metalloproteinase-12 enzymatically degrade elastin in alveolar septa, leading to emphysema [78]. However, ALI associated with massive inflammation does not usually result in emphysema, suggesting that inflammation may not be sufficient by itself for the development of emphysema. Oxidant stress and immunological injury also play important roles in pathogenesis of emphysema [78]. Recent investigations have highlighted a role of pulmonary cell apoptosis in the pathogenesis of emphysema [41, 45, 62] (see Chapter 26). Lung tissue from patients with emphysema displayed increased apoptosis of both epithelial and ECs in the alveolar septa [38, 45], and decreased expression of lung VEGF and VEGFR-2 [45]. In addition, inhibition of VEGF signaling through either chronic inhibition of VEGFR-2 or by genetic deletion of lung VEGF caused alveolar septal cell apoptosis and emphysema; effects that were prevented by caspase inhibition [41, 62]. Similar to VEGF signaling blockade, intra-tracheal instillation of C_{12} ceramide triggered apoptosis of ECs and type II alveolar epithelial cells, as well as emphysema-like disease in mice; effects which were prevented by pretreatment with a pancaspase inhibitor [76]. Furthermore, intratracheal instillation of the active form of caspase-3 caused alveolar wall cell apoptosis and emphysema in rats [81]. Recent studies have shown that targeted lung EC apoptosis results in enhanced oxidative stress, influx of macrophages, upregulation of ceramide, and development of emphysema, suggesting a central role of alveolar EC apoptosis in the development of emphysema [82]. In support of the idea that apoptosis is important in the pathogenesis of emphysema, studies have demonstrated protection against apoptosis-dependent emphysema. Petrache *et al.* demonstrated that the anti-protease AAT inhibits pulmonary microvascular EC apoptosis by a direct interaction with caspase-3, and that overexpression of AAT inhibits lung endothelial apoptosis and emphysema induced by caspase-3 instillation [83] and by VEGF signaling blockade [84]. In addition, Nana-Sinkam *et al.* found that prostacyclin synthase expression is decreased in human emphysematous lung tissues, while overexpression of prostacyclin synthase blunts lung endothelial apoptosis after cigarette smoke exposure [85].

In summary, these results suggest that lung EC apoptosis may be a critical step in the pathogenesis of emphysema (see Chapter 26). However, Wickenden *et al.* reported that cigarette smoke condensate induced necrosis, but inhibited apoptosis of cultured epithelial cells and ECs [86], suggesting that necrosis may play a role in the development of emphysema. In addition, Chen *et al.* have demonstrated that autophagy is significantly increased in lung tissue from chronic obstructive pulmonary disease patients and that cigarette smoke-induced autophagy may contribute to lung cell apoptosis and pathogenesis of emphysema [87]. Thus, lung EC necrosis and autophagy may also be important in the pathogenesis of emphysema.

Pulmonary Endothelial Apoptosis and ALI/Acute Respiratory Distress Syndrome

Extensive pulmonary EC apoptosis has been observed in patients with severe acute respiratory distress syndrome (ARDS) [88]. Recently, it has been reported that lung EC apoptosis is directly linked to the severity of lung injury resulting from allogeneic bone marrow transplantation [89]. The Fas/FasL system represents an important receptor-mediated, extrinsic pathway of apoptosis. Fas, a 45-kDa type I membrane protein, is expressed on the surface of various lung cells, including ECs, alveolar and bronchial epithelial cells, Clara cells, alveolar macrophages, and myofibroblasts. FasL, a 37-kDa type II protein, is expressed in neutrophils and lymphocytes. Increased expression of Fas and FasL has been observed in pulmonary edema fluid and lung tissue of patients with ALI and ARDS [90]. FasL can be cleaved to sFasL, a soluble form of FasL, by metalloproteinases. sFasL is also increased in bronchoalveolar lavage (BAL) of patients with ARDS [91]. Moreover, the BAL from these ARDS patients caused apoptosis of cultured lung epithelial cells; an effect inhibited by blocking the Fas/FasL system [91]. Whether BAL of ARDS patients has similar apoptosis-inducing effect on lung ECs is unknown. These results suggest that sFasL is released as a death-inducing mediator capable of inducing lung cell apoptosis during ALI by interaction with Fas. This notion is supported by a recent investigation demonstrating that Fas/FasL-deficient mice have lesser degrees of ALI when challenged with intrapulmonary deposition of IgG immune complexes; an effect associated with less lung cell apoptosis [92]. In addition, inhibition of caspase activity blunts neutrophil-induced ALI in wild-type mice [92]. Collectively, these results suggest that Fas/FasL-mediated lung cell apoptosis contributes to ALI and ARDS. However, the role of the Fas/FasL-induced lung EC apoptosis in ALI/ARDS is not yet clear.

LPS can induce ALI. Pulmonary EC apoptosis was observed during ALI in mice exposed to LPS [93]. Moreover, a broad-spectrum caspase inhibitor significantly attenuated pulmonary EC apoptosis and prolonged the survival of mice exposed to LPS [94]. These results suggest that pulmonary EC apoptosis plays an important role in LPS-induced ALI [93, 94]. Of note, in addition to alveolar EC, apoptosis of bronchial and alveolar epithelial cells, as well as interstitial inflammatory cells, also occurred during Fas/FasL- and LPS-induced ALI. Therefore, apoptosis of these cells may also contribute to ALI.

Pulmonary Cell Apoptosis and Pulmonary Fibrosis

Pulmonary fibrosis begins with alveolitis, followed by excess collagen deposition and destruction of the normal lung architecture. Epithelial apoptosis and necrosis are increased in lungs of patients with idiopathic pulmonary fibrosis (IPF) [95]. Apoptosis has been thought to be a noninflammatory means of removing injurious cells thus facilitating lung repair. However, there is increasing evidence that Fas/FasL-mediated lung epithelial apoptosis results in release of proinflammatory cytokines (such as TNF-α and TGF-β1), leading to inflammation and progression from ARDS to fibrosis [96]. FasL was upregulated in inflammatory cells in BAL from patients with IPF [97]. In addition, Fas expression was also elevated in alveolar and bronchiolar epithelial cells from patients with IPF [97], and sFasL was significantly enhanced in both serum and BAL from IPF patients [98]. Thus, it is possible that Fas/FasL-mediated epithelial cell apoptosis may contribute to pulmonary fibrosis. Recently, it has been demonstrated that an early, transient wave of epithelial apoptosis is essential for TGF-β1-induced pulmonary fibrosis [99, 100]. Whether pulmonary EC apoptosis occurs during initiation or progression of pulmonary fibrosis is unknown. Recently, Zhang *et al.* have demonstrated that the level of inhibitor of differentiation 1 (Id1) was upregulated in the lungs of mice challenged with bleomycin [101]. Id1-deficient mice displayed increased pulmonary EC apoptosis and enhanced lung collagen accumulation and fibrogenesis upon exposure to bleomycin, suggesting that Id1 may promote EC survival and attenuate fibrosis [101]. These results also suggest that pulmonary EC apoptosis may contribute to pulmonary fibrosis.

Pulmonary Endothelial Apoptosis and Lung Ischemia–Reperfusion Injury

Ischemia–reperfusion can cause organ injury in circumstances in which blood flow is interrupted and then restored, such as with the relief of vascular occlusion and after organ transplantation (see Chapter 17). Ischemia–reperfusion injury is an early complication of organ transplantation and is related to the duration of cold storage prior to transplantation [102]. Ischemia–reperfusion is characterized by increased ROS, inflammatory cell accumulation, and activation of neutral sphingomyelinase, resulting in ceramide accumulation [103]. Using a rat model of lung transplantation, Quadri *et al.* have reported that pulmonary EC apoptosis occurred during ischemia–reperfusion injury, and that caspase inhibition decreased apoptosis and enhanced the function of transplanted lungs [102]. Pulmonary EC apoptosis is associated with worsening of lung function upon lung ischemia–reperfusion [104].

Pulmonary thromboendarterectomy (PTE) is a surgical procedure to restore blood flow to the lung in chronic thromboembolism syndrome in which proximal pulmonary artery is obstructed by blood clot (see Chapter 29). ALI is a major complication of PTE that can lead to pulmonary edema and persistent pulmonary hypertension due to ischemia–reperfusion lung injury. Pulmonary artery EC apoptosis in PTE piglets was significantly increased five weeks after pulmonary artery ligation (chronic ischemia), dramatically increased 2 days after reperfusion, and returned to normal within 5 weeks after reperfusion [105]. These results suggest that lung ischemia–reperfusion-induced pulmonary EC apoptosis may contribute to the development of vascular dysfunction after PTE.

Pulmonary Endothelial Apoptosis and Pulmonary Hypertension

Voelkel *et al.* have proposed that the development of severe angioproliferative pulmonary artery hypertension (PAH) is associated with initial EC apoptosis followed by the emergence of hyperproliferative, apoptosis-resistant ECs and vascular smooth muscle cells (VSMCs) [106–108] (see Chapter 27). This notion is supported by studies demonstrating that development of monocrotaline-induced pulmonary hypertension was associated with increased lung EC apoptosis [109, 110]. Attenuation of PAH by SD208 (an inhibitor of TGF-β1 receptor ALK5) [110] or by pravastatin (a 3-hydroxy-3-methylglutaryl-coenzyme A reductase inhibitor) [109] correlated with reduction in lung EC apoptosis. Mutations in bone morphogenetic protein receptor-2 (BMPR-2) have been found in patients with familial PAH. Recently, Teichert-Kuliszewska *et al.* have shown that loss of BMPR-2 causes apoptosis of cultured human pulmonary artery ECs (PAECs) [111]. They also found that BMP attentuated serum deprivation-induced apoptosis in circulating endothelial progenitor cells (EPCs) from normal subjects, but not in EPCs isolated from patients with idiopathic PAH [111]. These results suggest that loss-of-function mutations in BMPR-2 may promote initial pulmonary EC apoptosis, thus contributing to pathogenesis of PAH. In support of this idea, greater EC apoptosis has been documented in heterozygous BMPR-2 mutant mice challenged with monocrotaline combined with intratracheal instillation of replication-deficient adenovirus expressing 5-lipoxygenase, which caused much severe pulmonary hypertension, compared to wild type mice [112].

In contrast, irreversible pulmonary hypertension is strongly associated with increased expression of the antiapoptotic protein Bcl-2 in vascular ECs from patients with pulmonary hypertension associated with congenital heart disease [113]. Consistently, PAECs derived from lungs of patients with idiopathic PAH showed a greater proliferation rate and decreased apoptosis in tissue culture, which was associated with persistent activation of the cell survival regulator, signal transducer and activator of transcription 3, and increased expression of its downstream antiapoptotic target, Mcl-1 [114]. Simvastatin has been shown to attenuate severe PH by inducing apoptosis of pulmonary microvascular ECs [115] and neointimal smooth muscle cells [116].

In summary, increased EC apoptosis at an early stage and decreased EC apoptosis at later stages of the disease may contribute to development of PAH. On the other hand, apoptosis may be an outcome of other causes of pulmonary hypertension. For example, acute pulmonary venous hypertension induces apoptosis of capillary ECs [117].

CONCLUSIONS AND PERSPECTIVES

Apoptosis occurs through the death receptor-mediated extrinsic pathway, mitochondria-mediated intrinsic pathway, and ER stress-mediated UPR pathway, which are tightly regulated by cell survival and adaptation signaling. Although cell death of different types of lung cells may have distinct effects on pathogenesis, pulmonary EC apoptosis appears to significantly contribute to the development of emphysema and ALI/ARDS. It may play a role in pulmonary fibrosis. Increased lung EC apoptosis may initiate PAH while resistance to apoptosis of alveolar ECs may contribute to progression to severe PAH. We hypothesize a central role for pulmonary EC apoptosis in the development of lung diseases is depicted in Figure 16.5. Caspase inhibitors have been successfully used to inhibit lung cell apoptosis and vascular injury

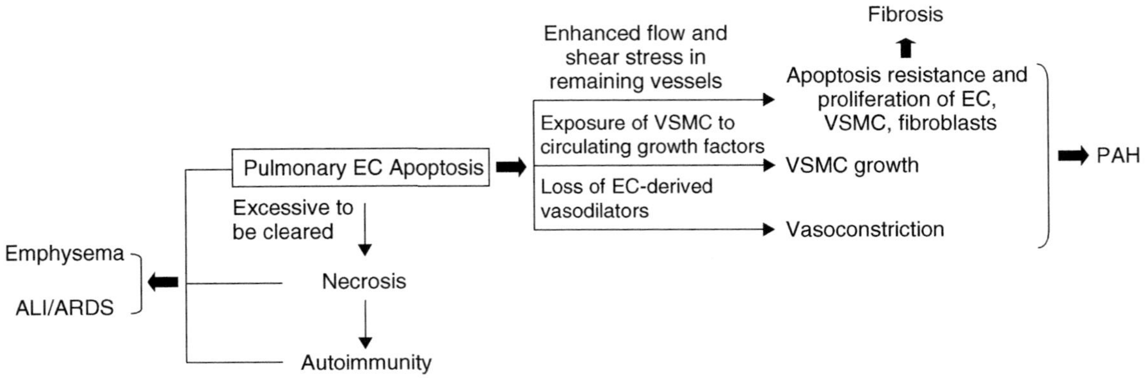

Figure 16.5 Available evidence suggests that apoptosis may play a role in emphysema, pulmonary fibrosis, and PAH.

in animal models. However, use of such drugs to treat apoptosis-associated lung diseases may be problematic due to breakdown of tissue homeostasis and potential activation of necrosis and autophagy, thus drugs acting locally are needed. Complementary approaches are necessary to precisely assess apoptosis in both cell culture and lung tissue.

Therapeutic potential of drugs that modulate cell death is dependent upon cell type-specific, tissue-specific, and vascular bed-specific actions. Areas where research is needed include:

- Relative apoptosis-resistance of the different lung ECs (conduit artery versus microvascular versus progenitor).
- The role of lung EC apoptosis due to loss of matrix contacts.
- Pulmonary EC resistance to apoptosis induced by viral infection.
- Role of lung EC apoptosis in pulmonary fibrosis.
- The insults and pathways leading to EC apoptosis in lung diseases.
- Apoptosis of specific lung cells in initiation and/or progression of lung diseases.
- The relative contributions of apoptosis, necrosis, and autophagy to pathogenesis of lung diseases.

References

1. Kerr, J., Wyllie, A., and Currie, A. (1972) Apoptosis: a basic biological phenomenon with wide-ranging implications in tissue kinetics. *British Journal of Cancer*, **26**, 239–57.
2. Henson, P.M. and Tuder, R.M. (2008) Apoptosis in the lung: induction, clearance and detection. *American Journal of Physiology: Lung Cellular and Molecular Physiology*, **294**, L601–11.
3. Huynh, M.L., Fadok, V.A., and Henson, P.M. (2002) Phosphatidylserine-dependent ingestion of apoptotic cells promotes TGF-beta1 secretion and the resolution of inflammation. *The Journal of Clinical Investigation*, **109**, 41–50.
4. Savill, J., Dransfield, I., Gregory, C., and Haslett, C. (2002) A blast from the past: clearance of apoptotic cells regulates immune responses. *Nature Reviews: Immunology*, **2**, 965–75.
5. Golpon, H.A., Fadok, V.A., Taraseviciene-Stewart, L. *et al.* (2004) Life after corpse engulfment: phagocytosis of apoptotic cells leads to VEGF secretion and cell growth. *The FASEB Journal*, **18**, 1716–18.
6. Gaipl, U.S., Kuhn, A., Sheriff, A. *et al.* (2006) Clearance of apoptotic cells in human SLE. *Current Directions in Autoimmunity*, **9**, 173–87.
7. Henriquez, M., Armisén, R., Stutzin, A., and Quest, A.F. (2008) Cell death by necrosis, a regulated way to go. *Current Molecular Medicine*, **8**, 187–206.
8. Sadoshima, J. (2008) The role of autophagy during ischemia/reperfusion. *Autophagy*, **4**, 402–3.
9. Gustafsson, A.B. and Gottlieb, R.A. (2008) Recycle or die: the role of autophagy in cardioprotection. *Journal of Molecular and Cellular Cardiology*, **44**, 654–61.
10. Lemasters, J.J., Nieminen, A.L., Qian, T. *et al.* (1998) The mitochondrial permeability transition in cell death: a common mechanism in necrosis, apoptosis and autophagy. *Biochimica et Biophysica Acta*, **1366**, 177–96.
11. Hung, R. and Chow, A. (2004) Dissecting the "end game": clinical relevance, molecular mechanisms and laboratory assessment of apoptosis. *Clinical and Investigative Medicine*, **27**, 324–44.
12. Slee, E., Harte, M., Kluck, R. *et al.* (1999) Ordering the cytochrome *c*-initiated caspase cascade: hierarchical activation of caspase-2, -3, -6, -7, and -10 in a caspase-9-dependent manner. *The Journal of Cell Biology*, **144**, 281–92.
13. Zou, H., Li, Y., Liu, X., and Wang, X. (1999) An APAF-1.cytochrome *c* multimeric complex is a functional apoptosome that activates procaspase-9. *The Journal of Biological Chemistry*, **274**, 11549–56.
14. Du, C., Fang, M., Li, Y. *et al.* (2000) Smac, a mitochondrial protein that promotes cytochrome c-dependent caspase activation by eliminating IAP inhibition. *Cell*, **102**, 33–42.
15. Yang, Q.R.C.-H., Ren, J., Newton, M., and Du, C. (2003) Omi/HtrA2 catalytic cleavage of inhibitor of apoptosis protein (IAP) irreversibly inactivates IAPs and facilitates caspase activity in apoptosis. *Genes and Development*, **17**, 1487–96.
16. Susin, S., Lorenzo, H., Zamzami, N. *et al.* (1999) Molecular characterization of mitochondrial apoptosis-inducing factor. *Nature*, **397**, 441–46.
17. Li, L., Luo, X., and Wang, X. (2001) Endonuclease G is an apoptotic DNase when released from mitochondria. *Nature*, **412**, 95–99.
18. Boldin, M., Goncharov, T., Goltsev, Y., and Wallach, D. (1996) Involvement of MACH, a novel MORT/FADD-interacting protease, in Fas/APO-1 and TNF receptor-induced cell death. *Cell*, **85**, 803–15.
19. Li, H., Zhu, H., Xu, C., and Yuan, J. (1998) Cleavage of BID by caspase 8 mediates the mitochondrial damage in the Fas-pathway of apoptosis. *Cell*, **94**, 491–501.

20. Ruchko, M., Gorodnya, O., LeDoux, S.P. *et al.* (2005) Mitochondrial DNA damage triggers mitochondrial dysfunction and apoptosis in oxidant-challenged lung endothelial cells. *American Journal of Physiology: Lung Cellular and Molecular Physiology*, **288**, L530–35.
21. Wang, H.L., Akinci, I.O., Baker, C.M. *et al.* (2007) The intrinsic apoptotic pathway is required for lipopolysaccharide-induced lung endothelial cell death. *Journal of Immunology*, **179**, 1834–41.
22. Lee, A.S. (2001) The glucose-regulated proteins: stress induction and clinical applications. *Trends in Biochemical Sciences*, **26**, 504–10.
23. Schroder, M. and Kaufman, R.J. (2005) The mammalian unfolded protein response. *Annual Review of Biochemistry*, **74**, 739–89.
24. Szegezdi, E., Logue, S.E., Gorman, A.M., and Samali, A. (2006) Mediators of endoplasmic reticulum stress-induced apoptosis. *EMBO Reports*, **7**, 880–85.
25. Kim, S.J., Zhang, Z., Hitomi, E. *et al.* (2006) Endoplasmic reticulum stress-induced caspase-4 activation mediates apoptosis and neurodegeneration in INCL. *Human Molecular Genetics*, **15**, 1826–34.
26. Szegezdi, E., Fitzgerald, U., and Samali, A. (2003) Caspase-12 and ER-stress-mediated apoptosis: the story so far. *Annals of the New York Academy of Sciences*, **1010**, 186–94.
27. Groenendyk, J. and Michalak, M. (2005) Endoplasmic reticulum quality control and apoptosis. *Acta Biochimica Polonica*, **52**, 381–95.
28. Di Sano, F., Ferraro, E., Tufi, R. *et al.* (2006) Endoplasmic reticulum stress induces apoptosis by an apoptosome-dependent but caspase 12-independent mechanism. *The Journal of Biological Chemistry*, **281**, 2693–700.
29. Obeng, E.A. and Boise, L.H. (2005) Caspase-12 and caspase-4 are not required for caspase-dependent endoplasmic reticulum stress-induced apoptosis. *The Journal of Biological Chemistry*, **280**, 29578–87.
30. Lin, J.H., Li, H., Yasumura, D. *et al.* (2007) IRE1 signaling affects cell fate during the unfolded protein response. *Science*, **318** (5852), 944–49.
31. Rutkowski, D.T., Arnold, S.M., Miller, C.N. *et al.* (2006) Adaptation to ER stress is mediated by differential stabilities of pro-survival and pro-apoptotic mRNAs and proteins. *PLoS Biology*, **4**, e374.
32. Adams, J. and Cory, S. (1998) The Bcl-2 protein family: arbiters of cell survival. *Science*, **281**, 1322–26.
33. Thome, M., Schneider, P., Hofmann, K. *et al.* (1997) Viral FLICE-inhibitory proteins (FLIPs) prevent apoptosis induced by death receptors. *Nature*, **386**, 517–21.
34. Aoudjit, F. and Vuori, K. (2001) Matrix attachment regulates Fas-induced apoptosis in endothelial cells: a role for c-FLIP and implications for anoikis. *The Journal of Cell Biology*, **152**, 633–43.
35. Verhagen, A., Coulson, E., and Vaux, D. (2001) Inhibitor of apoptosis proteins and their relatives: IAPs and other BIRPs. *Genome Biology*, **2**, REVIEW3009.
36. Tamatani, M., Che, Y., Matsuzaki, H. *et al.* (1999) Tumor necrosis factor induces Bcl-2 and Bcl-x expression through NFkappaB activation in primary hippocampal neurons. *The Journal of Biological Chemistry*, **274**, 8531–38.
37. Wang, C., Mayo, M., Korneluk, R. *et al.* (1998) NF-kappaB antiapoptosis: induction of TRAF1 and TRAF2 and c-IAP1 and c-IAP2 to suppress caspase-8 activation. *Science*, **281**, 1680–83.
38. Imai, K., Mercer, B.A., Schulman, L.L. *et al.* (2005) Correlation of lung surface area to apoptosis and proliferation in human emphysema. *The European Respiratory Journal*, **25**, 250–58.
39. Vermes, I., Haanen, C., Steffens-Nakken, H. *et al.* (1995) A novel assay for apoptosis. Flow cytometric detection of phosphatidylserine expression on earlier apoptotic cells using fluorescein labelled. *Journal of Immunological Methods*, **184**, 39–51.
40. Reers, M., Smith, T.W., and Chen, L.B. (1991) J-aggregate formation of a carbocyanine as a quantitative fluorescent indicator of membrane potential. *Biochemistry*, **30**, 4480–86.
41. Kasahara, Y., Tuder, R.M., Taraseviciene-Stewart, L. *et al.* (2000) Inhibition of VEGF receptors causes lung cell apoptosis and emphysema. *The Journal of Clinical Investigation*, **106**, 1311–19.
42. Lu, Q. (2008) Transforming growth factor-β1 protects against pulmonary artery endothelial cell apoptosis via ALK5. *American Journal of Physiology: Lung Cellular and Molecular Physiology*, **295**, L123–33.
43. Bellas, R.E., Harrington, E.O., Sheahan, K.L. *et al.* (2002) FAK blunts adenosine-homocysteine-induced endothelial cell apoptosis: requirement for PI3-kinase. *American Journal of Physiology: Lung Cellular and Molecular Physiology*, **282**, L1135–42.
44. Walker, P.R., Weaver, V.M., Lach, B. *et al.* (1993) Endonuclease activities associated with high molecular weight and internucleosomal DNA fragmentation in apoptosis. *Experimental Cell Research*, **213**, 100–6.

45. Kasahara, Y., Tuder, R.M., Cool, C.D. *et al.* (2001) Endothelial cell death and decreased expression of vascular endothelial growth factor and vascular endothelial growth factor receptor 2 in emphysema. *American Journal of Respiratory and Critical Care Medicine*, **163**, 737–44.
46. Bien, A., Seidenbecher, C.I., Böckers, T.M. *et al.* (1999) Apoptotic versus necrotic characteristics of retinal ganglion cell death after partial optic nerve injury. *Journal of Neurotrauma*, **16**, 153–63.
47. Harrington, E.O., Lu, Q., and Rounds, S. (2007) Endothelial cell apoptosis, in *Endothelial Biomedicine* (ed. W.C. Aird), Cambridge University Press, New York, pp. 1081–97.
48. Dubyak, G. and el-Moatassim, C. (1993) Signal transduction via P2-purinergic receptors for extracellular ATP and other nucleotides. *American Journal of Physiology: Cell Physiology*, **265**, C577–606.
49. Dawicki, D.D., Chatterjee, D., Wyche, J., and Rounds, S. (1997) Extracellular ATP and adenosine cause apoptosis of pulmonary artery endothelial cells. *American Journal of Physiology: Lung Cellular and Molecular Physiology*, **273**, L485–94.
50. Rounds, S., Yee, W., Dawicki, D.D. *et al.* (1998) Mechanism of extracellular ATP- and adenosine-induced apoptosis of cultured pulmonary artery endothelial cells. *American Journal of Physiology: Lung Cellular and Molecular Physiology*, **275**, L379–88.
51. Kramer, K., Harrington, E.O., Lu, Q. *et al.* (2003) Isoprenylcysteine carboxyl methyltransferase activity modulates endothelial cell apoptosis. *Molecular Biology of the Cell*, **14**, 848–57.
52. Dai, Q., Choy, E., Chiu, V. *et al.* (1998) Mammalian prenylcysteine carboxyl methyltransferase is in the endoplasmic reticulum. *The Journal of Biological Chemistry*, **273**, 15030–34.
53. Romano, J.D., Schmidt, W.K., and Michaelis, S. (1998) The *Saccharomyces cerevisiae* prenylcysteine carboxyl methyltransferase Ste14p is in the endoplasmic reticulum membrane. *Molecular Biology of the Cell*, **9**, 2231–47.
54. Lu, Q., Harrington, E.O., Newton, J. *et al.* (2007) Inhibition of ICMT induces endothelial cell apoptosis through GRP94. *American Journal of Respiratory Cell and Molecular Biology*, **37**, 20–30.
55. Nicchitta, C.V. and Reed, R.C. (2000) The immunological properties of endoplasmic reticulum chaperones: a conflict of interest? *Essays in Biochemistry*, **36**, 15–25.
56. Bruneel, A., Labas, V., Mailloux, A. *et al.* (2005) Proteomics of human umbilical vein endothelial cells applied to etoposide-induced apoptosis. *Proteomics*, **5**, 3876–84.
57. Dey, A., Kessova, I.G., and Cederbaum, A.I. (2006) Decreased protein and mRNA expression of ER stress proteins GRP78 and GRP94 in HepG2 cells over-expressing CYP2E1. *Archives of Biochemistry and Biophysics*, **447**, 155–66.
58. Bando, Y., Katayama, T., Aleshin, A.N. *et al.* (2004) GRP94 reduces cell death in SH-SY5Y cells perturbated calcium homeostasis. *Apoptosis*, **9**, 501–8.
59. Reddy, R.K., Lu, J., and Lee, A.S. (1999) The endoplasmic reticulum chaperone glycoprotein GRP94 with Ca^{2+}-binding and antiapoptotic properties is a novel proteolytic target of calpain during etoposide-induced apoptosis. *The Journal of Biological Chemistry*, **274**, 28476–83.
60. Bando, Y., Katayama, T., Kasai, K. *et al.* (2003) GRP94 (94kDa glucose-regulated protein) suppresses ischemic neuronal cell death against ischemia/reperfusion injury. *The European Journal of Neuroscience*, **18**, 829–40.
61. Voelkel, N.F., Vandivier, R.W., and Tuder, R.M. (2006) Vascular endothelial growth factor in the lung. *American Journal of Physiology: Lung Cellular and Molecular Physiology*, **290**, L209–21.
62. Tang, K., Rossiter, H.B., Wagner, P.D., and Breen, E.C. (2004) Lung-targeted VEGF inactivation leads to an emphysema phenotype in mice. *Journal of Applied Physiology*, **97**, 1559–66.
63. Pollman, M.J., Naumovski, L., and Gibbons, G.H. (1999) Vascular cell apoptosis: cell type-specific modulation by transforming growth factor-β1 in endothelial cells versus smooth muscle cells. *Circulation*, **99**, 2019–26.
64. Tsukada, T., Eguchi, K., Migita, K. *et al.* (1995) Transforming growth factor-β1 induces apoptotic cell death in cultured human umbilical vein endothelial cells with down-regulated expression of bcl-2. *Biochemical and Biophysical Research Communications*, **210**, 1076–82.
65. Yan, Q. and Sage, E.H. (1998) Transforming growth factor-β1 induces apoptotic cell death in cultured retinal endothelial cells but not pericytes: association with decreased expression of $p21^{waf1/cip1}$. *Journal of Cellular Biochemistry*, **70**, 70–83.
66. Mauro, M., Kim, J., Costello, C., and Laurence, J. (2001) Role of transforming growth factor-β1 in microvascular endothelial cell apoptosis associated with thrombotic thrombocytopenic purpura and hemolytic–uremic syndrome. *American Journal of Hematology*, **66**, 12–22.

67. Solovyan, V.T. and Keski-Oja, J. (2005) Apoptosis of human endothelial cells is accompanied by proteolytic processing of latent TGF-β binding proteins and activation of TGF-β. *Cell Death and Differentiation*, **12**, 815–26.
68. Vinals, F. and Pouyssegur, J. (2001) Transforming growth factor-β1 (TGF-β1) promotes endothelial cell survival during *in vitro* angiogenesis via an autocrine mechanism implicating TGF-alpha signaling. *Molecular Cell Biology*, **21**, 7218–30.
69. Li, C., Issa, R., Kumar, P. *et al.* (2003) CD105 prevents apoptosis in hypoxic endothelial cells. *Journal of Cell Science*, **116**, 2677–85.
70. Lu, Q., Patel, B., Harrington, E.O. *et al.* (2009) Transforming growth factor-β1 causes pulmonary microvascular endothelial cell apoptosis via ALK5. *American Journal of Physiology: Lung Cellular and Molecular Physiology*, **296**, L825–838.
71. Bulut, K., Pennartz, C., Felderbauer, P. *et al.* (2006) Vascular endothelial growth factor (VEGF164) ameliorates intestinal epithelial injury *in vitro* in IEC-18 and Caco-2 monolayers via induction of TGF-β release from epithelial cells. *Scandinavian Journal of Gastroenterology*, **41**, 687–92.
72. Li, Z.D., Bork, J.P., Krueger, B. *et al.* (2005) VEGF induces proliferation, migration, and TGF-β1 expression in mouse glomerular endothelial cells via mitogen-activated protein kinase and phosphatidylinositol 3-kinase. *Biochemical and Biophysical Research Communications*, **334**, 1049–60.
73. Bostrom, K., Zebboudj, A.F., Yao, Y. *et al.* (2004) Matrix GLA protein stimulates VEGF expression through increased transforming growth factor-β1 activity in endothelial cells. *The Journal of Biological Chemistry*, **279**, 52904–13.
74. Shih, S.C., Ju, M., Liu, N. *et al.* (2003) Transforming growth factor-β1 induction of vascular endothelial growth factor receptor 1: mechanism of pericyte-induced vascular survival *in vivo*. *Proceedings of the National Academy of Sciences of the United States of America*, **100**, 15859–64.
75. Ferrari, G., Pintucci, G., Seghezzi, G. *et al.* (2006) VEGF, a prosurvival factor, acts in concert with TGF-beta1 to induce endothelial cell apoptosis. *Proceedings of the National Academy of Sciences of the United States of America*, **103**, 17260–65.
76. Petrache, I., Natarajan, V., Zhen, L. *et al.* (2005) Ceramide upregulation causes pulmonary cell apoptosis and emphysema-like disease in mice. *Nature Medicine*, **11**, 491–98.
77. Martin, T.R., Hagimoto, N., Nakamura, M., and Matute-Bello, G. (2005) Apoptosis and epithelial injury in the lungs. *Proceedings of the American Thoracic Society*, **2**, 214–20.
78. Taraseviciene-Stewart, L. and Voelkel, N.F. (2008) Molecular pathogenesis of emphysema. *The Journal of Clinical Investigation*, **118**, 394–402.
79. Anonymous (1997) Alpha 1-antitrypsin deficiency: memorandum from a WHO meeting. *Bulletin of the World Health Organization*, **75**, 397–415.
80. Pastor, L.M., Sánchez-Gascón, F., Girona, J.C. *et al.* (2006) Morphogenesis of rat experimental pulmonary emphysema induced by intratracheally administered papain: changes in elastic fibres. *Histology and Histopathology*, **21**, 1309–19.
81. Aoshiba, K., Yokohori, N., and Nagai, A. (2003) Alveolar wall apoptosis causes lung destruction and emphysematous changes. *American Journal of Respiratory Cell and Molecular Biology*, **28**, 555–62.
82. Giordano, R.J., Lahdenranta, J., Zhen, L. *et al.* (2008) Targeted induction of lung endothelial cell apoptosis causes emphysema-like changes in the mouse. *The Journal of Biological Chemistry*, **283**, 29447–60.
83. Petrache, I., Fijalkowska, I., Medler, T.R. *et al.* (2006) Alpha-1 antitrypsin inhibits caspase-3 activity, preventing lung endothelial cell apoptosis. *American Journal of Pathology*, **169**, 1155–66.
84. Petrache, I., Fijalkowska, I., Zhen, L. *et al.* (2006) A novel antiapoptotic role for alpha1-antitrypsin in the prevention of pulmonary emphysema. *American Journal of Respiratory and Critical Care Medicine*, **173**, 1222–28.
85. Nana-Sinkam, S.P., Lee, J.D., Sotto-Santiago, S. *et al.* (2007) Prostacyclin prevents pulmonary endothelial cell apoptosis induced by cigarette smoke. *American Journal of Respiratory and Critical Care Medicine*, **175**, 676–85.
86. Wickenden, J.A., Clarke, M.C., Rossi, A.G. *et al.* (2003) Cigarette smoke prevents apoptosis through inhibition of caspase activation and induces necrosis. *American Journal of Respiratory Cell and Molecular Biology*, **29**, 562–70.
87. Chen, Z.H., Kim, H.P., Sciurba, F.C. *et al.* (2008) Egr-1 regulates autophagy in cigarette smoke-induced chronic obstructive pulmonary disease. *PLoS One*, **3**, e3316.
88. Abadie, Y., Bregeon, F., Papazian, L. *et al.* (2005) Decreased VEGF concentration in lung tissue and vascular injury during ARDS. *The European Respiratory Journal*, **25**, 139–46.

89. Gerbitz, A., Nickoloff, B.J., Olkiewicz, K. *et al.* (2004) A role for tumor necrosis factor-alpha-mediated endothelial apoptosis in the development of experimental idiopathic pneumonia syndrome. *Transplantation*, **78**, 494–502.
90. Albertine, K.H., Soulier, M.F., Wang, Z. *et al.* (2002) Fas and fas ligand are up-regulated in pulmonary edema fluid and lung tissue of patients with acute lung injury and the acute respiratory distress syndrome. *American Journal of Pathology*, **161**, 1783–96.
91. Matute-Bello, G., Liles, W.C., Steinberg, K.P. *et al.* (1999) Soluble Fas ligand induces epithelial cell apoptosis in humans with acute lung injury (ARDS). *Journal of Immunology* **163**, 2217–25.
92. Neff, T.A., Guo, R.-F., Neff, S.B. *et al.* (2005) Relationship of acute lung inflammatory injury to Fas/FasL system. *American Journal of Pathology*, **166**, 685–94.
93. Fujita, M., Kuwano, K., Kunitake, R. *et al.* (1998) Endothelial cell apoptosis in lipopolysaccharide-induced lung injury in mice. *International Archives of Allergy and Immunology*, **117**, 202–8.
94. Kawasaki, M., Kuwano, K., Hagimoto, N. *et al.* (2000) Protection from lethal apoptosis in lipopolysaccharide-induced acute lung injury in mice by a caspase inhibitor. *American Journal of Pathology*, **157**, 597–603.
95. Uhal, B.D., Joshi, I., Hughes, W.F. *et al.* (1998) Alveolar epithelial cell death adjacent to underlying myofibroblasts in advanced fibrotic human lung. *American Journal of Physiology: Lung Cellular and Molecular Physiology*, **275**, L1192–99.
96. Chapman, H.A. (1999) A Fas pathway to pulmonary fibrosis. *The Journal of Clinical Investigation*, **104**, 1–2.
97. Kuwano, K., Miyazaki, H., Hagimoto, N. *et al.* (1999) The involvement of Fas-Fas ligand pathway in fibrosing lung diseases. *American Journal of Respiratory Cell and Molecular Biology*, **20**, 53–60.
98. Kuwano, K., Kawasaki, M., Maeyama, T. *et al.* (2000) Soluble form of Fas and Fas ligand in BAL fluid from patients with pulmonary fibrosis and bronchiolitis obliterans organizing pneumonia. *Chest*, **118**, 451–58.
99. Lee, C.G., Cho, S.J., Kang, M.J. *et al.* (2004) Early growth response gene 1-mediated apoptosis is essential for transforming growth factor-β1-induced pulmonary fibrosis. *The Journal of Experimental Medicine*, **200**, 377–89.
100. Lee, C.G., Kang, H.R., Homer, R.J. *et al.* (2006) Transgenic modeling of transforming growth factor-β1: role of apoptosis in fibrosis and alveolar remodeling. *Proceedings of the American Thoracic Society*, **3**, 418–23.
101. Zhang, H., Lawson, W.E., Polosukhin, V.V. *et al.* (2007) Inhibitor of differentiation 1 promotes endothelial survival in a bleomycin model of lung injury in mice. *American Journal of Pathology*, **171**, 1113–26.
102. Quadri, S.M., Segall, L., de Perrot, M. *et al.* (2005) Caspase inhibition improves ischemia–reperfusion injury after lung transplantation. *American Journal of Transplantation*, **5**, 292–99.
103. Levade, T., Augé, N., Veldman, R.J. *et al.* (2001) Sphingolipid mediators in cardiovascular cell biology and pathology. *Circulation Research*, **89**, 957–68.
104. Ng, C.S., Wan, S., Yim, A.P., and Respir, J. (2005) Pulmonary ischaemia–reperfusion injury: role of apoptosis. *The European Respiratory Journal*, **25**, 356–63.
105. Sage, E., Mercier, O., Van den Eyden, F. *et al.* (2008) Endothelial cell apoptosis in chronically obstructed and reperfused pulmonary artery. *Respiratory Research*, **9**, 19.
106. Sakao, S., Taraseviciene–Stewart, L., Cool, C.D. *et al.* (2007) VEGF-R blockade causes endothelial cell apoptosis, expansion of surviving $CD34^+$ precursor cells and transdifferentiation to smooth muscle-like and neuronal-like cells. *The FASEB Journal* **21**, 3640–52.
107. Sakao, S., Taraseviciene-Stewart, L., Lee, J.D. *et al.* (2005) Initial apoptosis is followed by increased proliferation of apoptosis-resistant endothelial cells. *The FASEB Journal*, **19**, 1178–80.
108. Sakao, S., Taraseviciene-Stewart, L., Wood, K. *et al.* (2006) Apoptosis of pulmonary microvascular endothelial cells stimulates vascular smooth muscle cell growth. *American Journal of Physiology: Lung Cellular and Molecular Physiology*, **291**, L362–68.
109. Guerard, P., Rakotoniaina, Z., Goirand, F. *et al.* (2006) The HMG-CoA reductase inhibitor, pravastatin, prevents the development of monocrotaline-induced pulmonary hypertension in the rat through reduction of endothelial cell apoptosis and overexpression of eNOS. *Naunyn-Schmiedeberg's Archives of Pharmacology*, **373**, 401–14.
110. Zaiman, A.L., Podowski, M., Medicherla, S. *et al.* (2008) Role of TGF-β/ALK5 kinase in monocrotaline-induced pulmonary hypertension.

American Journal of Respiratory and Critical Care Medicine, **177**, 896–905.

111. Teichert-Kuliszewska, K., Kutryk, M.J., Kuliszewski, M.A. *et al.* (2006) Bone morphogenetic protein receptor-2 signaling promotes pulmonary arterial endothelial cell survival: implications for loss-of-function mutations in the pathogenesis of pulmonary hypertension. *Circulation Research*, **98**, 209–17.
112. Song, Y., Coleman, L., Shi, J. *et al.* (2008) Inflammation, endothelial injury, and persistent pulmonary hypertension in heterozygous BMPR2-mutant mice. *American Journal of Physiology: Heart and Circulatory Physiology*, **295**, H677–90.
113. Lévy, M., Raisky, O., Vouhé, P.R. *et al.* (2007) Apoptotic deregulation of endothelial cells and intimal proliferation in pulmonary hypertension of congenital heart disease. *Archives des Maladies du Coeur et des Vaisseaux*, **100**, 365–72.
114. Masri, F.A., Xu, W., Comhair, S.A. *et al.* (2007) Hyperproliferative apoptosis-resistant endothelial cells in idiopathic pulmonary arterial hypertension. *American Journal of Physiology: Lung Cellular and Molecular Physiology*, **293**, L548–54.
115. Taraseviciene-Stewart, L., Scerbavicius, R., Choe, K.H. *et al.* (2006) Simvastatin causes endothelial cell apoptosis and attenuates severe pulmonary hypertension. *American Journal of Physiology: Lung Cellular and Molecular Physiology*, **291**, L668–76.
116. Nishimura, T., Vaszar, L.T., Faul, J.L. *et al.* (2003) Simvastatin rescues rats from fatal pulmonary hypertension by inducing apoptosis of neointimal smooth muscle cells. *Circulation*, **108**, 1640–45.
117. Gotoh, N., Kambara, K., Jiang, X.W. *et al.* (2000) Apoptosis in microvascular endothelial cells of perfused rabbit lungs with acute hydrostatic edema. *Journal of Applied Physiology*, **88**, 518–26.

17

Oxidant-Mediated Signaling and Injury in Pulmonary Endothelium

Kenneth E. Chapman, Shampa Chatterjee and Aron B. Fisher

Institute for Environmental Medicine, University of Pennsylvania, Philadelphia, PA, USA

INTRODUCTION

The concept of oxidant stress associated with the generation of O_2-based radicals – reactive oxygen species (ROS) – as a cause of cellular injury has gained increasing attention in recent years and has become a well-recognized mechanism associated with many disease processes. More recently, a parallel concept of nitrosative stress has arisen based on the generation of nitrogen-based radicals – reactive nitrogen species (RNS). As an interesting contrast, ROS were initially studied as toxic agents, but are now known to be involved in physiologic processes (signaling). Conversely, RNS were first described as physiologic (signaling) agents, but are now know to be potentially toxic. Lungs, by virtue of their physiological function and anatomical site, come into contact with a broad spectrum of oxidizing species. First, lungs are the oxygen source for all other organs and consequently are exposed to the highest pO_2 that is normally present in the body. Second, the lungs are the internal organ of initial contact with pollutant oxidants in the inhaled air or drug oxidants injected intravenously or absorbed through the skin. Third, oxidants are produced endogenously by the lung and by the circulating cells such as polymorphonuclear neutrophils (PMNs) that accumulate in the lung. Although all lung cells are at risk for oxidant stress, the pulmonary endothelium, which comprises about 30% of the total endothelium in the body, forms a major target. Blood-borne oxidants reach the pulmonary endothelium directly, while O_2 and other diffusible inhaled species readily pass through the epithelial barrier to reach endothelial cells (ECs). This chapter will present the role of oxidants, specifically ROS and RNS, respectively, as both physiologic mediators and as injurious agents in the pulmonary endothelium. The next section presents oxidant–antioxidant balance (i.e., sources of oxidants and antioxidant defenses associated with the lung endothelium), followed by presentation of ROS-mediated signaling, the cellular manifestations of oxidative stress, the pathophysiologic mechanisms that lead to EC oxidative stress, and the role of ROS/RNS in lung disorders.

OXIDANT–ANTIOXIDANT BALANCE

Oxygen Free Radicals/ROS

A free radical is an atom or molecule that possesses one or more unpaired electrons in its outer orbital. Unpaired electrons are generally unstable and are therefore highly reactive with a variety of substrates. Interestingly, oxygen has two unpaired electrons in its outermost orbital and is thus a di-radical. However, these electrons have parallel spins that prevent reaction with most molecules. This characteristic is called the triplet state, indicated by the superscripted "3" in $^3O_2$, and represents the ground state of molecular O_2. Thus, the normally respired O_2 is a relatively unreactive molecule.

Ground-state oxygen can be "activated" by mechanisms associated with high energy input (e.g., high-intensity light), so that the spins of the electrons in the outermost orbital become antiparallel, thereby generating the highly reactive singlet state ($^1O_2$). This is a rare occurrence in biology, but might be associated with photodynamic therapy where an excited photosensitizer can transfer its energy to ground state O_2 [1]. Another mechanism for transformation into a reactive state is through the acceptance of a single electron (i.e., chemical reduction) that pairs with one of the outer orbital electrons (see Figure 17.1). The

The Pulmonary Endothelium: Function in health and disease Editors Norbert F. Voelkel, Sharon Rounds

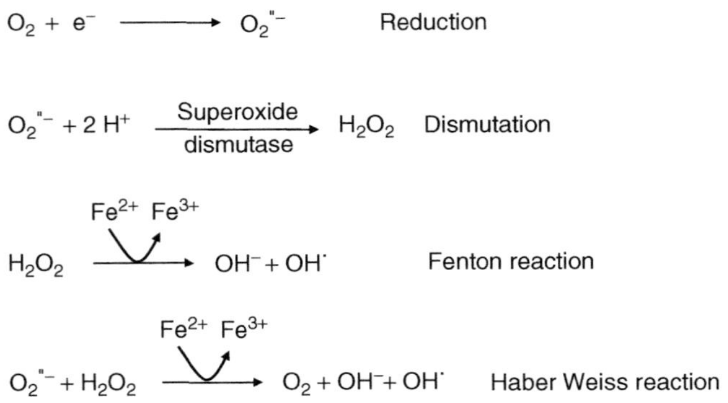

Figure 17.1 Electron flow for the generation of ROS. The source of the electron for initial reduction of O_2 can be from an enzymatic (e.g., NOX) or nonenzymatic (e.g., auto-oxidation of mitochondrial ubisemiquinone) source.

resultant species is called superoxide anion ($O_2^{\bullet-}$). In this molecule, there is now a single unpaired electron so it is by definition a free radical (the excess electron is indicated by the superscript "dot"). Superoxide can undergo monovalent reduction to produce a peroxide (O_2^{2-}), which in biological systems is neutralized by two protons to form hydrogen peroxide (H_2O_2). In the presence of a reduced transition metal such as iron (Fe^{2+}), H_2O_2 can dismute to hydroxyl ion (OH^-) and the highly reactive hydroxyl radical ($^{\bullet}OH$) – a process that is accelerated by the presence of $O_2^{\bullet-}$ [2]; the latter serves to re-reduce the oxidized transition metal (e.g., $Fe^{3+} \rightarrow Fe^{2+}$) to the reactive form (Haber–Weiss reaction, Figure 17.1). Of the O_2-derived products described above, $O_2^{\bullet-}$ and $^{\bullet}OH$ are free radicals, while $^1O_2$ and H_2O_2 are not. They all, though, show increased reactivity with biomolecules compared to ground state O_2 and are collectively called "ROS", ROS (especially $^{\bullet}OH$) can react with lipids (or proteins or sugars) to form the peroxyl radical $ROO^{\bullet}$ ($^{\bullet}OOH$ = hydroperoxyl, $LOO^{\bullet}$ = lipid peroxyl) and generally are included in the definition of ROS, as is hypochlorous acid (HOCl) that is generated in phagocytes by the myeloperoxidase enzyme. RNS, defined similarly to ROS, include nitric oxide ($^{\bullet}NO$), peroxynitrite ($ONOO^-$), nitrite radical ($NO_2^{\bullet}$), and other potential species that will be discussed further below.

Sources of ROS in the Pulmonary Endothelium

ROS are produced by pulmonary endothelium through enzymatic and nonenzymatic pathways. The former generate either $O_2^{\bullet-}$ or H_2O_2 by transfer of one or two electrons, respectively. Several of these pathways have been relatively well studied, but the majority of ROS-generating enzymes are not well known and are probably of minor importance with respect to the total oxidant burden. ROS generation by nonenzymatic pathways is due to the auto-oxidation of various compounds to generate $O_2^{\bullet-}$. In some cases, for example, the electron transport chain (ETC) or exogenous paraquat, the auto-oxidizable form of the compound can be regenerated by enzymatic reduction resulting in a cycle of oxidation/reduction and continuous O_2 production. The K_m for O_2 for enzymatic reactions that generate $O_2^{\bullet-}$ is generally low (in the physiological range) so that an elevated tissue pO_2 has little effect on these reactions. However, auto-oxidation generally is proportional to O_2 concentration so that the rate of $O_2^{\bullet-}$ generation is pO_2-dependent.

Generation of ROS from Endogenous Enzymes

NADPH Oxidase The enzyme NADPH oxidase (NOX) is a multiprotein complex that was first found in PMNs, but has since been described in a broad representation of cell types, including pulmonary endothelium. The prototype, now called NOX2, represents a family of proteins which, upon activation, utilize NADPH to reduce molecular O_2 to $O_2^{\bullet-}$. Phagocytic cells such as PMNs and macrophages utilize the $O_2^{\bullet-}$ generated by this pathway to kill microbes. The other major role of this pathway is to generate $O_2^{\bullet-}$ as a signaling molecule – a function that has been described relatively recently. NOX2 consists of both membrane and cytosolic protein components (see Figure 17.2). The former comprises two integral membrane protein subunits (the flavoprotein $gp91^{phox}$ and $p22^{phox}$) which together constitute flavocytochrome (b_{558}), the catalytic site of the oxidase. The cytosolic components include at least four proteins ($p47^{phox}$, $p67^{phox}$, $p40^{phox}$, and rac-1 or -2) and possibly others [3, 4]. For activity, the NOX2 complex requires assembly which in pulmonary endothelium (as well as in most cells) appears to be similar to

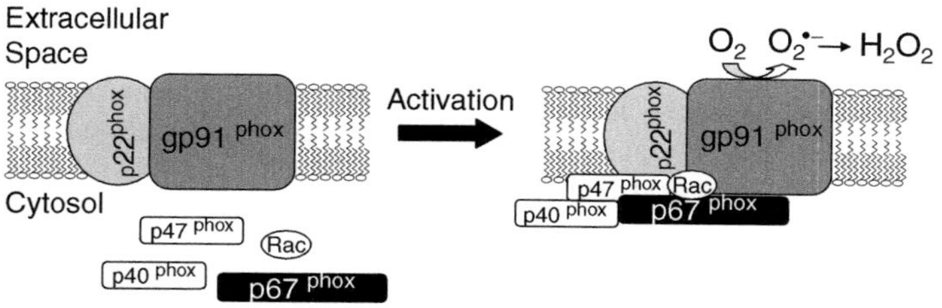

Figure 17.2 ROS generation by NOX. The complex consists of cytoplasmic and membrane bound components, shown in (a). The enzyme requires assembly (b) by translocation of the cytoplasmic components to the cell membrane. The enzyme generates $O_2^{\bullet-}$ in the extracellular space where it can dismute to H_2O_2.

that described for PMNs [5]. Assembly is initiated by activation of one of the small G-proteins (rac-1 in the pulmonary endothelium) resulting in phosphorylation of the cytosolic components which causes their translocation to the plasma membrane (Figure 17.2). A variety of physiological stimuli have been shown to activate endothelial NOX, including hormones and agonists such as angiotensin II and thrombin, inflammatory mediators such as tumor necrosis factor-α and interleukin (IL)-1, and mechanotransduction due to altered shear stress [6, 7]. The NOX family contains four other isoforms (as well as two related dual oxidase, DuOx, proteins) which differ from NOX2 in their flavoprotein and possibly some of the cytosolic components that are required for activation [3, 4]. Endothelium contains NOX4 and low levels of NOX1 in addition to NOX2. While NOX2 is localized primarily to the plasma membrane and possibly endosomes, NOX4 may be present on the membrane of intracellular organelles [8, 9]. In some (nonpulmonary) ECs, constitutively preassembled NOX2 has been demonstrated in a perinuclear cytoskeletal location, although its functional significance has not been determined [10].

Xanthine Oxidase Xanthine oxidase, a cytosolic metalloflavoprotein containing molybdenum, is a major source of oxygen-free radicals during re-oxygenation of hypoxic tissues. Under resting conditions, tissue xanthine dehydrogenase uses the reduced form of NAD (NADH) as the reductant to metabolize hypoxanthine to uric acid; xanthine dehydrogenase can be converted to an oxidase intracellularly by proteases that are activated by elevated Ca^{2+} levels. This truncated enzyme uses molecular O_2 as an electron acceptor to generate $O_2^{\bullet-}$. This enzyme has been demonstrated in pulmonary ECs [11].

Cyclooxygenases/Lipoxygenases The physiologic role of these enzymes is primarily in the generation of bioactive eicosanoids from arachidonic acid. This fatty acid essentially is "stored" within cell membranes where it is esterified to glycerol in phospholipids. Free arachidonate is released through action of the cytoplasmic phospholipase A_2 upon stimulation with a variety of agonists. Metabolism of free arachidonate to the active products is carried out by cyclooxygenases (COXs) and lipoxygenasess (LOX) – non-heme-containing dioxygenases that oxidize polyunsaturated fatty acids to hydroperoxy fatty acid derivatives. The biosynthesis of prostaglandins, prostacyclins, and thromboxanes occurs via the enzyme prostaglandin H synthase that possesses both COX and hydroperoxide peroxidase activities. In the initial step, O_2 is incorporated into arachidonic acid by COX converting it to hydroperoxy endoperoxide which is then reduced to the corresponding alcohol. The biosynthesis of leukotrienes occurs via 5-LOX activity which converts arachidonic acid to 5-hydroperoxy eicosatetraenoic acid followed by reduction to leukotriene A_4, the precursor of leukotrienes B_4, C_4, and D_4. Other fatty acids which are substrates for the COX group of enzymes are γ-linolenic and dihomo-γ-linolenic acids, eicosapentaenoic acid, docosahexaenoic acid, and stearidonic acid. The hydroperoxy fatty acid derivatives formed in the COX and LOX reactions are reactive, and result in formation of $O_2^{\bullet-}$ or H_2O_2.

NO Synthase NO synthase (nitric oxide synthase NOS) under normal conditions generates $^{\bullet}NO$ from arginine. Under conditions of tissue acidosis or with deficiency of the important cofactor, tetrahydrobiopterin, NOS can generate $O_2^{\bullet-}$ instead of $^{\bullet}NO$.

Others A variety of other oxidases such as amino acid oxidases, aldehyde oxidases, urate oxidase, acyl coenzyme A oxidase, L-gulonolactone oxidase, and dihydroorotate dehydrogenase may produce ROS, but their role in oxidant stress and specifically in endothelium is not well defined.

Generation of ROS from Nonenzymatic Sources

Mitochondrial ETC Under normal physiological conditions, around 1–3% of electrons carried by the mitochondrial ETC "leak out" of the pathway and pass directly to oxygen, generating $O_2^{\bullet -}$. Auto-oxidation of ubisemiquinone, formed at complex III by transfer of one electron from ubiquinol to cytochrome *c* of the ETC, appears to be major site for $O_2^{\bullet -}$ production; complex I (NADH dehydrogenase) contributes to $O_2^{\bullet -}$ generation through auto-oxidation of the semiquinone that is formed by one electron transfer from reduced flavin mononucleotide (FMN) [12]. These reactions are shown in Figure 17.3. As for auto-oxidation in general, the rate of $O_2^{\bullet -}$ production from ubisemiquinone increases in proportion to tissue O_2 concentration. Downstream inhibition of the ETC (e.g., by inhibition of cytochrome oxidase) results in reduction of chain components, which is reflected in an increased ubisemiquinone pool size and a consequent increase in the rate of $O_2^{\bullet -}$ production. The presence of antioxidants in the mitochondria ensures that the basal level of $O_2^{\bullet -}$ formed during normal electron transport is kept at a level whereby it does not damage proteins of the mitochondrial matrix. However, mitochondrial damage with oxidative stress can result with the increased production of $O_2^{\bullet -}$ under pathological conditions [13].

Cytochrome P450 Pathway of the Endoplasmic Reticulum "Leakage" of electrons also can occur from electron transport in the endoplasmic reticulum. This pathway involves sequential electron transfer from reduced flavin to the cytochrome P450–substrate complex and then to O_2. Auto-oxidation of the partially reduced cytochrome P450 can result in $O_2^{\bullet -}$ production. The expression of cytochrome P450 in pulmonary endothelium is relatively low (compared to Clara and alveolar type 2 cells) and in general this pathway appears to be less important than the mitochondrial ETC in cellular ROS production.

Free Iron During hypoxia, metabolites such as lactic acid accumulate causing a decrease in intracellular pH. Acidosis inhibits the binding of iron (Fe) and other transition metals to carrier proteins (transferrin, aconitase) thereby resulting in the accumulation of free intracellular Fe^{2+} [14]. Fe^{2+} also can be released from intracellular enzymes by the action of $O_2^{\bullet -}$. Free iron (Fe^{2+}) can catalyze the generation of $OH^{\bullet}$ radical from H_2O_2 and $O_2^{\bullet -}$ (Figure 17.1). Other transition metals such as Mn^{2+} and Cu^{+} also can catalyze these reactions.

Quinones The generation of $O_2^{\bullet -}$ from auto-oxidation of ubisemiquinone following its production in the

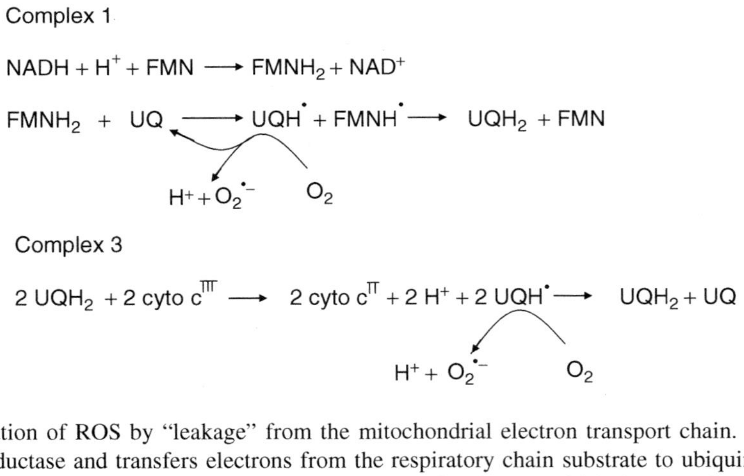

Figure 17.3 Generation of ROS by "leakage" from the mitochondrial electron transport chain. Complex 1 is NADH : ubiquinone oxidoreductase and transfers electrons from the respiratory chain substrate to ubiquinone (UQ, also called coenzyme Q). The initial step is a two-electron transfer to the associated FMN which then transfers one electron at a time to UQ to generate ubiquinol (UQH_2). The short-lived intermediate (one-electron reduction product) is ubisemiquinone, which can auto-oxidize to generate $O_2^{\bullet -}$. Complex 3 (ubiquinone : cytochrome *c* reductase) represents the site where protons are transported from the cell matrix to the mitochondrial inner membrane space through operation of the coenzyme Q cycle so the overall reactions are more complex than shown here. The net result is that the electrons from UQH_2 are passed to cytochrome *c* with the formation of a short-lived $UQH^{\bullet}$ intermediate. Auto-oxidation of $UQH^{-\bullet}$ generates $O_2^{\bullet -}$ as described for complex 1. cyto c^{III}, oxidized cytochrome *c* (Fe^{3+}); cyto c^{II}, reduced cytochrome *c* (Fe^{2+}).

mitochondrial ETC is shown in Figure 17.3. A variety of reductive enzymes, including microsomal NADPH-cytochrome P450 reductase, can metabolize quinones by one-electron reduction reactions [15]. Like ubisemiquinone, the resultant unstable semiquinones can autoxidize when molecular oxygen is present, causing a reformation of the quinone, with the concomitant generation of $O_2^{\bullet-}$ [16].

Xenobiotics Exposure to xenobiotics can result in oxidative stress due to the generation of ROS formed during their P450-mediated metabolism or by auto-oxidation following their reduction as described above for quinones [17]. An example is paraquat which is readily reduced to the paraquat radical ($PQ^{\bullet+}$) by a variety of diaphorases that transfer an electron from NAD(P)H [18, 19]. Auto-oxidation results in production of $O_2^{\bullet-}$, which can be continuous based on the cycle of reduction and auto-oxidation. Other xenobiotics that show similar redox cycling include the therapeutic agents bleomycin, streptonigrin, nitrofurantoin, and related compounds.

Extraendothelial Sources of ROS

In addition to these intracellular sources, ROS that target endothelium can be generated at extraendothelial sites. Activated inflammatory cells that adhere to endothelium in the pulmonary vasculature and migrate into the interstitium produce ROS during phagocytosis or following their stimulation by cytokines [20, 21]. As another source, pulmonary endothelium has the capacity to reduce circulating quinones such as coenzyme Q_1 by the transmembrane transport of electrons [22]. The enzymes responsible for quinone reduction are several NAD(P)H quinone : oxidoreductases. In addition to these non-endothelial sources, it should be noted that NOX2 of the endothelial plasma membrane actually generates $O_2^{\bullet-}$ on the extracellular side of the plasma membrane. ROS generated extracellularly by any of these pathways can react directly with the EC plasma membrane or, after conversion to H_2O_2, diffuse intracellularly.

RNS

Like ROS, RNS are formed constantly in the body. RNS comprise $^{\bullet}NO$ and products formed by its reactions with ROS. $^{\bullet}NO$ is a gaseous free radical that has a role in diverse biological functions, including host defense, vasodilation, and apoptosis (see Chapter 6). $^{\bullet}NO$ can react with $O_2^{\bullet-}$ to form $ONOO^-$, a powerful oxidizing and nitrating agent, and can function as an antimicrobial compound [23–25]. $ONOO^-$ may decompose to $^{\bullet}OH$ and $NO_2^{\bullet}$ – both potent oxidants that react with a variety of biomolecules [26–28].

The synthesis of $^{\bullet}NO$ occurs via the enzymatic oxidation of the terminal guanidine-nitrogen of L-arginine, thereby generating $^{\bullet}NO$ and L-citrulline. The enzymes involved in catalyzing this reaction are termed NOSs. There are three isoforms of these enzymes, one of which is specifically identified with ECs (endothelial nitric oxide synthase eNOS). An inducible form of the enzyme (inducible nitric oxide synthase iNOS) is largely expressed by PMNs, macrophages and related cell types. The NOS isoforms are multimeric complexes that require multiple cofactors such as FAD, heme, calmodulin, and tetrahydrobiopterin for activity. eNOS (as well as the neuronal NOS, neuronal nitric oxide synthase nNOS) are activated by Ca^{2+}/calmodulin binding and are therefore controlled by Ca^{2+} flux, while iNOS is regulated at the transcriptional level and is at most weakly affected by changes in intracellular Ca^{2+} levels. Thus, $^{\bullet}NO$ production by eNOS is transient and localized, whereas $^{\bullet}NO$ production by iNOS is greater and often sustained [23, 29, 30].

Antioxidant Defenses

The first line of defense against the toxic effects of ROS is the scavenging of $O_2^{\bullet-}$ and H_2O_2. There are no specific cellular defenses against $^{\bullet}OH$ as this potent electrophile can react with almost any adjacent molecule at a diffusion limited rate. Like the sources of ROS, cellular antioxidant defenses also can be classified as either enzymatic or nonenzymatic.

Enzymatic Defenses

Enzymatic defenses of the pulmonary endothelium against ROS include superoxide dismutases (SODs), catalase, glutathione (GSH) peroxidases (glutathione peroxidase GPxs), and the peroxiredoxins (Prdxs). Both GPx and Prdx ultimately use the reducing power of NAD(P)H to "neutralize" the oxidizing species; thus, the pathways for generating the reduced nucleotide, that is, primarily the pentose shunt pathway, play an important supportive role.

SOD SOD plays an important role in cellular defense against oxidative stress by converting $O_2^{\bullet-}$ to the less bioreactive H_2O_2. Three distinct mammalian SODs exist: cytosolic copper–zinc SOD (copper–zinc superoxide dismutase CuZnSOD), mitochondrial manganese SOD (manganese superoxide dismutase MnSOD) and extracellular CuZN SOD (extracellular copper–zinc superoxide dismutase ECSOD). The activities of CuZnSOD and MnSOD are lower in the lung than in several other organs such as kidney, liver, and heart [31]. In contrast, ECSOD activity has been shown to be remarkably high in the lungs with greatest expression in the matrix of vessels, airways and alveolar septa [31, 32].

Catalase This enzyme converts H_2O_2 to H_2O and O_2, a dismutation reaction that does not require a cofactor. Catalase activity is confined to H_2O_2 and other hydroperoxides are not substrates for this enzyme. Catalase is predominantly a peroxisomal enzyme. As an interesting mode of regulation, catalase mRNA stability is increased under oxidative stress conditions [33] as contrasted with other antioxidant enzymes that are transcriptionally regulated [34].

GPx The GPx family of seleno-enzymes has peroxidase activity that reduces H_2O_2 and other short chain hydroperoxides at the expense of GSH. The resultant oxidized form of GSH (GSSG) is restored to the reduced state (GSH) by GSH reductase utilizing NADPH:

$$2GSH + H_2O_2 \rightarrow GSSG + 2H_2O$$
$$GSSG + NADPH + H^+ \rightarrow 2GSH + NADP^+.$$

GPx1 is a ubiquitous enzyme that is present in the cytosolic and mitochondrial compartments. GPx2 and 3 have restricted distributions and are probably not present in pulmonary endothelium. GPx4 can reduce phospholipid hydroperoxides in addition to H_2O_2 and short-chain hydroperoxides; this activity is important for restoration of oxidized membrane phospholipids. Cytosolic GPx (GPx1) does not reduce phospholipid hydroperoxides [35, 36]. The activity of GPx4 is relatively low in the whole lung, while the activity specifically in pulmonary endothelium is not known. Although both GPx and catalase can degrade H_2O_2 to water, GPx has a significantly lower K_m (higher affinity) for H_2O_2. Owing to this and their respective intracellular distributions, GPx appears to have a greater role than catalase in the total cellular removal of H_2O_2.

Prdx The Prdxs differ from GPxs in having cysteine rather than selenocysteine as the active site. There are six known mammalian Prdx enzymes (I–VI), all of which find expression in the human lung. However, only PrdxI and VI appear to be expressed in the pulmonary endothelium and these are present at relatively low levels [37]. PrdxI–V metabolize peroxides (H_2O_2) similar to GPx, but with thioredoxin instead of GSH as the electron acceptor. The physiological role of these enzymes is still unclear as the rate constant for the peroxidase activity of Prdxs is at least an order of magnitude less than that of GPx. A primary role in regulating H_2O_2-mediated signal transduction has been suggested because of the relatively high affinity of Prdxs for H_2O_2 substrate. PrdxVI differs from the other mammalian Prdx enzymes in using GSH rather than thioredoxin as the cofactor and in its ability to catalyze reduction of phospholipid hydroperoxides [37, 38]. This property is of potential importance to reverse oxidant-mediate membrane lipid peroxidation and may account for the protective effect of this enzyme against lung oxidative stress [39, 40].

Nonenzymatic Antioxidants

GSH GSH, a tripeptide sulfhydryl compound (Cys–Gly–Glu), is a major constituent of cellular defenses against oxidative stress. This tripeptide is the most abundant nonprotein thiol of the cell and is present in most cells at 1–10 mM. GSH can interact directly, albeit slowly, with H_2O_2 [41] although its more important function is to serve as the reducing agent for GPx or PrdxVI activity. The antioxidant function of GSH results in oxidation of the thiol group of its cysteine residue with formation of a disulfide (GSSG), which in turn is reduced catalytically back to the thiol form (GSH) by GSH reductase using the reducing power of NADPH. Since generation of NADPH is through activity of the pentose phosphate shunt pathway, the provision of reducing equivalents is ultimately tied to the intermediary metabolism of the cell. GSH along with the GSH-utilizing peroxidases serve to detoxify H_2O_2 to water and molecular O_2. GSH also helps to maintain protein cysteinyl-thiols (R-CH_2-SH) in the reduced state, a possible requisite for their enzymatic function.

Tocopherols Tocopherols represent a family of compounds with multiple functions. Vitamin E (α-tocopherol) is a dietary constituent with high lipid solubility that consequently localizes in cell membranes. This compound can effectively compete with lipids for oxidation by lipid or lipid hydroperoxy radicals, thus serving to terminate the chain reaction of lipid peroxidation. The resultant tocopheryl radical is relatively stable, but can be reduced by radical scavengers, especially ascorbic acid (vitamin C).

Ascorbate (Vitamin C) and β-Carotene (Vitamin A Precursor) Ascorbate can function in antioxidant defense by reducing the α-tocopherol radical or by interaction with $O_2^{\bullet-}$ to generate H_2O_2. Although the latter reaction is relatively slow, the high concentration of ascorbate in tissues could allow this reaction to be of physiological importance. β-carotene is a terpenoid that is the precursor to vitamin A. This compound *in vitro* is a very effective quencher of singlet O_2. The product, all-*trans* β-carotene, can revert to the normal isomer with the gradual release of heat.

Blood-Borne Antioxidants

The endothelium is in contact with the blood so that blood-associated components can function in antioxidant protection of the pulmonary endothelium. Antioxidants in the serum such as vitamins A, C and E, and β-carotene, in addition to red blood cell enzymes such as SOD, catalase, and GPx provide antioxidant protection of the luminal pulmonary EC membrane from the effects of blood-derived oxidants.

Ancillary Antioxidants

A variety of additional enzymes, macromolecules, and small compounds can contribute in some way to antioxidant defense and may play important roles in special circumstances. The list includes ROS scavenging agents, reagents that support the role of the primary antioxidants, and enzymes that participate in repair of oxidatively damaged cells. Thus, the list can be quite extensive. Examples include the acute-phase reactant enzyme heme oxygenase that generates the ROS scavenger, bilirubin; glucose-6-phosphate dehydrogenase and other enzymes of the pentose shunt pathway that are responsible for generating NADPH required for enzymatic reductive reactions; GSH reductase that regenerates the reduced form of the tripeptide after its oxidation to GSSG; DNA repair enzymes to reverse single strand breaks; and thioredoxin that serves as the reductant cofactor for PrdxI–V.

PHYSIOLOGICAL ROLE OF ROS

ROS as Signaling Molecules

Although ROS are conventionally regarded as cytotoxic and mutagenic agents that induce oxidative damage and cell death, ROS at low levels are now known to function in cell signaling. To function in signaling, a molecule should be (i) present at low levels, (ii) diffusible, (iii) relatively short lived, (iv) present in the proper location, and (v) have effects that are reversible. H_2O_2 fulfills these requirements and is thus considered as a potential signaling molecule. This compound is present normally at around 10^{-8} M in cytosol, it is as diffusible as H_2O, it can be degraded rapidly by intracellular enzymes, and it can reversibly oxidize protein cysteine sulfhydryl groups. $O_2^{\bullet-}$ being a charged anion, is generally considered unlikely to initiate intracellular signaling due to its poor permeability through a phospholipid bilayer [42] and because its rapid dismutation to H_2O_2 results in a short diffusion distance [43]. However, recent evidence has indicated a discrete signaling role for $O_2^{\bullet-}$ that is different from H_2O_2 [44]. For example, $O_2^{\bullet-}$, but not H_2O_2, added to the cell medium leads to increased intracellular Ca^{2+} release and apoptosis in pulmonary ECs [45]. This differential effect in a mammalian cell is in line with observations in bacteria where the transcription factor OxyR has been shown to be sensitive to H_2O_2 while SoxR is sensitive to $O_2^{\bullet-}$ [46]. Using a fluorophore trap, $O_2^{\bullet-}$ has been shown to penetrate ECs through a chloride channel (ClC-3) and activate an inositol-3 phosphate receptor-mediated pathway [47]. Compared to $O_2^{\bullet-}$ and H_2O_2, $^{\bullet}OH$ is extremely short lived and reacts nonselectively with a broad spectrum of biomolecules in a diffusion limited manner; thus, it is totally unsuited to serve as a signaling agent.

ROS-Mediated Signaling Pathways

Signaling relies on enzymatically generated ROS, the best studied pathway being $O_2^{\bullet-}$ generation by NOX [4]. These pathways can be physiologically regulated unlike the pathways that produce ROS through auto-oxidation reactions. A broad spectrum of basic cellular functions is responsive to ROS-mediated regulation including cell division and proliferation [7, 9, 48–51], and its counterpart – apoptosis [52]. ROS promote tubular morphogenesis by endothelium compatible with a role in angiogenesis and neovascularization [53].

Mitogen-Activated Protein Kinases and Related Proteins

The mitogen-activated protein kinases (MAPKs) represent a large family of serine/threonine kinases that function in a signaling cascade. The three terminal kinases in these pathways are p38, extracellular signal-regulated kinase (ERK), and c-Jun N-terminal kinase (JNK). Activation of the MAPKs requires the phosphorylation of tyrosine and threonine of a ThrXTyr motif. The upstream kinases that result in MAPK phosphorylation (activation) and initiation of signaling are redox regulated and sensitive to ROS [54, 55]. ROS also can regulate MAPK activity by inactivation of the dephosphorylating enzyme (MAPK phosphatase); this occurs by oxidation of a highly reactive cysteine residue that is critical for catalytic function [56]. Other enzymes involved in signaling that have been shown to be redox sensitive include protein kinase C [57], several protein tyrosine kinases [58, 59], myosin light chain kinase [60], phospholipases A_2, C, and D [61–63], and various other phosphorylated protein phosphatases [64]. The products of these reactions have widespread effects on cellular metabolic pathways and function.

Transcription Factors

ROS mediate signaling via activation of several transcription factors, which in turn regulate genes that control many cellular functions, including antioxidant enzyme expression [34, 65, 66]. Activation of these transcription factors by ROS has been shown in pulmonary endothelium [67].

Nuclear Factor-κB Nuclear factor-κB (NF-κB) is an important transcription factor due to its rapid response, the wide range of genes that it controls (functional binding sites are present in the promoters of many genes), and its role in inflammatory and immunological processes. The activity of NF-κB is regulated through an inhibitory protein called IκB. An increase of intracellular oxidants such as ROS derived from NOX can lead to phosphorylation of IκB which is then released from NF-κB allowing the transcription factor to enter the nucleus and interact with DNA [68]. In the lung, NF-κB activation in response to LPS and *Pseudomonas aeruginosa* insult is impaired in NOX ($p47^{phox-/-}$ or $gp91^{phox-/-}$) gene-targeted mice [69, 70]. In other models of oxidant stress as well, the activation of the RelA/p65 subunit of NF-κB in mouse lungs is decreased upon inhibition of NOX activity [67, 71]. ROS-dependent activation of NF-κB induces the expression of genes transcribing various cytokines and chemotactic factors such as IL-6, IL-8, monocyte chemotactic protein-1, and matrix metalloproteinases (MMPs) such as MMP-2, -9, and -12, which have a profound effect on the process of inflammation.

Nuclear Factor-Erythroid 2-Related Factor Nuclear factor erythroid 2-related factor (Nrf2) is a redox-sensitive basic leucine zipper transcription factor which mediates the expression of a variety of genes, especially those related to antioxidant defense. This transcription factor binds to a DNA sequence called the antioxidant response element (ARE). Normally, cytosolic Nrf2 is bound to Kelch-like enoyl-coenzyme A hydratase-associated protein 1 (Keap1) that negatively regulates Nrf2 activity by targeting it to the proteasome for degradation. In response to oxidative stress, Nrf2 dissociates from Keap1 and enters the nucleus where it binds to an ARE in the upstream sequence of target genes. Binding generally requires the association of several other proteins such as small Maf proteins, c-Jun and cAMP-response element-binding protein in order to mediate gene transcription. Nrf2 null mice are more susceptible to oxidant stress as compared with wild-type, most likely related to impaired induction of antioxidant enzymes [65, 72, 73]. Chemical activation of Nrf2 reduced the oxidant injury associated with inflammation [74].

RNS-Mediated Signaling

•NO is a highly reactive free radical that rapidly diffuses and permeates cell membranes. In animals, •NO is implicated in diverse physiological processes, including neurotransmission, vascular smooth muscle relaxation (vasodilation, see Chapter 12), and regulation of platelet aggregation. It also functions as a messenger in immune responses and in the regulation of receptors (e.g., the cardiac ryanodine receptor), enzymes (e.g., activation of soluble guanylate cyclase), and transcription factors (e.g., the inactivation of zinc finger transcription factors and the *S*-nitrosylation/inactivation of NF-κB/IκB) [23, 24, 75]. Low •NO concentrations synthesized by constitutively expressed NOSs act on several signaling pathways activating transcription factors such as NF-κB or activated protein (AP)-1, thereby influencing gene expression. Recent evidence has implicated protein nitrosylation through conversion of protein cysteine moieties to an SNO compound as a mechanism for regulation of enzymatic activity [76]. An important recent observation is the identification of an enzyme capable of reversing this reaction [77]. The biological activities and regulation of lung endothelial •NO are discussed further (see Chapter 6).

CELLULAR MANIFESTATIONS OF OXIDATIVE STRESS

Although oxidant stress was once considered a relatively rare phenomenon, it is now known to be pervasive, associated with modest derangement of a broad range of normal cellular processes. This section will describe the cellular effects of oxidative stress in endothelium. Conversely, analysis of these alterations can be used in the diagnosis of oxidative stress.

Oxidation and Nitrosation of Biomolecules

Oxidative stress is determined by the balance between oxidant generation and antioxidant defenses. ROS production in excess of the relatively low levels associated with physiologic signaling can overwhelm the intrinsic antioxidant defenses. Excess ROS/RNS interact with cellular components, including lipids, proteins, and DNA, altering and modifying those by oxidation or nitration, thereby

Initiation	$L_1 + OH^{\bullet} \longrightarrow L_1^{\bullet} + H_2O$
Peroxidation	$L_1^{\bullet} + O_2 \longrightarrow L_1OO^{\bullet}$
Propagation	$L_2 + L_1OO^{\bullet} \longrightarrow L_2^{\bullet} + L_1OOH^{\bullet}$
Termination	$L_1^{\bullet} + L_1^{\bullet} \longrightarrow L_1 - L_1$
Termination by Vit E (α-tocopherol)	$L_1OO^{\bullet} + \text{Vit E} \longrightarrow \text{Vit E}^{\bullet} + L_1OOH$

Figure 17.4 Mechanism for the chain reaction of lipid peroxidation. Reaction of lipid with a free radical occurs in three steps. Initiation results in the formation of lipid radical, which peroxidizes in the presence of O_2, propagation occurs as the lipid peroxy radical abstracts an electron from another lipid (L_2): α-tocopherol (vitamin E) suppress the lipid peroxidation chain reaction by forming a stable product after accepting an electron from $LOO^{\bullet}$.

triggering biochemical, cellular, morphologic, and physiologic changes in the tissue [78–80]. The most potent electrophiles are $^{\bullet}OH$, $^1O_2$, and $ONOO^-$, and these members of the ROS/RNS family have extremely short diffusion distances as they rapidly react with adjacent organic molecules. The reactions of $O_2^{\bullet-}$ and H_2O_2 are more selective with specific targets for their role as signaling molecules.

Lipid Peroxidation

Oxidative stress may lead to peroxidation of lipids in cellular membranes. The interaction of a radical (e.g., $^{\bullet}OH$) with lipids results in hydrogen abstraction with the formation of a lipid radical that forms a lipid hydroperoxy radical in the presence of O_2 (see Figure 17.4). This reactive molecule can react with an adjacent lipid, abstracting hydrogen to form a stable lipid hydroperoxide but generating another lipid radical. Thus, these events cause a chain reaction of lipid peroxidation within the membrane. In addition to $^{\bullet}OH$, ferryl complexes and singlet O_2 can initiate lipid peroxidation; $O_2^{\bullet-}$ is much less potent (around six orders of magnitude) in that regard [81]. Lipid peroxidation in tissues can be measured by the generation of malondialdehyde, estimated as thiobarbituric acid-reactive substances, or the spectroscopic determination of lipid conjugated dienes [82, 83]. The double bonds in lipids are particularly susceptible to oxidation. Increased peroxidation of lipids in the whole lung has been demonstrated in animal models of hyperoxia, paraquat administration, and lung ischemia–reperfusion [39, 40, 83–85]. A recently developed method permits imaging of the sites of lipid peroxidation in lungs by using a fluorophore that specifically interacts with phospholipids hydroperoxides and has been used to show endothelial lipid peroxidation with ischemia [86]. Vitamin E (α-tocopherol) is considered to be the major chain-breaking antioxidant as it can intercept the electron from the lipid hydroperoxy radical to form a stable compound, thereby terminating the chain reaction of lipid peroxidation. Repair of peroxidized phospholipids in membranes is possible through direct reduction of lipid hydroperoxides by PrdxVI or Gpx4 [87, 36]. An alternate mechanism is the sequential activity of a phospholipase A_2 and an acyl transferase, although this pathway has been estimated as four orders of magnitude less efficient than direct reduction [88, 89].

Protein Oxidation

Oxidant stress can lead to the oxidation of proteins although this has been less well studied than lipid peroxidation. Protein cysteine thiol groups (-SH) can be oxidized to thiyl radicals ($-S^{\bullet}$), disulfides (-SS-), and sulfenic (-SOH), sulfinic (SO_2H), or sulfonic ($-SO_3H$) acid derivatives. Cysteines also can reversibly form mixed disulfides by their interaction with GSH. In addition to cysteines, methionine residues can be oxidized to sulfoxides. Oxidative attack on the side-chains of lysine, arginine, proline, or threonine residues results in the formation of carbonyl groups; proteins containing a metal cofactor are particularly vulnerable. Protein carbonyl formation can be detected by a colorimetric assay or by the use of carbonyl-specific antibodies [85]. These oxidative changes disrupt protein structure and function, and can also result in protein cleavage [90]. Proteins irreversibly inactivated by formation of methionine sulfones, cysteine sulfinic or sulfonic acids, and carbonyl derivatives cannot be repaired but are recognized and degraded by cellular proteolytic processes. Some oxidized and inactivated proteins may change conformation and become partially folded and aggregated. Accumulation of protein aggregates in the cell has been linked to the initiation and progression of numerous diseases and aging [91–93]. Lung protein oxidation has been shown in animal models of oxidant stress [39, 40, 83, 84] but not specifically in endothelium.

DNA Oxidation

Interaction of DNA with ROS can result in a variety of lesions including base damage, sites of base loss (abasic sites), and single-strand breaks that may contain modified 3′-ends and apurinic/apyrimidinic sites. A major oxidation product of ROS–DNA interaction is 8-oxoguanine which can be detected in body fluids as an index of DNA oxidation. Other products of oxidation include thymine glycol and some alkylated residues [94, 95]. Less than 1% of single strand breaks become converted to DNA double-strand breaks, predominantly at the time of DNA replication, while the overwhelming majority are repaired by usually error-free mechanisms. Repair to damaged nuclear DNA (and single-strand breaks) occurs through activity of poly(ADP-ribose) polymerase, which is activated with oxidant stress [96, 97]. Double-strand breaks are more difficult to repair and can lead to permanent chromosomal damage. Due to formation of oxo products, base mispairing can occur during DNA replication. Thus, these lesions can be mutagenic. In addition to point mutations, DNA–DNA and DNA–protein crosslinks and sister chromatid exchanges have been reported to occur due to interaction of ROS or RNS with DNA [98]. DNA modifications with experimental oxidative stress have been demonstrated in the vascular endothelial growth factor promoter in pulmonary artery ECs, in particular in those sequences of the gene that are the sites for binding of the AP-1 and hypoxia-inducible factor-1 transcription factors [99]. Thus, these DNA base modifications could have important functional consequences. Oxidative DNA damage in ECs has been studied further in mitochondrial DNA where the damage is more difficult to repair compared to nuclear DNA and can result in apoptosis [100]. The repair process for mitochondrial DNA of pulmonary venous endothelium has been shown to function less effectively than that for pulmonary arterial and microvascular endothelium [101].

GSH Oxidation

The role of GSH as a scavenger of H_2O_2 and the reductant for activity of GPxs has been described on p. 266. GSH participates in a cycle of oxidation to form GSSG followed by its reduction back to GSH by GSH reductase using NADPH. The cytoplasm of most cells is normally in a relatively reduced state with an intracellular GSH/GSSG ratio maintained at 10–100. With oxidative stress, GSH is utilized by peroxidases and converted to GSSG at an increased rate. Further, oxidative stress results in oxidation of protein cysteine moieties which then can interact with GSH to form mixed disulfides (Protein-SH + GSH $\rightarrow$ Protein-SSG). The net result is decreased GSH and increased GSSG resulting in a decrease in the GSH/GSSG. Cell membranes are impermeable to the normally charged GSH, but are modestly permeable to the uncharged GSSG. Thus, the increased generation of GSSG can result in its leakage from the cell which might be detected in the extracellular fluid. The total GSH pool (GSH + GSSG) frequently is decreased with oxidant stress because of the formation of protein disulfides and GSSG "leakage." Thus, oxidant stress is manifested by decreased total GSH, decreased GSH to GSSG ratio, and increased mixed GSH : protein disulfides.

Protein Nitration

While $^{\bullet}NO$ is generally short-lived and shows relatively little reactivity with biomolecules, its interaction with $O_2^{\bullet}$ generates the peroxynitrite anion ($ONOO^-$). Although the precise chemistry is not yet clear, $ONOO^{\bullet}$ is a strong oxidant, possibly through its decomposition to $^{\bullet}OH$ [23]. $ONOO^-$ is also a powerful protein-nitrating agent [23, 102–104]. Tyrosine residues are especially susceptible with the formation of nitrotyrosine. This reaction can be detected by chemical assay for nitrotyrosine or by use of an antibody that is specific for this moiety. Nitration of lung protein tyrosines has been demonstrated in an animal model of ischemia/reperfusion [105], but not applied specifically to endothelium. Nitration of proteins may lead to augmented or depressed enzymatic function [106–109] that may be related to either the prevention or stimulation of protein phosphorylation [110–113]. *S*-Nitrosylation of protein cysteine residues has been shown recently to be a reversible reaction that can play an important role in cellular signaling [77]. Nitration of other amino acid residues such as tyrosine does not appear to be reversible and these nitrated proteins are presumably degraded. Recent studies have indicated that lipids (fatty acids) also may be nitrosated and may play a role in cell signaling [114], although the physiologic significance of these reactions is not yet clear.

Alterations of Endothelial Function

The major functions of endothelium are to serve as a barrier between the plasma and tissue and also to serve as a platform for uptake of biomolecules and local function of enzymes. Oxidative stress can compromise these functions. For example, oxidative stress in the lung may alter the ability of endothelium to transform bioactive molecules, such as 5-hydroxytryptamine (5-HT or serotonin) or angiotensin I, while more severe stress can damage ECs with a resultant increase in barrier permeability and interstitial edema.

Uptake of Vasoactive Amines

The pulmonary endothelium modulates the plasma levels of vasoactive compounds including 5-HT and norepinephrine. These compounds are taken up by the pulmonary ECs through receptor-mediated processes and transformed to inactive metabolites [115, 116]. Depression of the uptake of 5-HT has been used as an index of oxidative damage to the pulmonary endothelium

[117–119]. The mechanism for this effect has been ascribed to damage to the EC membrane although alteration of the distribution of the pulmonary capillary blood flow can also alter the rate of amine clearance (see Chapter 7).

Angiotensin-Converting Enzyme

Angiotensins are biologically active peptides that are involved in regulation of vascular tone. Angiotensin I is a decapeptide that has activity as a vasoconstrictor following its cleavage to the octapeptide angiotensin II by angiotensin-converting enzyme (ACE). This enzyme is expressed on the cell membrane of ECs [120, 121]. Owing to its size and extent of vascularity, the lung has approximately 30% of the total body activity of this enzyme and is the major organ for generation of angiotensin II [122, 123]. Shedding of cell membrane-associated ACE from the pulmonary endothelium has been utilized as an index of oxidative stress [124–126] and has been shown in a variety of experimental models of oxidant stress including lung ischemia–reperfusion and infusion of H_2O_2 [125, 127] (see Chapter 7).

Lung Permeability and Edema

The major function of the pulmonary endothelium is to provide a barrier separating the fluid of the blood from the surrounding tissue (see Chapters 8 and 24). The width of the interstitial space between pulmonary alveolar epithelial cells and ECs is especially crucial to minimize the O_2 diffusion distance for blood oxygenation. Data from models of increased ROS exposure (lung ischemia–reperfusion or the addition of ROS to EC monolayers) indicate that cellular injury is associated with impairment of the pulmonary endothelial barrier function [128–131]. The initial effects of ROS on ECs that lead to barrier dysfunction may result from modulation of protein kinases or phosphatases and generation of intracellular second messengers that lead to the loss of normal cell–cell contacts. For example, ROS can alter the cytoskeletal organization [132, 133] via activation of focal adhesion kinases [134]. These non-receptor tyrosine kinases are involved in the structure and function of focal adhesions and are critical for promoting cellular integrity by maintaining cell–cell and cell–matrix interactions. More severe effects reflect cell damage resulting in loss of EC integrity manifested by disruption of cell–cell contacts or gaps in the barrier due to cell death.

Recruitment of PMNs and other Inflammatory Cells

ROS-mediated oxidant stress has been shown to result in the recruitment of inflammatory cells to the lungs [135–137]. The initial step in this process is related to a local increase in the concentration of cytokines and chemokines that act as cytoattractants for PMNs and other cells associated with inflammation. Although there does not appear to be specificity to oxidant-mediated injury for cell chemotaxis, oxidant stress appears to specifically increase the retention of these cells in the lungs. PMNs in the vasculature contact the lung tissue through their binding to ECs followed by their transmigration through the pulmonary endothelial barrier (see Chapter 10). Oxidant stress can lead to induction of adhesion molecules such as E- and P-selectins, vascular cell adhesion molecule-1, and intercellular cell adhesion molecule-1 (ICAM-1) which promote cell attachment. The mechanisms for increased expression of adhesion molecules by ECs may include ROS-mediated conformational changes, increased secretion from intracellular pools, or increased transcription [135, 138, 139]. However, the biochemical basis for these effects requires further investigation.

Oxidant stress during hyperoxia has resulted in differential sites of recruitment of PMNs to pulmonary endothelium. P-selectin was induced in pulmonary arteriolar endothelium and resulted in increased PMNs rolling, while induction of ICAM-1 was seen in pulmonary capillaries and venules resulting in increased PMNs adherence [140]. Thus, increased PMNs infiltration would be expected in the alveolar regions. Transmigration appears to be dependent chiefly on expression of vascular endothelial (VE)-cadherin and platelet-endothelial cell adhesion molecule-1 (CD31)–expression of the latter may be specifically involved in transendothelial migration associated with oxidative stress [141].

Cell Death

Cell death in response to oxidant stress can occur by either apoptosis or necrosis (see Chapter 16). In general, necrosis is thought to result from more severe insults associated with oxidation of cellular membranes and other components that lead to loss of cellular integrity. Apoptosis represents programmed cell death and is the manifestation of a signaling response, presumably initiated during an earlier stage of oxidant stress. $O_2^{\bullet-}$ and H_2O_2 can act as intracellular second messengers to activate and/or inhibit signal transduction pathways that alter expression patterns of stress response genes. ROS-initiated pathways of signal transduction, such as those involving MAPKs or the transcription factors, NF-κB and AP-1, eventually determine the course of cellular apoptosis and regeneration [142–144]. Apoptosis of pulmonary ECs has been seen with exposure to hyperoxia [145] and oxidants associated with wood or cigarette smoke extracts [146, 147]. Exogenous $O_2^{\bullet-}$ generated by activated macrophages has been shown to initiate apoptosis in a pulmonary microvascular EC cell line by initiating intracellular Ca^{2+} release and subsequent mitochondrial $O_2^{\bullet-}$ generation; this pathway was blocked by inhibiting the intracellular influx of $O_2^{\bullet-}$ through the anion transporter, ClC-3 [45]. Factors that determine the balance between signaling for apoptosis and the more severe injury that results in necrotic cell death are not fully understood.

PATHOPHYSIOLOGIC MECHANISMS FOR OXIDATIVE STRESS

A spectrum of diseases may be associated with oxidative stress as an aggravating factor. The source of ROS/RNS in these diseases may be exogenous (environmental) or endogenous associated with increased activity of pathways that are normally associated with physiological signaling or host defense.

Hyperoxia

Hyperoxia is defined as an inspired oxygen concentration greater than the normal atmospheric level [0.21 atmosphere absolute (ATA)]. Normobaric hyperoxia refers elevated fractional oxygen concentration (below 0.21 ATA) delivered at ambient pressure (1 ATA) while hyperbaric hyperoxia requires exposure in a hyperbaric chamber. Oxygen at elevated partial pressures can be cytotoxic as it causes the increased generation of ROS, probably through multiple sources that have not been fully delineated and may vary with the cell type. In isolated rat lungs, hyperoxia resulted in increased capillary endothelial ROS generation, initially from mitochondrial sources [132]. Activation of enzymatic pathways for ROS generation (i.e., NOX) may occur as a later event [132, 148]. Increased $O_2^{\bullet-}$ promotes release of free Fe^{2+} from intracellular stores, which potentiates the harmful effects of ROS. With the usually achievable levels of hyperoxygenation, the rates of ROS generation marginally exceed the antioxidant defenses resulting in a slowly evolving oxidative damage to tissue biomolecules. This accounts for the relatively prolonged initiation phase of O_2 toxicity; in rodents, increased ROS generation is observed immediately on hyperoxic exposure [132] but more than 48 h of exposure to 1 ATA O_2 is required before the earliest signs of lung injury are evident [39, 149].

Although nearly all bodily cells are susceptible to injury from elevated partial pressures of oxygen, lung cells appear to be especially at risk as they are exposed to the highest O_2 concentrations in the body. Lung injury associated with oxygen toxicity was first described experimentally by J. Lorrain Smith [150] more than a century ago and came to clinical relevance when treatment of patients with O_2-enriched gases or hyperbaric O_2 became feasible. Rats exposed to O_2 at 1 ATA showed significant damage to the pulmonary endothelium by morphologic analysis and by reduction in the clearance of 5-HT [117, 118, 149]. Cellular injury with subsequent cytokine release results in an influx of inflammatory cells which can amplify the endothelial injury, although this effect appears to play a relatively minor role in the manifestations of hyperoxia [151, 152]. Lung endothelium in rats shows significantly greater injury than epithelium, but in other species including primates, the injury to pulmonary endothelial and epithelial cells is similar [149]. Continued exposure to O_2 leads to increased pulmonary permeability, pulmonary edema, and cell death [151, 153]. With prolonged exposure, the number of functioning capillaries can be markedly diminished prior to death of the animal [149]. Discontinuation of exposure at a prelethal stage leads to proliferation of ECs and other cells (especially fibroblasts) that can regenerate a near normal lung or eventuate in lung fibrosis.

Inflammation

As described above, oxidant stress activates the recruitment of PMNs and other inflammatory cells to the lung. At the same time, production of ROS by recruited PMNs increases the oxidant load. ROS generation by inflammatory cells such as PMNs, eosinophils, and macrophages plays a fundamental role in the mammalian immune response to contain invading microbial pathogens. These ROS facilitate microbicidal activity of the cells. The "respiratory burst" representing the increased oxygen consumption that these cells demonstrate following phagocytosis arises from the activation of NOX (it was in PMNs that classical NOX was first discovered). This enzyme is dormant in resting cells but can be activated by chemoattractant peptides and chemokines that bind to membrane receptors as well as by stimuli following microbial phagocytosis. PMNs also can promote tissue destruction due to the secretion of various proteases. Thus, these cells serve to amplify the initial tissue injury associated with increased lung ROS generation. In the case of systemic sepsis where the lung is not directly involved initially, the recruitment of PMNs in the intravascular or interstitial spaces can lead to severe lung damage as manifested by the acute respiratory distress syndrome (ARDS) [21]. Although an important role of PMNs has been postulated for several other conditions such as hyperoxic lung injury, current evidence indicates that injury is not appreciably diminished in animal models lacking PMNs infiltration [152, 154].

Reoxygenation after Anoxia

Tissue anoxia (or hypoxia) generally reflects the consequences of compromised oxygen delivery (see Chapter

18). The usual cause in systemic organs is the impairment of blood flow [155], although that is not the case with the lung. In that organ, tissue hypoxia does not result from altered pulmonary perfusion but rather is associated with altered inspired gas composition. Acute hypoxia has been shown to result in the increased generation of ROS by mitochondria (due to inhibition of the terminal oxidase of the electron transport pathway), although this is controversial [156, 157]. If true, hypoxia-mediated ROS generation may contribute to tissue injury, but the greater insult with hypoxia is related to failure of oxidative phosphorylation resulting in an energy (ATP) deficit and tissue acidosis.

Compared to hypoxia, oxidative stress plays a much greater role during the reoxygenation period associated with restoration of the blood flow [158]. Anoxia in systemic organs (heart, brain, kidneys, etc.) followed by reoxygenation (i.e., ischemia–reperfusion) results in overproduction of ROS that can cause oxidation of cellular components with eventual cell death [159]. ROS generation is initiated within the first few minutes of reperfusion indicating that the return of O_2 to anoxic tissues is a critical event [158, 160, 161]. ROS generation in this syndrome has been attributed to the xanthine oxidase pathway, which is activated during anoxia by proteases (see Figure 17.5). Thus, anoxia results in ATP breakdown

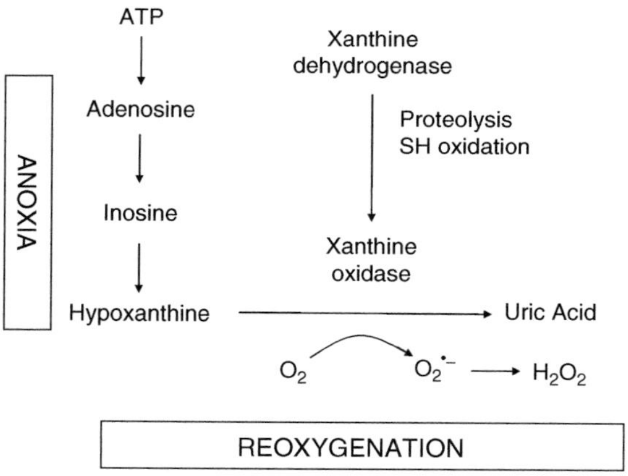

Figure 17.5 Mechanism for generation of ROS during reoxygenation following a hypoxic period. ROS production is postulated to occur during reperfusion by the xanthine oxidase pathway. Hypoxia results in ATP breakdown leading to the increased production of hypoxanthine – a substrate for xanthine oxidase. Xanthine oxidase is generated from xanthine dehydrogenase (a form of the enzyme that uses NADH as the electron acceptor) by Ca^{2+}-activated proteolysis. Xanthine oxidase generates $O_2^{\bullet-}$ in the presence of O_2 from the metabolism of hypoxanthine.

leading to increased cellular concentrations of hypoxanthine, a major substrate for the enzyme xanthine oxidase. The reintroduction of O_2 provides the electron acceptor for activity of this enzyme leading to $O_2^{\bullet-}$ generation. The physiological role for $O_2^{\bullet-}$ generation in these circumstances is not known.

Experimentally, anoxia in isolated rat lungs has been produced by ventilation with N_2 followed by reoxygenation [84]. This protocol resulted in evidence of oxidative stress that was prevented by pretreatment with allopurinol, an inhibitor of xanthine oxidase. Thus, the lung appears to show a response similar to systemic organs. Despite the theoretical lack of hypoxia with lung ischemia, increased oxidant stress and lung injury has been demonstrated experimentally with lung reperfusion [162, 163]. In some cases, the experimental model included occlusion of the bronchus which could result in atelectasis and tissue hypoxia during the ischemic period. Inflammation associated with the surgical procedures also could play a role. Nonetheless, the vigor of the reperfusion response in the lung appears to be significantly less than that observed in systemic organs.

Signaling Associated with Altered Mechanical Forces

Endothelial cells in situ are normally subjected to physical forces including shear stress and distension associated with increased intravascular pressure (see Chapter 20). The response of cells to physical forces is called mechanotransduction and refers to the mechanism by which cells convert a mechanical stimulus into a biochemical signal. Mechanical forces are increasingly recognized as important regulators of cell physiology [164–166]. The pulmonary ECs, like similar cells in systemic organs, are subjected constantly to the stimulus of blood flow and they are hence an important site for mechanotransduction. Inflation of the lung associated with respiration also gives rise to mechanical stimulation associated with "stretch." Mechanical forces are now known to activate various signal transduction pathways and generate a variety of second messengers depending upon the cell type and the characteristics of the physical forces. The signaling pathways in response to altered mechanotransduction are mediated through ROS.

Altered Shear Stress (Ischemia)

Ischemia is the loss of blood flow to an organ and, in systemic organs, leads to tissue hypoxia. However, oxygenation is maintained in the lung following vascular obstruction despite cessation of blood flow since the alveolar gas is the O_2 source. Thus, continued ventilation of the lung maintains adequate tissue oxygenation during the

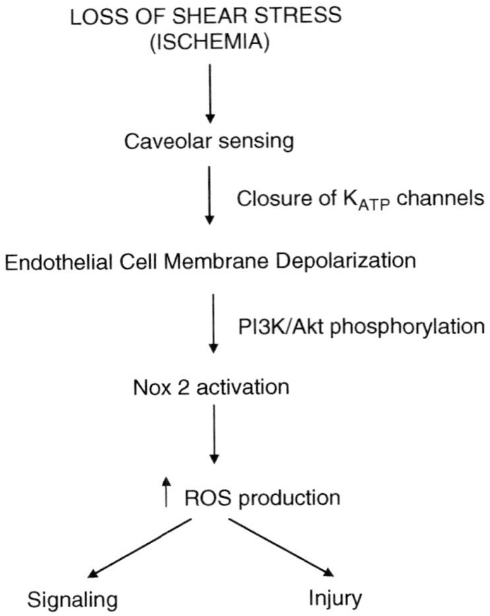

Figure 17.6 The mechanism for endothelial generation of ROS with lung ischemia. Endothelial ROS generation occurs with alteration of the mechanical stimulus of shear stress (mechanotransduction), and can initiate either signaling or injury depending on the level of production and antioxidant defenses of the cell.

ischemic period (at least until secondary manifestations develop such as atelectasis due to surfactant deficiency). Another significant effect of the loss of blood flow is an alteration (loss) of the normal shear to which the luminal endothelium is constantly exposed. Reduction of shear in lung ischemia is sensed by the pulmonary endothelium (altered mechanotransduction) leading to activation of signaling pathways that generate ROS (see Figure 17.6). Cessation of flow is initially sensed by structures on the endothelial membrane such as lipid rich membrane domains (caveolae) or perhaps by the cytoskeleton [167]. The signal is transmitted to cell membrane-localized ATP-sensitive K^+ channels (K_{ATP}) which normally maintain the pulmonary EC resting membrane potential [168]. Shear stress is required to maintain these channels in the open configuration, although whether by a direct or indirect mechanism is not yet clear. Loss of shear results in a decrease of the K_{ATP} channel open probability and a decreased (i.e., less negative) EC plasma membrane potential [5, 6, 168]. The decrease in EC plasma membrane potential in the intact lung with ischemia has been estimated as 20–30 mV [5, 169]. As expected, this change occurs immediately upon cessation of flow. Endothelial depolarization in turn triggers the activation of the plasma membrane NOX [5, 170] resulting in the production of ROS. Studies using pulmonary microvascular EC models of altered shear stress have demonstrated ROS-dependent activation of MAPKs and several transcription factors resulting in cell proliferation [67]. The change in membrane potential also leads to opening of voltage-gated calcium channels and Ca^{2+} influx into the cell [171] with activation of eNOS activity and •NO generation [172]. Acutely increased shear stress also has been shown to result in ROS generation, apparently by a mechanism similar to that described for ischemia [173]. This has led to the concept that any change from the "set-point" of flow adaptation activates the cell signaling pathway.

Cell Distention (Stretch)

Whereas endothelial responses to shear stress have been moderately well studied, the responses to circumferential vascular stretch are as yet poorly defined. Circumferential stretch in pulmonary microvessels is largely determined by the pressure gradient, and hence is determined by both vascular perfusion and alveolar ventilation pressures. The best-studied example is the response to mechanical ventilation at high lung volume that can result in lung injury from mechanical disruption of alveoli. The focus of hyperinflation studies has been primarily on the alveolar epithelium, with limited study of pulmonary endothelium [174, 175]. Overinflation of the alveoli can "stretch" the alveolar epithelium as well as the associated endothelium, although the extent of actual stretch *in situ* (versus simple unfolding) is difficult to calculate for either cell type. Vascular distension associated with increased pulmonary capillary pressure (e.g., during acute heart failure) also might lead to cellular stretch. Mechanical stretch in confluent pulmonary artery ECs in culture triggers ROS generation, possibly through mitochondrial pathways or via NOX activation similar to that seen with altered shear stress [176].

PULMONARY SYNDROMES ASSOCIATED WITH ENDOTHELIAL OXIDATIVE STRESS

Since ROS/RNS are known to exert effects on cell function ranging from subtle to powerful, a potential role for oxidant-mediated injury has been suggested for a variety of lung diseases. However, the number of disease conditions with a definitive link between oxidant stress and pulmonary endothelial dysfunction and injury is relatively limited. As discussed, any disease associated with lung inflammation, reoxygenation, or altered mechanical stresses could show evidence of oxidant-mediated injury. With some chronic lung diseases such as pulmonary fibrosis, oxidant stress appears to play a role, but the involvement of endothelium in their pathogenesis has not been demonstrated. This section will focus on diseases (syndromes) where oxidant stress to endothelium appears to play a major pathophysiologic role.

Oxygen Toxicity

The toxic effects of high concentrations of O_2 in experimental animals has been described in the section on Hyperoxia. Clinically, O_2 poisoning is a potential risk during treatment of patients with O_2 in the Intensive Care Unit [177]. Inhalation of O_2 at concentrations up to 0.6 ATA is considered safe, but only for relatively short periods (several hours), and inspired O_2 is generally maintained at concentrations below this level for patients on long-term therapy. However, direct evidence for oxygen poisoning in a clinical setting has been difficult to obtain because of the widespread appreciation of its toxic potential and the usually severe underlying lung disease that generated the need for supplemental O_2. The mechanism for lung injury during exposure to elevated pO_2 is the toxic effects of hyperoxia perhaps aggravated by the associated inflammation.

Chemical Poisoning

Oxidative injury as a consequence of chemical poisoning can be due to the inhalation or systemic administration of electrophiles, either inadvertently or for chemotherapy.

Environmental Toxins

A variety of chemicals and other environmental toxins have the ability to generate ROS. These can reach the lung endothelium indirectly by passage through the epithelium after inhalation or directly through the pulmonary circulation after intravenous injection, absorption through the skin, or after passage through the portal circulation following ingestion. A good example is paraquat – an agent that is used widely as an herbicide in agricultural applications. Ingestion, injection, or inhalation of this chemical leads to severe lung epithelial cell injury as it is specifically accumulated by these cells through polyamine transport pathways [178]. Damage to the endothelium also occurs with paraquat poisoning but is probably less severe than epithelial injury. The mechanism of ROS production by paraquat involves cyclic reduction and auto-oxidation, as described above. Experimentally endothelial injury after exposure to paraquat has been demonstrated *in vitro* by lactate dehydrogenase release from a pulmonary artery EC line [179] and *in vivo* by reduced uptake of 5-HT after an intraperitoneal injection of paraquat [119]. An intracellular superoxide dismutase mimetic decreased lactate dehydrogenase release in the cell line model of paraquat poisoning providing evidence that $O_2^{\bullet-}$ is involved in the injury [179]. Quinones such as menadione (2-methyl-1,4-naphthoquinone, vitamin K_3) also produce ROS by redox cycling. Menadione-induced damage related to oxidative stress includes the disruption of calcium homeostasis, depletion of cellular thiol levels, increases in lipid peroxidation, DNA strand breaks, and cell death [180].

Chemotherapeutic Agents

Bleomycin, often used as a component of multidrug chemotherapy for cancer, has been linked to pulmonary endothelial injury/dysfunction [181]. Enzymatic deactivation of bleomycin occurs in tissues expressing the enzyme, bleomycin hydrolase [182], and damage following exposure occurs in tissues such as the lung that do not express this enzyme [181, 183]. Bleomycin, a large hydrophobic protein, is administered by intravenous injection, binds copper in the blood stream, and is transported across the pulmonary EC membrane by an unknown mechanism. Intracellularly, the Cu^{2+} is replaced with free Fe^{2+} if available. Oxidation of Fe^{2+} to Fe^{3+} transfers the electron to molecular O_2 and creates a multicomponent, peroxide complex [184]. This complex of activated bleomycin is capable of single- or double-strand DNA breaks or it can decompose releasing ROS, possibly $^{\bullet}OH$ [184, 185]. Treatment of pulmonary artery ECs with bleomycin resulted in increased expression of γ-glutamylcysteine synthase, one of the enzymes in the GSH synthesis pathway, compatible with oxidant stress [183]. Adriamycin (doxorubicin) may exert effects by a similar mechanism [186, 187], although this compound has not been as well studied as bleomycin. Photodynamic therapy may result in production of singlet O_2. This agent can cause oxidation of biomolecules in a relatively discrete localization because of the short diffusion distances from the photosensitizer [1]. Possible effects specifically on pulmonary endothelium have not been studied.

Acute Lung Injury/ARDS

Endothelial injury is a hallmark of the pathology of acute lung injury (ALI) and its more severe manifestation, the ARDS (see Chapters 21 and 24). This lung syndrome most commonly follows sepsis, shock, or severe trauma, and its etiology has been attributed in large part to oxidant stress associated with activation of PMNs and their accumulation in the lung. Attraction of these cells to the lung is stimulated by inflammatory mediators as described in "Inflammation". Accumulation of activated neutrophils in the lung vasculature with binding of neutrophils to the pulmonary microvascular endothelium through ICAM-1 induces ROS production [188]. Oxidant stress due to ROS generation by PMNs can be amplified by therapeutic administration of O_2. A possible confounding factor in the pathophysiology of the syndrome is ROS generation by endothelium due to altered mechanotransduction associated with focal ischemia or mechanical distention. Although the oxidant injury hypothesis for pathogenesis of ALI is attractive, attempts at therapy using antioxidants have had decidedly mixed results [189]. Specific targeting of antioxidants to pulmonary endothelium could provide an improved therapeutic regimen (see Chapter 22).

Pulmonary Hypertension

Primary pulmonary hypertension (i.e., not secondary to left heart failure) is a progressive disease resulting from increased pulmonary vasoconstriction, thrombosis, and remodeling of the pulmonary arterial bed, leading to right heart failure. There are both adult and pediatric forms of the disease presumably due to different (but unknown) etiologies. Increased ROS generation has been shown in animal models of pulmonary hypertension of the newborn [190–192] and adult [193, 194]. There is increasing evidence to indicate that ROS generation also occurs in patients with pulmonary hypertension. Increased urinary excretion of isoprostanes indicating oxidative stress and extensive lung nitrotyrosine staining compatible with increased generation of $ONOO^-$ have been demonstrated in patients with severe pulmonary hypertension [195, 196]. These results suggest that ROS and RNS participate in the endothelial dysfunction of pulmonary hypertension and that vascular remodeling – a prominent part of the pathophysiology of these disorders – is preceded by endothelial injury (see Chapter 27). The endothelial proliferative response that results either as a response to cellular injury or from ROS-mediated signaling [50] can distort the pulmonary vascular bed and accentuate the alterations in pulmonary vascular resistance. Possible mechanisms for increased ROS generation in pulmonary hypertension are the increased shear stress due to the increased blood flow associated with the increased pulmonary vascular resistance or decreased shear stress due to pulmonary vascular obstruction. ROS associated with inflammation can add to the injury.

Lung Transplantation

Lung transplantation results in ischemia–reperfusion as the removal of the donor lung involves a brief period of no flow (ischemia) followed by restoration of flow (reperfusion) with the recipient's blood upon transplantation. ROS generation associated with the ischemic and reperfusion periods can cause direct damage as well as indirect damage through secondary inflammation. The solution used for storage of the donor lung also could promote ROS production due to the usually high K^+ content in the preservation solutions. High K^+ promotes EC membrane depolarization and has been shown to result in NOX2 activation in pulmonary microvascular endothelium [5, 6]. To minimize transplant injury, strategies have been adopted to prevent ischemia–reperfusion and to block release of cytokines with specific antibodies. ROS generation associated with ischemia is minimized by continuous ventilation and perfusion during preservation of the donor lung at 4 °C [197]. Donor lung storage in a low K^+ preservation solution also has been reported to improve viability of experimental lung transplantation [198].

CONCLUSIONS AND PERSPECTIVES

Although the toxicity of O_2 has been known for approximately 200 years, the role of ROS in tissue injury was suggested only 50 years ago and its rightful place in the universe of cell stresses has been appreciated only in the past 20 years. However, by this time, suggesting a role for ROS in various pathologies appears to have become a fad, although further research may indeed confirm their ubiquitous role. Clearly, ROS can oxidize tissue biomolecules resulting in cell injury and their role in oxygen toxicity, various chemical poisonings, and probably the ALI syndrome seems assured. We have made considerable progress in understanding the sources of ROS in these syndromes. These successes have raised the possibility of specific antioxidant therapy, although to date this approach has had limited success. There is an obvious need for continued rigorous study and, especially in the context of this chapter, for the specific role of ROS in injury to the pulmonary endothelium.

References

1. Dougherty, T.J., Gomer, C.J., Henderson, B.W. *et al.* (1998) Photodynamic therapy. *Journal of the National Cancer Institute*, **90**, 889–905.
2. Nunoshiba, T., Obata, F., Boss, A.C. *et al.* (1999) Role of iron and superoxide for generation of hydroxyl radical, oxidative DNA lesions, and mutagenesis in *Escherichia coli*. *The Journal of Biological Chemistry*, **274**, 34832–37.
3. Griendling, K.K. (2004) Novel NAD(P)H oxidases in the cardiovascular system. *Heart*, **90**, 491–93.
4. van der Vliet, A. (2008) NADPH oxidases in lung biology and pathology: host defense enzymes, and more. *Free Radical Biology and Medicine*, **44**, 938–55.

5. Zhang, Q., Matsuzaki, I., Chatterjee, S., and Fisher, A.B. (2005) Activation of endothelial NADPH oxidase during normoxic lung ischemia is K_{ATP} channel dependent. *American Journal of Physiology: Lung Cellular and Molecular Physiology*, **289**, L954–61.
6. Al-Mehdi, A.B., Zhao, G., Dodia, C. *et al.* (1998) Endothelial NADPH oxidase as the source of oxidants in lungs exposed to ischemia or high K^+. *Circulation Research*, **83**, 730–37.
7. Griendling, K.K., Sorescu, D., and Ushio-Fukai, M. (2000) NAD(P)H oxidase: role in cardiovascular biology and disease. *Circulation Research*, **86**, 494–501.
8. Van Buul, J.D., Fernandez-Borja, M., Anthony, E.C., and Hordijk, P.L. (2005) Expression and localization of NOX2 and NOX4 in primary human endothelial cells. *Antioxidants and Redox Signaling*, **7**, 308–17.
9. Petry, A., Djordjevic, T., Weitnauer, M. *et al.* (2006) NOX2 and NOX4 mediate proliferative response in endothelial cells. *Antioxidants and Redox Signaling*, **8**, 1473–84.
10. Li, J.M. and Shah, A.M. (2002) Intracellular localization and preassembly of the NADPH oxidase complex in cultured endothelial cells. *The Journal of Biological Chemistry*, **277**, 19952–60.
11. Phan, S.H., Gannon, D.E., Varani, J. *et al.* (1989) Xanthine oxidase activity in rat pulmonary artery endothelial cells and its alteration by activated neutrophils. *American Journal of Pathology*, **134**, 1201–11.
12. Esterhazy, D., King, M.S., Yakovlev, G., and Hirst, J. (2008) Production of reactive oxygen species by complex I (NADH: ubiquinone oxidoreductase) from *Escherichia coli* and comparison to the enzyme from mitochondria. *Biochemistry*, **47**, 3964–71.
13. Kowaltowski, A.J. and Vercesi, A.E. (1999) Mitochondrial damage induced by conditions of oxidative stress. *Free Radical Biology and Medicine*, **26**, 463–71.
14. Hernandez, L.A., Grisham, M.B., and Granger, D.N. (1987) A role for iron in oxidant-mediated ischemic injury to intestinal microvasculature. *American Journal of Physiology:Gastrointestinal and Liver Physiology*, **253**, G49–53.
15. Iyanagi, T. and Yamazaki, I. (1970) One-electron-transfer reactions in biochemical systems. V. Difference in the mechanism of quinone reduction by the NADH dehydrogenase and the NAD(P)H dehydrogenase (DT-diaphorase). *Biochimica et Biophysica Acta*, **216**, 282–94.
16. Tatsumi, H., Nakase, H., Kano, K., and Ikeda, T. (1998) Mechanistic study of the autoxidation of reduced flavin and quinone compounds. *Journal of Electroanalytical Chemistry*, **443**, 236–42.
17. Drechsel, D. and Patel, M. (2008) Role of reactive oxygen species in the neurotoxicity of environmental agents implicated in Parkinson's disease. *Free Radical Biology and Medicine*, **44**, 1873–86.
18. Bus, J.S., Aust, S.D., and Gibson, J.E. (1974) Superoxide- and singlet oxygen-catalyzed lipid peroxidation as a possible mechanism for paraquat (methyl viologen) toxicity. *Biochemical and Biophysical Research Communications*, **58**, 749–55.
19. Cocheme, H.M. and Murphy, M.P. (2008) Complex I is the major site of mitochondrial superoxide production by paraquat. *The Journal of Biological Chemistry*, **283**, 1786–98.
20. Krishnadasan, B., Naidu, B.V., Byrne, K. *et al.* (2003) The role of proinflammatory cytokines in lung ischemia–reperfusion injury. *The Journal of Thoracic and Cardiovascular Surgery*, **125**, 261–72.
21. Sherwood, E.R. and Toliver-Kinsky, T. (2004) Mechanisms of the inflammatory response. *Best Practice and Research: Clinical Anaesthesiology*, **18**, 385–405.
22. Merker, M.P., Audi, S.H., Lindemer, B.J. *et al.* (2007) Role of mitochondrial electron transport complex I in coenzyme Q1 reduction by intact pulmonary arterial endothelial cells and the effect of hyperoxia. *American Journal of Physiology: Lung Cellular and Molecular Physiology*, **293**, L809–19.
23. Beckman, J.S. and Koppenol, W.H. (1996) Nitric oxide, superoxide, and peroxynitrite: the good, the bad, and ugly. *American Journal of Physiology: Cell Physiology*, **271**, C1424–37.
24. Nathan, C. and Shiloh, M.U. (2000) Reactive oxygen and nitrogen intermediates in the relationship between mammalian hosts and microbial pathogens. *Proceedings of the National Academy of Sciences of the United States of America*, **97**, 8841–48.
25. Zaki, M.H., Akuta, T., and Akaike, T. (2005) Nitric oxide-induced nitrative stress involved in microbial pathogenesis. *Journal of Pharmacological Sciences*, **98**, 117–29.
26. Van der Vliet, A., Smith, D., O'Neill, C.A. *et al.* (1994) Interactions of peroxynitrite with human plasma and its constituents: oxidative damage and antioxidant depletion. *The Biochemical Journal*, **303**, 295–301.

27. Richeson, C.E., Mulder, P., Bowry, V.W., and Ungold, K.U. (1998) The complex chemistry of peroxynitrite decomposition: new insights. *Journal of the American Chemical Society*, **120**, 7211–19.
28. Szabo, C. (2003) Multiple pathways of peroxynitrite cytotoxicity. *Toxicology Letters*, **140-141**, 105–12.
29. Guzik, T.J., Korbut, R., and Adamek-Guzik, T. (2003) Nitric oxide and superoxide in inflammation and immune regulation. *Journal of Physiology and Pharmacology*, **54**, 469–87.
30. Aktan, F. (2004) iNOS-mediated nitric oxide production and its regulation. *Life Sciences*, **75**, 639–53.
31. Marklund, S.L. (1984) Extracellular superoxide dismutase in human tissues and human cell lines. *The Journal of Clinical Investigation*, **74**, 1398–403.
32. Fattman, C.L., Enghild, J.J., Crapo, J.D. *et al.* (2000) Purification and characterization of extracellular superoxide dismutase in mouse lung. *Biochemical and Biophysical Research Communications*, **275**, 542–48.
33. Clerch, L.B., Iqbal, J., and Massaro, D. (1991) Perinatal rat lung catalase gene expression: influence of corticosteroid and hyperoxia. *American Journal of Physiology: Lung Cellular and Molecular Physiology*, **260**, L428–33.
34. Cho, H.Y., Reddy, S.P., and Kleeberger, S.R. (2006) Nrf2 defends the lung from oxidative stress. *Antioxidants and Redox Signaling*, **8**, 76–87.
35. Michiels, C., Raes, M., Toussaint, O., and Remacle, J. (1994) Importance of Se-glutathione peroxidase, catalase, and Cu/Zn-SOD for cell survival against oxidative stress. *Free Radical Biology and Medicine*, **17**, 235–48.
36. Fisher, A.B., Dodia, C., Manevich, Y. *et al.* (1999) Phospholipid hydroperoxides are substrates for non-selenium glutathione peroxidase. *The Journal of Biological Chemistry*, **274**, 21326–34.
37. Schremmer, B., Manevich, Y., Feinstein, S., and Fisher, A.B. (2007). Peroxiredoxins in the lung with emphasis on peroxiredoxin VI, in *Peroxiredoxin Systems* (eds L. Flohe and J.R. Harris), Springer, New York, pp. 317–44.
38. Manevich, Y. and Fisher, A.B. (2005) Peroxiredoxin 6, a 1-Cys peroxiredoxin, functions in antioxidant defense and lung phospholipid metabolism. *Free Radical Biology and Medicine*, **38**, 1422–32.
39. Wang, Y., Feinstein, S.I., Manevich, Y. *et al.* (2004) Lung injury and mortality with hyperoxia are increased in peroxiredoxin 6 gene-targeted mice. *Free Radical Biology and Medicine*, **37**, 1736–43.
40. Wang, Y., Feinstein, S.I., Manevich, Y. *et al.* (2006) Peroxiredoxin 6 gene-targeted mice show increased lung injury with paraquat-induced oxidative stress. *Antioxidants and Redox Signaling*, **8**, 229–37.
41. Utzinger, G.E., Strait, L.A., and Tuck, L.D. (1963) The oxidation of glutathione with formic acid and hydrogen peroxide. *Experientia*, **19**, 324–27.
42. Tanabe, S., Wang, X., Takahashi, N. *et al.* (2005) HCO_3^--independent rescue from apoptosis by stilbene derivatives in rat cardiomyocytes. *FEBS Letters*, **579**, 517–22.
43. Finkel, T. (2003) Oxidant signals and oxidative stress. *Current Opinion in Cell Biology*, **15**, 247–54.
44. Madesh, M. and Hajnoczky, G. (2001) VDAC-dependent permeabilization of the outer mitochondrial membrane by superoxide induces rapid and massive cytochrome *c* release. *The Journal of Cell Biology*, **155**, 1003–15.
45. Madesh, M., Hawkins, B.J., Milovanova, T. *et al.* (2005) Selective role for superoxide in $InsP_3$ receptor-mediated mitochondrial dysfunction and endothelial apoptosis. *The Journal of Cell Biology*, **170**, 1079–90.
46. D'Autreaux, B. and Toledano, M.B. (2007) ROS as signalling molecules: mechanisms that generate specificity in ROS homeostasis. *Nature Reviews: Molecular Cell Biology*, **8**, 813–24.
47. Hawkins, B.J., Madesh, M., Kirkpatrick, C.J., and Fisher, A.B. (2007) Superoxide flux in endothelial cells via the chloride channel-3 mediates intracellular signaling. *Molecular Biology of the Cell*, **18**, 2002–12.
48. Stone, J.R. and Collins, T. (2002) The role of hydrogen peroxide in endothelial proliferative responses. *Endothelium: Journal of Endothelial Cell Research*, **9**, 231–38.
49. Luczak, K., Balcerczyk, A., Soszynski, M., and Bartosz, G. (2004) Low concentration of oxidant and nitric oxide donors stimulate proliferation of human endothelial cells *in vitro*. *Cell Biology International*, **28**, 483–86.
50. Milovanova, T., Chatterjee, S., Manevich, Y. *et al.* (2006) Lung endothelial cell proliferation with decreased shear stress is mediated by reactive oxygen species. *American Journal of Physiology: Cell Physiology*, **290**, C66–76.
51. Nakamura, Y., Patrushev, N., Inomata, H. *et al.* (2008) Role of protein tyrosine phosphatase 1B in

vascular endothelial growth factor signaling and cell–cell adhesions in endothelial cells. *Circulation Research*, **102**, 1182–91.

52. Matkar, S.S., Wrischnik, L.A., and Hellmann-Blumberg, U. (2008) Production of hydrogen peroxide and redox cycling can explain how sanguinarine and chelerythrine induce rapid apoptosis. *Archives of Biochemistry and Biophysics*. **477**, 43–52.
53. Shono, T., Ono, M., Izumi, H. *et al.* (1996) Involvement of the transcription factor NF-kappaB in tubular morphogenesis of human microvascular endothelial cells by oxidative stress. *Molecular and Cellular Biology*, **16**, 4231–39.
54. Torres, M. (2003) Mitogen-activated protein kinase pathways in redox signaling. *Frontiers in Bioscience*, **8**, d369–91.
55. Nguyen, A., Chen, P., and Cai, H. (2004) Role of CaMKII in hydrogen peroxide activation of ERK1/2, p38 MAPK, HSP27 and actin reorganization in endothelial cells. *FEBS Letters*, **572**, 307–13.
56. Kamata, H., Honda, S., Maeda, S. *et al.* (2005) Reactive oxygen species promote TNFalpha-induced death and sustained JNK activation by inhibiting MAP kinase phosphatases. *Cell*, **120**, 649–61.
57. Gopalakrishna, R. and Anderson, W.B. (1989) Ca^{2+}- and phospholipid-independent activation of protein kinase C by selective oxidative modification of the regulatory domain. *Proceedings of the National Academy of Sciences of the United States of America*, **86**, 6758–62.
58. Hardwick, J.S. and Sefton, B.M. (1995) Activation of the Lck tyrosine protein kinase by hydrogen peroxide requires the phosphorylation of Tyr-394. *Proceedings of the National Academy of Sciences of the United States of America*, **92**, 4527–31.
59. Gozin, A., Franzini, E., Andrieu, V. *et al.* (1998) Reactive oxygen species activate focal adhesion kinase, paxillin and $p130^{cas}$ tyrosine phosphorylation in endothelial cells. *Free Radical Biology and Medicine*, **25**, 1021–32.
60. Lopez-Ongil, S., Torrecillas, G., Perez-Sala, D. *et al.* (1999) Mechanisms involved in the contraction of endothelial cells by hydrogen peroxide. *Free Radical Biology and Medicine*, **26**, 501–10.
61. Shasby, D.M., Yorek, M., and Shasby, S.S. (1988) Exogenous oxidants initiate hydrolysis of endothelial cell inositol phospholipids. *Blood*, **72**, 491–99.
62. Chakraborti, S., Gurtner, G.H., and Michael, J.R. (1989) Oxidant-mediated activation of phospholipase A_2 in pulmonary endothelium. *American Journal of Physiology: Lung Cellular and Molecular Physiology*, **257**, L430–37.
63. Natarajan, V., Taher, M.M., Roehm, B. *et al.* (1993) Activation of endothelial cell phospholipase D by hydrogen peroxide and fatty acid hydroperoxide. *The Journal of Biological Chemistry*, **268**, 930–37.
64. Whisler, R.L., Goyette, M.A., Grants, I.S., and Newhouse, Y.G. (1995) Sublethal levels of oxidant stress stimulate multiple serine/threonine kinases and suppress protein phosphatases in Jurkat T cells. *Archives of Biochemistry and Biophysics*, **319**, 23–35.
65. Mitchell, J., Jiang, H., Berry, L., and Meyrick, B. (1996) Effect of antioxidants on lipopolysaccharide-stimulated induction of mangano superoxide dismutase mRNA in bovine pulmonary artery endothelial cells. *Journal of Cellular Physiology*, **169**, 333–40.
66. Yao, H., Yang, S.R., Kode, A. *et al.* (2007) Redox regulation of lung inflammation: role of NADPH oxidase and NF-kappaB signalling. *Biochemical Society Transactions*, **35**, 1151–55.
67. Wei, Z., Costa, K., Al-Mehdi, A.B. *et al.* (1999) Simulated ischemia in flow-adapted endothelial cells leads to generation of reactive oxygen species and cell signaling. *Circulation Research*, **85**, 682–89.
68. Gloire, G., Legrand-Poels, S., and Piette, J. (2006) NF-kappaB activation by reactive oxygen species: fifteen years later. *Biochemical Pharmacology*, **72**, 1493–505.
69. Koay, M.A., Christman, J.W., Segal, B.H. *et al.* (2001) Impaired pulmonary NF-kappaB activation in response to lipopolysaccharide in NADPH oxidase-deficient mice. *Infection and Immunity*, **69**, 5991–96.
70. Sadikot, R.T., Zeng, H., Yull, F.E. *et al.* (2004) $p47^{phox}$ deficiency impairs activation and host defense in Pseudomonas pneumonia. *Journal of Immunology*, **172**, 1801–8.
71. Yoshida, M., Korfhagen, T.R., and Whitsett, J.A. (2001) Surfactant protein D regulates NF-kappa B and matrix metalloproteinase production in alveolar macrophages via oxidant-sensitive pathways. *Journal of Immunology*, **166**, 7514–19.
72. Iizuka, T., Ishii, Y., Itoh, K. *et al.* (2005) Nrf2-deficient mice are highly susceptible to cigarette smoke-induced emphysema. *Genes to Cells*, **10**, 1113–25.
73. Rangasamy, T., Guo, J., Mitzner, W.A. *et al.* (2005) Disruption of Nrf2 enhances susceptibility

to severe airway inflammation and asthma in mice. *The Journal of Experimental Medicine*, **202**, 47–59.

74. Thimmulappa, R.K., Scollick, C., Traore, K. *et al.* (2006) Nrf2-dependent protection from LPS induced inflammatory response and mortality by CDDO-imidazolide. *Biochemical and Biophysical Research Communications*, **351**, 883–89.
75. Kroncke, K.D. (2003) Nitrosative stress and transcription. *Biological Chemistry*, **384**, 1365–77.
76. Kang-Decker, N., Cao, S., Chatterjee, S. *et al.* (2007) Nitric oxide promotes endothelial cell survival signaling through *S*-nitrosylation and activation of dynamin-2. *Journal of Cell Science*, **120**, 492–501.
77. Benhar, M., Forrester, M.T., Hess, D.T., and Stamler, J.S. (2008) Regulated protein denitrosylation by cytosolic and mitochondrial thioredoxins. *Science*, **320**, 1050–54.
78. de Zwart, L.L., Vermeulen, N.P., Hermanns, R.C. *et al.* (1999) Urinary excretion of biomarkers for radical-induced damage in rats treated with NDMA or diquat and the effects of calcium carbimide co-administration. *Chemico-Biological Interactions*, **117**, 151–72.
79. Droge, W. (2002) Free radicals in the physiological control of cell function. *Physiological Reviews*, **82**, 47–95.
80. Li, C., Wright, M.M., and Jackson, R.M. (2002) Reactive species mediated injury of human lung epithelial cells after hypoxia-reoxygenation. *Experimental Lung Research*, **28**, 373–89.
81. Bielski, B.H., Arudi, R.L., and Sutherland, M.W. (1983) A study of the reactivity of HO_2/O_2^- with unsaturated fatty acids. *The Journal of Biological Chemistry*, **258**, 4759–61.
82. Ayene, I.S., Dodia, C., and Fisher, A.B. (1992) Role of oxygen in oxidation of lipid and protein during ischemia/reperfusion in isolated perfused rat lung. *Archives of Biochemistry and Biophysics*, **296**, 183–89.
83. al-Mehdi, A.B., Dodia, C., Jain, M.K., and Fisher, A.B. (1993) A phospholipase A_2 inhibitor decreases generation of thiobarbituric acid reactive substance during lung ischemia–reperfusion. *Biochimica et Biophysica Acta*, **1167**, 56–62.
84. Zhao, G., al-Mehdi, A.B., and Fisher, A.B. (1997) Anoxia–reoxygenation versus ischemia in isolated rat lungs. *American Journal of Physiology: Lung Cellular and Molecular Physiology*, **273**, L1112–17.
85. Al-Mehdi, A.B., Song, C., Tozawa, K., and Fisher, A.B. (2000) Ca^{2+}- and phosphatidylinositol 3-kinase-dependent nitric oxide generation in lung endothelial cells in situ with ischemia. *The Journal of Biological Chemistry*, **275**, 39807–10.
86. Matot, I., Manevich, Y., Al-Mehdi, A.B. *et al.* (2003) Fluorescence imaging of lipid peroxidation in isolated rat lungs during nonhypoxic lung ischemia. *Free Radical Biology and Medicine*, **34**, 785–90.
87. Schnurr, K., Belkner, J., Ursini, F. *et al.* (1996) The selenoenzyme phospholipid hydroperoxide glutathione peroxidase controls the activity of the 15-lipoxygenase with complex substrates and preserves the specificity of the oxygenation products. *The Journal of Biological Chemistry*, **271**, 4653–58.
88. Sevanian, A., Muakkassah-Kelly, S.F., and Montestruque, S. (1983) The influence of phospholipase A_2 and glutathione peroxidase on the elimination of membrane lipid peroxides. *Archives of Biochemistry and Biophysics*, **223**, 441–52.
89. Zhao, L., Wang, H.P., Zhang, H.J. *et al.* (2003) L-PhGPx expression can be suppressed by antisense oligodeoxynucleotides. *Archives of Biochemistry and Biophysics*, **417**, 212–18.
90. Dalle-Donne, I., Aldini, G., Carini, M. *et al.* (2006) Protein carbonylation, cellular dysfunction, and disease progression. *Journal of Cellular and Molecular Medicine*, **10**, 389–406.
91. Berlett, B.S. and Stadtman, E.R. (1997) Protein oxidation in aging, disease, and oxidative stress. *The Journal of Biological Chemistry*, **272**, 20313–16.
92. Petropoulos, I. and Friguet, B. (2005) Protein maintenance in aging and replicative senescence: a role for the peptide methionine sulfoxide reductases. *Biochimica et Biophysica Acta*, **1703**, 261–66.
93. Costa, V., Quintanilha, A., and Moradas-Ferreira, P. (2007) Protein oxidation, repair mechanisms and proteolysis in *Saccharomyces cerevisiae*. *IUBMB Life*, **59**, 293–98.
94. Devasagayam, T.P., Subramanian, M., Pradhan, D.S., and Sies, H. (1993) Prevention of singlet oxygen-induced DNA damage by lipoate. *Chemico-Biological Interactions*, **86**, 79–92.
95. Budworth, H. and Dianov, G.L. (2003) Mode of inhibition of short-patch base excision repair by thymine glycol within clustered DNA lesions. *The Journal of Biological Chemistry*, **278**, 9378–81.
96. Jones, C.B., McIntosh, J., Huang, H. *et al.* (2001) Regulation of bleomycin-induced DNA breakage and chromatin structure in lung endothelial cells by integrins and poly(ADP-ribose) polymerase. *Molecular Pharmacology*, **59**, 69–75.

97. Pagano, A., Metrailler-Ruchonnet, I., Aurrand-Lions, M. *et al.* (2007) Poly(ADP-ribose) polymerase-1 (PARP-1) controls lung cell proliferation and repair after hyperoxia-induced lung damage. *American Journal of Physiology: Lung Cellular and Molecular Physiology*, **293**, L619–29.
98. De Bont, R. and van Larebeke, N. (2004) Endogenous DNA damage in humans: a review of quantitative data. *Mutagenesis*, **19**, 169–85.
99. Grishko, V., Solomon, M., Breit, J.F. *et al.* (2001) Hypoxia promotes oxidative base modifications in the pulmonary artery endothelial cell VEGF gene. *The FASEB Journal*, **15**, 1267–69.
100. Ruchko, M., Gorodnya, O., LeDoux, S.P. *et al.* (2005) Mitochondrial DNA damage triggers mitochondrial dysfunction and apoptosis in oxidant-challenged lung endothelial cells. *American Journal of Physiology: Lung Cellular and Molecular Physiology*, **288**, L530–35.
101. Grishko, V., Solomon, M., Wilson, G.L. *et al.* (2001) Oxygen radical-induced mitochondrial DNA damage and repair in pulmonary vascular endothelial cell phenotypes. *American Journal of Physiology: Lung Cellular and Molecular Physiology*, **280**, L1300–8.
102. Radi, R., Beckman, J.S., Bush, K.M., and Freeman, B.A. (1991) Peroxynitrite oxidation of sulfhydryls. The cytotoxic potential of superoxide and nitric oxide. *The Journal of Biological Chemistry*, **266**, 4244–50.
103. Alvarez, B., Rubbo, H., Kirk, M. *et al.* (1996) Peroxynitrite-dependent tryptophan nitration. *Chemical Research in Toxicology*, **9**, 390–96.
104. Alvarez, B., Ferrer-Sueta, G., Freeman, B.A., and Radi, R. (1999) Kinetics of peroxynitrite reaction with amino acids and human serum albumin. *The Journal of Biological Chemistry*, **274**, 842–48.
105. Ischiropoulos, H., al-Mehdi, A.B., and Fisher, A.B. (1995) Reactive species in ischemic rat lung injury: contribution of peroxynitrite. *American Journal of Physiology: Lung Cellular and Molecular Physiology*, **269**, L158–64.
106. Ischiropoulos, H. (1998) Biological tyrosine nitration: a pathophysiological function of nitric oxide and reactive oxygen species. *Archives of Biochemistry and Biophysics*, **356**, 1–11.
107. Gole, M.D., Souza, J.M., Choi, I. *et al.* (2000) Plasma proteins modified by tyrosine nitration in acute respiratory distress syndrome. *American Journal of Physiology: Lung Cellular and Molecular Physiology*, **278**, L961–67.
108. Greenacre, S.A. and Ischiropoulos, H. (2001) Tyrosine nitration: localisation, quantification, consequences for protein function and signal transduction. *Free Radical Research*, **34**, 541–81.
109. Balafanova, Z., Bolli, R., Zhang, J. *et al.* (2002) Nitric oxide (NO) induces nitration of protein kinase Cepsilon (PKCepsilon), facilitating PKCepsilon translocation via enhanced PKCepsilon–RACK2 interactions: a novel mechanism of no-triggered activation of PKCepsilon. *The Journal of Biological Chemistry*, **277**, 15021–27.
110. Gow, A.J., Duran, D., Malcolm, S., and Ischiropoulos, H. (1996) Effects of peroxynitrite-induced protein modifications on tyrosine phosphorylation and degradation. *FEBS Letters*, **385**, 63–66.
111. Kong, S.K., Yim, M.B., Stadtman, E.R., and Chock, P.B. (1996) Peroxynitrite disables the tyrosine phosphorylation regulatory mechanism: lymphocyte-specific tyrosine kinase fails to phosphorylate nitrated cdc2(6–20)NH_2 peptide. *Proceedings of the National Academy of Sciences of the United States of America*, **93**, 3377–82.
112. MacMillan-Crow, L.A., Greendorfer, J.S., Vickers, S.M., and Thompson, J.A. (2000) Tyrosine nitration of c-SRC tyrosine kinase in human pancreatic ductal adenocarcinoma. *Archives of Biochemistry and Biophysics*, **377**, 350–56.
113. Mallozzi, C., Di Stasi, A.M., and Minetti, M. (2001) Nitrotyrosine mimics phosphotyrosine binding to the SH2 domain of the src family tyrosine kinase lyn. *FEBS Letters*, **503**, 189–95.
114. Freeman, B.A., Baker, P.R.S., Schopfer, F.J. *et al.* (2008) Nitro-fatty acid formational and signaling. *The Journal of Biological Chemistry*, **283**, 15515–19.
115. Strum, J.M. and Junod, A.F. (1972) Radioautographic demonstration of 5-hydroxytryptamine-^{3}H uptake by pulmonary endothelial cells. *The Journal of Cell Biology*, **54**, 456–67.
116. Gillis, C.N., Cronau, L.H., Mandel, S., and Hammond, G.L. (1979) Indicator dilution measurement of 5-hydroxytryptamine clearance by human lung. *Journal of Applied Physiology*, **46**, 1178–83.
117. Block, E.R. and Fisher, A.B. (1977) Hyperoxia and lung serotonin clearance: a new observation. *Chest*, **71**, 289–91.
118. Block, E.R. and Fisher, A.B. (1977) Prevention of hyperoxic-induced depression of pulmonary serotonin clearance by pretreatment with superoxide dismutase. *The American Review of Respiratory Disease*, **116**, 441–47.
119. Block, E.R. and Schoen, F.J. (1982) Depression of serotonin uptake by rat lungs exposed to paraquat.

The Journal of Pharmacology and Experimental Therapeutics, **221**, 254–60.

120. Ng, K.K. and Vane, J.R. (1967) Conversion of angiotensin I to angiotensin II. *Nature*, **216**, 762–66.
121. Ryan, U.S., Ryan, J.W., Whitaker, C., and Chiu, A. (1976) Localization of angiotensin converting enzyme (kininase II). II. Immunocytochemistry and immunofluorescence. *Tissue and Cell*, **8**, 125–45.
122. Metzger, F.F., Bohle, R., Kerkman, L. *et al.* (1997) Angiotensin-1-converting enzyme (CD143) on endothelial cells in normal and in pathological conditions, in *Leukocyte Typing VI. White Cell Differentiation Antigens* (eds T. Kishimoto, H. Kikutani, A.E.G.Kr. von dem Borne *et al.*), Garland, New York, pp. 749–51.
123. Danilov, S.M., Gavrilyuk, V.D., Franke, F.E. *et al.* (2001) Lung uptake of antibodies to endothelial antigens: key determinants of vascular immunotargeting. *American Journal of Physiology: Lung Cellular and Molecular Physiology*, **280**, L1335–47.
124. Muzykantov, V.R. and Danilov, S.M. (1991) A new approach to the investigation of oxidative injury to the pulmonary endothelium: use of angiotensin-converting enzyme as a marker. *Biomedical Science*, **2**, 11–21.
125. Atochina, E.N., Muzykantov, V.R., Al-Mehdi, A.B. *et al.* (1997) Normoxic lung ischemia/reperfusion accelerates shedding of angiotensin converting enzyme from the pulmonary endothelium. *American Journal of Respiratory and Critical Care Medicine*, **156**, 1114–19.
126. Orfanos, S.E., Armaganidis, A., Glynos, C. *et al.* (2000) Pulmonary capillary endothelium-bound angiotensin-converting enzyme activity in acute lung injury. *Circulation*, **102**, 2011–18.
127. Atochina, E.N., Balyasnikova, I.V., Danilov, S.M. *et al.* (1998) Immunotargeting of catalase to ACE or ICAM-1 protects perfused rat lungs against oxidative stress. *American Journal of Physiology: Lung Cellular and Molecular Physiology*, **275**, L806–17.
128. McQuaid, K.E., Smyth, E.M., and Keenan, A.K. (1996) Evidence for modulation of hydrogen peroxide-induced endothelial barrier dysfunction by nitric oxide *in vitro*. *European Journal of Pharmacology*, **307**, 233–41.
129. Dodd-o, J.M., Hristopoulos, M.L., Faraday, N., and Pearse, D.B. (2003) Effect of ischemia and reperfusion without airway occlusion on vascular barrier function in the *in vivo* mouse lung. *Journal of Applied Physiology*, **95**, 1971–78.
130. Usatyuk, P.V., Vepa, S., Watkins, T. *et al.* (2003) Redox regulation of reactive oxygen species-induced p38 MAP kinase activation and barrier dysfunction in lung microvascular endothelial cells. *Antioxidants and Redox Signaling*, **5**, 723–30.
131. Neumann, P., Gertzberg, N., Vaughan, E. *et al.* (2006) Peroxynitrite mediates TNF-alpha-induced endothelial barrier dysfunction and nitration of actin. *American Journal of Physiology: Lung Cellular and Molecular Physiology*, **290**, L674–84.
132. Brueckl, C., Kaestle, S., Kerem, A. *et al.* (2006) Hyperoxia-induced reactive oxygen species formation in pulmonary capillary endothelial cells *in situ*. *American Journal of Respiratory Cell and Molecular Biology*, **34**, 453–63.
133. Moldovan, L., Mythreye, K., Goldschmidt-Clermont, P.J., and Satterwhite, L.L. (2006) Reactive oxygen species in vascular endothelial cell motility. Roles of NAD(P)H oxidase and Rac1. *Cardiovascular Research*, **71**, 236–46.
134. Vepa, S., Scribner, W.M., Parinandi, N.L. *et al.* (1999) Hydrogen peroxide stimulates tyrosine phosphorylation of focal adhesion kinase in vascular endothelial cells. *American Journal of Physiology: Lung Cellular and Molecular Physiology*, **277**, L150–58.
135. Patel, K.D., Zimmerman, G.A., Prescott, S.M. *et al.* (1991) Oxygen radicals induce human endothelial cells to express GMP-140 and bind neutrophils. *The Journal of Cell Biology*, **112**, 749–59.
136. Lo, S.K., Janakidevi, K., Lai, L., and Malik, A.B. (1993) Hydrogen peroxide-induced increase in endothelial adhesiveness is dependent on ICAM-1 activation. *American Journal of Physiology: Lung Cellular and Molecular Physiology*, **264**, L406–12.
137. Gaboury, J.P., Anderson, D.C., and Kubes, P. (1994) Molecular mechanisms involved in superoxide-induced leukocyte–endothelial cell interactions *in vivo*. *American Journal of Physiology: Heart and Circulatory Physiology*, **266**, H637–42.
138. Roebuck, K.A., Rahman, A., Lakshminarayanan, V. *et al.* (1995) H_2O_2 and tumor necrosis factor-alpha activate intercellular adhesion molecule 1 (ICAM-1) gene transcription through distinct *cis*-regulatory elements within the ICAM-1 promoter. *The Journal of Biological Chemistry*, **270**, 18966–74.
139. Kokura, S., Wolf, R.E., Yoshikawa, T. *et al.* (1999) Molecular mechanisms of neutrophil–endothelial cell adhesion induced by redox imbalance. *Circulation Research*, **84**, 516–24.

140. Nishio, K., Suzuki, Y., Aoki, T. *et al.* (1998) Differential contribution of various adhesion molecules to leukocyte kinetics in pulmonary microvessels of hyperoxia-exposed rat lungs. *American Journal of Respiratory and Critical Care Medicine*, **157**, 599–609.
141. Scalia, R. and Lefer, A.M. (1998) *In vivo* regulation of PECAM-1 activity during acute endothelial dysfunction in the rat mesenteric microvasculature. *Journal of Leukocyte Biology*, **64**, 163–69.
142. Tacchini, L., Radice, L., and Bernelli-Zazzera, A. (1999) Differential activation of some transcription factors during rat liver ischemia, reperfusion, and heat shock. *Journal of Cellular Physiology*, **180**, 255–62.
143. Frangogiannis, N.G., Smith, C.W., and Entman, M.L. (2002) The inflammatory response in myocardial infarction. *Cardiovascular Research*, **53**, 31–47.
144. Kaminski, K.A., Bonda, T.A., Korecki, J., and Musial, W.J. (2002) Oxidative stress and neutrophil activation – the two keystones of ischemia/reperfusion injury. *International Journal of Cardiology*, **86**, 41–59.
145. Wang, X., Wang, Y., Kim, H.P. *et al.* (2007) Carbon monoxide protects against hyperoxia-induced endothelial cell apoptosis by inhibiting reactive oxygen species formation. *The Journal of Biological Chemistry*, **282**, 1718–26.
146. Liu, P.L., Chen, Y.L., Chen, Y.H. *et al.* (2005) Wood smoke extract induces oxidative stress-mediated caspase-independent apoptosis in human lung endothelial cells: role of AIF and EndoG. *American Journal of Physiology: Lung Cellular and Molecular Physiology*, **289**, L739–49.
147. Nana-Sinkam, S.P., Lee, J.D., Sotto-Santiago, S. *et al.* (2007) Prostacyclin prevents pulmonary endothelial cell apoptosis induced by cigarette smoke. *American Journal of Respiratory and Critical Care Medicine*, **175**, 676–85.
148. Usatyuk, P.V., Romer, L.H., He, D. *et al.* (2007) Regulation of hyperoxia-induced NADPH oxidase activation in human lung endothelial cells by the actin cytoskeleton and cortactin. *The Journal of Biological Chemistry*, **282**, 23284–95.
149. Crapo, J.D. (1986) Morphologic changes in pulmonary oxygen toxicity. *Annual Review of Physiology*, **48**, 721–31.
150. Smith, J.L. (1899) The pathological effects due to increase of oxygen tension in air breathed. *The Journal of Physiology*, **24**, 19–35.
151. Fisher, A.B., Forman, H.J., and Glass, M. (1984) Mechanisms of pulmonary oxygen toxicity. *Lung*, **162**, 255–59.
152. Perkowski, S., Scherpereel, A., Murciano, J.C. *et al.* (2006) Dissociation between alveolar transmigration of neutrophils and lung injury in hyperoxia. *American Journal of Physiology: Lung Cellular and Molecular Physiology*, **291**, L1050–58.
153. Tang, P.S., Mura, M., Seth, R., and Liu, M. (2008) Acute lung injury and cell death: how many ways can cells die? *American Journal of Physiology: Lung Cellular and Molecular Physiology*, **294**, L632–41.
154. Das, D.K., Bandyopadhyay, D., Hoory, S., and Steinberg, H. (1988) Role of polymorphonuclear leukocytes in hyperoxic lung injury. Prevention of neutrophil influx into the lung endothelium during oxygen exposure by ibuprofen. *Biomedica Biochimica Acta*, **47**, 1023–36.
155. Malone, P.C. (1987) The physiology of intestinal oxygenation and the pathophysiology of intestinal ileus. *Medical Hypotheses*, **22**, 111–57.
156. Wang, X., Tong, M., Chinta, S. *et al.* (2006) Hypoxia-induced reactive oxygen species down-regulate ETB receptor-mediated contraction of rat pulmonary arteries. *American Journal of Physiology: Lung Cellular and Molecular Physiology*, **290**, L570–78.
157. Pastukh, V., Ruchko, M., Gorodnya, O. *et al.* (2007) Sequence-specific oxidative base modifications in hypoxia-inducible genes. *Free Radical Biology and Medicine*, **43**, 1616–26.
158. McCord, J.M. (1985) Oxygen-derived free radicals in postischemic tissue injury. *The New England Journal of Medicine*, **312**, 159–63.
159. Bayir, H. and Kagan, V.E. (2008) Bench-to-bedside review: mitochondrial injury, oxidative stress and apoptosis – there is nothing more practical than a good theory. *Critical Care*, **12**, 206.
160. Meerson, F.Z., Kagan, V.E., Kozlov, Y.P. *et al.* (1982) The role of lipid peroxidation in pathogenesis of ischemic damage and the antioxidant protection of the heart. *Basic Research in Cardiology*, **77**, 465–85.
161. Zweier, J.L., Flaherty, J.T., and Weisfeldt, M.L. (1987) Direct measurement of free radical generation following reperfusion of ischemic myocardium. *Proceedings of the National Academy of Sciences of the United States of America*, **84**, 1404–7.
162. Eppinger, M.J., Deeb, G.M., Bolling, S.F., and Ward, P.A. (1997) Mediators of ischemia–reperfusion injury of rat lung. *American Journal of Pathology*, **150**, 1773–84.

163. Georgieva, G.S., Kurata, S., Ikeda, S. *et al.* (2006) Prevention of ischemia reperfusion injury by positive pulmonary venous pressure in isolated rat lung. *Shock*, **25**, 66–72.
164. Davies, P.F. (1995) Flow-mediated endothelial mechanotransduction. *Physiological Reviews*, **75**, 519–60.
165. Ernstrom, G.G. and Chalfie, M. (2002) Genetics of sensory mechanotransduction. *Annual Review of Genetics*, **36**, 411–53.
166. Lehoux, S. and Tedgui, A. (2003) Cellular mechanics and gene expression in blood vessels. *Journal of Biomechanics*, **36**, 631–43.
167. Milovanova, T., Chatterjee, S., Hawkins, B.J. *et al.* (2008) Caveolae are an essential component of the pathway for endothelial cell signaling associated with abrupt reduction of shear stress. *Biochimica et Biophysica Acta*, **1783**, 1866–75.
168. Chatterjee, S., Levitan, I., Wei, Z., and Fisher, A.B. (2006) KATP channels are an important component of the shear-sensing mechanism in the pulmonary microvasculature. *Microcirculation*, **13**, 633–44.
169. Song, C., Al-Mehdi, A.B., and Fisher, A.B. (2001) An immediate endothelial cell signaling response to lung ischemia. *American Journal of Physiology: Lung Cellular and Molecular Physiology*, **281**, L993–1000, corrigenda, 1282: preceding L1167, 2002.
170. Zhang, Q., Chatterjee, S., Wei, Z. *et al.* (2008) Rac and PI3 kinase mediate endothelial cell-reactive oxygen species generation during normoxic lung ischemia. *Antioxidants and Redox Signaling*, **10**, 679–89.
171. Wei, Z., Manevich, Y., Al-Mehdi, A.B. *et al.* (2004) Ca^{2+} flux through voltage-gated channels with flow cessation in pulmonary microvascular endothelial cells. *Microcirculation*, **11**, 517–26.
172. Wei, Z., Al-Mehdi, A.B., and Fisher, A.B. (2001) Signaling pathway for nitric oxide generation with simulated ischemia in flow-adapted endothelial cells. *American Journal of Physiology: Heart and Circulatory Physiology*, **281**, H2226–32.
173. Hsieh, H.J., Cheng, C.C., Wu, S.T. *et al.* (1998) Increase of reactive oxygen species (ROS) in endothelial cells by shear flow and involvement of ROS in shear-induced c-fos expression. *Journal of Cellular Physiology*, **175**, 156–62.
174. Dreyfuss, D., Soler, P., Basset, G., and Saumon, G. (1988) High inflation pressure pulmonary edema. Respective effects of high airway pressure, high tidal volume, and positive end-expiratory pressure. *The American Review of Respiratory Disease*, **137**, 1159–64.
175. Dreyfuss, D. and Saumon, G. (1998) Ventilator-induced lung injury: lessons from experimental studies. *American Journal of Respiratory and Critical Care Medicine*, **157**, 294–323.
176. Mata-Greenwood, E., Grobe, A., Kumar, S. *et al.* (2005) Cyclic stretch increases VEGF expression in pulmonary arterial smooth muscle cells via TGF-beta1 and reactive oxygen species: a requirement for NAD(P)H oxidase. *American Journal of Physiology: Lung Cellular and Molecular Physiology*, **289**, L288–89.
177. Altemeier, W.A. and Sinclair, S.E. (2007) Hyperoxia in the intensive care unit: why more is not always better. *Current Opinion in Critical Care*, **13**, 73–78.
178. Smith, P. and Heath, D. (1975) The pathology of the lung in paraquat poisoning. *Journal of Clinical Pathology, Supplement (Royal College of Pathologists)*, **9**, 81–93.
179. Day, B.J., Shawen, S., Liochev, S.I., and Crapo, J.D. (1995) A metalloporphyrin superoxide dismutase mimetic protects against paraquat-induced endothelial cell injury, *in vitro*. *The Journal of Pharmacology and Experimental Therapeutics*, **275**, 1227–32.
180. Bianchet, M.A., Erdemli, S.B., and Amzel, L.M. (2008) Structure, function, and mechanism of cytosolic quinone reductases. *Vitamins and Hormones*, **78**, 63–84.
181. Sleijfer, S. (2001) Bleomycin-induced pneumonitis. *Chest*, **120**, 617–24.
182. Sebti, S.M., Mignano, J.E., Jani, J.P. *et al.* (1989) Bleomycin hydrolase: molecular cloning, sequencing, and biochemical studies reveal membership in the cysteine proteinase family. *Biochemistry*, **28**, 6544–48.
183. Day, R.M., Suzuki, Y.J., Lum, J.M. *et al.* (2002) Bleomycin upregulates expression of gamma-glutamylcysteine synthetase in pulmonary artery endothelial cells. *American Journal of Physiology: Lung Cellular and Molecular Physiology*, **282**, L1349–57.
184. Chen, J. and Stubbe, J. (2005) Bleomycins: towards better therapeutics. *Nature Reviews: Cancer*, **5**, 102–12.
185. Caspary, W.J., Lanzo, D.A., and Niziak, C. (1982) Effect of deoxyribonucleic acid on the production of reduced oxygen by bleomycin and iron. *Biochemistry*, **21**, 334–38.
186. Sarvazyan, N. (1996) Visualization of doxorubicin-induced oxidative stress in isolated cardiac myocytes. *American Journal of*

Physiology: Heart and Circulatory Physiology, **271**, H2079–85.

187. Wolf, M.B. and Baynes, J.W. (2006) The anti-cancer drug, doxorubicin, causes oxidant stress-induced endothelial dysfunction. *Biochimica et Biophysica Acta*, **1760**, 267–71.
188. Wang, Q. and Doerschuk, C.M. (2000) Neutrophil-induced changes in the biomechanical properties of endothelial cells: roles of ICAM-1 and reactive oxygen species. *Journal of Immunology*, **164**, 6487–94.
189. Rotstein, O.D. (2001) Oxidants and antioxidant therapy. *Critical Care Clinics*, **17**, 239–47.
190. Brennan, L.A., Steinhorn, R.H., Wedgwood, S. *et al.* (2003) Increased superoxide generation is associated with pulmonary hypertension in fetal lambs: a role for NADPH oxidase. *Circulation Research*, **92**, 683–91.
191. Wedgwood, S., Steinhorn, R.H., Bunderson, M. *et al.* (2005) Increased hydrogen peroxide downregulates soluble guanylate cyclase in the lungs of lambs with persistent pulmonary hypertension of the newborn. *American Journal of Physiology: Lung Cellular and Molecular Physiology*, **289**, L660–66.
192. Jancova, E., Vankova, Z., Honsova, E. *et al.* (2008) Prevalence and risk of hypertension in renal disease – data from the Czech Registry of Renal Biopsies. *Kidney and Blood Pressure Research*, **31**, 135–42.
193. Chen, M.J., Chiang, L.Y., and Lai, Y.L. (2001) Reactive oxygen species and substance P in monocrotaline-induced pulmonary hypertension. *Toxicology and Applied Pharmacology*, **171**, 165–73.
194. Hoshikawa, Y., Ono, S., Suzuki, S. *et al.* (2001) Generation of oxidative stress contributes to the development of pulmonary hypertension induced by hypoxia. *Journal of Applied Physiology*, **90**, 1299–306.
195. Cracowski, J.L., Cracowski, C., Bessard, G. *et al.* (2001) Increased lipid peroxidation in patients with pulmonary hypertension. *American Journal of Respiratory and Critical Care Medicine*, **164**, 1038–42.
196. Bowers, R., Cool, C., Murphy, R.C. *et al.* (2004) Oxidative stress in severe pulmonary hypertension. *American Journal of Respiratory and Critical Care Medicine*, **169**, 764–69.
197. Haniuda, M., Dresler, C.M., Mizuta, T. *et al.* (1995) Free radical-mediated vascular injury in lungs preserved at moderate hypothermia. *Annals of Thoracic and Cardiovascular Surgery*, **60**, 1376–81.
198. Kelly, R.F., Murar, J., Hong, Z. *et al.* (2003) Low potassium dextran lung preservation solution reduces reactive oxygen species production. *Annals of Thoracic and Cardiovascular Surgery*, **75**, 1705–10.

18

Hypoxia and the Pulmonary Endothelium

Matthew Jankowich, Gaurav Choudhary and Sharon Rounds

Vascular Research Laboratory, Alpert Medical School of Brown University, Providence VA Medical Center, Providence, RI, USA

INTRODUCTION

Cellular responses to oxygen are critical for normal energy metabolism, mediator release, proliferation, and survival. The lung has three sources of oxygen – from inspired gas, from the bronchial circulation, and from systemic venous blood returned to the lung via the right ventricle. The endothelia of conduit pulmonary arteries and veins are not exposed to oxygen in alveoli and the endothelium of small pulmonary blood vessels does not benefit from the bronchial circulation. The lung has unique responses to hypoxia – arterial vasoconstriction (hypoxic pulmonary vasoconstriction, see Chapter 12), vascular remodeling (see Chapters 11 and 27), and increased fluid flux into tissues (pulmonary edema, see Chapters 8, 21, and 24). Owing to these unique pulmonary physiologic responses to hypoxia, in this chapter we focus on the effects of hypoxia on pulmonary microvascular and arterial endothelium. Less is known about effects of hypoxia on pulmonary venous endothelium and endothelium of bronchial vessels (see Chapter 14).

Hypoxia is generally defined as a pO_2 less than 60 torr or blood oxygen saturation less than 90%, based on the sigmoid shape of the oxyhemoglobin desaturation curve. However, in the lung, endothelium of large pulmonary arteries is "normally" exposed to oxygen from mixed venous blood with a pO_2 about 40 torr, while microvascular endothelial cells (ECs) are exposed to both mixed venous oxygen and alveolar pO_2 of about 100 torr at sea level. Thus, it is not surprising that there is heterogeneity in the response of lung vascular endothelium to hypoxia, depending upon the type of blood vessel.

Studies utilizing cultured pulmonary ECs have been very important in understanding responses to hypoxia. However, studies of cultured cells are confounded by the fact that tissue culture media do not have the same oxygen carrying capacity as hemoglobin, oxygen can diffuse through tissue culture plastic, and long-term studies of hypoxia may necessitate intermittent return of cultures to room air conditions when the medium is changed. Indeed, while intact lungs display hypoxic vasoconstriction with ventilation by gas of F_IO_2 of 12%, it may be necessary to expose cultured ECs to oxygen concentrations of 3% or less to achieve a tissue culture media pO_2 of less than 60 torr. In addition, it is likely that intermittent hypoxia has more profound effects on reactive oxygen species (ROS) than sustained hypoxia [1]. Thus, interpretation of cultured cell studies requires careful consideration of experimental conditions.

HYPOXIA AND PULMONARY EC METABOLISM, VIABILITY, AND PROLIFERATION

In an early study from Una Ryan's laboratory, Cummiskey *et al.* compared responses to hypoxia of bovine pulmonary artery ECs (BPAECs) and bovine aortic ECs (BAECs) with respect to bioenergetic enzyme activities (pyruvate kinase, phosphofructokinase, and cytochrome oxidase) [2]. They noted increased glycolytic enzyme activity upon exposure to pO_2 15 torr for 48–96 h, but found no differences between the two cultured cell types. They noted increased glycolytic enzyme activity in freshly isolated intimal strips from bovine pulmonary artery when compared to aorta strips, suggesting that increased glycolysis is also seen under hypoxic conditions *in vivo*.

Lee and Fanburg reported that BPAECs exposed to 3 or 0% oxygen for up to 72 h displayed decreased cell proliferation and increased lactate release, but no change in ATP content, indicating a capacity to respond

The Pulmonary Endothelium: Function in health and disease Editors Norbert F. Voelkel, Sharon Rounds

to hypoxia with glycolysis [3]. Tretyakov and Farber compared hypoxia-tolerant BPAECs to immortalized opossum renal tubular ECs, which are more sensitive to hypoxia [4]. They found that the pulmonary artery cells exposed to 0% oxygen for up to 18 h were not damaged, displayed increased adenosine and guanosine uptake, and did not decrease cell ATP levels over 18 h hypoxic exposure. Farber *et al.* have further demonstrated that hypoxia-tolerant cultured main pulmonary artery endothelial cells express a specific set of stress proteins [5], including glyceraldehyde 3-phosphate dehydrogenase [6], non-neuronal enolase [7], and protein disulfide isomerase [8]. Thus, it is apparent that PAECs that are exposed to low environmental oxygen *in vivo* are tolerant of hypoxia and can upregulate enzymes that enhance glycolytic capacity and activity of the transcription factor hypoxia-inducible factor (HIF)-1α [9].

Farber *et al.* have demonstrated that BPAECs and BAECs are both capable of proliferation and retain responsiveness to hypoxic stimuli when cultured long-term (5 days to 16 weeks) under hypoxic conditions [4, 10, 11]. However, the rate of cell proliferation is slowed by hypoxia. Interestingly, lung microvascular ECs have recently been reported to have a proproliferative and vasculogenic phenotype [12, 13]. Since ECs from lungs of patients with pulmonary artery hypertension also replicate rapidly and display enhanced glycolytic capacity [14], it will be interesting to determine if there is a correlation between EC proliferative capacity and bioenergetics.

In summary, ECs from conduit pulmonary arteries are tolerant of hypoxia, and are able to enhance glycolysis and proliferate under hypoxic conditions. Further research is needed to determine if there is heterogeneity in these responses among ECs from different parts of the pulmonary vasculature (see Chapter 9).

HYPOXIA SENSOR(S)

The pulmonary EC sensor(s) for hypoxia are not well described. The pulmonary microvascular EC is appropriately positioned to sense alveolar hypoxia, thereby stimulating hypoxic pulmonary vasoconstriction of precapillary vessels of 60–100 μm internal diameter. However, it is now generally accepted that pulmonary vascular smooth muscle is the primary sensor cell for hypoxic vasoconstriction, while the EC is capable of modulating the vasoconstrictor response by mediator release (see Chapters 6 and 12) [15, 16]. Nevertheless, it is useful to review the various hypoxia sensors that have been proposed since it is possible that these sensors also function in lung ECs (Table 18.1).

Ward has categorized putative oxygen sensors as bioenergetic oxygen sensing mechanisms and biosynthetic oxygen sensing mechanisms [17]. Among the bioenergetic sensors are mitochondrial ROS production, ATP production, and redox state (see Chapter 17). There is controversy as to whether hypoxia is sensed via increased mitochondrial ROS production from electron transport [18] or via decreased mitochondrial ROS production resulting in a more reduced redox state and inhibition of O_2-sensitive K_v channels [16]. Previously investigators used chemical inhibitors of oxidative phosphorylation to assess the role of ATP production in oxygen sensing [19]. However, the moderate degrees of hypoxia that elicit physiologically significant responses are not sufficient to suppress mitochondrial ATP production. Thus, mitochondrial ATP production is probably not an important sensor of hypoxia *in vivo*.

Table 18.1 Candidates for hypoxia sensors

Bioenergetic sensing mechanisms
Mitochondrial ROS
ATP production
Redox state
Biosynthetic sensing mechanisms
ROS from NOXs
CO from heme oxygenases
H_2S from cystathione β-synthase and cystathione γ-ligase
Cytochrome P450 monooxygenases
HIF-1α

Among the biosynthetic sensing mechanisms are NADPH oxidases (NOXs), inhibition of which could result in decreased ROS production. However, mice deficient in the gp91^{phox}-containing NOX, NOX2, had decreased ROS production, but preserved pulmonary hypoxic vasoconstriction [16]. Pulmonary EC NOXs are similar to phagocyte NOXs and have been shown to play a role in ROS production in a variety of circumstances, such as inflammation and ischemia–reperfusion injury (see Chapter 17). However, there is no evidence that EC NOXs are important in sensing of hypoxia.

Heme oxygenases, HO-1, -2, and -3, have been suggested as oxygen sensors since they degrade heme to CO and biliverdin and Fe(II) in the presence of oxygen and NADPH [17], and since HO-1 and -2 are expressed in pulmonary arteries [20]. In rat PAECs, HO-1 has been localized to plasma membrane caveolae in association with caveolin-1 [21]. Thus, EC caveolae may act as a functional unit for HO-1 activity with modulation by caveolin-1. It is possible that HOs modulate pulmonary vasoconstrictor hypoxic responses via the product CO stimulating production of vasoconstrictor, endothelin [20]. However, knockdown or inhibition of HO-1 and -2 did not prevent hypoxic vasoconstriction of pulmonary arteries [20].

Cytochrome P450 monooxygenases include a large number of oxygen sensitive enzymes. Most attention has been paid to those metabolizing arachidonic acid. Among the products of ω-hydroxylases are hydroxyeicosatetraenoic acids (19- and 20-hydroxyeicosatetraenoic acid-HETE) and of epoxygenases are *cis*-epoxyeicosatrienoic acids (EETs). However, arachidonic acid availability, rather than oxygen tension may be rate-limiting for these enzymes. In elegant studies from the laboratory of Elizabeth Jacobs, cytochrome P450 4A (CYP4A) protein and mRNA have been localized in PAECs which also possess the capacity to synthesize the pulmonary vasodilator, 20-HETE [22].

Hydrogen sulfide is another possible oxygen sensor [23]. Like nitric oxide (NO) and CO, it is a gaseous molecule, soluble in tissues, and it is enzymatically generated in blood vessels in an oxygen-dependent manner. H_2S is generated from cysteine via cystathione β-synthase and cystathione γ-lyase. H_2S is a systemic vasodilator, like NO [24]. The effects of H_2S may be mediated by ATP-sensitive K^+ channels, by interaction with heme proteins such as cyclooxygenase, or by interactions with NO [25]. Pulmonary artery ECs respond to H_2S generation with increased NOX activity [26], suggesting that the pulmonary endothelium is capable of responding to H_2S.

HYPOXIA AND GENE TRANSCRIPTION

The transcription factor HIF-1α induces expression of genes involved in erythropoiesis, angiogenesis, and ion channel expression [27]. The mechanism of oxygen sensing by HIF-1α involves oxygen control of degradation of HIF-1α. HIF-1α is ubiquinated and degraded in proteosomes when bound to von Hippel–Lindau tumor suppressor protein, which requires proline hydroxylation. Pro564 and Pro402 of HIF-1α are hydroxylated by oxygen-dependent prolyl-hydroxylase-1 to -3 with K_m for O_2 slightly above atmospheric concentrations [28]. The Asp803 of HIF-1α is hydroxylated also in an O_2-dependent manner by factor-inhibiting HIF-1. Thus, hypoxia prevents degradation of HIF-1α and thereby facilitates gene transcription.

Via HIF-1α action, hypoxia induces endothelial gene expression of vasoactive and angiogenic factors, including endothelin [29], platelet-derived growth factor (PDGF) [30], inducible (type II) NO synthase (nitric oxide synthaseNOS) [31], and thrombospondin [32]. Among the angiogenic factors are vascular endothelial growth factor (VEGF), angiopoietin-1 and -2, placental growth factor, and PDGFβ. Manalo *et al.* have investigated gene expression (transcriptome) induced by hypoxia and/or by overexpression of HIF-1α in PAECs [9]. Remarkably, they found that more than 2% of all genes in human ECs are regulated by HIF-1α. The induced genes included oxidoreductases, collagens/modifying enzymes, cytokines/growth factors, receptors, signal transduction proteins, and transcription factors. Genes suppressed by hypoxia in PAECs included those involved with cell proliferation, RNA binding and metabolism, and protein ubiquitination and proteosomal degradation.

Using serial analysis of gene expression (SAGE), Choi *et al.* have assessed the effects of short-term hypoxia (1% oxygen for 8 and 24 h) on human pulmonary artery and aortic ECs derived from a single donor and maintained in tissue culture under identical conditions [33]. They found that hypoxia increased expression of stress-response genes, proapoptotic genes, and genes encoding extracellular matrix factors. Surprisingly, hypoxia increased expression of genes encoding antiproliferative factors in pulmonary artery endothelium. SAGE analysis demonstrated differences between human aortic and PAEC responses to hypoxia. For example, hypoxia decreased expression of pulmonary endothelial genes encoding proteins involved in oxidative energy production, such as ATP synthase, and decreased transcription of a transcriptional regulator of glycolytic genes. This is consistent with studies indicating increased glycolysis in hypoxic PAECs described above.

OTHER EFFECTS OF HYPOXIA ON PULMONARY ENDOTHELIUM

Hypoxic exposure changes the cellular morphology of pulmonary ECs. Bernal *et al.* have reported that rat pulmonary microvascular ECs contract reversibly when exposed to anoxic gas which reduced the medium pO_2 to 13 ± 2 torr [34]. These results suggest that the EC cytoskeleton contracts in response to acute hypoxia and that this contractility may contribute to hypoxic constriction of partially muscularized or nonmuscularized small pulmonary vessels.

Exposure of PAECs to more sustained hypoxia (1.5% v/v oxygen for 4 days) caused enlargement (megalocytosis) of cultured PAECs with enlargement of the Golgi [35]. These changes were accompanied by the loss of cell surface endothelial NOS (endothelial nitric oxide synthaseeNOS) and appearance of eNOS in the cytoplasmic compartment in Golgi and endoplasmic reticulum, and loss of NO production at the cell surface. Furthermore, eNOS colocalized with Golgi tethers and SNARES. Similar changes were seen with senescent cultured ECs and with cells treated with monocrotaline – an agent causing pulmonary hypertension in animal models. Similarly, Murata *et al.* described loss of eNOS from the cell membrane in "atrophied" PAECs from rats exposed to 1 week of hypoxia [36]. Owing to these changes and reported ultrastructural changes in pulmonary ECs in pulmonary hypertension, it has been suggested that dysfunctional intracellular trafficking of eNOS in pulmonary ECs might contribute to the pathogenesis of pulmonary hypertension [37]. Since optimal function of eNOS and vasodilator NO production requires appropriate protein–protein interactions (see Chapter 6), it is possible that reduced NO synthesis by hypoxic pulmonary ECs is due to effects of hypoxia on eNOS intracellular trafficking.

Farber *et al.* have described interesting differences in responses to hypoxia between cultured systemic ECs and PAECs. ECs from bovine systemic arteries responded to exposure to 10% oxygen (pO_2 85 torr) and 3% oxygen (pO_2 51 torr) with secretion of lipid-derived neutrophil chemoattractant activity, while main PAECs were less sensitive, requiring 0% oxygen (pO_2 32 torr) [38]. Similarly, PAECs were less responsive to hypoxia than aortic ECs in induction of lipid bodies [39]. Lipid bodies are non-membrane-bound, lipid-rich cytoplasmic inclusions that are an intracellular store of fatty acids and may be a nonmembrane site of eicosanoid formation. Finally, Farber has reported that cultured PAECs are slower than aortic ECs in synthesis of prostacyclin and thromboxane in response to acute hypoxia [40]. These studies suggested that main PAECs (that are exposed to lower pO_2 *in vivo*) are less responsive to hypoxic stimuli than ECs from the systemic vasculature, supporting the concept of heterogeneity of endothelium, depending upon the vascular bed (see Chapter 9).

Hypoxia regulates production of polyamines by PAECs [41]. The polyamines, putrescine, spermidine, and spermine, are low-molecular-weight compounds that are required for cell growth and differentiation, and may modulate other cell activities. Lung polyamine contents are increased in hypoxia. PAECs increase polyamine uptake with hypoxic exposure, although there is a decrease in the activity of the rate-limiting enzyme in polyamine synthesis, ornithine decarboxylase.

Hypoxia also modulates the production of heparan sulfates by PAECs [42, 43]. Heparan sulfates are cell surface-associated proteoglycans that help maintain an antithrombotic EC surface by catalyzing thrombin inactivation by antithrombin III. Karlinsky *et al.* reported that hypoxic exposure (3% oxygen for 24 h) decreased heparan sulfate production by both pulmonary artery and aortic ECs [43].

INTERMITTENT VERSUS SUSTAINED HYPOXIA AND PULMONARY ENDOTHELIAL CELLS

Sustained hypoxia complicates high-altitude exposure and lung diseases, such as chronic obstructive pulmonary disease and interstitial pulmonary fibrosis. Chronic intermittent hypoxia is seen in the common condition, obstructive sleep apnea, in which brief apneas or hypopneas during sleep result in frequent, intermittent decreases in oxygen saturation. Sustained hypoxia causes pulmonary hypertension and right ventricular failure, but does not increase systemic blood pressure. On the other hand, intermittent hypoxia results in more modest degrees of pulmonary hypertension, but sustained systemic hypertension, myocardial ischemia, and neuronal injury [1]. Studies of non-ECs indicate that the degree of oxidative stress and inflammation may be greater with intermittent hypoxia, as compared to sustained hypoxia [1]. Studies of gene transcription in rat lungs showed that intermittent hypoxia induced genes involved in ion transport and homeostasis, neurological processes, and steroid hormone receptor activity [44], while sustained hypoxia induced genes principally participating in immune responses. Transcriptional responses to chronic intermittent hypoxia [45] and post-translational protein modifications during chronic intermittent hypoxia [46] are just beginning to be understood. For example, intermittent hypoxia has been shown to increase HIF-1α phosphorylation in cultured ECs via protein kinase A [47]. Little is known regarding effects of intermittent hypoxia on pulmonary ECs.

HYPOXIA AND PULMONARY VASCULAR PERMEABILITY

Pulmonary ECs can modulate vasoconstriction and the proliferation of adjacent vascular smooth muscle. The effects of the hypoxic pulmonary endothelium on vasoreactivity are described in Chapter 12, while effects on pulmonary vascular remodeling are described in Chapter 11. In this chapter we focus on hypoxia effects on pulmonary endothelium that result in changes in lung vascular permeability.

The effect of hypoxia on permeability of the pulmonary endothelial barrier has been a topic of controversy for decades. A variety of experimental models, ranging from *in vivo* animal studies to isolated perfused lung models, to studies of cultured endothelial monolayer permeability, have attempted to address the question of whether hypoxia alone directly alters pulmonary endothelial barrier function. This question is most directly relevant to the study of the pathogenesis of high-altitude pulmonary edema (HAPE) – the most common situation in which global alveolar hypoxia occurs and a condition in which altered vascular permeability is implicated. In addition, in 1942, Madeline Warren and Cecil Drinker, pioneers in the study of hypoxic pulmonary vascular permeability, postulated that pulmonary edema caused by regional hypoxia could be conceived to contribute to "a vicious circle" of regional hypoxia leading to localized pulmonary edema, resulting in further impairment of gas exchange and worsening of hypoxemia [48].

EFFECTS OF HYPOXIA ON CULTURED PULMONARY EC MONOLAYER PERMEABILITY

Understanding of molecular pathways involved in endothelial permeability has expanded tremendously in recent years. *In vitro* studies using cultured ECs have demonstrated alterations in endothelial monolayer permeability under controlled hypoxic conditions and tentative elucidation of the mechanisms involved.

Alterations in endothelial monolayer permeability with hypoxia were initially demonstrated *in vitro* using cultured BAECs [49]. In this study, permeability of the endothelial monolayer to radiolabeled macromolecules was increased after 24 h in hypoxia. The relative increase in permeability was dependent on both the duration and the degree of hypoxia, and was reversible within 48 h of restoration of normoxia. The permeability changes were associated histologically with the formation of intercellular gaps and alterations in the actin cytoskeleton (see Chapter 8). A mild increase in monolayer permeability to albumin was demonstrated after only 90 min of exposure to a similar level of hypoxia in another study using BPAECs [50]. In this study, reoxygenation worsened barrier function, an effect prevented by antioxidants. Increased monolayer permeability to dextran was seen within 1 h of exposure to hypoxia in experiments with porcine PAECs [51]. Other work utilizing bovine pulmonary microvascular ECs demonstrated that ECs derived from the pulmonary microcirculation also responded to hypoxia with increased permeability after 4 h of hypoxia, associated with the formation of intercellular gaps and stress fiber formation. However, after 24 h of hypoxic incubation there was restoration of barrier function and resolution of intercellular gaps [52]. In this study, the oxygen content of the tissue culture medium at 24 h was greater than at 4 h, raising the question of whether the improvement in permeability with more prolonged hypoxic EC incubation was related to the apparent increase in available environmental oxygen. Pulmonary ECs derived from animals exposed to chronic hypoxia after birth displayed increased monolayer permeability even under normoxic conditions, suggesting that chronic hypoxic exposure induced persistent effects on endothelial permeability [51]. In summary, studies of cultured pulmonary ECs have established that endothelial monolayer barrier function is impaired by hypoxia alone in a dose–response relationship and that monolayer permeability changes following acute hypoxia were generally reversible following a return to normoxia. These principles derived from tissue culture experiments are helpful in interpreting the results of *in vivo* experiments of pulmonary vascular permeability using widely varying levels of hypoxia and conducted over various time courses.

CELL SIGNALING AND PULMONARY ENDOTHELIAL PERMEABILITY

Molecular transport across the endothelial barrier can occur via paracellular and transcellular pathways [53] (see Chapter 8). Most attention in hypoxia-induced endothelial permeability signaling has focused on paracellular transport involving signaling pathways which cause cell rounding and intercellular junction disassembly via regulation of the actin–myosin apparatus and cell junction stability. Morphologic changes in the actin cytoskeleton are seen following exposure of pulmonary arterial and microvascular ECs to hypoxia, with disassembly of the cortical actin band and formation of intracellular stress fibers [51, 52, 54], mediating changes in EC shape during hypoxia. Intercellular junctions are dispersed during hypoxia [51], allowing intercellular gaps to form [54]. These cytoskeletal rearrangements, well recognized following endothelial exposure to other permeability enhancing agonists such as thrombin, allow for increased paracellular permeability of the EC monolayer to small and large molecules under hypoxic conditions.

Multiple intracellular signaling pathways influence endothelial barrier maintenance and permeability, including signaling via cAMP, small GTPases, p38 mitogen-activated protein kinase (MAPK) and ROS; many of these systems have been demonstrated to influence endothelial permeability in hypoxia. Hypoxia-induced BPAEC monolayer permeability was associated with decreases in cAMP and adenylate cyclase activity, and cAMP analogs or activators of adenylate cyclase could restore barrier function [54]. Dexamethasone prevented the increase in monolayer permeability if given before or at the time of exposure to hypoxia, and prevented the decrease in cAMP seen with hypoxia exposure, but could not completely restore barrier function if given after exposure to hypoxia for 12 h or more [54]. In homogenized lung tissue preparations exposed to hypoxia, no decrease in cAMP content was observed compared with normoxic lung preparations, but hypoxic perfused lung preparations showed decreased ability to synthesize cAMP in response to terbutaline, as measured by lung perfusate cAMP levels [55]. These results support a role for cAMP second messenger signaling in the maintenance of the pulmonary vascular barrier in normoxia, whereas decreases in adenylate cyclase activity and secondarily cAMP result in hypoxia-induced alterations in barrier permeability.

The mechanisms leading from hypoxia to altered adenylate cyclase activity in hypoxia likely involve Ca^{2+} – an inhibitor of selected adenylate cyclase isoforms (Figure 18.1). Hypoxia leads to a transient spike in intracellular calcium content in BPAECs, followed by a higher baseline calcium level [56]. Hypoxia induces an increase in cytosolic calcium in human umbilical vein ECs (HUVECs) as well [57]. Increases in intracellular calcium have been shown to inhibit BPAEC cAMP production; BPAECs express a Ca^{2+}-inhibitable isoform of adenyl cyclase [58]. Intracellular Ca^{2+} levels and cAMP activity are inversely related in ECs, and normoxic monolayer permeability results from increased intracellular Ca^{2+} via decreased cAMP [58]. Intracellular Ca^{2+} concentration changes induced by hypoxia are likely involved in mediating the decrease in EC adenylate cyclase activity observed in hypoxia (see Chapters 5 and 9 regarding pulmonary EC calcium).

Regulators of actin have been implicated in hypoxia-induced increases in endothelial monolayer permeability. The p38 MAPK is activated in hypoxic rat microvascular ECs [110]. A substrate of p38, MK2, a protein kinase activated by hypoxia, appears to regulate actin redistribution in hypoxic pulmonary microvascular ECs [111]. Inhibition of p38 MAPK attenuates the permeability changes induced by hypoxia in both micro- and macrovascular pulmonary ECs [59]. Overexpression of the p38 substrate MK2 leads to analogous cytoskeletal changes to those seen in hypoxia and expression of dominant-negative MK2 blocks hypoxia-induced actin reorganization. Heat shock protein HSP27 appears to mediate the interaction between MK2 and the actin cytoskeleton [111]. Thus, the p38 pathway appears to regulate cytoskeletal alterations mediating microvascular endothelial monolayer permeability in hypoxia (Figure 18.1).

Rho GTPases are also among the key regulators of the actin cytoskeleton [60]. In hypoxia, activity of the small GTPase Rac1 falls while conversely RhoA activity increases in PAECs [61]. (Figure 18.1) Inhibitors of RhoA and its downstream effector, RhoA kinase, prevent actin redistribution seen with hypoxia, while Rac1 inhibitors prevent recovery of barrier function following reoxygenation, suggesting differential roles of these interrelated small GTPases in barrier regulation [61]. ROS produced via the NOX pathway appear to be critical regulators of small GTPase activity in lung ECs [61]. The role of small GTPases in regulating the cytoskeletal response of ECs to hypoxia is analogous to their role in regulating cytoskeletal rearrangements leading to permeability changes induced by inflammatory stimuli.

The role of ROS in hypoxia-induced signaling cascades associated with endothelial monolayer permeability changes is incompletely understood. Antioxidants can prevent the increase in permeability of monolayers of HUVECs associated with hypoxia [62] as well as reoxygenation [50]. Endothelial-derived interleukin (IL)-6, via autocrine and paracrine pathways, acts downstream of ROS to effect changes in HUVEC monolayer permeability in a finely tuned mechanism sensitive to interleukin-6 levels [62]. However, IL-6 production seems unlikely to represent the sole effector mechanism in permeability changes induced by ROS in hypoxia. There is in fact contradictory evidence regarding the effects of hypoxia on free radical production in ECs. In ECs derived from porcine pulmonary arteries, ROS are decreased in the setting of hypoxia (3% O_2) of 1 h duration [51], whereas in HUVECs, ROS formation is increased by hypoxia (1% O_2) within 2 h [62]. Both decreased ROS production and increased ROS production have been implicated in initiating different intracellular signaling pathways involved in endothelial barrier function changes. These differing observations may be related to species differences, EC vascular bed/tissue origin differences, or the specific experimental conditions and techniques employed. Further work is needed to better comprehend ROS signaling as related to endothelial permeability changes in hypoxia.

Potential extracellular stimulants of hypoxia-induced pulmonary vascular permeability include VEGF and inflammatory cytokines such as tumor necrosis factor (TNF)-α (Figure 18.1) TNF is a well-recognized vascular permeability agonist [53]. Hypoxia induces TNF-α production by pulmonary ECs, especially microvascular ECs, which may result in autocrine or paracrine effects on endothelial permeability [59], amplifying the permeability effect of hypoxia. As noted above, IL-6 is another proinflammatory cytokine that has been implicated in hypoxia-induced permeability and ROS produced by inflammatory cells recruited to hypoxic endothelium may provoke endothelial permeability alterations as well. There is considerable overlap between the EC molecular and phenotypic responses to hypoxia and to inflammation, and shared signaling pathways are likely involved in the increased vascular permeability seen in both conditions. There is emerging evidence that the principal transcriptional pathways in inflammation, governed by nuclear factor-κB (NF-κB), and in hypoxia, governed by HIF-1α, are linked by molecular cross-talk [63]. Evidence from non-EC models suggests a dependence of HIF-1α transcription on NF-κB [64] as well the ability of HIF-1α to induce NF-κB expression [65, 66]. If these findings are extended to ECs, this would help explain the characteristic induction of inflammatory responses, including permeability alterations, by hypoxic ECs.

VEGF also increases vascular permeability [67] through effects on endothelial barrier function [68].

Indeed, VEGF was originally called "vascular permeability factor". VEGF signaling has been implicated in the increased vascular permeability edema seen in ischemia–reperfusion lung injury [69]. VEGF expression is upregulated by hypoxia via HIF-1α in many cell types, including cultured HUVECs and lung epithelial cells, and *in vivo* [70–72]. This suggests the possibility of paracrine stimulation of lung ECs by VEGF leading to increased vascular permeability in the setting of hypoxia. The role of autocrine stimulation has been challenged as lung ECs were not seen to produce VEGF *in vivo* [73], although the capability of lung microvascular ECs to produce VEGF has been demonstrated [74]. The relevance of VEGF to increased vascular permeability *in vivo* is questionable, as hypoxia upregulates VEGF receptor (vascular endothelial growth factor receptorVEGFR)-1 expression in lung ECs [73], but VEGFR-2 signaling seems most relevant to vascular permeability, with VEGFR-2 stimulation resulting in alterations in the integrity of adherens junctions [75]. *In vivo*, serum venous [76] and capillary [77] VEGF levels do not increase in hypobaric hypoxia, even in human subjects with altitude sickness, though these values may not reflect local lung expression levels of VEGF. While VEGF is a plausible candidate molecule for affecting vascular permeability changes in hypoxia, the role played by VEGF signaling in inducing increased lung endothelial permeability in hypoxia is at present uncertain.

Bradykinin induces vascular permeability through pathways not involving Rho GTPase or myosin light chain kinase [53]. Bradykinin does not potentiate aortic endothelial monolayer permeability induced by hypoxia [78]. However, neprilysin, an enzyme present in the lung which degrades bradykinin, is downregulated by hypoxia in rats; hypoxic exposure of rats was associated with increased lung vascular permeability; the increased lung vascular permeability of hypoxia correlated with the decrease in neprilysin expression [79]. This suggests a possible role for unopposed bradykinin and/or substance P (a related neuropeptide also degraded by neprilysin) signaling in hypoxia-induced lung vascular permeability (Figure 18.1).

In summary, exposure to environmental hypoxia induces alterations in cytoskeletal arrangement and intracellular junction disassembly in lung and other ECs, leading to increased paracellular permeability to small and large molecules. This suggests that increased permeability pulmonary edema *in vivo*, induced by hypoxia alone, is indeed plausible. Mechanisms involved in mediating monolayer permeability changes have been examined in some detail; parallels to signaling pathways involved in vascular permeability induced by other agonists have been noted. The requirement for prolonged duration of hypoxia (hours) suggests that increased monolayer permeability requires protein expression.

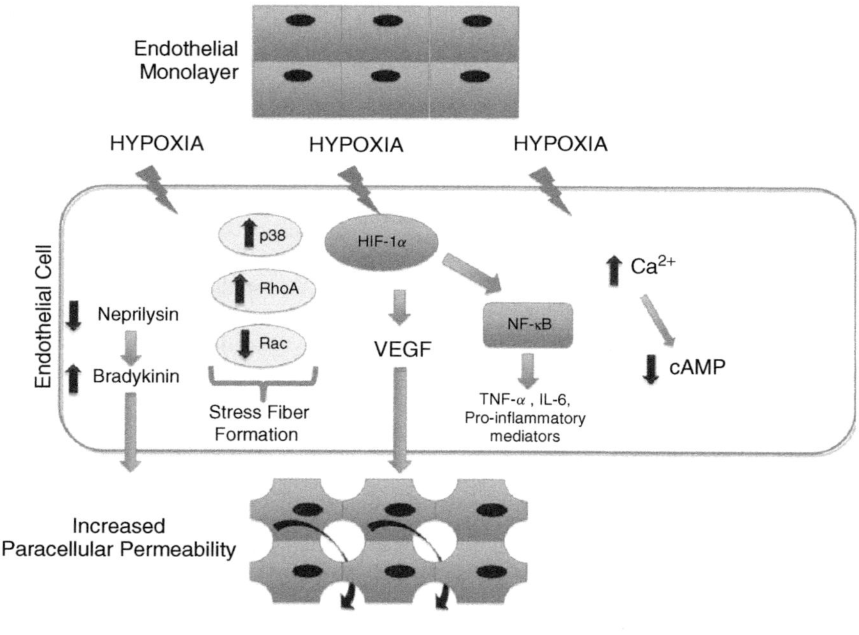

Figure 18.1 Signaling of hypoxia-induced increases in pulmonary endothelial paracellular permeability.

EFFECT OF HYPOXIA ON VASCULAR PERMEABILITY OF ISOLATED PERFUSED LUNGS

Isolated perfused lung models of hypoxia-induced vascular permeability edema are helpful in that the isolated perfused lung model allows manipulation of the perfusing pressure – a factor which confounds most *in vivo* studies given the hypoxic vasoconstrictive response. Most [55, 80, 81], but not all [82, 83], have demonstrated alterations in pulmonary vascular permeability or lung extravascular water content in isolated perfused lung preparations exposed to alveolar hypoxia. Kinasewitz *et al.* utilized isolated blood-perfused canine lungs, encased in a water impermeable membrane, and measured fluid and protein filtration into the artificial pleural space created by the membrane in the presence and absence of hypoxia. They utilized a calcium channel blocker to prevent hypoxic vasoconstriction; hydrostatic pressures were similar in the hypoxic and normoxic lung preparations. In this study, the hydraulic conductivity doubled and the diffusional permeability of protein tripled under hypoxic conditions (0% O_2). Greater protein concentration was measured in the pleural fluid collected from hypoxic preparations, consistent with increased permeability of the pleural capillary endothelium [80]. In this study, xanthine oxidase inhibition prevented the increased permeability associated with hypoxia, implying a role for free radicals in inducing the permeability change [80]. Parker *et al.* demonstrated an increase in the pulmonary capillary filtration coefficient in hypoxic isolated perfused dog lungs; in this study, perfusion pressures were maintained constant and no increase in capillary hydrostatic pressures occurred during hypoxia [81]. The authors attributed the increase in filtration coefficient to increased vascular permeability, as an increase in surface area for exchange seemed unlikely in the constant pressure system. Dehler *et al.* utilized isolated perfused rat lungs perfused at constant pressures and exposed to varying levels of oxygen (1.5–35%). They measured lung edema formation by changes in weight, and observed an earlier weight gain in lungs exposed to hypoxia [55]. Bronchoalveolar lavage of hypoxic lung preparations demonstrated 2.5-fold greater protein content in the bronchoalveolar lavage fluid in hypoxia compared with normoxia. The authors interpreted their findings as being consistent with an increase in vascular permeability caused by hypoxia.

In contrast, a study by Aarseth *et al.* demonstrated no change in the water content of hypoxic isolated rat lung preparations compared with control lungs exposed to normoxia [82]. The contradictory results may be related to the brief (4-min) hypoxic exposures utilized by Aarseth *et al.*, which, based on *in vitro* and *in vivo* data, may have been too short to allow for permeability changes and significant increased fluid filtration to occur.

Overall, the data from isolated lung preparations are consistent with the notion that hypoxia increases lung vascular permeability and causes lung edema. These studies are consistent with studies of cultured pulmonary endothelial monolayers which also display increased permeability under hypoxic conditions. Isolated lung preparations are helpful, in that perfusate hydrostatic forces can be maintained at constant levels, thus allowing the examination of permeability changes in isolation from the hypoxia-induced changes in hydrostatic forces which occur *in vivo*. Isolated lung preparations may be limited in that isolation of the heart and lungs is associated with some delay in perfusion which may cause tissue injury via ischemia–reperfusion and potentially accentuate the effects of subsequent exposure to hypoxia.

EFFECT OF HYPOXIA ON LUNG VASCULAR PERMEABILITY IN ANIMALS

Animal models have generated the most controversy in the study of pulmonary vascular permeability in hypoxia. Initial studies in anesthetized, ventilated dogs by Warren and Drinker utilized the rate of lymphatic outflow from the lungs as a surrogate for the measurement of lung fluid filtration. They demonstrated rapid increases in lymphatic flow from the lungs following exposure to hypoxia (8.6% O_2), concluding that "the pulmonary capillaries are peculiarly susceptible to oxygen lack as a cause of increased permeability" [48]. Their hemodynamic data were limited, although in a subsequent study they demonstrated a fall in cardiac output with hypoxia, concluding that increased flow was not a cause of the increased lymphatic production observed during hypoxia [84]. Many animal studies examining lung permeability changes in hypoxia would follow Warren and Drinker's seminal work, with conflicting and confusing results.

A number of studies have demonstrated no alteration in pulmonary vascular barrier function in hypoxia [85–87]; other studies suggested that hypoxia only produced or exacerbated pulmonary edema due to an increase in hydrostatic forces and did not increase permeability *per se* [88, 89]. These studies raise the question of whether increased permeability edema due to hypoxia exists *in vivo*. However, many other experiments in animals have demonstrated an increase in pulmonary vascular permeability with hypoxia, including experiments in puppies [90], dogs [91], and rats [79, 92]. Rats clearly develop histological evidence of pulmonary edema with hypoxia, with initial perivascular edema after exposures of less than 3 h [93], which then progresses to frank alveolar edema accompanied by inflammation with longer exposure times [94]. Furthermore, pulmonary edema occurs in humans at altitude in the setting of hypobaric hypoxia and is associated with increased permeability of the pulmonary vascular barrier [95], although other factors, including altered hydrostatic forces, are clearly involved in this disease [96]. The balance of evidence suggests that in some species, including humans, exposure to hypoxia is associated with increased pulmonary vascular permeability and pulmonary edema.

The animal studies that have not shown evidence of hypoxia-induced increased permeability may be due in part to genetically determined species differences. For example, sheep are particularly resistant to vascular permeability changes caused by hypoxia, whereas rats appear more vulnerable [92]. This is plausible, given that different species, such as domestic cattle and yaks, have genetically determined differences in pulmonary circulatory responses to hypoxia, as well as a different morphology of ECs [97]. The exposure time in experiments that have failed to demonstrate pulmonary edema with hypoxia *in vivo* may have been too short; in rats, hypoxia-induced pulmonary edema is most prominent after 16 h of exposure [94]. This is intriguing, in that most humans with HAPE develop symptoms 12 h or more after ascent to altitude. Alterations in endothelial barrier function are not necessarily immediate, in some experiments taking 4 h [52] or more [49] to develop. This suggests that short exposure times [86, 87] may have been too brief to allow significant alterations in endothelial barrier function to occur. However, it is also likely that some species are resistant to the development of pulmonary edema with hypoxia even after relatively prolonged exposure times. For example, Bland *et al.* exposed three sheep to 10% O_2 for 48 h without finding any evidence of pulmonary edema on postmortem examination [85].

While a number of studies support the concept that hypoxia can alter pulmonary vascular permeability *in vivo*, there is a relative paucity of data to exclude the hemodynamic consequences of hypoxia as a cause of this increase in permeability. Stelzner *et al.* was able to show that a short term elevation of the pulmonary artery pressure caused by hypoxic pulmonary vasoconstriction did not affect the protein leak index in rats, whereas the measured protein leak index as well as gravimetric lung water did increase after 24–48 h of exposure to hypoxia [92]. In this study, dexamethasone reduced transvascular protein leak without affecting pulmonary hemodynamics, while adrenalectomy exacerbated the pulmonary vascular permeability. The authors concluded that increased hydrostatic pressures alone do not explain the vascular permeability induced by hypoxia. This is in keeping with the observations of hypoxia-induced increased vascular permeability in isolated perfused lung models in which perfusion pressures were kept constant [80, 81]. Therefore hemodynamic forces are not likely to be the sole determinants of increased pulmonary vascular permeability in hypoxia.

In summary, the balance of evidence from animal models, coupled with observations of pulmonary edema due to hypobaric hypoxia in humans, suggests that hypoxia can stimulate pulmonary edema formation in at least some species. The evidence from cultured cell studies and experiments with isolated perfused lung models coupled with observations in animal models demonstrates that, in susceptible species, hypoxia induces alterations in endothelial barrier function, even in the absence of alterations in hydrostatic forces, which lead to increased paracellular protein and solute leak, manifesting as increased permeability pulmonary edema. This paracellular leak does not manifest immediately, as demonstrated *in vitro*, requiring hours to occur and being seen *in vivo* following several hours of exposure to hypoxia.

HAPE AND ALTERED LUNG VASCULAR PERMEABILITY IN HYPOXIC HUMANS

Lung histology and protein content of bronchoalveolar lavage indicate that HAPE is an increased permeability type of pulmonary edema [95, 98, 99]. As the main site of hypoxic pulmonary vasoconstriction is known to be the precapillary arterioles, relating pulmonary edema to altered hydrostatic forces at the capillary level was conceptually difficult. Early hemodynamic data did not support elevation of the pulmonary capillary wedge pressure in patients afflicted with HAPE [100]. However, lowering of elevated pulmonary arterial pressures using vasodilator therapy improves oxygenation in this condition [101], suggesting a role for hydrostatic pressures in the pathogenesis of the pulmonary edema. Elevated pulmonary capillary pressures may occur in HAPE-susceptible subjects as measured by pulmonary artery pressure decay curves, even in the absence of elevations in the pulmonary capillary wedge pressure [102]. Reconciling precapillary vasoconstriction, which would protect the pulmonary capillaries from elevated hydrostatic pressures, with the pulmonary edema that occurs in HAPE has been accomplished through the hypothesis of heterogeneous vasoconstriction as initially postulated by Hultgren, discussed in Bartsch 96. Heterogeneous pulmonary vasoconstriction in response to hypoxia would cause regional elevations in pulmonary capillary pressures in vessels not protected by vasoconstriction. It is possible that capillary mechanical stress failure subsequently occurs in those unprotected capillaries, thereby explaining the increased permeability edema seen in this disorder [96]. Ischemia–reperfusion injury could also potentially occur in this setting as regions of lung with low perfusion subsequently become reperfused as regional vasoconstriction lessens. Defects in alveolar fluid clearance have also been proposed as an adjunctive mechanism, as heterogeneous pulmonary vasoconstriction alone may be insufficient to induce this disorder [103].

A role for altered endothelial permeability due to hypoxia in the pathogenesis of HAPE, regulated by cytoskeletal changes occurring at the EC, is attractive for several reasons. Increased vascular permeability caused by hypoxia takes hours to occur in most cultured cell and *in vivo* models, consistent with the observed delay in onset of HAPE for hours or even days after exposure to hypobaric hypoxia. In contrast, capillary stress failure due to increased pressures occurs within a few minutes of exposure to elevated hydrostatic pressures [104, 113]. Other conditions in which pulmonary capillary stress failure has been postulated to occur, such as neurogenic pulmonary edema and the pulmonary edema occurring with extreme exercise [113], are of rapid onset, consistent with the time course of capillary stress failure. The transmural pressures associated with capillary stress failure in animal models [104] are much higher than the presumed transmural forces suggested by the capillary pressures recorded in humans with HAPE [102]. As reviewed above (Section "Effects of Hypoxia on Cultured Pulmonary EC Monolayer Permeability" and "Cell Signaling and Pulmonary Endothelial Permeability"), altered endothelial barrier function induced by hypoxia is rapidly reversible upon exposure to normoxia, consistent with the reversibility of HAPE with oxygen or return to lower altitudes. Ready reversibility seems incompatible with the tissue breaks observed in animals exposed to high transmural pressures resulting in capillary stress failure. However, Elliot *et al.* have shown that exposure to low transmural pressures following exposure to high transmural pressures did result in fewer apparent stress breaks, suggesting that stress failure is reversible [105]. Disruptions of the alveolar-capillary barrier have been demonstrated in an animal model of HAPE [113], although it is not clear that the ultrastructural changes seen in this model are incompatible with the occurrence of increased permeability due to regulated cell–cell junction and membrane alterations in cells induced by hypoxia.

In summary, current data does not exclude a role for altered endothelial permeability due to hypoxia in HAPE; the time course of altered endothelial permeability regulated by cytoskeletal and junctional changes induced by hypoxia is more consistent with the time course observed in the development of HAPE in humans than the stress failure hypothesis. Hypoxia-induced, cytoskeletally regulated endothelial permeability changes would potentially explain the occurrence of HAPE in humans at relatively low capillary pressures and hypoxia-induced edema in isolated perfused lung models under conditions of controlled hydrostatic pressures (see Chapter 20 for pressure/flow-induced changes in pulmonary endothelial function).

Altered endothelial permeability due to hypoxia is compatible with the finding that lowering pulmonary artery pressures results in improvements in clinical parameters in HAPE. Increases in either intravascular hydrostatic forces and/or membrane permeability favor fluid filtration out of the vascular space per the Starling equation, and improvements in either or both of these parameters would result in decreased fluid filtration across the alveolar–capillary barrier. Vasoactive agents including inhaled NO [101] and the phosphodiesterase inhibitor tadalafil [106] are effective in treatment or prevention of HAPE, confirming, but not proving, a role for elevated hydrostatic forces in the formation of HAPE. Intriguingly, increases in cGMP mediated by NO and sildenafil may decrease endothelial barrier dysfunction induced by hypoxia *in vitro*, suggesting that these agents may have vasomotor tone-independent *in vivo* [107]. Dexamethasone, not conventionally regarded as a vasoactive agent, is effective in prophylaxis against the respiratory symptoms of acute mountain sickness [108] and in preventing HAPE [106]. Dexamethasone minimized the increase in pulmonary artery pressures occurring with exposure to hypobaric hypoxia [106] and has other effects *in vivo*, including the regulation of gene expression, anti-inflammatory properties, and effects on barrier function. The mechanism of action of dexamethasone in preventing HAPE remains incompletely understood; alterations in cGMP levels, inflammatory mediators, and vascular barrier function are all possible [106].

Agents effective at preventing HAPE may have pleiotropic effects in addition to their beneficial effects on pulmonary hemodynamics that contribute to their usefulness in this condition. Improved understanding of the mechanisms of altered endothelial permeability in HAPE holds the potential for novel prophylactic agents and treatments of HAPE that may contribute to the role of vasoactive agents in this condition.

CONCLUSIONS AND PERSPECTIVES

Pulmonary ECs respond to hypoxia and these responses may be important in modulating lung vascular responses to hypoxia. Although the EC sensor(s) for hypoxia are poorly defined, it is likely that they are similar to those demonstrated in other tissues, including pulmonary vascular smooth muscle. Since there are differences between ECs from pulmonary conduit arteries and systemic arteries, it is likely that ECs within the lung circulation will also differ in response to hypoxia, dependent upon the

pO_2 of the blood to which the EC is exposed. Little is known regarding the regional heterogeneity of lung EC responses to hypoxia or of the effects of intermittent hypoxia on lung ECs.

A wealth of data has accumulated implicating hypoxia as a stimulus for altered pulmonary vascular barrier function. The data suggest that hypoxia causes increased pulmonary endothelial permeability in cultured cell monolayers and *in vivo* in some species, including humans. Plausible cell signaling mechanisms have been explored in hypoxia-induced endothelial barrier alteration, with intriguing, but perhaps not unexpected, parallels to signaling pathways involved in cytoskeletal dynamics and barrier regulation in inflammation.

HIF-1α has emerged as a central mechanism in cellular responses to hypoxia, and recent evidence has linked HIF-1α regulation to NF-κB, one of the central transcriptional signaling mechanisms in inflammation, albeit in non-endothelial-based experimental systems. Coordination between hypoxia-induced gene regulation mediated by HIF-1α and the generation of inflammatory cytokines via NF-κB transcriptional control would help to explain the parallels seen in endothelial permeability alterations in inflammatory states and in hypoxia. Demonstrating evidence of coordination between the inflammatory and hypoxic responses in pulmonary endothelial models would provide intriguing new targets for ameliorating hypoxic lung injury, and would help to explain the role of nonvasomotor drugs such as dexamethasone in the prevention of HAPE. *In vivo*, inflammatory cytokine production, in response to hypoxia, would lead to leukocyte recruitment and interactions between leukocytes and endothelium; as discussed in Section "Other Effects of Hypoxia on Pulmonary Endothelium", Farber *et al.* have demonstrated the production of a neutrophil chemoattractant factor by hypoxic ECs. Such interactions between leukocytes and the endothelium add further complexity to the signaling milieu of the hypoxic endothelium.

The role of other cell signaling pathways in endothelial permeability induced by hypoxia, such as the unfolded protein response induced by endoplasmic reticulum stress, is unknown. Finally, the potential usefulness of barrier-protective mechanisms, such as sphingosine 1-phosphate [109], in the prevention or amelioration of hypoxia-induced endothelial permeability changes remains to be explored (see Chapter 21).

In summary, further understanding of the mechanisms involved in hypoxia-induced endothelial barrier dysfunction may contribute to the effective management of specific lung conditions associated with global and regional hypoxia.

References

1. Douglas, R.M. and Haddad, G.G. (2008) Can O_2 dysregulation induce premature aging? *Physiology (Bethesda)*, **23**, 333–49.
2. Cummiskey, J.M., Simon, L.M., Theodore, J. *et al.* (1981) Bioenergetic alterations in cultivated pulmonary artery and aortic endothelial cells exposed to normoxia and hypoxia. *Experimental Lung Research*, **2**, 155–63.
3. Lee, S.L. and Fanburg, B.L. (1987) Glycolytic activity and enhancement of serotonin uptake by endothelial cells exposed to hypoxia/anoxia. *Circulation Research*, **60**, 653–58.
4. Tretyakov, A.V. and Farber, H.W. (1995) Endothelial cell tolerance to hypoxia. Potential role of purine nucleotide phosphates. *The Journal of Clinical Investigation*, **95**, 738–44.
5. Zimmerman, L.H., Levine, R.A., and Farber, H.W. (1991) Hypoxia induces a specific set of stress proteins in cultured endothelial cells. *The Journal of Clinical Investigation*, **87**, 908–14.
6. Graven, K.K., Troxler, R.F., Kornfeld, H. *et al.* (1994) Regulation of endothelial cell glyceraldehyde-3-phosphate dehydrogenase expression by hypoxia. *The Journal of Biological Chemistry*, **269**, 24446–53.
7. Aaronson, R.M., Graven, K.K., Tucci, M. *et al.* (1995) Non-neuronal enolase is an endothelial hypoxic stress protein. *The Journal of Biological Chemistry*, **270**, 27752–57.
8. Graven, K.K., Molvar, C., Roncarati, J.S. *et al.* (2002) Identification of protein disulfide isomerase as an endothelial hypoxic stress protein. *American Journal of Physiology: Lung Cellular and Molecular Physiology*, **282**, L996–L1003.
9. Manalo, D.J., Rowan, A., Lavoie, T. *et al.* (2005) Transcriptional regulation of vascular endothelial cell responses to hypoxia by HIF-1. *Blood*, **105**, 659–69.
10. Farber, H.W. and Rounds, S. (1990) Effect of long-term hypoxia on cultured aortic and pulmonary arterial endothelial cells. *Experimental Cell Research*, **191**, 27–36.
11. Tucci, M., Hammerman, S.I., Furfaro, S. *et al.* (1997) Distinct effect of hypoxia on endothelial cell proliferation and cycling. *American Journal of Physiology: Cell Physiology*, **272**, C1700–8.
12. Alvarez, D.F., Huang, L., King, J.A. *et al.* (2008) Lung microvascular endothelium is enriched with progenitor cells that exhibit vasculogenic capacity.

American Journal of Physiology: Lung Cellular and Molecular Physiology, **294**, L419–30.
13. Solodushko, V. and Fouty, B. (2007) Proproliferative phenotype of pulmonary microvascular endothelial cells. *American Journal of Physiology: Lung Cellular and Molecular Physiology*, **292**, L671–77.
14. Xu, W., Koeck, T., Lara, A.R. *et al.* (2007) Alterations of cellular bioenergetics in pulmonary artery endothelial cells. *Proceedings of the National Academy of Sciences of the United States of America*, **104**, 1342–47.
15. Gurney, A.M. (2002) Multiple sites of oxygen sensing and their contributions to hypoxic pulmonary vasoconstriction. *Respiratory Physiology and Neurobiology*, **132**, 43–53.
16. Moudgil, R., Michelakis, E.D., and Archer, S.L. (2005) Hypoxic pulmonary vasoconstriction. *Journal of Applied Physiology*, **98**, 390–403.
17. Ward, J.P. (2008) Oxygen sensors in context. *Biochimica et Biophysica Acta*, **1777**, 1–14.
18. Waypa, G.B., Chandel, N.S., and Schumacker, P.T. (2001) Model for hypoxic pulmonary vasoconstriction involving mitochondrial oxygen sensing. *Circulation Research*, **88**, 1259–66.
19. Rounds, S. and McMurtry, I.F. (1981) Inhibitors of oxidative ATP production cause transient vasoconstriction and block subsequent pressor responses in rat lungs. *Circulation Research*, **48**, 393–400.
20. Zhang, F., Kaide, J.I., Yang, L. *et al.* (2004) CO modulates pulmonary vascular response to acute hypoxia: relation to endothelin. *American Journal of Physiology: Heart and Circulatory Physiology*, **286**, H137–44.
21. Kim, H.P., Wang, X., Galbiati, F. *et al.* (2004) Caveolae compartmentalization of heme oxygenase-1 in endothelial cells. *The FASEB Journal*, **18**, 1080–89.
22. Zhu, D., Zhang, C., Medhora, M., and Jacobs, E.R. (2002) CYP4A mRNA, protein, and product in rat lungs: novel localization in vascular endothelium. *Journal of Applied Physiology*, **93**, 330–37.
23. Olson, K.R. (2008) Hydrogen sulfide and oxygen sensing: implications in cardiorespiratory control. *The Journal of Experimental Biology*, **211**, 2727–34.
24. Yang, G., Wu, L., Jiang, B. *et al.* (2008) H_2S as a physiologic vasorelaxant: hypertension in mice with deletion of cystathionine gamma-lyase. *Science*, **322**, 587–90.
25. Koenitzer, J.R., Isbell, T.S., Patel, H.D. *et al.* (2007) Hydrogen sulfide mediates vasoactivity in an O_2-dependent manner. *American Journal of Physiology: Heart and Circulatory Physiology*, **292**, H1953–60.
26. Muzaffar, S., Jeremy, J.Y., Sparatore, A. *et al.* (2008) H_2S-donating sildenafil (ACS6) inhibits superoxide formation and gp91^{phox} expression in arterial endothelial cells: role of protein kinases A and G. *British Journal of Pharmacology*, **155**, 984–94.
27. Walshe, T.E. and D'Amore, P.A. (2008) The role of hypoxia in vascular injury and repair. *Annual Review of Pathology*, **3**, 615–43.
28. Semenza, G.L. (2007) Hypoxia-inducible factor 1, in *Endothelial Biomedicine* (ed. W.C. Aird), Cambridge University Press, New York, pp. 236–45.
29. Kourembanas, S., Marsden, P.A., McQuillan, L.P., and Faller, D.V. (1991) Hypoxia induces endothelin gene expression and secretion in cultured human endothelium. *The Journal of Clinical Investigation*, **88**, 1054–57.
30. Kourembanas, S., Hannan, R.L., and Faller, D.V. (1990) Oxygen tension regulates the expression of the platelet-derived growth factor-B chain gene in human endothelial cells. *The Journal of Clinical Investigation*, **86**, 670–74.
31. Palmer, L.A., Semenza, G.L., Stoler, M.H., and Johns, R.A. (1998) Hypoxia induces type II NOS gene expression in pulmonary artery endothelial cells via HIF-1. *American Journal of Physiology: Lung Cellular and Molecular Physiology*, **274**, L212–19.
32. Faller, D.V. (1999) Endothelial cell responses to hypoxic stress. *Clinical and Experimental Pharmacology and Physiology*, **26**, 74–84.
33. Peters, D.G., Ning, W., Chu, T.J. *et al.* (2006) Comparative SAGE analysis of the response to hypoxia in human pulmonary and aortic endothelial cells. *Physiological Genomics*, **26**, 99–108.
34. Bernal, P.J., Leelavanichkul, K., Bauer, E. *et al.* (2008) Nitric-oxide-mediated zinc release contributes to hypoxic regulation of pulmonary vascular tone. *Circulation Research*, **102**, 1575–83.
35. Sehgal, P.B. and Mukhopadhyay, S. (2007) Dysfunctional intracellular trafficking in the pathobiology of pulmonary arterial hypertension. *American Journal of Respiratory Cell and Molecular Biology*, **37**, 31–37.
36. Murata, T., Sato, K., Hori, M. *et al.* (2002) Decreased endothelial nitric-oxide synthase (eNOS) activity resulting from abnormal interaction between eNOS and its regulatory proteins in hypoxia-induced pulmonary hypertension. *The Journal of Biological Chemistry*, **277**, 44085–92.
37. Mukhopadhyay, S., Xu, F., and Sehgal, P.B. (2007) Aberrant cytoplasmic sequestration of

eNOS in endothelial cells after monocrotaline, hypoxia, and senescence: live-cell caveolar and cytoplasmic NO imaging. *American Journal of Physiology: Heart and Circulatory Physiology*, **292**, H1373–89.

38. Farber, H.W., Center, D.M., and Rounds, S. (1987) Effect of ambient oxygen on cultured endothelial cells from different vascular beds. *American Journal of Physiology: Heart and Circulatory Physiology*, **253**, H878–83.
39. Scarfo, L.M., Weller, P.F., and Farber, H.W. (2001) Induction of endothelial cell cytoplasmic lipid bodies during hypoxia. *American Journal of Physiology: Heart and Circulatory Physiology*, **280**, H294–301.
40. Farber, H.W. and Barnett, H.F. (1991) Differences in prostaglandin metabolism in cultured aortic and pulmonary arterial endothelial cells exposed to acute and chronic hypoxia. *Circulation Research*, **68**, 1446–57.
41. Babal, P., Ruchko, M., Ault-Ziel, K. *et al.* (2002) Regulation of ornithine decarboxylase and polyamine import by hypoxia in pulmonary artery endothelial cells. *American Journal of Physiology: Lung Cellular and Molecular Physiology*, **282**, L840–46.
42. Humphries, D.E., Lee, S.L., Fanburg, B.L., and Silbert, J.E. (1986) Effects of hypoxia and hyperoxia on proteoglycan production by bovine pulmonary artery endothelial cells. *Journal of Cellular Physiology*, **126**, 249–53.
43. Karlinsky, J.B., Rounds, S., and Farber, H.W. (1992) Effects of hypoxia on heparan sulfate in bovine aortic and pulmonary artery endothelial cells. *Circulation Research*, **71**, 782–89.
44. Wu, W., Dave, N.B., Yu, G. *et al.* (2008) Network analysis of temporal effects of intermittent and sustained hypoxia on rat lungs. *Physiological Genomics*, **36**, 24–34.
45. Nanduri, J., Yuan, G., Kumar, G.K. *et al.* (2008) Transcriptional responses to intermittent hypoxia. *Respiratory Physiology and Neurobiology*, **164**, 277–81.
46. Kumar, G.K. and Prabhakar, N.R. (2008) Post-translational modification of proteins during intermittent hypoxia. *Respiratory Physiology and Neurobiology*, **164**, 272–6.
47. Toffoli, S., Feron, O., Raes, M., and Michiels, C. (2007) Intermittent hypoxia changes HIF-1alpha phosphorylation pattern in endothelial cells: unravelling of a new PKA-dependent regulation of HIF-1alpha. *Biochimica et Biophysica Acta*, **1773**, 1558–71.
48. Warren, M.F. and Drinker, C.K. (1942) The flow of lymph from the lungs of the dog. *American Journal of Physiology*, **136**, 207–21.
49. Ogawa, S., Koga, S., Kuwabara, K. *et al.* (1992) Hypoxia-induced increased permeability of endothelial monolayers occurs through lowering of cellular cAMP levels. *American Journal of Physiology: Cell Physiology*, **262**, C546–54.
50. Inauen, W., Payne, D.K., Kvietys, P.R., and Granger, D.N. (1990) Hypoxia/reoxygenation increases the permeability of endothelial cell monolayers: role of oxygen radicals. *Free Radical Biology and Medicine*, **9**, 219–23.
51. Wojciak-Stothard, B., Tsang, L.Y., and Haworth, S.G. (2005) Rac and Rho play opposing roles in the regulation of hypoxia/reoxygenation-induced permeability changes in pulmonary artery endothelial cells. *American Journal of Physiology: Lung Cellular and Molecular Physiology*, **288**, L749–60.
52. Partridge, C.A. (1995) Hypoxia and reoxygenation stimulate biphasic changes in endothelial monolayer permeability. *American Journal of Physiology: Lung Cellular and Molecular Physiology*, **269**, L52–58.
53. Mehta, D. and Malik, A.B. (2006) Signaling mechanisms regulating endothelial permeability. *Physiological Reviews*, **86**, 279–367.
54. Ogawa, S., Gerlach, H., Esposito, C. *et al.* (1990) Hypoxia modulates the barrier and coagulant function of cultured bovine endothelium. Increased monolayer permeability and induction of procoagulant properties. *The Journal of Clinical Investigation*, **85**, 1090–98.
55. Dehler, M., Zessin, E., Bartsch, P., and Mairbaurl, H. (2006) Hypoxia causes permeability oedema in the constant-pressure perfused rat lung. *The European Respiratory Journal*, **27**, 600–6.
56. Hampl, V., Cornfield, D.N., Cowan, N.J., and Archer, S.L. (1995) Hypoxia potentiates nitric oxide synthesis and transiently increases cytosolic calcium levels in pulmonary artery endothelial cells. *The European Respiratory Journal*, **8**, 515–22.
57. Michiels, C., Arnould, T., and Remacle, J. (2000) Endothelial cell responses to hypoxia: initiation of a cascade of cellular interactions. *Biochimica et Biophysica Acta*, **1497**, 1–10.
58. Stevens, T., Nakahashi, Y., Cornfield, D.N. *et al.* (1995) Ca^{2+}-inhibitable adenylyl cyclase modulates pulmonary artery endothelial cell cAMP content and barrier function. *Proceedings of the National Academy of Sciences of the United States of America*, **92**, 2696–700.

59. Irwin, D.C., Tissot van Patot, M.C., Tucker, A., and Bowen, R. (2005) Direct ANP inhibition of hypoxia-induced inflammatory pathways in pulmonary microvascular and macrovascular endothelial monolayers. *American Journal of Physiology: Lung Cellular and Molecular Physiology*, **288**, L849–59.
60. Dudek, S.M. and Garcia, J.G. (2001) Cytoskeletal regulation of pulmonary vascular permeability. *Journal of Applied Physiology*, **91**, 1487–500.
61. Wojciak-Stothard, B., Tsang, L.Y., Paleolog, E. *et al.* (2006) Rac1 and RhoA as regulators of endothelial phenotype and barrier function in hypoxia-induced neonatal pulmonary hypertension. *American Journal of Physiology: Lung Cellular and Molecular Physiology*, **290**, L1173–82.
62. Ali, M.H., Schlidt, S.A., Chandel, N.S. *et al.* (1999) Endothelial permeability and IL-6 production during hypoxia: role of ROS in signal transduction. *American Journal of Physiology: Lung Cellular and Molecular Physiology*, **277**, L1057–65.
63. Taylor, C.T. (2008) Interdependent roles for hypoxia inducible factor and nuclear factor-kappaB in hypoxic inflammation. *Journal of Physiology*, **586**, 4055–59.
64. Rius, J., Guma, M., Schachtrup, C. *et al.* (2008) NF-kappaB links innate immunity to the hypoxic response through transcriptional regulation of HIF-1alpha. *Nature*, **453**, 807–11.
65. Scortegagna, M., Cataisson, C., Martin, R.J. *et al.* (2008) HIF-1alpha regulates epithelial inflammation by cell autonomous NFkappaB activation and paracrine stromal remodeling. *Blood*, **111**, 3343–54.
66. Walmsley, S.R., Print, C., Farahi, N. *et al.* (2005) Hypoxia-induced neutrophil survival is mediated by HIF-1alpha-dependent NF-kappaB activity. *The Journal of Experimental Medicine*, **201**, 105–15.
67. Senger, D.R., Galli, S.J., Dvorak, A.M. *et al.* (1983) Tumor cells secrete a vascular permeability factor that promotes accumulation of ascites fluid. *Science*, **219**, 983–85.
68. Hippenstiel, S., Krull, M., Ikemann, A. *et al.* (1998) VEGF induces hyperpermeability by a direct action on endothelial cells. *American Journal of Physiology: Lung Cellular and Molecular Physiology*, **274**, L678–84.
69. Godzich, M., Hodnett, M., Frank, J.A. *et al.* (2006) Activation of the stress protein response prevents the development of pulmonary edema by inhibiting VEGF cell signaling in a model of lung ischemia–reperfusion injury in rats. *The FASEB Journal*, **20**, 1519–21.
70. Namiki, A., Brogi, E., Kearney, M. *et al.* (1995) Hypoxia induces vascular endothelial growth factor in cultured human endothelial cells. *The Journal of Biological Chemistry*, **270**, 31189–95.
71. Pham, I., Uchida, T., Planes, C. *et al.* (2002) Hypoxia upregulates VEGF expression in alveolar epithelial cells *in vitro* and *in vivo*. *American Journal of Physiology: Lung Cellular and Molecular Physiology*, **283**, L1133–42.
72. Shweiki, D., Itin, A., Soffer, D., and Keshet, E. (1992) Vascular endothelial growth factor induced by hypoxia may mediate hypoxia-initiated angiogenesis. *Nature*, **359**, 843–45.
73. Marti, H.H. and Risau, W. (1998) Systemic hypoxia changes the organ-specific distribution of vascular endothelial growth factor and its receptors. *Proceedings of the National Academy of Sciences of the United States of America*, **95**, 15809–14.
74. Sakao, S., Taraseviciene-Stewart, L., Wood, K. *et al.* (2006) Apoptosis of pulmonary microvascular endothelial cells stimulates vascular smooth muscle cell growth. *American Journal of Physiology: Lung Cellular and Molecular Physiology*, **291**, L362–68.
75. Esser, S., Lampugnani, M.G., Corada, M. *et al.* (1998) Vascular endothelial growth factor induces VE-cadherin tyrosine phosphorylation in endothelial cells. *Journal of Cell Science*, **111**, 1853–65.
76. Pavlicek, V., Marti, H.H., Grad, S. *et al.* (2000) Effects of hypobaric hypoxia on vascular endothelial growth factor and the acute phase response in subjects who are susceptible to high-altitude pulmonary oedema. *European Journal of Applied Physiology*, **81**, 497–503.
77. Walter, R., Maggiorini, M., Scherrer, U. *et al.* (2001) Effects of high-altitude exposure on vascular endothelial growth factor levels in man. *European Journal of Applied Physiology*, **85**, 113–17.
78. Nan, Y.S., Feng, G.G., Hotta, Y. *et al.* (2004) Neuropeptide Y enhances permeability across a rat aortic endothelial cell monolayer. *American Journal of Physiology: Heart and Circulatory Physiology*, **286**, H1027–33.
79. Carpenter, T.C. and Stenmark, K.R. (2001) Hypoxia decreases lung neprilysin expression and increases pulmonary vascular leak. *American Journal of Physiology: Lung Cellular and Molecular Physiology*, **281**, L941–48.
80. Kinasewitz, G.T., Groome, L.J., Marshall, R.P. *et al.* (1986) Effect of hypoxia on permeability of

pulmonary endothelium of canine visceral pleura. *Journal of Applied Physiology*, **61**, 554–60.
81. Parker, R.E., Granger, D.N., and Taylor, A.E. (1981) Estimates of isogravimetric capillary pressures during alveolar hypoxia. *American Journal of Physiology: Heart and Circulatory Physiology*, **241**, H732–39.
82. Aarseth, P., Bjertnaes, L., and Karlsen, J. (1980) Changes in blood volume and extravascular water content in isolated perfused rat lungs during ventilation hypoxia. *Acta Physiologica Scandinavica*, **109**, 61–67.
83. Goodale, R.L., Goetzman, B., and Visscher, M.B. (1970) Hypoxia and iodoacetic acid and alveolocapillary barrier permeability to albumin. *American Journal of Physiology*, **219**, 1226–30.
84. Warren, M.F., Peterson, D.K., and Drinker, C.K. (1942) The effects of heightened negative pressure in the chest, together with further experiments upon anoxia in increasing the flow of lung lymph. *American Journal of Physiology*, **137**, 641–48.
85. Bland, R.D., Demling, R.H., Selinger, S.L., and Staub, N.C. (1977) Effects of alveolar hypoxia on lung fluid and protein transport in unanesthetized sheep. *Circulation Research*, **40**, 269–74.
86. Homik, L.A., Bshouty, Z., Light, R.B., and Younes, M. (1988) Effect of alveolar hypoxia on pulmonary fluid filtration in *in situ* dog lungs. *Journal of Applied Physiology*, **65**, 46–52.
87. Landolt, C.C., Matthay, M.A., Albertine, K.H. *et al.* (1983) Overperfusion, hypoxia, and increased pressure cause only hydrostatic pulmonary edema in anesthetized sheep. *Circulation Research*, **52**, 335–41.
88. Courtice, F.C. and Korner, P.I. (1952) The effect of anoxia on pulmonary oedema produced by massive intravenous infusions. *The Australian Journal of Experimental Biology and Medical Science*, **30**, 511–26.
89. Hansen, T.N., Hazinski, T.A., and Bland, R.D. (1984) Effects of asphyxia on lung fluid balance in baby lambs. *The Journal of Clinical Investigation*, **74**, 370–76.
90. Boonyaprakob, U., Taylor, P.M., Watson, D.W. *et al.* (1969) Hypoxia and protein clearance from the pulmonary vascular beds of adult dogs and pups. *American Journal of Physiology*, **216**, 1013–19.
91. Martin, D.J., Grimbert, F.A., Baconnier, P., and Benchetrit, G. (1983) Effect of acute hypoxia on lung transvascular filtration in anaesthetized dogs. *Bulletin Europeen de Physiopathologie Respiratoire*, **19**, 7–11.
92. Stelzner, T.J., O'Brien, R.F., Sato, K., and Weil, J.V. (1988) Hypoxia-induced increases in pulmonary transvascular protein escape in rats. Modulation by glucocorticoids. *The Journal of Clinical Investigation*, **82**, 1840–47.
93. Whayne, T.F. Jr. and Severinghaus, J.W. (1968) Experimental hypoxic pulmonary edema in the rat. *Journal of Applied Physiology*, **25**, 729–32.
94. Rassler, B., Marx, G., Reissig, C. *et al.* (2007) Time course of hypoxia-induced lung injury in rats. *Respiratory Physiology and Neurobiology*, **159**, 45–54.
95. Schoene, R.B., Swenson, E.R., Pizzo, C.J. *et al.* (1988). The lung at high altitude: bronchoalveolar lavage in acute mountain sickness and pulmonary edema. *Journal of Applied Physiology* **64**, 2605–13.
96. Bartsch, P., Mairbaurl, H., Maggiorini, M., and Swenson, E.R. (2005) Physiological aspects of high-altitude pulmonary edema. *Journal of Applied Physiology*, **98**, 1101–10.
97. Durmowicz, A.G., Hofmeister, S., Kadyraliev, T.K. *et al.* (1993) Functional and structural adaptation of the yak pulmonary circulation to residence at high altitude. *Journal of Applied Physiology*, **74**, 2276–85.
98. Arias-Stella, J. and Kruger, H. (1963) Pathology of high altitude pulmonary edema. *Archives of Pathology*, **76**, 147–57.
99. Marticorena, E., Tapia, F.A., Dyer, J. *et al.* (1964) Pulmonary edema by ascending to high altitudes. *Diseases of the Chest*, **45**, 273–83.
100. Hultgren, H.N., Lopez, C.E., Lundberg, E., and Miller, H. (1964) Physiologic studies of pulmonary edema at high altitude. *Circulation*, **29**, 393–408.
101. Scherrer, U., Vollenweider, L., Delabays, A. *et al.* (1996) Inhaled nitric oxide for high-altitude pulmonary edema. *The New England Journal of Medicine*, **334**, 624–29.
102. Maggiorini, M., Melot, C., Pierre, S. *et al.* (2001) High-altitude pulmonary edema is initially caused by an increase in capillary pressure. *Circulation*, **103**, 2078–83.
103. Sartori, C., Allemann, Y., and Scherrer, U. (2007) Pathogenesis of pulmonary edema: learning from high-altitude pulmonary edema. *Respiratory Physiology and Neurobiology*, **159**, 338–49.
104. Tsukimoto, K., Mathieu-Costello, O., Prediletto, R. *et al.* (1991) Ultrastructural appearances of pulmonary capillaries at high transmural pressures. *Journal of Applied Physiology*, **71**, 573–82.
105. Elliott, A.R., Fu, Z., Tsukimoto, K. *et al.* (1992) Short-term reversibility of ultrastructural changes

in pulmonary capillaries caused by stress failure. *Journal of Applied Physiology*, **73**, 1150–58.

106. Maggiorini, M., Brunner-La Rocca, H.P., Peth, S. *et al.* (2006) Both tadalafil and dexamethasone may reduce the incidence of high-altitude pulmonary edema: a randomized trial. *Annals of Internal Medicine*, **145**, 497–506.

107. Kolluru, G.K., Tamilarasan, K.P., Rajkumar, A.S. *et al.* (2008) Nitric oxide/cGMP protects endothelial cells from hypoxia-mediated leakiness. *European Journal of Cell Biology*, **87**, 147–61.

108. Ellsworth, A.J., Larson, E.B., and Strickland, D. (1987) A randomized trial of dexamethasone and acetazolamide for acute mountain sickness prophylaxis. *The American Journal of Medicine*, **83**, 1024–30.

109. McVerry, B.J. and Garcia, J.G. (2005) In vitro and *in vivo* modulation of vascular barrier integrity by sphingosine 1-phosphate: mechanistic insights. *Cellular Signalling*, **17**, 131–39.

110. Kayyali, U.S., Donaldson, C., Huang, H. *et al.* (2001) Phosphorylation of xanthine dehydrogenase/oxidase in hypoxia. *The Journal of Biological Chemistry*, **276**, 14359–65.

111. Kayyali, U.S., Pennella, C.M., Trujillo, C. *et al.* (2002) Cytoskeletal changes in hypoxic pulmonary endothelial cells are dependent on MAPK-activated protein kinase MK2. *The Journal of Biological Chemistry*, **277**, 42596–602.

112. West, J.B., Colice, G.L., Lee, Y.J. *et al.* (1995) Pathogenesis of high-altitude pulmonary oedema: direct evidence of stress failure of pulmonary capillaries. *The European Respiratory Journal*, **8**, 523–29.

113. West, J.B., Tsukimoto, K., Mathieu-Costello, O., and Prediletto, R. (1991) Stress failure in pulmonary capillaries. *Journal of Applied Physiology*, **70**, 1731–42.

19

Viral Infection and Pulmonary Endothelial Cells

Norbert F. Voelkel

Victoria Johnson Center for Pulmonary Obstructive Disease Research, Pulmonary and Critical Care Medicine Division, Virginia Commonwealth University, Richmond, VA, USA

INTRODUCTION

The topic of viral infections and lung diseases is potentially a large area of interest and research in the context of acute lung injury (ALI) and pulmonary hypertension. For example, Hanta virus infection causes ALI [1] and HIV infection is associated with pulmonary arterial hypertension (PAH) [2]. In this chapter the focus is on the lung endothelial cell (EC) as a target of viral infections.

ALI associated with a viral infection (e.g., Hanta virus or influenza virus) is the result of a massive, overwhelming, acute infection, whereas the development of PAH requires a chronic or latent viral infection. The overall state of the art and knowledge of viral infection of lung ECs is almost entirely based on data derived from *in vitro* infection of cultured ECs; *in vivo* or *in situ* data linking lung EC infection to disease are rare. It is safe to say that lung EC virology is very much in its infancy and is a wide open field for seminal investigations.

ACUTE VIRAL INFECTIONS

Acute and fatal Hanta virus infections occurred several years ago as an epidemic in New Mexico; the deer mouse living at an elevation of 1300–2300 m in New Mexico was identified as the carrier of the virus and ALI was a common presentation [1, 3]. Other Hanta viruses cause hemorrhagic fever and renal disease [4]. Mechanistically the increase in vascular permeability is important and it has been recognized that Hanta viruses infect ECs [5]. It is now known that Hanta viruses enhance endothelial permeability 2–3 days postinfection, associated with impairment of the $\alpha_v\beta_3$ integrin that regulates the permeability-enhancing effect of vascular endothelial growth factor (VEGF). This was demonstrated by inhibition of Hanta virus-induced permeability in the presence of VEGF receptor-2 antibodies [6].

Similarly, there is now evidence that the avian influenza virus, H5N1, can infect EC and replicate in human EC and induce their apoptosis [7]. The H5N1 virus binds to sialic acid receptors present on EC [8]. Not only can influenza viruses infect EC, they can also induce tissue factor expression and induce a procoagulant EC phenotype [9], and stimulate the production of interleukin (IL)-6 [10]. Thus, avian influenza virus infection may produce its devastating effects importantly because of its endotheliotrophism, especially affecting lung ECs [11].

Another more recently discovered virus, the corona virus, the agent responsible for the severe acute respiratory syndrome (SARS) epidemic, also targets EC and damages small pulmonary vessel ECs [12]. Interestingly, SARS has been associated with the generation of anti-EC antibodies [13].

CHRONIC VIRAL INFECTIONS

Chronic viral infections are of interest because they can drive angiogenesis [14, 15], change the phenotype of cells – including that of ECs, and affect cell growth. One example of viral impact on EC is the immortalization of human umbilical vein ECs by the human papilloma virus (HPV)-16, E6 and E7 genes [16], and EC growth stimulation by HPV-16-infected keratinocytes [17]. Another example is neoplastic transformation of EC in Kaposi sarcoma triggered by human herpes virus (HHV)-8 infection [18].

The Pulmonary Endothelium: Function in health and disease Editors Norbert F. Voelkel, Sharon Rounds

HIV-RELATED PULMONARY HYPERTENSION

HIV infection is associated with infectious and noninfectious complications, and causes severe angioproliferative pulmonary hypertension in 1/200 patients with AIDS [19]. The first case of HIV infection-associated pulmonary hypertension was reported in 1987 [20]. Mehta *et al.* [21] reviewed 131 cases of HIV-associated pulmonary hypertension. Histologically, HIV-associated pulmonary hypertension is indistinguishable from other forms of severe pulmonary hypertension, including idiopathic sporadic and familial PAH. Early investigators attempted to localize the virus in the lung vasculature, but these attempts failed [22]. Humbert *et al.* [23] searched for expression of the HIV *gag* gene in pulmonary vessels from HIV-infected pulmonary hypertensive patients but were unable to detect this gene. The molecular pathogenesis of HIV-related pulmonary hypertension remained unclear until recently. Zuber *et al.* [24] reported that antiretroviral therapy in patients with HIV-related pulmonary hypertension did not affect the development of pulmonary hypertension and it was found that the HIV Nef protein was associated with the complex pulmonary vascular lesions in monkeys infected with the simian immune insufficiency virus which had been engineered such that it contained the *nef* gene isolated from a human AIDS patient [25]. Using immunohistochemistry, Marecki *et al.* [25] showed that ECs expressed the Nef protein and similarly that ECs in vascular lesions of a human patient with HIV-associated PAH expressed the Nef protein [26]. Thus, it is possible that pulmonary EC can be infected with the HIV, but it is also possible that the Nef protein is being shed by other infected cells(e.g., lymphocytes) and taken up by the EC. To assess whether the human Nef affects human ECs, aortic ECs were transfected with an adenovirus expressing the human *nef* gene. At 24 h after transfection, the ECs underwent apoptosis that was inhibited by a caspase inhibitor; without caspase inhibition the transfected EC became hyperproliferative at 48 and 72 h post-transfection (Marecki *et al.*, unpublished data). In the aggregate these data suggest that a mutated *nef* gene (Flores *et al.*, unpublished), when expressed in EC, can turn the angiogenic switch. It has been known for some time that proline-rich motifs in the HIV-1 Nef bind to Src homology-3 domains of Src kinases [27] and that Nef stimulates glomerular podocyte proliferation via Src-dependent Stat3 and mitogen-activated protein-1 and 2 pathways [28]. In recent years, HIV has been phylogenetically subclassified and it is possible that the different subtypes [29] (subtype A is prevalent in Eastern Europe and Central Africa, subtype B in America and Western Europe, and subtype C in India and South Africa) also display different degrees of EC trophism.

A categorically different mechanism of action of HIV on EC health is the mechanism of molecular mimicry (i.e., epitope sharing between host and virus). In this scenario antibodies developed against mutated Nef sequences may recognize pulmonary EC epitopes. For example, a Nef peptide has been incriminated in HIV-1 related immune thrombocytopenia [30].

In addition to HIV-related angioproliferative pulmonary vascular disease, HIV may be associated with pulmonary emphysema. Possibly the HIV-1 Tat protein, via production of ceramide, induces EC apoptosis and emphysema [31].

HHV-8 INFECTION

HHV-8 (also known as Kaposi sarcoma virus) is an oncogenic virus implicated in the pathogenesis of several malignancies. HHV-8 is expressed in Kaposi sarcoma, primary effusion lymphoma, and multicentric Castleman lymphoma [32]. HHV-8 infection of ECs causes transformation of the ECs and the virus-specific IL-6 causes angiogenesis [33, 34] via VEGF. ECs in some plexiform lesions (see also Chapter 27) have the appearance of spindle cells and patients with idiopathic PAH show signs of immune system deregulation, and a report of a patient with Castleman's lymphoma (known to be caused by HHV-8 infection [35] led to the investigation of plexiform lesions from patients with severe pulmonary hypertension and described, using immunehistochemistry, the expression of the latency-associated nuclear antigen-1 in plexiform lesion EC in patients with idiopathic but not secondary, forms of pulmonary hypertension [36]. Although these findings were not replicated by a number of studies from Germany, Israel, and Japan, the initial description of the findings by the Colorado group have not been invalidated and – together with the well-accepted notion that HIV causes pulmonary hypertension – have raised the question whether other latent virus infections, like HHV-8 and hepatitis viruses, could either cause the development of angioproliferative pulmonary hypertension or participate as cofactors in the pathobiology of severe pulmonary hypertension. Whereas the rationale for this hypothesis is valid, data are mostly lacking.

In the case of HHV-8 infection, it is known that this virus is EC-trophic. For example, Caselli *et al.* [37] have shown that HHV-8 induces expression of nuclear factor-κB in infected ECs with the subsequent release of monocyte chemoattractant protein-1, tumor necrosis factor-α, and RANTES. Fonsato *et al.* [38], concluded that HHV-8 in ECs may express the PAx2 oncogene which activates an angiogenic program. Infection with

HHV-8 of pulmonary microvascular ECs [39] caused significant changes in EC gene expression pattern. Among other genes, bone morphogenic protein gene (BMP4) expression was decreased in latently infected cells, whereas the expression of matrix metalloproteinase genes matrix metalloproteinase (MMP)-1, 2, and 10 was increased. Of particular interest, the latently infected EC were resistant to camptothecin-induced apoptosis. BMP4 expression, also shown to be decreased in HHV-8-infected dermal microvascular ECs [40], is likely also a protein which controls vascular homeostasis.

CYTOMEGALOVIRUS

Cytomegalovirus (CMV, also known as HHV-5) infection has been associated with atherosclerosis, transplant vascular sclerosis, and coronary artery restenosis, and anti-human CMV antibodies are prevalent in patients with scleroderma. CMV infects ECs and macrophages, and a shared pathogenetic scheme is EC apoptosis, vascular and perivascular infiltrates. Human CMV binds to β_1 and β_3 integrins and to the epithelial growth factor receptor [14]. A significant portion of patients with the limited form of systemic sclerosis (scleroderma) develop severe angioproliferative pulmonary hypertension (see Chapter 28), but whether CMV infection plays a role in the development of the pulmonary vascular disease in these patients is unknown.

HEPATITIS VIRUS INFECTION

Severe angioproliferative pulmonary hypertension is a recognized complication of chronic liver disease, and is also called porto-pulmonary hypertension . Many of these patients are infected with a hepatitis virus and whether the pulmonary hypertension is secondary to the liver cirrhosis and portal hypertension or due to the viral infection has not been resolved [41]. A patient with hepatitis B virus infection and angioproliferative pulmonary hypertension has been reported by Cool (personal communication). Introduction of angiogenesis by the hepatitis B virus X protein via stabilization of hypoxia-inducible factor (HIF)-1α has been described [42] and conceptually X protein-induced HIF-1α protein stabilization could be part of a pulmonary angioproliferation program.

CONCLUSIONS AND PERSPECTIVES

Viral infections of pulmonary endothelial cells are of great interest in the context of massive and acute infections that lead to severe lung tissue damage as observed in influenza, Hanta, and corona virus infections with destruction of lung microvascular ECs and fatal increased permeability lung edema. A second interest is severe PAH and the question of whether chronic latent viral infections of the lung can cause or contribute to the development of severe angioproliferative pulmonary hypertension. Whereas it has been accepted that the AIDS virus infection is associated with the development of severe PAH, the mechanism of HIV-related pulmonary plexiform lesion formation is less clear. The work by Marecki [25] suggests a role for a mutated *nef* HIV gene as described in "HIV-Related Pulmonary Hypertension". HHV-8 infection, without accompanying HIV infection, has been associated with Castleman lymphoma and pulmonary hypertension [35], and in a HIV/HHV-8 dual-infected patient with Castleman lymphoma [43]. With the availability of precise molecular virology tools and knowledge of viral gene sequences, it is now possible to search for the presence of viral genes in lung tissue and lung ECs. Whether viral genes detected in such a way contribute to EC activation, cause EC apoptosis by triggering an immune response based on molecular mimicry, and stimulate antiendothelial antibodies will be more difficult to establish [44].

References

1. Duchin, J.S., Koster, F.T., Peters, C.J. *et al.* The Hantavirus Study Group (1994) Hantavirus pulmonary syndrome: a clinical description of 17 patients with a newly recognized disease. *The New England Journal of Medicine*, **330** (14), 949–55.
2. Sitbon, O. (2008) HIV-related pulmonary arterial hypertension: clinical presentation and management. *AIDS*, **22** (Suppl 3), S55–62.
3. Eisen, R.J., Glass, G.E., Eisen, L. *et al.* (2007) A spatial model of shared risk for plague and hantavirus pulmonary syndrome in the southwestern United States. *The American Journal of Tropical Medicine and Hygiene*, **77** (6), 999–1004.
4. Tai, P.W., Chen, L.C., and Huang, C.H. (2005) Hanta hemorrhagic fever with renal syndrome: a case report and review. *The Journal of Microbiology Immunology & Infection*, **38** (3), 221–3.
5. Geimonen, E., Neff, S., Raymond, T. *et al.* (2002) Pathogenic and nonpathogenic hantaviruses differentially regulate endothelial cell responses. *Proceedings of the National Academy of Sciences of the United States of America*, **15** (99), 13837–42.
6. Gavrilovskaya, I.N., Gorbunova, E.E., Mackow, N.A., and Mackow, E.R. (2008) Hantaviruses direct endothelial cell permeability by sensitizing cells to the vascular permeability factor VEGF, while angiopoietin 1 and sphingosine 1-phosphate inhibit hantavirus-directed permeability. *Journal of Virology*, **82** (12), 5797–806.

7. Sumikoshi, M., Hashimoto, K., Kawasaki, Y. *et al.* (2008) Human influenza virus infection and apoptosis induction in human vascular endothelial cells. *Journal of Medical Virology*, **80** (6), 1072–78.
8. Yao, L., Korteweg, C., Hsueh, W., and Gu, J. (2008) Avian influenza receptor expression in H5N1-infected and noninfected human tissues. *The FASEB Journal*, **22** (3), 733–40.
9. Visseren, F.L., Bouwman, J.J., Bouter, K.P. *et al.* (2000) Procoagulant activity of endothelial cells after infection with respiratory viruses. *Thrombosis and Haemostasis*, **84** (2), 319–24.
10. Visseren, F.L., Verkerk, M.S., van der Bruggen, T. *et al.* (2002) Iron chelation and hydroxyl radical scavenging reduce the inflammatory response of endothelial cells after infection with *Chlamydia pneumoniae* or influenza A. *European Journal of Clinical Investigation*, **32** (Suppl 1), 84–90.
11. Feldmann, A., Schafer, M.K., Garten, W., and Klenk, H.D. (2000) Targeted infection of endothelial cells by avian influenza virus A/FPV/Rostock/34 (H7N1) in chicken embryos. *Journal of Virology*, **74** (17), 8018–27.
12. Ye, J., Zhang, B., Xu, J. *et al.* (2007) Molecular pathology in the lungs of severe acute respiratory syndrome patients. *American Journal of Pathology*, **170** (2), 538–45.
13. Yang, Y.H., Huang, Y.H., Chuang, Y.H. *et al.* (2005) Autoantibodies against human epithelial cells and endothelial cells after severe acute respiratory syndrome (SARS)-associated coronavirus infection. *Journal of Medical Virology*, **77**(1), 1–7.
14. Bentz, G.L. and Yurochko, A.D. (2008) Human CMV infection of endothelial cells induces an angiogenic response through viral binding to EGF receptor and beta1 and beta3 integrins. *Proceedings of the National Academy of Sciences of the United States of America*, **105** (14), 5531–36.
15. Mazzanti, R., Messerini, L., Comin, C.E. *et al.* (2007) Liver angiogenesis as a risk factor for hepatocellular carcinoma development in hepatitis C virus cirrhotic patients. *World Journal of Gastroenterology*, **13** (37), 5009–14.
16. Rhim, J.S., Tsai, W.P., Chen, Z.Q. *et al.* (1998) A human vascular endothelial cell model to study angiogenesis and tumorigenesis. *Carcinogenesis*, **19** (4), 673–81.
17. Bequet-Romero, M. and López-Ocejo, O. (2000) Angiogenesis modulators expression in culture cell lines positives for HPV-16 oncoproteins. *Biochemical and Biophysical Research Communications*, **277** (1), 55–61.
18. Douglas, J.L., Gustin, J.K., Dezube, B. *et al.* (2007) Kaposi's sarcoma: a model of both malignancy and chronic inflammation. *Panminerva Medica*, **49** (3), 119–38.
19. Humbert, M., Sitbon, O., Chaouat, A. *et al.* (2006) Pulmonary arterial hypertension in France: results from a national registry. *The American Journal of Respiratory and Critical Care Medicine*, **173** (9), 1023–30.
20. Kim, K.K. and Factor, S.M. (1987). Membranoproliferative glomerulonephritis and plexogenic pulmonary arteriopathy in a homosexual man with acquired immunodeficiency syndrome. *Human Pathology*, **18** (12), 1293–6.
21. Mehta, N.J., Khan, I.A., Mehta, R.N., and Sepkowitz, D.A. (2000) HIV-related pulmonary hypertension: analytic review of 131 cases. *Chest*, **118** (4), 1133–41.
22. Mette, S.A., Palevsky, H.I., Pietra, G.G. *et al.* (1992) Primary pulmonary hypertension in association with human immunodeficiency virus infection. A possible viral etiology for some forms of hypertensive pulmonary arteriopathy. *The American Review of Respiratory Disease*, **145** (5), 1196–200.
23. Humbert, M., Monti, G., Fartoukh, M. *et al.* (1998) Platelet-derived growth factor expression in primary pulmonary hypertension: comparison of HIV seropositive and HIV seronegative patients. *The European Respiratory Journal*, **11** (3), 554–59.
24. Zuber, J.P., Calmy, A., Evison, J.M. *et al.* (2004) Pulmonary arterial hypertension related to HIV infection: improved hemodynamics and survival associated with antiretroviral therapy. Clinical Infectious Diseases, **38** (8), 1178–85.
25. Marecki, J.C., Cool, C.D., Parr, J.E. *et al.* (2006) HIV-1 Nef is associated with complex pulmonary vascular lesions in SHIV-*nef*-infected macaques. *The American Journal of Respiratory and Critical Care Medicine*, **174** (4), 437–45.
26. Voelkel, N.F., Cool, C.D., and Flores, S. (2008) From viral infection to pulmonary arterial hypertension: a role for viral proteins? *AIDS*, Suppl 3, S49–53.
27. Saksela, K., Cheng, G., and Baltimore, D. (1995) Proline-rich (PxxP) motifs in HIV-1 Nef bind to SH3 domains of a subset of Src kinases and are required for the enhanced growth of Nef^{+} viruses but not for down-regulation of CD4. *The EMBO Journal*, **14** (3), 484–91.
28. He, J.C., Husain, M., Sunamoto, M. *et al.* (2004) Nef stimulates proliferation of glomerular podocytes through activation of Src-dependent Stat3 and MAPK1,2 pathways. *The Journal of Clinical Investigation*, **114** (5), 643–51.

29. Taylor, B.S., Sobieszczyk, M.E., McCutchan, F.E., and Hammer, S.M. (2008) The challenge of HIV-1 subtype diversity. *The New England Journal of Medicine*, **358** (15), 1590–602.
30. Li, Z., Nardi, M.A., and Karpatkin, S. (2005) Role of molecular mimicry to HIV-1 peptides in HIV-1-related immunologic thrombocytopenia. *Blood*, **106** (2), 572–6.
31. Petrache, I., Diab, K., Knox, K.S. *et al.* (2008) HIV associated pulmonary emphysema: a review of the literature and inquiry into its mechanism. *Thorax*, **63** (5), 463–69.
32. Sullivan, R.J., Pantanowitz, L., Casper, C. *et al.* (2008) HIV/AIDS: epidemiology, pathophysiology, and treatment of Kaposi sarcoma-associated herpesvirus disease: Kaposi sarcoma, primary effusion lymphoma, and multicentric Castleman disease. *Clinical Infectious Diseases*, **47** (9), 1209–15.
33. Mead, M.B. and Medveczky, P.G. (2004) Kaposi's sarcoma-associated herpesvirus-encoded viral interleukin-6 secreted and modified differently than human interleukin-6: evidence for a unique autocrine signaling mechanism. *Journal of Biological Chemistry*, **279** (50), 51793–803.
34. Liu, C., Okruzhnov, Y., Li, H., and Nicholas, J. (2001) Human herpesvirus 8 (HHV-8)-encoded cytokines induce expression of and autocrine signaling by vascular endothelial growth factor (VEGF) in HHV-8-infected primary-effusion lymphoma cell lines and mediate VEGF-independent antiapoptotic effects. *Journal of Virology*, **75** (22), 10933–40.
35. Bull, T.M., Cool, C.D., Serls, A.E. *et al.* (2003) Primary pulmonary hypertension, Castleman's disease and human herpesvirus-8. *The European Respiratory Journal*, **22** (3), 403–7.
36. Cool, C.D., Rai, P.R., Yeager, M.E. *et al.* (2003) Expression of human herpesvirus 8 in primary pulmonary hypertension. *The New England Journal of Medicine*, **349** (12), 1113–22.
37. Caselli, E., Florentini, S., Amici, C. *et al.* (2007) Human herpesvirus 8 acute infection of endothelial cells induces monocyte chemoattractant protein 1-dependent capillary-like structure formation: role of the IKK/NF-kappaB pathway. *Blood*, **109** (7), 2718–726.
38. Fonsato, V., Buttiglieri, S., Deregibus, M.C. *et al.* (2008) PAX2 expression by HHV-8-infected endothelial cells induced a proangiogenic and proinvasive phenotype. *Blood*, **111** (5), 2806–15.
39. Bull, T.M., Meadows, C.A., Coldren, C.D. *et al.* (2008) Human herpesvirus-8 infection of primary pulmonary microvascular endothelial cells. *American Journal of Respiratory Cell and Molecular Biology*, **39** (6), 706–16.
40. Poole, L.J., Yu, Y., Kim, P.S. *et al.* (2002) Altered patterns of cellular gene expression in dermal microvascular endothelial cells infected with Kaposi's sarcoma-associated herpesvirus. *Journal of Virology*, **76** (7), 3395–420.
41. Moorman, J., Saad, M., Kosseifi, S., and Krishnaswamy, G. (2005) Hepatitis C virus and the lung: implications for therapy. *Chest*, **128** (4), 2882–92.
42. Moon, E.J., Jeong, C.H., Jeong, J.W. *et al.* (2004) Hepatitis B virus X protein induces angiogenesis by stabilizing hypoxia-inducible factor-1alpha. *The FASEB Journal*, **18** (2), 382–84.
43. Montani, D., Achouh, L., Marcelin, A.G. *et al.* (2005) Reversibility of pulmonary arterial hypertension in HIV/HHV8-associated Castleman's disease. *The European Respiratory Journal*, **26** (5), 969–72.
44. Lunardi, C., Bason, C., Corrocher, R., and Puccetti, A. (2005) Induction of endothelial cell damage by hCMV molecular mimicry. *Trends in Immunology*, **26** (1), 19–24.

20

Effects of Pressure and Flow on the Pulmonary Endothelium

Wolfgang M. Kuebler[1,2]

[1]Charité–Universitaetsmedizin Berlin, Lung and Circulatory Research Laboratory, Institute of Physiology, Berlin, Germany

[2]The Keenan Research Centre at the Li Ka Shing Knowledge Institute of St. Michael's Hospital, Toronto, Ontario, Canada

INTRODUCTION

As a result of lung perfusion and ventilation, the pulmonary endothelium is constantly exposed to a unique combination of biomechanical forces. Under physiological conditions, some of these forces act continuously while others follow oscillatory, chaotic, or even random patterns. Excessive increases in mechanical forces result in structural damage primarily at the alveolo-capillary barrier and may cause a subsequent loss of compartmentalization [1]. Smaller changes activate endothelial responses which are triggered by a variety of mechanotransduction cascades and may initiate or promote inflammatory processes and edema formation in disorders such as high-altitude pulmonary edema (HAPE) or ventilator-induced lung injury (VILI). Chronic exposure to mechanical stress causes structural and functional adaptations of the endothelial phenotype with consequences for lung function and pulmonary hemodynamics. This chapter reviews the biomechanical forces acting upon the pulmonary endothelium, and the cellular mechanisms and pathophysiological relevance of endothelial cell (EC) dysfunction and injury as a consequence of acute or chronic exposure to excess forces.

MECHANICAL FORCES ACTING ON THE PULMONARY ENDOTHELIUM

The pulmonary endothelium is typically subjected to two major types of mechanical stress –shear and stretch (Figure 20.1). Shear stress is the result of blood flow through the pulmonary circulation that continuously exerts a viscous drag on the luminal endothelial surface. For laminar flow of a Newtonian fluid, shear stress (τ) is given by:

$$\tau = 4\bar{v} \cdot \frac{\eta}{r_\mathrm{i}} \qquad (20.1)$$

with $\bar{v}$ representing the mean flow velocity, η the viscosity, and r_i the inner vessel radius. In non-Newtonian fluids like blood, η varies in dependence of flow velocity and vessel diameter (Fåhraeus–Lindqvist effect) [2]. When η is assumed as 0.02 Poise (g/cm/s) [3], τ in pulmonary microvessels can be estimated based on our published blood flow velocities *in vivo* [4, 5]. In alveolar capillaries and venules, these estimates yield τ of 5–10 dyn/cm^2, which is comparable to data from the systemic circulation [6], while τ estimates in pulmonary arterioles are in the range of 3–4 dyn/cm^2 and thus almost a magnitude smaller than in respective vessel segments of the systemic circulation [3]. Additional effects of turbulence on shear stress in pulmonary microvessels can be considered negligible since the relatively low flow velocities and small characteristic lengths result in small Reynolds numbers. Flow and thus shear stress in pulmonary microvessels is yet not steady, but subjected to a combination of oscillatory patterns of varying frequencies. Pulsations of pressure and flow attributable to the cardiac rhythm are not only prominent in pulmonary arteries and arterioles, but also transmitted into the alveolar capillary bed [7, 8]. Furthermore, cyclic changes of flow due to the respiratory cycle affect all segments of the pulmonary microvasculature. Even under constant flow conditions and in the absence of respiratory movements, blood flow switches

The Pulmonary Endothelium: Function in health and disease Editors Norbert F. Voelkel, Sharon Rounds

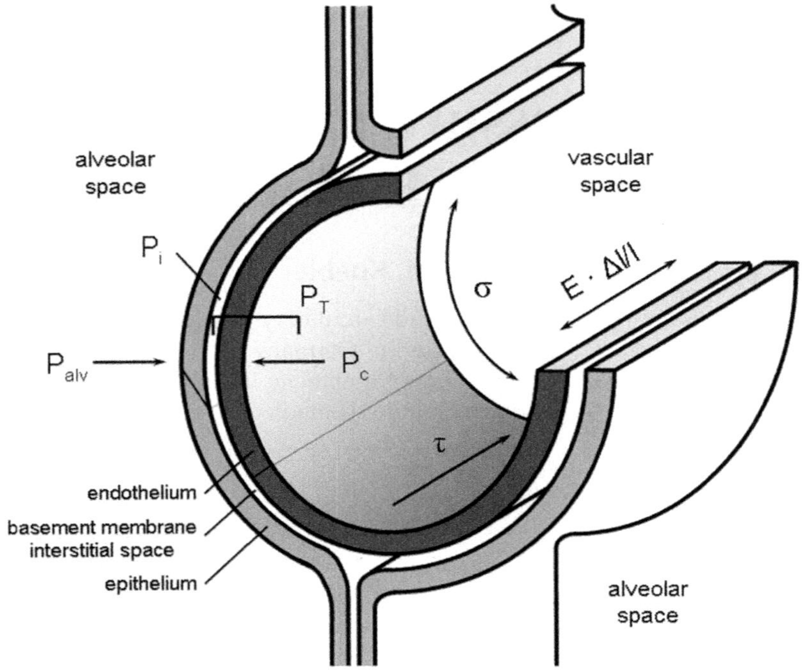

Figure 20.1 Mechanical forces acting on the pulmonary endothelium. Schematic cut of the blood–gas barrier shows a tubular alveolar capillary surrounded by alveolar spaces. The transmural pressure across the capillary wall (P_T) is given by the difference between the capillary (P_c) and the interstitial (P_i) pressure. Shear stress (τ), circumferential wall stress (σ), and longitudinal wall stress, given as the product of longitudinal strain ($\Delta l/l$) and Young's modulus E, act directly on the pulmonary endothelium.

continuously between different perfusion pathways of the alveolar capillary network in a nonrandom pattern with a fractal dimension near 1.0 [9]. The fact that this switching of capillaries is dependent on the actual hematocrit suggests that the continually varying size of plasma gaps between individual red blood cells may play a critical role in the opening and closure of individual capillary segments [10]. The constant recruitment and derecruitment of capillary segments results in a continuous switch between conditions of high and zero shear stress acting on the pulmonary capillary endothelium. It is conceivable that these chaotic oscillations have profound effects on endothelial signaling and cell function, but the physiological significance of this intrinsic phenomenon is still obscure.

Endothelial stretch, on the other hand, is the result of acute distensions of the inner vascular diameter or of the length of individual vascular segments. The term "stretch" is in fact not precisely defined in as far as it is frequently used as a synonym for "strain" (i.e., relative elongation), but equally applied to describe the concomitant stress, (i.e., the force acting upon a surface divided by the respective area). Circumferential (or longitudinal) strain (ε) is defined as relative increase in radius (or length):

$$\varepsilon = \frac{\Delta r}{r} \tag{20.2}$$

The respective change in stress will be $E \cdot \varepsilon$, where E is Young's modulus describing the material's resistance to extension and compression [11]. In case of the endothelium, E is an intrinsic measure of the cell's tensile–compressive elasticity. Unless excessive distension causes structural disintegration, the stress in tubular structures equals the circumferential wall stress (or hub stress; σ) as described by the Young–Laplace equation:

$$\sigma = P_T \cdot \frac{r_i}{h} \tag{20.3}$$

with P_T representing the transmural pressure, r_i the inner radius, and h the wall thickness of the vessel segment.

Elevation of lung microvascular pressure causes rapid distension, and thus endothelial strain in pulmonary capillary and venular segments. Direct intravital microscopic assessment in isolated perfused lungs demonstrated a linear pressure–diameter relation over a range from 0 to 15 mmHg [12, 13]. Calculation of vascular distensibility D, defined as strain over pressure increment (ΔP):

$$D = \frac{\varepsilon}{\Delta P} \tag{20.4}$$

revealed a distensibility of approximately 3.1 ± 0.2% per mmHg in pulmonary arterioles and 1.8 ± 0.2% per mmHg in pulmonary venules [13, 14]. These data are essentially comparable to total vascular distensibility in intact murine lungs which has been reported as 3.2 ± 1.1% per mmHg [15]. In larger pulmonary blood vessels, distensibility is slightly lower and has been reported as 2–2.5% per mmHg for pulmonary arteries of rats and dogs [16, 17], and as 1.2% per mmHg for canine pulmonary veins [18]. Importantly, distensibilities are approximately a magnitude higher in the lung as compared to the vascular segments of the systemic circulation, for which distensibilities are approximately 0.05–0.25% per mmHg in larger arteries [19], 0.1–0.2% per mmHg in capillaries [20, 21], and 0.3–0.8% in larger veins [19]. Hence, small increases in transmural pressure will result in almost 10-fold larger distension and thus endothelial strain in pulmonary as compared to systemic microvessels.

Endothelial strain may not only result from increased hydrostatic pressure, but also occurs during lung expansion in normal and mechanical ventilation. Importantly, lung inflation imposes competing vascular stresses on different microvascular segments [22]. At the alveolar level, the majority of capillaries embedded within the alveolar wall is compressed during inflation by the expansion of adjoining alveoli [23]. Using intravital microscopy of ventilated rabbit lungs, we found that an increase in airway pressure from 8 to 12 mmHg reduces functional capillary density in alveolar networks by 31 ± 3% (unpublished data), demonstrating the effective collapse of almost a third of the previously perfused capillaries. In extra-alveolar microvessels in contrast, transmural pressure increases during lung inflation due to a decrease in interstitial pressure [24]. As a consequence, extra-alveolar lung microvessels as well as pulmonary arteries and veins distend resulting in circumferential strain of the vascular endothelium [25, 26]. In addition to determining radial distension and thus circumferential strain, lung inflation also causes axial elongation of pulmonary blood vessels resulting in longitudinal strain of the vascular endothelium [27, 28]. In intravital microscopic observations, we determined a longitudinal elongation of small pulmonary arterioles and venules by 8.9 ± 2.1% when airway pressure was raised from 8 to 12 mmHg [29]. Thus, lung inflation modulates endothelial strain in two different axes. Circumferential and longitudinal strain may occur in parallel, as in extra-alveolar lung vessels during inspiration. Yet, opposing strain changes may simultaneously occur in different axes in the alveolar microvessels, which necessitates a rapid and complex reorientation of lung endothelial microfilaments and organelles.

MECHANOTRANSDUCTION IN ECs

As a dynamic interface between the vascular compartment and the extravascular space, the endothelium can sense shear and stretch, and respond to these mechanical stresses by rapid adaptations in shape and function. However, how ECs sense mechanical forces is still far from clear. The quest for the endothelial mechanosensor has been hampered by the traditional difficulty to differentiate between actual mechanosensation and early downstream signals. A considerable number of different mechanotransduction mechanisms in ECs have been proposed and will be discussed in this chapter, yet it should be kept in mind that none of these structures or pathways may actually present the mechanosensor itself. Moreover, different modes of mechanotransduction and subsequent signaling pathways seem to exist between different EC types as well as in response to different mechanical stimuli [30].

Different cellular structures have been involved into the sensing of mechanical forces by the endothelium including cytoskeletal components, cell–cell and cell–matrix interactions, ion channels, caveolae, or the endothelial surface layer (ESL). Work in this area has largely been based on two seemingly opposing models –the "centralized" notion of a localized sensing of mechanical forces at their site of action (i.e., the plasma membrane) or the "decentralized" assumption of a rapid dissemination of mechanical forces via the cytoskeleton, which in theory could place the mechanosensor anywhere in the cell [30].

According to the latter "decentralization" model, cellular responses occur as a result of spatial integration of molecular signaling events as well as internal and peripheral force transmission throughout the cell [31, 32]. This force transmission is primarily achieved via cytoskeletal elements which couple distant molecules in the extracellular matrix, the cytoplasm, and the nucleus to form a mechanical continuum [32]. ECs contain three major cytoskeletal networks composed of actin microfilaments, vimentin intermediate filaments, and tubulin microtubules [31]. F-actin microfilaments serve as tension bearing elements that resist the greatest amount of intracellular stress at small strains [33]. At larger strains, both the microfilament and microtubule networks rupture, while intermediate filaments can still retain their connected structure [34], and thus maintain the mechanical integrity of the cell during force adaptation [31]. Actin filaments are anchored in association with focal adhesions and integrins at the basal membrane, intercellular adhesion proteins like cadherins at the lateral membrane, the ESL and caveolae at the apical membrane, and at the nuclear membrane (Figure 20.2), thus allowing for simultaneous mechanotransduction at different locations and rapid spread of mechanical stresses [35]. In accordance with this notion, activation of integrin-mediated signaling at the basolateral membrane induces the formation of focal adhesions [36, 37], and subsequent stimulation of cytosolic, cytoskeletal, and nuclear responses [38]. Focal adhesion kinase (FAK) plays a central role in this mechanotransduction cascade, but whether FAK is directly mechanosensitive or presents a critical initial target of a cellular mechanotransducer remains to be elucidated [30]. Similarly, intercellular adherens junctions [39, 40], caveolae [41, 42], or the endothelial glycocalyx [43, 44], which are all anchored to the actin microfilamentous network, may sense mechanical forces and disseminate these signals according to the decentralization model.

Centralized models, on the other hand, postulate localized mechanosensors in or at the cell membrane, and focus in particular on the potential role of mechanosensitive ion channels and protein kinases [30]. Membrane K^+ channels have been implicated in the endothelial response to changes in shear stress [45], since plasma membrane permeability to K^+ increases with shear [46, 47]. The rapid influx of Ca^{2+} into ECs under shear stress furthermore suggests the existence of a Ca^{2+}-selective channel or a nonselective cation channel [48, 49]. In cell-attached membrane patches on aortic ECs, Lansman *et al.* identified a stretch-sensitive cation channel which mediates Ca^{2+} entry and activates Ca^{2+}-dependent downstream signaling cascades [50]. Subsequent patch-clamp analyses by Hoyer *et al.* demonstrated the existence of a stretch-sensitive cation channel that is permeable to K^+, Na^+, and Ca^{2+} at a ratio of 1 : 0.98 : 0.23, and upon activation allowed for sufficient Ca^{2+} influx to activate neighboring Ca^{2+}-sensitive K^+ channels [51–53]. In addition, these authors identified a K^+-selective stretch-activated channel with a $K^+ : Na^+$ permeability ratio of 10.9 : 1 [52]. The functional integrity of stretch-dependent Ca^{2+} channels was demonstrated in fluorescence microscopic measurements by Naruse and Sokabe [54, 55]. In human umbilical ECs cultured on silicon membranes they observed an increase in the endothelial Ca^{2+} concentration ($[Ca^{2+}]_i$) in response to stretch that was blocked by both removal of extracellular Ca^{2+} or addition of the trivalent lanthanide gadolinium –an unspecific inhibitor of mechanosensitive cation channels [56].

Recently, our understanding of the molecular nature of mechanosensitive ion channels and their regulation

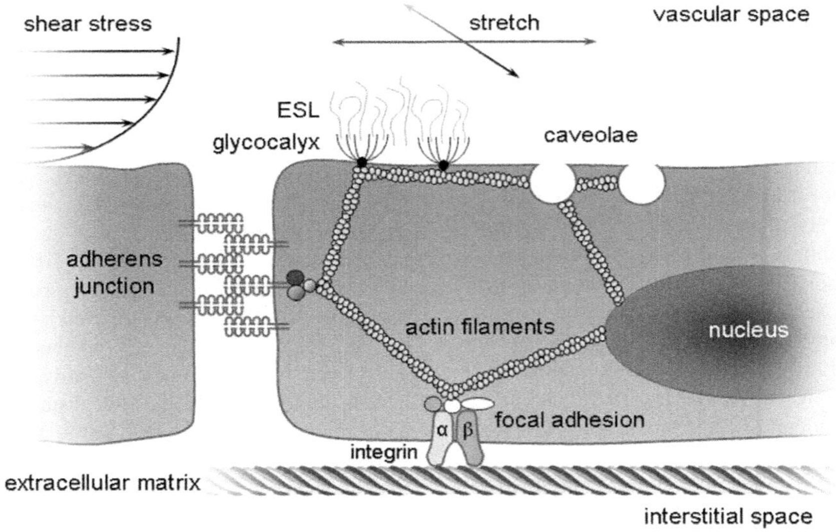

Figure 20.2 Decentralized model of mechanotransduction in the pulmonary endothelium. ECs may sense mechanical forces like shear stress and stretch at various locations of the plasma membrane, the nucleus or even the cytosol. Actin filaments link mechanosensitive structures like the glycocalyx, the ESL, and caveolae at the apical cell surface with the nuclear membrane, adherens junctions at the lateral and focal adhesions at the basal membrane. This spatial arrangement allows for rapid transmission of mechanical forces from one part of the cell to another and simultaneous mechanotransduction at different subcellular localizations.

has been propelled by the identification of the transient receptor potential (TRP) superfamily of ion channels, which comprises a group of channel proteins with multiple sensory roles including mechanosensation (Table 20.1) (see Chapters 5 and 9). Most notable in this respect are members of the transient receptor potential vanilloid (TRPV) subfamily of channels that exhibit largely conserved sequences in species as different as *Homo sapiens*, *Caenorhabditis elegans*, and *Drosophila melanogaster* [57]. The TRPV subfamily comprises a group of currently six members which –with the possible exception of TRPV5 and TRPV6 that mediate Ca^{2+} absorption in kidney and intestine [58] –each fulfill a variety of sensory functions by responding to multiple modal stimulations. In lung ECs, TRPV4 expression has been demonstrated in alveolar capillaries and –although not consistently –in extra-alveolar vessels [59], while TRPV1 and TRPV2 are at least expressed on the mRNA level in pulmonary artery endothelium [60]. While all four sensory TRPV channels (TRPV1–TRPV4) have been implicated in thermosensing at different temperatures, TRPV1, TRPV2, and TRPV4 are also activated by changes in osmolarity indicating their mechanosensitive properties [57, 61]. By use of pharmacological interventions and a knockout mouse model, TRPV4 was recently shown to mediate shear stress-induced Ca^{2+} entry in rat carotid artery ECs and to trigger nitric oxide (NO) and endothelium-derived hyperpolarizing factor-dependent vasodilatory responses both *in situ* and *in vivo* [62, 63]. As will be discussed later, recent data from the lab of Mary Townsley and our own group has also implicated TRPV4 in the endothelial Ca^{2+} response to mechanical stretch [64, 65]. While Ca^{2+} influx via stretch or shear activated TRPV4 is currently studied intensely, other members of the TRPV family may contribute to mechanosensation in ECs. In TRPV2-expressing murine aortic myocytes, both hypotonic swelling and membrane stretch have been shown to activate a Ca^{2+} influx that could be blocked by the TRPV channel inhibitor ruthenium red and an antisense nucleotide against TRPV2 [66]. A mechanosensory role has also been proposed for the TRPV1 channel based on the finding that the stretch-evoked release of ATP and NO from urothelial strips is significantly decreased in TRPV1 knockout mice [67]. This notion is supported by recent data demonstrating that the volume-evoked rise in contractile amplitude in isolated rat bladders is effectively inhibited by ruthenium red and the TRPV1-selective antagonist capsazepine [68]. In addition to TRPV channels, a mechanosensitive function has also been proposed for the members of the polycystin TRP subfamily (transient receptor potential polycystinTRPP). In renal embryonic kidney cells, both TRPP1 and TRPP2 localize to primary cilia [69, 70], which are sensitive to fluid flow [71]. Expression of TRPP1 in ECs has been demonstrated both by reverse-transcription polymerase chain reaction and immunohistochemistry, while conflicting data have been reported on the endothelial expression of TRPP2 [72]. Other TRP channels from the TRPA (ANKTM1) and the TRPN (NOMPC) subfamilies also have mechanosensitive properties, and play a role in sound sensation in zebrafish [73, 74], but lack expression in ECs or mammalian homologs, respectively [57].

In addition to ion channels, activation of protein kinases has been implicated in local mechanotransduction [30]. Phosphorylation of the mitogen-activated protein kinases (MAPKs), extracellular signal-regulated kinase (ERK) 1/2, constitutes an early endothelial response to shear stress, but is itself mediated via tyrosine kinase signaling pathways [76], which may be activated via

Table 20.1 Activation modes and tissue distribution of TRP channel subfamilies with mechanosensitive function in vertebrates.

TRP channel	Modes of mechanical activation	EC expression	Detection method
TRPV1	hypotonicity, stretch	+	RT, IC
TRPV2	hypotonicity, stretch	+	RT, IHC
TRPV3	none	–	RT
TRPV4	hypotonicity, shear stress, stretch	+	RT, NB, IHC
TRPV5	none	–	RT
TRPV6	none	–	RT
TRPP1	shear stress (?)	+	RT, IC, IHC
TRPP2	shear stress (?)	+/–	IHC/IHC, WB
TRPA1	shear stress (in hair cells)	?	–

EC expression of TRP channels was tested for by immunostaining in cultured cells (IC), immunohistochemistry (IHC), Northern blot (NB), reverse transcription polymerase chain reaction (RT), or Western blot (WB).
Data compiled from Inoue *et al.* [75], Liedtke and Kim [61], O'Neil and Heller [57], and Yao and Garland [72].

mechanosensitive receptors or redox-sensitive pathways [77, 78]. Caveolae located in the endothelial plasma membrane have been proposed as possible sites of mechanosensitive protein kinase activity, because disruption of caveolae by mild detergent and anti-caveolin-1 antibodies prevents the activation of ERK by shear stress [79]. Importantly, caveolae have also been proposed as possible entry ports for extracellular Ca^{2+} [80]. Mice deficient in caveolin-1 show impaired endothelial Ca^{2+} fluxes in response to acetylcholine and Ca^{2+} entry is rescued in caveolin-1 knockout mice by reconstitution with endothelium-specific caveolin-1 [81]. Upon stimulation with bradykinin, the canonical TRP channel TRPC6 was demonstrated to translocate to caveolin-1-rich areas of the EC membrane where it mediates Ca^{2+} influx [82]. Physical and functional interaction with caveolin-1 was also demonstrated for TRPC1 in human pulmonary artery ECs [83], thus further substantiating the role for caveolae in endothelial Ca^{2+} influx and possibly mechanotransduction.

Due to their anchoring to the endothelial cytoskeleton [42], caveolae may link localized and disseminated mechanosensitive responses, and thus integrate centralized and decentralized concepts of mechanotransduction. Modulation of caveolae expression (e.g., by recruitment and derecruitment of caveolin-1 to the plasma membrane [84]) may thus provide a yet unexplored mechanism how cells can actively regulate their mechanosensitivity. In ECs, shear stress recruits caveolin-1 to the apical plasma membrane [85] where it is localized to newly formed caveolae [86] and may thus establish an intrinsic feedback loop to allow for an individual adaptation of cellular mechanosensation. This notion is supported by recent findings demonstrating that shear preconditioning modulates the phosphorylation of several downstream targets including endothelial NO synthase (endothelial nitric oxide synthase eNOS), Akt, caveolin-1, and ERK1/2 in response to an acute step in laminar shear stress [86, 87].

EFFECTS OF ACUTE PRESSURE STRESS ON THE PULMONARY ENDOTHELIUM

In 1748, Ippolito Francesco Albertini (1662–1738), a scholar of Marcello Malpighi at the University of Bologna, published a treatise entitled "Animadversiones super quibusdam difficilis respirationis vitiis a laesa cordis et praecordiorum structura pendentibus [Observations on certain diseases that produce difficulty in breathing and are caused by structural damage of the heart and precordia]," which gave the first scientific account of hydrostatic lung edema [88, 89]. The landmark work of Ernest Starling in the late nineteenth century outlined the role of hydrostatic forces in regulating fluid shifts across the capillary membrane [90], and its pathophysiological relevance in lung edema was confirmed in experiments by Gaar *et al.* who demonstrated a linear relationship between fluid content and capillary pressure in isolated perfused dog lungs [91].

Electron microscopic analyses of isolated rabbit lungs which had been perfused at high capillary pressures of around 29 mmHg provided first evidence that the formation of pulmonary edema cannot be explained solely by uniform membrane models of fluid exchange [92]. Instead, endothelial (and epithelial) lesions were demonstrated to result in distinct barrier leaks [93]. Subsequent work by West *et al.* showed that elevated hydrostatic pressure causes breaks and discontinuities in endothelial and epithelial membranes of the blood–gas barrier (Figures 20.3 and 20.4), and introduced the term "stress failure" of pulmonary capillaries [94, 95]. In rabbit lungs, they identified stress failure at capillary transmural pressures of 40 mmHg or higher, corresponding to a circumferential wall tension of around 25 dyn/cm [95]. With higher pressures the number of breaks per endothelial length increased while the average break lengths did not change [94]. In the intact lung, capillaries bulging into the alveolar space are stabilized by the surface tension of the alveolar lining layer [95]. This notion was elegantly confirmed by experiments in saline-filled isolated rabbit lungs in which abolition of the alveolar gas–liquid surface tension increased the number of breaks in the alveolo-capillary membrane at high transmural pressure [96]. Remarkably, the majority of breaks in the blood–gas barrier resealed after a transient exposure to high transmural pressures for 1 min. The rapid reversibility of capillary stress failure suggests that ECs can move along their underlying matrix by rapid disengagement and reattachment of cell adhesion molecules, causing breaks to open or close within minutes [97]. This view provides a mechanistic morphometric basis for the well-described reversibility of pressure-induced increases in pulmonary capillary permeability once pressure is reset to normal values [98, 99]. Ultrastructural changes in the capillary wall allow for extravasation of macromolecules and blood cells and thus, explain the presence of high concentrations of protein and cells in the bronchoalveolar lavage (BAL) fluid of isolated rabbit lungs at high transmural capillary pressures [94]. Increased levels of protein are also evident in edema fluid from patients with cardiogenic lung edema [100], confirming the notion that high transmural capillary pressures result in a high-permeability form of lung edema.

The earliest disruptions of the capillary endothelium were found to occur at capillary transmural pressures as low as 24 mmHg [101]. While physiological capillary pressures at rest are approximately 7 mmHg, pressures may exceed 25 mmHg not only in pathological conditions

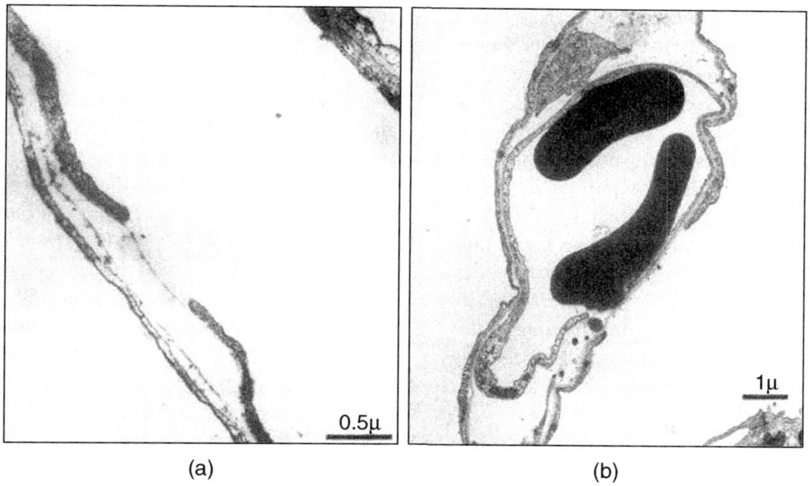

Figure 20.3 Electron micrographs of stress failure at raised capillary pressure in rabbit lungs. (a) Capillary endothelium is disrupted, but alveolar epithelium and basement membranes are intact. Capillary transmural pressure was 52.5 ± 2.5 cmH_2O. (b) Alveolar epithelium and capillary endothelium are disrupted, but basement membrane is intact. Capillary transmural pressure was 72.5 ± 2.5 cmH_2O. Reproduced from [95], with permission of the American Physiological Society.

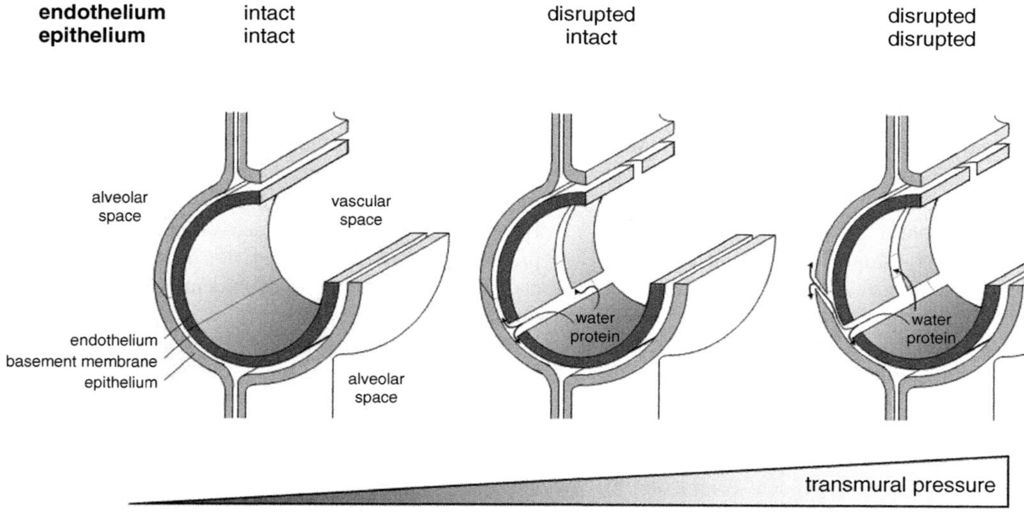

Figure 20.4 Schematic model of stress failure in an alveolar septal capillary. With increasing transmural pressure, the endothelial barrier disrupts and fluid and protein leak through the breaks into the interstitial space (center). Concomitant disruption of the alveolar epithelial layer (right) allows fluid, protein, and even red blood cells to enter the alveolar space.

such as left-sided heart disease [102], HAPE [103], or neurogenic pulmonary edema (NPE) [104], but also during strenuous exercise in healthy subjects [105]. After an uphill sprint at maximal effort, BAL of elite competition cyclists revealed red blood cells and increased protein concentrations suggestive of capillary stress failure [106]. Indeed, measurements of mean pulmonary arterial wedge pressure during severe exercise yielded values of up to 20 mmHg, which will result in capillary pressures greater than 25 mmHg at the lung base [107]. Probably the most prominent example of exercise-induced stress failure is the frequent occurrence of pulmonary hemorrhage in highly trained thoroughbreds after a race. During treadmill galloping these horses develop pulmonary arterial and left atrial pressures of 120 and 70 mmHg, respectively [108], and subsequent ultrastructural analyses yielded evidence for stress failure of pulmonary capillaries [109]. Further studies showed that the threshold pressure for inducing capillary breaks in the horse lung is very high (above 100 mmHg) due to its relatively thick blood–gas barrier [110], which may constitute a physiological protection mechanism against stress failure in this species. Conversely, in newborn rabbits with a very thin blood–gas barrier the threshold pressure was found to be as low as around 11 mmHg [111].

In addition to inflicting structural damage on the capillary barrier increased transmural pressure evokes active endothelial responses which contribute pivotally to pressure-induced lung pathology. This notion first arose from the observations of Bhattacharya *et al.*, who recorded weight changes in isolated perfused canine lung lobes at different venous outflow pressures [112]. While considerations based on the Starling principle or the stress failure concept would have predicted either a linear increase or a step increment in lung weight upon pressure elevation, the authors actually observed an exponential increase suggestive of a progressive deterioration of the lung capillary barrier. The notion of a pressure-induced dysregulation of the endothelial barrier function was subsequently confirmed in studies by Parker and Ivey, who detected a marked increase in the filtration coefficient (K_f) of isolated perfused rat lungs when venous outflow pressure was raised from 11 to 23 and 32 mmHg, respectively [113]. Extravasation of red blood cells was evident at 32 mmHg, but not at 23 mmHg, suggesting that the increase in K_f was not attributable to capillary stress failure. In contrast, partial attenuation of the permeability increase by administration of the β-adrenergic agonist isoproterenol indicated a regulated cellular mechanism. Parker and Ivey speculated that an isoproterenol-induced increase in adenosine 3′,5′-cAMP may counteract a Ca^{2+} and myosin light chain kinase-dependent contraction of endothelial cytoskeletal myofibrils [113].

By real-time imaging of endothelial second messenger responses [114], experiments from our group confirmed the notion of an endothelial Ca^{2+} response to pressure stress [12]. In the isolated perfused rat lung preparation, acute elevation of microvascular pressure resulted in two distinct and independent endothelial $[Ca^{2+}]_i$ responses, namely an endothelial Ca^{2+} influx via gadolinium-inhibitable cation channels and a concomitant Ca^{2+} release from intracellular stores which amplified endothelial $[Ca^{2+}]_i$ oscillations (Figure 20.5). Endothelial Ca^{2+} transients were induced by pressure elevations of as little as 4 mmHg and increased linearly in magnitude with vascular pressure over a range of 4–15 mmHg. Since pressure increments and microvascular distension correlate linearly over this pressure range [13], this finding indicates a directly proportional activation of Ca^{2+} entry channels by endothelial strain.

Recent evidence from the group of Mary Townsley and our own laboratory demonstrates that the endothelial Ca^{2+} response to pressure stress critically depends upon the mechanosensitive Ca^{2+} channel TRPV4 [64, 65]. Blocking of TRPV channels by ruthenium red or a genetic loss-of-function of TRPV4 results in an almost complete inhibition of the pressure-induced endothelial $[Ca^{2+}]_i$ increase (Figure 20.6). Ca^{2+} influx via TRPV4 may play a critical role in lung barrier failure and the formation of pulmonary edema because pharmacological activation of TRPV4 was shown to increase lung microvascular permeability [59]. This view is confirmed by recent data demonstrating that pharmacological inhibition or genetic deficiency of TRPV4 attenuates the pressure-induced increase in lung endothelial permeability and reduces lung edema formation [64, 65]. Hence, activation of TRPV4 is critical for endothelial mechanotransduction in response to circumferential stretch and stimulates downstream signaling cascades which contribute to pressure-induced lung pathology. Yet, it remains to be elucidated whether the Ca^{2+} channel itself is directly sensitive to strain. TRPV4 activity is regulated by various signaling molecules including epoxyeicosatrienoic acids [115], guanosine 3′,5′-cyclic monophosphate (Figure 20.7), or PACSIN 3, a protein implicated in vesicular trafficking and endocytosis [116]. Hence it is conceivable that activation of TRPV4 simply presents an early and critical step in the signaling cascade downstream from a yet unidentified strain-sensitive mechanosensor.

In addition to regulating microvascular permeability, Ca^{2+} influx activates a series of endothelial responses which are relevant for the pulmonary pathology at high vascular pressure. Real-time imaging of vesicular trafficking in pulmonary ECs demonstrated that pressure elevation triggers the exocytosis of endothelial vesicles [117]. Microinfusion of the styryl dye FM1-43, a fluorescent marker of exocytotic fusion pores [118], into lung venular capillaries reveals discrete fluorescent spots that cluster mainly at vessel bifurcations (Figures 20.8 and 20.9). While spots are relatively sparse at baseline, elevation of microvascular pressure increases the frequency of exocytotic spots per vessel wall surface. A characteristic feature is that the fluorescent spots are short-lived and within the same image, decay of fluorescent spots in one region co-occurs with the appearance of new fluorescent spots in adjacent regions (Figures 20.8 and 20.9). Pressure-induced exocytotic

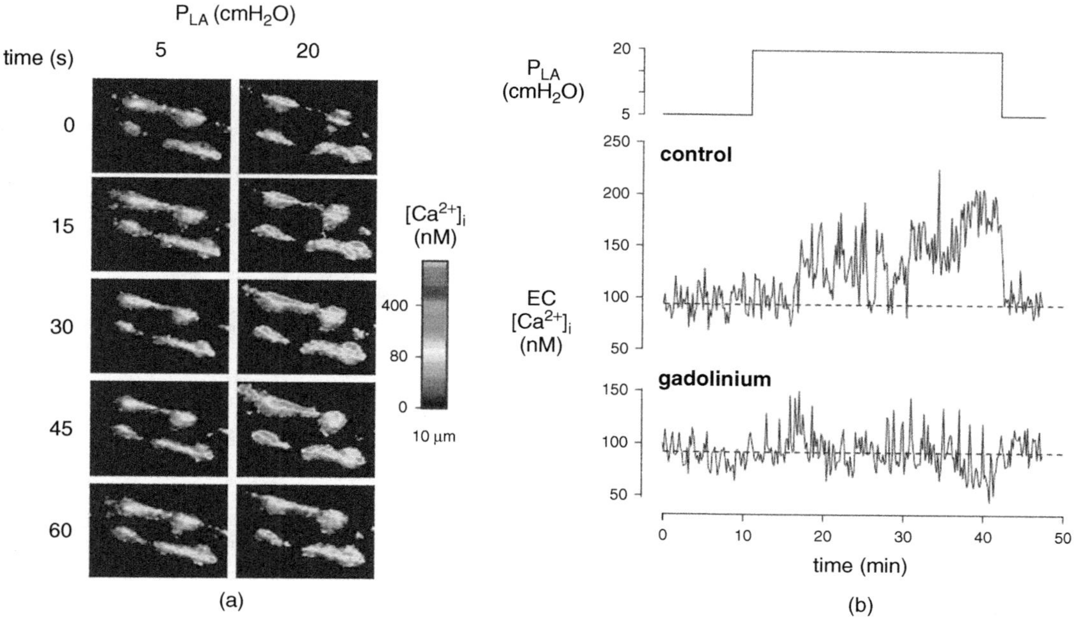

Figure 20.5 Endothelial $[Ca^{2+}]_i$ response to increased left atrial pressure (P_{LA}). (a) Sequential ratiometric images of Fura-2-loaded ECs in a lung venular capillary are color coded for endothelial $[Ca^{2+}]_i$. Images obtained in 15-s intervals at P_{LA} of 5 and 20 cmH_2O show $[Ca^{2+}]_i$ oscillations and pressure-induced increase in mean endothelial $[Ca^{2+}]_i$. (b) Representative $[Ca^{2+}]_i$ profiles in single ECs of lung venular capillaries in the absence (top) or presence (bottom) of gadolinium –an unspecific inhibitor of mechanosensitive cation channels. $[Ca^{2+}]_i$ is determined at baseline (P_{LA}, 5 cmH_2O), during 30 min of P_{LA} elevation to 20 cmH_2O and for 5 min after return to baseline P_{LA}. Gadolinium blocks the pressure-induced increase in mean endothelial $[Ca^{2+}]_i$, but does not affect $[Ca^{2+}]_i$ oscillations that originate from Ca^{2+} release from intracellular stores. A color version of this figure appears in the plate section of this volume.

events colocalize with the microvascular expression of P-selectin, identifying the exocytosed vesicles as endothelial Weibel–Palade bodies [117]. In the resting endothelium, these large vesicles serve as intracellular storage pools for P-selectin, von Willebrand factor, and interleukin-8 [119–121]. Pressure-induced exocytosis of Weibel–Palade bodies results in the expression of P-selectin on the microvascular endothelium where it initiates rolling and subsequent adhesion of circulating inflammatory cells and platelets [122]. Gadolinium effectively blocks pressure-induced Weibel–Palade body exocytosis (Figure 20.8) and P-selectin expression in lung microvessels, identifying Ca^{2+} influx via TRPV4 as direct trigger of this proinflammatory response [117]. This endothelial signaling cascade stimulates the recruitment of inflammatory cells into the lung as shown by Ichimura *et al.* [123] and data from our own group (Figure 20.10) demonstrating an accumulation of white blood cells in lung microvessels at elevated vascular pressure. A blocking anti-P-selectin antibody and the L- and P-selectin inhibitor fucoidin each inhibited the accumulation of leukocytes. Thus, the endothelial Ca^{2+} response to pressure initiates an inflammatory response in lung microvessels that is likely to underlie or at least to contribute to the alveolar influx of neutrophils and the upregulation of inflammatory mediators in clinical and experimental hydrostatic lung disease [94, 100, 124].

Pressure stress not only promotes barrier failure and inflammation, but also stimulates endothelial NO production in lung microvessels (Figure 20.11). eNOS activity is regulated via different signaling pathways, notably the binding of Ca^{2+}/calmodulin and the phosphorylation of a serine residue in the reductase domain (Ser^{1177}), which are generally considered to act independently [125] (see Chapter 6). Interestingly, activation of lung endothelial NO production by elevated vascular pressure could be blocked by either removal of extracellular Ca^{2+} or inhibition of phosphatidylinositol 3-kinase (PI3K), suggesting that mechano-induced activation of eNOS in lung ECs requires both Ca^{2+} influx and PI3K-dependent phosphorylation of the enzyme [126, 127]. Pressure-induced endothelial NO production occurs in lung microvessels

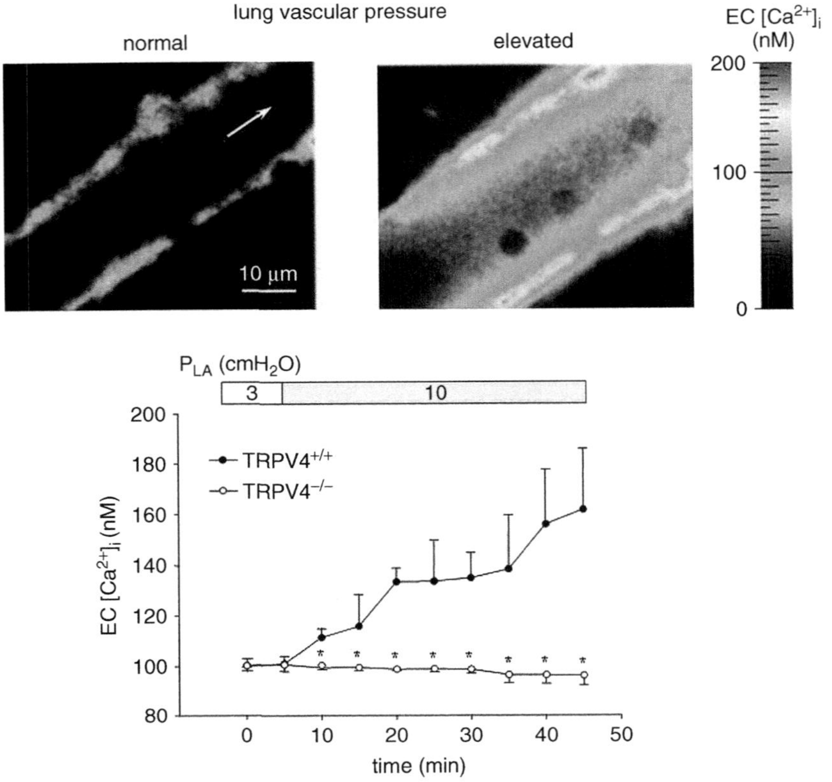

Figure 20.6 Role of TRPV4 in the pulmonary endothelial $[Ca^{2+}]_i$ response to acute pressure stress. Fluorescence microscopic images of a murine lung venular capillary show Fura-2-loaded ECs grayscale coded for $[Ca^{2+}]_i$ at baseline left atrial pressure (P_{LA}) of 3 cmH_2O (left) and 30 minutes after P_{LA} elevation to 10 cmH_2O (right). Note vessel distension at elevated P_{LA} indicating endothelial stretch, and the rise in fluorescence signal representing an increase in endothelial $[Ca^{2+}]_i$. Group data of endothelial $[Ca^{2+}]_i$ (EC $[Ca^{2+}]_i$) in lungs of TRPV4$^{-/-}$ and wild-type (TRPV4$^{+/+}$) mice are shown as 5-min averages at baseline left atrial pressure ($P_{LA} = 3$ cmH_2O) and over 40 min of P_{LA} elevation to 10 cmH_2O. The pressure-induced increase of EC $[Ca^{2+}]_i$ in TRPV4$^{+/+}$ is absent in TRPV4$^{-/-}$ mice. $*p < 0.05$ versus TRPV4$^{+/+}$.

of less than 30 µm diameter which lack smooth muscle cells and hence, vascular tone [128]. Endothelial-derived NO does therefore not induce vasodilation in these vessel segments, but may nevertheless play an important role in the pathophysiology of hydrostatic lung edema. In recent studies we could demonstrate that endothelial-derived NO attenuates endothelial Ca^{2+} influx via mechanosensitive TRPV4 channels by a cGMP-dependent mechanism. Hence, pressure-induced and Ca^{2+}-dependent activation of eNOS inhibits the endothelial Ca^{2+} influx in a negative feedback loop (Figure 20.11) and thus, limits the increase in microvascular permeability [65]. These findings would suggest that endothelial-derived NO may attenuate hydrostatic lung edema. Yet, in isolated mouse lungs perfused at elevated microvascular pressures, water content was actually increased after pharmacological inhibition of NO synthase or in lungs of eNOS knockout mice [126]. These seemingly contradictory findings are explained by the inhibitory effect of NO on alveolar fluid clearance, an ion transport-driven mechanism by which the alveolar epithelium absorbs fluid from the alveolar space to counteract lung edema formation [129]. Pressure-induced simulation of endothelial NO synthesis blocks this intrinsic rescue mechanisms and thus, promotes flooding of the alveolar space [126]. Hydrostatic lung edema thus presents another example of the well recognized Janus face of NO in pathophysiological scenarios, in that NO regulates precapillary vessel tone and strengthens the endothelial barrier but simultaneously promotes edema formation by inhibiting epithelial fluid absorption (Figure 20.12).

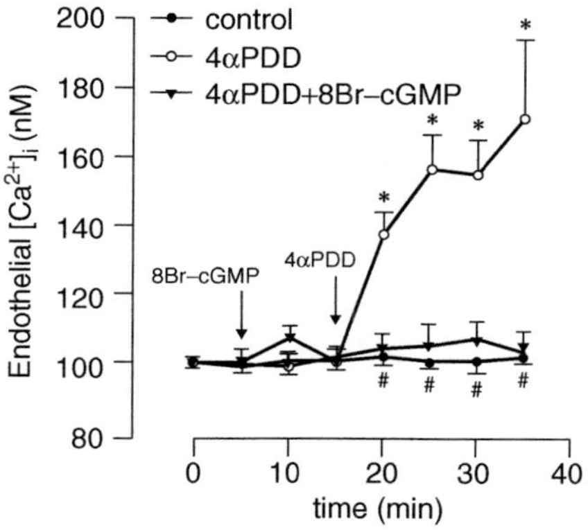

Figure 20.7 Regulation of TRPV4 by cGMP. Group data of endothelial $[Ca^{2+}]_i$ determined by Fura-2 ratiometric imaging in isolated-perfused rat lungs demonstrate the endothelial $[Ca^{2+}]_i$ increase in response to TRPV4 activation by 4α-phorbol 12,13-didecanoate (4α-PDD) (10 μmol/l) which is completely blocked by pretreatment with the cell-permeable cGMP analog 8Br-cGMP (100 μmol/l). $*p < 0.05$ versus control, $\#p < 0.05$ versus 4α-PDD.

Intercompartmental signaling between the vascular and the alveolar space by endothelial-derived NO may also account for other epithelial responses to increased vascular pressure, such as the release of surfactant [130, 131].

EFFECTS OF CHRONIC PRESSURE STRESS ON THE PULMONARY ENDOTHELIUM

Chronic elevation of lung microvascular pressures typically occurs as a consequence of left sided heart disease and can be detected in more than 60% of patients with heart failure of New York Heart Association classification class II–IV [132]. Chronic pressure stress results in structural and functional adaptations of lung ECs which determine the pulmonary pathology in heart failure. Morphometric analyses of lungs from a guinea-pig chronic heart failure model revealed a marked thickening of ECs at the alveolo-capillary membrane [133, 134] (Figure 20.13). Thickening and proliferation of ECs may reduce the luminal space of lung microvessels and thus, contribute to the characteristic hourglass-shaped vascular narrowings that have been identified by intravital microscopy in lungs from rats with chronic heart failure (Figure 20.14).

Another consistent clinical and experimental finding in heart failure is a dysfunction of the pulmonary endothelium, characterized by an impaired NO availability and increased expression of endothelin. The imbalanced release of endothelial-derived vasoactive mediators results in an increase of vascular smooth muscle tone and promotes pulmonary vascular remodeling [126, 135, 136]. The resulting rise in pulmonary vascular resistance further increases right ventricular afterload, limits right ventricular output, and may ultimately cause fatal right ventricular failure [132].

In rats with congestive heart failure due to supracoronary aortic banding, lung eNOS protein expression is not diminished, suggesting that lung endothelial dysfunction results from an impaired post-translational activation of the enzyme (Figure 20.15). In pulmonary artery segments from rats with chronic left ventricular failure following myocardial infarction, Ontkean *et al.* found that the vasodilator response to acetylcholine was impaired, whereas the response to the Ca^{2+} ionophore A23187 was normal [137]. Preliminary data from our own group confirmed that administration of A23187 reconstitutes endothelial NO production in lungs of heart failure rats [138]. The notion that endothelial Ca^{2+} signaling may be impaired in chronic pressure stress is also supported by data from the group of Mary Townsley, who showed that the endothelial permeability increase in response to various stimuli including histamine, angiotensin II, acute pressure elevation, or stimulation of capacitative Ca^{2+} entry with thapsigargin is virtually abolished in lungs of dogs with pacing induced heart failure [139–142]. In contrast, the Ca^{2+} ionophore A23187 increased permeability in both control and heart failure lungs, indicating again that lung endothelial Ca^{2+} signaling is impaired in chronic pressure stress [140, 143]. Mechanisms underlying the general lack of endothelial Ca^{2+} responses in heart failure are still unclear, but may involve downregulation of store-operated [143] as well as mechanosensitive (Figure 20.15) TRP channels. Such a fundamental impairment in endothelial second messenger responses will at first seem surprising. Yet, together with the above-mentioned endothelial thickening [133, 134], it may constitute an important protective mechanism by which the lung limits fluid filtration from pulmonary microvessels under conditions of chronically elevated vascular pressure and thus, prevents the formation of severe pulmonary edema [144, 145]. Endothelial dysfunction may additionally contribute to this protective effect, since the lack of endothelial NO production reconstitutes alveolar fluid clearance and thus, further counteracts edema formation [126].

EFFECTS OF FLOW ON THE PULMONARY ENDOTHELIUM

Over the past decade, the regulation of gene expression and cell function of ECs by blood flow and resulting shear stress has been a subject of intense research activities. These efforts have created new insights into important

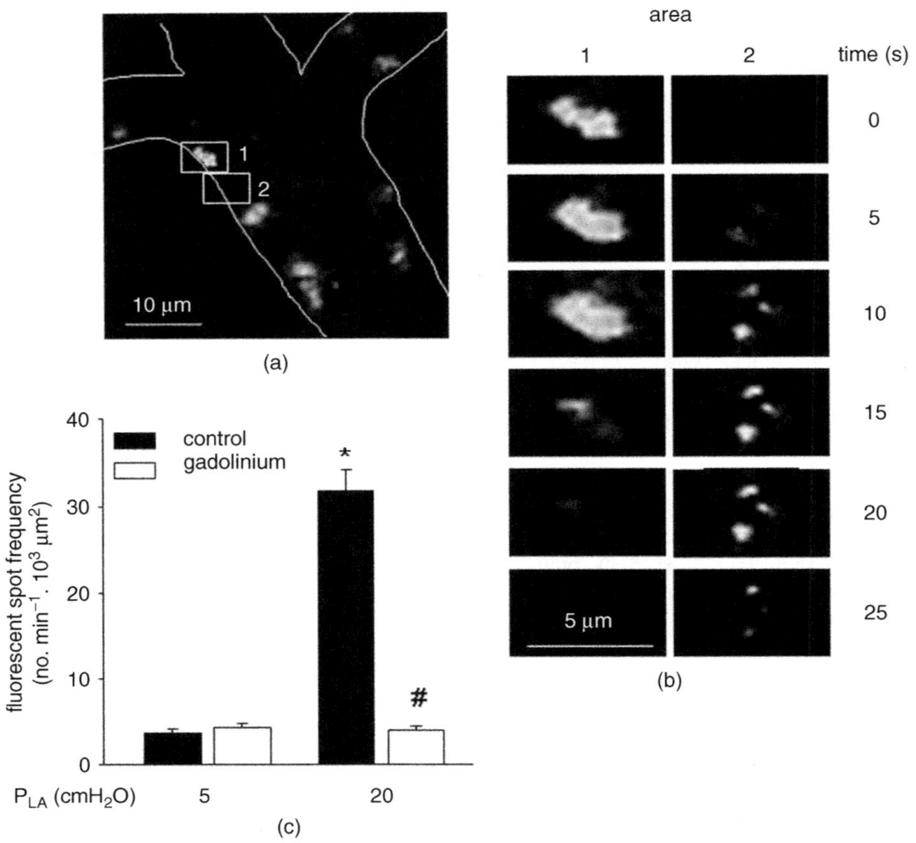

Figure 20.8 Formation of exocytotic fusion pores in lung venular capillaries. Image of a lung venular capillary perfused with the styryl dye FM1-43 shows several fluorescent spots representing exocytotic fusion pores (a). Vessel margins are depicted by line sketches. Temporal sequences of enlarged sections from the identical capillary reveal appearance and decay of large and small fluorescent spots at different time points in adjacent areas (b). Elevation of left atrial pressure (P_{LA}) from 5 to 20 cmH_2O increases the frequency of fusion pore formation but this effect is completely blocked by the unspecific inhibitor of mechanosensitive cation channels, gadolinium (c). $*p < 0.05$ versus $P_{LA} = 5$ cmH_2O, $\#p < 0.05$ versus control.

vascular mechanisms which underlie physiological regulations, such as in angioadaptation as well as pathophysiological processes such as atherosclerosis [146–148]. In contrast, studies focusing specifically on the effects of flow and shear stress in the pulmonary circulation are relatively scarce. Similar to the effects of pressure and stretch described before, the available data demonstrate both structural and functional changes to flow and shear stress in lung ECs, but the (patho-)physiological relevance of these changes is so far unclear.

Birukov *et al.* exposed bovine and human pulmonary artery endothelial monolayers in static culture to physiological relevant laminar shear stresses of 10 dyn/cm^2 and observed a rapid (less than 15 min) cytoskeletal reorganization with increased stress fiber formation in random orientation [149]. Prolonged exposure to shear stress over 24 h resulted in cell reorientation in the direction of flow and re-establishment of the prominent cortical actin ring. Inhibition of these adaptive responses by dominant-negative Rac1 identified a critical role for this small GTPase in the shear stress-induced cytoskeletal rearrangement of pulmonary ECs.

Functional endothelial responses to an abrupt cessation of flow (i.e., ischemia) were outlined in a series of elegant experiments by the group of Aron Fisher using lung surface fluorometry and intravital microscopy. In these studies the authors showed that ischemia in the normoxic lung leads to plasma membrane depolarization, an influx of extracellular Ca^{2+}, and the generation of reactive oxygen species (ROS) by the vascular NAD(P)H oxidase isoform

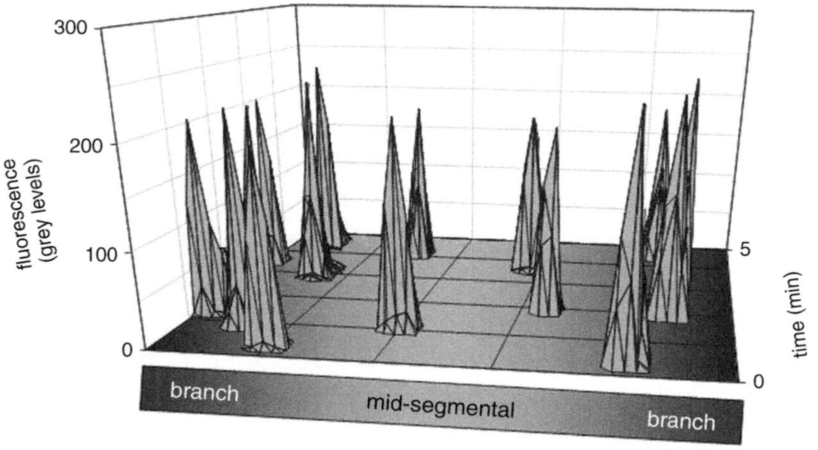

Figure 20.9 Temporospatial analysis of exocytotic fusion pore formation in a lung venular capillary at elevated microvascular pressure. Fluorescence intensity determined in a 5-μm band between two adjacent branch points is displayed over time. Note rapid appearance and decay of fluorescence spots that are clustered at the branch points (branch) as compared to midsegmental locations.

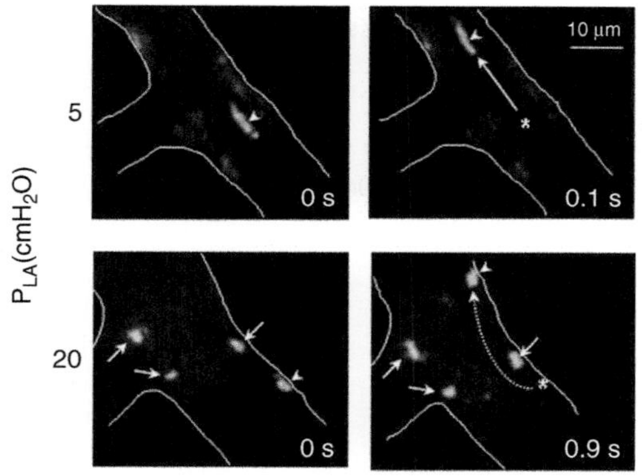

Figure 20.10 Pressure-induced leukocyte margination. Images show rhodamine 6G-labeled leukocytes in a lung venular capillary. (Top) Images taken at baseline left atrial pressure (P_{LA}) of 5 cmH_2O in 0.1-s intervals show the passage of a freely flowing leukocyte (arrowhead) in the vessel center stream. (Bottom) Images taken at elevated P_{LA} of 20 cmH_2O in 0.9-s intervals show three adherent leukocytes (arrows) and a rolling leukocyte (arrowhead) in the same lung microvessel which is now dilated by the increased hydrostatic pressure. Vessel margins are depicted by line sketches.

NOX2 [150–152]. The formation of intracellular ROS plays a critical role in the ischemia-induced activation of endothelial transcription factors [153] and MAPKs [154] as demonstrated by inhibition of NAD(P)H oxidases or addition of antioxidants and is likely to contribute to the increased peroxidation of lipids in non-hypoxic lung ischemia [150, 151].

Ischemia also results in endothelial NO synthesis which was blocked by removal of extracellular Ca^{2+} as well as by inhibitors of calmodulin or PI3K [155]. Hence, the pulmonary endothelial NO response to altered shear stress is strikingly similar to the reaction to elevated pressure and endothelial stretch, in that it depends on both endothelial Ca^{2+} entry and PI3K activation. Yet, the functional relevance of shear-dependent NO release in lung microvessels lacking smooth muscle remains to be determined.

Because the ischemia-induced increase in endothelial $[Ca^{2+}]_i$ was blocked by the K^+ channel agonist cromaglakim, Fisher *et al.* postulated the involvement of a flow-sensitive K^+ channel, which becomes inactivated in ischemia, thus causing membrane depolarization and subsequent Ca^{2+} influx via voltage-dependent Ca^{2+} channels [152, 155]. Due to the fact that ischemia-induced NO synthesis was blocked by the cholesterol-binding reagents filipin and cyclodextrin Wei *et al.* furthermore suggested plasma membrane cholesterol, possibly as a component of caveolae, as an additional shear stress sensor [154].

Based on these findings, it is tempting to hypothesize a role for the mechanosensitive TRPV channels in the pulmonary endothelial response to flow. As discussed before, TRP channels have been linked to caveolae [72] and several studies have demonstrated the sensitivity of TRPV4 to fluid flow and shear stress [57, 63].

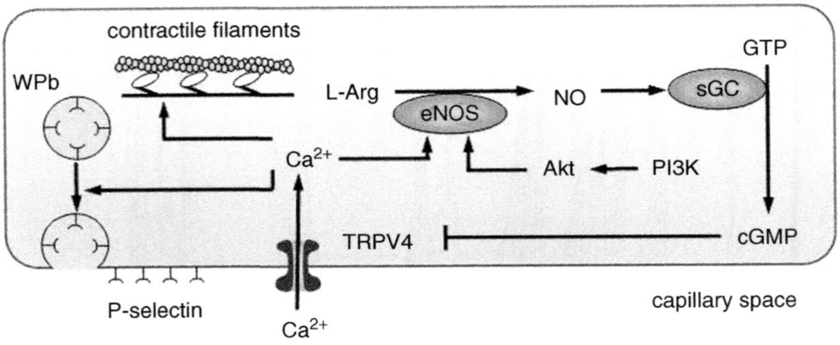

Figure 20.11 Pulmonary endothelial response to acute pressure stress. Schematic drawing shows pressure-induced Ca^{2+} entry into an EC via mechanosensitive TRPV4 channels and stimulation of the following downstream signaling cascades: (i) activation of contractile filaments with a subsequent increase in endothelial permeability, (ii) exocytosis of Weibel-Palade bodies (WPb) and surface expression of the pro-inflammatory adhesion molecule P-selectin, and (iii) Ca^{2+}- and PI3K-dependent activation of eNOS. The resulting formation of NO from L-arginine limits the endothelial $[Ca^{2+}]_i$ response by blocking TRPV4 channels in a negative, cGMP-regulated feedback loop. sGC, soluble guanylate cyclase.

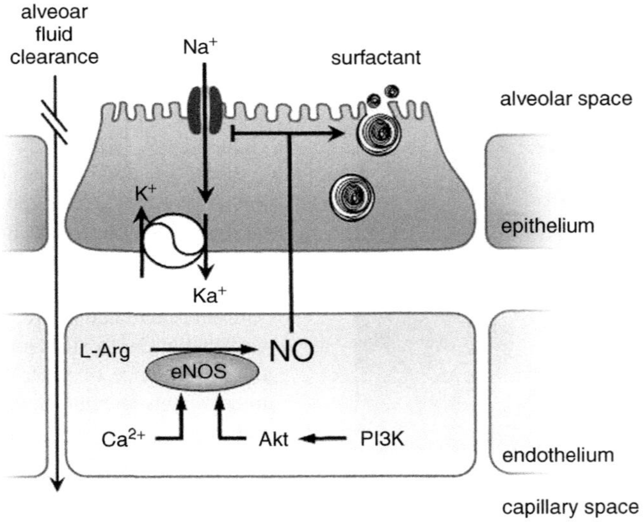

Figure 20.12 Intercompartmental signaling at the alveolo-capillary barrier. Increased vascular pressure blocks alveolar fluid clearance by inhibition of epithelial Na^+ transport and stimulates surfactant secretion from lamellar bodies. Ca^{2+} and PI3K-dependent activation of eNOS increases endothelial production of NO that acts as intercompartmental signal between the capillary and the alveolar space.

Ca^{2+} entry via activated TRPV4 increases endothelial permeability and may thus provide a mechanistic basis for lung edema formation under conditions of increased pulmonary blood flow [156]. This notion may shed new light into the long-standing and yet unresolved controversy whether changes in lung blood flow can cause edema formation –a problem that is traditionally complex due to the experimental difficulty to separate the effects of changes in flow, pressure, and surface area in the intact lung [157, 158].

MECHANICAL INJURY TO THE ENDOTHELIUM IN LUNG DISEASE

Effects of pressure and flow on the pulmonary endothelium appear to play a major role in several pathological states.

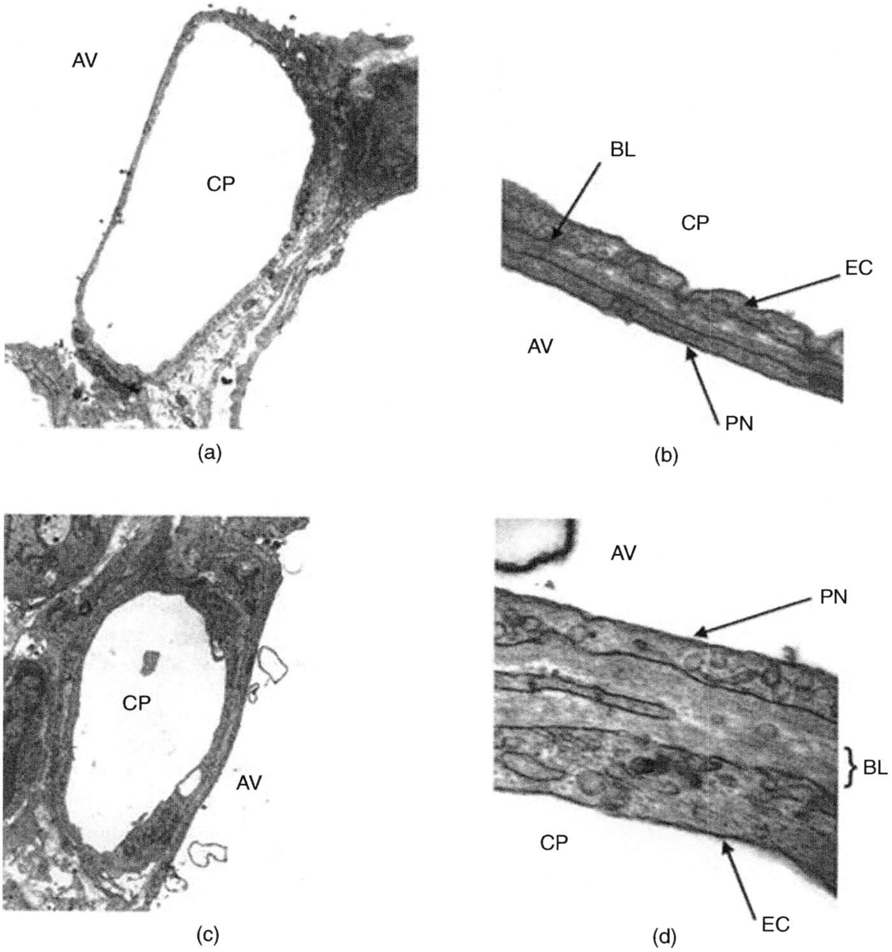

Figure 20.13 Photomicrographs of (a) low-power magnification (×7625) of normal lung showing orientation of capillary (CP) and alveolus (AV), (b) high-power magnification (×76 710) of normal lung, showing detail of basal laminae (BL), (c) low-power magnification of heart failure lung, to show orientation of capillary and alveolus, and (d) high-power magnification of heart failure lung to show detail of basal laminae. Note the thickening of the basal laminae, cellular infiltration and increased cell size in the heart failure lung. PN, type I pneumocyte. Reproduced from [133], with permission of Elsevier Science.

Cardiogenic Pulmonary Edema

Left heart failure results in increased vascular pressures in all segments of the pulmonary vasculature and subsequent formation of hydrostatic lung edema. In acute heart failure, pressure-induced increases in lung endothelial permeability [100] and simultaneous inhibition of alveolar fluid clearance [159] are likely to contribute considerably to lung edema formation. Concomitantly, endothelial activation seems to initiate proinflammatory responses which are reflected by increased cytokine levels [124] and the recruitment of inflammatory cells into the alveolar space [100]. The resulting parenchymal inflammation may be functionally relevant in as far as it may injure the alveolo-capillary barrier and thus account for the vulnerability of these patients to recurrent pulmonary fluid accumulation [160, 161]. In chronic heart failure, impairment of cellular second messenger signaling may dampen the endothelial response to pressure stress and thus, help to adapt the lung microvasculature to chronic pressure stress, while endothelial dysfunction at the same time promotes pulmonary hypertension and right ventricular failure [136].

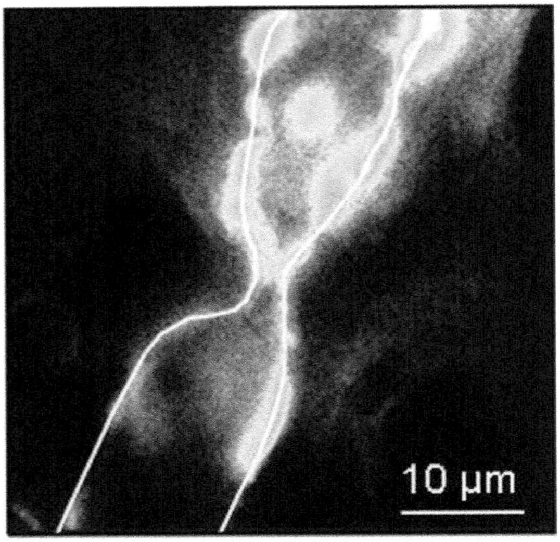

Figure 20.14 Image of Fura-2-loaded ECs in a lung venular capillary shows characteristic hourglass-shaped vascular narrowing in the lung from a rat with chronic heart failure nine weeks after aortic banding. Vessel margins are depicted by line sketches.

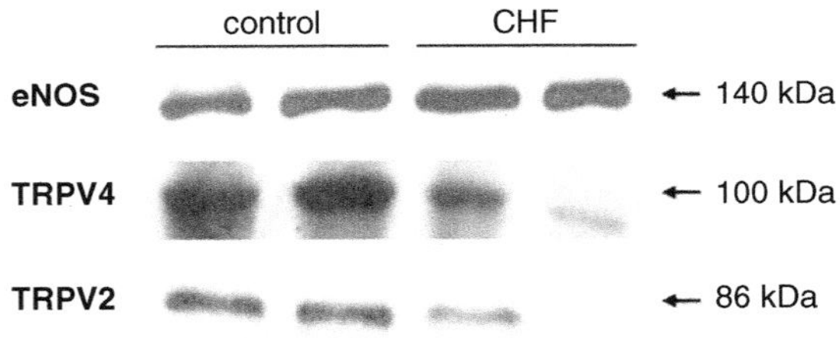

Figure 20.15 (Top) Representative Western blots of eNOS expression in freshly isolated ECs from lungs of untreated control rats or rats with chronic heart failure (CHF) 9 weeks after aortic banding. (Bottom) Representative Western blots of TRPV2 and TRPV4 expression in lungs of untreated control rats or chronic heart failure rats.

High-altitude Pulmonary Edema (HAPE)

Advanced HAPE shares many features of high-permeability type of edema in that it is characterized by a proteinaceous edema fluid and increased cytokine levels and neutrophil numbers in the BAL [162, 163] (see Chapter 18). However, there is a convincing body of evidence that the early stage of HAPE is hydrostatic edema. Individual susceptibility to HAPE has been linked to an exaggerated hypoxic pulmonary vasoconstrictive response [164, 165] and genetic factors such as polymorphisms in the genes encoding eNOS, angiotensin-converting enzyme, or endothelin-1 may contribute to this effect [166, 167]. The abnormal rise in pulmonary arterial pressure is accompanied by an increased pulmonary capillary pressure of 20–25 mmHg in HAPE-susceptible as compared to an average of 16 mmHg in nonsusceptible individuals [103]. The notion of a critical elevation in lung capillary pressure is supported by experimental data from Madison strain Sprague-Dawley rats which show a brisk pulmonary pressure response to acute hypoxia and are susceptible to HAPE. Ultrastructural lung examination after hypobaric hypoxia showed evidence of stress failure of pulmonary capillaries, such as disruption of the capillary endothelial layer, and red blood cells in the interstitial and alveolar spaces [168]. In humans, pulmonary capillary pressure correlates well with the radiographic features of HAPE and a concomitant decline in arterial oxygenation suggesting a causal relationship [103].

The mechanisms accounting for increased capillary pressures in HAPE are still under discussion, but a number of different hypotheses have been put forward. (i) Not only pulmonary arterioles, but also pulmonary venules constrict in response to hypoxia and thus increase pulmonary capillary pressure [169, 170]. (ii) In man, lung capillaries do not solely originate from small precapillaries, but frequently directly branch off from arterioles larger than 100 μm in diameter [171] (i.e., prior to the main resistance site of the pulmonary microvasculature [172]). These capillaries are therefore directly exposed to elevated pulmonary arterial pressures during hypoxia [173]. (iii) Increased capillary pressure has also been proposed to result from regional differences in hypoxic pulmonary vasoconstriction [174]. In areas with the least arterial vasoconstriction, capillaries will then be exposed to relatively higher pressures as compared to areas with a marked constrictive response. The notion of a spatial heterogeneity is supported by experimental data from pigs and dogs demonstrating non-uniform distributions of pulmonary blood flow in hypoxia [175, 176]. In a recent study in humans using functional magnetic resonance imaging, pulmonary blood flow heterogeneity was also found to be higher in HAPE-susceptible subjects exposed to hypoxia as compared to HAPE-resistant subjects [177]. Regional heterogeneities in blood flow have also been proposed to contribute directly to the formation of HAPE [178, 179]. This hypothesis is based on the notion that regional overperfusion in areas with low vasoconstriction will result in an increase in capillary pressure that is required to overcome pulmonary venous resistance at high flow [180]. Yet, the resulting pressure increase is probably relatively low. By use of double-occlusion and blue dextran elution techniques in isolated perfused rat lungs [181, 182], we determined that a step increment in perfusion rate by 70% increases lung vascular surface area by around 50%, but elevates lung capillary pressure by only around 7% (Figure 20.16). Regional overperfusion may nevertheless contribute importantly to the pathophysiology of HAPE by activation of mechanosensitive TRP channels. As discussed before, TRPV4, which is expressed in lung ECs and mediates lung edema formation [59], is not only responsive to mechanical stretch but similarly to shear stress [57, 63]. The notion of a role for shear stress in HAPE is also in agreement with the clinical observation that in many cases, particularly at lower elevations, exercise may be the essential component in the pathogenesis of this disease [179].

Neurogenic Pulmonary Edema (NPE)

NPE may develop in individuals with head trauma or seizures and is considered to have a hydrostatic basis due to the severe degree of pulmonary hypertension that occurs [183, 184]. "Blast injury" has been proposed as the underlying pathogenetic mechanism and refers to a sudden increase in intracranial pressure which triggers a transient, yet dramatic, α-adrenergic vasoconstrictive response in both the systemic and the pulmonary circulation [185]. The formation of NPE appears to be promoted in many cases by an increased microvascular permeability in the lung as suggested from animal studies demonstrating high interstitial or alveolar protein concentrations [186, 187]. Similarly, several clinical studies detected a proteinaceous edema fluid in NPE suggestive of high-permeability type of edema [188, 189]. The mechanisms underlying the permeability increase are incompletely understood but may comprise direct endothelial effects of sympathetic neurotransmitters such as neuropeptide Y [190], inflammatory mechanisms [185], or pressure-dependent endothelial injury [191]. In support of the latter theory, elevated pulmonary artery wedge pressures have been observed in a few cases in humans [184, 192]. In a relatively large group of 12 patients, Smith and Matthay found that in the majority of cases the initial alveolar edema fluid to plasma protein concentration was 0.65 or less, suggesting an underlying hydrostatic mechanism [193]. None of these patients had cardiac failure or intravascular volume overload, indicating that mechanisms underlying the increase in lung capillary pressure may be similar to those discussed for HAPE. Thus, NPE seems to be at least in part attributable to mechanical injury to the pulmonary endothelium.

Ventilator-induced Lung Injury (VILI)

Mechanical ventilation with high tidal volumes results in rapid and diverse endothelial responses which promote inflammatory processes and edema formation and may thus play a central role in the pathophysiology of VILI. At plateau airway pressures above 35 mmHg, baro- and/or volutrauma can result in stress failure of endothelial and epithelial barriers with subsequent hemorrhage into the airspace and recruitment of inflammatory cells [1, 194, 195]. Lower inflation pressures of 30 mmHg increase lung microvascular permeability, and this effect can be fully inhibited by the mechanosensitive cation channel blocker gadolinium [196]. Interestingly, the permeability increases similarly in alveolar and extra-alveolar vessels [197] that, as discussed in "Mechanical Forces Acting on the Pulmonary Endothelium", show opposing changes in circumferential but parallel changes in longitudinal

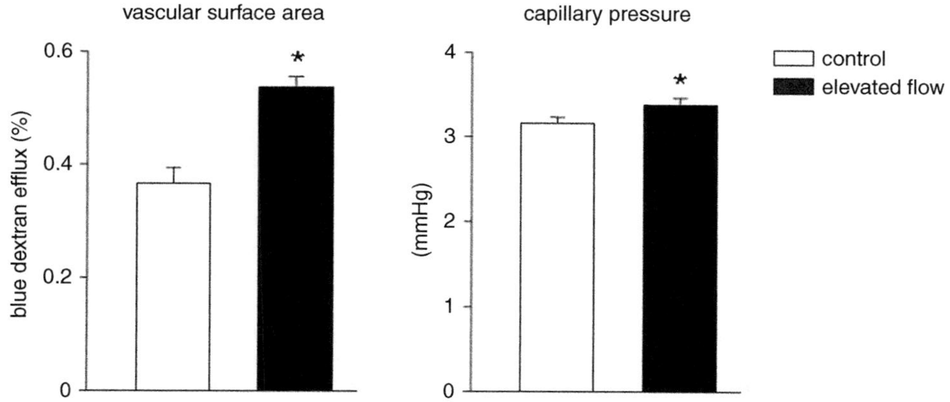

Figure 20.16 Effect of increased lung perfusion on capillary recruitment and pressure. Group data from isolated-perfused rat lungs show increases in vascular surface area and capillary pressure following a 70% flow increase. $*p < 0.05$ versus control.

strain. Thus, ventilation-induced elongation of lung blood vessel may be the predominant trigger for transcapillary fluid and protein leak. Recently, Hamanaka *et al.* could demonstrate that lung distention causes endothelial Ca^{2+} entry in isolated mouse lungs, thus providing a mechanistic basis for the observed gadolinium-sensitive increase in filtration coefficient K_f [198]. More importantly, both the K_f increase and the Ca^{2+} influx were absent in the presence of the TRPV4 inhibitor ruthenium red or in lungs of $TRPV4^{-/-}$ mice. Thus, VILI shares common pathophysiological characteristics with the effects of acute vascular pressure elevation in that TRPV4 mediates an endothelial Ca^{2+} response and the subsequent permeability change. This parallelism also applies to the endothelial NO response to overventilation, which is again dependent on PI3K and likely contributes to the impairment of alveolar fluid absorption in VILI [199] by similar mechanisms as identified in cardiogenic lung edema [126].

Furthermore, high-tidal volume ventilation with 12 ml/kg body weight causes a structural remodeling of the endothelial barrier as demonstrated by an increased formation of focal adhesions and tyrosine phosphorylation of focal adhesion proteins [200]. The concomitant increase in endothelial P-selectin may be relevant for the rapid and massive recruitment of leukocytes to the lung [29] and appears to be amplified by the interaction of the endothelium with circulating inflammatory cells [201]. Thus, endothelial responses to ventilation-dependent vascular stretch appear to play a major role in the initiation of the pathophysiological and clinical hallmarks of VILI (i.e., edema and inflammation).

CONCLUSIONS AND PERSPECTIVES

The pulmonary endothelium is continuously exposed to mechanical forces exerted by vascular and airspace pressures and hemodynamic flow. Importantly, these forces are not static in nature, but oscillate with cardiac and respiratory movements, resulting in continuous and superimposed changes in shear stress, and circumferential and longitudinal endothelial stretch. Excessive mechanical forces either cause ultrastructural damage or even physical disruption of ECs resulting in stress failure of the vascular barrier, or activate endothelial mechanosensors and downstream signaling pathways which promote inflammatory responses and pulmonary edema formation. Consequently, endothelial mechanotransduction constitutes a critical pathophysiological mechanism in a variety of lung diseases including cardiogenic, neurogenic, and high-altitude pulmonary edema. VILI as a iatrogenic disease constitutes a particular challenge in this context, and strategies to minimize mechanical stress such as high-frequency oscillatory ventilation need to be further developed, refined, and implemented into clinical routine.

Substantial and comprehensive research efforts are required to improve our understanding of endothelial mechanosensing and mechanotransduction pathways and to integrate the different existing concepts. Recently, seminal work in prokaryotes and invertebrates has led to the identification of a group of mechanosensitive TRP ion channels, of which TRPV4 has been recognized in particular to play a central role in the activation of endothelial Ca^{2+} signaling and the regulation of microvascular permeability in lung pathology following mechanical stress. A series of specific TRPV4 channels blockers currently undergo preclinical assessment, and may provide new therapeutic tools for the prevention of lung edema and inflammation caused by excess hemodynamic or respiratory forces. Concomitantly, we need to elucidate the regulation of TRPV4 and understand whether (and if so, how) this channel recognizes mechanical forces itself, or rather is downstream of a structurally interacting, close or potentially even distant primary mechanosensor yet to be identified.

Chronic pressure stress in the pulmonary vasculature results in lung endothelial dysfunction which promotes vasoconstriction and smooth muscle cell hypertrophy of lung resistance vessels, thus contributing critically to pulmonary hypertension in patients with atrial, valvular, or ventricular left heart disease. Cellular mechanisms underlying endothelial dysfunction in pulmonary hypertension with left heart disease remain to be elucidated, but seem to involve a unique impairment in endothelial Ca^{2+} signaling. Further insights into this process may not only provide a better understanding of how ECs adapt to mechanical stress, but shed new lights on basic principles of Ca^{2+} homeostasis and signaling in the vascular wall.

Currently, there is no specific treatment for pulmonary hypertension with left heart disease. While additional treatment options for this large patient population are desperately in need, it should be considered that lung vascular adaptation to chronic pressure stress constitutes an important rescue mechanism: while endothelial dysfunction promotes pulmonary hypertension, it concomitantly brings about a decrease in vascular permeability and an increase in alveolar fluid absorption, thus providing critical protection from hydrostatic lung edema. New therapeutic strategies will thus have to walk a tightrope in aiming to reduce right ventricular afterload without aggravating the risk for pulmonary edema.

The recognition that ECs respond actively to pressure and flow, and thus contribute significantly to lung pathology in scenarios where respiratory or hemodynamic forces are altered, poses new challenges both for basic scientists and physicians, yet also provides new and exciting opportunities to gain novel insights into cellular mechanotransduction and pulmonary vascular regulation, and to devise new treatment strategies for old clinical problems.

ACKNOWLEDGMENTS

The author's cited research work was supported by grants from the Deutsche Forschungsgemeinschaft (Ku1218/1, Ku1218/4, Ku1218/5 and GRK 865); the European Commission under the Sixth Framework Program (contract LSHM-CT-2005-018725, PULMOTENSION); Pfizer GmbH, Karlsruhe, Germany; and the Kaiserin-Friedrich Foundation, Berlin, Germany. I am indebted to Julia Hoffmann and Stephanie Kaestle for help in preparation of the manuscript, and to Jahar Bhattacharya, Jun Yin, Wolfgang Liedtke, and Ning Yin for their valuable contributions to the presented data.

References

1. Uhlig, S. (2002) Ventilation-induced lung injury and mechanotransduction: stretching it too far? *American Journal of Physiology: Lung Cellular and Molecular Physiology*, **282**, L892–96.
2. Fahraeus, R. (1929) The suspension stability of the blood. *Physiological Reviews*, **9**, 241–71.
3. Frame, M.D. and Sarelius, I.H. (1996) Endothelial cell dilatory pathways link flow and wall shear stress in an intact arteriolar network. *Journal of Applied Physiology*, **81**, 2105–14.
4. Kuebler, W.M., Kuhnle, G.E., Groh, J., and Goetz, A.E. (1994) Leukocyte kinetics in pulmonary microcirculation: intravital fluorescence microscopic study. *Journal of Applied Physiology*, **76**, 65–71.
5. Kuebler, W.M., Kuhnle, G.E., Groh, J., and Goetz, A.E. (1997) Contribution of selectins to leucocyte sequestration in pulmonary microvessels by intravital microscopy in rabbits. *The Journal of Physiology*, **501**, 375–86.
6. Hudlicka, O., Brown, M.D., May, S. *et al.* (2006) Changes in capillary shear stress in skeletal muscles exposed to long-term activity: role of nitric oxide. *Microcirculation*, **13**, 249–59.
7. Bergel, D.H., Caro, C.G., and Seed, W.A. (1967) Asymmetrical transmission of pressure waves in the pulmonary vascular system of the dog. *The Journal of Physiology*, **188**, 41P–42P.
8. Presson, R.G. Jr., Baumgartner, W.A. Jr., Peterson, A.J. *et al.* (2002) Pulmonary capillaries are recruited during pulsatile flow. *Journal of Applied Physiology*, **92**, 1183–90.
9. Wagner, W.W. Jr., Todoran, T.M., Tanabe, N. *et al.* (1999) Pulmonary capillary perfusion: intra-alveolar fractal patterns and interalveolar independence. *Journal of Applied Physiology*, **86**, 825–31.
10. Baumgartner, W.A. Jr., Peterson, A.J., Presson, R.G. Jr. *et al.* (2004) Blood flow switching among pulmonary capillaries is decreased during high hematocrit. *Journal of Applied Physiology*, **97**, 522–26.
11. Yu, Q., Zhou, J., and Fung, Y.C. (1993) Neutral axis location in bending and Young's modulus of different layers of arterial wall. *American Journal of Physiology: Heart and Circulatory Physiology*, **265**, H52–60.
12. Kuebler, W.M., Ying, X., and Bhattacharya, J. (2002) Pressure-induced endothelial Ca^{2+} oscillations in lung capillaries. *American Journal of Physiology: Lung Cellular and Molecular Physiology*, **282**, L917–23.
13. Sadurski, R., Tsukada, H., Ying, X. *et al.* (1994) Diameters of juxtacapillary venules determined by oil-drop method in rat lung. *Journal of Applied Physiology*, **77**, 718–25.
14. Hillier, S.C., Godbey, P.S., Hanger, C.C. *et al.* (1993) Direct measurement of pulmonary microvascular distensibility. *Journal of Applied Physiology*, **75**, 2106–11.
15. Spohr, F., Cornelissen, A.J., Busch, C. *et al.* (2005) Role of endogenous nitric oxide in endotoxin-induced alteration of hypoxic pulmonary vasoconstriction in mice. *American Journal of Physiology: Heart and Circulatory Physiology*, **289**, H823–31.
16. al-Tinawi, A., Madden, J.A., Dawson, C.A. *et al.* (1991) Distensibility of small arteries of the dog lung. *Journal of Applied Physiology*, **71**, 1714–22.
17. Molthen, R.C., Karau, K.L., and Dawson, C.A. (2004) Quantitative models of the rat pulmonary arterial tree morphometry applied to hypoxia-induced arterial remodeling. *Journal of Applied Physiology*, **97**, 2372–84.
18. al-Tinawi, A., Clough, A.V., Harder, D.R. *et al.* (1992) Distensibility of small veins of the dog lung. *Journal of Applied Physiology*, **73**, 2158–65.
19. Eiken, O. and Kolegard, R. (2004) Comparison of vascular distensibility in the upper and lower extremity. *Acta Physiologica Scandinavica*, **181**, 281–87.
20. Kassab, G.S., Le, K.N., and Fung, Y.C. (1999) A hemodynamic analysis of coronary capillary blood flow based on anatomic and distensibility data. *American Journal of Physiology: Heart and Circulatory Physiology*, **277**, H2158–66.

21. Kindig, C.A. and Poole, D.C. (1999) Effects of skeletal muscle sarcomere length on *in vivo* capillary distensibility. *Microvascular Research*, **57**, 144–52.
22. Howell, J.B., Permutt, S., Proctor, D.F., and Riley, R.L. (1961) Effect of inflation of the lung on different parts of pulmonary vascular bed. *Journal of Applied Physiology*, **16**, 71–76.
23. Marini, J.J. (2004) Microvasculature in ventilator-induced lung injury: target or cause? *Minerva Anestesiologica*, **70**, 167–73.
24. Lai-Fook, S.J. (1982) Perivascular interstitial fluid pressure measured by micropipettes in isolated dog lung. *Journal of Applied Physiology*, **52**, 9–15.
25. Albert, R.K., Lamm, W.J., Rickaby, D.A. *et al.* (1993) Lung inflation distends small arteries (<1mm) in excised dog lungs. *Journal of Applied Physiology*, **75**, 2595–601.
26. Wanner, A., Zarzecki, S., and Sackner, M.A. (1975) Effects of lung inflation on pulmonary arterial blood volume in intact dogs. *Journal of Applied Physiology*, **38**, 675–80.
27. Macklin, C.C. (1946) Evidence of increase in the capacity of the pulmonary arteries and veins of dogs, cats and rabbits, during inflation of the freshly excised lung. *Revue Canadienne de Biologie*, **5**, 199–232.
28. Sun, R.Y., Nieman, G.F., Hakim, T.S., and Chang, H.K. (1987) Effects of lung volume and alveolar surface tension on pulmonary vascular resistance. *Journal of Applied Physiology*, **62**, 1622–26.
29. Kuebler, W.M. (2005) Inflammatory pathways and microvascular responses in the lung. *Pharmacological Reports*, **57** (Suppl), 196–205.
30. Ali, M.H. and Schumacker, P.T. (2002) Endothelial responses to mechanical stress: where is the mechanosensor? *Critical Care Medicine*, **30**, S198–S206.
31. Helmke, B.P. and Davies, P.F. (2002) The cytoskeleton under external fluid mechanical forces: hemodynamic forces acting on the endothelium. *Annals of Biomedical Engineering*, **30**, 284–96.
32. Ingber, D.E. (1997) Tensegrity: the architectural basis of cellular mechanotransduction. *Annual Review of Physiology*, **59**, 575–99.
33. Wang, N., Butler, J.P., and Ingber, D.E. (1993) Mechanotransduction across the cell surface and through the cytoskeleton. *Science*, **260**, 1124–27.
34. Janmey, P.A., Euteneuer, U., Traub, P., and Schliwa, M. (1991) Viscoelastic properties of vimentin compared with other filamentous biopolymer networks. *The Journal of Cell Biology*, **113**, 155–60.
35. Davies, P.F. and Tripathi, S.C. (1993) Mechanical stress mechanisms and the cell. An endothelial paradigm. *Circulation Research*, **72**, 239–45.
36. Chen, K.D., Li, Y.S., Kim, M. *et al.* (1999) Mechanotransduction in response to shear stress. Roles of receptor tyrosine kinases, integrins, and Shc. *The Journal of Biological Chemistry*, **274**, 18393–400.
37. Takahashi, M., Ishida, T., Traub, O. *et al.* (1997) Mechanotransduction in endothelial cells: temporal signaling events in response to shear stress. *Journal of Vascular Research*, **34**, 212–19.
38. Sastry, S.K. and Burridge, K. (2000) Focal adhesions: a nexus for intracellular signaling and cytoskeletal dynamics. *Experimental Cell Research*, **261**, 25–36.
39. Ko, K.S., Arora, P.D., and McCulloch, C.A. (2001) Cadherins mediate intercellular mechanical signaling in fibroblasts by activation of stretch-sensitive calcium-permeable channels. *The Journal of Biological Chemistry*, **276**, 35967–77.
40. Shasby, D.M. (2007) Cell–cell adhesion in lung endothelium. *American Journal of Physiology: Lung Cellular and Molecular Physiology*, **292**, L593–607.
41. Kawamura, S., Miyamoto, S., and Brown, J.H. (2003) Initiation and transduction of stretch-induced RhoA and Rac1 activation through caveolae: cytoskeletal regulation of ERK translocation. *The Journal of Biological Chemistry*, **278**, 31111–17.
42. van Deurs, B., Roepstorff, K., Hommelgaard, A.M., and Sandvig, K. (2003) Caveolae: anchored, multifunctional platforms in the lipid ocean. *Trends in Cell Biology*, **13**, 92–100.
43. Pries, A.R. and Kuebler, W.M. (2006) Normal endothelium, in *Handbook of Experimental Pharmacology: The Vascular Endothelium* (eds S. Moncada and A. Higgs), Springer, Berlin, pp. 1–40.
44. Weinbaum, S., Zhang, X., Han, Y. *et al.* (2003) Mechanotransduction and flow across the endothelial glycocalyx. *Proceedings of the National Academy of Sciences of the United States of America*, **100**, 7988–95.
45. Olesen, S.P., Clapham, D.E., and Davies, P.F. (1988) Haemodynamic shear stress activates a K^+ current in vascular endothelial cells. *Nature*, **331**, 168–70.
46. Alevriadou, B.R., Eskin, S.G., McIntire, L.V., and Schilling, W.P. (1993) Effect of shear stress on $^{86}Rb^+$ efflux from calf pulmonary artery endothelial cells. *Annals of Biomedical Engineering*, **21**, 1–7.

47. Hutcheson, I.R. and Griffith, T.M. (1994) Heterogeneous populations of K^+ channels mediate EDRF release to flow but not agonists in rabbit aorta. *American Journal of Physiology: Heart and Circulatory Physiology*, **266**, H590–96.
48. Schwarz, G., Callewaert, G., Droogmans, G., and Nilius, B. (1992) Shear stress-induced calcium transients in endothelial cells from human umbilical cord veins. *The Journal of Physiology*, **458**, 527–38.
49. Schwarz, G., Droogmans, G., and Nilius, B. (1992) Shear stress induced membrane currents and calcium transients in human vascular endothelial cells. *Pflugers Archiv*, **421**, 394–96.
50. Lansman, J.B., Hallam, T.J., and Rink, T.J. (1987) Single stretch-activated ion channels in vascular endothelial cells as mechanotransducers? *Nature*, **325**, 811–13.
51. Hoyer, J., Kohler, R., Haase, W., and Distler, A. (1996) Up-regulation of pressure-activated Ca^{2+}-permeable cation channel in intact vascular endothelium of hypertensive rats. *Proceedings of the National Academy of Sciences of the United States of America*, **93**, 11253–58.
52. Hoyer, J., Kohler, R., and Distler, A. (1997) Mechanosensitive cation channels in aortic endothelium of normotensive and hypertensive rats. *Hypertension*, **30**, 112–19.
53. Kohler, R., Schonfelder, G., Hopp, H. *et al.* (1998) Stretch-activated cation channel in human umbilical vein endothelium in normal pregnancy and in preeclampsia. *Journal of Hypertension*, **16**, 1149–56.
54. Naruse, K. and Sokabe, M. (1993) Involvement of stretch-activated ion channels in Ca^{2+} mobilization to mechanical stretch in endothelial cells. *American Journal of Physiology: Cell Physiology*, **264**, C1037–44.
55. Naruse, K., Yamada, T., and Sokabe, M. (1998) Involvement of SA channels in orienting response of cultured endothelial cells to cyclic stretch. *American Journal of Physiology: Heart and Circulatory Physiology*, **274**, H1532–38.
56. Yang, X.C. and Sachs, F. (1989) Block of stretch-activated ion channels in Xenopus oocytes by gadolinium and calcium ions. *Science*, **243**, 1068–71.
57. O'Neil, R.G. and Heller, S. (2005) The mechanosensitive nature of TRPV channels. *Pflugers Archiv*, **451**, 193–203.
58. Hoenderop, J.G., Nilius, B., and Bindels, R.J. (2005) Calcium absorption across epithelia. *Physiological Reviews*, **85**, 373–422.
59. Alvarez, D.F., King, J.A., Weber, D. *et al.* (2006) Transient receptor potential vanilloid 4-mediated disruption of the alveolar septal barrier: a novel mechanism of acute lung injury. *Circulation Research*, **99**, 988–95.
60. Fantozzi, I., Zhang, S., Platoshyn, O. *et al.* (2003) Hypoxia increases AP-1 binding activity by enhancing capacitative Ca^{2+} entry in human pulmonary artery endothelial cells. *American Journal of Physiology: Lung Cellular and Molecular Physiology*, **285**, L1233–45.
61. Liedtke, W. and Kim, C. (2005) Functionality of the TRPV subfamily of TRP ion channels: add mechano-TRP and osmo-TRP to the lexicon *Cellular and Molecular Life Sciences*, **62**, 2985–3001.
62. Hartmannsgruber, V., Heyken, W.T., Kacik, M. *et al.* (2007) Arterial response to shear stress critically depends on endothelial TRPV4 expression. *PLoS One*, **2**, e827.
63. Kohler, R., Heyken, W.T., Heinau, P. *et al.* (2006) Evidence for a functional role of endothelial transient receptor potential V4 in shear stress-induced vasodilatation. *Arteriosclerosis, Thrombosis, and Vascular Biology*, **26**, 1495–502.
64. Jian, M.Y., King, J.A., Al-Mehdi, A.B. *et al.* (2008) High vascular pressure-induced lung injury requires P450 epoxygenase-dependent activation of TRPV4. *American Journal of Respiratory Cell and Molecular Biology*, **38**, 386–92.
65. Yin, J., Hoffmann, J., Kaestle, S.M. *et al.* (2008) Negative-feedback loop attenuates hydrostatic lung edema via a cGMP-dependent regulation of transient receptor potential vanilloid 4. *Circulation Research*, **102**, 966–74.
66. Muraki, K., Iwata, Y., Katanosaka, Y. *et al.* (2003) TRPV2 is a component of osmotically sensitive cation channels in murine aortic myocytes. *Circulation Research*, **93**, 829–38.
67. Birder, L.A., Nakamura, Y., Kiss, S. *et al.* (2002) Altered urinary bladder function in mice lacking the vanilloid receptor TRPV1. *Nature Neuroscience*, **5**, 856–60.
68. Gevaert, T., Vandepitte, J., Hutchings, G. *et al.* (2007) TRPV1 is involved in stretch-evoked contractile changes in the rat autonomous bladder model: a study with piperine, a new TRPV1 agonist. *Neurourology and Urodynamics*, **26**, 440–50.
69. Nauli, S.M., Alenghat, F.J., Luo, Y. *et al.* (2003) Polycystins 1 and 2 mediate mechanosensation in the primary cilium of kidney cells. *Nature Genetics*, **33**, 129–37.
70. Yoder, B.K., Hou, X., and Guay-Woodford, L.M. (2002) The polycystic kidney disease proteins, polycystin-1, polycystin-2, polaris, and cystin, are

co-localized in renal cilia. *Journal of the American Society of Nephrology*, **13**, 2508–16.

71. Praetorius, H.A. and Spring, K.R. (2001) Bending the MDCK cell primary cilium increases intracellular calcium. *The Journal of Membrane Biology*, **184**, 71–79.
72. Yao, X. and Garland, C.J. (2005) Recent developments in vascular endothelial cell transient receptor potential channels. *Circulation Research*, **97**, 853–63.
73. Corey, D.P., Garcia-Anoveros, J., Holt, J.R. *et al.* (2004) TRPA1 is a candidate for the mechanosensitive transduction channel of vertebrate hair cells. *Nature*, **432**, 723–30.
74. Sidi, S., Friedrich, R.W., and Nicolson, T. (2003) NompC TRP channel required for vertebrate sensory hair cell mechanotransduction. *Science*, **301**, 96–99.
75. Inoue, R., Jensen, L.J., Shi, J. *et al.* (2006) Transient receptor potential channels in cardiovascular function and disease. *Circulation Research*, **99**, 119–31.
76. Tseng, H., Peterson, T.E., and Berk, B.C. (1995) Fluid shear stress stimulates mitogen-activated protein kinase in endothelial cells. *Circulation Research*, **77**, 869–78.
77. Jin, Z.G., Wong, C., Wu, J., and Berk, B.C. (2005) Flow shear stress stimulates Gab1 tyrosine phosphorylation to mediate protein kinase B and endothelial nitric-oxide synthase activation in endothelial cells. *The Journal of Biological Chemistry*, **280**, 12305–9.
78. Tai, L.K., Okuda, M., Abe, J. *et al.* (2002) Fluid shear stress activates proline-rich tyrosine kinase via reactive oxygen species-dependent pathway. *Arteriosclerosis, Thrombosis, and Vascular Biology*, **22**, 1790–96.
79. Park, H., Go, Y.M., Darji, R. *et al.* (2000) Caveolin-1 regulates shear stress-dependent activation of extracellular signal-regulated kinase. *American Journal of Physiology: Heart and Circulatory Physiology*, **278**, H1285–93.
80. Isshiki, M. and Anderson, R.G. (2003) Function of caveolae in Ca^{2+} entry and Ca^{2+}-dependent signal transduction. *Traffic*, **4**, 717–23.
81. Murata, T., Lin, M.I., Stan, R.V. *et al.* (2007) Genetic evidence supporting caveolae microdomain regulation of calcium entry in endothelial cells. *The Journal of Biological Chemistry*, **282**, 16631–43.
82. Fleming, I., Rueben, A., Popp, R. *et al.* (2007) Epoxyeicosatrienoic acids regulate Trp channel dependent Ca^{2+} signaling and hyperpolarization in endothelial cells. *Arteriosclerosis, Thrombosis, and Vascular Biology*, **27**, 2612–18.
83. Kwiatek, A.M., Minshall, R.D., Cool, D.R. *et al.* (2006) Caveolin-1 regulates store-operated Ca^{2+} influx by binding of its scaffolding domain to transient receptor potential channel-1 in endothelial cells. *Molecular Pharmacology*, **70**, 1174–83.
84. Conrad, P.A., Smart, E.J., Ying, Y.S. *et al.* (1995) Caveolin cycles between plasma membrane caveolae and the Golgi complex by microtubule-dependent and microtubule-independent steps. *The Journal of Cell Biology*, **131**, 1421–33.
85. Sun, R.J., Muller, S., Zhuang, F.Y. *et al.* (2003) Caveolin-1 redistribution in human endothelial cells induced by laminar flow and cytokine. *Biorheology*, **40**, 31–39.
86. Rizzo, V., Morton, C., DePaola, N. *et al.* (2003) Recruitment of endothelial caveolae into mechanotransduction pathways by flow conditioning *in vitro*. *American Journal of Physiology: Heart and Circulatory Physiology*, **285**, H1720–29.
87. Boyd, N.L., Park, H., Yi, H. *et al.* (2003) Chronic shear induces caveolae formation and alters ERK and Akt responses in endothelial cells. *American Journal of Physiology: Heart and Circulatory Physiology*, **285**, H1113–22.
88. Albertini, H.F. (1748) Animadversiones super quibusdam difficilis respirationis vitiis a laesa cordis et praecordiorum structura pendentibus. *De Bononiensi Scientiarum et Artium Instituto Atque Academia Comentarii*, **1**, 382–404.
89. Jarcho, S. (1972) The Animadversiones of Ippolito Francesco Albertini (1748) on cardiac and precordial diseases which cause dyspnea. *Proceedings of the American Philosophical Society*, **116**, 163–81.
90. Starling, E.H. (1896) On the absorption of fluids from the connective tissue spaces. *The Journal of Physiology*, **19**, 312–26.
91. Gaar, K.A. Jr., Taylor, A.E., Owens, L.J., and Guyton, A.C. (1967) Effect of capillary pressure and plasma protein on development of pulmonary edema. *American Journal of Physiology*, **213**, 79–82.
92. Bachofen, H., Schurch, S., Michel, R.P., and Weibel, E.R. (1993) Experimental hydrostatic pulmonary edema in rabbit lungs: morphology. *The American Review of Respiratory Disease*, **147**, 989–96.
93. Bachofen, H., Schurch, S., and Weibel, E.R. (1993) Experimental hydrostatic pulmonary edema in rabbit lungs. Barrier lesions. *The American Review of Respiratory Disease*, **147**, 997–1004.

94. Tsukimoto, K., Yoshimura, N., Ichioka, M. *et al.* (1994) Protein, cell, and LTB4 concentrations of lung edema fluid produced by high capillary pressures in rabbit. *Journal of Applied Physiology*, **76**, 321–27.
95. West, J.B., Tsukimoto, K., Mathieu-Costello, O., and Prediletto, R. (1991) Stress failure in pulmonary capillaries. *Journal of Applied Physiology*, **70**, 1731–42.
96. Namba, Y., Kurdak, S.S., Fu, Z. *et al.* (1995) Effect of reducing alveolar surface tension on stress failure in pulmonary capillaries. *Journal of Applied Physiology*, **79**, 2114–21.
97. Elliott, A.R., Fu, Z., Tsukimoto, K. *et al.* (1992) Short-term reversibility of ultrastructural changes in pulmonary capillaries caused by stress failure. *Journal of Applied Physiology*, **73**, 1150–58.
98. Nicolaysen, G., Waaler, B.A., and Aarseth, P. (1979) On the existence of stretchable pores in the exchange vessels of the isolated rabbit lung preparation. *Lymphology*, **12**, 201–7.
99. Rippe, B., Townsley, M., Thigpen, J. *et al.* (1984) Effects of vascular pressure on the pulmonary microvasculature in isolated dog lungs. *Journal of Applied Physiology*, **57**, 233–39.
100. Nakos, G., Pneumatikos, J., Tsangaris, I. *et al.* (1997) Proteins and phospholipids in BAL from patients with hydrostatic pulmonary edema. *American Journal of Respiratory and Critical Care Medicine*, **155**, 945–51.
101. West, J.B. (2000) Invited review: pulmonary capillary stress failure. *Journal of Applied Physiology*, **89**, 2483–89.
102. Givertz, M.M., Slawsky, M.T., Moraes, D.L. *et al.* (2001) Noninvasive determination of pulmonary artery wedge pressure in patients with chronic heart failure. *The American Journal of Cardiology*, **87**, 1213–15.
103. Maggiorini, M., Melot, C., Pierre, S. *et al.* (2001) High-altitude pulmonary edema is initially caused by an increase in capillary pressure. *Circulation*, **103**, 2078–83.
104. Sarnoff, S.J., Berglund, E., and Sarnoff, L.C. (1953) Neurohemodynamics of pulmonary edema. III. Estimated changes in pulmonary blood volume accompanying systemic vasoconstriction and vasodilation. *Journal of Applied Physiology*, **5**, 367–74.
105. West, J.B. (1998) Left ventricular filling pressures during exercise: a cardiological blind spot? *Chest*, **113**, 1695–97.
106. Hopkins, S.R., Schoene, R.B., Henderson, W.R. *et al.* (1997) Intense exercise impairs the integrity of the pulmonary blood–gas barrier in elite athletes. *American Journal of Respiratory and Critical Care Medicine*, **155**, 1090–94.
107. Wagner, P.D., Gale, G.E., Moon, R.E. *et al.* (1986) Pulmonary gas exchange in humans exercising at sea level and simulated altitude. *Journal of Applied Physiology*, **61**, 260–70.
108. Hiraga, A., Hobo, S., Birks, E.K. *et al.* (1999) Changes in left ventricular dynamics during graded exercise. *Equine Veterinary Journal: Supplement*, **30**, 122–25.
109. West, J.B., Mathieu-Costello, O., Jones, J.H. *et al.* (1993) Stress failure of pulmonary capillaries in racehorses with exercise-induced pulmonary hemorrhage. *Journal of Applied Physiology*, **75**, 1097–109.
110. Birks, E.K., Mathieu-Costello, O., Fu, Z. *et al.* (1997) Very high pressures are required to cause stress failure of pulmonary capillaries in thoroughbred racehorses. *Journal of Applied Physiology*, **82**, 1584–92.
111. Fu, Z., Heldt, G.P., and West, J.B. (2003) Increased fragility of pulmonary capillaries in newborn rabbit. *American Journal of Physiology: Lung Cellular and Molecular Physiology*, **284**, L703–9.
112. Bhattacharya, J., Nakahara, K., and Staub, N.C. (1980) Effect of edema on pulmonary blood flow in the isolated perfused dog lung lobe. *Journal of Applied Physiology*, **48**, 444–49.
113. Parker, J.C. and Ivey, C.L. (1997) Isoproterenol attenuates high vascular pressure-induced permeability increases in isolated rat lungs. *Journal of Applied Physiology*, **83**, 1962–67.
114. Kuebler, W.M., Parthasarathi, K., Lindert, J., and Bhattacharya, J. (2007) Real-time lung microscopy. *Journal of Applied Physiology*, **102**, 1255–64.
115. Watanabe, H., Vriens, J., Prenen, J. *et al.* (2003) Anandamide and arachidonic acid use epoxyeicosatrienoic acids to activate TRPV4 channels. *Nature*, **424**, 434–38.
116. D'Hoedt, D., Owsianik, G., Prenen, J. *et al.* (2008) Stimulus-specific modulation of the cation channel TRPV4 by PACSIN 3. *The Journal of Biological Chemistry*, **10**, 6272–80.
117. Kuebler, W.M., Ying, X., Singh, B. *et al.* (1999) Pressure is proinflammatory in lung venular capillaries. *The Journal of Clinical Investigation*, **104**, 495–502.
118. Betz, W.J. and Bewick, G.S. (1992) Optical analysis of synaptic vesicle recycling at the frog neuromuscular junction. *Science*, **255**, 200–3.

119. McEver, R.P., Beckstead, J.H., Moore, K.L. *et al.* (1989) GMP-140, a platelet alpha-granule membrane protein, is also synthesized by vascular endothelial cells and is localized in Weibel–Palade bodies. *The Journal of Clinical Investigation*, **84**, 92–99.
120. Wagner, D.D., Olmsted, J.B., and Marder, V.J. (1982) Immunolocalization of von Willebrand protein in Weibel–Palade bodies of human endothelial cells. *The Journal of Cell Biology*, **95**, 355–60.
121. Wolff, B., Burns, A.R., Middleton, J., and Rot, A. (1998) Endothelial cell "memory" of inflammatory stimulation: human venular endothelial cells store interleukin 8 in Weibel–Palade bodies. *The Journal of Experimental Medicine*, **188**, 1757–62.
122. Kuebler, W.M. (2006) Selectins revisited: the emerging role of platelets in inflammatory lung disease. *The Journal of Clinical Investigation*, **116**, 3106–8.
123. Ichimura, H., Parthasarathi, K., Issekutz, A.C., and Bhattacharya, J. (2005) Pressure-induced leukocyte margination in lung postcapillary venules. *American Journal of Physiology: Lung Cellular and Molecular Physiology*, **289**, L407–12.
124. De Pasquale, C.G., Arnolda, L.F., Doyle, I.R. *et al.* (2003) Prolonged alveolocapillary barrier damage after acute cardiogenic pulmonary edema. *Critical Care Medicine*, **31**, 1060–67.
125. Fleming, I. and Busse, R. (2003) Molecular mechanisms involved in the regulation of the endothelial nitric oxide synthase. *American Journal of Physiology: Regulatory, Integrative and Comparative Physiology*, **284**, R1–12.
126. Kaestle, S.M., Reich, C.A., Yin, N. *et al.* (2007) Nitric oxide-dependent inhibition of alveolar fluid clearance in hydrostatic lung edema. *American Journal of Physiology: Lung Cellular and Molecular Physiology*, **293**, L859–69.
127. Kuebler, W.M., Uhlig, U., Goldmann, T. *et al.* (2003) Stretch activates nitric oxide production in pulmonary vascular endothelial cells *in situ*. *American Journal of Respiratory and Critical Care Medicine*, **168**, 1391–98.
128. Tabuchi, A., Mertens, M., Kuppe, H. *et al.* (2008) Intravital microscopy of the murine pulmonary microcirculation. *Journal of Applied Physiology*, **104**, 338–46.
129. Song, W. and Matalon, S. (2007) Modulation of alveolar fluid clearance by reactive oxygen–nitrogen intermediates. *American Journal of Physiology: Lung Cellular and Molecular Physiology*, **293**, L855–58.
130. Kaestle, S.M., Kuppe, H., and Kuebler, W.M. (2004) Compartmentalisation of the inflammatory response, in *Anaesthesia, Pain, Intensive Care and Emergency Medicine* (ed. A. Gullo), Springer, Milan, pp. 639–54.
131. Wang, P.M., Fujita, E., and Bhattacharya, J. (2002) Vascular regulation of type II cell exocytosis. *American Journal of Physiology: Lung Cellular and Molecular Physiology*, **282**, L912–16.
132. Ghio, S., Gavazzi, A., Campana, C. *et al.* (2001) Independent and additive prognostic value of right ventricular systolic function and pulmonary artery pressure in patients with chronic heart failure. *Journal of the American College of Cardiology*, **37**, 183–88.
133. Huang, W., Kingsbury, M.P., Turner, M.A. *et al.* (2001) Capillary filtration is reduced in lungs adapted to chronic heart failure: morphological and haemodynamic correlates. *Cardiovascular Research*, **49**, 207–17.
134. Kingsbury, M.P., Huang, W., Donnelly, J.L. *et al.* (2003) Structural remodelling of lungs in chronic heart failure. *Basic Research in Cardiology*, **98**, 295–303.
135. Hentschel, T., Yin, N., Riad, A. *et al.* (2007) Inhalation of the phosphodiesterase-3 inhibitor milrinone attenuates pulmonary hypertension in a rat model of congestive heart failure. *Anesthesiology*, **106**, 124–31.
136. Moraes, D.L., Colucci, W.S., and Givertz, M.M. (2000) Secondary pulmonary hypertension in chronic heart failure: the role of the endothelium in pathophysiology and management. *Circulation*, **102**, 1718–23.
137. Ontkean, M., Gay, R., and Greenberg, B. (1991) Diminished endothelium-derived relaxing factor activity in an experimental model of chronic heart failure. *Circulation Research*, **69**, 1088–96.
138. Kuebler, W.M., Kerem, A., Kaestle, S. *et al.* (2004) Mechanism of pulmonary endothelial dysfunction in congestive heart failure. *The FASEB Journal*, **18**, A714.
139. Dyess, D.L., Hunter, J.L., Lakey, J.R. *et al.* (2000) Attenuation of histamine-induced endothelial permeability responses after pacing-induced heart failure: role for endogenous catecholamines. *Microcirculation*, **7**, 307–15.
140. Ivey, C.L., Roy, B.J., and Townsley, M.I. (1998) Ablation of lung endothelial injury after pacing-induced heart failure is related to alterations in Ca^{2+} signaling. *American Journal of Physiology: Heart and Circulatory Physiology*, **275**, H844–51.
141. Roy, B.J., Pitts, V.H., and Townsley, M.I. (1996) Pulmonary vascular response to angiotensin II

in canine pacing-induced heart failure. *American Journal of Physiology: Heart and Circulatory Physiology*, **271**, H222–27.

142. Townsley, M.I., Fu, Z., Mathieu-Costello, O., and West, J.B. (1995) Pulmonary microvascular permeability. Responses to high vascular pressure after induction of pacing-induced heart failure in dogs. *Circulation Research*, **77**, 317–25.
143. Alvarez, D.F., King, J.A., and Townsley, M.I. (2005) Resistance to store depletion-induced endothelial injury in rat lung after chronic heart failure. *American Journal of Respiratory and Critical Care Medicine*, **172**, 1153–60.
144. Davies, S.W., Wilkinson, P., Keegan, J. *et al.* (1991) Pulmonary microvascular permeability in patients with severe mitral stenosis. *British Heart Journal*, **65**, 89–93.
145. Davies, S.W., Bailey, J., Keegan, J. *et al.* (1992) Reduced pulmonary microvascular permeability in severe chronic left heart failure. *American Heart Journal*, **124**, 137–42.
146. Barakat, A. and Lieu, D. (2003) Differential responsiveness of vascular endothelial cells to different types of fluid mechanical shear stress. *Cell Biochemistry and Biophysics*, **38**, 323–43.
147. Brooks, A.R., Lelkes, P.I., and Rubanyi, G.M. (2004) Gene expression profiling of vascular endothelial cells exposed to fluid mechanical forces: relevance for focal susceptibility to atherosclerosis. *Endothelium*, **11**, 45–57.
148. Zakrzewicz, A., Secomb, T.W., and Pries, A.R. (2002) Angioadaptation: keeping the vascular system in shape. *News in Physiological Sciences*, **17**, 197–201.
149. Birukov, K.G., Birukova, A.A., Dudek, S.M. *et al.* (2002) Shear stress-mediated cytoskeletal remodeling and cortactin translocation in pulmonary endothelial cells. *American Journal of Respiratory Cell and Molecular Biology*, **26**, 453–64.
150. Al-Mehdi, A.B., Zhao, G., Dodia, C. *et al.* (1998) Endothelial NADPH oxidase as the source of oxidants in lungs exposed to ischemia or high K^+. *Circulation Research*, **83**, 730–37.
151. Al-Mehdi, A.B., Shuman, H., and Fisher, A.B. (1997) Intracellular generation of reactive oxygen species during nonhypoxic lung ischemia. *American Journal of Physiology: Lung Cellular and Molecular Physiology*, **272**, L294–300.
152. Tozawa, K., Al-Mehdi, A.B., Muzykantov, V., and Fisher, A.B. (1999) In situ imaging of intracellular calcium with ischemia in lung subpleural microvascular endothelial cells. *Antioxidants and Redox Signaling*, **1**, 145–54.
153. Wei, Z., Costa, K., Al-Mehdi, A.B. *et al.* (1999) Simulated ischemia in flow-adapted endothelial cells leads to generation of reactive oxygen species and cell signaling. *Circulation Research*, **85**, 682–89.
154. Wei, Z., Al-Mehdi, A.B., and Fisher, A.B. (2001) Signaling pathway for nitric oxide generation with simulated ischemia in flow-adapted endothelial cells. *American Journal of Physiology: Heart and Circulatory Physiology*, **281**, H2226–32.
155. Al-Mehdi, A.B., Song, C., Tozawa, K., and Fisher, A.B. (2000) Ca^{2+}- and phosphatidylinositol 3-kinase-dependent nitric oxide generation in lung endothelial cells in situ with ischemia. *The Journal of Biological Chemistry*, **275**, 39807–10.
156. Sniderman, A., Burdon, T., Homan, J., and Salerno, T.A. (1984) Pulmonary blood flow. A potential factor in the pathogenesis of pulmonary edema. *The Journal of Thoracic and Cardiovascular Surgery*, **87**, 130–35.
157. Ehrhart, I.C., McCloud, L.L., Orfanos, S.E. *et al.* (1994) Effect of high blood flow on pulmonary vascular permeability to protein. *Journal of Applied Physiology*, **76**, 2342–47.
158. Tancredi, R.G. and Yipintsoi, T. (1980) Interrelationships of flow, intravascular pressure, and tissue perfusion in the measurement of capillary permeability to sodium in isolated dog lung lobes. *Circulation Research*, **46**, 669–80.
159. Verghese, G.M., Ware, L.B., Matthay, B.A., and Matthay, M.A. (1999) Alveolar epithelial fluid transport and the resolution of clinically severe hydrostatic pulmonary edema. *Journal of Applied Physiology*, **87**, 1301–12.
160. Kramer, K., Kirkman, P., Kitzman, D., and Little, W.C. (2000) Flash pulmonary edema: association with hypertension and reoccurrence despite coronary revascularization. *American Heart Journal*, **140**, 451–55.
161. Light, R.W. and George, R.B. (1983) Serial pulmonary function in patients with acute heart failure. *Archives of Internal Medicine*, **143**, 429–33.
162. Kubo, K., Hanaoka, M., Yamaguchi, S. *et al.* (1996) Cytokines in bronchoalveolar lavage fluid in patients with high altitude pulmonary oedema at moderate altitude in Japan. *Thorax*, **51**, 739–42.
163. Schoene, R.B., Swenson, E.R., Pizzo, C.J. *et al.* (1988) The lung at high altitude: bronchoalveolar lavage in acute mountain sickness and pulmonary edema. *Journal of Applied Physiology* **64**, 2605–13.

164. Hultgren, H.N., Lopez, C.E., Lundberg, E., and Miller, H. (1964) Physiologic studies of pulmonary edema at high altitude. *Circulation*, **29**, 393–408.

165. Hultgren, H.N., Grover, R.F., and Hartley, L.H. (1971) Abnormal circulatory responses to high altitude in subjects with a previous history of high-altitude pulmonary edema. *Circulation*, **44**, 759–70.

166. Charu, R., Stobdan, T., Ram, R.B. *et al.* (2006) Susceptibility to high altitude pulmonary oedema: role of ACE and ET-1 polymorphisms. *Thorax*, **61**, 1011–12.

167. Droma, Y., Hanaoka, M., Ota, M. *et al.* (2002) Positive association of the endothelial nitric oxide synthase gene polymorphisms with high-altitude pulmonary edema. *Circulation*, **106**, 826–30.

168. West, J.B., Colice, G.L., Lee, Y.J. *et al.* (1995) Pathogenesis of high-altitude pulmonary oedema: direct evidence of stress failure of pulmonary capillaries. *The European Respiratory Journal*, **8**, 523–29.

169. Gao, Y. and Raj, J.U. (2005) Role of veins in regulation of pulmonary circulation. *American Journal of Physiology: Lung Cellular and Molecular Physiology*, **288**, L213–26.

170. Hillier, S.C., Graham, J.A., Hanger, C.C. *et al.* (1997) Hypoxic vasoconstriction in pulmonary arterioles and venules. *Journal of Applied Physiology*, **82**, 1084–90.

171. Horsfield, K. (1978) Morphometry of the small pulmonary arteries in man. *Circulation Research*, **42**, 593–97.

172. Bhattacharya, S., Glucksberg, M.R., and Bhattacharya, J. (1989) Measurement of lung microvascular pressure in the intact anesthetized rabbit by the micropuncture technique. *Circulation Research*, **64**, 167–72.

173. Goetz, A.E., Kuebler, W.M., and Peter, K. (1996) High-altitude pulmonary edema. *The New England Journal of Medicine*, **335**, 206–7.

174. Hultgren, H.N. and Hegnauer, A.H. (eds) (1969) High altitude pulmonary edema, in *Biomedicine Problems of High Terrestrial Altitude* (ed. A.H. Hegnauer), Springer, New York, pp. 131–41.

175. Hlastala, M.P., Lamm, W.J., Karp, A. *et al.* (2004) Spatial distribution of hypoxic pulmonary vasoconstriction in the supine pig. *Journal of Applied Physiology*, **96**, 1589–99.

176. Lamm, W.J., Starr, I.R., Neradilek, B. *et al.* (2004) Hypoxic pulmonary vasoconstriction is heterogeneously distributed in the prone dog. *Respiratory Physiology and Neurobiology*, **144**, 281–94.

177. Hopkins, S.R., Garg, J., Bolar, D.S. *et al.* (2005) Pulmonary blood flow heterogeneity during hypoxia and high-altitude pulmonary edema. *American Journal of Respiratory and Critical Care Medicine*, **171**, 83–87.

178. Bartsch, P., Mairbaurl, H., Maggiorini, M., and Swenson, E.R. (2005) Physiological aspects of high-altitude pulmonary edema. *Journal of Applied Physiology*, **98**, 1101–10.

179. Maggiorini, M. (2006) High altitude-induced pulmonary oedema. *Cardiovascular Research*, **72**, 41–50.

180. Younes, M., Bshouty, Z., and Ali, J. (1987) Longitudinal distribution of pulmonary vascular resistance with very high pulmonary blood flow. *Journal of Applied Physiology*, **62**, 344–58.

181. Townsley, M.I., Korthuis, R.J., Rippe, B. *et al.* (1986) Validation of double vascular occlusion method for $P_{c,i}$ in lung and skeletal muscle. *Journal of Applied Physiology*, **61**, 127–32.

182. Nelin, L.D., Roerig, D.L., Rickaby, D.A. *et al.* (1992) Influence of flow on pulmonary vascular surface area inferred from blue dextran efflux data. *Journal of Applied Physiology*, **72**, 874–80.

183. Maron, M.B. (1985) A canine model of neurogenic pulmonary edema. *Journal of Applied Physiology*, **59**, 1019–25.

184. Wray, N.P. and Nicotra, M.B. (1978) Pathogenesis of neurogenic pulmonary edema. *The American Review of Respiratory Disease*, **118**, 783–86.

185. Baumann, A., Audibert, G., McDonnell, J., and Mertes, P.M. (2007) Neurogenic pulmonary edema. *Acta Anaesthesiologica Scandinavica*, **51**, 447–55.

186. Kowalski, M.L., Didier, A., and Kaliner, M.A. (1989) Neurogenic inflammation in the airways. I. Neurogenic stimulation induces plasma protein extravasation into the rat airway lumen. *The American Review of Respiratory Disease*, **140**, 101–9.

187. Minnear, F.L., Kite, C., Hill, L.A., and van der Zee, H. (1987) Endothelial injury and pulmonary congestion characterize neurogenic pulmonary edema in rabbits. *Journal of Applied Physiology*, **63**, 335–41.

188. Fein, A., Grossman, R.F., Jones, J.G. *et al.* (1979) The value of edema fluid protein measurement in patients with pulmonary edema. *American Journal of Medicine*, **67**, 32–8.

189. Sprung, C.L., Rackow, E.C., Fein, I.A. *et al.* (1981) The spectrum of pulmonary edema: differentiation of cardiogenic, intermediate, and noncardiogenic forms of pulmonary edema. *The American Review of Respiratory Disease*, **124**, 718–22.

190. Hirabayashi, A., Nishiwaki, K., Shimada, Y., and Ishikawa, N. (1996) Role of neuropeptide Y and its receptor subtypes in neurogenic pulmonary edema. *European Journal of Pharmacology*, **296**, 297–305.

191. Novitzky, D., Wicomb, W.N., Rose, A.G. *et al.* (1987) Pathophysiology of pulmonary edema following experimental brain death in the chacma baboon. *The Annals of Thoracic Surgery*, **43**, 288–94.

192. Carlson, R.W., Schaeffer, R.C. Jr., Michaels, S.G., and Weil, M.H. (1979) Pulmonary edema following intracranial hemorrhage. *Chest*, **75**, 731–34.

193. Smith, W.S. and Matthay, M.A. (1997) Evidence for a hydrostatic mechanism in human neurogenic pulmonary edema. *Chest*, **111**, 1326–33.

194. Dreyfuss, D. and Saumon, G. (1998) Ventilator-induced lung injury: lessons from experimental studies. *American Journal of Respiratory and Critical Care Medicine*, **157**, 294–323.

195. Uhlig, S. and Uhlig, U. (2004) Pharmacological interventions in ventilator-induced lung injury. *Trends in Pharmacological Sciences*, **25**, 592–600.

196. Parker, J.C., Ivey, C.L., and Tucker, J.A. (1998) Gadolinium prevents high airway pressure-induced permeability increases in isolated rat lungs. *Journal of Applied Physiology*, **84**, 1113–18.

197. Parker, J.C. and Yoshikawa, S. (2002) Vascular segmental permeabilities at high peak inflation pressure in isolated rat lungs. *American Journal of Physiology: Lung Cellular and Molecular Physiology*, **283**, L1203–9.

198. Hamanaka, K., Jian, M.Y., Weber, D.S. *et al.* (2007) TRPV4 initiates the acute calcium-dependent permeability increase during ventilator-induced lung injury in isolated mouse lungs. *American Journal of Physiology: Lung Cellular and Molecular Physiology*, **293**, L923–32.

199. Frank, J.A., Pittet, J.F., Lee, H. *et al.* (2003) High tidal volume ventilation induces NOS2 and impairs cAMP-dependent air space fluid clearance. *American Journal of Physiology: Lung Cellular and Molecular Physiology*, **284**, L791–98.

200. Bhattacharya, S., Sen, N., Yiming, M.T. *et al.* (2003) High tidal volume ventilation induces proinflammatory signaling in rat lung endothelium. *American Journal of Respiratory Cell and Molecular Biology*, **28**, 218–24.

201. Yiming, M.T., Parthasarathi, K., Issekutz, A.C., and Bhattacharya, S. (2005) Sequence of endothelial signaling during lung expansion. *American Journal of Respiratory Cell and Molecular Biology*, **33**, 549–54.

21

Therapeutic Strategies to Limit Lung Endothelial Cell Permeability

Rachel K. Wolfson[1], Gabriel Lang[2], Jeff Jacobson[3] and Joe G. N. Garcia[4]

[1]Section of Critical Care, Department of Pediatrics, Pritzker School of Medicine, University of Chicago, Chicago, IL, USA
[2]Section of Pulmonary and Critical Care Medicine, Department of Medicine, Pritzker School of Medicine, University of Chicago, Chicago, IL, USA
[3]Section of Pulmonary and Critical Care Medicine, Department of Medicine, Pritzker School of Medicine, University of Chicago, Chicago, IL, USA
[4]Department of Medicine, Pritzker School of Medicine, University of Chicago, Chicago, IL, USA

INTRODUCTION

The vascular endothelium is a primary cellular target and key participant in the profound physiologic derangement which accompanies acute (inflammatory) lung injury (ALI). A key expression of this involvement is the substantial vascular hyperpermeability that results in parenchymal accumulation of leukocytes and extravascular lung water – defining features of ALI that contribute to the increased morbidity and mortality of this devastating syndrome. Consequently, there is substantial interest in the development and utilization of clinically effective agents that either interfere with or prevent lung vascular endothelial cell (EC) barrier dysfunction, restore EC barrier integrity, reduce alveolar flooding, and improve respiratory mechanics. Complementing other chapters in this volume targeting aspects of EC barrier regulation (see Chapter 8), this chapter focuses on molecular mechanisms of both homeostatic and pathobiologic lung vascular barrier regulation and barrier restoration processes, and the translation of this information into novel permeability-resolving therapeutic strategies.

As noted throughout this volume, there is striking diversity in the mechanisms by which bioactive and biophysical stimuli alter lung EC function including a predilection toward increased permeability (Figure 21.1). Equally striking, however, are the commonalities that exist in the signaling pathways elicited by novel barrier-protective agents (Figure 21.2) within the unique microenvironment of the membrane–cytoskeleton interface. A "conversation" elicited by ligation of G-protein-coupled receptors takes place between transmembrane components (such as large and small GTPases) and cytoskeletal proteins in membrane domains such as caveolin-enriched lipid rafts and Rac GTPase-dependent lamellipodia [1, 2]. These pathways induce EC cytoskeletal rearrangement resulting in enhanced junctional linkages between ECs as well as increased linkage of the cytoskeleton with the underlying extracellular matrix [3]. These events provide the conceptual underpinning for the molecular targeting of these permeability-reducing therapeutic strategies.

SIGNALS TO THE EC CYTOSKELETON AS TARGETS FOR BARRIER RESTORATION

The regulation of pulmonary endothelial barrier function is a complex process defined by the relative balance of adhesive and contractile forces mediated by the intimate linkage between the actin cytoskeleton and various intercellular junctional proteins localized to tight junctions (TJs, e.g., claudins), adherens junctions (AJs) (e.g., VE-cadherin, α, β, and γ catenins) (see Chapter 3), and the focal adhesion complex (e.g.,

The Pulmonary Endothelium: Function in health and disease Editors Norbert F. Voelkel, Sharon Rounds

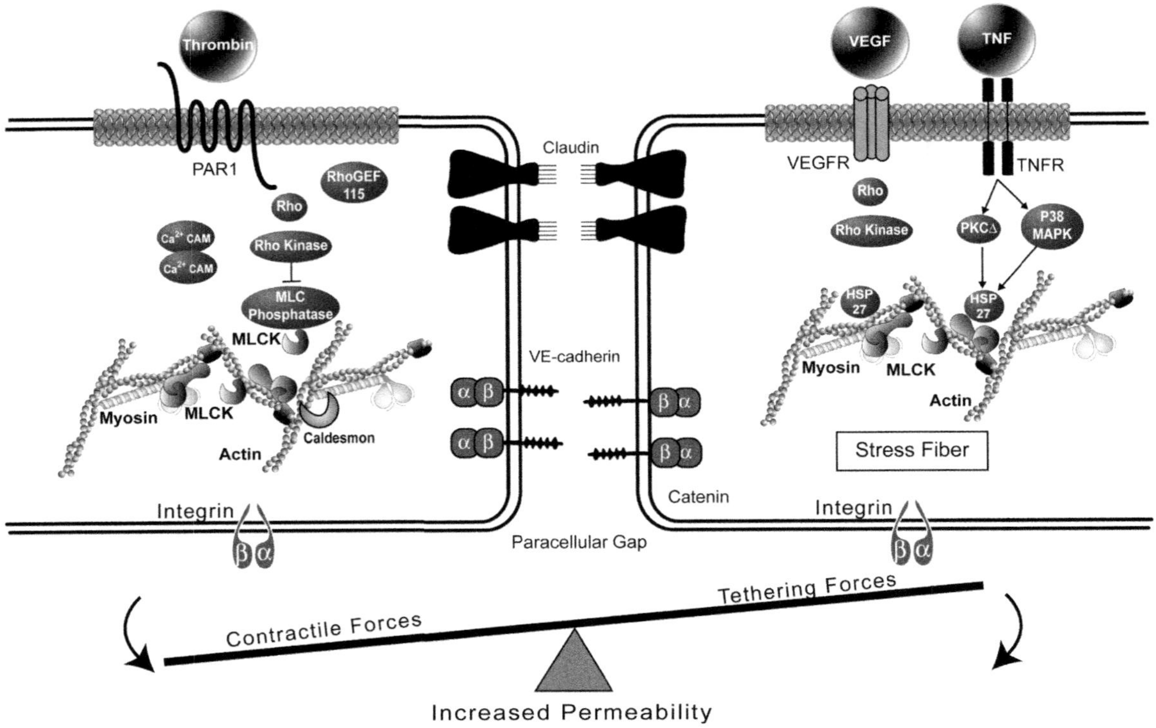

Figure 21.1 Intracellular signaling pathways that are evoked by EC edemagenic agents and contractile agonists, and result in paracellular gap formation.

paxillin, α-actinin) (see Chapter 4) (Figure 21.2) [3, 4]. It has become clear that actin polymerization has profound functional consequences for barrier regulation that are highly dependent on the exact spatial location of this actin rearrangement occurring as either barrier-disrupting cytosolic stress fibers (Figure 21.1) or as a barrier-enhancing thickened cortical actin ring (Figure 21.2). When actin polymerization produces transcytoplasmic stress fibers (Figure 21.1), the cortical actin ring disassembles and contractile tension is produced, a function of the abundant tension-generating molecular machinery (actin and myosin comprise ~16% of total EC protein). A key initial event in transducing endothelial paracellular permeability-enhancing signals to the actin cytoskeleton is an increase in intracellular calcium (Ca^{2+}) that follows inflammatory mediators binding to their respective cognate receptors [such as vascular endothelial growth factor (VEGF), tumor necrosis factor (TNF), and protease-activated receptor (PAR)-1] [5] (see Chapters 5 and 9). The source and site of the increased intracellular Ca^{2+} is key to the ultimate Ca^{2+} effects on EC barrier function [6] with increased intracellular Ca^{2+} a stimulus to many signaling processes, including physical interaction of Ca^{2+} with the Ca^{2+}-binding protein, calmodulin, thereby activating Ca^{2+}/calmodulin-dependent enzymes such as calmodulin-dependent kinase II [7–9] and the nonmuscle (nm) isoform of myosin light chain (MLC) kinase (myosin light chain kinase MLCK). The nmMLCK isoform is a key regulator of the EC cytoskeleton initially cloned by the Garcia lab [10], and is essential for the generation of EC centripetal tension as nmMLCK activation results in increased MLC phosphorylation, enhanced actin–myosin interaction, and increased actomyosin motor activity [3]. The resultant contractile force increases intracellular tension and produces paracellular gaps which increase paracellular permeability [3]. Neutrophil migration into lung tissues across the endothelium also requires changes in EC shape and the formation of paracellular gaps and as a result, nmMLCK activity exerts a major gatekeeper function during acute lung inflammation [11].

Given the essential role for post-translational modifications in driving cytoskeletal dynamics, it is not surprising that serine/threonine phosphatases are also critical to EC barrier regulation [12, 13], with EC contraction and barrier dysfunction significantly augmented by decreases in the activity of the MLC phosphatase known as myosin light chain phosphatase MYPT1 – a key regulatory enzyme of cell contractile activity [14,

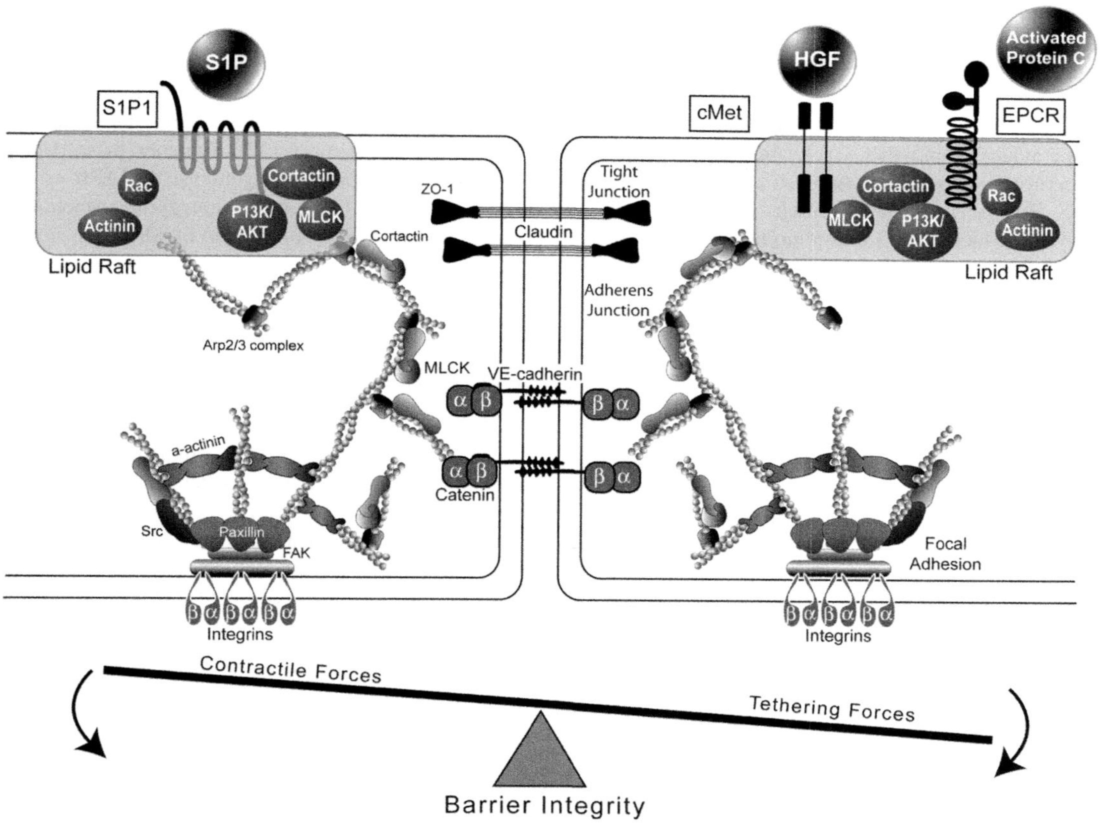

Figure 21.2 Intracellular signals elicited by barrier-protective agonists with cortical cytoskeletal linkage to target junctional effectors and integrin-linked matrix focal adhesion components.

15]. Decreased MYPT1 activity reduces MLC dephosphorylation, thereby increasing the cellular content of phosphorylated MLC, enhancing actin–myosin interaction and contractile force generation [3]. Activation of the small GTPase, Rho, by edemagenic agents such as VEGF [16], TNF-α, and thrombin, leads to rapid increases in the activity of Rho kinase with subsequent MYPT1 phosphorylation on Thr686 and Thr850 [17]. This post-translational modification is critically important for MYPT1 inhibition, highlighting the coordinated activities of several signaling pathways to achieve optimal increase in intracellular tension and paracellular gap formation via targeting of the endothelial cytoskeleton (Figure 21.1) [3, 18–20]. Thus, attractive therapeutic targets for attenuating lung vascular permeability, particularly during profound inflammation, include strategies designed to reduce nmMLCK activity [3] and Rho kinase activity [17, 20, 21]. These pharmacologic approaches reduce overall endothelial contractile apparatus activity and reduce thrombin-induced MLC phosphorylation, paracellular gap formation and EC barrier dysfunction [22] as well as TNF-α-induced EC apoptosis [23]. Furthermore, nmMLKC knockout mice are protected from lipopolysaccharide (LPS)-induced ALI and subsequent ventilator-induced lung injury (VILI) with small-molecule nmMLCK inhibitors similarly protective against ALI [24–26].

MLCK-independent pathways also regulate EC contraction and barrier dysfunction, and despite a lack of increased nmMLCK activity, these pathways generally involve rearrangement of the endothelial cytoskeleton. For example, protein kinase C (PKC) is a central signaling molecule with many isotype-specific functions, and is well appreciated as a key determinant of endothelial barrier integrity, cytoskeletal rearrangement, and alterations in vascular permeability [27]. Isotype-specific increases in PKC activity directly increase levels of MLC phosphorylation but with phosphorylation at MLC serine/threonine residues Ser1, Ser2, and Thr9 [28, 29], which are distinct from nmMLCK-mediated sites at Ser19 and Thr18. PKC actually produces time-dependent MLC dephosphorylation of Ser19 and Thr18 concurrent with rearrangement of actin stress fibers into a grid-like network [30]. The potent edemagenic agent, thrombin, induces rapid PKCδ activation that is essential for actin stress fiber formation and barrier disruption [22, 31]. However, PKCδ inhibition does not attenuate thrombin-induced MLC phosphorylation, suggesting PKCδ is downstream of Rho GTPase activation. Consistent with this relationship, Rho kinase inhibition attenuates thrombin-induced PKCδ translocation to the plasma membrane with thrombin-induced stress fiber formation attenuated by PKCδ inhibition [small interfering RNA (siRNA)]. PKCζ has an opposite effect on barrier regulation and is critical to barrier recovery after thrombin, consistent with a proinflammatory role for PKCδ and an anti-inflammatory role for PKCζ in EC barrier regulation. Thus, PKC-mediated increases in EC permeability involve activation of the EC contractile apparatus, but do not require substantial increases in nmMLCK activity [30]. A potential PKC-dependent mechanism leading to cytoskeletal rearrangement and increased endothelial permeability is via phosphorylation of caldesmon – an actin- and myosin-binding protein that acts as a molecular switch for regulation of actomyosin contractile activities [32–34]. PKC inhibition has been identified as a potential therapeutic strategy to decrease EC permeability [35, 36] both *in vitro* as well as *in vivo* in hypochlorite-induced rabbit lung injury [37]. These results suggest PKCδ and PKCζ serve as useful molecular targets in addressing the increased permeability responses observed in acute inflammatory lung syndromes.

Inflammatory mediators such as TNF-α also induce MLCK-independent EC barrier disruption [23] via coordinated increases in PKC and p38 mitogen-activated protein kinase (MAPK) activation [38], exerting dual barrier-disrupting effects through downstream phosphorylation of heat shock protein 27 (HSP27) – an actin-binding protein that stabilizes actin in its dephosphorylated state. Activation of p38 MAPK results in increased phosphorylation of caldesmon and the activation of MAPK-activated protein kinase-2/3, which in turn phosphorylates HSP27 [39]. Phosphorylated HSP27 does not stabilize actin, thereby contributing to the formation of stress fibers and paracellular gap formation [3]. The effects of p38 MAPK are independent of MLCK, supported by the fact that both pharmacologic and dominant-negative inhibition of p38 MAPK did not affect diphosphorylation of MLC, and by the observation of different time courses of MLCK and p38 MAPK activation after a thrombin challenge [40]. p38 MAPK activity is critically important to the lung EC barrier disruption induced by pertussis toxin (PTX) and PTX-mediated decreases in transmonolayer electrical resistance (TER) were related temporally to the phosphorylation of downstream targets of p38 MAPK, including HSP27 and caldesmon [41]. In addition, p38 MAPK participates in EC permeability augmentation by microtubule disassembly [42]. Together these data underscore the rationale for targeting p38 MAPK as a key barrier-regulatory therapeutic strategy for ameliorating hyperpermeability states observed in ALI.

RECEPTOR-MEDIATED BARRIER PROTECTION: ROLE OF RAC GTPASE AND CORTICAL ACTIN REMODELING

Prior to the last decade, permeability-reducing strategies primarily consisted of cAMP augmentation, producing only modest barrier enhancement [43–46]. More recently, a number of barrier-promoting agents have been identified that share common signal transduction mechanisms that are distinct from cAMP signals and target the endothelial actin cytoskeleton to facilitate barrier-restorative processes. We have conceptualized a paradigm whereby barrier recovery after edemagenic agonists involves development of a cortical actin ring to anchor cellular junctions and a carefully choreographed (but poorly understood) gap-closing process via formation of Rac GTPase-dependent lamellipodial protrusions into the paracellular space between activated ECs (Figure 21.3). Within these lamellipodia, signals are transduced to actin-binding proteins (nmMLCK and cortactin) and phosphorylated MLCs in spatial-specific cellular locations (Figure 21.2). Lamellipodia also require formation of focal adhesions (regulated by the cytoskeleton) critical to establishment of linkage of the actin cytoskeleton to target effectors that restore cell–cell adhesion and cell–matrix adhesion. This process is essential to the barrier restoration as we have previously described following stimulation with sphingosine 1-phosphate (S1P), hepatocyte growth factor (HGF), simvastatin, activated protein C (APC), ATP, oxidized phospholipids, and hyaluronan [1, 247–51]. Central to these events is the activation of small GTPases, Rac and Cdc42 [52] which follows ligation of barrier-protective receptors and drives cortical actin remodeling and lamellipodia formation (Figure 21.3). In addition to lamellipodia, there is increased actin polymerization at the cell periphery (i.e., the cortical actin ring) that occurs with increased force driven by the actin-binding proteins, cortactin and nmMLCK, which also translocate to this spatially defined region. Like lamellipodia formation, Rac GTPase-dependent increases in cortical actin follow exposure to multiple barrier-enhancing levels of shear stress or to potent barrier-enhancing agonists [48–50]

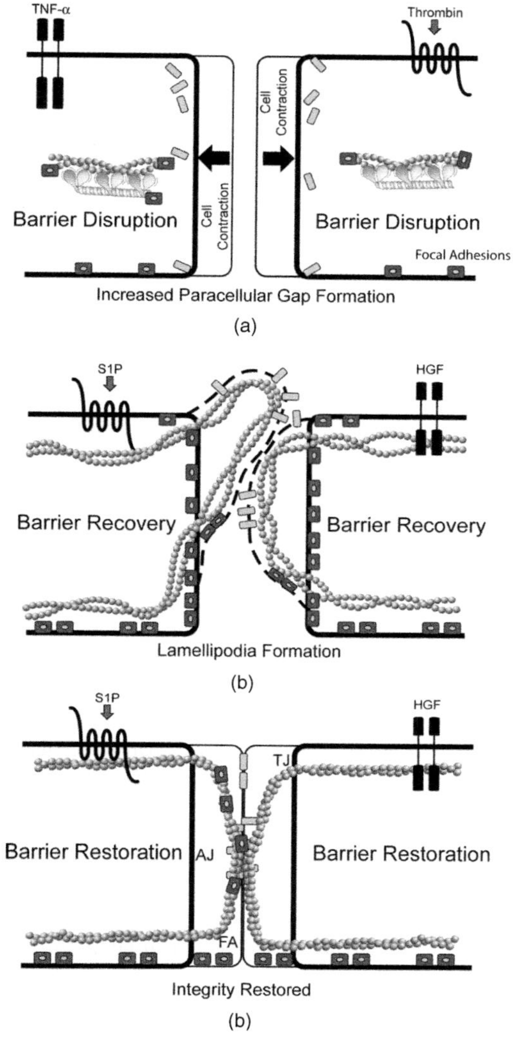

Figure 21.3 Paradigm for resolution of inflammation-mediated EC paracellular gaps. Shown in (a–c) is a sequence of lamellipodia-mediated closure of the paracellular gap induced by thrombin or other edemagenic stimuli (like TNF) (a) with closure of gaps, which is greatly facilitated by HGF or S1P that induce lamellipodial membrane protrusion into the gap (b) driven by actin polymerization. Closure of the paracellular gap occurs anchored by focal contacts (c).

including S1P [1, 53], HGF [47] (Figure 21.2), ATP [50], simvastatin [48], APC [49], prostaglandin (PG) E_2 [54], and oxidized phospholipid OxPAPC (oxidized 1-palmitoyl-2-arachidonoyl-*sn*-glycero-3-phosphochlorine) [51]. These observations serve to highlight the importance of the cellular location of cytoskeletal proteins in maintaining or enhancing EC barrier function with cortactin directly interacting with nmMLCK, an association which is increased by $p60^{Src}$ tyrosine phosphorylation of either cortactin or nmMLCK [55]. Rac activation is in conjunction with Akt-mediated phosphorylation events known to be involved in EC proliferation and migration [56] and EC barrier enhancement. Akt-induced phosphorylation of the S1P1 receptor is important in barrier enhancement produced by high-molecular-weight hyaluronan [2, 57].

Discussion of EC barrier enhancement would be incomplete without attention to the cell junction proteins, notably focal adhesions and AJs, and their relationship with the underlying EC cytoskeleton (Figure 21.2) (see Chapter 4). Focal adhesions bridge the intracellular and extracellular space, and are composed of extracellular matrix proteins, transmembrane proteins, and cytoplasmic focal adhesion plaques [4]. Focal adhesions facilitate communication between the actin cytoskeleton and the extracellular space. In comparison, AJs are composed of cadherins which in turn interact through their intracellular component with catenins, which attach to the actin cytoskeleton. Both protein complexes are critical to maintaining EC barrier function and signal transduction from within each cell to its surrounding matrix and neighboring cells.

S1P AND CLOSELY RELATED ANALOGS

S1P is a sphingolipid resulting from the phosphorylation of sphingosine, a product of sphingomyelinase catabolism of sphingomyelin, catalyzed by sphingosine kinase (Figure 21.4) [58]. S1P ligates a family of receptors known as S1P receptors (also termed endothelial differentiation gene receptors) with prominent effects on the vasculature, promoting EC mitogenesis, chemotaxis, and angiogenesis. Our earlier studies were the first to link the angiogenic factor S1P and S1P receptor ligation [59] to vascular barrier regulation and demonstrated that physiologic doses of S1P induce EC activation, marked cytoskeletal rearrangement, and stabilization of lung EC barrier function *in vitro* [1]. This novel function for S1P was of particular relevance to clinical medicine as thrombocytopenia is well known to be associated with increased vascular leak [60], and while the mechanism of this effect was unknown, we demonstrated that activated platelets are an important source of S1P and directly enhance barrier function via S1P1 ligation [61]. Platelets contain significant levels of sphingosine kinase but reduced levels of sphingosine lyase thereby serving as enriched sources for the barrier-promoting S1P [61].

Subsequent mechanistic studies linked the specific S1P receptor S1P1 receptor (also known as Edg1) [1, 58, 62, 63] and PTX-sensitive $G_{i\alpha}$-protein-coupled

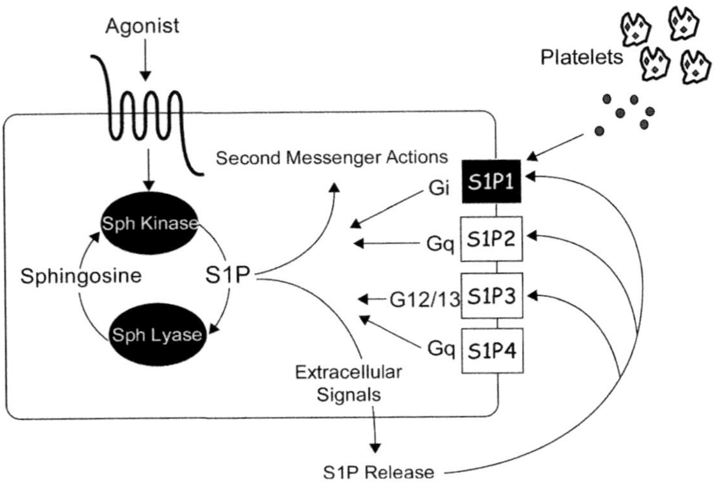

Figure 21.4 Schematic depiction of S1P formation, release, and ligation of S1P receptors by S1P with S1P1 having the primary barrier-enhancing properties.

signaling, to increases in Rac GTPase activity [1] and endoplasmic reticulum-derived cytosolic calcium [64] – processes critical to S1P-mediated barrier enhancement. Subsequent studies emphasized that the aggregation of a number of barrier-regulatory signaling components into caveolin-rich lipid rafts was critical to the signaling cascade driving rearrangement of the actin cytoskeleton and S1P-mediated barrier enhancement [1]. S1P-recruited lipid raft components include the Rac GTPase target, p21-associated serine/threonine kinase (PAK), and its downstream target, cofilin – an actin-binding protein [65]. PAK and cofilin allow polymerization–depolymerization cycling to occur and thus facilitate rearrangement of actin from primarily transcytoplasmic to primarily cortical in a spatially distinct organization as a cortical actin cellular ring, findings which are integral to EC barrier function [1]. Consistent with the conceptual framework that barrier regulation is intimately linked to the cytoskeleton, changes in the actin cytoskeleton were essential for S1P-mediated barrier enhancement as cytochalasin B, an actin depolymerizing agent, and latrunculin B, which inhibits actin polymerization, each prevent the barrier-enhancing effects of S1P [3].

While increases in MLC phosphorylation within stress fibers are critical to barrier disruption (Figure 21.1), MLC phosphorylation is also a key element in S1P-mediated barrier enhancement and occurs in a peripheral distribution within the cortical actin ring [1], providing strength to this spatially directed scaffolding force and enhancing cell–cell tethering as we described via atomic force microscopy [66]. Immunofluorescence studies demonstrated that overexpressed Green Fluorescent Protein–nmMLCK distributes along cytoplasmic actin fibers, but rapidly translocates to the cortical regions of the cell after S1P treatment, rapidly catalyzing MLC phosphorylation. In addition, confocal microscopy studies show EC challenged with S1P demonstrate colocalization of nmMLCK with the key actin-binding and EC barrier-regulatory protein, cortactin [53] (Figure 21.5). Cortactin is involved in stimulating actin polymerization [55], cortical actin rearrangement [67], and tyrosine phosphorylation of cortactin is seen after stimuli which cause cytoskeletal rearrangement [67]. The C-terminal Src homology-3 (SH3) region of phosphorylated cortactin directly interacts with nmMLCK at higher rates than nonphosphorylated cortactin [55], and the interaction of cortactin and nmMLCK decreases cortactin-stimulated actin polymerization [53, 55] and is essential to S1P barrier protection. A cortactin-blocking peptide that competitively blocks the cortactin SH3 site, and therefore nmMLCK interaction, did not affect S1P-induced cortactin translocation or cortical actin ring formation, but significantly attenuated S1P-induced barrier enhancement [53]. Immunofluorescence showed that S1P as well as other barrier-enhancing agents (such as HGF) produced rapid translocation of cortactin to the EC periphery, an effect not seen when EC are treated with the barrier-disrupting agent, thrombin [53]. Thus, tyrosine phosphorylation of cortactin is not necessary for peripheral translocation of cortactin after S1P but is necessary for S1P-induced barrier enhancement [53]. $p60^{Src}$ is not involved in this pathway, but other tyrosine kinases such as c-Abl, are likely involved [53]. Thus, cortactin, nmMLCK, and the intracellular location of phosphorylated MLC are all critically important in the barrier enhancement that results from S1P.

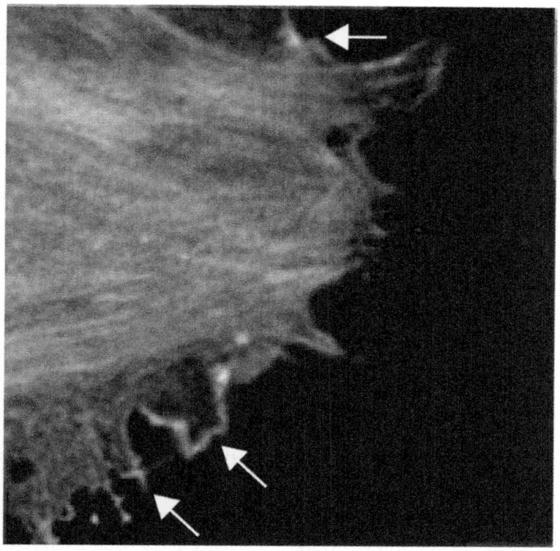

Figure 21.5 Colocalization of lamellipodia stress fibers. High-magnification photomicrograph of cortactin and MLCK within the lamellipodia protruding out to close paracellular gaps. Shown is nmMLCK (green) and cortactin (red), which are colocalized (yellow) within lamellipodia. A color version of this figure appears in the plate section of this volume.

S1P-induced Rac activation and cytoskeletal rearrangement produce increased linkage of actin to VE-cadherin and β-catenin, both important AJ components, as well as S1P-induced phosphorylation of focal adhesion-related proteins paxillin and focal adhesion kinase (FAK), with translocation of these proteins to the EC periphery, further implicating S1P-induced cell–cell adhesive changes as part of the mechanism of S1P-induced barrier enhancement [64, 68]. Coimmunoprecipitation studies revealed the increased association of VE-cadherin with FAK and paxillin in S1P-challenged human pulmonary artery EC (human pulmonary artery endothelial cell HPAEC) monolayers. Furthermore, S1P-induced enhancement of VE-cadherin interaction with α-catenin and β-catenin was associated with the increased formation of FAK–β-catenin protein complexes. Depletion of β-catenin using specific siRNA resulted in complete loss of S1P effects on VE-cadherin association with FAK and paxillin rearrangement. These results demonstrate that S1P-induced endothelial barrier enhancement involves β-catenin-linked AJ/focal adhesion interaction.

In addition to the abundant *in vitro* data describing the EC barrier-enhancing effect of S1P, the potential utility of S1P in restoring lung water balance in patients with inflammatory injury was underscored in studies involving small and large animal models of ALI in which S1P provided dramatic attenuation of LPS-mediated lung inflammation and permeability [58, 69]. Mice treated with S1P had significantly less histologic evidence of inflammatory changes/lung injury, with decreased neutrophil alveolitis on bronchoalveolar lavage (BAL) and decreased lung myeloperoxidase (MPO) activity [69]. Interestingly, mice treated with S1P after intratracheal LPS also showed an attenuated renal inflammatory response compared to control, measured by tissue MPO activity and Evan's blue dye extravasation as a measure of capillary leak. S1P also protected against intrabronchial LPS-induced ALI and concomitant VILI in a canine model, with decreased shunt fraction, decreased BAL protein, decreased extravascular lung water, and improved oxygenation [70]. Use of a large animal canine model allowed investigation of regional lung changes in ALI and the effect of S1P on these changes. Computed tomography scans of animals subjected to LPS/VILI found that animals treated with S1P had dramatic improvement in alveolar air content (with decreased edema) in all lung regions [70]. Additional *in vivo* studies found that S1P protects against VILI in a murine model as assessed by Evan's blue dye extravasation [70].

We have also evaluated a potential role for S1P in ameliorating lung ischemia–reperfusion injury, a common complication of lung transplantation, which is characterized by alveolar damage, edema, and inflammation in donor lungs and is a significant cause of transplant failure. Utilizing a rat model of ischemia–reperfusion injury (pulmonary artery ligation and reperfusion), we determined that rats pretreated with S1P exhibited reduced lung vascular permeability and inflammation compared to controls [71]. Lung MPO activity, an index of parenchymal leukocyte infiltration, and levels of interleukins IL-6, -1β, and -2 were also attenuated in S1P-treated animals exposed to ischemia–reperfusion injury [71]. Together, these findings suggest that S1P may serve as an effective permeability reducing agent in diverse conditions which share an element of lung inflammatory burden.

Despite the profound attractiveness of S1P as a therapeutic agent which targets the endothelium in high permeability states, S1P has several attributes which limit its potential utility as a permeability-reducing strategy. With an affinity for ligation of the $S1P_3$ receptor, intratracheal S1P has been implicated as a cause of pulmonary edema via endothelial/epithelial barrier disruption [72]. S1P also causes bradycardia via ligation of cardiac $S1P_3$ receptors [73]. These findings generated increased interest in FTY720, a derivative of the natural immunosuppressant myriocin [74], and a recently described immunosuppressive agent that causes peripheral lymphopenia by inhibiting cellular egress from lymphoid tissues. FTY720 is structurally similar (but not identical) to S1P and is phosphorylated by sphingosine kinase to FTY720-phosphate, which is an agonist at S1P receptors [74]. This characteristic prompted investigation of the effect of FTY720 on EC barrier function. FTY720 did not have superior efficacy compared to mycophenolate mofetil in preventing renal transplant rejection [75], but it is in phase III clinical trials as an immunosuppressant in multiple sclerosis patients. The clinical availability of FTY720 makes it attractive as a potential mediator of EC barrier function in patients with ALI. An *in vivo* study demonstrated that intraperitoneal FTY720 protected against intratracheal LPS in a murine model of ALI, as measured by Evan's blue dye extravasation [69]. Like S1P, FTY720 causes increased TER measurements in pulmonary ECs, an effect which is abolished by PTX (implicating G_i-coupled receptor activation), and requires the generation of signaling components within membrane lipid rafts [76]. Interestingly, however, the mechanism of FTY720-induced EC barrier enhancement diverges from the mechanism described for S1P in several ways including the delayed kinetics of the rise in TER compared to S1P [76]. Decreased expression of the S1P1 receptor prevented S1P-induced increase in TER, but did not affect FTY720-induced TER increases. Unlike S1P, FTY720 did not result in threonine phosphorylation of the S1P1 receptor, nor did inhibition of phosphatidylinositol 3-kinase (PI3K) prevent FTY720-induced EC barrier enhancement [76]. Furthermore, FTY720 did not cause the increased intracellular calcium, the MLC phosphorylation, or the cytoskeletal rearrangement seen in response to S1P [76]. Downregulation of Rac or cortactin using siRNAs attenuated the barrier enhancing effect of S1P, but not that of FTY720 [76]. Although FTY720 is an S1P receptor agonist, its mechanism of barrier enhancement is distinct from that of S1P and does not require the S1P1 receptor.

SIMVASTATIN

Another class of prominent barrier-protective agonists under intense scrutiny are the statin family of compounds known as 3-hydroxy-3-methylglutaryl coenzyme A (HMG-CoA) reductase inhibitors [77]. These drugs inhibit cholesterol synthesis in the liver, are commonly used in clinical practice as lipid-lowering agents, and prevent acute coronary events. A plethora of reports have now demonstrated that the benefits of statin therapy cannot be entirely attributed to decreased serum cholesterol. This led to the investigation of the effect of statins on endothelial function, which plays a central role in the development of atherosclerosis and acute coronary events. Interest in the effect of statins on endothelial function in ALI is also drawn from an ever-growing literature that demonstrates improved outcomes in patients with sepsis who are treated with statins. Early work found decreased mortality in bacteremic patients admitted to the hospital while on statin therapy compared to patients who were not taking statins [78]. Subsequent animal studies found dramatically improved survival in mice treated with simvastatin prior to initiation of sepsis by cecal ligation and puncture compared to mice which were not pretreated with simvastatin. The improved survival time in simvastatin-treated mice was due to preservation of hemodynamics [79]. Further experiments found that statin treatment after the onset of sepsis can similarly rescue mice from derangements in hemodynamic status [80]. A retrospective study in human patients with multiple organ dysfunction syndrome found that those receiving statins had significantly lower 28-day mortality and hospital mortality compared to matched controls not receiving statin therapy [81]. Clearly, the mechanism of action of these drugs and their effects on the endothelium need to be understood, and many experiments have sought to accomplish this task.

Early EC studies found that simvastatin attenuated thrombin-induced stress fiber formation, paracellular gap formation, and barrier dysfunction [48, 77]. Coincubation with mevalonate (the product of HMG-CoA reductase activity) eliminated the protective effect of simvastatin against thrombin-induced EC permeability, indicating this effect is due to HMG-CoA reductase inhibition. The effect of simvastatin on EC permeability did not involve either intracellular increased cAMP levels or increased levels of endothelial nitric oxide synthase. Statins inhibit geranylgeranylation of small GTPases, essential for GTPase interaction with cell membranes [48], and translocation of the small GTPases Rac and Rho to the plasma membrane. EC pretreatment with simvastatin prevented thrombin-induced translocation of Rho to the plasma membrane [77] and simvastatin was found to confer greater protection against thrombin-induced barrier dysfunction than Rho inhibition alone. Rac inhibition may be protective via decreased activation of NADPH oxidase and resultant superoxides that induce barrier dysfunction, and this was also found to be important in simvastatin-induced EC barrier protection [82]. Simvastatin pretreatment resulted in reduced diphosphorylated MLC, reduced stress fibers, increased Rac GTPase activation [48], cortactin translocation to the EC periphery [48], and increased cortical actin and decreased paracellular gap formation after thrombin (Figure 21.6). Unlike S1P, simvastatin does not cause an increased baseline TER [48].

Simvastatin pretreatment of EC for greater than 1 h was required for attenuation of the decrease in endothelial integrity (TER) caused by thrombin and more rapid recovery to baseline TER values [48] suggesting the

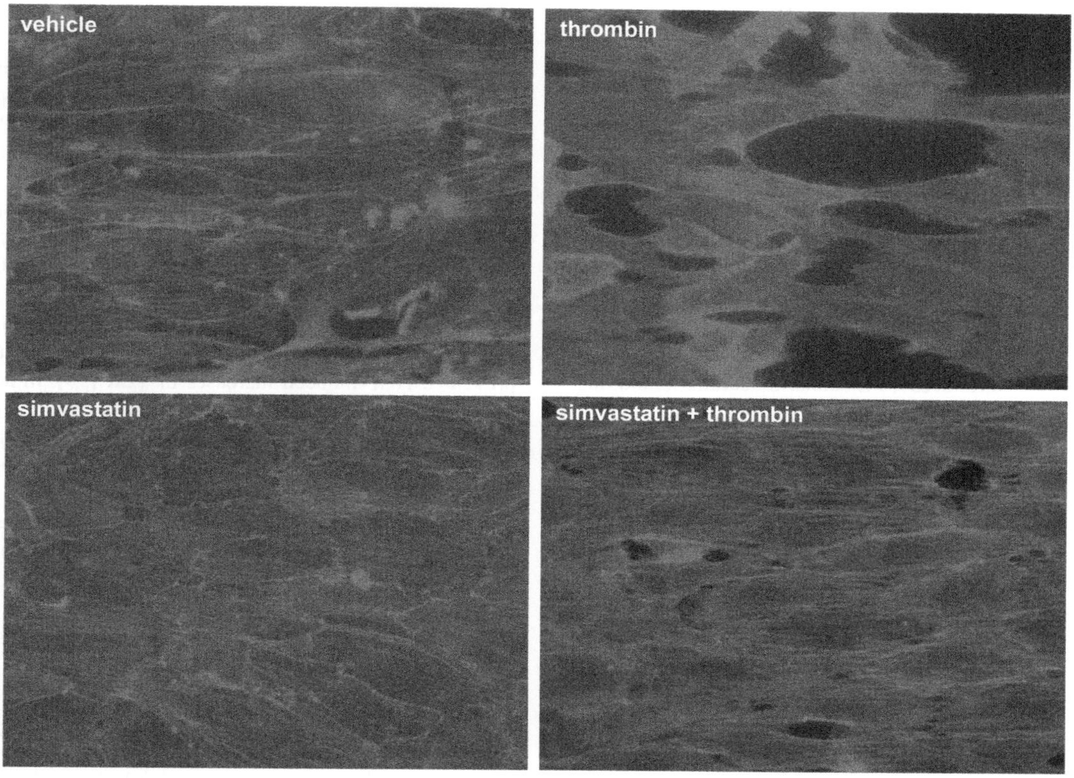

Figure 21.6 Simvastatin-mediated inhibition of thrombin and 18% cyclic stretch-induced human lung EC gap formation.

possibility that simvastatin elicits changes in gene expression or protein translation. This was investigated using Affymetrix microarray analysis from EC after 24 h of exposure to simvastatin, and yielded intriguing downregulation of caldesmon and the thrombin receptor PAR-1, as well as upregulation of integrin β_4 (known to function in cell–cell adhesion), Rac 1, and guanine exchange factors (GEFs), which may regulate Rho GTPase activity [48]. The importance of new protein synthesis to the barrier protective effect of simvastatin was established by the elimination of the protective effect by coincubation of EC with simvastatin and the protein synthesis inhibitor cycloheximide [48].

Initial *in vivo* data from an intratracheal-LPS murine model of ALI supports the *in vitro* finding that simvastatin is protective of EC barrier function. Mice were pretreated with 20 mg/kg intraperitoneal simvastatin 24 h prior to and again at the time of intratracheal LPS instillation. Simvastatin protected against markers of inflammatory lung injury compared to control, with decreased BAL neutrophil count and MPO activity, as well as decreased BAL protein and Evan's blue dye extravasation. Mice pretreated with simvastatin showed marked reduction of inflammatory histology changes after thrombin challenge as compared to controls [83]. Investigation of gene expression in lung tissue of mice pretreated with simvastatin in this LPS-induced model of ALI found that simvastatin caused differential regulation of several families of genes, including inflammatory and immune response genes, as well as nuclear factor-κB regulation and cell adhesion genes [83]. Although the mice in this model were pretreated with simvastatin before the ALI-inducing event (intratracheal LPS), simvastatin may prove to be clinically relevant in treating ALI, as ALI typically has a prolonged course, and treatment with simvastatin along the trajectory of the illness may be beneficial. To this end, a blinded, randomized controlled clinical trial of simvastatin in ALI is currently underway.

ATP

ATP is found in abundance in the EC microenvironment and participates in EC barrier regulation with constitutive release of ATP across the EC apical membrane in basal conditions [84, 85]. ATP reduced EC albumin permeability in a concentration-dependent manner in ECs from a variety of origins, including porcine aorta and pulmonary artery, bovine aorta, and human umbilical vein ECs [86]. The mechanism of ATP-induced EC barrier enhancement involves G_i/G_o proteins [84], but does not involve adenosine receptors [86], increased PKC activity, or increases in cGMP [86]. However, ATP-induced decreases in EC permeability were found to involve the phospholipase C (PLC) signaling pathway [86], as well as alterations in EC MLC phosphorylation [87, 88]. We have also demonstrated that ATP produces Ca^{2+}- and p42/44 MAPK-independent increases in cell–cell interfaces (VE-cadherin staining), and increased thickness and continuity of zona occluden-1 in TJs [84], mediated in part via cAMP-independent activation of PKA. We also noted that ATP produced a biphasic effect on MLC phosphorylation, with an initial increase followed by a decrease in levels of phospho-MLC. However, the delayed decrease in phospho-MLC was prevented by phosphatase inhibitors, emphasizing the importance of G-protein-mediated phosphatase activity in the ATP-induced decrease in MLC phosphorylation and ATP-induced barrier enhancement [84]. Similar to S1P (as well as HGF, APC, etc.) ATP-mediated barrier enhancement required Rac-dependent cytoskeletal rearrangement with decreased central actin stress fibers, increased cortical distribution of actin, peripheral MLC phosphorylation, and cortactin translocation to the cortical actin ring [50]. In addition, a rapid, transient increase in MLC diphosphorylation was observed after ATP with phosphorylated MLC localized at the cell periphery – a stark contrast to the central, stress fiber-associated phosphorylated MLC seen in EC treated with thrombin [50].

Extending these *in vitro* studies, the effect of purinergic stimulation was assessed in a murine model of ALI with intratracheal LPS. As ATP is rapidly degraded intravascularly, the nonhydrolyzable analog ATPγS was used for *in vivo* studies. Mice given intravenous ATPγS concomitant with intratracheal LPS were protected from LPS-induced ALI compared to controls as assessed by neutrophil infiltration and MPO activity [89]. ATPγS also attenuated the lung microvascular permeability elicited by LPS, with decreased BAL protein and decreased Evan's blue albumin extravasation in mice treated with ATPγS compared to controls [89]. ATPγS-treated animals were also protected from the LPS-induced decrease in body weight that was seen in control mice [89]. In addition, *in vitro* studies found that ATPγS alone produced increased TER in ECs and also showed delayed protection against the reduction in TER caused by LPS [89].

HGF

Alterations in vascular permeability are requisite steps in the angiogenic process [1, 59]. We were the first to report that HGF, a well-known angiogenic factor, like S1P, is a potent EC barrier-protective agonist [47]. HGF signals via a tyrosine kinase receptor, c-Met, and serves to recruit CD44v10, a key transactivated receptor for CD44, into caveolin-enriched microdomains (CEMs) or lipid rafts [90]. In experiments using siRNA, both c-Met and CD44 were found to be important in HGF-induced increases in EC TER [90]. Furthermore, pretreatment of ECs with the CEM-interfering compound methyl-β-cyclodextran also prevented HGF-induced increases in TER [90]. In addition, Rac activation by HGF was found to require CEM formation, c-Met, CD44, Tiam1, and dynamin-2 [90]. In a mouse model of LPS-induced ALI, HGF was protective against markers of lung inflammation, an effect not noted in CD44 knockout mice [90]. The signaling mechanism involved in HGF-induced EC barrier enhancement is complex, with important roles for c-Met, CD44, and CEM formation.

Stabilization of the EC actin cytoskeleton is critical to HGF-induced EC barrier enhancement, as EC pretreatment with actin-disrupting cytochalasin B prevented HGF-induced increases in TER. HGF produced Rac-dependent increases in cortical actin, cortactin translocation, and cortical levels of phosphorylated MLC [47]. Further mechanistic studies found that HGF-induced EC barrier enhancement critically involves PI3K activity, distinguishing the mechanism of HGF-induced barrier enhancement from that of S1P [47] with important roles for MAPKs [extracellular signal-regulated kinase (ERK) and p38] and PKC in HGF-induced EC barrier enhancement [47]. Attention to the role of improved cell–cell or cell–matrix adhesion elicited by HGF found that HGF produced increased β-catenin localization to the EC periphery alongside cortical actin and increased association of β-catenin with VE-cadherin [47]. The cell signaling effectors of HGF (PI3K, ERK, p38, PKC) were found to converge at phosphorylation of glycogen synthase kinase-3β, which regulates the association of β-catenin and cadherin, thereby controlling cell–cell adhesion [47]. Additional work found that HGF pretreatment protected EC from the barrier-disrupting effect of thrombin as measured by TER and stress fiber and paracellular gap formation [91] – effects mediated by attenuation of thrombin-induced Rho activation and by increased Rac activation involving the Rac-specific GEF Tiam1. Decreased Rho activation was associated with decreased MLC phosphorylation and activation of the Tiam1 and Rac pathways, and increased Rac activation was associated with increased PAK-1 phosphorylation [91].

APC

APC is a serine protease that modulates coagulation and inflammation. In 2001, the US Food and Drug Administration approved Xigris® [recombinant human (rh) APC, also known as drotrecogin alfa (activated)] for treatment of severe sepsis in adults after a randomized trial found a 28-day survival benefit in treated patients [92]. As severe sepsis involves ALI and systemic increased vascular permeability, the effect of APC on pulmonary EC permeability is intriguing. Interest in the effect of the anticoagulant APC on EC permeability is also related to the well-described role of the procoagulant thrombin in EC barrier disruption. Furthermore, the mechanism of the survival benefit imparted by treatment with rhAPC is unclear, as APC given to human subjects in the setting of endotoxin infusion improved hemodynamics but did not have an anti-inflammatory or antithrombotic effect [93], suggesting that a different mechanism may be involved.

We demonstrated that APC prevented and was able to reverse thrombin-induced increased permeability [49]. APC also increased MLC phosphorylation and actin at the EC periphery, and decreased central stress fibers. The barrier-enhancing effect of APC was found to be mediated by Rac1 activation, similar to the barrier-enhancing effect of S1P, simvastatin, and HGF [49]. The endothelial protein C receptor (EPCR) is critical to APC-induced barrier enhancement and MLC phosphorylation. Furthermore, EPCR-mediated transactivation of the S1P1 receptor via PI3K is essential, and involves direct interaction between EPCR and S1P1 receptor [49]. This novel pathway for APC-induced EC barrier enhancement may contribute significantly to the survival benefit offered by rhAPC in patients with severe sepsis.

More recent work has focused on APC in animal models of ALI. Using a rat model of intestinal ischemia/reperfusion injury-induced ALI, investigators found that APC treatment just prior to reperfusion attenuated subsequent pulmonary edema, which was accompanied by fewer neutrophils on histology and a marked improvement in histologic appearance compared to animals that did not receive APC [94]. In addition, rats treated with APC prior to intestinal reperfusion had lower serum levels of TNF-α, IL-6, and D-dimer compared to controls [94]. Investigation of APC in a mouse model of VILI found that APC pretreatment was protective against VILI caused by high tidal volume ventilation with mice pretreated with APC exhibiting significant reductions in BAL protein and Evan's blue dye extravasation compared to controls (Finigan, Garcia, and Hassoun, unpublished observations).

OXIDIZED PHOSPHOLIPIDS

Oxidized phospholipids are derived from oxidized low-density-lipoproteins and have been the focus of much investigation in the areas of vascular injury and inflammation [95] with increased levels noted in ALI [96]. Oxidized phospholipids resulting from the OxPAPC activate MAPKs ERK1/2 and c-Jun N-terminal kinase, but not p38 or its downstream target, HSP27 [95], and increased the activity of both PKC and PKA [95] and Src kinases, processes involved in OxPAPC-mediated EC barrier enhancement, whereas Rho, Rho kinase, Erk-1,2, p38, and PI3K were not involved [97]. Furthermore, OxPAPC resulted in phosphorylation of the actin-binding protein cofilin as well as phosphorylation of the focal adhesion proteins FAK and paxillin, indicating that OxPAPC may affect the EC actin cytoskeleton and cell–cell adhesions [95].

OxPAPC protects against EC barrier dysfunction *in vitro* with increased TER in HPAECs, reflecting enhanced EC barrier function [51]. OxPAPC increased TER and protected against thrombin-induced decreases in TER in lung microvessel ECs in a dose-dependent manner [98]. In addition, ECs treated with OxPAPC after treatment with barrier-disrupting doses of LPS showed attenuated barrier disruption as measured by TER [99]. OxPAPC alone accentuated peripheral F-actin in a unique, zip-like configuration [51], and OxPAPC subsequent to LPS pretreatment attenuated LPS-induced actin stress fiber and paracellular gap formation [99]. OxPAPC also resulted in continuous focal adhesions (noted via staining for vinculin – a focal adhesion protein) and caused accumulation of β-catenin, which links AJs to the cytoskeleton, at cell–cell interfaces [99]. Thus, OxPAPC has significant EC barrier-enhancing effects. The barrier-protective effects of OxPAPC have been compared to those of other oxidized phospholipids, including oxidized phosphoserine and oxidized glycerophosphate, and the polar head groups of oxidized phospholipids are important in mitigating the barrier-protective effects of these compounds [100].

Significant mechanistic investigations have defined the signaling pathways involved in OxPAPC-mediated endothelial barrier protection. Initial work found that the barrier protective effect of OxPAPC was mediated by the small GTPases Rac and Cdc42, as inhibition of Rac and Cdc42 attenuated OxPAPC-induced increases in TER [51]. OxPAPC also induced translocation of Rac, Cdc42, and the Rac effector PAK-1 from the cytosol to the cell membrane [51]. Furthermore, concomitant Rac and Cdc42 activation resulted in increased peripheral F-actin similar to the effect caused by OxPAPC, and transfection of EC with dominant-negative constructs of Rac and Cdc42 inhibited the augmented peripheral F-actin caused by OxPAPC [51]. Having established the importance of Rac and Cdc42 in OxPAPC-mediated EC barrier enhancement, further studies examined the upstream Rac/Cdc42-specific GEFs Tiam1 and βPIX (PAK-interacting exchange factor), and found OxPAPC led to membrane translocation of Tiam1, βPIX, Cdc42, and Rac [101]. In addition, decreased expression of Tiam1 and βPIX resulted in inhibition of OxPAPC-mediated Rac activation, EC barrier enhancement measured by TER, peripheral F-actin enhancement, and membrane translocation of the actin-binding proteins cortactin and Arp3 [101]. Lastly, OxPAPC was found to cause a novel interaction between focal adhesion and AJ complexes, a process mediated by association of paxillin and β-catenin and dependent upon Rac and Cdc42 [102].

The effect of intravenous OxPAPC in a rat model of ALI utilizing aerosolized LPS resulted in significant attenuation of LPS-induced increases in BAL cell counts, tissue neutrophils, BAL protein content, and BAL levels of IL-6 and -1β. These results suggest a potential role for oxidized phospholipids in protecting from inflammation and vascular leak in ALI [99]. In a rat model of VILI, treatment with OxPAPC at the onset of mechanical ventilation was protective against VILI caused by high tidal volume ventilation measured by several parameters, including BAL cell count, BAL neutrophils, BAL protein, Evan's blue extravasation, and lung injury and inflammation evaluated by histology [98]. The mechanism of OxPAPC-induced attenuation of VILI was studied *in vitro* using EC exposed to pathologic repetitive cyclic stretch (18%). EC treated with OxPAPC during 18% cyclic stretch were protected from cyclic stretch-induced cytoskeletal changes, with enhanced peripheral F-actin staining and decreased central stress fibers [98]. Furthermore, pretreatment of EC with OxPAPC in thrombin-treated cells subsequently challenged with 18% cyclic stretch was protective against the dramatic paracellular gap and stress fiber formation otherwise found after such treatment [98]. In addition, EC pretreated with OxPAPC in thrombin-treated cells prior to 18% cyclic stretch had decreased thrombin-induced Rho activation. OxPAPC increased Rac activation in EC exposed to both physiologic (5%) and pathologic (18%) cyclic stretch, but after thrombin stimulation, Rac activity was high for the 5% cyclic stretch EC, not for the pathologically strained EC [98]. Thus, OxPAPC protects EC from mechanical stress-induced injury via cytoskeletal rearrangements and changes in Rho and Rac activation, and remains a potential therapy for the profound pulmonary edema associated with inflammatory states.

PGs

PGs are produced by the arachidonic acid pathways and are released by many cell types, including ECs, and play a complex role in EC barrier function [54]. Beraprost, a PGI_2 analog, is used clinically to treat pulmonary hypertension and has been described to have antiedemagenic effects in a model of ALI [103]. Iloprost, another clinically available PGI_2 analog, and PGE_2 have been shown to decrease EC permeability *in vitro* [104]. Investigation of the mechanism of this barrier-protective effect found that both PGE_2 and beraprost produce a rapid and sustained increase in baseline TER in HPAECs, an effect associated with increased membrane-related cortactin and β-catenin, decreased central stress fibers and increased peripheral F-actin, and increased AJs as seen by VE-cadherin immunofluorescence [54]. The cytoskeletal changes caused by PGE_2 were found to be Rac-dependent [54]. Another similarity to the mechanisms of other barrier enhancing agents was rapid activation of Rac GTPase, although the PGs did not affect Rho activation [54]. The PGs also resulted in membrane translocation of the GEF Tiam1, similar to the effect of HGF, and also caused phosphorylation of the Rac effector PAK-1, as is described for S1P [54]. Additional mechanistic studies found that PGs caused increased EC PKA activity, increased cAMP, but not increased cGMP, and PKA activity was required for PG-induced barrier enhancement as measured by TER [54].

PGs were also found to be protective against a thrombin challenge *in vitro*, as EC pretreatment with PGE_2 or beraprost attenuated both the thrombin-induced decrease in TER and increase in EC dextran permeability, and protected against thrombin-induced paracellular gap and stress fiber formation. These effects in EC were accompanied by a decrease in thrombin-induced Rho activation and MLC phosphorylation [54]. Furthermore, intraperitoneal beraprost was protective in a murine model of VILI, with treated mice having decreased indices of inflammatory lung injury (BAL protein and cell counts) after high tidal volume ventilation compared to controls [54].

METHYLNALTREXONE

Methylnaltrexone (MNTX) is a peripherally restricted mu opioid receptor (mOP-R) antagonist recently approved by the US Food and Drug Administration for the treatment of postoperative ileus, and also recently found to work synergistically with 5-fluorouracil and bevacizumab to inhibit VEGF-induced pulmonary EC proliferation and migration [105]. Antagonists of mOP-R are of interest as potential EC barrier enhancing agents because of the barrier-disruptive properties of the mOP-R agonist, morphine [106]. Pretreatment of human pulmonary microvascular ECs with 0.1 μM MNTX was found to protect against the decrease in TER caused by mOP-R agonists morphine and the synthetic opiod peptide DAMGO ([D-Ala^2, *N*-MePhe^4, Gly-ol]-enkephalin), and also protected against the barrier-disruptive effects of thrombin and LPS, which act independently of mOP-R [107]. MNTX augments the barrier-enhancing effect of S1P [107]. EC pretreatment with naloxone, a charged mOP-R antagonist, protected against morphine and DAMGO-induced barrier disruption, but was not protective against barrier disruption caused by thrombin or LPS. This data, together with the observation that siRNA targeting mOP-R had minimal effect of MNTX-induced protection against thrombin and LPS, suggests that the protective effect of MNTX cannot be attributed to mOP-R antagonism alone [107]. Further experiments found that MNTX confers its barrier-protective effect by inhibiting the association of the RhoA-activating GEF, $p115^{RhoGEF}$, with the $S1P_3$ receptor and resultant RhoA activation that is caused by barrier-disrupting agents [107]. Complimentary *in vivo* experiments found that intravenous MNTX given after ALI was established via intratracheal LPS was protective against ALI at 24 h, as assessed by histology and BAL protein and TNF-α levels [107].

CONCLUSIONS AND PERSPECTIVES

As can readily be seen by the studies described, a number of agents have been added to the scientific armory for alleviation of uncontrolled lung vascular leakage and alveolar flooding. Nearly each of these agents has been successfully evaluated in preclinical models of ALI which increase enthusiasm for the potential administration of these agents in the critically ill. One agent, FTY720, is in phase III trials and two agents, APC and MNTX, are currently approved by the US Food and Drug Administration for other medical conditions. Thus, the prospects for the rapid translation of these lung vascular barrier-protective strategies are extremely high. Looking beyond clinical trials of the agents described above in the treatment of ALI, there is much work that remains in order to optimally treat these patients and reduce the tremendous morbidity and mortality associated with ALI. Our group currently focuses on the use of genomics to define candidate genes that place different patient populations at risk for ALI. In-depth study of each of these genes illuminates a variety of cell signaling pathways that may be exploited in the treatment of ALI. We believe that the use of powerful genomic strategies will ultimately allow the determination of which barrier-enhancing therapies will yield the best result for individual patients based on individual genetic polymorphisms and the clinical scenario at hand.

References

1. Garcia, J.G., Liu, F., Verin, A.D. *et al.* (2001) Sphingosine 1-phosphate promotes endothelial cell barrier integrity by Edg-dependent cytoskeletal rearrangement. *The Journal of Clinical Investigation*, **108**, 689–701.
2. Singleton, P.A., Dudek, S.M., Ma, S.F., and Garcia, J.G. (2006) Transactivation of sphingosine 1-phosphate receptors is essential for vascular barrier regulation. Novel role for hyaluronan and CD44 receptor family. *The Journal of Biological Chemistry*, **281**, 34381–93.
3. Dudek, S.M. and Garcia, J.G. (2001) Cytoskeletal regulation of pulmonary vascular permeability. *Journal of Applied Physiology*, **91**, 1487–500.
4. Romer, L.H., Birukov, K.G., and Garcia, J.G. (2006) Focal adhesions: paradigm for a signaling nexus. *Circulation Research*, **98**, 606–16.
5. Tiruppathi, C., Minshall, R.D., Paria, B.C. *et al.* (2002) Role of Ca^{2+} signaling in the regulation of endothelial permeability. *Vascular Pharmacology*, **39**, 173–85.
6. Garcia, J.G., Schaphorst, K.L., Shi, S. *et al.* (1997) Mechanisms of ionomycin-induced endothelial cell barrier dysfunction. *American Journal of Physiology: Lung Cellular and Molecular Physiology*, **273**, L172–84.
7. Borbiev, T., Verin, A.D., Shi, S. *et al.* (2001) Regulation of endothelial cell barrier function by calcium/calmodulin-dependent protein kinase II. *American Journal of Physiology: Lung Cellular and Molecular Physiology*, **280**, L983–90.
8. Borbiev, T., Verin, A.D., Birukova, A. *et al.* (2003) Role of CaM kinase II and ERK activation in thrombin-induced endothelial cell barrier dysfunction. *American Journal of Physiology: Lung Cellular and Molecular Physiology*, **285**, L43–L54.
9. Cai, H., Liu, D., and Garcia, J.G. (2008) CaM kinase II-dependent pathophysiological signalling in endothelial cells. *Cardiovascular Research*, **77**, 30–34.
10. Garcia, J.G., Lazar, V., Gilbert-McClain, L.I. *et al.* (1997) Myosin light chain kinase in endothelium: molecular cloning and regulation. *American Journal of Respiratory Cell and Molecular Biology*, **16**, 489–94.

11. Garcia, J.G., Verin, A.D., Herenyiova, M., and English, D. (1998) Adherent neutrophils activate endothelial myosin light chain kinase: role in transendothelial migration. *Journal of Applied Physiology*, **84**, 1817–21.
12. Verin, A.D., Cooke, C., Herenyiova, M. *et al.* (1998) Role of Ca^{2+}/calmodulin-dependent phosphatase 2B in thrombin-induced endothelial cell contractile responses. *American Journal of Physiology: Lung Cellular and Molecular Physiology*, **275**, L788–99.
13. Wang, P., Verin, A.D., Birukova, A. *et al.* (2001) Mechanisms of sodium fluoride-induced endothelial cell barrier dysfunction: role of MLC phosphorylation. *American Journal of Physiology: Lung Cellular and Molecular Physiology*, **281**, L1472–83.
14. Verin, A.D., Csortos, C., Durbin, S.D. *et al.* (2000) Characterization of the protein phosphatase 1 catalytic subunit in endothelium: involvement in contractile responses. *Journal of Cellular Biochemistry*, **79**, 113–25.
15. Verin, A.D., Wang, P., and Garcia, J.G. (2000) Immunochemical characterization of myosin-specific phosphatase 1 regulatory subunits in bovine endothelium. *Journal of Cellular Biochemistry*, **76**, 489–98.
16. Liu, F., Verin, A.D., Wang, P. *et al.* (2001) Differential regulation of sphingosine-1-phosphate- and VEGF-induced endothelial cell chemotaxis. Involvement of $G_{i\alpha2}$-linked Rho kinase activity. *American Journal of Respiratory Cell and Molecular Biology*, **24**, 711–19.
17. Birukova, A.A., Smurova, K., Birukov, K.G. *et al.* (2004) Role of Rho GTPases in thrombin-induced lung vascular endothelial cells barrier dysfunction. *Microvascular Research*, **67**, 64–77.
18. Verin, A.D., Patterson, C.E., Day, M.A., and Garcia, J.G. (1995) Regulation of endothelial cell gap formation and barrier function by myosin-associated phosphatase activities. *American Journal of Physiology: Lung Cellular and Molecular Physiology*, **269**, L99–108.
19. Csortos, C., Kolosova, I., and Verin, A.D. (2007) Regulation of vascular endothelial cell barrier function and cytoskeleton structure by protein phosphatases of the PPP family. *American Journal of Physiology: Lung Cellular and Molecular Physiology*, **293**, L843–54.
20. Kawkitinarong, K., Linz-McGillem, L., Birukov, K.G., and Garcia, J.G. (2004) Differential regulation of human lung epithelial and endothelial barrier function by thrombin. *American Journal of Respiratory Cell and Molecular Biology*, **31**, 517–27.
21. Garcia, J.G., Verin, A.D., Schaphorst, K. *et al.* (1999) Regulation of endothelial cell myosin light chain kinase by Rho, cortactin, and $p60^{src}$. *American Journal of Physiology: Lung Cellular and Molecular Physiology*, **276**, L989–98.
22. Garcia, J.G., Davis, H.W., and Patterson, C.E. (1995) Regulation of endothelial cell gap formation and barrier dysfunction: role of myosin light chain phosphorylation. *Journal of Cellular Physiology*, **163**, 510–22.
23. Petrache, I., Birukov, K., Zaiman, A.L. *et al.* (2003) Caspase-dependent cleavage of myosin light chain kinase (MLCK) is involved in TNF-alpha-mediated bovine pulmonary endothelial cell apoptosis. *The FASEB Journal*, **17**, 407–16.
24. Wainwright, M.S., Rossi, J., Schavocky, J. *et al.* (2003) Protein kinase involved in lung injury susceptibility: evidence from enzyme isoform genetic knockout and *in vivo* inhibitor treatment. *Proceedings of the National Academy of Sciences of the United States of America*, **100**, 6233–38.
25. Behanna, H.A., Watterson, D.M., and Ranaivo, H.R. (2006) Development of a novel bioavailable inhibitor of the calmodulin-regulated protein kinase MLCK: a lead compound that attenuates vascular leak. *Biochimica et Biophysica Acta*, **1763**, 1266–74.
26. Rossi, J.L., Velentza, A.V., Steinhorn, D.M. *et al.* (2007) MLCK210 gene knockout or kinase inhibition preserves lung function following endotoxin-induced lung injury in mice. *American Journal of Physiology: Lung Cellular and Molecular Physiology*, **292**, L1327–34.
27. Verin, A.D., Liu, F., Bogatcheva, N. *et al.* (2000) Role of ras-dependent ERK activation in phorbol ester-induced endothelial cell barrier dysfunction. *American Journal of Physiology: Lung Cellular and Molecular Physiology*, **279**, L360–70.
28. Ikebe, M., Hartshorne, D.J., and Elzinga, M. (1986) Identification, phosphorylation, and dephosphorylation of a second site for myosin light chain kinase on the 20,000-dalton light chain of smooth muscle myosin. *The Journal of Biological Chemistry*, **261**, 36–39.
29. Ikebe, M., Hartshorne, D.J., and Elzinga, M. (1987) Phosphorylation of the 20,000-dalton light chain of smooth muscle myosin by the calcium-activated, phospholipid-dependent protein kinase. Phosphorylation sites and effects of phosphorylation. *The Journal of Biological Chemistry*, **262**, 9569–73.

30. Bogatcheva, N.V., Verin, A.D., Wang, P. *et al.* (2003) Phorbol esters increase MLC phosphorylation and actin remodeling in bovine lung endothelium without increased contraction. *American Journal of Physiology: Lung Cellular and Molecular Physiology*, **285**, L415–26.
31. Tinsley, J.H., Teasdale, N.R., and Yuan, S.Y. (2004) Involvement of PKCdelta and PKD in pulmonary microvascular endothelial cell hyperpermeability. *American Journal of Physiology: Cell Physiology*, **286**, C105–11.
32. Stasek, J.E. Jr., Patterson, C.E., and Garcia, J.G. (1992) Protein kinase C phosphorylates caldesmon77 and vimentin and enhances albumin permeability across cultured bovine pulmonary artery endothelial cell monolayers. *Journal of Cellular Physiology*, **153**, 62–75.
33. Bogatcheva, N.V., Birukova, A., Borbiev, T. *et al.* (2006) Caldesmon is a cytoskeletal target for PKC in endothelium. *Journal of Cellular Biochemistry*, **99**, 1593–605.
34. Mirzapoiazova, T., Kolosova, I.A., Romer, L. *et al.* (2005) The role of caldesmon in the regulation of endothelial cytoskeleton and migration. *Journal of Cellular Physiology*, **203**, 520–28.
35. Lynch, J.J., Ferro, T.J., Blumenstock, F.A. *et al.* (1990) Increased endothelial albumin permeability mediated by protein kinase C activation. *The Journal of Clinical Investigation*, **85**, 1991–98.
36. Siflinger-Birnboim, A., Goligorsky, M.S., Del Vecchio, P.J., and Malik, A.B. (1992) Activation of protein kinase C pathway contributes to hydrogen peroxide-induced increase in endothelial permeability. *Laboratory Investigation*, **67**, 24–30.
37. Hammerschmidt, S., Vogel, T., Jockel, S. *et al.* (2007) Protein kinase C inhibition attenuates hypochlorite-induced acute lung injury. *Respiratory Medicine*, **101**, 1205–11.
38. Chiang, E.T., Persaud-Sawin, D.A., Kulkarni, S. *et al.* (2006) Bluetongue virus and double-stranded RNA increase human vascular permeability: role of p38 MAPK. *Journal of Clinical Immunology*, **26**, 406–16.
39. Guay, J., Lambert, H., Gingras-Breton, G. *et al.* (1997) Regulation of actin filament dynamics by p38 MAP kinase-mediated phosphorylation of heat shock protein 27. *Journal of Cell Science*, **110**, 357–68.
40. Borbiev, T., Birukova, A., Liu, F. *et al.* (2004) p38 MAP kinase-dependent regulation of endothelial cell permeability. *American Journal of Physiology: Lung Cellular and Molecular Physiology*, **287**, L911–18.
41. Garcia, J.G., Wang, P., Schaphorst, K.L. *et al.* (2002) Critical involvement of p38 MAP kinase in pertussis toxin-induced cytoskeletal reorganization and lung permeability. *The FASEB Journal*, **16**, 1064–76.
42. Birukova, A.A., Birukov, K.G., Gorshkov, B. *et al.* (2005) MAP kinases in lung endothelial permeability induced by microtubule disassembly. *American Journal of Physiology: Lung Cellular and Molecular Physiology*, **289**, L75–L84.
43. Stelzner, T.J., Weil, J.V., and O'Brien, R.F. (1989) Role of cyclic adenosine monophosphate in the induction of endothelial barrier properties. *Journal of Cellular Physiology*, **139**, 157–66.
44. Patterson, C.E., Lum, H., Schaphorst, K.L. *et al.* (2000) Regulation of endothelial barrier function by the cAMP-dependent protein kinase. *Endothelium*, **7**, 287–308.
45. Liu, F., Verin, A.D., Borbiev, T., and Garcia, J.G. (2001) Role of cAMP-dependent protein kinase A activity in endothelial cell cytoskeleton rearrangement. *American Journal of Physiology: Lung Cellular and Molecular Physiology*, **280**, L1309–17.
46. Birukova, A.A., Liu, F., Garcia, J.G., and Verin, A.D. (2004) Protein kinase A attenuates endothelial cell barrier dysfunction induced by microtubule disassembly. *American Journal of Physiology: Lung Cellular and Molecular Physiology*, **287**, L86–L93.
47. Liu, F., Schaphorst, K.L., Verin, A.D. *et al.* (2002) Hepatocyte growth factor enhances endothelial cell barrier function and cortical cytoskeletal rearrangement: potential role of glycogen synthase kinase-3beta. *The FASEB Journal*, **16**, 950–62.
48. Jacobson, J.R., Dudek, S.M., Birukov, K.G. *et al.* (2004) Cytoskeletal activation and altered gene expression in endothelial barrier regulation by simvastatin. *American Journal of Respiratory Cell and Molecular Biology*, **30**, 662–70.
49. Finigan, J.H., Dudek, S.M., Singleton, P.A. *et al.* (2005) Activated protein C mediates novel lung endothelial barrier enhancement: role of sphingosine 1-phosphate receptor transactivation. *The Journal of Biological Chemistry*, **280**, 17286–93.
50. Jacobson, J.R., Dudek, S.M., Singleton, P.A. *et al.* (2006) Endothelial cell barrier enhancement by ATP is mediated by the small GTPase Rac and cortactin. *American Journal of Physiology: Lung Cellular and Molecular Physiology*, **291**, L289–95.
51. Birukov, K.G., Bochkov, V.N., Birukova, A.A. *et al.* (2004) Epoxycyclopentenone-containing oxidized phospholipids restore endothelial barrier function via Cdc42 and Rac. *Circulation Research*, **95**, 892–901.

52. Broman, M.T., Mehta, D., and Malik, A.B. (2007) Cdc42 regulates the restoration of endothelial adherens junctions and permeability. *Trends in Cardiovascular Medicine*, **17**, 151–56.
53. Dudek, S.M., Jacobson, J.R., Chiang, E.T. *et al.* (2004) Pulmonary endothelial cell barrier enhancement by sphingosine 1-phosphate: roles for cortactin and myosin light chain kinase. *The Journal of Biological Chemistry*, **279**, 24692–700.
54. Birukova, A.A., Zagranichnaya, T., Fu, P. *et al.* (2007) Prostaglandins PGE_2 and PGI_2 promote endothelial barrier enhancement via PKA- and Epac1/Rap1-dependent Rac activation. *Experimental Cell Research*, **313**, 2504–20.
55. Dudek, S.M., Birukov, K.G., Zhan, X., and Garcia, J.G. (2002) Novel interaction of cortactin with endothelial cell myosin light chain kinase. *Biochemical and Biophysical Research Communications*, **298**, 511–19.
56. Morales-Ruiz, M., Fulton, D., Sowa, G. *et al.* (2000) Vascular endothelial growth factor-stimulated actin reorganization and migration of endothelial cells is regulated via the serine/threonine kinase Akt. *Circulation Research*, **86**, 892–96.
57. Miyahara, T., Hamanaka, K., Weber, D.S. *et al.* (2007) Phosphoinositide 3-kinase, Src, and Akt modulate acute ventilation-induced vascular permeability increases in mouse lungs. *American Journal of Physiology: Lung Cellular and Molecular Physiology*, **293**, L11–21.
58. McVerry, B.J. and Garcia, J.G. (2005) *In vitro* and *in vivo* modulation of vascular barrier integrity by sphingosine 1-phosphate: mechanistic insights. *Cellular Signalling*, **17**, 131–39.
59. English, D., Welch, Z., Kovala, A.T. *et al.* (2000) Sphingosine 1-phosphate released from platelets during clotting accounts for the potent endothelial cell chemotactic activity of blood serum and provides a novel link between hemostasis and angiogenesis. *The FASEB Journal*, **14**, 2255–65.
60. Gimbrone, M.A. Jr., Aster, R.H., Cotran, R.S. *et al.* (1969) Preservation of vascular integrity in organs perfused *in vitro* with a platelet-rich medium. *Nature*, **222**, 33–36.
61. Schaphorst, K.L., Chiang, E., Jacobs, K.N. *et al.* (2003) Role of sphingosine-1 phosphate in the enhancement of endothelial barrier integrity by platelet-released products. *American Journal of Physiology: Lung Cellular and Molecular Physiology*, **285**, L258–67.
62. English, D., Kovala, A.T., Welch, Z. *et al.* (1999) Induction of endothelial cell chemotaxis by sphingosine 1-phosphate and stabilization of endothelial monolayer barrier function by lysophosphatidic acid, potential mediators of hematopoietic angiogenesis. *Journal of Hematotherapy and Stem Cell Research*, **8**, 627–34.
63. McVerry, B.J. and Garcia, J.G. (2004) Endothelial cell barrier regulation by sphingosine 1-phosphate. *Journal of Cellular Biochemistry*, **92**, 1075–85.
64. Mehta, D., Konstantoulaki, M., Ahmmed, G.U., and Malik, A.B. (2005) Sphingosine 1-phosphate-induced mobilization of intracellular Ca^{2+} mediates rac activation and adherens junction assembly in endothelial cells. *The Journal of Biological Chemistry*, **280**, 17320–28.
65. Guo, Y., Singleton, P.A., Rowshan, A. *et al.* (2007) Quantitative proteomics analysis of human endothelial cell membrane rafts: evidence of MARCKS and MRP regulation in the sphingosine 1-phosphate-induced barrier enhancement. *Molecular and Cellular Proteomics*, **6**, 689–96.
66. Arce, F.T., Whitlock, J.L., Birukova, A.A. *et al.* (2008) Regulation of the micromechanical properties of pulmonary endothelium by S1P and thrombin: role of cortactin. *Biophysical Journal*, **95**, 886–94.
67. Weed, S.A. and Parsons, J.T. (2001) Cortactin: coupling membrane dynamics to cortical actin assembly. *Oncogene*, **20**, 6418–34.
68. Shikata, Y., Birukov, K.G., and Garcia, J.G. (2003) S1P induces FA remodeling in human pulmonary endothelial cells: role of Rac, GIT1, FAK, and paxillin. *Journal of Applied Physiology*, **94**, 1193–203.
69. Peng, X., Hassoun, P.M., Sammani, S. *et al.* (2004) Protective effects of sphingosine 1-phosphate in murine endotoxin-induced inflammatory lung injury. *American Journal of Respiratory and Critical Care Medicine*, **169**, 1245–51.
70. McVerry, B.J., Peng, X., Hassoun, P.M. *et al.* (2004) Sphingosine 1-phosphate reduces vascular leak in murine and canine models of acute lung injury. *American Journal of Respiratory and Critical Care Medicine*, **170**, 987–93.
71. Moreno-Vinasco, L., Jacobson, J.R., Bonde, P. *et al.* (2008) Attenuation of ischemia–reperfusion lung injury by sphingosine 1-phosphate in rodents. *Journal of Organ Dysfunction*, **4**, 106–14.
72. Gon, Y., Wood, M.R., Kiosses, W.B. *et al.* (2005) S1P3 receptor-induced reorganization of epithelial tight junctions compromises lung barrier integrity and is potentiated by TNF. *Proceedings of the National Academy of Sciences of the United States of America*, **102**, 9270–75.

73. Forrest, M., Sun, S.Y., Hajdu, R. *et al.* (2004) Immune cell regulation and cardiovascular effects of sphingosine 1-phosphate receptor agonists in rodents are mediated via distinct receptor subtypes. *The Journal of Pharmacology and Experimental Therapeutics*, **309**, 758–68.
74. Chiba, K. (2005) FTY720, a new class of immunomodulator, inhibits lymphocyte egress from secondary lymphoid tissues and thymus by agonistic activity at sphingosine 1-phosphate receptors. *Pharmacology and Therapeutics*, **108**, 308–19.
75. Tedesco-Silva, H., Szakaly, P., Shoker, A. *et al.* (2007) FTY720 versus mycophenolate mofetil in *de novo* renal transplantation: six-month results of a double-blind study. *Transplantation*, **84**, 885–92.
76. Dudek, S.M., Camp, S.M., Chiang, E.T. *et al.* (2007) Pulmonary endothelial cell barrier enhancement by FTY720 does not require the S1P1 receptor. *Cellular Signalling*, **19**, 1754–64.
77. van Nieuw Amerongen, G.P., Vermeer, M.A., Negre-Aminou, P. *et al.* (2000) Simvastatin improves disturbed endothelial barrier function. *Circulation*, **102**, 2803–9.
78. Liappis, A.P., Kan, V.L., Rochester, C.G., and Simon, G.L. (2001) The effect of statins on mortality in patients with bacteremia. *Clinical Infectious Diseases*, **33**, 1352–57.
79. Merx, M.W., Liehn, E.A., Janssens, U. *et al.* (2004) HMG-CoA reductase inhibitor simvastatin profoundly improves survival in a murine model of sepsis. *Circulation*, **109**, 2560–65.
80. Merx, M.W., Liehn, E.A., Graf, J. *et al.* (2005) Statin treatment after onset of sepsis in a murine model improves survival. *Circulation*, **112**, 117–24.
81. Schmidt, H., Hennen, R., Keller, A. *et al.* (2006) Association of statin therapy and increased survival in patients with multiple organ dysfunction syndrome. *Intensive Care Medicine*, **32**, 1248–51.
82. Chen, W., Pendyala, S., Natarajan, V. *et al.* (2008) Endothelial cell barrier protection by simvastatin: GTPase regulation and NADPH oxidase inhibition. *American Journal of Physiology: Lung Cellular and Molecular Physiology*, **295**, L575–83.
83. Jacobson, J.R., Barnard, J.W., Grigoryev, D.N. *et al.* (2005) Simvastatin attenuates vascular leak and inflammation in murine inflammatory lung injury. *American Journal of Physiology: Lung Cellular and Molecular Physiology*, **288**, L1026–32.
84. Kolosova, I.A., Mirzapoiazova, T., Adyshev, D. *et al.* (2005) Signaling pathways involved in adenosine triphosphate-induced endothelial cell barrier enhancement. *Circulation Research*, **97**, 115–24.
85. Schwiebert, L.M., Rice, W.C., Kudlow, B.A. *et al.* (2002) Extracellular ATP signaling and P2X nucleotide receptors in monolayers of primary human vascular endothelial cells. *American Journal of Physiology: Cell Physiology*, **282**, C289–C301.
86. Noll, T., Holschermann, H., Koprek, K. *et al.* (1999) ATP reduces macromolecule permeability of endothelial monolayers despite increasing $[Ca^{2+}]i$. *American Journal of Physiology*, **276**, H1892–901.
87. Noll, T., Schafer, M., Schavier-Schmitz, U., and Piper, H.M. (2000) ATP induces dephosphorylation of myosin light chain in endothelial cells. *American Journal of Physiology: Cell Physiology*, **279**, C717–23.
88. Gunduz, D., Hirche, F., Hartel, F.V. *et al.* (2003) ATP antagonism of thrombin-induced endothelial barrier permeability. *Cardiovascular Research*, **59**, 470–78.
89. Kolosova, I.A., Mirzapoiazova, T., Moreno-Vinasco, L. *et al.* (2008) Protective effect of purinergic agonist ATPgammaS against acute lung injury. *American Journal of Physiology: Lung Cellular and Molecular Physiology*, **294**, L319–24.
90. Singleton, P.A., Salgia, R., Moreno-Vinasco, L. *et al.* (2007) CD44 regulates hepatocyte growth factor-mediated vascular integrity. Role of c-Met, Tiam1/Rac1, dynamin 2, and cortactin. *The Journal of Biological Chemistry*, **282**, 30643–57.
91. Birukova, A.A., Alekseeva, E., Mikaelyan, A., and Birukov, K.G. (2007) HGF attenuates thrombin-induced endothelial permeability by Tiam1-mediated activation of the Rac pathway and by Tiam1/Rac-dependent inhibition of the Rho pathway. *The FASEB Journal*, **21**, 2776–86.
92. Bernard, G.R., Vincent, J.L., Laterre, P.F. *et al.* (2001) Efficacy and safety of recombinant human activated protein C for severe sepsis. *The New England Journal of Medicine*, **344**, 699–709.
93. Kalil, A.C., Coyle, S.M., Um, J.Y. *et al.* (2004) Effects of drotrecogin alfa (activated) in human endotoxemia. *Shock*, **21**, 222–29.
94. Teke, Z., Sacar, M., Yenisey, C. *et al.* (2008) Activated protein C attenuates intestinal reperfusion-induced acute lung injury: an experimental study in a rat model. *American Journal of Surgery*, **195**, 861–73.
95. Birukov, K.G., Leitinger, N., Bochkov, V.N., and Garcia, J.G. (2004) Signal transduction pathways activated in human pulmonary endothelial cells

by OxPAPC, a bioactive component of oxidized lipoproteins. *Microvascular Research*, **67**, 18–28.

96. Chabot, F., Mitchell, J.A., Gutteridge, J.M., and Evan's, T.W. (1998) Reactive oxygen species in acute lung injury. *The European Respiratory Journal*, **11**, 745–57.
97. Birukova, A.A., Chatchavalvanich, S., Oskolkova, O. *et al.* (2007) Signaling pathways involved in OxPAPC-induced pulmonary endothelial barrier protection. *Microvascular Research*, **73**, 173–81.
98. Nonas, S., Birukova, A.A., Fu, P. *et al.* (2008) Oxidized phospholipids reduce ventilator-induced vascular leak and inflammation *in vivo*. *Critical Care*, **12**, R27.
99. Nonas, S., Miller, I., Kawkitinarong, K. *et al.* (2006) Oxidized phospholipids reduce vascular leak and inflammation in rat model of acute lung injury. *American Journal of Respiratory and Critical Care Medicine*, **173**, 1130–38.
100. Birukova, A.A., Fu, P., Chatchavalvanich, S. *et al.* (2007) Polar head groups are important for barrier-protective effects of oxidized phospholipids on pulmonary endothelium. *American Journal of Physiology: Lung Cellular and Molecular Physiology*, **292**, L924–35.
101. Birukova, A.A., Malyukova, I., Mikaelyan, A. *et al.* (2007) Tiam1 and betaPIX mediate Rac-dependent endothelial barrier protective response to oxidized phospholipids. *Journal of Cellular Physiology*, **211**, 608–17.
102. Birukova, A.A., Malyukova, I., Poroyko, V., and Birukov, K.G. (2007) Paxillin-beta-catenin interactions are involved in Rac/Cdc42-mediated endothelial barrier-protective response to oxidized phospholipids. *American Journal of Physiology: Lung Cellular and Molecular Physiology*, **293**, L199–211.
103. Ueno, Y., Koike, H., Annoh, S., and Nishio, S. (1997) Anti-inflammatory effects of beraprost sodium, a stable analogue of PGI_2, and its mechanisms. *Prostaglandins*, **53**, 279–89.
104. Farmer, P.J., Bernier, S.G., Lepage, A. *et al.* (2001) Permeability of endothelial monolayers to albumin is increased by bradykinin and inhibited by prostaglandins. *American Journal of Physiology: Lung Cellular and Molecular Physiology*, **280**, L732–38.
105. Singleton, P.A., Garcia, J.G., and Moss, J. (2008) Synergistic effects of methylnaltrexone with 5-fluorouracil and bevacizumab on inhibition of vascular endothelial growth factor-induced angiogenesis. *Molecular Cancer Therapeutics*, **7**, 1669–79.
106. Liu, H.C., Anday, J.K., House, S.D., and Chang, S.L. (2004) Dual effects of morphine on permeability and apoptosis of vascular endothelial cells: morphine potentiates lipopolysaccharide-induced permeability and apoptosis of vascular endothelial cells. *Journal of Neuroimmunology*, **146**, 13–21.
107. Singleton, P.A., Moreno-Vinasco, L., Sammani, S. *et al.* (2007) Attenuation of vascular permeability by methylnaltrexone: role of mOP-R and S1P3 transactivation. *American Journal of Respiratory Cell and Molecular Biology*, **37**, 222–31.

22

Targeted Delivery of Biotherapeutics to the Pulmonary Endothelium

Vladimir R. Muzykantov

Department of Pharmacology and Program in Targeted Therapeutics of the Institute for Translational Medicine Therapeutics, University of Pennsylvania School of Medicine, Philadelphia, PA, USA

INTRODUCTION – PULMONARY ENDOTHELIUM: A THERAPEUTIC TARGET

Drug targeting to selected compartments of the pulmonary endothelium may optimize diagnosis, treatment, and modeling of many disease conditions that involve this tissue. This chapter (i) overviews drug delivery means to achieve optimal addressing of drugs in the endothelium, (ii) describes endothelial determinants potentially useful for such targeting, and (iii) gives examples of targeted delivery of biotherapeutics to the pulmonary endothelium. It also briefly discusses current challenges in translation of these approaches from animal to human studies.

The pulmonary endothelium functions include (i) blood gas exchange, (ii) blood filtering, (iii) conversion of vasoactive compounds, (iv) control of vascular permeability, and (v) control of leukocyte recruitment into the lung tissue [1]. Pulmonary endothelium is involved in the development of pulmonary inflammation, edema, tumor growth, and hypertension, and thus is an important target for diagnostic, prophylactic, and therapeutic interventions. As most drugs have no particular affinity to endothelial cells (ECs), therapeutic interventions do not optimally affect the endothelium. Inadequate delivery to a desirable site of action is especially important for the utility of biotherapeutics – enzymes, proteins, genetic materials, and other biological compounds. EC-targeted drugs require precise subcellular delivery to exert the needed activity. The need for targeted delivery of therapeutics to the pulmonary endothelium is clear and compelling. Attempts to achieve this goal were initiated three decades ago by groups exploring thrombomodulin (TM) and angiotensin-converting enzyme (ACE) as targets for drug delivery to the pulmonary endothelium [2–4]. Many new target determinants and delivery means have since evolved. This chapter gives a brief overview of this area of research, with a focus on approaches that showed in animal studies a promise of utility for the treatment of pulmonary inflammation, thrombosis, oxidative stress, and hypertension.

CARRIERS OPTIMIZING VASCULAR PHARMACOKINETICS OF DRUGS

The bloodstream is the natural route for endothelial drug delivery. However, drugs are rapidly eliminated from blood by renal clearance, and uptake by the liver and the reticuloendothelial system [5]. Drug carriers are being designed in order to overcome this obstacle, and to (i) optimize drug pharmacokinetics, and protect drugs against inactivation and premature activity *en route* to the target, (ii) control drug release kinetics, (iii) deliver drugs to target cells, and (iv) optimize their subcellular delivery.

Pharmacokinetics can be improved by conjugating polyethylene glycol (PEG) to drugs or their carriers. PEG forms a hydrated shell that enhances solubility of carriers or drugs themselves (e.g., PEG-coated insulin), and hinders interaction with clearance and defense systems in the body, thereby prolonging drug circulation [6]. PEG-coated liposomal drug vehicles circulate in the vasculature for hours [7]. Polymersomes, the polymer analog of liposomes assembled from biocompatible polymers such as polylactic/polyglycolic acid conjugated

The Pulmonary Endothelium: Function in health and disease Editors Norbert F. Voelkel, Sharon Rounds

with PEG, are more robust and circulate for days [8]. Solid PEG-copolymer nanoparticles formed by disintegration of a polymer emulsion can be loaded by either hydrophobic or hydrophilic drugs incorporated into the polymer matrix or into the internal aqueous space, respectively [9]. Therapeutic proteins encapsulated in polymer nanoparticles are protected against proteolysis [10]. In general, nanocarriers with diameters within a range of 50–500 nm (i.e., a size permitting circulation through capillaries and delivery into the ECs [11]) have been employed for targeted drug and gene delivery to the vasculature.

However, these formulations do not accumulate in the lungs. Microspheres with diameters greater than 20 μm and cationic liposomes forming large aggregates with DNA lodge in the pulmonary capillaries after intravenous (IV) injection [12], but afford neither selective delivery of cargoes to cells of interest nor control over subcellular localization of the delivered cargoes. Blood flow eliminates cargoes released from carriers retained in the lumen, compromising endothelial delivery. More specific and effective means for targeting drugs and drug carriers to the pulmonary endothelium are needed.

VASCULAR TARGETING TO THE PULMONARY ENDOTHELIUM: TARGET DEFINITION

This goal can be achieved by using vascular targeting (i.e., coupling of drugs or drug carriers with molecules possessing affinity to endothelial determinants, providing targeted delivery to the pulmonary vasculature) [13–16]. Drugs and their carriers can be conjugated with such affinity moieties (antibodies, their fragments, peptides, aptamers) [15] or fused with targeting peptides using recombinant techniques [17].

Identification of candidate target determinants employs standard immunological techniques, such as staining of tissue sections and fluorescence-activated cell sorting analysis of isolated cells, as well as functional genomics, proteomics, phage display, and whole-body imaging, among other approaches. Comparing mRNA of ECs obtained from different areas of the vasculature [18] characterizes a relative abundance of messages for a plethora of proteins [19]. Proteomics of endothelial plasmalemma obtained from vasculature of organs of interest using perfusion of silica beads [20], reveals distribution of membrane proteins in endothelia in these organs, including the lungs [21, 22]. Separation of plasmalemma domains including caveoli [23], lipid rafts, and cellular junctions reveals endothelial "microcartography" [22, 24]. Phage-display selection techniques (repetitive cycles of injections of phages encoding stochastic peptides followed by identification of specific phages homing into selected tissues and identification of the homing determinants [25]) help to identify the binding sites accessible from the circulation in selected normal and pathological vascular areas [26, 27]. This method allows the identification of endothelial surface determinants in tumors and vascular lesions [28–30] that are being explored for vascular targeting of imaging probes and drugs [31]. Tracing of the organ distribution of injected carriers of a given size coated with affinity ligands to these determinants provides the ultimate test of their accessibility to circulation [22].

The last parameter is critical – a target should be present on the pulmonary endothelium lumen, accessible to circulating carriers. Some determinants are masked by the glycocalyx or hidden within plasmalemma structures. For example, caveolar determinants support lung-selective targeting of antibodies and small peptides, but not carriers larger than 100 nm (the caveolar neck diameter is < 100 nm, "Antibodies Directed to Specific Endothelial Domains"). The surface density of certain endothelial molecules (e.g., ACE and TM, see following section) is reduced by disease conditions.

In order to be useful for endothelial drug delivery, an accessible determinant must meet criteria of safety and precision of targeting. First, unless designed for such a goal, targeting should not cause harmful side-effects to ECs. Binding of targeted drugs may activate shedding or/and internalization of target determinants, or affect their function. Ideally, binding of an antibody–drug complex to a target antigen should cause therapeutic effects. Second, docking to a surface determinant should provide a proper subcellular localization of a drug. Depending on the therapeutic goal, a drug should be either retained on the cell surface or traffic to a proper subcellular compartment.

The pulmonary vasculature is the major capillary network containing around 30% of the endothelial surface in the body and receiving the entire cardiac output of venous blood. As a result, agents with an endothelial affinity accumulate in the lungs after intravascular injection, even if their target determinants are relatively evenly distributed throughout all types of ECs in the body [32].

Targeting drugs to either normal or pathological ECs can be useful for either prophylaxis or therapy [33, 34]. Targeting to pathological endothelium might permit more specific therapeutic interventions. These goals can be attained by directing drugs to constitutive and inducible endothelial targets. The following sections provide examples of candidate endothelial determinants for drug targeting to the pulmonary endothelium.

CONSTITUTIVELY EXPRESSED TRANSMEMBRANE GLYCOPROTEINS ENRICHED IN THE PULMONARY ENDOTHELIUM: ACE AND TM

Initial studies of drug targeting to the pulmonary endothelium focused on the constitutive endothelial proteins enriched in the pulmonary vessels – ACE [2, 3, 35, 36] and TM [4, 37].

ACE, a transmembrane glycoprotein expressed on the endothelial luminal surface, converts angiotensin I into angiotensin II – a vasoactive peptide that exerts vasoconstricting, pro-oxidant, prothrombotic, and proinflammatory activities [38, 39] (see Chapter 7). Some ACE antibodies (anti-ACE) do not affect its activity, while other antibodies inhibit ACE and cause its shedding [35, 40]. This "side-effect" of ACE targeting may be beneficial in treatment conditions associated with hypertension, oxidative stress, and inflammation [16, 33, 41]. However, ACE also inactivates substance P and bradykinin; elevated levels of these peptides due to ACE inhibition may cause adverse effects including hypotension, cough, and edema.

The pulmonary vasculature is enriched in ACE: around 100% versus less than 15% ECs are ACE-positive in the alveolar as compared with the extrapulmonary capillaries [36]. ECs internalize anti-ACE and anti-ACE conjugates, which thereby deliver drugs intracellularly [33, 42]. Oxidants, cytokines, and other pathological agents suppress ACE expression, and thus inhibit targeting to ACE [43, 44]. Monoclonal ACE antibodies with species specificity, including rat, mouse, cat, primate, and human ACE, have been produced, and showed effective and selective accumulation in the pulmonary vasculature after IV injection in these species [2, 36, 45, 46]. Pilot safety tests did not reveal overt harmful effects of injection of anti-ACE in normal animals and humans [35, 47]. ACE targeting is being explored for delivery of imaging, antioxidant, genetic materials, and other agents (see "Drug Targeting to the Pulmonary Endothelium: Specific Applications"). Figure 22.1 illustrates immunotargeting to ACE.

TM, a transmembrane glycoprotein expressed on the luminal surface of the endothelium, converts thrombin into an antithrombotic and anti-inflammatory enzyme [48]. TM antibodies accumulate in the lungs after IV injection, and have been employed in animals for targeting of liposomes, genetic materials, model enzymes, and isotopes to the pulmonary endothelium [49]. However, interventions that affect TM activity may compromise important endothelial functions and provoke thrombosis [50, 51].

CONSTITUTIVELY EXPRESSED CELL ADHESION MOLECULES

Platelet-endothelial cell adhesion molecule (PECAM)-1 and intercellular adhesion molecule (ICAM)-1 are constitutive transmembrane glycoproteins expressed by ECs throughout the vasculature (see Chapter 10). Their ligands (e.g. anti-ICAM and anti-PECAM, or conjugates of these antibodies) infused locally in a conduit artery accumulate in the downstream vascular areas including cardiac [52], cerebral [53], and mesentery [54, 55] vasculature. Similarly to other pan-endothelial antibodies, anti-PECAM and anti-ICAM tend to accumulate in the lungs after IV injection, and may be used for drug targeting to either normal and/or pathologically altered pulmonary endothelium [13]. These two determinants have been extensively explored for pulmonary drug delivery in animal studies that showed promise for the medical utility of this strategy.

PECAM-1

PECAM (CD31) is predominantly localized in intercellular borders of ECs [56]. Platelets and leukocytes also express PECAM, but at levels that are orders of magnitude lower than ECs, and bind a minor fraction of circulating anti-PECAM. PECAM is involved in mechanisms of cellular recognition, adhesion, and transendothelial migration of leukocytes [57] (see Chapter 10). Therefore, drug targeting to PECAM may cause side-effects (e.g., inhibit leukocyte trafficking) [58]. Leukocyte pulmonary infiltration is generally viewed as an injurious factor [59]. Consistent with this notion, anti-PECAM strategies attenuate inflammation in animal studies [60]. Clinical studies aimed to achieve such effects using cell adhesion molecule (CAM)-blocking agents have yet to provide a therapeutic benefit [13] due to redundancy of leukocyte-migratory mechanisms [58, 61]. However, attenuation of leukocyte infiltration by PECAM targeting would be a bonus in treatment of lung inflammation.

Engagement of CAMs by antibodies may also activate endothelium [62–64]. Of note, multivalent engagement of endothelial vascular cell adhesion molecule (VCAM)-1, but not PECAM-1, caused generation of oxidants via NADPH oxidase activation [65]. Animal studies using electron microscopy analysis of the lung tissue and functional tests of vascular permeability showed that even multivalent anti-PECAM conjugates and polymeric drug carriers coated by anti-PECAM caused no acute or delayed endothelial abnormalities [66, 67] or detectable adverse effects in the lungs and other organs of mice [68–71]. Anti-PECAM single-chain variable fragment (scFv)/urokinase caused no injurious effects and was

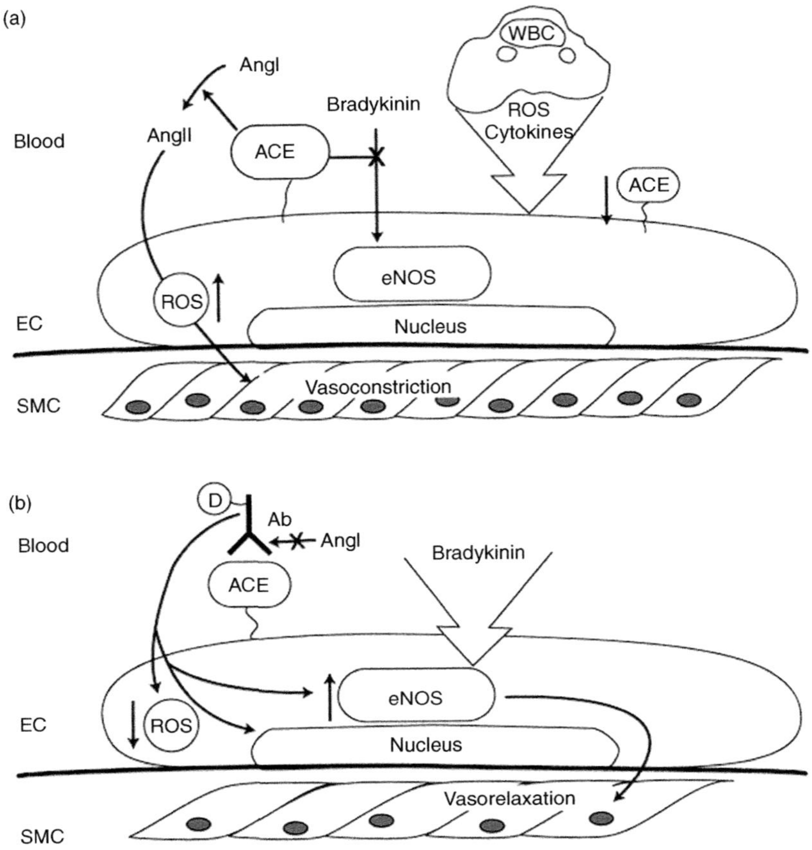

Figure 22.1 Immunotargeting to ACE. (a) ACE is normally expressed on the endothelial surface (left), but inflammatory factors (e.g., ROS and cytokines released from activated leukocytes) downregulate the density of ACE molecules on the endothelial surface (right). (b) Anti-ACE delivers drugs to the endothelium and undergoes internalization. This paradigm can be utilized for delivery of antioxidant enzymes (to intercept intracellular ROS) or genetic materials. In addition, anti-ACE inhibits ACE. This attenuates angiotensin I conversion, protects bradykinin (both of these effects lead to vasorelaxation), and suppresses pro-oxidative effects of angiotensin II. SMCs, smooth muscle cells; WBCs, white blood cells; Ab, monoclonal antibody (anti-ACE) conjugated with drugs (D). Reproduced with permission from [33], © 2002 by Kluwer Academic Publishers.

protective in a model of cerebrovascular thrombosis in mice [53].

PECAM is stably expressed on endothelium at a level of a million copies per cell, which permits a robust PECAM-targeted drug delivery to either normal or pathologically altered vasculature, for either prophylaxis or therapy. Multivalent binding of high-affinity anti-PECAM/nanocarriers to the endothelium further enhances pulmonary targeting [36, 72]. Diverse reporter [73], enzymatic [52, 74, 75], and genetic materials [76] conjugated to anti-PECAM accumulate and display their functional activity in the pulmonary endothelium as soon as 10 min after IV injection in mice and pigs.

ECs bind anti-PECAM without internalization, but internalize multivalent anti-PECAM complexes with a diameter of 100–500 nm [72, 73], regardless of the chemistry of formation of such multivalent conjugates [15]. Cross-linking of PECAM by multivalent anti-PECAM conjugates induces signaling via the PECAM-1 cytosolic tail [66], leading to endocytosis [77]. In contrast, monomeric anti-PECAM and anti-PECAM scFv cause no such effects [72, 77]. This feature permits anti-PECAM utility as an affinity moiety for targeting drugs either to the endothelial surface or intracellularly. Figure 22.2 illustrates some characteristics of vascular immunotargeting to PECAM.

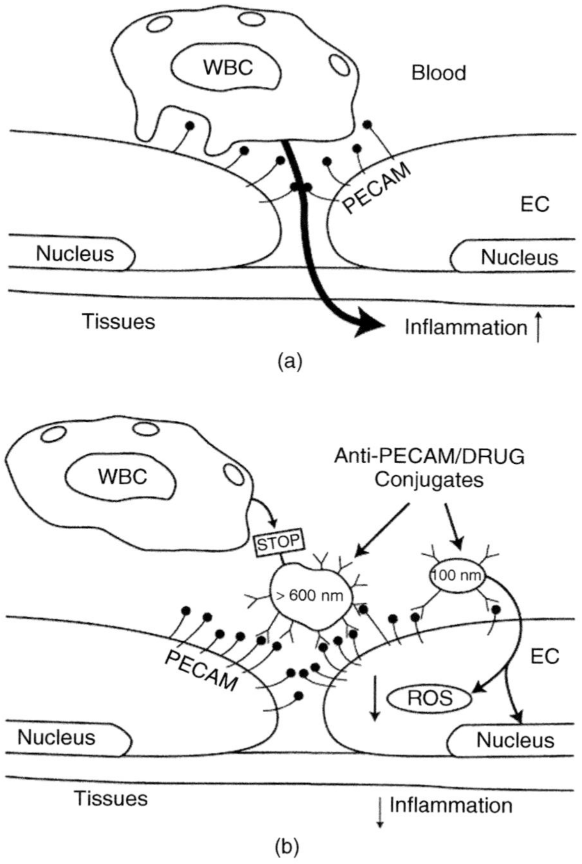

Figure 22.2 Immunotargeting to PECAM-1. (a) PECAM-1 is constitutively expressed by endothelium, predominantly in the intercellular contacts. It supports white blood cell (WBC) transmigration and facilitates inflammation in the tissues. (b) Large (>600 nm diameter) anti-PECAM conjugates poorly internalize in the ECs, whereas smaller counterparts (100–300 nm diameter) enter the cells, and can be used for intracellular delivery of genetic materials and antioxidant enzymes to detoxify ROS. In addition, anti-PECAM conjugates may suppress inflammation via blocking white blood cell transmigration. Reproduced with permission from [33], © 2002 by Kluwer Academic Publishers.

ICAM-1

ICAM-1 (CD54) is another immunoglobulin superfamily surface glycoprotein with cytoplasmic, transmembrane, and extracellular domains. Quiescent confluent ECs in culture do not express appreciable amounts of ICAM-1, but expression if increased 20–50 times after cytokine treatment [70]. In rat and mouse vasculature, however, ICAM-1 is normally expressed by ECs at a surface density that, by various estimates, ranges from modest to relatively high (2×10^4–2×10^5 surface copies per cell) [78]. Other cell types also express ICAM, yet the blood-accessible ICAM is located predominantly to ECs that represent the main target for anti-ICAM. A robust and specific binding of ICAM antibodies and anti-ICAM conjugates to vascular endothelium after IV administration has been documented in animals; a major fraction of the injected anti-ICAM accumulates in the lungs [70, 79].

Pathological stimuli, such as oxidants, cytokines, and abnormal shear stress, stimulate *de novo* synthesis and surface expression of ICAM by ECs [80] and thereby facilitate anti-ICAM endothelial targeting [70, 79, 81, 82]. Conjugation of anti-ICAM to therapeutics [70, 83], liposomes [84], or polymer carriers [55] providing multivalent binding to endothelium further enhances drug delivery.

ICAM supports leukocyte adhesion to the endothelium [85]. ICAM antibodies inhibit this process, providing protective effects in animal models of inflammation, ischemia/reperfusion, and oxidative stress [86–88]. ICAM may also serve for docking of pathogens [89] and as a signaling molecule [90]. Adverse effects of ICAM blocking in animal and human studies have been rare and confined to aggravated airway infection by anti-ICAM delivered via airways and an adverse reaction to heterologous anti-ICAM used to inhibit inflammation in stroke [13, 86, 91].

Internalization of anti-ICAM follows the same pattern as described for PECAM: ECs internalize multivalent anti-ICAM conjugates within 100–500 nm diameter, but not monomeric anti-ICAM [77], via a unique internalization pathway, CAM-mediated endocytosis that involves PECAM or ICAM clustering by multivalent anti-CAM complexes triggering signaling for the fast formation of actin stress fibers, resulting in formation of endocytic vacuoles and internalization of cell-bound conjugates [11, 77, 92].

Therefore, monomolecular anti-ICAM carriers anchor drugs to the endothelial surface [70], whereas polyvalent anti-ICAM conjugates deliver drugs intracellularly [11, 13]. After internalization, ICAM-1 dissociates from anti-ICAM carriers and recycles to the plasma membrane, supporting multiple cycles of intracellular delivery [67]. The intracellular traffic of cargoes internalized via PECAM-1 and ICAM-1 can be regulated by auxiliary pharmacological agents. For example, the alteration of proton–sodium balance in endosomal–lysosomal compartments decelerates degradation of proteolysis-susceptible cargoes, and permits their recycling to the plasma membrane [67, 92]. Disruption of endothelial microtubules blocks lysosomal traffic and prolongs the duration of activity of internalized drugs [92].

A list of ICAM-targeted drug delivery systems includes immunoliposomes [84, 93], polymer nanocarriers [55, 67, 94], protein conjugates [70, 95], imaging agents [82, 96], and acoustic micron-size bubbles used as a contrast for ultrasound imaging [83]. These systems have been characterized in cell cultures [67, 84, 93, 94] and animals [70, 82, 83, 96]. Figure 22.3 illustrates several modes of potential application of drug targeting to ICAM-1.

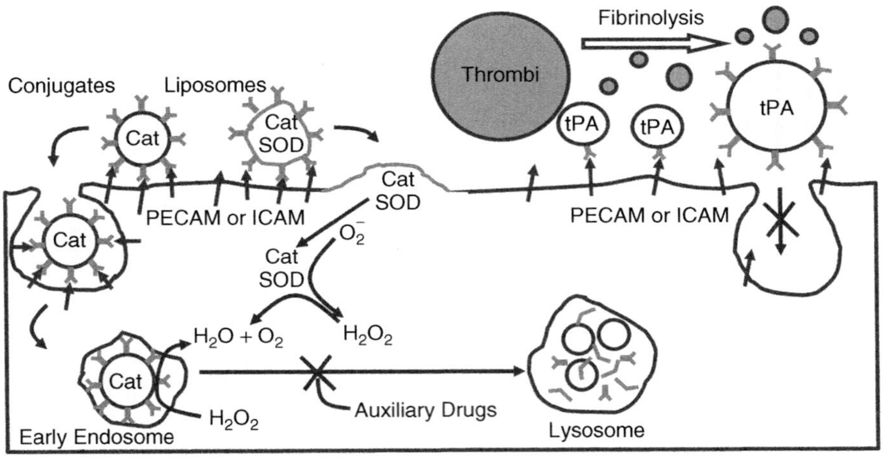

Figure 22.3 Vascular immunotargeting of antioxidant and fibrinolytic enzymes to constitutive EC adhesion molecules (CAM). (Left) containment of vascular oxidative stress. Multivalent conjugates of 100–300 nm diameter, consisting of AOEs (catalase and SOD) and carrier PECAM-1 or ICAM-1 antibodies, bind to ECs, enter cells via CAM-mediated endocytosis, and detoxify ROS in the cell interior. Auxiliary drugs decelerating lysosomal trafficking and/or degradation can be employed to prolong the duration of the protective effects. A fraction of the conjugates transiently bound to the endothelial surface intercepts external ROS and inhibits leukocytes adhesion via blocking of ICAM-1 or PECAM-1 (not shown). As an alternative strategy, anti-CAM stealth immunoliposomes bound to the endothelium can fuse with the plasma membrane, thus delivering their AOE payload into the cytosol; this tentative strategy is suitable for delivery of SOD/catalase tandem for orchestrated detoxification of both superoxide anion and hydrogen peroxide ROS. (Right) boosting of fibrinolytic activity on the endothelial luminal surface. Endothelial cells poorly internalize both small monomolecular and large micron-size anti-CAM conjugates. Plasminogen activators (e.g., tPA) delivered in the form of such conjugates are retained on the surface of ECs and facilitate fibrinolysis of clots lodged in the vasculature. Reproduced from [13] with permission from Bentham Science Publishers.

INDUCIBLE ENDOTHELIAL CAMs: SELECTINS AND VCAM-1

These molecules are normally absent on the vascular lumen, but become exposed on pathologically activated endothelium. For example, pathological mediators cause mobilization of intracellular P-selectin to the endothelial surface within 10–30 min [97], and induce *de novo* synthesis and surface expression of E-selectin [85] and VCAM-1 [56] within several hours. Selectins and VCAM-1 facilitate adhesion of leukocytes to ECs [98] (see Chapter 10).

Selectin and VCAM ligands represent attractive affinity moieties for the delivery of diagnostics and therapeutic agents to activated endothelium. Attenuation of leukocyte adhesion is a potential secondary benefit of this approach. Experiments in cell cultures and animals show that anti-selectins permit drug targeting to cytokine-activated endothelium [99, 100]. ECs constitutively internalize selectins via clathrin-coated pits [101–103], permitting entry into ECs of anti-E-selectin targeted liposomes [104], anti-inflammatory drugs [104, 105], and genetic materials [106].

However, even at the activation peak, selectins and VCAM-1 are exposed at a surface density lower than PECAM-1 and ICAM-1; hence, robustness of the targeting may be suboptimal for therapies requiring delivery of large doses of drugs. P-selectin-targeted compounds also bind to activated platelets [107]. Interestingly, E-selectin and VCAM-1 seem to be more readily expressed in activated endothelia of nonpulmonary vasculature origin (e.g., in arteries and in the skin microvasculature) [108]. These determinants seem to be useful for diagnostic visualization of activated endothelium in inflammation to deliver conjugated isotopes [109] or ultrasound contrasts [107, 110].

ANTIBODIES DIRECTED TO SPECIFIC ENDOTHELIAL DOMAINS

Specific domains of the endothelial plasmalemma are enriched in certain molecules (see Chapter 15). For example, rat glycoprotein GP85 is predominantly localized to the luminal surface of the plasmalemma domain that belongs to a thin part of EC body that lacks main organelles, and separates the alveolar and vascular compartments [111, 112]. Anti-GP85 accumulates in rat pulmonary vasculature without internalization and delivers conjugated cargoes into the pulmonary vasculature [113]. A human counterpart of this antigen could be an interesting candidate for drug delivery to the surface of alveolar capillaries.

Phage display libraries selection *in vivo* yielded several candidate peptides binding to and providing homing of phages to the pulmonary [e.g., peptides with an affinity to aminopeptidase P (APP)] and other endothelia including tumor endothelium [e.g., peptides with an affinity to aminopeptidase N (APN)] [25, 114]. Endothelial proteomics showed that APP is enriched in the pulmonary vasculature and represents an interesting model target determinant in animal studies [22]. This enzyme, however, plays important physiological roles including degradation of bradykinin [115]; hence, its inhibition may cause side-effects including edema. APN is involved in angiogenesis and its inhibition may provide a beneficial side-effect in the context of anticancer therapy [116].

The pulmonary endothelium contains cholesterol-rich plasma membrane flask-shaped invaginations named caveoli [117–119]. Antibodies of determinants localized to caveoli including APP and glycoproteins GP60 and GP90 accumulate in the pulmonary vasculature in rats after IV injection, enter endothelial intracellular vesicles, and traverse endothelial barriers [120]. Caveoli-mediated endocytosis and transcytosis are involved in endothelial transport functions [118, 121–125]. Interaction of a protein ligand (e.g., antibodies) leading to receptor clustering in caveoli further activates endocytosis and transcytosis [126, 127]. Caveoli seem to be involved in transendothelial transport of albumin [128, 129] – a process further stimulated by albumin nitration that may take place during oxidant stress [130].

Caveolar transcytosis is envisioned as a pathway for transcellular delivery of drugs targeted to caveolae-located receptors (see Chapter 8). For example, after IV injection, tracers conjugated with antibodies directed against specific antigen GP90 localized in pulmonary endothelial caveolae undergo transport through the pulmonary endothelium [120]. Recently, very rapid transendothelial transport of caveoli-targeted antibodies and affinity peptides identified by a phage display has been demonstrated in mice [131]. The caveoli neck diameter is around 50 nm; therefore, the size of the drug carriers that can employ this pathway is limited to less than 100 nm: phage nanoparticles targeted to these caveolar determinants do not accumulate in the lungs because of the steric hindrances [131]. Caveoli represent endothelial transport and signaling organelles; therefore, side-effects of targeting caveolar determinants must be rigorously characterized.

DRUG TARGETING TO THE PULMONARY ENDOTHELIUM: SPECIFIC APPLICATIONS

As recently as 5 years ago, this chapter would have been limited to a description of candidate determinants and the design of drug delivery systems. However, animal studies published in the last few years describe targeting to the pulmonary endothelium of agents that can find application in medicine. This section gives a brief overview of these studies.

Imaging Agents

Visualization of vascular and other compartments in lung tissue holds promise to improve the diagnosis of pulmonary diseases. Modalities applicable for lung imaging include gamma-scintigraphy, computer tomography (CT), inert gas-contrasted magnetic resonance imaging (MRI), positron emission tomography (PET), single-photon emission CT (single photon emission computed tomography SPECT), and their combinations.

For example, ACE antibodies labeled with radioactive ^{125}I and ^{99}Tc isotopes accumulate in the lungs after IV injection in diverse animal species including primates [36, 46, 47]. The pulmonary accumulation of anti-ACE has been visualized in real time in a γ-camera in these animal species [2] and humans [47].

Reduction of the pulmonary uptake of isotope labeled anti-ACE is a sensitive indicator of endothelial disturbance in the lungs in rat models of endotoxemia, edema, and ischemia–reperfusion [44, 132, 133]. Limited clinical studies using γ-camera thoracic imaging revealed that pulmonary uptake of isotope-labeled anti-ACE is reduced in the patients with sarcoidosis, in comparison with healthy volunteers [47]. Similarly, pulmonary accumulation of [^{125}I]anti-TM was reduced to 50% of basal level in mice exposed to hyperoxic lung injury [134], consistent with loss of TM in the pathologically altered pulmonary endothelium [135].

^{125}I-labeled anti-ICAM accumulates preferentially in the lungs, the mesentery, and to a lesser extent in other highly vascularized organs after IV injection in rats [79] and mice [136]. In contrast to anti-ACE and anti-TM, isotope-labeled anti-ICAM show elevated vascular accumulation in oxidative stress and inflammation. Enhanced pulmonary uptake of [^{125}I]anti-ICAM has been shown in lungs of rats and mice challenged intratracheally with immune complexes [81], bacterial endotoxin lipopolysaccharide (LPS) [80], cytokines [136, 137], as well as challenged with hyperoxia [138] and LPS systemically [70, 79, 139] and in a model of chronic hypertension [136]. Gamma-camera imaging showed that the lungs represent the main target organ after injection of ^{111}In-labeled anti-ICAM in rats [82]. Pulmonary uptake of [^{111}In]anti-ICAM was enhanced in allograft rat lungs during acute rejection [140], rat models of acute lung injury caused by oleic acid injury [141], and bleomycin-induced lung injury [142]. Expression of other CAMs in the pulmonary vasculature in mice has also been probed using radiolabeled antibodies. For example, elevated pulmonary accumulation of ^{125}I-labeled antibody to P-selectin has been reported within few minutes in mice treated with histamine [143], several hours after treatment with LPS [143], and 1 day after exposure to hyperoxia [144].

Polymer nanoparticles labeled with positron emission isotopes including ^{64}Cu and Cd^{125m}Te, targeted to anti-ICAM [96] and anti-TM [145], respectively, have been employed to visualize the pulmonary vasculature in real time in mice. Consistent with other studies that employed labeled anti-ICAM, pulmonary accumulation of anti-ICAM/^{64}Cu-labeled polymer nanocarriers was markedly enhanced in mice treated with bacterial LPS, reflecting enhanced exposure of ICAM-1 on the endothelial lumen [96]. The combination of PET or SPECT with micro-CT augmented topographical localization of isotope signals in the chest in these studies.

Monoclonal antibodies, recombinant derivatives of these antibodies, and high-affinity peptides binding to endothelial determinants localized in the caveoli accumulate within minutes in the pulmonary vasculature after IV injection in rats and permit lung imaging using γ-scintigraphy [146]. For example, a monoclonal antibody to APP, ^{125}I-labeled anti-APP provided discernable lung images in rats within minutes after IV injection and retained a strong signal in the lungs without major cardiac signal in the thorax for ensuing 48 h [131]. ^{99}Tc-labeled anti-APP provided lung imaging using whole-body SPECT/CT in rats, while fluorescently labeled anti-APP provided a high-resolution real-time microscopy imaging of its permeation across endothelium in rat lungs [131].

Phage display-defined affinity peptides to pulmonary endothelial determinants including APP also permit targeting to and visualization of the lung vasculature (e.g., by optical imaging of injected peptide-decorated quantum dot nanocrystals) [147]. Hybrid viral vectors decorated by vascular cell-addressing peptides offer new modalities for targeted interventions and imaging capitalizing on the local expression of enzymes converting latent agents into imaging probes [148].

Vascular imaging utilizing reporter probes targeted to specific endothelial determinants represents an exponentially developing area of modern biomedicine.

Antitumor Agents, Radioisotope Therapies, and Toxic Compounds (Glucose Oxidase)

Some drug delivery systems, in particular, anti-TM, have been utilized in animals to model effects of injurious actions of toxic agents targeted to endothelium. Outcomes of these studies include both design of new models of pulmonary vascular injury and the understanding of sequelae of injurious side-effects of endothelial targeting. Furthermore, at least in theory, targeting of toxic compounds to proper endothelial determinants may help eradicate tumor vascular endothelium and tumors.

Liposomes targeted to TM [149, 150] have been used in this context, as a model to study lung tumor eradication. Anti-TM/liposomes loaded with prodrugs deliver their payloads to ECs in culture and to the pulmonary endothelium after IV injection in mice, thus permitting local formation of a toxic drug derivative in the vasculature in a mouse model of lung metastases [151]. Although clinical application of this delivery system is unlikely due to side-effects, it provided interesting correlations between parameters of the pulmonary drug targeting and effects.

Kennel *et al.* studied pulmonary targeting and effects of anti-TM labeled with close-range emitting isotopes (e.g., ^{213}Bi and ^{225}Ac α-emitters) [152]. [^{213}Bi]anti-TM accumulates in mice lungs after IV injection [153], alleviates tumor burden in mouse models of lung carcinoma, and thereby prolongs survival [153, 154]. Interestingly, despite the fact that ECs rapidly internalize TM [50], [^{213}Bi]anti-TM, and [^{225}Ac]anti-TM apparently remain in the lung tissue for a prolonged time (a half-life reported as 30 and 49 h, respectively [152, 153]). A prolonged retention in the target is likely consistent with effective tumor eradication.

Unfortunately, animals treated with [^{225}Ac]anti-TM died within few days because of vascular radiotoxicity [152]. [^{213}Bi]anti-TM eradicated tumors and thus prolonged survival of tumor-bearing mice, but 3 months after treatment surviving animals succumbed to fatal lung fibrosis that developed likely due to vascular injury caused by targeted isotope, cytokine release, and TM inhibition [153, 155]. Even injection of naked anti-TM caused transient disturbance of the pulmonary endothelium detectable by transmission electron microscopy, leading to enhanced susceptibility of the pulmonary vasculature in mice to colonization by circulating tumor cells [156].

These data, taken together with known key protective functions of TM in the vasculature, reduce enthusiasm for safety and potential clinical utility of TM targeting. However, vascular toxicities inflicted by anti-TM conjugates represent an interesting opportunity to study pathological pathways involved in specific types of vascular injury and provide models for testing new treatments. For example, glucose oxidase (GOX), an enzyme producing H_2O_2 from glucose, has been conjugated with anti-TM [134] and other carrier antibodies directed to endothelial determinants [157, 158] including anti-ACE [159] and anti-PECAM [160]. These GOX conjugates accumulate in the pulmonary vasculature after IV injection and cause dose-dependant oxidative stress in the lungs [49], providing a tool to model this pathological condition. Anti-TM/GOX is more toxic and causes a more profound pulmonary thrombosis than anti-PECAM/GOX, likely due to TM inhibition [49]. The level of anti-TM/GOX toxicity is dependent on the level of lung oxygenation providing secondary GOX substrate oxygen needed to oxidize glucose [134]. A combined anti-TM/GOX and hyperoxia treatment represent an interesting model of double-hit acute lung injury. A model of acute pulmonary oxidative stress caused by anti-TM/GOX in mice has been used for the initial testing of the protective effects of anti-PECAM/catalase conjugate in mice (see next section) [71].

Antioxidants

The lung is a vulnerable target for reactive oxygen species (ROS; superoxide anion) and a more diffusible and stable product of its dismutation, H_2O_2 (see Chapter 17). ROS sources may be localized in the alveolar, interstitial, and vascular compartments. In particular, both activated leukocytes (e.g., in inflammation) and ECs (in inflammation, ischemia, and hyperoxia) can produce ROS in the pulmonary vasculature via enzymatic systems including mitochondrial respiratory chain components, xanthine oxidase and NADPH oxidase. In many instances, the pulmonary endothelium represents both the source and the target of oxidative stress. Unfortunately, insufficient potency and delayed time window of the effect restrict the utility of existing antioxidant formulations (e.g., oxidant scavengers and glutathione donors) and inducers of antioxidant enzymes (AOEs) for treatment of acute oxidative stress. Potent AOEs [e.g., superoxide dismutase, (SOD) and catalase] do not have medical utility, either, because of unfavorable pharmacokinetics and inadequate delivery to endothelium.

Diverse chemical modifications of catalase and SOD have been designed to prolong the circulation time and improve its delivery, including coupling with PEG and encapsulation into liposomes [161–163] (Figure 22.3). Some of these derivatives and SOD mimetic showed enhanced potency in animal models of systemic and focal oxidative stress [161, 164]. Considerable efforts have also been devoted to optimize SOD delivery to ECs. For example, lecithinized SOD was shown to have increased affinity to ECs [165, 166]. Soluble SOD mutants with high affinity to heparan sulfates localized on the endothelial surface [167, 168], and SOD fused with a scFv antibody to lung tumor have been synthesized and shown to maintain enzymatic activity and affinity moieties [169]. The pharmacokinetics and pulmonary targeting of these compounds remain to be characterized in animal studies. However, endothelial uptake of these monomolecular SOD derivatives, needed for interception of intracellular superoxide [170], may be suboptimal. In theory, delivery of AOEs into endothelium (e.g., via targeting endocytic pathways described in "Constitutively Expressed Cell

Adhesion Molecules") may boost antioxidant defense in the pulmonary vasculature [34].

For example, radiolabeled catalase and SOD conjugated with anti-ACE, anti-PECAM, or anti-ICAM accumulate in the pulmonary vasculature after IV injection in mice and rats [72, 171]. These conjugates also accumulated in and protected perfused rat lungs against oxidative injury induced by infused H_2O_2 [68, 95]. Anti-PECAM/catalase conjugates bind to and protect ECs against H_2O_2 toxicity in cell cultures [72, 172]. Anti-PECAM/SOD conjugates also bind to ECs and protect against oxidative stress caused by either extracellularly or intracellularly produced superoxide anion [134].

Anti-PECAM/catalase protected the pulmonary vasculature against acute oxidant stress in mice by local production of H_2O_2 in the lungs and in a model of lung transplantation in rats [68, 71]. Consistent with these reports, anti-ACE/catalase injected in rats protected against pulmonary ischemia–reperfusion *in situ*, providing improved blood oxygenation, reduced edema, decreased serum level of endothelin-1, and lung level of inducible nitric oxide synthase (iNOS) mRNA [173].

Studies by Muro *et al.* revealed that ECs internalize within 30 min multivalent anti-ICAM and anti-PECAM complexes carrying catalase [92]. Since H_2O_2 easily diffuses through cellular membranes, catalase conjugates trafficking within endosomal vesicles retain activity and protect cells against oxidative stress [92]. However, within 3 h after internalization catalase gets degraded in lysosomes [174], which terminates its protective effect [67]. Use of auxiliary pharmacological agents affecting lysosomal traffic and degradation prolongs the protective effects of anti-ICAM/catalase formulations by many hours [67, 174]. An alternative approach uses biocompatible polymer nanocarriers selectively permeable for H_2O_2, but not for proteases. Dziubla *et al.* have recently developed approaches allowing encapsulation of active catalase into such protecting nanocarriers with controlled size and shape (e.g., nanospheres with a diameter < 500 nm [10]) applicable for vascular delivery into ECs [175]. Catalase-loaded polymer nanocarriers targeted to PECAM-1 deliver active cargo to the pulmonary vasculature after IV injection in animals and protect ECs against H_2O_2-induced injury for a prolonged time [69].

Ongoing studies produce encouraging pilot data regarding the protective effects of targeting catalase and SOD to the pulmonary endothelium for treatment of oxidative stress in diverse animal models, including lung transplantation in pigs (G. Pressler and R. Wiewrodt, unpublished data). Pulmonary transplantation represents an especially attractive setting for using vascular immunotargeting to endothelium: (i) time for injection of conjugates into a donor is well defined (just prior to organ procurement) and (ii) targeted agents will be metabolized within the graft with minimal systemic side-effects. Treatment of acute lung injury and hyperoxic lung injury represent additional attractive areas of potential application of endothelial targeting of antioxidants.

Enzyme Replacement Therapies

Lysosomes represent the final destination of anti-ICAM and anti-PECAM conjugates and nanocarriers. Muro *et al.* proposed that lysosomal targeting provides an ideal natural mechanism for intracellular delivery of enzyme replacement therapies for lysosomal storage diseases – a group of disorders caused by genetic deficit of lysosomal enzymes [176]. Some of these diseases, such as type B Niemann–Pick disease (NPD) caused by a genetic deficiency of acid sphingomyelinase (ASM) leading to accumulation of sphingomyelin and cholesterol in the cellular vesicles [177], involve the pulmonary endothelium and are associated with lung inflammation [178, 179]. Poor delivery into such cells limits the therapeutic effect of infused recombinant ASM. Furthermore, the pulmonary endothelium is an especially challenging target, because only a minor fraction of recombinant ASM accumulates in the lungs [179, 180]. ICAM-1-directed targeting of nanocarriers loaded with ASM seems especially attractive for targeting of ASM and other cargoes to the lysosomal compartment of the pulmonary endothelium, because endothelial expression of ICAM-1 is upregulated and CAM-mediated endocytosis bypasses endocytic pathways affected in these disease conditions [176].

In support of this hypothesis, Muro *et al.* designed ASM-loaded polymer nanocarriers targeted to ICAM-1 and showed that anti-ICAM/ASM/carriers, but not free ASM or nontargeted ASM/carriers, bind to and enter ECs and NPD-affected cells, thereby delivering active ASM into the lysosomes, where ASM corrects the storage disorder [176]. In a recent study, this group has demonstrated that anti-ICAM/ASM/nanocarriers, but not free ASM or untargeted ASM/nanocarriers, accumulate in the lungs of naïve and NPD disease model mice [54]. It is tempting to predict that this promising targeting strategy, rationally capitalizing on specific biological features of the disease, will eventually improve treatment of lysosomal storage diseases and that LSD involving pulmonary endothelium may be especially amenable to this type of intervention.

Genetic Materials

The pulmonary vasculature is an anatomical filter of venous blood and, therefore, a natural site for local delivery of genetic materials associated with positively charged polyplexes and other delivery systems forming micron-size aggregates that get entrapped in the microvasculature downstream from the injection site [12]. The effectiveness of transfection in the pulmonary vasculature and toxic effects can be modulated by control of the chemical content, charge, and size of polyplexes [181, 182]. Combined administration of liposomal polyplexes with adenoviral vectors enhances the pulmonary transfection [183]. Animal studies reported beneficial effects of transfection of the pulmonary vasculature by genes encoding protective proteins. Thus, injection of DNA/polyplexes provided transfection of the pulmonary tissue with a gene encoding indoleamine 2,3-dioxygenase a cytosolic enzyme with antioxidant and other protective features, thereby attenuating subsequent ischemia–reperfusion injury in rat model of lung transplantation [184]. Transfection of heat shock protein HSP70 and nitric oxide synthase NOS using IV injected adenovirus also provides protection in this model [185, 186]. Of note, non-EC types (e.g., smooth muscle cells) also get transfected by DNA released from vehicles in addition to their intended cellular targets, by not fully understood mechanisms including potentially harmful transendothelial transport of genetic materials [187].

Targeting genetic materials to endothelial determinants enhances the specificity of the cell type recognition and intracellular delivery via active endocytosis constitutively involving these determinants (e.g., E-selectin) or caused by their cross-linking (e.g., PECAM-1). Endothelial delivery of plasmid DNA and transfection of cultured cells has been accomplished by targeting to E-selectin [106, 188], transferrin receptor [188], ACE [45], PECAM-1 [73], and other determinants using antibodies conjugated with nonviral (e.g., liposomal) [73, 106, 188] and viral gene delivery means [45].

Diverse viruses including adenoviruses and adeno-associated viruses [189–191] have been modified genetically and chemically to enhance endothelial selectivity of gene transfer. Recombinant insertion of peptides with endothelial affinity in viral coat proteins facilitates specific homing and inhibits non-specific uptake by non-target cells such as hepatocytes [189, 191]. Insertion of endothelium-specific promoters such as utilized by genes encoding TM [192] or VEGF receptors (Flk-1) enhances endothelial specificity of gene therapies [28, 193, 194]. Analysis of *in vitro* studies and studies of the nonpulmonary vasculature is beyond the scope of this chapter [195–197]. It should be noted, however, that both viral and nonviral means for endothelial delivery of genetic materials can harm or/and activate ECs [198–200]. The immune response to the pulmonary gene therapies is also of a significant concern [201, 202].

Gene delivery targeted to the endothelium may help reduce the gene dose and side-effects. Gene delivery using liposomes targeted to TM (proof-of-principle studies) demonstrated enhanced efficacy of transfection versus nontargeted means in mice [203, 204]. Anti-PECAM/liposomes provided enhanced transfection of reporter gene products in the lungs after IV injection in mice [76]. Furthermore, hetero-conjugates that consist of anti-ACE and antibody directed to a viral fiber protein-recognizing cellular receptor attain dual function: they attenuate natural viral tropism to liver and non-target organs and redirect it toward ECs [205]. Such anti-ACE/antivirus conjugates have been used for retargeting of viral gene therapy to ECs in culture [46] and pulmonary endothelium in rats [189, 205]. Combining targeted delivery provided by anti-ACE/antivirus conjugates with insertion of endothelium-specific promoter in the genetic construct carried by viral envelope augments the pulmonary specificity of transgene expression by several orders of magnitude [189]. Using this approach for transfection of pulmonary endothelium by viral gene delivery of genes encoding NOS and bone marrow morphogenetic protein type 2 receptor showed a reduction of spontaneous pulmonary hypertension [206] and hypoxic pulmonary hypertension [207].

Antithrombotic Agents

The pulmonary vasculature is vulnerable in situ thrombosis. Pulmonary pathologies are associated with a high risk of pulmonary thrombosis in part due to reduction of the natural antithrombotic features of the endothelium (e.g., loss of TM level in response to oxidants and cytokines [134]). Effective and safe management of thrombosis is a challenging and still elusive goal, because most antithrombotic agents such as anticoagulants, platelet inhibitors, and fibrinolytics pose a danger of bleeding and other side-effects, whereas their therapeutic effects are restricted by inadequate delivery to the sites of thrombosis and embolism. In theory, targeting these drugs to and retention on the luminal surface in a vasculature prone to thrombosis or embolism could enhance the antithrombotic potential of the endothelium. Experimental studies in animal models showing that transfection of ECs with antithrombotic proteins including TM and plasminogen activators [e.g., tissue type plasminogen activator (tPA) and urokinase plasminogen activator (uPA)] helps alleviate intravascular thrombosis [208–211]. However, the utility of gene delivery in acute clinical settings looks uncertain. Direct anchoring of antithrombotic agents on the endothelial lumen may offer more expedited and localized interventions.

Early attempts to deliver antithrombotic agents to injured ECs in culture cannot be translated to animal studies due to lack of identified determinants useful for such a targeting [212]. Targeting antithrombotic drugs to selectins is being tested in cultured ECs challenged by cytokines and in a pilot animal study [99, 100], but the robustness of the delivery and drug retention on the luminal surface are restricted by modest and transient expression of these internalizable determinants. Functionally active conjugate anti-ACE/tPA was synthesized and shown to accumulate in the lungs after IV injection in rats [213]. Alternative immunoconjugates have been designed and showed promising accumulation and activity in rat lungs [214]. However, antithrombotic effect of drugs targeted to internalizable determinants is likely limited [42]. Anchoring to noninternalizable endothelial determinants, such as glycoprotein GP85, expressed on the lumen of the rat pulmonary ECs, seems to be a more logical choice [113]. In this context, targeting antithrombotic agents to ICAM-1 and PECAM-1 is of interest, as these molecules do not support internalization of monomolecular antibodies.

Indeed, radiolabeled tPA chemically conjugated with anti-ICAM accumulates in the pulmonary vasculature after IV injection in mice and rats and facilitates subsequent dissolution of fibrin emboli entrapped in the isolated perfused rat lungs [70]. This study demonstrated that targeting to ICAM-1 provides anchoring of an active tPA to the endothelial luminal surface. In further development of this paradigm, urokinase (uPA) fused (using recombinant technique) with a single-chain antigen-binding fragment (scFv) of anti-PECAM (anti-PECAM scFv/uPA) accumulated in mouse lungs after IV injection in wild-type, but not PECAM$^{-/-}$ mice, which provided a gold standard control of the specificity of targeting [17]. Anti-PECAM scFv/uPA was retained in the lungs for several hours and afforded enhanced dissolution of fibrin emboli subsequently injected in intact animals [17]. Figure 22.4 illustrates this strategy.

Furthermore, using a mutant uPA in which a natural plasmin-sensitive activation site was replaced by a thrombin-sensitive activation site (uPA-T), provided a prodrug anti-PECAM scFv/uPA-T [215]. This fusion product was targeted to the pulmonary endothelium in mice similarly to its plasmin-sensitive prototype analog, but maintained resistance to plasma inhibitors, remained latent, and caused no fibrinogen consumption, yet was activated by thrombin in the mouse models of pulmonary thrombosis caused by tissue factor and ischemia–reperfusion, and thereby afforded a more effective and durable antithrombotic effect than plasmin-sensitive anti-PECAM scFv/uPA [215]. Of note, the scFv-fusion format is modular and conducive to industrial production of diverse iterations of this class of proteins, directed to diverse endothelial determinants and carrying diverse antithrombotic protein cargoes.

CONCLUSIONS AND PERSPECTIVES

The pulmonary vasculature is an unusual but important target for drug delivery. Traditionally, drug delivery strategies are designed to deliver toxic agents to eradicate tumors. In contrast, the pulmonary vasculature represents a preferable site for therapeutic or/and prophylactic action of noninjurious agents. Design of delivery systems for targeting nontoxic drugs involves the rigorous consideration of potential side-effects that can be unintentionally inflicted by engaging, cross-linking, or blocking the functions of target determinants. In contrast to many other target organs, where drugs are needed to be delivered beyond the vascular wall, pulmonary ECs represent an important therapeutic site that binds a major fraction of the drugs, even if they are targeted to common endothelial determinants.

Targeting of active reporter and therapeutic cargoes including enzymes and genes has been achieved in intact animals and animal models of diseases. Recent animal studies showed that determinants including ACE, APP, ICAM, and PECAM have a potential utility when used for drug targeting to the pulmonary endothelium. The functions of these molecules are fairly well understood, which helps to avoid unintentional side-effects. Targeting caveoli provides an avenue for the intracellular and transcellular delivery to the pulmonary vasculature. Careful selection of targets and modulation of features of the antibody–drug conjugates like valency and size provide powerful tools for the control of intracellular uptake and trafficking of cargoes.

The scaling-up of synthesis and the quality control of targeted drug delivery systems with a standard, US Food and Drug Administration-acceptable level of homogeneity represent a significant challenge for their industrial development and clinical utility. Recombinant fusion of protein drugs and prodrugs with protein affinity moieties provides homogeneous and relatively easy to scale-up therapeutic agents. Recombinant design of these constructs permits the deletion of unnecessary parts of molecules or insertion of point mutations endowing products with novel, favorable pharmacokinetics, and/or functional features. Clinical testing of biotherapeutics targeted to ECs outside of the pulmonary vasculature have been recently conducted and showed promising levels of safety and efficacy. It is reasonable to hope that in the next decade targeted interventions addressed to the pulmonary ECs will be translated into medical practice.

ACKNOWLEDGMENTS

This work was supported by National Institutes of Health grants HL71175, HL078785, HL087036, and HL73940,

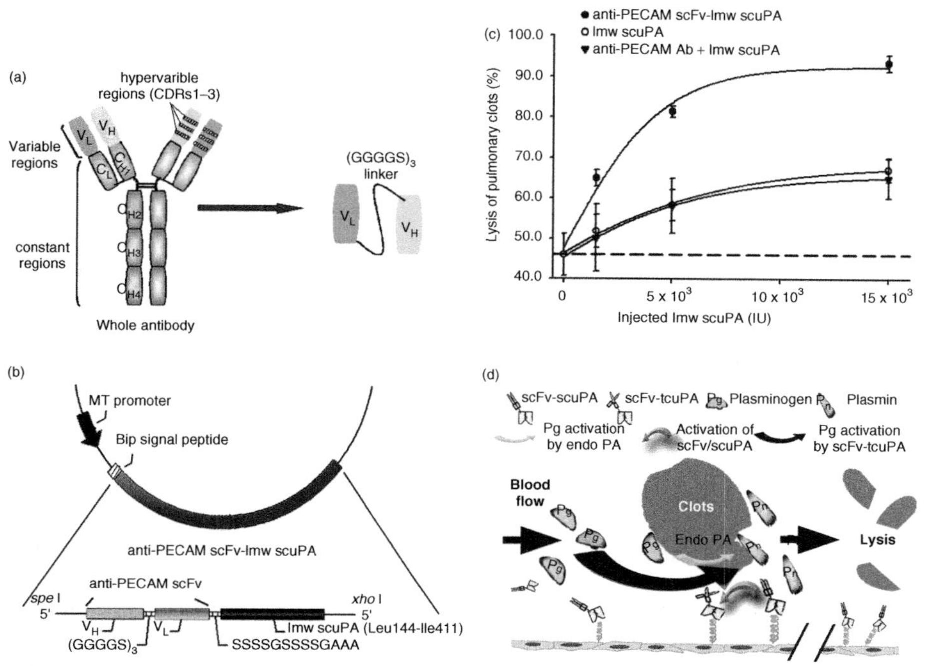

Figure 22.4 Vascular immunotargeting of recombinant fusion protein combining a scFv of anti-PECAM and uPA. (a) Schematic representation of different antibody formats. Left image shows generic structure of a whole IgG molecule, comprising two heavy and light chains linked by disulfide bonds. Hypervariable regions (complementarity determining regions) forming antigen-binding sites are indicated. Right image shows structure of a scFv in which variable domains of light chain and heavy chain are covalently linked by a flexible interchain linker. As depicted, the size of scFv (30 kDa) is sixfold smaller than whole IgG molecule (180 kDa). (b) Schematic diagram describing the cloning strategy for the fusion construct scFv/uPA. Variable domains of antibody heavy chain and light chain were linked and then fused to the N-terminus of low-molecular-weight (lmw) single-chain uPA (scuPA) by a $(Ser_4Gly)_2Ala_3$ linker. (c) Pulmonary thrombolysis in mice by anti-PECAM scFv/uPA. The graph shows the dose–response curve of dissolution of pulmonary thrombi by bolus injection of equal doses anti-PECAM scFv/uPA versus nontargeted uPA in mice. Thrombolytic potency was expressed as percent of fibrinolysis versus dose. Dashed line indicates spontaneous lysis. (d) Simplified schema of a proposed strategy for thromboprophylaxis using vascular immunotargeting of genetically engineered scFv/uPA. ScFv/uPA circulates in a form of a prodrug, binds to PECAM-1, and remains anchored on the luminal surface of endothelium for at least several hours. *In situ* thrombosis or embolism induces initial local conversion of plasminogen (Pg) into plasmin (Pn) by endogenous plasminogen activators (Endo-PA). Plasmin (and perhaps other enzymes) formed in the vicinity of the clot converts the endothelium-bound scFv/uPA into enzymatically active two-chain (tcuPA), which in turn amplifies local formation of plasmin, reinforcing local thrombolysis, preventing clot extension, and reocclusion. Reproduced with permission from Bi-Sen Ding *et al.*, (2006) [175], Copyright © 2006, American Society for Pharmacology and Experimental Therapeutics.

and a pilot grant from the Transdisciplinary Awards Program in Translational Medicine and Therapeutics/PENN. The author thanks his collaborators, Drs Silvia Muro (University of Pennsylvania), and Thomas Dziubla (University of Kentucky) for their invaluable contributions to the joint previous studies and publications that provided a framework for this chapter and numerous stimulating discussions.

References

1. Orlandi, P.A. and Fishman, P.H. (1998) Filipin-dependent inhibition of cholera toxin: evidence for toxin internalization and activation through caveolae-like domains. *The Journal of Cell Biology*, **141**, 905–15.
2. Danilov, S.M., Martynov, A.V., Klibanov, A.L. *et al.* (1989) Radioimmunoimaging of lung

vessels: an approach using indium-111-labeled monoclonal antibody to angiotensin-converting enzyme. *Journal of Nuclear Medicine*, **30**, 1686–92.

3. Danilov, S.M., Muzykantov, V.R., Martynov, A.V. *et al.* (1991) Lung is the target organ for a monoclonal antibody to angiotensin-converting enzyme. *Laboratory Investigation*, **64**, 118–24.
4. Kennel, S.J., Lee, R., Bultman, S., and Kabalka, G. (1990) Rat monoclonal antibody distribution in mice: an epitope inside the lung vascular space mediates very efficient localization. *International Journal of Radiation Applications and Instrumentation, Part B*, **17**, 193–200.
5. Moghimi, S.M. and Szebeni, J. (2003) Stealth liposomes and long circulating nanoparticles: critical issues in pharmacokinetics, opsonization and protein-binding properties. *Progress in Lipid Research*, **42**, 463–78.
6. Abuchowski, A., McCoy, J.R., Palczuk, N.C. *et al.* (1977) Effect of covalent attachment of polyethylene glycol on immunogenicity and circulating life of bovine liver catalase. *The Journal of Biological Chemistry*, **252**, 3582–86.
7. Discher, D.E. and Eisenberg, A. (2002) Polymer vesicles. *Science*, **297**, 967–73.
8. Discher, B.M., Won, Y.Y., Ege, D.S. *et al.* (1999) Polymersomes: tough vesicles made from diblock copolymers. *Science*, **284**, 1143–46.
9. Dziubla, T.D. and Muzykantov, V. (2006) Synthetic carriers for the delivery of protein therapeutics, *Biotechnology and Genetic Engineering Reviews*, **22**, 267–98.
10. Dziubla, T.D., Karim, A., and Muzykantov, V.R. (2005) Polymer nanocarriers protecting active enzyme cargo against proteolysis. *Journal of Controlled Release*, **102**, 427–39.
11. Muro, S., Koval, M., and Muzykantov, V. (2004) Endothelial endocytic pathways: gates for vascular drug delivery. *Current Vascular Pharmacology*, **2**, 281–99.
12. Muller, D.W., Gordon, D., San, H. *et al.* (1994) Catheter-mediated pulmonary vascular gene transfer and expression. *Circulation Research*, **75**, 1039–49.
13. Muro, S. and Muzykantov, V.R. (2005) Targeting of antioxidant and anti-thrombotic drugs to endothelial cell adhesion molecules. *Current Pharmaceutical Design*, **11**, 2383–401.
14. Muzykantov, V. and Danilov, S. (2006) Delivery of therapeutic and genetic materials to vascular endothelium, in *Encyclopedia of the Microvasculature* (eds D. Sherpo and J. Garcia), Elsevier, New York, pp. 1053–59.
15. Muro, S., Muzykantov, V.R., and Murciano, J.C. (2004) Characterization of endothelial internalization and targeting of antibody–enzyme conjugates in cell cultures and in laboratory animals. *Methods in Molecular Biology*, **283**, 21–36.
16. Muzykantov, V.R. (1998) Immunotargeting of drugs to the pulmonary vascular endothelium as a therapeutic strategy. *Pathophysiology*, **5**, 15–33.
17. Ding, B.S., Gottstein, C., Grunow, A. *et al.* (2005) Endothelial targeting of a recombinant construct fusing a PECAM-1 single-chain variable antibody fragment (scFv) with prourokinase facilitates prophylactic thrombolysis in the pulmonary vasculature. *Blood*, **106**, 4191–98.
18. Passerini, A.G., Polacek, D.C., Shi, C. *et al.* (2004) Coexisting proinflammatory and antioxidative endothelial transcription profiles in a disturbed flow region of the adult porcine aorta. *Proceedings of the National Academy of Sciences of the United States of America*, **101**, 2482–87.
19. St Croix, B., Rago, C., Velculescu, V. *et al.* (2000) Genes expressed in human tumor endothelium. *Science*, **289**, 1197–202.
20. Jacobson, B.S., Schnitzer, J.E., McCaffery, M., and Palade, G.E. (1992) Isolation and partial characterization of the luminal plasmalemma of microvascular endothelium from rat lungs. *European Journal of Cell Biology*, **58**, 296–306.
21. Durr, E., Yu, J., Krasinska, K.M. *et al.* (2004) Direct proteomic mapping of the lung microvascular endothelial cell surface *in vivo* and in cell culture. *Nature Biotechnology*, **22**, 985–92.
22. Oh, P., Li, Y., Yu, J. *et al.* (2004) Subtractive proteomic mapping of the endothelial surface in lung and solid tumours for tissue-specific therapy. *Nature*, **429**, 629–35.
23. Jacobson, B.S., Stolz, D.B., and Schnitzer, J.E. (1996) Identification of endothelial cell-surface proteins as targets for diagnosis and treatment of disease. *Nature Medicine*, **2**, 482–84.
24. Schnitzer, J.E. (1998) Vascular targeting as a strategy for cancer therapy. *The New England Journal of Medicine*, **339**, 472–74.
25. Rajotte, D., Arap, W., Hagedorn, M. *et al.* (1998) Molecular heterogeneity of the vascular endothelium revealed by *in vivo* phage display. *The Journal of Clinical Investigation*, **102**, 430–37.
26. Pasqualini, R., Arap, W., and McDonald, D.M. (2002) Probing the structural and molecular diversity of tumor vasculature. *Trends in Molecular Medicine*, **8**, 563–71.
27. Pasqualini, R., McDonald, D.M., and Arap, W. (2001) Vascular targeting and antigen presentation. *Nature Immunology*, **2**, 567–68.

28. Nicklin, S.A., White, S.J., Watkins, S.J. *et al.* (2000) Selective targeting of gene transfer to vascular endothelial cells by use of peptides isolated by phage display. *Circulation*, **102**, 231–37.
29. Houston, P., Goodman, J., Lewis, A. *et al.* (2001) Homing markers for atherosclerosis: applications for drug delivery, gene delivery and vascular imaging. *FEBS Letters*, **492**, 73–77.
30. Smith, J., Kontermann, R.E., Embleton, J., and Kumar, S. (2005) Antibody phage display technologies with special reference to angiogenesis. *The FASEB Journal*, **19**, 331–41.
31. Kelly, K.A., Allport, J.R., Tsourkas, A. *et al.* (2005) Detection of vascular adhesion molecule-1 expression using a novel multimodal nanoparticle. *Circulation Research*, **96**, 327–36.
32. Muzykantov, V.R. (2005) Biomedical aspects of targeted delivery of drugs to pulmonary endothelium. *Expert Opinion on Drug Delivery*, **2**, 909–26.
33. Muzykantov, V.R. (2002) Targeting pulmonary endothelium, in *Biomedical Aspects of Drug Targeting*, (eds V.R. Muzykantov and V.P. Torchilin), Kluwer, Boston, MA, pp. 129–48.
34. Muzykantov, V.R. (2001) Delivery of antioxidant enzyme proteins to the lung. *Antioxidants and Redox Signaling*, **3**, 39–62.
35. Danilov, S., Atochina, E., Hiemisch, H. *et al.* (1994) Interaction of mAb to angiotensin-converting enzyme (ACE) with antigen *in vitro* and *in vivo*: antibody targeting to the lung induces ACE antigenic modulation. *International Immunology*, **6**, 1153–60.
36. Danilov, S.M., Gavrilyuk, V.D., Franke, F.E. *et al.* (2001) Lung uptake of antibodies to endothelial antigens: key determinants of vascular immunotargeting. *American Journal of Physiology: Lung Cellular and Molecular Physiology*, **280**, L1335–47.
37. Kennel, S.J., Falcioni, R., and Wesley, J.W. (1991) Microdistribution of specific rat monoclonal antibodies to mouse tissues and human tumor xenografts. *Cancer Research*, **51**, 1529–36.
38. Erdos, E.G. (1990) Angiotensin I converting enzyme and the changes in our concepts through the years. Lewis K. Dahl Memorial Lecture. *Hypertension*, **16**, 363–70.
39. Heitsch, H., Brovkovych, S., Malinski, T., and Wiemer, G. (2001) Angiotensin-(1–7)-stimulated nitric oxide and superoxide release from endothelial cells. *Hypertension*, **37**, 72–76.
40. Balyasnikova, I.V., Karran, E.H., Albrecht, R.F. II, and Danilov, S.M. (2002) Epitope-specific antibody-induced cleavage of angiotensin-converting enzyme from the cell surface. *The Biochemical Journal*, **362**, 585–95.
41. Muzykantov, V.R. (2001) Targeting of superoxide dismutase and catalase to vascular endothelium. *Journal of Controlled Release*, **71**, 1–21.
42. Muzykantov, V.R., Atochina, E.N., Kuo, A. *et al.* (1996) Endothelial cells internalize monoclonal antibody to angiotensin-converting enzyme. *American Journal of Physiology: Lung Cellular and Molecular Physiology*, **270**, L704–13.
43. Watanabe, K., Lam, G., Keresztes, R.S., and Jaffe, E.A. (1992) Lipopolysaccharides decrease angiotensin converting enzyme activity expressed by cultured human endothelial cells. *Journal of Cellular Physiology*, **150**, 433–39.
44. Atochina, E.N., Hiemisch, H.H., Muzykantov, V.R., and Danilov, S.M. (1992) Systemic administration of platelet-activating factor in rat reduces specific pulmonary uptake of circulating monoclonal antibody to angiotensin-converting enzyme. *Lung*, **170**, 349–58.
45. Balyasnikova, I.V., Metzger, R., Visintine, D.J. *et al.* (2005) Selective rat lung endothelial targeting with a new set of monoclonal antibodies to angiotensin I-converting enzyme. *Pulmonary Pharmacology and Therapeutics*, **18**, 251–67.
46. Balyasnikova, I.V., Sun, Z.L., Metzger, R. *et al.* (2006) Monoclonal antibodies to native mouse angiotensin-converting enzyme (CD143): ACE expression quantification, lung endothelial cell targeting and gene delivery. *Tissue Antigens*, **67**, 10–29.
47. Muzykantov, V.R. and Danilov, S.M. (1995) Targeting of radiolabeled monoclonal antibody against ACE to the pulmonary endothelium, in *Targeted Delivery of Imaging Agents* (ed. V. Torchilin), CRC Press, Roca Baton, FL, pp. 465–85.
48. Esmon, C.T. (1989) The roles of protein C and thrombomodulin in the regulation of blood coagulation. *The Journal of Biological Chemistry*, **264**, 4743–46.
49. Christofidou-Solomidou, M., Kennel, S., Scherpereel, A. *et al.* (2002) Vascular immunotargeting of glucose oxidase to the endothelial antigens induces distinct forms of oxidant acute lung injury: targeting to thrombomodulin, but not to PECAM-1, causes pulmonary thrombosis and neutrophil transmigration. *American Journal of Pathology*, **160**, 1155–69.
50. Brisson, C., Archipoff, G., Hartmann, M.L. *et al.* (1992) Antibodies to thrombomodulin induce receptor-mediated endocytosis in human

saphenous vein endothelial cells. *Thrombosis and Haemostasis*, **68**, 737–43.

51. Guermazi, S., Mellouli, F., Trabelsi, S. *et al.* (2004) Anti-thrombomodulin antibodies and venous thrombosis. *Blood Coagulation and Fibrinolysis*, **15**, 553–58.
52. Scherpereel, A., Wiewrodt, R., Christofidou-Solomidou, M. *et al.* (2001) Cell-selective intracellular delivery of a foreign enzyme to endothelium *in vivo* using vascular immunotargeting. *The FASEB Journal*, **15**, 416–26.
53. Danielyan, K., Ding, B.S., Gottstein, C. *et al.* (2007) Delivery of anti-platelet-endothelial cell adhesion molecule single-chain variable fragment–urokinase fusion protein to the cerebral vasculature lyses arterial clots and attenuates postischemic brain edema. *The Journal of Pharmacology and Experimental Therapeutics*, **321**, 947–52.
54. Garnacho C., Dhami R., Simone E. *et al.* (2008) Delivery of acid sphingomyelinase in normal and Niemann–Pick disease mice using ICAM-1-targeted polymer nanocarriers. *The Journal of Pharmacology and Experimental Therapeutics*, **325**, 400–8.
55. Muro, S., Dziubla, T., Qiu, W. *et al.* (2006) Endothelial targeting of high-affinity multivalent polymer nanocarriers directed to intercellular adhesion molecule 1. *The Journal of Pharmacology and Experimental Therapeutics*, **317**, 1161–69.
56. Albelda, S.M. (1991) Endothelial and epithelial cell adhesion molecules. *American Journal of Respiratory Cell and Molecular Biology*, **4**, 195–203.
57. Newman, P.J. (1997) The biology of PECAM-1. *The Journal of Clinical Investigation*, **99**, 3–8.
58. Murohara, T., Delyani, J.A., Albelda, S.M., and Lefer, A.M. (1996) Blockade of platelet endothelial cell adhesion molecule-1 protects against myocardial ischemia and reperfusion injury in cats. *Journal of Immunology*, **156**, 3550–57.
59. Doerschuk, C.M. (2001) Mechanisms of leukocyte sequestration in inflamed lungs. *Microcirculation*, **8**, 71–88.
60. Vaporciyan, A.A., DeLisser, H.M., Yan, H.C. *et al.* (1993) Involvement of platelet-endothelial cell adhesion molecule-1 in neutrophil recruitment *in vivo*. *Science*, **262**, 1580–82.
61. Tasaka, S., Qin, L., Saijo, A. *et al.* (2003) Platelet endothelial cell adhesion molecule-1 in neutrophil emigration during acute bacterial pneumonia in mice and rats. *American Journal of Respiratory and Critical Care Medicine*, **167**, 164–70.
62. Doerschuk, C.M. (2000) Leukocyte trafficking in alveoli and airway passages. *Respiratory Research*, **1**, 136–40.
63. Wang, Q., Yerukhimovich, M., Gaarde, W.A. *et al.* (2005) MKK3 and -6-dependent activation of p38alpha MAP kinase is required for cytoskeletal changes in pulmonary microvascular endothelial cells induced by ICAM-1 ligation. *American Journal of Physiology: Lung Cellular and Molecular Physiology*, **288**, L359–69.
64. Wang, Q., Pfeiffer, G.R. II, Stevens, T., and Doerschuk, C.M. (2002) Lung microvascular and arterial endothelial cells differ in their responses to intercellular adhesion molecule-1 ligation. *American Journal of Respiratory and Critical Care Medicine*, **166**, 872–77.
65. Matheny, H.E., Deem, T.L., and Cook-Mills, J.M. (2000) Lymphocyte migration through monolayers of endothelial cell lines involves VCAM-1 signaling via endothelial cell NADPH oxidase. *Journal of Immunology*, **164**, 6550–59.
66. Garnacho, C., Shuvaev, V., Thomas, A. *et al.* (2008) RhoA activation and actin reorganization involved in endothelial CAM-mediated endocytosis of anti-PECAM carriers: critical role for tyrosine 686 in the cytoplasmic tail of PECAM-1. *Blood*, **111**, 3024–33.
67. Muro, S., Gajewski, C., Koval, M., and Muzykantov, V.R. (2005) ICAM-1 recycling in endothelial cells: a novel pathway for sustained intracellular delivery and prolonged effects of drugs. *Blood*, **105**, 650–58.
68. Kozower, B.D., Christofidou-Solomidou, M., Sweitzer, T.D. *et al.* (2003) Immunotargeting of catalase to the pulmonary endothelium alleviates oxidative stress and reduces acute lung transplantation injury. *Nature Biotechnology*, **21**, 392–98.
69. Dziubla, T.D., Shuvaev, V.V., Hong, N.K. *et al.* (2008) Endothelial targeting of semi-permeable polymer nanocarriers for enzyme therapies. *Biomaterials*, **29**, 215–27.
70. Murciano, J.C., Muro, S., Koniaris, L. *et al.* (2003) ICAM-directed vascular immunotargeting of antithrombotic agents to the endothelial luminal surface. *Blood*, **101**, 3977–84.
71. Christofidou-Solomidou, M., Scherpereel, A., Wiewrodt, R. *et al.* (2003) PECAM-directed delivery of catalase to endothelium protects against pulmonary vascular oxidative stress. *American Journal of Physiology: Lung Cellular and Molecular Physiology*, **285**, L283–92.
72. Muzykantov, V.R., Christofidou-Solomidou, M., Balyasnikova, I. *et al.* (1999) Streptavidin facilitates internalization and pulmonary targeting of an anti-endothelial cell antibody (platelet-endothelial cell adhesion molecule 1): a strategy for vascular

immunotargeting of drugs. *Proceedings of the National Academy of Sciences of the United States of America*, **96**, 2379–84.

73. Wiewrodt, R., Thomas, A.P., Cipelletti, L. *et al.* (2002) Size-dependent intracellular immunotargeting of therapeutic cargoes into endothelial cells. *Blood*, **99**, 912–22.
74. Christofidou-Solomidou, M., Pietra, G.G., Solomides, C.C. *et al.* (2000) Immunotargeting of glucose oxidase to endothelium *in vivo* causes oxidative vascular injury in the lungs. *American Journal of Physiology: Lung Cellular and Molecular Physiology*, **278**, L794–805.
75. Scherpereel, A., Rome, J.J., Wiewrodt, R. *et al.* (2002) Platelet-endothelial cell adhesion molecule-1-directed immunotargeting to cardiopulmonary vasculature. *The Journal of Pharmacology and Experimental Therapeutics*, **300**, 777–86.
76. Li, S., Tan, Y., Viroonchatapan, E. *et al.* (2000) Targeted gene delivery to pulmonary endothelium by anti-PECAM antibody. *American Journal of Physiology: Lung Cellular and Molecular Physiology*, **278**, L504–11.
77. Muro, S., Wiewrodt, R., Thomas, A. *et al.* (2003) A novel endocytic pathway induced by clustering endothelial ICAM-1 or PECAM-1. *Journal of Cell Science*, **116**, 1599–609.
78. Almenar-Queralt, A., Duperray, A., Miles, L.A. *et al.* (1995) Apical topography and modulation of ICAM-1 expression on activated endothelium. *American Journal of Pathology*, **147**, 1278–88.
79. Panes, J., Perry, M.A., Anderson, D.C. *et al.* (1995) Regional differences in constitutive and induced ICAM-1 expression *in vivo*. *American Journal of Physiology: Heart and Circulatory Physiology*, **269**, H1955–64.
80. Beck-Schimmer, B., Schimmer, R.C., Warner, R.L. *et al.* (1997) Expression of lung vascular and airway ICAM-1 after exposure to bacterial lipopolysaccharide. *American Journal of Respiratory Cell and Molecular Biology*, **17**, 344–52.
81. Vaporciyan, A.A., Mulligan, M.S., Warren, J.S. *et al.* (1995) Up-regulation of lung vascular ICAM-1 in rats is complement dependent. *Journal of Immunology*, **155**, 1442–49.
82. Sasso, D.E., Gionfriddo, M.A., Thrall, R.S. *et al.* (1996) Biodistribution of indium-111-labeled antibody directed against intercellular adhesion molecule-1. *Journal of Nuclear Medicine*, **37**, 656–61.
83. Villanueva, F.S., Jankowski, R.J., Klibanov, S. *et al.* (1998) Microbubbles targeted to intercellular adhesion molecule-1 bind to activated coronary artery endothelial cells. *Circulation*, **98**, 1–5.
84. Bloemen, P.G., Henricks, P.A., van Bloois, L. *et al.* (1995) Adhesion molecules: a new target for immunoliposome-mediated drug delivery. *FEBS Letters*, **357**, 140–44.
85. Kishimoto, T.K. and Rothlein, R. (1994) Integrins, ICAMs, and selectins: role and regulation of adhesion molecules in neutrophil recruitment to inflammatory sites. *Advances in Pharmacology*, **25**, 117–69.
86. Rothlein, R., Mainolfi, E.A., and Kishimoto, T.K. (1993) Treatment of inflammation with anti-ICAM-1. *Research in Immunology*, **144**, 735–39; discussion 754–762.
87. Kavanaugh, A.F., Davis, L.S., Jain, R.I. *et al.* (1996) A phase I/II open label study of the safety and efficacy of an anti-ICAM-1 (intercellular adhesion molecule-1; CD54) monoclonal antibody in early rheumatoid arthritis. *The Journal of Rheumatology*, **23**, 1338–44.
88. Lefer, D.J., Flynn, D.M., Anderson, D.C., and Buda, A.J. (1996) Combined inhibition of P-selectin and ICAM-1 reduces myocardial injury following ischemia and reperfusion. *American Journal of Physiology: Heart and Circulatory Physiology*, **271**, H2421–29.
89. Staunton, D.E., Merluzzi, V.J., Rothlein, R. *et al.* (1989) A cell adhesion molecule, ICAM-1, is the major surface receptor for rhinoviruses. *Cell*, **56**, 849–53.
90. Zimmerman, G.A., McIntyre, T.M., and Prescott, S.M. (1996) Adhesion and signaling in vascular cell–cell interactions. *The Journal of Clinical Investigation*, **98**, 1699–702.
91. Zivin, J.A. (1998) Factors determining the therapeutic window for stroke. *Neurology*, **50**, 599–603.
92. Muro, S., Cui, X., Gajewski, C. *et al.* (2003) Slow intracellular trafficking of catalase nanoparticles targeted to ICAM-1 protects endothelial cells from oxidative stress. *American Journal of Physiology: Cell Physiology*, **285**, C1339–47.
93. Mastrobattista, E., Storm, G., van Bloois, L. *et al.* (1999) Cellular uptake of liposomes targeted to intercellular adhesion molecule-1 (ICAM-1) on bronchial epithelial cells. *Biochimica et Biophysica Acta*, **1419**, 353–63.
94. Zhang, N., Chittasupho, C., Duangrat, C. *et al.* (2008) PLGA nanoparticle – peptide conjugate effectively targets intercellular cell-adhesion molecule-1. *Bioconjugate Chemistry*, **19**, 145–52.
95. Atochina, E.N., Balyasnikova, I.V., Danilov, S.M. *et al.* (1998) Immunotargeting of catalase to ACE

or ICAM-1 protects perfused rat lungs against oxidative stress. *American Journal of Physiology: Lung Cellular and Molecular Physiology*, **275**, L806–17.

96. Rossin, R., Muro, S., Welch, M.J. *et al.* (2008) *In vivo* imaging of ^{64}Cu-labeled polymer nanoparticles targeted to the lung endothelium. *Journal of Nuclear Medicine*, **49**, 103–11.
97. Doerschuk, C.M., Quinlan, W.M., Doyle, N.A. *et al.* (1996) The role of P-selectin and ICAM-1 in acute lung injury as determined using blocking antibodies and mutant mice. *Journal of Immunology*, **157**, 4609–14.
98. Springer, T.A. (1990) Adhesion receptors of the immune system. *Nature*, **346**, 425–34.
99. Spragg, D.D., Alford, D.R., Greferath, R. *et al.* (1997) Immunotargeting of liposomes to activated vascular endothelial cells: a strategy for site-selective delivery in the cardiovascular system. *Proceedings of the National Academy of Sciences of the United States of America*, **94**, 8795–800.
100. Kiely, J.M., Cybulsky, M.I., Luscinskas, F.W., and Gimbrone, M.A. Jr. (1995) Immunoselective targeting of an anti-thrombin agent to the surface of cytokine-activated vascular endothelial cells. *Arteriosclerosis, Thrombosis, and Vascular Biology*, **15**, 1211–18.
101. Kuijpers, T.W., Raleigh, M., Kavanagh, T. *et al.* (1994) Cytokine-activated endothelial cells internalize E-selectin into a lysosomal compartment of vesiculotubular shape. A tubulin-driven process. *Journal of Immunology*, **152**, 5060–69.
102. Straley, K.S. and Green, S.A. (2000) Rapid transport of internalized P-selectin to late endosomes and the TGN: roles in regulating cell surface expression and recycling to secretory granules. *The Journal of Cell Biology*, **151**, 107–16.
103. von Asmuth, E.J., Smeets, E.F., Ginsel, L.A. *et al.* (1992) Evidence for endocytosis of E-selectin in human endothelial cells. *European Journal of Immunology*, **22**, 2519–26.
104. Kessner, S., Krause, A., Rothe, U., and Bendas, G. (2001) Investigation of the cellular uptake of E-selectin-targeted immunoliposomes by activated human endothelial cells. *Biochimica et Biophysica Acta*, **1514**, 177–90.
105. Everts, M., Kok, R.J., Asgeirsdottir, S.A. *et al.* (2002) Selective intracellular delivery of dexamethasone into activated endothelial cells using an E-selectin-directed immunoconjugate. *Journal of Immunology*, **168**, 883–89.
106. Harari, O.A., Wickham, T.J., Stocker, C.J. *et al.* (1999) Targeting an adenoviral gene vector to cytokine-activated vascular endothelium via E-selectin. *Gene Therapy*, **6**, 801–7.
107. Lindner, J.R., Song, J., Christiansen, J. *et al.* (2001) Ultrasound assessment of inflammation and renal tissue injury with microbubbles targeted to P-selectin. *Circulation*, **104**, 2107–12.
108. Tsourkas, A., Shinde-Patil, V.R., Kelly, K. *et al.* (2005) *In vivo* imaging of activated endothelium using an anti-VCAM-1 magneto-optical probe. *Bioconjugare Chemistry*, **16**, 576–81.
109. Keelan, E.T., Harrison, A.A., Chapman, P.T. *et al.* (1994) Imaging vascular endothelial activation: an approach using radiolabeled monoclonal antibodies against the endothelial cell adhesion molecule E-selectin. *Journal of Nuclear Medicine*, **35**, 276–81.
110. Lindner, J.R., Klibanov, A.L., and Ley, K. (2003) Targeting inflammation, in *Biomedical Aspects of Drug Targeting* (eds V.R. Muzykantov and V.P. Torchilin), Kluwer, Boston, MA, pp. 149–72.
111. Ghitescu, L., Jacobson, B.S., and Crine, P. (1999) A novel, 85 kDa endothelial antigen differentiates plasma membrane macrodomains in lung alveolar capillaries. *Endothelium*, **6**, 241–50.
112. Ghitescu, L.D., Crine, P., and Jacobson, B.S. (1997) Antibodies specific to the plasma membrane of rat lung microvascular endothelium. *Experimental Cell Research*, **232**, 47–55.
113. Murciano, J.C., Harshaw, D.W., Ghitescu, L. *et al.* (2001) Vascular immunotargeting to endothelial surface in a specific macrodomain in alveolar capillaries. *American Journal of Respiratory and Critical Care Medicine*, **164**, 1295–302.
114. Kolonin, M.G., Sun, J., Do, K.A. *et al.* (2006) Synchronous selection of homing peptides for multiple tissues by *in vivo* phage display. *The FASEB Journal*, **20**, 979–81.
115. Skidgel, R.A. (1992) Bradykinin-degrading enzymes: structure, function, distribution, and potential roles in cardiovascular pharmacology. *Journal of Cardiovascular Pharmacology*, **20** (Suppl 9), S4–S9.
116. Rangel, R., Sun, Y., Guzman-Rojas, L. *et al.* (2007) Impaired angiogenesis in aminopeptidase N-null mice. *Proceedings of the National Academy of Sciences of the United States of America*, **104**, 4588–93.
117. Chun, M., Liyanage, U.K., Lisanti, M.P., and Lodish, H.F. (1994) Signal transduction of a G protein-coupled receptor in caveolae: colocalization of endothelin and its receptor with caveolin. *Proceedings of the National Academy of Sciences of the United States of America*, **91**, 11728–32.

118. Schnitzer, J.E. (2001) Caveolae: from basic trafficking mechanisms to targeting transcytosis for tissue-specific drug and gene delivery *in vivo*. *Advanced Drug Delivery Reviews*, **49**, 265–80.
119. Stevens, T., Garcia, J.G., Shasby, D.M. *et al.* (2000) Mechanisms regulating endothelial cell barrier function. *American Journal of Physiology: Lung Cellular and Molecular Physiology*, **279**, L419–22.
120. McIntosh, D.P., Tan, X.Y., Oh, P., and Schnitzer, J.E. (2002) Targeting endothelium and its dynamic caveolae for tissue-specific transcytosis *in vivo*: a pathway to overcome cell barriers to drug and gene delivery. *Proceedings of the National Academy of Sciences of the United States of America*, **99**, 1996–2001.
121. Riezman, H., Woodman, P.G., van Meer, G., and Marsh, M. (1997) Molecular mechanisms of endocytosis. *Cell*, **91**, 731–38.
122. Predescu, D. and Palade, G.E. (1993) Plasmalemmal vesicles represent the large pore system of continuous microvascular endothelium. *American Journal of Physiology: Heart and Circulatory Physiology*, **265**, H725–33.
123. Stan, R.V. (2002) Structure and function of endothelial caveolae. *Microscopy Research and Technique*, **57**, 350–64.
124. Minshall, R.D., Tiruppathi, C., Vogel, S.M., and Malik, A.B. (2002) Vesicle formation and trafficking in endothelial cells and regulation of endothelial barrier function. *Histochemistry and Cell Biology*, **117**, 105–12.
125. Schnitzer, J.E., McIntosh, D.P., Dvorak, A.M. *et al.* (1995) Separation of caveolae from associated microdomains of GPI-anchored proteins. *Science*, **269**, 1435–39.
126. Minshall, R.D., Tiruppathi, C., Vogel, S.M. *et al.* (2000) Endothelial cell-surface gp60 activates vesicle formation and trafficking via G_i-coupled Src kinase signaling pathway. *The Journal of Cell Biology*, **150**, 1057–70.
127. Vogel, S.M., Easington, C.R., Minshall, R.D. *et al.* (2001) Evidence of transcellular permeability pathway in microvessels. *Microvascular Research*, **61**, 87–101.
128. John, T.A., Vogel, S.M., Tiruppathi, C. *et al.* (2003) Quantitative analysis of albumin uptake and transport in the rat microvessel endothelial monolayer. *American Journal of Physiology: Lung Cellular and Molecular Physiology*, **284**, L187–96.
129. Tiruppathi, C., Song, W., Bergenfeldt, M. *et al.* (1997) Gp60 activation mediates albumin transcytosis in endothelial cells by tyrosine kinase-dependent pathway. *The Journal of Biological Chemistry*, **272**, 25968–75.
130. Predescu, D., Predescu, S., and Malik, A.B. (2002) Transport of nitrated albumin across continuous vascular endothelium. *Proceedings of the National Academy of Sciences of the United States of America*, **99**, 13932–37.
131. Oh, P., Borgstrom, P., Witkiewicz, H. *et al.* (2007) Live dynamic imaging of caveolae pumping targeted antibody rapidly and specifically across endothelium in the lung. *Nature Biotechnology*, **25**, 327–37.
132. Muzykantov, V.R., Puchnina, E.A., Atochina, E.N. *et al.* (1991) Endotoxin reduces specific pulmonary uptake of radiolabeled monoclonal antibody to angiotensin-converting enzyme. *Journal of Nuclear Medicine*, **32**, 453–60.
133. Atochina, E.N., Muzykantov, V.R., Al-Mehdi, A.B. *et al.* (1997) Normoxic lung ischemia/reperfusion accelerates shedding of angiotensin converting enzyme from the pulmonary endothelium. *American Journal of Respiratory and Critical Care Medicine*, **156**, 1114–19.
134. Shuvaev, V.V., Christofidou-Solomidou, M., Scherpereel, A. *et al.* (2007) Factors modulating the delivery and effect of enzymatic cargo conjugated with antibodies targeted to the pulmonary endothelium. *Journal of Controlled Release*, **118**, 235–44.
135. Perkowski, S., Sun, J., Singhal, S. *et al.* (2003) Gene expression profiling of the early pulmonary response to hyperoxia in mice. *American Journal of Respiratory Cell and Molecular Biology*, **28**, 682–96.
136. Komatsu, S., Panes, J., Russell, J.M. *et al.* (1997) Effects of chronic arterial hypertension on constitutive and induced intercellular adhesion molecule-1 expression *in vivo*. *Hypertension*, **29**, 683–89.
137. Mulligan, M.S., Vaporciyan, A.A., Miyasaka, M. *et al.* (1993) Tumor necrosis factor alpha regulates *in vivo* intrapulmonary expression of ICAM-1. *American Journal of Pathology*, **142**, 1739–49.
138. Perkowski, S., Scherpereel, A., Murciano, J.C. *et al.* (2006) Dissociation between alveolar transmigration of neutrophils and lung injury in hyperoxia. *American Journal of Physiology: Lung Cellular and Molecular Physiology*, **291**, L1050–58.
139. Panes, J., Perry, M.A., Anderson, D.C. *et al.* (1996) Portal hypertension enhances endotoxin-induced intercellular adhesion molecule 1 up-regulation in the rat. *Gastroenterology*, **110**, 866–74.

140. Amano, J., Hiroe, M., Ohta, Y. *et al.* (1996) Uptake of indium-111-anti-intercellular adhesion molecule-1 monoclonal antibody in the allografted rat lung during acute rejection. *The Journal of Heart and Lung Transplantation*, **15**, 1027–33.
141. Weiner, R.E., Sasso, D.E., Gionfriddo, M.A. *et al.* (2001) Early detection of oleic acid-induced lung injury in rats using ^{111}In-labeled anti-rat intercellular adhesion molecule-1. *Journal of Nuclear Medicine*, **42**, 1109–15.
142. Weiner, R.E., Sasso, D.E., Gionfriddo, M.A. *et al.* (1998) Early detection of bleomycin-induced lung injury in rat using indium-111-labeled antibody directed against intercellular adhesion molecule-1. *Journal of Nuclear Medicine*, **39**, 723–28.
143. Eppihimer, M.J., Wolitzky, B., Anderson, D.C. *et al.* (1996) Heterogeneity of expression of E- and P-selectins *in vivo*. *Circulation Research*, **79**, 560–69.
144. Perkowski, S., Scherpereel, A., Murciano, J.C. *et al.* (2006) Dissociation between alveolar transmigration of neutrophils and lung injury in hyperoxia. *American Journal of Physiology: Lung Cellular and Molecular Physiology*, **291**, L1050–58.
145. Woodward, J.D., Kennel, S.J., Mirzadeh, S. *et al.* (2007) *In vivo* SPECT/CT imaging and biodistribution using radioactive Cd^{125m}Te/ZnS nanoparticles. *Nanotechnology*, **18**, 175103.
146. Valadon, P., Garnett, J.D., Testa, J.E. *et al.* (2006) Screening phage display libraries for organ-specific vascular immunotargeting *in vivo*. *Proceedings of the National Academy of Sciences of the United States of America*, **103**, 407–12.
147. Akerman, M.E., Chan, W.C., Laakkonen, P. *et al.* (2002) Nanocrystal targeting *in vivo*. *Proceedings of the National Academy of Sciences of the United States of America*, **99**, 12617–21.
148. Hajitou, A., Trepel, M., Lilley, C.E. *et al.* (2006) A hybrid vector for ligand-directed tumor targeting and molecular imaging. *Cell*, **125**, 385–98.
149. Maruyama, K., Takizawa, T., Yuda, T. *et al.* (1995) Targetability of novel immunoliposomes modified with amphipathic poly(ethylene glycol)s conjugated at their distal terminals to monoclonal antibodies. *Biochimica et Biophysica Acta*, **1234**, 74–80.
150. Hughes, B.J., Kennel, S., Lee, R., and Huang, L. (1989) Monoclonal antibody targeting of liposomes to mouse lung *in vivo*. *Cancer Research*, **49**, 6214–20.
151. Mori, A., Kennel, S.J., van Borssum Waalkes, M. *et al.* (1995) Characterization of organ-specific immunoliposomes for delivery of 3′,5′-*O*-dipalmitoyl-5-fluoro-2′-deoxyuridine in a mouse lung-metastasis model. *Cancer Chemotherapy and Pharmacology*, **35**, 447–56.
152. Kennel, S.J., Chappell, L.L., Dadachova, K. *et al.* (2000) Evaluation of ^{225}Ac for vascular targeted radioimmunotherapy of lung tumors. *Cancer Biotherapy and Radiopharmaceuticals*, **15**, 235–44.
153. Kennel, S.J. and Mirzadeh, S. (1998) Vascular targeted radioimmunotherapy with ^{213}Bi – an alpha-particle emitter. *Nuclear Medicine and Biology*, **25**, 241–46.
154. Kennel, S.J., Boll, R., Stabin, M. *et al.* (1999) Radioimmunotherapy of micrometastases in lung with vascular targeted ^{213}Bi. *British Journal of Cancer*, **80**, 175–84.
155. Davis, I.A. and Kennel, S.J. (1999) Radioimmunotherapy using vascular targeted ^{213}Bi: the role of tumor necrosis factor alpha in the development of pulmonary fibrosis. *Clinical Cancer Research*, **5**, 3160s–3164s.
156. Kennel, S.J., Lankford, T.K., Ullrich, R.L., and Jamasbi, R.J. (1988) Enhancement of lung tumor colony formation by treatment of mice with monoclonal antibodies to pulmonary capillary endothelial cells. *Cancer Research*, **48**, 4964–68.
157. Muzykantov, V.R., Trubetskaya, O.V., Puchnina, E.A. *et al.* (1990) Cytotoxicity of glucose oxidase conjugated with antibodies to target cells: killing efficiency depends on the conjugate internalization. *Biochimica et Biophysica Acta*, **1053**, 27–31.
158. Muzykantov, V.R., Sakharov, D.V., Sinitsyn, V.V. *et al.* (1988) Specific killing of human endothelial cells by antibody-conjugated glucose oxidase. *Analytical Biochemistry*, **169**, 383–89.
159. Muzykantov, V.R., Martynov, A.V., Puchnina, E.A., and Danilov, S.M. (1989) *In vivo* administration of glucose oxidase conjugated with monoclonal antibodies to angiotensin-converting enzyme. The tissue distribution, blood clearance, and targeting into rat lungs. *The American Review of Respiratory Disease*, **139**, 1464–73.
160. Gow, A.J., Branco, F., Christofidou-Solomidou, M. *et al.* (1999) Immunotargeting of glucose oxidase: intracellular production of H_2O_2 and endothelial oxidative stress. *American Journal of Physiology: Lung Cellular and Molecular Physiology*, **277**, L271–81.
161. Gaspar, M.M., Boerman, O.C., Laverman, P. *et al.* (2007) Enzymosomes with surface-exposed superoxide dismutase: *in vivo* behaviour and therapeutic activity in a model of adjuvant arthritis. *Journal of Controlled Release*, **117**, 186–95.
162. Fujita, T. and Ushiki, T. (1992) Scanning electron microscopic observations of the immunodefensive

systems with special reference to the surface morphology of the non-lymphoid cells. *Archives of Histology and Cytology*, **55** (Suppl), 105–13.

163. Freeman, B.A., Young, S.L., and Crapo, J.D. (1983) Liposome-mediated augmentation of superoxide dismutase in endothelial cells prevents oxygen injury. *The Journal of Biological Chemistry*, **258**, 12534–42.

164. Duann, P., Datta, P.K., Pan, C. *et al.* (2006) Superoxide dismutase mimetic preserves the glomerular capillary permeability barrier to protein. *The Journal of Pharmacology and Experimental Therapeutics*, **316**, 1249–54.

165. Igarashi, R., Hoshino, J., Ochiai, A. *et al.* (1994) Lecithinized superoxide dismutase enhances its pharmacologic potency by increasing its cell membrane affinity. *The Journal of Pharmacology and Experimental Therapeutics*, **271**, 1672–77.

166. Koo, D.D., Welsh, K.I., West, N.E. *et al.* (2001) Endothelial cell protection against ischemia/reperfusion injury by lecithinized superoxide dismutase. *Kidney International*, **60**, 786–96.

167. Bonder, C.S., Knight, D., Hernandez-Saavedra, D. *et al.* (2004) Chimeric SOD2/3 inhibits at the endothelial–neutrophil interface to limit vascular dysfunction in ischemia–reperfusion. *American Journal of Physiology: Gastrointestinal and Liver Physiology*, **287**, G676–84.

168. Gao, B., Flores, S.C., Leff, J.A. *et al.* (2003) Synthesis and anti-inflammatory activity of a chimeric recombinant superoxide dismutase: SOD2/3. *American Journal of Physiology: Lung Cellular and Molecular Physiology*, **284**, L917–25.

169. Lu, M., Gong, X., Lu, Y. *et al.* (2006) Molecular cloning and functional characterization of a cell-permeable superoxide dismutase targeted to lung adenocarcinoma cells. Inhibition cell proliferation through the Akt/p27^{kip1} pathway. *The Journal of Biological Chemistry*, **281**, 13620–27.

170. Jiang, J., Kurnikov, I., Belikova, N.A. *et al.* (2007) Structural requirements for optimized delivery, inhibition of oxidative stress, and antiapoptotic activity of targeted nitroxides. *The Journal of Pharmacology and Experimental Therapeutics*, **320**, 1050–60.

171. Muzykantov, V.R., Atochina, E.N., Ischiropoulos, H. *et al.* (1996) Immunotargeting of antioxidant enzyme to the pulmonary endothelium. *Proceedings of the National Academy of Sciences of the United States of America*, **93**, 5213–18.

172. Sweitzer, T.D., Thomas, A.P., Wiewrodt, R. *et al.* (2003) PECAM-directed immunotargeting of catalase: specific, rapid and transient protection against hydrogen peroxide. *Free Radical Biology and Medicine*, **34**, 1035–46.

173. Nowak, K., Weih, S., Metzger, R. *et al.* (2007) Immunotargeting of catalase to lung endothelium via anti-angiotensin-converting enzyme antibodies attenuates ischemia–reperfusion injury of the lung *in vivo*. *American Journal of Physiology: Lung Cellular and Molecular Physiology*, **293**, L162–69.

174. Muro, S., Mateescu, M., Gajewski, C. *et al.* (2006) Control of intracellular trafficking of ICAM-1-targeted nanocarriers by endothelial Na^+/H^+ exchanger proteins. *American Journal of Physiology: Lung Cellular and Molecular Physiology*, **290**, L809–17.

175. Ding, B.S., Dziubla, T., Shuvaev, V.V. *et al.* (2006) Advanced drug delivery systems that target the vascular endothelium. *Molecular Interventions*, **6**, 98–112.

176. Muro, S., Schuchman, E.H., and Muzykantov, V.R. (2006) Lysosomal enzyme delivery by ICAM-1-targeted nanocarriers bypassing glycosylation- and clathrin-dependent endocytosis. *Molecular Therapy*, **13**, 135–41.

177. Schuchman, E.H. and Desnick, R.J. (2000) Niemann–Pick disease types A and B: acid sphingomyelinase deficiencies, in *The Metabolic and Molecular Bases of Inherited Disease*, 8th edn (eds C. Scriver, A. Beaudet, W. Sly *et al.*), McGraw-Hill, New York, pp. 3589–610.

178. Desnick, R.J. and Schuchman, E.H. (2002) Enzyme replacement and enhancement therapies: lessons from lysosomal disorders. *Nature Reviews: Genetics*, **3**, 954–66.

179. Miranda, S.R., He, X., Simonaro, C.M. *et al.* (2000) Infusion of recombinant human acid sphingomyelinase into Niemann–Pick disease mice leads to visceral, but not neurological, correction of the pathophysiology. *The FASEB Journal*, **14**, 1988–95.

180. He, X., Miranda, S.R., Xiong, X. *et al.* (1999) Characterization of human acid sphingomyelinase purified from the media of overexpressing Chinese hamster ovary cells. *Biochimica et Biophysica Acta*, **1432**, 251–64.

181. Uyechi, L.S., Gagne, L., Thurston, G., and Szoka, F.C. Jr. (2001) Mechanism of lipoplex gene delivery in mouse lung: binding and internalization of fluorescent lipid and DNA components. *Gene Therapy*, **8**, 828–36.

182. Ma, Z., Zhang, J., Alber, S. *et al.* (2002) Lipid-mediated delivery of oligonucleotide to pulmonary endothelium. *American Journal of*

Respiratory Cell and Molecular Biology, **27**, 151–59.

183. Ma, Z., Mi, Z., Wilson, A. *et al.* (2002) Redirecting adenovirus to pulmonary endothelium by cationic liposomes. *Gene Therapy*, **9**, 176–82.
184. Liu, H., Liu, L., and Visner, G.A. (2007) Nonviral gene delivery with indoleamine 2,3-dioxygenase targeting pulmonary endothelium protects against ischemia–reperfusion injury. *American Journal of Transplantation*, **7**, 2291–300.
185. Hiratsuka, M., Mora, B.N., Yano, M. *et al.* (1999) Gene transfer of heat shock protein 70 protects lung grafts from ischemia–reperfusion injury. *The Annals of Thoracic Surgery*, **67**, 1421–27.
186. Suda, T., Mora, B.N., D'Ovidio, F. *et al.* (2000) *In vivo* adenovirus-mediated endothelial nitric oxide synthase gene transfer ameliorates lung allograft ischemia–reperfusion injury. *The Journal of Thoracic and Cardiovascular Surgery*, **119**, 297–304.
187. Goula, D., Becker, N., Lemkine, G.F. *et al.* (2000) Rapid crossing of the pulmonary endothelial barrier by polyethylenimine/DNA complexes. *Gene Therapy*, **7**, 499–504.
188. Tan, P.H., Manunta, M., Ardjomand, N. *et al.* (2003) Antibody targeted gene transfer to endothelium. *The Journal of Gene Medicine*, **5**, 311–23.
189. Reynolds, P.N., Nicklin, S.A., Kaliberova, L. *et al.* (2001) Combined transductional and transcriptional targeting improves the specificity of transgene expression *in vivo*. *Nature Biotechnology*, **19**, 838–42.
190. Rancourt, C., Robertson, M.W. III, Wang, M. *et al.* (1998) Endothelial cell vehicles for delivery of cytotoxic genes as a gene therapy approach for carcinoma of the ovary. *Clinical Cancer Research*, **4**, 265–70.
191. Trepel, M., Grifman, M., Weitzman, M.D., and Pasqualini, R. (2000) Molecular adaptors for vascular-targeted adenoviral gene delivery. *Human Gene Therapy*, **11**, 1971–81.
192. Weiler-Guettler, H., Aird, W.C., Husain, M. *et al.* (1996) Targeting of transgene expression to the vascular endothelium of mice by homologous recombination at the thrombomodulin locus. *Circulation Research*, **78**, 180–87.
193. White, S.J., Nicklin, S.A., Buning, H. *et al.* (2004) Targeted gene delivery to vascular tissue *in vivo* by tropism-modified adeno-associated virus vectors. *Circulation*, **109**, 513–19.
194. Nicklin, S.A., Von Seggern, D.J., Work, L.M. *et al.* (2001) Ablating adenovirus type 5 fiber-CAR binding and HI loop insertion of the SIGYPLP peptide generate an endothelial cell-selective adenovirus. *Molecular Therapy*, **4**, 534–42.
195. Rade, J.J., Cheung, M., Miyamoto, S., and Dichek, D.A. (1999) Retroviral vector-mediated expression of hirudin by human vascular endothelial cells: implications for the design of retroviral vectors expressing biologically active proteins. *Gene Therapy*, **6**, 385–92.
196. Waehler, R., Russell, S.J., and Curiel, D.T. (2007) Engineering targeted viral vectors for gene therapy. *Nature Reviews: Genetics*, **8**, 573–87.
197. Heistad, D.D. (2006) Gene therapy for vascular disease. *Vascular Pharmacology*, **45**, 331–33.
198. Maus, U., Rosseau, S., Mandrakas, N. *et al.* (1999) Cationic lipids employed for antisense oligodeoxynucleotide transport may inhibit vascular cell adhesion molecule-1 expression in human endothelial cells: a word of caution. *Antisense and Nucleic Acid Drug Development*, **9**, 71–80.
199. Godbey, W.T., Wu, K.K., and Mikos, A.G. (2001) Poly(ethylenimine)-mediated gene delivery affects endothelial cell function and viability. *Biomaterials*, **22**, 471–80.
200. Kader, K.N., Sweany, J.M., and Bellamkonda, R.V. (2002) Cationic lipid-mediated transfection of bovine aortic endothelial cells inhibits their attachment. *Journal of Biomedical Materials Research*, **60**, 405–10.
201. Li, S., Wu, S.P., Whitmore, M. *et al.* (1999) Effect of immune response on gene transfer to the lung via systemic administration of cationic lipidic vectors. *American Journal of Physiology: Lung Cellular and Molecular Physiology*, **276**, L796–804.
202. Rodman, D.M., San, H., Simari, R. *et al.* (1997) *In vivo* gene delivery to the pulmonary circulation in rats: transgene distribution and vascular inflammatory response. *American Journal of Respiratory Cell and Molecular Biology*, **16**, 640–49.
203. Gref, R., Minamitake, Y., Peracchia, M.T. *et al.* (1994) Biodegradable long-circulating polymeric nanospheres. *Science*, **263**, 1600–3.
204. Wilson, A., Zhou, W., Champion, H.C. *et al.* (2005) Targeted delivery of oligodeoxynucleotides to mouse lung endothelial cells *in vitro* and *in vivo*. *Molecular Therapy*, **12**, 510–18.
205. Reynolds, P.N., Zinn, K.R., Gavrilyuk, V.D. *et al.* (2000) A targetable, injectable adenoviral vector for selective gene delivery to pulmonary endothelium *in vivo*. *Molecular Therapy*, **2**, 562–78.
206. Miller, W.H., Brosnan, M.J., Graham, D. *et al.* (2005) Targeting endothelial cells with adenovirus expressing nitric oxide synthase prevents elevation of blood pressure in stroke-prone spontaneously hypertensive rats. *Molecular Therapy*, **12**, 321–27.

207. Reynolds, A.M., Xia, W., Holmes, M.D. *et al.* (2007) Bone morphogenetic protein type 2 receptor gene therapy attenuates hypoxic pulmonary hypertension. *American Journal of Physiology: Lung Cellular and Molecular Physiology*, **292**, L1182–92.
208. Waugh, J.M., Kattash, M., Li, J. *et al.* (1999) Gene therapy to promote thromboresistance: local overexpression of tissue plasminogen activator to prevent arterial thrombosis in an *in vivo* rabbit model. *Proceedings of the National Academy of Sciences of the United States of America*, **96**, 1065–70.
209. Waugh, J.M., Li-Hawkins, J., Yuksel, E. *et al.* (2000) Thrombomodulin overexpression to limit neointima formation. *Circulation*, **102**, 332–37.
210. Waugh, J.M., Yuksel, E., Li, J. *et al.* (1999) Local overexpression of thrombomodulin for *in vivo* prevention of arterial thrombosis in a rabbit model. *Circulation Research*, **84**, 84–92.
211. Dichek, D.A., Anderson, J., Kelly, A.B. *et al.* (1996) Enhanced *in vivo* antithrombotic effects of endothelial cells expressing recombinant plasminogen activators transduced with retroviral vectors. *Circulation*, **93**, 301–9.
212. de Bono, D.P. and Pringle, S. (1991) Local inhibition of thrombosis using urokinase linked to a monoclonal antibody which recognises damaged endothelium. *Thrombosis Research*, **61**, 537–45.
213. Muzykantov, V.R., Barnathan, E.S., Atochina, E.N. *et al.* (1996) Targeting of antibody-conjugated plasminogen activators to the pulmonary vasculature. *The Journal of Pharmacology and Experimental Therapeutics*, **279**, 1026–34.
214. Ding, B.S., Zhou, Y.J., Chen, X.Y. *et al.* (2003) Lung endothelium targeting for pulmonary embolism thrombolysis. *Circulation*, **108**, 2892–98.
215. Ding, B.S., Hong, N., Murciano, J.C. *et al.* (2008) Prophylactic thrombolysis by thrombin-activated latent prourokinase targeted to PECAM-1 in the pulmonary vasculature. *Blood*, **111**, 1999–2006.

SECTION III:

PULMONARY ENDOTHELIUM IN DISEASE

23

Endothelial Regulation of the Pulmonary Circulation in the Fetus and Newborn

Yuansheng Gao[1] and J. Usha Raj[2]

[1]Department of Physiology and Pathophysiology, Peking University Health Science Center, Beijing, China

[2]Department of Pediatrics, University of Illinois at Chicago, Chicago, IL, USA

INTRODUCTION

High vascular resistance and low blood flow characterize the pulmonary circulation in the fetus. This is due to multiple factors: fluid-filled lungs that result in high intraluminal pressures [1, 2], low oxygen tension *in utero* [3, 4], and low shear stress resulting from low blood flow, all of which may contribute to low vasodilator activity coupled with high vasoconstrictor activity [5–8]. At birth, with the onset of breathing, pulmonary vascular resistance decreases more than 10-fold and pulmonary blood flow increases 8- to 10-fold. The change from a liquid- to an air-filled organ, an increase in oxygen tension and shear stress, and the surge of vasodilator activity together with suppression of vasoconstrictor activity and many other factors contribute to this dramatic hemodynamic change. This reduction in pulmonary vascular resistance and the accompanying increase in blood flow are essential for the lungs to take over the gas-exchange function from the placenta and for the successful postnatal adaptation of the newborn [9–11].

The endothelium consists of a single layer of cells which is strategically located between blood and vascular smooth muscle. There is overwhelming evidence to indicate that the endothelium plays a critical role in regulating vasomotor tone in the pulmonary circulation of the fetus and newborn. This is achieved largely by the release of various vasoactive agents that act on the underlying smooth muscle cells (SMCs; see Chapter 12). In addition, the endothelium plays a role in regulating smooth muscle growth and proliferation, and thereby regulates vascular structure to some extent [12, 13]. The role of some of the more important vasoactive agents, in particular nitric oxide (NO), prostanoids, and endothelin (ET), in regulation of pulmonary vasomotor tone in the fetus and during the normal transition at birth will be discussed in this chapter. The altered balance in the production of vasodilators and constrictors, that leads to abnormally high pulmonary vascular resistance after birth and failure of postnatal adaptation resulting in the clinical condition of persistent pulmonary hypertension of the newborn (PPHN), will also be discussed.

ENDOTHELIUM AND THE FETAL PULMONARY CIRCULATION

Human lung development starts from the lung bud, which appears as a ventral diverticulum of the foregut during the fourth week of gestation (see Chapter 1). Studies suggest that intrapulmonary arteries and veins are formed *de novo* from the endothelial cells (ECs) of the lung bud. At about 34 days of gestation, a continuous pulmonary circulation appears, with the artery extending from the outflow tract of the heart and the vein connecting to the prospective left atrium. Between these arteries and veins lies a mesenchymal capillary plexus [12, 14–16].

In early gestation, there are fewer small pulmonary arteries and thus a smaller cross-sectional area, which contributes to the high pulmonary vascular resistance. In the second half of the gestation, the number of pulmonary vessels increases dramatically with increased pulmonary blood flow [10, 17]. In fetal lambs, the proportion of cardiac output distributed to the lung is 3–4% from 0.4 to 0.7 gestation. It increases progressively to 8–10% at term [9].

The Pulmonary Endothelium: Function in health and disease Editors Norbert F. Voelkel, Sharon Rounds

In the human fetus, the proportion of cardiac output distributed to the lung increases from 13 to 25% from 20 to 30 weeks of gestation and then remains unchanged to term [18]. Despite increased vascular cross-section area, pulmonary pressures increase progressively with gestational age. In ovine lungs, the pressure increases from around 30 mmHg at 0.4 gestation to about 50 mmHg at term [9]. This is due to the fact that pulmonary vessels become more responsive with advancing gestation, to both constrictors and dilators, with constrictor activity being predominant [9–11, 19]. The consensus in the literature is that the predominance of constrictor influences is primarily mediated by the low oxygen tension in the fetal environment [20] and is also in part due to the developmental characteristics of the fetal pulmonary vascular smooth muscle *per se*, such as the intrinsic activities of myosin light chain kinase and myosin light chain phosphatase (MLCP) [21], Rho kinase [22], cGMP and cAMP pathways [23], and potassium channels [24, 25]. The endothelium exerts a critical role in the regulation of fetal pulmonary vasomotor tone, predominantly through various agents it releases, in particular NO, prostacyclin, and ETs. While the mechanisms underlying the role of the endothelium-derived vasoactive agents in regulating the fetal pulmonary vasoactivity are not well understood, it is generally recognized that the balance between these agents is toward high pulmonary vasocontractility [6, 9, 12, 13, 23].

ENDOTHELIUM-DERIVED VASOCONSTRICTORS IN THE FETAL PULMONARY CIRCULATION

ET

ET is a family of bicyclic 21-amino-acid peptides composed of three isoforms, ET-1, -2, and -3. ET-1 is the major isoform that has prominent vasoactive properties. ET-1 is predominantly produced in ECs. It is not stored in secretory granules within cells and hence can be secreted within minutes after an appropriate stimulus. ET-1 is synthesized first as a 203-amino-acid prepro-ET-1 peptide, which is cleaved to a 38-amino-acid peptide (big ET-1) and to ET-1 by furin convertase and endothelin-converting enzyme (ECE)-1, respectively. The actions of ET-1 are mediated via two receptor types, ET_A and ET_B receptors. The binding of ET-1 to ET_A and ET_B receptors in SMCs causes vasoconstriction. ET-1 may also cause vasodilation through endothelium-derived nitric oxide (EDNO) and prostacyclin by activation of endothelial ET_B receptors. ET_B receptors also mediate the pulmonary clearance of circulating ET-1 and the reuptake of ET-1 by ECs [26–28].

In human lungs, enzyme-linked immunosorbent assay analysis showed an increase in ET-1 expression during midterm gestation, but a decrease in infants. ET_A expression is strong throughout gestation and remains stable after birth. In contrast, ET_B receptor expression is weak in the canalicular stage of lung development, increases markedly during the saccular and alveolar stages, and remains stable after birth [29]. In fetal ovine lungs ET protein content was highest at around 0.9 gestation, but decreases before birth in the fetal lamb lung. ET_A receptor mRNA expression and ET_B receptor mRNA increases from 0.6 to 0.9 gestation [30]. In fetal lambs of late gestation, intrapulmonary infusion of ET_A receptor agonists causes pulmonary vasoconstriction or vasodilation, while ET_A blockers cause moderate change in pulmonary vascular tone [8, 30–36]. Acute blockade of the ET_B receptor has no effect on basal fetal ovine pulmonary vascular tone at any gestational age, indicating that the ET_B receptor does not exert a significant role in resting vascular tone [8, 37]. However, prolonged ET_A receptor blockade decreases pulmonary artery pressure, right ventricular hypertrophy, and distal muscularization of small pulmonary arteries of the ovine fetus [36]. Chronic treatment with an ET_B receptor blocker in fetal lambs increases pulmonary arterial pressure and pulmonary vascular resistance. The animals show greater right ventricular hypertrophy, muscularization of small pulmonary arteries, and elevated lung ET-1 levels [37]. The mechanisms underlying the effects of chronic blockade of ET_A or ET_A receptor are not clear. It appears at least in part to result from their effects on proliferation of pulmonary SMCs [38].

Platelet Activating Factor

Ibe *et al.* demonstrated, by using specific platelet-activating factor (PAF) receptor antagonists infused into fetal lambs *in vivo*, that PAF contributes significantly to maintenance of high tone in the pulmonary circulation *in utero* [39]. Consistent with the observed high pulmonary vasomotor tone, they found very high circulating levels of PAF in the fetus [39–41]. Hypoxia increases PAF synthesis so that in the hypoxic environment of fetal lungs, there is more PAF available for binding to its receptor in pulmonary vessels. Ibe *et al.* also reported that PAF receptor (platelet-activating factor receptor, PAF-R) gene mRNA expression as well as PAF-R density is high in fetal lungs [42]. Hypoxia significantly upregulates PAF-R binding and PAF-R protein level [43]. PAF is inactivated by acetylhydrolase (platelet-activating factor acetylhydrolase, PAF-Ah) [44, 45]. In fetal lamb lungs, activity of PAF-Ah is significantly attenuated by hypoxia [41, 43, 46] suggesting that enzymatic degradation of PAF in fetal pulmonary vasculature is low, resulting in a high level of PAF in fetal lungs.

Therefore, the combination of high PAF synthesis and low PAF catabolism by PAF-Ah should mean that a high PAF level will be available for binding to the receptor. These factors all account for this unique role for PAF as an important endogenous mediator of increased tone in pulmonary vasculature of the fetus. PAF is synthesized by platelets, vascular SMCs, and endothelium in response to endogenous and exogenous stimuli. In fetal lambs, Ibe *et al.* show that the amount of PAF synthesized by pulmonary arterial SMCs is at least 400-fold more than that by the ECs, calculated based on equal number of cells. Considering the total number of SMCs in the lung is much greater than that of the ECs, endothelium-derived PAF is likely to be minimal. Moreover, the production of PAF is augmented by hypoxia in SMCs, but not in ECs (Figure 23.1) [41]. Therefore, PAF from pulmonary vascular SMCs and sources other than ECs may play a more important role in maintaining the high vessel tone of fetal lungs.

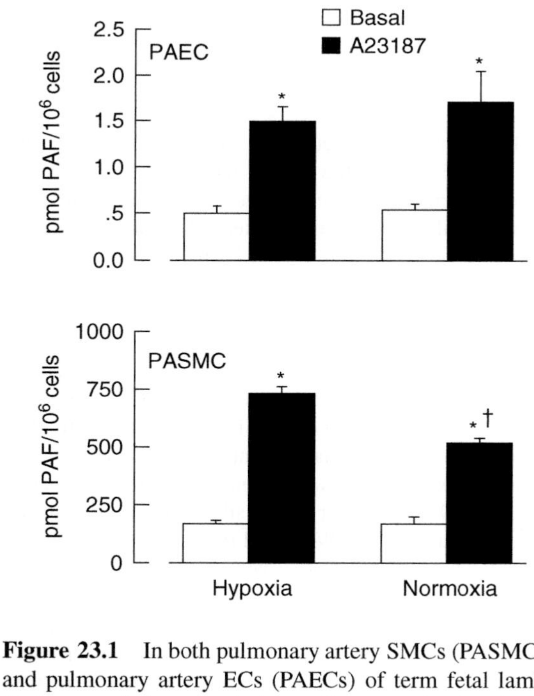

Figure 23.1 In both pulmonary artery SMCs (PASMCs) and pulmonary artery ECs (PAECs) of term fetal lambs PAF synthesis increased when cells were stimulated with the calcium ionophore A23187 (10^{-6} M). Hypoxia augmented PAF synthesis in pulmonary artery SMCs, but not in pulmonary artery ECs. Data are means $\pm$ standard error, $n = 4$ different cell preparations: *, significantly different from basal; †, significantly different from normoxia plus A23187 cells ($p < 0.05$). Modified from [41], with permission.

SUPPRESSED ENDOTHELIUM-DEPENDENT VASODILATOR ACTIVITY IN FETAL PULMONARY CIRCULATION

EDNO

EDNO is a major endogenous agent in regulating both basal and stimuli-induced vasodilation of the pulmonary vasculature (see Chapter 6). EDNO is synthesized by endothelial NO synthase (endothelial nitric oxide synthase-eNOS) by the conversion of L-arginine to L-citrulline. Three distinct isoforms of NOS have been identified: neuronal NO synthase (neuronal nitric oxide synthase-eNOS), inducible NO synthase (inducible nitric oxide synthaseeNOS), and eNOS. They are all present in the lung. In the pulmonary vasculature, eNOS is expressed primarily in ECs, while nNOS is expressed in the neurons of blood vessel walls. The expression of iNOS is induced by cytokines and is present predominantly in SMCs [47–49].

In the human fetus, immunohistochemical labeling shows that eNOS immunoreactivity appears in the cells of the 14-day fetal lung and increases as gestation proceeds. These cells coalesce to form an endothelial layer of pulmonary vessels [50]. Immunohistochemical study of the lungs shows that eNOS is strong expressed in the canalicular as well as in the saccular stages, with comparable intensity. This falls sharply in the alveolar stage of development and further decreases after birth [29]. In the baboon lung from 125 to 140 days of gestation (term = 175 days) a marked increase was noticed in total NOS activity and in the expression of eNOS and nNOS, whereas iNOS expression and activity were minimal. From 140 day of gestation to term, total NOS activity remains constant, eNOS and nNOS fall dramatically, but iNOS rises sharply [51]. In sheep lungs, eNOS mRNA expression increases from low levels at 70 days of gestation to peak at 113 days and remains high for the rest of fetal life (term = 150 days). Lung eNOS protein expression and activity in the fetus rises and peaks at 118 days of gestation but decreases before birth [52]. In the rat, both mRNA and protein of eNOS are detectable in 16-day fetal lung and increases to maximal levels at 20 days of gestation (term = 22 days). In contrast, nNOS protein increases while its mRNA abundance declined during late fetal life. These findings suggest that the regulation of pulmonary eNOS may primarily involve alterations in transcription or mRNA stability, whereas nNOS expression in the late gestation also involves post-transcriptional modifications [53].

In near-term and term fetal lambs, infusion of nitro-L-arginine decreases pulmonary artery blood flow and increases pulmonary pressure. Pulmonary vasodilation to the endothelium-dependent stimulus acetylcholine is also attenuated by nitro-L-arginine [54]. These results suggest that EDNO is functional and may exert an inhibitory effect on fetal pulmonary vasoactivity. Since oxygen is the necessary substrate for NO synthesis and fetal pulmonary ECs are in a low oxygen environment (17–20 mmHg), the role of EDNO in opposing pulmonary vasoconstriction is likely to be limited in the fetus. In a study in near-term lambs, fetal arterial oxygen tension (p_aO_2) was raised from 25 to 55 torr by making the pregnant ewes breath hyperbaric 100% oxygen at 3 atm pressure and the proportion of right ventricular output distributed to the fetal lungs increased from 8 to 59% [55]. However, when in very immature lambs (94–101 days of gestation), fetal p_aO_2 was increased from 27 to 174 torr, there was no change in the proportion of right ventricular output distributed to the lungs, indicating that EDNO is not functional at this early stage of gestation. Several studies also demonstrate that heterogeneity exists along the pulmonary vascular tree. In term fetal pigs, N^{ω}-nitro-L-arginine methyl ester, an inhibitor of eNOS, attenuated relaxation to the endothelium-dependent vasodilator bradykinin by 50% in conduit pulmonary arteries, but almost completely abolished relaxation in resistant pulmonary arteries [56].

Although many studies have demonstrated that EDNO is predominantly produced by eNOS, the other NOS isoforms may be also involved in the regulation of fetal pulmonary vasoreactivity. In chronically prepared fetal lambs (around 128 days of gestation), selective inhibition of nNOS with 7-nitroindazole resulted in increased basal pulmonary vascular resistance by 37%. Western blot analysis detected nNOS protein in the fetal lung and large pulmonary vessels. Since nNOS is detected in intact and endothelium-denuded vessels, the enzyme may be present in the medial or adventitial layer [7]. The iNOS isoform is constitutively expressed predominately in airway epithelium and vascular smooth muscle in the late-gestation ovine fetal lung. Intrapulmonary infusions of selective iNOS antagonists (aminoguanidine and *S*-ethylisothiourea) increase basal pulmonary vascular resistance in late-gestation fetal lambs and attenuate shear stress-induced pulmonary vasodilation caused by acute compression of the ductus arteriosus, whereas nonselective blockade with nitro-L-arginine completely blocked this response [57–59].

NO induces pulmonary vasodilation mainly by increasing the intracellular level of cGMP resulting from activation of soluble guanylyl cyclase (sGC). sGC is a heterodimer consisting of α and β subunits, and the predominant sGC isoform in vascular system is $\alpha_1\beta_1$ [60]. Reaction of NO with the heme moiety of sGC induces a conformational change leading to a several hundred-fold increase in production of cGMP from GTP [61]. In near-term fetal lambs, sGC immunostaining is more pronounced in small pulmonary arteries than in large ones; in veins, however, sGC immunostaining is more pronounced in large than in small vessels [62]. Stimulation of sGC causes marked pulmonary vasodilation of near-term fetal lambs [63]. Abundant mRNA and protein of α_1 and β_1 subunit of sGC have also been found in lungs of late-gestation fetal and neonatal rats, with markedly reduced levels detected in adult lungs. Pulmonary sGC activity stimulated with sodium nitroprusside is approximately sevenfold greater in 1- and 8-day-old rats than in adult rats [64]. The effects of cGMP are mediated through activation of cGMP-dependent protein kinase (PKG), nucleotide-gated ion channels, and cGMP-regulated phosphodiesterases (PDEs). In perinatal ovine pulmonary vessels, our studies show that cGMP-mediated relaxation is largely mediated by PKG [65–67]. cGMP-mediated relaxation of ovine pulmonary arteries is less in fetal than in newborn and adult sheep [68]. However, PKG protein expression and activity are developmentally downregulated (Gao and Raj, unpublished observations). The underling mechanism is not clear.

cGMP is degraded by PDEs. Among the 11 PDE subtypes that have been identified, the type 5 PDE (PDE5) specifically hydrolyzes cGMP and is found to be abundant in lung tissues [69]. RNA blot hybridization shows that PDE5 mRNA is detectable in fetal lung tissue as early as 18.5 days of the 22-day term gestation fetal rat and reaches maximal levels in neonatal rat lungs. mRNA levels in adult rat lungs are markedly less than the those levels measured in lungs of newborn rats [70]. In term fetal lambs, inhibition of PDE5 with E4021 causes significant relaxation of intrapulmonary arteries. The effect is blocked by inhibition of NOS [71]. When treated with sildenafil, a specific PDE5 inhibitor, pulmonary vascular resistance of term fetal lambs is lower during maternal O_2 inhalation than that of control lambs. Furthermore, the drop in pulmonary vascular resistance during acute ductus arteriosus compression, which causes "shear stress" and induces eNOS, is greater in the sildenafil group than in the control lambs [72]. We found that rates of hydrolysis of cGMP in ovine pulmonary vessels are greater in fetal than in newborn lambs and that rates of hydrolysis of cGMP are greater in pulmonary arteries than in veins. A higher PDE5 activity may contribute to the greater contractility of fetal pulmonary vessels, particularly in the arteries [73].

Prostanoids

The prostaglandins PGI_2 (prostacyclin) and PGE_2 are potent dilator prostanoids of pulmonary vessels of the fetus and newborn. They are produced mainly from the endothelium, with the production of PGI_2 being dominant [74–77]. PG are synthesized from arachidonic acid (AA) released from cell membrane following the activation of phospholipase A_2 by calcium. Released AA is converted by cyclooxygenases (COXs) to 15-OH-prostaglandin-9,11-endoperoxide (prostaglandin H_2), which is further converted to PGs and thromboxanes (Txs) by their respective synthases. COXs are the rate-limiting enzymes for the production of prostanoids. The enzymes are present as two types, the constitutive and the inducible, termed COX-1 and COX-2, respectively. Although the expression of COX-2 is induced by inflammatory factors, studies suggest that it seems to be present also under normal or developmental conditions [78].

Vasodilator effects of PGs are very low in fetal lungs. In isolated resistance pulmonary arteries and veins of term fetal lambs, the COX inhibitor, indomethacin, has no effect on arteries pre-equilibrated at a low pO_2 (around 21 mmHg) but induces contraction in arteries exposed to an intermediate (~40 mmHg) or high (~70 mmHg) pO_2 [79]. In intrapulmonary arteries from 110 day ovine fetuses to 4-week-old newborn lambs, basal and stimulated PGI_2 and PGE_2 rises in an age-dependent manner, accompanied with increased expression of COX-1 mRNA and protein, indicating that the synthesis of dilator prostanoids is developmentally regulated, in part by the changes in COX-1 expression [76, 80].

TRANSITION OF THE PULMONARY CIRCULATION AT BIRTH: A SHIFT IN THE BALANCE OF VASODILATORS AND VASOCONSTRICTORS FROM THE ENDOTHELIUM

At birth, with the onset of breathing, pulmonary vascular resistance decreases greater than 10-fold and pulmonary blood flow increases 8- to 10-fold. The decrease in pulmonary vascular resistance consists of two components, a rapid decrease during the first 30 s of ventilation and a slower decline through the first 2 h. Thereafter, the decrease in pulmonary artery pressure occurs more gradually, and it is reduced to approximately 50% of mean systemic arterial pressure by 24 h after birth and reaches adult values by 2–6 weeks. These dramatic hemodynamic changes at birth are critical for the lung to replace the placenta as the organ of gas exchange. Among the various mechanisms which contribute to the postnatal adaptation of pulmonary circulation, endothelium-derived vasoactive agents, in particular EDNO, may play a vital role for a successful transition [10, 81–83].

IMPORTANCE OF THE EDNO–cGMP PATHWAY IN THE TRANSITIONAL PULMONARY CIRCULATION

In the late-gestation ovine fetus, infusion of the inhibitor of NOS not only increases pulmonary arterial pressure and decreases pulmonary blood flow induced by acetylcholine but also reduces the rise in pulmonary blood flow at birth [54, 84]. In near-term fetal lambs, the increase in pulmonary blood flow with an increase in pO_2 in the pulmonary arterial blood using hyperbaric oxygen given to the ewe is entirely due to an increase in EDNO as the increase was reversed by inhibition of NOS [85], indicating that vasodilation of the fetal pulmonary circulation at birth is most likely to be mediated by oxygen-induced EDNO. Oxygenation may acutely increase EDNO production by acting as an essential substrate for eNOS, and oxygen may also increase eNOS activity and thus NO production by upregulating eNOS expression through transcriptional and post-transcriptional mechanisms [86, 87]. These mechanisms may be involved in the sustained reduction in pulmonary vascular resistance. Besides oxygenation, the increase in fluid shear stress at birth, resulting from increased pulmonary blood flow, also increases NO production by phosphorylation of eNOS and upregulation of eNOS expression [88, 89]. One study showed that shear stress-induced activation of signaling pathways leading to the phosphorylation of c-Jun, Akt, and eNOS occurs in pulmonary arterial ECs from fetal lambs but not from adult sheep, indicating that the ECs within the fetal pulmonary circulation are primed to respond immediately to shear stress [89]. In addition to eNOS, studies suggest that the other isoforms of NOSs, nNOS, and iNOS, may also contribute to birth-related increase in pulmonary blood flow [7, 58, 59, 90].

The downstream enzymes for EDNO action, sGCs and PKG, are also upregulated after birth. A maturational increases in sGC protein levels has been found to be associated with the augmented pulmonary vasodilation in the newborn rat and piglet [91, 92]. In pulmonary arteries of newborn (3–18 h of age) and 2-week-old piglets, the expression of sGC β_1 subunit increases with postnatal age, both at mRNA and protein levels, which correlates with increased vasorelaxant responses to NO and to sGC activator YC-1 [93]. Our studies show that relaxation of pulmonary arteries and veins of term fetal lambs to 8-Br-cGMP, a cell-permeable cGMP analog, are greater after exposure for 4 h to normoxia (pO_2, 140 mmHg) compared to hypoxia (pO_2, 30 mmHg). The decreased relaxation of pulmonary veins to cGMP in hypoxia may result from reduced expression of PKG protein and mRNA as well as post-transcriptional modification of PKG by peroxynitrite. However, the suppressed response to cGMP of pulmonary arteries in hypoxia seems largely due to a PKG independent mechanism (Figure 23.2) [67, 94]. cGMP is predominantly inactivated by PDE5, the cGMP-specific PDE. In ovine and mouse lungs, PDE5 activity, protein, and mRNA levels decrease within 1 h following birth. The decreased PDE5 activity would lead to higher cGMP levels and greater pulmonary vasodilation during postnatal transition [95]. In pig pulmonary arteries, however, it was found that PDE5 may not be responsible for the maturational increase in NO-mediated response during the first days of extrauterine life [91].

Increased Production of Dilator PGs

PGI_2 synthesis in ovine pulmonary artery is greater in newborn than in fetal vessels. The increase occurs both in ECs and vascular SMCs, and is caused by upregulation of COX-1 [96]. In fetal goats and lambs, ventilation-induced pulmonary vasodilation is associated with increased production of PGI_2. These changes are prevented by the inhibition of COX with indomethacin, suggesting that PGI_2 is involved in the decrease in pulmonary vascular resistance at birth [75, 97]. The decrease in pulmonary vascular resistance caused by PGI_2 at birth is modest in comparison to that induced by EDNO and PGI_2-induced relaxation may be in part mediated by EDNO [6, 98].

PGs are thought to induce vasodilation by elevating the intracellular level of cAMP through adenylyl cyclase. In ovine lungs, direct activation of adenylyl cyclase by forskolin induced greater relaxation of pulmonary veins of newborn compared to those of fetal lambs, whereas the cAMP analog induced similar responses in all vessel groups. Furthermore, the stimulated activity of adenylyl cyclase is greater in newborn than in fetal veins [99]. When comparing the relative sensitivities of

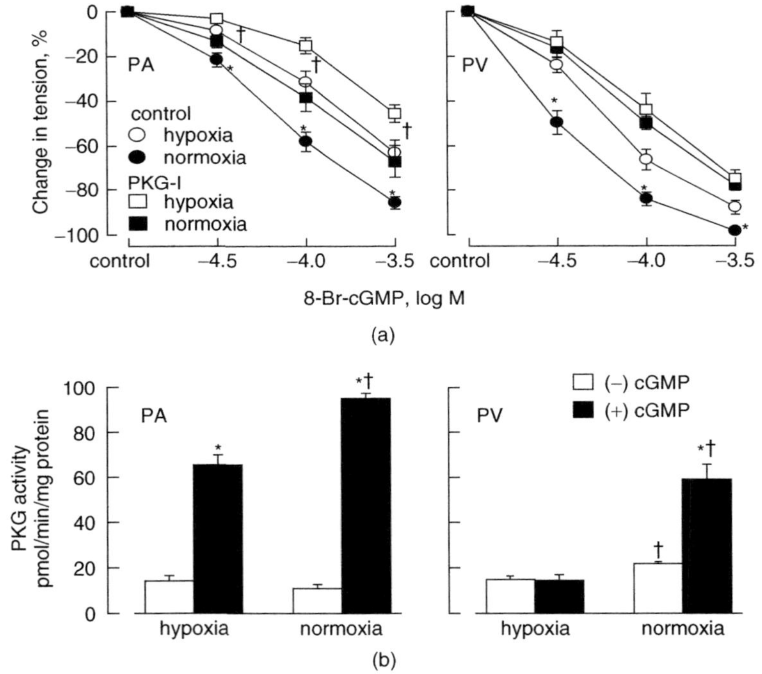

Figure 23.2 (a) Relaxations of pulmonary arteries (PA) and veins (PV) of fetal lambs to 8-Br-cGMP after a 4-h incubation under hypoxic and normoxic conditions (pO_2: 30 and 140 mmHg, respectively). Vessels were preconstricted to a similar tension with endothelin-1. Data are shown as means $\pm$ SE; $n = 6$ for each group: *, significant difference between vessels incubated under hypoxic and normoxic conditions (no inhibitor added); †, significant difference between vessels incubated under hypoxia and normoxia in the presence of Rp-8-Br-PET-cGMPS (PKG-I; 3×10^{-5} M) ($p < 0.05$). (b) PKG activity in pulmonary vessels after a 4-hour incubation under hypoxia and normoxia (pO_2: 30 and 140 mmHg, respectively). (–) cGMP, without cGMP; (+) cGMP, in the presence of cGMP at 5×10^{-6} M. Data are shown as means $\pm$ standard error; $n = 4$ for each group: *, significantly different from those without cGMP; †, significantly different from those incubated under hypoxia ($p < 0.05$). Reproduced from [94], used with permission of The American Physiological Society.

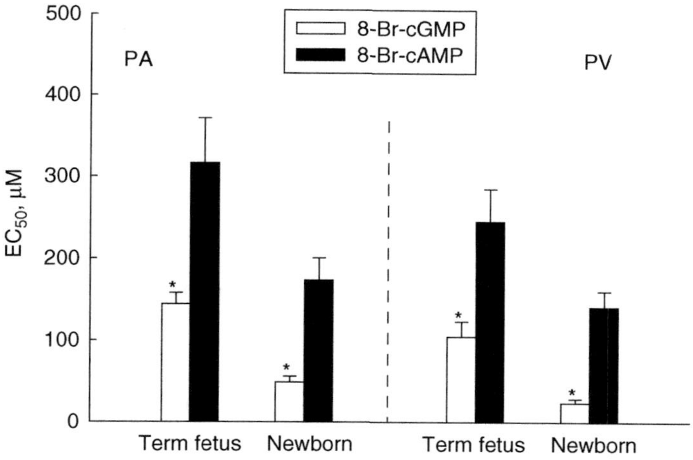

Figure 23.3 The concentrations of 8-Br-cGMP and 8-Br-cAMP that elicit 50% of maximum vasodilation (EC_{50}) of ovine pulmonary arteries (PA) and veins (PV) of the term fetus and newborns (6–13 days). Active tension of vessels was first raised to a similar level by ET-1 (3×10^{-9} to 6×10^{-9} M). Data are shown as means ± standard error; $n = 5$ for each group: *, significant different from those treated with 8-Br-cAMP ($p < 0.05$).

intrapulmonary arteries and veins of near-term term fetal and newborn lambs to cAMP and cGMP, we found that pulmonary arteries and veins of term fetal and newborn lambs were much more sensitive to the cGMP analog than to the cAMP analog (Figure 23.3). Therefore, it appears that the NO–cGMP–PKG pathway may play a more important role in regulating the relaxation responses of pulmonary arteries and veins in the perinatal period than the agents which act through cAMP pathway.

Increased ET_B Receptor-Mediated Vasodilation and Decreased ET_A Receptor-Mediated Vasoconstriction

Plasma ET-1 levels are high in the fetus and at birth, and decrease gradually [100–103]. In near-term fetal lambs, blockade of ET_B receptors does not affect pulmonary vascular resistance at baseline, but attenuates the reduction in pulmonary vascular resistance induced by ventilation and oxygenation, indicating that ET_B receptor stimulation contributes to pulmonary vasodilation at birth [104]. The effect of ET_B receptor seems to be mediated, in part, by EDNO and by ATP-sensitive potassium channels [105, 106].

In contrast to that of ET_B receptors, activation of ET_A receptors causes contraction of pulmonary vessels. In rabbit pulmonary resistance arteries and pig pulmonary veins, studies show a postnatal decrease in ET-1 induced a contractile response and an increase in ET-1 dilator response in pulmonary veins [107, 108]. Studies also show that in some species including humans, the threshold for ET-1-induced contraction is lower in pulmonary veins than in pulmonary arteries [79, 109, 110].

Decreased Vasoconstrictor Action of PAF

In term fetal lambs infusion of the PAF-R antagonist caused marked reduction of pulmonary vascular resistance, indicating that PAF contributes significantly to maintenance of high tone in the pulmonary circulation *in utero* [39]. The plasma levels of PAF are high in the fetus and fell dramatically after birth with the onset of oxygenation. PAF is inactivated by acetylhydrolase, its activity in lamb lungs are upregulated in the immediate newborn period, thereby facilitating the fall in postnatal PAF levels [40, 46, 111]. PAF-R gene mRNA expression as well as PAF-R density decrease within 2 h after birth, returning to a new intermediate level after a few weeks of life [42, 43, 112]. These findings suggest that PAF may play an important role in the postnatal change of pulmonary circulation.

PULMONARY ENDOTHELIUM IN PPHN

PPHN is a clinical syndrome due to the failure of pulmonary vascular resistance to decrease at birth so that venous blood is shunted away from the lungs through the ductus arteriosus and/or foramen ovale. This markedly diminishes oxygenation of venous blood and leads to severe systemic arterial hypoxemia. PPHN affects between one and two infants per 1000 live births. Despite the advances of treatment, the mortality is still about 10–20%. Infants who survive PPHN face increased risks for serious and long-term sequelae such as chronic lung disease, seizures, and neurodevelopmental problems [13, 113, 114].

Mechanisms for PPHN are not well understood and likely to be multifactorial, which include elevated vascular reactivity, smooth muscle remodeling, and impaired angiogenesis [13, 113, 114]. Among these possible mechanisms, EC dysfunction, in particular decreased production of EDNO and increased production of ET-1, may play a key role in the development of PPHN [13, 36, 115, 116]. It should be pointed out that, in addition to the altered endothelial function, the responsiveness of the pulmonary vascular smooth muscle also changes favoring myocyte contraction, which constitutes the fundamental impairments of PPHN [13, 117].

Decreased eNOS expression, reduced release of EDNO, and impaired pulmonary vasodilation have been found in PPHN in human infants and several animal models [13, 115, 118, 119]. In addition to the decreased eNOS expression, the decreased release of EDNO may also be due to scavenging of endogenous NO by increased production of superoxide, which can be from NADPH oxidase in pulmonary arteries [120, 121], xanthine oxidase [122], and uncoupled eNOS [121]. Superoxide reacts readily with NO to form peroxynitrite. A recent study show that, in PPHN ovine model, peroxynitrite may decrease the interactions between heat shock protein 90 (HSP90) and eNOS through tyrosine nitration of HSP90, which leads to uncoupling of eNOS, further increase in superoxide production, and decrease in NO production [123].

The production of EDNO can be inhibited by an endogenous inhibitor of eNOS, asymmetric dimethylarginine (ADMA; [124]). ADMA levels are high in fetal blood and gradually decline to undetectable after birth [125]. ADMA levels are elevated in patients with PPHN, which may contribute to the reduced EDNO production [125, 126]. ADMA is metabolized to citrulline by dimethyl-arginine dimethylaminohydrolase (DDAH). In the newborn pig with pulmonary hypertension, a study suggests that the suppressed DDAH type 2 may be accountable for the elevated ADMA levels [127].

In addition to its role in vasodilation, EDNO is also critical for vascular development of the fetal and postnatal lung. In eNOS-deficient mice pulmonary arterial muscularity was greater [128], which contributes to hypertensive remodeling [115]. Exposure to mild hypoxia in the neonatal period led to a failure of capillary and alveolar growth in $eNOS^{-/-}$ mice that was not seen in normal mice, suggesting that EDNO preserves normal distal lung growth during hypoxic stress, perhaps through preservation of vascular endothelial growth factor receptor-1/2 signaling [129].

In PPHN, the downstream enzymes of EDNO signaling may also be impaired, including decreased activity of sGC [130] and increased PDE5 activity [95, 131]. In fetal lambs exposed to chronic high altitude hypoxia, the medial wall thickness of pulmonary arteries is significantly increased. In these animals, PKG-dependent relaxation of pulmonary arteries was attenuated and associated with decreased PKG-specific activity [132, 133]. These studies would suggest that augmentation of the impaired NO pathway may be of therapeutic benefit in PPHN. The management of PPHN includes NO inhalation therapy, specific PDE5 inhibitors, and sGC stimulators and activators [13].

There is a paucity of studies on the role of PGI_2 in PPHN. In pulmonary arteries of newborn pigs, prolonged hypoxia reduces the production of PGI_2 but does not affect that of TxA_2. This leads to an increased ratio of TxA_2 to PGI_2 [134]. TxA_2 is a potent constrictor prostanoid. Neonatal pulmonary arterial myocytes show increased sensitivity and reactivity to the Tx agonist after prolonged hypoxia. It has been found that hypoxia increases the affinity of Tx receptors for TxA_2, which may be due to reduced phosphorylation of Tx receptors by cAMP-dependent protein kinase (PKA), possibly due to decreased dilator PGs. In addition, hypoxia also induces the expression of Tx receptors in neonatal resistance pulmonary arteries, which would also contribute to the increased pulmonary arterial pressure observed in PPHN [135, 136].

cAMP is believed to be the primary mediator for relaxation induced by vasodilator PGs. In newborn lambs the elevation of cAMP as well as relaxation of pulmonary arteries to PGE_2 are markedly potentiated by EDNO, resulting from the inhibition of degradation of cAMP by activating the cGMP-inhibitory phosphodiesterase, PDE3 [137]. Based on this, it may be that the combination use of vasodilator prostanoids with EDNO or PDE3 inhibitors may be a useful alternative in the treatment of PPHN. Indeed, intravenous milrinone, a PDE3 inhibitor, significantly shortened the onset, prolonged the duration and degree of pulmonary vasodilation produced by PGI_2 in newborn lambs with pulmonary hypertension [138].

In patients with PPHN, plasma levels of ET-1 are elevated and have been shown to correlate with disease severity and to decline with clinical improvement [139–141]. In fetal lambs, prolonged ET_A receptor blockade attenuates chronic pulmonary hypertension [36] while prolonged ET_B receptor blockade causes pulmonary hypertension [37]. A genetic rat model of ET_B receptor deficiency, after 3 weeks of severe hypoxia, develops exaggerated pulmonary hypertension characterized by elevated pulmonary arterial pressure, diminished cardiac output, and increased total pulmonary resistance. Although mRNA for prepro-ET-1 in the lungs is not different from the control rat, mRNA for ECE-1 of the lungs and plasma ET-1 level are greater than in controls [142]. These findings suggest that the activation of ET_A receptor promotes PPHN, while activation of ET_B receptors protects against PPHN. Also, in rats of ET_B receptor deficiency, the elevated ET-1 levels may in part result from increased activity of ECE-1 and decreased clearance [143]. Although it is believed that the ET_A receptor is the principal subtype for ET-1-induced pulmonary vasoconstriction, ET_B receptors may also play a significant role in mediating ET-1-induced constriction of intrapulmonary conduit and resistance arteries [144]. The vasoconstrictor actions of the ET_B receptor may become more pronounced in the pathologic setting of pulmonary hypertension [145]. This would explain the findings that dual blockade is necessary to maximize the inhibition of ET-1-induced pulmonary vasoconstriction in humans [143, 144].

The elevated ET-1 level in PPHN may also result from decreased production of EDNO and vasodilator PGs, as these dilators can negatively regulate ET-1 production by inhibition of prepro-ET-1 transcription [140–143]. The elevated ET-1 level may in turn further decrease the production of EDNO through downregulation of the expression of eNOS, and thus may further potentiate ET-1-mediated pulmonary vasoconstriction and smooth muscle proliferation [116, 143].

Activation of RhoA–Rho kinase (ROCK) is a key mechanism for contraction induced by ET-1. Following activation, ROCK phosphorylates myosin phosphatase target subunit 1 (MYPT1), the regulatory subunit of MLCP, which leads to decreased activity of MLCP, increased Ca^{2+} sensitivity of the contractile filaments, and augmented contractility [146]. In pulmonary hypertension, ROCK activity has been found to be preferentially augmented. Moreover, ROCK inhibitors have been shown to be effective in preventing and reversing pulmonary hypertension in animal models and in treatment of pulmonary hypertension in humans [147, 148]. In newborn ovine pulmonary arteries, chronic hypoxia augments both the expression and activity of ROCK. The increased ROCK activity causes greater phosphorylation of MYPT1 at Thr696 and Thr850, and counteracts PKG action on MYPT1, leading to reduced relaxation of pulmonary arteries in response to cGMP [133].

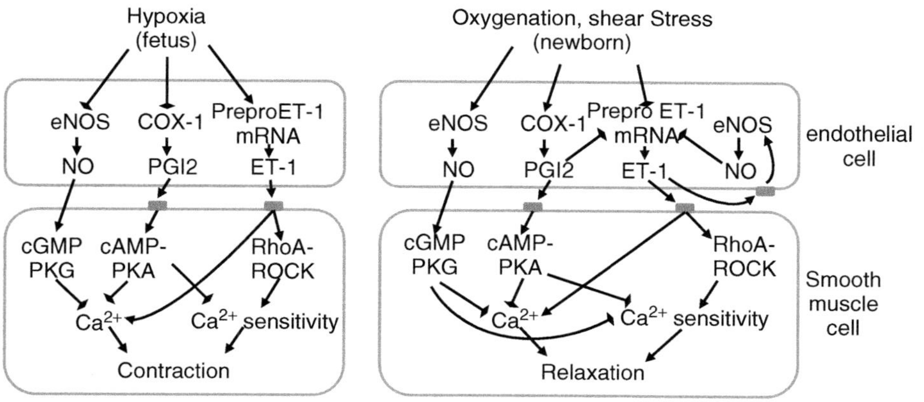

Figure 23.4 Possible mechanisms involved in endothelial modulation of perinatal pulmonary vasoactivity. NO synthesized by eNOS, PGI_2 synthesized through constitutive COX-1, and ET-1 are the major endothelium-derived vasoactive agents involved in modulating the vasoactivity of perinatal lungs. NO and PGI_2 act via cGMP–PKG and cAMP–PKA pathways, respectively. Activation of these pathways leads to reduced intracellular Ca^{2+} level, and decreased sensitivity of myofilaments to Ca^{2+} and thus vasodilation. ET-1 promotes vasoconstriction by elevating intracellular Ca^{2+} levels and increasing sensitivity of myofilaments to Ca^{2+} via activation of the RhoA–ROCK pathway. In the fetus (a), the hypoxic environment that the pulmonary vasculature encounters suppresses the activities of eNOS and COX-1, but augments the synthesis of propro-ET-1 mRNA and thereby renders the vessels more constricted. After birth (b), oxygenation and increased shear stress are potent stimulators of eNOS and COX-1 synthesis which promote vasodilation. Oxygenation and shear stress are also potent inhibitors of ET-1 synthesis. The increased production of NO and PGI_2 may exert an inhibitory effect on ET-1 production, while ET-1 acts through ET_B receptors on ECs to further stimulate the production of NO. All these actions contribute to the dramatic reduction of pulmonary vascular resistance that occurs after birth. PKA, cAMP-dependent protein kinase. The normal arrowheads represent stimulatory, while the blunted arrowheads represent inhibitory actions.

CONCLUSIONS AND PERSPECTIVES

The regulation of perinatal pulmonary circulation is an integrated activity. Among the multiple and complex interactions, the endothelium plays a pivotal a role in modulating the activity of the underlying SMCs, primarily by releasing NO, PGI_2, and ET-1 (Figure 23.4). In perinatal lungs, NO and PGI_2 exert their actions predominantly via cGMP–PKG and cAMP–PKA pathways, respectively, which lead to vasodilation by decreasing the intracellular Ca^{2+} level and decreasing the sensitivity of myofilaments toward Ca^{2+}. ET-1 promotes vasoconstriction through elevating intracellular Ca^{2+} level and increasing sensitivity of myofilaments to Ca^{2+}. The Ca^{2+} sensitization is mainly through the inhibition of MLCP by RhoA–ROCK signaling, while Ca^{2+} desensitization is mainly through the activation of MLCP by PKG and PKA. In fetal lungs, the production and actions of NO and PGI_2 are suppressed, while those of ET-1 are augmented due to the hypoxic environment, promoting contraction of the fetal pulmonary vasculature. The roles of endothelium-derived NO and PGI_2 in the fetal pulmonary circulation are relatively well understood, but the role of ET-1 is still not fully elucidated. Its vasoconstrictive role may result from "priming" of the SMCs to be more contractile and proliferative [38, 149]. At birth, oxygenation and increased shear stress are two major factors that stimulate the synthesis of NO and PGI_2. NO production can also be stimulated by ET-1 through ET_B receptors on ECs. The synthesis of ET-1 is inhibited not only by oxygenation and increased shear stress, but also by NO and PGI_2. All these mechanisms contribute to the marked reduction in pulmonary vascular resistance that is vital for postnatal adaptation of the lungs to a gas-exchanging organ occurred after birth (Figure 23.4). It should be pointed out that many other nonendothelial factors are also importantly involved in the regulation of perinatal vasoactivity. For example, PAF derived from pulmonary SMCs and other sources may contribute importantly to hypoxic vasoconstriction in the fetal lungs. Also, the intrinsic myogenic characteristics of vascular smooth muscle and RhoA–ROCK pathway may be important factors of the high vasoconstrictivity of fetal lungs. Finally, there may be differential roles for microvascular versus macrovascular endothelium in the regulation of the pulmonary circulation, as ECs in these two regions have some differential characteristics.

References

1. Hooper, S.B. (1998) Role of luminal volume changes in the increase in pulmonary blood flow at birth in sheep. *Experimental Physiology*, **83**, 833–42.
2. Vilos, G.A. and Liggins, G.C. (1982) Intrathoracic pressures in fetal sheep, *Journal of Developmental Physiology*, **4**, 247–256.
3. Iwamoto, H.S., Teitel, D.F., and Rudolph, A.M. (1993) Effects of lung distension and spontaneous fetal breathing on hemodynamics in sheep. *Pediatric Research*, **33**, 639–44.
4. Teitel, D.F., Iwamoto, H.S., and Rudolph, A.M. (1990) Changes in the pulmonary circulation during birth-related events. *Pediatric Research*, **27**, 372–78.
5. Cassin, S. (1993) The role of eicosanoids and endothelium-dependent factors in regulation of the fetal pulmonary circulation. *Journal of Lipid Mediators*, **6**, 477–85.
6. Heymann, M.A. (1999) Control of the pulmonary circulation in the fetus and during the transitional period to air breathing. *Obstetrics and Gynecology*, **84**, 127–32.
7. Rairigh, R.L., Storme, L., Parker, T.A. *et al.* (2000) Role of neuronal nitric oxide synthase in regulation of vascular and ductus arteriosus tone in the ovine fetus. *American Journal of Physiology: Lung Cellular and Molecular Physiology*, **278**, L105–10.
8. Wong, J., Fineman, J.R., and Heymann, M.A. (1994) The role of endothelin and endothelin receptor subtypes in regulation of fetal pulmonary vascular tone. *Pediatric Research*, **35**, 664–70.
9. Rudolph, A.M. (1979) Fetal and neonatal pulmonary circulation. *Annual Review of Physiology*, **41**, 383–95.
10. Fineman, J.R., Soifer, S.J., and Heymann, M.A. (1995) Regulation of pulmonary vascular tone in the perinatal period. *Annual Review of Physiology*, **57**, 115–34.
11. Kiserud, T. (2005) Physiology of the fetal circulation. *Seminars in Fetal and Neonatal Medicine*, **10**, 493–503.
12. Wojciak-Stothard, B. and Haworth, S.G. (2006) Perinatal changes in pulmonary vascular endothelial function. *Pharmacology and Therapeutics*, **109**, 78–91.
13. Abman, S.H. (2007) Recent advances in the pathogenesis and treatment of persistent pulmonary hypertension of the newborn. *Neonatology*, **91**, 283–90.
14. Hall, S.M., Hislop, A.A., Pierce, C.M., and Haworth, S.G. (2000) Prenatal origins of human intrapulmonary arteries: formation and smooth muscle maturation. *American Journal of Respiratory Cell and Molecular Biology*, **23**, 194–203.

15. Hall, S.M., Hislop, A.A., Pierce, C.M., and Haworth, S.G. (2002) Origin, differentiation, and maturation of human pulmonary veins. *American Journal of Respiratory Cell and Molecular Biology*, **26**, 333–40.
16. Haworth, S.G. and Hislop, A.A. (2003) Lung development – the effects of chronic hypoxia. *Seminars in Neonatology*, **8**, 1–8.
17. Levin, D.L., Rudolph, A.M., Heymann, M.A., and Phibbe, R.H. (1976) Morphological development of the pulmonary vascular bed in fetal lambs. *Circulation*, **53**, 144–51.
18. Rasanen, J., Wood, D.C., Weiner, S. *et al.* (1996) Role of the pulmonary circulation in the distribution of human fetal cardiac output during the second half of pregnancy. *Circulation*, **94**, 1068–73.
19. Lewis, A.B., Heymann, M.A., and Rudolph, A.M. (1976) Gestational changes in pulmonary vascular responses in fetal lambs *in utero*. *Circulation Research*, **39**, 536–41.
20. Storme, L., Rairigh, R.L., Parker, T.A. *et al.* (1999) *In vivo* evidence for a myogenic response in the fetal pulmonary circulation. *Pediatric Research*, **45**, 425–31.
21. Belik, J., Kerc, E., and Pato, M.D. (2006) Rat pulmonary arterial smooth muscle myosin light chain kinase and phosphatase activities decrease with age. *American Journal of Physiology: Lung Cellular and Molecular Physiology*, **290**, L509–16.
22. Parker, T.A., Roe, G., Grover, T.R., and Abman, S.H. (2006) Rho kinase activation maintains high pulmonary vascular resistance in the ovine fetal lung. *American Journal of Physiology: Lung Cellular and Molecular Physiology*, **291**, L976–82.
23. Gao, Y. and Raj, J.U. (2005) Role of veins in regulation of the pulmonary circulation. *American Journal of Physiology: Lung Cellular and Molecular Physiology*, **288**, L213–26.
24. Cornfield, D.N., Saqueton, C.B., Porter, V.A. *et al.* (2000) Voltage-gated K^+-channel activity in ovine pulmonary vasculature is developmentally regulated. *American Journal of Physiology: Lung Cellular and Molecular Physiology*, **278**, L1297–304.
25. Rhodes, M.T., Porter, V.A., Saqueton, C.B. *et al.* (2001) Pulmonary vascular response to normoxia and K_{Ca} channel activity is developmentally regulated. *American Journal of Physiology: Lung Cellular and Molecular Physiology*, **280**, L1250–57.
26. Yanagisawa, M., Kurihara, H., Kimura, S. *et al.* (1988) A novel potent vasoconstrictor peptide produced by vascular endothelial cells. *Nature*, **332**, 411–15.
27. Masuda, Y., Miyazaki, H., Kondoh, M. *et al.* (2004) The endothelin system in pulmonary arterial hypertension. *Cardiovascular Research*, **61**, 227–37.
28. Schneider, M.P., Boesen, E.I., and Pollock, D.M. (2007) Contrasting actions of endothelin ET_A and ET_B receptors in cardiovascular disease. *Annual Review of Pharmacology and Toxicology*, **47**, 731–59.
29. Levy, M., Maurey, C., Chailley-Heu, B. *et al.* (2005) Developmental changes in endothelial vasoactive and angiogenic growth factors in the human perinatal lung. *Pediatric Research*, **57**, 248–53.
30. Ivy, D.D., le Cras, T.D., Parker, T.A. *et al.* (2000) Developmental changes in endothelin expression and activity in the ovine fetal lung. *American Journal of Physiology: Lung Cellular and Molecular Physiology*, **278**, L785–93.
31. Cassin, S., Kristova, V., Davis, T. *et al.* (1991) Tone-dependent responses to endothelin in the isolated perfused fetal sheep pulmonary circulation *in situ*. *Journal of Applied Physiology*, **70**, 1228–34.
32. Chatfield, B.A., McMurtry, I.F., Hall, S.L., and Abman, S.H. (1991) Hemodynamic effects of endothelin-1 on ovine fetal pulmonary circulation. *American Journal of Physiology: Renal Physiology*, **261**, R182–87.
33. Tod, M.L. and Cassin, S. (1992) Endothelin-1-induced pulmonary arterial dilation is reduced by *N* omega-nitro-*L*-arginine in fetal lambs. *Journal of Applied Physiology*, **72**, 1730–34.
34. Ivy, D.D., Kinsella, J.P., and Abman, S.H. (1994) Physiologic characterization of endothelin A and B receptor activity in the ovine fetal pulmonary circulation. *The Journal of Clinical Investigation*, **93**, 2141–48.
35. Ivy, D.D., Kinsella, J.P., and Abman, S.H. (1996) Endothelin blockade augments pulmonary vasodilation in the ovine fetus. *Journal of Applied Physiology*, **81**, 2481–87.
36. Ivy, D.D., Parker, T.A., Ziegler, J.W. *et al.* (1997) Prolonged endothelin A receptor blockade attenuates chronic pulmonary hypertension in the ovine fetus. *The Journal of Clinical Investigation*, **99**, 1179–86.
37. Ivy, D.D., Parker, T.A., and Abman, S.H. (2000) Prolonged endothelin B receptor blockade causes pulmonary hypertension in the ovine fetus. *American Journal of Physiology: Lung Cellular and Molecular Physiology*, **279**, L758–65.
38. Barman, S.A. and Marrero, M.B. (2005) Mechanism of endothelin-1 activation of MAP kinases

in neonatal pulmonary vascular smooth muscle. *Lung*, **183**, 425–39.

39. Ibe, B.O., Hibler, S., and Raj, J.U. (1998) Platelet activating factor modulates pulmonary vasomotor tone in the perinatal lamb. *Journal of Applied Physiology*, **85**, 1079–85.
40. Ibe, B.O., Sander, F.C., and Raj, J.U. (2000) Platelet activating factor acetylhydrolase activity in lamb lungs is up-regulated in the immediate newborn period. *Molecular Genetics and Metabolism*, **69**, 46–55.
41. Ibe, B.O., Pham, H.H., Kääpä, P., and Raj, J.U. (2001) Maturational changes in ovine pulmonary metabolism of platelet-activating factor: implications for postnatal adaptation. *Molecular Genetics and Metabolism*, **74**, 385–95.
42. Ibe, B.O., Sander, F.C., and Raj, J.U. (2000) Platelet-activating factor receptors in lamb lungs are down-regulated immediately after birth. *American Journal of Physiology: Heart and Circulatory Physiology*, **278**, H1168–76.
43. Ibe, B.O., Portugal, A.M., and Raj, J.U. (2002) Metabolism of platelet activating factor by intrapulmonary vascular smooth muscle cells. Effect of oxygen on phospholipase A_2 protein expression and activities of acetyl-CoA acetyltransferase and cholinephosphotransferase. *Molecular Genetics and Metabolism*, **77**, 237–48.
44. Snyder, F. (1995) Platelet-activating factor: the biosynthetic and catabolic enzymes. *The Biochemical Journal*, **305**, 689–705.
45. Montrucchio, G., Alloatti, G., and Camussi, G. (2000) Role of platelet-activating factor in cardiovascular pathophysiology. *Physiological Reviews*, **80**, 1669–99.
46. Salva, A.M.L., Ibe, B.O., Cliborn, E. *et al.* (1996) Hypoxia attenuates metabolism of platelet activating factor by fetal and newborn lamb lungs. *Journal of Lipid Research*, **37**, 783–89.
47. Barnes, P.J. and Liu, S.F. (1995) Regulation of pulmonary vascular tone. *Pharmacological Reviews*, **47**, 87–131.
48. Forstermann, U. and Kleinert, H. (1995) Nitric oxide synthase: expression and expressional control of the three isoforms. *Naunyn-Schmiedeberg's Archives of Pharmacology*, **352**, 351–64.
49. Coggins, M.P. and Bloch, K.D. (2007) Nitric oxide in the pulmonary vasculature. *Arteriosclerosis, Thrombosis, and Vascular Biology*, **27**, 1877–85.
50. Xue, C., Reynolds, P.R., and Johns, R.A. (1996) Developmental expression of NOS isoforms in fetal rat lung: implications for transitional circulation and pulmonary angiogenesis. *American Journal of Physiology: Lung Cellular and Molecular Physiology*, **270**, L88–100.
51. Shaul, P.W., Afshar, S., Gibson, L.L. *et al.* (2002) Developmental changes in nitric oxide synthase isoform expression and nitric oxide production in fetal baboon lung. *American Journal of Physiology: Lung Cellular and Molecular Physiology*, **283**, L1192–99.
52. Parker, T.A., le Cras, T.D., Kinsella, J.P., and Abman, S.H. (2000) Developmental changes in endothelial nitric oxide synthase expression and activity in ovine fetal lung. *American Journal of Physiology: Lung Cellular and Molecular Physiology*, **278**, L202–8.
53. North, A.J., Star, R.A., Brannon, T.S. *et al.* (1994) Nitric oxide synthase type I and type III gene expression are developmentally regulated in rat lung. *American Journal of Physiology: Lung Cellular and Molecular Physiology*, **266**, L635–41.
54. Abman, S.H., Chatfield, B.A., Hall, S.L., and McMurtry, I.F. (1990) Role of endothelium-derived relaxing factor during transition of pulmonary circulation at birth. *American Journal of Physiology: Heart and Circulatory Physiology*, **259**, H1921–27.
55. Morin, F.C., Egan, E.A., Ferguson, W., and Lundgren, C.E.G. 3rd. (1988) development of pulmonary vascular response to oxygen. *American Journal of Physiology: Heart and Circulatory Physiology*, **254**, H542–46.
56. Boels, P.J., Deutsch, J., Gao, B., and Haworth, S.G. (2001) Perinatal development influences mechanisms of bradykinin-induced relaxations in pulmonary resistance and conduit arteries differently. *Cardiovascular Research*, **51**, 140–50.
57. Sherman, T.S., Chen, Z., Yuhanna, I.S. *et al.* (1999) Nitric oxide synthase isoform expression in the developing lung epithelium. *American Journal of Physiology: Lung Cellular and Molecular Physiology*, **276**, L383–90.
58. Rairigh, R.L., Le Cras, T.D., Ivy, D.D. *et al.* (1998) Role of inducible nitric oxide synthase in regulation of pulmonary vascular tone in the late-gestation ovine fetus. *The Journal of Clinical Investigation*, **101**, 15–21.
59. Rairigh, R.L., Storme, L., Parker, T.A. *et al.* (1999) Inducible NO synthase inhibition attenuates shear stress-induced pulmonary vasodilation in the ovine fetus. *American Journal of Physiology: Lung Cellular and Molecular Physiology*, **276**, L513–21.
60. Friebe, A. and Koesling, D. (2003) Regulation of nitric oxide-sensitive guanylyl cyclase. *Circulation Research*, **93**, 96–105.

61. Ballou, D.P., Zhao, Y., Brandish, P.E., and Marletta, M.A. (2002) Revisiting the kinetics of nitric oxide (NO) binding to soluble guanylate cyclase: the simple NO-binding model is incorrect. *Proceedings of the National Academy of Sciences of the United States of America*, **99**, 12097–101.
62. D'Angelis, C.A., Nickerson, P.A., Steinhorn, R.H., and Morin, F.C. III (1998) Heterogeneous distribution of soluble guanylate cyclase in the pulmonary vasculature of the fetal lamb. *The Anatomical Record*, **250**, 62–69.
63. Deruelle, P., Grover, T.R., and Abman, S.H. (2005) Pulmonary vascular effects of nitric oxide-cGMP augmentation in a model of chronic pulmonary hypertension in fetal and neonatal sheep. *American Journal of Physiology: Lung Cellular and Molecular Physiology*, **289**, L798–806.
64. Bloch, K.D., Filippov, G., Sanchez, L.S. *et al.* (1997) Pulmonary soluble guanylate cyclase, a nitric oxide receptor, is increased during the perinatal period. *The American Journal of Physiolog: Lung Cellular and Molecular Physiology y*, **272**, L400–6.
65. Gao, Y., Dhanakoti, S., Tolsa, J.-F., and Raj, J.U. (1999) Role of protein kinase G in nitric oxide and cGMP-induced relaxation of newborn ovine pulmonary veins. *Journal of Applied Physiology*, **87**, 993–98.
66. Dhanakoti, S., Gao, Y., Nguyen, M.Q., and Raj, J.U. (2000) Involvement of cGMP-dependent protein kinase in the relaxation of ovine pulmonary arteries to cGMP and cAMP. *Journal of Applied Physiology*, **88**, 1637–42.
67. Negash, S., Gao, Y., Zhou, W. *et al.* (2007) Regulation of cGMP-dependent protein kinase (PKG1) mediated vasodilation by hypoxia-induced reactive species in ovine fetal pulmonary veins. *American Journal of Physiology: Lung Cellular and Molecular Physiology*, **293**, L1012–20.
68. Kolber, K.A., Gao, Y., and Raj, J.U. (2000) Maturational changes in endothelium-derived nitric oxide-mediated relaxation of ovine pulmonary arteries. *Biology of the Neonate*, **77**, 123–30.
69. Corbin, J.D., Beasley, A., Blount, M.A., and Francis, S.H. (2005) High lung PDE5: a strong basis for treating pulmonary hypertension with PDE5 inhibitors. *Biochemical and Biophysical Research Communications*, **334**, 930–38.
70. Sanchez, L.S., de la Monte, S.M., Filippov, G. *et al.* (1998) Cyclic-GMP-binding, cyclic-GMP-specific phosphodiesterase (PDE5) gene expression is regulated during rat pulmonary development. *Pediatric Research*, **43**, 163–68.
71. Dukarm, R.C., Russell, J.A., Morin, F.C. III *et al.* (1999) The cGMP-specific phosphodiesterase inhibitor E4021 dilates the pulmonary circulation. *American Journal of Respiratory and Critical Care Medicine*, **160**, 858–65.
72. Jaillard, S., Larrue, B., Deruelle, P. *et al.* (2006) Effects of phosphodiesterase 5 inhibitor on pulmonary vascular reactivity in the fetal lamb. *The Annals of Thoracic Surgery*, **81**, 935–42.
73. Okogbule-Wonodi, A.C., Ibe, B.O., Yue, B.W. *et al.* (1998) Phosphodiesterase activity in intrapulmonary arteries and veins of perinatal lambs. *Molecular Genetics and Metabolism*, **65**, 229–37.
74. Cassin, S., Tyler, T., Leffler, C., and Wallis, R. (1979) Pulmonary and systemic vascular responses of perinatal goats to prostaglandins E_1 and E_2. *American Journal of Physiology: Heart and Circulatory Physiology*, **236**, H828–32.
75. Leffler, C.W., Hessler, J.R., and Terragno, N.A. (1980) Ventilation-induced release of prostaglandin like material from fetal lungs. *American Journal of Physiology: Heart and Circulatory Physiology*, **238**, H282–86.
76. Brannon, T.S., North, A.J., Wells, L.B., and Shaul, P.W. (1994) Prostacyclin synthesis in ovine pulmonary artery is developmentally regulated by changes in cyclooxygenase-1 gene expression. *The Journal of Clinical Investigation*, **93**, 2230–35.
77. Gao, Y., Zhou, H., Ibe, B.O., and Raj, J.U. (1996) Prostaglandins E_2 and I_2 cause greater relaxations in pulmonary veins than in arteries of newborn lambs. *Journal of Applied Physiology*, **81**, 2534–39.
78. Fitzpatrick, F.A. and Soberman, R. (2001) Regulated formation of eicosanoids. *The Journal of Clinical Investigation*, **107**, 1347–51.
79. Wang, Y. and Coceani, F. (1992) Isolated pulmonary resistance vessels from fetal lambs. Contractile behavior and responses to indomethacin and endothelin-1. *Circulation Research*, **71**, 320–30.
80. Brannon, T.S., MacRitchie, A.N., Jaramillo, M.A. *et al.* (1998) Ontogeny of cyclooxygenase-1 and cyclooxygenase-2 gene expression in ovine lung. *American Journal of Physiology: Lung Cellular and Molecular Physiology*, **274**, L66–71.
81. Leffler, C.W., Tyler, T.L., and Cassin, S. (1978) Effect of indomethacin on pulmonary vascular response to ventilation of fetal goats. *American Journal of Physiology: Heart and Circulatory Physiology*, **234**, H346–51.
82. Iwamoto, H.S., Teitel, D., and Rudolph, A.M. (1987) Effects of birth-related events on blood flow distribution. *Pediatric Research*, **22**, 634–40.

83. Fineman, J.R., Wong, J., Morin, F.C. III *et al.* (1994) Chronic nitric oxide inhibition *in utero* produces persistent pulmonary hypertension in newborn lambs. *The Journal of Clinical Investigation*, **93**, 2675–83.
84. Moore, P., Velvis, H., Fineman, J.R. *et al.* (1992) EDRF inhibition attenuates the increase in pulmonary blood flow due to oxygen ventilation in fetal lambs. *Journal of Applied Physiology*, **73**, 2151–57.
85. Tiktinsky, M.H. and Morin, F.C. III (1993) Increasing oxygen tension dilates fetal pulmonary circulation via endothelium-derived relaxing factor. *American Journal of Physiology: Heart and Circulatory Physiology*, **265**, H376–80.
86. Liao, J.K., Zulueta, J.J., Yu, F. *et al.* (1995) Regulation of bovine endothelial constitutive nitric oxide synthase by oxygen. *The Journal of Clinical Investigation*, **96**, 2661–66.
87. North, A.J., Lau, K.S., Brannon, T.S. *et al.* (1996) Oxygen upregulates nitric oxide synthase gene expression in ovine fetal pulmonary artery endothelial cells. *American Journal of Physiology: Lung Cellular and Molecular Physiology*, **270**, L643–49.
88. Fleming, I. and Busse, R. (2003) Molecular mechanisms involved in the regulation of the endothelial nitric oxide synthase. *American Journal of Physiology: Renal Physiology*, **284**, R1–12.
89. Wedgwood, S., Mitchell, C.J., Fineman, J.R., and Black, S.M. (2003) Developmental differences in the shear stress-induced expression of endothelial NO synthase: changing role of AP-1. *American Journal of Physiology: Lung Cellular and Molecular Physiology*, **284**, L650–62.
90. Rairigh, R.L., Parker, T.A., Ivy, D.D. *et al.* (2001) Role of inducible nitric oxide synthase in the pulmonary vascular response to birth-related stimuli in the ovine fetus. *Circulation Research*, **88**, 721–26.
91. Moreno, L., Losada, B., Cogolludo, A. *et al.* (2004) Postnatal maturation of phosphodiesterase 5 (PDE5) in piglet pulmonary arteries: activity, expression, effects of PDE5 inhibitors, and role of the nitric oxide/cyclic GMP pathway. *Pediatric Research*, **56**, 563–70.
92. Chicoine, L.G., Paffett, M.L., Girton, M.R. *et al.* (2007) Maturational changes in the regulation of pulmonary vascular tone by nitric oxide in neonatal rats. *American Journal of Physiology: Lung Cellular and Molecular Physiology*, **293**, L1261–70.
93. Moreno, L., Gonzalez–Luis, G., Cogolludo, A. *et al.* (2005) Soluble guanylyl cyclase during postnatal porcine pulmonary maturation. *American Journal of Physiology: Lung Cellular and Molecular Physiology*, **288**, L125–30.
94. Gao, Y., Dhanakoti, S., Trevino, E.M. *et al.* (2003) Effect of oxygen on cyclic GMP-dependent protein kinase-mediated relaxation in ovine fetal pulmonary arteries and veins. *American Journal of Physiology: Lung Cellular and Molecular Physiology*, **285**, L611–18.
95. Hanson, K.A., Burns, F., Rybalkin, S.D. *et al.* (1998) Developmental changes in lung cGMP phosphodiesterase-5 activity, protein, and message. *American Journal of Respiratory and Critical Care Medicine*, **158**, 279–88.
96. Shaul, P.W., Pace, M.C., Chen, Z., and Brannon, T.S. (1999) Developmental changes in prostacyclin synthesis are conserved in cultured pulmonary endothelium and vascular smooth muscle. *American Journal of Respiratory Cell and Molecular Biology*, **20**, 113–21.
97. Leffler, C.W. and Hessler, J.R. (1981) Perinatal pulmonary prostaglandin production. *American Journal of Physiology: Heart and Circulatory Physiology*, **241**, H756–59.
98. Zenge, J.P., Rairigh, R.L., Grover, T.R. *et al.* (2001) NO and prostaglandin interactions during hemodynamic stress in the fetal ovine pulmonary circulation. *American Journal of Physiology: Lung Cellular and Molecular Physiology*, **281**, L1157–63.
99. Gao, Y., Tolsa, J-F., Botello, M., and Raj, J.U. (1998) Developmental change in isoproterenol-mediated relaxation of pulmonary veins of fetal and newborn lambs. *Journal of Applied Physiology*, **84**, 1535–39.
100. Nakamura, T., Kasai, K., Konuma, S. *et al.* (1990) Immunoreactive endothelin concentrations in maternal and fetal blood. *Life Sciences*, **46**, 1045–50.
101. Malamitsi–Puchner, A., Economou, E., Sevastiadou, S. *et al.* (1993) Endothelin 1-21 plasma levels on the first and fourth postpartum day in normal full-term neonates. *Developmental Pharmacology and Therapeutics*, **20**, 195–98.
102. MacDonald, P.D., Paton, R.D., and Logan, R.W. (1994) Neonatal endothelin-1 concentrations in term infants. *Archives of Disease in Childhood*, **70**, F223–24.
103. Levy, M., Tulloh, R.M., Komai, H. *et al.* (1995) Maturation of the contractile response and its endothelial modulation in newborn porcine intrapulmonary arteries. *Pediatric Research*, **38**, 25–29.

104. Ivy, D.D., Lee, D.S., Rairigh, R.L. *et al.* (2004) Endothelin B receptor blockade attenuates pulmonary vasodilation in oxygen-ventilated fetal lambs. *Biology of the Neonate*, **86**, 155–59.

105. Pinheiro, J.M. and Malik, A.B. (1993) Mechanisms of endothelin-1-induced pulmonary vasodilation in neonatal pigs. *Journal of Physiology*, **469**, 739–52.

106. Wong, J., Vanderford, P.A., Fineman, J.R. *et al.* (1993) Endothelin-1 produces pulmonary vasodilation in the intact newborn lamb. *American Journal of Physiology: Heart and Circulatory Physiology*, **265**, H1318–25.

107. Docherty, C. and MacLean, M.R. (1998) Development of endothelin receptors in perinatal rabbit pulmonary resistance arteries. *British Journal of Pharmacology*, **124**, 1165–74.

108. Schindler, M.B., Hislop, A.A., and Haworth, S.G. (2007) Postnatal changes in pulmonary vein responses to endothelin-1 in the normal and chronically hypoxic lung. *American Journal of Physiology: Lung Cellular and Molecular Physiology*, **292**, L1273–79.

109. Brink, C., Gillard, V., Roubert, P. *et al.* (1991) Effects and specific binding sites of endothelin in human lung preparations. *Pulmonary Pharmacology*, **4**, 54–59.

110. Toga, H., Ibe, B.O., and Raj, J.U. (1992) *In vitro* responses of ovine intrapulmonary arteries and veins to endothelin-1. *American Journal of Physiology: Lung Cellular and Molecular Physiology*, **263**, L15–21.

111. Ibe, B.O., Pham, H.H., Kaapa, P., and Raj, J.U. (2001) Maturational changes in ovine pulmonary metabolism of platelet-activating factor: implications for postnatal adaptation. *Molecular Genetics and Metabolism*, **74**, 385–95.

112. Ibe, B.O., Portugal, A.M., Chaturvedi, S., and Raj, J.U. (2005) Oxygen-dependent PAF receptor binding and intracellular signaling in ovine fetal pulmonary vascular smooth muscle. *American Journal of Physiology: Lung Cellular and Molecular Physiology*, **288**, L879–86.

113. Geggel, R.L. and Reid, L.M. (1984) The structural basis of PPHN. *Clinics in Perinatology*, **11**, 525–49.

114. Dakshinamurti, S. (2005) Pathophysiologic mechanisms of persistent pulmonary hypertension of the newborn. *Pediatric Pulmonology*, **39**, 492–503.

115. Shaul, P.W., Yuhanna, I.S., German, Z. *et al.* (1997) Pulmonary endothelial NO synthase gene expression is decreased in fetal lambs with pulmonary hypertension. *American Journal of Physiology: Lung Cellular and Molecular Physiology*, **16**, L1005–12.

116. Wedgwood, S. and Black, S.M. (2005) Endothelin-1 decreases endothelial NOS expression and activity through ET_A receptor- mediated generation of hydrogen peroxide. *American Journal of Physiology: Lung Cellular and Molecular Physiology*, **288**, L480–87.

117. Cool, C.D., Groshong, S.D., Oakey, J., and Voelkel, N.F. (2005) Pulmonary hypertension: cellular and molecular mechanisms. *Chest*, **128**, 565S–571S.

118. Villanueva, M.E., Zaher, F.M., Svinarich, D.M., and Konduri, G.G. (1998) Decreased gene expression of endothelial nitric oxide synthase in newborns with persistent pulmonary hypertension. *Pediatric Research*, **44**, 338–43.

119. Shehata, S.M., Sharma, H.S., Mooi, W.J., and Tibboel, D. (2006) Pulmonary hypertension in human newborns with congenital diaphragmatic hernia is associated with decreased vascular expression of nitric-oxide synthase. *Cell Biochemistry and Biophysics*, **44**, 147–55.

120. Brennan, L.A., Steinhorn, R.H., Wedgwood, S. *et al.* (2003) Increased superoxide generation is associated with pulmonary hypertension in fetal lambs: a role for NADPH oxidase. *Circulation Research*, **92**, 683–91.

121. Grobe, A.C., Wells, S.M., Benavidez, E. *et al.* (2006) Increased oxidative stress in lambs with increased pulmonary blood flow and pulmonary hypertension: role of NADPH oxidase and endothelial NO synthase. *American Journal of Physiology: Lung Cellular and Molecular Physiology*, **290**, L1069–77.

122. Jankov, R.P., Kantores, C., Pan, J., and Belik, J. (2008) Contribution of xanthine oxidase-derived superoxide to chronic hypoxic pulmonary hypertension in neonatal rats. *American Journal of Physiology: Lung Cellular and Molecular Physiology*, **294**, L233–45.

123. Konduri, G.G., Bakhutashvili, I., Eis, A., and Pritchard, K. Jr. (2007) Oxidant stress from uncoupled nitric oxide synthase impairs vasodilation in fetal lambs with persistent pulmonary hypertension. *American Journal of Physiology: Heart and Circulatory Physiology*, **292**, H1812–20.

124. Vallance, P. and Leiper, J. (2004) Cardiovascular biology of the asymmetric dimethylarginine: dimethylarginine dimethylaminohydrolase pathway. *Arteriosclerosis, Thrombosis, and Vascular Biology*, **24**, 1023–30.

125. Pierce, C.M., Krywawych, S., and Petros, A.J. (2004) Asymmetric dimethyl arginine and symmetric dimethyl arginine levels in infants with persistent pulmonary hypertension of the newborn. *Pediatric Critical Care Medicine*, **5**, 517–20.
126. Pierce, C.M., Krywawych, S., and Petros, A.J. (2005) Elevated levels of asymmetric di-methylarginine in neonates with congenital diaphragmatic hernia. *European Journal of Pediatrics*, **164**, 248–49.
127. Arrigoni, F.I., Vallance, P., Haworth, S.G., and Leiper, J.M. (2003) Metabolism of asymmetric dimethylarginines is regulated in the lung developmentally and with pulmonary hypertension induced by hypobaric hypoxia. *Circulation*, **107**, 1195–201.
128. Miller, A.A., Hislop, A.A., Vallance, P.J., and Haworth, S.G. (2005) Deletion of the eNOS gene has a greater impact on the pulmonary circulation of male than female mice. *American Journal of Physiology: Lung Cellular and Molecular Physiology*, **289**, L299–306.
129. Balasubramaniam, V., Tang, J.R., Maxey, A. *et al.* (2003) Mild hypoxia impairs alveolarization in the endothelial nitric oxide synthase-deficient mouse. *American Journal of Physiology: Lung Cellular and Molecular Physiology*, **284**, L964–71.
130. Tzao, C., Nickerson, P.A., Russell, J.A. *et al.* (2001) Pulmonary hypertension alters soluble guanylate cyclase activity and expression in pulmonary arteries isolated from fetal lambs. *Pediatric Pulmonology*, **31**, 97–105.
131. Hanson, K.A., Ziegler, J.W., Rybalkin, S.D. *et al.* (1998) Chronic pulmonary hypertension increases fetal lung cGMP phosphodiesterase activity. *American Journal of Physiology: Lung Cellular and Molecular Physiology*, **275**, L931–41.
132. Bixby, C.E., Ibe, B.O., Abdallah, M.F. *et al.* (2007) Role of platelet-activating factor in pulmonary vascular remodeling associated with chronic high altitude hypoxia in ovine fetal lambs. *American Journal of Physiology: Lung Cellular and Molecular Physiology*, **293**, L1475–82.
133. Gao, Y., Portugal, A.D., Negash, S. *et al.* (2007) Role of Rho kinases in PKG-mediated relaxation of pulmonary arteries of fetal lambs exposed to chronic high altitude hypoxia. *American Journal of Physiology: Lung Cellular and Molecular Physiology*, **292**, L678–84.
134. Fike, C.D., Kaplowitz, M.R., Zhang, Y., and Pfister, S.L. (2005) Cyclooxygenase-2 and an early stage of chronic hypoxia-induced pulmonary hypertension in newborn pigs. *Journal of Applied Physiology*, **98**, 1111–18.
135. Hinton, M., Mellow, L., Halayko, A.J. *et al.* (2006) Hypoxia induces hypersensitivity and hyperreactivity to thromboxane receptor agonist in neonatal pulmonary arterial myocytes. *American Journal of Physiology: Lung Cellular and Molecular Physiology*, **290**, L375–84.
136. Hinton, M., Gutsol, A., and Dakshinamurti, S. (2007) Thromboxane hypersensitivity in hypoxic pulmonary artery myocytes: altered TP receptor localization and kinetics. *American Journal of Physiology: Lung Cellular and Molecular Physiology*, **292**, L654–63.
137. Gao, Y., Tolsa, J.-F., Shen, H., and Raj, J.U. (1998) Effect of selective phosphodiesterase inhibitors on the responses of ovine pulmonary veins to prostaglandin E_2. *Journal of Applied Physiology*, **84**, 13–18.
138. Rashid, N., Morin, F.C. III, Swartz, D.D. *et al.* (2006) Effects of prostacyclin and milrinone on pulmonary hemodynamics in newborn lambs with persistent pulmonary hypertension induced by ductal ligation. *Pediatric Research*, **60**, 624–29.
139. Rosenberg, A.A., Kennaugh, J., Koppenhafer, S.L. *et al.* (1993) Elevated immunoreactive endothelin-1 levels in newborn infants with persistent pulmonary hypertension. *The Journal of Pediatrics*, **123**, 109–14.
140. Christou, H., Adatia, I., Van Marter, L.J. *et al.* (1997) Effect of inhaled nitric oxide on endothelin-1 and cyclic guanosine 5′-monophosphate plasma concentrations in newborn infants with persistent pulmonary hypertension. *The Journal of Pediatrics*, **130**, 603–11.
141. Endo, A., Ayusawa, M., Minato, M. *et al.* (2001) Endogenous nitric oxide and endothelin-1 in persistent pulmonary hypertension of the newborn. *European Journal of Pediatrics*, **160**, 217–22.
142. Ivy, D.D., Yanagisawa, M., Gariepy, C.E. *et al.* (2002) Exaggerated hypoxic pulmonary hypertension in endothelin B receptor-deficient rats. *American Journal of Physiology: Lung Cellular and Molecular Physiology*, **282**, L703–12.
143. Channick, R.N., Sitbon, O., Barst, R.J. *et al.* (2004) Endothelin receptor antagonists in pulmonary arterial hypertension. *Journal of the American College of Cardiology*, **43**, 62S–67S.
144. McCulloch, K.M. and MacLean, M.R. (1995) Endothelin B receptor-mediated contraction of human and rat pulmonary resistance arteries and the effect of pulmonary hypertension on endothelin responses in the rat. *Journal of Cardiovascular Pharmacology*, **26**, S169–76.
145. Dupuis, J., Jasmin, J.F., Prie, S., and Cernacek, P. (2000) Importance of local production of

endothelin-1 and of the ET_B receptor in the regulation of pulmonary vascular tone. *Pulmonary Pharmacology and Therapeutics*, **13**, 135–40.

146. Somlyo, A.P. and Somlyo, A.V. (2003) Ca^{2+} sensitivity of smooth muscle and nonmuscle myosin II: modulated by G proteins, kinases, and myosin phosphatase. *Physiological Reviews*, **83**, 1325–58.
147. Fukumoto, Y., Matoba, T., Ito, A. *et al.* (2005) Acute vasodilator effects of a Rho-kinase inhibitor, fasudil, in patients with severe pulmonary hypertension. *Heart*, **91**, 391–92.
148. Shimokawa, H. and Takeshita, A. (2005) Rho-kinase is an important therapeutic target in cardiovascular medicine. *Arteriosclerosis, Thrombosis, and Vascular Biology*, **25**, 1767–75.
149. Shimoda, L.A., Sham, J.S., Liu, Q., and Sylvester, J.T. (2002) Acute and chronic hypoxic pulmonary vasoconstriction: a central role for endothelin-1? *Respiratory Physiology and Neurobiology*, **132**, 93–106.

24

Genetic Insights into Endothelial Barrier Regulation in the Acutely Inflamed Lung

Sumegha Mitra, Daniel Turner Lloveras, Shwu-Fan Ma and Joe G. N. Garcia

Section of Pulmonary and Critical Care Medicine, Department of Medicine, Pritzker School of Medicine, University of Chicago, Chicago, IL, USA

INTRODUCTION

Acute lung injury (ALI) is a devastating syndrome of diffuse alveolar damage that develops via a variety of local and systemic insults such as sepsis, trauma, pneumonia, and aspiration [1]. Deranged alveolar capillary permeability, profound inflammation, and extravasation of edema fluid into the alveolar spaces are critical elements of ALI, reflecting the substantial surface area of the pulmonary vasculature needed for alveolar gas exchange. ALI, together with its most severe form, acute respiratory distress syndrome (ARDS), accounts for approximately 190 000 cases per year in the United States of America. However, with a mortality rate of 35–50%, only a subset of individuals exposed to potential ALI-inciting insults develop the disorder and the severity of the disease varies from complete resolution to death [2, 3]. In addition, ALI susceptibility and severity are also affected by ethnicity [4] as evident by the higher mortality rate in African-American ALI patients than those belonging to other ethnic groups in the United States [4]. Moreover, marked differences in strain-specific ALI responses to inflammatory and injurious agents are observed in preclinical animal models [5]. Together, these observations strongly indicate genetic components to be involved in the pathogenesis of ALI.

The role that genetics plays in determining ALI risk or subsequent severity of outcome is one of the many unanswered questions regarding ALI pathogenesis and epidemiology. While understanding of the pathogenesis of ALI continues to evolve, the identification of genes contributing to ALI would potentially provide a better understanding of ALI pathobiology and yield novel biomarkers that identify individuals or populations at risk/severity proving useful for the development of novel and individualized therapies. However, genetic studies of ALI are challenging due to the tremendous phenotypic variance in critically ill patients, diversity in the lung injury evoking stimuli, presence of varied comorbid illnesses, complex gene–environment interactions, potentially incomplete gene penetrance, and a high likelihood for locus heterogeneity [6, 7]. Moreover, the traditional genetic studies using family linkage mapping are not feasible given the sporadic nature of ALI and the necessity of an extreme environmental insult. Nevertheless, much progress has been made in the post-Human Genome Project era with the utilization of sophisticated bioinformatics and high-throughput methodologies. These tools are now linked to escalating knowledge of the molecular mechanisms of lung endothelial permeability, a hallmark of ALI and an attractive target for the design of novel therapies, to identify candidate genes whose variants are potentially involved in ALI susceptibility. Genome-wide searches in animal models have identified a number of quantitative trait loci that associate with ALI susceptibility [8]. In this chapter, we utilize a system biology approach combining cellular signaling pathway analysis with population-based association studies to evaluate established and suspected candidate genes that contribute to dysfunction of endothelial cell (EC) barrier integrity and ALI susceptibility (Figure 24.1). Integrating high-throughput gene expression profiling in preclinical models of ALI with bioinformatics has led to the identification of differentially expressed genes in response to ALI whose variants are potentially involved in ALI susceptibility and severity. Filtering these results for unidirectional and significant changes in gene expression confirmed long-suspected ALI candidate genes, such as angiotensin-converting

The Pulmonary Endothelium: Function in health and disease Editors Norbert F. Voelkel, Sharon Rounds

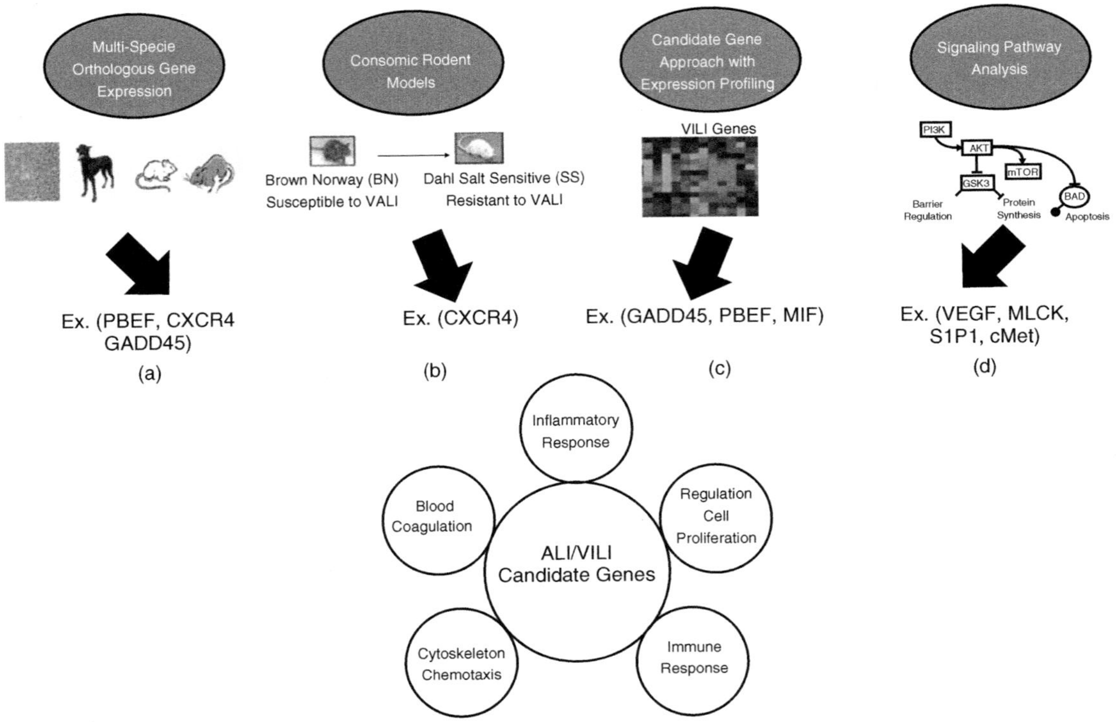

Figure 24.1 Representative novel approaches to identify ALI-implicated genes. Our multispecies orthologous gene approach (a) in human (ECs), rat, mouse, and canine models of VALI exhibits expression of common ALI-implicated evolutionarily conserved genes (orthologs) across the species. The genes with unidirectional 1.3-fold change ($p > 0.05$) are found to reside at high density on rat chromosomes 13 and 16 – the chromosomal loci used to develop the consomic rodent model (b). Together, these approaches identified novel ALI genes like PBEF, CXCR4, and GADD45. The differential gene expression between lung apex/base regions as well as between gravitationally dependent/nondependent regions of the lung base in the canine model of VALI identified ALI-implicated genes in response to local stress within the lung. This approach (c) identified the already established ALI gene MIF, and novel genes like GADD45 and PBEF. (d) Intervening in the prospective pathways involved in endothelial permeability and correlation with these differentially expressed genes in VALI models identified the most putative ALI genes like MLCK, S1P1, c-Met, and VEGF.

enzyme (ACE), tumor necrosis factor (TNF)-α, and interleukin (IL-6), but more importantly identified novel genes not previously implicated in ALI [9–12] (Figure 24.2 and Table 24.1). Increasing knowledge of the molecular mechanisms of endothelial barrier-regulatory pathways has also enhanced the ability to find novel ALI candidate genes. The analysis of the molecular pathways involving the cytoskeletal scaffolding and the dynamic cytoskeletal changes in cell shape, a key feature of cell permeability [13], has identified additional genes, such as myosin light chain (MLC) kinase (myosin light chain kinaseMLCK) contributing to the development and severity of ALI, thereby providing novel therapeutic targets in this devastating illness. Genes encoding proinflammatory cytokines, growth factors and mediators (vascular endothelial growth factor (VEGF), TNF-α, ACE, receptors for barrier-regulatory agonists [sphingosine 1-phosphate (S1P) and hepatocyte growth factor (HGF)], and mechanical stress-sensitive genes expressed in endothelium which regulate inflammatory responses (such as growth arrest DNA damage-inducible (GADD) 45α, macrophage-migration inhibitory factor (MIF), VEGF, pre-B cell colony-enhancing factor (PBEF), chemokine receptor (CXCR) 4] also serve as attractive ALI candidate genes, and are representative of the diverse but fertile areas of exploration for candidate single nucleotide polymorphisms (SNPs) affecting ALI susceptibility and severity. These approaches are likely to further our

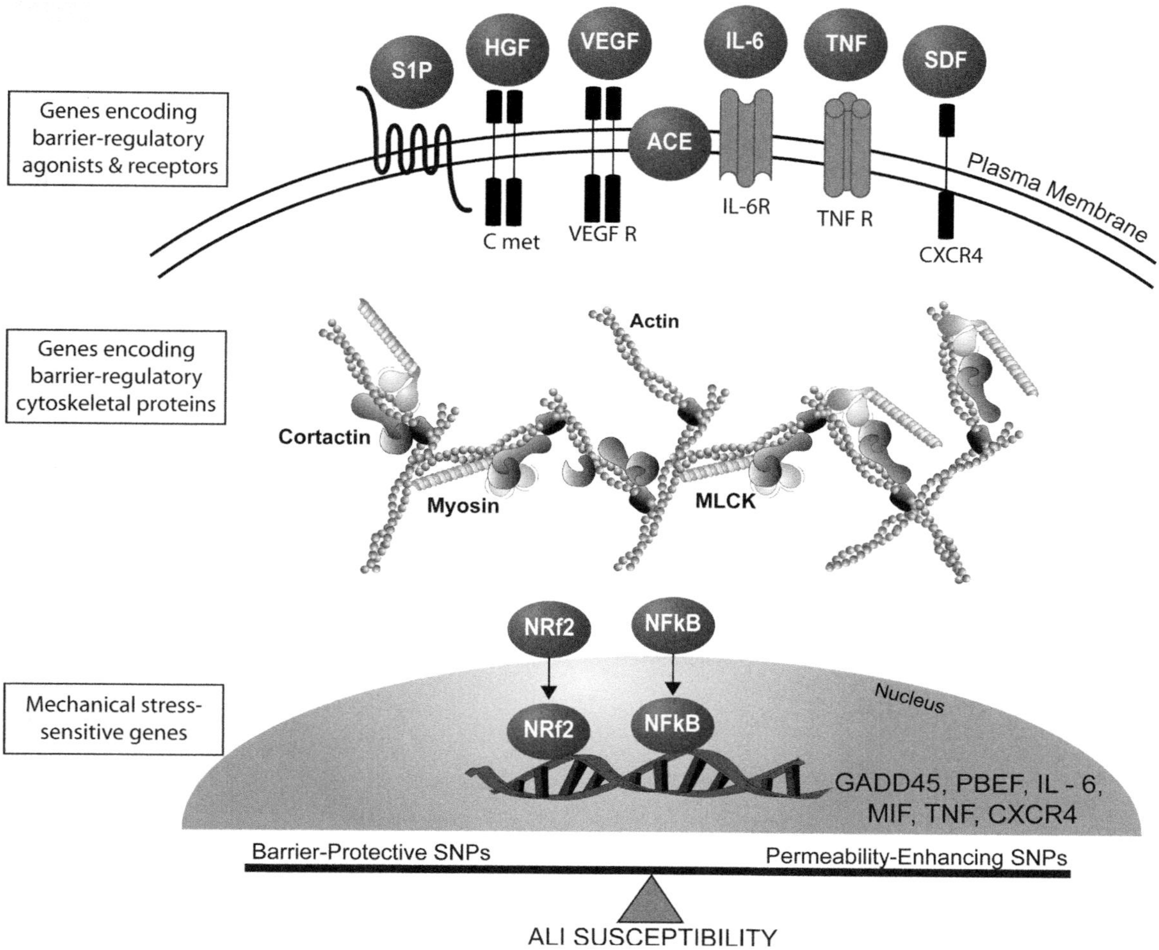

Figure 24.2 Representative EC barrier-regulatory genes. The genes encoding cytokines and cytokines receptors like IL-6, IL-6R, TNF, VEGF, and CXCR4 that play important roles in leukocytes trafficking and inflammatory diseases; S1P1 and c-Met, the receptors for barrier function-enhancing agonists; cytoplasmic proteins like MLCK, cortactin and MIF that play a crucial role in actin–myosin interaction and endothelial cytoskeleton regulation; mechanosensitive proteins like GADD45α and PBEF that are differentially expressed in ALI models are the significant genes involved in endothelial barrier permeability. The barrier-protective and disruptive polymorphisms in these genes direct toward ALI susceptibility.

understanding of vascular endothelial barrier regulation as well as elucidate novel ALI targets.

VASCULAR BARRIER REGULATORY CYTOKINES, GROWTH FACTORS, AND MEDIATORS

ACE

ACE is a significant member of the renin–angiotensin system (RAS), balancing the levels of angiotensin I and II, with profound expression in lung vascular endothelium as compared to other vascular beds (Figure 24.2) [14] (see Chapter 7). RAS is considered to be an important regulator of inflammation that contributes to ALI by altering vascular permeability, vascular tone, fibroblast activation, and endothelial–epithelial cell survival [15–17]. For example, angiotensin II activates inflammatory processes by upregulating proinflammatory cytokines and chemokines via type I and type II angiotensin II receptors that subsequently activate the nuclear factor-κB (NF-κB) pathway [18, 19]. The RAS is also involved in the fibrotic

Table 24.1 Genes with significant differential expression in multispecies models of ALI and number of PubMatrix citations.

Genes	Gene symbol	Fold change ($p > 0.05$)	No. PubMatrix[a] citations					
			ALI	Endothelium	Endothelial permeability	Sepsis	Inflammation	Lung diseases
Interleukin-1β	IL-1β	1.53	4	22	1	20	133	23
Interleukin-6	IL-6	1.84	344	1098	104	1788	8854	1444
Tissue factor/thromboplastin	F3	1.52	68	946	32	437	753	513
Plasminogen activator inhibitor type 1	PAI-1	1.47	57	1091	20	175	718	252
Cyclooxygenase II	COX2	1.79	16	272	10	59	700	107
Interleukin-13	IL-13	1.30	14	49	7	23	633	554
Aquaporin-1	AQP-1	1.30	13	68	51	1	17	16
Plasminogen activator, urokinase receptor	PLAUR	1.47	4	157	6	13	92	77
Interleukin -1 receptor antagonist	CCL2	2.00	25	70	4	170	926	137
Pre-B cell colony enhancing factor	PBEF	2.82	16	12	2	10	43	14
Chemokine receptor 4	CXCR4	1.62	1	297	7	0	7	3
Growth arrest DNA damage-inducible α	GADD45α	1.71	0	6	0	0	3	0

[a]PubMatrix: a tool for multiplex literature mining (http://pubmatrix.grc.nia.nih.gov/). Analysis performed August 2008.

response to ALI via inducing transforming growth factor (TGF) expression [20]; however, the compelling evidence for RAS involvement in ALI has come from the effective attenuation of ALI pathobiology by ACE inhibitors or angiotensin receptor blocking drugs [21, 22] and ACE knockout mice in preclinical models of ALI [23].

An intronic insertion (I) or deletion (D) of a 287-bp Alu repeat sequence in the human ACE gene, located on chromosome 17q35, has been associated with changed ACE levels and activity in serum [24, 25]. The D allele possesses a higher enzyme activity the parallels the higher gene expression in individuals with the DD genotype [26]. The initial association of the DD genotype in the ACE gene with increased ALI mortality [17] provided the impetus for subsequent studies to more firmly establish a genetic basis of ALI and to identify ALI candidate genes. Caucasian patients with ARDS show significantly higher frequencies of the DD genotype and the D allele as compared to either ventilated intensive care unit (ICU) patients without ARDS, patients after coronary artery bypass surgery, or healthy controls. Moreover, ARDS patients with DD genotype show markedly higher mortality (54%) in comparison with the II genotype (11%) or ID genotype (28%) [17]. The higher mortality rate in ARDS patients with DD or ID genotype as compared to II genotype has subsequently been confirmed in Han Chinese patients in Taiwan, although the frequency of the D allele is significantly lower in the Chinese population as compared to Western populations [27]. Compared to Caucasians, a higher frequency of D allele has been reported among Africans (Nigerian and African-American populations) [28, 29], potentially contributing to the observed disparity in ALI associated higher mortality rates in African-Americans [4]. However, to date, no association study of ACE polymorphisms and lung injury has been performed in African-Americans. In contrast, Mexican and Amerindian populations have slightly lower allelic frequencies of the D allele [28]. Thus, ACE represents a highly viable endothelial candidate gene and attractive target in acute inflammatory lung disease.

TNF

TNF-α, an early mediator of ALI development [1], is a potent proinflammatory cytokine that dramatically increases EC permeability, cytokine production, and a variety of cytotoxic and proinflammatory compounds, leading to subsequent vascular leakage and disturbed lung water balance. Both TNF-α and TNF-β subtypes appear in the circulation, in bronchoalveolar lavage (BAL) and in pulmonary edema during the onset of lung injury. The elevated levels of TNF and its soluble receptors are commonly used as a marker of inflammation, and are associated with morbidity and mortality in ALI patients [30]. Both the TNF-α and TNF-β genes lie in close proximity within the major histocompatibility complex, with several polymorphisms described in this region. The −308G/A promoter polymorphism in the TNF-α gene and the *Nco*I restriction fragment length polymorphism in the TNF-β gene appear to influence the expression of TNF-α. The carriers of the −308A allele (–308A) and homozygotes for the TNF-β 2 allele (22 genotype) exhibit increased TNF-α expression, and are associated with increased susceptibility and mortality in sepsis [31, 32]. The allele −308A is also associated with increased 60-day mortality in ARDS patients, with the strongest association found among younger patients [33]. However, in ARDS patients with direct or indirect pulmonary injury, these SNPs are associated with alterations in ALI susceptibility (TNF-α −308G/A SNP only in the direct pulmonary injury group and TNF-β *Nco*I only in the indirect pulmonary injury group). Due to the extent of linkage disequilibrium in the region, it remains unclear as to whether these are regulatory SNPs or if the TNF protein level is modulated by a third locus or a haplotype [34]. Two other promoter SNPs of TNF-α gene, −238G/A and −857C/T, along with −308 G/A SNP, have been associated with inflammatory bowel disease [35]. Thus, the role of TNF variants in inflammatory disorders is apparent and indicates a need for further study of other TNF variants in association with ALI.

IL-6

IL-6 is an acute-phase response cytokine that plays a key role in the activation of B and T cells. Inflammatory cytokines, including IL-6, are essential for the immune system homeostasis, but their exaggerated productions show destructive effects of inflammation. Increased levels of IL-6 are observed in inflammatory lung disorders including ALI [36, 37]. ALI-related increased levels of IL-6 have been established in the BAL of critically ill patients with ARDS, sepsis, and trauma [38, 39]. Increased levels of IL-6 are also implicated with ALI outcome [40] and development of multisystem organ failure [41]. We observed significantly higher expression of IL-6 and the IL-6 receptor (interluekin-6 receptorIL-6R) genes across multiple species ALI models and in human lung endothelium exposed to ventilator-induced mechanical stress as well as in differential region-specific expression in lungs of the canine ALI model [9, 10, 12]. Based on these evidences, the IL-6 gene constitutes an excellent candidate gene to understand the genetic basis underlying ALI. A functional polymorphism in the IL-6 gene promoter region at the −174 position (G–174C) has been associated with gene expression and IL-6 levels. Allele −174C is associated with lower circulating IL-6 concentrations and lower mortality rate in patients with acute respiratory failure admitted to the ICU [36]. The contrasting correlation between G174C alleles and circulating IL-6 levels has also been reported [37]. The haplotype involving promoter SNP (–174G/C) and two other IL-6 gene SNPs, 1753C/G and 2954G/C, is associated with higher mortality, and other secondary clinical outcomes, in a cohort of septic patients of European descent [42]. We further evaluated 14 IL-6 gene-tagging SNPs covering the entire gene in sepsis and ALI patients of European descent (Figure 24.3) [37]. No single SNP was identified as significantly associated with ALI; however, a common haplotype comprised

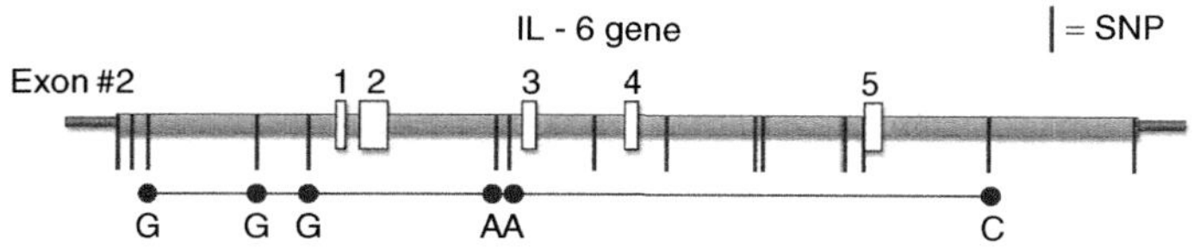

Figure 24.3 Illustration of IL-6 SNPs with minor allele frequency above 5%. The black dots represent the haplotype from position −1363 to 4835 that demonstrated the strongest association ($p > 0.05$) with ALI [37]. Boxes represent the exons and vertical lines represent the position of SNPs in the gene.

of −1363G/–572G/–174G/1208A/1305A/4835C with a frequency of 63% in cases and 49.4% in controls shows association with ALI susceptibility. Homozygote carriers of the risk haplotype are approximately threefold more frequent in ALI cases (44.8%) than in controls (22.9%) accounting for a significant odds ratio for developing ALI (odds ratio = 2.73; 95% confidence interval, 1.39–5.37; $p = 0.003$). This haplotype comprises the entire IL-6 gene including the G allele at position −174, a risk allele for susceptibility to ALI as shown in other studies. These data support the association of the IL-6 gene with ALI susceptibility and illustrate the value of haplotype analysis as a robust approach in association studies.

VEGF

VEGF is an EC-specific mitogen that regulates angiogenesis, migration, and cell permeability [43]. VEGF plays an important role in several organs by directly regulating vascular permeability to water and proteins. Lung overexpression of VEGF induces increased pulmonary vascular permeability resulting in marked pulmonary edema [44] and plasma VEGF levels are significantly elevated in ALI patients [45]. A number of studies have reported the association of low levels of VEGF with the severity of ARDS and elevated levels with the recovery from ARDS, indicating a role for VEGF in the repair process of lung injury [46]. Several polymorphisms have been described in the VEGF gene, primarily in association with cancer susceptibility and severity. The C/T SNP at position 936 of the 3′-untranslated region (UTR) of the gene has been associated with higher VEGF plasma levels in healthy subjects [47]. Recently, C936T SNP in VEGF gene has been associated with ARDS susceptibility and severity (increased mortality) in subjects of European descent [48, 49]. The haplotype TCT at position C–460T, C+405G, and C+936T was significantly associated with a higher rate of mortality in ARDS patient and higher plasma levels of VEGF [49]. These studies highlight VEGF gene as an attractive barrier-regulatory ALI candidate gene and molecular target in ALI therapeutic strategies.

CXCR4

The CXCR4 is an α-CXCR specific for stromal-derived-factor (SDF-1; also known as CXCL12) that plays an important role in cell migration, inflammation, B lymphocyte development, angiogenesis, and HIV infection (HIV co-receptor) [50–52]. CXCRs are G-protein-coupled receptors that trigger diverse signaling cascades including activation of G-proteins, and the phosphatidylinositol 3-kinase, Jak/STAT, Rho/p160-ROCK (a serine/threonine protein kinase), and mitogen-activated protein kinase signaling pathways [53]. The activation of these signaling pathways is often accompanied by the internalization of CXCRs and their trafficking back to the plasma membrane. This intracellular turnover determines the leukocyte responsiveness to chemokines [54]. Non-muscle myosin II A is a molecular motor that binds with the cytoplasmic tail of CXCR4 and CCR5 [52], and participates in the SDF-1-dependent endocytosis of CXCR4 via dynamic interaction with α-arrestin, a key component of the CXCR4 internalization pathway [52].

CXCR4 gene was identified as a novel candidate gene for ALI as it survived two filtering strategies dedicated to identifying ALI susceptibility genes. Our orthologous gene approach determined ALI-specific gene ontologies – coagulation, inflammation, chemotaxis/cell motility, and immune response [10] involving already recognized genes highly likely to play a role in ALI [IL-6, aquaporin (AQP)-1, plasminogen activator inhibitor (PAI)-1], as well as novel genes not previously known to be mechanistically involved in ALI including CXCR4 [10]. We also utilized a consomic rodent approach with introgression of rat chromosomes 2, 13, 16, and 17, which contained the highest density of these ventilator-associated lung injury (VALI)-response genes. Introgression of the VALI-sensitive Brown Norway rat chromosome 13, containing several genes including CXCR4 (Table 24.1 and Table 24.2), into the VALI-resistant Dahl salt-sensitive (SS) rat (Figure 24.1b), resulted in alterations in the phenotype of SS consomic rats to a VALI-sensitive phenotype [11]. Surface expression of CXCR4 is downregulated by IL-4, IL-13, and granulocyte-macrophage colony-stimulating factor, and upregulated by IL-10 and TGF-β [55]. CXCR4 may also play a role in the fibrotic response to ALI via TGF-β signaling.

Polymorphisms in the CXCR4 gene have not yet been reported; however, SNP in the 3′-UTR of the SDF1 gene (G801A) is associated with susceptibility to AIDS and type I diabetes [56, 57]. We are currently exploring CXCR4 as a potential ALI-associated candidate gene as suggested by the number of articles in the PubMatrix citing CXCR4 in relation to inflammation ($n = 1151$), endothelium ($n = 297$), ALI ($n = 28$), and endothelial permeability ($n = 7$) (Table 24.2).

RECEPTORS FOR BARRIER-REGULATORY AGONISTS

S1P Receptor 1

The bioactive sphingolipid metabolite S1P is an important platelet-derived lipid mediator that enhances EC barrier function *in vivo* and *in vitro* by ligating the S1P receptor, S1P1, which is encoded by the endothelial

Table 24.2 Significant ALI candidate genes residing on rat chromosomes 13 and 16 and number of PubMatrix citations.

Genes	No. PubMatrix citations						
	ALI	Endothelium	Inflammation	Sepsis	Cell permeability	Mechanical ventilation	Coagulation
Chromosome 13 genes							
coagulation factor 5	207	4575	2737	786	3725	103	13 840
prostaglandin-endoperoxide synthase 2	41	712	2527	112	105	6	66
troponin T2, cardiac	10	23	91	23	4	10	17
laminin, $\gamma 1$	1	33	21	2	4	0	3
regulator of G-protein signaling 2	0	18	18	0	6	0	3
chemokine (CXC motif) receptor 4	1	297	7	0	0	0	0
Chromosome 16 genes							
plasminogen activator, tissue	65	1597	848	199	6272	30	4598
tropomyosin 4	0	10	19	2	611	0	3
mitochondrial tumor suppressor 1	0	7	17	1	728	0	0
ADP-ribosylation factor 4	1	11	8	2	688	0	0

differentiation gene *Edg1* (Figure 24.2) [58, 59]. The signaling events evoked by S1P are reviewed in detail in Chapter 21. Briefly, S1P1 is a pertussis toxin-sensitive G_i-coupled receptor which induces Rac GTPase-dependent substantial increases in cortical actin polymerization critical to EC barrier enhancement [59, 60]. S1P1 activation enhances the organization and redistribution of VE-cadherin and β-catenin in junctional complexes in endothelium by phosphorylation of cadherin as well as p120-catenin and inducing the formation of cadherin/catenin/actin complexes. Understanding the role of S1P in enhancing EC barrier function makes it an important molecule for therapeutic applications that reverse the loss of EC barrier integrity. *In vivo* administration of selective S1P1 competitive antagonists induces a dose-dependent disruption of barrier integrity in pulmonary endothelium [61, 62], whereas S1P1 agonists, SEW2871 and FTY720, promote vascular endothelial barrier function [63–65]. A compelling argument for S1P1 as an attractive ALI candidate gene is not only its ability to transduce signals that restore barrier integrity, but that S1P1 is the target for transactivation by receptors for other potent barrier-protective agonists. These include endothelial activated protein C receptor [66], c-Met (receptor for HGF) [67], CD44 (receptor for high-molecular-weight hyaluronan) [68], and the ATP receptor [69]. We recently resequenced S1P1 gene (14 African and 13 European-Americans) to search for common variations in the *Edg1* gene and identified 39 SNPs in the *Edg1* gene. These SNPs are currently being explored to identify polymorphism(s) which is associated with inflammatory lung disorders such as asthma and ALI.

c-Met (HGF)

The role of HGF and its tyrosine kinase receptor, c-Met, has been investigated in lung development, inflammation, and repair [70], as well as in neoplastic processes such as cellular transformation, neoplastic invasion, and metastasis [71, 72]. SNPs causing underexpression of c-Met have been associated with autism and c-Met SNPs/mutations appear to be linked to lung cancer disparities in different ethnic groups. These include a N375S mutation in the HGF-binding domain of c-Met, a R988C SNP/mutation in the juxtamembrane domain, and an activating M1268T mutation in the tyrosine kinase domain (exon 19), all linked to development of solid tumors such as lung cancer, renal cancer, gastric cancer, and hepatocellular carcinoma [72].

HGF influences morphogenesis in epithelial cells from a variety of organs including the lungs, where HGF antisense oligonucleotides blocked alveolar and branching morphogenesis [73]. HGF expression and activity increase after 3–6 h of lung injury with intratracheal hydrochloric acid, suggesting that HGF plays a role in reparative response to lung injury [74]. .c-Met expression on type II pneumocyte is likely involved in increased type II pneumocyte proliferation and restoration of an intact alveolar epithelium [75]. c-Met is composed of a 50-kDa extracellular α-subunit and a 145-kDa transmembrane β-subunit [76] which contains tyrosine kinase domains, tyrosine phosphorylation sites, and tyrosine-docking sites [77]. We demonstrated that HGF-mediated c-Met phosphorylation and c-Met recruitment to caveolin-enriched microdomains (CEMs) protects against the lipopolysaccharide (LPS)-induced pulmonary vascular hyperpermeability that is regulated by high-molecular-weight hyaluronan (CD44 ligand) [78]. Our novel findings indicate that HGF/c-Met-mediated, CD44-regulated CEM signaling promotes Tiam1 (a Rac1 exchange factor)/dynamin-2-dependent Rac1 activation and peripheral recruitment of cortactin (an actin cytoskeletal regulator) – processes essential for EC barrier integrity (see Chapter 21). Understanding the mechanism(s) by which HGF/c-Met promotes increased EC barrier function may lead to novel treatments for diseases involving vascular barrier disruption, including inflammation, tumor angiogenesis, atherosclerosis, and ALI. However, on the contrary, the higher mortality rate in ALI patients with increased levels of HGF in BAL fluids [79] and in pulmonary edematous fluids [80] indicate more severe injury and inflammation in response to increased HGF levels. It is now interestingly clear that HGF plays an important role in normal and injured lung, and may have a therapeutic potential in lung diseases.

ENDOTHELIAL CYTOSKELETAL PROTEIN TARGETS WITH BARRIER-REGULATORY PROPERTIES

MLCK

MLCK is an enzyme that phosphorylates regulatory MLCs that allows interaction with F-actin (Figure 24.2). Consequently, the contraction of the actomyosin complex generates a stronger centripetal force that overcomes the force keeping the adjacent EC tethered (Figure 24.2) leading to endothelial retraction, and decreased intercellular adhesion and vascular permeability [81, 82] (see Chapter 21). Further, the nonmuscle MLCK isoform (nonmuscle myosin light chain kinasenmMLCK) knockout mice, which retain the smooth muscle MLCK isoform (smooth muscle myosin light chain kinasesmMLCK), are less susceptible to LPS and high mechanical ventilation-induced ALI [83]. The treatment with MLCK inhibitor prior to LPS exposure in the wild-type mice shows attenuated inflammation and EC barrier dysfunction [83].

The *MLCK* gene, which encodes for MLCK protein, is an excellent candidate gene for ALI identified by our multispecies gene expression profiling approach (twofold increase) and analysis of pathways involving cytoskeleton in vascular barrier regulation, as discussed in the Introduction. Since initial cloning of the highly expressed nmMLCK in endothelium in our laboratory [84], we have identified substantial roles of nmMLCK in cytoskeleton rearrangement of ECs regulating vascular barrier function [13, 81], angiogenesis, and leukocyte diapedesis [85], consistent with a potential mechanistic role for MLCK in the elaboration of ALI.

The human *MYLK* gene is located on chromosome 3q21 and encodes three proteins, including both nmMLCK and smMLCK, and telokin. We sequenced exons, exon–intron boundaries, and 2 kb of 5′-UTR of *MYLK* in healthy, sepsis-alone, and sepsis-associated ALI patients of European and African-American descent [86], and identified 51 SNPs (10 exonic, 31 intronic, nine in the 5′-UTR, and one in the noncoding exon 1) of which 28 SNPs were chosen for further linkage disequilibrium studies. Five of the 10 coding *MYLK* SNPs confer an amino acid change (Pro21His, Pro147Ser, Val261Ala, Ser1341Pro, and Arg1450Gln) in MLCK. Subsequently, association analysis of both single SNPs and haplotypes demonstrated very strong associations in both the population groups [86]. In European-Americans, the rs3845915A/MYLK_037C haplotype is associated with more than a fivefold increase in the risk of developing ALI and sepsis. In contrast, the haplotype MYLK_021G/MYLK_022G/MYLK_011T confers specific risk for ALI, but not sepsis [86]. The 5′ haplotype of *MYLK* gene also confers ALI-specific risk in both European and African descent subjects; however, the 3′ region haplotype is associated with ALI only in African descent subjects. In African-Americans, the haplotype hcv1602689C/MYLK_037A/rs11707609G is prevalent in ALI (11%) as compared to sepsis (1%) [86]. This CAG haplotype is not found in European-Americans, highlighting a potential genetic contribution to the observed ethnicity specific differences in ALI/ARDS prevalence and susceptibility [4]. Similar findings were noted in association studies involving a cohort of trauma-induced ALI [87]. We evaluated the association of 17 *MYLK* genetic variants with severe asthma in both European-American and African-American populations and identified a SNP highly associated with severe asthma in African-Americans [88]. Thus, the chromosomal locus of *MYLK* (3q21.1) has not only been linked with ALI and sepsis, but also with asthma and asthma-related phenotypes, as identified by a number of genome-wide studies [88]. Taken together, these data strongly implicate *MYLK* genetic variants as risk variants in inflammatory lung disorders, such as ALI and asthma. We are currently exploring the potential association of *MYLK* SNPs with other chronic inflammatory disorders such as inflammatory bowel disease and ischemia reperfusion injuries.

MIF

MIF is another recognized ALI candidate gene and biomarker, initially discovered as a soluble product of activated T cells and named for its role in inhibiting random macrophage migration [89]. MIF is a proinflammatory cytokine that binds to CD44 and CD74, and is produced by many cell types, including monocytes/macrophages, pituitary cells, vascular endothelium, and respiratory epithelium [90, 91]. MIF may serve as a delicate regulator of the cytokine balance between immunity and inflammation as MIF counter-regulates the immunosuppressive effects of glucocorticoids [92]. The role of MIF as an endogenous prosurvival factor has been demonstrated *in vitro*. LPS-mediated induction of Flice-like inhibitory protein by MIF confers resistance to LPS mediated EC death [93]. Suppression of MIF by RNA interference induces cell death and sensitivity to apoptotic stimuli [93]. In addition, MIF interacts with the multidimensional nmMLCK [94] isoform that regulates TNF-mediated apoptosis in addition to its potent effects on endothelial barrier dysfunction as discussed in the previous section [81, 82]. This implicates the role of MIF in regulation of nonmuscle cytoskeletal dynamics and vascular pathophysiology, which is evident from the enhanced MIF levels in the serum, BAL fluid and alveolar endothelium of patients with ARDS as compared to other critically ill patients [89, 91, 95]. We found significant increases in MIF transcript and protein levels in murine and canine models of ALI (using high mechanical ventilation and endotoxin exposure) [95] and in human lung endothelium cells exposed to 48 h of cyclic stretch [96]. MIF deficiency or immuno-neutralization appears to protect mice or rats from fatal endotoxic shock or other inflammatory diseases [97] although these results are not without controversy [98]. MIF also upregulates the expression of AQP-1, the water channels expressed in alveolar endothelial and epithelial cells, perhaps modulating fluid movement into alveolar spaces – a hallmark of ALI [99].

To extend the likelihood that MIF serves as a putative candidate gene in ALI and sepsis, we studied the association of eight MIF polymorphisms, including the most-studied MIF promoter G/C SNP at position −173, in a sepsis-induced ALI cohort ($n = 506$) of African and European descent cases [95]. No individual SNP showed a significant association with either ALI or sepsis; however, the carriers of the CC genotype (rs755622) and the TT genotype (rs2070767) showed more than twofold increased risk to develop sepsis and ALI, respectively. This association was lost, however, after age and gender adjustment in a logistic regression model. In contrast, MIF haplotypes at the 3′ region of the gene display strong association with ALI and sepsis, conferring both protection as well as susceptibility to ALI, in European and African populations [95]. Furthermore, the haplotype at 5′ promoter region of the gene involving a short tandem repeat at position −794, (CATT)5, the and −173G allele show significant association with both ALI and trauma [95], however, no association is found between promoter region haplotypes and MIF levels. Rheumatoid arthritis patients with the −173C allele have higher levels of MIF in the serum and synovial fluid than the carriers of the G allele and have the higher probability of developing idiopathic arthritis [100]. Thus, given these diverse MIF functions, MIF remains an attractive target in inflammatory diseases.

MECHANOSENSITIVE GENES WITH ENDOTHELIAL BARRIER-REGULATORY EFFECTS

PBEF

Two genes that highlight the power of the genomic approach in search of novel disease-susceptibility genes and potentially novel biomarkers are GADD45α and PBEF (Figure 24.2). PBEF is an obscure gene first identified by [123] as a protein secreted by activated lymphocytes in bone marrow stromal cells that stimulate early stage B cell formation in conjugation with stem cell factor and IL-7. We first identified marked upregulation of PBEF in microarray analyses of various models of murine and canine ALI, and increased gene/protein expression in BAL fluid and serum samples from critically ill ICU patients with ALI and sepsis [101]. With only eight papers in PubMed at that time, but the impressive data supporting the role of PBEF as a novel candidate gene for ALI, we next performed direct sequencing of PBEF gene in 36 subjects with ALI, sepsis, and healthy controls, and a PBEF SNP-based association study in ALI subjects of European and African-American descent [101]. We identified 11 SNPs in the PBEF gene (Figure 24.4), of these two promoter SNPs T–1001G and C–1543T, the former in the proximal promoter region and the latter in the distal, were associated with ALI and sepsis. Genotyping of PBEF C–1543T and T–1001G SNPs revealed significant associations of sepsis and ALI with the strongest association found with the −1543C/–1001G haplotype. In a univariate analysis carriers of G allele (T1001G) are found to have 2.75-fold higher risk of developing ALI as compared to controls ($p = 0.002$) [101]. These results were subsequently confirmed in a comparable but separate ALI population [102]. Interestingly, the −1543G/–1001C

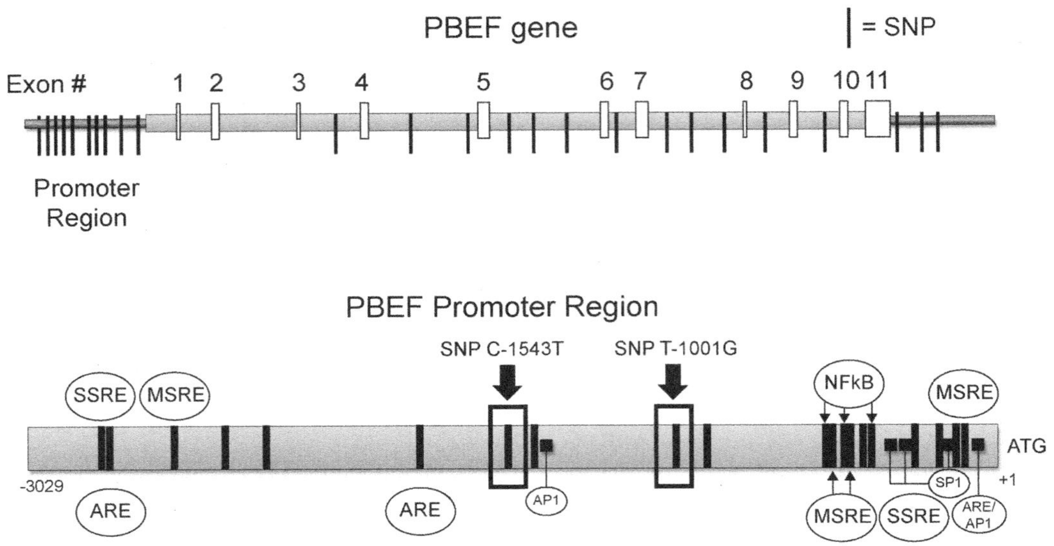

Figure 24.4 Illustration of human PBEF gene SNPs with minor allele frequency above 5%. The promoter SNP C–154T lies in the proximity of the AP-1 transcription factor binding site and SNP T–1001G shows association with ALI. Boxes represent the exons and vertical lines represent the position of SNPs in the gene.

haplotype was also associated with increased ICU mortality while the −1543T/–1001T haplotype was associated with fewer ventilator days and decreased ICU mortality [102].

A key challenge in genomic explorations is the ability to confirm the contribution of a specific gene to a disease process. Additional reports have highlighted the capacity for PBEF gene to influence far beyond any B cell regulatory function with a key role in regulating vascular permeability [103] as well as inhibiting neutrophil apoptosis [104]. To further explore mechanistic participation of PBEF in ALI and ventilator-induced lung injury (VILI), we focused on the contribution of PBEF to endothelial function. Our prior immunohistochemical staining of canine-injured lung tissues localized PBEF expression to vascular ECs, in addition to infiltrating neutrophils and type 2 alveolar epithelial cells [101]. Our *in vitro* studies showed that expression of PBEF in pulmonary artery ECs increases thrombin-mediated vascular permeability [101], suggesting that enhanced PBEF expression may mediate the early increase in vascular permeability that is characteristic of ALI. Neutrophils harvested from the circulation of septic and ALI patients show marked inhibition of the apoptotic process in association with the evidence of enhanced respiratory burst capacity [105, 106], with both activities largely restored with administration of PBEF antisense oligonucleotides. Our initial *in vitro* studies further demonstrated recombinant human (rh) PBEF as a direct rat neutrophil chemotactic factor with *in vivo* studies demonstrating marked increases in BAL polymorphonuclear neutrophils (PMNs) following intratracheal injection in C57BL/6J mice [107]. These changes were accompanied by increased BAL levels of PMN chemoattractants (cytokine-induced neutrophil chemoattractant KC and macrophage inflammatory protein-2) and modest increases in lung vascular and alveolar permeability. We also noted synergism between rhPBEF challenge and a model of limited VILI, and observed dramatic increases in BAL PMNs, BAL protein, and cytokine levels (IL-6, TNF-α, and KC) compared with either challenge alone. Gene expression profiling identified induction of ALI- and VILI-associated gene modules (NF-κB, leukocyte extravasation, apoptosis, and Toll-receptor pathways). Heterozygous $PBEF^{+/-}$ mice were significantly protected (reduced BAL protein, BAL IL-6 levels, and peak inspiratory pressures) when exposed to a model of severe VILI (4 h, 40 ml/kg tidal volume) and exhibited significantly reduced gene expression of VILI-associated modules. Finally, strategies to reduce PBEF availability (neutralizing antibody) resulted in significant protection from VILI [107]. PBEF is now recognized as associated with modestly increased risk of type 2 diabetes and elevated levels of acute-phase proteins [108], and a C–948G SNP has been associated with an increased diastolic blood pressure in obese children [109]. These studies implicate PBEF, now associated with a number of inflammatory disorders such as inflammatory bowel disease, multiple sclerosis, cystic fibrosis, and asthma [110–112], as a key inflammatory mediator intimately involved in both the development and severity of ventilator-induced ALI.

*GADD45*α

The other gene acquired in response to mechanical stress is *GADD45*α, as noted in Figure 24.1. GADD45, a member of evolutionarily conserved gene family, is implicated as stress sensors that modulate the response of mammalian cells to genotoxic or physiological stress [113, 114]. GADD45α is a small 18-kDa predominantly nuclear protein that interacts with other proteins implicated in stress responses, including proliferating cell nuclear antigen, p21, Cdc2/cyclin B1, MAPK kinase kinase 4, and p38 kinase [115, 116]. GADD45α induces cell cycle arrest and apoptosis in most of the cells as well as promoting DNA repair functions and survival [114]. GADD45α also maintains genomic stability in a p53-responsive manner [117]. Despite its multiple known functions, the role of GADD45α in ALI, endothelial/epithelial barrier dysfunction, or repair of injured lung is unknown [10]. GADD45 exhibited differential expression in orthologous global gene expression profiling, in multispecies ALI models [10], and in region-specific lung tissue expression profiling [12], and was markedly upregulated in response to the VILI [118]. We explored the mechanistic involvement of GADD45α in endotoxin (LPS)-induced injury and VILI by comparing multiple biochemical and genomic parameters of inflammatory lung injury in wild-type C57Bl/6 and GADD45α$^{-/-}$ knockout mice exposed to high tidal volume ventilation (VILI) or intratracheal LPS [119]. GADD45α$^{-/-}$ mice were modestly susceptible to LPS-induced injury and profoundly susceptible to VILI, demonstrating increased inflammation and increased microvascular permeability. VILI-exposed GADD45α$^{-/-}$ mice manifested striking neutrophilic alveolitis with increased BAL fluid levels of protein, IgG, and inflammatory cytokines. Expression profiling of lung homogenates revealed strong dysregulation in the B cell receptor signaling pathway in GADD45α$^{-/-}$ mice, suggesting the involvement of PI3K/Akt signaling components. Western blots confirmed a threefold reduction in Akt protein and phospho-Akt levels observed in GADD45α$^{-/-}$ lungs. Electrical resistance measurements across human lung EC monolayers transfected with small interfering RNAs to reduce GADD45α or Akt expression revealed significant potentiation of LPS-induced endothelial barrier dysfunction that was attenuated by overexpression of a constitutively active Akt1 transgene. Whereas prior studies did not demonstrate a role for GADD45 in hyperoxic lung injury [120, 121], these studies validate GADD45α as a novel inflammatory lung injury candidate gene and a significant participant in vascular barrier regulation via effects on Akt-mediated endothelial signaling [122] Thus, both Akt and GADD45 are extremely attractive ALI candidate genes. The human GADD45α gene (Figure 24.5) contains 25 validated SNPs (National Center for Biotechnology Information SNP database) whose role in the ALI pathogenesis is completely unknown [113]. We are currently pursuing further characterization of the role of GADD45α and its association of genetic variants with sepsis and ALI.

CONCLUSIONS AND PERSPECTIVES

ALI is an enigmatic syndrome with limited insights into the full spectrum of ALI pathobiology. Given the unacceptably high mortality rate of 35–50% observed in ALI,

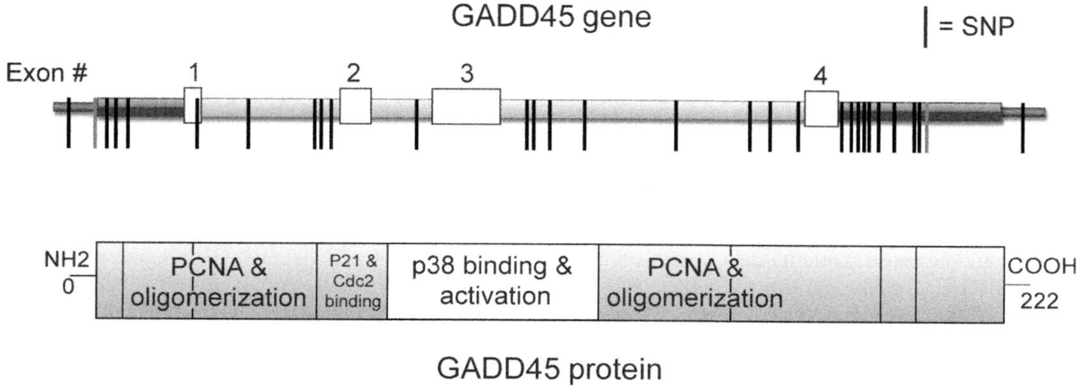

Figure 24.5 (a) Human GADD45α gene with exons (boxes) and Single Nucleotide Polymorphism database SNPs (vertical lines). (b) GADD45α protein with characterized domains corresponding to the gene.

and the paucity of novel therapies and biomarkers, it is essentially important to recognize the molecular targets associated with ALI in order to identify individuals at risk and to develop novel therapeutic targets. Although studies depicting genetic contribution to ALI pathogenesis and severity are still in a nascent phase, in this chapter we have highlighted how global gene expression profiling in multispecies ALI models served to broaden our net knowledge of ALI implicated genes and provide a basis for hope that increased insights and therapies may be forthcoming. For example, PBEF represents a powerful example of the utility of integrating global genomic and candidate gene approaches in identifying viable candidate genes and functional polymorphisms (historically considered to be a daunting task with the use of linkage or association analyses alone) and, hence, translates into not only a novel clinical biomarker, but a novel and genomically derived therapy for VILI [107]. As genotyping becomes more rapid and easily accessed, combining advanced bioinformatics techniques with high-throughput methodologies will be the future practice of personalizing barrier-reducing strategies, pressure support, mode of ventilation, and supportive strategies such as selection of appropriate antibiotics. Continued challenges will be the gene–gene and gene–environment interactions that add complexity to our understanding of the genome. These novel genetic approaches may prove exceptionally useful in ushering toward personalized medicine for the critically ill individuals.

References

1. Ware, L.B. and Matthay, M.A. (2000) The acute respiratory distress syndrome. *The New England Journal of Medicine*, **342** (18), 1334–49.
2. Flores, C., Ma, S.F., Maresso, K. *et al.* (2006) Genomics of acute lung injury. *Seminars in Respiratory and Critical Care Medicine*, **27** (4), 389–95.
3. Pepe, P.E., Potkin, R.T., Reus, D.H. *et al.* (1982) Clinical predictors of the adult respiratory distress syndrome. *American Journal of Surgery*, **144** (1), 124–30.
4. Moss, M. and Mannino, D.M. (2002) Race and gender differences in acute respiratory distress syndrome deaths in the United States: an analysis of multiple-cause mortality data (1979–1996). *Critical Care Medicine*, **30** (8), 1679–85.
5. Savov, J.D., Whitehead, G.S., Wang, J. *et al.* (2004) Ozone-induced acute pulmonary injury in inbred mouse strains. *American Journal of Respiratory Cell and Molecular Biology*, **31** (1), 69–77.
6. Kamp, R., Sun, X., and Garcia, J.G. (2008) Making genomics functional: deciphering the genetics of acute lung injury. *Proceedings of the American Thoracic Society*, **5** (3), 348–53.
7. Meyer, N.J. and Garcia, J.G. (2007) Wading into the genomic pool to unravel acute lung injury genetics. *Proceedings of the American Thoracic Society*, **4** (1), 69–76.
8. Bauer, A.K., Malkinson, A.M., and Kleeberger, S.R. (2004) Susceptibility to neoplastic and non-neoplastic pulmonary diseases in mice: genetic similarities. *American Journal of Physiology: Lung Cellular and Molecular Physiology*, **287** (4), L685–703.
9. Grigoryev, D.N., Finigan, J.H., Hassoun, P., and Garcia, J.G. (2004) Science review: searching for gene candidates in acute lung injury. *Critical Care*, **8** (6), 440–47.
10. Grigoryev, D.N., Ma, S.F., Irizarry, R.A. *et al.* (2004) Orthologous gene-expression profiling in multi-species models: search for candidate genes. *Genome Biology*, **5** (5), R34.
11. Nonas, S.A., Moreno–Vinasco, L., Ma, S.F. *et al.* (2007) Use of consomic rats for genomic insights into ventilator-associated lung injury. *American Journal of Physiology: Lung Cellular and Molecular Physiology*, **293** (2), L292–L302.
12. Simon, B.A., Easley, R.B., Grigoryev, D.N. *et al.* (2006) Microarray analysis of regional cellular responses to local mechanical stress in acute lung injury. *American Journal of Physiology: Lung Cellular and Molecular Physiology*, **291** (5), L851–61.
13. Dudek, S.M. and Garcia, J.G. (2001) Cytoskeletal regulation of pulmonary vascular permeability. *Journal of Applied Physiology*, **91** (4), 1487–500.
14. Ng, K.K. and Vane, J.R. (1968) Fate of angiotensin I in the circulation. *Nature*, **218** (5137), 144–50.
15. Dimmeler, S., Rippmann, V., Weiland, U. *et al.* (1997) Angiotensin II induces apoptosis of human endothelial cells. Protective effect of nitric oxide. *Circulation Research*, **81** (6), 970–76.
16. Kiely, D.G., Cargill, R.I., Wheeldon, N.M. *et al.* (1997) Haemodynamic and endocrine effects of type 1 angiotensin II receptor blockade in patients with hypoxaemic cor pulmonale. *Cardiovascular Research*, **33** (1), 201–8.
17. Marshall, R.P., Webb, S., Bellingan, G.J. *et al.* (2002) Angiotensin converting enzyme insertion/deletion polymorphism is associated with susceptibility and outcome in acute respiratory distress syndrome. *American Journal of Respiratory and Critical Care Medicine*, **166** (5), 646–50.

18. Esteban, V., Lorenzo, O., Ruperez, M. *et al.* (2004) Angiotensin II, via AT1 and AT2 receptors and NF-kappaB pathway, regulates the inflammatory response in unilateral ureteral obstruction. *Journal of the American Society of Nephrology*, **15** (6), 1514–29.
19. Suzuki, Y., Ruiz-Ortega, M., Lorenzo, O. *et al.* (2003) Inflammation and angiotensin II. *The International Journal of Biochemistry and Cell Biology*, **35** (6), 881–900.
20. Marshall, R.P., Gohlke, P., Chambers, R.C. *et al.* (2004) Angiotensin II and the fibroproliferative response to acute lung injury. *American Journal of Physiology: Lung Cellular and Molecular Physiology*, **286** (1), L156–64.
21. Wang, R., Zagariya, A., Ibarra-Sunga, O. *et al.* (1999) Angiotensin II induces apoptosis in human and rat alveolar epithelial cells. *American Journal of Physiology: Lung Cellular and Molecular Physiology*, **276** (5), L885–89.
22. Wang, R., Ibarra-Sunga, O., Verlinski, L. *et al.* (2000) Abrogation of bleomycin-induced epithelial apoptosis and lung fibrosis by captopril or by a caspase inhibitor. *American Journal of Physiology: Lung Cellular and Molecular Physiology*, **279** (1), L143–51.
23. Imai, Y., Kuba, K., Rao, S. *et al.* (2005) Angiotensin-converting enzyme 2 protects from severe acute lung failure. *Nature*, **436** (7047), 112–16.
24. Tiret, L., Rigat, B., Visvikis, S. *et al.* (1992) Evidence, from combined segregation and linkage analysis, that a variant of the angiotensin I-converting enzyme (ACE) gene controls plasma ACE levels. *American Journal of Human Genetics*, **51** (1), 197–205.
25. Rigat, B., Hubert, C., Alhenc-Gelas, F. *et al.* (1990) An insertion/deletion polymorphism in the angiotensin I-converting enzyme gene accounting for half the variance of serum enzyme levels. *The Journal of Clinical Investigation*, **86** (4), 1343–46.
26. Costerousse, O., Allegrini, J., Lopez, M., and Alhenc-Gelas, F. (1993) Angiotensin I-converting enzyme in human circulating mononuclear cells: genetic polymorphism of expression in T-lymphocytes. *The Biochemical Journal*, **290** (Pt 1), 33–40.
27. Jerng, J.S., Yu, C.J., Wang, H.C. *et al.* (2006) Polymorphism of the angiotensin-converting enzyme gene affects the outcome of acute respiratory distress syndrome. *Critical Care Medicine*, **34** (4), 1001–6.
28. Vargas-Alarcon, G., Hernandez-Pacheco, G., Rodriguez-Perez, J.M. *et al.* (2003) Angiotensin-converting enzyme gene (ACE) insertion/deletion polymorphism in Mexican populations. *Human Biology*, **75** (6), 889–96.
29. Mathew, J., Basheeruddin, K., and Prabhakar, S. (2001) Differences in frequency of the deletion polymorphism of the angiotensin-converting enzyme gene in different ethnic groups. *Angiology*, **52** (6), 375–79.
30. Parsons, P.E., Matthay, M.A., Ware, L.B., and Eisner, M.D. (2005) Elevated plasma levels of soluble TNF receptors are associated with morbidity and mortality in patients with acute lung injury. *American Journal of Physiology: Lung Cellular and Molecular Physiology*, **288** (3), L426–31.
31. Maier, L.A., Sawyer, R.T., Bauer, R.A. *et al.* (2001) High beryllium-stimulated TNF-alpha is associated with the −308 TNF-alpha promoter polymorphism and with clinical severity in chronic beryllium disease. *American Journal of Respiratory and Critical Care Medicine*, **164** (7), 1192–99.
32. Tang, G.J., Huang, S.L., Yien, H.W. *et al.* (2000) Tumor necrosis factor gene polymorphism and septic shock in surgical infection. *Critical Care Medicine*, **28** (8), 2733–36.
33. Gong, M.N., Zhou, W., Williams, P.L. *et al.* (2005) –308GA and TNFB polymorphisms in acute respiratory distress syndrome. *The European Respiratory Journal*, **26** (3), 382–89.
34. Hajeer, A.H. and Hutchinson, I.V. (2001) Influence of TNFalpha gene polymorphisms on TNFalpha production and disease. *Human immunology*, **62** (11), 1191–99.
35. Ferguson, L.R., Huebner, C., Petermann, I. *et al.* (2008) Single nucleotide polymorphism in the tumor necrosis factor-alpha gene affects inflammatory bowel diseases risk. *World Journal of Gastroenterology*, **14** (29), 4652–61.
36. Marshall, R.P., Webb, S., Hill, M.R. *et al.* (2002) Genetic polymorphisms associated with susceptibility and outcome in ARDS. *Chest*, **121** (Suppl 3), 68S–69S.
37. Flores, C., Ma, S.F., Maresso, K. *et al.* (2008) IL6 gene-wide haplotype is associated with susceptibility to acute lung injury. *Translational Research*, **152** (1), 11–17.
38. Park, W.Y., Goodman, R.B., Steinberg, K.P. *et al.* (2001) Cytokine balance in the lungs of patients with acute respiratory distress syndrome. *American Journal of Respiratory and Critical Care Medicine*, **164** (10Pt 1), 1896–903.
39. Takala, A., Jousela, I., Takkunen, O. *et al.* (2002) A prospective study of inflammation markers in patients at risk of indirect acute lung injury. *Shock*, **17** (4), 252–57.

40. Meduri, G.U., Headley, S., Kohler, G. *et al.* (1995) Persistent elevation of inflammatory cytokines predicts a poor outcome in ARDS. Plasma IL-1 beta and IL-6 levels are consistent and efficient predictors of outcome over time. *Chest*, **107** (4), 1062–73.
41. Ranieri, V.M., Giunta, F., Suter, P.M., and Slutsky, A.S. (2000) Mechanical ventilation as a mediator of multisystem organ failure in acute respiratory distress syndrome. *The Journal of the American Medical Association*, **284** (1), 43–44.
42. Sutherland, A.M., Walley, K.R., Manocha, S., and Russell, J.A. (2005) The association of interleukin 6 haplotype clades with mortality in critically ill adults. *Archives of Internal Medicine*, **165** (1), 75–82.
43. Mura, M., dos Santos, C.C., Stewart, D., and Liu, M. (2004) Vascular endothelial growth factor and related molecules in acute lung injury. *Journal of Applied Physiology*, **97** (5), 1605–17.
44. Kaner, R.J., Ladetto, J.V., Singh, R. *et al.* (2000) Lung overexpression of the vascular endothelial growth factor gene induces pulmonary edema. *American Journal of Respiratory Cell and Molecular Biology*, **22** (6), 657–64.
45. Thickett, D.R., Armstrong, L., Christie, S.J., and Millar, A.B. (2001) Vascular endothelial growth factor may contribute to increased vascular permeability in acute respiratory distress syndrome. *American Journal of Respiratory and Critical Care Medicine*, **164** (9), 1601–5.
46. Abadie, Y., Bregeon, F., Papazian, L. *et al.* (2005) Decreased VEGF concentration in lung tissue and vascular injury during ARDS. *The European Respiratory Journal*, **25** (1), 139–46.
47. Renner, W., Kotschan, S., Hoffmann, C. *et al.* (2000) A common 936C/T mutation in the gene for vascular endothelial growth factor is associated with vascular endothelial growth factor plasma levels. *Journal of Vascular Research*, **37** (6), 443–48.
48. Medford, A.R., Keen, L.J., Bidwell, J.L., and Millar, A.B. (2005) Vascular endothelial growth factor gene polymorphism and acute respiratory distress syndrome. *Thorax*, **60** (3), 244–48.
49. Zhai, R., Gong, M.N., Zhou, W. *et al.* (2007) Genotypes and haplotypes of the VEGF gene are associated with higher mortality and lower VEGF plasma levels in patients with ARDS. *Thorax*, **62** (8), 718–22.
50. Berger, E.A., Murphy, P.M., and Farber, J.M. (1999) Chemokine receptors as HIV-1 coreceptors: roles in viral entry, tropism, and disease. *Annual Review of Immunology*, **17**, 657–700.
51. Muller, A., Homey, B., Soto, H. *et al.* (2001) Involvement of chemokine receptors in breast cancer metastasis. *Nature*, **410** (6824), 50–56.
52. Rey, M., Valenzuela-Fernandez, A., Urzainqui, A. *et al.* (2007) Myosin IIA is involved in the endocytosis of CXCR4 induced by SDF-1alpha. *Journal of Cell Science*, **120** (Pt 6), 1126–33.
53. Ganju, R.K., Brubaker, S.A., Meyer, J. *et al.* (1998) The alpha-chemokine, stromal cell-derived factor-1alpha, binds to the transmembrane G-protein-coupled CXCR-4 receptor and activates multiple signal transduction pathways. *The Journal of Biological Chemistry*, **273** (36), 23169–75.
54. Fan, G.H., Lapierre, L.A., Goldenring, J.R. *et al.* (2004) Rab11-family interacting protein 2 and myosin Vb are required for CXCR2 recycling and receptor-mediated chemotaxis. *Molecular Biology of the Cell*, **15** (5), 2456–69.
55. Wang, J., Guan, E., Roderiquez, G. *et al.* (2001) Role of tyrosine phosphorylation in ligand-independent sequestration of CXCR4 in human primary monocytes-macrophages. *The Journal of Biological Chemistry*, **276** (52), 49236–43.
56. Dubois-Laforgue, D., Hendel, H., Caillat-Zucman, S. *et al.* (2001) A common stromal cell-derived factor-1 chemokine gene variant is associated with the early onset of type 1 diabetes. *Diabetes*, **50** (5), 1211–13.
57. Winkler, C., Modi, W., Smith, M.W. *et al.* (1998) Genetic restriction of AIDS pathogenesis by an SDF-1 chemokine gene variant. ALIVE Study, Hemophilia Growth and Development Study (HGDS), Multicenter AIDS Cohort Study (MACS), Multicenter Hemophilia Cohort Study (MHCS), San Francisco City Cohort (SFCC). *Science*, **279** (5349), 389–93.
58. Singleton, P.A., Dudek, S.M., Chiang, E.T., and Garcia, J.G. (2005) Regulation of sphingosine 1-phosphate-induced endothelial cytoskeletal rearrangement and barrier enhancement by S1P1 receptor, PI3 kinase, Tiam1/Rac1, and alpha-actinin. *The FASEB journal*, **19** (12), 1646–56.
59. Garcia, J.G., Liu, F., Verin, A.D. *et al.* (2001) Sphingosine 1-phosphate promotes endothelial cell barrier integrity by Edg-dependent cytoskeletal rearrangement. *The Journal of Clinical Investigation*, **108** (5), 689–701.
60. Arce, F.T., Whitlock, J.L., Birukova, A.A. *et al.* (2008) Regulation of the micromechanical properties of pulmonary endothelium by S1P and thrombin: role of cortactin. *Biophysical Journal*, **95** (2), 886–94.

61. Sanna, M.G., Wang, S.K., Gonzalez-Cabrera, P.J. *et al.* (2006) Enhancement of capillary leakage and restoration of lymphocyte egress by a chiral S1P1 antagonist *in vivo*. *Nature Chemical Biology*, **2** (8), 434–41.
62. Foss, F.W. Jr., Snyder, A.H., Davis, M.D. *et al.* (2007) Synthesis and biological evaluation of gamma-aminophosphonates as potent, subtype-selective sphingosine 1-phosphate receptor agonists and antagonists. *Bioorganic and Medicinal Chemistry*, **15** (2), 663–77.
63. Peng, X., Hassoun, P.M., Sammani, S. *et al.* (2004) Protective effects of sphingosine 1-phosphate in murine endotoxin-induced inflammatory lung injury. *American Journal of Respiratory and Critical Care Medicine*, **169** (11), 1245–51.
64. Rosen, H., Sanna, M.G., Cahalan, S.M., and Gonzalez-Cabrera, P.J. (2007) Tipping the gatekeeper: S1P regulation of endothelial barrier function. *Trends in Immunology*, **28** (3), 102–7.
65. Sammani, S., Mirzapoiazova, T., Moreno-Vinasco, L., and Garcia, J. (2008) Effect of S1P1 receptor agonists on murine lung airway function. *Journal of Investigative Medicine*, **56**, 659.
66. Finigan, J.H., Dudek, S.M., Singleton, P.A. *et al.* (2005) Activated protein C mediates novel lung endothelial barrier enhancement: role of sphingosine 1-phosphate receptor transactivation. *The Journal of Biological Chemistry*, **280** (17), 17286–93.
67. Liu, F., Verin, A.D., Wang, P. *et al.* (2001) Differential regulation of sphingosine-1-phosphate- and VEGF-induced endothelial cell chemotaxis. Involvement of $G_{i\alpha2}$-linked Rho kinase activity. *American Journal of Respiratory Cell and Molecular Biology*, **24** (6), 711–19.
68. Singleton, P.A., Dudek, S.M., Ma, S.F., and Garcia, J.G. (2006) Transactivation of sphingosine 1-phosphate receptors is essential for vascular barrier regulation. Novel role for hyaluronan and CD44 receptor family. *The Journal of Biological Chemistry*, **281** (45), 34381–93.
69. Jacobson, J.R., Dudek, S.M., Singleton, P.A. *et al.* (2006) Endothelial cell barrier enhancement by ATP is mediated by the small GTPase Rac and cortactin. *American Journal of Physiology: Lung Cellular and Molecular Physiology*, **291** (2), L289–95.
70. Brinkmann, V., Foroutan, H., Sachs, M. *et al.* (1995) Hepatocyte growth factor/scatter factor induces a variety of tissue-specific morphogenic programs in epithelial cells. *The Journal of Cell Biology*, **131** (6Pt 1), 1573–86.
71. Lesko, E. and Majka, M. (2008) The biological role of HGF–MET axis in tumor growth and development of metastasis. *Frontiers in Bioscience*, **13**, 1271–80.
72. Ma, P.C., Kijima, T., Maulik, G. *et al.* (2003) c-MET mutational analysis in small cell lung cancer: novel juxtamembrane domain mutations regulating cytoskeletal functions. *Cancer Research*, **63** (19), 6272–81.
73. Itakura, A., Kurauchi, O., Morikawa, S. *et al.* (1997) Involvement of hepatocyte growth factor in formation of bronchoalveolar structures in embryonic rat lung in primary culture. *Biochemical and Biophysical Research Communications*, **241** (1), 98–103.
74. Yanagita, K., Matsumoto, K., Sekiguchi, K. *et al.* (1993) Hepatocyte growth factor may act as a pulmotrophic factor on lung regeneration after acute lung injury. *The Journal of Biological Chemistry*, **268** (28), 21212–17.
75. Panos, R.J., Patel, R., and Bak, P.M. (1996) Intratracheal administration of hepatocyte growth factor/scatter factor stimulates rat alveolar type II cell proliferation *in vivo*. *American Journal of Respiratory Cell and Molecular Biology*, **15** (5), 574–81.
76. Hammond, D.E., Carter, S., and Clague, M.J. (2004) Met receptor dynamics and signalling. *Current Topics in Microbiology and Immunology*, **286**, 21–44.
77. Kermorgant, S. and Parker, P.J. (2005) c-Met signalling: spatio-temporal decisions. *Cell Cycle*, **4** (3), 352–55.
78. Singleton, P.A., Salgia, R., Moreno-Vinasco, L. *et al.* (2007) CD44 regulates hepatocyte growth factor-mediated vascular integrity. Role of c-Met, Tiam1/Rac1, dynamin 2, and cortactin. *The Journal of Biological Chemistry*, **282** (42), 30643–57.
79. Stern, J.B., Fierobe, L., Paugam, C. *et al.* (2000) Keratinocyte growth factor and hepatocyte growth factor in bronchoalveolar lavage fluid in acute respiratory distress syndrome patients. *Critical Care Medicine*, **28** (7), 2326–33.
80. Verghese, G.M., McCormick-Shannon, K., Mason, R.J., and Matthay, M.A. (1998) Hepatocyte growth factor and keratinocyte growth factor in the pulmonary edema fluid of patients with acute lung injury. Biologic and clinical significance. *American Journal of Respiratory and Critical Care Medicine*, **158** (2), 386–94.
81. Garcia, J.G., Davis, H.W., and Patterson, C.E. (1995) Regulation of endothelial cell gap formation and barrier dysfunction: role of myosin light chain phosphorylation. *Journal of Cellular Physiology*, **163** (3), 510–22.

82. Garcia, J.G. and Schaphorst, K.L. (1995) Regulation of endothelial cell gap formation and paracellular permeability. *Journal of Investigative Medicine*, **43** (2), 117–26.
83. Wainwright, M.S., Rossi, J., Schavocky, J. *et al.* (2003) Protein kinase involved in lung injury susceptibility: evidence from enzyme isoform genetic knockout and *in vivo* inhibitor treatment. *Proceedings of the National Academy of Sciences of the United States of America*, **100** (10), 6233–38.
84. Garcia, J.G., Lazar, V., Gilbert-McClain, L.I. *et al.* (1997) Myosin light chain kinase in endothelium: molecular cloning and regulation. *American Journal of Respiratory Cell and Molecular Biology*, **16** (5), 489–94.
85. Garcia, J.G., Verin, A.D., Herenyiova, M., and English, D. (1998) Adherent neutrophils activate endothelial myosin light chain kinase: role in transendothelial migration. *Journal of Applied Physiology*, **84** (5), 1817–21.
86. Gao, L., Grant, A., Halder, I. *et al.* (2006) Novel polymorphisms in the myosin light chain kinase gene confer risk for acute lung injury. *American Journal of Respiratory Cell and Molecular Biology*, **34** (4), 487–95.
87. Christie, J.D., Ma, S.F., Aplenc, R. *et al.* (2008) Variation in the MYLK gene is associated with development of acute lung injury following major trauma. *Critical Care Medicine*, Excepted.
88. Flores, C., Ma, S.F., Maresso, K. *et al.* (2007) A variant of the myosin light chain kinase gene is associated with severe asthma in African Americans. *Genetic Epidemiology*, **31** (4), 296–305.
89. Donnelly, S.C., Haslett, C., Reid, P.T. *et al.* (1997) Regulatory role for macrophage migration inhibitory factor in acute respiratory distress syndrome. *Nature Medicine*, **3** (3), 320–23.
90. Baugh, J.A. and Bucala, R. (2002) Macrophage migration inhibitory factor. *Critical Care Medicine*, **30** (Suppl 1), S27–S35.
91. Lai, K.N., Leung, J.C., Metz, C.N. *et al.* (2003) Role for macrophage migration inhibitory factor in acute respiratory distress syndrome. *The Journal of Pathology*, **199** (4), 496–508.
92. Calandra, T., Bernhagen, J., Metz, C.N. *et al.* (1995) MIF as a glucocorticoid-induced modulator of cytokine production. *Nature*, **377** (6544), 68–71.
93. Damico, R.L., Chesley, A., Johnston, L. *et al.* (2008) Macrophage migration inhibitory factor governs endothelial cell sensitivity to LPS-induced apoptosis. *American Journal of Respiratory Cell and Molecular Biology*, **39** (1), 77–85.
94. Wadgaonkar, R., Dudek, S.M., Zaiman, A.L. *et al.* (2005) Intracellular interaction of myosin light chain kinase with macrophage migration inhibition factor (MIF) in endothelium. *Journal of Cellular Biochemistry*, **95** (4), 849–58.
95. Gao, L., Flores, C., Fan-Ma, S. *et al.* (2007) Macrophage migration inhibitory factor in acute lung injury: expression, biomarker, and associations. *Translational Research*, **150** (1), 18–29.
96. Gao, L., Ye, S., Maloney, J. *et al.* (2003) Role of macrophage migration inhibitory factor (MIF) in human and animal models of acute lung injury (ALI) and sepsis: association of a promoter polymorphism and increased gene expression. *American Journal of Respiratory and Critical Care Medicine*, **167**, A162.
97. Tohyama, S., Onodera, S., Tohyama, H. *et al.* (2008) A novel DNA vaccine–targeting macrophage migration inhibitory factor improves the survival of mice with sepsis. *Gene Therapy*, **15** (23), 1513–22.
98. Korsgren, M., Kallstrom, L., Uller, L. *et al.* (2000) Role of macrophage migration inhibitory factor (MIF) in allergic and endotoxin-induced airway inflammation in mice. *Mediators of Inflammation*, **9** (1), 15–23.
99. King, L., Barnes, K., Ashworth, R. *et al.* (2003) Aquaporin-1: a candidate gene in sepsis and lung injury. *American Journal of Respiratory and Critical Care Medicine*, **167**, A662.
100. Donn, R.P., Shelley, E., Ollier, W.E., and Thomson, W. (2001) A novel 5′-flanking region polymorphism of macrophage migration inhibitory factor is associated with systemic-onset juvenile idiopathic arthritis. *Arthritis and Rheumatism*, **44** (8), 1782–85.
101. Ye, S.Q., Simon, B.A., Maloney, J.P. *et al.* (2005) Pre-B-cell colony-enhancing factor as a potential novel biomarker in acute lung injury. *American Journal of Respiratory and Critical Care Medicine*, **171** (4), 361–70.
102. Bajwa, E.K., Yu, C.L., Gong, M.N. *et al.* (2007) Pre-B-cell colony-enhancing factor gene polymorphisms and risk of acute respiratory distress syndrome. *Critical Care Medicine*, **35** (5), 1290–95.
103. Ye, S.Q., Zhang, L.Q., Adyshev, D. *et al.* (2005) Pre-B-cell-colony-enhancing factor is critically involved in thrombin-induced lung endothelial cell barrier dysregulation. *Microvascular Research*, **70** (3), 142–51.
104. Jia, S.H., Li, Y., Parodo, J. *et al.* (2004) Pre-B cell colony-enhancing factor inhibits neutrophil apoptosis in experimental inflammation and clinical

sepsis. *The Journal of Clinical Investigation*, **113** (9), 1318–27.

105. Taneja, R., Parodo, J., Jia, S.H. *et al.* (2004) Delayed neutrophil apoptosis in sepsis is associated with maintenance of mitochondrial transmembrane potential and reduced caspase-9 activity. *Critical Care Medicine*, **32** (7), 1460–69.
106. Jimenez, M.F., Watson, R.W., Parodo, J. *et al.* (1997) Dysregulated expression of neutrophil apoptosis in the systemic inflammatory response syndrome. *Archives of Surgery*, **132** (12), 1263–69; discussion 69–70.
107. Hong, S.B., Huang, Y., Moreno-Vinasco, L. *et al.* (2008) Essential role of pre-B-cell colony enhancing factor in ventilator-induced lung injury. *American Journal of Respiratory and Critical Care Medicine*, **178** (6), 605–17.
108. Zhang, Y.Y., Gottardo, L., Thompson, R. *et al.* (2006) A visfatin promoter polymorphism is associated with low-grade inflammation and type 2 diabetes. *Obesity (Silver Spring)*, **14** (12), 2119–26.
109. Korner, A., Bottcher, Y., Enigk, B. *et al.* (2007) Effects of genetic variation in the visfatin gene (PBEF1) on obesity, glucose metabolism, and blood pressure in children. *Metabolism*, **56** (6), 772–77.
110. Vandenbroeck, K., Fiten, P., Heggarty, S. *et al.* (2002) Chromosome 7q21–22 and multiple sclerosis: evidence for a genetic susceptibility effect in vicinity to the protachykinin-1 gene. *Journal of Neuroimmunology*, **125** (1-2), 141–48.
111. Koppelman, G.H., Stine, O.C., Xu, J. *et al.* (2002) Genome-wide search for atopy susceptibility genes in Dutch families with asthma. *The Journal of Allergy and Clinical Immunology*, **109** (3), 498–506.
112. Satsangi, J., Parkes, M., Louis, E. *et al.* (1996) Two stage genome-wide search in inflammatory bowel disease provides evidence for susceptibility loci on chromosomes 3, 7 and 12. *Nature Genetics*, **14** (2), 199–202.
113. Lal, A. and Gorospe, M. (2006) *Egad*, more forms of gene regulation: the *gadd45* α story. *Cell Cycle*, **5** (13), 1422–25.
114. Liebermann, D.A. and Hoffman, B. (2007) Gadd45 in the response of hematopoietic cells to genotoxic stress. *Blood Cells, Molecules and Diseases*, **39** (3), 329–35.
115. Fornace, A.J. Jr., Jackman, J., Hollander, M.C. *et al.* (1992) Genotoxic-stress-response genes and growth-arrest genes. *gadd*, MyD, and other genes induced by treatments eliciting growth arrest. *Annals of the New York Academy of Sciences*, **663**, 139–53.
116. Liebermann, D.A. and Hoffman, B. (2002) Myeloid differentiation (MyD)/growth arrest DNA damage (GADD) genes in tumor suppression, immunity and inflammation. *Leukemia*, **16** (4), 527–41.
117. Hollander, M.C., Philburn, R.T., Patterson, A.D. *et al.* (2005) Genomic instability in Gadd45$\alpha^{-/-}$ cells is coupled with S-phase checkpoint defects. *Cell Cycle*, **4** (5), 704–9.
118. Dolinay, T., Szilasi, M., Liu, M., and Choi, A.M. (2004) Inhaled carbon monoxide confers antiinflammatory effects against ventilator-induced lung injury. *American Journal of Respiratory and Critical Care Medicine*, **170** (6), 613–20.
119. Meyer, N., Huang, Y., Sammani, S. *et al.* (2008) GADD45α protects against ventilator induced lung injury. *Journal of Investigative Medicine*, **56**, 639.
120. Roper, J.M., Gehen, S.C., Staversky, R.J. *et al.* (2005) Loss of Gadd45a does not modify the pulmonary response to oxidative stress. *American Journal of Physiology: Lung Cellular and Molecular Physiology*, **288** (4), L663–71.
121. O'Reilly, M.A., Staversky, R.J., Watkins, R.H. *et al.* (2000) p53-independent induction of GADD45 and GADD153 in mouse lungs exposed to hyperoxia. *American Journal of Physiology: Lung Cellular and Molecular Physiology*, **278** (3), L552–59.
122. Altemeier, W.A., Matute-Bello, G., Gharib, S.A. *et al.* (2005) Modulation of lipopolysaccharide-induced gene transcription and promotion of lung injury by mechanical ventilation. *Journal of Immunology*, **175** (5), 3369–76.
123. Samal, B., Sun, Y., Stearns, G. *et al.* (1994) Cloning and characterization of the cDNA encoding a novel human pre-B-cell colony-enhancing factor. *Molecular and Cellular Biology* **14** (2), 1431–7.

25

Interactions of Pulmonary Endothelial Cells with Immune Cells and Platelets: Implications for Disease Pathogenesis

Mark R. Nicolls[1], Rasa Tamosiuniene[2], Ashok N. Babu[3] and Norbert F. Voelkel[4]

[1]Divisions of Pulmonary and Critical Care Medicine, Immunology and Rheumatology, VA Palo Alto Health Care System, Stanford University School of Medicine, Palo Alto, CA, USA

[2]Palo Alto Institute of Research Education, VA Palo Alto Health Care System, Stanford University, Palo Alto, CA, USA

[3]Department of Medicine, University of Colorado Health Sciences Center, Aurora, CO, USA

[4]Victoria Johnson Center for Pulmonary Obstructive Disease Research, Pulmonary and Critical Care Medicine Division, Virginia Commonwealth University, Richmond, VA, USA

INTRODUCTION

This chapter focuses on pulmonary vascular endothelial cells (ECs) and cells of the immune system. The usefulness of considering these cell types together is that their interaction may form the basis of many pulmonary diseases. Jordan Pober, a Yale researcher who has studied the endothelium extensively, recently noted that microvascular ECs at inflammatory sites are "both active participants in and regulators of inflammatory processes" [1]. The compartmentalization of scientific disciplines has previously divided vascular biologists from immunologists, but improved dialogue promises to provide major insights into lung pathophysiology. "Immune cells" are the cellular constituents of inflammatory responses, and can be broadly divided into innate and adaptive arms (Figure 25.1). This chapter is divided according to this scheme.

An important caveat to the theme of this chapter is that the proximity of immune cells to pulmonary endothelium does not necessarily indicate that immune cells are damaging the endothelium. This ambiguous proximity is largely responsible for the occasional difficulty in distinguishing perivascular inflammation from true vasculitis. (The terms vasculitis, arteritis, and angiitis are used interchangeably, and imply inflammatory damage to the intimal, muscular and/or adventitial layer of blood vessels.) To help focus this chapter, we will concentrate on the interplay between immune cells and the intimal (or EC-rich) layer. We will proceed with the understanding that it is not always possible to separate which lung conditions feature nonpathological perivascular immune cells from those where the immune response is directly injurious to the pulmonary endothelium (and are therefore rightly called vasculitides).

A helpful way to broadly consider immune cell interactions with the pulmonary endothelium is to remember that (i) innate responses generally precede and influence adaptive responses, (ii) inflammation is quelled by restoration of normal tissue or by introduction of changed tissue (e.g., fibrosis), and (iii) without resolution, chronic inflammation may persist for extended periods. This chapter will emphasize how interaction between immune cells and the pulmonary endothelium is not only a common finding, but also may be pivotally important in the initiation of multiple lung diseases [e.g., emphysema, pulmonary arterial hypertension (PAH), transplant rejection and the bronchiolitis obliterans syndrome (BOS)].

The Pulmonary Endothelium: Function in health and disease Editors Norbert F. Voelkel, Sharon Rounds

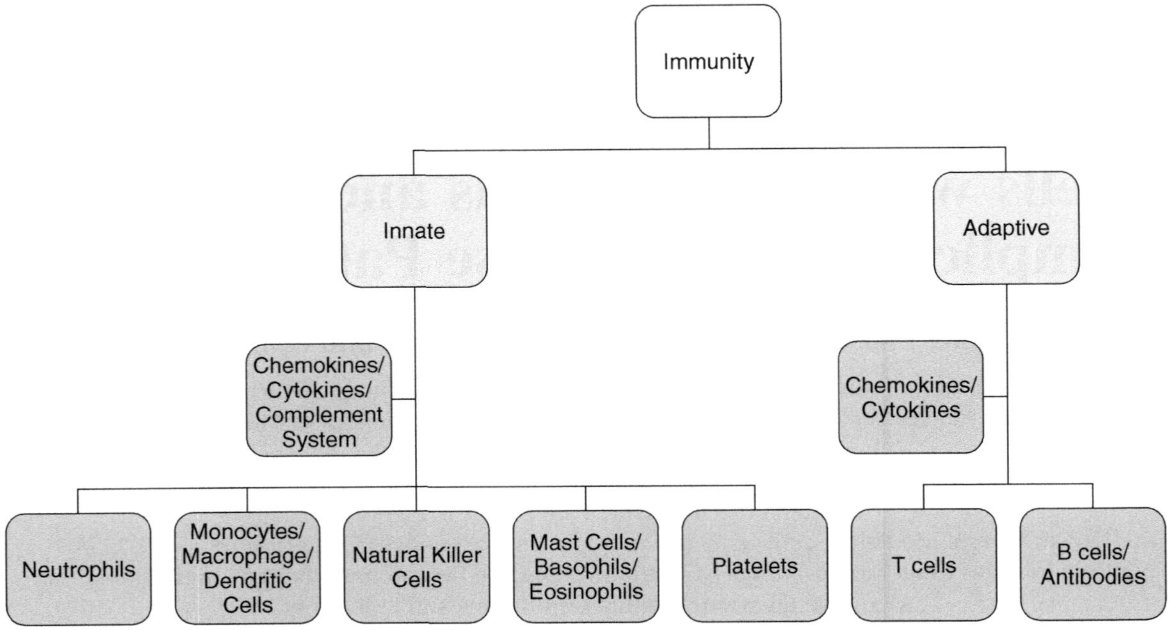

Figure 25.1 Innate and adaptive immunity.

The Quiescent and Inflamed Endothelium

In noninflamed tissues, vascular endothelium promotes blood fluidity, regulates vessel wall permeability, and leaves passing leukocytes unmolested [1]. The endothelium promotes blood fluidity by several mechanisms including expression of heparan sulfate proteoglycans (binds antithrombin III), tissue factor pathway inhibitors (blocks coagulation initiation), and thrombomodulin (promotes anticoagulant activator of protein C). Quiescent vascular endothelium normally does not express P-selectin, E-selectin, vascular cell adhesion molecule (VCAM)-1, and intercellular adhesion molecule (ICAM)-1, and therefore does not interact with leukocytes. Additionally, ECs sequester leukocyte-attracting chemokines in the absence of inflammatory stimuli (see Chapter 10).

In acute inflammation, during neutrophil recruitment, the resting endothelium becomes activated. Rapid responses, referred to as type I activation, are independent of new gene expression whereas Type II activation is a slower process that does depend on new gene expression [2]. Type I activation is mediated by ligands that engage the extracellular domains of G-protein-coupled receptors (e.g., histamine H1 receptors) initiating a signaling cascade that culminates in an elevation of cytosolic free Ca^{2+} and activation of Rho kinase. $Ca^{2+}$and Rho kinase pathway activation results in vasodilation and vascular leakiness of plasma proteins. Subsequent neutrophil interaction with the endothelium is described further in "Neutrophils" (see also Chapter 10).

If inflammation does not resolve, chronic inflammation results, and has a unique molecular and cellular phenotype distinguishing it from acute inflammation. Frequently, the persistence of foreign antigen drives this process and so, not surprisingly, the role of ECs as antigen-presenting cells (APCs) may be critical to subsequent resolution or maladaptive tissue remodeling [3]. ECs express both class I and class II major histocompatibility complex (MHC) molecules as well as costimulatory ligands for CD2, OX40, 4-1BB, and inducible T cell costimulator [4]. Thus, ECs share important characteristics with more "professional" APCs and may accordingly skew T cell responses. If the combined innate and adaptive immune responses fail to clear persistent foreign antigen, increased angiogenesis and tertiary lymphoid development may occur. Angiogenesis occurs by migration of venule ECs into the new matrix of remodeled tissue in proximity to the affected vessels [5]. Angiogenic factors such as vascular endothelial growth factor (VEGF) A, (basic) fibroblast growth factor (FGF)-2, and angiopoietin-1 and -2, are likely provided by T cells and mononuclear phagocytes [6]. Thus, a quiescent endothelium becomes, with appropriate stimuli, an active participant in the immune response.

INNATE IMMUNITY

The innate immune system comprises the cells and molecules that defend the host from injury in a nonspecific manner. This means that the cells of the innate system recognize and respond to pathogens in a generic way, and, unlike adaptive immune cells, innate responses do not confer long-lasting antigen-specific memory. Given the exposure of lung tissue to the external environment, innate immune responses are considerably important in protecting pulmonary parenchyma. The major functions of the innate immune system include: (i) recruiting immune cells to sites of inflammation, through the production of chemokines and cytokines, (ii) activating the complement cascade to promote clearance of dead cells or antibody complexes, (iii) identifying and removing foreign substances via phagocytic leukocytes, (iv) acting as "professional" APCs to the adaptive immune system, and (v) initiating potent anti-inflammatory pathways to quell leukocyte activity. The latter property can lead to resolution of injury or may contribute to pathological fibrosis. The following section focuses on the known interactions between immune cells of the innate immunity arm and the pulmonary endothelium.

Neutrophils

Neutrophils are the most abundant type of white blood cells and figure centrally in innate immune responses. During the acute phase of inflammation, neutrophils leave the pulmonary vasculature through the pulmonary endothelium (diapedesis) (see Chapter 10) and migrate toward the site of pulmonary inflammation along chemotactic gradients provided by chemokines and cytokines expressed by activated endothelium, mast cells, and macrophages. As highly motile cells, neutrophils are usually the "first responders" congregating at a focus of lung inflammation. With EC activation, leaking plasma proteins enter into tissues and assemble into a provisional matrix that supports the docking, migration, and survival of penetrating neutrophils [7, 8].

Neutrophils are associated by proximity to pulmonary endothelium in inflammatory states. This is clearly evident as neutrophils undergo diapedesis in multiple inflammatory lung conditions, such as transfusion-related acute lung injury (ALI) [9]. Neutrophils are also in close proximity to proliferating ECs in an animal model of chronic pulmonary inflammation, such as when infected with *Mycoplasma pulmonis* [10]. However, the extent to which neutrophils are responsible for directly damaging the pulmonary endothelium is less well described. Thus, the issue of delineating immune cell proximity and immune cell-mediated EC damage is raised again here as it will be in the examination of other types of leukocytes.

As described in "The Quiescent and Inflamed Endothelium," type I (fast) and type II (slow) activation of quiescent endothelium can lead to neutrophil/leukocyte recruitment. With the rise in intracellular Ca^{2+}, there is an exocytosis of specialized secretory vesicles, known as Weibel–Palade bodies, which brings P-selectin to the luminal EC surface. Simultaneously, platelet-activating factor (PAF) is displayed on the EC surface (see Chapter 10) and, with P-selectin, helps to mediate neutrophil extravasation between adjacent ECs [11–13]. Type II activation of ECs occurs after type I activation and leads to a more sustained response. Activated leukocytes arriving at a site of inflammation secrete tumor necrosis factor (TNF)-α and interleukin (IL)-1 [2], which bind TNF receptor-1 and the IL-1 receptor-1, respectively. Signaling from these two activated pathways leads to an increased transcription of chemokines, E-selectin, ICAM-1, VCAM-1, and cycloxygenase (COX)-2 [1]. The upregulation of COX-2 facilitates the conversion of arachidonic acid to prostaglandin I_2. The sum effect of this increased transcriptional activity is that leukocyte recruitment is enhanced, blood flow is increased, and vascular permeability is promoted.

Leukocyte Transmigration across Vascular Endothelium

Leukocyte migration through the EC barrier occurs in about 2–5 min and penetrating the EC basement membrane takes longer than 5–15 min. Whether transmigration itself routinely damages endothelium with each occurrence is not known. The process can be broadly considered to consist of three steps. (i) The cells loosely adhere to the vascular endothelium via the interaction between P-, E-, and L-selectins and their carbohydrate ligands on various glycoproteins such as P-selectin glycoprotein (PSGL)-1, CD34, cutaneous lymphocyte-associated antigen, E-selectin ligand-1, and CD44 [14–18]. The pulmonary vascular leak syndrome, which does appear to involve true EC injury, is greatly attenuated in the absence of one of these ligands, CD44 [18]. Given the loose nature of this attachment, the leukocytes are dragged along as the blood flows through the vasculature resulting in a rolling movement of the cells. (ii) The leukocytes become firmly attached to the endothelium via integrins (usually β_2 integrins) which can be upregulated by the bacterial peptide formyl methionyl-leucyl-phenylalanine [19, 20]. (iii) Leukocytes diapedese through the endothelium utilizing the platelet-endothelial cell adhesion molecule (PECAM)-1 and CD99 [19, 21]. In this manner, neutrophils (and other leukocytes) are routinely and intimately associated with the pulmonary vascular endothelium (see Chapter 10).

Monocytes/Macrophages/Dendritic Cells

Monocytes, macrophages, and dendritic cells (DCs) are an important component of innate immune responses, and facilitate adaptive immune responses through antigen presentation. Monocytes are produced by the bone marrow from hematopoietic stem cell precursors that circulate for approximately 1–3 days and then typically move into tissues throughout the body where they mature into macrophages and DCs. Monocytes can also kill infected host cells via antibody, in a process called antibody-mediated cellular cytotoxicity. Monocytes patrol blood vessels by crawling on resting endothelium mediated by lymphocyte function-associated antigen (LFA)-1 and the chemokine receptor CX_3CR1 [22]. CC chemokine ligand 2 (CCL2) is synthesized by vascular ECs and can stimulate monocyte/macrophage migration and smooth muscle cell proliferation. Macrophages have been noted in proximity to injured vascular endothelium in both experimental [23] and clinical [24] pathological specimens of lungs with PAH. Patients with idiopathic PAH have elevated CCL2 protein levels in plasma and lung tissue as well as elevated CCL2 release by pulmonary ECs. These results suggest that CCL2 overproduction may be a feature of the abnormal pulmonary EC phenotype in PAH, and contributes to the inflammatory process and to pulmonary vascular remodeling seen in this disease [25] (see Chapter 27).

DCs are key cells of the innate immune system that (i) may arise from monocytes or from plasma-like cells, (ii) rapidly recognize, interrogate, and directly eliminate microbial pathogens and transformed cells, (iii) induce acute inflammation, and (iv) initiate, polarize, and regulate adaptive immune responses [26]. Early DC recruitment to the lung depends in part on the CCL2 receptor chemokine CCR2 to transverse the vascular endothelium [27]. To effectively activate T cells, DCs must mature into a fully activated state [28]. Human vascular endothelial growth inhibitor (VEGI, TNF superfamily 15), an EC-produced antiangiogenic cytokine, has been shown to induce DC maturation [29]. VEGI is able to facilitate the induction of maturation of DCs, the early activation of key DC maturation signaling molecules, the reorganization of cytoskeleton, and the formation of dendrites. VEGI enhances expression of DC maturation markers, decreases antigen endocytosis, increases cell surface translocation of MHC class II, promotes the secretion of IL-12 and TNF, and facilitates the activation of $CD4^+$ T cells. There is an intimate relationship between DCs and ECs as illustrated by the transdifferentiation of DCs to ECs [30]. The proximity between DCs and ECs allows close monitoring by DCs of blood-borne antigens and endothelium-derived cytokines and chemokines (see Chapter 11).

In summary, monocytes, macrophages, and DCs all have close *in vivo* relationships with the vascular endothelium. Vascular ECs appear to have integral importance in providing a surface for monocytes to patrol for inflammation as well as secreting chemokines that summon these cells. ECs help to mature DCs to optimize their capacity as professional APCs. The activated endothelium may subsequently contribute to the development of pulmonary diseases by its participation in an exuberant inflammatory response.

Natural Killer Cells

Natural killer (NK) cells are important effector cells in the innate immune response, being critical in the defense against both viral infection and well as cells that have been malignantly transformed [31]. NK cells are normally inhibited by recognition of self-MHC class I molecules and provide protection for target cells that express normal levels of class I molecules [32]. There is good evidence indicating that NK cells are directly cytotoxic to vascular endothelium in certain biological contexts. Xenogeneic NK cells lyse porcine ECs [33]. NK–EC interactions may play a fundamental role in the rejection of noncompatible tissues because of their importance to the chemotaxis of leukocytes to sites of inflammation and necrosis. How NK cells interact with pulmonary vascular endothelium has not been well studied to date.

Mast Cells, Eosinophils, and Basophils

Mast cells, eosinophils, and basophils are important granulocytic leukocytes that release multiple factors that impact the vascular endothelium. Mast cells, while generally considered part of the innate immune response, do have a well-known adaptive immunity component in their ability to bind IgE through high-affinity Fc receptors. IgE molecules, like all antibodies, are specific to one particular antigen [34]. However, as a cell typically resident in tissues, mast cell functions are generally attributed to innate (and thus less antigen specific) functions. Importantly, mast cells are not simply effector responses but also greatly facilitate both adaptive and innate immune responses [35]. Mast cells are generally located around blood vessels. Vascular EC adherence to mast cells promotes mast cell survival via membrane-bound stem cell factor/c-kit and VCAM-1/very late antigen-4 interactions on human mast cells and ECs [36]. Mast cells can impact vascular EC permeability, blood flow, and coagulation via multiple mediators; most prominently, histamine, leukotriene C_4, chymase, and heparin. The pioneering work of Stephen Galli's group at Stanford has demonstrated that mast cells can impact disease both negatively (as when globally exacerbating experimental asthma [37]) but also, perhaps surprisingly, in a positive manner (as when enhancing resistance to snake and honeybee venoms) [38]. In the latter case, mast cells do this by releasing carboxypeptidase A and other proteases that degrade the venom. Thus, while mast cell activation is ordinarily associated with increased vascular permeability, inflammation, and abnormalities of the clotting and fibrinolytic systems [39, 40], it appears to paradoxically counter the effects normally produced by these venoms. Of note, mast cells promote the degradation of the potent vasoconstrictor endothelin (ET)-1 that is produced by ECs [41]. Thus, it is possible that increases in pulmonary mast cells in PAH [42] reflects efforts to counter-regulate inappropriately elevated ET-1 levels.

Eosinophils are granulocytes implicated in allergy and asthma. Nitric oxide (NO) derived from pulmonary vascular ECs is involved in the extravasation of eosinophils from the circulation into the lung tissue [43]. Chronic endothelial NO synthase overexpression contributes to the suppression of allergic inflammation by reducing the production of eosinophil-derived eotaxin in the airspaces as well as the expression of adhesion molecules in the vascular endothelium [44]. Antigen-activated peripheral blood mononuclear cells from atopic asthmatics induce eosinophil transmigration across vascular EC in a CCR3-dependent fashion [45]. The effect of transendothelial migration on eosinophils appears to increase the surface expression of CD69, HLA-DR, and ICAM-1 as well as increase the eosinophil's oxidative burst and promote eosinophil survival [46]. Cumulatively, these results suggest that transmigration through vascular endothelium alters the phenotype of airway eosinophils.

Basophils are the least common type of granulocytic leukocyte and less is known about this cell type's specific interaction with vascular endothelium. It is presumably important in allergic airway pathology and release of substances like histamine likely affect EC permeability in pulmonary conditions involving allergic inflammation. IL-3 promotes the selective activation of basophil adherence to ECs [47]. The adhesion molecules P-selectin and β_1 integrins CD49d, CD49e, and CD49f, and the CCR7 chemokine receptor are involved in the recruitment of basophils on IL-3-activated endothelium [48]. In summary, the granulocytic leukocytes (i.e., mast cells, eosinophils, and basophils) have an intimate relationship with vascular endothelium, and these interactions are likely important in asthma and allergic pulmonary conditions. They may have great, and as yet undiscovered, roles in other lung diseases where autoimmune injury occurs (e.g., PAH).

Platelets

Beyond their integral role in hemostasis and thrombosis, platelets are also characterized by their key functions in assisting and modulating immune responses. More recently, it has become apparent that platelets are a linking element between hemostasis and tissue repair, and they play a major role in inflammation [9]. Although not always included as members of innate immune responses, platelets do, in fact, serve as classical nonadaptive inflammatory cells. They undergo chemotaxis, contain and release adhesive proteins, activate other inflammatory cells, release vasoactive substances, and have the capacity to express or release proinflammatory mediators. There is growing evidence that platelets play a crucial role in the pathogenesis of ALI and the acute respiratory distress syndrome, postischemic reperfusion injury following lung transplantation, cystic fibrosis, and asthma [49–52]. Greater attention is given to this cell type in this chapter because platelets, as immune cells, routinely interact with the pulmonary endothelium and, in this capacity, are implicated in pulmonary disease.

Platelets are produced in the bone marrow, but also in peripheral blood and in the pulmonary circulation [53]. The pulmonary circulation may be seen as the birthplace of the platelet because megakaryocytes entering the lung release their platelets there. As a result, pulmonary venous blood has a much larger number of platelets than pulmonary arterial blood. Coupled with the fact that the pulmonary circulation is the major reservoir for marginated polymorphonuclear neutrophils (PMNs), there is ample opportunity for platelet–PMN interactions to affect lung health. Platelets normally circulate without attaching to the endothelium, but do so when ECs become activated, and platelet adherence triggers inflammation [54]. Therefore, entrapped platelets may promote activation and recruitment of leukocytes at the site of injury and aggravate pulmonary ECs damage.

Although platelets are anuclear fragments of megakaryocytes, they possess cellular components enabling their interaction with the ECs. The molecular pairs allowing adhesion of platelets to endothelium include PSGL-1/P-selectin, GPIbα/von Willebrand factor, GPIbα/P-selectin, GPIIb/IIIa/fibrinogen, and LFA-1/CAM-1 respectively [55–57]. Recently, EC-derived fractalkine has also been show to contribute to platelet activation and adhesion [58]. Activated platelets produce massive amounts of proinflammatory mediators and activate a variety of cells; in turn, platelets are activated by EC-derived proinflammatory substances binding to cognate receptors on the platelets' surface. Platelet activation is regarded as an important contributing factor in pulmonary vascular remodeling and hypertension [59, 60].

Platelets are activated via traditional pathways [thrombin, ADP, thromboxane A_2 (TxA_2)], but they can also be stimulated by antigens, antigen–antibody complexes, microorganisms, and bacterial endotoxins, including lipopolysaccharide (LPS) from *Pseudomonas aeruginosa* [61, 62]. Release of mediators stored in platelet granules and *de novo* platelet production of other mediators enhance the inflammatory response. Platelet mediators are stored in α-granules and dense body systems [63], and are promptly released upon platelet activation, including histamine, serotonin, TxA_2, oxygen radicals, PAF, platelet factor (PF)-4, prostaglandins E_2 and D_2, transforming growth factor-β, platelet-derived growth factor (PDGF), multiple chemokines [RANTES (regulated upon activation, normal T cell expressed and secreted), epithelia-derived neutrophil-activating 78 (ENA-78), monocyte chemotactic protein (MCP)-3, growth-related oncogene-α, and macrophage-inflammatory protein (MIP)-1α], IL-1β, and thrombocidins, all of which target immune cells [55, 64–73] (Figure 25.2). Of the inflammatory molecules listed herein, four of the most potent secreted by platelets are the chemokines PF-4, RANTES, MIP-1α, and ENA-78 [74]. PF-4 facilitates macrophage differentiation, and recruits and activates monocytes. RANTES is also a powerful chemoattractant, drawing monocytes and memory T lymphocytes. Once secreted, RANTES is deposited by platelets on the endothelial surface, enabling mononuclear cells to be tethered to the disrupted vascular wall. In addition, RANTES directly stimulates genes that control inflammatory pathways in monocytes, provoking the synthesis of more inflammatory mediators, such as IL-8, MCP-1, MIP-1α, and TNF-α. Activated platelets not only secrete MIP-1α, a monocyte chemoattractant and macrophage activator, but also induce its production by ECs. The last of the four chemokines produced by platelets, ENA-78, is probably less well known, but it is equally potent. The ENA-78 induces β_2 integrin signaling, which greatly increases neutrophil adhesion to the endothelium. It is also synthesized by ECs in response to platelet expression of IL-1β. Il-1β is produced when platelets are activated and its expression on the platelet membrane triggers the production not only of ENA-78, but also of E-selectin and IL-8, all of which encourage EC adhesiveness [74] (Figure 25.3).

The molecular determinants orchestrating leukocyte-dependent platelet adhesion are being elucidated. P-selectin (CD62P) [75–77] plays a critical role in mediating platelet adhesion to endothelium, and is an important adhesion molecule for PSGL-1 as it mediates adhesion of activated platelets to monocytes, neutrophils, and lymphocytes, resulting in the formation of platelet/leukocyte complexes and, vice versa, supports leukocyte rolling and arrest on surface-adherent platelets. As noted above, P-selectin is stored in α-granules of platelets and in Weibel–Palade bodies of ECs [78]. Upon activation of these cells, P-selectin is expressed on the cell surface within seconds and, once expressed, P-selectin binds to leukocytes via its major ligand PSGL-1 [79, 80] and mediates rolling of leukocytes along the inflamed endothelium. P-selectin initiates the adhesion of platelets by establishing reversible bonds that transform tethering into rolling and subsequently allow firm arrest along the endothelial surface which

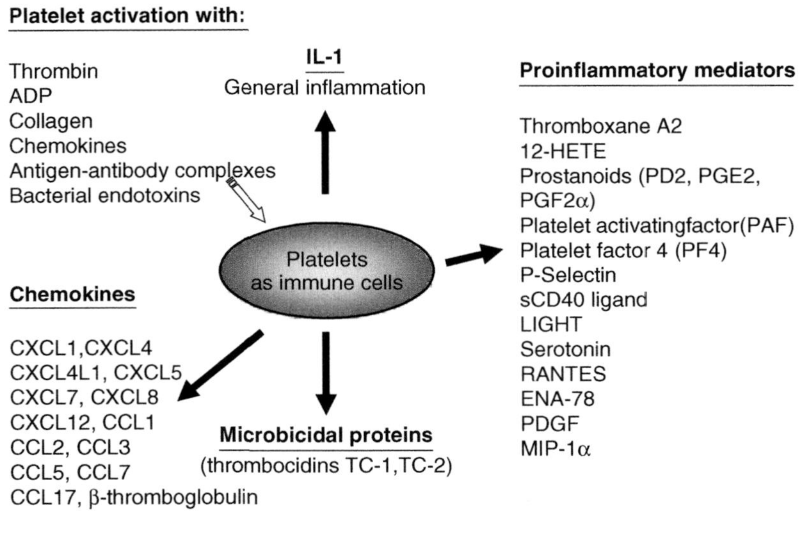

Figure 25.2 Platelets as immune cells. The activation of platelets results in the release of multiple diverse and soluble mediators with miscellaneous functions in inflammation.

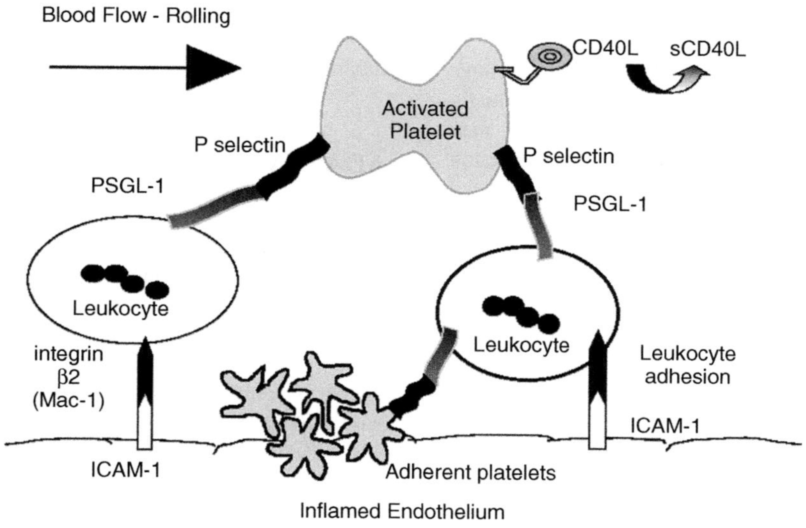

Figure 25.3 Platelets and leukocyte trafficking. Scheme for the role of platelets on leukocyte rolling on activated vascular endothelium and formation of platelet-leukocyte aggregates via interactions between P-selectin and PSGL-1.

is also the result of upregulation of integrin molecules, particularly integrin $\alpha_M\beta_2$ (Mac-1, CD11b/CD18) and adhesion molecules such as ICAM-1 [77]. Once adhered, platelets create a platform onto which a leukocyte can roll and adhere firmly through leukocyte-expressed PSGL-1 and platelet-expressed P-selectin [81] (see Chapter 10).

Platelets are crucial for leukocyte rolling on vascular endothelium and platelet–leukocyte aggregates help to amplify the recruitment of leukocytes to sites of vascular injury or inflammation. Although endothelial expression of P-selectin alone can lead to leukocyte rolling, this process (which is a necessary precursor to firm attachment and diapedesis) is much more efficient in the presence of platelet P-selectin, in part due to formation of platelet–leukocyte aggregates, which amplify the ability of leukocytes to be recruited to the endothelial surface by cross-linking. Thus, platelet activation with expression of P-selectin and release of chemoattractants enhances leukocyte recruitment in pulmonary vessels.

In addition to adhesion molecules, other mechanisms that mediate EC–leukocyte–platelet interactions rely on chemokines or the CD40/CD40 ligand (CD40L) pathway. Various experimental models demonstrate that release of platelet-stored chemokines that adhere to ECs allows binding and retention of monocytes or lymphocytes [68, 82]. In addition to mediating EC–platelet and EC–T cell binding, T cell-associated and platelet CD40L upregulates the density of CD40 expression on vascular ECs *in vivo* [83] which likely have significant immunomodulatory and proinflammatory implications.

Platelets produce membrane-bound and soluble CD40L (soluble CD40 ligandsCD40L), which engages CD40 on the surface of ECs, leading to adhesion molecule upregulation, chemokine secretion, and leukocyte recruitment [54]. The platelet surface-expressed CD40L is cleaved and shed from the platelet surface in a time-dependent manner as sCD40L. In this regard, activated platelets mimic the action of activated T cells, which express and release CD40L [84]. In doing so, platelets modulate the immune response by establishing a link between innate and adaptive immunity [85, 86]. Finally, CD40 ligation by platelet CD40L not only promotes immune activation and inflammation, but also tissue factor induction and blood coagulation [81].

Platelet-induced modulation of inflammation involves platelet expression of ligands in the TNF superfamily such as CD40L, plus Fas ligand and LIGHT ("lymphotoxins-like inducible protein that competes with glycoprotein D for herpesvirus entry mediator on T lymphocytes") [85, 87]. It is estimated that more than 95% of the circulating soluble sCD40L is derived from platelets. Raised plasma levels of sCD40L in PAH significantly correlate with the prothrombin fragment F1+2, further suggesting that platelet activation is associated with ongoing thrombus formation in this disorder. Recently, much attention has been focused on the role of platelet-derived CD40L and LIGHT in the inflammatory loop between platelets and ECs. Evident findings indicates that platelet-derived CD40L and LIGHT display prothrombotic properties by inducing tissue factor expression and plasminogen activator inhibitor (PAI)-1 and decreased thrombomodulin levels in ECs and also by stabilizing arterial thrombi by an integrin-dependent mechanism [88]. These findings suggest that such platelet-driven mechanisms potentially promote both inflammation and thrombosis, and may be operating in PAH possibly through an interaction between platelets and ECs involving chemokine-related mechanisms.

Studies provide strong evidence for an important role for activated, P-selectin surface-positive platelets in the recruitment of inflammatory cells into the lungs of patients with asthma. Pitchford *et al.*, in a series of elegant experiments' have demonstrated the importance of platelets and platelet-derived P-selectin in the recruitment of leukocytes into the lungs in allergic asthma [51]. Platelet–leukocyte aggregates, also known as heterotypic aggregates, are caused by binding of platelet P-selectin to leukocyte PSGL-1 and are a more sensitive measure of platelet activation than is measurement of P-selectin expression alone. In the blood of human subjects with allergic asthma, increased numbers of platelet–leukocyte aggregates were observed compared with healthy control subjects. In addition, platelets from these patients augmented adhesion of human PMNs to cultured vascular ECs, suggesting that platelets participate in cell recruitment and transmigration into tissue.

Recent intravital microscopic data suggest that activated platelets may attach to lung capillaries via platelet-derived P-selectin. Intrapulmonary causes of ALI, such as pneumonia or acid aspiration, can be expected to activate the vascular endothelium [89] and activated platelets may attach to lung capillaries via platelet-derived P-selectin [90] preceding the interaction of platelets with neutrophils in this event. Systemic inflammatory stimuli seem to act primarily on the lung microvascular endothelium, demonstrating that endotoxemia-induced neutrophil accumulation is dependent on endothelial, but not leukocytic, expression of the LPS receptor Toll-like receptor-4 [91].

Zarbock *et al.* [92] have convincingly integrated into the pathophysiology of ALI the concept of neutrophil–platelet interactions and platelet–endothelial interactions facilitate the secondary capture of neutrophils and other leukocytes. They identified the eicosanoid TxA_2 (mainly derived by platelets via COX) [71] as an important proinflammatory signal released by activated platelet–neutrophil aggregates, which mediates firm neutrophil adhesion by inducing the expression of endothelial adhesion molecules such as ICAM-1. However, this concept may extend even further to other inflammatory respiratory lung diseases. In summary, while not typically considered to be immune cells, platelets demonstrate many properties of innate immune cells and contribute significantly to perivascular inflammatory events.

ADAPTIVE IMMUNITY

T Cells

T cells are key components of the adaptive immune response and can be distinguished from other lymphocyte types, such as B cells and NK T cells by the presence of a special receptor on their cell surface called the T cell receptor (TCR). The TCR engages the MHC on APCs which is complexed with an antigenic peptide. Thus, with appropriate T cell affinity for the MHC–peptide complex, T cells become activated. Vascular ECs constitutively express both class I and class II MHC, and can effectively present antigen to circulating T cells. Antigen presentation via vascular endothelium in turn influences circulating T cells, indicating a bidirectional relationship between ECs and T cells. Jordan Pober has written extensively about T cell/EC antigen-specific encounters and has detailed major regulatory functions of vascular ECs that will be outlined below, including regulation of blood vessel formation and remodeling, permselectivity, blood flow and fluidity, hemostasis, and EC-mediated T cell activation and differentiation (reviewed in [3]).

T Cells and Blood Vessel Formation and Remodeling

ECs orchestrate the growth and remodeling of new blood vessels as well as new lymphatics. These processes are known as vasculogenesis (isolated endothelial precursor cells form into tubular aggregates), angiogenesis (outgrowth of previously established vessels into new vessels), and lymphangiogenesis (vascular precursor ECs give rise to lymphatic ECs). VEGF isoforms are key growth factors for these processes. While vessel formation normally ceases at birth, several stimuli, including inflammation, can trigger new vascular growth. While other mononuclear cells are known to be sources of classical vascular factors such as VEGFs, angiopoietins or PDGFs, T cells can synthesize FGF-2, a prominent mediator of angiogenesis in tissue repair [93], and the angiogenic heparin-binding epidermal-like factor [94] and TNF [95]. Conversely, T cells produce the potentially antiangiogenic cytokines such as interferon (IFN)-γ [96]. Pober *et al.* have postulated that the ability of T cells to produces these angio-reactive proteins in concert with the ability of T cells to induce EC apoptosis [97] gives them a potential role in the angiogenesis that occurs during inflammation [3].

T Cells and Permselectivity, Blood Flow, Blood Fluidity, and Hemostasis

Pober postulates that T cell-produced cytokines likely have protean impact on the vascular endothelium. For example, normal permselectivity, which is the ability to regulate macromolecules that traverse the tight junctions separating ECs is a function of ECs, and can be lost in the presence of certain T-cell derived cytokines including IFN-γ, TNF, and IL-1 [98–100]. T cells produce IL-2, which promotes the vascular leak syndrome [101] and can increase permeability through a contact-dependent non-cytokine-dependent mechanism [102]. T cell cytokines, including TNF and IFN-γ, have differential impacts on the NO pathways that directly affect vascular smooth muscle tone (reviewed in [3]). Thus, it is possible that activated T cells in the proximity of vascular endothelium have an impact on blood flow that is increased with inflammation. As described above, ECs have several mechanisms that limit in situ coagulation (e.g., binding and activating antithrombin III). T cell-derived cytokines can convert ECs into a prothrombotic microenvironment. For example, TNF can stimulate ECs to synthesize procoagulant proteins such as PAI-1 and tissue factor [103].

EC -Mediated Activation of T cells

Human ECs basally express both MHC class I and class II molecules [104], and the ability of ECs to serve as APCs has been extensively studied. Cultured ECs acquire protein antigens, which are subsequently processed and presented as peptides in the context of MHC [105, 106]. *In vitro* data have implicated the vascular EC as a potential APC [107, 108]; more recently, *in vivo* evidence suggests that vascular ECs may function as antigenic targets in experimental heart transplantation for both CD4 and CD8 T cells [109, 110]. Surprisingly little is known about alloantigen presentation by vascular ECs in lung transplantation, but it is highly likely that the endothelium can serve as APCs in this clinical setting. One study has shown increased expression of HLA class II antigens in vascular ECs in transplant recipients even though increased expression did not correlate with acute rejection or obliterative bronchiolitis [111]. The ability of class II MHC expressing human lung parenchymal cells to present alloantigen to $CD4^+$ T lymphocytes is not as effective as lung microvascular ECs [112].

ECs share many ligand–receptor pairs with the so-called "professional" antigen presenting DCs (Figure 25.4). ECs form an effective immune synapse, which is the physical structure of the interacting surface of the T cell and EC. The immune synapse typically consists of an external ring of LFA-1 that primarily interacts with ICAM-1 surrounding a central TCR-rich area [113–115]. The architecture of the synapse appears to favor docking stabilization of the T cell against the APC such that antigen-specific T cells are activated through costimulatory signals. EC interactions with T cells are thoroughly reviewed in [3].

Pulmonary Diseases with Disordered T Cell/EC Interactions

T cells are implicated in vascular pathology in PAH, lung transplant rejection, and emphysema – conditions in which the vascular endothelium may serve as an antigenic target for T cells. In PAH, a disease that notably impacts the pulmonary vascular endothelium, CD4 and CD8 cells cluster around plexiform lesions along with mast cells and macrophages [24, 42116–118]. Bosentan, an ET-1 receptor antagonist used to treat pulmonary hypertension, regulates the expression of adhesion molecules on circulating T cells in systemic sclerosis-associated pulmonary hypertension [119].

It is also possible that with pulmonary vascular EC injury, regulatory T cells may actually protect against destructive immune responses. Our group has recently found that animals lacking T cells develop severe PAH following vascular endothelial injury induced by a VEGF receptor blocker that induces EC apoptosis [23]. Indeed, a number of conditions associated with the development of PAH are significantly diminished lymphocyte populations, including the putative regulatory $CD4^+CD25^+$ cells [118] (see Chapter 27). Thus, PAH appears to be an important example of the complex interaction between immune cells and ECs; the presence of perivascular T cells in this setting may reflect either anti-inflammatory effects or may indicate T cells being directly cytotoxic to pulmonary vascular cells. A model of how immune cells may contribute to the development of PAH is presented in Figure 25.5.

Lung transplantation perhaps represents the least ambiguous setting in which endothelium may be an antigenic target. However, even this area is mired in some controversy as questions of proximity and direct toxicity still remain (i.e., T cells are present around the vessels, but it is unclear whether they are always harming vascular endothelium). The very definition of lung transplantation rejection begins with perivascular lymphocyte accumulations. However, in A1 (minimal) rejection, this infiltration has no clear clinical relevance. When T cell accumulation is more severe (A2–A4 rejection), it is likely that vascular endothelial injury is occurring. As with most coordinated adaptive immune responses, vascular injury in rejection likely involves the coordinated attack of T cells and antibodies. The differences in rejection pathology may reflect the gradient of how dominant the T cell response is versus how strong the antibody

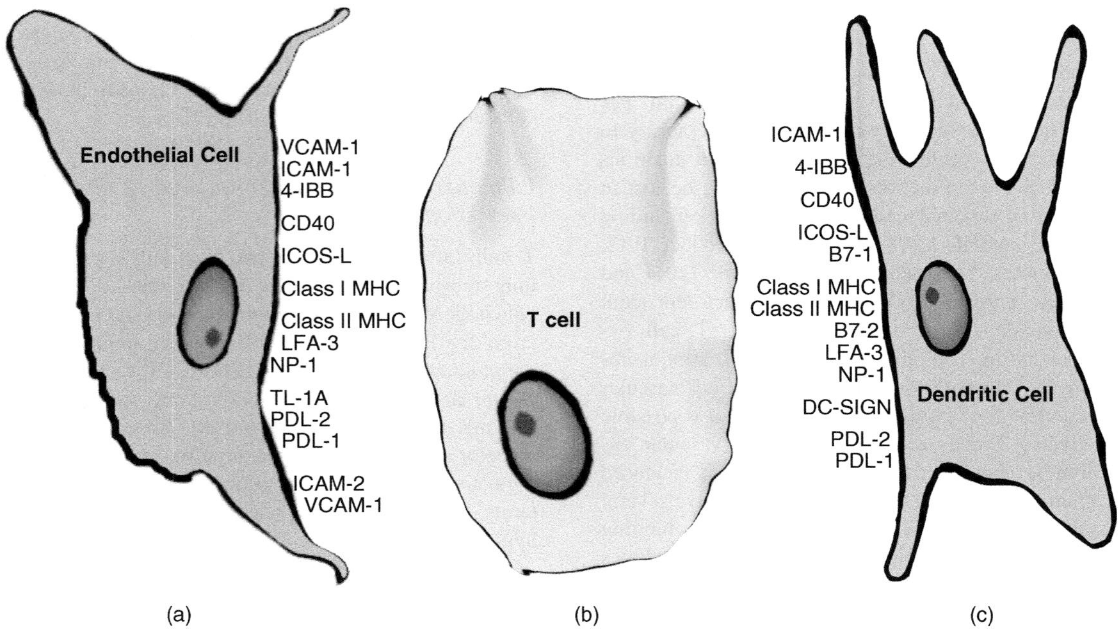

Figure 25.4 Comparing the immune synapse between vascular ECs/T cells and DCs/T cells. ECs share a number of T cell ligand pairs with DCs. Adapted from Choi *et al.* [3].

response. Our group has recently explored the mechanisms of alloimmune inflammation leading to microvascular destruction and loss of perfusion to the transplant [120]. This study examined the relationship between alloimmune airway inflammation and the development of airway fibrosis with specific regard to alterations in microvascular function. Utilizing orthotopic airway grafts, it was determined that transplants were reperfused by connection of recipient vessels to donor vessels at the surgical anastomosis site. Microcirculation through the newly-formed vascular anastomoses appeared partially dependent on VEGF receptor-2 and CXCR2 pathways. In the absence of immunosuppression, the microvasculature of rejecting allografts exhibited intravascular complement deposition, diminished endothelial CD31 expression, and absent perfusion prior to the onset of fibroproliferation. When rejecting grafts, which are characterized by a significant CD4, CD8, and macrophage accumulations, also develop extensive EC injury, they become refractory to immunotherapy. This was the first study to closely relate allograft microvascular injury and a loss of tissue perfusion to immunotherapy-resistant rejection. These results provide one look at how a T cell-mediated process impacts vascular endothelium and how this might influence the course of a pulmonary disease.

The idea that a loss of a functional microvasculature identifies lung transplant airways destined for fibrotic occlusion (i.e., the BOS) is consistent with recent clinical findings by Luckraz *et al.* [121] who found in an autopsy study of 99 lung transplant patients that there is a drop-off in the presumably pulmonary artery-derived microvasculature in "normal" lung tissue adjacent to BOS lung. This was interpreted to mean that the microvasculature is lost prior to the development of BOS. Neovascularization does occur in advanced fibrotic lungs, but these appear to be composed of vessels of a significantly smaller gauge. Further, in lung transplantation, the bronchial artery circulation is sacrificed at the time of transplantation and so the source of the microvasculature is presumed to be from the low O_2 pulmonary artery circulation. Thus, it is possible that even normally functioning lung transplants are hypoxic at baseline and may be especially vulnerable to fibrotic wound healing responses in the face of unmitigated, O_2-consuming inflammation. These ideas are summarized in Figure 25.6.

Finally, another pulmonary disease where vascular endothelial/T cell interactions may be important for disease pathogenesis is emphysema (see Chapter 26). Emphysema patients have a significantly reduced capillary length and length density [122]. In chronic obstructive pulmonary disease (COPD), increased numbers of T lymphocytes are known to contribute to the inflammatory

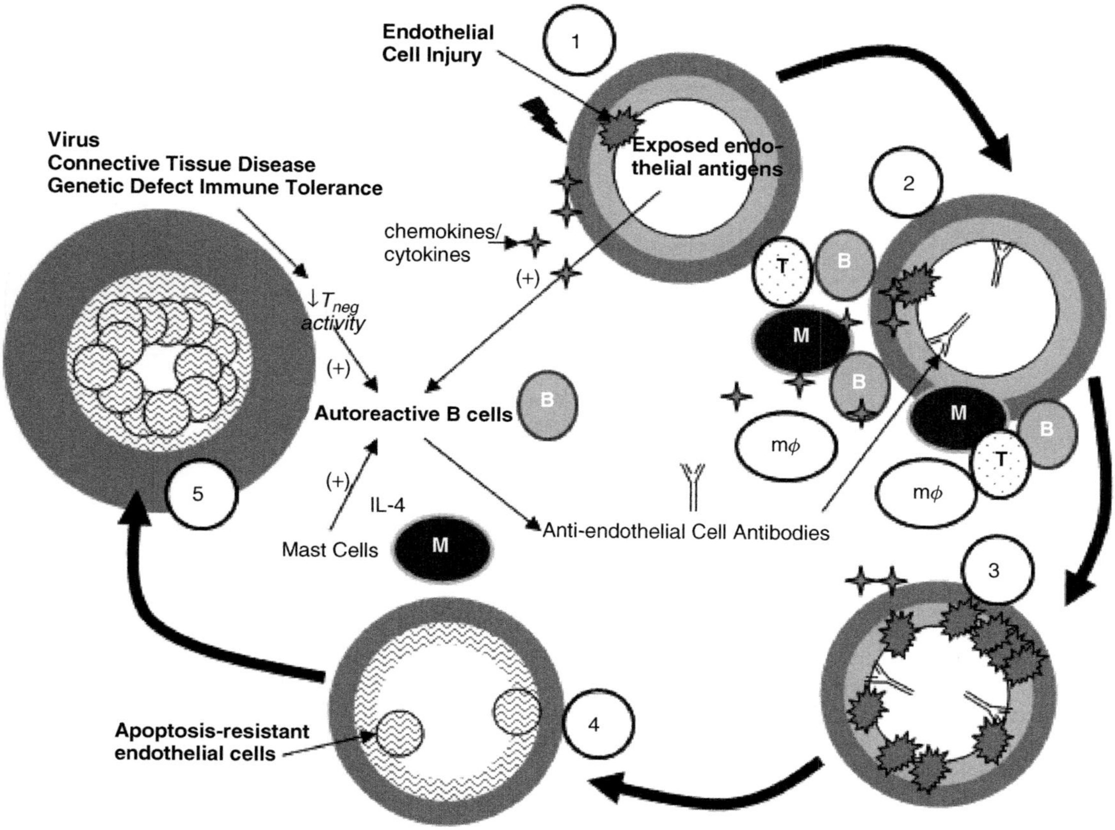

Figure 25.5 Inflammation in the evolution of PAH: a hypothetical model of disease progression (1) Inflammatory injury to vascular endothelium exposes endothelial antigens and increases local chemokine/cytokine concentrations. When the "two hits" of vascular injury and diminished peripheral immune tolerance occur simultaneously, this may lead to a loss of normal control normally exerted on autoreactive B cells. (2) Clinical PAH is characterized by B cells, T cells, macrophages (mØ), and mast cells (M) infiltrating plexiform lesions, and antibody-complement deposits that are located in the pulmonary arteries of patients with PAH. (3) Antibody deposition may contribute to ongoing endothelial apoptosis. (4) As a tissue repair response, ongoing endothelial apoptosis results in the generation of apoptosis-resistant ECs that have a malignant phenotype (5) Apoptosis-resistant ECs become "heaped-up" and begin to occlude the lumen of the vessel, and there is thickening of the vessel wall. The resulting vascular remodeling leads vascular occlusion, an increased vascular resistance and worsening of PAH.

infiltrate in airways [123] and may be involved in pulmonary vascular destruction. Experimental work from Norbert Voelkel's group has demonstrated that immunization of rats with human umbilical vein ECs induces antibody responses against the pulmonary vasculature culminating in emphysema [124]. This is a T cell-dependent process because athymic rats that are T cell-deficient did not develop emphysema. Further, adoptive transfer of $CD4^+$ cells from affected animals is sufficient to transfer disease to immunologically naïve animals. Unpublished data from Norbert Voelkel's group suggests that this experimental model of emphysema is amenable to immune tolerance strategies (e.g., intrathymic inoculation of antigen) similar to other autoimmune experimental models (e.g., the nonobese diabetic mouse for type I diabetes). These results suggest that immunomodulation may benefit a subset of COPD patients.

BCells/Antibodies

B cells are lymphocytes that play a large role in the adaptive immune response by generating antibodies (i.e., humoral immunity) against antigens. B and T cell responses

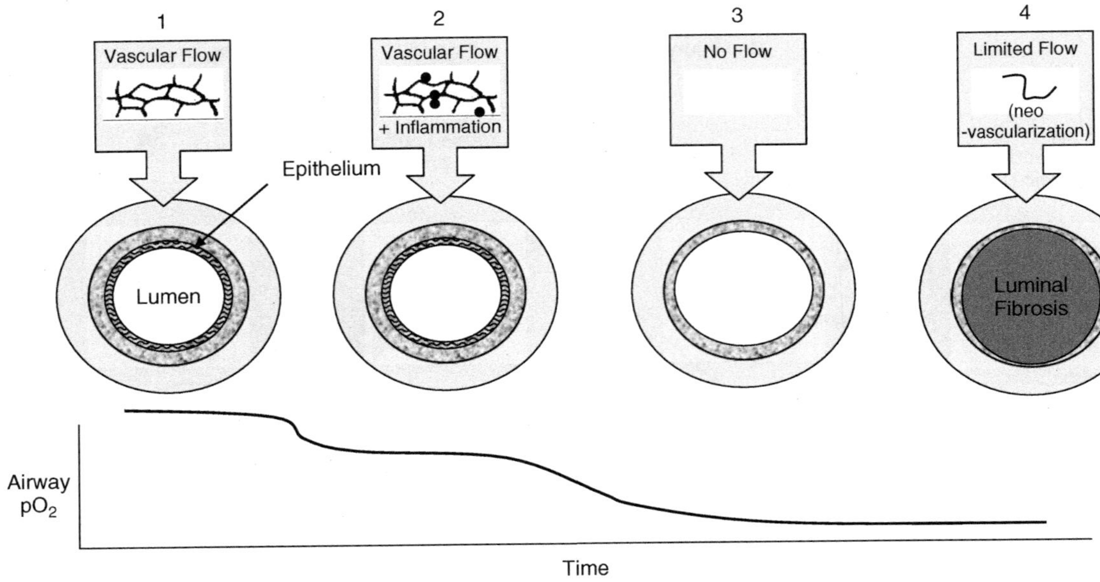

Figure 25.6 The fate of the microvasculature in airway allograft rejection. This figure, based on experimental findings [120] and clinical observations [121], is a putative model of how alloimmune-mediated loss of microvasculature in lung transplantation may influence the course of chronic rejection. (1) Normal airway without inflammation has normal microvasculature, good vascular flow, and normal oxygenation. (2) With early airway graft rejection, vascular flow is maintained but pO_2 falls likely due to the O_2 consumption of invading leukocytes. During this time, intravascular complement deposition occurs. (3) With ongoing inflammation, blood flow to the graft ceases and the overlying epithelium is sloughed. Anti-inflammatory molecules (e.g., indoleamine 2,3-dioxygenase) are upregulated and leukocytic infiltration diminishes. (4) The apparent consequence of inflammation, ischemia, and tissue hypoxia is a fibrotic wound-healing response leading to airway-occluding collagen deposition (BOS).

are frequently well-coordinated but occasionally humoral immune responses can actually overshadow cellular immune responses (i.e., T cell-mediated) to cause pulmonary pathology. The extent to which B cell/antibody engagement of pulmonary vascular endothelium impacts lung disease is currently an area of high scientific and clinical interest. Again, PAH, lung transplantation, and emphysema are three particular areas where evidence is mounting that this arm of the adaptive immune response contributes to disease pathobiology.

A notable example of antibody engagement of the pulmonary vascular endothelium is Goodpasture's syndrome in which autoantibodies specific for Goodpasture's antigens in glomerular basement membranes (GBMs) cross-react with alveolar cell proteins and lead to pulmonary hemorrhage. Anti-GBM membrane antibodies from all Goodpasture's patients recognize the same component of GBM, known as the Goodpasture antigen [125, 126]. This antigen has been identified as the NC1 domain of the α_3-chain of type IV collagen [α_3(IV)NC1] – a collagen present in lung vascular ECs [127–130]. Presumably, complement-fixing anti-GBM antibodies lead to a loss of vascular integrity resulting in both glomerulonephritis and pulmonary hemorrhage. Rarely, anti-GBM antibodies leading to pulmonary manifestations without glomerular involvement occurs, which indicates that antivascular antibodies are sufficient to cause disease in only the lung [131]. The α_3(IV)NC1 epitope is normally hidden within the $\alpha_3 \cdot \alpha_4 \cdot \alpha_5$(IV) protomer, and exposure to environmental agents, such as hydrocarbons or cigarette smoke, may be required in order to reveal these epitopes and allow binding of the anti-GBM antibody [132]. Finally, B cells from Goodpasture's patients may receive T cell help because autoreactive T cells, also specific for α_3(IV)NC1, have been found to be higher in patients with Goodpasture's syndrome and decline with treatment of the disease [133].

As described above, pathogenic antibodies with endothelial specificity [anti-EC antibodies (anti-endothelial cell antibodyAECAs)] have been implicated in the pathogenesis of PAH. AECAs have been described in scleroderma, anti-phospholipid syndrome, mixed connective tissue disease, Behçet disease, lupus and Schönlein–Henoch purpura [134–139], but it remains to be determined whether AECAs are actually pathogenic. *Ex vivo* assays of anti-EC IgG or IgM antibodies show that these antibodies engage ECs, activate complement pathways and platelet binding [140], and induce EC apoptosis [141–145].

Several studies have established a link between serum AECAs and the severity of disease activity in various autoimmune states, including lupus and scleroderma [146–149] (see Chapter 28). For example, patients with scleroderma and lupus with IgG-specific AECAs had a higher incidence of PAH compared with patients with no detectable antibodies. Based on the positive correlation between the degree of PAH and the staining intensity of the assay, it has already been posited that AECAs may trigger PAH [146]. In one study, 76 patients with systemic scleroderma and 50 matched healthy control subjects were examined with respect to AECAs, ANA, rheumatoid factor, and Scl-70. Patients with AECAs had a significantly higher incidence of digital infarct, gangrene, and PAH than those without these antibodies. Furthermore, in lupus and Sjögren's syndrome, antibody and complement deposits have been localized in the walls of pulmonary arteries of patients with PAH [150, 151]. These results in conjunction with strong experimental evidence implicating autoimmunity in the pathogenesis of PAH have led to the development of a National Institutes of Health-funded clinical trial scheduled for 2009 that will investigate the efficacy of rituximab, an anti-CD20 monoclonal antibody which depletes B cells, for the treatment of systemic sclerosis-associated PAH.

The contribution of B cells and AECAs to lung transplant rejection is a growing, if somewhat contentious, area of clinical research. While most experts in the field agree that humoral rejection must occur (as it does for other solid organ transplants), the prevalence and identification of this condition is a matter of considerable debate. Cynthia Magro, of Cornell University, has been a proponent of the idea that humoral immunity occurs relatively commonly in lung transplantation. Her group has demonstrated that humoral rejection can occur in the absence of antibodies with HLA specificity and that the antigenic targets may be of EC origin [152]. Further, AECAs may play a role in the development of chronic rejection, with BOS lung biopsies exhibiting antibody and complement deposition in the microvasculature [153]. Pulmonary capillaritis may be another pathological manifestation of humoral-mediated rejection and can be responsive to plasmapheresis when high-dose corticosteroids fail [154, 155]. It is important to note that not all pathologists agree with the evidence presented for the presence of humoral rejection [156]; thus, this remains an open and highly interesting question in the field of lung transplantation.

Finally, COPD may be affected by the B cell/antibody arm of adaptive immune responses. As noted above (in previous section), an experimental model employing human umbilical vein EC injections into rats that induces emphysema has demonstrated that AECAs are present and, like $CD4^+$ cells in this model, sufficient to transfer disease to immunologically naïve animals [124]. A recent study demonstrates that COPD patients have a high prevalence of AECAs (in addition to AECAs) [157]. Thus, there is growing interest in how the timing of the appearance of these antibodies correlates with pathogenesis, and how these factors may act in concert with other inflammatory mediators (including T cells) to begin or exacerbate this lung disease.

CONCLUSIONS AND PERSPECTIVES

This chapter has highlighted how immune cells and ECs interact with a special view to this engagement in the pulmonary circulation. The innate and adaptive arms of immunity have significant contact with pulmonary ECs. The meaning of immune cell proximity to the vascular endothelium *in vivo* can be ambiguous; this includes cell–cell contact reflecting activation of the endothelium from the quiescent to an activated state, transmigration through the endothelium, a direct attack on the vasculature, or even an anti-inflammatory response. Neutrophils are early responders to sites of inflammation and readily migrate across EC barriers. Monocytes, macrophages, and DCs arrive at inflamed sites later, and have important roles in antigen presentation and resolving inflammation – roles that ECs likely facilitate. NK cells appear to have a key role in vascular rejection in transplantation. Granulocytic leukocytes, including mast cells, eosinophils, and basophils, have significant EC interactions, and in addition to having numerous functions in allergic and asthmatic-related inflammation, mast cells, in particular, may have important protective effects that can limit endothelial damage while promoting angiogenesis. Platelets are not typically considered to be immune cells and yet have highly important immune functions in the lungs that are likely important in a number of lung conditions, including PAH. As key players in adaptive immune responses, T cells may have importance to the vascular endothelium by regulating blood vessel growth, altering EC permeability, and affecting blood flow and hemostasis. Similarly, B cells and their antibody products are likely involved in a number of deleterious endothelial-directed responses that may culminate in disease pathology. In turn, ECs are not passive participants in these processes and have significant effects on the immune cells with which they have contact.

Important questions remain to be answered. How does the nature of antigen and antigen presentation between ECs and T cells affect various disease manifestations? In situations where the EC–immune interface leads to disease, what immunological targets for therapy are appropriate? Is it possible that *all* pulmonary diseases begin with inflammation and, if so, how important are EC–immune interactions at the inception of these processes? Does successful intervention with immunotherapy require administration during a putative "threshold" period that precedes a terminal non-immunotherapy-responsive period? The study of lung immunology is burgeoning. The anatomical nexus of the endothelial–immune cell interface is beckoning clinical and basic researchers alike to this emerging field. The research performed thus far indicates that there is a high likelihood that new therapies that address the relationship between immune cells and ECs could have a striking benefit to patients with lung disease.

References

1. Pober, J.S. and Sessa, W.C. (2007) Evolving functions of endothelial cells in inflammation. *Nature Reviews: Immunology*, **7**, 803–15.
2. Pober, J.S. and Cotran, R.S. (1990) The role of endothelial cells in inflammation. *Transplantation*, **50**, 537–44.
3. Choi, J., Enis, D.R., Koh, K.P. *et al.* (2004) T lymphocyte–endothelial cell interactions. *Annual Review of Immunology*, **22**, 683–709.
4. Shiao, S.L., Mcniff, J.M., and Pober, J.S. (2005) Memory T cells and their costimulators in human allograft injury. *Journal of Immunology*, **175**, 4886–96.
5. Phillips, G.D., Whitehead, R.A., and Knighton, D.R. (1991) Initiation and pattern of angiogenesis in wound healing in the rat. *The American Journal of Anatomy*, **192**, 257–62.
6. Monaco, C., Andreakos, E., Kiriakidis, S. *et al.* (2004) T-cell-mediated signalling in immune, inflammatory and angiogenic processes: the cascade of events leading to inflammatory diseases. *Current Drug Targets: Inflammation and Allergy*, **3**, 35–42.
7. De Caterina, R., Libby, P., Peng, H.B. *et al.* (1995) Nitric oxide decreases cytokine-induced endothelial activation. Nitric oxide selectively reduces endothelial expression of adhesion molecules and proinflammatory cytokines. *The Journal of Clinical Investigation*, **96**, 60–68.
8. Frangogiannis, N.G. (2006) Targeting the inflammatory response in healing myocardial infarcts. *Current Medicinal Chemistry*, **13**, 1877–93.
9. Sachs, U.J. (2007) Pathophysiology of TRALI: current concepts. *Intensive Care Medicine*, **33** (Suppl 1), S3–S11.
10. Ezaki, T., Baluk, P., Thurston, G. *et al.* (2001) Time course of endothelial cell proliferation and microvascular remodeling in chronic inflammation. *American Journal of Pathology*, **158**, 2043–55.
11. Prescott, S.M., Zimmerman, G.A., and Mcintyre, T.M. (1984) Human endothelial cells in culture produce platelet-activating factor (1-alkyl-2-acetyl-*sn*-glycero-3-phosphocholine) when stimulated with thrombin. *Proceedings of the National Academy of Sciences of the United States of America*, **81**, 3534–38.
12. Lorant, D.E., Patel, K.D., Mcintyre, T.M. *et al.* (1991) Coexpression of GMP-140 and PAF by endothelium stimulated by histamine or thrombin: a juxtacrine system for adhesion and activation of neutrophils. *The Journal of Cell Biology*, **115**, 223–34.
13. Marchesi, V.T. (1961) The site of leucocyte emigration during inflammation. *Quarterly Journal of Experimental Physiology and Cognate Medical Sciences*, **46**, 115–18.
14. Butcher, E.C. (1991) Leukocyte–endothelial cell recognition: three (or more) steps to specificity and diversity. *Cell*, **67**, 1033–36.
15. Mcever, R.P. and Cummings, R.D. (1997) Role of PSGL-1 binding to selectins in leukocyte recruitment. *The Journal of Clinical Investigation*, **100**, S97–S103.
16. Bruehl, R.E., Moore, K.L., Lorant, D.E. *et al.* (1997) Leukocyte activation induces surface redistribution of P-selectin glycoprotein ligand-1. *Journal of Leukocyte Biology*, **61**, 489–99.
17. Alon, R., Rossiter, H., Wang, X. *et al.* (1994) Distinct cell surface ligands mediate T lymphocyte attachment and rolling on P and E selectin under physiological flow. *The Journal of Cell Biology*, **127**, 1485–95.
18. Rafi-Janajreh, A.Q., Chen, D., Schmits, R. *et al.* (1999) Evidence for the involvement of CD44 in endothelial cell injury and induction of vascular leak syndrome by IL-2. *Journal of Immunology*, **163**, 1619–27.
19. Wagner, J.G. and Roth, R.A. (2000) Neutrophil migration mechanisms, with an emphasis on the pulmonary vasculature. *Pharmacological Reviews*, **52**, 349–74.
20. Mackarel, A.J., Russell, K.J., Brady, C.S. *et al.* (2000) Interleukin-8 and leukotriene-B_4, but not formylmethionyl leucylphenylalanine, stimulate CD18-independent migration of neutrophils

across human pulmonary endothelial cells *in vitro*. *American Journal of Respiratory Cell and Molecular Biology*, **23**, 154–61.

21. Katayama, Y., Hidalgo, A., Chang, J. *et al.* (2005) CD44 is a physiological E-selectin ligand on neutrophils. *The Journal of Experimental Medicine*, **201**, 1183–89.
22. Auffray, C., Fogg, D., Garfa, M. *et al.* (2007) Monitoring of blood vessels and tissues by a population of monocytes with patrolling behavior. *Science*, **317**, 666–70.
23. Taraseviciene-Stewart, L., Nicolls, M.R., Kraskauskas, D. *et al.* (2007) Absence of T cells confers increased pulmonary arterial hypertension and vascular remodeling. *American Journal of Respiratory and Critical Care Medicine*, **175**, 1280–89.
24. Tuder, R.M., Marecki, J.C., Richter, A. *et al.* (2007) Pathology of pulmonary hypertension. *Clinics in Chest Medicine*, **28**, 23–42, vii.
25. Sanchez, O., Marcos, E., Perros, F. *et al.* (2007) Role of endothelium-derived CC chemokine ligand 2 in idiopathic pulmonary arterial hypertension. *American Journal of Respiratory and Critical Care Medicine*, **176**, 1041–47.
26. Banchereau, J. and Palucka, A.K. (2005) Dendritic cells as therapeutic vaccines against cancer. *Nature Reviews: Immunology*, **5**, 296–306.
27. Osterholzer, J.J., Ames, T., Polak, T. *et al.* (2005) CCR2 and CCR6, but not endothelial selectins, mediate the accumulation of immature dendritic cells within the lungs of mice in response to particulate antigen. *Journal of Immunology*, **175**, 874–83.
28. Figdor, C.G., De Vries, I.J., Lesterhuis, W.J., and Melief, C.J. (2004) Dendritic cell immunotherapy: mapping the way. *Nature Medicine*, **10**, 475–80.
29. Tian, F., Grimaldo, S., Fujita, M. *et al.* (2007) The endothelial cell-produced antiangiogenic cytokine vascular endothelial growth inhibitor induces dendritic cell maturation. *Journal of Immunology*, **179**, 3742–51.
30. Gottfried, E., Kreutz, M., Haffner, S. *et al.* (2007) Differentiation of human tumour-associated dendritic cells into endothelial-like cells: an alternative pathway of tumour angiogenesis. *Scandinavian Journal of Immunology*, **65**, 329–35.
31. Cerwenka, A. and Lanier, L.L. (2001) Natural killer cells, viruses and cancer. *Nature Reviews: Immunology*, **1**, 41–49.
32. Biassoni, R., Cantoni, C., Marras, D. *et al.* (2003) Human natural killer cell receptors: insights into their molecular function and structure. *Journal of Cellular and Molecular Medicine*, **7**, 376–87.
33. Kwiatkowski, B., Artrip, J.H., Michler, R.E., and Itescu, S. (1997) Human NK cells demonstrate increased adherence and lysis of porcine endothelium following activation with IL-2. *Annals of Transplantation*, **2**, 21–25.
34. Prussin, C. and Metcalfe, D.D. (2003) IgE, mast cells, basophils, and eosinophils. *The Journal of Allergy and Clinical Immunology*, **111**, S486–94.
35. Galli, S.J., Nakae, S., and Tsai, M. (2005) Mast cells in the development of adaptive immune responses. *Nature Immunology*, **6**, 135–42.
36. Mierke, C.T., Ballmaier, M., Werner, U. *et al.* (2000) Human endothelial cells regulate survival and proliferation of human mast cells. *The Journal of Experimental Medicine*, **192**, 801–11.
37. Yu, M., Tsai, M., Tam, S.Y. *et al.* (2006) Mast cells can promote the development of multiple features of chronic asthma in mice. *The Journal of Clinical Investigation*, **116**, 1633–41.
38. Metz, M., Piliponsky, A.M., Chen, C.C. *et al.* (2006) Mast cells can enhance resistance to snake and honeybee venoms. *Science*, **313**, 526–30.
39. Metcalfe, D.D., Baram, D., and Mekori, Y.A. (1997) Mast cells. *Physiological Reviews*, **77**, 1033–79.
40. Gurish, M.F. and Austen, K.F. (2001) The diverse roles of mast cells. *The Journal of Experimental Medicine*, **194**, F1–F5.
41. Maurer, M., Wedemeyer, J., Metz, M. *et al.* (2004) Mast cells promote homeostasis by limiting endothelin-1-induced toxicity. *Nature*, **432**, 512–16.
42. Heath, D. (1972) Pulmonary hypertension in pulmonary parenchymal disease. *Cardiovascular Clinics*, **4**, 79–96.
43. Feder, L.S., Stelts, D., Chapman, R.W. *et al.* (1997) Role of nitric oxide on eosinophilic lung inflammation in allergic mice. *American Journal of Respiratory Cell and Molecular Biology*, **17**, 436–42.
44. Kobayashi, K., Nishimura, Y., Yamashita, T. *et al.* (2006) The effect of overexpression of endothelial nitric oxide synthase on eosinophilic lung inflammation in a murine model. *International Immunopharmacology*, **6**, 1040–52.
45. Nagata, M., Yamamoto, H., Tabe, K., and Sakamoto, Y. (2001) Eosinophil transmigration across VCAM-1-expressing endothelial cells is upregulated by antigen-stimulated mononuclear cells. *International Archives of Allergy and Immunology*, **125** (Suppl 1), 7–11.
46. Yamamoto, H., Sedgwick, J.B., Vrtis, R.F., and Busse, W.W. (2000) The effect of transendothelial

migration on eosinophil function. *American Journal of Respiratory Cell and Molecular Biology*, **23**, 379–88.

47. Bochner, B.S., Mckelvey, A.A., Sterbinsky, S.A. *et al.* (1990) IL-3 augments adhesiveness for endothelium and CD11b expression in human basophils but not neutrophils. *Journal of Immunology*, **145**, 1832–37.
48. Lim, L.H., Burdick, M.M., Hudson, S.A. *et al.* (2006) Stimulation of human endothelium with IL-3 induces selective basophil accumulation *in vitro*. *Journal of Immunology*, **176**, 5346–53.
49. Heffner, J.E., Sahn, S.A., and Repine, J.E. (1987) The role of platelets in the adult respiratory distress syndrome. Culprits or bystanders? *The American Review of Respiratory Disease*, **135**, 482–92.
50. Okada, Y., Marchevsky, A.M., Zuo, X.J. *et al.* (1997) Accumulation of platelets in rat syngeneic lung transplants: a potential factor responsible for preservation–reperfusion injury. *Transplantation*, **64**, 801–6.
51. Pitchford, S.C., Yano, H., Lever, R. *et al.* (2003) Platelets are essential for leukocyte recruitment in allergic inflammation. *The Journal of Allergy and Clinical Immunology*, **112**, 109–18.
52. O'Sullivan, B.P., Linden, M.D., Frelinger, A.L. III *et al.* (2005) Platelet activation in cystic fibrosis. *Blood*, **105**, 4635–41.
53. Italiano, J.J. and Hartwig, J. (2002) Megakaryocyte development and platelet formation, in *Platelets* (ed. A.D. Michelson), Academic Press, San Diego, CA, pp. 21–35.
54. Weyrich, A.S. and Zimmerman, G.A. (2004) Platelets: signaling cells in the immune continuum. *Trends in Immunology*, **25**, 489–95.
55. Ruf, A. and Patscheke, H. (1995) Platelet-induced neutrophil activation: platelet-expressed fibrinogen induces the oxidative burst in neutrophils by an interaction with CD11C/CD18. *British Journal of Haematology*, **90**, 791–96.
56. Bevilacqua, M.P. and Nelson, R.M. (1993) Selectins. *The Journal of Clinical Investigation*, **91**, 379–87.
57. Tailor, A., Cooper, D., and Granger, D.N. (2005) Platelet–vessel wall interactions in the microcirculation. *Microcirculation*, **12**, 275–85.
58. Schafer, A., Schulz, C., Eigenthaler, M. *et al.* (2004) Novel role of the membrane-bound chemokine fractalkine in platelet activation and adhesion. *Blood*, **103**, 407–12.
59. Herve, P., Humbert, M., Sitbon, O. *et al.* (2001) Pathobiology of pulmonary hypertension. The role of platelets and thrombosis. *Clinics in Chest Medicine*, **22**, 451–58.
60. Tamosiuniene, R., Pidgeon, G.P., Chen, G. *et al.* (2004) Intravascular thrombosis after hypoxia-induced pulmonary hypertension: regulation by cyclooxygenase-2. *Circulation*, **110**, 2701–7.
61. Konig, B., Jaeger, K.E., Sage, A.E. *et al.* (1996) Role of Pseudomonas aeruginosa lipase in inflammatory mediator release from human inflammatory effector cells (platelets, granulocytes, and monocytes). *Infection and Immunity*, **64**, 3252–58.
62. Yeaman, M. and Bayer, A. (2002) Antimicrobial host defense, in *Platelets* (ed. A.D. Michelson), Academic Press, San Diego, CA, pp. 469–90.
63. Klinger, M.H. and Jelkmann, W. (2002) Role of blood platelets in infection and inflammation. *Journal of Interferon and Cytokine Research*, **22**, 913–22.
64. Leo, R., Pratico, D., Iuliano, L. *et al.* (1997) Platelet activation by superoxide anion and hydroxyl radicals intrinsically generated by platelets that had undergone anoxia and then reoxygenated. *Circulation*, **95**, 885–91.
65. Forde, R.C. and Fitzgerald, D.J. (1997) Reactive oxygen species and platelet activation in reperfusion injury. *Circulation*, **95**, 787–89.
66. Piccardoni, P., Evangelista, V., Piccoli, A. *et al.* (1996) Thrombin-activated human platelets release two NAP-2 variants that stimulate polymorphonuclear leukocytes. *Thrombosis and Haemostasis*, **76**, 780–85.
67. Deuel, T.F., Senior, R.M., Chang, D. *et al.* (1981) Platelet factor 4 is chemotactic for neutrophils and monocytes. *Proceedings of the National Academy of Sciences of the United States of America*, **78**, 4584–87.
68. von Hundelshausen, P., Weber, K.S., Huo, Y. *et al.* (2001) RANTES deposition by platelets triggers monocyte arrest on inflamed and atherosclerotic endothelium. *Circulation*, **103**, 1772–77.
69. Brunetti, M., Martelli, N., Manarini, S. *et al.* (2000) Polymorphonuclear leukocyte apoptosis is inhibited by platelet-released mediators, role of TGFbeta-1. *Thrombosis and Haemostasis*, **84**, 478–83.
70. Raiden, S., Schettini, J., Salamone, G. *et al.* (2003) Human platelets produce granulocyte-macrophage colony-stimulating factor and delay eosinophil apoptosis. *Laboratory Investigation*, **83**, 589–98.
71. Weber, A. (2004) Aspirin and activated platelets, in *The Eicosanoids* (ed. P. Curtis-Prior), John Wiley & Sons, Ltd, Chichester, pp. 373–85.
72. Watanabe, J., Marathe, G.K., Neilsen, P.O. *et al.* (2003) Endotoxins stimulate neutrophil adhesion followed by synthesis and release of

platelet-activating factor in microparticles. *The Journal of Biological Chemistry*, **278**, 33161–68.
73. Brandt, E., Ludwig, A., Petersen, F., and Flad, H.D. (2000) Platelet-derived CXC chemokines: old players in new games. *Immunological Reviews*, **177**, 204–16.
74. Vorchheimer, D.A. and Becker, R. (2006) Platelets in atherothrombosis. *Mayo Clinic Proceedings*, **81**, 59–68.
75. Bevilacqua, M.P. (1993) Endothelial–leukocyte adhesion molecules. *Annual Review of Immunology*, **11**, 767–804.
76. Hattori, R., Hamilton, K.K., Mcever, R.P., and Sims, P.J. (1989) Complement proteins C5b-9 induce secretion of high molecular weight multimers of endothelial von Willebrand factor and translocation of granule membrane protein GMP-140 to the cell surface. *The Journal of Biological Chemistry*, **264**, 9053–60.
77. Ma, Y.Q., Plow, E.F., and Geng, J.G. (2004) P-selectin binding to P-selectin glycoprotein ligand-1 induces an intermediate state of alphaMbeta2 activation and acts cooperatively with extracellular stimuli to support maximal adhesion of human neutrophils. *Blood*, **104**, 2549–56.
78. Stenberg, P.E., Mcever, R.P., Shuman, M.A. *et al.* (1985) A platelet alpha-granule membrane protein (GMP-140) is expressed on the plasma membrane after activation. *The Journal of Cell Biology*, **101**, 880–86.
79. Mcever, R.P. (2002) P-selectin/PSGL-1 and other interactions between platelets, leukocytes and endothelium, in *Platelets* (ed. A.D. Michelson), Academic Press, San Diego, CA, pp. 139–55.
80. Massberg, S., Enders, G., Leiderer, R. *et al.* (1998) Platelet-endothelial cell interactions during ischemia/reperfusion: the role of P-selectin. *Blood*, **92**, 507–15.
81. Slupsky, J.R., Kalbas, M., Willuweit, A. *et al.* (1998) Activated platelets induce tissue factor expression on human umbilical vein endothelial cells by ligation of CD40. *Thrombosis and Haemostasis*, **80**, 1008–14.
82. Danese, S., De La Motte, C., Sturm, A. *et al.* (2003) Platelets trigger a CD40-dependent inflammatory response in the microvasculature of inflammatory bowel disease patients. *Gastroenterology*, **124**, 1249–64.
83. Vowinkel, T., Wood, K.C., Stokes, K.Y. *et al.* (2006) Differential expression and regulation of murine CD40 in regional vascular beds. *American Journal of Physiology: Heart and Circulatory Physiology*, **290**, H631–39.
84. Danese, S. and Fiocchi, C. (2005) Platelet activation and the CD40/CD40 ligand pathway: mechanisms and implications for human disease. *Critical Reviews in Immunology*, **25**, 103–21.
85. Otterdal, K., Andreassen, A.K., Yndestad, A. *et al.* (2007) Raised LIGHT levels in pulmonary arterial hypertension – potential role in thrombus formation. *American Journal of Respiratory and Critical Care Medicine*, **177**, 202–7.
86. Elzey, B.D., Tian, J., Jensen, R.J. *et al.* (2003) Platelet-mediated modulation of adaptive immunity. A communication link between innate and adaptive immune compartments. *Immunity*, **19**, 9–19.
87. Damas, J.K., Otterdal, K., Yndestad, A. *et al.* (2004) Soluble CD40 ligand in pulmonary arterial hypertension: possible pathogenic role of the interaction between platelets and endothelial cells. *Circulation*, **110**, 999–1005.
88. Andre, P., Prasad, K.S., Denis, C.V. *et al.* (2002) CD40L stabilizes arterial thrombi by a beta3 integrin-dependent mechanism. *Nature Medicine*, **8**, 247–52.
89. Kuebler, W.M., Parthasarathi, K., Wang, P.M., and Bhattacharya, J. (2000) A novel signaling mechanism between gas and blood compartments of the lung. *The Journal of Clinical Investigation*, **105**, 905–13.
90. Kiefmann, R., Heckel, K., Schenkat, S. *et al.* (2006) Role of P-selectin in platelet sequestration in pulmonary capillaries during endotoxemia. *Journal of Vascular Research*, **43**, 473–81.
91. Andonegui, G., Bonder, C.S., Green, F. *et al.* (2003) Endothelium-derived Toll-like receptor-4 is the key molecule in LPS-induced neutrophil sequestration into lungs. *The Journal of Clinical Investigation*, **111**, 1011–20.
92. Zarbock, A., Singbartl, K., and Ley, K. (2006) Complete reversal of acid-induced acute lung injury by blocking of platelet-neutrophil aggregation. *The Journal of Clinical Investigation*, **116**, 3211–19.
93. Broadley, K.N., Aquino, A.M., Woodward, S.C. *et al.* (1989) Monospecific antibodies implicate basic fibroblast growth factor in normal wound repair. *Laboratory Investigation*, **61**, 571–75.
94. Blotnick, S., Peoples, G.E., Freeman, M.R. *et al.* (1994) T lymphocytes synthesize and export heparin-binding epidermal growth factor-like growth factor and basic fibroblast growth factor, mitogens for vascular cells and fibroblasts: differential production and release by $CD4^+$ and $CD8^+$ T cells. *Proceedings of the National Academy*

of Sciences of the United States of America, **91**, 2890–94.
95. Leibovich, S.J., Polverini, P.J., Shepard, H.M. *et al.* (1987) Macrophage-induced angiogenesis is mediated by tumour necrosis factor-alpha. *Nature*, **329**, 630–32.
96. Fathallah-Shaykh, H.M., Zhao, L.J., Kafrouni, A.I. *et al.* (2000) Gene transfer of IFN-gamma into established brain tumors represses growth by antiangiogenesis. *Journal of Immunology*, **164**, 217–22.
97. Zheng, L., Ben, L.H., Pober, J.S., and Bothwell, A.L. (2002) Porcine endothelial cells, unlike human endothelial cells, can be killed by human CTL via Fas ligand and cannot be protected by Bcl-2. *Journal of Immunology*, **169**, 6850–55.
98. Stolpen, A.H., Guinan, E.C., Fiers, W., and Pober, J.S. (1986) Recombinant tumor necrosis factor and immune interferon act singly and in combination to reorganize human vascular endothelial cell monolayers. *American Journal of Pathology*, **123**, 16–24.
99. Joris, I., Cuenoud, H.F., Doern, G.V. *et al.* (1990) Capillary leakage in inflammation. A study by vascular labeling. *American Journal of Pathology*, **137**, 1353–63.
100. Yi, E.S. and Ulich, T.R. (1992) Endotoxin, interleukin-1, and tumor necrosis factor cause neutrophil-dependent microvascular leakage in postcapillary venules. *American Journal of Pathology*, **140**, 659–63.
101. Cotran, R.S., Pober, J.S., Gimbrone, M.A. Jr. *et al.* (1988) Endothelial activation during interleukin 2 immunotherapy. A possible mechanism for the vascular leak syndrome. *Journal of Immunology*, **140**, 1883–88.
102. Damle, N.K. and Doyle, L.V. (1990) Ability of human T lymphocytes to adhere to vascular endothelial cells and to augment endothelial permeability to macromolecules is linked to their state of post-thymic maturation. *Journal of Immunology*, **144**, 1233–40.
103. Martin, N.B., Jamieson, A., and Tuffin, D.P. (1993) The effect of interleukin-4 on tumour necrosis factor-alpha induced expression of tissue factor and plasminogen activator inhibitor-1 in human umbilical vein endothelial cells. *Thrombosis and Haemostasis*, **70**, 1037–42.
104. Pober, J.S. (1999) Immunobiology of human vascular endothelium. *Immunologic Research*, **19**, 225–32.
105. Savage, C.O., Brooks, C.J., Harcourt, G.C. *et al.* (1995) Human vascular endothelial cells process and present autoantigen to human T cell lines. *International Immunology*, **7**, 471–79.
106. Savinov, A.Y., Wong, F.S., Stonebraker, A.C., and Chervonsky, A.V. (2003) Presentation of antigen by endothelial cells and chemoattraction are required for homing of insulin-specific $CD8^+$ T cells. *The Journal of Experimental Medicine*, **197**, 643–56.
107. Burger, D.R., Ford, D., Vetto, R.M. *et al.* (1981) Endothelial cell presentation of antigen to human T cells. *Human Immunology*, **3**, 209–30.
108. Geppert, T.D. and Lipsky, P.E. (1985) Antigen presentation by interferon-gamma-treated endothelial cells and fibroblasts: differential ability to function as antigen-presenting cells despite comparable Ia expression. *Journal of Immunology*, **135**, 3750–62.
109. Grazia, T.J., Pietra, B.A., Johnson, Z.A. *et al.* (2004) A two-step model of acute CD4T-cell mediated cardiac allograft rejection. *Journal of Immunology*, **172**, 7451–58.
110. Kreisel, D., Krupnick, A.S., Gelman, A.E. *et al.* (2002) Non-hematopoietic allograft cells directly activate $CD8^+$ T cells and trigger acute rejection: an alternative mechanism of allorecognition. *Nature Medicine*, **8**, 233–39.
111. Shreeniwas, R., Schulman, L.L., Narasimhan, M. *et al.* (1996) Adhesion molecules (E-selectin and ICAM-1) in pulmonary allograft rejection. *Chest*, **110**, 1143–49.
112. Cunningham, A.C., Zhang, J.G., Moy, J.V. *et al.* (1997) A comparison of the antigen-presenting capabilities of class II MHC-expressing human lung epithelial and endothelial cells. *Immunology*, **91**, 458–63.
113. Grakoui, A., Bromley, S.K., Sumen, C. *et al.* (1999) The immunological synapse: a molecular machine controlling T cell activation. *Science*, **285**, 221–27.
114. Lee, K.H., Holdorf, A.D., Dustin, M.L. *et al.* (2002) T cell receptor signaling precedes immunological synapse formation. *Science*, **295**, 1539–42.
115. Monks, C.R., Freiberg, B.A., Kupfer, H. *et al.* (1998) Three-dimensional segregation of supramolecular activation clusters in T cells. *Nature*, **395**, 82–86.
116. Tuder, R.M., Groves, B., Badesch, D.B., and Voelkel, N.F. (1994) Exuberant endothelial cell growth and elements of inflammation are present in plexiform lesions of pulmonary hypertension. *American Journal of Pathology*, **144**, 275–85.
117. Humbert, M., Morrell, N.W., Archer, S.L. *et al.* (2004) Cellular and molecular pathobiology of pulmonary arterial hypertension. *Journal of the American College of Cardiology*, **43**, 13S–24S.

118. Nicolls, M.R., Tarasevicience-Stewart, L., Rai, P.R. *et al.* (2005) Autoimmunity and pulmonary hypertension: a perspective. *The European Respiratory Journal*, **26**, 1110–18.
119. Iannone, F., Riccardi, M.T., Guiducci, S. *et al.* (2007) Bosentan regulates the expression of adhesion molecules on circulating T cells and serum soluble adhesion molecules in systemic sclerosis-associated pulmonary arterial hypertension. *Annals of the Rheumatic Diseases*, **67**, 1121–6.
120. Babu, A.N., Murakawa, T., Thurman, J.M. *et al.* (2007) Microvascular destruction identifies murine allografts that cannot be rescued from airway fibrosis. *The Journal of Clinical Investigation*, **117**, 3774–85.
121. Luckraz, H., Goddard, M., Mcneil, K. *et al.* (2006) Is obliterative bronchiolitis in lung transplantation associated with microvascular damage to small airways? *The Annals of Thoracic Surgery*, **82**, 1212–18.
122. Wiebe, B.M. and Laursen, H. (1998) Lung morphometry by unbiased methods in emphysema: bronchial and blood vessel volume, alveolar surface area and capillary length. *Acta Pathologica, Microbiologica, et Immunologica Scandinavica*, **106**, 651–56.
123. Hogg, J.C., Chu, F., Utokaparch, S. *et al.* (2004) The nature of small-airway obstruction in chronic obstructive pulmonary disease. *The New England Journal of Medicine*, **350**, 2645–53.
124. Taraseviciene-Stewart, L., Scerbavicius, R., Choe, K.H. *et al.* (2005) An animal model of autoimmune emphysema. *American Journal of Respiratory and Critical Care Medicine*, **171**, 734–42.
125. Pusey, C.D., Dash, A., Kershaw, M.J. *et al.* (1987) A single autoantigen in Goodpasture's syndrome identified by a monoclonal antibody to human glomerular basement membrane. *Laboratory Investigation*, **56**, 23–31.
126. Derry, C.J., Dunn, M.J., Rees, A.J., and Pusey, C.D. (1991) Restricted specificity of the autoantibody response in Goodpasture's syndrome demonstrated by two-dimensional western blotting. *Clinical and Experimental Immunology*, **86**, 457–63.
127. Wieslander, J., Kataja, M., and Hudson, B.G. (1987) Characterization of the human Goodpasture antigen. *Clinical and Experimental Immunology*, **69**, 332–40.
128. Saus, J., Wieslander, J., Langeveld, J.P. *et al.* (1988) Identification of the Goodpasture antigen as the alpha 3(IV) chain of collagen IV. *The Journal of Biological Chemistry*, **263**, 13374–80.
129. Hudson, B.G., Tryggvason, K., Sundaramoorthy, M., and Neilson, E.G. (2003) Alport's syndrome, Goodpasture's syndrome, and type IV collagen. *The New England Journal of Medicine*, **348**, 2543–56.
130. Boosani, C.S., Mannam, A.P., Cosgrove, D. *et al.* (2007) Regulation of COX-2 mediated signaling by alpha3 type IV noncollagenous domain in tumor angiogenesis. *Blood*, **110**, 1168–77.
131. Sethi, S., Lewin, M., Lopez, L., and Lager, D. (2007) Linear anti-glomerular basement membrane IgG but no glomerular disease: Goodpasture's syndrome restricted to the lung. *Nephrology, Dialysis, Transplantation*, **22**, 1233–35.
132. Donaghy, M. and Rees, A.J. (1983) Cigarette smoking and lung haemorrhage in glomerulonephritis caused by autoantibodies to glomerular basement membrane. *Lancet*, **2**, 1390–93.
133. Salama, A.D., Chaudhry, A.N., Ryan, J.J. *et al.* (2001) In Goodpasture's disease, $CD4^+$ T cells escape thymic deletion and are reactive with the autoantigen alpha3(IV)NC1. *Journal of the American Society of Nephrology*, **12**, 1908–15.
134. Renaudineau, Y., Dugue, C., Dueymes, M., and Youinou, P. (2002) Antiendothelial cell antibodies in systemic lupus erythematosus. *Autoimmunity Reviews*, **1**, 365–72.
135. Sasaki, N., Kurose, A., Inoue, H., and Sawai, T. (2002) A possible role of anti-endothelial cell antibody in the sera of MCTD patients on pulmonary vascular damage relating to pulmonary hypertension. *Ryumachi*, **42**, 885–94.
136. Shingu, M. and Hurd, E.R. (1981) Sera from patients with systemic lupus erythematosus reactive with human endothelial cells. *The Journal of Rheumatology*, **8**, 581–86.
137. Manoussakis, M.N., Constantopoulos, S.H., Gharavi, A.E., and Moutsopoulos, H.M. (1987) Pulmonary involvement in systemic sclerosis. Association with anti-Scl 70 antibody and digital pitting. *Chest*, **92**, 509–13.
138. Furuta, K., Hildebrandt, B., Matsuoka, S. *et al.* (1998) Immunological characterization of heterochromatin protein p25beta autoantibodies and relationship with centromere autoantibodies and pulmonary fibrosis in systemic scleroderma. *Journal of Molecular Medicine*, **76**, 54–60.
139. Sachsenberg-Studer, E.M., Prins, C., Saurat, J.H., and Salomon, D. (2000) Raynaud's phenomenon, anticentromere antibodies, and digital necrosis without sclerodactyly: an entity independent of scleroderma? *Journal of the American Academy of Dermatology*, **43**, 631–34.

140. Cines, D.B., Lyss, A.P., Reeber, M. *et al.* (1984) Presence of complement-fixing anti-endothelial cell antibodies in systemic lupus erythematosus. *The Journal of Clinical Investigation*, **73**, 611–25.
141. Dieude, M., Senecal, J.L., and Raymond, Y. (2004) Induction of endothelial cell apoptosis by heat-shock protein 60-reactive antibodies from anti-endothelial cell autoantibody-positive systemic lupus erythematosus patients. *Arthritis and Rheumatism*, **50**, 3221–31.
142. Worda, M., Sgonc, R., Dietrich, H. *et al.* (2003) *In vivo* analysis of the apoptosis-inducing effect of anti-endothelial cell antibodies in systemic sclerosis by the chorionallantoic membrane assay. *Arthritis and Rheumatism*, **48**, 2605–14.
143. Bordron, A., Dueymes, M., Levy, Y. *et al.* (1998) The binding of some human antiendothelial cell antibodies induces endothelial cell apoptosis. *The Journal of Clinical Investigation*, **101**, 2029–35.
144. Kaneko, K., Savage, C.O., Pottinger, B.E. *et al.* (1994) Antiendothelial cell antibodies can be cytotoxic to endothelial cells without cytokine pre-stimulation and correlate with ELISA antibody measurement in Kawasaki disease. *Clinical and Experimental Immunology*, **98**, 264–69.
145. Sgonc, R., Gruschwitz, M.S., Boeck, G. *et al.* (2000) Endothelial cell apoptosis in systemic sclerosis is induced by antibody-dependent cell-mediated cytotoxicity via CD95. *Arthritis and Rheumatism*, **43**, 2550–62.
146. Negi, V.S., Tripathy, N.K., Misra, R., and Nityanand, S. (1998) Antiendothelial cell antibodies in scleroderma correlate with severe digital ischemia and pulmonary arterial hypertension. *The Journal of Rheumatology*, **25**, 462–66.
147. Song, J., Park, Y.B., Lee, W.K. *et al.* (2000) Clinical associations of anti-endothelial cell antibodies in patients with systemic lupus erythematosus. *Rheumatology International*, **20**, 1–7.
148. Chan, T.M. and Cheng, I.K. (1996) A prospective study on anti-endothelial cell antibodies in patients with systemic lupus erythematosus. *Clinical Immunology and Immunopathology*, **78**, 41–46.
149. Navarro, M., Cervera, R., Font, J. *et al.* (1997) Anti-endothelial cell antibodies in systemic autoimmune diseases: prevalence and clinical significance. *Lupus*, **6**, 521–26.
150. Quismorio, F.P. Jr., Sharma, O., Koss, M. *et al.* (1984) Immunopathologic and clinical studies in pulmonary hypertension associated with systemic lupus erythematosus. *Seminars in Arthritis and Rheumatism*, **13**, 349–59.
151. Nakagawa, N., Osanai, S., Ide, H. *et al.* (2003) Severe pulmonary hypertension associated with primary Sjogren's syndrome. *Internal Medicine*, **42**, 1248–52.
152. Magro, C.M., Klinger, D.M., Adams, P.W. *et al.* (2003) Evidence that humoral allograft rejection in lung transplant patients is not histocompatibility antigen-related. *American Journal of Transplantation*, **3**, 1264–72.
153. Magro, C.M., Ross, P. Jr., Kelsey, M. *et al.* (2003) Association of humoral immunity and bronchiolitis obliterans syndrome. *American Journal of Transplantation*, **3**, 1155–66.
154. Badesch, D.B., Zamora, M., Fullerton, D. *et al.* (1998) Pulmonary capillaritis: a possible histologic form of acute pulmonary allograft rejection. *The Journal of Heart and Lung Transplantation*, **17**, 415–22.
155. Astor, T.L., Weill, D., Cool, C. *et al.* (2005) Pulmonary capillaritis in lung transplant recipients: treatment and effect on allograft function. *The Journal of Heart and Lung Transplantation*, **24**, 2091–97.
156. Wallace, W.D., Reed, E.F., Ross, D. *et al.* (2005) C4d staining of pulmonary allograft biopsies: an immunoperoxidase study. *The Journal of Heart and Lung Transplantation*, **24**, 1565–70.
157. Feghali-Bostwick, C.A., Gadgil, A.S., Otterbein, L.E. *et al.* (2008) Autoantibodies in patients with chronic obstructive pulmonary disease. *American Journal of Respiratory and Critical Care Medicine*, **177**, 156–63.

26

Role of the Endothelium in Emphysema: Emphysema – A Lung Microvascular Disease

Norbert F. Voelkel and Ramesh Natarajan

Victoria Johnson Center for Pulmonary Obstructive Disease Research, Pulmonary and Critical Care Medicine Division, Virginia Commonwealth University, Richmond, VA, USA

INTRODUCTION

Emphysema is defined as (alveolar) airspace enlargement [1], usually occurring in adults. It is distinguished from lung developmental abnormalities that, due to disturbed alveolarization, can result in a lung characterized by reduced number of enlarged alveoli. Although world-wide the most common cause of emphysema is cigarette smoking [2], adult human emphysema does also occur in nonsmokers and can be associated with HIV infection or hypersensitivity pneumonitis [3, 4]. Occupational exposure to cadmium and copper compounds have been reported to cause emphysema [5] as well as semi-starvation [6].

Until recently the tacit assumption has been that emphysema is a consequence of lung epithelial cell damage and death [7]. A vascular component as part of the pathobiology of emphysema was first discussed by A. Liebow at the occasion of the First Emphysema Conference in Aspen, Colorado in 1958 [8]. However, as recent as 10 years ago, the two words "endothelium" and "emphysema" did not appear in the same sentence. The traditional theory of emphysema pathogenesis posits that inflammatory cells in the lung, particularly neutrophils and macrophages [9, 10], generate and release proteolytic enzyme activity emanating in a gradient from clusters of activated inflammatory cells that is met by another gradient of antiproteolytic enzyme activity, hence the protease/antiprotease theory [11, 12].

As it has been recognized that endothelial cells (ECs) can express proteases [13, 14], a process of autoproteolysis and therefore loss of ECs within the framework of a protease/antiprotease imbalance can also be considered. Given the intimate functional interactions between alveolar septum cells and alveolar macrophages, and neutrophils and lymphocytes within the capillaries, a multicellular participation in emphysema formation is likely. However, this chapter will make an attempt to build a case for a high vulnerability of the alveolar septal ECs and a privileged role of these cells in the emphysema pathobiology. One of the easily understandable reasons for the early focus on airways in the emphysema pathogenesis is that, experimentally, proteases and protease containing tissue extracts had been delivered to the lung via airway aerolization. For example, neutrophil and alveolar macrophage extracts, when aerosolized into the airways, caused emphysema-like lung lesions, but not when added to the lung circulation [15].

As mentioned, emphysema is a pathohistological diagnosis and airspace enlargement leads to a loss of lung parenchyma which can be very severe, hence the term "vanishing lung syndrome." A reasonable estimate of the degree of loss of functional gas exchange units can be obtained by measuring the diffusing capacity for carbon monoxide. Thus, a significant decrease of the carbon monoxide diffusing capacity (DL_{CO}) reflects, in the absence of lung tissue fibrosis, the loss of alveolar capillaries and small vessels. It has been known for many years [16] that the lung capillaries are embedded in an extremely dense mesh of elastin fibers and perhaps for this reason unbridled elastolysis in emphysema leads to a loss of lung capillaries. Other causes or reasons that can explain capillary loss in emphysema are discussed below.

Whereas computed tomography of the lung is now being used to localize areas of emphysematous lung tissue

The Pulmonary Endothelium: Function in health and disease Editors Norbert F. Voelkel, Sharon Rounds

destruction and methods to quantitate the emphysematous lung tissue loss are being validated, lung tissue sections are examined morphometrically and emphysema is assessed histologically by measuring the "mean linear intercept" (the distance between alveolar septae is measured using a grid) or the alveolar surface area [17] or the volume of the airspaces.

EMPHYSEMA: A FAILURE OF THE LUNG STRUCTURE MAINTENANCE PROGRAM

Modern concepts of emphysema pathogenesis have been recently reviewed [18] and [19] discuss the molecular pathogenesis of emphysema in the context of a postulated adult lung structure maintenance program (LSMP). This concept is an extrapolation from the now well-studied lung development program with its succinct and demonstrates sequential phases of organ building [20] (see Chapter 1). This extrapolation appears to be justified since developmental arrest or impairment does result in airspace enlargement. Indeed, postnatal inhibition of angiogenesis causes airspace enlargement [21]. The adult LSMP is conceptualized as a vestigal homeostatic program that is endowed with a largely unexplored measure of plasticity and potential to reconstitute – likely from stem cells – injured parts of the lung as well as a compensatory lung growth potential – which is realized at least in rodents – following pneumonectomy [22]. The postulate is that this repair and growth potential (of stem cells) is reactive and functioning in response to a variety of challenges to the lung organ's structural integrity (see Chapter 13). However, it may be irresponsive and malfunctioning when confronted with challenges that overwhelm the homeostatic equilibrium. A further postulate is that the lung microvascular EC play a critically important role in the adult LSMP (see Chapter 9), and the third postulate is that vascular endothelial growth factor (VEGF) is the principal lung microvascular EC maintenance factor.

EC APOPTOSIS

Although apoptosis in human emphysema lung tissue samples has been described [23, 24] and confirmed [25–28], documentation of pulmonary EC apoptosis in emphysema (Figure 26.1a) is sparse [29] (see Chapter 16). More data are available in rodent models of emphysema [23, 30–32], and agents like the VEGF receptor blocker SU5416 and the lipid metabolism product ceramide, which have been associated with emphysema, have been shown to cause apoptosis of cultured ECs [31] (see Table 26.1). In addition, α_1-antitrypsin, an antiprotease and target of elastase that prevents experimental rodent emphysema development, also protects ECs against apoptosis [32]. The question whether alveolar septal EC apoptosis comes first or is a consequence of alveolar epithelial cell death or matrix destruction (anoikis), particularly pericapillary elastin digestion, has been raised, but not resolved. Mechanisms of vessel regression have recently been proposed [33]. Even the remaining lung EC which are not undergoing apoptosis present evidence of EC dysfunction and phenotype alterations, like the loss of the prostacyclin synthase gene and protein (Figure 26.1b) [34, 35]. A complete description of the EC phenotype in emphysematous lungs is lacking. In addition to the loss of prostacyclin, ECs in emphysematous lungs have decreased endothelial nitric oxide synthase [36] and nuclear factor erythroid 2-related factor 2 (Nrf2) expression [37]. The loss of the expression of the transcription factor Nrf2 may explain the loss of expression of the antioxidant enzyme HO-1 [38].

VEGF RECEPTOR BLOCKADE CAUSES EMPHYSEMA

A single subcutaneously implanted dose of the combined VEGF receptor (vascular endothelial growth factor receptor VEGFR)-1 and -2 blocker SU5416 (developed as an antiangiogenesis/anticancer drug) causes emphysema in rats, associated with a dramatic loss of lung vessels within 3 weeks of implantation. This drug causes an

Table 26.1 Agents and drugs inducing EC apoptosis and emphysema.

	EC apoptosis	Emphysema
SU5416	[99]	[23]
Methylprednisolone	not investigated	[100]
CSE	[34]	[48]
Acrolein	[101]	(Lee J. *et al.*, unpublished)
Caspase	[102]	[102]
Adenosine	[103]	[104]
Ceramide	[32, 105]	[106]

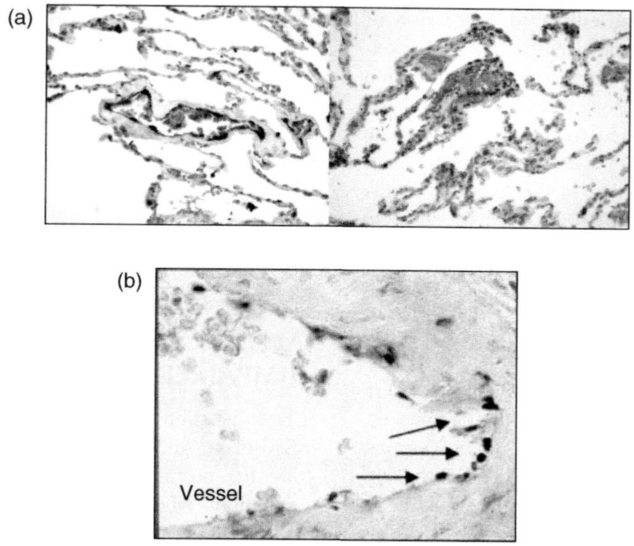

Figure 26.1 (a) Human lung tissue sections. The tissue was stained using an antibody directed against the enzyme (protein) prostacyclin synthase. The EC monolayer stains grown in the normal lung tissue section (left), whereas staining is absent in the emphysema lung section (right). This finding is indicative of an altered lung EC phenotype in emphysema. Reproduced from [34] with the permission of the American Thoracic Society. (b) Terminal deoxynucleotidyl transferase biotin-dUTP nick end-labeling staining of a lung vessel in a human emphysema lung section demonstrates apoptotic ECs within the vessel EC monolayer (arrows). Reproduced from [29] with the permission of the American Thoracic Society. A color version of this figure appears in the plate section of this volume.

impressive alveolar septal cell (including EC) apoptosis, which is functionally important because concomitant treatment of the animals with a broad-spectrum caspase antagonist prevents emphysema development [23]. Similar results (i.e., emphysema) were generated when mice received anti-VEGFR antibodies [32] or following a conditional knockout that abolished lung VEGF protein production [39].

In addition, Taraseviciene-Stewart *et al.* [4] immunized rats with human umbilical vein ECs (human umbilical vein endothelial cell HUVECs) – a strategy that produces anti-EC antibodies [40, 41] and emphysema in these immunized rats. Further, when plasma from immunized emphysematous rats is transferred to naïve mice, they also develop emphysema. Taken together, these experimental data provide evidence that decreased lung VEGF expression and inhibition of VEGFRs cause emphysema, and the HUVEC immunization experiments provide perhaps the strongest support of the concept that anti-EC antibodies and pathological T lymphocytes can attack lung ECs resulting in emphysema.

In summary, human emphysema is associated with septal cell-including EC apoptosis and reduced VEGF gene and protein expression and reduced VEGFR-2 [also known as kinase insert domain-containing receptor (KDR)] gene and protein expression. Experimental VEGFR blockade in rodents induces alveolar septal cell apoptosis and emphysema. In addition, cigarette smoke extract (CSE) induces lung microvascular EC apoptosis and inhibits VEGF expression *in vitro* [34], and intratracheal CSE instillation causes lung septal cell apoptosis, emphysema, and reduced lung tissue VEGF protein expression in rats [42].

VEGF: THE LYNCH PIN OF EMPHYSEMA PATHOGENESIS

It is apparent that studies of human lungs diseased with emphysema are descriptive. Nevertheless, the clinical data also raise questions which probe the role of VEGF in the pathobiology of emphysema. The first question is: why are expression of VEGF and KDR low in emphysematous lung tissue, whereas the expression of the ligand and receptor protein are high in lung tissue from asthma patients [43, 44]? This differential gene and protein expression might not only be related but actually causative for the high-angiogenesis state of asthma and the impaired angiogenesis state of emphysema [45]. Although both asthma and emphysema are characterized by lung tissue inflammation, interestingly the inflammation in emphysema does not cause enhanced expression of the pleiotropic VEGF and, paradoxically, neither hypoxemia nor lung tissue hypoxia effect increased expression in lung tissue VEGF levels. Both inflammation and tissue hypoxia would be expected to increase the release of this heparin-binding growth factor [46] and not decrease VEGF expression. Added to this paradox is the fact that lung microvascular ECs generate (at least *in vitro*) spontaneously large quantities of VEGF, leading to the speculation that this high microvascular EC VEGF production and secretion may serve a paracrine and even autocrine survival purpose [47]. A similar conclusion had recently been reached by Lee [48], distinguishing in a report in *Cell* between the role of cell-autonomous VEGF signaling in vascular homeostasis and its angiogenic role.

Translated to emphysemagenesis: lung microvascular ECs depend on their own high levels of VEGF for survival and are very vulnerable when VEGF signaling is impaired, resulting in microvascular EC apoptosis. If this scenario is correct, then indeed lung microvascular EC maintenance is critically VEGF-dependent and the microvascular ECs play a central role in emphysemagenesis. Since VEGF is also involved in the successful phagocyte removal of apoptotic cells (which limits inflammation) [49], decreased VEGF tissue levels and impaired VEGFR signaling may impair apoptotic cell removal in emphysematous lungs and contribute to persistence of inflammation or impaired resolution of inflammation. Whether an interplay or imbalance between VEGF and transforming growth factor (TGF)-β signaling causes EC apoptosis [50, 51], and whether, for example, a combination of decreased VEGF and decreased TGF-β expression would cause emphysema, are interesting questions to pursue [52].

WHY IS VEGF EXPRESSION REDUCED (IMPAIRED) IN EMPHYSEMA?

This question may point towards a fundamental problem of vessel homeostasis and we can simply ask: what are the mechanisms that prevent VEGF expression in situations where it would make teleological sense for VEGF to be highly expressed? One such situation is ischemia–reperfusion, the other is emphysema. In both cases the transcriptional stimuli provided by low oxygen and inflammatory cells are apparently ineffective, as are paracrine effects by activated macrophages and mast cells that secrete VEGF [53, 54] (Figure 26.2). VEGF

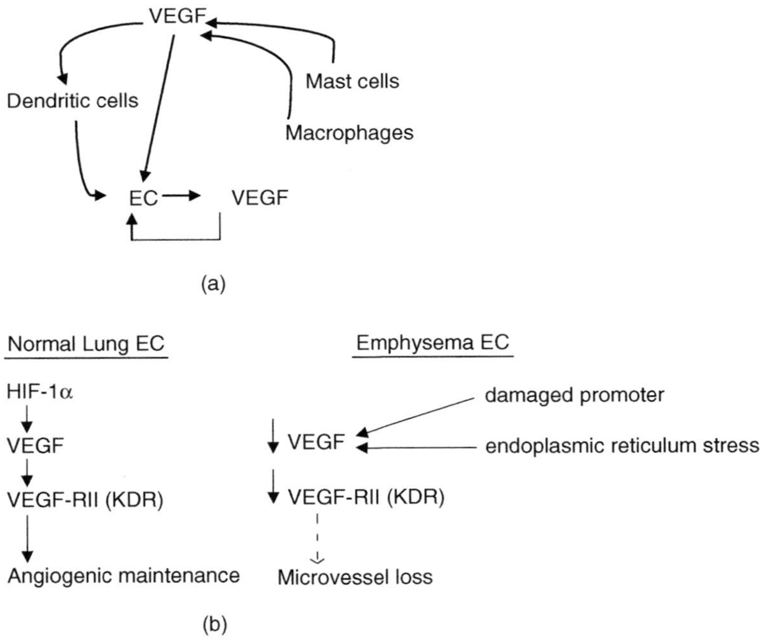

Figure 26.2 (a) The schematic depicts VEGF in the center of cell–cell interactions as well as an autocrine loop. VEGF is secreted by mast cells and macrophages, and acts on ECs. In addition, microvascular ECs secrete VEGF for their own maintenance (autocrine action). Dendritic cells have been shown to transdifferentiate into EC under the influence of VEGF. (b) Schematic comparison of the angiogenic maintenance in normal lung tissue and the microvessel loss attributable to reduced VEGF gene and protein expression and impaired VEGFR signaling.

gene expression is "sensitive" to oxidative stress, and it is known that hyperoxia decreases VEGF and KDR expression in the lungs from hyperoxic rats [55]. Gillespie's group investigated the effect of severe hypoxia on pulmonary EC VEGF gene expression and demonstrated oxidative base modifications in the hypoxia response element of the VEGF promoter [56, 57].

In addition to this proposed mechanism that results in impaired VEGF gene transcription, the VEGF protein, like many secreted proteins, is subject to endoplasmic reticulum (ER) stress or the unfolded protein response (UPR) 58-60 (see Chapter 16). The ER is a cellular organelle that directs folding of secretory and membrane proteins [61, 62].The ER senses oxidative stress, maintains calcium homeostasis, and triggers apoptotic signaling [60, 63]. Activation of this response would predictably result both in reduced VEGF transcription and in reduced VEGF protein secretion [64]. Indeed, cigarette smoke induced ER stress in rat lungs [65] and aqueous CSE induced the ER response in fibroblasts [66]. Importantly, excessive and prolonged ER stress triggers apoptotic cell death [59].

As shown in Figure 26.3, the UPR is mediated by at least three regulatory pathways; two involved in transcriptional regulation [activation of transcription factor (ATF)-6 and X-box binding protein (XBP)-1] and a third that controls protein translation [eukaryotic initiation factor (eIF)-2α] 67-69. ER stress is initially sensed by the primary proximal effectors of the UPR (see Figure 26.3), protein kinase R-like ER kinase (protein kinase R-like endoplasmic reticulum kinase PERK) [70], ATF-6 [71], and inositol-requiring enzyme (IRE)-1 [72]. Immunoglobulin heavy chain-binding protein (BiP, also known as Grp78), a central ER-resident chaperone, docks to these proteins, thereby preventing their activation. Increased numbers of misfolded proteins within the ER prompts BiP to "undock" from IRE-1, PERK, and ATF-6. ATF-6 subsequently undergoes proteolytic activation in the Golgi following which it translocates to the nucleus, forms homodimers or heterodimers with bZip transcription factors such as XBP-1 and regulates expression of ER stress response genes [73]. IRE-1 and PERK undergo oligomerization within ER membranes. IRE-1's unusual ribonuclease activity splices the mRNA that encodes for active XBP-1, a transcription factor inducing genes that restore protein folding (BiP) or degrade unfolded proteins [ER degradation-enhancing α-mannosidase-like protein (ER degradation-enhancing α-mannosidase-like protein EDEM)] [74]. Oligomerized IRE-1 also signals kinases that activate nuclear factor-κB and c-Jun (activating protein-1), driving host defense-associated genes. PERK oligomerization activates its intrinsic kinase activity, resulting in eIF-2α phosphorylation – an event which heralds suppressed mRNA translation and cell cycle arrest [58, 70, 75]. Following eIF-2α phosphorylation, only select mRNAs, such as ATF-4, are translated. ATF-4 expression promotes both prosurvival (early) and proapoptotic (late) transcriptional programs.

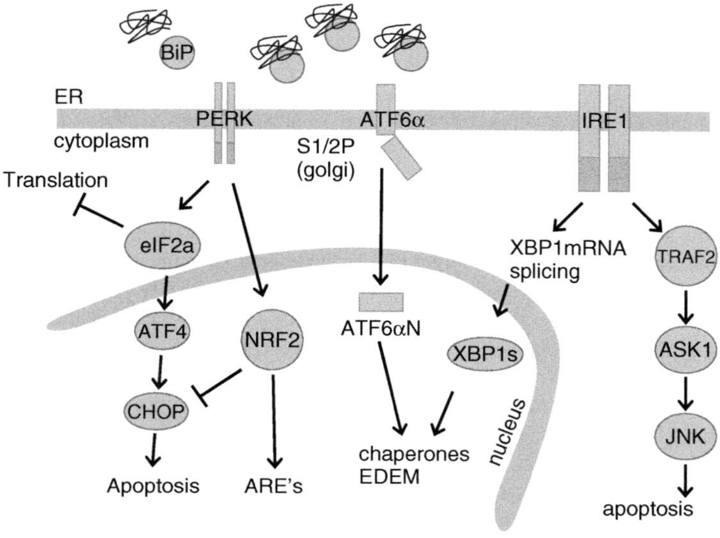

Figure 26.3 This schematic depicts some of the pathways which are activated during ER stress leading to the UPR. Secreted proteins are affected by this stress response and apoptosis can be a consequence. (See text for a detailed discussion of these pathways; TRAF2, tumor necrosis factor receptor-associated factor-2; ASK1, apoptosis signal-regulating kinase-1.) From [98].

Sustained ER stress, acting through ATF-4, leads to activation of the proapoptotic peptide growth arrest and DNA damage/CCAAT/enhancer-binding protein (CEBP) homology protein (CCAAT/enhancer-binding protein homology protein CHOP). CHOP overexpression induces apoptosis, through a Bcl-2-inhibitable mechanism [76, 77]. CHOP silencing (chop$^{-/-}$ mice) confers significant resistance to ER stress-induced renal injury (induced by tunicamycin) as well as resistance to brain injury resulting from cerebral artery occlusion. This suggests that CHOP plays a significant role in cell death associated with ER stress [78]. Thus it is conceivable that oxidant-mediated misfolding of proteins could not only impair the ER's ability to synthesize proteins (such as VEGF and KDR), but also trigger apoptosis in lung ECs and contribute to smoking-induced emphysematous tissue destruction.

IMMUNE MECHANISMS IN EMPHYSEMAGENESIS

One interesting aspect of the above mentioned autoimmune/HUVEC immunization model of emphysema is the transfer of the disease (i.e., emphysema) by spleen CD4^{+} T lymphocytes. The work of Jordan Pober has shown that ECs are antigen-presenting cells (APCs) [79, 80] and thus it is conceivable that pathogenic T cells in this autoimmune rat model attack the lung vessels, particularly the alveolar septal ECs (see Chapter 25). In addition, septal EC which undergo apoptosis may become antigenic after processing by macrophages/monocytes or phagocytic uptake by neighboring ECs. The phagocytic properties of ECs have been documented convincingly by studies conducted many years ago in Una Ryan's lab [81]. EC immunogenicity may also be generated or enhanced by altered extracellular matrix composition or altered three-dimensional spatial orientation [82, 83]. [84], demonstrated the presence of high titers of proteins which bind to ECs (likely anti-EC antibodies) in the sera from patients with severe emphysema. These proteins inhibited EC growth *ex vivo* in culture.

Dendritic cells are professional APCs, their role in emphysema has not been explored (see Chapter 25). Peripheral immature dendritic cells derived from monocytic precursors are recruited to sites of inflammation in the lung and thus EC/dendritic cell interactions are to be expected. Here again, VEGF might be of great importance since VEGF is a dendritic cell-suppressive cytokine [85]. In this case reduced VEGF levels and impaired VEGF signaling would favor the maturation of dendritic cell, s and possibly impede dendritic cell/EC transdifferentiation that has been reported *in vitro* by Sozzani *et al.* [86].

Finally, Toll-like receptors (TLRs) must be considered as part of a "danger and damage" reporting system; they can convert T cell autoreactivity into overt autoimmune disease [87] and appear to play a role in the adult LSMP [30]. The studies by Zhang *et al.* showed spontaneous emphysema development in TLR-4 knockout mice. This emphysema was indeed associated with lung EC apoptosis [30].

LUNG ECs AND REPAIR OF THE EMPHYSEMATOUS LUNG STRUCTURE

The introduction of the concept of a LSMP [19] fills a gap in our understanding of lung biology; it adds an active homeostatic component to the injury/repair spectrum, and directs our thoughts toward cell turnover and ongoing repair. Progressive emphysema is not only due to alveolar septal cell destruction, but also due to tissue repair failure. A VEGF-based view of emphysemagenesis implies that meaningful (i.e., functional) regeneration of the emphysematous lung [88] cannot be accomplished without normalization of VEGF tissue expression and the repair of VEGF receptor signaling. Such a strategy is now feasible. Contemporary models of lung injury repair operate with at least two components: bone marrow-derived precursor cells and lung-resident precursor cells [89, 90]. Whether bone marrow-derived precursor cells participate in the activity of the adult LSMP or in repair of the emphysematous lung is not known. The discovery of highly proliferative lung microvascular ECs in adult rats [91] makes one wonder whether these ECs participate in lung tissue repair. If so, one would hope that these resident lung EC progenitors can lay down the appropriate matrix [92] required for the meaningful repair of alveolar spaces (see Chapter 13).

CONCLUSIONS AND PERSPECTIVES

A compartment-oriented analysis of lung diseases, which posits that interstitial lung diseases begin and end within the interstitial compartment, has not stood the test of time. Given the intricately connected multicellular structure of the lung and the fact that the lung is not only a gas exchanger, but also a large metabolically active organ (to a large measure because of EC function), it is difficult to see how emphysema could only and strictly be an airway disease and can develop without significant involvement of lung ECs. Here, we argue that pulmonary ECs are not only involved, but indeed play a central role in the pathobiology of emphysema because of the critical dependence of their survival and regenerative growth on VEGF. Apoptosis of ECs by itself cannot explain emphysemagenesis because apoptosis *per se* begets EC growth [93], and apoptotic bodies enhance the number and differentiation of EC progenitor cells [94] – a process requiring VEGF. Indeed, VEGF can be seen as an important hub or master controller of angiogenesis and lung tissue repair [95], integrating bone marrow responses and affecting the behavior and function of cells of the innate and adaptive immune system.

In this chapter we have reviewed aspects of EC involvement in emphysemagenesis and it is apparent that many important questions have not been asked or answered. If some – or a considerable – part of the lung tissue maintenance and repair program relies on resident EC precursor cells it will be important to know whether their separative potential diminishes with age and chronic cigarette smoke exposure. Further, it is unclear whether such lung EC precursor cells are pluripotent (i.e., can give rise to vascular smooth muscle cells and fibroblasts) and whether they can repair destroyed matrix. Antioxidant strategies need to be discovered which protect against lung EC apoptosis without inhibiting cell growth. Such strategies may be of benefit and control EC–mesenchymal transformation as has been described in the myocardium [96]. We predict that a focus on the important role of lung EC in homeostasis will help explain the importance of angiogenesis and inhibition of angiogenesis in the development of both lung emphysema and fibrosis. Experimental strategies have now become available to inhibit ER stress resulting in hypoxia-inducible factor-1α-related preservation of tissue VEGF levels and normal VEGFR function (Natarajan *et al.*, unpublished). Such strategies may be successfully used to restore lung alveolar septal structures and repair the emphysematous lung because of the centrality of VEGF actions that also affect integrin (collagen receptor) expression [97].

References

1. Snider, G.L., Kleinerman, J., Thurlbeck, W.M., and Bengali, Z.H. (1985) The definition of emphysema. *The American Review of Respiratory Disease*, **132**, 182–85.
2. Rushton, L. (2007) Occupational causes of chronic obstructive pulmonary disease. *Reviews on Environmental Health*, **22**, 195–212.
3. Sahebjami, H. (1992) Emphysema-like changes in HIV. *Annals of Internal Medicine*, **116**, 876.
4. Taraseviciene-Stewart, L., Scerbavicius, R., Choe, K.H. *et al.* (2005) An animal model of autoimmune emphysema. *American Journal of Respiratory and Critical Care Medicine*, **171**, 734–42.
5. Kirschvink, N., Martin, N., Fievez, L. *et al.* (2006) Airway inflammation in cadmium-exposed rats is associated with pulmonary oxidative stress and emphysema. *Free Radical Research*, **40**, 241–50.
6. Coxson, H.O., Chan, I.H., Mayo, J.R. *et al.* (2004) Early emphysema in patients with anorexia nervosa. *American Journal of Respiratory and Critical Care Medicine*, **170**, 748–52.
7. Hodge, S., Hodge, G., Holmes, M., and Reynolds, P.N. (2005) Increased airway epithelial and T-cell apoptosis in COPD remains despite smoking cessation. *European Respiratory Journal*, **25**, 447–54.
8. Aspen Lung Conference (1959) Symposium of Emphysema and the "Chronic Bronchitis" Syndrome. *The American Review of Respiratory Disease*, **80**, 1–213.
9. Minematsu, N. and Shapiro, S.D. (2007) To live and die in the LA (lung airway): mode of neutrophil death and progression of chronic obstructive pulmonary disease. *American Journal of Respiratory Cell and Molecular Biology*, **37**, 129–30.
10. Houghton, A.M., Quintero, P.A., Perkins, D.L. *et al.* (2006) Elastin fragments drive disease progression in a murine model of emphysema. *The Journal of Clinical Investigation*, **116**, 753–59.
11. Hogg, J.C. and Senior, R.M. (2002) Chronic obstructive pulmonary disease–part 2: pathology and biochemistry of emphysema. *Thorax*, **57**, 830–34.
12. Kassim, S.Y., Fu, X., Liles, W.C. *et al.* (2005) NADPH oxidase restrains the matrix metalloproteinase activity of macrophages. *The Journal of Biological Chemistry*, **280**, 30201–5.
13. van Hinsbergh, V. and Koolwijk, P. (2008) Endothelial sprouting and angiogenesis: matrix metalloproteinases in the lead. *Cardiovascular Research*, **78**, 203–12.
14. Taraboletti, G., D'Ascenzo, S., Borsotti, P. *et al.* (2002) Shedding of the matrix metalloproteinases MMP-2, MMP-9, and MT1-MMP as membrane vesicle-associated components by endothelial cells. *American Journal of Pathology*, **160**, 673–80.
15. Kimbel, P. and Weinbaum, G. (1974) Role of leucoproteases in the genesis of emphysema, in *Lung Metabolism* (eds A.F. Junod and R. de Haller), Academic Press, New York, pp. 25–41.
16. Orsós, F. (1907) Über das elastische Gerüst der normalen und der emphysematösen Lunge. *Beitrag zur pathologischen Anatomie*, **41**, 95.
17. Chetty, A., Cao, G.J., Severgnini, M. *et al.* (2008) Role of matrix metalloprotease-9 in hyperoxic injury in developing lung. *American Journal of Physiology: Lung Cellular and Molecular Physiology*, **295**, L584–92.

18. Yoshida, T. and Tuder, R.M. (2007) Pathobiology of cigarette smoke-induced chronic obstructive pulmonary disease. *Physiological Reviews*, **87**, 1047–82.
19. Taraseviciene-Stewart, L. and Voelkel, N.F. (2008) Molecular pathogenesis of emphysema. *The Journal of Clinical Investigation*, **118**, 394–402.
20. Galambos, C. and Demello, D.E. (2008) Regulation of alveologenesis: clinical implications of impaired growth. *Pathology*, **40**, 124–40.
21. Jakkula, M., Le Cras, T.D., Gebb, S. *et al.* (2000) Inhibition of angiogenesis decreases alveolarization in the developing rat lung. *American Journal of Physiology: Lung Cellular and Molecular Physiology*, **279**, L600–7.
22. Fernandez, L.G., Le Cras, T.D., Ruiz, M. *et al.* (2007) Differential vascular growth in postpneumonectomy compensatory lung growth. *Journal of Thoracic and Cardiovascular Surgery*, **133**, 309–16.
23. Kasahara, Y., Tuder, R.M., Taraseviciene-Stewart, L. *et al.* (2000) Inhibition of VEGF receptors causes lung cell apoptosis and emphysema. *The Journal of Clinical Investigation*, **106**, 1311–19.
24. Imai, K., Mercer, B.A., Schulman, L.L. *et al.* (2005) Correlation of lung surface area to apoptosis and proliferation in human emphysema. *European Respiratory Journal*, **25**, 250–58.
25. Segura-Valdez, L., Pardo, A., Gaxiola, M. *et al.* (2000) Upregulation of gelatinases A and B, collagenases 1 and 2, and increased parenchymal cell death in COPD. *Chest*, **117**, 684–94.
26. Demedts, I.K., Demoor, T., Bracke, K.R. *et al.* (2006) Role of apoptosis in the pathogenesis of COPD and pulmonary emphysema. *Respiratory Research*, **7**, 53.
27. Calabrese, F., Giacometti, C., Beghe, B. *et al.* (2005) Marked alveolar apoptosis/proliferation imbalance in end-stage emphysema. *Respiratory Research*, **6**, 14.
28. Yokohori, N., Aoshiba, K., and Nagai, A. (2004) Increased levels of cell death and proliferation in alveolar wall cells in patients with pulmonary emphysema. *Chest*, **125**, 626–32.
29. Kasahara, Y., Tuder, R.M., Cool, C.D. *et al.* (2001) Endothelial cell death and decreased expression of vascular endothelial growth factor and vascular endothelial growth factor receptor 2 in emphysema. *American Journal of Respiratory and Critical Care Medicine*, **163**, 737–44.
30. Zhang, X., Shan, P., Jiang, G. *et al.* (2006) Toll-like receptor 4 deficiency causes pulmonary emphysema. *The Journal of Clinical Investigation*, **116**, 3050–59.
31. Petrache, I., Fijalkowska, I., Zhen, L. *et al.* (2006) A novel antiapoptotic role for alpha1-antitrypsin in the prevention of pulmonary emphysema. *American Journal of Respiratory and Critical Care Medicine*, **173**, 1222–28.
32. Petrache, I., Fijalkowska, I., Medler, T.R. *et al.* (2006) Alpha-1 antitrypsin inhibits caspase-3 activity, preventing lung endothelial cell apoptosis. *American Journal of Pathology*, **169**, 1155–66.
33. Im, E. and Kazlauskas, A. (2006) New insights regarding vessel regression. *Cell Cycle*, **5**, 2057–59.
34. Nana-Sinkam, S.P., Lee, J.D., Sotto-Santiago, S. *et al.* (2007) Prostacyclin prevents pulmonary endothelial cell apoptosis induced by cigarette smoke. *American Journal of Respiratory and Critical Care Medicine*, **175**, 676–85.
35. Klumpp, G., Schildknecht, S., Nastainczyk, W. *et al.* (2005) Prostacyclin in the cardiovascular system: new aspects and open questions. *Pharmacological Reports*, **57** (Suppl), 120–26.
36. Barberá, J.A., Peinado, V.I., Santos, S. *et al.* (2001) Reduced expression of endothelial nitric oxide synthase in pulmonary arteries of smokers. *American Journal of Respiratory and Critical Care Medicine*, **164**, 709–13.
37. Malhotra, D., Thimmulappa, R., Navas-Acien, A. *et al.* (2008) Decline in NRF2-regulation antioxidants in chronic obstructive pulmonary disease lungs due to loss of its positive regulator, DJ-1. *American Journal of Respiratory and Critical Care Medicine*, **178**, 592–604.
38. Maestrelli, P., Páska, C., Saetta, M. *et al.* (2003) Decreased haem oxygenase-1 and increased inducible nitric oxide synthase in the lung of severe COPD patients. *European Respiratory Journal*, **21**, 971–76.
39. Tang, K., Rossiter, H.B., Wagner, P.D., and Breen, E.C. (2004) Lung-targeted VEGF inactivation leads to an emphysema phenotype in mice. *Journal of Applied Physiology*, **97**, 1559–66.
40. Wei, Y.Q., Wang, Q.R., Zhao, X. *et al.* (2000) Immunotherapy of tumors with xenogeneic endothelial cells as a vaccine. *Nature Medicine*, **6**, 1160–66.
41. Scappaticci, F.A., Contreras, A., Boswell, C.A. *et al.* (2003) Polyclonal antibodies to xenogeneic endothelial cells induce apoptosis and block support of tumor growth in mice. *Vaccine*, **21**, 2667–77.
42. Lee, J.H., Hanaoka, M., Kitaguchi, Y., and Voelkel, N.F. (2008) Can ovalbumin immunization protect against the development of emphysema. *American Journal of Respiratory and Critical Care Medicine*, **77**, A658.

43. Kanazawa, H., Hirata, K., and Yoshikawa, J. (2003) Imbalance between vascular endothelial growth factor and endostatin in emphysema. *European Respiratory Journal*, **22**, 609–12.
44. Koyama, S., Sato, E., Haniuda, M. *et al.* (2002) Decreased level of vascular endothelial growth factor in bronchoalveolar lavage fluid of normal smokers and patients with pulmonary fibrosis. *American Journal of Respiratory and Critical Care Medicine*, **166**, 382–85.
45. Voelkel, N.F. (2008) Vascular endothelial growth factor and its role in emphysema and asthma, in *Chronic Obstructive Lung Diseases* (eds N. Voelkel and W. MacNee), BC Decker, Hamilton, ON, pp. 77–85.
46. Krilleke, D., DeErkenez, A., Schubert, W. *et al.* (2007) Molecular mapping and functional characterization of the VEGF164 heparin-binding domain. *The Journal of Biological Chemistry*, **282**, 28045–56.
47. Stevens, T., Kasper, M., Cool, C., and Voelkel, N. (2005) Pulmonary circulation and pulmonary hypertension, in *Endothelial Cells in Health and Disease* (ed. W.C. Aird), Taylor & Francis, Boca Raton, FL, pp. 467–505.
48. Lee, S., Chen, T.T., Barber, C.L. *et al.* (2007) Autocrine VEGF signaling is required for vascular homeostasis. *Cell*, **130**, 691–703.
49. Henson, P.M. and Tuder, R.M. (2008) Apoptosis in the lung: induction, clearance and detection. *American Journal of Physiology: Lung Cellular and Molecular Physiology*, **294**, L601–11.
50. Coomber, B.L. (1991) Cytoskeleton in TFG-beta- and bFGF-modulated endothelial monolayer repair. *Experimental Cell Research*, **194**, 42–47.
51. Ferrari, G., Pintucci, G., Seghezzi, G. *et al.* (2006) VEGF, a prosurvival factor, acts in concert with TGF-beta1 to induce endothelial cell apoptosis. *Proceedings of the National Academy of Sciences of the United States of America*, **103**, 17260–65.
52. Gauldie, J., Kolb, M., Ask, K. *et al.* (2006) Smad3 signaling involved in pulmonary fibrosis and emphysema. *Proceedings of the American Thoracic Society*, **3**, 696–702.
53. De Ponti, C., Carini, R., Alchera, E. *et al.* (2007) Adenosine A2a receptor-mediated, normoxic induction of HIF-1 through PKC and PI-3K-dependent pathways in macrophages. *Journal of Leukocyte Biology*, **82**, 392–402.
54. Dvorak, A.M. (2005) Mast cell-derived mediators of enhanced microvascular permeability, vascular permeability factor/vascular endothelial growth factor, histamine, and serotonin, cause leakage of macromolecules through a new endothelial cell permeability organelle, the vesiculo-vacuolar organelle. *Chemical Immunology and Allergy*, **85**, 185–204.
55. Kiekamp, J.G., Jarzecka, K., and Perkett, E.A. (1999) Exposure to hyperoxia decreases the expression of vascular endothelial growth factor and its receptors in adult rat lungs. *American Journal of Pathology*, **154**, 823–31.
56. Grishko, V., Solomon, M., Breit, J.F. *et al.* (2001) Hypoxia promotes oxidative base modifications in the pulmonary artery endothelial cell VEGF gene. *The FASEB Journal*, **15**, 1267–69.
57. Ziel, K.A., Grishko, V., Campbell, C.C. *et al.* (2005) Oxidants in signal transduction: impact on DNA integrity and gene expression. *The FASEB Journal*, **19**, 387–94.
58. Brewer, J.W. and Diehl, J.A. (2000) PERK mediates cell-cycle exit during the mammalian unfolded protein response. *Proceedings of the National Academy of Sciences of the United States of America*, **97**, 12625–30.
59. Szegezdi, E., Fitzgerald, U., and Samali, A. (2003) Caspase-12 and ER-stress-mediated apoptosis: the story so far. *Annals of the New York Academy of Sciences*, **1010**, 186–94.
60. Szabadkai, G. and Rizzuto, R. (2004) Participation of endoplasmic reticulum and mitochondrial calcium handling in apoptosis: more than just neighborhood? *FEBS Letters*, **567**, 111–15.
61. Kaufman, R.J. (1999) Stress signaling from the lumen of the endoplasmic reticulum: coordination of gene transcriptional and translational controls. *Genes and Development*, **13**, 1211–33.
62. Ron, D. (2002) Translational control in the endoplasmic reticulum stress response. *The Journal of Clinical Investigation*, **110**, 1383–88.
63. Xu, C., Bailly-Maitre, B., and Reed, J.C. (2005) Endoplasmic reticulum stress: cell life and death decisions. *The Journal of Clinical Investigation*, **115**, 2656–64.
64. Ozawa, K., Kondo, T., Hori, O. *et al.* (2001) Expression of the oxygen-regulated protein ORP150 accelerates wound healing by modulating intracellular VEGF transport. *The Journal of Clinical Investigation*, **108**, 41–50.
65. Kelsen, S.G., Duan, X., Ji, R. *et al.* (2008) Cigarette smoke induces an unfolded protein response in the human lung: a proteomic approach. *American Journal of Respiratory Cell and Molecular Biology*, **38**, 541–50.
66. Hengstermann, A. and Muller, T. (2008) Endoplasmic reticulum stress induced by aqueous extracts of cigarette smoke in 3T3 cells activates the

unfolded-protein-response-dependent PERK pathway of cell survival. *Free Radical Biology and Medicine*, **44**, 1097–107.

67. Schroder, M. and Kaufman, R.J. (2005) ER stress and the unfolded protein response. *Mutation Research*, **569**, 29–63.
68. Shen, X., Zhang, K., and Kaufman, R.J. (2004) The unfolded protein response: a stress signaling pathway of the endoplasmic reticulum. *Journal of Chemical Neuroanatomy*, **28**, 79–92.
69. Rao, R.V. and Bredesen, D.E. (2004) Misfolded proteins, endoplasmic reticulum stress and neurodegeneration. *Current Opinion in Cell Biology*, **16**, 653–62.
70. Shi, Y., Vattem, K.M., Sood, R. *et al.* (1998) Identification and characterization of pancreatic eukaryotic initiation factor 2 alphasubunit kinase, PEK, involved in translational control. *Molecular and Cellular Biology*, **18**, 7499–509.
71. Yoshida, H., Haze, K., Yanagi, H. *et al.* (1998) Identification of the cisacting endoplasmic reticulum stress response element responsible for transcriptional induction of mammalian glucose-regulated proteins. Involvement of basic leucine zipper transcription factors. *The Journal of Biological Chemistry*, **273**, 33741–49.
72. Cox, J.S., Shamu, C.E., and Walter, P. (1993) Transcriptional induction of genes encoding endoplasmic reticulum resident proteins requires a transmembrane protein kinase. *Cell*, **73**, 1197–206.
73. Ye, J., Rawson, R.B., Komuro, R. *et al.* (2000) ER stress induces cleavage of membrane-bound ATF6 by the same proteases that process SREBPs. *Molecules and Cells*, **6**, 1355–64.
74. Yoshida, H., Matsui, T., Yamamoto, A. *et al.* (2001) XBP1 mRNA is induced by ATF6 and spliced by IRE1 in response to ER stress to produce a highly active transcription factor. *Cell*, **107**, 881–91.
75. Harding, H.P., Zhang, Y., and Ron, D. (1999) Protein translation and folding are coupled by an endoplasmic-reticulum-resident kinase. *Nature*, **397**, 271–74.
76. Zinszner, H., Kuroda, M., Wang, X. *et al.* (1998) CHOP is implicated in programmed cell death in response to impaired function of the endoplasmic reticulum. *Genes and Development*, **12**, 982–95.
77. McCullough, K.D., Martindale, J.L., Klotz, L.O. *et al.* (2001) Gadd153 sensitizes cells to endoplasmic reticulum stress by downregulating Bcl2 and perturbing the cellular redox state. *Molecular and Cellular Biology*, **21**, 1249–59.
78. Oyadomari, S., Takeda, K., Takiguchi, M. *et al.* (2001) Nitric oxide-induced apoptosis in pancreatic beta cells is mediated by the endoplasmic reticulum stress pathway. *Proceedings of the National Academy of Sciences of the United States of America*, **98**, 10845–50.
79. Pober, J.S. and Sessa, W.C. (2007) Evolving functions of endothelial cells in inflammation. *Nature Reviews, Immunology*, **7**, 803–15.
80. Rothermel, A.L., Wang, Y., Schechner, J. *et al.* (2004) Endothelial cells present antigens *in vivo*. *BMC Immunology*, **5**, 5.
81. Ryan, U.S. (2001) Endothelial cell activation responses, in *Pulmonary Endothelium in Health and Disease* (ed. U.S. Ryan), Marcel Dekker, New York, pp. 3–34.
82. Methe, H., Hess, S., and Edelman, E.R. (2007) Endothelial immunogenicity – a matter of matrix microarchitecture. *Thrombosis and Haemostasis*, **98**, 278–82.
83. Methe, H., Groothuis, A., Sayegh, M.H., and Edelman, E.R. (2007) Matrix adherence of endothelial cells attenuates immune reactivity: induction of hyporesponsiveness in allo- and xenogeneic models. *The FASEB Journal*, **21**, 1515–26.
84. Taraseviciene-Stewart L., Kraskauskiene V., Burns N., and Voelkel N.F. (2008) Presence of anti-endothelial cell antibodies in patients with COPD. AJRCCM, A 177.
85. Gabrilovich, D.I., Chen, H.L., Girgis, K.R. *et al.* (1996) Production of vascular endothelial growth factor by human tumors inhibits the functional maturation of dendritic cells. *Nature Medicine*, **2**, 1096–103.
86. Sozzani, S., Rusnati, M., Riboldi, E. *et al.* (2007) Dendritic cell–endothelial cell cross-talk in angiogenesis. *Trends in Immunology*, **28**, 385–92.
87. Lang, K.S., Recher, M., Junt, T. *et al.* (2005) Toll-like receptor engagement converts T-cell autoreactivity into overt autoimmune disease. *Nature Medicine*, **11**, 138–45.
88. Weinberg, K.S. and Hayes, J.A. (1982) Elastase-induced emphysema: asynchronous bronchial, alveolar and endothelial cell proliferation during the acute response to injury. *The Journal of Pathology*, **136**, 253–64.
89. Rafii, S., Meeus, S., Dias, S. *et al.* (2002) Contribution of marrow-derived progenitors to vascular and cardiac regeneration. *Seminars in Cell and Developmental Biology*, **13**, 61–67.
90. Suratt, B.T., Cool, C.D., Serls, A.E. *et al.* (2003) Human pulmonary chimerism after hematopoietic stem cell transplantation. *American Journal of Respiratory and Critical Care Medicine*, **168**, 318–22.

91. Alvarez, D.F., Huang, L., King, J.A. *et al.* (2008) Lung microvascular endothelium is enriched with progenitor cells that exhibit vasculogenic capacity. *American Journal of Physiology: Lung Cellular and Molecular Physiology*, **294**, L419–30.
92. Divya, P., Sreerekha, P.R., and Krishnan, L.K. (2007) Growth factors upregulate deposition and remodeling of ECM by endothelial cells cultured for tissue-engineering applications. *Biomolecular Engineering*, **24**, 593–602.
93. Golpon, H.A., Fadok, V.A., Taraseviciene-Stewart, L. *et al.* (2004) Life after corpse engulfment: phagocytosis of apoptotic cells leads to VEGF secretion and cell growth. *The FASEB Journal*, **18**, 1716–18.
94. Hristov, M., Erl, W., Linder, S., and Weber, P.C. (2004) Apoptotic bodies from endothelial cells enhance the number and initiate the differentiation of human endothelial progenitor cells *in vitro*. *Blood*, **104**, 2761–66.
95. Siner, J.M., Jiang, G., Cohen, Z.I. *et al.* (2007) VEGF-induced heme oxygenase-1 confers cytoprotection from lethal hyperoxia *in vivo*. *The FASEB Journal*, **21**, 1422–32.
96. Zeisberg, E., Tarnavski, O., Zeisberg, M. *et al.* (2007) Endothelial-to-mesenchymal transition contributes to cardiac fibrosis. *Nature Medicine*, **13**, 952–61.
97. Senger, D.R., Claffey, K.P., Benes, J.E. *et al.* (1997) Angiogenesis promoted by vascular endothelial growth factor: regulation through alpha1beta1 and alpha2beta1 integrins. *Proceedings of the National Academy of Sciences of the United States of America*, **94**, 13612–17.
98. Zhao, L. and Ackerman, S.L. (2006) Endoplasmic reticulum stress in health and disease. *Current Opinion in Cell Biology*, **18**, 444–52.
99. Sakao, S., Taraseviciene-Stewart, L., Lee, J.D. *et al.* (2005) Initial apoptosis is followed by increased proliferation of apoptosis-resistant endothelial cells. *The FASEB Journal*, **19**, 1178–80.
100. Choe, K.H., Taraseviciene-Stewart, L., Scerbavicius, R. *et al.* (2003) Methylprednisolone causes matrix metalloproteinase-dependent emphysema in adult rats. *American Journal of Respiratory and Critical Care Medicine*, **167**, 1516–21.
101. Wu, C.C., Hsieh, C.W., Lai, P.H. *et al.* (2006) Upregulation of endothelial heme oxygenase-1 expression through the activation of the JNK pathway by sublethal concentrations of acrolein. *Toxicology and Applied Pharmacology*, **214**, 244–52.
102. Aoshiba, K., Yokohori, N., and Nagai, A. (2003) Alveolar wall apoptosis causes lung destruction and emphysematous changes. *American Journal of Respiratory Cell and Molecular Biology*, **28**, 555–62.
103. Lu, Q., Harrington, E.O., and Rounds, S. (2005) Apoptosis and lung injury. *The Keio Journal of Medicine*, **54**, 184–89.
104. Ma, B., Blackburn, M.R., Lee, C.G. *et al.* (2006) Adenosine metabolism and murine strain-specific IL-4-induced inflammation, emphysema, and fibrosis. *The Journal of Clinical Investigation*, **116**, 1274–83.
105. Medler, T.R., Petrusca, D.N., Lee, P.J. *et al.* (2008) Apoptotic sphingolipid signaling by ceramides in lung endothelial cells. *American Journal of Respiratory Cell and Molecular Biology*, **38**, 639–46.
106. Petrache, I., Natarajan, V., Zhen, L. *et al.* (2005) Ceramide upregulation causes pulmonary cell apoptosis and emphysema-like disease in mice. *Nature Medicine*, **11**, 491–98.

27

Pulmonary Endothelium and Pulmonary Hypertension

Rubin M. Tuder[1] and Serpil C. Erzurum[2]

[1]Program in Translational Lung Research, Division of Pulmonary Sciences and Critical Care Medicine, University of Colorado Denver School of Medicine, Aurora, CO, USA

[2]Department of Pathobiology and Respiratory Institute, The Cleveland Clinic Foundation, Cleveland, OH, USA

INTRODUCTION

There is an increasing realization that pulmonary hypertension is a critical indicator of outcome in several lung diseases, including chronic obstructive pulmonary diseases (COPDs) [1] and interstitial lung disease [2]. A wide range of systemic diseases are also associated with pulmonary arterial hypertension (PAH), including congenital heart malformations, collagen vascular disorders, hepatic cirrhosis, HIV infection, schistosomiasis, among others [termed associated diseases of PAH (APAH)]. Once all these associated conditions are clinically ruled out, the disease is classified as idiopathic PAH (IPAH) [3]. The elevation of pulmonary artery pressures above the upper limit of 25 mmHg imparts a progressive strain on right ventricular function. Pulmonary hypertension, in its more severe form in which the pulmonary artery pressures increase above 40–45 mmHg, carries a high mortality due to right ventricular failure. This unifying hemodynamic definition encompasses complex heterogeneous clinical and pathophysiological features, which are being progressively unraveled. Insights into this increasing body of knowledge will be central for the development of novel diagnostic and prognostic markers and mechanism-targeted therapies.

Even though the pathogenetic concepts underlying pulmonary hypertension have classically revolved around the role of vascular cells [i.e., endothelial cells (ECs) and smooth muscle cells (SMCs), and adventitial fibroblasts) [4], the recent focus on the potential role of inflammation and the lung cellular environment expand significantly our understanding of the complexity of the pathophysiology of pulmonary hypertension [5]. As the pulmonary endothelium is a critical integrator of these cellular and molecular processes, a leading pathogenetic role has been ascribed to lung ECs in pulmonary hypertension [6]. The multitude of pathogenetic processes involving pulmonary ECs has been recently reviewed [7]. In fact, many of these mechanistic insights provide the pathophysiological rationale of available therapies in the disease [8], including endothelin receptor blockers, phosphodiesterase inhibitors, and prostaglandin analogs. Despite these prominent advances, significant challenges remain as these therapies carry significant side-effects, and do not lead to resolution of the cellular and molecular processes underlying pulmonary hypertension.

The present chapter focuses on novel concepts of EC pathobiology in pulmonary hypertension. Based on data generated in the recent years, decade-old hypotheses can now be tested, which can significantly impact on our understanding of the disease and development of novel therapies. We emphasize new insights into the phenotypic shift of pulmonary ECs that allow them to acquire some of the features seen in neoplastic processes, particularly the potential role of hypoxia-inducible factor (HIF)-1α in the metabolic shift towards a more anaerobic metabolism, the potential involvement of bone marrow elements in the proangiogenic environment present in lung arteries and veins, and the rationale for cell-based therapies to reconstitute a normal pulmonary endothelium.

The Pulmonary Endothelium: Function in health and disease Editors Norbert F. Voelkel, Sharon Rounds

EXUBERANT EC PROLIFERATION AND MISGUIDED ANGIOGENESIS IN PULMONARY HYPERTENSION

Although IPAH has been recognized for more than a century [9], in the 1950s it was postulated that abnormal ECs might participate in some of the pathological vascular lesions, particularly plexiform and dilation lesions [10, 11]. Interest in the pathogenesis of these lesions has been tempered by belief that these lesions were "terminal events," given that they were often recognized in autopsied lungs [12, 13]. Notwithstanding these potential limitations, immunohistochemical markers confirm that ECs, with variable participation of myofibroblasts, account for the glomeruloid-like intimal obliteration of pulmonary arteries [14] (Figure 27.1). Moreover, these lesions contain actively proliferating vascular cells [15, 16] allied to a decreased expression of cyclin kinase inhibitors (Figure 27.1) [13].

The realization that these lesions have a tumor-like three-dimensional structure [16] led to studies that documented that plexiform lesions in lungs of patients with IPAH are composed of a monoclonal cell population, while similar lesions present in lungs of patients with APAH due to Eisenmenger's syndrome are uniformly polyclonal [17]. Of note, a similar observation also pertained to PAH caused by anorexigen use [18]. These aforementioned observations led to predictions concerning the pathobiology of severe pulmonary hypertension (particularly IPAH), which have been partly validated by recent research performed by multiple centers.

The finding of a clonal expansion in IPAH lungs raised the novel hypothesis that this disease shares common genetic, cellular, and molecular pathobiological underlying neoplastic processes [19, 20]. A recent review on this topic has rather focused on the expression of a set of markers in pulmonary hypertensive tissue that are also present in neoplastic process, yet they can also be present in reactive tissues [21]. The early focus on the tumor suppressor function of transforming growth factor (TGF)-β family signaling as a molecular underpinning of pulmonary hypertension [22] was subsequently validated by the documentation of somatic microsatellite instability in the TGF-β receptor-2 associated with loss of receptor expression [23] and the discovery of bone morphogenetic protein receptor (BMPR)-2 mutation as one of the genes

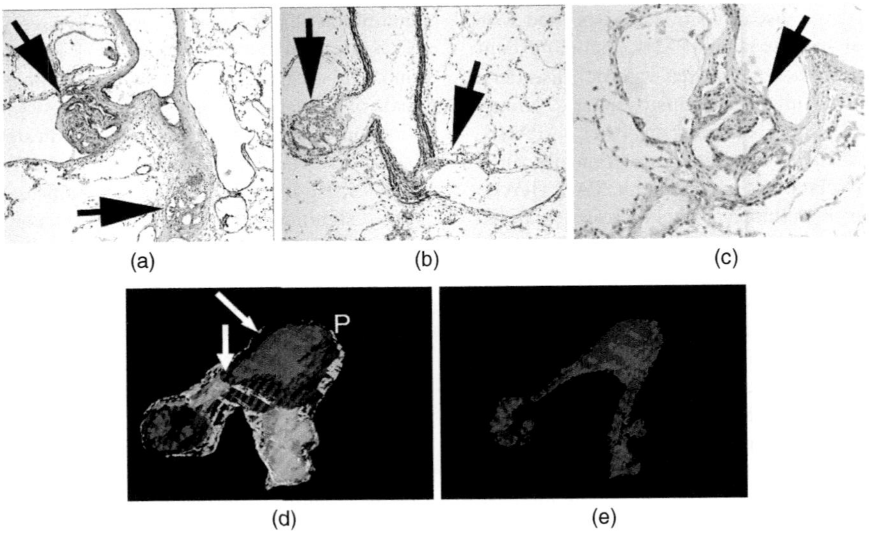

Figure 27.1 Plexiform lesions occurring along two branches of medium-sized pulmonary arteries (arrows). (a) ECs are highlighted by Factor VIII-related antigen immunohistochemistry. (b) Serial section stained by anti-smooth muscle α-actin immunohistochemistry. (c) Lack of expression of the TGF-β-dependent cell cycle inhibitor $p21^{Kip1}$ in the core of plexiform lesion ECs (arrow). (d) Three-dimensional reconstruction of plexiform lesion, showing marked intraluminal expansion of ECs and therefore creating a tumor-like appearance to the intraluminal growth (arrows). (e) The cell layers have been removed and only the cast of the pulmonary vascular lumen in the vascular segment compromised by the plexiform lesions is highlighted. A color version of this figure appears in the plate section of this volume. Reproduced from [16] by permission of the American Society for Investigative Pathology.

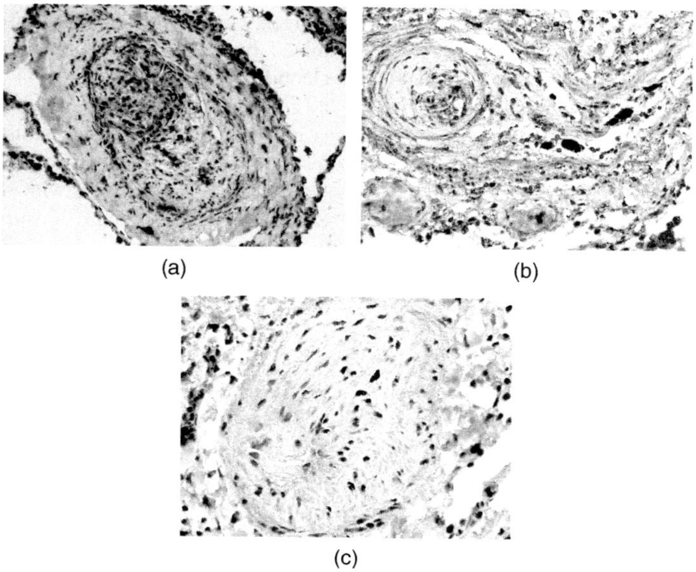

Figure 27.2 Expression of HIF-1α in a plexiform lesion (a) and in a concentric lesion (b), and of HIF-1β in a plexiform lesion (c). A color version of this figure appears in the plate section of this volume. Reproduced from [31] with permission from John Wiley & Sons, Ltd.

underlying familial IPAH [24, 25]. Furthermore, these findings suggested that the clonal growth of ECs occurs through an expansion of a stem-like/progenitor vascular cell population, arising from pulmonary arteries themselves or from extrapulmonary tissues, particularly the bone marrow [19]. In the setting of IPAH, a single cell acquires the ability to expand and form a lesion, while stimuli, such as shear stress, inflammation, or viral products, would recruit several vascular progenitor cells to form a polyclonal cell population in APAH seen in the setting of congenital heart malformations, collagen vascular diseases, and HIV infection. Stem-like/progenitor cancer cells are now leading candidates as the source of primary and metastatic cancers [26]. Despite evidence that there are increased numbers of circulating mature ECs in severe pulmonary hypertension [27], this hypothesis has not been addressed until recently (as discussed in the following section).

As observed in cancers, genetic alterations cause abnormal cell and molecular signaling, leading to cell expansion due to enhanced cell proliferation allied to decreased apoptosis. To overcome a strong adverse selection process, clonal ECs might become apoptosis-resistant. This postulate is supported by the evidence of remarkably low apoptosis seen in IPAH lungs [28], and animal and cell culture data that, following EC apoptosis, there is emergence of an apoptosis-resistant EC population [29, 30].

Although the precise cell signaling involved in the maintenance and expansion of a progenitor cell population in pulmonary hypertension remains undefined, *in situ* studies have delineated the increased expression of vascular endothelial growth factor (VEGF), VEGF receptors, and HIF-1α and -1β in plexiform lesions (Figure 27.2) [31], therefore suggesting that these lesions represent a manifestation of a process of "misguided angiogenesis."

The resemblance of EC proliferation in IPAH and APAH with neoplastic growth suggests that these processes may follow a two-hit hypothesis. It is tempting that somatic genetic events might indeed lead to loss of tumor suppressor or proapoptotic genes. We have previously documented that IPAH ECs in plexiform lesions show microsatellite instability, with somatic mutations in TGF receptor-2 and Bax [23]. However, there is not a somatic loss of the second copy of BMPR-2 in plexiform lesions [32] that might explain the reduced expression of the receptor in lungs of patients with IPAH [33].

The aggregate of these findings prompts one to address fundamental questions regarding the pathogenesis of pulmonary hypertension:

(i) Are there additional genetic or epigenetic events that allow for the expansion of a progenitor cell population in the diseased pulmonary arteries?
(ii) What is the contribution of the bone marrow precursors in the initiation and progression of pulmonary vascular lesions in pulmonary hypertension?
(iii) Can we identify a vascular progenitor cell population in pulmonary arteries that is the seed of the abnormal vascular lesions in uniquely predisposed patients?
(iv) Like in cancer cells, does hypoxia-related signaling lead to unique pulmonary vascular phenotypes that can be harnessed for diagnosis and, more importantly, for therapies? Are these genetic and signaling events a vital property pertaining to the physiological role of pulmonary vascular progenitors, which would allow for targeted manipulation in disease-predisposed individuals and in those with the disease?

The discussion that follows indicates that recent advances have provided some answers to these questions, and portends a bright and exciting future in the research of PAH.

VASCULAR PROGENITOR CELLS IN PULMONARY ARTERIES

The heterogeneity of ECs in normal pulmonary arteries was emphasized by the studies of Stevens *et al.*, in which they isolated and characterized large and microvascular ECs from rat pulmonary arteries based on differential lectin expression [34]. The aggregate of data obtained from rat microvascular ECs indicate that these cells have increase growth ability and tighter barrier function when compared to large pulmonary artery ECs [35] (see Chapter 9). The enhanced growth properties of microvascular ECs reside in the presence of EC progenitors in this population. Ingram *et al.* and Yoder *et al.* have redefined the concept of EC progenitors based on classic hematological concepts of clonogenic assays and repopulation studies [36, 37]. A late outgrowth population of ECs is capable of reconstituting an entire EC population after a single progenitor cell is plated. These studies also suggest that the predominant circulating proangiogenic population is comprised of angiogenic monocytic cells, which provide angiogenic signals for native endothelial and pulmonary vascular cells (see Chapter 13).

The isolation and culture of pulmonary artery ECs from IPAH lungs has provided the opportunity for a parallel line of investigation to identify biologic and molecular differences among ECs derived from IPAH and healthy lungs [38]. The study of ECs obtained from IPAH lungs shows that they have greater proliferation as determined by bromodeoxyuridine incorporation and Ki-67 nuclear antigen expression, and decreased apoptosis as determined by caspase 3 activation and terminal deoxynucleotidyl transferase biotin-dUTP nick end-labeling assay as compared to control cells [39]. The cell proliferation is dependent upon VEGF, Janus kinase, and the signal transducer and activator of transcription (STAT)-3 pathways, which is consistent with the current paradigm of STAT-3 being the central prosurvival molecular signaling pathway for ECs and a primary regulator of angiogenesis [40–43]. The identification of phosphorylated STAT-3 in ECs within IPAH lesions *in vivo* supports a role for STAT-3 activation in the genesis of the proliferative vascular lesions in IPAH lungs [44] (Figure 27.3). Furthermore, hypertensive ECs appear to have abnormal Golgi sorting of proteins and intracellular transport of crucial signaling molecules [45], potentially leading to abnormal intracellular STAT-3 localization and signaling [46]. The differences in proliferation and apoptosis between IPAH and control pulmonary artery ECs provide evidence for the concept of phenotypically altered ECs within IPAH vascular lesions [14], and support the notion of apoptosis-resistant and hyperproliferative vascular ECs in the origin of pulmonary hypertension [29, 30]. A recent study indicates that there are greater numbers of $CD34^+$ endothelial progenitor cells among the IPAH cells in culture as compared to control ECs, which likely accounts for the greater proliferative potential [47].

METABOLIC SHIFT OF HYPERTENSIVE PULMONARY VASCULAR CELLS IN PULMONARY HYPERTENSION

The earlier observations of expression of angiogenic molecules, including HIF-1α in plexiform lesions in IPAH lungs [31], were subsequently expanded in a series of elegant animal models of pulmonary hypertension by the groups of Yuan, Michelakis, and Archer. These studies linked the downregulation of voltage-dependent potassium channels (K_v channels) in pulmonary artery SMCs from IPAH lungs [48] with the resistance to apoptosis and mitochondrial hyperpolarization in hypertensive SMCs [49]. They showed that the altered phenotype of cells is related to abnormal expression and localization of the inhibitor of apoptosis survivin [50], and a metabolic shift toward a more anaerobic metabolism, which is amenable to inhibition by dichloroacetate (an inhibitor of pyruvate dehydrogenase kinase) [51]. More recently, this paradigm came full circle with the experimental observation that HIF-1α expression results in a cascade of events including ion channel abnormalities and alterations in mitochondrial redox, metabolic regulation,

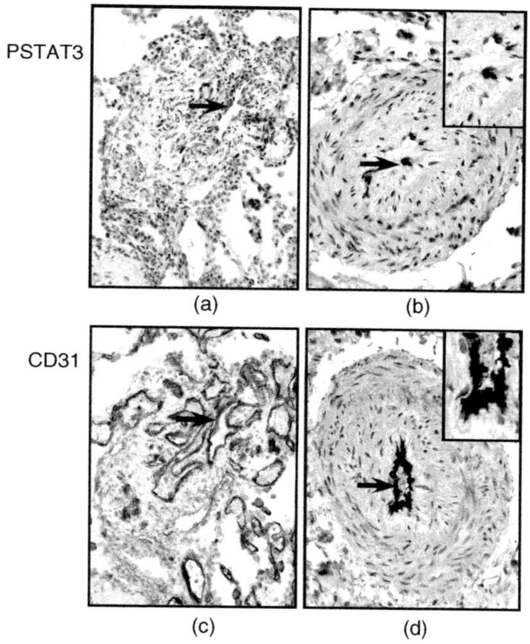

Figure 27.3 Cellular localization of phospho-STAT-3 by immunohistochemical staining in IPAH lung. Strong positive immunoreactivity of phospho-STAT-3 is present in endothelium in a plexiform lesion (arrowhead; a) and a concentric intimal lesion (arrowhead; b). (c and d) Immunohistochemical staining for CD31 in the laminar lining cells (arrowheads) of the vessels confirms the endothelial phenotype of cells. A color version of this figure appears in the plate section of this volume. Reproduced from [39] with permission from The American Physiological Society.

and apoptotic machinery, which dictate a growth-prone, apoptosis-resistant pulmonary SMC [52].

Deficiency in the vasodilator nitric oxide (NO) has been identified in the pathogenesis of pulmonary hypertension [53–60] (see Chapter 6). Pulmonary and total body NO are lower in IPAH patients as compared to healthy controls [38, 59, 61, 62], and NO production by endothelial NO synthase (endothelial nitric oxide synthase eNOS) is lower than normal in IPAH ECs *in vitro* [38, 44]. In addition to effects on vascular tone, NO regulates cellular bioenergetics through effects on glycolysis, oxygen consumption by mitochondria, and mitochondrial biogenesis [63–65]. The eNOS-deficient mice, which develop more pronounced pulmonary hypertension under hypoxia [66], have reduced mitochondria content in a wide range of tissues associated with significantly lower oxygen consumption and ATP content [63–65]. Similarly, in human IPAH cells with reduced NO production, cellular metabolic energy pathways are characterized by lower oxygen consumption of mitochondria and significantly higher glycolytic rate [44]. Greater than normal glucose uptake in the lungs of IPAH patients is detectable by nuclear imaging using [^{18}F]fluoro-deoxy-D-glucose positron emission tomography (PET), which indicates that the proliferative ECs in the pulmonary vasculature also have greater reliance on the glycolytic pathway *in vivo* [44] (Figure 27.4).

The switch to energy derived from primarily glycolytic metabolism in human pulmonary hypertension parallels the metabolic findings identified in avian and rodent pulmonary hypertension [52, 67–70], which are analogs to the alterations in cancer cell metabolism. Tumor cells often exhibit this combination of alterations in cellular energy production – a phenomenon first described over 80 years ago and known as the Warburg effect [71]. The condition of aerobic glycolysis is not a unique feature of tumor cells, but is also found in nontransformed rapidly proliferating cells when sufficient glucose is available [72]. Thus, the occurrence of aerobic glycolysis, or the Warburg effect, in IPAH ECs is consistent with the increased proliferative capacity of these cells [39].

Despite many years of investigation, it is still unclear as to what regulatory mechanisms transition proliferating cells from oxidative glucose metabolism to anaerobic glycolysis. However, in IPAH pulmonary artery ECs, the decreased cellular respiration and greater glycolysis are due in large part to low mitochondrial numbers, which is related to reduced NO/cGMP-dependent mitochondrial numbers [44] (Figure 27.5). In contrast, studies of pulmonary hypertension in avian and rodent species identify intrinsic deficiencies in mitochondrial function, rather than numbers [73]. Nevertheless, the cumulative data indicate that decreased mitochondrial function, whether related to lower numbers of mitochondria and/or an intrinsic impairment of function, is a pathologic hallmark of pulmonary hypertension.

MORE THAN JUST PULMONARY VASCULAR CELLS IN PULMONARY HYPERTENSION: ROLE OF BONE MARROW

Pathological studies recognized that inflammatory cells cluster in the vicinity of remodeled pulmonary arteries in IPAH [14, 74]. Although these inflammatory cells exhibit markers of mature lymphocytes, dendritic cells, mast cells, and macrophages, a contribution of more immature bone marrow progenitors is feasible. Experimentally, Stenmark *et al.* showed that pulmonary arteries of hypoxic cows, rats, and mice exhibit infiltration by CD45^{+} collagen-producing fibrocytic cells, which play a

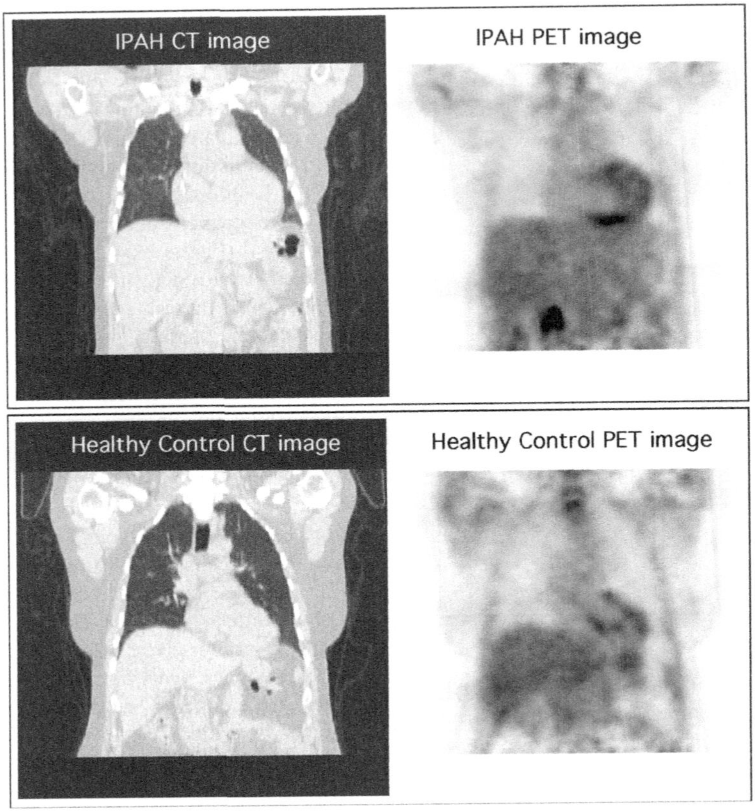

Figure 27.4 PET and computed tomography (CT) images of IPAH and healthy control subject. Reproduced with permission from Xu *et al.*, (2007) [44]. Copyright (2007) National Academy of Sciences, U.S.A.

contributory role in the development of pulmonary hypertension in these models [75] (see Chapter 11).

The concept of "vasoformative" cells arriving from the blood stream that contribute to the formation of plexiform lesions was postulated nearly three decades ago by Smith and Heath [76]. Recent studies now clearly support a role for circulating bone marrow-derived progenitors in remodeling of pulmonary arteries [47, 77, 78] (see Chapter 13). The current model for repair and/or remodeling of blood vessels proposes essential interactions among ECs within vessels and several types of stem cells, some of which are bone marrow-derived and others resident in the blood vessel wall. The bone marrow-derived proangiogenic precursor cells in the peripheral blood circulation are particularly enriched within the $CD34^+$ and $CD133^+$ subsets, and are thus designated as $CD34^+$ $CD133^+$ endothelial progenitor cells [79, 80]. The $CD34^+$ $CD133^+$ cells are bone marrow-derived $CD45^+$ mononuclear cells and are proangiogenic, but not true ECs [81].

Although the exact mechanisms of action are unknown [81], there is growing evidence that $CD34^+$ $CD133^+$ progenitor cells contribute to the formation of new blood vessels in a paracrine manner, possibly by disruption of mature ECs lining the vessel walls or by interaction with high proliferative true endothelial stem cells in the vascular wall [81, 82]. $CD34^+$ $CD133^+$ progenitor cells arrive first at sites of injury, facilitating vascular repair by recruitment and activation of resident endothelial stem cells, and thus providing a trophic effect on neovascularization [81].

Circulating $CD34^+$ $CD133^+$ cells are increased in the blood of IPAH patients and have significantly enhanced angioproliferative potential [47]. *In vivo* subcutaneous inoculation of IPAH $CD34^+$ $CD133^+$ endothelial progenitor cells into immune-deficient severe combined immunodeficiency (SCID) mice shows that there are more proliferative precursors circulating in IPAH patients. The mobilization of proangiogenic bone marrow-derived cells may be a normal physiologic repair response to ongoing pulmonary vascular shear stress and endothelial injury of IPAH. In support of this, progenitor cell mobilization is intrinsic to hypoxic conditions [83] and

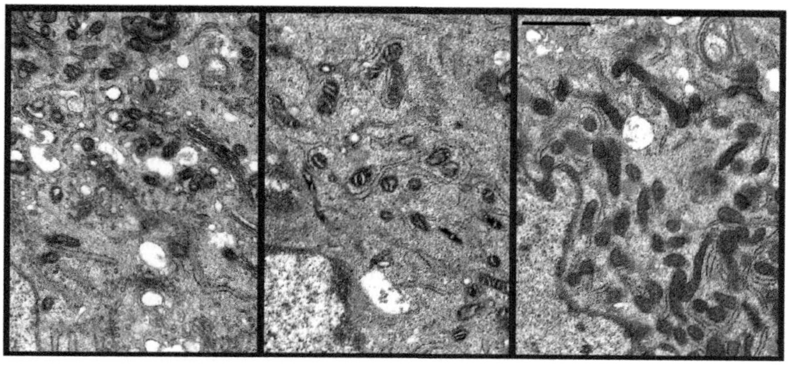

Figure 27.5 Ultrastructure detail of mitochondria in untreated IPAH pulmonary artery ECs (b), healthy control (a), and IPAH exposed to the NO donor DETA NONOate (c) (bar = 1 μm). Reproduced with permission from Xu *et al.*, (2007) [44]. Copyright (2007) National Academy of Sciences, U.S.A.

increased bone marrow progenitors have been reported in hypoxia-induced animal models of PAH [84, 85]. However, even if the greater number of circulating cells is part of a reparative response, circulating proangiogenic bone marrow-derived cells may contribute to the development of proliferative vascular lesions in IPAH by disruption and activation of the endothelium through release of matrix metalloproteinases (MMPs) and other angiogenic factors. IPAH endothelial progenitors produce much more MMP-2 – a protease that plays crucial roles in vascular regeneration [86]. In contrast, the circulating nonproliferative $CD14^+$ monocyte cells from IPAH patients that are being used in human clinical trials [87] are not proliferative when inoculated into nonobese diabetic/SCID mice, confirming that these cells are unlikely to contribute to hyperplastic endothelial lesions [47].

Irrespective of the multiple types of bone marrow-derived and resident vascular wall proangiogenic precursors, mobilization of these cells appears to be a characteristic of IPAH patients and offers an opportunity for future mechanistic studies using human circulating cells as opposed to lung vascular tissues.

EC DYSFUNCTION: MORE THAN JUST IPAH

Although much of the research involving the role of the endothelium in pulmonary hypertension has focused in IPAH, the vast majority of pulmonary hypertension occurs in the setting of left ventricular dysfunction and in COPDs. The understanding of the role of a dysfunctional pulmonary endothelium in the former is lacking. However, recent insights provided evidence that some of the alterations pertaining to IPAH ECs also apply to cigarette smoke induced vascular dysfunction. The group of Barbera *et al.* documented that human pulmonary arteries of smokers are dysfunctional, with reduced ability to promote NO dependent vasodilation [88]. Experimentally, cigarette smoke exposure induces expression of VEGF and endothelin, while it decreases expression of nitric oxide synthaseNOS [1].

The aforementioned examples of EC production of growth factors for SMCs is ever more pertinent as a prototypic growth factor for SMCs, serotonin, is also secreted by pulmonary ECs [89]. Serotonin transporter polymorphisms and overexpression have been documented in IPAH and COPD-associated pulmonary hypertension [90]. The prominence of the role of ECs in the pathogenesis of pulmonary hypertension is therefore anchored on the combined evidence of decreased production of vasodilating/antiproliferative agents, such as prostacyclin and NO [10], allied to overexpression of vasoconstrictive/proliferative molecules.

However, there are no EC proliferative lesions in pulmonary hypertension associated with COPD, interstitial lung disease, or sleep apnea. This categorical distinction suggests that proliferation of ECs or mutational events in pulmonary vascular cells are relatively rare or infrequent events, and that EC dysfunction *per se* or pulmonary vascular remodeling with SMCs is not sufficient to cause a "neoplastic-like" proliferation of pulmonary artery ECs.

What then sets the different forms of pulmonary hypertension apart? Short of the finding of clonal expansion of ECs in IPAH, the presence of plexiform lesions, and mutations in the TGF-β family genes, there are no clear molecular signatures that differentiate the forms of severe versus milder forms of pulmonary hypertension. Is it possible that the differences in clinical severity are determined by an early inciting event? Unfortunately, little is known about how and when the disease starts. Based on our studies of severe pulmonary hypertension caused by the combination of chronic hypoxia and VEGF receptor inhibition with SU5416 in the rat [30], we have proposed that early EC apoptosis leads to severe pulmonary hypertension, associated with proliferated ECs (see Chapter 16).

In agreement with a multiple hit hypothesis, injuries promoted by autoimmune or viral agents might trigger the pathobiological events associated with pulmonary hypertension (see Chapter 19). Human herpesvirus (HHV)-8 is potentially linked to the proliferative switch of pulmonary ECs, as demonstrated in Kaposi's sarcoma [91]. The initial demonstration of the presence of HHV-8 in IPAH lungs [92] has been questioned by other studies [93, 94]. However, a recent study lends support that herpes viruses might be linked to experimental pulmonary hypertension [95]. Furthermore, monkeys infected with a mutant *nef* gene in the simian immunodeficiency virus demonstrate development of pulmonary vascular remodeling, including plexiform lesions – a finding paralleled by the demonstration of the HIV *nef* in HIV-associated pulmonary hypertension [96].

CONCLUSIONS AND PERSPECTIVES

Pulmonary ECs have a lead role in the pathobiology of pulmonary hypertension. This chapter attempts to capture the increasing complexity of the alterations present in ECs in the different forms of pulmonary hypertension. Dysregulation of expression of mediators of vascular tone and vascular cell proliferation, potential for mutational events, and abnormalities in EC metabolism underlie the molecular heterogeneity of pulmonary hypertension as a disease. Thus, successful strategies to re-establish vascular patency and reverse the occlusive pulmonary panvasculopathy will require approaches that target the dysfunctional ECs.

The future of pulmonary hypertension research is promising, with a "second revolution" in novel therapies and diagnostic tools in the horizon. These are the result of the conceptual advances in the past 20 years, validated by solid and extensive experimentation. Testimony to these accomplishments, imatinib (Gleevec™) [97] and sorafenib [98], which have been used in cancer research based on their profile of tyrosine kinase inhibition, have made their way to treat patients with pulmonary hypertension. However, this is clearly not enough. We still lack a refined understanding of the inciting event(s) in the pathogenesis of the different forms of pulmonary hypertension. There is still a pending conflict if the pulmonary vasoconstriction versus vascular remodeling is key for preventing or treating the disease. Accordingly, the former will lead to investigations focused on SMC contraction, while the latter will focus on the molecular cross-talk between endothelial and SMCs leading to abnormal cell growth. Furthermore, inflammatory cells and cytokines, mediated largely by pulmonary ECs, may have a significant role not yet fully understood or harnessed for future therapies. The role of circulating precursors originated from the pulmonary vessels or bone marrow awaits further elucidation, pending the development of suitable cell markers for detection and relevant models to test their potential pathogenic or therapeutic functions. The finding of EC precursors in pulmonary arteries may lead to a revolution in our current paradigms to explain, diagnose, and treat the disease. Finally, pulmonary hypertension is not a uniform entity, both pathologically, pathobiologically, or clinically, although, therapeutic developments are used to treat different forms of pulmonary hypertension as they are a single uniform entity. Studies dedicated to the unique aspects of the pathobiology of the different presentations of pulmonary hypertension will be central to continue the accomplishments in the management of the disease.

ACKNOWLEDGMENTS

This work was supported by the grants HL60917 (SCE) and Cardiovascular Medical Research and Education Fund (CMREF) (RMT and SCE).

References

1. Wright, J.L., Levy, R.D., and Churg, A. (2005) Pulmonary hypertension in chronic obstructive pulmonary disease: current theories of pathogenesis and their implications for treatment. *Thorax*, **60**, 605–9.
2. Nathan, S.D., Noble, P.W., and Tuder, R.M. (2007) Idiopathic pulmonary fibrosis and pulmonary hypertension: connecting the dots. *American Journal of Respiratory and Critical Care Medicine*, **175**, 875–80.
3. Simonneau, G., Galie, N., Rubin, L.J. *et al.* (2004) Clinical classification of pulmonary hypertension. *Journal of the American College of Cardiology*, **43**, S5–S12.
4. Rubin, L.J. (1998) Primary pulmonary hypertension. *Chest*, **104**, 236–50.
5. Mouthon, L., Guillevin, L., and Humbert, M. (2005) Pulmonary arterial hypertension: an autoimmune disease? *European Respiratory Journal*, **26**, 986–88.
6. Loscalzo, J. (1992) Endothelial dysfunction in pulmonary hypertension. *The New England Journal of Medicine*, **327**, 70–75.

7. Budhiraja, R., Tuder, R.M., and Hassoun, P.M. (2004) Endothelial dysfunction in pulmonary hypertension. *Circulation*, **109**, 159–65.
8. Humbert, M., Sitbon, O., and Simonneau, G. (2004) Treatment of pulmonary arterial hypertension. *The New England Journal of Medicine*, **351**, 1425–36.
9. Zaiman, A., Fijalkowska, I., Hassoun, P.M., and Tuder, R.M. (2005) One hundred years of research in the pathogenesis of pulmonary hypertension. *American Journal of Respiratory Cell and Molecular Biology*, **33**, 425–31.
10. Tuder, R.M., Marecki, J.C., Richter, A. *et al.* (2007) Pathology of pulmonary hypertension. *Clinics in Chest Medicine*, **28**, 23–42, vii.
11. Heath, D. and Edwards, J.E. (1958) The pathology of pulmonary hypertensive disease. A description of six grades of structural changes in the pulmonary arteries with special reference to congenital cardiac septal changes. *Circulation*, **18**, 533–47.
12. Eddahibi, S., Morrell, N., d'Ortho, M.P. *et al.* (2002) Pathobiology of pulmonary arterial hypertension. *European Respiratory Journal*, **20**, 1559–72.
13. Yamaki, S. and Wagenvoort, C.A. (1985) Comparison of primary plexogenic arteriopathy in adults and children. A morphometric study in 40 patients. *British Heart Journal*, **54**, 428–34.
14. Tuder, R.M., Groves, B.M., Badesch, D.B., and Voelkel, N.F. (1994) Exuberant endothelial cell growth and elements of inflammation are present in plexiform lesions of pulmonary hypertension. *American Journal of Pathology*, **144**, 275–85.
15. Cool, C.D., Kennedy, D., Voelkel, N.F., and Tuder, R.M. (1997) Pathogenesis and evolution of plexiform lesions in pulmonary hypertension associated with scleroderma and human immunodeficiency virus infection. *Human Pathology*, **28**, 434–42.
16. Cool, C.D., Stewart, J.S., Werahera, P. *et al.* (1999) Three-dimensional reconstruction of pulmonary arteries in plexiform pulmonary hypertension using cell specific markers: evidence for a dynamic and heterogeneous process of pulmonary endothelial cell growth. *American Journal of Pathology*, **155**, 411–19.
17. Lee, S.D., Shroyer, K.R., Markham, N.E. *et al.* (1998) Monoclonal endothelial cell proliferation is present in primary but not secondary pulmonary hypertension. *The Journal of Clinical Investigation*, **101**, 927–34.
18. Tuder, R.M., Radisavljevic, Z., Shroyer, K.R. *et al.* (1998) Monoclonal endothelial cells in appetite suppressant-associated pulmonary hypertension. *American Journal of Respiratory and Critical Care Medicine*, **158**, 1999–2001.
19. Tuder, R.M., Lee, S.D., and Cool, C.D. (1998) Histopathology of pulmonary hypertension. *Chest*, **114**, 1S–6S.
20. Voelkel, N.F., Cool, C.D., Lee, S.D. *et al.* (1999) Primary pulmonary hypertension between inflammation and cancer. *Chest*, **114**, 225S–230S.
21. Rai, P.R., Cool, C.D., King, J.A.C. *et al.* (2008) The cancer paradigm of severe pulmonary arterial hypertension. *American Journal of Respiratory and Critical Care Medicine*, **178**, 558–64.
22. Yeager, M.E., Voelkel, N.F., and Tuder, R.M. (1999) Mutational analysis of endothelial cell TGF-β receptor type II in plexiform lesions of patients with primary pulmonary hypertension. *Circulation*, **100**, I-587.
23. Yeager, M.E., Halley, G.R., Golpon, H.A. *et al.* (2001) Microsatellite instability of endothelial cell growth and apoptosis genes within plexiform lesions in primary pulmonary hypertension. *Circulation Research*, **88**, e8–e11.
24. Deng, Z., Morse, J.H., Slager, S.L. *et al.* (2000) Familial primary pulmonary hypertension (gene PPH1) Is caused by mutations in the bone morphogenetic protein receptor-II gene. *American Journal of Human Genetics*, **67**, 737–44.
25. Lane, K.B., Machado, R.D., Pauciulo, M.W. *et al.* (2000) Heterozygous germline mutations in BMPR2 encoding a TGF-B receptor cause familiar pulmonary hypertension. *Nature Genetics*, **26**, 81–84.
26. Tan, B.T., Park, C.Y., Ailles, L.E., and Weissman, I.L. (2006) The cancer stem cell hypothesis: a work in progress. *Laboratory Investigation*, **86**, 1203–7.
27. Bull, T.M., Golpon, H., Hebbel, R.P. *et al.* (2003) Circulating endothelial cells in pulmonary hypertension. *Thrombosis and Haemostasis*, **90**, 688–703.
28. Kasahara, Y., Tuder, R.M., Cool, C.D. *et al.* (2001) Endothelial cell death and decreased expression of vascular endothelial growth factor and vascular endothelial growth factor receptor 2 in emphysema. *American Journal of Respiratory and Critical Care Medicine*, **163**, 737–44.
29. Sakao, S., Taraseviciene-Stewart, L., Lee, J.D. *et al.* (2005) Initial apoptosis is followed by increased proliferation of apoptosis-resistant endothelial cells. *The FASEB Journal*, **19**, 1178–180.
30. Taraseviciene-Stewart, L., Kasahara, Y., Alger, L. *et al.* (2001) Inhibition of the VEGF receptor 2 combined with chronic hypoxia causes cell death-dependent pulmonary endothelial cell proliferation and severe pulmonary hypertension. *The FASEB Journal*, **15**, 427–38.
31. Tuder, R.M., Chacon, M., Alger, L.A. *et al.* (2001) Expression of angiogenesis-related molecules in

plexiform lesions in severe pulmonary hypertension: evidence for a process of disordered angiogenesis. *Journal of Pathology*, **195**, 367–74.

32. Machado, R.D., James, V., Southwood, M. *et al.* (2005) Investigation of second genetic hits at the BMPR2 locus as a modulator of disease progression in familial pulmonary arterial hypertension. *Circulation*, **111**, 607–13.
33. Atkinson, C., Stewart, S., Upton, P.D. *et al.* (2002) Primary pulmonary hypertension is associated with reduced pulmonary vascular expression of type II bone morphogenetic protein receptor. *Circulation*, **105**, 1672–78.
34. Lowe, K., Alvarez, D., King, J., and Stevens, T. (2007) Phenotypic heterogeneity in lung capillary and extra-alveolar endothelial cells. Increased extra-alveolar endothelial permeability is sufficient to decrease compliance. *The Journal of Surgical Research*, **143**, 70–77.
35. Stevens, T., Kasper, M., Cool, C., and Voelkel, N.F. (2005) Pulmonary circulation and pulmonary hypertension, in *Endothelial Cells in Health and Disease*, 1st edn (ed. W.C. Aird), Taylor & Francis, Boca Raton, FL, pp. 417–38.
36. Ingram, D.A., Mead, L.E., Tanaka, H. *et al.* (2004) Identification of a novel hierarchy of endothelial progenitor cells using human peripheral and umbilical cord blood. *Blood*, **104**, 2752–60.
37. Yoder, M.C., Mead, L.E., Prater, D. *et al.* (2007) Redefining endothelial progenitor cells via clonal analysis and hematopoietic stem/progenitor cell principals. *Blood*, **109**, 1801–9.
38. Xu, W., Kaneko, F.T., Zheng, S. *et al.* (2004) Increased arginase II and decreased NO synthesis in endothelial cells of patients with pulmonary arterial hypertension. *The FASEB Journal*, **18**, 1746–48.
39. Masri, F.A., Xu, W., Comhair, S.A. *et al.* (2007) Hyperproliferative apoptosis-resistant endothelial cells in idiopathic pulmonary arterial hypertension. *American Journal of Physiology: Lung Cellular and Molecular Physiology*, **293**, L548–54.
40. Bartoli, M., Gu, X., Tsai, N.T. *et al.* (2000) Vascular endothelial growth factor activates STAT proteins in aortic endothelial cells. *The Journal of Biological Chemistry*, **275**, 33189–92.
41. Niu, G., Wright, K.L., Huang, M. *et al.* (2002) Constitutive Stat3 activity up-regulates VEGF expression and tumor angiogenesis. *Oncogene*, **21**, 2000–8.
42. Shibata, R., Kai, H., Seki, Y. *et al.* (2003) Inhibition of STAT3 prevents neointima formation by inhibiting proliferation and promoting apoptosis of neointimal smooth muscle cells. *Human Gene Therapy*, **14**, 601–10.
43. Yahata, Y., Shirakata, Y., Tokumaru, S. *et al.* (2003) Nuclear translocation of phosphorylated STAT3 is essential for vascular endothelial growth factor-induced human dermal microvascular endothelial cell migration and tube formation. *The Journal of Biological Chemistry*, **278**, 40026–31.
44. Xu, W., Koeck, T., Lara, A.R. *et al.* (2007) Alterations of cellular bioenergetics in pulmonary artery endothelial cells. *Proceedings of the National Academy of Sciences of the United States of America*, **104**, 1342–47.
45. Sehgal, P.B. and Mukhopadhyay, S. (2007) Pulmonary arterial hypertension: a disease of tethers, SNAREs and SNAPs? *American Journal of Physiology: Heart and Circulatory Physiology*, **293**, H77–85.
46. Mukhopadhyay, S., Shah, M., Xu, F. *et al.* (2007) Cytoplasmic provenance of STAT3 and PY-STAT3 in endolysosomal compartments in pulmonary arterial endothelial and smooth muscle cells: implications in PAH. *American Journal of Physiology: Lung Cellular and Molecular Physiology*, **294**, L449–68.
47. Asosingh, K., Aldred, M.A., Vasanji, A. *et al.* (2008) Circulating angiogenic precursors in idiopathic pulmonary arterial hypertension. *American Journal of Pathology*, **172**, 615–27.
48. Yuan, J.X., Aldinger, A.M., Juhaszova, M. *et al.* (1998) Dysfunctional voltage-gated K^+ channels in pulmonary artery smooth muscle cells of patients with primary pulmonary hypertension. *Circulation*, **98**, 1400–6.
49. Krick, S., Platoshyn, O., McDaniel, S.S. *et al.* (2001) Augmented K^+ currents and mitochondrial membrane depolarization in pulmonary artery myocyte apoptosis. *American Journal of Physiology: Lung Cellular and Molecular Physiology*, **281**, L887–94.
50. McMurtry, M.S., Archer, S.L., Altieri, D.C., Bonnet, S. *et al.* (2005) Gene therapy targeting survivin selectively induces pulmonary vascular apoptosis and reverses pulmonary arterial hypertension. *The Journal of Clinical Investigation*, **115**, 1479–91.
51. McMurtry, M.S., Bonnet, S., Wu, X. *et al.* (2004) Dichloroacetate prevents and reverses pulmonary hypertension by inducing pulmonary artery smooth muscle cell apoptosis. *Circulation Research*, **95**, 830–40.
52. Michelakis, E.D., Porter, C.J., Bonnet, S. *et al.* (2006) An abnormal mitochondrial-hypoxia inducible factor-1 alpha-K_v channel pathway disrupts oxygen sensing and triggers pulmonary arterial

hypertension in fawn hooded rats – similarities to human pulmonary arterial hypertension. *Circulation*, **113**, 2630–41.

53. Archer, S. and Rich, S. (2000) Primary pulmonary hypertension: a vascular biology and translational research "work in progress". *Circulation*, **102**, 2781–91.
54. Azeka, E., Costa, A.J. Jr., Kajita, L. *et al.* (2002) Effects of low doses of inhaled nitric oxide combined with oxygen for the evaluation of pulmonary vascular reactivity in patients with pulmonary hypertension. *Pediatric Cardiology*, **23**, 20–26.
55. Champion, H.C., Bivalacqua, T.J., Greenberg, S.S. *et al.* (2002) Adenoviral gene transfer of endothelial nitric-oxide synthase (eNOS) partially restores normal pulmonary arterial pressure in eNOS-deficient mice. *Proceedings of the National Academy of Sciences of the United States of America*, **99**, 13248–53.
56. Cooke, J.P. and Dzau, V.J. (1997) Nitric oxide synthase: role in the genesis of vascular disease. *Annual Review of Medicine*, **48**, 489–509.
57. Ignarro, L.J. (1993) Nitric oxide-mediated vasorelaxation. *Thrombosis and Haemostasis*, **70**, 148–51.
58. Jeffery, T.K. and Morrell, N.W. (2002) Molecular and cellular basis of pulmonary vascular remodeling in pulmonary hypertension. *Progress in Cardiovascular Diseases*, **45**, 173–202.
59. Kaneko, F.T., Arroliga, A.C., Dweik, R.A. *et al.* (1998) Biochemical reaction products of nitric oxide as quantitative markers of primary pulmonary hypertension. *American Journal of Respiratory and Critical Care Medicine*, **158**, 917–23.
60. Moncada, S. (1999) Nitric oxide: discovery and impact on clinical medicine. *Journal of the Royal Society of Medicine*, **92**, 164–69.
61. Girgis, R.E., Champion, H.C., Diette, G.B. *et al.* (2005) Decreased exhaled nitric oxide in pulmonary arterial hypertension: response to bosentan therapy. *American Journal of Respiratory and Critical Care Medicine*, **172**, 352–57.
62. Ozkan, M., Dweik, R.A., Laskowski, D. *et al.* (2001) High levels of nitric oxide in individuals with pulmonary hypertension receiving epoprostenol therapy. *Lung*, **179**, 233–43.
63. Momken, I., Fortin, D., Serrurier, B. *et al.* (2002) Endothelial nitric oxide synthase (NOS) deficiency affects energy metabolism pattern in murine oxidative skeletal muscle. *The Biochemical Journal*, **368**, 341–47.
64. Nisoli, E., Clementi, E., Paolucci, C. *et al.* (2003) Mitochondrial biogenesis in mammals: the role of endogenous nitric oxide. *Science*, **299**, 896–99.
65. Nisoli, E., Falcone, S., Tonello, C. *et al.* (2004) Mitochondrial biogenesis by NO yields functionally active mitochondria in mammals. *Proceedings of the National Academy of Sciences of the United States of America*, **101**, 16507–12.
66. Fagan, K.A., Fouty, B.W., Tyler, R.C. *et al.* (1998) The pulmonary circulation of mice with either homozygous or heterozygous disruption of endothelial nitric oxide synthase is hyperresponsive to chronic mild hypoxia. *The Journal of Clinical Investigation*, **1-3**, 291–99.
67. Cawthon, D., Iqbal, M., Brand, J. *et al.* (2004) Investigation of proton conductance in liver mitochondria of broilers with pulmonary hypertension syndrome. *Poultry Science*, **83**, 259–65.
68. Iqbal, M., Cawthon, D., Wideman, R.F., and Bottje, W.G. Jr. (2001) Lung mitochondrial dysfunction in pulmonary hypertension syndrome. I. Site-specific defects in the electron transport chain. *Poultry Science*, **80**, 485–95.
69. Michelakis, E.D., McMurtry, M.S., Wu, X.C. *et al.* (2002) Dichloroacetate, a metabolic modulator, prevents and reverses chronic hypoxic pulmonary hypertension in rats: role of increased expression and activity of voltage-gated potassium channels. *Circulation*, **105**, 244–50.
70. Tang, Z., Iqbal, M., Cawthon, D., and Bottje, W.G. (2002) Heart and breast muscle mitochondrial dysfunction in pulmonary hypertension syndrome in broilers (Gallus domesticus). *Comparative Biochemistry and Physiology, Part A: Molecular and Integrative Physiology*, **132**, 527–40.
71. Warburg, O., Posener, K., and Negelein, E. (1924) Ueber den stoffwechsel der carcinomzelle. *Biochemische Zeitschrift*, **152**, 319–44.
72. Wang, T., Marquardt, C., and Foker, J. (1976) Aerobic glycolysis during lymphocyte proliferation. *Nature*, **261**, 702–5.
73. Bonnet, S., Michelakis, E.D., Porter, C.J., ndrade-Navarro, M.A.,, Thebaud, B., Bonnet, S., Haromy, A., Harry, G., Moudgil, R., McMurtry, S., Weir, E.K. and Archer, S.L. (2006) An abnormal mitochondrial-hypoxia inducible factor-1 alpha-Kv channel pathway disrupts oxygen sensing and triggers pulmonary arterial hypertension in fawn hooded rats-Similarities to human pulmonary arterial hypertension. *Circulation*, **113**, 2630–2641.
74. Caslin, A.W., Heath, D., Madden, B. *et al.* (1990) The histopathology of 36 cases of plexogenic pulmonary arteriopathy. *Histopathology*, **16**, 9–19.
75. Frid, M.G., Brunetti, J.A., Burke, D.L. *et al.* (2006) Hypoxia-induced pulmonary vascular remodeling

requires recruitment of circulating mesenchymal precursors of a monocyte/macrophage lineage. *American Journal of Pathology*, **168**, 659–69.

76. Smith, P. and Heath, D. (1979) Electron microscopy of the plexiform lesion. *Thorax*, **34**, 177–86.
77. Stenmark, K.R., Fagan, K.A., and Frid, M.G. (2006) Hypoxia-induced pulmonary vascular remodeling: cellular and molecular mechanisms. *Circulation Research*, **99**, 675–91.
78. Strieter, R.M., Gomperts, B.N., and Keane, M.P. (2007) The role of CXC chemokines in pulmonary fibrosis. *The Journal of Clinical Investigation*, **117**, 549–56.
79. Gehling, U.M., Ergun, S., Schumacher, U. *et al.* (2000) *In vitro* differentiation of endothelial cells from AC133-positive progenitor cells. *Blood*, **95**, 3106–12.
80. Pesce, M., Orlandi, A., Iachininoto, M.G. *et al.* (2003) Myoendothelial differentiation of human umbilical cord blood-derived stem cells in ischemic limb tissues. *Circulation Research*, **93**, e51–e62.
81. Prater, D.N., Case, J., Ingram, D.A., and Yoder, M.C. (2007) Working hypothesis to redefine endothelial progenitor cells. *Leukemia*, **21**, 1141–49.
82. Hur, J., Yoon, C.H., Kim, H.S. *et al.* (2004) Characterization of two types of endothelial progenitor cells and their different contributions to neovasculogenesis. *Arteriosclerosis, Thrombosis, and Vascular Biology*, **24**, 288–93.
83. Ceradini, D.J., Kulkarni, A.R., Callaghan, M.J. *et al.* (2004) Progenitor cell trafficking is regulated by hypoxic gradients through HIF-1 induction of SDF-1. *Nature Medicine*, **10**, 858–64.
84. Davie, N.J., Crossno, J.T., Frid, M.G. *et al.* (2004) Hypoxia-induced pulmonary artery adventitial remodeling and neovascularization: contribution of progenitor cells. *American Journal of Physiology: Lung Cellular and Molecular Physiology*, **286**, L668–78.
85. Hayashida, K., Fujita, J., Miyake, Y. *et al.* (2005) Bone marrow-derived cells contribute to pulmonary vascular remodeling in hypoxia-induced pulmonary hypertension. *Chest*, **127**, 1793–98.
86. Cheng, X.W., Kuzuya, M., Nakamura, K. *et al.* (2007) Mechanisms underlying the impairment of ischemia-induced neovascularization in matrix metalloproteinase 2-deficient mice. *Circulation Research*, **100**, 904–13.
87. Teichert-Kuliszewska, K., Kutryk, M.J., Kuliszewski, M.A. *et al.* (2006) Bone morphogenetic protein receptor-2 signaling promotes pulmonary arterial endothelial cell survival: implications for loss-of-function mutations in the pathogenesis of pulmonary hypertension. *Circulation Research*, **98**, 209–17.
88. Peinado, V.I., Barbera, J.A., Ramirez, J. *et al.* (1998) Endothelial dysfunction in pulmonary arteries of patients with mild COPD. *American Journal of Physiology: Lung Cellular and Molecular Physiology*, **18**, L908–13.
89. Eddahibi, S., Guignabert, C., Barlier-Mur, A.M. *et al.* (2006) Cross talk between endothelial and smooth muscle cells in pulmonary hypertension: critical role for serotonin-induced smooth muscle hyperplasia. *Circulation*, **113**, 1857–64.
90. Eddahibi, S., Humbert, M., Fadel, E. *et al.* (2001) Serotonin transporter overexpression is responsible for pulmonary artery smooth muscle hyperplasia in primary pulmonary hypertension. *The Journal of Clinical Investigation*, **108**, 1141–50.
91. Grisotto, M.G., Garin, A., Martin, A.P. *et al.* (2006) The human herpesvirus 8 chemokine receptor vGPCR triggers autonomous proliferation of endothelial cells. *The Journal of Clinical Investigation*, **116**, 1264–73.
92. Cool, C.D., Rai, M.D., Yeager, M.E. *et al.* (2003) Expression of human herpesvirus 8 in primary pulmonary hypertension. *The New England Journal of Medicine*, **349**, 1113–22.
93. Laney, A.S., De, M., Peters, J.S. *et al.* (2005) Kaposi sarcoma-associated herpesvirus and primary and secondary pulmonary hypertension. *Chest*, **127**, 762–67.
94. Nicastri, E., Vizza, C.D., Carletti, F. *et al.* (2005) Human herpesvirus 8 and pulmonary hypertension. *Emerging Infectious Diseases*, **11**, 1480–82.
95. Spiekerkoetter, E., Alvira, C.M., Kim, Y.M. *et al.* (2008) Reactivation of gammaHV68 induces neointimal lesions in pulmonary arteries of S100A4/Mts1-overexpressing mice in association with degradation of elastin. *American Journal of Physiology: Lung Cellular and Molecular Physiology*, **294**, L276–89.
96. Marecki, J.C., Cool, C.D., Parr, J.E. *et al.* (2006) HIV-1 *nef* is associated with complex pulmonary vascular lesions in SHIV-*nef*-infected macaques. *American Journal of Respiratory and Critical Care Medicine*, **174**, 437–45.
97. Schermuly, R.T., Dony, E., Ghofrani, H.A. *et al.* (2005) Reversal of experimental pulmonary hypertension by PDGF inhibition. *The Journal of Clinical Investigation*, **115**, 2811–21.
98. Moreno-Vinasco, L., Gomberg-Maitland, M., Maitland, M.L. *et al.* (2008) Genomic assessment of a multikinase inhibitor, sorafenib, in a rodent model of pulmonary hypertension. *Physiological Genomics*, **33**, 278–91.

28

Collagen Vascular Diseases and Pulmonary Endothelium

Pradeep R. Rai[1] and Carlyne D. Cool[2,3]

[1]Division of Pulmonary and Critical Care Medicine, University of Colorado Health Sciences Center, Aurora, CO, USA
[2]Department of Pathology, National Jewish Health, Denver, CO, USA
[3]Department of Pathology, University of Colorado Health Sciences Center, Aurora, CO, USA

INTRODUCTION

Of the many collagen vascular diseases, systemic sclerosis (SSc), or scleroderma, is perhaps the best known of the diseases that affect the endothelium. Scleroderma derives from the Greek – *skleros* (meaning hard or indurated) and *derma* (skin). Although scleroderma has been described since the time of Hippocrates, it was Robert H. Goetz who, in 1945, introduced the concept of scleroderma as a progressive disease and introduced the term progressive SSc [1]. SSc affects the connective tissue and the vasculature of many organs, including the lungs, kidneys, and skin [2, 3]. There is extensive damage to the microvessels. Specific pulmonary manifestations include interstitial fibrosis, pulmonary arterial hypertension (PAH), constriction of the chest wall due to skin thickening, and chronic aspiration due to esophageal dysfunction [4]. Patients with SSc who develop pulmonary complications, especially PAH, frequently die from their pulmonary disease [4–6].

Spontaneous cases of PAH are considered "idiopathic PAH (IPAH)," cases with a familial background are designated "familial PAH," and cases attributable to collagen vascular diseases, congenital systemic to pulmonary shunts, portal hypertension, HIV, drug exposure and others, are now categorized as "associated PAH" [7] (see Chapter 27).

Although all forms of collagen vascular disease have been associated with the development of PAH, there is a wide range of risk depending on the clinical classification of collagen vascular disease. CREST (calcinosis, Raynaud's phenomenon, esophageal dysfunction, sclerodactyly, telangiectasia), which is the limited form of SSc, has a very high incidence of pulmonary hypertension – up to 60% patients are affected [8–11]. The incidence of pulmonary hypertension in patients with diffuse SSc is variable, ranging from 6 to 33% [8–11]. Pulmonary hypertension can be seen in SSc patients both with and without interstitial lung disease. Over 50% of SSc patients will have pathologic changes of pulmonary vessels at autopsy [3].

Patients with mixed connective tissue disease (MCTD), a disease that shares clinical features of scleroderma, systemic lupus erythematosus (SLE), and polymyositis/dermatomyositis (PM/DM), have similar incidences of PAH to SSc [12–16]. The high incidence of pulmonary hypertension in patients with MCTD is likely related to the scleroderma/SSc component of the disease.

SLE has a somewhat lower risk, with approximately 4–14% of individuals affected [17–20]. Other connective tissue diseases, including Sjögren's disease, rheumatoid arthritis, and PM/DM, rarely develop associated pulmonary hypertension [21–25].

MECHANISMS OF ENDOTHELIAL CELL INJURY IN COLLAGEN VASCULAR DISEASE-ASSOCIATED PAH

Vascular Lesions

There are three main components of pulmonary arteries – the intima, the media, and the adventitia. The

The Pulmonary Endothelium: Function in health and disease Editors Norbert F. Voelkel, Sharon Rounds

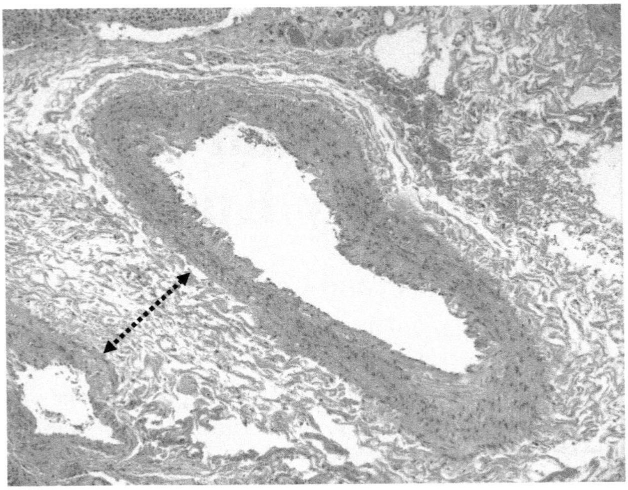

Figure 28.1 Pulmonary artery from a patient with diffuse scleroderma/SSc showing a marked thickening of the adventitial collagen (double arrow). The intima and media, however, show minimal change. A color version of this figure appears in the plate section of this volume.

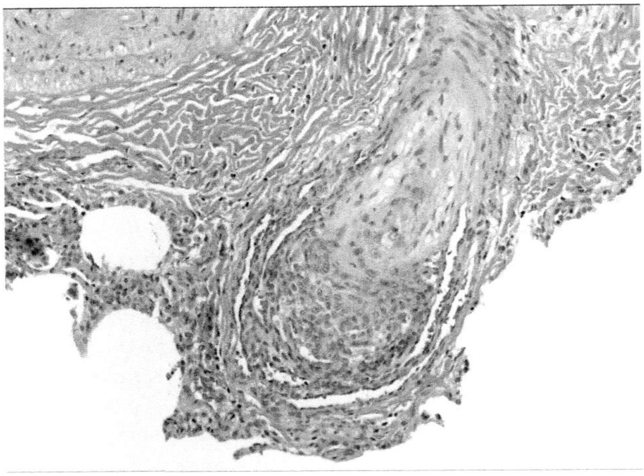

Figure 28.2 This plexiform lesion (P) from a patient with severe PAH demonstrates the proliferative, lumen-obliterating appearance of the ECs. A color version of this figure appears in the plate section of this volume.

pulmonary vascular remodeling of collagen vascular disease-associated pulmonary hypertension can affect any and all of these layers. The adventitia can be markedly thickened, particularly in scleroderma (Figure 28.1). The smooth muscle of the pulmonary artery can be variably hypertrophied in disease, but smooth muscle hypertrophy is not well correlated with clinically significant PAH. The lesion that has the best correlation with clinically severe PAH is the plexiform lesion – an abnormal proliferation of predominantly endothelial cells (ECs) (Figure 28.2). There are variants of this lesion, including concentric and dilatation or angiomatoid lesions [26] (Figure 28.3). Patients with scleroderma have a preponderance of the concentric, or "onionskin," lesions [27]. All of the endothelial-based lesions cause progressive narrowing and obstruction of the vascular lumen, which leads to severe PAH, subsequent right heart failure, and death. The concept of EC growth as a feature of plexiform lesions was first described by Tuder *et al.* [28]. Subsequent work has focused on the concept that PAH is an angioproliferative, possibly neoplastic disease, not simply a reparative process [29] (see Chapter 27).

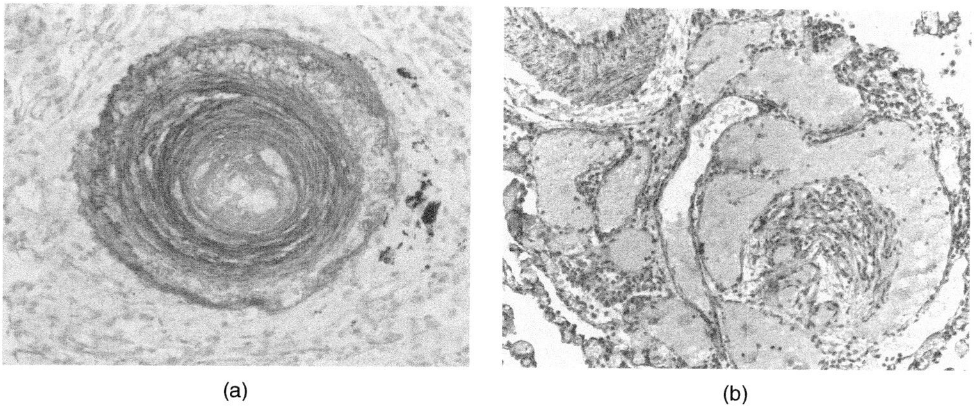

Figure 28.3 (a) Pulmonary artery obliterated by a concentric, "onionskinning," proliferation of ECs, highlighted by immunohistochemical stain for ECs (Factor VIII-related antigen). (b) Dilatation lesion at the distal end of a plexiform lesion. Immunohistochemical stain for EC marker, CD31. A color version of this figure appears in the plate section of this volume.

Although formerly thought to be solely the result of vasoconstriction, the most likely initiating event in the EC proliferative lesions of severe PAH is injury to the endothelium. The initiating injurious event may be mechanical (e.g., shear stress), hypoxia, toxins, drugs, infections, and/or immunologic factors. The Spanish toxic oil syndrome, which was a scleroderma-like disease caused by ingestion of adulterated rapeseed oil [30], resulted from direct EC injury by the toxic agent, which likely triggered severe PAH in the susceptible individuals. In HIV-associated PAH, while the ECs are not directly infected, the virus must cross the EC barrier to infect tissue and, in doing so, activates the ECs [31] (see Chapter 19). The specific location of plexiform lesions at sites just distal to bifurcation sites of pulmonary arteries suggests that shear stress-induced EC damage could lead to a proliferating EC [32] (Figure 28.4). Sakao *et al.* have demonstrated in an artificial capillary system that shear

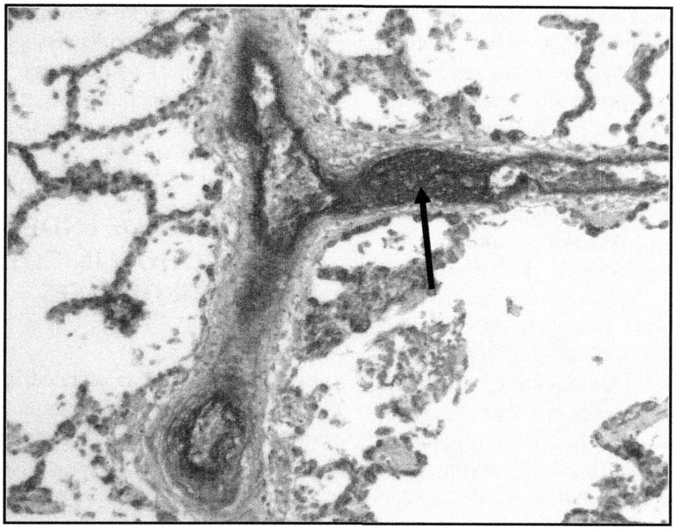

Figure 28.4 Bifurcating pulmonary artery from a patient with CREST and severe PAH. This immunohistochemical stain for an EC marker (Factor VIII-related antigen) highlights the proliferation of ECs that occludes the vascular lumen just distal to the bifurcation (arrow). A color version of this figure appears in the plate section of this volume.

stress causes apoptosis and subsequent proliferation of ECs [33]. It may be that the initial shear stress-induced apoptosis determines the degree of subsequent cell proliferation (see Chapter 16).

ECs within plexiform lesions in all forms of PAH show a decrease in the expression of antiproliferative and vasodilator factors, and an increase in the expression of angiogenic and mitogenic factors, as well as significant alterations in growth and apoptosis key regulatory genes [28, 32, 34–42]. In aggregate, these studies show that the cells that comprise the plexiform lesions are phenotypically altered, with loss of tumor suppressor proteins and overexpression of proapoptotic proteins (see Chapter 27).

Immune Dysfunction

Immune dysfunction is a prominent feature of scleroderma, as well as other collagen vascular diseases, and involves changes in immunoregulation and autoimmunity [43] (see Chapter 25). Data suggest that T lymphocytes in patients with scleroderma are activated and unusually adherent to endothelium [44, 45]. Perivascular mononuclear inflammatory cell infiltrates of arteries affected by plexiform growth has been described in SSc/scleroderma-related PAH [27, 46]. These inflammatory cells express cytokines, including interleukin (IL)-1 and IL-6, as well as growth factors, such as vascular endothelial growth factor (VEGF), transforming growth factor (TGF)-β and platelet-derived growth factor (PDGF). IL-1β increases the expression of hypoxia inducible factor (HIF)-1α, which in turn increases VEGF expression [47]. IL-6 protects ECs against apoptosis [48].

Anti-EC autoantibodies (AECAs) have been identified in SSc [49]. These antibodies are a heterogeneous group of autoantibodies that specifically recognize EC proteins and molecules present on the EC surface. AECAs have been reported in 40% of patients with SSc and in 13% of patients with CREST [50]. Patients with AECAs have a higher incidence of pulmonary hypertension and digital infarcts [50]. AECAS are not specific to SSc, though, and can be found in the sera of patients with idiopathic PAH, SLE, rheumatoid arthritis, and Wegener's granulomatosis [49, 51]. Interestingly, incubation of ECs with AECA-containing sera causes increased production of cytokines (including IL-1 and -6), enhanced expression of adhesion molecules, and initiation of EC apoptosis [52–54]. *In vitro* experiments demonstrate that autoantibodies from patients with collagen vascular diseases cause upregulation of intercellular adhesion molecule (ICAM-1)-1, endothelium leukocyte adhesion molecule-1, and major histocompatibility complex class II molecules on the surface of human ECs [55]. Thus, autoimmunity and/or active inflammation could lead to a proliferative pulmonary vasculopathy.

Antifibroblast antibodies have been reported not only in SSc patients [56–58], but also in patients with PAH [59]. The antifibroblast antibodies in the SSc patients induce fibroblast activation as well as a proadhesive, proinflammatory phenotype. Activation of the fibroblasts could lead to production of cytokines and upregulation of ICAM-1 on the fibroblast surface [60]. The antifibroblast antibodies may, in part, explain why SSc patients often have marked adventitial collagenous thickening (see Figure 28.1).

There have been numerous reports regarding antibody patterns in patients with PAH and collagen vascular diseases. SSc patients often have additional autoantibodies – including antifibrillarin (anti-U3-nucleolar ribonucleoprotein), anticentromere, antihistone and, in the presence of the human leukocyte antigen HLA-B35, antitopoisomerase IIα antibodies – which are more frequently seen in patients with PAH [61–63]. SLE patients with pulmonary hypertension have antifibrillarin antibodies and antiphospholipid antibodies even in the absence of thromboembolic disease [64]. Patients with CREST have antibodies to fibrin-bound tissue-type plasminogen activator [65, 66].

Apoptosis and Collagen Vascular Diseases

As mentioned in the previous section, AECAs from sera of patients with SSc can initiate apoptosis of human ECs. Previous work has demonstrated that EC apoptosis is the primary pathogenetic event in SSc skin lesions [67, 68]. There is a body of evidence suggesting that EC apoptosis may be a primary pathogenetic event in severe PAH and that the initial apoptotic event may be followed by the development of an apoptosis-resistant, proliferative EC phenotype [33, 42, 69, 70].

MEDIATORS OF ENDOTHELIAL DYSFUNCTION IN COLLAGEN VASCULAR DISEASES

Endothelin-1

Endothelin (ET)-1 is secreted in limited amounts by the normal pulmonary vasculature. Patients with PAH, however, have markedly increased levels of serum and vascular ET-1, and those levels are directly correlated with the severity of the disease as well as the extent of the development of the plexiform lesions [37]. Although ET-1 is a potent vasoconstrictor, it is also linked to vascular remodeling because it causes an increase in the expression of serotonin 1B receptors in vascular smooth muscle cells and acts as a smooth muscle mitogen [71]. ET-1 promotes vasoconstriction and results in proliferation and elevated production of the key profibrotic factors TGF-β and PDGF. Patients with both diffuse and limited SSc have been shown to have increased serum ET-1 levels [72, 73]. ET-1 is increased in both the fibroblasts and ECs of SSc patients [74, 75]. Patients with SLE-associated PAH show higher serum levels of ET-1 than in non-PAH SLE patients [76]. ET-1 increases extracellular matrix production and increases adhesion molecule expression, thereby facilitating leukocyte–fibroblast interactions [77]. The presumed importance of ET-1 in pulmonary vascular remodeling of PAH has led to the use of ET-1 receptor antagonists, bosentan and sitaxsentan, in treatment of this disease [78, 79].

Nitric Oxide

Patients with SSc have defective endothelial-dependent vasodilation, which may be related to reduced endothelial nitric oxide (NO) synthase (eNOS) [80]. Reduced expression of eNOS has been found in lung tissue from patients with PAH [81]. Additionally, there is decreased NO production in both SSc and PAH patients [80, 82, 83], the extent of which is directly correlated with the degree of vascular resistance. Reduction of NO promotes smooth muscle hypertrophy and vasoconstriction (see Chapter 6). More recent evidence suggests that the plexiform lesions in patients with PAH associated with congenital heart disease have increased expression of eNOS by immunohistochemistry, suggesting a possible angiogenic role, or perhaps illustrating a defective, nonfunctioning form of the enzyme [40, 84].

Prostacyclin

The ECs in PAH demonstrate decreased prostacyclin synthase levels in both autoimmune and idiopathic disease [41]. Prostacyclin synthase is responsible for the production of prostacyclin, a potent vasodilator and platelet aggregator. Prostacyclin and prostacyclin analogs have become a central part of treatment in patients with IPAH and in collagen vascular disease-associated PAH [85–89]. Prostacyclin treatment in SSc patients has also been shown to improve some of the skin lesions.

CIRCULATING EC IN COLLAGEN VASCULAR DISEASE

The presence of circulating ECs (CECs) in vascular disorders likely provides direct evidence of endothelial injury, even though it is unknown whether the circulating cells correlate with the extent of endothelial lesions. Circulating EC progenitors (CEPs) increase in response to ischemia or cytokine stimulation, and then home to sites of angiogenesis and/or vascular damage, contributing to neovascularization and wound-healing processes [90–92] (see Chapter 13). CECs have been reported in patients with PAH [93]. Increased levels of CECs and CEPs in patients with SSc have also been reported, including those with scleroderma-associated pulmonary hypertension [94–96]. The increased levels of CECs might correlate with the degree of EC damage in these patients, but the levels of CECs did not vary between limited or diffuse variants of SSc. Some theorize that CECs populate pulmonary arterioles/arteries, where they implant, proliferate, and eventually obliterate the vascular lumen. Damaged and dysfunctional endothelium plays a critical role in the initiation and progression of pulmonary hypertension in SSc and other secondary forms of pulmonary hypertension, and therefore detection of increased circulating levels of CEC may aid in early detection of disease, as well as monitoring of disease activity and therapy efficacy.

CONCLUSIONS AND PERSPECTIVES

It has long been known that patients with collagen vascular diseases are at high risk for the development of PAH. The vascular lesions in collagen vascular disease-associated PAH and IPAH are similar – both show marked endothelial abnormalities. Although injury to the endothelium can occur with a variety of mechanisms, the pathophysiology of injury in collagen vascular diseases is undoubtedly related to dysregulation of immunity. AECAs have been found in not only SSc and CREST, but in other autoimmune diseases as well. Other antibodies to specific lung vascular components may also play a role in the development of EC dysregulation in the lung. AECAs can initiate apoptosis in ECs: apoptosis can lead to later development of apoptosis-resistant, proliferative ECs. Although the mediators of EC dysfunction, including ET-1, NO, and prostacyclin, have been extensively studied and have proved useful in the development of therapeutic interventions, there remains much to be understood about the pathophysiology of endothelial dysfunction in scleroderma and other collagen vascular diseases associated with PAH. Whether cause or consequence, the endothelium is at the center of PAH, as any factor that causes injury to endothelium causes downstream effects including release of vasoactive agents and change in vascular tone (see Chapter 12). Although there is still much work to be done in discovering the key mechanisms of the complex cellular and molecular interactions that lead to PAH, the recognition of the EC's central role in the process has done much to advance the field in recent years.

References

1. Goetz, R. (1945) Pathology of progressive systemic sclerosis with special reference to changes in the viscera. *Clinical Proceedings*, **4**, 337–92.
2. al-Sabbagh, M.R., Steen, V.D., Zee, B.C. *et al.* (1989) Pulmonary arterial histology and morphometry in systemic sclerosis: a case-control autopsy study. *The Journal of Rheumatology*, **16**, 1038–42.
3. Young, R.H. and Mark, G.J. (1978) Pulmonary vascular changes in scleroderma. *The American Journal of Medicine*, **64**, 998–1004.
4. Minai, O.A., Dweik, R.A., and Arroliga, C. (1998) Manifestations of scleroderma pulmonary disease. *Clinics in Chest Medicine*, **19**, 713–31.
5. Arroliga, A.C., Podell, D.N., and Matthay, R.A. (1992) Pulmonary manifestations of scleroderma. *Journal of Thoracic Imaging*, **7**, 30–45.
6. Koh, E.T., Lee, P., Gladman, D.D. *et al.* (1996) Pulmonary hypertension in systemic sclerosis: an analysis of 17 patients. *British Journal of Rheumatology*, **35**, 989–93.
7. Third World Symposium on Pulmonary Arterial Hypertension. (2004) A review of the Third World Symposium on Pulmonary Arterial Hypertension. *Journal of the American College of Cardiology*, **43** (12 Suppl S), 1S–90S.
8. Salerni, R., Rodnan, G.P., Leo, D.F., and Shaver, J.A. (1977) Pulmonary hypertension in the CREST syndrome variant of progressive systemic sclerosis (scleroderma). *Annals of Internal Medicine*, **86**, 394–99.
9. MacGregor, A.J., Canavan, R., Knight, C. *et al.* (2001) Pulmonary hypertension in systemic sclerosis: risk factors for progression and consequences for survival. *Rheumatology (Oxford, England)*, **40**, 453–59.
10. Ungerer, R.G., Tashkin, D.P., Furst, D. *et al.* (1983) Prevalence and clinical correlates of pulmonary arterial hypertension in progressive systemic sclerosis. *The American Journal of Medicine*, **75**, 65–74.
11. Stupi, A.M., Steen, V.D., Owens, G.R. *et al.* (1986) Pulmonary hypertension in the CREST syndrome variant of systemic sclerosis. *Arthritis and Rheumatism*, **29**, 515–24.
12. Esther, J.H., Sharp, G.C., Agia, G. *et al.* (1981) Pulmonary hypertension in patients with connective tissue disease and antibody to nuclear ribonucleoprotein. *Arthritis and Rheumatism*, **24**, S105.
13. Wiener–Kronish, J.P., Solinger, A.M., Warnock, M.L. *et al.* (1981) Severe pulmonary involvement in mixed connective tissue disease. *The American Review of Respiratory Disease*, **124**, 499–503.
14. Hosoda, Y., Suzuki, Y., Takano, M. *et al.* (1987) Mixed connective tissue disease with pulmonary hypertension: a clinical and pathological study. *The Journal of Rheumatology*, **14**, 826–30.
15. Prakash, U.B. (1998) Respiratory complications in mixed connective tissue disease. *Clinics in Chest Medicine*, **19**, 733–46.
16. Burdt, M.A., Hoffman, R.W., Deutscher, S.L. *et al.* (1999) Long-term outcome in mixed connective tissue disease: longitudinal clinical and serologic findings. *Arthritis and Rheumatism*, **42**, 899–909.
17. Asherson, R.A., Higenbottam, T.W., Dinh Xuan, A.T. *et al.* (1990) Pulmonary hypertension in a lupus clinic: experience with twenty-four patients. *The Journal of Rheumatology*, **17**, 1292–98.
18. Li, E.K. and Tam, L.S. (1999) Pulmonary hypertension in systemic lupus erythematosus: clinical association and survival in 18 patients. *The Journal of Rheumatology*, **26**, 1923–29.
19. Shen, J.Y., Chen, S.L., Wu, Y.X. *et al.* (1999) Pulmonary hypertension in systemic lupus erythematosus. *Rheumatology International*, **18**, 147–51.
20. Tanaka, E., Harigai, M., Tanaka, M. *et al.* (2002) Pulmonary hypertension in systemic lupus erythematosus: evaluation of clinical characteristics and response to immunosuppressive treatment. *The Journal of Rheumatology*, **29**, 282–87.
21. Dawson, J.K., Goodson, N.G., Graham, D.R. *et al.* (2000) Raised pulmonary artery pressures measured with Doppler echocardiography in rheumatoid arthritis patients. *Rheumatology (Oxford, England)*, **39**, 1320–25.
22. Denbow, C.E., Lie, J.T., Tancredi, R.G. *et al.* (1979) Cardiac involvement in polymyositis: a clinicopathologic study of 20 autopsied patients. *Arthritis and Rheumatism*, **22**, 1088–92.
23. Nakagawa, N., Osanai, S., Hiroshi, I.D.E. *et al.* (2003) Severe pulmonary hypertension associated with primary Sjogren's syndrome. *Internal Medicine*, **42**, 1248–52.
24. Hunninghake, G.W. and Fauci, A.S. (1979) Pulmonary involvement in the collagen vascular diseases. *The American Review of Respiratory Disease*, **119**, 471–503.
25. Goldman, J. and Edwards, W.D. (1994) Plexogenic pulmonary hypertension in systemic lupus erythematosus: report of two cases and review of the literature. *Cardiovascular Pathology*, **3**, 65–69.

26. Tuder, R.M., Marecki, J.C., Richter, A. *et al.* (2007) Pathology of pulmonary hypertension. *Clinics in Chest Medicine*, **28**, 23–42.
27. Cool, C.D., Kennedy, D., Voelkel, N.F., and Tuder, R.M. (1997) Pathogenesis and evolution of plexiform lesions in pulmonary hypertension associated with scleroderma and human immunodeficiency virus infection. *Human Pathology*, **28**, 434–42.
28. Tuder, R.M., Groves, B.M., Badesch, D.B. *et al.* (1994) Exuberant endothelial cell growth and elements of inflammation are present in plexiform lesions of pulmonary hypertension. *American Journal of Pathology*, **144**, 275–85.
29. Voelkel, N.F., Cool, C., Lee, S.D. *et al.* (1998) Primary pulmonary hypertension between inflammation and cancer. *Chest*, **114**, 225S–230S.
30. Gomez-Sanchez, M.A., Saenz de la Calzada, C., Gomez-Pajuelo, F.J. *et al.* (1991) Clinical and pathologic manifestations of pulmonary vascular disease in the toxic oil syndromes. *Journal of the American College of Cardiology*, **122**, 851–55.
31. Hofman, F.M., Wright, A.D., Dohadwala, M.M. *et al.* (1993) Exogenous tat protein activates human endothelial cells. *Blood*, **82**, 2774–80.
32. Cool, C.D., Stewart, J.S., Werahera, P. *et al.* (1999) Three-dimensional reconstruction of pulmonary arteries in plexiform pulmonary hypertension using cell-specific markers. Evidence for a dynamic and heterogeneous process of pulmonary endothelial cell growth. *American Journal of Pathology*, **155**, 411–19.
33. Sakao, S., Taraseviciene-Stewart, L., Lee, J.D. *et al.* (2005) Initial apoptosis is followed by increased proliferation of apoptosis-resistant endothelial cells. *The FASEB Journal*, **19**, 1178–80.
34. Tuder, R.M., Chacon, M., Alger, L. *et al.* (2001) Expression of angiogenesis-related molecules in plexiform lesions in severe pulmonary hypertension: evidence for a process of disordered angiogenesis. *The Journal of Pathology*, **15**, 367–74.
35. Du, L., Sullivan, C.C., Chuy, D. *et al.* (2003) Signaling molecules in nonfamilial pulmonary hypertension. *The New England Journal of Medicine*, **348**, 500–9.
36. Wright, L., Tuder, R.M., Wang, J. *et al.* (1998) 5-lipoxygenase and 5-lipoxygenase activating protein (FLAP) immunoreactivity in lungs from patients with primary pulmonary hypertension. *American Journal of Respiratory and Critical Care Medicine*, **157**, 219–29.
37. Giaid, A., Yanagisawa, M., Langleben, D. *et al.* (1993) Expression of endothelin-1 in lungs of patients with pulmonary hypertension. *The New England Journal of Medicine*, **328**, 1732–39.
38. Golpon, H.A., Geraci, M.W., Moore, M.D. *et al.* (2001) HOX genes in human lung: altered expression in primary pulmonary hypertension and emphysema. *American Journal of Pathology*, **158**, 955–66.
39. Dorfmuller, P., Zarka, V., Durand-Gasselin, I. *et al.* (2002) Chemokine RANTES in severe pulmonary arterial hypertension. *American Journal of Respiratory and Critical Care Medicine*, **165**, 534–39.
40. Mason, N.A., Springall, D.R., Burke, M. *et al.* (1998) High expression of endothelial nitric oxide synthase in plexiform lesions of pulmonary hypertension. *The Journal of Pathology*, **185**, 313–18.
41. Tuder, R.M., Cool, C.D., Geraci, M.W. *et al.* (1999) Prostacyclin synthase expression is decreased in lungs from patients with severe pulmonary hypertension. *American Journal of Respiratory and Critical Care Medicine*, **159**, 1925–32.
42. Ameshima, S., Golpon, H., Cool, C.D. *et al.* (2003) Peroxisome proliferator-activated receptor gamma (PPARγ) expression is decreased in pulmonary hypertension and affects endothelial cell growth. *Circulation Research*, **92**, 1162–69.
43. Murrell, D.F. (1993) A radical proposal for the pathogenesis of scleroderma. *Journal of the American Academy of Dermatology*, **28**, 78–85.
44. LeRoy, E.C. (1992) A brief overview of the pathogenesis of scleroderma (systemic sclerosis). *Annals of the Rheumatic Diseases*, **51**, 286–88.
45. Kahaleh, M.B. and LeRoy, E.C. (1989) The immune basis for human fibrotic disease, especially scleroderma (systemic sclerosis). *Clinical Aspects of Immunology*, **3**, 19–29.
46. Tuder, R.M. and Voelkel, N.F. (1998) Pulmonary hypertension and inflammation. *The Journal of Laboratory and Clinical Medicine*, **132**, 16–24.
47. Jung, Y.J., Isaacs, J.S., Lee, S. *et al.* (2003) IL-1 beta-mediated up-regulations of HIF-1alpha via an NFkappaB/COS-2 pathway identifies HIF-1 as a critical link between inflammation and oncogenesis. *The FASEB Journal*, **17**, 2115–17.
48. Waxman, A.B., Mahboubi, K., Knickelbein, R.G. *et al.* (2003) Interleukin-11 and interleukin-6 protect cultured human endothelial cells from H_2O_2-induced cell death. *American Journal of Respiratory Cell and Molecular Biology*, **29**, 513–22.
49. Renaudineau, Y., Revelen, R., Levy, Y. *et al.* (1999) Anti-endothelial cell antibodies in systemic sclerosis. *Clinical and Diagnostic Laboratory Immunology*, **6**, 156–60.
50. Negi, V.S., Tripathy, N.K., Msra, R. *et al.* (1998) Antiendothelial cell antibodies in scleroderma correlate with severe digital ischemia and pulmonary

arterial hypertension. *The Journal of Rheumatology*, **25**, 461–66.

51. Tamby, M., Chanseaud, Y., Humbert, M. *et al.* (2005) Antiendothelial cell antibodies in idiopathic and systemic sclerosis associated pulmonary arterial hypertension. *Thorax*, **60**, 765–72.
52. Carvalho, D., Savage, C.O., Black, C.M., and Pearson, J.D. (1997) IgG antiendothelial cell autoantibodies from scleroderma patients induce leukocyte adhesion to human vascular endothelial cells *in vitro*. Induction of adhesion molecule expression and involvement of endothelium-derived cytokines. *The Journal of Clinical Investigation*, **97**, 111–19.
53. Bordron, A., Dueymes, M., Levy, Y. *et al.* (1998) The binding of human antiendothelial cell antibodies induces endothelial cell apoptosis. *The Journal of Clinical Investigation*, **101**, 2029–35.
54. Belizna, C., Duijvestijn, A., Hamidou, M., and Cohen Tervaert, J.W. (2006) Antiendothelial cell antibodies in vasculitis and connective tissue disease. *Annals of the Rheumatic Diseases*, **65**, 1545–50.
55. Okawa-Takatsuji, M., Aotsuka, S., Fujinami, M. *et al.* (1999) Up-regulation of intercellular adhesion molecule-1 (ICAM-1), endothelial leuocoycte adhesion molecule-1 (ELAM-1) and class II MHC molecules on pulmonary arterial endothelial cells by antibodies against U1-ribonucleoprotein. *Clinical and Experimental Immunology*, **116**, 174–80.
56. Chizzolini, C., Raschi, E., Rezzonico, R. *et al.* (2002) Autoantibodies to fibroblasts induce a proadhesive and proinflammatory fibroblast phenotype in patients with systemic sclerosis. *Arthritis and Rheumatism*, **46**, 1602–13.
57. Mouthon, L., Guillevin, L., and Humbert, M. (2005) Pulmonary arterial hypertension: an autoimmune disease? *The European Respiratory Journal*, **26**, 986–88.
58. Henault, J., Tremblay, M., Clement, I. *et al.* (2004) Direct binding of anti-DNA topoisomerase I antibodies to the cell surface of fibroblasts in patients with systemic sclerosis. *Arthritis and Rheumatism*, **50**, 3265–74.
59. Tamby, M.C., Humbert, M., Guilpain, P. *et al.* (2006) Antibodies to fibroblasts in idiopathic and scleroderma-associated pulmonary hypertension. *The European Respiratory Journal*, **28**, 799–807.
60. Ihn, H., Sato, S., Fulimoto, M. *et al.* (1997) Circulating ICAM-1 in the sera of patients with systemic sclerosis: enhancement by inflammatory cytokines. *British Journal of Rheumatology*, **36**, 1270–75.
61. Torney, V.J., Bunn, C.C., Denton, C.P. *et al.* (2001) Anti-fibrillarin antibodies in systemic sclerosis. *Rheumatology (Oxford, England)*, **40**, 1157–62.
62. Martin, L., Pauls, J.D., Ryan, P. *et al.* (1993) Identification of a subset of patients with scleroderma with severe pulmonary and vascular disease by the presence of autoantibodies to centromere and histone. *Annals of the Rheumatic Diseases*, **52**, 780–84.
63. Grigolo, B., Mazzetti, I., Meliconi, R. *et al.* (2000) Anti-topoisomerase II alpha autoantibodies in systemic sclerosis – association with pulmonary hypertension and HLA-B35. *Clinical and Experimental Immunology*, **121**, 539–43.
64. Asherson, R.A. and Oakley, C.M. (1986) Pulmonary hypertension and systemic lupus erythematosus. *The Journal of Rheumatology*, **13**, 1–5.
65. Fritzler, M.J., Hart, D.A., Wilson, D. *et al.* (1994) Antibodies to fibrin bound tissue type plasminogen activator in systemic sclerosis. *The Journal of Rheumatology*, **22**, 1688–93.
66. Morse, J.H., Barst, R.J., Fotino, M. *et al.* (1997) Primary pulmonary hypertension, tissue plasminogen activator antibodies, and HLA-DQ7. *American Journal of Respiratory and Critical Care Medicine*, **155**, 274–78.
67. Sgonc, R., Gruschwitz, M.S., Dietrich, H. *et al.* (1996) Endothelial cell apoptosis is a primary pathogenetic event underlying skin lesions in avian and human scleroderma. *The Journal of Clinical Investigation*, **98**, 785–92.
68. Bordron, A., Dueymes, M., Levy, Y. *et al.* (1998) The binding of some human antiendothelial cell antibodies induces endothelial cell apoptosis. *The Journal of Clinical Investigation*, **101**, 2029–35.
69. Taraseviciene-Stewart, L., Kasahara, Y., Alger, L. *et al.* (2001) Inhibition of the VEGF receptor 2 combined with chronic hypoxia causes cell-dependent pulmonary endothelial cell proliferation and severe pulmonary hypertension. *The FASEB Journal*, **15**, 427–38.
70. Voelkel, N.F., Cool, C., Taraseviciene-Stewart, L. *et al.* (2002) Janus face of vascular endothelial growth factor: the obligatory survival factor for lung vascular endothelium controls precapillary artery remodeling in severe pulmonary hypertension. *Critical Care Medicine*, **30**, S251–56.
71. Rondelet, B., van Beneden, R., Kerbaul, F. *et al.* (2003) Expression of the serotonin 1b receptor in experimental pulmonary hypertension. *The European Respiratory Journal*, **22**, 408–12.
72. Morelli, S., Ferri, C., Polettini, E. *et al.* (1995) Plasma endothelin-1 levels, pulmonary hypertension, and lung fibrosis in patients with systemic sclerosis. *The American Journal of Medicine*, **99**, 255–60.

73. Yamane, K. (1994) Endothelin and collagen vascular disease: a review with special reference to Raynaud's phenomenon and systemic sclerosis. *Internal Medicine*, **33**, 579–82.
74. Kawaguchi, Y., Suzuki, K., Hara, M. *et al.* (1994) Increased endothelin-1 production in fibroblasts derived from patients with systemic sclerosis. *Annals of the Rheumatic Diseases*, **53**, 506–10.
75. Tabata, H., Yamakage, A., and Yamazaki, S. (1997) Cutaneous localization of endothelin-1 in patients with systemic sclerosis: immunoelectron microscopic study. *International Journal of Dermatology*, **36**, 272–75.
76. Shen, J.Y., Chen, S.L., Wu, Y.X. *et al.* (1999) Pulmonary hypertension in systemic lupus erythematosus. *Rheumatology International*, **18**, 147–51.
77. Fernandez-Patron, C., Zouki, C., Whittal, R. *et al.* (2001) Matrix metalloproteinases regulate neutrophil–endothelial cell adhesion through generation of endothelin-1. *The FASEB Journal*, **15**, 2230–40.
78. Channick, R.N., Simonneau, G., Sitbon, O. *et al.* (2001) Effects of the dual endothelin-receptor antagonist bosentan in patients with pulmonary hypertension: a randomized, placebo-controlled study. *Lancet*, **358**, 1119–23.
79. Rubin, L.J., Badesch, D.B., Barst, R.J. *et al.* (2002) Bosentan therapy for pulmonary arterial hypertension. *The New England Journal of Medicine*, **346**, 896–903.
80. Romero, L.I., Zhang, D.N., Cooke, J.P. *et al.* (2000) Differential expression of nitric oxide by dermal microvascular endothelial cells from patients with scleroderma. *Vascular Medicine*, **5**, 147–58.
81. Giaid, A. and Saleh, D. (1995) Reduced expression of endothelial nitric oxide synthase in the lungs of patients with pulmonary hypertension. *The New England Journal of Medicine*, **333**, 214–21.
82. Kharitonov, S.A., Cailes, J.B., Black, C.M. *et al.* (1997) Decreased nitric oxide in the exhaled air of patients with systemic sclerosis with pulmonary hypertension. *Thorax*, **52**, 1051–55.
83. Rolla, G., Colagrande, P., Scappaticci, E. *et al.* (2000) Exhaled nitric oxide in systemic sclerosis. Relationships with lung involvement and pulmonary hypertension. *The Journal of Rheumatology*, **27**, 1693–99.
84. Berger, R., Geiger, R., Hess, J. *et al.* (2001) Altered arterial expression patterns of inducible and endothelial nitric oxide synthase in pulmonary plexogenic arteriopathy caused by congenital heart disease. *American Journal of Respiratory and Critical Care Medicine*, **163**, 1493–99.
85. Barst, R.J., Rubin, L.J., Long, W.A. *et al.* (1996) A comparison of continuous intravenous epoprostenol (prostacyclin) with conventional therapy for primary pulmonary hypertension. The Primary Pulmonary Hypertension Study Group. *The New England Journal of Medicine*, **334**, 296–302.
86. McLaughlin, V.V., Genthner, D.E., Panella, M.M. *et al.* (1999) Compassionate use of continuous prostacyclin in the management of secondary pulmonary hypertension: a case series. *Annals of Internal Medicine*, **130**, 740–43.
87. Menon, N., McAlpine, L., Peacock, A.J., and Madhok, R. (1998) The acute effects of prostacyclin on pulmonary hemodynamics in patients with pulmonary hypertension secondary to systemic sclerosis. *Arthritis and Rheumatism*, **41**, 466–69.
88. Robbins, I.M., Gaine, S.P., Schilz, R. *et al.* (2000) Epoprostenol for treatment of pulmonary hypertension in patients with systemic lupus erythematodes. *Chest*, **117**, 14–18.
89. Badesch, D.B., Tapson, V.F., McGoon, M.D. *et al.* (2000) Continuous intravenous epoprostenol for pulmonary hypertension due to the scleroderma spectrum of disease. *Annals of Internal Medicine*, **132**, 425–34.
90. Takahashi, T., Kalka, C., Masuda, H. *et al.* (1999) Ischemia- and cytokine-induced mobilization of bone marrow-derived endothelial progenitor cells for neovascularization. *Nature Medicine*, **5**, 434–38.
91. Rafii, S. (2000) Circulating endothelial precursors: mystery, reality, and promise. *The Journal of Clinical Investigation*, **105**, 17–19.
92. Luttun, A., Carmeliet, G., and Carmeliet, P. (2002) Vascular progenitors: from biology to treatment. *Trends in Cardiovascular Medicine*, **12**, 88–96.
93. Bull, T.M., Golpon, H., Hebbel, R.P. *et al.* (2003) Circulating endothelial cells in pulmonary hypertension. *Thrombosis and Haemostasis*, **90**, 698–703.
94. Del Papa, N., Quirici, N., Soligo, D. *et al.* (2006) Bone marrow endothelial progenitors are defective in systemic sclerosis. *Arthritis and Rheumatism*, **54**, 2605–15.
95. Allanore, Y., Batteux, F., Avouac, J. *et al.* (2007) Levels of circulating endothelial progenitor cells in systemic sclerosis. *Clinical and Experimental Rheumatology*, **25**, 60–66.
96. Del Papa, N., Colombo, G., Fracchiolla, N. *et al.* (2004) Circulating endothelial cells as a marker of ongoing vascular disease in system sclerosis. *Arthritis and Rheumatism*, **50**, 1296–304.

29

Pulmonary Endothelium in Thromboembolism

Irene M. Lang

Division of Cardiology, Medical University of Vienna, Vienna, Austria

INTRODUCTION

Abnormal thrombus formation and resolution occur in a majority of vascular disorders, including venous thromboembolism, stroke, hypertension, and myocardial infarction. There is increasing evidence for a link between venous and arterial thrombosis. The two vascular complications share several risk factors, such as age, obesity, diabetes mellitus, blood hypertension, hypertriglyceridemia, and metabolic syndrome. Moreover, there are many examples of conditions accounting for both venous and arterial thrombosis, such as the antiphospholipid antibody syndrome, hyperhomocysteinemia, malignancies, infections, and hormone treatment. Furthermore, recent trials have demonstrated that patients with venous thromboembolism are at increased risk of arterial thrombosis.

Venous thromboembolism is a continuum including deep vein thrombosis, thrombus in transit, acute pulmonary embolism (PE), and chronic thromboembolic pulmonary hypertension (CTEPH) (Figure 29.1). Venous thromboembolism is frequent (~500–800 cases/million/year, ~15 000 cases/year in Austria), and carries a high morbidity and mortality, leading to sudden death in about 10% of patients, accounting for around 300 000 yearly clinical episodes and 50 000 deaths in the United States of America [1, 2]. Still, pulmonary thromboemboli resolve in the majority of cases with restoration of normal pulmonary hemodynamics [3–5]. Resolution occurs by mechanical fragmentation [6], through organization of the thromboembolus by invasion of capillary buds and fibroblasts leading to recanalization, or via endogenous thrombolysis. Only in a small percentage of cases do venous thromboemboli fail to lyse, thereby resulting in CTEPH, or in occlusion of the deep veins.

The generation of thrombin from its precursor prothrombin is the central event of blood coagulation, which is essential to normal hemostasis and pathological thrombosis. It is a highly regulated, dynamic, and rapid process. By contrast, the timely removal of thrombus results from the concerted action of plasma fibrinolysis and a complex vascular remodeling process [7], which is time-consuming (Figure 29.2), and involves circulating cells and cells within the vessel wall [8]. Thrombosis itself may be considered the first event in this timeline, which takes between 15 s and 120 min. In addition to mechanisms of thrombus formation [9, 10], a new view of thrombosis will address the vascular biology and gene expression of thrombus resolution.

THE ENDOTHELIUM AND COAGULATION

The blood vessel wall plays a major role in the pathophysiology of thrombosis, in addition to blood flow and plasmatic coagulation (also famously known as Virchow's triad).

A crucial physiologic function of the endothelium is to facilitate blood flow by providing an antithrombotic surface that inhibits platelet adhesion and clotting [11]. The intact endothelium is a barrier separating platelets from adhesive substrates in the subendothelial matrix. Disruption of the integrity of the vessel wall by mechanical or functional trauma allows circulating platelets to come in contact with and adhere to the thrombogenic subendothelial matrix. After disturbance of the integrity of the endothelial monolayer, endothelial cells (ECs) undergo programmatic biochemical changes that result in their transformation to a prothrombotic surface [11]. Thrombus formation ensues once a critical mass of fibrinogen

The Pulmonary Endothelium: Function in health and disease Editors Norbert F. Voelkel, Sharon Rounds

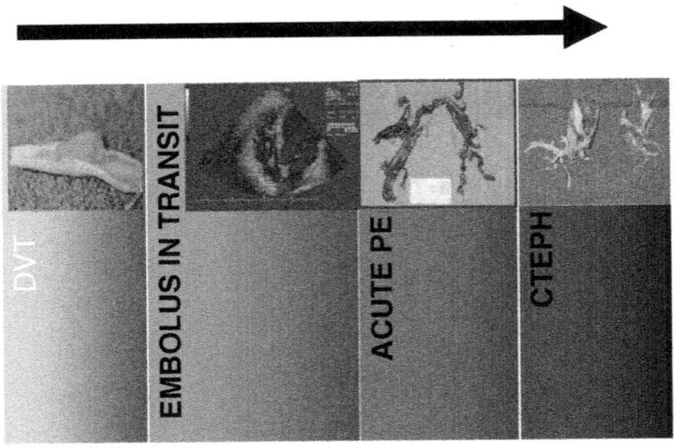

Figure 29.1 Spectrum of venous thromboembolism. DVT, deep venous thrombosis.

is cleaved and serves to protect organ integrity by limiting vascular damage. Once the procoagulant stimulus has disappeared, the injured endothelium can often return to its unperturbed state.

THE ENDOTHELIUM AND THROMBOSIS

Blood coagulation and hemostasis are essential defense mechanisms against bleeding. After clot formation, ECs proliferate and migrate to the site of endothelial damage. This has been demonstrated *in vivo* several hours after acute PE using antibodies against proliferating cell nuclear antigen [7]. This process is accelerated by high concentrations of growth factors that are released by activated platelets. Local high expression of plasminogen activator inhibitor (PAI)-1 in the endothelium immediately adjacent to the thrombus may serve to immobilize the thrombus on the vascular wall and permits a sequence of events eventually resulting in resolution/organization (Figure 29.2). Where in this process the thrombus resolution process succumbs to thrombus persistence is unknown. Thrombus persistence may be driven by inflammation, autoimmunity, phospholipids, and infection [12]. In a resting state, the endothelial surface is profibrinolytic and helps to maintain the blood's fluid state [11]. The contribution of ECs to coagulation/fibrinolysis varies with their metabolic state (i.e., quiescent or activated), their organ-specific functions, and the concentration of other hemostatically active molecules in the local plasma milieu [11].

Binding of tissue-type plasminogen activator (tPA) to an EC promotes its fibrinolytic activity and stimulates cell proliferation [13, 14]. ECs produce abundant PAI-1 that is associated primarily with the extracellular matrix, resulting in stabilization of its activity [15]. PAI-1 is a serine protease inhibitor and its main function is to inhibit tPA and urokinase type plasminogen activator (uPA). The liver is the major source of plasma PAI-1 and its synthesis is stimulated by thrombin, endotoxin, various cytokines, Lp(a) lipoprotein, oxidized low-density lipoprotein, and other mediators [16]. Quiescent EC express little or no PAI-1 [11], but after exposure to inflammatory stimuli, the expression of PAI-1 is highly upregulated, which results in impaired fibrinolytic function [17].

Binding of thrombin to thrombomodulin (TM) accelerates its capacity to activate a protein known as thrombin-activatable fibrinolysis inhibitor (TAFI). When activated, TAFI cleaves basic C-terminal residues within fibrin and other proteins. This results in the loss of plasminogen/plasmin and tPA binding sites on fibrin such that fibrinolysis is retarded [18]. By regulating the expression of TM, EC decrease the rate of intravascular fibrinolysis [11].

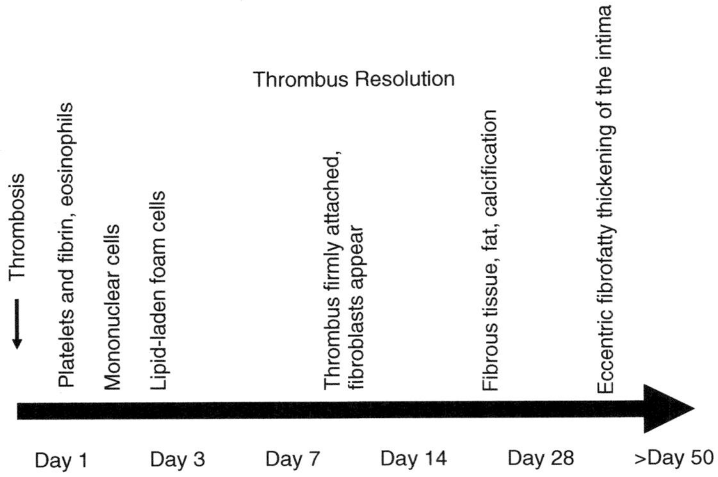

Figure 29.2 Timelines of thrombosis and thrombus resolution. While thrombosis may occur within seconds, the process of resolution is time-consuming and is subject to numerous disturbances (e.g., infection, inflammation, autoimmunity).

CHARACTERISTICS OF LUNG ENDOTHELIUM

For the regulation of vascular tone, the pulmonary endothelium is endowed with a well-adjusted balance of endothelial vasodilators and vasoconstrictors [19]. Although few data exist regarding the differential expression patterns of normal pulmonary ECs as a distinct pulmonary vascular compartment, there is evidence for a differential regulation and expression of transforming growth factor (TGF)-β – a family of multifunctional cytokines controlling cell growth, differentiation, and extracellular matrix deposition in the lung [20]. Furthermore, hypoxia-induced hemoxygenase regulation differs in systemic and lung ECs [21]. Genetic profiling of ECs in pulmonary hypertension has disclosed an increase of 5-lipoxygenase, the enzyme regulating the synthesis of leukotriene A_4 [22], and of endothelin (ET) [23], in the presence of a decrease of prostacyclin synthase [24]. Studies of the fibrinolytic capacity of pulmonary when compared to aortic ECs have disclosed increased tPA activity in unstimulated main pulmonary arterial ECs compared with those recovered simultaneously from the aorta of transplant donors at the time of organ donation [25].

ENDOTHELIUM OF THE DEEP VEINS

In humans, venous thromboembolism originates in the deep veins. Increased venous pressure, hemodynamic consequences of arteriovenous communications, valvular incompetence, and primary connective tissue abnormalities have emerged as the principal pathogenic theories of primary varicose vein formation. Endothelial nitric oxide (NO) synthase (endothelial nitric oxide synthaseeNOS) is involved in the regulation of resting and stimulated arterial tone by producing NO from the guanidino nitrogen of L-arginine (see Chapter 6). To investigate venous endothelial dysfunction in veins, the expression of eNOS, inducible NO synthase (inducible nitric oxide synthaseiNOS) and tPA was analyzed in veins from 24 patients undergoing elective vein resection (mean age 56.1 ± 4 years, 14 females/10 males) and compared with segments of greater saphenous veins from 12 patients undergoing coronary artery bypass grafting (CABG, mean age 56.4 ± 3.5 years, six females/six males, unpublished data). In addition, EC lysates freshly scraped from the vessel walls were used to determine the nitric oxide synthaseNOS enzyme activity by the conversion of L-[^{14}C]arginine to L-[^{14}C]citrulline. Analysis of the specimens revealed severe intimal and medial thickening in all patients undergoing elective vein resections and in two of nine veins harvested from CABG patients. eNOS expression in the endothelium of varicose veins was dramatically reduced compared with controls both by protein measurement (1.5 ± 0.4 versus 6.1 ± 1.1, $p = 0.0001$) or mRNA expression (unpublished data). In contrast, iNOS expression was not different between the groups. Analysis of EC scrapings demonstrated increased endothelial PAI-1 in varicose veins versus normal veins (PAI-1*von Willebrand factor activity %/100, 1.3 ± 0.3 versus 0.2 ± 0.07) and a decreased eNOS activity in varicose compared to control veins. These data illustrate that the endothelium of normal leg veins expresses high basal levels of eNOS. In contrast, eNOS disappears in varicose veins from the luminal endothelium, in parallel with an upregulation of PAI-1 and downregulation of tPA, resulting in loss of vasomotor function. These alterations can further thrombosis

ROLE OF LUNG ENDOTHELIUM IN NORMAL CLOT CLEARANCE – EC-DEPENDENT FIBRINOLYSIS

Since pulmonary emboli generally resolve within 1 year, it has been assumed that the pulmonary circulation harbors remarkable fibrinolytic capacity [26]. Proteases of the fibrinolytic system are crucial to the degradation of pulmonary thrombi [27]. In accord with this paradigm, tPA is expressed in ECs and smooth muscle cells of human main pulmonary arteries. These cells may become a significant source of luminal plasminogen activator activity. In combination with a relative decrease of whole vessel wall PAI-1 expression, an increased net fibrinolytic activity is the consequence. Within hours after pulmonary thromboembolism a sequential upregulation of fibrinolytic genes occurs in the wall of the mainstem pulmonary artery [7]. One may speculate that the capacity to efficiently lyse pulmonary emboli may be developmentally required in humans to compensate for an increased rate of thromboembolic episodes due to an erect posture. An elevated tPA/PAI-1 ratio reflects decreased steady-state expression of PAI-1, at the mRNA and protein level, with free tPA rapidly accessible for thrombolysis in the pulmonary artery vessel wall (Figure 29.3A). The adventitia is an important source for plasminogen activator activity, but does not account for differential fibrinolytic gene expression (Figure 29.3B).

Although shear stress was shown to increase EC tPA mRNA and tPA protein release [28], tPA steady-state expression is not physiologically different between the aorta and the pulmonary artery. However, more efficient macrophage recruitment in this vascular compartment [29] could contribute to differential fibrinolytic gene expression and enhanced spontaneous thrombus degradation.

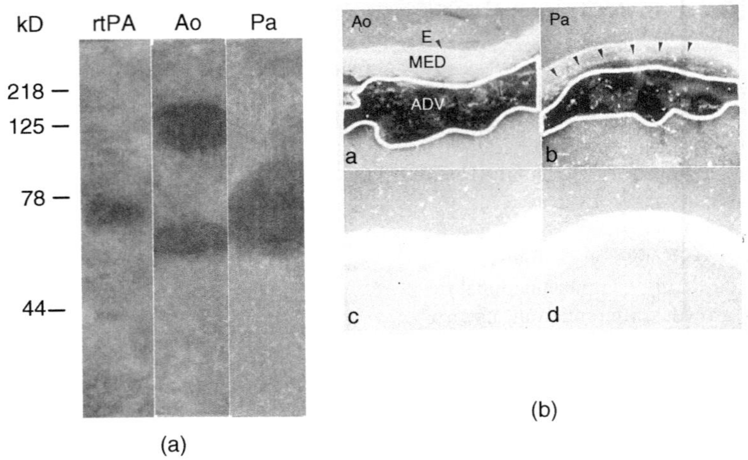

Figure 29.3 Fibrolinytic activity is increased in the normal pulmonary artery. Direct fibrin autography (A) and *in situ* zymography (B) of a representative pair of aorta (Ao) and pulmonary artery (Pa) from a human transplant donor. Dark areas represent lysis zones. (Bc and Bd) represent *in situ* zymographies in the absence of plasminogen. E, EC layer; MED, medial smooth muscle cell layer; ADV, adventitia.

RATE AND SEQUENCE OF THROMBUS ORGANIZATION IN THE DEEP VEINS

Thrombosis of the veins of the lower extremities usually occurs without a requirement of a primary inflammatory stimulus [30]. However, histological evidence suggests that inflammatory cells appear at the thrombus attachment sites to the endothelium [7]. To clarify the natural history of thrombus resolution we have utilized a mouse model of stagnant flow vena cava thrombosis. In brief, a stenosis is produced in the vein by tying a silk suture around the inferior vena cava and a stagnant flow venous thrombosis ensues.

In this model, thrombi resolve to approximately one sixth of their original size by 28 days after inferior vena cava ligation. Histologically, and by gene array analyses (unpublished data), procollagen type I, fibronectin 1, procollagen type XVI α1, latent TGF-β-binding protein 2, thrombospondin 2, and matrix metalloproteinase (MMP)-12 and -13 predominate by day 7 after ligation, indicating matrix protein accumulation. Tropomyosin 2β, fast skeletal troponin T3, tropomodulin 4, fast troponin C2, actinin α3, titin, phosphoglycerate mutase 2, histidine-rich calcium-binding protein, and muscle creatine kinase appear in parallel with ingrowth of smooth muscle cells (Figure 29.4). Inflammatory cells represented by genes such as calgranulins A and B, CC chemokines and CXC chemokines, and interleukin 1 beta are present by day 14, and then disappear.

VASCULAR REMODELING IN PULMONARY EMBOLISM

Based on the knowledge of a key role of the fibrinolytic system in thrombus organization, human tissues from patients who had died from acute PE have been investigated. Several important observations were made. First, evidence was found for a staged process of thrombosis. Different stages of thrombus organization were found in a single individual specimen. Local expression of PAs and PAI-1 were observed in distinct patterns. tPA antigen was primarily detected in regions containing an intact endothelial lining, whereas uPA expression was initially restricted to monocytic cells within thromboemboli and subsequently in cells that appeared to be migrating from the vessel wall toward the thrombus. The distribution of uPA is in accord with published data on uPA expression in monocytes/macrophages [31] and with the proposed role of uPA in supporting cell migration [32]. High PAI-1 concentrations were detected in EC directly in contact with fibrin, which is in good agreement with data showing PAI-1 induction by thrombin [33], and by TGF β, a polypeptide growth factor released from platelets [34]. PAI-1 expression induced by picomolar concentrations of TGF-β occurs within 2–4 h [35]. In another model system, PAI-1 expression was observed within neo-ECs overgrowing vascular thrombi, as soon as 1 week after thrombosis [36]. An embolizing thrombus may cause local EC injury by impinging on the EC surface. Direct contact of EC with fibrin has been shown to modulate a

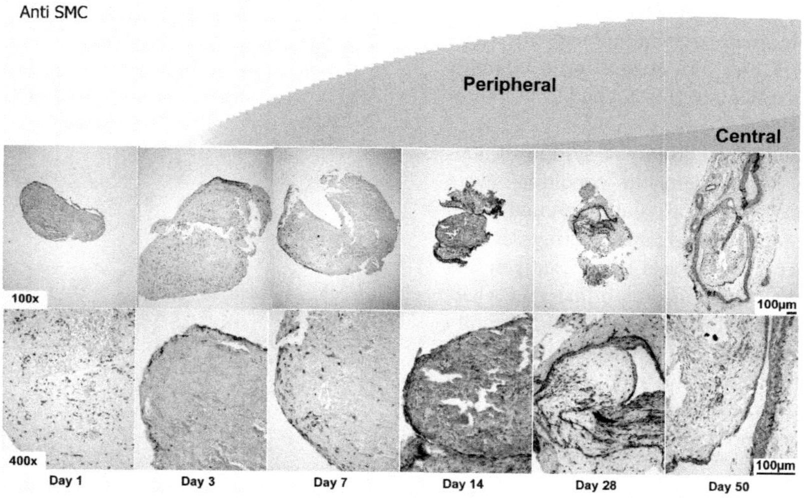

Figure 29.4 Characterization of cells in a vena cava thrombus of the mouse. Representative histological sections are shown. Standard immunohistochemical staining with anti-smooth muscle cell (SMC) antibody M851 (Dako, Glostrup, Denmark) was performed. Immunoreactivity in the periphery of the thrombus is shown in blue and in the center of the thrombus in grey. Note, that the immunoreactivity increases over time, indicating a relative increase in the smooth muscle cell number. A color version of this figure appears in the plate section of this volume.

number of phenotypic properties, including loss of organization, severing of cell–cell contacts, and cell retraction [37]. In addition to enhancing the production of PAI-1 in the single layer of ECs lining the vessel wall, the deposition of a thromboembolus within pulmonary vessel results in the generation of a new interface, which is subsequently penetrated by cells that are involved in the organization process. These cells are characterized by the concomitant production of proteases and protease inhibitors, initiating the entry of macrophages, and facilitating the degradation of fibrin and capillary sprouting [32]. Furthermore, local hypoxia, which has been shown to be a trigger for PAI-1 upregulation [38] could be a factor responsible for the elevation of PAI-1 in the remodeling/organization regions within the thrombi, and for the induction of connective tissue growth factor (CTGF) via a hypoxia-responsive element in the CTGF promotor [39].

Recent data have shown that genetic ablation of the bone morphogenetic protein receptor (BMPR)-2 gene in pulmonary endothelium induces *in situ* thrombosis [40].

ROLE OF LUNG ENDOTHELIUM IN ABNORMAL CLOT CLEARANCE AND MECHANISMS OF IN SITU THROMBOSIS

There exists no evidence that thrombus organization in the pulmonary circulation is regulated differently than that in the deep veins. We investigated the EC-associated fibrinolytic system in major pulmonary vessels free of thrombus of patients with CTEPH. CTEPH is characterized by predominantly major-vessel obstructions [41] resulting in increased pulmonary vascular resistance. The molecular mechanisms underlying thrombus persistence are unknown [42, 43]. The natural history of acute pulmonary thromboemboli is to undergo almost complete resolution within 6 months [44]. However, in 0.1–5% of survived acute events [45–47] thromboemboli undergo an organization process leading to permanent fibrotic obstruction of the pulmonary vascular bed. CTEPH is largely understood as of thromboembolic origin. However, patients with CTEPH lack classic plasma thromboembolic risk factors [48], such as antithrombin III deficiency, alterations in proteins C and S, and Factor V Leiden deficiency [49]. Furthermore, neither systemic [50] nor local [51] imbalances of fibrinolytic proteins in the pulmonary arterial wall have been detected. In addition, it is virtually impossible to induce the disease in animal models by repeated embolizations [52], suggesting alternative nonthromboembolic hypotheses [53]. The difficulty to induce CTEPH by repeated release of preformed clots from the inferior vena cava of dogs [52] was resolved by a thorough biochemical dissection of factors contributing to increased vascular fibrinolytic activity in these animals [54]. It was found that high plasma levels of uPA activity are present. Furthermore, uPA is associated with canine platelets and mediates rapid clot lysis.

In recent years, it has been recognized that major-vessel remodeling and classic small-vessel pulmonary arteriopathy coexist in CTEPH [42, 55], suggesting a complex remodeling process involving factors beyond local thrombosis.

To dissect the endothelial fibrinolytic system in nonresolving pulmonary thromboemboli, conditions were established to culture ECs from unthrombosed main pulmonary arteries of patients obtained during surgical pulmonary endarterectomies. Levels of tPA antigen and PAI-1 activity in media conditioned by primary ECs harvested from areas free of thrombus were not significantly different between patients with chronic thromboemboli and organ donors. Cultured patient pulmonary arterial ECs increased the secretion of tPA and PAI-1 in response to thrombin similar to donor pulmonary arterial EC [51].

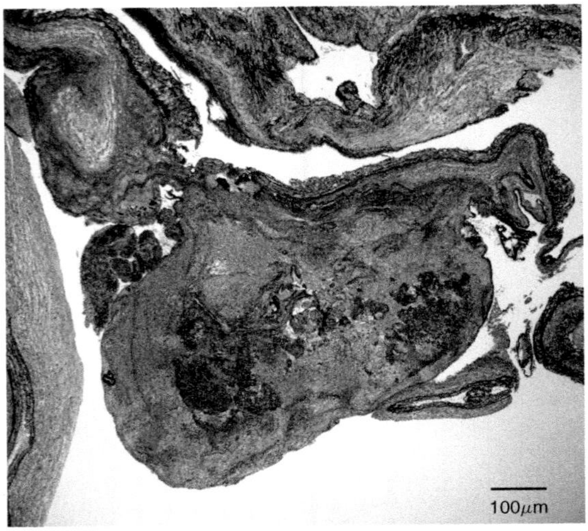

Figure 29.5 Representative histological section of a chronic pulmonary embolus, illustrating an area with *in situ* thrombosis. A trichrome stain is shown. The red color represents areas of fresh fibrin within the organized, collagen-containing thrombus tissue. A color version of this figure appears in the plate section of this volume.

CANDIDATE GENE EXPRESSION ANALYSIS IN NONRESOLVING THROMBI

A very different observation was derived from the analyses of pulmonary arterial obstructions in patients with nonresolving thromboemboli. Red, fibrin-rich thrombi within thromboendarterectomy specimens lined by a single layer of ECs exhibited high levels of PAI-1 antigen and an *in situ* hybridization signal in the ECs lining fresh thrombi in comparison to the signal present in the ECs from noninvolved areas of patients' pulmonary arteries ($p < 0.001$) [56]. Yellowish-white thrombi were composed of smooth muscle cells and ECs in numerous vessels that stained prominently for PAI-1 antigen. Both types of cells within the highly organized tissues also exhibited elevated PAI-1 mRNA levels in comparison to patient pulmonary artery specimens that were free of thrombus ($p < 0.02$). The prevalence of PAI-1 expression within pulmonary thromboemboli suggests that this inhibitor may play a role in the stabilization of vascular thrombi and provide grounds for *in situ* thrombosis (Figure 29.5) [56]. Another molecule of interest is Factor VIII (FVIII). Apart from its role as a plasma marker in patients with recurrent venous thromboembolism [57] that is present in 40% of patients with CTEPH [58], FVIII expression is high in organized pulmonary vascular obstructions of CTEPH (Figure 29.6), thus providing another molecular explanation for the phenomenon of *in situ* thrombosis.

FVIII antisense

Anti-FVIII

Anti-SMC

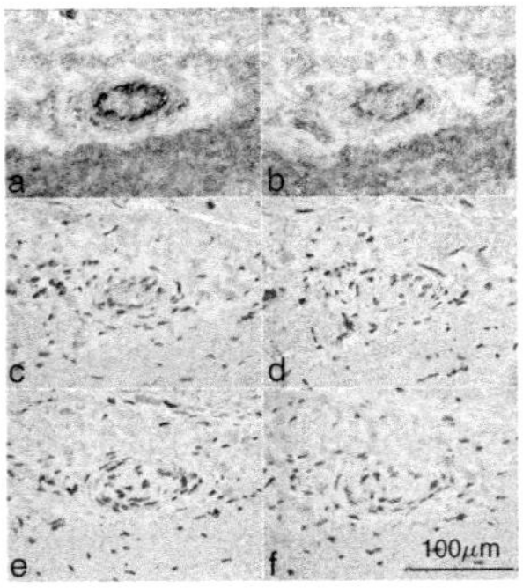

FVIII sense

control

control

Figure 29.6 Elevated expression of FVIII on thrombus surfaces favors *in situ* thrombosis. *In situ* hybridization, and immunohistochemical labeling of von Willebrand factor mRNA and protein in a small vessel of an organizing thrombus. Panels (b), (d), and (f) demonstrate that FVIII is expressed and that smooth muscle cells (SMC) are present. A color version of this figure appears in the plate section of this volume.

One hypothesis emerging from these studies is that deep venous thrombi undergo extensive organization in the vascular compartment of the deep femoral and pelvic veins. When such thrombi are embolized even well-functioning machineries of fibrinolysis and thrombolysis do not suffice to restore vascular patency. In other studies, the expression of a potent inhibitor of Factor IXa and Factor XIa, protease nexin-2/myloid β-protein precursor has been demonstrated in the organized vascular occlusions harvested from patients with this disease [59]. Clot vessel hemorrhage is a feature of CTEPH thrombus histology and it is speculated to be a powerful stimulator of angiogenesis.

ANGIOGENESIS IN THROMBUS RESOLUTION

Angiogenesis is a key histologic feature of thrombus growth and organization (Figure 29.7).

Two growth factors, basic fibroblast growth factor and vascular endothelial cell growth factors (VEGFs), are required for the initiation of vascular development [60]. VEGF interacts with specific tyrosine kinase receptors, and stimulates receptor autophosphorylation, EC replication, and migration. In the mouse, Flk-1 protein (corresponding to human VEGF receptor-2) is upregulated during thrombus organization (Figure 29.8). Contact of angioblasts and EC is essential for the expansion of the vascular bed. Distinct cell surface receptors, including platelet-endothelial cell adhesion molecule-1 (CD31) and VE-cadherin, mediate important cell–cell adhesions [11, 61, 62]. Interestingly, the use of constitutive (CD31) or conditional (Flk-1) knockout mice in the vena cava thrombosis model lead to very similar failures of thrombus resolution. These data confirm an integral role for angiogenesis in the organization process of vascular thrombi.

SERIAL ANALYSIS OF GENE EXPRESSION IN VENOUS THROMBI

In contrast to genetic pulmonary hypertension [63, 64] with mutations in the genes for BMPR-2 and activin-like kinase, two members of the TGF-beta receptor superfamily, CTEPH does not demonstrate a heritable trait. Therefore, gene expression studies at the tissue level are justified.

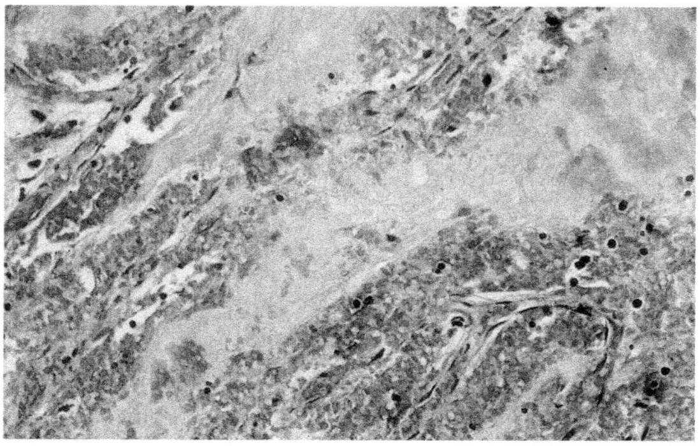

Figure 29.7 Trichrome stain of a histological section of a thrombus from a patient with chronic thromboembolic pulmonary hypertension. Numerous capillary structures are shown invading blood islands within the thrombus. A color version of this figure appears in the plate section of this volume.

mRNA display of hundreds of genes comparing pulmonary arteries and pulmonary arterial thromboemboli of CTEPH patients has demonstrated a remarkable loss of the gene expression repertoire in chronic clots in comparison with the expression profile of the parent pulmonary artery. After confirmation of clones by reverse Northern analysis, 67 different clones were identified. From these, nine were unknown and 58 were molecules found in a published database. As expected, most of the clones had been previously isolated. Among the genes in the database were a number of genes involved in lipid metabolism, for example, lipoprotein lipase, low-density lipoprotein-C, a brefeldin A-sensitive peripheral Golgi protein required for normal Golgi function, apoER2 [65], and LRP-6 [66], a member of the low-denisty lipoprotein receptor proteins. Dysregulation of lipid metabolism is a key mechanism underlying atherosclerosis. As atherosclerosis is associated with vascular remodeling, an adaptive process that is aimed at the preservation of a patent lumen [67], a shut-down of this process may result in progressive occlusion.

THROMBOSIS, THE ENDOTHELIUM, AND REGULATION OF VASCULAR TONE

The most important EC-dependent vasodilators are NO, prostaglandin I_2, and endothelium-derived hyperpolarization factor [68]. ET and platelet-activating factor (PAF) are potent vasoconstrictors.

NO is a heterodiatomic free radical product, which is generated through the oxidation of L-arginine to L-citrulline by NOSs [69] (see Chapter 6). It has several important effects on the vasculature. First, NO maintains basal tone by relaxing vascular smooth muscle cells through binding of NO to the heme prosthetic group of guanylyl cyclase [70]. NO also inhibits platelet adhesion, activation, secretion, and aggregation, also promoting platelet disaggregation [11, 71]. NO also suppresses the conformational change in the heterodimeric integrin glycoprotein $\alpha_{IIb}\beta_3$ (GPIIb–IIIa), which is required for fibrinogen binding [72]. NO inhibits leukocyte adhesion to the endothelium [73, 74], and migration and proliferation of smooth muscle cells [75, 76].

ET-1, the most potent vasoconstrictor identified to date, is synthesized by EC [11, 77]. ET-1 is formed after stimulation by hypoxia, shear stress, or ischemia [11]. Binding to the ET_A receptor on vascular smooth muscle cells results in an increased intracellular calcium concentration and increases vascular smooth muscle cell tone [78]. ET-1 potentiates the vasoconstrictor actions of catecholamines, which, in turn, potentiate the actions of ET-1 [11]. The concentration of bioactive NO is reduced in states of endothelial dysfunction, such as atherosclerosis. This results in vasoconstriction and smooth muscle

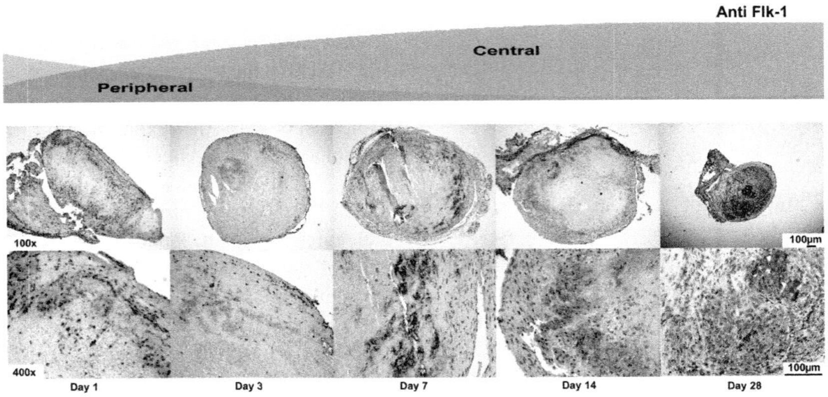

Figure 29.8 Characterization of cells in a vena cava thrombus of the mouse. Representative histological sections are shown. Standard immunohistochemical staining with anti-Flk-1 antibody (ab2349 rabbit polyclonal to VEGF receptor-2; Abcam, Cambridge, UK) was performed. Immunoreactivity in the periphery of the thrombus is shown in blue and in the center of the thrombus in grey. Note that the immunoreactivity increases over time, possibly indicating a relative increase of vascular structures. A color version of this figure appears in the plate section of this volume.

cell proliferation due to relatively unopposed actions of ET-1 [79].

In an *in vitro* vasoconstriction model, bolus injection of 300 μl homogenized venous thrombus material led to an increase in vessel wall tension of 65.5 (56.1–73.3)% of the maximal contractile capacity lasting for approximately 1 h. ET specificity of the vasoconstriction was examined by preincubation of porcine coronary artery rings with 100 nM of the dual ET receptor blocker tezosentan (ACT-050089A; provided by Actelion, Allschwil, Switzerland). These experiments suggest that thrombus-bound ET could exert a microcirculatory constriction after embolization [95].

CROSS-TALK BETWEEN ECS AND PLATELETS IN VENOUS THROMBOSIS

From a mechanistic standpoint, platelets and ECs communicate on multiple levels. Cross-talk may occur over a distance (paracrine signaling), via transient interactions, or through receptor-mediated cell–cell adhesion [80, 81]. Platelets and the endothelium influence each other in different ways (see Chapters 10 and 25). Platelets release interleukin-1β [82], TGF-β, platelet-derived growth factor, and VEGF, each of which may trigger signal transduction pathways in the endothelium. In the other direction, ECs express cell surface receptors or soluble mediators that either inhibit platelet function (e.g., nucleoside triphosphate diphosphohydrolases, prostaglandin I_2, or NO) or promote platelet activation (e.g., PAF) [81]. Studies have demonstrated a critical role for the CD40–CD40 ligand (CD40L) system in mediating reciprocal interactions between platelets and ECs [81, 83]. Platelet activation results in increased expression of CD40 and CD40L [84]. GPIIb–IIIa-dependent adhesion of platelets to the endothelium results in CD40L-induced activation of EC with secondary induction of TF [84], cytokines, adhesion molecules [85], MMPs, uPA, tPA, and urokinase receptor [83]. Thus, platelets coordinate indirectly (via the endothelium) changes in coagulation, leukocyte trafficking, and extracellular matrix modeling/turnover. At the same time, interaction between platelets and EC results in GPIIb–IIIa-mediated outside-in signaling with secondary induction of CD40L [86], and P-selectin expression on the platelet surface [87]. In addition, soluble trimeric CD40L, released from activated platelets, may engage platelet CD40 in an autocrine or paracrine manner, resulting in shape change, and dense granule and α-granule release [81, 86]. A recent study suggests a role for CD40L in the pathogenesis of pulmonary hypertension [88]. However, the role of thrombosis in this process is unclear.

Platelet adhesion causes the secretion of inflammatory factors that stimulate EC and thereby recruit additional platelets to the growing thrombus [89]. Endothelial selectins mediate platelet rolling, which is defined as the first loose contact between circulating platelets and vascular endothelium [87]. In response to inflammatory stimuli, P-selectin is rapidly expressed on the endothelial surface by translocation from membranes of the Weibel–Palade bodies to the plasma membrane. Inflamed ECs also express E-selectin, which promotes a loose contact between platelets and the endothelium [90].

In vivo studies in mice lacking P- and/or E-selectin showed that platelets from these knockout mice roll as efficiently as wild-type platelets. Thus, platelet rolling does not require previous platelet activation, which is in accordance with the concept of endothelial activation as a trigger for platelet accumulation [91]. Recent data have demonstrated that high P-selectin plasma levels independently predict venous thromboembolism in cancer patients [92]. Whether cancers alter the underlying mechanisms of thrombosis via P-selectin expression is unclear. Recently, homozygosity in the single nucleotide polymorphism Ser128Arg in the E-selectin gene has been found associated with recurrent venous thromboembolism [93]. However, in patients with nonresolving thromboemboli this polymorphism was not more frequent than in the general population (unpublished data).

Functional analyses of platelets from patients with CTEPH have recently illustrated a state of activation. Circulating heterotypic monocytes–platelet aggregates (MPAs) measured by $CD14^+/CD41^+\%$ fluorescence-activated cell sorting and platelet–leukocyte aggregates (PLA) formation assessed by coexpression of $CD45^+/CD41^+\%$ were significantly higher in patients with nonresolving venous thrombi. After activation with thrombin receptor-activating peptide-6, MPA and PLA were increased in CTEPH. In addition, platelet surface coverage and average size of aggregates measured by the cone and platelet analyzer Impact-R were significantly higher in CTEPH patients (unpublished data).

CONCLUSIONS AND PERSPECTIVES

In conclusion, while thrombosis is increasingly understood and utilized for targeted treatments, the importance of the vascular remodeling process underlying thrombus resolution/organization has not been appropriately appreciated. CTEPH serves as a model disease, as it provides clinical data and surgical thrombus specimens, thus permitting translational insights in mechanisms underlying nonresolution of thrombus. For example, the observation that splenectomy is a condition increasing the risk of CTEPH at least 10-fold above background risk [94] has led to the concept that phospholipids from recycling cell membranes that cannot be deposited in the spleen may alter the propensity for vascular thrombosis. Furthermore, the observation that infected intravenous lines increase the risk for CTEPH has led to the discovery of *Staphylococcus aureus* promoting thrombosis and thrombus persistence [12]. *In situ* thrombosis is associated with abnormal gene expression, that is, increased PAI-1 [56], increased von Willebrand factor (unpublished data), inadequate expression of antagonists of thrombosis [59], abnormal TGF-β receptor expression [40] but also high numbers of circulating microparticles [96] and slow flow. Key mechanisms that need to be addressed in the future are angiogenesis, leukocyte–EC interactions, immune mechanisms, EC regeneration, and the effects of endothelial shear stress in the context of thrombosis/thrombus organization. Research is also needed to better define the mechanisms underlying the vascular remodeling processes following thrombosis.

References

1. Lilienfeld, D.E., Chan, E., Ehland, J. *et al.* (1990) Mortality from pulmonary embolism in the United States: 1962 to 1984. *Chest*, **98** (5), 1067–72.
2. Anderson, F.A. Jr., Wheeler, H.B., Goldberg, R.J. *et al.* (1991) A population-based perspective of the hospital incidence and case-fatality rates of deep vein thrombosis and pulmonary embolism. The Worcester DVT Study. *Archives of Internal Medicine*, **151** (5), 933–38.
3. Paraskos, J.A., Adelstein, S.J., Smith, R.E. *et al.* (1973) Late prognosis of acute pulmonary embolism. *The New England Journal of Medicine*, **289** (2), 55–58.
4. Dalen, J.E. and Dexter, L. (1969) Pulmonary embolism. *The Journal of the American Medical Association*, **207** (8), 1505–7.
5. Murphy, M.L. and Bulloch, R.T. (1968) Factors influencing the restoration of blood flow following pulmonary embolization as determined by angiography and scanning. *Circulation*, **38** (6), 1116–26.
6. Wagenvoort, C.A. (1995) Pathology of pulmonary thromboembolism. *Chest*, **107** (Suppl 1), 10S–17S.
7. Lang, I.M., Moser, K.M., and Schleef, R.R. (1998) Elevated expression of urokinase-like plasminogen activator and plasminogen activator inhibitor type 1 during the vascular remodeling associated with pulmonary thromboembolism. *Arteriosclerosis, Thrombosis, and Vascular Biology*, **18** (5), 808–15.
8. Modarai, B., Burnand, K.G., Sawyer, B. *et al.* (2005) Endothelial progenitor cells are recruited into resolving venous thrombi. *Circulation*, **111** (20), 2645–53.
9. Cushman, M. (2007) Epidemiology and risk factors for venous thrombosis. *Seminars in Hematology*, **44** (2), 62–69.
10. Nossent, A.Y., Eikenboom, J.C., and Bertina, R.M. (2007) Plasma coagulation factor levels in venous thrombosis. *Seminars in Hematology*, **44** (2), 77–84.

11. Cines, D.B., Pollak, E.S., Buck, C.A. *et al.* (1998) Endothelial cells in physiology and in the pathophysiology of vascular disorders. *Blood*, **91** (10), 3527–61.
12. Bonderman, D., Jakowitsch, J., Redwan, B. *et al.* (2008) Role for staphylococci in misguided thrombus resolution of chronic thromboembolic pulmonary hypertension. *Arteriosclerosis, Thrombosis, and Vascular Biology*, **28** (4), 678–84.
13. Barnathan, E.S., Kuo, A., Van der Keyl, H. *et al.* (1988) Tissue-type plasminogen activator binding to human endothelial cells. Evidence for two distinct binding sites. *The Journal of Biological Chemistry*, **263** (16), 7792–99.
14. Welling, T.H., Huber, T.S., Messina, L.M. *et al.* (1996) Tissue plasminogen activator increases canine endothelial cell proliferation rate through a plasmin-independent, receptor-mediated mechanism. *The Journal of Surgical Research*, **66** (1), 36–42.
15. Levin, E.G. and Santell, L. (1987) Association of a plasminogen activator inhibitor (PAI-1) with the growth substratum and membrane of human endothelial cells. *The Journal of Cell Biology*, **105** (6Pt 1), 2543–49.
16. Loskutoff, D.J., Sawdey, M., and Mimuro, J. (1989) Type 1 plasminogen activator inhibitor. *Progress in Hemostasis and Thrombosis*, **9**, 87–115.
17. Sawdey, M.S. and Loskutoff, D.J. (1991) Regulation of murine type 1 plasminogen activator inhibitor gene expression *in vivo*. Tissue specificity and induction by lipopolysaccharide, tumor necrosis factor-alpha, and transforming growth factor-beta. *The Journal of Clinical Investigation*, **88** (4), 1346–53.
18. Bajzar, L., Morser, J., and Nesheim, M. (1996) TAFI, or plasma procarboxypeptidase B, couples the coagulation and fibrinolytic cascades through the thrombin-thrombomodulin complex. *The Journal of Biological Chemistry*, **271** (28), 16603–8.
19. Kinlay, S., Libby, P., and Ganz, P. (2001) Endothelial function and coronary artery disease. *Current Opinion in Lipidology*, **12** (4), 383–89.
20. Zhao, Y., Young, S.L., McIntosh, J.C. *et al.* (2000) Ontogeny and localization of TGF-beta type I receptor expression during lung development. *American Journal of Physiology: Lung Cellular and Molecular Physiology*, **278** (6), L1231–39.
21. Hartsfield, C.L., Alam, J., and Choi, A.M. (1999) Differential signaling pathways of HO-1 gene expression in pulmonary and systemic vascular cells. *American Journal of Physiology: Lung Cellular and Molecular Physiology*, **277** (6), L1133–41.
22. Wright, L., Tuder, R.M., Wang, J. *et al.* (1998) 5-Lipoxygenase and 5-lipoxygenase activating protein (FLAP) immunoreactivity in lungs from patients with primary pulmonary hypertension. *American Journal of Respiratory and Critical Care Medicine*, **157** (1), 219–29.
23. Giaid, A., Yanagisawa, M., Langleben, D. *et al.* (1993) Expression of endothelin-1 in the lungs of patients with pulmonary hypertension. *The New England Journal of Medicine*, **328** (24), 1732–39.
24. Tuder, R.M., Cool, C.D., Geraci, M.W. *et al.* (1999) Prostacyclin synthase expression is decreased in lungs from patients with severe pulmonary hypertension. *American Journal of Respiratory and Critical Care Medicine*, **159** (6), 1925–32.
25. Rosenhek, R., Korschineck, I., Gharehbaghi-Schnell, E. *et al.* (2003) Fibrinolytic balance of the arterial wall: pulmonary artery displays increased fibrinolytic potential compared with aorta. *Laboratory Investigation*, **83** (6), 871–76.
26. Todd, A.S. (1972) Endothelium and fibrinolysis. *Atherosclerosis*, **15** (2), 137–40.
27. Carmeliet, P., Mulligan, R.C., Collen, D. *et al.* (1994) Transgenic animals as tools for the study of fibrinolysis *in vivo* Physiological consequences of loss of plasminogen activator gene function in mice. *Journal of Internal Medicine*, **236** (4), 455–59.
28. Kawai, Y., Matsumoto, Y., Watanabe, K. *et al.* (1996) Hemodynamic forces modulate the effects of cytokines on fibrinolytic activity of endothelial cells. *Blood*, **87** (6), 2314–21.
29. Schwenke, D.C. (1997) Comparison of aorta and pulmonary artery: II. LDL transport and metabolism correlate with susceptibility to atherosclerosis. *Circulation Research*, **81** (3), 346–54.
30. Stein, P.D. and Evans, H. (1967) An autopsy study of leg vein thrombosis. *Circulation*, **35** (4), 671–81.
31. Samad, F., Schneiderman, J., and Loskutoff, D. (1997) Expression of fibrinolytic genes in tissues from human atherosclerotic aneurysms and from obese mice. *Annals of the New York Academy of Sciences*, **811**, 350–58; discussion 358–360.
32. Pepper, M.S., Montesano, R., Mandriota, S.J. *et al.* (1996) Angiogenesis: a paradigm for balanced extracellular proteolysis during cell migration and morphogenesis. *Enzyme and Protein*, **49** (1–3), 138–62.
33. Fearns, C., Samad, F., and Loskutoff, D.J. (1995) Synthesis and localization of PAI-1 in the vessel wall, in *Vascular Control of Hemostasis* (ed. V.W.M. van Hinsbergh), Harwood, Amsterdam, pp. 207–26.
34. Fujii, S., Hopkins, W.E., and Sobel, B.E. (1991) Mechanisms contributing to increased synthesis of

plasminogen activator inhibitor type 1 in endothelial cells by constituents of platelets and their implications for thrombolysis. *Circulation*, **83** (2), 645–51.
35. Sawdey, M., Podor, T.J., and Loskutoff, D.J. (1989) Regulation of type 1 plasminogen activator inhibitor gene expression in cultured bovine aortic endothelial cells. Induction by transforming growth factor-beta, lipopolysaccharide, and tumor necrosis factor-alpha. *The Journal of Biological Chemistry*, **264** (18), 10396–401.
36. Sawa, H., Fujii, S., and Sobel, B.E. (1992) Augmented arterial wall expression of type-1 plasminogen activator inhibitor induced by thrombosis. *Arteriosclerosis and Thrombosis*, **12** (12), 1507–15.
37. Kadish, J.L., Butterfield, C.E., and Folkman, J. (1979) The effect of fibrin on cultured vascular endothelial cells. *Tissue and Cell*, **11** (1), 99–108.
38. Gertler, J.P., Perry, L., L'Italien, G. *et al.* (1993) Ambient oxygen tension modulates endothelial fibrinolysis. *Journal of Vascular Surgery*, **18** (6), 939–45; discussion 945–936.
39. Hong, K.H., Yoo, S.A., Kang, S.S. *et al.* (2006) Hypoxia induces expression of connective tissue growth factor in scleroderma skin fibroblasts. *Clinical and Experimental Immunology*, **146** (2), 362–70.
40. Hong, K.H., Lee, Y.J., Lee, E. *et al.* (2008) Genetic ablation of the BMPR2 gene in pulmonary endothelium is sufficient to predispose to pulmonary arterial hypertension. *Circulation*, **118** (7), 722–30.
41. Moser, K.M., Auger, W.R., and Fedullo, P.F. (1990) Chronic major-vessel thromboembolic pulmonary hypertension. *Circulation*, **81** (6), 1735–43.
42. Lang, I.M. (2004) Chronic thromboembolic pulmonary hypertension – not so rare after all. *The New England Journal of Medicine*, **350** (22), 2236–38.
43. Tapson, V.F. and Humbert, M. (2006) Incidence and prevalence of chronic thromboembolic pulmonary hypertension: from acute to chronic pulmonary embolism. *Proceedings of the American Thoracic Society*, **3** (7), 564–67.
44. Menendez, R., Nauffal, D., and Cremades, M.J. (1998) Prognostic factors in restoration of pulmonary flow after submassive pulmonary embolism: a multiple regression analysis. *The European Respiratory Journal*, **11** (3), 560–64.
45. Ribeiro, A., Lindmarker, P., Johnsson, H. *et al.* (1999) Pulmonary embolism: one-year follow-up with echocardiography doppler and five-year survival analysis. *Circulation*, **99** (10), 1325–30.
46. Pengo, V., Lensing, A.W., Prins, M.H. *et al.* (2004) Incidence of chronic thromboembolic pulmonary hypertension after pulmonary embolism. *The New England Journal of Medicine*, **350** (22), 2257–64.
47. Fedullo, P.F., Auger, W.R., Kerr, K.M. *et al.* (2001) Chronic thromboembolic pulmonary hypertension. *The New England Journal of Medicine*, **345** (20), 1465–72.
48. Wolf, M., Boyer-Neumann, C., Parent, F. *et al.* (2000) Thrombotic risk factors in pulmonary hypertension. *The European Respiratory Journal*, **15** (2), 395–99.
49. Lang, I.M., Klepetko, W., and Pabinger, I. (1996) No increased prevalence of the factor V Leiden mutation in chronic major vessel thromboembolic pulmonary hypertension (CTEPH). *Thrombosis and Haemostasis*, **76** (3), 476–77.
50. Olman, M.A., Marsh, J.J., Lang, I.M. *et al.* (1992) Endogenous fibrinolytic system in chronic large-vessel thromboembolic pulmonary hypertension. *Circulation*, **86** (4), 1241–48.
51. Lang, I.M., Marsh, J.J., Olman, M.A. *et al.* (1994) Parallel analysis of tissue-type plasminogen activator and type 1 plasminogen activator inhibitor in plasma and endothelial cells derived from patients with chronic pulmonary thromboemboli. *Circulation*, **90** (2), 706–12.
52. Marsh, J.J., Konopka, R.G., Lang, I.M. *et al.* (1994) Suppression of thrombolysis in a canine model of pulmonary embolism. *Circulation*, **90** (6), 3091–97.
53. Egermayer, P. and Peacock, A.J. (2000) Is pulmonary embolism a common cause of chronic pulmonary hypertension? Limitations of the embolic hypothesis. *The European Respiratory Journal*, **15** (3), 440–48.
54. Lang, I.M., Marsh, J.J., Konopka, R.G. *et al.* (1993) Factors contributing to increased vascular fibrinolytic activity in mongrel dogs. *Circulation*, **87** (6), 1990–2000.
55. Moser, K.M. and Bloor, C.M. (1993) Pulmonary vascular lesions occurring in patients with chronic major vessel thromboembolic pulmonary hypertension. *Chest*, **103** (3), 685–92.
56. Lang, I.M., Marsh, J.J., Olman, M.A. *et al.* (1994) Expression of type 1 plasminogen activator inhibitor in chronic pulmonary thromboemboli. *Circulation*, **89** (6), 2715–21.
57. Kyrle, P.A., Minar, E., Hirschl, M. *et al.* (2000) High plasma levels of factor VIII and the risk of recurrent venous thromboembolism. *The New England Journal of Medicine*, **343** (7), 457–62.

58. Bonderman, D., Turecek, P.L., Jakowitsch, J. *et al.* (2003) High prevalence of elevated clotting factor VIII in chronic thromboembolic pulmonary hypertension. *Thrombosis and Haemostasis*, **90** (3), 372–76.
59. Lang, I.M., Moser, K.M., and Schleef, R.R. (1996) Expression of Kunitz protease inhibitor – containing forms of amyloid beta-protein precursor within vascular thrombi. *Circulation*, **94** (11), 2728–34.
60. Beck, L. Jr. and D'Amore, P.A. (1997) Vascular development: cellular and molecular regulation. *The FASEB Journal*, **11** (5), 365–73.
61. Albelda, S.M., Muller, W.A., Buck, C.A. *et al.* (1991) Molecular and cellular properties of PECAM-1 (endoCAM/CD31): a novel vascular cell-cell adhesion molecule. *The Journal of Cell Biology*, **114** (5), 1059–68.
62. Risau, W. and Flamme, I. (1995) Vasculogenesis. *Annual Review of Cell and Developmental Biology*, **11**, 73–91.
63. Deng, Z., Morse, J.H., Slager, S.L. *et al.* (2000) Familial primary pulmonary hypertension (gene PPH1) is caused by mutations in the bone morphogenetic protein receptor-II gene. *American Journal of Human Genetics*, **67** (3), 737–44.
64. Harrison, R.E., Flanagan, J.A., Sankelo, M. *et al.* (2003) Molecular and functional analysis identifies ALK-1 as the predominant cause of pulmonary hypertension related to hereditary haemorrhagic telangiectasia. *Journal of Medical Genetics*, **40** (12), 865–71.
65. Kim, D.H., Iijima, H., Goto, K. *et al.* (1996) Human apolipoprotein E receptor 2. A novel lipoprotein receptor of the low density lipoprotein receptor family predominantly expressed in brain. *The Journal of Biological Chemistry*, **271** (14), 8373–80.
66. Brown, S.D., Twells, R.C., Hey, P.J. *et al.* (1998) Isolation and characterization of LRP6, a novel member of the low density lipoprotein receptor gene family. *Biochemical and Biophysical Research Communications*, **248** (3), 879–88.
67. Glagov, S., Weisenberg, E., Zarins, C.K. *et al.* (1987) Compensatory enlargement of human atherosclerotic coronary arteries. *The New England Journal of Medicine*, **316** (22), 1371–75.
68. Garland, C.J., Plane, F., Kemp, B.K. *et al.* (1995) Endothelium-dependent hyperpolarization: a role in the control of vascular tone. *Trends in Pharmacological Sciences*, **16** (1), 23–30.
69. Stamler, J.S., Singel, D.J., and Loscalzo, J. (1992) Biochemistry of nitric oxide and its redox-activated forms. *Science*, **258** (5090), 1898–902.
70. Loscalzo, J. and Welch, G. (1995) Nitric oxide and its role in the cardiovascular system. *Progress in Cardiovascular Diseases*, **38** (2), 87–104.
71. Mendelsohn, M.E., O'Neill, S., George, D. *et al.* (1990) Inhibition of fibrinogen binding to human platelets by *S*-nitroso-*N*-acetylcysteine. *The Journal of Biological Chemistry*, **265** (31), 19028–34.
72. Michelson, A.D., Benoit, S.E., Furman, M.I. *et al.* (1996) Effects of nitric oxide/EDRF on platelet surface glycoproteins. *American Journal of Physiology: Heart and Circulatory Physiology*, **270** (5Pt 2), H1640–48.
73. Kubes, P., Suzuki, M., and Granger, D.N. (1991) Nitric oxide: an endogenous modulator of leukocyte adhesion. *Proceedings of the National Academy of Sciences of the United States of America*, **88** (11), 4651–55.
74. De Caterina, R., Libby, P., Peng, H.B. *et al.* (1995) Nitric oxide decreases cytokine-induced endothelial activation. Nitric oxide selectively reduces endothelial expression of adhesion molecules and proinflammatory cytokines. *The Journal of Clinical Investigation*, **96** (1), 60–68.
75. Marks, D.S., Vita, J.A., Folts, J.D. *et al.* (1995) Inhibition of neointimal proliferation in rabbits after vascular injury by a single treatment with a protein adduct of nitric oxide. *The Journal of Clinical Investigation*, **96** (6), 2630–38.
76. Garg, U.C. and Hassid, A. (1989) Nitric oxide-generating vasodilators and 8-bromo-cyclic guanosine monophosphate inhibit mitogenesis and proliferation of cultured rat vascular smooth muscle cells. *The Journal of Clinical Investigation*, **83** (5), 1774–77.
77. Levin, E.R. (1996) Endothelins as cardiovascular peptides. *American Journal of Nephrology*, **16** (3), 246–51.
78. Simonson, M.S. and Dunn, M.J. (1990) Cellular signaling by peptides of the endothelin gene family. *The FASEB Journal*, **4** (12), 2989–3000.
79. Lopez, J.A., Armstrong, M.L., Piegors, D.J. *et al.* (1990) Vascular responses to endothelin-1 in atherosclerotic primates. *Arteriosclerosis*, **10** (6), 1113–18.
80. Warkentin, T.E., Aird, W.C., and Rand, J.H. (2003) Platelet–endothelial interactions: sepsis, HIT, and antiphospholipid syndrome. *Hematology American Society of Hematology: Education Program*, 497–519.
81. Siegel-Axel, D.I. and Gawaz, M. (2007) Platelets and endothelial cells. *Seminars in Thrombosis and Hemostasis*, **33** (2), 128–35.
82. Lindemann, S., Tolley, N.D., Dixon, D.A. *et al.* (2001) Activated platelets mediate inflammatory

signaling by regulated interleukin 1beta synthesis. *The Journal of Cell Biology*, **154** (3), 485–90.

83. May, A.E., Kalsch, T., Massberg, S. *et al.* (2002) Engagement of glycoprotein IIb/IIIa (alpha(IIb)beta3) on platelets upregulates CD40L and triggers CD40L-dependent matrix degradation by endothelial cells. *Circulation*, **106** (16), 2111–17.
84. Slupsky, J.R., Kalbas, M., Willuweit, A. *et al.* (1998) Activated platelets induce tissue factor expression on human umbilical vein endothelial cells by ligation of CD40. *Thrombosis and Haemostasis*, **80** (6), 1008–14.
85. Gawaz, M., Neumann, F.J., Dickfeld, T. *et al.* (1998) Activated platelets induce monocyte chemotactic protein-1 secretion and surface expression of intercellular adhesion molecule-1 on endothelial cells. *Circulation*, **98** (12), 1164–71.
86. Inwald, D.P., McDowall, A., Peters, M.J. *et al.* (2003) CD40 is constitutively expressed on platelets and provides a novel mechanism for platelet activation. *Circulation Research*, **92** (9), 1041–48.
87. Gawaz, M., Langer, H., and May, A.E. (2005) Platelets in inflammation and atherogenesis. *The Journal of Clinical Investigation*, **115** (12), 3378–84.
88. Damas, J.K., Otterdal, K., Yndestad, A. *et al.* (2004) Soluble CD40 ligand in pulmonary arterial hypertension: possible pathogenic role of the interaction between platelets and endothelial cells. *Circulation*, **110** (8), 999–1005.
89. Gawaz, M. (2004) Role of platelets in coronary thrombosis and reperfusion of ischemic myocardium. *Cardiovascular Research*, **61** (3), 498–511.
90. Frenette, P.S., Moyna, C., Hartwell, D.W. *et al.* (1998) Platelet-endothelial interactions in inflamed mesenteric venules. *Blood*, **91** (4), 1318–24.
91. Massberg, S., Enders, G., Matos, F.C. *et al.* (1999) Fibrinogen deposition at the postischemic vessel wall promotes platelet adhesion during ischemia–reperfusion *in vivo*. *Blood*, **94** (11), 3829–38.
92. Ay, C., Simanek, R., Vormittag, R. *et al.* (2008) High plasma levels of soluble P-selectin are predictive of venous thromboembolism in cancer patients: results from the Vienna Cancer and Thrombosis Study (CATS). *Blood*, **112** (7), 2703–8.
93. Jilma, B., Kovar, F.M., Hron, G. *et al.* (2006) Homozygosity in the single nucleotide polymorphism Ser128Arg in the E-selectin gene associated with recurrent venous thromboembolism. *Archives of Internal Medicine*, **166** (15), 1655–59.
94. Bonderman, D., Wilkens, H., Wakounig, S. *et al.* (2009) Risk factors for chronic thromboembolic pulmonary hypertension. *The European Respiratory Journa*, **33** (2), 325–31.
95. Adlbrecht, C., Bonderman, D., Plass, C. *et al.* (2006) Thrombus bound endothelin and leukocytes extracted by thrombectomy in acute myocardial infarction correlate with ST-segment resolution. *Circulation*, **114** (Suppl II), II-458.
96. Mackman, N. (2009) On the trail of microparticles. Circ Res. *Circ Res*, **104** (8), 925–7.

30

Pulmonary Endothelium and Malignancies

Abu-Bakr Al-Mehdi

Department of Pharmacology, University of South Alabama College of Medicine, Mobile AL, USA

INTRODUCTION

The entire cardiac output of the right ventricle first passes through the lung circulation, lined by pulmonary endothelial cells. Pulmonary ECs are actively involved in the arrest and growth of hematogenous spread of metastases to the lungs. Although some large cancer cells (12–15 μm in diameter) fail to negotiate the capillary tunnels (5–7 μm in diameter) of the pulmonary circulation and thereby get trapped by size restriction, many other cell types "home" in to the endothelium of precapillary arterioles (20–80 μm in diameter) as they attach to the ECs via specific molecular mechanisms. It is increasingly recognized that the cellular basis of organotropism of metastases of different cancer types might rest mainly with the ECs in the vessels of the target organ.

LUNG METASTATIC TUMORS AND THE PULMONARY ENDOTHELIUM

Organotropism of Metastasis is Endotheliotropism of Metastasis

The lung, along with the liver, bone, and brain, are the most frequent targets of metastasis of different cancers. Cancers that commonly metastasize to lung include breast cancer, renal cell cancer, melanoma, sarcomas, lymphomas and leukemias, germ cell tumors, ovarian cancer, contralateral primary lung cancer, and head and neck cancer. In contrast, prostate cancer usually metastasizes to the bone and lung involvement is rarely seen. On the other hand, gastrointestinal tumors most frequently metastasize to the liver [1], but rarely to the lung.

This relatively selective organotropism brings back to life the original "seed and soil" hypothesis of Paget [2]. The presence of organotropism suggests that there is a role for the "soil" in metastatic efficiency. Traditionally, because of the prevalence of the notion of extravascular development of metastasis, the "soil" has been thought to be the parenchyma of the host organ. However, metastatic cell proliferation does not require special permissiveness of a host site. Cancer cells directly injected into sites where they normally would not metastasize are not prevented from growing into tumors. This suggests that the factors that determine the specificity of attachment of circulating cancer cells to the vascular bed of a particular organ would essentially function as the determinant "soil" for organospecificity of metastasis. The notion of organ-specific and vascular segment-specific EC heterogeneity and the intravascular nature of the origin of metastasis support the endotheliospecificity of metastatic cell homing. It appears that the specific attachment of circulating cancer cells to a particular vascular segment determines the organospecificity of metastasis. It follows that this specificity of attachment should depend upon the expression of complementary types of cell adhesion molecules on a particular cancer cell type and on the target organ ECs that serve as their anchorage sites for new growth.

Pulmonary Endothelium Cell Adhesion Molecules as the Basis for Organotropism of Pulmonary Metastasis

Specificity of metastasis to the lung can be explained by the presence of pulmonotropic cell adhesion molecules on the cancer cells and their counter-receptors or ligands on the pulmonary endothelium. Since the EC is the common denominator, it seemed logical to explore the genomic profile of lung-homing cancer cell populations in search

The Pulmonary Endothelium: Function in health and disease Editors Norbert F. Voelkel, Sharon Rounds

of genes that confer lung organotropism. It was found that the expression of matrix metalloproteinase-2 (gelatinase A), the cell adhesion molecule SPARC ("secreted protein acidic and rich in cysteine"), the interleukin-13 decoy receptor IL-13Rα2, and the cell adhesion receptor vascular cell adhesion molecule (VCAM)-1 were generally restricted to aggressive lung metastatic populations of a breast cancer cell line [3].

Pulmonary ECs express a variety of cell adhesion molecules that take part in cancer cell attachment. Among the selectin family of adhesion molecules, pulmonary ECs express both E- and P-selectins. Selectins recognize fucosylated, sialylated, and sulfated ligands on scaffold glycoproteins (see Chapter 10 for discussion of leukocyte/endothelial adhesion molecules). The tumor cell ligands for E-selectin are sialyl-Lewis^a and sialyl-Lewis^x that are found in colon and renal cell carcinomas among others [4–7]. Cancer cells bind to P-selectin of ECs via the P-selectin glycoprotein ligand-1 and mucin CD24. Among the immunoglobulin superfamily of cell adhesion molecules, the ECs have been shown to express intercellular cell adhesion molecule (ICAM)-1, VCAM-1, and mucosal addressin cell adhesion molecule-1 which mediate cancer cell attachment via α_4, β_2, and $\alpha_4\beta_1$ integrins on the surface of cancer cells [8]. The integrins α_6, β_1, and $\alpha_v\beta_3$ expressed on pulmonary ECs may mediate cancer cell adhesion via unknown counter-receptors on cancer cells [9, 10]. The β integrins are expressed on both ECs and cancer cells. Fibronectin, vitronectin, fibrinogen, VCAM-1, ICAM-1, and platelet-endothelial cell adhesion molecule-1 can serve as ligands for β integrins. We have shown that tumor cell $\alpha_3\beta_1$ integrin and vascular laminin-5 mediate pulmonary attachment and metastasis in a mouse breast cancer model [11]. An antibody against melanoma cellular adhesion molecule (MelCAM or MUC18) inhibited spontaneous pulmonary metastasis of osteosarcoma, indicating a role of MelCAM for organospecific targeting [12].

Among other adhesion molecules, thrombospondin, heparan sulfate, and neural cell adhesion molecule have been implicated in cancer cell attachment to lung endothelium. The truly lung endothelium-specific cell adhesion molecule implicated in cancer metastasis is dipeptidyl peptidase (DPP)-IV (CD26) expressed on pulmonary ECs [13–16]. Its ligand on cancer cells is cell surface polymeric fibronectin [17]. DPP-IV can serve as the homing mechanism for lung specific metastasis. Chemokine receptors such as CXCR4 and CCR7 on cancer cells and their ligands CXCL12 and CCL21 have been implicated in breast cancer cell metastasis to the lung [18]. Although it has been thought that the cell adhesion molecules that are constitutively expressed on ECs serve as a common denominator of the homing mechanism, recently the notion of primary tumors preparing the "soil" for an impending metastasis is challenging this concept of adhesion molecule expression constancy [19–21]. Therefore, the absence of homing cell adhesion molecules in a particular vascular bed under normal conditions may not mean that organospecific targeting of cancer cannot take place, because the primary tumor itself may induce upregulation of ligands/receptors that create a premetastatic niche in a particular vascular segment.

Cancer Cell–EC Interaction during Attachment under Flow

The pulmonary endothelium receives the entire cardiac output and all the circulating metastatic cells. The circulating cells could potentially be trapped by size restriction in the sieve of the pulmonary capillaries. However, it has been demonstrated that in an experimental metastasis model, where tumor cells were directly injected into the tail vein of mice, about two-thirds of the attached cells were found in precapillary arterioles with lumen diameters far exceeding the size of the tumor cells [22]. The attachment of cancer cells to ECs under flow is often modeled after polymorphonuclear neutrophil (PMN) interaction with the endothelium (see Chapter 10). Leukocytes in blood with laminar flow tend to occupy the center of the vessel lumen where velocity is highest and hydrodynamic pressure is lowest. However, when a gradient of chemoattractants originating in the vascular wall is encountered by PMNs, the cells respond by moving toward the signal and away from the center of the vessel while their forward velocity is not significantly altered (margination phase). Approximation of PMNs to the vascular lining progressively slows them down due to transient binding to ECs sequentially mediated by L-, P-, and E-selectins (rolling phase). Integrins, ICAM-1, VCAM, and other adhesion molecules then firmly arrest the PMNs on the surface of the EC (attachment phase). The arrested PMNs then penetrate the vascular wall through interendothelial junctions or intraendothelial fenestrae and end up in the interstitium (transmigration or extravasation phase).

Although this model provides a suitable frame of reference for EC–tumor cell interaction, the particulars of this interaction are different for a cancer cell. The major difference is in the shortened rolling phase for the cancer cells. PMNs roll for 30–90 s before they exhibit no apparent movement for more than 30 s, when they are considered attached. In contrast, cancer cells in the pulmonary circulation took less than 10 s of rolling time before stopping in the isolated, perfused rat lung. Another difference is the intravascular division of the attached cell, indicating a role for sustained EC–cancer cell communication in the metastatic efficiency. Only up to 2% of attached tumor cells were shown to transmigrate in to

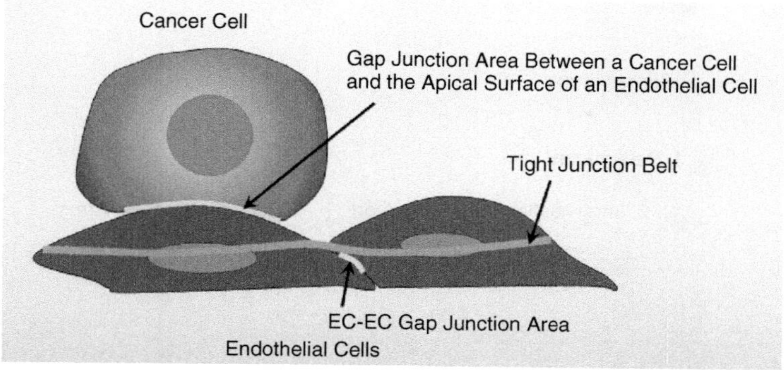

Figure 30.1 An apical localization of EC–cancer cell gap junctions. Although gap junctions are normally localized on the lateral surfaces of the ECs below the tight junction belt, attachment of cancer cells to pulmonary ECs *in vivo* leads to formation of functional gap junctions between them that play a critical role in the metastatic efficiency [26]. This type of close interaction via gap junctions between ECs and cancer cells is not only able to enhance attachment of tumor cells in the lung, but also assist in the vascularization of metastatic tumors from the proliferation of interacting ECs.

the lung interstitium or the alveolar space, from where they were cleared within 24–48 h [22]. This suggests that the lung interstitium and the alveolar space are hostile spaces for metastatic cancer cells. In order to elicit the role of selectins and integrins in syngeneic breast adenocarcinoma cell attachment to the pulmonary endothelium, both P- and E-selectin homogenous "knockout" mice were used. The attachment efficiency was blocked significantly in both "knockout" mice (54 ± 8 for P-selectin and 52 ± 10 for E-selectin knockout versus 100 in control mice; $p < 0.05$; $n = 6$ for each). An integrin β_1 antibody was similarly effective in reducing adhesion in control mice (51 ± 9; $p < 0.05$; $n = 6$), but had no additive effect in the P-selectin "knockout" mice. An integrin β_3 antibody was ineffective in blocking cancer cell attachment to the lung endothelium.

Recently, nonjunctional roles for connexins with signaling-type functions have received considerable attention [23, 24] (see Chapter 3). Adhesion-mediated functional gap-junctional intercellular communication (GJIC) between lung-metastatic B16F10 melanoma cells and endothelium was dependent on the expression of Cx43 in both cell types [25]. We have shown that gap junctional communication via Cx43 facilitates metastatic homing by increasing attachment efficiency of cancer cells to the lung endothelium [26]. Cancer cells with a dominant-negative G138R mutation in the Cx43 gene that allows formation of gap-junction plaques, but not of functionally competent gap junctions, led to significant reduction in the number of adherent tumor cell in the lungs. This suggests that the formation of functional gap junctions between a cancer cell and the EC is critical for tumor cell adhesion to the pulmonary endothelium. Generally, gap junction plaques are found in the lateral membranes of cells below the belt of the tight junctions. Formation of heterologous gap junction between a cancer cell and the apical surface of an EC would require the presence of connexins at the apical surface of ECs (Figure 30.1). We have observed marked upregulation of Cx43 in tumor cell–EC contact areas, whether in pre-existing "homing" vessels or in newly formed tumor vessels [26]. However, as a cancer cell progresses through the steps of intravascular metastasis, its level of interaction with the neighboring cells is highly variable (Figure 30.2). First, detachment of a metastatic cell from the primary tumor would necessitate downregulation of intercellular junctions, followed by their upregulation during the intravasation step as the cell negotiates its entry between or through the ECs. During the transport of cancer cells in the bloodstream, the requirement for their intercellular junctions could be minimal, followed by an increased cell–cell interaction during the endotheliospecific attachment step. Again, intravascular division of the metastatic cell minimizes intercellular interactions, followed by an upregulation in cell–cell junctions during the growth and formation of the secondary tumor.

PULMONARY EC MALIGNANCIES

In addition to interacting with circulating malignant cells, ECs themselves can undergo malignant transformation. Systemically, EC malignancy may manifest in the form of hemangioblastomas, hemangioendotheliomas, von Hippel–Lindau disease, and angiosarcomas [27–30]. In the lung, the primary pulmonary endothelial malignancies described are pulmonary capillary hemangiomatosis [31–35], pulmonary sclerosing hemangioma

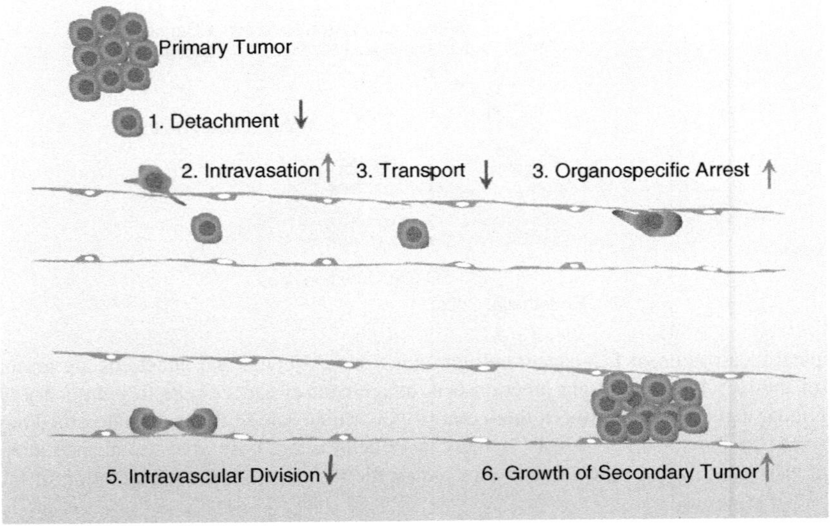

Figure 30.2 Variable requirement of intercellular interaction during the steps of hematogenous metastasis. A cancer cell has to undergo variable interactions with ECs and other neighboring cells during the distinct steps of the metastatic process. Depicted is the intravascular model of metastasis, in which metastatic tumors develop from endothelium-attached tumor cells without the requirement for extravasation [22]. Up arrows indicate increased requirement of intercellular interaction, down arrows indicate decreased requirement of cell–cell interaction. EC–cancer cell interaction plays a crucial role in most steps of the metastatic process.

[36–39], malignant angioendotheliomatosis [40–42], and epithelioid hemangioendothelioma [43–46]. Pulmonary capillary hemangiomatosis is uncontrolled proliferation of capillaries in the alveolar septae, airway and venous walls, pleura, and regional lymph nodes [47]. EC proliferation at the capillary level is a major hallmark of many of these malignancies. *In vivo*, ECs are quiescent cell types with low metabolic and proliferative activity. However, *in vitro*, pulmonary microvascular ECs share some characteristics of cancer cells – they grow in culture in multiple layers [48], they can dedifferentiate back into mesenchymal cells [49], and they contribute to the formation of intravascular, multicellular plexiform lesions in pulmonary hypertension [50]. These characteristics might explain occurrence of EC tumors at the capillary level in contrast to development of sarcomas at the pulmonary artery and vein levels. The malignant transformation of pulmonary microvascular cells *in vivo* may involve creation of *in vitro*-like microenvironmetal conditions due to inflammation or metabolic alterations.

CONCLUSIONS AND PERSPECTIVES

Organotropism of pulmonary metastasis may be a manifestation of preferential interaction between cancer cells and lung ECs. The pulmonary ECs express a multitude of cell adhesion molecules that assist in the "homing" of metastatic cells. ECs and the attached cancer cells also establish heterologous functional gap junctions that enhance metastatic efficiency and promote vascularization of the metastatic tumor.

Since the mortality and morbidity of cancer stems from metastasis and since organotropism of metastasis is determined by EC–cancer cell interaction, a dissection of the nature of this interaction with the aim of preventing it should be the focus of future research efforts. The pulmonary EC stands as a clear target for this approach because of the prevalence of metastasis to the lungs in a variety of cancer types. We hypothesize that metastatic organotropism can be explained by metastatic endotheliotropism. An important future direction for research is extensive segmental characterization of EC surface heterogeneity in major target organs, such as the lung, liver, brain, and bone. We need to answer the question: Do organotropic genes exist in target organ ECs? Further characterization of the role of organotropism genes associated with particular cancer cell types will be also required. Prevention of organotropic dissemination of cancer may be the key to reducing the impact of metastasis. The pulmonary endothelium, with its structural heterogeneity, functional diversity, and large dimension is a promising target for future research.

References

1. Hess, K.R., Varadhachary, G.R., Taylor, S.H. *et al.* (2006) Metastatic patterns in adenocarcinoma. *Cancer*, **106** (7), 1624–33.
2. Paget, S. (1889) The distribution of secondary growths in cancer of the breast. *Lancet*, **1**, 571–73.
3. Minn, A.J., Gupta, G.P., Siegel, P.M. *et al.* (2005) Genes that mediate breast cancer metastasis to lung. *Nature*, **436** (7050), 518–24.
4. Kajiwara, H., Yasuda, M., Kumaki, N. *et al.* (2005) Expression of carbohydrate antigens (SSEA-1, sialyl-Lewis X, DU-PAN-2 and CA19-9) and E-selectin in urothelial carcinoma of the renal pelvis, ureter, and urinary bladder. *The Tokai Journal of Experimental and Clinical Medicine*, **30** (3), 177–82.
5. Tozawa, K., Okamoto, T., Kawai, N. *et al.* (2005) Positive correlation between sialyl Lewis X expression and pathologic findings in renal cell carcinoma. *Kidney International*, **67** (4), 1391–96.
6. Kannagi, R. (2004) Molecular mechanism for cancer-associated induction of sialyl Lewis X and sialyl Lewis A expression – the Warburg effect revisited. *Glycoconjugate Journal*, **20** (5), 353–64.
7. Tomlinson, J., Wang, J.L., Barsky, S.H. *et al.* (2000) Human colon cancer cells express multiple glycoprotein ligands for E-selectin. *International Journal of Oncology*, **16** (2), 347–53.
8. Orr, F.W., Wang, H.H., Lafrenie, R.M. *et al.* (2000) Interactions between cancer cells and the endothelium in metastasis. *The Journal of Pathology*, **190** (3), 310–29.
9. Legler, D.F., Johnson-Leger, C., Wiedle, G. *et al.* (2004) The alpha v beta 3 integrin as a tumor homing ligand for lymphocytes. *European Journal of Immunology*, **34** (6), 1608–16.
10. Virtanen, I., Laitinen, A., Tani, T. *et al.* (1996) Differential expression of laminins and their integrin receptors in developing and adult human lung. *American Journal of Respiratory Cell and Molecular Biology*, **15** (2), 184–96.
11. Wang, H., Fu, W., Im, J.H. *et al.* (2004) Tumor cell alpha3beta1 integrin and vascular laminin-5 mediate pulmonary arrest and metastasis. *The Journal of Cell Biology*, **164** (6), 935–41.
12. McGary, E.C., Heimberger, A., Mills, L. *et al.* (2003) A fully human antimelanoma cellular adhesion molecule/MUC18 antibody inhibits spontaneous pulmonary metastasis of osteosarcoma cells *in vivo*. *Clinical Cancer Research*, **9** (17), 6560–66.
13. Abdel-Ghany, M., Cheng, H., Levine, R.A., and Pauli, B.U. (1998) Truncated dipeptidyl peptidase IV is a potent anti-adhesion and anti-metastasis peptide for rat breast cancer cells. *Invasion and Metastasis*, **18** (1), 35–43.
14. Cheng, H.C., Abdel-Ghany, M., Elble, R.C., and Pauli, B.U. (1998) Lung endothelial dipeptidyl peptidase IV promotes adhesion and metastasis of rat breast cancer cells via tumor cell surface-associated fibronectin. *The Journal of Biological Chemistry*, **273** (37), 24207–15.
15. Yamaguchi, N., Plant, C., Biancone, L. *et al.* (1996) *In vivo* modulation of CD26 (dipeptidyl peptidase IV) in the mouse: effects of polyreactive and monoreactive antibodies. *Transplantation*, **62** (7), 973–85.
16. Cheng, H.C., Abdel-Ghany, M., Zhang, S., and Pauli, B.U. (1999) Is the Fischer 344/CRJ rat a protein-knock-out model for dipeptidyl peptidase IV-mediated lung metastasis of breast cancer? *Clinical and Experimental Metastasis*, **17** (7), 609–15.
17. Cheng, H.C., Abdel-Ghany, M., and Pauli, B.U. (2003) A novel consensus motif in fibronectin mediates dipeptidyl peptidase IV adhesion and metastasis. *The Journal of Biological Chemistry*, **278** (27), 24600–7.
18. Muller, A., Homey, B., Soto, H. *et al.* (2001) Involvement of chemokine receptors in breast cancer metastasis. *Nature*, **410** (6824), 50–56.
19. Hiratsuka, S., Watanabe, A., Aburatani, H., and Maru, Y. (2006) Tumour-mediated upregulation of chemoattractants and recruitment of myeloid cells predetermines lung metastasis. *Nature Cell Biology*, **8** (12), 1369–75.
20. Kaplan, R.N., Psaila, B., and Lyden, D. (2006) Bone marrow cells in the "pre-metastatic niche": within bone and beyond. *Cancer Metastasis Reviews*, **25** (4), 521–29.
21. Kaplan, R.N., Rafii, S., and Lyden, D. (2006) Preparing the "soil": the premetastatic niche. *Cancer Research*, **66** (23), 11089–93.
22. Al-Mehdi, A.B., Tozawa, K., Fisher, A.B. *et al.* (2000) Intravascular origin of metastasis from the proliferation of endothelium-attached tumor cells: a new model for metastasis. *Nature Medicine*, **6** (1), 100–2.
23. Stout, C., Goodenough, D.A., and Paul, D.L. (2004) Connexins: functions without junctions. *Current Opinion in Cell Biology*, **16** (5), 507–12.
24. Evans, W.H., De Vuyst, E., and Leybaert, L. (2006) The gap junction cellular internet: connexin hemichannels enter the signalling limelight. *The Biochemical Journal*, **397** (1), 1–14.
25. el-Sabban, M.E., and Pauli, B.U. (1994) Adhesion-mediated gap junctional communication between lung-metastatatic cancer cells and

endothelium. *Invasion and Metastasis*, **14** (1–6), 164–76.

26. Elzarrad, M.K., Haroon, A., Willecke, K. *et al.* (2008) Connexin-43 upregulation in micrometastases and tumor vasculature and its role in tumor cell attachment to pulmonary endothelium. *BMC Medicine*, **6**, 20.
27. Shuin, T., Yamasaki, I., Tamura, K. *et al.* (2006) Von Hippel–Lindau disease: molecular pathological basis, clinical criteria, genetic testing, clinical features of tumors and treatment. *Japanese Journal of Clinical Oncology*, **36** (6), 337–43.
28. Bauland, C.G., van Steensel, M.A., Steijlen, P.M. *et al.* (2006) The pathogenesis of hemangiomas: a review. *Plastic and Reconstructive Surgery*, **117** (2), 29e–35e.
29. Bell, C.D. (2003) Endothelial cell tumors. *Microscopy Research and Technique*, **60** (2), 165–70.
30. Larbcharoensub, N., Singhsnaeh, A., and Leelachaikul, P. (2004) Primary cardiac angiosarcoma with systemic metastases; a case report and review of the literature. *Journal of the Medical Association of Thailand*, **87** (9), 1116–22.
31. Sullivan, A., Chmura, K., Cool, C.D. *et al.* (2006) Pulmonary capillary hemangiomatosis: an immunohistochemical analysis of vascular remodeling. *European Journal of Medical Research*, **11** (5), 187–93.
32. Ginns, L.C., Roberts, D.H., Mark, E.J. *et al.* (2003) Pulmonary capillary hemangiomatosis with atypical endotheliomatosis: successful antiangiogenic therapy with doxycycline. *Chest*, **124** (5), 2017–22.
33. Varnholt, H. and Kradin, R. (2004) Pulmonary capillary hemangiomatosis arising in hereditary hemorrhagic telangiectasia. *Human Pathology*, **35** (2), 266–68.
34. Assaad, A.M., Kawut, S.M., Arcasoy, S.M. *et al.* (2007) Platelet-derived growth factor is increased in pulmonary capillary hemangiomatosis. *Chest*, **131** (3), 850–55.
35. Kradin, R., Matsubara, O., and Mark, E.J. (2005) Endothelial nitric oxide synthase expression in pulmonary capillary hemangiomatosis. *Experimental and Molecular Pathology*, **79** (3), 194–97.
36. Lin, D., Zou, S., Lu, N. *et al.* (2002) Thyroid transcription factor-1 in the histogenesis of pulmonary sclerosing hemangioma. *Zhonghua Zhong Liu Za Zhi*, **24** (4), 384–87.
37. Illei, P.B., Rosai, J., and Klimstra, D.S. (2001) Expression of thyroid transcription factor-1 and other markers in sclerosing hemangioma of the lung. *Archives of Pathology and Laboratory Medicine*, **125** (10), 1335–39.
38. Alvarez-Fernandez, E., Carretero-Albinana, L., and Menarguez-Palanca, J. (1989) Sclerosing hemangioma of the lung. An immunohistochemical study of intermediate filaments and endothelial markers. *Archives of Pathology and Laboratory Medicine*, **113** (2), 121–24.
39. Williams, L.J. Jr., Greenberg, S.D. and Mace, M.L. Jr. (1982) Pulmonary sclerosing hemangioma: sometimes a form of pulmonary hamartoma. *Southern Medical Journal*, **75** (6), 665–70.
40. Curtis, J.L., Warnock, M.L., Conrad, D.J. *et al.* (1991) Intravascular (angiotropic) large-cell lymphoma ("malignant angioendotheliomatosis") with small vessel pulmonary vascular obstruction and hypercalcemia. *The Western Journal of Medicine*, **155** (1), 72–76.
41. Pellicone, J.T. and Goldstein, H.B. (1990) Pulmonary malignant angioendotheliomatosis. Presentation with fever and syndrome of inappropriate antidiuretic hormone. *Chest*, **98** (5), 1292–94.
42. Yousem, S.A. and Colby, T.V. (1990) Intravascular lymphomatosis presenting in the lung. *Cancer*, **65** (2), 349–53.
43. Schattenberg, T., Pfannschmidt, J., Herpel, E. *et al.* (2007) Bilateral surgical resection in pulmonary epitheloid hemangioendothelioma. *The Thoracic and Cardiovascular Surgeon*, **55** (3), 199–200.
44. Ergun, E.L. and Lim, E. (2006) Increased FDG uptake in pulmonary epitheloid hemangioendothelioma. *Revista Espanola de Medicina Nuclear*, **25** (3), 188–92.
45. Carretero, A., Elmberger, P.G., Skold, C.M., and Collins, B.T. (2006) Pulmonary epithelioid hemangioendothelioma: report of a case with fine needle aspiration biopsy. *Acta Cytologica*, **50** (4), 455–59.
46. Massera, F., Delfanti, R., Rocco, G. *et al.* (2004) Pulmonary epithelioid haemangioendothelioma mimicking bronchogenic carcinoma. *The Journal of Cardiovascular Surgery (Torino)*, **45** (4), 397–98.
47. Yi, E.S. (2004) Tumors of the pulmonary vasculature. *Cardiology Clinics*, **22** (3), 431–40, vi–vii.
48. Gebb, S. and Stevens, T. (2004) On lung endothelial cell heterogeneity. *Microvascular Research*, **68** (1), 1–12.
49. Arciniegas, E., Neves, C.Y., Carrillo, L.M. *et al.* (2005) Endothelial–mesenchymal transition occurs during embryonic pulmonary artery development. *Endothelium*, **12** (4), 193–200.
50. Tuder, R.M., Cool, C.D., Yeager, M. *et al.* (2001) The pathobiology of pulmonary hypertension. Endothelium. *Clinics in Chest Medicine*, **22** (3), 405–18.

Epilogue

Norbert F. Voelkel

Victoria Johnson Center for Pulmonary Obstructive Disease Research, Pulmonary and Critical Care Medicine Division, Virginia Commonwealth University, Richmond, VA, USA

This volume has been written and edited not only for the endothelial cell (EC) biologist with an interest in the lung, but also for the clinician with an interest in lung diseases and the translation of problems of lung endothelial pathobiology. The goal is to improve the outcomes of acute and chronic lung diseases. For the majority of pulmonary scientists, investigation of lung epithelial, but not endothelial, cells continues to be a focal point of research endeavors. This is not surprising if one thinks of the lung as an organ that interacts with the environment. However, the lung should be considered in the context of integrated systems as not only an air-intake organ, but also as a blood filter. The recent acceptance of the concept of "EC dysfunction" mostly benefits the systemic circulation [1] because we think that we can thus understand the systemic vasculopathies of diabetes, the metabolic syndrome, and atherosclerosis [2]. That the lung endothelium is likely also "sick," in some chronic lung disorders and in congestive heart failure, has neither been firmly proven, nor has such a concept taken hold in the minds of many lung researchers. Pulmonary hypertension researchers know that the anorexigen fenfluramine is not inhaled, but arrives in the lung via the blood stream. Few researchers invested in interstitial lung diseases consider the lung vessels to be importantly involved in the pathobiology of idiopathic pulmonary fibrosis, although injected bleomycin causes both acute lung injury and pulmonary fibrosis. The overwhelming majority of asthma researchers accept the hypervascularity of the airway mucosa as a part of the chronic airway remodeling in asthma, but we do know that the asthmatic lung parenchyma is also hypervascular (D.M. Hyde, personal communication and unpublished data).

Investigators such as Wolfgang Kuebler (Chapter 20) will be able to tell us in the future whether ischemia–reperfusion biology is part of chronic progressive lung diseases and whether it is a highly regional effect. It is possible that ischemia–reperfusion plays a role in conditions other than lung transplantation and post-thrombendarterectomy syndrome. We are learning that it is the organ context, the microenvironment that drives the fate of stem cells to differentiate into endothelial or neural cells [3]. Sakao *et al.* [4] recently demonstrated that lung microvascular (likely precursor) ECs, under conditions of stress, can *in vitro* turn into neuronal cells, which raises the question of EC–nerve cell interactions in the small lung vessels [5, 6].

Research strategies employed by Jan Schnitzer (Chapter 15) and Renata Pasqualini's group (based on phage libraries) [7] can eventually provide a complete list of lung EC membrane proteins and a list of surface epitopes. We wonder whether the work of Norm Gillis [8] and Chris Dawson [9] needs to be revisited and re-evaluated – after all, the development of a clinical test that quantifies lung endothelial (metabolic) function remains a desirable goal still today. We think that we now know that ECs affect lung vascular tone – via nitric oxide (NO) and other mechanisms described by Resta *et al.* in Chapter 12. There is also a concept of a functional endothelial/smooth muscle syncytium about which much needs to be learned. A largely uninvestigated hypothesis is that one adaptive response of the lung resistance vessel ECs to high pressure and vasoconstriction is to fortify the vessel wall by transdifferentiation into smooth muscle cells [10]. On the lumenal side, endothelial–lymphocyte interactions may also contribute to the outcome of lung vascular remodeling [11, 12], as described in Chapter 25.

Asked what the clinically most important aspect of lung vessel pathobiology might be, this author has decided that it is the clinically prevalent setting of left heart failure associated with pulmonary hypertension and

The Pulmonary Endothelium: Function in health and disease Editors Norbert F. Voelkel, Sharon Rounds

subsequent right heart failure [13], exemplified by mitral stenosis and formerly termed "mitral lung." The pathology of the lung vessels in mitral stenosis was described more than 50 years ago. It was recognized by the contemporaries of Paul Wood [14] that irreversible pulmonary hypertension in the setting of mitral valve disease is unresponsive to infused acetylcholine – an activator of NO synthase (nitric oxide synthaseNOS) as we know today. Therefore, irreversible pulmonary hypertension was associated with lung EC dysfunction. Possibly both pulmonary venous hypertension – because of left ventricular failure – and the so-called cor pulmonale (heart disease secondary to chronic lung disease) can only be understood after examination of heart–lung interactions. Given our modern concept of the role of the endothelium in the maintenance of the microcirculation [15], it is now perhaps intuitive that the "bad humors" released from a chronically inflamed lung will impact the microcirculation of the myocardium, and that left ventricular pump failure will affect the venous circulation of the lung and subsequently the performance of the right ventricle. It is surprising that the role of the lung circulation in chronic heart failure has not been an area of intense research. Instead, heart failure management focuses on the pump performance of the left ventricle.

A combination of diastolic and contractile cardiac dysfunction develops in both "pure" diffuse lung tissue damage (e.g., emphysema) or "pure" mechanical cardiac stress under chronic conditions. Deliberately, the focus of this discussion is the lung circulation and the lung vessel EC damage and dysfunction and the critical loss of myocardial microvessels due to EC apoptosis and perhaps transforming growth factor-β-triggered EC–fibroblast transdifferentiation [16]. The destroyed emphysematous lung, because of its impaired microvessel EC function [17], will have lost much of its filter function, while congestive pulmonary venous hypertension in primary left ventricle pump failure likely will impair the lung EC phagocyte (macrophage-like) [18] function. In both cases, we speculate that lung effluent may be "toxic" to cardiac microvessel EC in part by turning off the expression and secretion of the EC survival factor vascular endothelial growth factor [19, 20].

At the present time, there are no known genetic abnormalities that directly determine the conditions and responses of the lung endothelium. Thus, a more productive approach may be to investigate epigenetic influences on the lung ECs. Epigenetic factors that might modify the responses of lung ECs are subject to experimental manipulations, such as dietary or hormonal factors. For example, we already know that chronic cigarette smoke exposure causes loss of small lung vessel expression of the enzymes prostaglandin I_2 synthase [17] and endothelial NOS [21]. It is not far fetched to postulate that diabetes, an antioxidant-rich Mediterranean diet, or high or low estrogen metabolite levels will also affect the behavior of lung ECs. Such epigenetic factors may prove to be of significance in the pathogenesis of emphysema and pulmonary hypertension. For example, the monocrotaline model of pulmonary hypertension, caused by the monocrotaline pyrrole metabolite, which is generated via cytochrome P450 in the liver, has for many years been called a model of "dietary pulmonary hypertension.". In addition, copper intestinal absorption failure, as in the "blotchy mouse" [22], and a Cu^{2+}-reduced diet cause emphysema (unpublished data).

The horizons of lung EC biology and pathobiology are wide, and reach from EC–matrix interactions to sphingolipid/ceramide metabolism [23] and pulmonary angiogenesis in the hepatopulmonary syndrome [24]. What we must not forget is that the lung is constructed around two "tube systems," one of which is the airways, and the other a very intricate and complex circulation, but the first and principal locus of blood-born information processing is the lung endothelium. The lung endothelium "decides" what to do with this information and "pronounces its decision" – in the language of the lung ECs: it is that language that we have yet to learn.

References

1. Kanda, T., Brown, J.D., Orasanu, G. *et al.* (2009) PPARgamma in the endothelium regulates metabolic responses to high-fat diet in mice. *The Journal of Clinical Investigation*, **119**, 110–24.
2. Eickhoff, P., Valipour, A., Kiss, D. *et al.* (2008) Determinants of systemic vascular function in patients with stable chronic obstructive pulmonary disease. *American Journal of Respiratory and Critical Care Medicine*, **178**, 1211–18.
3. Gualandris, A., Noghero, A., Geuna, M. *et al.* (2009) Microenvironment drives the endothelial or neural fate of differentiating embryonic stem cells coexpressing neuropilin-1 and Flk-1. *The FASEB Journal*, **23**, 68–78.
4. Sakao, S., Taraseviciene-Stewart, L., Cool, C.D. *et al.* (2007) VEGF-R blockade causes endothelial cell apoptosis, expansion of surviving CD34$^+$ precursor cells and transdifferentiation to smooth muscle-like and neuronal-like cells. *The FASEB Journal*, **21**, 3640–52.
5. Erlinge, D. and Burnstock, G. (2008) P2 receptors in cardiovascular regulation and disease. *Purinergic Signalling*, **4**, 1–20.
6. Burnstock, G. (2008) Dual control of vascular tone and remodelling by ATP released from nerves and endothelial cells. *Pharmacological Reports*, **60**, 12–20.

7. Giordano, R.J., Lahdenranta, J., Zhen, L. *et al.* (2008) Targeted induction of lung endothelial cell apoptosis causes emphysema-like changes in the mouse. *The Journal of Biological Chemistry*, **283**, 29447–60.
8. Riggs, D., Havill, A.M., Pitt, B.R., and Gillis, C.N. (1988) Pulmonary angiotensin-converting enzyme kinetics after acute lung injury in the rabbit. *Journal of Applied Physiology*, **64**, 2508–16.
9. Roerig, D.L., Audi, S.H., Linehan, J.H. *et al.* (1999) Detection of changes in lung tissue properties with multiple-indicator dilution. *Journal of Applied Physiology*, **86**, 1866–80.
10. Voelkel, N.F. and Tuder, R.M. (2000) Hypoxia-inducible pulmonary vascular remodeling: a model for what human disease? *The Journal of Clinical Investigation*, **106**, 733–38.
11. Daley, E., Emson, C., Guignabert, C. *et al.* (2008) Pulmonary arterial remodeling induced by a T_h2 immune response. *The Journal of Experimental Medicine*, **205**, 361–72.
12. Taraseviciene-Stewart, L., Nicolls, M.R., Kraskauskas, D. *et al.* (2007) Absence of T cells confers increased pulmonary arterial hypertension and vascular remodeling. *American Journal of Respiratory and Critical Care Medicine*, **175**, 1280–89.
13. Ghio, S., Gavazzi, A., Campana, C. *et al.* (2001) Independent and additive prognostic value of right ventricular systolic function and pulmonary artery pressure in patients with chronic heart failure. *Journal of the American College of Cardiology*, **37**, 183–88.
14. Wood, P. (1959) Pulmonary hypertension with special reference to the vasoconstrictive factor. *British Heart Journal*, **21**, 557–70.
15. Aird, W.C. (2005) *Endothelial Cells in Health and Disease*, Taylor & Francis, Boca Raton, FL.
16. Zeisberg, E.M., Tarnavski, O., Zeisberg, M. *et al.* (2007) Endothelial-to-mesenchymal transition contributes to cardiac fibrosis. *Nature Medicine*, **13**, 952–61.
17. Nana-Sinkam, S.P., Lee, J.D., Sotto-Santiago, S., *et al.* (2007) Prostacyclin prevents pulmonary endothelial cell apoptosis induced by cigarette smoke. *American Journal of Respiratory and Critical Care Medicine*, **175**, 676–85.
18. Ryan, U.S. (1987) Endothelial cell activation response, in *Lung Biology in Health Disease*, vol. **32** (eds U.S. Ryan and M. Dekker), Dekker, New York, pp. 3–33.
19. Friehs, I., Barillas, R., Vasilyev, N.V. *et al.* (2006) Vascular endothelial growth factor prevents apoptosis and preserves contractile function in hypertrophied infant heart. *Circulation*, **114** (Suppl), 1290–95.
20. Lee, S., Chen, T.T., Barber, C.L. *et al.* (2007) Autocrine VEGF signaling is required for vascular homeostasis. *Cell*, **130**, 691–703.
21. Barberá, J.A., Peinado, V.I., Santos, S. *et al.* (2001) Reduced expression of endothelial nitric oxide synthase in pulmonary arteries of smokers. *American Journal of Respiratory and Critical Care Medicine*, **164**, 709–13.
22. Fisk, D.E. and Kuhn, C. (1976) Emphysema-like changes in the lungs of the blotchy mouse. *The American Review of Respiratory Disease*, **113**, 787–97.
23. Berdyshev, E.V., Gorshkova, I., Skobeleva, A. *et al.* (2009) FTY720 inhibits ceramide synthases and upregulates dihydrosphingosine-1-phosphate formation in human lung endothelial cells. *The Journal of Biological Chemistry*, **284**, 5467–77.
24. Zhang, J., Luo, B., Tang, L. *et al.* (2009) Pulmonary angiogenesis in rat model of experimental hepatopulmonary syndrome. *Gastroenterology*, **136**, 1070–80.

Index

The Pulmonary Endothelium: Function in health and disease Editors Norbert F. Voelkel, Sharon Rounds

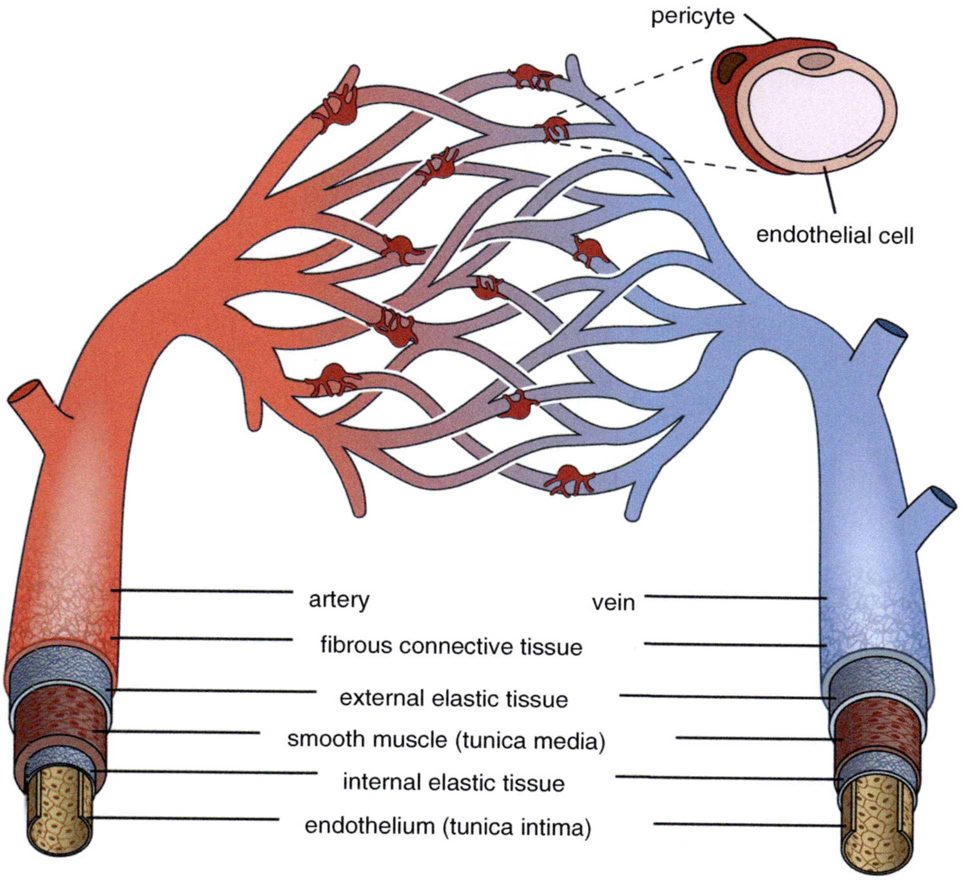

Plate 1.2 Fundamental architecture of blood vessels.

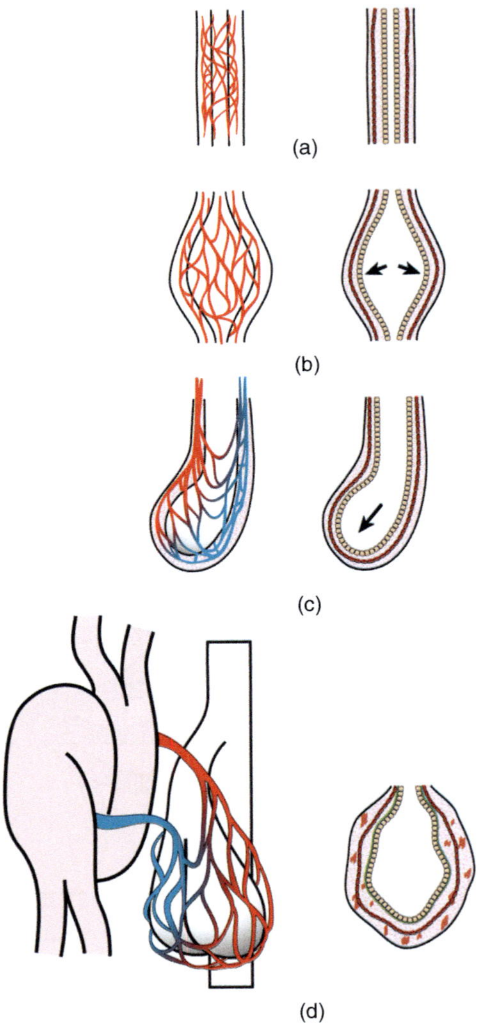

Plate 1.4 Proposal for the sequential progression of lung vascular development.

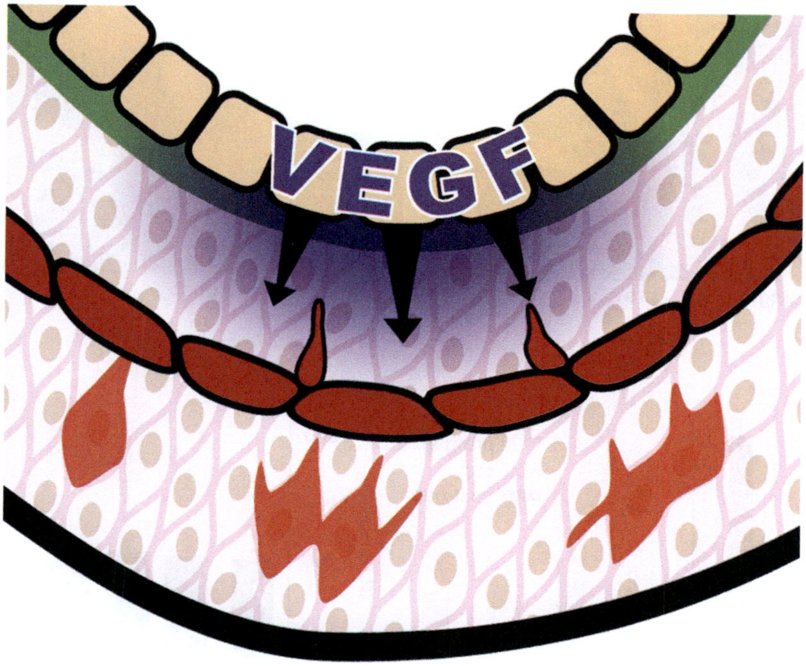

Plate 1.5 Vascular remodeling and establishment of intervascular connections is in part due to the interactions between epithelial VEGF gradients, the vasculogenic pools, and angiogenic extensions from the growing lung plexus.

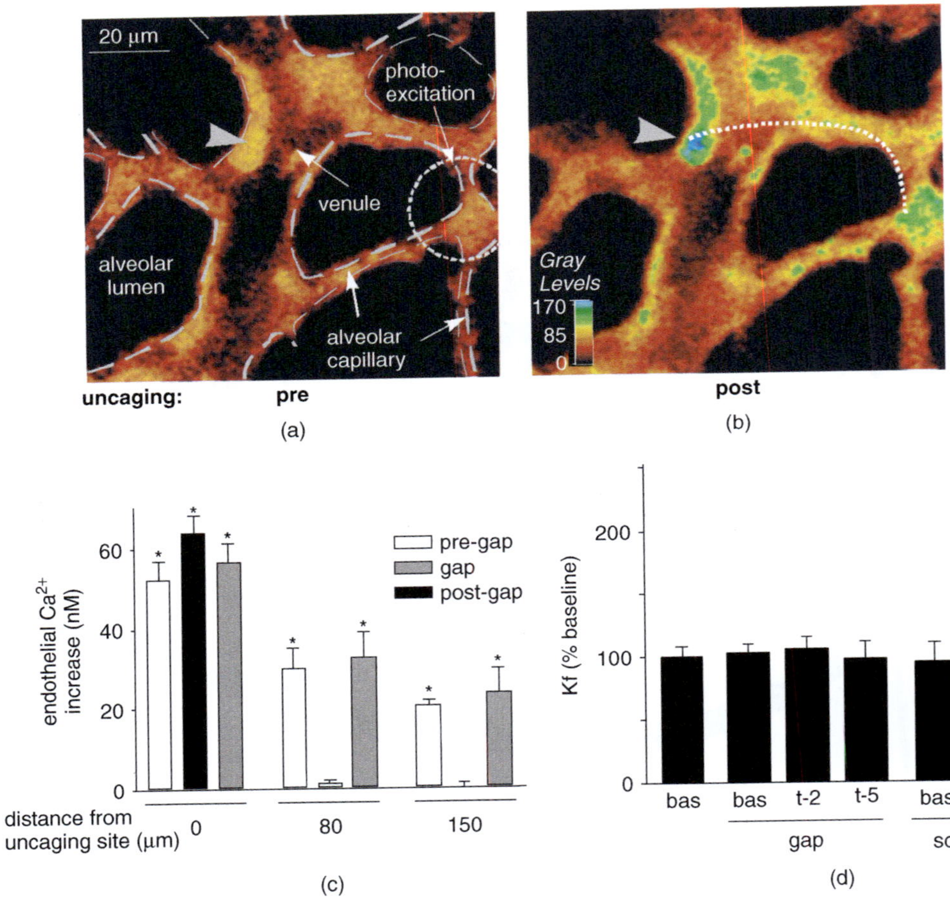

Plate 3.4 GJ-dependent responses in lung microvessels. Reproduced from Parthasarathi *et al.* (2006), *The Journal of Clinical Investigation*, **116**, 2193–200. by permission of the American Society for Clinical Investigation.

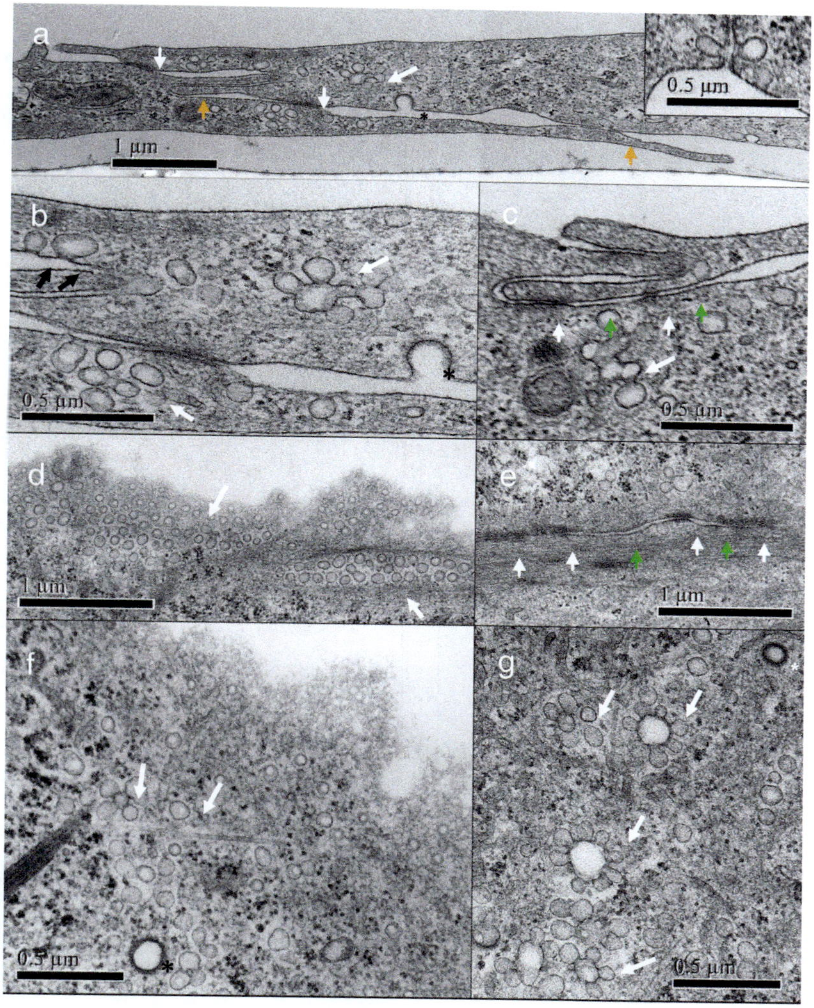

Plate 8.3 TEMs showing cell–cell junctions and abundance of caveolae in human lung microvascular ECs.

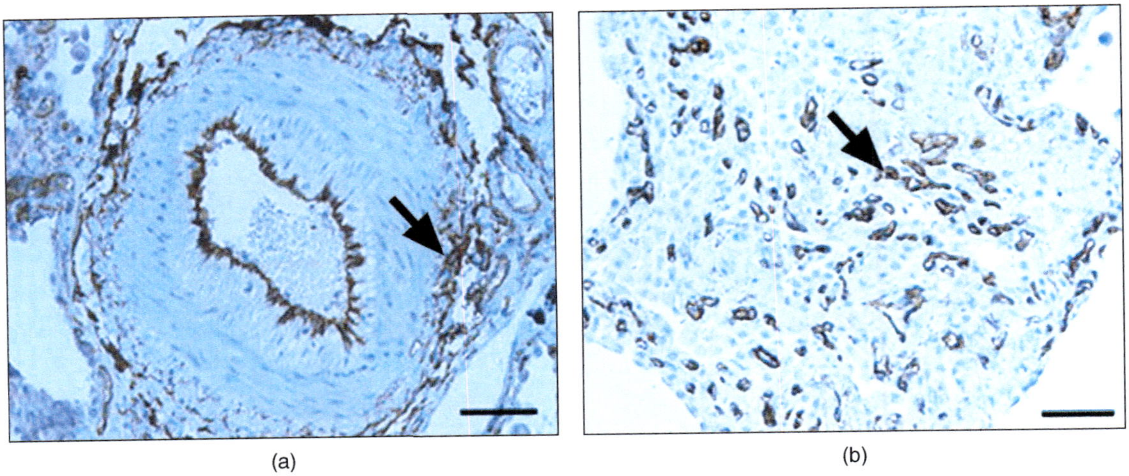

Plate 11.1 Angiogenic responses in the perivascular region of a patient with pulmonary fibrosis and associated pulmonary hypertension.

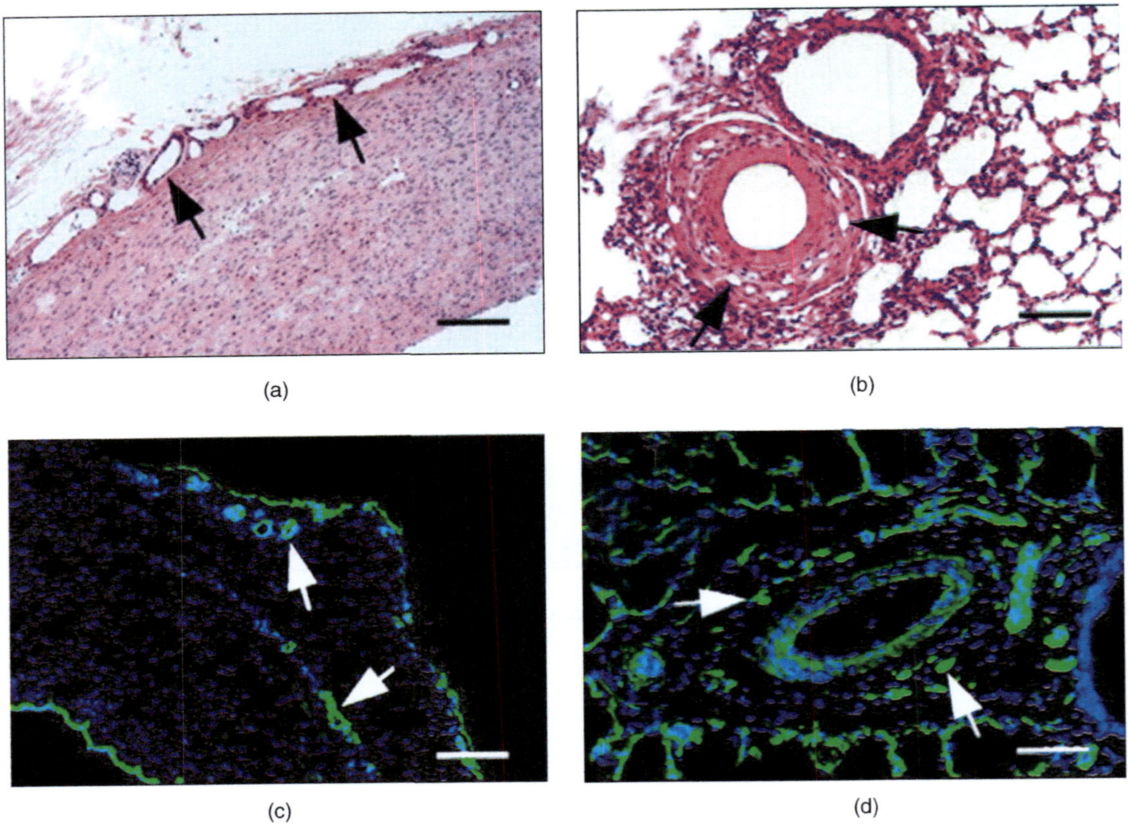

Plate 11.2 Angiogenic responses in the pulmonary arteries of calves with severe hypoxia-induced pulmonary hypertension.

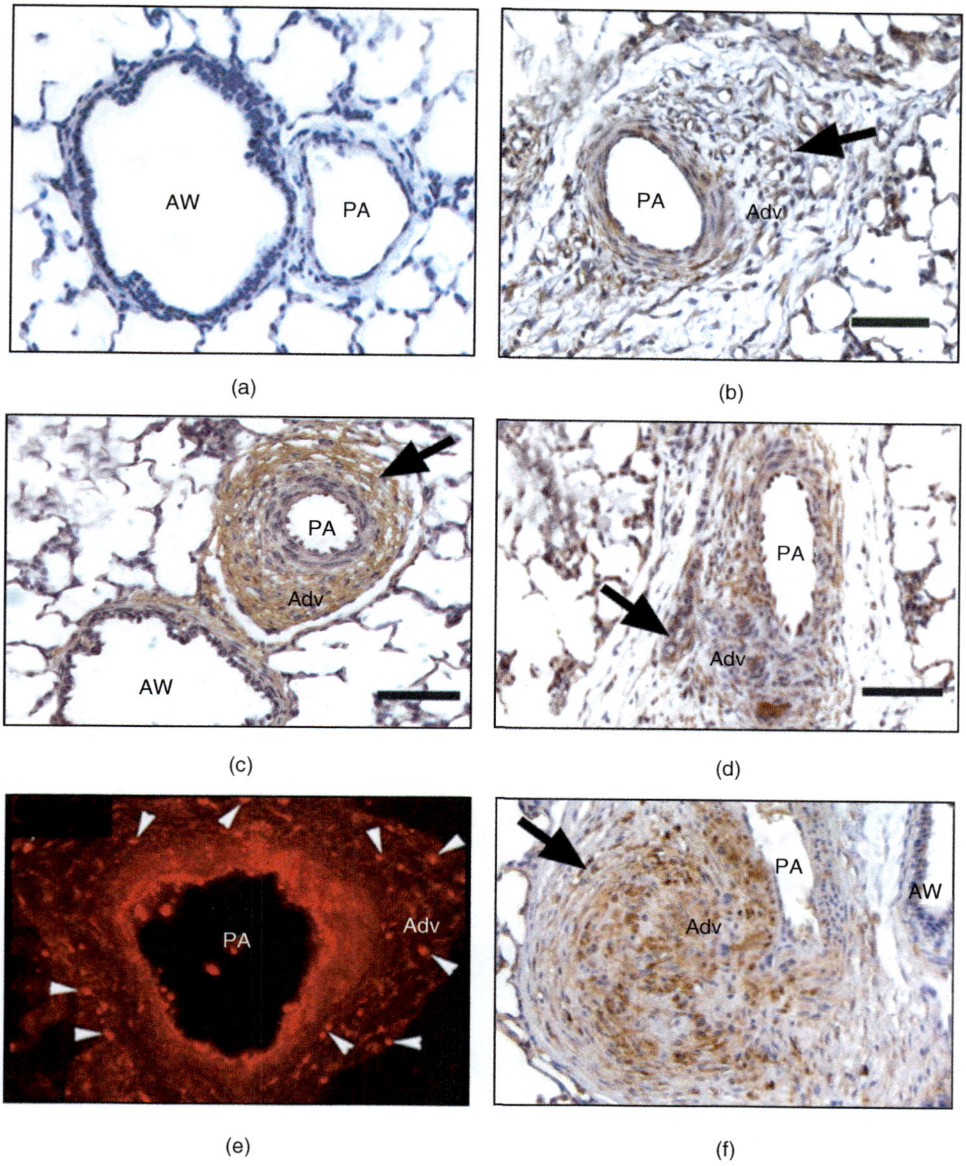

Plate 11.6 Proteins described as having proangiogenic potential [VEGF (b), fibronectin (c), thrombin (d), TGF-β1 (e), and S100A4 (f)], are all expressed in the remodeled adventitia of neonatal calves with severe hypoxia-induced pulmonary hypertension.

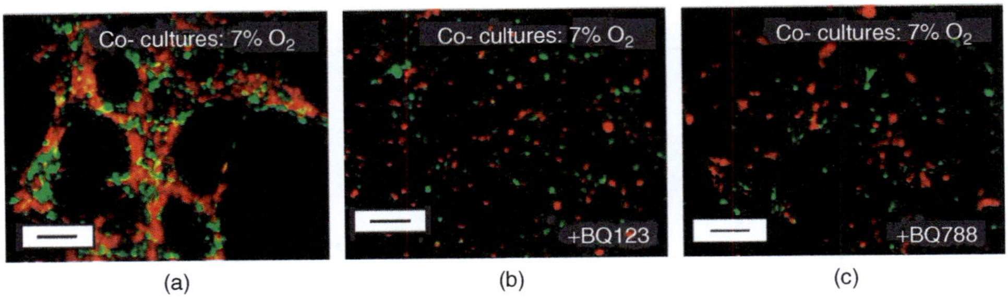

Plate 11.7 Fluorescence microscopy showing that cord-like networks, formed in hypoxic VVEC-AdvFBs Matrigel co-cultures (a), were markedly attenuated when cells were incubated with either the ET_A receptor antagonist BQ123 (b) or the ET_B receptor antagonist BQ788 (c).

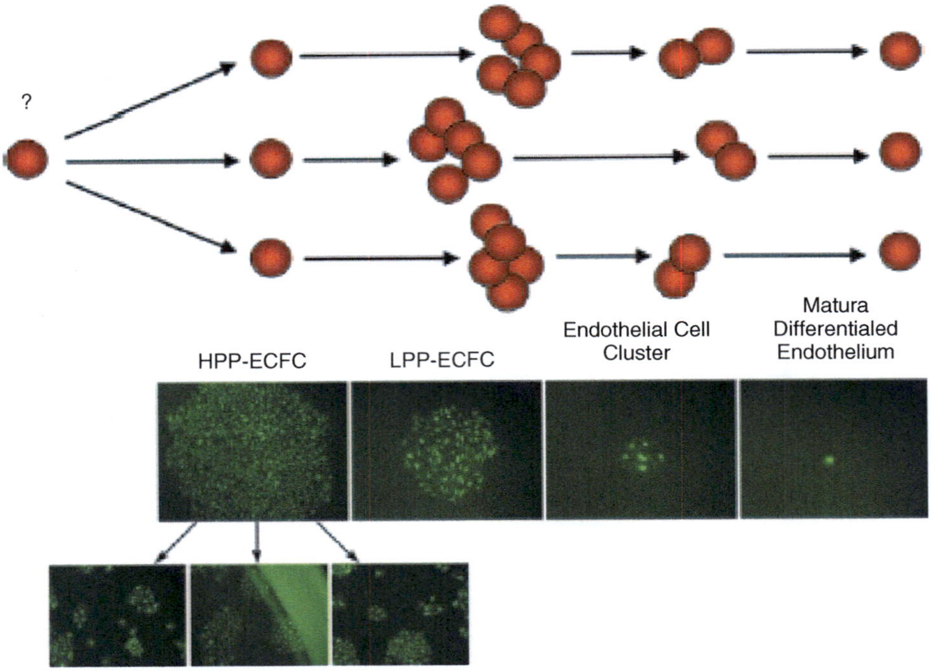

Plate 13.1 Model of an EPC hierarchy based on the proliferative and clonogenic potential of discrete populations of progenitor cells.

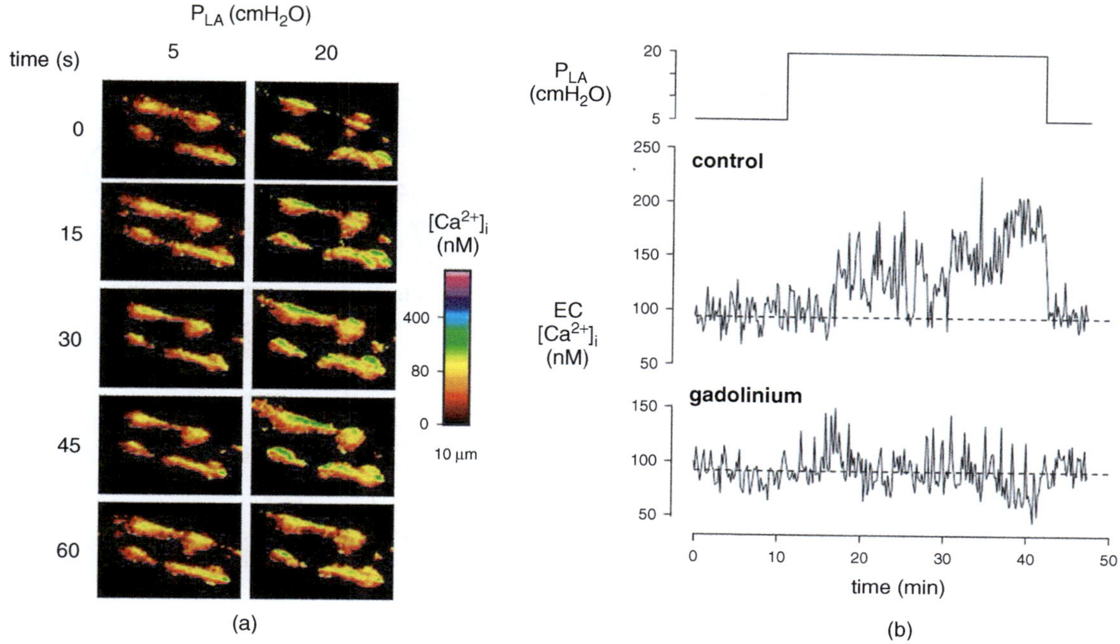

Plate 20.5 Endothelial [Ca^{2+}]i response to increased left atrial pressure (P_{LA}).

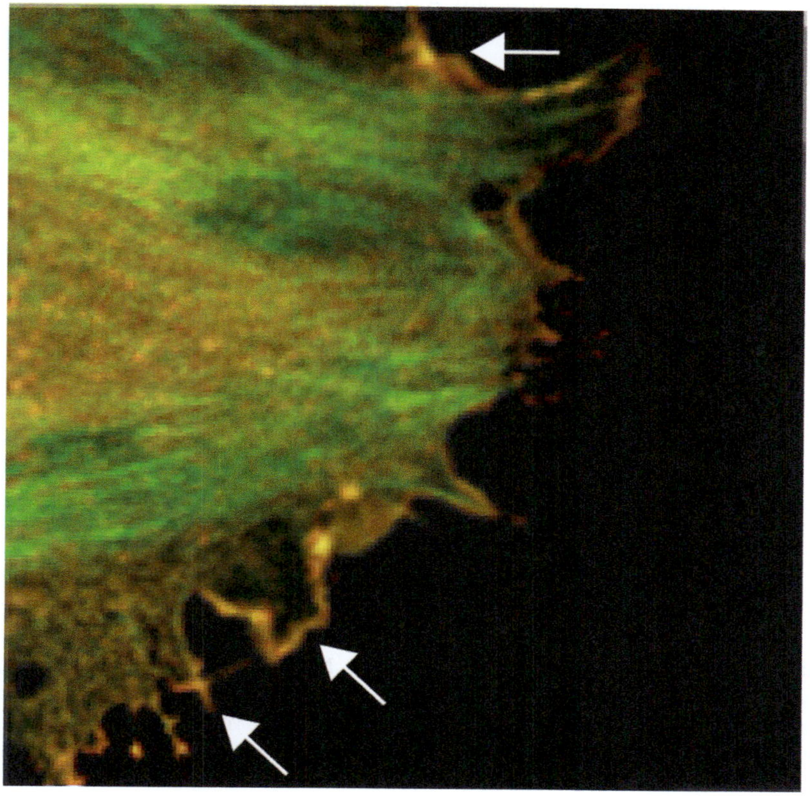

Plate 21.5 Colocalization of lamellipodia stress fibers.

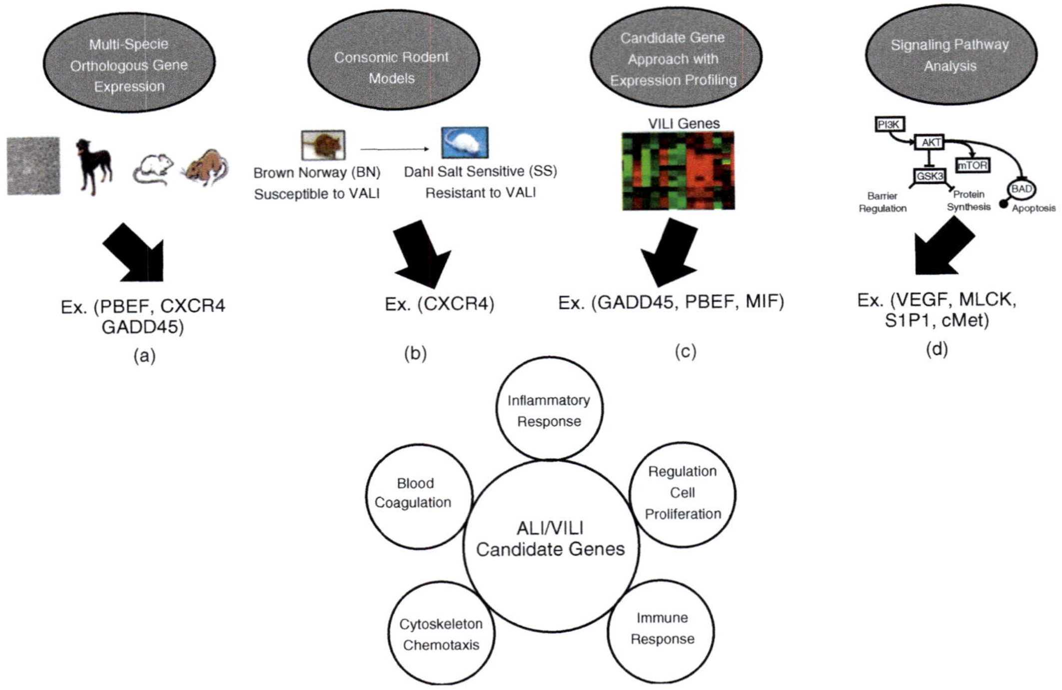

Plate 24.1 Representative novel approaches to identify ALI-implicated genes.

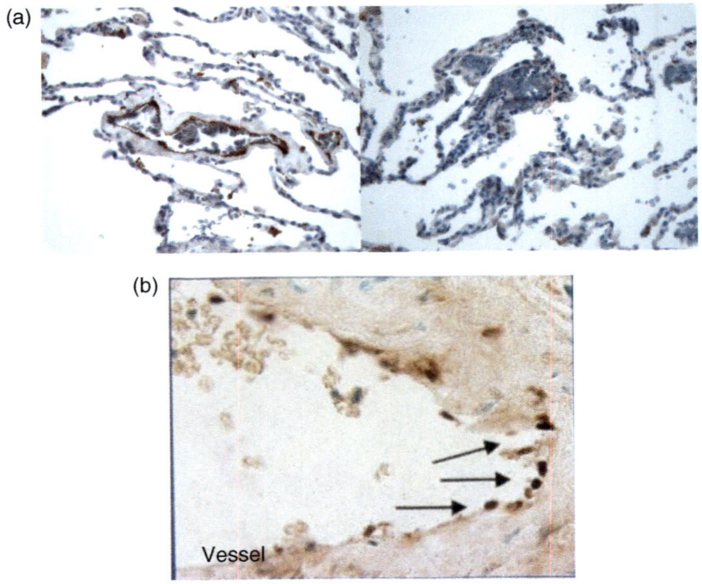

Plate 26.1 (a) Human lung tissue sections. Reproduced from Nana-Sinkam *et al.* (2007), *American Journal of Respiratory and Critical Care Medicine*, **175**, 676–85 with the permission of the American Thoracic Society. (b) Terminal deoxynucleotidyl transferase biotin-dUTP nick end-labeling staining of a lung vessel in a human emphysema lung section demonstrates apoptotic ECs within the vessel EC monolayer (arrows). Reproduced from Kasahara *et al.* (2001) *American Journal of Respiratory and Critical Care Medicine*, **163**, 737–44 with the permission of the American Thoracic Society.

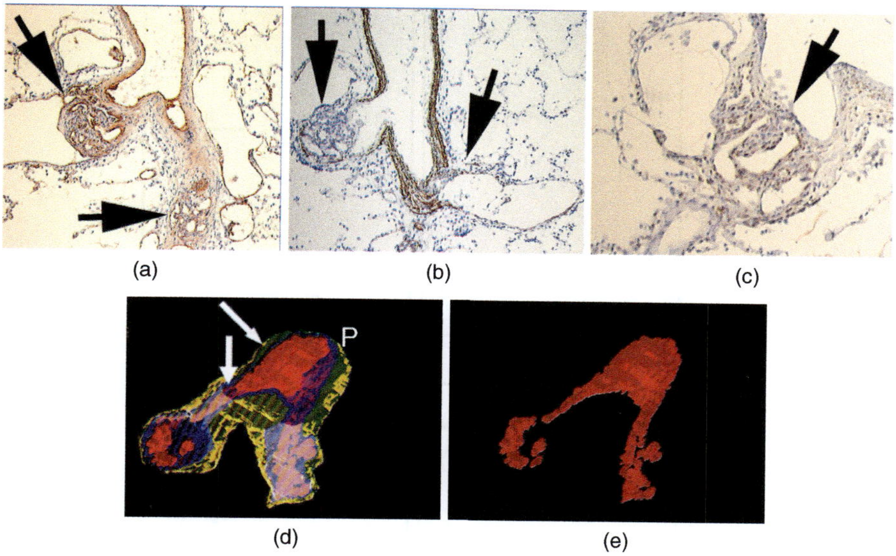

Plate 27.1 Plexiform lesions occurring along two branches of medium-sized pulmonary arteries (arrows). Reproduced from Cool *et al.* (1999) *American Journal of Pathology*, **155**, 411–19, by permission of the American Society for Investigative Pathology.

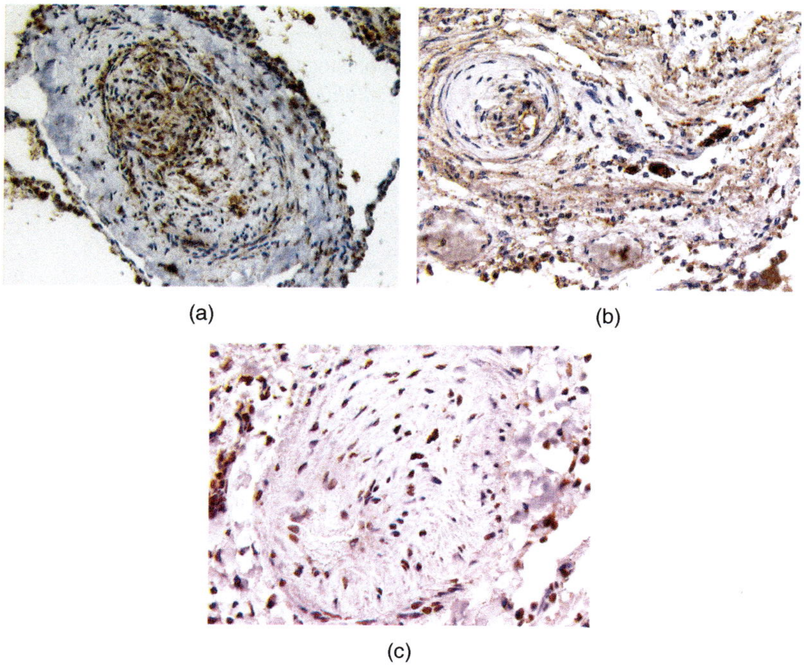

Plate 27.2 Expression of HIF-1α in a plexiform lesion (a) and in a concentric lesion (b), and of HIF-1β in a plexiform lesion (c). Reproduced from Tuder *et al.* (2001) *Journal of Pathology*, **195**, 367–74 with permission from John Wiley & Sons, Ltd.

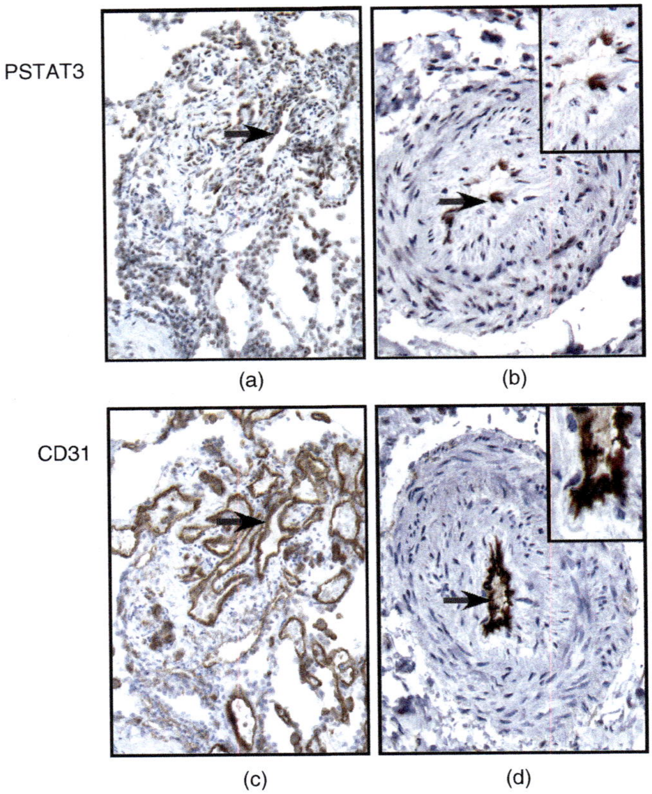

Plate 27.3 Cellular localization of phospho-STAT-3 by immunohistochemical staining in IPAH lung. Reproduced from Masri *et al*. (2007), *American Journal of Physiology: Lung Cellular and Molecular Physiology*, **293**, L548–54, with permission from The American Physiological Society.

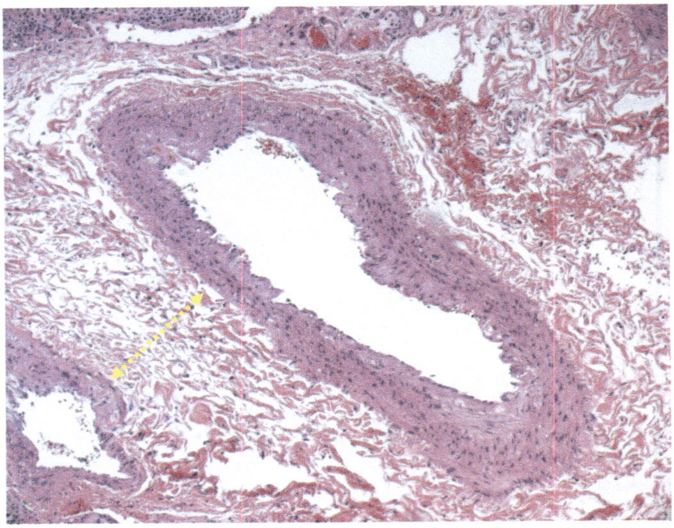

Plate 28.1 Pulmonary artery from a patient with diffuse scleroderma/SSc showing a marked thickening of the adventitial collagen (double arrow).

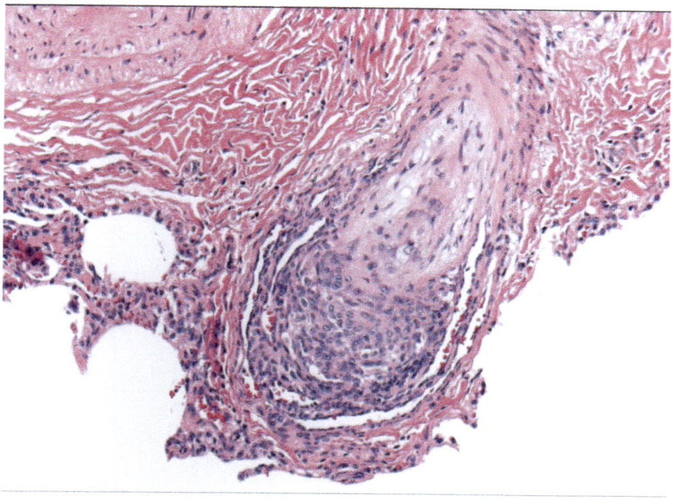

Plate 28.2 This plexiform lesion (P) from a patient with severe PAH demonstrates the proliferative, lumen-obliterating appearance of the ECs.

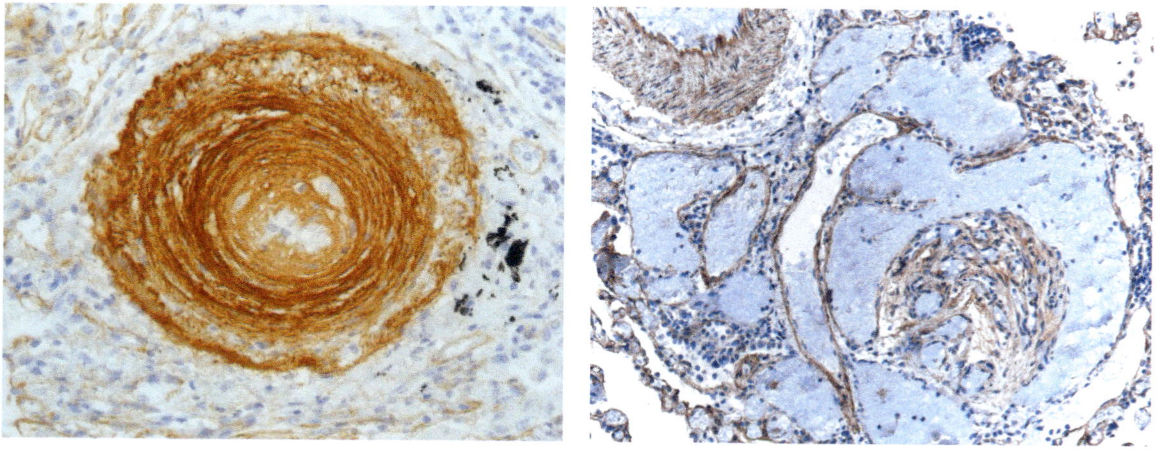

(a)

(b)

Plate 28.3 (a) Pulmonary artery obliterated by a concentric, "onionskinning," proliferation of ECs, highlighted by immunohistochemical stain for ECs (Factor VIII-related antigen). (b) Dilatation lesion at the distal end of a plexiform lesion. Immunohistochemical stain for EC marker, CD31.

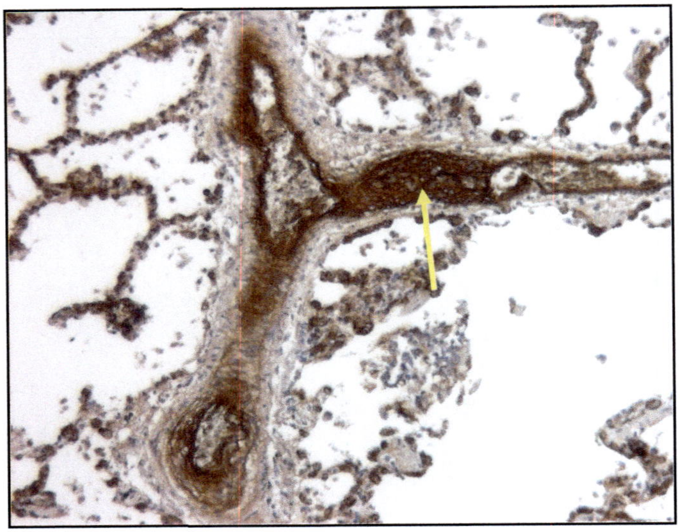

Plate 28.4 Bifurcating pulmonary artery from a patient with CREST and severe PAH.

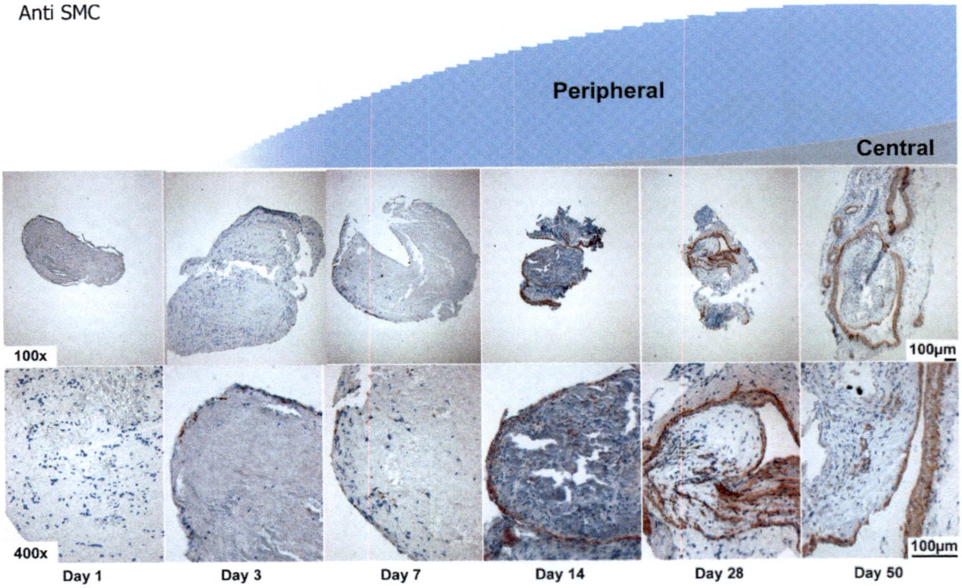

Plate 29.4 Characterization of cells in a vena cava thrombus of the mouse.

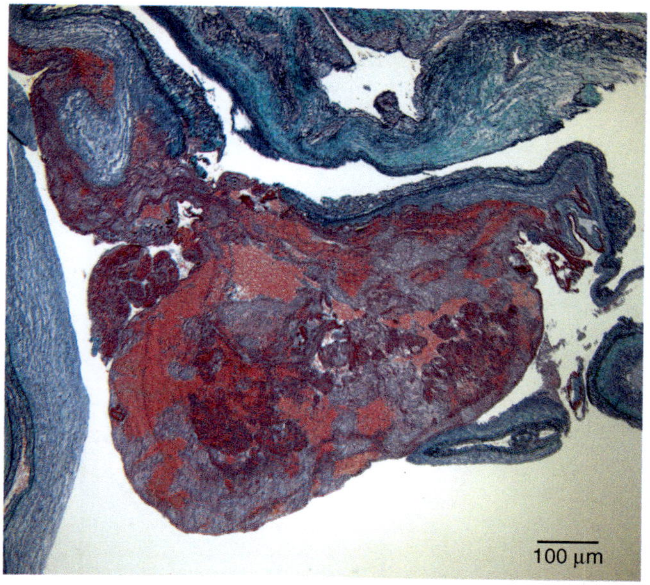

Plate 29.5 Representative histological section of a chronic pulmonary embolus, illustrating an area with *in situ* thrombosis.

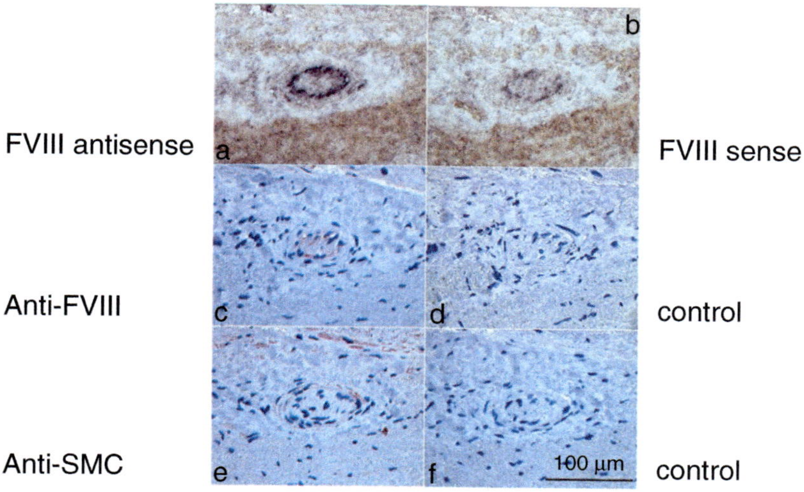

Plate 29.6 Elevated expression of FVIII on thrombus surfaces favors *in situ* thrombosis.

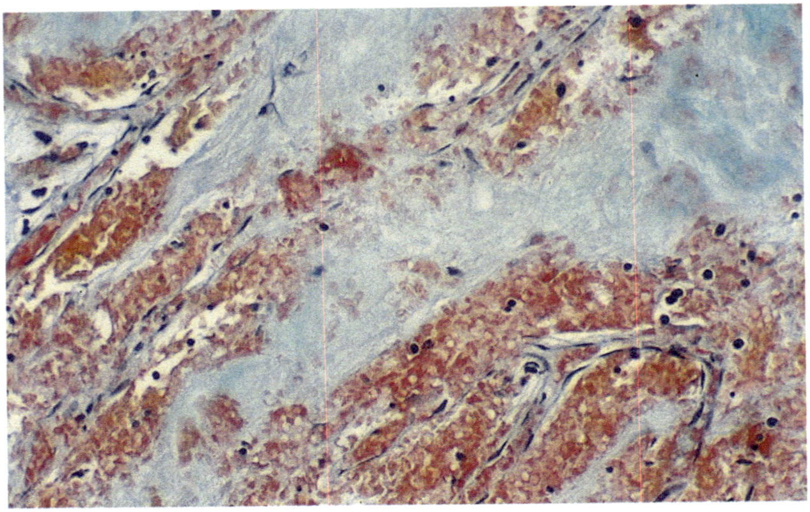

Plate 29.7 Trichrome stain of a histological section of a thrombus from a patient with chronic thromboembolic pulmonary hypertension.

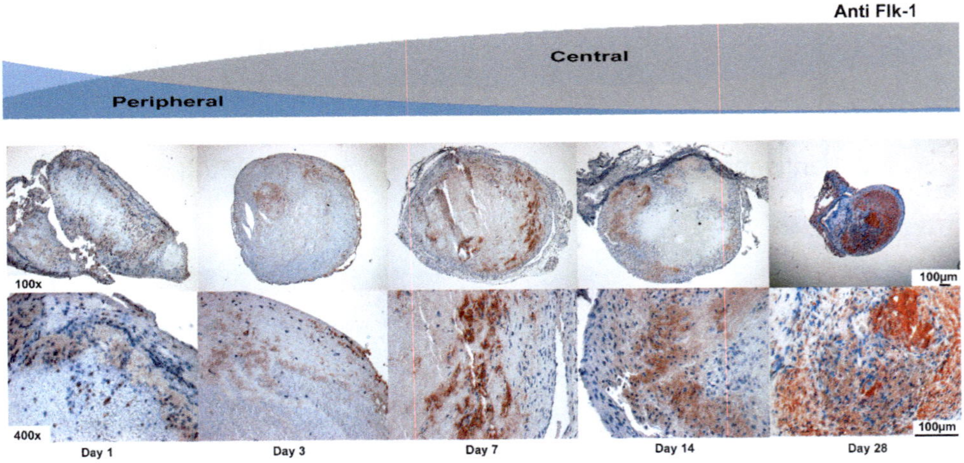

Plate 29.8 Characterization of cells in a vena cava thrombus of the mouse.

Printed and bound by CPI Group (UK) Ltd, Croydon, CR0 4YY
30/09/2026
14973388-0002